Major Biological Events

Millions of Years Ago

Plants	Millions of Years Ago	Animals
	0.135 — Modern *Homo sapiens* arises. 2 — Genus *Homo* arises. 4 — Australopithecines present.	
Gymnosperms, angiosperms widespread. Temperate grasslands, forests expand.	65 — Mammals diversify. Primates arise.	
Angiosperms arise and diversify.		Major extinction event. Most large reptiles, ancient birds extinct.
		Teleost fish diversify. Dinosaurs dominant. Modern crustaceans common.
Gymnosperms, ferns dominant.		Dinosaur ancestors common. First mammals, birds.
Conifers appeared.		Mammallike reptiles common. Major extinction of invertebrates, amphibia.
Forests widespread. Coal deposits form.		First reptiles. Amphibia diversify. Major extinction event.
First forests. Vascular plants and seeds present.		First insects, sharks, amphibians. Fish diversify.
Green, red, brown algae common.		First land arthropods. Jawed fish arise.
First vascular land plants probably appeared.		Second major extinction event. Jawless fish diversify; large invertebrates present; mollusks diversify. First tracks left by land animals.
Algae dominant.		First major extinction event. Trilobites common; onychophorans, first jawless fish at end of period. Evolution of many phyla.
Algae abundant. Multicellular organisms: algae, fungi. Cyanobacteria diversified. Eukaryotes present: green algae, protists.		Wormlike animals, cnidarians present.
Photosynthetic cells liberate oxygen.		
Origin of life. Prokaryotic heterotrophs. Chemical evolution.		

THE NATURE OF LIFE

ABOUT THE AUTHORS

JOHN H. POSTLETHWAIT, Professor of Biology at the University of Oregon, holds a B.A. degree from Purdue University and a Ph.D. from Case Western Reserve University, and he trained as a post-doctoral fellow at Harvard University. He was Visiting Research Scientist for a year each at the Institut für Molekular Biologie of the Austrian Academy of Sciences in Salzburg, Austria; the Laboratoire de Génétique Moléculaire, Faculté de Médecine, in Strasbourg, France; and the Imperial Cancer Research Fund in Oxford, England. From 1971 to the present, his research in developmental and molecular genetics has been funded by grants from the National Institutes of Health, the American Heart Association, and the U.S. Department of Agriculture, and he has published the results of his work in more than one hundred research articles. He has taught general biology to majors and nonmajors since 1971 at the University of Oregon, where he received the Ersted Award for Distinguished Teaching. His current efforts include work funded by the National Science Foundation program in biology education, developing a workshop-oriented approach in the teaching of non-biology majors that emphasizes open-ended investigation and the application of biological concepts to solving problems in society.

JANET L. HOPSON is a lecturer for the Science Communication Program, University of California at Santa Cruz, as well as a freelance science writer. She has also taught writing courses at the University of California at Berkeley and Mills College. She holds B.A. and M.S. degrees from Southern Illinois University and the University of Missouri. Coauthor of three other biology textbooks for McGraw-Hill, *Biology* and *Essentials of Biology* with Norman K. Wessells, and *Biology! Bringing Science to Life* with John H. Postlethwait and Ruth C. Veres, she has also written a trade book on the human sense of smell. She won the Russell L. Cecil Award for magazine writing from the Arthritis Foundation and has published dozens of articles in national magazines and newspapers, including *Smithsonian, Psychology Today, Science News, Science Digest, Outside,* and others. Her biography is included in the first edition of *Who's Who in Science and Engineering*.

SECOND EDITION

THE NATURE OF LIFE

JOHN H. POSTLETHWAIT **UNIVERSITY OF OREGON**

JANET L. HOPSON **UNIVERSITY OF CALIFORNIA, SANTA CRUZ**

McGRAW-HILL, INC.
NEW YORK ST. LOUIS SAN FRANCISCO AUCKLAND BOGOTÁ CARACAS HAMBURG LISBON
LONDON MADRID MEXICO MILAN MONTREAL NEW DELHI PARIS SAN JUAN
SÃO PAULO SINGAPORE SYDNEY TOKYO TORONTO

With appreciation for the love and support of my mother, Sara M. Postlethwait, and my father, Samuel N. Postlethwait, a wonderful inspiration for students and teachers of biology. —J.H.P.

For Samuel F. Smith and Charles R. Billings, with my deepest appreciation and affection. —J.L.H.

The Nature of Life

1 2 3 4 5 6 7 8 9 0 VNH VNH 9 0 9 8 7 6 5 4 3 2

ISBN 0-07-050633-7

Library of Congress Cataloging-in-Publication Data

Postlethwait, John H.
    The nature of life/John H. Postlethwait, Janet L. Hopson.—2nd ed.
        p.   cm.
    Includes bibliographical references and index.
    ISBN 0-07-050633-7
    1. Biology. 2. Life (Biology)   I. Hopson, Janet L.   II. Title.
    [DNLM: 1. Biology.   QH 308.2 P858n]
QH308.2.P67   1992
574—dc20
DNLM/DLC   91-25800
for Library of Congress   CIP

Sponsoring Editors: Denise Schanck, June Smith
Senior Associate Editor: Mary Eshelman
Senior Editing Supervisor: Alice Mace Nakanishi
Assistant Production Manager: Pattie Myers
Designer: BB&K Design
Art Coordinator: Marian Hartsough
Developmental Art Consultants: Iris Martinez Kane, Cherie Wetzel, Arthur Ciccone
Illustrators: Martha Blake, Wayne Clark, Cecile Duray-Bito, JAK Studio, Paula McKenzie, Linda McVay, Elizabeth Morales-Denney, Victor Royer, Carla Simmons, Tek-Nēk´ Inc., John Waller, Judith Waller, Cyndie C.H.-Wooley
Photo Researchers: Darcy Lanham, Monica Suder
Copyeditor: Janet Greenblatt
Proofreader: Sarah Miller
Indexer: Barbara Littlewood
Compositor: Graphic Typesetting Service
Color Separator: Black Dot Graphics
Printer and Binder: Von Hoffmann Press
Cover Photos: Rusty parrotfish (*Scarus ferrugineus*) (inset) and rusty parrotfish scales (background) by Jeff Rotman
Cover Photo Researcher: Natalie Goldstein
Cover Color Separator and Printer: New England Book Components
Production assistance by: Betsy Dilernia, Brian Jones, Jane Moorman, Tralelia Twitty

Figure 2.21c: Adapted with permission of Procter & Gamble.

Figure 12.15: Adapted from "The Human Gene Map," compiled by Victor A. McKusick, 1986. Reprinted with permission of Dr. Victor A. McKusick, University Professor of Medical Genetics, Johns Hopkins University, and Cold Spring Harbor Laboratory.

Figure 33.28: Modified from *Horses: The Story of the Horse Family in the Modern World and Through Sixty Million Years of History* by G. G. Simpson. Copyright © 1951 by Oxford University Press, Inc.; renewed 1979 by G. G. Simpson. Reprinted by permission.

Credits are continued on pages C-1 to C-6.

CONTENTS IN BRIEF

C O N T E N T S

PART ONE ■ Life's Fundamentals

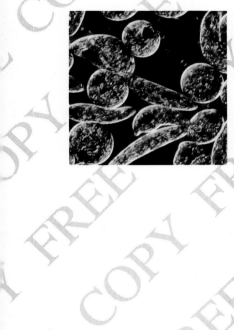

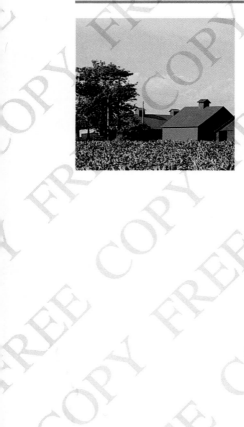

PART TWO   Perpetuation of Life

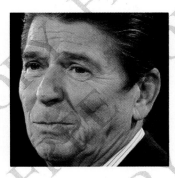

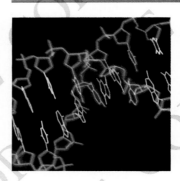

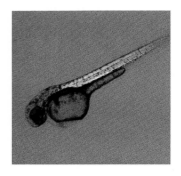

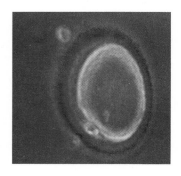

PART THREE ■ Life's Variety

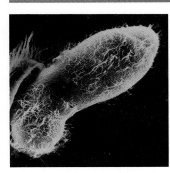

PART FOUR ■ How Animals Survive

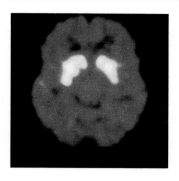

PART FIVE ▪ How Plants Survive

PART SIX ■ Interactions: Organisms and Environment

CHAPTER 36 ■ Ecosystems: Webs of Life and the
Physical World 730

CHAPTER 37 ■ The Biosphere: Earth's Thin
Film of Life 750

PART SEVEN ▪ Behavior and the Future

Modern biology is often one of the most popular undergraduate science courses at colleges and universities. Many students enroll because they'd like to learn more about human physiology, genetics, and the environment. Some sign up because science is required and they hope that biology will be easier than chemistry or physics. And a handful mainly need a M–W–F at 10:40.

Reasons aside, this will be the first and last exposure many nonmajors will have to a science course during their college years, and it presents their best opportunity to become biologically literate citizens who can make wise decisions about nutrition, exercise, health care, drugs, alcohol, smoking, sexually transmitted diseases, consumer products, habits that impact the environment, and important issues before the electorate.

While introductory biology has obvious relevance to a student's life, the course can nevertheless be challenging—for some students, even formidable—because of the sheer number of facts and terms that biologists use in their work and that students have traditionally been required to learn. The well-documented trend toward science illiteracy among Americans is partially caused by the burgeoning of facts and terms in all scientific fields, biology included. But the expansion of science is just one factor—and considering the higher levels of science literacy in a dozen other industrialized nations, perhaps a minor one. More to the point, say many science educators, is the way we have customarily presented science from elementary school through graduate school.

The National Science Foundation has established a grant program to encourage new approaches to the teaching of undergraduate science courses. The collective innovations proposed by professors at dozens of colleges and universities represent an emerging consensus that emphasizes:

- Concepts over details
- Issues and student-relevant topics over traditional vocabulary and facts
- Hands-on, self-generated laboratory projects over teacher demonstrations and "cookbook" exercises
- Writing and speaking skills over memorization and standard exams
- Cooperative learning by small groups in addition to attendance at large lectures

Overall, life science professors want their students to experience the excitement and to see the applications of modern biology; to carry away an understanding of basic concepts, not a memory full of transitory facts; to appreciate science as a way of knowing about the world around them; and to apply biological concepts to the health, reproductive, and environmental choices they will inevitably face.

Features from the First Edition
Our approach to the first edition of *The Nature of Life* reflected a number of these innovations and was enthusiasti-

cally received in many colleges and universities. To motivate and excite nonmajors, we generated 38 stories, one per chapter, around which to pose and answer central questions about the natural world and to organize basic biological concepts. We mapped out three bookwide unifying themes to help orient readers:

1. Living things take in *energy,* needed to maintain their internal order and organization.
2. Living things undergo *reproduction,* enabling the species to continue after the individual ceases to exist.
3. Specialized means of acquiring energy and characteristic patterns of reproduction arise by *evolution,* allowing living organisms to adapt to changing environments.

The first edition included chapterwide themes, as well, to help establish the relationship of parts to the whole and to guide students to what is most significant, and why. What's more, we created an art program full of orienting icons, unique process diagrams, and colorful photos, and closely coordinated them to the text so that readers could verbalize and visualize biological structures and activities simultaneously. We incorporated current issues and topics of student relevance throughout the chapters and the more than 50 boxed essays. Finally, we added several special study features—underlined take-home messages, integrating end-of-chapter reflections called Connections, Highlights in Review, Key Terms, Study Questions, and Further Readings—to help the student learn biology most effectively. Our first edition provided a guided tour of the features in a typical chapter, and an updated version appears in this new edition, beginning on page xxx.

As in the first edition of *The Nature of Life,* this new revision takes a hierarchical approach to the study of biology. An introductory chapter discusses the main themes of the book as well as the process of science. In Part One, we consider how molecules (Chapter 2), cells (Chapter 3), and cellular activities (Chapter 4) provide the common groundwork for life. We end Part One with a clear analysis of how cells obtain and use energy (Chapters 5 and 6), the book's first theme.

From this cellular foundation, we move on to the principles of reproduction, the book's second theme, and the general subject of Part Two. Chapters 7 through 10 discuss how cells and organisms pass on to their offspring the hereditary units that cause "like to beget like." A short, very up-to-date chapter on the exciting world of biotechnology and recombinant DNA techniques (Chapter 11) leads in to the fascinating subject of human genetics (Chapter 12). This part of the book ends with an analysis of how egg cells decode instructions stored in DNA to build a fish, fly, frog, or person (Chapters 13 and 14).

Equipped with a thorough knowledge of the cellular and genetic features that unify life, we survey the diverse range of life forms in Part Three. We take an evolutionary approach (the book's third theme) to the questions of life's origins (Chapter 15). Then we examine the characteristics of organisms from each kingdom of life (Chapters 16 through 19), ending with an in-depth discussion of the human species' own evolutionary descent.

Parts Four and Five investigate how plants and animals maintain their bodies against the inevitable disorganization that occurs over time (a reprise of our first theme). Chapter 20 orients students by presenting the themes that will recur throughout the study of anatomy and physiology. Chapters 21 through 28 analyze individual physiological systems in animals and include dozens of examples that we hope will captivate readers and allow them to understand the biological bases for their own body functions. We include a complete discussion of the immune system (Chapter 22), an in-depth treatment of the nervous system, brain, and behavior (Chapters 27 and 28), and a unique chapter that focuses on exercise physiology (Chapter 29) and that shows how body systems work together as a dynamic whole. This chapter promises to have great appeal for physically active college students.

Part Five builds on the discussion of plant diversity from Chapter 17 with three more plant chapters: one on plant structure (Chapter 30), one on the regulation of plant growth (Chapter 31), and one that deals with how plants function (Chapter 32). These chapters incorporate the many new applications of genetic engineering to plant science, and they emphasize the importance of plants to students' everyday lives.

With Part Six we return to our evolutionary theme, devoting Chapter 33 to the science of evolution, and Chapters 34 through 37 to ecology, how organisms interact with each other and the world around them. We give special emphasis to the ecological and environmental issues that affect our current quality of life and that of future generations.

The book ends with the engrossing subject of behavior (Part Seven, Chapter 38), viewed from evolutionary and ecological perspectives. We believe this is a fitting conclusion because human behavior will shape the world of the future— a world our students must prepare to lead as professionals and to help preserve as responsible citizens.

New to This Edition
Although we retain this basic organization in our current revision of *The Nature of Life,* we made numerous changes, reflecting both our reviewers' helpful advice and new ideas. Our goal was to continue motivating and exciting introductory biology students in ways that support both traditional and innovative courses, but to do so in an even more effective manner:

- We cut detail in many discussions, simplifying the overall presentation and strengthening the concepts and take-home messages.
- We updated the science where appropriate in the text and boxed essays.

- We significantly increased our coverage of environmental science and added the preservation of the environment as a new bookwide theme.
- We enlarged and strengthened our section on the physical evidence for evolution.
- We introduced more human relevance, new societal issues, and new opening examples in many chapters.
- We streamlined dozens of diagrams to improve their readability, and we replaced or enlarged many photos to integrate the visual with the verbal more successfully.
- We sharpened our presentation of the scientific process.
- We enhanced the end-of-chapter questions and bibliographies.
- We redesigned the book's graphic elements to make the book easier to read, clearer, and more inviting.

We hope that the second edition of *The Nature of Life* will be a useful educational tool. We have tried to preserve the best features of the first edition while improving upon them in innovative ways that will intrigue nonmajors and help them learn the basics of biological science and then apply that knowledge to their own lives.

This second edition is the designated textbook to accompany a planned college-level biology telecourse and an eight-part prime-time series coming to public television, produced by WGBH Boston.

SUPPLEMENTARY MATERIALS

A comprehensive and completely integrated package of supplementary materials accompanies *The Nature of Life,* Second Edition.

- ***Instructor's Manual and Resource Guide***
  Dennis Todd, University of Oregon
- ***Test Bank***
  Dennis Todd, University of Oregon
- ***Critical Thinking Workbook to Accompany* The Nature of Life**
  Gail Patt, Boston University
- ***Hands-On Biology*** (a laboratory manual for introductory biology) and ***Preparator's Guide to Accompany* Hands-On Biology**
  Theodore Taigen, University of Connecticut, Storrs
  Thomas Terry, University of Connecticut, Storrs
  David Wagner, University of Connecticut, Storrs
  Eileen Jokinen, University of Connecticut, Storrs
  Andris Indars, University of Connecticut, Storrs
- Computerized Instructor's Manual (available in IBM, Macintosh, Apple)
- Computerized Test Bank (available in IBM, Macintosh, Apple)
- ***BioPartner*** (Computerized Study Guide; available in IBM, Macintosh)
- Videos

- Biology Slides and Acetate Package
- Videodisk
- HyperMedia Software

For further information regarding the supplements available, please contact your local McGraw-Hill representative.

ACKNOWLEDGMENTS

This revision grew out of the original material we presented in the first edition, upon the insightful comments of our many reviewers and marketing consultants, and finally upon our new manuscript and illustrations so skillfully handled by the McGraw-Hill team. We wish, therefore, to thank those central to the inception and development of *The Nature of Life:* our former sponsoring editor, Eirik Børve; our former developmental editor, Ruth Veres; our former developmental art consultant, Peter Veres; our seven original contributors, Charles L. Aker, The National Autonomous University of Nicaragua; Russ Fernald, University of Oregon; Craig Heller, Stanford University; Kent Holsinger, University of Connecticut; V. Pat Lombardi, Stanford University; Christopher Stringer, British Museum, London; and Daniel Udovic, University of Oregon; and our general consultant, the late Howard Schneiderman, whose counsel, support, and enthusiasm were so important over the years. We also extend special thanks to Dennis Todd from the University of Oregon for his design of Study Questions and material in both editions.

We have sought the advice of hundreds of instructors around the country to help us create a textbook that would meet the unique needs of the introductory biology market. Our sincere thanks are extended to the following individuals who responded to our market questionnaires:

Dr. Laura Adamkewicz, George Mason University; *Olukemi Adewusi,* Ferris State University; *Kraig Adler,* Cornell University; *Dr. John U. Aliff,* Glendale Community College; *Joanna T. Ambron,* Queensborough Community College; *Steven Austad,* Harvard University; *Robert J. Baalman,* California State University, Hayward; *Stuart S. Bamforth,* Tulane University; *Sarah F. Barlow,* Middle Tennessee State University; *R. J. Barnett,* California State University, Chico; *Joseph A. Beatty,* Southern Illinois University, Carbondale; *Nancy Benchimol,* Nassau Community College; *Dr. Rolf W. Benseler,* California State University, Hayward; *Gerald Bergtrom,* University of Wisconsin, Milwaukee; *Dr. Dorothy B. Berner,* Temple University; *Dr. A. K. Boateng,* Florida Community College, Jacksonville; *William S. Bradshaw,* Brigham Young University; *Jonathan Brosin,* Sacramento City College; *Howard E. Buhse, Jr.,* University of Illinois, Chicago; *John Burger,* University of New Hampshire; *F. M. Butterworth,* Oakland University; *Guy Cameron,* University of Houston; *Ian M. Campbell,* University of Pittsburgh; *John L. Caruso,* University of Cincinnati; *Brenda Casper,* University of Pennsylvania; *Doug Cheeseman,* De Anza College; *Dr. Gregory Cheplick,* University of Wisconsin; *Dr. Joseph P.*

Chinnici, Virginia Commonwealth University; *Carl F. Chuey,* Youngstown State University; *Dr. Simon Chung,* Northeastern Illinois University; *Norman S. Cohn,* Ohio University; *Paul Colinvaux,* Ohio State University; *Scott L. Collins,* University of Oklahoma; *Dr. August J. Colo,* Middlesex County College; *G. Dennis Cooke,* Kent State University; *Jack D. Cote,* College of Lake County; *Gerald T. Cowley,* University of South Carolina; *Louis Crescitelli,* Bergen Community College; *Orlando Cuellar,* University of Utah; *Thomas Daniel,* University of Washington; *J. Michael DeBow,* San Joaquin Delta College; *Loren Denny,* Southwest Missouri State University; *Ron DePry,* Fresno City College; *Dr. Kathryn Dickson,* California State University, Fullerton; *Patrick J. Doyle,* Middle Tennessee State University; *Dr. David W. Eldridge,* Baylor University; *Lynne Elkin,* California State University, Hayward; *Paul R. Elliott,* Florida State University; *Eldon Enger,* Delta College; *Gauhari Farooka,* University of Nebraska, Omaha; *Marvin Fawley,* North Dakota State University; *Ronald R. Fenstermacher,* Community College of Philadelphia; *Edwin Franks,* Western Illinois University; *C. E. Freeman,* University of Texas, El Paso; *Lawrence D. Friedman,* University of Missouri, St. Louis; *Dr. Ric A. Garcia,* Clemson University; *Wendell Gauger,* University of Nebraska, Lincoln; *Dr. S. M. Gittleson,* Fairleigh Dickinson University; *E. Goudsmit,* Oakland University; *John S. Graham,* Bowling Green State University; *Shirley Graham,* Kent State University; *Thomas Gregg,* Miami University; *Alan Groeger,* Southwest Texas State University; *Thaddeus A. Grudzien,* Oakland University; *James A. Guikema,* Kansas State University; *Robert W. Hamilton,* Loyola University of Chicago; *Richard C. Harrel,* Lamar University; *T. P. Harrison,* Central State University; *Maurice E. Hartman,* Palm Beach Community College; *Dr. Karl H. Hasenstein,* University of Southwestern Louisiana; *Martin A. Hegyi,* Fordham University; *Dr. John J. Heise,* Georgia Institute of Technology; *H. T. Hendrickson,* University of North Carolina, Greensboro; *T. R. Hoage,* Sam Houston State University; *Kurt G. Hofter,* Florida State University; *Dr. Rhodes B. Holliman,* Virginia Polytechnic Institute and State University; *Harry L. Holloway,* University of North Dakota; *E. Bruce Holmes,* Western Illinois University; *Jerry H. Hubschman,* Wright State University; *Hadar Isseroff,* State University of New York College, Buffalo; *Dr. Ira James,* California State University, Long Beach; *Dr. Wilmar B. Jansma,* University of Northern Iowa; *Dr. Margaret Jefferson,* California State University, Los Angeles; *Dr. Ira Jones,* California State University, Long Beach; *Dr. Patricia P. Jones,* Stanford University; *Dr. Craig T. Jordan,* University of Texas, San Antonio; *Maurice C. Kalb,* University of Wisconsin, Whitewater; *Bonnie Kalison,* Mesa College; *Judy Kandel,* California State University, Fullerton; *Arnold Karpoff,* University of Louisville; *L. G. Kavaljian,* California State University, Sacramento; *Donald R. Kirk,* Shasta College; *R. Koide,* Pennsylvania State University; *Mark Konikoff,* University of Southwestern Louisiana; *Barbara S. Lake,* Central Piedmont Community College; *Jim des Lauvérs,* Chaffey College; *Tami*

Levitt-Gilmarr, Pennsylvania State University; *Daniel Linzer*, Northwestern University; *J. R. Loewenberg*, University of Wisconsin, Milwaukee; *Dr. Robert Lonard*, University of Texas, Pan American; *Sharon R. Long*, Stanford University; *Carmita E. Love*, Community College of Philadelphia; *C. E. Ludwig*, California State University, Sacramento; *Dr. Ann S. Lumsden*, Florida State University; *Dr. Bonnie Lustigman*, Montclair State College; *Edward B. Lyke*, California State University, Hayward; *Douglas Lyng*, Indiana University–Purdue University, Ft. Wayne; *George L. Marchin*, Kansas State University; *Philip M. Mathis*, Middle Tennessee State University; *Mrs. Margaret L. May*, Virginia Commonwealth University; *Edward McCrady*, University of North Carolina, Greensboro; *Bruce McCune*, Oregon State University; *Dr. John O. Mecom*, Richland College; *Tekié Mehary*, University of Washington; *Richard L. Miller*, Temple University; *Phyllis Moore*, University of Arkansas, Little Rock; *Carl Moos*, State University of New York, Stony Brook; *Doris Morgan*, Middlesex County College; *Donald B. Morzenti*, Milwaukee Area Technical College; *Steve Murray*, California State University, Fullerton; *Robert Neill*, University of Texas, Arlington; *Paul Nollen*, Western Illinois University; *Kenneth Nuss*, University of Northern Iowa; *William D. O'Dell*, University of Nebraska, Omaha; *Dr. Joyce K. Ono*, California State University, Fullerton; *James T. Oris*, Miami University; *Clark L. Ovrebo*, Central State University, Edmond; *Charles Page*, El Camino College; *Kay Pauling*, Foothill College; *Dr. Chris E. Petersen*, College of DuPage; *Richard Petersen*, Portland State University; *Jeffrey Pommerville*, Glendale Community College; *David I. Rasmussen*, Arizona State University; *Daniel Read*, Central Piedmont Community College; *Dr. Don Reinhardt*, Georgia State University; *Louis Renaud*, Prince George's Community College; *Jackie Reynolds*, Richland College; *Jennifer H. Richards*, Florida International University; *Thomas L. Richards*, California State Polytechnic University; *Tom Rike*, Glendale Community College, California; *C. L. Rockett*, Bowling Green State University; *Hugh Rooney*, J. S. Reynolds Community College; *Wayne C. Rosing*, Middle Tennessee State University; *Frederick C. Ross*, Delta College; *A. H. Rothman*, California State University, Fullerton; *Mary Lou Rottman*, University of Colorado, Denver; *Dr. Donald J. Roufa*, Kansas State University; *Dr. Michael Rourke*, Bakersfield College; *Chester E. Rufh*, Youngstown State University; *Mariette Ruppert*, Clemson University; *Charles L. Rutherford*, Virginia Polytechnic Institute and State University; *Dr. Milton Saier*, University of California, San Diego; *Lisa Sardinia*, San Francisco State University; *A. G. Scarbrough*, Towson State University; *Dan Scheirer*, Northeastern University; *Randall Schietzelf*, Harper College; *Robert W. Schuhmacher*, Kean College of New Jersey; *Joel S. Schwartz*, College of Staten Island; *Roger S. Sharpe*, University of Nebraska, Omaha; *Stanley Shostak*, University of Pittsburgh; *J. Kenneth Shull, Jr.*, Appalachian State University; *C. Steven Sikes*, University of South Alabama; *Christopher C. Smith*, Kansas State University; *John O. Stanton*, Monroe Community College; *D. R. Starr*, Mt. Hood Community College; *Dr. Ruth B. Thomas*, Sam Houston State University;

Nancy C. Tuckman, Loyola University of Chicago; *Dr. Spencer Jay Turkel*, New York Institute of Technology; *William A. Turner*, Wayne State University; *C. L. Tydings*, Youngstown State University; *John Tyson*, Virginia Polytechnic Institute and State University; *Richard R. Vance*, University of California, Los Angeles; *Harry van Keulen*, Cleveland State University; *Roy M. Ventullo*, University of Dayton; *Judith A. Verbeke*, University of Illinois, Chicago; *Dr. Ronald B. Walter*, Southwest Texas State University; *Stephen Watts*, University of Alabama, Birmingham; *Dr. Joel D. Weintraub*, California State University, Fullerton; *Marion R. Wells*, Middle Tennessee State University; *James White*, New York City Technical College; *Joe Whitesell*, University of Arkansas, Little Rock; *Fred Whittaker*, University of Louisville; *Roberta Williams*, University of Nevada, Las Vegas; *Chuck Wimpee*, University of Wisconsin, Milwaukee; *Mala Wingerd*, San Diego State University; *Richard Wise*, Bakersfield College; *Gary Wisehart*, San Diego City College; *Dan Wivagg*, Baylor University; *Richard P. Wurst*, Central Connecticut State University; *Edward K. Yeargers*, Georgia Institute of Technology; *Linda Yasui*, Northern Illinois University.

In addition, many users of the first edition graciously provided us with feedback, which allowed us to focus on areas that could be enhanced or modified. They are:

Dean A. Adkins, Marshall University; *Leslie Drew*, Texas Tech University; *Douglas J. Eder*, Southern Illinois University, Edwardsville; *Richard Haas*, Cal State University, Fresno; *Earl L. Hanebrink*, Arkansas State University; *Dr. John P. Harley*, Eastern Kentucky University; *Marcia Harrison*, Marshall University; *Frank Heppner*, University of Rhode Island; *B. Hunnicutt*, Seminole Community College; *Ursula Jando*, Washburn University of Topeka; *Norma G. Johnson*, University of North Carolina, Chapel Hill; *Clyde Jones*, Texas Tech University; *Thomas L. Keefe*, Eastern Kentucky University; *Robin C. Kennedy*, University of Missouri, Columbia; *Eugene C. Perri*, Bucks County Community College; *Eugene C. Perri*, Bucks County Community College; *Joel B. Piperberg*, Millersville University of Pennsylvania; *William D. Rogers*, Ball State University; *Dr. Fred Schreiber*, California State University, Fresno; *Dr. Jane R. Shoup*, Purdue University, Calumet; *Joseph D. Stogner*, Ferrum College; *Bert Tribbey*, California State University, Fresno; *John Twente*, University of Missouri, Columbia; *Leonard S. Vincent*, Fullerton College; *T. Weaver*, Montana State University; *Kenneth A. Wilson*, California State University, Northridge; *Thomas Wolf*, Washburn University of Topeka; *Paul Wright*, Western Carolina University.

We express our sincere appreciation to reviewers who carefully reviewed the first-edition textbook or drafts of the second-edition manuscript and provided extensive comments and advice. They are:

Aimée H. Bakken, University of Washington; *Jack Bostrack*, University of Wisconsin, River Falls; *Charlotte Clark*, Fullerton College; *David Darda*, Central Washington University; *Kathryn A. Dickson*, California State University, Fullerton;

Thomas Dolan, Butler University; *Helen Dunlap,* Millersville University of Pennsylvania; *Grace Gagliardi,* Bucks County Community College; *Gregory Grove,* Pennsylvania State University; *Madeline M. Hall,* Cleveland State University; *Robert N. Hurst,* Purdue University; *Jerry Kaster,* University of Wisconsin, Milwaukee—Center for Great Lakes Studies; *Robin C. Kennedy,* University of Missouri, Columbia; *Eliot Krause,* Seton Hall University; *Elmo A. Law,* University of Missouri, Kansas City; *James Luken,* Northern Kentucky University; *Gail Patt,* Boston University; *David Polcyn,* California State University, San Bernardino; *Michael Pollock,* Mount Royal College, Canada; *Deborah D. Ross,* Indiana University–Purdue University, Ft. Wayne; *Erik P. Scully,* Towson State University; *Guy Steucek,* Millersville University of Pennsylvania; *Raymond Tampari,* Northern Arizona University; *William A. Turner,* Wayne State University; *Linda Van Thiel,* Wayne State University; *C. David Vanicek,* California State University, Sacramento; *Thomas Wolf,* Washburn University of Topeka.

Finally, we are enormously grateful to the fine professionals at McGraw-Hill who directed our work so smoothly— sponsoring editors Denise Schanck and June Smith and senior associate editor Mary Eshelman—and who polished and produced our manuscript so artfully—Alice Mace Nakanishi, senior editing supervisor; Marian Hartsough, art coordinator; Iris Martinez Kane, Cherie Wetzel, and Arthur Ciccone, art reviewers; Darcy Lanham and Monica Suder, photo researchers; Pattie Myers, assistant production manager; Janet Greenblatt, copyeditor; Sarah Miller, proofreader; Tess Joseph, typist; and Lesley Walsh, office manager.

If this edition of *The Nature of Life* inspires and informs the undergraduate students of the early 1990s, it will be in large measure because the above-mentioned contributed their ideas and energy so generously to help us improve this new version.

John H. Postlethwait and Janet L. Hopson

A Guided Tour to THE NATURE OF LIFE

Central Example

▶ ▶ ▶ ▶ ▶ ▶ ▶ ▶ ▶ ▶ ▶ ▶ ▶ ▶ ▶ ▶ ▶ ▶ ▶ ▶

The student is led through an intriguing real-world narrative to the main chapter concepts. This popular feature from the first edition has been written in a style that motivates the student to read further.

C H A P T E R ▪ 1 0

How Genes Work: From DNA to RNA to Protein

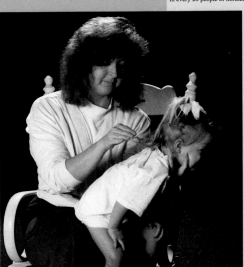

Cystic Fibrosis: A Case Study in Gene Action

A tiny infant receives a parent's tender kiss on the cheek, and the mother or father receives a piece of disquieting information: The baby tastes so *salty*. Why? This exchange may be the first indication that the newborn has cystic fibrosis, the most common lethal genetic disease among Caucasians. One in every 20 people of northern European ancestry carries one copy of the recessive mutation which leads to the disease, but shows no symptoms. Only a child who inherits one copy from each parent (and is thus a homozygous recessive like a pea plant with a short stem or white flowers) will develop this life-threatening illness.

Cystic fibrosis is essentially a disease of clogged ducts; the recessive gene and the faulty protein it encodes (described shortly) lead to a buildup of sticky mucus in lung passages, pancreas ducts, sweat glands, and sperm ducts. As a result, the individual with cystic fibrosis tends to have difficulty breathing, as well as dangerous bacterial infections in the lungs, stomachaches due to poor absorption of essential fats in the diet, a salty secretion on the skin, and, in adult males, usually sterility.

Right now, doctors must treat the symptoms of cystic fibrosis one by one rather than correcting the genetic defect itself. As we will see later in the chapter, that basic genetic correction may be possible in the not-so-distant future and is one of the most exciting medical prospects for the early 1990s. The current treatments, however, include mainly "percussion sessions" during which a parent thumps the child firmly on the back to dislodge gummy mucus and allow easier breathing for a while (Figure 10.1); a special diet; powerful antibiotics to fight the lung infections; and pills containing certain digestive

FIGURE 10.1 ▪ **Manual Therapy for a Child with Cystic Fibrosis.** A "percussion session," during which a mother thumps her child's back vigorously, to loosen and help expel the sticky mucus clogging the victim's lungs. This therapy decreases the infections, scarring, and gradual loss of lung function that characterize cystic fibrosis. A mutated gene leads to this devastating condition.

CHAPTER 10   HOW GENES WORK: FROM DNA TO RNA TO PROTEIN   209

DNA
mRNA
Protein
Transcription
Translation

FIGURE 10.2 ▪ **Path of Information Flow in a Cell: DNA to RNA to Protein.** Enzymes transcribe DNA into a messenger RNA molecule; ribosomes translate the mRNA into the polypeptide chain of a protein.

enzymes to replace those blocked up in the pancreas by clogged ducts. Despite these symptomatic treatments, nearly half of the affected individuals fail to survive to age 20. That's why the prospect of gene therapy for cystic fibrosis is so important.

How does a mutation in a single gene cause defects in so many systems (the lungs, pancreas, sweat glands, sperm ducts)? What makes one human gene out of approximately 100,000 so important? And how can physicians combat the effects of the deadly gene more effectively? The answers to these questions are crucial to the health of about 50,000 young people in the United States and are based on the principles of gene action discussed in this chapter.

We know that each gene specifies the synthesis of one polypeptide (see Chapter 9), so it is not surprising that a disease like cystic fibrosis—the result of a single defective gene—results from errors in the synthesis of a single protein. In cystic fibrosis, the protein is called CFTR (for cystic fibrosis transmembrane regulator). This protein is embedded in the cell's plasma membrane and probably transports chloride ions (Cl⁻) out of the cell. Sufferers of cystic fibrosis make a faulty CFTR protein, and so chloride transport is abnormal. Our broad goal in this chapter is to discuss the links between genes and proteins, whether normal or abnormal, and the cause of devastating inherited diseases such as cystic fibrosis.

Three unifying themes will emerge as we follow the processes by which a mutation in a single gene results in a defective protein and ultimately in a range of defects in the individual organism. The first theme is that information flows from DNA to RNA to protein (Figure 10.2) and from proteins to the building of an organism's phenotype. The gene for the CFTR protein in a healthy person, for example, is normally transcribed into an RNA molecule, which is then translated into CFTR, and this, in turn, regulates normal ion passage through cell membranes, and the person does not suffer the symptoms of cystic fibrosis. A second unifying theme is that protein synthesis requires a large expenditure of energy, and cells have evolved ways that minimize that energy cost by regulating *gene expression*—the translation of genetic information into proteins. A final theme is that basic genetic mechanisms are essentially universal: All creatures, from bacteria to field mice to flowering dogwood, share the same approach to protein synthesis.

This chapter will answer several important questions about how genes function, controlling protein shape and activity and ultimately the activities of living things:

■ How does information flow from DNA to RNA to proteins?
■ How can even slight alterations in DNA lead to diseases like cystic fibrosis?
■ What determines when and where the information in a gene will be used in a cell?

Unifying Chapter Themes

◀ ◀ ◀ ◀ ◀ ◀ ◀ ◀ ◀ ◀ ◀ ◀ ◀ ◀ ◀ ◀ ◀

Three or four thematic statements in the central example introduce the student to the main topics of the chapter.

Advance Organizer

◀ ◀ ◀ ◀ ◀ ◀ ◀ ◀ ◀ ◀ ◀ ◀ ◀ ◀ ◀ ◀

The chapter's specific objectives are framed as questions to stimulate critical thinking skills.

BOX 9.1   PERSONAL IMPACT

THE HUMAN GENOME PROJECT

In October of 1990, the journal *Science* included a large foldout wall chart printed with figures that resemble military decorations and candy-striped worms. The poster was, in fact, an up-to-date map of the human genome, placing genes of known function and location, regions where the nucleotide sequence has been worked out, and other genetic data in their correct positions on blown-up, diagrammatic versions of the human's 22 chromosomes plus *X* and *Y* (Figure 1). By looking at the bands of yellow, orange, blue, and magenta, one can interpret how complete—or more accurately, how incomplete—the mapping effort was at that time.

Although molecular geneticists had by then already spent over five years and $100 million of federal research funds in what will eventually be the largest scientific project ever undertaken, there are far more empty spaces on the human gene map than data to plug in. Some have labeled the Human Genome Project the ultimate measure of humankind—an effort that could revolutionize medicine, biology, and psychology. Noting that only a tiny fraction of the map is completed, however, critics wonder whether the benefits can possibly outweigh the tremendous costs.

First suggested in 1985, the Human Genome Project will cost more and take longer than building the atom bomb or landing people on the moon. Hundreds of researchers will labor simultaneously (1) to create a low-resolution physical map that places the 1000 or so currently known human genes in the proper positions on the 23 pairs of chromosomes; (2) to locate and identify the 50,000 to 100,000 additional genes (some 4400 of which occur on an average chromosome); and (3) to determine the nucleotide sequences of all 3 billion bases in the human DNA—even sequences that don't code for proteins or that simply repeat each other.

The project will cost at least $3 billion and won't be completed before the year 2005. The result will be a 500-volume encyclopedia of human genetics spelling out the detailed instructions for every protein in the body, and with them, a high-resolution view of how these proteins function normally day to day and perhaps participate in mental and physical illness. Collectively, humans have more than 4000 hereditary diseases, including sickle-cell anemia, cystic fibrosis, and Huntington's disease. These three and a few hundred others can be traced to single-gene mutations. Most diseases, however, including cancer, diabetes, Alzheimer's, and heart disease, probably have roots in multiple genes, and the interactions between them will be challenging to work out, even with a genome map.

To some critics—including 23 Harvard professors who published a dissenting view of the Human Genome Project in *Science* shortly before the 1990 map appeared—the promised high-resolution view is a mixed blessing. Knowing the complete nucleotide sequence, wrote Bernard Davis and colleagues, would be "like viewing a painting through a microscope," with researchers having to "plow through 1 to 2 million 'junk' bases

FIGURE 1 ■ **The Human Genome Map.**

before encountering an interesting sequence"* and then trying to discover its unknown function. Even senior statesman of molecular biology Sydney Brenner (see Box 7.2 on page 155) jokes that working out nucleotide sequences is so tedious that a penal colony could hand out sequencing projects as prison sentences. And everyone from the critics to Nobelist James Watson, who is heading those parts of the project overseen by the National Institutes of Health, agree that only 3 percent of the bases in the total sequence code for proteins of any kind.

Other major criticisms are financial and ethical. The last decade or so has seen major cuts in funding for biomedical research. The money allocated for Genome Project work in 1991 alone could provide sizable ($200,000) research grants to over 750 investigators. Critics also worry that a human gene map could lead to genetic discrimination, since doctors, insurance companies, and employers could "read" your inherited tendencies toward heart disease, cancer, or dementia, let's say, even though you have no symptoms now and may never have them.

Despite these legitimate concerns, many scientists are convinced that the benefits will far exceed the risks and point to the wealth of basic information already unlocked from DNA sequencing in viruses, bacteria, plants, and other animals. James Watson, probably the Genome Project's most eloquent spokesman, sees it this way: "When finally interpreted, the genetic message encoded within our DNA molecules will provide the ultimate answers to the chemical underpinnings of human existence."†

---

*Bernard D. Davis et al., "The Human Genome and Other Initiatives," *Science* 249 (1990): 342–343.
†James D. Watson, "The Human Genome Project: Past, Present, and Future," *Science* 248 (1990): 44–49.

Boxed Essays

◀ ◀ ◀ ◀ ◀ ◀ ◀ ◀ ◀ ◀ ◀ ◀

These stimulating accounts explore the scientific search for answers, the people behind the results, and how biology influences our own lives and our environment. There are four kinds of boxes: How Do We Know?; Biology: A Human Endeavor; Personal Impact; and Focus on the Environment.

Emphasis on the Environment

▶ ▶ ▶ ▶ ▶ ▶ ▶ ▶ ▶ ▶ ▶ ▶ ▶ ▶ ▶

An increased coverage of environmental science helps the student make a connection between the principles of biology and the world in which he or she lives. Here, both text and photo demonstrate how anaerobic respiration relates to the balance of nature.

FIGURE 5.11 ■ **Environmental Carbon Cycling: Anaerobic Respiration Plays a Key Role.** Hikers pausing for a rest by a clear mountain lake see only the sparkling upper layers of water, where photosynthesis in the tissues of algae and green plants fixes carbon and releases oxygen into the water. When water plants and oxygen-breathing fish die, their bodies sink to the lake bottom. These deep layers are colder and soon become depleted of oxygen, as oxygen-utilizing microorganisms begin to decompose the debris. Once the oxygen is fully depleted, the microorganisms switch to anaerobic metabolism and complete the decomposition of the organic matter to carbon dioxide and often methane gas, which hikers can sometimes see bubbling up to the surface. Without this anaerobic recycling, most of the carbon would be tied up in dead plants and animals.

During fermentation in muscle cells, yeast cells, and certain other kinds of microorganisms, the pyruvate molecules are degraded to wastes such as alcohol, carbon dioxide, and lactic acid, while the NADH is recycled to NAD^+. The energy yields of these two metabolic pathways may seem meager—just two ATPs per glucose. However, anaerobic metabolism (energy breakdown in the absence of oxygen) is crucial to the global recycling of carbon and the stability of the environment.

Organic matter from dead leaves, dead microorganisms, and other sources often sinks into an environment devoid of oxygen, such as the soft layers at the bottom of lakes or oceans (Figure 5.11). If it weren't for anaerobic decomposers—organisms capable of breaking down organic matter via anaerobic metabolism—most of the world's carbon would eventually be locked up in undecomposed organic material in these oxygen-poor environments, and deeper and deeper layers would build up. As a result, there would be too little carbon dioxide available as a raw material for photosynthesis, plants would be unable to generate new glucose molecules, and neither plants nor animals would survive.

Although simple and microscopic, anaerobic decompos-

ers clearly play an enormous role in the balance of nature. The vast majority of life forms, however, expire fairly quickly without oxygen. But their dependence on oxygen is based on a form of metabolism that provides an energy harvest 18 times greater than that provided by anaerobic metabolism.

AEROBIC RESPIRATION: THE BIG ENERGY HARVEST

While muscle cells as well as yeasts and other decomposers have the ability to metabolize sugars even when oxygen is unavailable (for a while, at least), the vast majority of living things, from euglenas to elephants, roses to redwoods, are made up of cells that require oxygen for their major metabolic pathway. Known as aerobic respiration, this pathway shunts the products of glycolysis through the Krebs cycle and harvests energy in the electron transport chain. Being 18 times more productive than glycolysis alone, this extended pathway can provide the huge amounts of ATP an active cell needs, and its superiority is clearly demonstrated: One out of every 500 yeast cells has a mutation, or change, in its DNA that prevents it from carrying out aerobic respiration even when oxygen is present. These so-called *petite* yeasts get by solely on glycolysis and fermentation, but they grow two to three times more slowly than their normal counterparts.

The reason for the greater efficiency of aerobic respiration is revealed by the overall equation for aerobic respiration:

$$C_6H_{12}O_6 + 6\ O_2 + 36\ ADP + 36\ P_i \rightarrow 6\ CO_2 + 6\ H_2O + 36\ ATP$$

The initial glucose molecule is broken down completely to inorganic waste products that are not energy-rich; thus, most of the energy residing in the molecular bonds of the sugar is released and a large proportion of it stored as ATP—36 ATPs per glucose, to be precise.

Aerobic respiration has two phases that use the products of glycolysis. The first phase is known as the Krebs cycle, named after the scientist who worked out its reactions, Sir Hans Krebs. The **Krebs cycle** of reactions cleaves the carbons from pyruvate, releases them as carbon dioxide, and stores energy in reduced carrier molecules. The second phase, the **electron transport chain**, then strips the electrons from the reduced carriers (Figure 5.12). The flow of electrons down this transport chain creates a current that is then used to build the cellular currency ATP.

The Krebs Cycle: Metabolic Clearinghouse

As in fermentation, the two pyruvate molecules produced from each glucose during glycolysis are the raw material for the Krebs cycle. Unlike fermentation, however, the important events of the Krebs cycle take place not in the cytoplasm but in the mitochondria of eukaryotic cells (or on the plasma membranes of bacteria). Recall from Chapter 3 that the mito-

Complex Topics Made Relevant

Many topics, such as cell and molecular biology, are made relevant through the use of health and ecological applications. Here, both text and photo demonstrate the relationship between the endoplasmic reticulum and human health.

FIGURE 3.18 ■ **Some Bedouin Women Have a Smooth ER Problem.** Because this woman's clothing leaves little or no skin exposed to sunlight, her smooth ER may not be able to make enough of the vitamin D necessary to maintain strong, healthy bones.

African women of one Bedouin tribe who wear dark, full-length garments get very little exposure to sunlight. As a result, the regions of smooth ER in their cells are often unable to convert enough cholesterol to vitamin D, and their bones grow soft and weak (Figure 3.18).

Another part of the endoplasmic reticulum is studded with ribosomes, looks rough under the microscope, and is called the **rough ER.** This region is involved in the synthesis of certain proteins. For example, the rough ER helps produce the enzymes that digest ice cream and most of the other foods you eat. Tracing the production and transport of an enzyme through this membrane system helps us understand how the ER functions. Let's focus on the rough ER in cells within a person's pancreas, a cucumber-shaped organ that generates digestive enzymes and secretes them into a duct leading to the intestine, where food is broken down and digested.

Within a pancreatic cell, the nucleus makes a special RNA called *messenger RNA* that carries genetic information (in this case, information for the digestive enzyme) out of the nucleus and into the cytoplasm (see Figure 3.15). Once in the cytoplasm, the messenger RNA joins the small, beadlike ribosomes—biochemical anvils on which protein molecules will be forged. The ribosomes then stick to the surface of the rough ER—the reason, in fact, that it appears rough. Proteins like these digestive enzymes that are assembled on rough ER enter the ER cavity, are modified as they move along through the channels, and are eventually pinched off in little sacs, or vesicles.

Most of the sacs pinched off from the endoplasmic reticulum enter another membrane system, the **Golgi apparatus,** where the digestive enzymes or other proteins in the sacs are further modified. The modified proteins then leave the Golgi apparatus headed either for the cell's plasma membrane, for tiny digestive sacs within the cell (the lysosomes), or for export from the cell (see Figure 3.17). When exported from the pancreatic cells, the digestive enzymes move down a duct and into the intestine. There they can break down the fats, sugars, and other nutrients present in the ice cream.

The Golgi apparatus is a key component of the membranous organelles along the enzyme secretion pathway. Named after the Italian cell biologist who first spotted it, the Golgi apparatus is like a packaging department for the eukaryotic cell. In 1898, Camillo Golgi observed this apparatus in the nerve cells of a barn owl, but it was not until the availability of electron microscopes over 40 years later that biologists could see the structure clearly. Each Golgi apparatus is a stack of saucer-shaped, baglike membranes surrounded by small, round, membranous containers, or **vesicles** (see Figure 3.17c). A cell can have just one Golgi apparatus or many thousands. Golgi-packaged proteins and lipids repair the plasma membrane itself when it is damaged; and in plants, the Golgi apparatus (called a *dictyosome* by botanists) packages for export the precursors to the cellulose that forms the outer cell wall.

■ **Vesicles in the Cytoplasm** While many vesicles that pinch off a Golgi apparatus leave the cell, two main types of vesicles—lysosomes and microbodies—take up permanent residence in the cytoplasm. **Lysosomes** are spherical vesicles

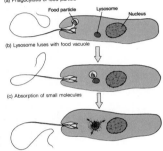

(a) Phagocytosis of food particle

(b) Lysosome fuses with food vacuole

(c) Absorption of small molecules

FIGURE 3.19 ■ *Euglena* **Engulfing a Food Particle.** (a) Phagocytosis. When a *Euglena* cell happens upon a food particle of the right size and composition, the cell membrane forms a pocket around it and engulfs it. (b) Fusion. A lysosome then fuses with the food vacuole, and digestive enzymes break down the food particle. (c) Absorption. Small nutrient molecules then pass through the lysosome membrane into the cytoplasm and nourish the cell.

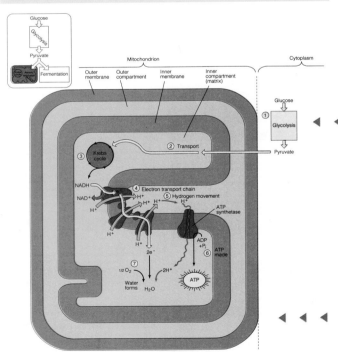

FIGURE 5.16 ■ **Mitochondrion: Site of the Krebs Cycle and Electron Transport Chain.** This diagram shows where the phases of aerobic respiration take place in the mitochondrion. Pyruvate molecules (1) generated during glycolysis in the cytoplasm are transported through the outer and inner mitochondrial membranes to the inner compartment (matrix) (2), where the Krebs cycle takes place (3). Then electrons from energy carriers are passed down the electron transport chain (4), and proteins use this released energy to pump hydrogen ions across the inner membrane (5). Finally, these hydrogen ions flow through the protein ATP synthetase back across the inner mitochondrial membrane, and the protein traps energy in ATP (6) much as a turbine captures the energy of water flowing over a dam and converts it to electricity. The final acceptor of the electrons is oxygen (7), which joins hydrogen ions and electrons and forms water.

Extensive Use of Icons

Orienting diagrams place structures and processes into a physical context for the student.

Extensive Use of Process Diagrams

These figures depict sequential biological events with individual steps numbered and keyed to step-by-step discussions in the text or figure legend.

226　　　　PART TWO　PERPETUATION OF LIFE

systems for engineered genes, but before long, medical workers will be applying the basic biology of transcription, translation, and gene regulation described in this chapter to diseases like cystic fibrosis and sickle-cell anemia and helping to improve the quality of life for millions of sufferers around the world.

CONNECTIONS

In a very real sense, the possibility of helping people with cystic fibrosis and other genetic diseases stems directly from life's basic unity at the level of genes and molecules. Mendel initially discovered genes in plants; later experimenters learned that genes are made of DNA by studying viruses; research on fungi supported the one gene–one enzyme concept; and studies with bacteria elucidated the mechanisms of transcription, translation, and gene regulation. All these principles are being applied now to helping people with genetic diseases, and the future uses of this knowledge are potentially limitless. Few genetic applications would be possible, however, without our modern techniques for manipulating genes, which grew out of a remarkable episode in scientific history that began in the late 1970s. This ability led to a revolution in biological engineering that is still going on and is the subject of Chapter 11.

Highlights in Review

1. Genes work by specifying the base sequence of mRNA, which specifies the amino acid sequence of a polypeptide. Information in DNA is copied to RNA through a process called transcription, and information in RNA is transferred to protein through a process called translation.
2. Messenger RNA (mRNA), transfer RNA (tRNA), and ribosomal RNA (rRNA) differ from DNA in that they are made of ribonucleotide subunits rather than deoxyribonucleotide subunits. A single oxygen atom on the sugar ribose differentiates the two kinds of nucleotides. RNA consists of only a single strand of nucleotides, not a double strand, as in DNA; and the RNA base uracil replaces the DNA base thymine. RNA molecules are also generally much shorter than DNA molecules because they tend to code for only one or a few genes rather than for the whole genome.
3. Like replication, transcription is directed by base pairing and carried out by a polymerase enzyme. But in transcription, ribonucleotide bases pair with the deoxyribonucleotide bases on the DNA strand, and the enzyme is RNA polymerase rather than DNA polymerase. During transcription, a few genes may be copied thousands of times, while during replication, the whole genome is copied just once. The newly formed RNA strand separates from the DNA molecule in transcription, instead of joining with it as does a newly replicated DNA strand.
4. The genetic code, transfer RNA, and ribosomes together bring about translation. The genetic code is identical in nearly all organisms. A group of three bases is a codon. Except for the start codon and the three stop codons (which tell RNA polymerase where to begin and end translation), each codon specifies one amino acid. Several different codons may specify the same amino acid, but no codon specifies more than one amino acid.
5. The tRNAs translate the mRNA codons into an amino acid sequence. At one end of each tRNA molecule is an anticodon; at the other end is a specific amino acid. The anticodon pairs up with the mRNA codon, so that the amino acids are ordered according to the mRNA codon sequence.

6. Ribosomes hold mRNAs, tRNAs, and amino acids in place until the amino acids can be joined together into a polypeptide. Each mRNA may have several ribosomes running along it at once, all translating polypeptides off the same mRNA.
7. A change in the base sequence, a mutation, may consist of a base substitution, insertion, or deletion. Usually, DNA repair enzymes detect and fix altered DNA; if these enzymes fail, a permanent mutation occurs. A change in the sequence of DNA bases results in a change in the sequence of RNA bases, which can result in a change in the sequence of amino acids—an altered polypeptide.
8. Mutations in the sequence of base pairs in DNA can result in genetic defects or cancer.
9. Cells limit the production of unnecessary proteins by regulating gene activity. They can accomplish this by regulating transcription, modifying the mRNA after transcription, regulating translation, modifying the polypeptides after translation, and adjusting protein activity itself. The most efficient level at which to regulate—and the most common—is transcription.
10. A classic example of transcription-level control in prokaryotes is the lactose operon of *E. coli*. In the absence of lactose, a repressor protein binds to an operator sequence (a special site on the DNA) and so prevents transcription of genes for lactose-digesting enzymes. In the presence of lactose, the repressor protein binds to the lactose instead of to DNA. RNA polymerase then transcribes the structural genes that code for enzymes that digest lactose.
11. Prokaryotes change the kinds of proteins they make both often and rapidly, which accommodates changes in their environment. They achieve flexible gene regulation dependent on environmental conditions by using transcription-level control, by transcribing and translating simultaneously, and by destroying mRNAs almost as fast as they are made.
12. All cells in eukaryotic organisms have the same genes, but some cells use one set of genes while others use another set. These differences arise through the developmental process called cell differentiation. How eukaryotes achieve this cell differentiation is one of the great mysteries of biology.

Connections

◄ ◄ ◄ ◄ ◄ ◄ ◄ ◄ ◄ ◄ ◄

A short reflection on key themes and chapter concepts ties these to the chapter that follows.

Highlights in Review

◄ ◄ ◄ ◄ ◄ ◄ ◄ ◄ ◄ ◄

A point-by-point recap of the chapter's main points aids review.

CHAPTER 10　HOW GENES WORK: FROM DNA TO RNA TO PROTEIN　　　227

13. Unlike prokaryotic genes, eukaryotic genes that are regulated simultaneously are not necessarily located near each other; they may even be on different chromosomes. Like prokaryotic genes, eukaryotic genes may be flanked by regulatory sequences of DNA that are responsive to gene regulators made of protein.
14. In eukaryotes, genes are clustered according to their historical relatedness in multigene families. The hemoglobin gene family in humans encodes several structurally similar globins, some of which are expressed at different times in development.
15. Our increasing understanding of genetics gives us hope that we may one day be able to help victims of genetic defects.

Key Terms

anticodon, 214	mutation, 216
chromosomal mutation, 217	operon, 222
codon, 212	protein synthesis, 215
elongation, 216	reading frame, 213
exon, 225	ribonucleotide, 210
gene mutation, 217	ribosomal RNA (rRNA), 210
gene regulation, 219	RNA polymerase, 211
genetic code, 212	start codon, 213
initiation, 216	stop codon, 213
intron, 225	termination, 216
message, 211	transcription, 210
messenger RNA	transfer RNA (tRNA), 210
(mRNA), 210	translation, 210

Study Questions

Review What You Have Learned

1. What is the difference between transcription and translation?
2. How do RNA and DNA differ?
3. During transcription, what will the order of the bases in mRNA be if the base sequence in DNA is CTAGCT?
4. Refer to the codon dictionary in Figure 10.6 and give the possible codons for the following amino acids: lysine; phenylalanine; glycine.
5. Explain why the genetic code is identical in humans, bacteria, and other organisms.
6. Assume that there is a CAG anticodon on a tRNA. Which mRNA codon does the anticodon pair up with? Which amino acid is attached to this tRNA?

7. What are three causes of gene mutation?
8. Explain the functioning of a bacterial operon.
9. Explain the statement, "You can take the cell out of the liver, but you can't take the liver out of the cell."
10. Compare the basic features of gene regulation in prokaryotes and eukaryotes.

Apply What You Have Learned

1. How could a scientist use the Ames test to determine whether the chemicals in a new hair dye are carcinogenic?
2. Although Jacob and Monod received the Nobel Prize for their elucidation of the operon model, that model does not apply to a horse or a maple tree. Why not?
3. A mutation occurs in a human cell that changes the anticodon of a transfer RNA from AUG to AUU. The tRNA carries tyrosine. What will be the likely consequences?

For Further Reading

ALBERTS, B., D. BRAY, J. LEWIS, M. RAFF, K. ROBERTS, and J. D. WATSON. *Molecular Biology of the Cell.* 2d ed. New York: Garland, 1989.

DAVIES, K. "The Search for the Cystic Fibrosis Gene." *New Scientist,* October 21, 1989, pp. 54–58.

FLINT, J. A., V. S. HILL, D. K. BOWDEN, S. J. OPPENHEIMER, P. R. SILL, S. W. SERJEANTSON, J. BANA-KOIRI, K. BHATIA, M. P. ALPERS, A. J. BOYCE, D. J. WEATHERALL, and J. B. CLEGG. "High Frequencies of Alpha-Thalassaemia Are the Result of Natural Selection by Malaria." *Nature* 321 (1986): 744–750.

ROBERTS, L. "Cystic Fibrosis Corrected in Lab." *Science* 249 (1990): 1503–1504.

SCHULMAN, L. H., and J. ABELSON. "Recent Excitement in Understanding Transfer RNA Identity." *Science* 240 (1988): 1591–1592.

STEPHENS, J. C., M. L. CAVANAUGH, M. I. GRADLE, M. L. MADOR, and K. K. KIDD. "Mapping the Human Genome: Current Status." *Science* 250 (1990): 237–244.

WEISS, R. "Cystic Fibrosis Treatments Promising." *Science News* 139 (1991): 132.

WEISS, R. "Upping the Antisense Ante: Scientists Bet on Profits from Reverse Genetics." *Science News* 139 (1991): 108–109.

Key Terms

▶ ▶ ▶ ▶ ▶ ▶ ▶ ▶ ▶ ▶ ▶ ▶ ▶ ▶ ▶ ▶ ▶ ▶

A list of the chapter's boldface vocabulary terms provides page numbers for easy reference and review.

Study Questions

▶ ▶ ▶ ▶ ▶ ▶ ▶ ▶ ▶ ▶ ▶ ▶ ▶ ▶ ▶ ▶ ▶ ▶

- *Review What You Have Learned*
  Short-answer questions aid in reviewing the chapter's main concepts.
- *Apply What You Have Learned*
  Thought-provoking questions encourage the student to relate the concepts of the chapter to other potential applications.

For Further Reading

▶ ▶ ▶ ▶ ▶ ▶ ▶ ▶ ▶ ▶ ▶ ▶ ▶ ▶ ▶ ▶ ▶ ▶

Up-to-date references encourage reading beyond the text material.

THE NATURE OF LIFE

The Nature of Life: An Introduction

The Courtship of a Scarlet Frog

Several times a year, in the lush, misty cloud forests of Central America, an unusual mating ritual takes place. The partners are male and female strawberry frogs—small, flamboyantly scarlet animals with deadly poisonous skin glands (Figure 1.1). As a result of their union, they produce and care for a new generation of strawberry frogs.

The courtship begins when the male issues a harsh, tinny love song from a perch such as a moss-covered tree root or a fallen log. Most of the world's frogs are well camouflaged in greens and browns. The brilliantly colored strawberry frog, however, sits out in plain view, uttering his "tick-ticking" call, loudly advertising his whereabouts and availability to friend and foe alike. His crooning attracts a nearby female with ripe eggs in her body and brings her hopping slowly toward him through the leaf litter. He jumps down from his perch and, like a scarlet pied piper, leads her to a rolled-up dry leaf, chirping loudly whenever she lags behind. Eventually, he crawls into the leafy boudoir, and his mate follows.

For nearly half an hour, the female rubs the male's head and chin with her snout while he calls softly. Periodically, she wheels about and presents her vent (urogenital opening) to him. Finally, the male spins around too, their vents touch, and he releases sperm onto the now moistened leaf as the female deposits half a dozen eggs in a mass of clear jelly. The ritual complete, the male hops onto a large green leaf and calls loudly as the female disappears quietly into the forest.

FIGURE 1.1 ▪ **Strawberry Frog in a Cloud Forest.** *Dendrobates pumilio,* a tiny resident of the Central American jungle, has flaming red skin. The frog carries a newly hatched tadpole on its back to a pool of water trapped in a leaf. This behavior ultimately helps ensure the survival of new generations of strawberry frogs.

■ **Organization and Reproduction Combat the Tendency Toward Disorder and the Certainty of Death.** A sugar maple seedling, pushing upward through dying fallen leaves, symbolizes the collection of energy, the maintenance of order, and the reproduction of the species despite the universal tendency toward disorder and the eventuality of death.

For the next 10 to 12 days, the gaudy male returns daily and wets the fertilized eggs, a behavior that keeps the eggs from drying out. Soon the eggs hatch, and dark, wriggling tadpoles—all head, mouth, and slowly whipping tail—emerge with yolk from the egg still filling much of their bodies.

The mother now returns, and the tadpoles wriggle up her legs onto a sticky patch of mucus on her back. The vibrant red female then delivers her tadpoles to a pool of clear water trapped in a tropical plant and returns weekly to feed her young. At her arrival, each tadpole communicates its presence by stiffening its body and rapidly vibrating its tail. To feed her young, the mother deposits a few infertile eggs, and the tadpoles consume their contents. After six weeks, the strawberry tadpoles sprout arms and legs, lose their tails, and emerge from the water as brilliant red adult frogs, ready to mate, brood, and care for new young.

Each distinguishing trait of the strawberry frog is unusual among frogs: the dazzling coloration; the piggyback transport of tadpoles; the feeding of tadpoles with sterile eggs; the communication between mother and young; and the protective poison glands in the skin. Yet these uncommon traits can also exemplify some of the universal themes that we will encounter again and again as we explore **biology,** the science of life. This chapter introduces the five most important biological themes: the role of energy in life; the perpetuation of life through reproduction and development; evolution, or biological change over time; the interaction of organisms and their environment; and the process of science as a way of learning about the natural world. We will see in this first chapter how these themes provide a framework for understanding biology. These same themes will reappear throughout the book.

We will also discuss fascinating historical experiments as well as ongoing explorations of still puzzling phenomena to discover not just what biologists know, but also how they learn about living systems. We will consider many of the challenges now facing our planet and its organisms, and we will show how biological science contributes to medicine, agriculture, industry, and preservation of the environment. Finally, we will see how life science enriches people's health, well-being, and appreciation for the world around them.

Chapter 1 and each chapter that follows will address several questions, which we will list first to show the general organization of the chapter at hand. Here are the questions for this chapter:

■ How do the themes of energy, reproduction, evolution, and environment help us understand the nature of life?
■ What characteristics define a living thing?
■ How do living things use energy in maintaining their complex organization?
■ What are some general features of reproduction?
■ How do species evolve new ways to obtain and use energy and to reproduce?
■ How do people learn about the living world and the functioning of individual organisms?
■ How can biological science help people solve some of the world's current problems?

TO BE ALIVE: UNIFYING THEMES

What is a living thing? When is it alive or dead? What, in fact, is the nature of life? These questions have never been more important, because today's biologists and physicians have unprecedented abilities to sustain the human body and individual organs on life-support machines; to freeze human embryos for later use; and to change and merge hereditary traits of microbes, plants, and animals. Perhaps, one day, this list will extend to generating life in the test tube and to creating hybrids between machines and living things. In the broadest sense, biological science probes the question, *What is life?*, and it takes this entire book to provide a definition.

As this fascinating story unfolds, the five themes of the book—energy, perpetuation, evolution, environment, and the process of science—will occur again and again. In this chapter, the themes help organize a set of nine observable characteristics that together define life (Table 1.1). The first four

TABLE 1.1   Characteristics of Life

Themes and Life Characteristics	Explanation
Energy and Organization	
Order	Each structure or activity is in a specific relationship to all other structures and activities
Metabolism	Organized chemical steps break down molecules and convert them into products that build body parts or make energy available
Motility	Using their own power, organisms propel themselves or their body parts through space
Responsiveness	Organisms perceive the environment and react to it
Perpetuation of Life	
Reproduction	Organisms give rise to others of the same type
Development	Ordered sequences of progressive changes result in increased complexity
Genes	Organisms have units of inheritance that control physical, chemical, and behavioral traits
Evolution	
Evolution	Through evolution, species acquire new ways to survive, to obtain and use energy, and to reproduce
Adaptations	Specific structures and behaviors suit life form to its own environment

traits relate to the use of energy and the maintenance of organization. The energy-related characteristics of life include *order,* a precise arrangement of structural units and activities; *metabolism,* the chemical breakdown, conversion, and use of energy-rich compounds; *motility,* the self-generated motion of an individual or its parts; and *responsiveness,* the tendency of a living thing to sense and react to its surroundings. Without these trademarks, most life forms would perish rapidly.

The next three life characteristics relate to the perpetuation theme, the persistence of groups of like organisms over time, even though individuals die. Parents give rise to offspring in the life characteristic we call *reproduction;* each offspring undergoes *development,* an orderly sequence of physical and behavioral changes during that organism's life cycle; and each living thing has units of inheritance called *genes,* passed from parent to offspring, that control many daily functions.

Finally, living things are adapted, and they evolve. *Adaptations* are structures and activities that allow an organism to make better use of its environment. And populations of living things *evolve,* or adjust to environmental variations through biological changes over time. Ultimately, evolution is the central unifying theme in all biology, and it is closely tied to the interactions between organisms and their environment. Evolution provides new ways for organisms to use energy, maintain order, and perpetuate themselves. But always, these processes must operate within the resources and restrictions of the physical setting.

Many nonliving entities have one or more of these nine characteristics of life; for example, waves move, flames use energy, and crystals grow. Only living organisms, however, display *all* the characteristics at some point during their individual life cycle or species history. Throughout this chapter and the rest of the book, you will find hundreds more examples of how living things display the collective traits that distinguish a rock from a rockrose and water from a water lily.

Now let us consider the defining characteristics of life one by one and see how they allow organisms to maintain their internal organization and survive.

LIFE TRAITS INVOLVING ENERGY AND ORGANIZATION

Living things are highly ordered. Strawberry frogs, for example, have two large, dark eyes, muscles in certain places with certain dimensions that allow the legs to paddle and hop, and a dazzling red skin that covers the animal's exterior. As time passes, some of this organization starts to deteriorate. While climbing a tree one day in search of ants, a frog may slip and crash to the forest floor, puncturing its skin, bruising its legs, and becoming more disorganized.

This kind of wear and tear is not unique to life: Rocks, stars, rivers, and mountain ranges all tend to become more disordered as time goes by. You need only look as far as the

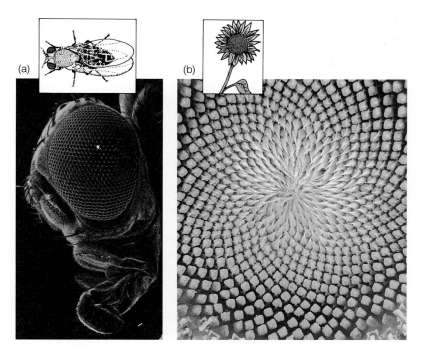

FIGURE 1.2 ■ **Order in Life and Nonlife.** (a) A fruit fly's eye and (b) the tightly packed seeds of a sunflower form geometric arrays that reveal the inherent order in living things.

nearest heap of papers on your desk and the accumulation of dirty clothes and dust balls in the corner to witness the tremendous tendency toward disorder. To combat a housekeeping disaster, you must expend energy. Likewise, in an injured frog, broken skin is replaced and wounded muscle heals when the animal's body extracts energy and materials from the insects it catches and eats, then uses the energy and materials to generate new skin and healthy muscle. In general, living things counteract the disorder that comes with time by taking energy and materials from their environment and employing them for maintaining order, growing, and other survival activities.

Life Characteristic 1: Order Within Biological Systems

The workings of a Swiss clock are marvelously intricate, and the parts of a race car are beautifully engineered for high performance. But living organisms possess a degree of **order**— a structural and behavioral complexity and regularity—far greater than anything in the nonliving world. The eye of a fly, for example, and the spiral-packed seeds of a sunflower head consist of highly organized units repeated and arranged in precise geometric arrays (Figure 1.2). In fact, the most highly organized structure yet discovered in the universe is the human brain, such as the one allowing you to read and understand this page.

While life's organization is obvious in structures such as eyes or flowers, it occurs at the microscopic level as well and at the level of large living groups in their environments. Consider, for example, the African savanna (Figure 1.3). The levels of biological order and organization it represents reveal a

FIGURE 1.3 ■ **Order Reigns at Every Level in the Living World.** An elephant herd in the shadow of Mt. Kilimanjaro on the African savanna symbolizes organization within organization. The elephants' bodies are made up of highly ordered cell parts, cells, tissues, organs, and organ systems that function together smoothly. The herd members are part of a population within a diverse community, including the birds picking insects off their thick hides and the grasses they graze and trample. The community, in turn, is part of the savanna ecosystem, with its expansive plain, snowcapped mountain range, and arid climate.

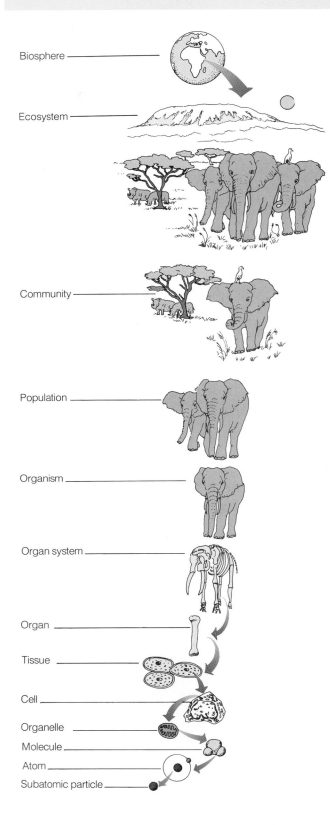

Biosphere

Ecosystem

Community

Population

Organism

Organ system

Organ

Tissue

Cell

Organelle

Molecule

Atom

Subatomic particle

FIGURE 1.4 ■ The Hierarchy of Life and Its Levels of Organization. As the text explains, elephants, as whole organisms, lie midway on a continuum of life's inherent hierarchy of order, stretching from subatomic particles to the biosphere.

fascinating *hierarchy of life* (Figure 1.4). The savanna is a collection of elephants, egrets, acacia trees, tussock grasses, clouds, and arid plains. Although it is vast and complex, it is but a small portion of the *biosphere:* our entire planet earth and all its living inhabitants. The biosphere consists of many *ecosystems:* living things in particular regions and their non-living physical surroundings. The savanna ecosystem includes elephants, the egrets that pick insects off their skin, and the coarse grass they chew and trample, as well as the water in the clouds, the sandy soil underfoot, and the hot African sunshine. The living part of an ecosystem is a *community:* an assemblage of interacting organisms living in a particular place—in this case, the plants, animals, and other organisms of the savanna.

Communities are made up of *populations:* groups of individuals of a particular type that live in the same area and actively interbreed with one another, such as an elephant herd or a field of grass. Populations are made up of *organisms,* the next lowest level of organization. Organisms are independent individuals that express life's characteristics. A tuskless, old female elephant, the matriarch and leader in this particular population, is a single organism (see Figure 1.3). Each organism, in turn, is made up of *organ systems:* various body parts arranged to carry out a particular general function within the organism. The skeletal system, for example, supports the elephant's body.

Moving further down the hierarchy, organ systems are composed of *organs:* units that perform specialized functions, such as a single bone that supports part of an elephant's leg or a single blade of grass that produces sugars for the grass plant.

Each organ, in turn, is made up of *tissues:* groups of similar cells. *Cells* are the simplest entities in the hierarchy that have all the properties of life listed in Table 1.1; they are the least complicated units that we can say are truly alive. Within cells are *organelles:* structures that perform specialized functions for the cell, such as the information-storing cell nucleus or the energy-harvesting mitochondrion. Organelles are made up of *molecules,* which are clusters of atoms; and *atoms* are the smallest units of matter that still have distinct chemical properties. Atoms, finally, are composed of *subatomic particles,* fundamental units of energy and matter.

The hierarchy of life is important because it identifies the various levels at which we can examine and understand biology. For example, to understand the significance of a strawberry frog's fiery coloration (organism level), a biologist wants to know how the skin makes its brilliant colors (organ system level and cellular level); how the bright colors affect the frog's reproductive activities (population level); and how these colors help the frog avoid being eaten (community level). Biologists delight in discovering new phenomena like the strawberry frog's piggyback tadpole transport or intricate mating ritual, and they enjoy discovering the origin and utility of such phenomena at various levels in the hierarchy of life.

Life's levels of organization provide a general outline for the inquiry in this book. Part One probes the nature of atoms and molecules, because all living things are built of them and are governed by the same rules of matter and energy. Part Two discusses the rules of heredity and development common to all organisms. Part Three describes the full range of living things, from bacteria to animals. Parts Four and Five explain how those organisms function. And Parts Six and Seven cover the highest levels of organization, the interactions of living things with one another and with their environment.

Life Characteristic 2: Metabolism

The intricate order within the body of a frog or the trunk of an acacia tree can only be maintained through the gathering and expenditure of energy. Once an organism has taken in energy-containing food molecules or has manufactured its own food using sunlight and gases, an organized series of chemical steps called *metabolism* breaks down the molecules. Other metabolic processes then rearrange these simpler molecules into products that are useful for repairing old structures or building new ones or unlocking energy necessary to transport and organize materials within the organism's body. Light shining on a palm leaf (Figure 1.5), for example, fuels a series of metabolic steps that causes the plant to construct high-energy compounds, move organelles around inside cells, construct new leaves, and break down worn-out cells and organs.

Life Characteristic 3: Motility

Energy unlocked by the steps of metabolism allows the movement of materials to areas where they are used to replace worn parts, help increase size, and generally combat disorganization. Self-propelled motion, or **motility,** is a diagnostic feature of life. Even organisms as simple as bacteria can move on their own toward concentrations of food particles. Organisms such as plants, which may not seem to move about, nonetheless have a constant streaming rotation of fluid within each cell. What's more, whole plant organs, such as leaves,

FIGURE 1.6 ■ Poetry in Motion: Animal Movement Can Be Swift and Elegant. These impalas, residents of savanna and woodland, cavort in the African sunshine.

usually track the sun's path through the sky each day. Animals, of course, have elevated movement to an art form in their pursuit of food, their escape from enemies, and their social relations (Figure 1.6).

Life Characteristic 4: Responsiveness

To maintain a well-ordered body, an organism must be responsive to its environment. It must exhibit **responsiveness,** detecting the presence of food or enemies, sun or cold, water or dry land, and then reacting in ways that help maintain body organization. Responsiveness can be instantaneous, as when a small moth hears the high-pitched whine of a swooping, hungry bat, or it can be gradual and seasonal, as when trumpeter swans detect the shortening days of autumn and respond by flying to wetlands in warmer southern regions. Plants, too, must respond to their environments, conserving water during times of drought and capturing sunlight when it is most readily available. And some plants even show instantaneous responses, reacting to external signals rapidly enough to trap an unsuspecting fly and digest it (Figure 1.7).

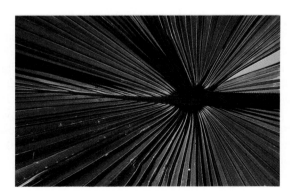

FIGURE 1.5 ■ Metabolism: Conversion of Energy and Materials. The spoked leaf of a fan palm acts as a solar collector. Leaf cells take in the sun's energy and trap it in a chemical form that metabolic processes employ for growth, repair, and movement—activities that help maintain order within the plant.

FIGURE 1.7 ■ Responsiveness: The Life Characteristics of Sensing and Reacting. This Venus flytrap has reacted to the featherlight step of a housefly by closing rapidly and imprisoning the hapless insect.

The life characteristics of order, metabolism, motility, and responsiveness involve energy and allow individuals to survive. Nevertheless, disorganization inevitably gets the upper hand as time passes. Individuals appear to be programmed to age and die, and eventually all do. Since life has continued to exist for billions of years, however, this tendency toward death has clearly been overcome by a second set of life characteristics.

LIFE CHARACTERISTICS RELATED TO PERPETUATION

Among the millions of species alive today, human beings, with our typical life span of three score and ten years, are fairly long-lived. A few other species, however, live for much longer periods. The classic examples are the 4000- to 5000-year-old bristlecone pine trees standing stalwart but gnarled with the wear and tear of millennia on windswept limestone ridges in the White Mountains of California and Nevada (Figure 1.8). Despite their tenure throughout much of human his-

FIGURE 1.8 ■ **Methuselah Tree. One of the earth's oldest living organisms stands twisted and tortured-looking on a slope in the White Mountains.**

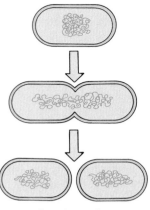

Daughter cells

FIGURE 1.9 ■ **Some Single-Celled Organisms Reproduce by Simple Division. Bacteria divide in two, and the daughter cells grow into new bacteria.**

tory, even these astonishing survivors will eventually die. To overcome the inevitability of dying, living entities generate similar copies of themselves. When metabolism, motility, and the other characteristics of life cease in the original, they continue in the "copies."

Life Characteristic 5: Reproduction

The mechanism by which organisms give rise to others of the same type is called **reproduction.** Reproduction can be asexual, as in the fission of a single-celled bacterium into two daughter cells (Figure 1.9). More complex organisms, however, usually undergo sexual reproduction, during which two specialized cells called the egg and sperm unite and form the first cell of a new individual. Many sexually reproducing organisms have curious reproductive adaptations. A male strawberry frog, for example, or a dazzling golden toad increases his chances of reproducing if he advertises his location and amorous intent by a distinctive song attractive to females of his species. In another example, tube-shaped scarlet gilias produce stores of energy-rich nectar, and both the flaming red flowers and the nectar attract hummingbirds (Figure 1.10). Each iridescent visitor solves its energy needs this way and inadvertently assists in the flower's reproduction. Its feathers pick up sticky pollen grains, which contain the flower's sperm, and when the bird flies to the next plant, these grains can reach the eggs of another flower and lead to the formation of a new plant generation.

Life Characteristic 6: Development

When an organism reproduces, the young always start out smaller and usually much simpler in form than the parent(s). The offspring then grow in size, increase in complexity, and

FIGURE 1.10 ■ **Botanical Want Ad: The Scarlet Gilia's Brilliant Red Tube and Sweet Nectar Attract Hummingbirds and Ensure the Plant's Reproduction.** This *Gilia aggregata* is visited by a broad-tailed hummingbird. The animal laps up nectar with a long tongue, gains energy, and inadvertently carries pollen grains to the next flower it samples.

become parents themselves in a process called **development,** a key characteristic of life.

Although the details vary considerably from one species to the next, consider the sequence of events in the development of a common frog. With the union of the father's sperm and mother's egg, a new individual begins as a single, very large cell (Figure 1.11a). The newly fertilized cell divides and redivides into many smaller cells, which rapidly organize themselves into an embryo and then a tadpole with eyes, internal organs, and tail (Figure 1.11b). The tadpole hatches and commences its main job—eating—then grows larger, sprouts legs, resorbs its tail, and finally matures into an adult frog (Figure 1.11c). Although an adult frog must also eat to survive, its job, in the broadest sense, is to reproduce, and various adaptations, including its voice, sex organs, and innate behaviors for mating and parental care, aid the reproductive process.

One of the most intriguing questions in all biology is how a fertilized egg cell develops into the millions of cells of various types that function as a viable organism. In other words, what makes an egg develop into a strawberry frog instead of a strawberry? The answer is found in the remarkable molecule of inheritance called DNA.

Life Characteristic 7: Heritable Units of Information, the Genes

Identical twins make a convincing argument that some kind of information must direct the development of each individual in a fairly precise manner so that two very similar organisms

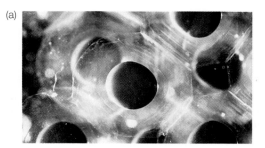

FIGURE 1.11 ■ **Development: A Sequence of Growth, a Consequence of Reproduction.** (a) A European common frog (*Rana temporaria*) begins as a fertilized egg cell. (b) After a few hours, divisions in that original cell produce an embryo. Development continues during metamorphosis, producing a tadpole, and the physical transformation is completed by the time (c) the adult forms.

result. When we contrast a set of twins with their different-looking brothers and sisters, however, we can see that there are also variations in hereditary information. The units of inheritance that guide heredity and control the development of specific physical, chemical, or behavioral traits in a living organism are the **genes.** Specific genes, inherited from parents, determine whether a strawberry frog has big, bright red splashes of color or small, pale pink patches, whether a scarlet gilia produces a lot of nectar or only a little, and whether a person has blue, brown, or hazel eyes.

Genes are made of a remarkable and beautiful molecule

called DNA (Figure 1.12). Each of your cells contains 46 of these twisting, threadlike molecules, and each DNA molecule contains on average about 2000 genes. While a DNA molecule may be transmitted from parent to offspring without change, variations occasionally arise in its structure. Such changes are called *mutations,* and the most famous mutation in history probably occurred in Queen Victoria of England about 150 years ago.

Normally, people possess a gene that helps blood to clot after a cut or bruise. Some of Queen Victoria's eggs, however, contained an altered gene that would not allow blood to clot and that led to a disease called hemophilia in males who inherited the gene. By intermarriage, Queen Victoria's mutated gene was passed throughout the royal houses of Europe, and it caused ten young nobles to bleed to death.

As in hemophilia, many mutations in DNA are harmful. But occasionally, mutations occur that allow an organism to combat disorganization and death in some new and better way. The gradual accumulation of such mutations over many generations tends to change organisms, and the amassing of such changes is called **evolution.**

LIFE CHARACTERISTICS RELATED TO EVOLUTION AND ENVIRONMENT

Life Characteristic 8: Living Things Evolve

Living organisms change over time. We know this partly from the fossilized imprints of early organisms (Figure 1.13). The older a fossil, the less similar it is to present-day forms, and this provides good evidence of the constant change in living species over millennia. Using fossils, analysis of DNA, and other kinds of evidence, biologists can trace the history of a group of organisms. All groups alive today are like the youngest branches of a family tree. For example, the strawberry frog (see Figure 1.1) and the "play-actor" frog (see Figure 19.11a) are closely related to each other and somewhat more distantly related to common frogs (see Figure 1.11c). Reaching farther

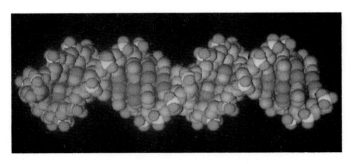

FIGURE 1.12 ■ **DNA and Genes. Multicolor computer graphic of the DNA molecule. DNA is the repository of genetic information for reproduction and cell function in all living cells.**

back in time, frogs and salamanders would share a common ancestor, and even farther back in the history of this lineage, frogs would be related to reptiles, fish, and marine worms. Tracing the evolutionary tree of life to ever more remote periods, all organisms would eventually share a common ancestor in the ancient past. This family tree concept helps explain the origin of species: All living things are descended from a common ancestor and arose as the result of genetic modification in species that lived before them in a process of change called *evolution.*

Life Characteristic 9: Living Things Are Adapted

As groups of organisms evolve, genetic modifications arise that allow them to cope with their environments in special ways. Different species, for example, have different ways of extracting energy and materials from their surroundings. Since these different methods adapt the organisms to their own special ways of life, they are called **adaptations.** One adaptation in an adult frog, for example, is a long gluey tongue folded over at the tip (Figure 1.14). When an ant or fly ventures near, the frog quickly unfurls its tongue, mires the victim in a sticky coating, retracts its tongue, and devours the morsel.

(a)

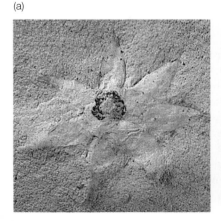

(b)

FIGURE 1.13 ■ **Floral Ancestor in Stone: Fossil Gentian Compared with Modern Flower.** (a) This 50-million-year-old fossil of a seven-petaled flower bears a strong resemblance to (b) a modern gentian. Biologists use such comparisons in the study of evolution.

FIGURE 1.14 ■ **Sticky Tongues: Feeding Adaptation in Frogs.** An adult frog has a long, sticky tongue that it flips forward with lightning speed to entrap insects. This adaptation allows the frog to extract energy and materials from its environment.

Not all adaptations relate to the intake of energy and materials. Some adaptations improve the organism's ability to grow; others, to reproduce, move, live in a group, or attract a mate. Some adaptations are truly wondrous and bizarre. Certain flowers, for instance, so closely resemble female wasps that male wasps "copulate" with them. Pollen clings to the insects' bodies, and as they move on to "conquer" new "females," they carry the pollen from one flower to the next. We will encounter many strange adaptations in this book. But no matter how strange they may seem, all adaptations evolve only over many generations and increase the individual's chances of combating disorganization or improve the species' chance of perpetuating into the future.

Adaptations and the broader theme of evolution so pervade the way biologists think about, understand, and explain life that an evolutionary perspective will appear in virtually every chapter that follows. The next section provides an overview of evolution, which can serve as a useful framework until we explore the science of evolution in greater detail in Chapter 33.

EVOLUTION: BIOLOGY'S CENTRAL THEME

Plunge with a mask and snorkel into the warm, blue waters off Hawaii and count the varieties of fish you see in just a few minutes (Figure 1.15)—perhaps butterfly fish, bird

FIGURE 1.15 ■ **Diversity Under the Waves: Rainbow of Fish on a Tropical Reef.** This coral reef is home to lavishly colored parrot fish, striped sergeant majors, and several other bright species. These represent only a few of earth's more than 30,000 fish species—a stunning variety that is itself only a small fraction of our planet's 5 to 10 million species of living things.

wrasses, striped Moorish idols with streaming back fins, or intensely turquoise parrot fishes like the one on the book cover. This same dazzling variety is evident in the colorful songbirds of eastern forests, the butterflies of midwestern weed fields, and the wildflowers of the Rocky Mountains. The earth, in fact, is covered with immensely varied communities of living species, all products of evolution.

Organizing Life's Dazzling Diversity

To help make sense of this vast array of different organisms—this *diversity of life*—biologists have created a system for categorizing organisms that separates living things into groups according to their similarities. Strawberry frogs, for example, are more similar to bullfrogs than to elephants, but frogs and elephants are more similar to each other than to tussock grass. To make sense of life's wonderful diversity, biologists divide organisms into *species,* sets of structurally similar individuals that all descend from the same initial group and that have the potential to successfully breed with one another. Several related and similar species may make up a *genus* (plural, genera). Just as a person has a family name and a given name, biologists refer to each species by its genus name followed by its species name. For example, the strawberry frog is called *Dendrobates pumilio,* while the closely related play-actor frog is called *Dendrobates histrionicus* (see Figure 19.11a); its color markings are different and it does not interbreed with *D. pumilio.*

Just as species are grouped into genera, genera are grouped into *families,* families into *orders,* orders into *classes,* and classes into *phyla* or *divisions.* In each successive grouping, similar organisms are arranged by broader and broader criteria. Ultimately, biologists assign the dozens of divisions and phyla—and the millions of life forms they include—into *kingdoms.*

The ancient Greeks divided all life into just two kingdoms—plants and animals. Biologists in the mid-nineteenth century added a third kingdom, the single-celled life forms. During the 1930s, biologists split the single-celled kingdom into monerans and protists. The kingdom *Monera* includes only simple, single-celled, microscopic organisms such as the bacteria that turn milk into yogurt or cause strep throat. Members of the kingdom *Protista* (some now call it Protoctista) are more complicated than bacteria and include amoebas and the protozoa that swim around in drops of pond water. In the late 1950s, taxonomists added yet another kingdom, *Fungi,* which includes some single-celled forms like yeasts and some multicellular forms like mushrooms, but which all absorb their energy and materials after decomposing other living or dead organisms or their parts.

The designation of the newer kingdoms allowed biologists to hone their definitions of plants and animals: Members of the kingdom *Plantae,* such as ferns and maple trees, are usually green and multicellular, and they generate their own food

from air, water, and sunlight. Members of the largest kingdom, *Animalia,* including insects, earthworms, snails, and people, are multicellular and get their energy by ingesting other organisms, alive or dead. Throughout this book, we will see hundreds of examples from the five kingdoms and explore how these organisms evolved, function, and survive.

This five-kingdom system remained the standard until the mid-1980s, when microbiologist Carl Woese of the University of Illinois proposed a sixth kingdom of very primitive bacteria-like organisms, which he called Archaebacteria. Included in this kingdom are species that live in a cow's gut and digest grass and others that survive in extremely hot, salty, or acidic environments, such as salt flats or smoldering mine tailings. By 1990, Woese had collected enough biochemical and genetic information on the major divisions of living things to propose a taxonomic category above the kingdom, called the *domain,* and to divide the entire tree of life into three domains, or branches: The first domain, Bacteria, includes many members of the old kingdom Monera, which Woese divides into a new fifth kingdom, Eubacteria, and a sixth kingdom, Archaebacteria. He places all archaebacteria into a new second domain, Archaea. The third domain, Eucarya, includes the complex organisms in the other four kingdoms, the plants, animals, protists, and fungi.

Throughout this book, we will use the new three-domain system and the lobed icon shown in Figure 1.16. We'll discuss Bacteria and Archaea in Chapters 15 and 16, and we'll break the third domain, Eucarya, into its four kingdoms and cover them in Chapters 17 through 19. There is so much biochemical diversity in Bacteria and Archaea that Carl Woese predicts they will eventually be divided into at least eight or more separate kingdoms. Significantly, he sees the simple, ancient Archaea as genetically more similar to plants and animals.

All these kingdoms, domains, and other categories are helpful for understanding evolutionary relationships between the various types of organisms. Nevertheless, the great number of genetic similarities among the members of all three branches strongly suggests that all living things on earth arose from a single group of ancestral cells present at the dawn of life. These genetic similarities stand as powerful evidence for the *unity of life,* a unity both of origin and of basic day-to-day functioning. We will see dozens of examples in the chapters to come of common life mechanisms at the molecular level, and we will see how these unite our planet's incredible diversity of life forms at a very fundamental level.

Natural Selection: A Mechanism of Evolution

Another major guiding principle for the study of life is **natural selection,** the evolutionary mechanism that brings about the diversity of new species from ancestral lines. Consider, for a moment, the graceful, towering giraffe, one of nature's magnificent products. This animal's extremely long neck and legs have been objects of human curiosity for several centuries.

(d) Fungi

(e) Plantae

(f) Animalia

(c) Protista

(a) Eubacteria

(b) Archaebacteria

Domain Eucarya

Fungi

Plants

Animals

Protists

Domain Bacteria

Domain Archaea

(Eubacteria)   Monera   (Archaebacteria)

FIGURE 1.16 ■ The Family Tree of Life. (a) These newly divided *Streptococcus* cells (microscopic bacteria, magnification 20,750×) represent Eubacteria (part of the old kingdom Monera). (b) This *Methanobacterium ruminantium* cell (magnification 30,000×), which inhabits a cow's intestines, is from the Archaebacteria. (c) The kingdom Protista also contains microscopic single-celled organisms, but they are generally much larger and more complex than bacteria. This *Paramecium* cell, with its exterior covering of beating hairlike organelles, is a protist. (d) Fungi, such as this *Hygrophorus conicus* mushroom decomposing leaf litter on the forest floor, make up a fourth kingdom. (e) The plant kingdom includes a large group of many-celled organisms that generate their own food molecules and are often conspicuous and graceful, such as this pine. (f) The largest of the five kingdoms contains over a million species of animals. Most of these lack backbones, but some, like this arctic loon, have an internal skeleton and complex behavior patterns.

Early naturalists noted that the giraffe's long neck allows it to browse leaves from high branches that are inaccessible to wildebeests, zebras, elephants, and other inhabitants of the African savanna. We know today that these adaptations help the giraffe overcome disorder by collecting energy and materials and that the world's tallest animal evolved from shorter ancestors. At the turn of the nineteenth century, however, curious biologists could only speculate about how the giraffe got its wondrous neck and legs.

In 1809, French naturalist Jean Baptiste Lamarck suggested that early giraffes must have stretched their necks trying to graze on the leaves of high branches and that the long neck that an individual acquired through such stretching was passed along to its young. Experiments eventually showed that Lamarck's theory, the so-called inheritance of acquired characteristics, is incorrect and is not the mechanism underlying evolution. Consider some examples that clearly disprove Lamarck's idea: Many human families have for thousands of years and hundreds of generations removed the foreskins of infant males. Despite this deliberate alteration, however, every baby boy in each new generation is born with a foreskin—clear evidence that the acquired characteristic is not inherited. Other alterations of the body, including neck and lip stretching and foot binding, can also be carried out generation after generation without the traits being inherited. It remained for two English naturalists working in the mid-1800s, Charles Darwin and Alfred Russel Wallace, to devise an alternative explanation for how living things evolve.

In the 1830s, young Charles Darwin sailed around the world, investigating nature's diversity. (An account of his voyage and a description of Wallace's very similar scientific work appear in Box 1.1.) Darwin agreed with Lamarck and others that evolution occurred, but remained dissatisfied with existing explanations for how species change over time. Eventually, Darwin drew together two indisputable facts based on his own observations and synthesized a far-reaching conclusion:

Fact 1   Individuals in a population vary in many ways, and some of these variations are inheritable.

Fact 2   Populations have the inherent ability to produce many more offspring than the environment's food, space, and other assets can possibly support. As a consequence, individuals of the same population struggle among each other to survive on the limited resources.

Darwin's conclusion   Individuals equipped with traits that allow them to cope efficiently with the local environment leave more offspring than individuals with less-adaptive traits. As a result, certain heritable variations become more common in succeeding generations.

Darwin used the term *natural selection* to describe the greater reproductive success displayed by those individuals with adaptive characteristics as compared to members of the same

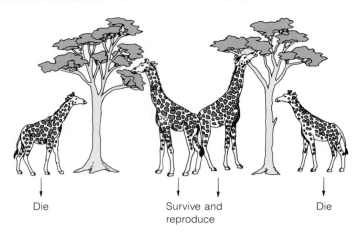

Die   Survive and   Die
                              reproduce

FIGURE 1.17 ■ Natural Selection on the Savanna: Giraffes with Genetically Determined Long Necks Have Advantages for Survival and Reproduction.   Amid tall trees and long-necked neighbors, short-necked giraffes would have died and left few offspring, while long-necked ones would have survived, reproduced more often, and left more long-neck genes in the population.

species lacking the adaptations. He chose this term because nature "selects" the parents for the next generation.

The principle of natural selection, now widely accepted as a main mechanism behind evolution in nature, can explain adaptations such as the long neck of giraffes and the brilliant color of strawberry frogs. If we begin with a population of giraffes browsing on trees in the savanna many thousands of years ago (Figure 1.17), we can envision how some of the giraffes would have long necks and others short necks, just as some people have longer necks than others. The hereditary units called genes help determine neck length in both giraffes and people. Now, if on the savanna there were too few low-hanging leaves to feed all the giraffes in the population, then the long-necked giraffes could reach more food, harvest more energy and materials, and survive to produce more offspring—many, like their parents, possessing the genes for long necks. With proportionately more long-neck genes around, the average neck length in the giraffe population would increase over time to present-day lengths. It is important to note, however, that natural selection is not the cause of variations within the population. Short necks, long necks, and other variations preexist in the population as a result of random changes in genes, called *mutations*. Natural selection simply chooses the best-adapted, best-competing individuals to be parents for the next generation.

Natural selection can also account for the color of strawberry frogs. The Central American jungle lacks enough food and nesting sites to support all the strawberry tadpoles that hatch, since a single female can lay 600 eggs a year. Thus, only a few strawberry frogs from each clutch of eggs survive to reproduce, and this natural selection of survivors is probably based on adaptations such as crimson skin color and the

BOX 1.1   BIOLOGY: A HUMAN ENDEAVOR

DARWIN, WALLACE, AND EVOLUTION BY NATURAL SELECTION

History has not dealt equally with Charles Darwin and Alfred Russel Wallace. Yet these two men, living in the same country during the same era, proposed remarkably similar theories of evolution by natural selection—a central theme of this book and a unifying concept in all of biological science. To understand this historical oversight, as well as the stunning achievement evolutionary theory represents, one must consider both the social and scientific tenor of the mid-nineteenth century.

Darwin and Wallace lived in Victorian England, an era of courtly manners, rigid social roles, and inflexible stratification by social class. In contrast to the social constraints of the day, there was excitement and progressive thinking in many areas of science, including a growing belief that the natural world is in a constant state of change.

For hundreds of years prior to this time, most scientists believed that all species originated about 6000 years ago and remained unchanged in form throughout history. With the careful study of fossils in the late eighteenth and early nineteenth centuries, however, biologists Georges Cuvier, Georges Louis Leclerc de Buffon, Jean Baptiste Lamarck, and others began to realize that extinct organisms were unlike those alive in the 1800s and that the form of plants and animals seemed to change, or evolve, over the years. Simultaneously, geologists such as James Hutton and Charles Lyell were investigating rock layers and other evidence of mountain building, erosion, and volcanic eruptions and suggested that our planet must be millions—not thousands—of years old and must be undergoing slow but continual change. Within this social and scientific context, Darwin and Wallace independently made their own observations and separately drew nearly identical conclusions about the mechanisms of evolutionary change within the organisms on our changing planet.

Darwin was a wealthy upper-class gentleman who had studied for the clergy at Cambridge University but was passionately interested in natural history (Figure 1). Upon graduating, he agreed to serve as a social companion to Captain Fitzroy of the HMS *Beagle* on a five-year voyage (1831–1836) to map the coastline of South America. Throughout the voyage, Darwin had ample opportunity to explore the jungles, plains, and mountains of South America. One of the most seminal stops on the trip was the Galápagos, a cluster of black volcanic islands 1000 kilometers (600 miles)* west of Ecuador with a small but unique set of plants and animals.

Throughout his travels, Darwin was continually impressed by the variable shapes and colors that the members of a single species often showed, and he believed that this variation must have a fundamental role in evolution. He returned to England with the idea of natural selection already half formed and notebooks full of observations to support the contention, but he continued studying variation in domesticated species for nearly 20 years.

Darwin had written a short summary of his theory in 1842, but it remained unpublished as late as 1858. That changed abruptly, however, when Alfred Russel Wallace's work in the

FIGURE 1 ■
Darwin (1809–1882).

FIGURE 2 ■
Wallace (1823–1913).

same arena came to light (Figure 2). Younger than Darwin, Wallace had sailed to South America himself in 1848 to collect plants, insects, and other natural specimens. Later, he visited Singapore and Borneo to make similar collections. Like Darwin, Wallace had read the works of Malthus and Lyell and had noticed on his travels the striking variations among populations of individual species. And like Darwin, Wallace had returned and devised a theory of evolution by natural selection to explain what he saw (even choosing the term *natural selection* independently!). Unlike Darwin, however, Wallace was a working-class man, and perhaps because he had far less to lose and few fears of public disapproval, he was more eager to promote his ideas.

By 1858, Wallace had published four papers containing portions of a theory of evolution that included the mechanism of natural selection, the notion of descent with modification, and the idea of survival of the fittest. Wallace sent Darwin his fourth paper in 1858, and Darwin—realizing that his life's work was being scooped—rushed to prepare an abstract. Friends read this abstract before the Linnean Society later that year along with Wallace's published work. Then the following year, Darwin released his monumental and well-documented book, *The Origin of Species by Means of Natural Selection.*

As biologist William V. Mayer points out,[†] the simultaneous publication of nearly identical theories in today's highly competitive scientific world would probably lead to "recrimination and lawsuits." But products of the gracious Victorian era, Darwin and Wallace instead approached their situation with courtesy and generosity, each fully acknowledging the other's contribution at every opportunity. Darwin and Wallace, in fact, became personal friends, Darwin arranging for the impoverished Wallace to receive a government pension, and Wallace serving as a pallbearer at Darwin's funeral.

The intellectual stimulation each scientist gave the other provided generations of biologists with their broadest conceptual framework for understanding the living world. Now, over 130 years after the independent contributions of Darwin and Wallace and the thorough documentation of evolutionary principles, their insights still foster lively experimental inquiry into the mechanisms of life.

*km (mi); see the abbreviations chart on the inside back cover.

[†]William V. Mayer, "Wallace and Darwin," *The American Biology Teacher* 49 (1987): 406–410.

(a)

(b)

FIGURE 1.18 ■ **Natural Variation in the Strawberry Frog.** As these photos show, not all strawberry frogs are a vivid scarlet. (a) Some have red bodies and large dark spots; and (b) some have yellowish bodies and small spots. Natural selection acts on such variation, "selecting" those individuals best equipped to survive in the local environment as parents for the next generation.

skin's content of powerful poison (Figure 1.18). (Such frog poisons serve as a hunting resource for South American Indians; see Figure 1.19.) After a bird or snake takes one disagreeable nibble on a poison strawberry frog, most learn to leave the frogs alone. The more brilliantly colored the frog, the more easily an enemy will recognize it as poisonous. Over time, more frogs belonging to brightly colored families would survive and thereby produce more offspring, and genes for bright color would increase in frequency in the population along with genes for poison production.

Darwin argued that evolution by natural selection, working over long stretches of time, could have produced the millions of living forms we now observe in all their splendid diversity (see Box 1.2 on page 18 for another example). Natural selection fine-tunes each species to its own environment by selecting genes that, in the broadest sense, help individual organisms overcome disorder and death. Evolution by natural selection is so grand an organizing principle for all biology that it will resurface time and again in this book, in the many magazine and newspaper articles you may read on life science, and in any future biology courses you may take.

THE PROCESS OF SCIENCE: A MAJOR SYSTEM FOR LEARNING ABOUT LIFE

A science like biology is sometimes presented as a compendium of facts to memorize. This approach is not just dull, it's incomplete, because science is a human endeavor, a lively, ongoing process for observing the world around us and gaining new knowledge about it. Before we begin to explore biology in the chapters that follow, it is useful to consider how scientists have coaxed, prodded, and pried the secrets of nature from living organisms. It is also important to keep in mind that behind every fact and concept we present, there were people in laboratories or field stations engaged in the often tedious and frustrating but sometimes joyful and exciting pursuit of information—of truth—about living things. So let's look more closely at the fundamental principles behind the scientific process, as well as the general method for problem solving that scientists employ and that, in fact, we all use daily.

Fundamental Principles: Causality and Uniformity

A lightning bolt flashes in a cloud-darkened sky. A man eating in a restaurant becomes enraged and abusive when the waitress tells him they are out of apple pie. Modern scientists assume that events like these are due to natural causes, a principle they call **causality.** The ancient Greeks, on the other hand, believed that thunderbolts arose when the god Zeus hurled them at the earth and that a mentally disturbed person

FIGURE 1.19 ■ **Poison-Tipped Dart and Blowgun.** At some point in their cultural history, Indian tribes living in the jungles of Peru, in South America, learned that strawberry frogs and similar tropical species make powerful poisons that will kill animals. One type of frog is so potent that the poison from a single individual can kill 20 adult people! An Indian catches a frog with poison glands, rubs the point of a dart several times along the animal's back, then inserts the venomous dart into a long blowgun and moves quietly through the forest, tracking monkeys, wild pigs, birds, and other animals to shoot for food. Since such poisons block the function of nerve and muscle cells, researchers also use them to investigate how nerves and muscles do their jobs.

was possessed. Today's scientists may not yet fully understand the natural causes of a phenomenon like mental illness, but they firmly believe that by applying the scientific process, they someday will.

A second fundamental principle of science is the **uniformity** of phenomena in time and space. Scientists contend that the laws of nature operate the same way today in Columbus, Ohio, as they did in Olduvai Gorge, in East Africa, 1 million years ago or in the cloud of dust and gases that existed before our solar system formed billions of years ago. The principle of uniformity is important to biologists because events that led to life's origin and diversity occurred long before people lived to observe them. Yet a biologist can feel confident that the same natural laws operating today functioned in the same way at the dawn of time, during the emergence of life, and all during its evolution on the planet.

The Power of Scientific Reasoning

Resting on the twin pillars of causality and uniformity is the power of scientific reasoning. We have seen that Darwin stated two facts from which he drew a grand conclusion: that natural selection is a mechanism of evolution. This kind of thinking is called **inductive reasoning,** or generalizing from specific cases to arrive at broad principles. This is a particularly creative, intuitive, and exciting part of science. It's like creating a new sonnet or sculpture or sonata. The instant when the mind leaps from previously isolated facts to a broad, unifying generalization is as thrilling, some scientists say, as starting down the steepest slope of a roller coaster!

The counterpart to inductive reasoning is **deductive reasoning,** or analyzing specific cases on the basis of preestablished general principles. Let's see how inductive reasoning and deductive reasoning differ in application. A person who knows, for example, that bright red strawberry frogs and red and yellow coral snakes (Figure 1.20a) are poisonous might conclude through inductive reasoning that all reddish organisms are dangerous. Such a tentative generalization is called a **hypothesis.** If that same person then encounters a flashy pink and black lizard called a Gila monster (Figure 1.20b), the person might, using deductive reasoning, assume that the Gila, too, is poisonous and avoid it.

While inductive and deductive reasoning help shape how scientists think, the final and equally crucial element in the scientific process is the method scientists use to try to disprove their generalizations.

Testing Generalizations

The person who generalized from strawberry frogs and coral snakes and formed the hypothesis that all red organisms are poisonous might one day see a friend eating a bowl of red raspberries and start waving his or her arms at the confused diner. If the friend survives, however, this will *disprove* the hypothesis, or show it to be false, and a new hypothesis will be needed—something like: Bright red animals (not all organisms) are poisonous. But then, what about cardinals (Figure 1.20c)? A quick test could show that these flashy scarlet birds are also nonpoisonous, requiring still further modification of the hypothesis.

(a) Coral snake

(b) Gila monster

(c) Cardinal

FIGURE 1.20 ■ Brightly Colored Animals and the Scientific Process. Like strawberry frogs, coral snakes (a) are poisonous. One could generalize that all brightly colored animals are poisonous, and from this deduce that brightly colored Gila monsters (b) are also poisonous—which, in fact, they are. But this fact, while consistent with the hypothesis, does not prove it. Tests can only *disprove* hypotheses. Since a test can show that cardinals (c) are bright red but not poisonous, the hypothesis is clearly incorrect. It can, however, be modified to this: Certain brightly colored reptiles and amphibians are poisonous.

BOX 1.2   *HOW DO WE KNOW?*

EVOLUTIONARY ODDITIES PROVE THE RULE

Evolution by natural selection can clearly account for the long neck of the tree-feeding giraffe and the crimson "advertisements" and protective poison glands of the strawberry frog. But, wrote Harvard professor Stephen Jay Gould in 1980,* "Ideal design is a lousy argument for evolution," because it suggests that engineers could have designed the organism from scratch. "Odd arrangements and funny solutions," Gould continues, "are the proof of evolution," showing that organisms can evolve only in the ways their particular past history allows. Such an odd evolutionary arrangement is the giant panda's thumb.

The handsome, thick-coated panda lives only in remote, remnant bamboo forests along the Tibetan plateau in eastern China (Figure 1). Giant pandas once inhabited much of China, but pelt hunters and the steady encroachment of human farms and settlements have caused panda populations to fall to about 1000 animals surviving in just six small regions with a total land area less than the size of West Virginia. Pandas eat primarily bamboo stalks—up to one-third of their body weight daily. But the most curious of the panda's traits is their front paws: Instead of having five digits like people and many other mammals, they alone have *six* on each paw. The panda's five "fingers" plus one opposable "thumb" are fully adapted for holding bamboo stalks, their favorite food.

Biologist D. Dwight Davis and others have studied the anatomy of this "thumb" in great detail and have found that it is not a true thumb. It is, instead, an enlarged wrist bone (sesamoid bone) with attached muscles and tendons (Figure 2). This arrangement enables the "thumb" to grasp the bamboo powerfully against the other five, more fingerlike digits and in this way to substitute for the panda's fifth digit—its true thumb—which is already occupied.

Sometime in the early evolutionary history of the giant panda, mutations occurred giving some individuals an enlarged wrist bone with muscles and tendons properly connected for opposable movement. These mutants must have been able to gather more bamboo, survive in greater numbers than their "thumbless" cohorts, and leave more offspring, because today, all surviving pandas have the opposable sixth digit. It is not a true thumb—and not the graceful innovation of some skilled bioengineer. But it functions like a thumb, and its gradual origin from a preexisting structure is a good demonstration of how evolution generally proceeds in the natural world.

*Stephen Jay Gould, *The Panda's Thumb* (New York: Norton, 1980), pp. 20–21.

FIGURE 1 ■ **Giant Panda (*Ailuropoda melanoleuca*).**

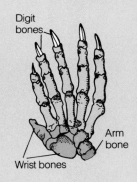

Digit bones

Arm bone

Wrist bones

FIGURE 2 ■ **The Panda's Thumb.** An enlarged wrist bone serves as an armature moved by muscles and tendons.

A good scientist is always skeptical about hypotheses and ready to discard or modify them when data from actual tests disprove them. Some hypotheses, however, survive all attempts at disproof. A broad general hypothesis that is repeatedly tested but never disproved comes to be accepted as a **theory,** a general truth about the natural world, like the theory of gravity or the theory of evolution.

To accurately describe and understand nature, scientists combine the kinds of creative thinking, reasoning, and testing we have been discussing into a series of steps called the **scientific method:**

1. They ask a question or identify a problem to be solved based on observations of the natural world.
2. They propose a hypothesis, a possible answer to the question or a potential solution to the problem.

3. They make a **prediction** of what they will observe in a specific situation if the hypothesis is correct.
4. They test the prediction by performing an experiment.

While these steps may sound very regimented, they are really little more than an organized commonsense approach—one you use regularly in your own life. Let's say that one evening you observe that your desk lamp isn't working. You would pose a question (step 1): "What made my desk lamp go out?" And you would probably create a hypothesis (step 2): "Maybe the light bulb burned out." Next you would make a prediction (step 3): "If the bulb burned out, then when I replace it with a working bulb, the lamp should light." Finally, you would perform an experiment (step 4): You would remove a bulb from a floor lamp that works, screw it into your desk lamp, and watch what happened.

When you performed the test, you automatically included what scientists call a **control,** a check of the experiment based on keeping all factors the same except for the one in question. Here, the control was the borrowed bulb that worked in the floor lamp. If the borrowed bulb failed to work in the desk lamp, you could conclude, based on your control, that a burned-out bulb was not the problem. You would next discard the faulty-bulb hypothesis and ask new questions: "Is the lamp itself broken? Is something wrong with the wiring to the wall socket?" You could then make new hypotheses, new predictions, and perform new tests until you discovered why your lamp went out.

No tool is more powerful for understanding the natural world than the scientific method. It does not apply, however, to matters of religion, politics, culture, ethics, or art. These valuable ways of approaching the world and its problems proceed along different lines of inquiry and experience. Nevertheless, many of the world's complex problems have underlying biological bases, and thus they demand mainly biological solutions.

HOW BIOLOGICAL SCIENCE CAN HELP SOLVE WORLD PROBLEMS

Anyone who reads a daily newspaper is well acquainted with world problems: overpopulation, famine, war, crime, drug addiction, AIDS, cancer, heart disease, pollution, ozone depletion, acid rain, changes in climate. While these problems tend to have multiple roots, their biological bases may yield to biological solutions, and these solutions, in turn, will help ease the associated pressures on society. For this reason, citizens in all fields must understand the biological bases of the world's problems.

Many of our most vexing problems stem from our species' enormous and burgeoning population. Five billion people are currently straining our planet's environmental resources, and by the year 2000, we will number 6 billion. This immense population is accumulating because the human species, like all others, is finely honed to obtain energy and materials and to reproduce. The resulting human adaptations—the abilities to reason, communicate, and manipulate the physical world—have been so successful that our single species is busily exhausting the limited resources that support all life on our small planet. Many observers believe that our future security and quality of life are threatened less by war among nations than by the burden we place on natural systems and resources with the sheer crush of humanity.

Take, for example, the plight of Madagascar, a large island off the southeast coast of Africa. In the past 35 years, half of Madagascar's forests have been leveled to provide fuel and to uncover farmland for the impoverished and rapidly growing population. Heavy rains, however, cause the cleared hills to erode, bleeding the rich red topsoil into the rivers and destroying the land's ability to grow crops (Figure 1.21). Farmers plant clove tree seedlings on the cleared slopes and then wait seven years before the first full crop of the pungent clove buds are ready for harvest and export, simply hoping that the slope will not wash away before harvest. Whole villages depend on the developed world's appetite for vanilla ice cream, cola drinks, and other foods that contain extracts from clove buds. Yet their own children go hungry because cloves are not an adequate food source.

FIGURE 1.21 ■ **Deforestation in Madagascar: Landscapes Effaced, Healing Plants Lost. Hills in Madagascar's central plateau lie denuded of their original forests, with precious topsoil washing away each time it rains. The jungles were cleared for firewood and for the creation of new farmland, but after a few years, the land's productivity plummeted, and farmers cleared new jungle. Thousands of species may be lost to this deforestation process.**

The exploitation of Madagascar's forests and wildlife began long ago, when the human population was much smaller. In the process, many species became extinct, including the "elephant bird," an astonishing animal that stood 3 meters (10 feet) tall, weighed 500 kilograms (half a ton), and laid 10-kilogram (20-pound) eggs.* The destruction has since accelerated, and with it has come the loss of hundreds of species, including plants that produce potentially life-saving drugs. Residents of Madagascar export a pink-petaled periwinkle flower to pharmaceutical companies in Europe and North America that use the petals to make drugs for children with leukemia. Observers fear that many such plants with potential uses in medicine and agriculture will perish along with Madagascar's tropical forests before they are ever studied and applied. And worse, they fear that Madagascar's own people are not even benefiting substantially from the wholesale clearing of forested lands.

Two decades ago, one might have predicted a similar fate for the Central American nation of Costa Rica. Although this small country's national debt is one of the highest in the world per citizen, the leaders are committed to sustainable development based on the application of modern biological principles and practices. Among other things, Costa Ricans are replanting deforested hillsides with fast-growing tropical hardwood trees. They are also growing nutritionally improved food crops, attempting to set aside up to 15 percent of their land as natural preserves for their tropical species, and instituting up-to-date health care and family-planning practices whenever possible. Costa Rica's success is a hopeful sign for the future: biological solutions for social problems of biological origin.

In the chapters that follow, you will find many discussions of how biology is helping to solve world problems. We are, in fact, in the midst of a revolution in the biological sciences, with exciting new information surfacing weekly in the fight against cancer, heart disease, AIDS, infertility, and obesity. Researchers are making rapid advances in gene manipulation to create new drugs, crops, and farm animals; in exercise physiology to improve human performance; in the diagnosis of genetic diseases; and in the transplantation of organs, including brain tissue. Across all frontiers of biological science, at all levels of life's organization—from molecules to the biosphere—scientists are learning the most profound secrets of how living things use energy to overcome disorganization and reproduce to overcome death. In studying the living world, you are embarking on an adventure of discovery that will not only excite your imagination and enrich your appreciation of the natural world, but will provide a basis on which you can contribute intelligently to the difficult choices society must make in the future.

CONNECTIONS

All living things share a mutual interdependence, a common origin, and two fundamental problems of survival: the tendencies toward disorder and death. Each organism must conquer the challenge of disorganization by extracting energy and materials from its environment and using them for growth, repair, and continued activity. While a strawberry frog captures ants and recycles the energy stored in the insects' muscles, a periwinkle plant captures energy directly from the sun and absorbs materials from the air and soil. As different as these solutions might be, animals, plants, and all other organisms share a basic set of characteristics that allow them to combat the challenges to survival, and these attest to the common origin and to the unity of all living things.

Because life's characteristics and challenges are common to all species, people, as living organisms, are powerless to escape all of its consequences. To save ourselves and our world from its current ecological problems, biologists and citizens in all fields must learn about life processes and the interactions of organisms with their environments. The underlying principles of biological science begin with the laws of chemistry and physics and how they govern the atoms and molecules that make up living things—the subject of our next chapter.

Highlights in Review

1. The strawberry frog, with its bright red skin and particular sequence of courting and brooding behaviors, illustrates five main themes in biology: the use of energy to maintain order, reproduction to overcome the death of individuals, the evolution of new traits, the interaction of organisms with the environment, and the utility of the scientific process in learning about the natural world.
2. Biologists have compiled a list of characteristics common to all living things. Four of these traits (order, metabolism, mo-

tility, and responsiveness) relate to an organism's systems for overcoming disorder; three of them (reproduction, development, and genes) relate to perpetuation; and two of them (evolution and adaptation) relate to interactions with the environment.
3. Living things exhibit a hierarchy of biological organization that includes the biosphere, ecosystems, communities, populations, individual organisms, organ systems, organs, tissues, cells, organelles, molecules, atoms, and finally, subatomic particles.

---

*m (ft), kg (ton), kg (lb); see the abbreviations chart on the inside back cover.

4. Living things exhibit metabolism, or the breaking down, rearrangement, and use of energy compounds.

5. Self-propelled movement is a diagnostic feature of life, even in plants and other organisms that appear to remain stationary.

6. All organisms are responsive; they detect information about their surroundings and react in appropriate ways.

7. All organisms reproduce, allowing a particular lineage to be perpetuated, despite the eventual death of the individual members.

8. Development, or change and growth during the life cycle, is another key characteristic of life.

9. Living things have genes, or units of heritable information, that pass from one generation to the next and control specific physical, chemical, and behavioral activities within the organism.

10. All species evolve, and this change over time accounts for both the unity of life and its amazing diversity.

11. Adaptations, or particular innovations of structure and function, suit organisms to survive in their particular environments.

12. Natural selection is a major mechanism leading to evolutionary change. Nature selects as the parents of the next generation those individuals best adapted to coping with the local environment and conditions. Since those "selected" parents leave more offspring, their genes become more frequent in the population.

13. Two additional qualities characterize life as a whole: unity, manifest by shared molecular-level characteristics, and diversity, the fantastic variety of living things, all related by descent from common ancient progenitors.

14. Biologists categorize each living thing into a genus and a species, and into increasingly inclusive groups, including families, orders, classes, phyla or divisions, and six main kingdoms arranged in a family tree with three main branches or domains. The kingdoms are Eubacteria and Archaebacteria (together, old Monera), Protista, Fungi, Plantae, and Animalia.

15. Biologists apply the process of science to their study of the natural world. This includes the principles of causality and uniformity as well as inductive and deductive reasoning and the testing of hypotheses. The formal scientific method involves asking a question about some puzzling phenomenon, proposing a hypothesis, predicting an observable result, then testing the prediction through experimentation.

16. Biological research and applications have the potential to help solve many of the world's current problems, a number of which stem from overpopulation and the strain that our sheer numbers place on the environment.

17. We live amidst a revolution in the biological sciences, and citizens in all fields must become informed so that they can contribute to societal choices.

Key Terms

adaptation, 10	motility, 7
biology, 3	natural selection, 12
causality, 16	order, 5
control, 19	prediction, 19
deductive reasoning, 17	reproduction, 8
development, 9	responsiveness, 7
evolution, 10	scientific method, 18
gene, 9	theory, 18
hypothesis, 17	uniformity, 17
inductive reasoning, 17	

Study Questions

Review What You Have Learned

1. Which life characteristics are related to overcoming disorder? To overcoming death?

2. Unity and diversity are two additional qualities that characterize life. How do they relate to evolution?

3. Define organism.

4. All organisms must carry out one basic activity to counteract the tendency toward disorder. What is it?

5. List the levels of hierarchy of biological organization on earth.

6. Once an organism takes in energy and materials, what basic process breaks them down and rearranges them into useful products?

7. What are the two basic types of reproduction?

8. What do genes do in organisms? What is the result of a mutation?

9. What is the most important mechanism leading to evolutionary change?

10. Define adaptation.

11. What is the difference between inductive and deductive reasoning?

Apply What You Have Learned

1. A biology student wonders whether her philodendron plants really need their weekly dose of nitrogen fertilizer. Using your knowledge of the scientific method, explain how this student can satisfy her curiosity.

2. Poisonous butterfly species are often brilliantly colored. Edible butterfly species may evolve that resemble poisonous species. Explain how natural selection might lead to this resemblance.

For Further Reading

DARWIN, C. *On the Origin of Species: A Facsimile of the First Edition.* Cambridge, MA: Harvard University Press, 1975.

GOULD, S. J. *The Panda's Thumb.* New York: Norton, 1980.

GREEN, G. M., and R. W. SUSSMAN. "Deforestation History of the Eastern Rain Forests of Madagascar from Satellite Images." *Science* 248 (1990): 212–215.

JOLLY, A. "Madagascar: A World Apart." *National Geographic,* February 1987, pp. 149–183.

KORMANDY, E. J. "Ethics and Values in the Biology Classroom." *The American Biology Teacher* 52 (1990): 403–407.

MYERS, C. W., and J. W. DALY. "Dart-Poison Frogs." *Scientific American,* February 1983, pp. 120–133.

SCHNEIDERMAN, H. A., and W. D. CARPENTER. "Planetary Patriotism: Sustainable Agriculture for the Future." *Environmental Science and Technology* 24 (1990): 466–473.

SUN, M. "Costa Rica's Campaign for Conservation." *Science* 239 (1988): 1366–1369.

WELLS, K. D. "Courtship and Parental Behavior in a Panamanian Poison-Arrow Frog (*Dendrobates auratus*)." *Herpetologica* 34 (1978): 148–155.

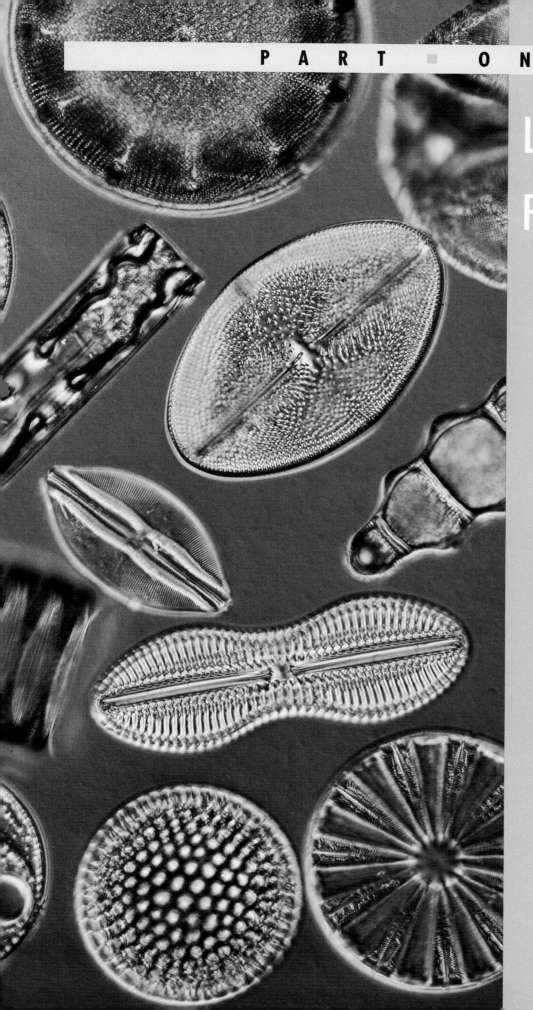

LIFE'S
FUNDAMENTALS

Atoms, Molecules, and Life

Water, Survival, and the Chemistry of Living Things

The vast interior of Australia—more than one-third of the continent's total landmass—is a desert of red sand, timeworn mountains, and sparse, scrubby plants that receive less than 25 centimeters (10 inches)* of rain each year. Australian aborigines have inhabited this parched terrain for thousands of years, surviving largely because they have devised a set of techniques for locating water (Figure 2.1). Traditional aborigines knew where to dig down in just the right part of a dry creek bed to find buried patches of damp sand to suck on for moisture. They gathered early morning dew on grasses and soaked up water from cavities in desert oaks with sponges made of dry grass. They knew a dozen plant species that store water in shallow roots, which they could tap in an emergency. And they even learned to locate desert frogs, encased in hard, deep mud, but alive and bloated with water. The aborigines would dig up the frogs and squeeze them like lemons for their precious moisture.

For desert inhabitants such as the aborigines, and in fact for all living organisms, life depends on water. A human drinks about 980 liters (265 gallons)† or more of water per year, in addition to imbibing another 2.7 L (3 quarts)‡ of moisture in food per day. An apple tree in summer uses 10 to 20 L (2.4 to 4.7 gal) of water or more in one day. Obviously, without sufficient water, organisms eventually die. So why is it necessary for life?

One reason life depends so heavily on water is that living things are made up mostly of water molecules (H_2O). A tree is about 50 percent water, most animals (including humans) are about 65 percent water, and a jellyfish is more than 90

FIGURE 2.1 ■ **Life Depends on Water.** And Australian aborigines know how to find it, even in the continent's parched desert areas.

*cm (in.); see the abbreviations chart on the inside back cover.

†L (gal); see the abbreviations chart on the inside back cover.

‡L (qt); see the abbreviations chart on the inside back cover.

percent water. Biologists believe that life arose in water early in the earth's history and that living things have been made up mostly of water ever since. Henry David Thoreau once called the pickerel fish swimming in Walden Pond "animalized water." And renowned essayist Loren Eiseley described people as "myriad little detached ponds" and "a way that water has of going about beyond the reach of rivers"—poetic sentiments, perhaps, but not far from the truth.

In short, the living state we recognize by means of movement, growth, use of energy, reproduction, and so on, is based on a set of interacting biochemical structures and processes that require the presence of water molecules. As we explore the chemical constituents of life, we will see the many roles of water in the living cell, including its importance in dissolving substances, transporting materials, and maintaining appropriate body temperature. We will also see how the chemistry of water relates to the chemistry of life.

This chapter lays the groundwork for understanding cells by discussing their fundamental building blocks—atoms and molecules—the majority of which, in any living thing, happen to be hydrogen and oxygen in the ratio of 2:1. The remaining 10 to 50 percent of an organism's compounds are equally important, and most of them contain the element carbon. As the chapter explains, both water and compounds containing carbon have unique chemical and physical properties that make life possible.

Our focus will first be on atoms and molecules, because, as we saw in Chapter 1, the hierarchy of organization in the biological world builds from these basic constituents of all matter to cells, tissues, organs, and whole organisms. We must understand chemistry to understand life. Why? The answer is a major unifying theme of this chapter: The behavior of atoms and molecules underlies and explains the behavior of living cells, whether those cells are breaking down food molecules, taking in materials, building new cell parts, dividing, or moving. Furthermore, a second unifying theme is related to the first: The physical structure of atoms and molecules determines their chemical properties and hence the roles they play in cells. The close relationship between structure and function will be evident throughout our study of biology, and significantly, it begins at the chemical level. That's why we begin the study of life by considering the chemical constituents of living things—atoms and molecules.

A fascinating idea—the concept of *emergent properties*—helps explain the connection between structure and function. In chemistry as well as biology, the whole (the whole atom, the whole molecule, the whole cell, the whole organism) is always more than the sum of its parts. The collective proper-

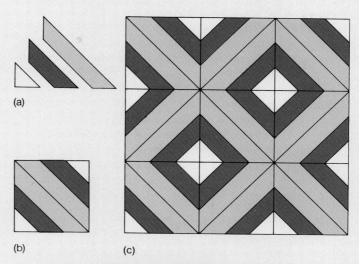

FIGURE 2.2 ■ Emergent Properties. The whole is more than the sum of its parts. In a quilt, small geometric units (a) can be combined into a block to form a pattern (b). When sets of these blocks are arranged correctly, a higher-order pattern emerges (c). In a similar way, the properties of a living thing emerge from the precise arrangement of component parts: atoms, molecules, cell parts, cells, and so on.

ties of the whole emerge not just from the properties of the individual parts, but from the precise way those parts are arranged and interact (Figure 2.2). For example, all of the 92 naturally occurring chemical elements are made up of the same three kinds of particles, but the properties of powdery yellow sulfur, let's say, are very different from the properties of metallic iron because of the way the particles are arranged in sulfur and iron atoms. Likewise, the body of a desert-dwelling aborigine contains the same six major chemical elements as the woody tissue in the eucalyptus tree he or she rests beneath for shade. But those six elements are arranged into different sorts of molecules in human and eucalyptus. Hence, the properties of person and tree emerge from both constituent parts and their precise arrangement.

In this chapter, we answer several questions about the chemistry of life:

■ What is the structure of atoms and molecules, and how do they form the units of all matter, including water, rocks, air, and living things?

■ How does the structure of the water molecule give it properties essential to life?

■ What special property of carbon compounds allows them to form the four major types of biological molecules?

ATOMS AND MOLECULES

The Nature of Matter: Atoms and Elements

Ancient Greek philosophers noticed that some materials, such as rocks, wood, and soil, are composed of more than one substance, while other materials, such as chunks of iron, gold, silver, aluminum, and sulfur, cannot be further decomposed into different constituents by chemical processes. Scientists later studied the pure substances and called them elementary substances, or **elements,** and named the various combinations of two or more elements **compounds.** The essential difference between elements and compounds was shown experimentally in the 1820s by the English chemist Michael Faraday. Faraday took common table salt, heated it to the boiling point, passed electricity through it, and produced the soft, sil-

very metal he called sodium (Na) plus the corrosive, greenish yellow gas chlorine (Cl). Since neither substance could be further decomposed, both were elements.

About the same time, the English chemist John Dalton concluded that each element is composed of identical particles. He called the particles **atoms** (meaning "indivisible"). Atoms are the smallest particles into which an element can be divided and yet still display the properties of that element—gold, for example, versus iron. Since Dalton's time, other chemists have confirmed his **atomic theory.** In addition, chemists have discovered 92 naturally occurring elements and have created 13 more in the laboratory. The properties of these elements are marvelously different. For example, in its pure state, helium is a colorless, odorless gas; lithium a soft, whitish metal; calcium a white powder; copper a reddish metal; sulfur a yellow solid; and iron a hard, dark, shiny metal.

(a) Composition of living things

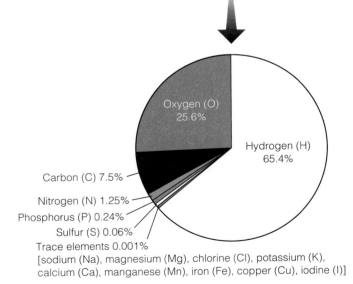

(b) Composition of the earth

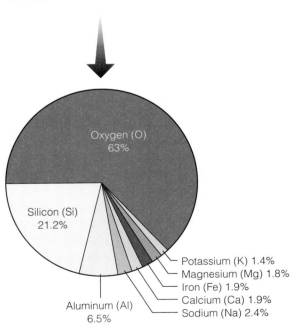

FIGURE 2.3 ■ **Elements in Earth and Living Organisms.** If analyzed at the level of atoms, our planet has a very different composition from the living organisms that inhabit it. The crust at our planet's surface, containing rock, soil, sand, and other materials, is mostly oxygen, silicon, and aluminum atoms in various compounds. The people, fish, trees, and other organisms that live on that crust are mostly hydrogen, oxygen, and carbon atoms. The unique properties of living things emerge from their combination of elements and the precise arrangement of those elements.

Although the earth contains 92 elements, scientists have found that living things contain just a few. In fact, earth and the life it supports have radically different proportions of elements. Geologists studying the elemental composition of the earth's crust, for example, have discovered that 98 percent of all the atoms in our planet's surface layer belong to the eight elements listed in Figure 2.3, headed by oxygen (O), silicon (Si), and aluminum (Al). Biologists, on the other hand, have found a very different set of elements in living things. In a typical organism, 99 percent of all the atoms belong to the six elements carbon (C), hydrogen (H), nitrogen (N), oxygen (O), phosphorus (P), and sulfur (S). In addition to these six basic elements, Figure 2.3 also names the nine other elements that occur in trace amounts in living organisms. In our discussion of water and carbon, we will see why living tissue is a unique and special form of matter and why it is so unlikely that living things could be based mainly on the common crustal elements silicon and aluminum.

The atomic theory that Dalton devised to explain the essential differences between common materials became a foundation for the modern sciences. In his day, however, Dalton did not have the means to probe still deeper into the nature of atoms and elements and discover the key to understanding them. This was left to later generations of chemists, who did in fact find that key: Properties of elements are based on the internal organization of their atoms. Let's see why that is true.

Organization Within the Atom

Chemists since Dalton have discovered that all atoms are composed of the same three types of subatomic particles: protons, neutrons, and electrons. A **proton** is a positively charged particle, and chemists have defined the mass of a proton (roughly speaking, how heavy it is) as about 1 atomic mass unit. A **neutron** is a neutral particle with no electrical charge; neutrons also have a mass of about 1 unit. **Electrons** are very small negatively charged particles with a mass 2000 times less than that of a proton or neutron.

Within each atom, these three types of particles are arranged a bit like an unimaginably small solar system. The sun at the center is the **atomic nucleus,** and it contains a set number of protons and neutrons that account for most of the atom's mass. A specific number of electrons, equal to the number of protons, orbit about the nucleus at a relatively great distance like planets encircling the sun. Figure 2.4 shows the structure of hydrogen, the simplest atom, with its single proton, single electron, and no neutron, as well as the structures of carbon and oxygen. If an atom—say, carbon—were the size of the Houston Astrodome, the nucleus would be a small marble on the 50-yard line. Of course, in reality, an atom is minute: About 1 million carbon atoms could sit side by side on the period ending this sentence.

Because the nucleus contains both positively charged protons and chargeless neutrons, it has an overall positive charge; by contrast, the orbiting electrons have a negative charge. The

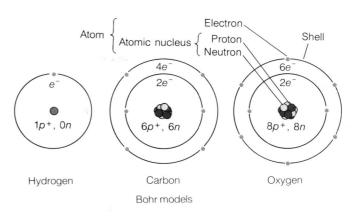

FIGURE 2.4 ■ Structure of the Atom. In each atom, negatively charged electrons (here, small gold dots) orbit the positively charged nucleus, which contains one proton (brown spheres) for each electron and a set of neutrons (tan spheres), usually very close to the number of protons. A balance between opposing physical forces keeps the electrons moving at a given distance around the nucleus. The system of static rings (to represent energy levels) and dots (to represent electrons around the nucleus) is called the Bohr model.

attraction between these positive and negative charges pulls the electrons toward the nucleus, but the movement of the rapidly orbiting electrons tends to throw them outward, away from the nucleus, not unlike a rock tied to a twirling string.

Just as it takes different amounts of energy to twirl rocks attached to different lengths of string, the electrons in an atom are organized in shells with different energies (see Figure 2.4). The shell nearest the nucleus holds two electrons and will be filled with two before any appear in the second shell. The second shell can accommodate eight electrons and will be filled before any electrons appear in the higher-energy third shell (which can also be stable with eight electrons). Subsequent shells also become filled with set numbers of electrons. These shell-filling rules are important for understanding how atoms join together to form molecules.

Since all atoms consist of the same three types of particles arranged in similar ways, what gives them different properties? What makes a chunk of the element carbon, with its billions of atoms, look black and solid, while the element oxygen is a clear gas? The answer is that each type of atom contains a unique number of protons in the nucleus. The carbon atom has six, for example, and the oxygen atom eight. This is the atom's **atomic number.** Because each proton and each neutron contribute one atomic mass unit, the combined number of protons and neutrons in the nucleus is the weight of the atom, known as its *atomic mass number,* or simply **atomic mass.**

An element often has an equal number of protons and neutrons, but not always. For example, the nucleus of the carbon atom contains 6 protons and 6 neutrons (for an atomic mass of 12), but phosphorus has 15 protons and 16 neutrons (and an atomic mass of 31). The number of electrons in an element equals the number of protons (and thus the atomic number): one for hydrogen, six for carbon, eight for oxygen, and so on. Box 2.1 shows how physicians have learned to use atomic structure in the diagnosis of major diseases.

Variations in Atomic Structure: Atomic Bombs and Nerve Impulses

Atomic bombs and the actions of your nerve cells both depend on slight exceptions to the standard structure of atoms. While the number of protons in the nucleus of a given atom always remains the same, the number of neutrons sometimes varies from the normal number. Atoms with the same number of protons but with different numbers of neutrons are called **isotopes** or *nuclides*. A good example of an isotope is uranium-235 (^{235}U), the explosive component of an atomic bomb. ^{235}U has three fewer neutrons in its nucleus than ^{238}U, or normal uranium. An isotope such as ^{235}U is **radioactive;** its nucleus is unstable and emits energy as it loses neutrons and changes to a more stable form.

^{235}U is an interesting isotope because when it breaks apart, it releases three neutrons, which can strike three more atoms

BOX 2.1   P E R S O N A L   I M P A C T

MAGNETIC RESONANCE IMAGING AND THE SECRETS IN HYDROGEN

Doctors are now using a powerful diagnostic tool called magnetic resonance imaging (MRI), based on the special properties of protons. Chemists discovered many decades ago that in certain atoms, such as hydrogen and phosphorus, the protons whirl like tops and create small magnetic fields in the nucleus that can be altered with external magnets and radio signals and then detected electronically. More recently, medical engineers created a diagnostic instrument out of a huge, ring-shaped magnet that can encompass the human body and generate a magnetic field 40,000 times greater than the earth's (Figure 1). The magnet is strong enough to jerk a 2 lb wrench from your hands, and yet the force fields it generates do not appear to harm the body in any way. The magnets cause protons in the body's hydrogen atoms (and in certain other types of atoms) to align and spin in the same direction. Then, when a second magnetic field is applied, the spinning protons flip back to their normal alignment; this flipping gives off a signal that can be recorded and used to reveal the presence and concentration of the atoms. Physicians can interpret the signal and use it to diagnose diseases.

Because MRI can detect protons in the nuclei of hydrogen atoms, doctors can now observe blood (which is 90 percent water) flowing through various tissues. This allows them to peer into the brain tissue of stroke victims to see which vessels were blocked during the stroke and which still allow the free flow of blood (Figure 2). Doctors are also using MRI to pinpoint the exact location of blockages caused by brain tumors, to locate sources of pain in the spine and limb joints, and to diagnose broken bones anywhere in the body, since blood seeps into the fracture. MRI may also replace some angiogram tests (which involve inserting tubes into the body) and reveal the fatty material blocking blood vessels in the heart.

MRI is part of a new generation of medical technology that is replacing the doctor's black bag and the medical laboratory. Other techniques include computed axial tomography, or CAT scans, which combine X-rays and computers, and positron emission tomography, or PET, which detects electrons given off by certain isotopes injected into the patient. While each of these life-saving techniques is unique, all three are based, ultimately, on our knowledge of the structure and properties of atoms.

FIGURE 1 ■ Medical Diagnostic Device That Employs MRI.

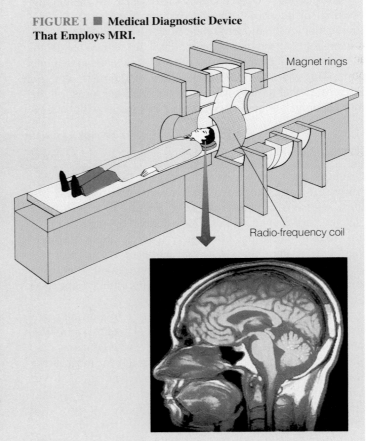

Magnet rings

Radio-frequency coil

FIGURE 2 ■ MRI Reveals Vivid Details of the Brain's Soft Tissue.

of ^{235}U, cause them to fly apart, send nine neutrons out toward nine more ^{235}U atoms, and carry on this way in a rapid-fire chain reaction that instantaneously and explosively releases an enormous amount of energy. In contrast, other isotopes, such as carbon-14 (^{14}C), with two extra neutrons, are useful to biologists in experiments to mark and trace molecules that contain them. We'll discuss such experiments in later chapters.

Ions are the second exception to an element's standard atomic structure. Ions are crucial in the normal functioning of nerves, including the nerves that are enabling you to read this sentence. **Ions** are atomic variants with a positive or negative charge because their electron number is not equal to their proton number. For example, if a normal sodium atom (Na), having 11 protons and 11 electrons, loses an electron, it will become a sodium ion with 10 electrons and 11 protons and an overall positive charge. It is designated Na$^+$ and has different properties from a sodium atom. The rapid movement of electrically charged sodium ions into and out of your nerve cells causes electrical signals that a researcher can detect with sensitive equipment and record (see the graph in Figure 2.5). The movement of sodium ions and hence the graphs of nerve action are slightly different between a person meditating and one not meditating.

Electrons in Orbit: Atomic Properties Emerge

With this picture of the atom and its variations in mind, we are ready to address the basic question, What is it about the structure of an atom that gives each element its unique properties: powdery, hard, or gaseous texture; green, yellow, white, black, or metallic color; and so on? Moreover, why do your nerve cells use sodium, not iron? What special feature of carbon atoms allows them to combine with other atoms into thousands of different chemicals in your body? What is so crucial about the oxygen atom that without sufficient quantities of it, one dies in a few minutes?

The answers lie mainly with the number of electrons in the atom's energy shells. As we saw in Figure 2.4, electrons fill these shells in a specified order—two in the first shell, eight in the second, and so on. For our purposes here, it is enough to know that electrons in the outermost shell are "in contact with the world," and their number and position largely determine the atom's chemical behavior. For example, atoms of the colorless, odorless gas neon have a filled second energy level (Figure 2.6a). Since the energy level is filled, this atom is very stable—so stable, in fact, that it is **inert**; it does not react easily with other atoms because it does not have to accept or eject electrons to achieve a complete outer shell. This stability is why an electric spark causes neon gas to emit energy in a certain way and glow in the dark instead of reacting chemically and forming compounds with other atoms.

In contrast to neon's full outer shell, the black, solid element carbon has a filled first energy level but is missing four electrons in its second energy level (see Figure 2.6b). As a

(a)

(b) Brain wave pattern of person meditating shows alpha waves

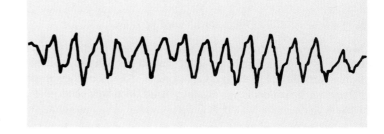

(c) Brain wave pattern of person resting

FIGURE 2.5 ■ Ions, Nerve Impulses, and Meditation. Ions have more or fewer orbiting electrons than normal atoms of the same element. The change gives the atom an overall positive or negative charge. Sodium ions (Na$^+$) are crucial to nerve cell functioning and electrical activity in the brain, which can be monitored and recorded. A person meditating with eyes half-closed (a) has a very distinctive brain wave pattern showing alpha waves (b; see Chapter 28 for more details). The brain wave pattern of the same person while not meditating is quite different and less relaxed (c). The brain waves shown here are actually from a Japanese Zen Buddhist priest, during and after meditating in a manner similar to the one this woman is using.

result of this "electron dearth," carbon tends to combine (share electrons) with many other elements in a huge number of compounds, including those that make up living cells, and to take part in the chemical activities of the life process. Hydrogen, which needs one electron to fill its outer level, and oxygen, which has two unfilled slots in the second energy level, also react and form compounds with each other (including water, H_2O) and with other elements. The soft, shiny metal sodium has stable, filled orbitals in the first and second energy levels, but has just a single electron in the third energy level (Figure 2.6c). It is thus highly unstable and extremely reactive, combining rapidly—almost explosively—with many other elements as it relinquishes the lone electron in the third energy level and becomes a sodium ion (Na^+), which, as we saw, is necessary for nerve impulses.

Appendix A.1 describes atomic orbitals in more detail.

We have seen that the properties of the elements emerge from both the structure of the parts and the way the parts are arranged. We can move on from here to examine the tendency of atoms, based on their outermost electrons, to react and combine with other atoms. This potential for combining into molecules is perhaps the most important aspect of atoms we will discuss, because the cells of living things are composed of molecules.

Molecules: Atoms Linked with "Energy Glue"

Molecules are groups of identical or different atoms that are linked by a kind of energy glue called **molecular bonds.** Bonds are not actual physical objects, like the couplings between railroad cars, nor are they solid, like hardened glue;

instead, they are *links of pure energy* usually based on shared or donated electrons. Bonds act like invisible springs; once a bond between two atoms is formed, it requires energy to pull the atoms apart or push them closer together.

■ **Covalent Bonds** The most common type of bond forms when a pair of electrons is shared by two or more atoms. A shared pair of electrons constitutes a **covalent bond.** Water molecules are a good example (Figure 2.7). Two hydrogen atoms joined with one oxygen atom form a water molecule. The formula for water, H_2O, shows this relationship. As the two hydrogen atoms approach the oxygen atom, each positively charged nucleus begins to attract electrons orbiting the other nuclei. Eventually, the electron orbits overlap and fuse. You can count the number of electrons shared in the outer shells in Figure 2.7b—eight for oxygen and two for each of the hydrogens—and see that all outer shells are filled.

This sharing of a pair of electrons creates a **single bond.** A single bond is often represented in a simple, two-dimensional way by a single dash. Water, with its two single bonds, is designated H—O—H. An atom of some elements, such as carbon, can share two pairs of electrons with the same atom, thus forming a **double bond.** A double bond is usually represented with two dashes, as in molecular carbon: C=C. Double bonds between the carbons in certain kinds of dietary fat seem to play a role in the development of heart disease. In addition, an unpaired electron orbiting a molecule can sometimes cause problems in the human body (see Box 2.2).

In most molecules, the electrons spend as much time orbiting one nucleus as the other. In a case like this, the electrical charge is evenly distributed about both ends, or *poles,* of the molecule. Because of this equal sharing, the molecule

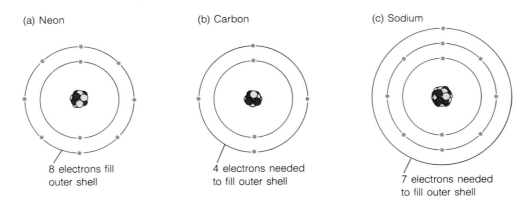

(a) Neon   (b) Carbon   (c) Sodium

8 electrons fill   4 electrons needed   7 electrons needed
outer shell   to fill outer shell   to fill outer shell

FIGURE 2.6 ■ **Why Atoms React or Remain Inert.** (a) If an atom's outermost energy shell is filled with a stable number of electrons, the atom does not need to accept or eject electrons to reach stability, so it remains inert, or unreactive with other atoms. Neon is inert because its first level is filled with two electrons and its second shell is filled with eight electrons. (b) Carbon, however, is reactive. Carbon has only four electrons in the second energy level, and so lacks the four additional electrons that would be needed to fill its outer shell. (c) Sodium is even more reactive than carbon. It is missing seven electrons in the third energy shell, having two and eight electrons filling the first and second shells, respectively. Sodium loses its lone electron very easily, leaving a filled second shell. For this reason, sodium is one of the most reactive chemical elements.

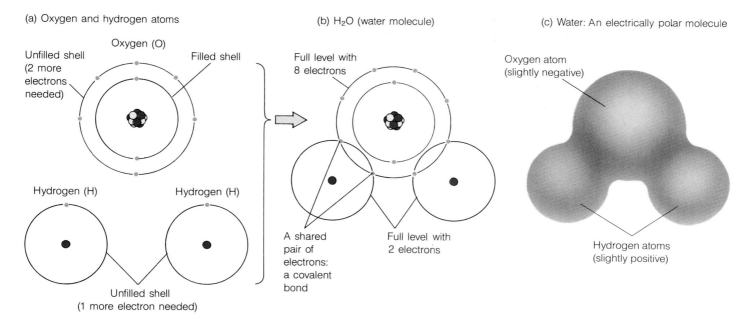

(a) Oxygen and hydrogen atoms

Oxygen (O)

Unfilled shell
(2 more
electrons
needed)

Filled shell

Hydrogen (H)   Hydrogen (H)

Unfilled shell
(1 more electron needed)

(b) H₂O (water molecule)

Full level with
8 electrons

A shared
pair of
electrons:
a covalent
bond

Full level with
2 electrons

(c) Water: An electrically polar molecule

Oxygen atom
(slightly negative)

Hydrogen atoms
(slightly positive)

FIGURE 2.7 ■ Covalent Bonds. **(a) The oxygen atom has only six electrons in its outer shell and needs two to be filled. Hydrogen atoms have one electron in their outer shell and need one to be filled. (b) In the H₂O molecule, the oxygen nucleus with its six protons attracts the two hydrogen electrons, the outer shells overlap, and the atoms share electron pairs in two covalent bonds. This sharing fills the outer shells of all three atoms. (c) The electrons in a water molecule spend more time orbiting the oxygen atom than the two hydrogens. This causes the oxygen atom to act as if it is slightly negatively charged (or, in chemical terms, *electronegative*) and the two hydrogen atoms to act as if they are slightly positively charged (*electropositive*). This kind of bonding is called a polar covalent bond, because regions (poles) of the molecule have slightly positive or negative charges.**

BOX 2.2   P E R S O N A L   I M P A C T

FREE RADICALS AND CIGARETTE SMOKING

Some molecules are dangerously reactive owing to a change in their electrons. Such a molecule, which usually contains oxygen, has a single unpaired electron orbiting a nucleus and hence is called a *free radical.* Free radicals are highly damaging to cells and tissues of living things because they react rapidly with any neighboring molecule and thus destroy it. In light of this fact, chemists recently discovered one more reason not to smoke: Cigarette smoking generates several kinds of free radicals.

Recall that in covalent bonds, two atoms share a pair of electrons. The heat of the smoldering cigarette alters the shared electron pairs of covalent bonds to produce free radicals. Since one of its electrons lacks a partner, a free radical grabs an electron from a stable compound. This stabilizes the former free radical but damages the new compound in the process. Tobacco smoke containing reactive free radicals is drawn into the lungs along with the hot gases. Here, chemists suspect, the free radicals remain active for several minutes after each inhalation,

altering compounds in nearby cells in the way just described. If the cellular compound is crucial to the cell's activity, then the cell may be injured or killed.

Some biologists believe that free radicals from cigarette smoke, from foods such as rancid fats and oils, from sunlight, and from other sources may play a major role in the process of aging and in causing cancer by damaging the genetic material in cells. These facts on free radicals help explain why the average two-pack-a-day smoker dies eight years earlier than his or her nonsmoking friends. The best way to decrease one's risk of cell damage from free radicals obviously is to avoid agents that create them, such as sunlight and cigarette smoking. But in addition, some compounds minimize free radical reactions, such as the fat-soluble vitamin E, the water-soluble vitamin C, and beta-carotene, the substance that gives carrots their orange color. Eating foods rich in these nutrients may help guard us against those free radicals we cannot easily avoid.

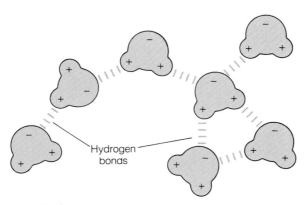

Water molecules

FIGURE 2.8 ■ Hydrogen Bonds. A hydrogen atom can be shared between the negative regions of two molecules to form a hydrogen bond. Such bonds form between the negative regions of water molecules. This kind of bond is weak and easily broken, but it figures prominently in the molecular interactions within living organisms.

is said to be **nonpolar.** In a molecule like water, however, the electrons spend more time orbiting the oxygen than the hydrogen (Figure 2.7c). This leaves the oxygen pole of the molecule with a slightly negative charge and the hydrogen pole of the molecule with a slightly positive charge. In a case like this, the molecule is said to be **polar.** The three-dimensional shape of the water molecule reflects this polarity; rather than the linear arrangement H—O—H, the two hydrogens are displaced toward one end, a bit like the corners of a triangle:

$$
\begin{array}{c}
O \\
/ \quad \backslash \\
H \qquad H
\end{array}
$$

This shape and polarity can lead to the formation of another kind of chemical bond—the hydrogen bond.

■ Hydrogen Bonds A polar molecule can interact with another polar molecule because of the slight charges at their poles. In liquid water, for example, the molecules form the structure shown in Figure 2.8, with a hydrogen from one water molecule forming a bond with an oxygen from a different water molecule, a result of the attraction of opposite charges. This sort of bond between a polar molecule with a negative pole and the slight positive charge on a hydrogen atom is called a **hydrogen bond.** Hydrogen bonds are much more easily broken and re-formed than are covalent bonds. Some of water's unusual properties (such as the tendency to form droplets) are based on hydrogen bonds, and some important biological molecules (such as the protein molecules that make up your hair) have millions of these bonds, contributing to the shape and function of the molecule.

Using a curling iron is a good demonstration of how easily hydrogen bonds can be broken and re-formed in protein molecules. The application of heat breaks the thousands of hydro-

gen bonds between the protein molecules that make up your hair—bonds that held it in its original shape (Figure 2.9). As the hair cools in its new curled shape, new hydrogen bonds form between the protein molecules and maintain—at least for a while—a different shape with curls or waves.

■ Ionic Bonds The third type of molecular bond depends on the complete transfer of electrons from one atom to another, rather than the sharing of electrons. This is most common when an atom's outermost orbital contains just one or two electrons or needs just one or two to be filled. Common table salt (sodium chloride, NaCl) is a good example (Figure 2.10a and b). To be stable, the sodium atom needs to lose one electron and the chlorine atom needs to gain one (Figure 2.10c). When these two atoms encounter each other, an electron leaves the sodium atom and begins to orbit the chlorine nucleus, creating two ions. The oppositely charged ions attract each other, rather like magnets, and an **ionic bond** forms between them that holds the atoms together as a sodium chloride molecule. These same attractions can cause the formation of a regular latticework of many molecules, called an *ionic crystal.* Ionic bonds are much stronger than hydrogen bonds, but not as strong as covalent bonds; when dissolved in water, ionic molecules often dissociate (fall apart) into ions once again.

Our discussion of bonding helps show how the properties of molecules emerge from the internal organization and elec-

FIGURE 2.9 ■ Hydrogen Bonds and Curling Hair. Hydrogen bonds are relatively weak and can be broken readily by heat. Each protein molecule in a person's hair is linked to the neighboring molecules by hydrogen bonds. When hair is wrapped around a curling iron, the heat breaks the original hydrogen bonds. When the curled hair cools, new hydrogen bonds form, linking new molecules together and giving the hair a curled, but more stressed, form. Unfortunately, these new hydrogen bonds slowly break under the strain—especially in damp weather—and allow the hair to return to its original form. Here, a young woman tightly curls one side of her long hair.

trical activities of their constituent atoms. Salt is an excellent example of such emergent properties. Ionic crystals emerge from the molecular structure of NaCl, and visible salt crystals emerge from the structure of submicroscopic ionic crystals. The water molecule is another example: This single remarkable chemical entity makes up most of the matter in most living organisms, and its structure and bonding properties make it uniquely suited to biological systems.

LIFE AND THE CHEMISTRY OF WATER

Our planet and its life forms are practically synonymous with water: Fully three-quarters of the earth's surface is covered by ocean. Life began in this aquatic realm, and hundreds of thousands of species still live in water today. Since water is transparent, light can penetrate for many meters and allow photosynthesis to take place, supporting much aquatic life. What's more, organisms themselves—whether in the sea or on land—are 50 to 90 percent water. What makes water so special chemically and so necessary to life?

The answer is that hydrogen bonding between water molecules and other substances makes life possible. Recall that water is a polar molecule with positive and negative ends that can attract the oppositely charged portions of other water molecules. This attraction has a profound effect on the properties of water.

Water, Temperature, and Life

Anyone who has ever had frostbite knows that water behaves differently at different temperatures and that these behaviors can have a significant impact on living cells. When a person is caught in a blizzard without gloves, the liquid water in his or her fingers turns to ice crystals, which can tear cells apart and cause pain, dysfunction, and even loss of the digits. It is fortunate for living things that water usually remains liquid, within the normal temperature range found on earth, instead of assuming one of its other physical forms—gas (water vapor) or solid (ice). If our planet were much hotter or much colder, this would not be the case.

Living things are also lucky that water is slow to heat; that is, the amount of heat needed to raise the temperature of a certain volume of water is greater than for most other liquids. This property is a direct result of water's hydrogen bonds. Much of the heat energy applied to water—say, the sunshine striking a small pond in the Australian desert—goes into stretching or breaking hydrogen bonds instead of raising the temperature. This is important to living things because it helps ensure the relatively constant external and internal environments they need. Oceans, large lakes, and rivers are all slow to change temperature, and thus aquatic life forms do not have to cope with rapid temperature changes in their surroundings. On land, too, an organism's temperature must remain fairly constant. The fact that living things are largely

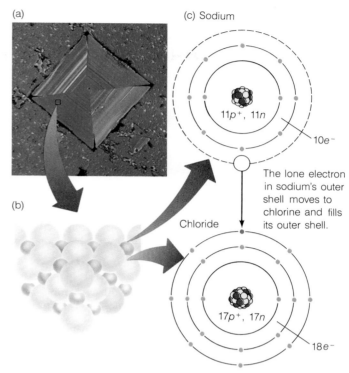

FIGURE 2.10 ■ Ionic Bonds. Oppositely charged ions attract each other like magnets and form ionic bonds. (a) The cubic structure of a salt crystal, enlarged 14 times in this photograph. (b) The regular three-dimensional latticework of sodium and chloride ions in a salt crystal results from ionic bonds, the electrical attraction of a positively charged sodium ion and a negatively charged chloride ion. (c) A sodium atom contains one lone electron in its third energy shell, while a chlorine atom lacks just one electron to complete its third shell. If the lone outer electron (here, colored red) leaves the sodium atom and joins the chlorine atom, then both will have filled outer shells. The result is a sodium ion with a positive charge (because it has one more proton than electrons) and a negatively charged chloride ion (having one more electron). These ions attract each other and form an ionic bond.

water means that body temperature, especially of large organisms like crocodiles or large sea bass, also changes slowly.

An Australian living in the great central desert and sweating under the hot sun, or a college student perspiring at the end of a vigorous tennis match, can appreciate another important temperature-related property of water: An unusually high amount of heat is required to turn liquid water to water vapor. Before a water molecule can *evaporate* (leave the liquid phase and enter the gaseous phase), it must begin to jostle about rapidly enough to fly off the surface of a water droplet. And before it can really jostle, the hydrogen bonds linking the molecule to its neighbors must be broken by heat energy. Thus, water must absorb a great deal of heat from its surroundings before evaporating, and in the process of absorbing heat, it cools its surroundings. For a living thing, this property provides a natural cooling system and explains the evolution of sweat glands in horses, people, and many other mammals.

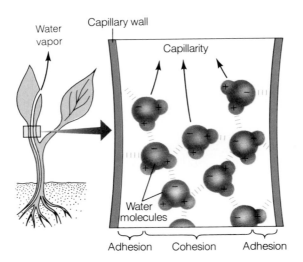

FIGURE 2.11 ■ **Water's Mechanical Properties: Cohesion,
Adhesion, and Capillarity.** As a result of hydrogen bonding,
water molecules cling to other water molecules—a property of
like bonding to like called cohesion. Water molecules also form
hydrogen bonds with unlike molecules, the property known as
adhesion. Finally, water tends to move upward through a narrow
space because water molecules adhere to the walls above and to
the other molecules below. This is called capillarity. Water movement in plants reflects all three mechanical properties. Most
plants have very fine, continuous transport "tubes" running from
the roots up through the stem to the leaves. Water enters the
roots and moves up the vessels partly because of cohesion and
partly because of adhesion to the vessel walls and capillarity.
Water molecules form long, unbroken chains with great tensile
strength, and these chains are drawn up link by link as water
evaporates from the leaves. Without the mechanical properties of
water, magnificent towering eucalyptus and redwood trees could
never have evolved—nor, in fact, could other medium-sized or tall
plants.

Envision an Australian aborigine hunting beneath the blazing
desert sun. Sweat pours from the glands in his skin and covers
parts of his body surface in a thin layer. As the sweat evaporates, large amounts of body heat transfer from the body to
the skin and break the hydrogen bonds in the water molecules.
This heat eventually dissipates, and the person's body is
cooled.

Mechanical Properties of Water

In addition to its properties relating to temperature, water has
several mechanical properties also based on hydrogen bonding, and these, too, are important to living things.

Water molecules exhibit **cohesion,** or the tendency of like
molecules to cling to each other, and **adhesion,** the tendency
of unlike molecules to cling to each other (e.g., water to paper,
soil, or glass) (Figure 2.11). Together, cohesion and adhesion
account for **capillarity,** the tendency of a liquid substance to

move upward against the pull of gravity through a narrow
space, such as between the fibers of a paper towel being used
to wipe up spilled milk, or up the inside walls of a narrow
glass tube, or into a wad of dry grass that an aborigine might
have used to soak up moisture from a hollow tree. An example of the importance of adhesion of water molecules is in the
transport of water in many kinds of plants (see Figure 2.11).
Without the adhesion of water molecules to each other, there
would be no tall trees such as redwoods and eucalyptus.

One mechanical property of water must be overcome the
minute a newborn draws its first breath. **Surface tension** is
the tendency of molecules at the surface of liquid water to
cohere to each other but not to the molecules of air above
them (Figure 2.12). The water molecules in a newborn's lungs
cohere so tightly that they could cause the lungs to collapse.
The body, however, produces special molecules called surfactants that function like detergent molecules and lower the surface tension of the water in the lungs, circumventing this collapse. Premature babies born before their lungs are capable of
generating surfactants sometimes die because of this collapse.

Water has yet another mechanical property with implications for biology: the tendency of ice (solid water) to float in
liquid water (Figure 2.13a). The hydrogen bonds in ice create

FIGURE 2.12 ■ **Surface Tension.** A baby's first breath is facilitated by a special coating called a surfactant, which the lungs
secrete. This material acts much like a detergent to decrease the
surface tension of the fluid layer lining the lungs. Without the
surfactant, hydrogen bonds in the water lining the small sacs of
the lungs would pull water molecules together so tightly that the
sacs would collapse. Newborns unable to produce enough surfactant can suffer collapsed lungs and die of respiratory failure.

(a) Why does ice float on water?

(b) Hydrogen bonds generate an open lattice in ice.

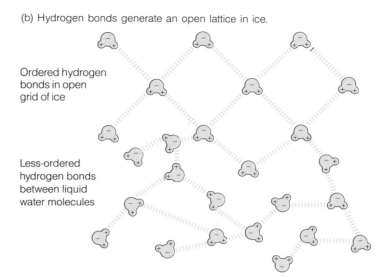

Ordered hydrogen bonds in open grid of ice

Less-ordered hydrogen bonds between liquid water molecules

FIGURE 2.13 ■ Why Ice Floats. (a) Many organisms would perish if it weren't for the simple fact that ice floats, insulating the liquid lower depths. (b) Frozen water is less dense than liquid water because the hydrogen bonds in ice create a rigid, open latticework. Hence, this less dense material floats in the more dense water.

a fairly rigid, open latticework that holds water molecules farther apart than the flexible bonds of the liquid form. Thus, ice is less dense than liquid water and floats in it (Figure 2.13b). This is a unique property; other substances become more dense when they freeze. If ice did not become less dense, lakes would freeze solid in winter, and only the top few centimeters would thaw in summer. Instead, with a frozen top layer insulating the lower depths, plants and animals can survive winter in the chilly—but liquid—water beneath the ice.

Chemical Properties of Water

While washing dishes is no fun, it can—if one is in the right frame of mind—serve as a reminder of the chemical properties of water so important to living cells. Dishwater—in fact, all water—is a **solvent,** a substance capable of dissolving other molecules. Dissolved substances are called **solutes.** Water can dissolve polar compounds, such as table sugar, and most kinds of ionic molecules, such as table salt. When polar sugar molecules—syrup on dirty plates, let's say, or glucose molecules inside cells—become surrounded by water molecules, hydrogen bonds form. With an ionic solute like salt, the component ions dissociate, and each becomes surrounded by an oriented cloud of water molecules (Figure 2.14a).

Compounds such as sugar and salt that dissolve readily in water are called **hydrophilic,** or "water-loving," compounds. In contrast, nonpolar compounds, such as cooking oils and animal fats on dirty dinner dishes, do not dissolve readily and are called **hydrophobic,** or "water-fearing," compounds. Instead of dissolving, they form a boundary, or *interface,* with oil molecules cohering to oil molecules in a film at the surface

of the dishwater (Figure 2.14b). (A detergent, of course, can help dissolve the grease.) Hydrophobic interactions are essential to life, because, as we shall see, every cell is a miniature pool of water and dissolved substances surrounded by an envelope of hydrophobic fatty compounds.

Water Dissociation: Acids and Bases

Water has yet another property—a slight tendency to fall apart—again with significant implications for all living things, particularly those organisms subjected to acid rain, which forms downwind from our huge industrialized regions. Just as NaCl tends to *ionize,* or dissociate, into positive and negative ions when dissolved, water molecules themselves ionize into a positively charged hydrogen ion (H^+) and a negatively charged hydroxide ion (OH^-):

$$H_2O \rightarrow H^+ + OH^-$$

This falling apart, however, is a relatively rare event in pure water; only about 2 out of every 10^9 water molecules tend to dissociate at a given time. By definition, an **acid** is any substance that gives off hydrogen ions when dissolved in water and thereby increases the H^+ concentration of the solution, while a **base** is any substance that accepts hydrogen ions when dissociated in water. The concentration of hydrogen ions is important to living cells because many of the chemical reactions that drive life processes—the digestion of foods, for example—depend on concentrations of these ions.

Biologists measure hydrogen ion concentration on the **pH scale,** which ranges from 1 to 14; on this scale, water has the

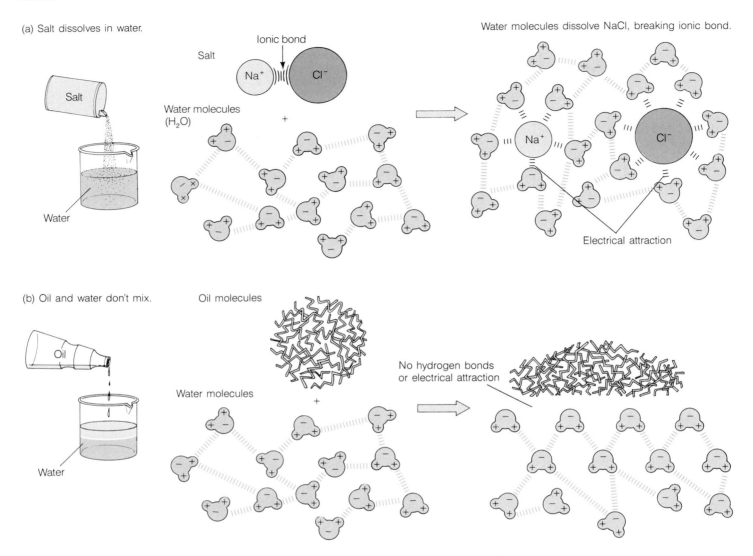

(a) Salt dissolves in water.

Ionic bond

Salt

Water molecules dissolve NaCl, breaking ionic bond.

Water molecules (H$_2$O)

Electrical attraction

(b) Oil and water don't mix.

Oil molecules

Water molecules

No hydrogen bonds or electrical attraction

FIGURE 2.14 ■ Hydrophilic Compounds Dissolve Readily, Hydrophobic Ones Do Not. (a) Salt is hydrophilic, or "water loving," because its charged ions form hydrogen bonds with water molecules. (b) Oil is hydrophobic, or "water fearing," because its chemical structure is such that hydrogen bonds do not usually form, and hence an oil/water boundary develops.

neutral value of 7, in the middle of the scale. Acidic solutions have pH values between 0 and 7. Basic solutions, by comparison, have pH values between 7 and 14. The pH inside most cells also stays fairly neutral, between about 6.5 and 7.5, and it is only within this narrow range that many vital cellular reactions take place at optimal speed (see Chapters 4 and 5). The pH scale is logarithmic (pH $= -\log_{10}[\text{H}^+]$, i.e., minus the log to the base ten of the hydrogen ion concentration), and thus a solution with pH 6 has a hydrogen ion concentration of 10^{-6} mole per liter,* which is ten times higher than a solution with a pH of 7 (10^{-7} mol/L).

Figure 2.15 shows the pH of common solutions. Notice that one of the entries at pH 4 is acid rain. This forms when oxides of sulfur given off by car engines and industrial smokestacks dissolve in water droplets in the atmosphere and form sulfuric acid. When this acid is carried to earth in rain or snow—now with a pH as low as 3 or 4 in areas downwind of the industrialized and heavily populated regions—the precipitation burns the leaves and tender twigs of trees (Figure 2.16), stunts fish, and can essentially eliminate most life forms from small lakes and streams. We'll return to the subject of acid rain in Chapter 36.

You may, at some time or other, have had a personal encounter with high levels of acid in the form of acid indigestion. This uncomfortable condition is caused when cells in

*mol/L; see the abbreviations chart on the inside back cover. A mole is 6.023 × 10^{23} particles of a substance, or 1 gram molecular weight, which is the weight in grams equal to the atomic weight of each atom multiplied by the number of those atoms.

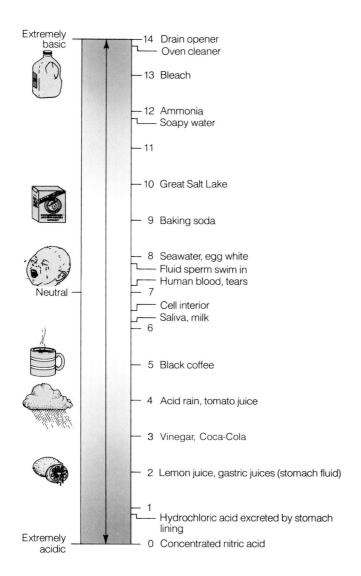

FIGURE 2.15 ■ **The pH Scale.** This scale is a representation of the hydrogen ion concentration in common biological and non-biological substances. The more hydrogen ions, the more acidic a substance; the fewer hydrogen ions, the more basic. In a solution, the concentration of H^+ times the concentration of OH^- is always the same, whether we are talking about battery acid or household cleaning ammonia. This constant value is 10^{-14}. In pure water, the concentrations of H^+ and OH^- are the same; they are both 10^{-7} (and recall that 10^{-7} times 10^{-7} is 10^{-14}). Because of this constant feature, H^+ and OH^- concentrations are at the opposite ends of a seesaw: Whenever the concentration of H^+ goes up, the concentration of OH^- must go down, and vice versa. So if the concentration of H^+ is 10^{-3}—the concentration you would find in a carbonated soft drink—the concentration of OH^- would be 10^{-11} (since 10^{-3} times 10^{-11} is 10^{-14}). To avoid dealing with cumbersome and sometimes confusing exponents, chemists have devised the pH scale to measure H^+ concentrations. *To get the pH from the concentration of H^+ is easy: Just take the exponent without the minus sign.* Thus, water has a pH of 7, since the concentration of hydrogen ions is 10^{-7}, and carbonated pop has a pH of 3, since the concentration of hydrogen ions is 10^{-3}.

your stomach secrete too much hydrochloric acid (HCl). While the acid speeds up the digestion of food, it can also irritate the stomach lining. When hydrochloric acid dissolves in water (in your stomach or elsewhere), most of the molecules dissociate and release H^+ into the solution:

$$HCl \rightarrow H^+ + Cl^-$$

Thus, HCl is a very strong acid.

When basic, or **alkaline,** substances dissociate in water, instead of giving off hydrogen ions, they combine with them, resulting in an excess of OH^-. For example, NaOH dissociates into sodium and hydroxide ions:

$$NaOH \rightarrow Na^+ + OH^-$$

The sodium ions dissolve in the water, and some of the freed hydroxide ions combine with free H^+ to form H—OH (or, more familiarly, H_2O), leaving fewer hydrogen ions than before NaOH was added to the water.

As an antidote to acid indigestion you may have taken some bicarbonate—the active ingredient in heartburn reme-

FIGURE 2.16 ■ **Acid Rain and a Dead Forest. Skeletal Fraser fir trees stand at the summit of Mount Mitchell, North Carolina, stark reminders of the environmental consequences of air pollution.**

dies. Bicarbonate can act as a **buffer.** When hydrogen ion concentrations are high, buffers bind to H^+; when hydrogen ion concentrations are low, buffers release H^+. Thus, buffers regulate pH by "soaking up" or "doling out" hydrogen ions as needed. The bicarbonate ion (HCO_3^-) also helps maintain blood pH at about 7.2 by soaking up hydrogen ions when the blood is too acidic and doling out hydrogen ions when the blood is too alkaline.

We have seen in this section that water—the sparkling substance of ocean waves, waterfalls, snow drifts, clouds, and icebergs—is also the major ingredient in living things. And we've also seen that the molecular properties of water, based on the atomic structures of oxygen and hydrogen and on the versatile hydrogen bond, make it possible for people to stay cool even in the desert, for plants to grow tall, and for life to occur in its myriad forms. Before we can truly understand the chemistry of life, however, we must consider another set of materials that makes up living things: compounds containing carbon.

THE STUFF OF LIFE: COMPOUNDS CONTAINING CARBON

Despite the preponderance of oxygen and hydrogen in living organisms, fully 18 percent of a person's weight comes from carbon atoms, and for a large tree, that figure can approach 50 percent. So prevalent and so important are carbon atoms to living things that the study of compounds containing carbon is called **organic** ("related to organisms") **chemistry,** leaving to **inorganic** ("lifeless") **chemistry** all other substances. Interestingly, of the two divisions, organic chemistry deals with a far larger number of compounds. Why? Because the structure of the carbon atom is such that carbon can form millions of different combinations with other atoms—ten times more than all the compounds formed by the dozens of other elements put together. This bonding versatility is carbon's key characteristic, and it explains why crustal elements such as silicon and aluminum, which form relatively few compounds, could never have formed the basis of life with all its many manifestations. Four classes of organic compounds—*carbohydrates, lipids, proteins,* and *nucleic acids*—are vital to the structure and function of living things and together with a few other materials account for the diverse shapes, colors, textures, and characteristics of organisms.

Carbon: Compounds and Characteristics

The carbon atom has six electrons, and as we saw in Figure 2.4, two of these fill the first energy level and four occupy half of the eight available spots in the second energy level. Because the carbon atom is "looking for" four more electrons, carbon can form covalent bonds with up to four atoms at a time. Methane gas (CH_4), sometimes called swamp gas,

is a simple example (Figure 2.17a). Many compounds in living cells, however, are larger and have a "backbone" of several carbon atoms bonded to each other in long straight chains, branched chains, or rings (Figure 2.17b).

While an organic molecule's backbone shape contributes to its role in a cell, small clusters of atoms called **functional groups** often hang from the backbone and impart specific chemical behaviors to the molecules to which they are attached. Table 2.1 on page 40 shows seven biologically important functional groups and their properties. Note that ring-shaped glucose molecules, such as the one shown in Figure 2.17b, have several hydroxyl (OH) functional groups.

Figure 2.18 shows the structure of an intriguing carbon compound and the functional groups it contains. Allin is the substance that is converted inside the garlic bulb into the strong odor we associate with that vegetable. Notice that it has several functional groups, including the amine group (NH_2), the hydroxyl group (OH), and the carboxyl group (COOH). The presence of these various groups and the way they are arranged on the carbon-sulfur backbone give garlic the characteristics we know so well.

Another characteristic of carbon compounds is closely tied to the presence of certain functional groups. Some of the most important classes of biological molecules, including carbohydrates and proteins, are long chains called **polymers** ("many parts"). Like strings of beads, polymers are made up of smaller subunits, called **monomers** ("single parts"), linked together at certain functional groups by covalent bonds. As an example, consider the carbohydrate polymer called *starch,* the main nutrient in potatoes. The subunits, or monomers, in starch are hundreds or thousands of molecules of the simple sugar glucose joined together into the long polymer chain (Figure 2.17b and c). Other biological polymers are constructed in a similar way. As we will see shortly, it is the chemical characteristics of the simpler component subunits that determine the distinctive biological functions of the various polymers and enable them to carry out very different tasks in the living cell. Let's start now to investigate the four main classes of large biological molecules: the carbohydrates, lipids, proteins, and nucleic acids.

Carbohydrates

Plants and fungi are mostly water and carbohydrates. Considering the earth's great forested expanses, its grassy plains, and all the aquatic plants of lake and sea, by far the most abundant carbon compounds in living organisms are carbohydrates, and they serve both as structural components of cells and as an energy reserve to fuel life processes. The term *carbohydrate* comes from "hydrate (or water) of carbon," and indeed, **carbohydrates** generally contain carbon, hydrogen, and oxygen in the ratio of 1:2:1 (CH_2O). For example, the sugar in grapes, called glucose, has the formula $C_6H_{12}O_6$. The monomers in carbohydrates are sugar molecules called **monosaccharides** ("single sugars") that are linked into two-unit molecules

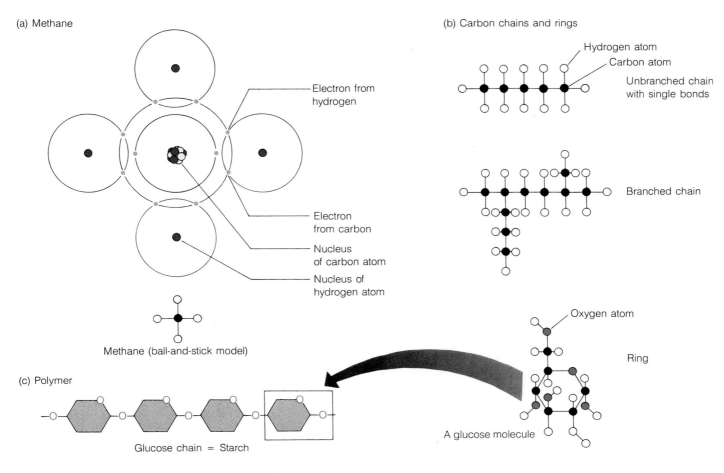

(a) Methane

Electron from hydrogen

Electron from carbon

Nucleus of carbon atom

Nucleus of hydrogen atom

Methane (ball-and-stick model)

(b) Carbon chains and rings

Hydrogen atom

Carbon atom

Unbranched chain with single bonds

Branched chain

Oxygen atom

Ring

A glucose molecule

(c) Polymer

Glucose chain = Starch

FIGURE 2.17 ■ **Carbon Compounds: The Stuff of Life.** (a) Carbon can form millions of organic compounds because each atom can form covalent bonds with up to four other atoms, such as in methane, CH_4. Throughout the remainder of this chapter, we will mainly use the two-dimensional ball-and-stick molecular models for clarity, with each colored ball representing an atom and each stick a covalent bond. Double or triple sticks represent double or triple bonds (that is, two or three shared pairs of electrons). (b) Many organic compounds have unbranched, branched, or circular (ring) carbon "backbones," with hydrogens or other atoms projecting. Many important biological molecules are hundreds of times larger than the simple molecules depicted here. For large molecules, the ball-and-stick model depicting every atom can become unwieldy. Thus, we can use an even simpler version, with the carbons of the ring omitted but implied by corners. (c) Here a polymer made up of four glucose subunits of monomers is shown. Compare these monomers to the glucose molecule in part b.

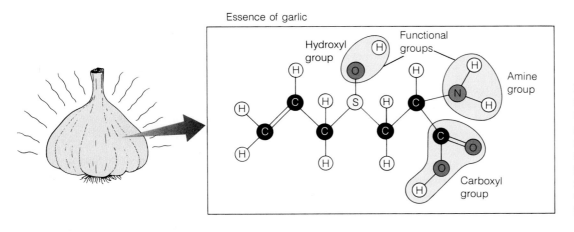

Essence of garlic

Hydroxyl group

Functional groups

Amine group

Carboxyl group

FIGURE 2.18 ■ **Functional Groups at Work in Organic Compounds.** One complex substance in garlic, called allin, has three functional groups that give the molecule (and the whole garlic bulb) the properties we recognize: a distinctive, strong smell with staying power on the breath.

TABLE 2.1   Several Common Functional Groups and the Properties They Impart to Compounds

Name, Symbol, and Structure of Group		Characteristics
Hydroxyl group (OH)	R—O—H	Polar, water-soluble, involved in hydrogen bond formation; found in alcohols
Carboxyl group (COOH)	R—C(=O)—O—H	Weak acid (H⁺ donor); gives molecules the properties of an acid; found in organic acids like vinegar
Amine group (NH₂)	R—N(H)(H)	Weak base (H⁺ acceptor); gives molecules the properties of a base; found in proteins and urine
Aldehyde group (COH)	R—C(=O)—H	Polar, water-soluble, common in sugar molecules; found in molecules exhaled on breath after drinking alcohol
Keto group (CO)	R,R—C=O	Polar, soluble, common in sugar molecules; found in molecules expelled in urine when on high-fat diet
Methyl group (CH₃)	R—C(H)(H)(H)	Hydrophobic (insoluble); gives molecules this property; found in fats and oils
Phosphate group (PO₄)	R—O—P(=O)(O—H)(O—H)	Acidic; gives molecules properties of an acid; found in molecules used in energy flow within organisms; occurs in DNA molecules

Legend: R = rest of molecule

Disaccharides are the common form in which sugars are transported inside plants and result when two simpler sugars combine; for example, glucose bonds to fructose to yield **sucrose,** or table sugar (Figure 2.19a). Sucrose is abundant in the saps of sugarcane, maple trees, and sugar beets—our major sources of sugar for refining (Figure 2.19b). Honey is also made up of glucose and fructose and has virtually no more nutrient value than refined sugar. The predominant

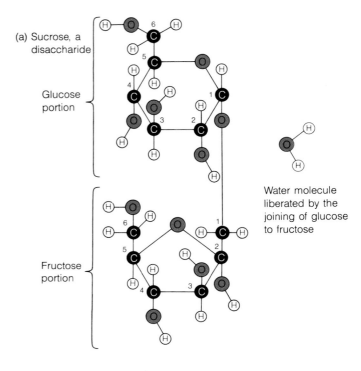

(a) Sucrose, a disaccharide

Glucose portion

Fructose portion

Water molecule liberated by the joining of glucose to fructose

(b) Sugar beet, a source of sucrose

FIGURE 2.19 ■ Simple Carbohydrates: Sugars. (a) Sucrose is a disaccharide that forms from the condensation reaction between two monosaccharides, glucose and fructose. **(b)** Sugar beets, the enlarged storage roots of the sugar beet plant, store large quantities of sucrose in special cells. Sugar beets are a major source of the millions of tons of sugar Americans consume each year.

called **disaccharides** (Figure 2.19) or into polymers called **polysaccharides** (''many sugars''; Figure 2.20).

The monosaccharides **glucose, fructose,** and **galactose** are the most important carbohydrate monomers, since those units make up the complex carbohydrates in starch, wood, and other biological materials. As sugars, some of these monomers have a sweet taste. Fructose, for example, is called fruit sugar because it is the compound that gives many kinds of fruit their sweet flavor. Glucose is the universal cellular fuel, broken down by virtually all living things with the release of energy stored in its bonds. Glucose and its conversion products will figure prominently in our discussions of energy in Chapter 5. Galactose is a common subunit of many other sugars, such as milk sugar. Glucose, galactose, and fructose share the molecular formula $C_6H_{12}O_6$, but the different properties of the three sugars emerge from their different arrangements of the same atoms.

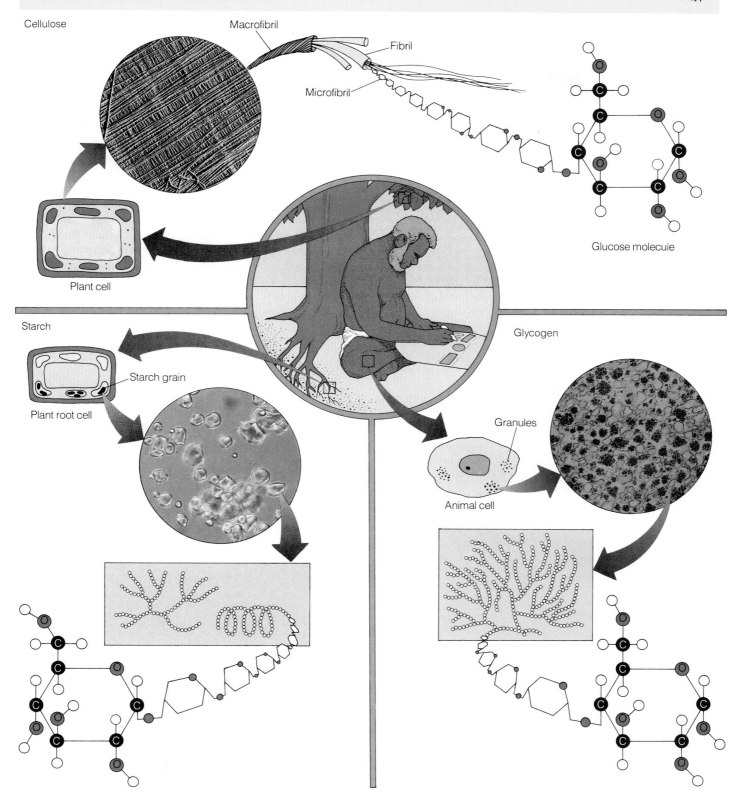

FIGURE 2.20 ■ **Complex Carbohydrates: Cellulose, Starch, and Glycogen.**   The roots of trees and other plants that a traditional aborigine might have harvested for food contain cells that store starch grains. These grains are made up of branched or coiled starch molecules, which are themselves long polymers of glucose subunits. The plant cell walls contain cellulose, a tough, fibrous structural material. Cellulose occurs in cablelike bundles, or macrofibrils, made up of smaller cords or fibrils that contain even smaller strands or microfibrils. Each microfibril strand is composed of cellulose molecules, large polymers of glucose subunits that do not coil or twist. The aborigine's muscles and liver cells contain granules of a third complex carbohydrate, glycogen, which is very highly branched and can be broken down rapidly and supply quick energy for muscular activity.

sugar in milk is **lactose,** galactose bonded to glucose. **Maltose**—two joined glucose subunits—gives barley seeds a sweet taste; beer brewers ferment the sugars in barley and other grains into alcohol.

Polysaccharides are carbohydrate polymers formed by the condensation of simple sugars into long chains, which may be straight or branched. The polysaccharides **starch** and **glycogen** are molecular storage bins that serve as the primary energy reserves of plants and animals, respectively. **Cellulose** and **chitin** are structural polysaccharides that give form and rigidity to plants, insects, and other oganisms.

Starch is a polysaccharide composed of long chains of glucose subunits. The bonds between the glucose subunits allow the chains to coil and form granules that are stored in plant tissues and seeds. Most humans consume a diet that is mainly starch gathered from the seeds of rice, wheat, corn, and other cereal plants grown agriculturally. People who lived by hunting and gathering food, such as the traditional Australian aborigines, also depended heavily on starchy roots and seeds (Figure 2.20).

In people and other animals, energy in the muscles and liver is stored in the form of highly branched glycogen molecules composed of many short chains, each containing 10 to 12 glucose units (see Figure 2.20). Some chemists think that this branching allows for faster liberation of the energy stored in the bonds between and within the glucose molecules. An animal suddenly confronted with a fierce and hungry predator needs such a surge of energy for its rapid escape.

Cellulose has a structure somewhat similar to that of starch: long chains of glucose units connected by bonds. But cellulose is a tough, fibrous structural material (the stuff of wood and paper), not a soft, water-soluble material like starch. The orientation of the bond in cellulose prevents twisting and imparts a rigidity to the chains that make them ideal components for the cell walls in crisp leaves and woody stems (see Figure 2.20). Since people cannot digest cellulose, it provides us with no calories. Food technologists have recently learned to process cellulose into a flour they call "fluffy cellulose." It has no calories and can be mixed with starch and baked into diet-conscious cakes and bread.

(a) Saturated fatty acids

Bacon

Saturated fatty acids

Glycerol portion

Acid group

Chain of carbon and hydrogen

Fatty acid portion

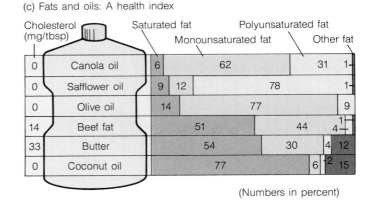

(c) Fats and oils: A health index

Cholesterol (mg/tbsp)		Saturated fat	Monounsaturated fat	Polyunsaturated fat	Other fat
0	Canola oil	6	62	31	1
0	Safflower oil	9	12	78	1
0	Olive oil	14	77		9
14	Beef fat	51	44	4	1
33	Butter	54	30	4	12
0	Coconut oil	77	6	2	15

(Numbers in percent)

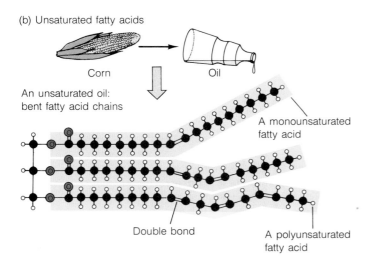

(b) Unsaturated fatty acids

Corn

Oil

An unsaturated oil: bent fatty acid chains

A monounsaturated fatty acid

Double bond

A polyunsaturated fatty acid

FIGURE 2.21 ■ **Triglycerides: Fat and Oils.** **(a)** A fat such as the solid white part of bacon contains mostly straight-chain fatty acids with no single bonds. The fatty acids are saturated with hydrogen. A diet high in saturated fats appears to contribute to diseases of the heart and blood vessels. The long straight chains can pack close together, generating the solid fat. **(b)** An oil such as this golden liquid corn oil generally contains unsaturated fatty acids. Some positions in the fatty acids have double bonds instead of single bonds. The double bonds cause the fatty acid molecules to bend and kink; thus, they cannot pack closely together, and the result is a slippery liquid. **(c)** Not all oils are equally healthful in the diet, since they contain different percentages of saturated and unsaturated fats and may or may not contain cholesterol. Monounsaturated fats have one double bond and lower your LDL, the "bad cholesterol" component in the blood. Polyunsaturated fats have several double bonds and lower the total count of cholesterol in the blood.

Starch and cellulose are excellent examples of the emergent properties concept: The very different characteristics of the two materials, one soft and edible, the other tough and fibrous, arise from the way their identical subunits are arranged.

The structural carbohydrate chitin is a major component of the hard outer shells of insects, crabs, lobsters, and other animals (see Chapter 18), as well as the cell walls of certain bacteria and the tissues of mushrooms and other fungi (see Chapters 16 and 17).

In later chapters, we will talk more about the great importance of carbohydrates—as structural materials in plant cell walls, as nutrient storage compounds in most kinds of organisms, and as a medium for energy exchange in all forms of life.

Lipids

The term *lipid* may not have a familiar ring, but some of the compounds classed as lipids probably do: **Lipids** include the **fats,** such as bacon fat, lard, and butter; the **oils,** such as corn oil, coconut oil, and olive oil; the **waxes,** like beeswax and earwax; the **phospholipids,** which are important components of cell membranes; and the **steroids,** including certain vitamins, some hormones, and cholesterol (the blood vessel clogger). Lipids can serve as energy storage molecules (like carbohydrates) or as waterproof coverings around cells. However, the chemical structures of lipids are quite different from those of carbohydrates, and these differences account for the unique properties of each type of lipid molecule.

■ **Triglycerides: Fats and Oils**   Fats and oils have similar basic structures and serve as rich energy storage molecules. Each contains a three-carbon **glycerol** molecule bonded to three **fatty acids,** long chains of nothing but carbon and hydrogen atoms with acidic functional groups on the ends (Figure 2.21). Because three fatty acids are joined to the three carbons of the glycerol, fats and oils are usually called **triglycerides.**

You have probably heard warnings of how saturated fats in the diet are linked to heart disease, and perhaps you have wondered about the difference between saturated and so-called polyunsaturated fats. The fatty acids of **saturated fats** are "saturated" with hydrogen; that is, all their carbon atoms are bonded to at least two hydrogen atoms (Figure 2.21a). Saturated fatty acids are straight-chain polymers, and they pack together tightly as in the dense white fat in a slab of bacon. In contrast, **unsaturated fats** have one or more positions along the polymer chain where two adjacent carbons share a double bond, and hence there are fewer hydrogens bonded to the carbons at this spot (Figure 2.21b). The presence of double bonds causes kinks along the polymer chains of the unsaturated fatty acids, and this kinking decreases the ability of fatty acids to pack together. Unsaturated fats are therefore liquid oils. Nutritional research suggests that it is much healthier to consume very little saturated fat (like butter

FIGURE 2.22 ■ Waxes: Lipids That Form Solid Masses and Layers. Waxes have long-chain fatty acids combined with long-chain alcohols rather than glycerol. This structure gives waxes a sticky, solid, waterproof character. Waxes give plums their delicate whitish blush and help keep citrus fruit juicy.

and lard) and instead to cook with oils such as canola, safflower, and olive oils, which are high in unsaturated fatty acids.

■ **Waxes**   The soft but solid substance called wax—in boot polish, honeycombs, and earwax—is part of the lipid family, but its structure is different from that of triglycerides (Figure 2.22). Wax molecules are insoluble in water and can pack together to form solid masses and layers that serve as important waterproof coatings on leaves, bark, some fruits, and feathers; as the structural materials in the honeycombs of beehives; and as protective coatings for the ear canal.

■ **Phospholipids**   Another class of lipids, the phospholipids, has properties essential to the living cell. A phospholipid molecule has two fatty acid chains attached to a glycerol molecule, rather than three as in fats and oils, but it has a phosphate-containing group in place of the third chain (Figure 2.23a). This group creates a water-soluble (hydrophilic) region on an otherwise insoluble molecule. The overall shape can be visualized as a soluble ball (the phosphate-containing group) on an insoluble stick (the fatty acid chains; Figure 2.23b). When added to water, phospholipids form single layers, double layers, or spheres, with the polar "balls" oriented toward the water and the nonpolar tails oriented away from it. This characteristic behavior is crucial to living things, for as we will see in Chapter 3, every cell is surrounded by a double layer of phospholipids (Figure 2.23c), and this vital barrier allows the cell to maintain its watery contents and integrity as a living unit while still exchanging materials with the fluid environment surrounding it.

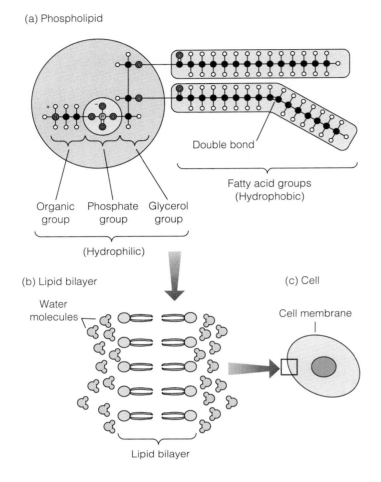

(a) Phospholipid

Double bond

Organic group | Phosphate group | Glycerol group

(Hydrophilic)

Fatty acid groups (Hydrophobic)

(b) Lipid bilayer

Water molecules

Lipid bilayer

(c) Cell

Cell membrane

FIGURE 2.23 ■ **Phospholipids as Waterproof Barriers in Living Organisms.** (a) Each phospholipid has a head (containing a phosphate group) and a tail (made up of two fatty acids). (b) The head is water-soluble (hydrophilic), but the tail is not. When phospholipid molecules are surrounded by water, they can form double layers (bilayers), with their heads oriented toward the water molecules and their tails oriented away from the water, toward the inside of the "lipid sandwich." (c) The membranes surrounding living cells contain a double layer of phospholipids.

■ **Steroids** The last major class of lipids is the steroids. Steroid molecules are insoluble in water, but they can dissolve in oils or in lipid membranes. Examples of steroids include various regulatory molecules that must pass in and out of cells easily, such as the sex hormones estrogen and testosterone (see Chapter 13) and vitamins A, D, and E (see Chapter 22). Figure 2.24a shows the common steroid cholesterol. Regularly manufactured by plant and animal cells, cholesterol is a necessary component of cell membranes that affects fluidity and thus the passage of materials. In some people, however, a diet high in saturated fats from dairy products, meat, and eggs can lead to a buildup of cholesterol deposits in blood vessels, which in turn can lead to higher blood pressure and the risk of strokes and heart attacks (Figure 2.24b; see Chapter 20).

Proteins: Key to Life's Diversity

The complexities of our bodies and the rich diversity of life—the millions of organisms of different shapes, colors, textures,

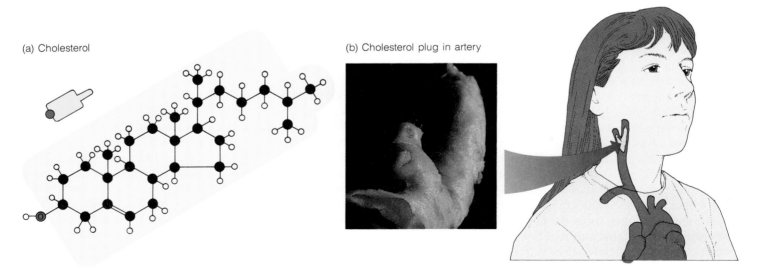

(a) Cholesterol

(b) Cholesterol plug in artery

FIGURE 2.24 ■ **Cholesterol and Plugged Arteries.** Steroids are a class of lipid made up of four interconnected rings of carbon atoms with functional groups attached. Steroids can pass through phospholipid layers; therefore, hormones—important chemical messengers and regulators—are often steroids. (a) Cholesterol is a common steroid that occurs within cell membranes. But cholesterol deposits can build up into solid clots and plugs that block arteries and lead to serious diseases of the heart and blood vessels. (b) A plug, or plaque, of solid cholesterol and other materials was removed from a patient's major neck artery to allow greater blood flow to the brain.

sizes, and life-styles—could never be accounted for by carbohydrates and lipids alone. These classes of compounds are important for cell structure and for energy storage and use, but they simply do not contain enough different types of molecules to account for life's vast diversity. **Proteins,** on the other hand, come in such a wide variety of forms (at least 10 to 100 million different kinds in the spectrum of the earth's organisms) that they can easily explain the myriad shapes and functions of living things.

There are several types of proteins. *Structural proteins* form cell parts, such as the keratin fibers in hair and the contractile proteins in muscles. *Regulatory proteins* control cell processes. The so-called *ras* proteins are regulators which, if altered slightly by mutation, cause cancer. Some proteins are *hormones,* or molecular messengers. Growth hormone, for instance, helps govern how tall a person will grow. *Transport proteins* carry other substances through the body. The hemoglobin protein in red blood cells, for example, transports oxygen. *Antibodies* are proteins that protect many types of animals from disease. And important proteins called *enzymes* facilitate virtually all the life processes that go on in cells. Moreover, the specialized shapes and functions of different cell types depend on protein. In other words, the cells in your eyes that detect light are different from the cells in your liver that detoxify poisons because eye cells and liver cells have different sets of proteins.

■ **Amino Acids: Protein Building Blocks**   Proteins are long chains of **amino acid** subunits, which have the general structure shown in Figure 2.25a. Each amino acid has three functional groups bound to one central carbon atom, and the term *amino acid* comes from two of the attached functional groups, the amine group (NH_2) and the acidic carboxyl group (COOH). Five of the 20 amino acids common in proteins are shown in Figure 2.25b. In a protein, amino acids are joined by covalent bonds in an order that is distinctive for each protein and determined by the organism's genes. Long chains of amino acid subunits are called **polypeptides.**

(a) Components of an amino acid

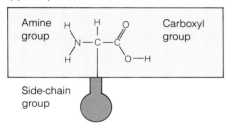

Amine group

Carboxyl group

Side-chain group

FIGURE 2.25 ■ Amino Acids: Protein Building Blocks. The term *amino acid* is based on the structure of these protein monomers. **(a) Each amino acid has an amine group (NH_2) on one end and an acidic carboxyl group (COOH) on the other end. The third "group" is always a hydrogen atom, shown here at the top of the molecule. Only the fourth group, the side-chain group, varies from one amino acid to another. The side-chain group can be as simple as the single hydrogen atom found in the amino acid glycine or as complex as a double-ring structure or chain. (b) Representative amino acids.**

(b) Representative amino acids

Glycine (Gly)	Phenylalanine (Phe)	Cysteine (Cys)	Glutamic Acid (Glu)	Arginine (Arg)
Glycine has the simplest side-chain group, a hydrogen atom. The animal protein collagen has many glycines.	Seven of the 20 most common amino acids have side-chain groups that are insoluble in water. Phenylalanine is such an amino acid.	Nine amino acids have side-chain groups that are soluble in water. Cysteine is such an amino acid. It is important in holding protein chains together.	Glutamic acid has a negatively charged side-chain group. Glutamic acid with a sodium ion is monosodium glutamate, the flavor enhancer.	Arginine is a positively charged amino acid because of its side-chain group. Arginine helps rid the body of dangerous nitrogen wastes.

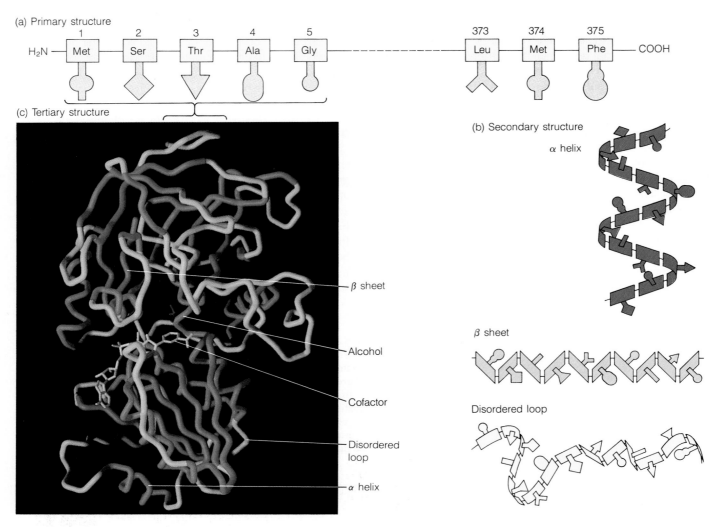

(a) Primary structure

(c) Tertiary structure

(b) Secondary structure

α helix

β sheet

Disordered loop

β sheet

Alcohol

Cofactor

Disordered loop

α helix

FIGURE 2.26 ■ **Primary, Secondary, and Tertiary Structure of the ADH Protein.** All proteins have at least three levels of structure, as does the alcohol dehydrogenase protein responsible for detoxifying alcohol in liver cells. (a) The primary structure is the unique linear sequence of amino acids in the polypeptide chain. (b) The secondary structure consists of regular folding patterns in the polypeptide. Areas of secondary structure are regions along the polypeptide chain folded in such a way that the maximum number of hydrogen bonds form. The α helix configuration is shown here in purple. The β pleated sheets are shown in blue, and the disordered loops are shown in yellow. (c) The tertiary structure is the unique three-dimensional shape of each protein molecule. ADH is a globular protein, and its rounded, ball-like shape depends on the positions of the disordered loop regions relative to the α-helix and β-sheet regions. Some proteins also have a quaternary structure, which is the way two or more polypeptide chains fit together, like the pieces of a three-dimensional puzzle. A cofactor and an alcohol molecule are shown here in light blue and red, respectively.

The 20 types of amino acids function as subunits in a biological alphabet, "spelling out" complex proteins much as the 26 letters of our alphabet can be combined into a nearly infinite array of words. And just as with a word, the "meaning" of a protein—its shape, properties, and functions—rests in the exact order of the amino acid "letters." Proteins usually contain from 100 to 10,000 amino acids; in even a short polypeptide, the potential number of unique amino acid sequences would equal the astonishingly huge number 20^{100}. (For comparison, 20^5 is 3.2 million.) Actually, not every potential sequence of amino acids occurs in nature, just as many random combinations of letters fail to form real words. Nevertheless, there are tens of millions of unique sequences in the living world, each spelling out a specific type of protein and together making possible life's remarkable diversity. The human body alone contains about 50,000 different proteins, each with a unique structure.

■ **The Structure of Proteins** Proteins have complex shapes based on four levels of structure, with each level based on the characteristics of the next simpler level. Each protein has a **primary structure** (the order of its amino acids), a **secondary structure** (regions of localized bending or pleating of amino acid sequences), and a **tertiary structure** (a three-dimensional folding of the entire polypeptide chain). Some proteins also have a **quaternary structure** (the fitting together of two or more chains in a three-dimensional configuration). Figures 2.26 and 2.28 explore these four levels of structure in two separate proteins.

The protein in Figure 2.26 is called alcohol dehydrogenase, or ADH, and it helps eliminate the poisonous effects of alcohol on the body. The ADH protein is a liver enzyme that helps break down alcohol, and its structure and activity may be relevant to the disease of alcoholism, which affects perhaps one out of ten adults who drink. Alcoholism researchers suspect there may be a genetic component to this serious condition, because children who have an alcoholic parent are predisposed to problem drinking, even if they are adopted and raised by nonalcoholic parents.

Some people's ADH appears to work more rapidly than others' in converting alcohol to a compound called acetaldehyde, which is even more toxic to the body than ethyl alcohol itself. In many Asians, the ADH enzyme converts alcohol to this toxin so rapidly that fairly soon after taking a drink, they feel flushed and nauseated and their hearts pound. This sensation acts as a strong negative reinforcement to excessive drinking, and thus the risk of alcoholism tends to be lower in people with fast-acting ADH. People with slow-acting ADH do not get this negative reinforcement and hence tend to be at higher risk for alcoholism. Researchers hope that understanding the structure of ADH may someday lead the way to a better understanding of alcoholism with its tragic consequences for the drinker, the family, and society in general, as well as leading to new methods of prevention and treatment.

We'll use the ADH molecule to illustrate the first three levels of protein structure. The *primary structure* of ADH, or any protein, is its linear sequence of amino acids (Figure 2.26a). The ADH protein is a long chain of 375 amino acids arranged in a unique and constant order. The first 5 of these 375 amino acids, for example, starting at the tip at the top of Figure 2.26c, are methionine, serine, threonine, alanine, and glycine.

As the long single chain twists and loops like a wad of stiff string, we can detect areas along the chain with regular folding patterns. One pattern is a coiled, springlike form (here, purple), which the chemist Linus Pauling discovered and called the *alpha helix* (α *helix*). Another pattern consists of parallel rows of the polypeptide chain folded into an accordion-like sheet called a *beta sheet* (β *sheet;* here, blue). A third pattern (shown here in yellow) is the *disordered loop,* which tends to lie between areas of α helices and β sheets. These three repeating structural motifs constitute the enzyme's *secondary structure* (Figure 2.26b).

The α helices and β sheets are held in their positions by weak hydrogen bonds. If a protein gets too hot, these hydrogen bonds break, and the protein loses its shape. That breakage, in fact, causes an egg white to solidify when you boil it; the secondary structure of the egg white protein is destroyed as a result of the broken hydrogen bonds.

Note that in the ADH molecule, the springlike α helices form rigid sides, while the parallel strands making up the β sheets look like a slightly twisted sheet of cardboard, and the disordered loops run between helical and sheet regions and hold things together. The specific way that the sheets, helices, and loops are ordered in space with respect to each other is the protein's *tertiary structure* (Figure 2.26c).

Different proteins have different primary structures—sequences of amino acids—and hence different secondary and tertiary structures. For example, *keratin,* the protein in hair, fingernails, horns, claws, and feathers, has many helical regions. The fact that weak hydrogen bonds produce the helix shape explains why hair can be stretched or curled when warm or wet (review Figure 2.9). Heat and water disrupt the bonds and allow the helix to unfold and the polypeptide chain to stretch out. When the keratin cools, helices re-form, and the hair takes on the new shape of the brush, roller, or curling iron it was coiled around. Permanent wave lotions induce new strong covalent bonds to form—hence the longer-lasting effect.

Fibroin, the protein in silk, occurs largely as pleated sheets. The many bonds cross-linking the polypeptide chains in fibroin result in a very strong fiber. The keratin in hair and fibroin in silk are *fibrous proteins,* and so is *collagen,* the most abundant protein in the animal kingdom. Collagen gives strength, flexibility, and shape to skin, tendons, ligaments, cartilage, and bones. Some physicians now inject collagen to "fill in" the scars, dents, and wrinkles left by acne, accidents, and aging (Figure 2.27).

(a)
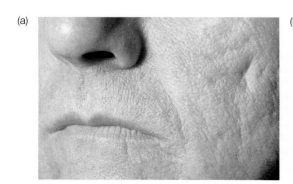

(b)
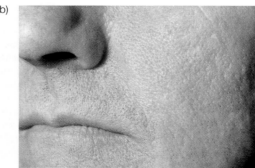

FIGURE 2.27 ■ Collagen Injections: Before and After. This man's facial scar (a) all but disappears (b) after cosmetic collagen injections.

Other proteins called *globular proteins* tend to ball up into rounded shapes. The alcohol dehydrogenase protein is a globular protein, and so is hemoglobin, the iron-containing protein that carries oxygen in the blood. Hemoglobin also illustrates the fourth level of protein structure, or *quaternary structure,* the way two or more folded polypeptide chains fit together like the pieces of a three-dimensional puzzle. The blood protein hemoglobin consists of four polypeptide chains, each balled up with its particular tertiary structure and fitting together into a larger, globular shape, as shown in Figure 2.28. When assembled this way, hemoglobin forms an active protein that is contained inside red blood cells and that transports oxygen from the lungs to tissues throughout the body.

As a final note to protein structure, notice that both ADH (Figure 2.26) and hemoglobin (Figure 2.28) have structures (here, light blue) that are not made up of amino acids and so are not links in the polypeptide chain. These added-on substances, called *cofactors,* help the proteins function. We will see later that the reason some vitamins are so essential in the diet is that they act as cofactors for certain enzymes.

In the years since biochemists discovered the four levels of protein structure, additional studies have revealed that proteins often assume their three-dimensional shapes automatically, given the right chemical environment. Scientists have also confirmed that the sequence of amino acids does indeed determine the higher-order foldings and that these, in turn, establish the individual properties of the millions of protein types. Thus, whether a protein serves a structural role, aids movement, transports oxygen, speeds chemical reactions, or attacks foreign invaders depends on its shape and solubility, and those biochemical "meanings" emerge from the language written in amino acids. What's more, the properties of each amino acid letter are determined by the chemical characteristics of the attached functional groups. The behavior of atoms and molecules indeed underlies the behavior of biological molecules and of living things.

Nucleic Acids: Information Storage and Energy Transfer

Among the outstanding members of the final class of biological molecules are the **nucleic acids,** which carry the chemical "code of life" and bear genetic information from one generation to the next. This last group also includes transport compounds involved in critical energy transformations in every cell. All these molecules are built of monomers called *nucleotides,* which are composed of a nitrogen-containing base, a five-carbon sugar (either ribose or deoxyribose), and a phosphate group. An example of a nucleotide is shown in Figure 2.29.

The nucleic acids DNA (for deoxyribonucleic acid) and RNA (for ribonucleic acid) are long polymers of nucleotide subunits. Each DNA molecule has an elegant double helix shape, while RNA molecules are smaller and simpler. Figure 2.29 shows these structures for DNA, and the legend explains how the order of the bases in DNA and RNA carries crucial

information for constructing and maintaining cells. What is important here is that the sequence of amino acids in proteins (which, as we saw, determines their shapes and properties) is coded for and assembled according to the sequence of nucleotides in genes made of either DNA or RNA.

Several other important biological molecules are based on modified nucleotide monomers, but aren't themselves polymers. One class, the **adenosine phosphates,** includes ATP, an energy-carrying molecule that acts as a form of "currency" whose "expenditure" (chemical breakdown) releases energy and enables the cell to accomplish most of its tasks. We'll discuss ATP and related compounds in detail in Chapters 4 and 5. Another member of the class, cyclic adenosine mono-

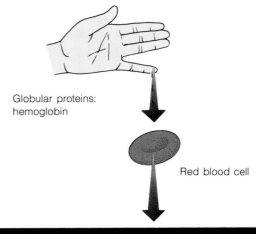

Globular proteins: hemoglobin

Red blood cell

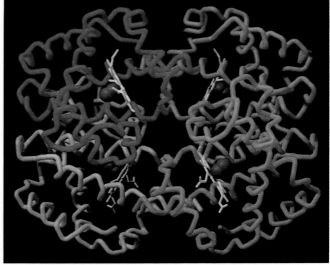

FIGURE 2.28 ■ **Quaternary Structure: A Fourth Level of Protein Organization.** Proteins with two or more separate polypeptide chains, each folded into its secondary and tertiary shapes, have a fourth level of organization, a quaternary structure, which determines the exact way these folded chains fit together. Hemoglobin, the protein that transports oxygen in the blood (shown as red spheres), has four such chains. Two are called α chains (here, two shades of blue), and two are called β chains (here, two shades of purple). Each is folded about a gray iron-containing heme group.

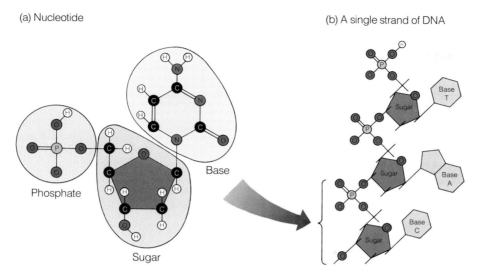

(a) Nucleotide

(b) A single strand of DNA

FIGURE 2.29 ■ **DNA: A Double Helix Polymer Made Up of Nucleotides.** (a) A nucleotide has a phosphate group, a ribose or deoxyribose sugar (here, we show deoxyribose), and a nitrogen-containing base. There are five types of bases: adenine, guanine, cytosine, thymine, and uracil—each with slightly different chemical compositions. (Here, cytosine is shown.) DNA monomers use all but uracil; RNA nucleotides use all but thymine. (b) Nucleotides join together, forming a single strand of DNA. A DNA molecule in a cell generally consists of two of the polymer chains twisted about each other in a double helix, with the outer coiled "rails" formed by deoxyribose sugars connected by phosphate "bridges," and the "ladder rungs" formed by pairs of bases (adenine and thymine, or guanine and cytosine) attached by hydrogen bonds (see Chapter 9). The precise order of the "rungs" carries information for the primary structure of enzymes and other proteins. In turn, enzymes act on raw materials to build all four kinds of biological molecules—carbohydrates, lipids, proteins, and nucleic acids—as well as the cell parts constructed of them. DNA is also responsible for the passing of heredity information from one cell generation to the next, as we will see later. RNA molecules, most of which are single chains of nucleotides, act as intermediaries in the building of proteins by transferring and translating the hereditary information contained in DNA to amino acid sequences (see Chapter 10).

phosphate (cAMP), carries chemical signals rather than energy and is involved in turning certain cellular activities on and off. Nucleotides called *coenzymes* are transport compounds necessary for energy harvest and the building of new cellular structures. In the alcohol dehydrogenase protein (Figure 2.26), the coenzyme is a nucleotide. We will become well acquainted with such cofactors in Chapters 5 and 6.

In summary, nucleotides provide the immediate energy source for most activities of living cells and join together in long chains to form nucleic acids, the reservoir of hereditary information in the cell.

CONNECTIONS

We have seen in this chapter that living things—from individual cells to indigenous peoples dwelling in the great Australian outback—are largely made up of water and depend utterly on this small molecule with its special chemical properties. We have also seen that most of the other compounds in a living organism are carbon-based. Moreover, all of the biologically important molecules, whether as simple as water or as complex as a protein, owe their characteristic shapes and

activities to the atoms within them and the properties arising from the unique arrangement of particles in each of the elements. Thus, the laws of chemistry and physics dictate how atoms behave, and moving up the ladder of biological organization, the way molecules, organelles, cells, organisms, and populations behave, as well.

Such ideas may seem abstract at first. But once you begin to think of living things as incredibly complex and precisely arranged collections of atoms and molecules, then the fundamental importance of water and carbon compounds, the relationships of structure to function, and the concept of emergent properties all unite to help explain a great deal that we see in our daily lives: How physicians can diagnose life-threatening diseases with giant magnetic instruments. How a person's liver helps detoxify the poisonous effects of alcohol. Why you can curl or straighten your hair. How sweating helps cool you off. How collagen treatments can fill out scars and wrinkles. Why organisms store energy as fats and oils. And so on.

We will encounter atoms and molecules again and again in our exploration of biology. And we will continue to see the relationships between the structure of these building blocks and the nature of the life process. Chapter 3 explains how the life process emerges from the arrangement of molecules in living cells.

Highlights in Review

1. Biological organization rests ultimately on atoms, and the properties of atoms in turn emerge from the precise arrangement of their constituent parts. Atoms of a given element have a specific number of protons and neutrons in the nucleus, as well as a specific number of orbiting electrons equal to the number of protons. Exceptions include isotopes, which have an uncommon number of neutrons, and ions, which have more or fewer electrons than normal.

2. The distribution of electrons orbiting around atoms helps us understand the shapes and properties of those atoms and why certain elements like neon are inert while others like sodium are violently reactive.

3. Because of the instability of unfilled outer shells, atoms tend to combine with each other and form molecules. They can share pairs of electrons in a covalent bond or lose or gain electrons in an ionic bond. Bonds can be single, double, or triple, and covalent bonds can be polar or nonpolar. All these bonding patterns influence molecular behavior.

4. Hydrogen bonds figure prominently in the behavior of water, contributing to its slowness to heat up and evaporate, its cohesion, adhesion, capillarity, and surface tension, as well as the tendency of ice to float. All these properties have important biological implications.

5. Water is the universal solvent and can dissolve other substances by surrounding and separating the component materials with a layer of oriented water molecules. Compounds that dissolve easily in water are hydrophilic; water-insoluble compounds are hydrophobic.

6. Water tends to ionize, acting as a weak acid whose pH is 7. Living cells are sensitive to pH and usually have an internal pH range of 6.5 to 7.5, often maintained by buffers.

7. Living things consist of four main classes of carbon-containing, or organic, compounds: carbohydrates, lipids, proteins, and nucleic acids. All are polymers made up of a particular family of monomers: sugars, fatty acids, amino acids, and nucleotides, respectively.

8. Carbohydrates include the simple sugars, or monosaccharides, such as glucose and galactose; the double sugars, or disaccharides, such as sucrose and lactose; and the long-chain polymers, or polysaccharides, such as starch, glycogen, cellulose, and chitin.

9. Lipids include the fats, oils, waxes, phospholipids, and steroids. Fats and oils are triglycerides; most have a glycerol molecule joined to three long-chain fatty acids. Single or double bonds in the long carbon chains determine the degree of saturation with hydrogen, and this, in turn, determines whether a lipid is a solid fat or a liquid oil.

10. Phospholipids have hydrophobic and hydrophilic regions and form single and double layers that are of great importance to the structure of cell membranes. Steroids have a multiple-ring structure and are soluble in fats and oils.

11. There are many classes of proteins, including structural proteins, regulatory proteins, hormones, transport proteins, antibodies, and enzymes. The amino acid sequence is a protein's primary structure; its secondary structure consists of helical, pleated sheet, or disordered loop configurations of the polypeptide chain; the tertiary structure is a three-dimensional folding of the entire chain; and a quaternary structure occurs in proteins with more than one polypeptide chain.

12. The fourth class of biological molecules includes the nucleic acids DNA (with its double helix shape) and RNA (with its single chains). The class also contains the adenosine phosphates, coenzymes, and other compounds based on nucleotide monomers containing a nitrogenous base, a five-carbon sugar, and a phosphate group.

Key Terms

acid, 35	isotope, 28
adenosine phosphate, 48	lactose, 42
adhesion, 34	lipid, 43
alkaline, 37	maltose, 42
amino acid, 45	molecular bond, 30
atom, 26	molecule, 30
atomic mass, 27	monomer, 38
atomic nucleus, 27	monosaccharide, 38
atomic number, 27	neutron, 27
atomic theory, 26	nonpolar, 32
base, 35	nucleic acid, 48
buffer, 38	oil, 43
capillarity, 34	organic chemistry, 38
carbohydrate, 38	pH scale, 35
cellulose, 42	phospholipid, 43
chitin, 42	polar, 32
cohesion, 34	polymer, 38
compound, 26	polypeptide, 45
covalent bond, 30	polysaccharide, 40
disaccharide, 40	primary structure, 47
double bond, 30	protein, 45
electron, 27	proton, 27
element, 26	quaternary structure, 47
fat, 43	radioactive, 28
fatty acid, 43	saturated fat, 43
fructose, 40	secondary structure, 47
functional group, 38	single bond, 30
galactose, 40	solute, 35
glucose, 40	solvent, 35
glycerol, 43	starch, 42
glycogen, 42	steroid, 43
hydrogen bond, 32	sucrose, 40
hydrophilic, 35	surface tension, 34
hydrophobic, 35	tertiary structure, 47
inert, 29	triglyceride, 43
inorganic chemistry, 38	unsaturated fat, 43
ion, 29	wax, 43
ionic bond, 32	

Study Questions

Review What You Have Learned

1. Although atoms of different elements contain the same basic particles (protons, neutrons, and electrons), they nevertheless have different properties. Explain.
2. How do atomic number and atomic mass differ?
3. What is the difference between a covalent bond and an ionic bond? Give an example of each.
4. While the atoms of an element normally have the same number of protons, neutrons, and electrons, there are exceptions. Name two types.
5. Explain why the pH of water is 7.
6. List the different types of carbohydrates, and describe how they differ from one another. Give an example of each.
7. Name the different classes of lipids and their characteristics.
8. Describe the four levels of protein structure.
9. How are nucleotides important to living things?
10. How do the carbohydrates, lipids, and amino acids differ in basic structure?

Apply What You Have Learned

1. A water strider can do something few other living things can do: walk on the surface of water. Which property of water allows this activity?

2. Cholesterol, which our cells manufacture, serves a structural role in cell membranes; yet a doctor may place a patient on a low-cholesterol diet. Why?
3. Explain the change taking place in the white part of the egg when you fry it.

For Further Reading

ATKINS, P. W. *Molecules.* New York: Scientific American, 1987.

CHANG, R. *Chemistry.* 2d ed. New York: Random House, 1984.

EHRIG, T., W. BOSRON, and T. K. LI. "Alcohol and Aldehyde Dehydrogenase." *Alcohol and Alcoholism* 25 (1990): 105–116.

RICHARDS, F. M. "The Protein Folding Problem." *Scientific American,* January 1991, pp. 54–63.

STRYER, L. *Biochemistry.* 3d ed. New York: Freeman, 1988.

Cells: The Basic Units of Life

Euglena: An Exemplary Living Cell

If you were to collect a water sample from almost any small pond, stream, swamp, or drainage ditch and then view drops of it under a microscope, you might see an organism that acts partially like a plant and partially like an animal—but is neither. The organism is a member of the genus *Euglena.* It belongs to neither the plant nor the animal kingdom, but is instead a single-celled life form in the kingdom Protista (see Chapter 1). The microscopic, spindle-shaped *Euglena* is a bright grassy green and traps energy from the sun just as a plant does. Yet it also moves about rapidly as if it were a miniature animal, propelled by a long, whiplike thread that spins around like a twirling lasso (Figure 3.1).

Euglena delighted and confounded generations of early biologists and is a good case study today. This versatile single cell can help answer fundamental questions about life's basic units: What are cells? How are they constructed? How do they carry out their functions? What traits does *Euglena* share with the 200 types of cells that make up the human body, as well as with the cells of millions of other types of living organisms? And what traits distinguish the various cell types from each other?

If you focused your microscope on a single *Euglena,* translucent as an emerald and cavorting about the province of a water droplet like a tiny sea creature, you would be following the traditions of one of history's greatest microscopists, Anton van Leeuwenhoek. This Dutch cloth merchant, in 1675, was the first to describe what was probably a *Euglena.* Using the term *animalcules* for the pond-water life forms he saw magnified in his small hand-held microscopes, he noted that some had

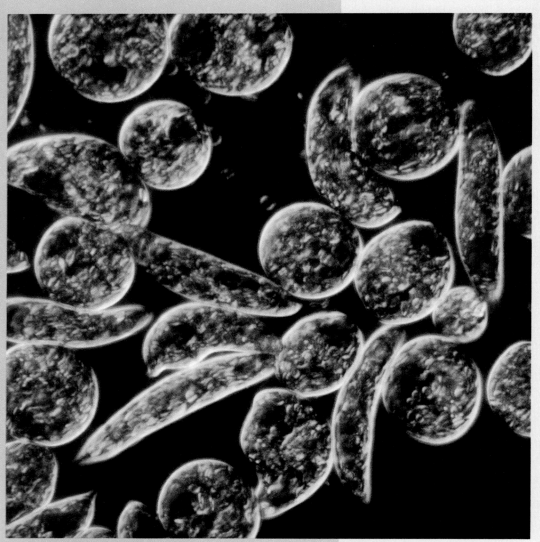

FIGURE 3.1 ■ *Euglena:* **A Versatile Single Cell.** Members of the genus *Euglena* are usually bright green due to chloroplasts, the green discs inside these cells.

"green and very glittering little scales, and a motion so swift, and so various, upwards, downwards, and round about, that 'twas wonderful to see."*

The *Euglena* studied by Leeuwenhoek and other early "voyeurs" of the microscopic realm is a rather complicated single cell and has all the characteristics of life (growth, metabolism, reproduction, and so on) that a larger being with many cells displays. Each *Euglena* species is unique and yet illustrates important unifying themes that relate to all organisms. First, all cells share certain basic functions responsible for keeping them alive. Separated from the world around them by protective enclosures—membranes and walls—cells must take in and convert energy to usable form; they must also build proteins, grow, reproduce, and control the extent and timing of their activities. To accomplish all this, all cells share certain physical structures that might collectively be called the machinery of life, including an outer envelope; a control center; a jellylike inner substance; and several other small internal structures discussed in this chapter. *Euglena* shares these structures (or equivalent ones) with every other living cell. The sharing of these structures is a strong argument for the *unity of life,* the evolutionary relatedness of all living things through their shared physical structures and the often identical ways in which they function.

Second, each cell's specialization—how it lives and what it can do—depends on its shape, its size, and the particular organelles, or complex internal bodies, it contains. *Euglena* is a single-celled organism that converts the energy in sunlight into the chemical bonds of carbohydrate molecules. In other words, like a plant, it makes its own food. Like a tiny animal, however, it moves about to satisfy its needs; with its streamlined spindle shape and its whiplike propeller, it darts about following the appropriate level of sunlight so that its rate of converting solar energy into food energy (*photosynthesis*) is maximized. It is, in short, a single cell specialized to pursue sunlight.

Both *Euglena*'s plantlike and animal-like characteristics are based on specific *organelles*. As Chapter 1 stated, organelles are individual cell structures with definite roles. *Euglena*'s specializations are based on green organelles called chloroplasts and whiplike organelles called flagella (both structures are discussed later in this chapter). Specializations like photosynthesis and mobility reflect life's enormous diversity: The thousands of cell types that make up the earth's single-celled and multicelled species owe their divergent structures and traits to the huge variety of proteins in nature (see Chapter 2) and to the spectrum of particular cell structures formed from proteins and other biological molecules.

Third, despite the importance of physical structures to the general and specialized functions of the cell, the parts of a cell are not themselves alive; the living state is an emergent property based on the arrangement of the cell's parts and their integrated activities. If you grew *Euglena* cells in a beaker and then disintegrated them in a blender, the whole and fragmented cell parts that remained would not re-form living cells. The order of the system—the organized structure and activity within the cell—would have been lost, and without this, there is no life. Cells are unique arrangements of matter in the universe in that their complex internal order must be passed down from previous generations of living cells and maintained through regulated activity. Thus, contemporary living things are part of an unbroken lineage that stretches back in time to the very first cells, early in our planet's history. How cells may have originated is the subject of Chapter 15. Here, we concentrate on what cells are, how they live, and why they are so important to the understanding of biology. This chapter will answer several questions:

- ■ How were cells discovered, and what basic theory describes them?
- ■ What are the main cell types, cell organization, and cell sizes and shapes?
- ■ What are the common functions carried out by all cells, and what are the physical structures that all cells share?
- ■ What are the specialized structures and functions that distinguish one type of cell from another?

---

*Science Digest, March 1982, p. 92.

THE DISCOVERY AND BASIC THEORY OF CELLS

While Leeuwenhoek was probably the first person ever to gaze upon the green and gently gliding *Euglena,* he was not the first to use microscopes or the first to discover cells. Robert Hooke, one of England's greatest scientists, had also been intrigued with the devices.

About 1663, Hooke directed an instrument maker to fashion a microscope of Hooke's own design based on imported ones (Figure 3.2a). Hooke began to focus his new instrument on everyday objects—the point of a pin, the edge of a razor, the surface of a nettle leaf, the body of a flea—and was astonished by the fine detail never before revealed.

When Hooke peered at a thin slice of cork through his microscope, he saw what he called "pores" or "cells" that reminded him of the little rooms monks inhabit (Figure 3.2b). We know today that Hooke was seeing not living cells but their remains—specifically, plant cell walls, composed of cellulose and other molecules deposited outside each plant cell and remaining even after the cell dies.

Leeuwenhoek was a contemporary of Hooke. Working in Delft, Holland, in the mid-1600s, he had far greater success at

(a)

(b)

FIGURE 3.3 ■ **Anton van Leeuwenhoek: Microscopist Extraordinaire.** **(a) Anton van Leeuwenhoek (1632–1723). Leeuwenhoek was a master at making ultrathin slices of living tissue for examination and at grinding lenses for his microscopes. A few of his original specimens were recently found still intact after 300 years. (b) Some of Leeuwenhoek's small, simple, hand-held microscopes are still capable of magnifying objects up to 500 times.**

(a) Hooke's microscope

(b) Hooke's drawing of cork

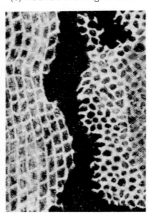

FIGURE 3.2 ■ **Discovery of the Cell.** **(a) Robert Hooke, a seventeenth-century British scientist, used a simple microscope to explore common objects in a new way. (b) Hooke observed and drew a slice of cork, with its small chambers. These subunits of a formerly living thing he called "cells," and they were the first cells to be discovered.**

seeing living cells such as *Euglena* in action, despite his smaller, simpler, hand-held microscopes (Figure 3.3). Considered the greatest early microscopist, Leeuwenhoek was unsurpassed in his day at grinding lenses and slicing and mounting specimens for viewing. Some of his 5 cm (2 in.) long instruments had lenses no bigger than a grain of sand but a magnification power of up to 500! And when some of his original specimens were recently found at the Royal Society of London, viewers realized that the 300-year-old ultrathin slices he had made of cork, feathers, the pith of the elder bush, and the optic nerve of a cow were still admirable specimens when viewed through a twentieth-century microscope. (Modern microscopes are described in Box 3.1 on page 62.)

Although Leeuwenhoek was the first to see and record the antics of living single-celled organisms, it took more than two centuries of study before biologists could extend the work of the early microscopists past simple observation and comprehend the true significance of cells to the living state. This significance is stated in the **cell theory,** a major doctrine of biology based on the work of three nineteenth-century scientists. In 1838, a German botanist named Matthias Schleiden drew an important conclusion based on years of studying plant tissues: All plants, he said, are made up of cells. The following year, he and German zoologist Theodor Schwann together published the even broader conclusion that *all* living things are composed of cells. Twenty years later, German physician Rudolf Virchow stated that living cells arise only from preexisting cells, making them a rare and marvelous form of matter—matter with a history.

According to modern cell theory:

1. All living things are made up of one or more cells.
2. Cells are the basic living units within organisms, and the chemical reactions of life take place within cells.
3. All cells arise from preexisting cells.

These tenets may seem obvious today, but in an era with few scientists, few microscopes, and millions of species to explore, they were profound revelations that led to a far greater understanding of the structures and processes that make up the living state.

THE UNITS OF LIFE: AN OVERVIEW

Saying that cells are the basic units of life does not tell much about the units themselves. What are the major kinds of cells? How are they arranged in living things? What shapes and sizes are most cells? What tasks do cells carry out, and how do they do it? Let's look at the answers to these questions.

The Two Major Kinds of Cells

■ **Prokaryote or Eukaryote**   As we saw in Chapter 1, biologists divide the millions of species of living things into three domains, or branches on the tree of life, based on genetic rela-tionships (review Figure 1.16). Two of the domains, Bacteria and Archaea, have very simple cells. The third domain, Eucarya, includes mushrooms, dogs, maple trees, and all the organisms we see in a normal day—all of which are made up of more complex cells. These complex cells contain a central membrane-enclosed organelle called the *nucleus*. Often roundish in shape, the nucleus directs the cell's activities and contains DNA and the cell's hereditary information. Cells containing a nucleus are called **eukaryotic cells** (*eu* means "true," *karyon* means "kernel"; Figure 3.4b). All members of the domain Eucarya are made up of eukaryotic cells.

In contrast, each member of the 3000 or so species in the other two domains—Bacteria and Archaea—is a single-celled organism composed of a **prokaryotic cell.** The term *prokaryotic* means "before nucleus," and perhaps the most obvious thing about a prokaryote (a prokaryotic organism) is that the cell lacks a distinct nucleus (Figure 3.4a). Prokaryotic cells possess DNA, but unlike the DNA of eukaryotic cells, it is a circular strand that is concentrated in an unbounded region called the **nucleoid.** While prokaryotic cells lack a membrane-enclosed nucleus, they do have a membrane forming an outer envelope around the cell (the *cell membrane, or plasma membrane*); and in some species, additional inpocket-ings of this membrane act as surfaces for important energy conversion reactions (see Chapter 5). Beyond this, the internal organization of prokaryotic cells tends to be fairly simple. And most prokaryotic cells are surrounded by a *cell wall* just

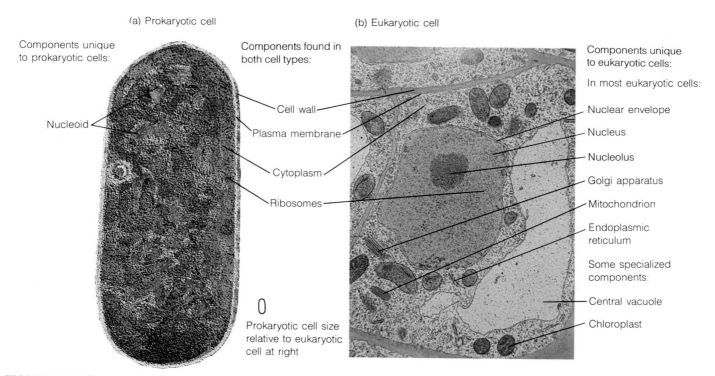

(a) Prokaryotic cell

(b) Eukaryotic cell

Components unique to prokaryotic cells:

Components found in both cell types:

Components unique to eukaryotic cells:

In most eukaryotic cells:

Nucleoid

Cell wall

Plasma membrane

Cytoplasm

Ribosomes

Nuclear envelope

Nucleus

Nucleolus

Golgi apparatus

Mitochondrion

Endoplasmic reticulum

Some specialized components:

Central vacuole

Chloroplast

Prokaryotic cell size relative to eukaryotic cell at right

FIGURE 3.4 ■ Prokaryotic and Eukaryotic Cells. Compare (a) a prokaryotic cell, such as this bacillus (*Clostridium perfringens*), with major cellular components labeled, with (b) a eukaryotic cell, such as this one from a plant in the mustard family (*Arabidopsis thaliana*), with its components labeled. Eukaryotic cells are usually over ten times bigger than the simpler prokaryotic cells.

outside the cell membrane that can be either rigid or flexible, but always gives shape and support to the cell. See Table 3.1 for the distinguishing traits of prokaryotes.

While the three domains are based on lineage (i.e., genetic relationships), the division of all organisms into eukaryotes and prokaryotes is based on form and function. All eukaryotic organisms are closely related. But Bacteria—the prokaryotes in one domain—are only distantly related to the prokaryotic members of the domain Archaea. Bacteria are noted for their common occurrence in soil, air, and water, as well as on other organisms, while archaebacteria inhabit inhospitable environments such as salt flats, hot springs, cow's intestines, and sewage treatment plants. The prokaryotes in both domains grow and reproduce rapidly and as a group use a huge spectrum of organic and inorganic materials as sources of energy. They are thus very diverse biochemically, and this diversity has

TABLE 3.1 Components of Prokaryotic, Plant, and Animal Cells

Structure	Function	Present in Prokaryotic Cells	Present in Eukaryotic Cells	
			Plant Cells	Animal Cells
Plasma membrane	Protection; communication; regulates passage of materials	✔	✔	✔
DNA	Contains genetic information	✔	✔	✔
Nuclear envelope	Surrounds genetic material		✔	✔
Several linear chromosomes	Contain genes that govern cell structure and activity		✔	✔
Cytoplasm	Gel-like interior of cell	✔	✔	✔
Cytoskeleton	Aids in support and movement in cell		✔	✔
Endoplasmic reticulum	Intracellular transport		✔	✔
Golgi apparatus	Package materials		✔	✔
Ribosomes	Manufacture proteins	✔	✔	✔
Lysosomes	Contain enzymes; aid in cell digestion; play a role in programmed cell death		✔	✔
Microbodies	Storage		✔	✔
Mitochondria	Provide cellular energy		✔	✔
Chloroplasts	Capture sunlight; produce energy for cell		✔	
Central vacuole	Maintains cell shape; stores materials and water		✔	
Flagella	Cell movement; only found in a few cells in each group	✔	✔	✔
Cilia	Cell movement; present only in certain cells			✔
Cell wall	Protects cell; maintains cell shape	✔	✔	
Glycocalyx	Surrounds and protects cell			✔
Intercellular links Pili Junctions	Mating Cell-to-cell communication; prevent fluid leakage; strengthen tissues	✔	✔	✔

allowed prokaryotes to exploit environments inaccessible to eukaryotes, with their more complex cells. Some archaebacteria, for example, thrive in the boiling waters of Yellowstone's hot springs, while bright orange bacterial cells grow in a slimy mat at the cooler but still steamy edges of the spring (Figure 3.5). Microbiologist T. D. Brock suggests that cells with nuclei cannot synthesize a nuclear membrane that remains functional at high temperatures.

Eukaryotic cells may fail to thrive in hot springs, but they make up the individual cells of multicellular plants, animals, and fungi as well as the single-celled protists (such as *Euglena*). Like prokaryotes, eukaryotic cells are surrounded by a *cell membrane* (or *plasma membrane*) (Figure 3.4b). But as we saw, eukaryotic cells have a membrane-bound nucleus. Eukaryotic cells also have other membrane-bound organelles that carry out specific functions, such as harvesting energy or recycling worn-out cell parts. Eukaryotes are generally 10 to 25 times larger than most prokaryotes, and this allows them to contain the numerous organelles. *Euglena,* for example, is about 30 times larger than a medium-sized bacterium. (See Table 3.1 for the eukaryotic organelles, many of which are labeled in Figure 3.4b.) Eukaryotic cells also have a highly organized *cytoplasm,* the clear, watery mass contained by the cell membrane and surrounding the nucleus. Crisscrossing this mass is an internal framework called the *cytoskeleton* that is important to the structural support and movement of the cell and its parts.

■ **Eukaryotic Cells Can Coexist and Cooperate in Multicellular Organisms**   Eukaryotes have evolved an important trait not shared with the simpler cells: a capacity to coexist and cooperate as subunits of multicellular organisms. Within a many-celled organism, eukaryotic cells are usually specialized for

FIGURE 3.5 ■ **Biochemical Versatility.   Prokaryotic cells can utilize a much broader spectrum of energy sources and environments than can eukaryotic cells. Archaebacteria thrive in the boiling waters of Yellowstone National Park's Prismatic Spring, while photosynthetic bacterial cells grow in a slimy orange mat at the spring's cooler edges.**

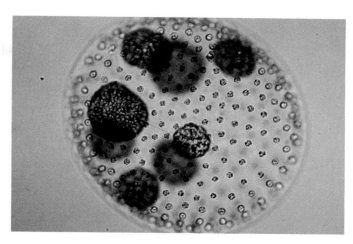

FIGURE 3.6 ■ **Colonial Organisms: Simple Division of Labor.** *Volvox* is a type of simple green alga that forms jewel-like spherical colonies. Each colony has dozens of regular colony cells as well as a few larger reproductive cells. The reproductive cells are capable of generating new cells within the colony or leaving to start a new colony elsewhere.

given chores. In fact, each multicellular organism depends on a *division of labor* in which groups of cells accomplish specific tasks. Thus, one subset of cells, usually in a tissue or organ, is involved in reproduction, another in digestion or photosynthesis, yet another in mechanical support, and so on.

A similar but simpler division of labor occurs in *colonial species*. These are groups of cells that are capable of living independently as single-celled organisms but that tend to live in clusters with certain cells specialized for reproduction. A relative of *Euglena* called *Volvox* is an example of a colonial form (Figure 3.6). The evolution of cellular cooperation and coexistence probably began in colonial organisms but reached a zenith in the multicellular organisms.

■ **Eukaryotic Cells: Generalized and Specialized**   Figure 3.7 shows generalized animal and plant cells with their internal organelles. A quick tour of these cells will help orient you for the more detailed discussions in the rest of the chapter. Notice that the two cell types share many structural features: The *nucleus* controls cell functions and houses the genetic material; *mitochondria* produce usable energy; *ribosomes* make proteins; the *endoplasmic reticulum* and *Golgi apparatus* process and transport the proteins; and the *cytoskeleton* helps maintain cell shape and allows cell movements. A common trait of eukaryotic cells, therefore, is the separation of many cell tasks into individual cell *compartments*. These compartments, or organelles, concentrate cell machinery needed for each task and isolate potentially interfering reactions from each other.

While plant, animal, and other eukaryotic cells share a set of compartmentalizing organelles, they also have obvious structural differences associated with different life-styles. For example, an animal cell is surrounded only by an outer or

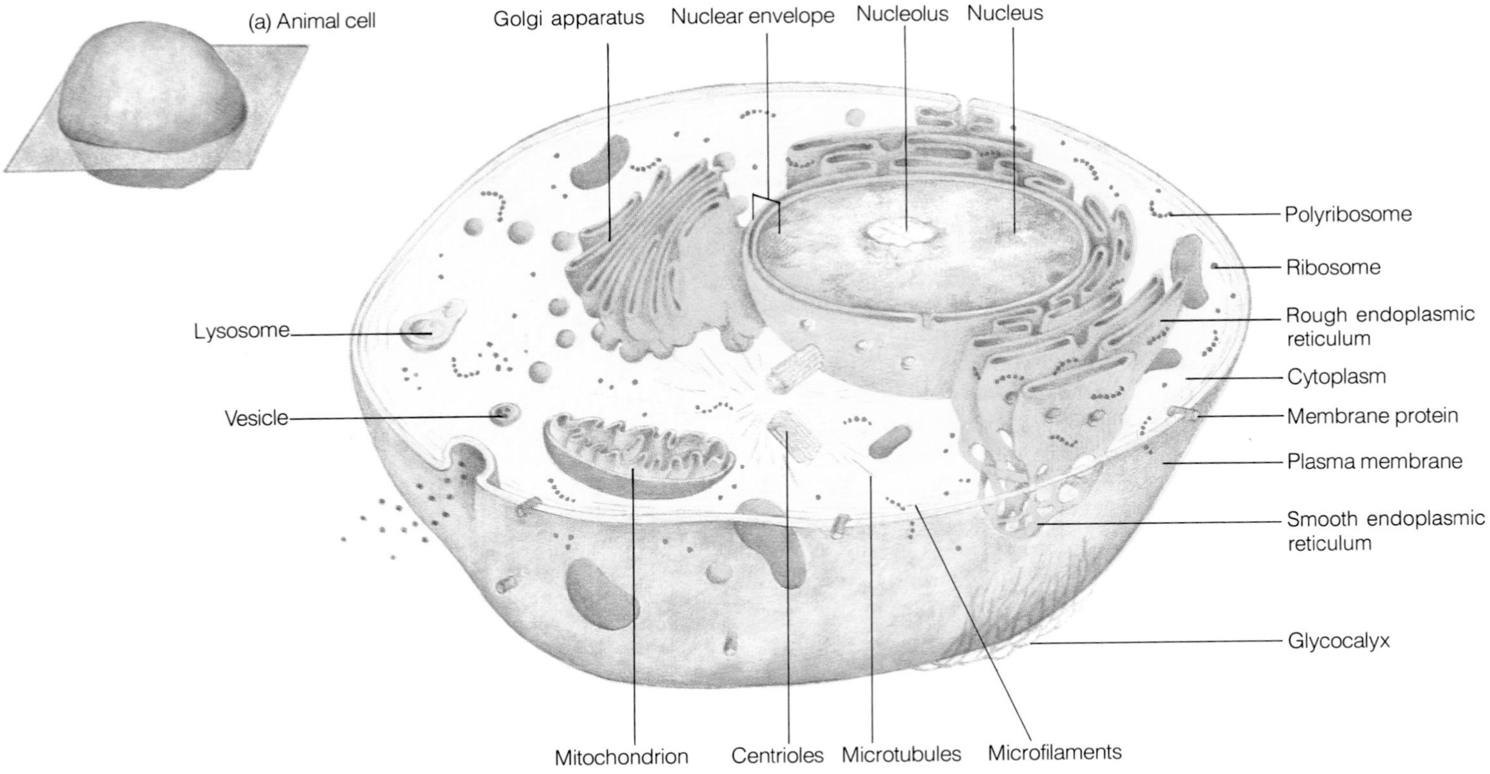

(a) Animal cell

Golgi apparatus   Nuclear envelope   Nucleolus   Nucleus

Polyribosome

Ribosome

Rough endoplasmic reticulum

Cytoplasm

Membrane protein

Plasma membrane

Smooth endoplasmic reticulum

Glycocalyx

Lysosome

Vesicle

Mitochondrion   Centrioles   Microtubules   Microfilaments

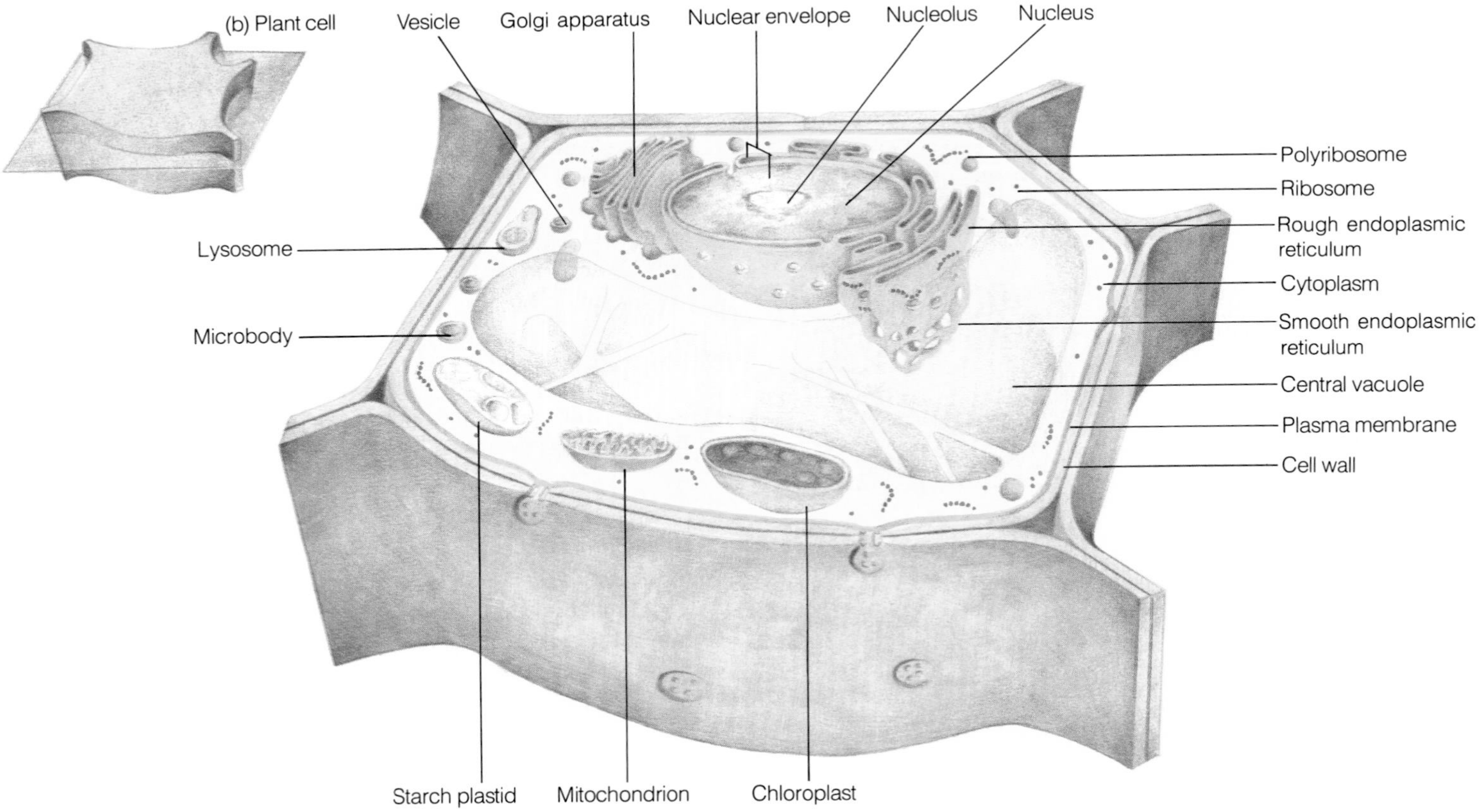

(b) Plant cell   Vesicle   Golgi apparatus   Nuclear envelope   Nucleolus   Nucleus

Polyribosome

Ribosome

Rough endoplasmic reticulum

Cytoplasm

Smooth endoplasmic reticulum

Central vacuole

Plasma membrane

Cell wall

Lysosome

Microbody

Starch plastid   Mitochondrion   Chloroplast

FIGURE 3.7 ■ Animal and Plant Cells Compared. These generalized drawings show the cellular components of (a) an animal cell and (b) a plant cell.

plasma membrane and can have tiny whips (flagella or cilia), which help the cell move. In a plant cell, however, the plasma membrane lies just inside a solid supportive *cell wall* made of cellulose and other fibers that prevent the cells from migrating within the organism. Even though both plant cells and pro-karyotic bacterial cells are surrounded by cell walls, the structures of these walls differ. Plant cells also have a large, usually clear storage organelle, the *vacuole,* filling most of the cell (as opposed to many smaller vacuoles, as in animal cells). Many plant cells also contain one or more bright green *chloroplasts.*

Most plant cells are rigid because of their cell wall and vacuole and can carry out photosynthesis because of their chloroplasts. Animal cells, on the other hand, are notable for their ability to ingest food particles and move about if placed in a Petri dish. However, in addition to these general cellular attributes, most plants, animals, and fungi have subsets of cells specialized by their shapes and organelles to carry out particular tasks. A large flowering plant—say, a magnolia tree—has about a dozen different types of cells. It has, for example, sets of rigid, tubelike cells known as *sieve tube elements* which are stacked end to end in long pipelines and are involved in the transport of sugars throughout the plant. A magnolia also has flattened *epidermal cells,* which serve as an outer "skin" on leaves and young stems (Figure 3.8a). An animal—a spiny anteater, let's say—has about 200 kinds of cells. They include the *motor neurons,* nerve cells that carry impulses between the central nervous system and the muscles, and the *striated muscle cells,* spindle-shaped cells that can contract thanks to an internal grid of movable filaments (Figure 3.8b). Specializations on the cellular level thus lead to specialization at the level of tissues, organs, and organisms and help explain the wonderful diversity of life.

The Numbers and Sizes of Cells

When Robert Hooke first discovered the basic units of life in 1663, he also noticed how very numerous they were in a given piece of tissue—over 1 billion, he calculated, in a "cubick inch" of cork. One encounters such astonishing numbers frequently in the study of cell biology. A newborn human baby contains 2 trillion cells; an adult, 60 trillion. Each of your eyes has a retina containing 125 million light-sensitive cells, and connected to these are 1 million nerve cells that carry visual information to the brain. When you donate blood, you give away 5.4 billion cells—and scarcely miss them. Each day, in fact, your body sloughs off and replaces 1 percent of its cells, or about 600 billion. (No wonder you don't miss a mere 5.4 billion!)

For a body to contain so many cells, the cells must be extremely small. But just how small? And more importantly, *why* are cells so small? Why don't we just have a few extremely large cells instead of trillions of microscopic ones?

Prokaryotes are the smallest cells, and some bacteria are no more than 0.2 micrometer* in length. For comparison, this

*μm, one-millionth of a meter; see the abbreviations chart on the inside back cover.

(a) Flowering magnolia tree

Sieve tube element within stems

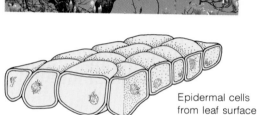

Epidermal cells from leaf surface

(b) Echidna (spiny anteater)

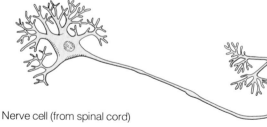

Muscle cell (from leg muscle)

Nerve cell (from spinal cord)

FIGURE 3.8 ■ Specialized Cells for Specialized Tasks. (a) Like all multicellular organisms, different parts of a magnolia tree have specialized functions that require cells equipped with specialized structures. The tree's trunk houses "pipelines" that carry water and materials to and from the roots via tubelike cells called sieve tube elements. The tree's green leaves are protected by a "skin" made up of tough, flattened epidermal cells. (b) Cells in an animal like an echidna, or spiny anteater, have specialized functions too. The animal's nervous system carries information back and forth in the body and contains long, spindly motor neurons, or nerve cells. Muscle cells allow the animal to move because they contain a movable internal grid of filaments and tubules.

page is 100 μm thick. Other prokaryotic cells, such as a typical bacterium like *Escherichia coli,* normally a harmless inhabitant of the human intestinal tract, have a volume about ten times greater than the smallest prokaryote and are about 2 μm long. Hundreds of *E. coli* cells can fit on the tip of a sharp

Portion of a chick egg cell (yolk): 3 cm diameter

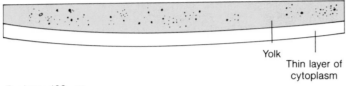

Yolk

Thin layer of cytoplasm

Euglena: 100 μm

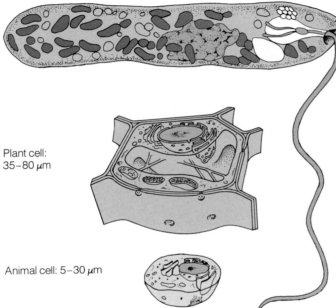

Plant cell: 35–80 μm

Animal cell: 5–30 μm

Escherichia coli: 2 μm

├───┤
10 μm

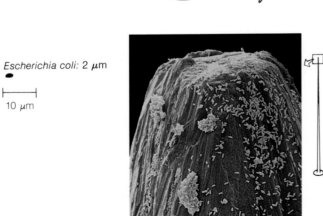

FIGURE 3.9 ■ Cell Size Can Differ Widely. If a typical bacterium such as *Escherichia coli* were the size of the drawing at the bottom of this figure, then the animal and plant cells would be considerably larger: A euglena would look positively whalelike, and the yolk of a chicken egg, a single cell 3 cm in diameter, would look like a small planet 50 m across! In the scanning electron micrograph, *E. coli* cells are dwarfed by the point of a pin.

(a) Doubling the linear dimensions of a cell

(b) A large cube made of small cubes

FIGURE 3.10 ■ Why Are Cells So Small? Surface-to-Volume Ratios. The life of a cell depends on the exchange of materials across its surface. The greater the cell volume, the more surface area required. Most cells are microscopic, and their surface-to-volume ratios are favorable. (a) Doubling the linear dimensions of a cube increases its volume eight times while increasing its surface area only four times. (b) A large cube made up of many small cubes still has the same surface-to-volume ratio as each individual cube. That's why a large organism can survive. It has more cells than a small organism, but the cells are roughly the same size.

pin (Figure 3.9). An average-sized animal cell at about 20 μm in length is ten times larger again (about five animal cells could fit across the thickness of this page), and a typical plant cell is about 35 μm across. Although cells are rarely more than 50 μm across, a very large *Euglena,* such as *Euglena ehrenbergii,* can be 200 μm long (still just the thickness of two pages here). And there are a few types of truly colossal cells, including the yolk of an ostrich egg, which can be more than 7 cm (70,000 μm) across, and the nerve cells that extend more than 3 m (9.7 ft) down a giraffe's leg or along its neck. These nerve cells appear to contradict the cells-must-be-small rule. But they function quite well, and here's why.

The critical dimensions of any cell are its *surface area* and its *volume.* The life of a cell depends on exchanges of materials (ions, gases, nutrients, and wastes) with the environment, and the inward and outward movement of all these materials takes place through the cell's surface. The larger a cell's volume, the greater the amount of material to be exchanged. Since this exchange depends on the cell's surface area, the **surface-to-volume ratio** imposes limits on cell size (Figure 3.10). An analogy is a pile of wet laundry. If left in a heap, it will take a long time to dry; but with each item separated and hung on a line, the exposed surface area increases, water evaporates to the environment, and the laundry dries rapidly.

Biological needs depending on surface area and volume thus determine how large a cell can be. A large cell with no special modifications would exchange materials with the environment too slowly to survive, but many small cells have enough surface area for the rapid exchanges that sustain life. It may sound surprising, but an elephant's liver cell is the same size as a mouse's; the elephant's liver simply has trillions more cells than the mouse's, not larger cells.

One way around this surface-to-volume constraint on size is through altered cell shape. A long, thin cell, such as a giraffe's nerve cell, or an extremely flat one, such as the thin layer of living cytoplasm that surrounds an egg yolk, can have the same volume as a round or cuboidal cell but a greatly expanded surface area. If you took a round balloon and pulled it into a long shape or squashed it into a flat one, the volume of air inside would not change, but the rubber would have stretched considerably and the surface area would expand.

Surface area can also be expanded through a series of slender, fingerlike extensions of the cell's outer membrane. Figure 3.11 shows the kind of cells that line the human small intestine. The numerous extensions, called microvilli, greatly expand the cell's surface area, allowing it to absorb nutrients with speed and efficiency.

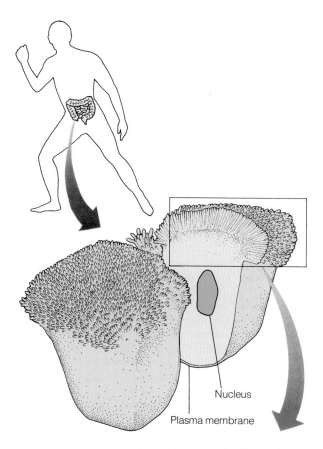

Nucleus

Plasma membrane

FIGURE 3.11 ■ One Way to Expand Cell Surface. High-powered magnification of the surface of intestinal cells reveals microscopic, fingerlike projections called microvilli. These expand the surface area, allowing the cells to absorb nutrients quickly.

THE COMMON FUNCTIONS AND STRUCTURES OF ALL CELLS

A cell is a compact, highly organized unit of matter that simultaneously carries out a number of "housekeeping" tasks, including growth, maintenance, and repair. The absence of any one of these tasks would lead to the disorganization and eventual death of the cell. Thus, the survival activities carried out by *all* living cells are essentially synonymous with the life process itself. An emerald green euglena, for example, is like every other living cell on earth in at least five ways: (1) It is separated from the surrounding environment by a boundary that maintains the appropriate pH and allows concentrations of cellular raw materials to be maintained inside the cell; (2) it takes in raw materials and expels wastes through that boundary; (3) it absorbs energy in some form—usually either as light or as the chemical bonds in molecules—and alters the energy to a form useful for powering cellular activities; (4) it builds biological molecules and cell parts for repair, growth, and reproduction; and (5) its myriad activities are closely regulated and coordinated.

All cells carry out these tasks and maintain a set of basic structures. In this section, we consider the structures one by one, starting at the cell's outer boundary, or plasma membrane, then moving inward to the most visible cell part, the nucleus, and finally considering the cytoplasm and all the various organelles within it.

During this survey of cellular structures, it would be good to bear in mind that something is always missing whenever one takes a *reductionist* approach—that is, whenever one reduces or breaks a system down into its components and describes each one separately. What's missing is the emergent property mentioned earlier—the next higher level of organization, the *life* of the cell, the thousands of simultaneous biochemical reactions that are received from previous generations and passed on to succeeding ones and that cannot be resumed by reassembling a set of static parts. So try to imagine the "music" of the life process humming quietly but continuously in the background as we focus on each cell part, each "molecular instrument," separately.

The Cell's Dynamic Boundary: The Plasma Membrane

Living things such as human beings are mostly water. Yet organisms don't dissolve into puddles or simply merge with the environment. For example, a euglena gliding about in a pond can be easily distinguished from the pond water itself, even though the cell's composition is nearly 80 percent water.

A cell's content is delineated from the world around it by a flexible sheet of fatty material called the **plasma membrane** or **cell membrane** (see Figure 3.7). This outer boundary keeps most of the unnecessary or harmful materials out of the cell, while sequestering most of the useful substances within. It is not, however, a tight seal around the cell. On the contrary, the membrane regulates a constant traffic of materials into and out of the cell, allowing water, ions, and certain organic molecules to *permeate,* or pass through, the boundary and allowing toxic or useless by-products of cellular metabolism to exit, at the same time keeping unneeded materials from entering and useful cell contents from oozing out. In a sense, the plasma membrane is like gate and gatekeeper combined. Besides its barrier activities, the plasma membrane also receives and generates signals that are an important part of the communication and coordination between cells.

The membrane responsible for these important delimiting and controlling actions is only 0.1 μm thick, or about one-thousandth the thickness of a sheet of paper, but it is both flexible and dynamic and varies in composition, depending on cell type. The matrix of the membrane—that is, its background material—is formed by the group of fatty organic compounds called phospholipids (see Figure 2.23). Each molecule of this type has a head and a tail. When such compounds are surrounded by water, they align in a characteristic two-layered sheet with the hydrophobic ("water-fearing") tails pointing inward, the hydrophilic ("water-loving") heads pointing outward, and water excluded from the middle (Fig-

BOX 3.1   HOW DO WE KNOW?

MICROSCOPES: TOOLS FOR STUDYING CELLS

How do we know that eukaryotic cells—which are usually far too small to see with the unaided eye—have a membrane-bound nucleus, chloroplasts, or any of the other organelles discussed in this chapter? Much of what we know about cells comes from biologists peering into microscopes of three basic types and witnessing the beauty inside and on the surfaces of the cells.

Light microscopes are instruments containing optical lenses that *refract,* or bend, light rays so that an object appears larger than it really is (Figure 1a). Because it can illuminate and magnify, biologists use the light microscope extensively to locate cells in tissues, to observe the behavior of living cells like *Euglena,* and to detect cell organelles that can be stained bright colors, such as the nucleus, chloroplasts, and mitochondria.

If a specimen is thin enough for light to pass through, a light microscope can magnify its physical details to more than 2000 times normal size. Colored stains can heighten the contrast between various structures in the specimen, making them easier to distinguish. Stains are usually used in combination with fixatives—agents that preserve the cells or tissues so that they remain unchanged. However, stains and fixatives kill cells, so to view living cells, biologists use *phase-contrast light microscopes,* which augment the differences in light refraction between unstained structures so that the contrast between them is bright even without stains.

In the light microscope, magnification is limited by **resolving power,** the ability of the human eye to distinguish adjacent objects as distinct and separate. Two dots brought closer and closer together will eventually be perceived as merging at a distance of about 0.1 millimeter;* that is the lower limit of the eye's resolving power. (An analogy is the way the headlights of an approaching car appear as a single spot when the car is far away, but as two individual lights when the vehicle gets closer.) With a good light microscope, resolving power can be extended some 500 times to about 0.2 μm, or 200 nanometers.† But this in itself represents a lower limit determined by the length of the

electromagnetic waves in visible light. Red light has wavelengths of about 750 nm, and violet light has wavelengths of about 400 nm. Objects with a diameter of less than 200 nm, or half the shortest wavelength of visible light, are no longer visible with the light microscope.

It was the desire to see finer details in cells that led to the invention of the electron microscope. **Electron microscopes (EMs)** use beams of electrons with wavelengths 100,000 times shorter than visible light to view objects, thus magnifying them that much more. These microscopes also use magnets rather than ground glass lenses to focus the electron beams, and the beams must be generated in a vacuum. To use a **transmission electron microscope (TEM)**, the microscopist must first cut an ultrathin slice of a specimen with a piece of broken glass or a diamond knife, then "stain" the specimen with heavy metal such as lead or uranium to increase contrast between the inner structures. Some electrons are scattered or absorbed by structures in the specimen while others pass through and strike a small fluorescent screen (Figure 1b). The result is a greatly enlarged image of the object, which can be photographed and studied for its fine detail.

A second type of electron microscope, the **scanning electron microscope (SEM)**, has a greater depth of field and thus allows scientists to see a specimen's outer surface in three dimensions. The specimen is coated with a thin layer of gold or other metal, and electrons are then aimed to scan it rapidly (Figure 1c). The surface atoms become excited and emit secondary electrons, which are detected by a screen to reveal the object's surface characteristics.

Electron microscopes opened a new and marvelous realm to biologists and have provided much of our current knowledge of cell structure. TEMs have revealed a level of complexity and detail in the cell never suspected until the middle of the twentieth century. And SEMs have shown us a beautiful and sometimes bizarre vision of the intricate shapes of living cells and organisms.

*mm; see the abbreviations chart on the inside back cover.

†nm; see the abbreviations chart on the inside back cover.

ure 3.12a). Since the cytoplasm is watery and since cells must be bathed in fluid on the outside (for reasons explored later), the phospholipids naturally assume this two-tiered configuration. And it is this **lipid bilayer** that is largely responsible for the membrane's barrier functions; many polar substances, as well as large hydrophilic substances such as sugars, cannot pass through it without the aid of special passageways. The membranes surrounding internal cell organelles have essentially the same structure.

While the plasma membrane excludes certain substances, it also allows ions, nutrients, organic wastes, and other molecules to pass in and out. What could possibly account for these well-regulated exchanges? The answer is that proteins embedded in the membrane form passageways for materials. Cell biologists have shown that the cell membrane is a selectively permeable lipid bilayer (one that allows passage of certain materials but not others) studded with proteins that regulate the flow of materials into and out of the cell. This concept is the **fluid mosaic model** of membrane structure (Figure 3.12b). The membrane has the consistency of butter on a warm day rather than the solidity of lard, and individual phospholipid molecules can move about fairly freely in the plane of the membrane. The term *mosaic* refers to the fact that the proteins are scattered about as in a tile mosaic and can

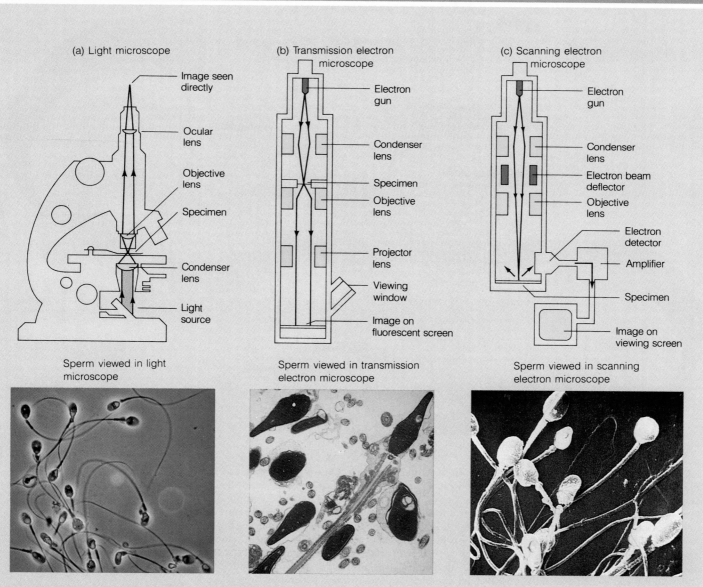

(a) Light microscope
Image seen directly
Ocular lens
Objective lens
Specimen
Condenser lens
Light source

(b) Transmission electron microscope
Electron gun
Condenser lens
Specimen
Objective lens
Projector lens
Viewing window
Image on fluorescent screen

(c) Scanning electron microscope
Electron gun
Condenser lens
Electron beam deflector
Objective lens
Electron detector
Amplifier
Specimen
Image on viewing screen

Sperm viewed in light microscope

Sperm viewed in transmission electron microscope

Sperm viewed in scanning electron microscope

FIGURE 1 ■ **Types of Microscopes and Their Images.**

move about the fluid plane like floating icebergs. Cholesterol molecules inserted in the membrane can alter its fluidity.

The proteins of the plasma membrane are large, irregularly shaped bodies, some embedded in the outer surface of the bilayer and projecting outward, some embedded in the inner surface and extending into the cell's interior, and some extending all the way from one side to the other. Some of these proteins recognize specific materials and allow them to pass through the membrane (or they transport them across). This explains why it is a *selectively* **permeable membrane.**

Defects in one of these transport proteins are responsible for cystic fibrosis, the most frequent serious genetic disease among Caucasian Americans. Chapter 10 explains cystic fibrosis in more detail. Other membrane proteins react to signals, such as those of hormones, or respond to the surfaces of other cells. Different types of cells contain different populations of membrane proteins, and each cell's subset endows it with specific capabilities for regulating the traffic of materials and communicating with outside elements. For example, your red blood cells have special membrane molecules that label

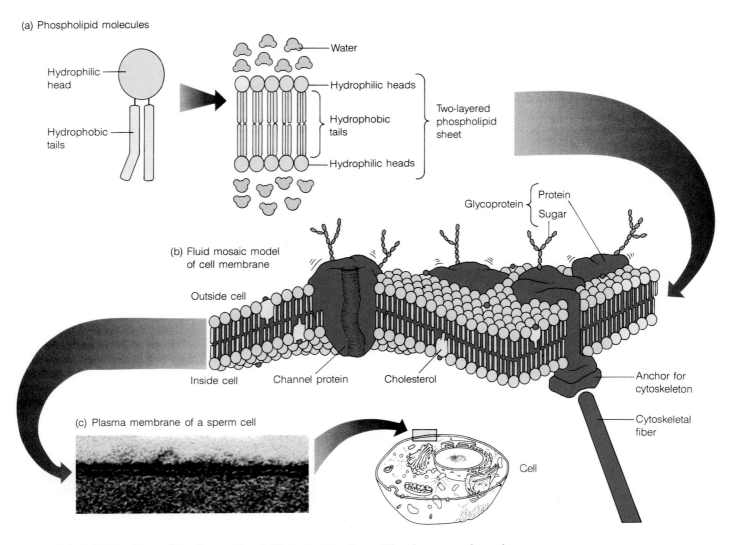

FIGURE 3.12 ■ **The Plasma Membrane: The Cell's Outer Boundary.** The plasma membrane is a flexible fatty sheet that protects the cell and controls the traffic of materials into and out of the cell. **(a)** The plasma membrane is a lipid bilayer, two tiers of phospholipid molecules with hydrophilic heads oriented outward toward the cell's watery interior and exterior and hydrophobic tails oriented inward toward the middle of the membrane. **(b)** Fluid mosaic model. The plasma membrane is pictured as a fluid plane with floating "icebergs"—proteins that can extend through the membrane and project from either or both sides. Glycoproteins, special cell surface molecules that look like antennae, are made up of sugars and proteins and act as cellular labels and as receptors for incoming chemical signals. Some membrane proteins have channels or pores through which ions or other substances may pass. Some membrane proteins attach to cytoskeletal fibers, as shown here. **(c)** Electron micrograph of a sperm cell's plasma membrane.

(a) Phagocytosis

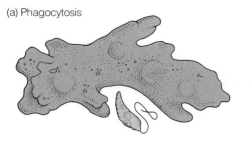
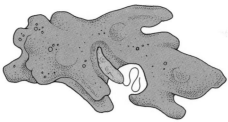
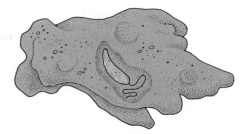

An amoeba ingests a *Euglena* cell (phagocytosis).

(b) Endocytosis and exocytosis

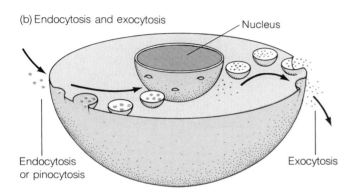

Nucleus

Endocytosis
or pinocytosis

Exocytosis

FIGURE 3.13 ■ Infolding and Outpocketing of the Cell Membrane. Material can be drawn into or expelled from the cell in a membranous pocket by means of *endocytosis* and *exocytosis.* (a) In one kind of membrane infolding, the cell membrane forms a pocket around a cluster of molecules or a food particle, or sometimes around an entire cell, in a process called *phagocytosis.* Here, an amoeba ingests a *Euglena* cell through phagocytosis. (b) When the cell takes in fluid, the process is called *pinocytosis.* Here, the membrane indents around the fluid and dissolved proteins (designated by blue dots), which in turn become trapped as the membrane pinches off a vesicle inside the cell. By *exocytosis,* a process that is the reverse of endocytosis, the cell can secrete pockets of material—indigestible wastes or products such as hormones for export to other parts of a multicellular organism. In exocytosis, tiny membranous spheres packed with the material to be jettisoned fuse with the plasma membrane and then open and release their contents to the outside. (c) In receptor-mediated endocytosis, specific proteins, such as hormones or egg yolk or iron-containing molecules, are sequestered inside the cell. The specific protein binds to a receptor and becomes concentrated in a pit. The pit folds into a vesicle, bringing the protein into the cell. In these photos, the protein is yolk being taken into an eggshell.

(c) Electron micrograph of endocytosis

1   2   3   4

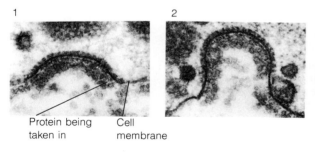
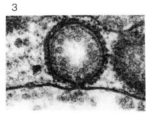
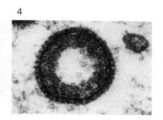

Protein being   Cell
taken in   membrane

the cell, giving it your A, B, AB, or O blood type. Other molecular labels lead to either acceptance or rejection of a transplanted kidney, heart, or other organ.

Surprisingly, there are two methods of transport that allow materials to enter or exit the cell without passing through the membrane at all. This apparent magic involves the infolding or outpocketing of the membrane itself and the taking in or exporting of materials. Figure 3.13 depicts and describes this membrane activity. Interestingly, in some people heart attacks are associated with the infolding of the cell membrane. These people have an inherited condition that slows down the rate at which cells take in cholesterol circulating in the blood. The fatty material thus builds up outside the cells, clogs the blood vessels, and can lead to heart attacks.

The Nucleus: The Cell's Control Center

In a eukaryotic cell, the largest organelle and often the most conspicuous when viewed through the microscope is the **nucleus,** the "brain" of the cellular operation. This roughly spherical structure contains life's master plan, the genetic information (in the form of DNA) that controls a cell's activities. It also makes and exports various forms of RNA, the nucleic acids that relay genetic orders to the cytoplasm. These orders are patterns for building proteins—sometimes functional proteins (enzymes) and sometimes structural proteins in cell walls, cell organelles, and so on. The enzymes, in turn, play an active role in procuring materials for the cell, in building needed cell parts, and in other vital tasks. Information thus

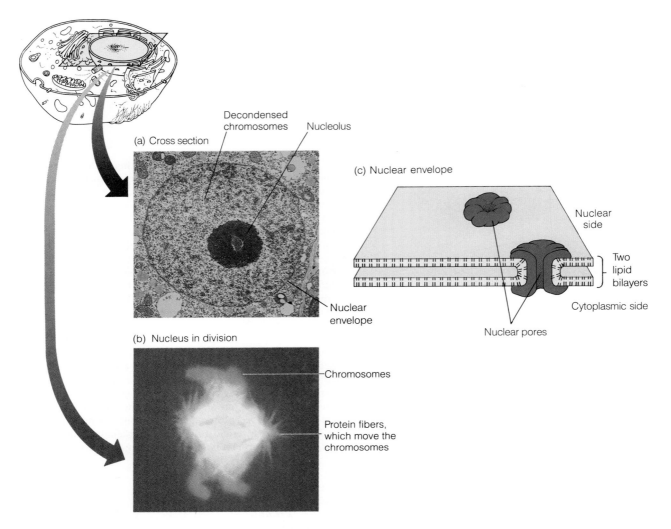

(a) Cross section

Decondensed chromosomes

Nucleolus

Nuclear envelope

(b) Nucleus in division

Chromosomes

Protein fibers, which move the chromosomes

(c) Nuclear envelope

Nuclear side

Two lipid bilayers

Cytoplasmic side

Nuclear pores

FIGURE 3.14 ■ The Nucleus: The Cell's Central Control. Most of a cell's daily activities are directed by information molecules that issue from the nucleus. (a) A cross section of a nucleus reveals a large area of decondensed chromosomes and a dark-staining area, the nucleolus, which is the site of ribosome production. (b) During cell division, the chromosomes condense into visible rods, and protein fibers appear that move the chromosomes into the forming daughter cells. The surface of the often spherical nucleus looks a bit like a golf ball with its double membrane perforated by dozens of pores, or channels made up of proteins. The pores regulate the passage of RNA and other substances. (c) The nuclear envelope is composed of two lipid bilayers with a space in between.

flows from the nucleus to RNA to the proteins, which carry out the work of the cell. Through its RNA messengers to the cytoplasm, the nucleus also controls the rate of cell growth, cell division, and energy acquisition and use. In other words, information molecules issuing from the nucleus direct most of a cell's activities on a day-to-day basis.

The nucleus is surrounded by a **nuclear envelope.** This envelope is made up of *two* lipid bilayer membranes separated by a space, and the entire envelope is perforated at dozens of points by **nuclear pores,** giving the organelle the appearance of a golf ball (Figure 3.14). Each pore is a cluster of proteins

that form a channel that somehow regulates the passage of RNA to the cell's cytoplasm.

Some biologists believe that the nuclear membrane helps keep the DNA molecule intact; the eukaryotic cell contains so much DNA that the vital molecule would become tangled or broken if it were left floating freely in the cell's interior. (In a prokaryotic cell, the much shorter single circular DNA molecule is coiled up in an unbounded region, from where it controls cellular activity more directly.)

Within the eukaryotic nucleus, proteins associate with DNA and form long fibrils, or threads. Whenever the cell is not actively dividing, the nucleus appears grainy because of the loose, extended nature of these threads. However, at the start of cell division, the DNA/protein threads compact into **chromosomes** (literally, "colored bodies"), structures that carry hereditary information (see Chapters 7 and 9).

Within the nucleus, there may be one or more dark-staining regions called **nucleoli** (or "little nuclei"; see Figure 3.14a). Each nucleolus makes a special type of RNA necessary for the synthesis of all proteins. Nucleoli, therefore, help set the pace for the productivity and growth of the cell. The nucleus of a very active cell, such as a growing egg cell primed for the rapid development of an embryo, can have as many as 1000 nucleoli. In a cell with little or no protein synthesis, such as a mature sperm cell, there is no nucleolus.

Cytoplasm and the Cytoskeleton: The Dynamic Background

The plasma membrane and nucleus account for only a small portion of the cell's mass. Most of the cell is made up of **cytoplasm,** a semifluid, highly organized ground substance that acts as a pool of raw materials. About 70 percent of the fluid portion of the cytoplasm consists of water molecules, and about 15 to 20 percent is made up of 10,000 different kinds of protein molecules, with 10 billion or so molecules occurring in an average cell! Many of the proteins are enzymes that speed biochemical reactions, while others are structural proteins that can be assembled into various cell components.

Suspended in the cytoplasm are numerous particles, fibers, and organelles. Thousands of beadlike clusters called **ribosomes** are embedded throughout the cytoplasm and participate in the synthesis of proteins (Figure 3.15). Based on the location of the ribosomes, protein synthesis goes on throughout the cytoplasm but not in the nucleus itself. Ribosomes are made up of a number of special proteins and three or four specific RNA molecules (called *ribosomal RNAs*) that are produced in the nucleolus, the dark-staining region within the nucleus. These associate in the cytoplasm with another form of RNA called messenger RNA, described later.

The electron microscope reveals another major feature of eukaryotic cells: a three-dimensional latticework of minute protein fibers called the **cytoskeleton** that permeates the cytoplasm like a spider's web, suspending the organelles in proper spatial relationships to each other and allowing precisely regulated movement of cell parts. Most of the movements of living things, such as the beating of hearts, the running, flying, or swimming of whole organisms, and the tiny movements of structures within the cell, are based on the activities of these fiberlike "engines" and "skeletal" parts. Protein fibers in the cytoskeleton also act as internal girders and cables that help maintain a cell's shape and transport cell organelles, much as skiers are pulled along on moving rope tows. A rod-shaped organelle called the *centriole* (see Figure 3.7) organizes certain cytoskeletal fibers. Figure 3.16 describes the cytoskeleton in more detail.

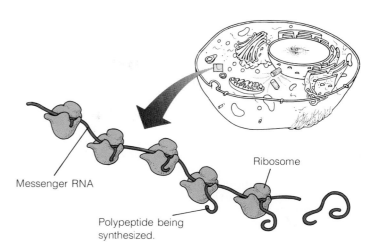

Messenger RNA

Ribosome

Polypeptide being synthesized.

FIGURE 3.15 ■ Ribosomes: Sites of Protein Manufacture. Ribosomes are dense, beadlike clusters of RNA and certain proteins. Ribosomes can be seen throughout the cell cytoplasm, associated with various membrane surfaces, or in chains called polyribosomes. In a polyribosome, several ribosomes become associated with a molecule of messenger RNA like beads on a string. From information in the RNA, each ribosome builds a protein.

In addition to the ribosome particles and cytoskeletal fibers, a eukaryotic cell's cytoplasm contains several more specialized organelles, each with a unique role in the cell's survival.

A System of Internal Membranes Providing Synthesis, Storage, and Export

As we have seen, cells take energy from the environment and use it to build proteins and other molecules, then transport those molecules to the proper regions of the cell where they can be stored or incorporated into new cell parts for repair, growth, or reproduction. In eukaryotic cells, these functions are accomplished by an interconnected system of membranous organelles.

The most conspicuous part of this membrane system is a network of flattened, hollow tubules, sheets, and cisterns (reservoirs) that form an interconnected set of channels throughout the cell. That system of channels is called the **endoplasmic reticulum** (literally, "network within the cell"), and its convoluted passageway extends from the nuclear envelope to the plasma membrane (Figure 3.17). Some areas of the endoplasmic reticulum (abbreviated ER) are folded into smooth sheets and tubes that biologists rather logically call **smooth ER.** The smooth ER manufactures lipids and also detoxifies poisonous chemicals. For example, when sunlight strikes the skin, enzymes in smooth ER convert molecules of the lipid cholesterol into vitamin D, a compound necessary for maintaining strong, healthy bones. Interestingly, North

FIGURE 3.16 ■ The Cytoskeleton: Cell Support and Movement. The cytoskeleton acts as an internal skeleton, supporting the cell's shape and activities. Three types of protein fibers make up the cytoskeleton: *microfilaments, microtubules,* and *intermediate filaments.* The photos here are analogous to X-rays of bones; they show only one particular cytoskeletal element at a time. However, all three types occur simultaneously in many cells.

(a) Microfilaments are cytoskeletal elements about 3 to 6 nm in diameter, made mostly of *actin* proteins. One end of the fiber is usually anchored to the inner surface of the plasma membrane, and along its length, it may associate with other proteins and cause a contracting force. Actin proteins can also rapidly assemble to or disassemble from the filament, and this dynamic activity causes the filament to lengthen or shorten. Actin filaments can pinch an animal cell in two via a drawstring-like action during cell division, and they are also critical elements in the contraction of muscles. If actin from a rabbit's muscle is mixed with a protein from a funguslike organism called a slime mold, the filaments contract forcefully, just as if they were in a muscle cell. Thus, actin and the related molecule myosin are widespread in life, suggesting a very ancient and common origin. Actin filaments can move organelles around inside of cells. The alga Chara has giant cells that rely on actin filaments to move organelles from one end of the cell to the other.

(b) Microtubules, a second cytoskeletal element, are hollow cylindrical tubes 20 to 25 nm in diameter composed of subunits of the globular protein *tubulin.* These subunits are stacked in a spiral form, and like actin, they assemble and disassemble rapidly to lengthen or shorten the microtubules. Microtubules, for example, move colored pigment granules around in the skin cells of reptiles and fish, allowing chameleons, for example, to blend in with the color of their surroundings. Microtubules also help establish and maintain the cell's shape, make up the *spindle* apparatus that separates the chromosomes during cell division (see Chapter 7), and are instrumental in generating cell movement.

(c) A third component of the cytoskeleton, the intermediate filaments (about 10 nm across), are made up of *keratin* and other proteins and are midway in size between microfilaments and microtubules. These filaments are believed to be involved in maintaining cell shape, acting like girders or tension-bearing cables that stabilize the cell's perimeter. Intermediate filament proteins provide the reinforcement that makes an animal's skin cells tough, and they make up most of the matter in skin, nails, hair, and feathers. They thus provide animals, such as this male peacock, with protection, camouflage, and sexual attraction.

(a) Microfilaments   (b) Microtubules   (c) Intermediate filaments

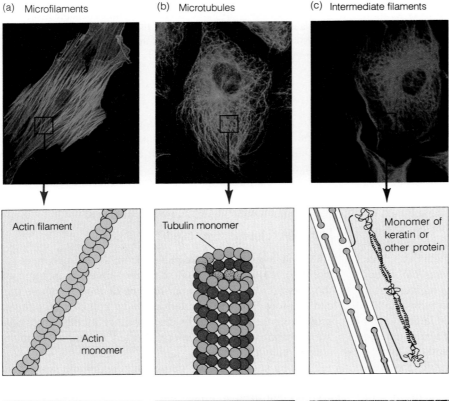

Actin filament

Actin monomer

Tubulin monomer

Monomer of keratin or other protein

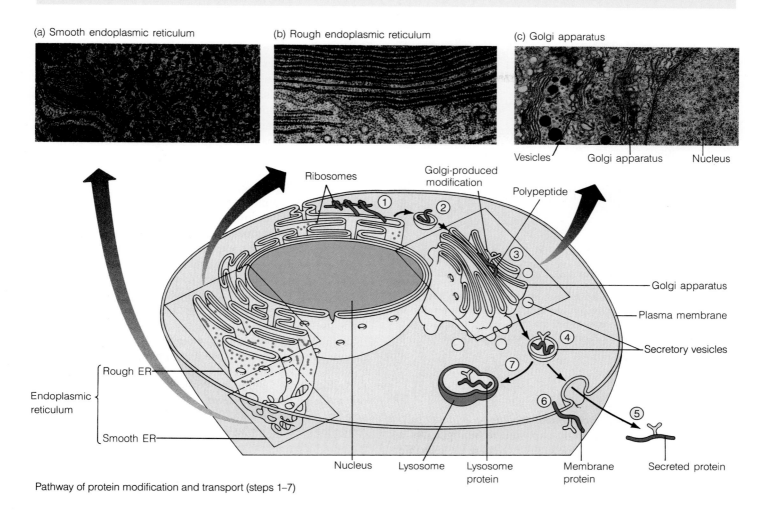

(a) Smooth endoplasmic reticulum   (b) Rough endoplasmic reticulum   (c) Golgi apparatus

Vesicles   Golgi apparatus   Nucleus

Ribosomes   Golgi-produced modification   Polypeptide

Golgi apparatus

Plasma membrane

Secretory vesicles

Rough ER

Endoplasmic reticulum

Smooth ER

Nucleus   Lysosome   Lysosome protein   Membrane protein   Secreted protein

Pathway of protein modification and transport (steps 1–7)

FIGURE 3.17 ■ The Internal Network of Channels and Sacs: Endoplasmic Reticulum and Golgi Apparatus.

(a) The endoplasmic reticulum (ER) is a network of membranes forming an interconnected set of channels throughout the cell. Certain areas of the ER have a smooth appearance and are called *smooth ER*. Smooth ER is most common in cells that produce, store, and secrete steroid hormones and other lipids. Prominent examples of smooth ER occur in the hormone-producing cells of the ovaries and testes and in the glands that produce oil on the face and hair.

(b) The exterior of some ER is studded with ribosomes, giving it a grainy appearance when viewed with the electron microscope. These areas make up the *rough ER,* which is especially prominent in cells exporting or secreting proteins that do their jobs outside the cell of origin. As the sketch shows, proteins assembled on rough ER ribosomes (step 1) pass into the space within the channels of the rough ER and are pinched off into transport vesicles (step 2). These vesicles move to the Golgi apparatus.

(c) A Golgi apparatus is a stack of saucer-shaped membranes surrounded by vesicles. Enzymes inside the Golgi sacs modify the newly made molecules—largely protein, fat, or steroid molecules in animal cells, and mainly proteins or complex carbohydrates in plant cells (step 3). The modifications can include the addition or removal of a functional group or a change in the molecule's basic structure. Finally, the cargo is packaged in new transport vesicles that pinch off from the outer edges of the Golgi sacs (step 4). These vesicles may move toward the plasma membrane, merge with it and release their contents through exocytosis (membrane outpocketing) (step 5), or they may insert a protein into the plasma membrane (step 6) or deliver it to a lysosome (a refuse-digesting organelle) or other cell part (step 7).

FIGURE 3.18 ■ Some Bedouin Women Have a Smooth ER Problem. Because this woman's clothing leaves little or no skin exposed to sunlight, her smooth ER may not be able to make enough of the vitamin D necessary to maintain strong, healthy bones.

African women of one Bedouin tribe who wear dark, full-length garments get very little exposure to sunlight. As a result, the regions of smooth ER in their cells are often unable to convert enough cholesterol to vitamin D, and their bones grow soft and weak (Figure 3.18).

Another part of the endoplasmic reticulum is studded with ribosomes, looks rough under the microscope, and is called the **rough ER.** This region is involved in the synthesis of certain proteins. For example, the rough ER helps produce the enzymes that digest ice cream and most of the other foods you eat. Tracing the production and transport of an enzyme through this membrane system helps us understand how the ER functions. Let's focus on the rough ER in cells within a person's pancreas, a cucumber-shaped organ that generates digestive enzymes and secretes them into a duct leading to the intestine, where food is broken down and digested.

Within a pancreatic cell, the nucleus makes a special RNA called *messenger RNA* that carries genetic information (in this case, information for the digestive enzyme) out of the nucleus and into the cytoplasm (see Figure 3.15). Once in the cytoplasm, the messenger RNA joins the small, beadlike ribosomes—biochemical anvils on which protein molecules will be forged. The ribosomes then stick to the surface of the rough ER—the reason, in fact, that it appears rough. Proteins like these digestive enzymes that are assembled on rough ER enter the ER cavity, are modified as they move along through the channels, and are eventually pinched off in little sacs, or vesicles.

Most of the sacs pinched off from the endoplasmic reticulum enter another membrane system, the **Golgi apparatus,** where the digestive enzymes or other proteins in the sacs are further modified. The modified proteins then leave the Golgi apparatus headed either for the cell's plasma membrane, for tiny digestive sacs within the cell (the lysosomes), or for export from the cell (see Figure 3.17). When exported from the pancreatic cells, the digestive enzymes move down a duct and into the intestine. There they can break down the fats, sugars, and other nutrients present in the ice cream.

The Golgi apparatus is a key component of the membranous organelles along the enzyme secretion pathway. Named after the Italian cell biologist who first spotted it, the Golgi apparatus is like a packaging department for the eukaryotic cell. In 1898, Camillo Golgi observed this apparatus in the nerve cells of a barn owl, but it was not until the availability of electron microscopes over 40 years later that biologists could see the structure clearly. Each Golgi apparatus is a stack of saucer-shaped, baglike membranes surrounded by small, round, membranous containers, or **vesicles** (see Figure 3.17c). A cell can have just one Golgi apparatus or many thousands. Golgi-packaged proteins and lipids repair the plasma membrane itself when it is damaged; and in plants, the Golgi apparatus (called a *dictyosome* by botanists) packages for export the precursors to the cellulose that forms the outer cell wall.

■ **Vesicles in the Cytoplasm**   While many vesicles that pinch off a Golgi apparatus leave the cell, two main types of vesicles—lysosomes and microbodies—take up permanent residence in the cytoplasm. **Lysosomes** are spherical vesicles

(a) Phagocytosis of food particle

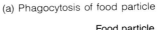

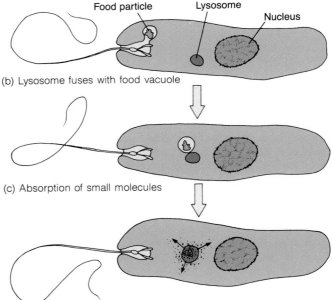

(b) Lysosome fuses with food vacuole

(c) Absorption of small molecules

FIGURE 3.19 ■ *Euglena* **Engulfing a Food Particle.** (a) Phagocytosis. When a *Euglena* cell happens upon a food particle of the right size and composition, the cell membrane forms a pocket around it and engulfs it. (b) Fusion. A lysosome then fuses with the food vacuole, and digestive enzymes break down the food particle. (c) Absorption. Small nutrient molecules then pass through the lysosome membrane into the cytoplasm and nourish the cell.

within the cell that contain powerful digestive enzymes. These enzymes can help recycle worn-out cell constituents. Biologists know that this recycling by lysosomes is crucial to human survival because babies born with defective lysosomes have Tay-Sachs disease. Lysosomes in patients' brain cells fail to recycle certain carbohydrates, allowing them to accumulate, block brain development, and result in mental retardation and death.

Besides carrying out recycling, lysosomes can act like minute cellular stomachs. When a *Euglena* cell, or even one of your white blood cells, engulfs a bacterium, lysosomes merge with the membrane-encapsulated "meal" and release their enclosed digestive enzymes, breaking down the particle to its constituent molecules (Figure 3.19). These molecules then pass through the lysosomal membrane into the cytoplasm, where they can be used as raw materials for producing energy or building new cell parts.

Lysosomes also sometimes act as "suicide bags," breaking open and spilling their contents and literally digesting entire damaged or aged cells from the inside out. This process is especially dramatic in insects that are transformed within cocoons from caterpillars to moths or butterflies. One by one, the caterpillar cells die and are digested, making way for the new adult moth cells.

It is still not clear why the enzymes in the suicide bags don't digest the bags themselves, but clearly they can't break down every type of material to constituent molecules. In fact, aging eukaryotic organisms often accumulate brown-pigmented granules in their lysosomes; these are believed to be undigested cellular "refuse" that must simply be stored until the cell dies. In an 80-year-old person, the material can occupy 10 percent of the volume of each heart cell. Older people often develop brownish "age spots" on their skin from the same pigments.

Microbodies are the other class of membranous vesicles that reside permanently in the cell. One type of microbody called a *peroxisome* contains the enzyme catalase. This enzyme breaks down the corrosive compound hydrogen peroxide that forms as a by-product of energy conversions and other cell functions. You can watch catalase at work if you cut yourself and then clean the wound with hydrogen peroxide. The bubbling you see results from the action of catalase from your cells breaking down the hydrogen peroxide into bubbles of oxygen and water. Peroxisomes within liver and kidney cells break down and detoxify fully half of the alcohol a person drinks, and other peroxisomes can produce a small amount of energy for the cell. However, such microbodies provide only a small percentage of the energy compounds used by the cell. The cell's real power plant is in the mitochondrion (Figure 3.20).

Mitochondria: The Cell's Powerhouse

Within eukaryotic cells, one to several hundred organelles called **mitochondria** provide chemical fuel for cellular activities that go on continually, such as the production, shuffling

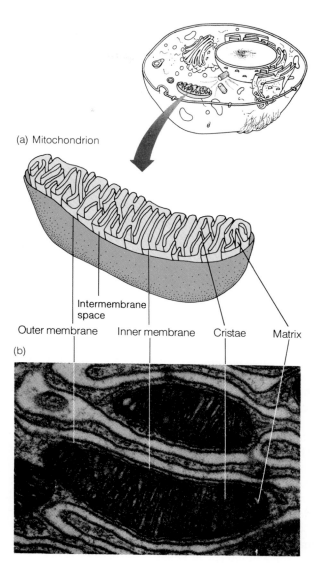

(a) Mitochondrion

Intermembrane space

Outer membrane   Inner membrane   Cristae   Matrix

(b)

FIGURE 3.20 ■ Mitochondria: Power for the Cell. (a) Mitochondria look like tiny grains of rice filled with threads. Mitochondria have a smooth *outer membrane* and an *inner membrane* folded into a complex of overlapping sheets, or *cristae*, like the teeth of enmeshed gears. Between the two membranes is the *intermembrane space*, a compartment crucial for the generation of ATP. Inside the inner membrane, there is another compartment, called the *matrix*, that contains a concentrated mixture of the genetic material DNA, ribosomes, and many different enzymes involved in producing ATP for energy. (b) The electron micrograph shows kidney cell mitochondria.

about, and export of proteins. Mitochondria break down small carbon-containing molecules into carbon dioxide and water and in the process release energy, which is stored in ATP molecules (see Chapter 2). These "energy packets" then diffuse throughout the cell and fuel the biochemical transactions of life processes. The chemical conversions that go on inside the mitochondria require oxygen and are collectively called **aerobic respiration.** The details of these important conversions are discussed in Chapter 5.

The activities of each mitochondrion are possible because of the organelle's unique structure. The term *mitochondrion* comes from the Greek words for "thread" and "grain," and indeed, the organelle can look like tiny grains of rice or, at low magnification, threads (see Figure 3.20). The shape can vary from small spheres to long, sausagelike bodies roughly the size of a bacterium. Figure 3.20 shows that the mitochondrion, like the nucleus, has two lipid bilayer membranes: a smooth **outer membrane** and an **inner membrane** thrown into folds called **cristae.** Production of ATP takes place on these folds. Prokaryotic cells lack mitochondria, and they convert energy on the inner surface of their plasma membranes.

While all eukaryotic cells have mitochondria, the characteristics and abundance of the organelles depend on the cell's activity level. Very active cells that expend a great deal of energy have larger, more numerous mitochondria with more extensive, highly folded inner membranes. A heart muscle cell, for example, can have thousands of large mitochondria. In addition, mitochondria are often most densely packed in regions of the cell with the highest energy demands. The mitochondria in a human sperm cell, for instance, are wrapped around the base of the whiplike tail that propels the cell.

Mitochondria have a semiautonomous existence in the cell: They have their own DNA that directs production of some, but not all, of their component proteins. Information to form the rest of the mitochondrion is encoded in the DNA in the nucleus. Mitochondria also have their own ribosomes, and mitochondria can divide in half and thus reproduce independently of the cell's normal cell division cycle. These facts, plus the bacteria-like size and shape of most mitochondria, have led some biologists to believe that the organelles originated millions of years ago when primitive bacteria were engulfed by larger cells. This hypothesis is called endosymbiosis, and we will return to it in Chapter 15.

Not only do mitochondria have their own DNA, but they are passed to an animal only from its mother, since mitochondria are present in eggs but not the sperm nucleus. Thus, people can trace their mitochondria back to their mothers, grandmothers, great-grandmothers, and so on. Analysis of mitochondrial DNA from people all over the world has allowed researchers to conclude that all people alive today are members of a lineage that began with one individual woman (or a small group of women) who lived (probably in Africa) about 200,000 years ago. Controversy still surrounds the exact time this woman (or women) lived, and whether she was anatomically modern. Nevertheless, she has caught the popular imagination as "mitochondrial Eve."

The endoplasmic reticulum, lysosomes, and mitochondria are present in almost all eukaryotic cells. In the next section, we will consider tasks and organelles found in some cells but not in others. Whether a given structure occurs in all cells or

only certain cells, one fact remains: It is the simultaneous, coordinated activities of *all* the cell parts that add up to a living unit. Separately, those parts may be marvels of biochemical organization, but they are still just individual instruments in the music of life.

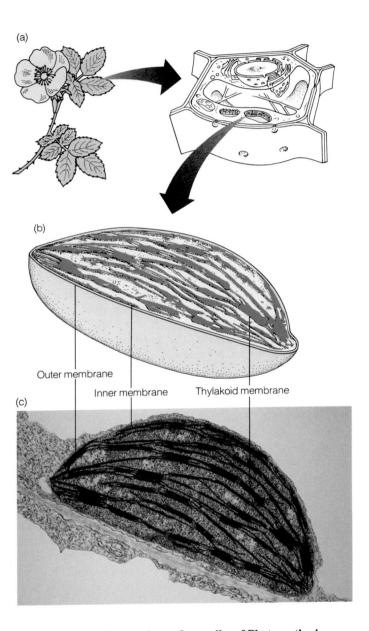

FIGURE 3.21 ■ **Chloroplasts: Organelles of Photosynthesis.** (a) The chloroplasts in plant cells are elongated organelles that contain green-colored pigments that trap the energy from sunlight. (b) Chloroplast pigments are embedded in membranes that arise from the inner membrane and form a stack of flattened disklike sacs called the *thylakoids.* These thylakoid membranes trap sunlight and transfer the energy to energy-carrying molecules that fuel the production of sugar and starch (see Chapter 6). (c) The electron micrograph of a chloroplast is magnified 40,000 times.

SPECIALIZED FUNCTIONS AND STRUCTURES IN CELLS

This chapter began with *Euglena,* a cell specialized to pursue sunlight by means of particular organelles that "outfit" the organism to move and to trap solar energy. Like *Euglena,* most cells have specialized organelles in addition to their nucleus, mitochondria, and other organelles of life support.

Plastids: Organelles of Photosynthesis and Storage

Plants and some protists have several types of oval organelles, collectively called **plastids,** that harvest solar energy and produce and store food. Like mitochondria, plastids have two lipid bilayer membranes, and the most important and widespread of the plastids, the **chloroplasts,** also carry out critical energy conversions. As the organelles of photosynthesis, chloroplasts trap the sun's energy in a chemical form (carbohydrate molecules) that is ultimately used by nearly every organism on earth to power the myriad activities of life. Chloroplasts contain the green, light-absorbing pigment **chlorophyll** and sometimes brownish and yellowish pigments as well. These pigments are embedded in chloroplast membranes, in a similar way to the anchoring of respiratory enzymes in mitochondrial membranes.

Figure 3.21 depicts the internal structure of chloroplasts. Through photosynthesis, plant cells manufacture their own organic nutrients instead of absorbing ready-made nutrients, as animal and fungal cells and most protozoa must do. Chapter 6 explains how, ultimately, those ready-made nutrients come from chloroplasts. *Euglena* chloroplasts contain their own DNA, and many biologists believe that they, like the mitochondria, evolved from bacteria that came to inhabit larger cells. Chapter 15 describes this theory.

Two other types of plastids are common in plant cells. **Chromoplasts** ("colored plastids") store yellow, orange, and red energy-trapping pigments and give color to many fruits and flower petals. **Leucoplasts** ("white plastids"), on the other hand, are colorless organelles. Some leucoplasts store starch granules, while others store proteins or lipids. Potatoes, for example, contain billions of starch-storing leucoplasts.

Vacuoles: Not-So-Empty Vesicles

Through the microscope, most plant cells look strangely hollow, with what appears to be empty space filling most of the cell. Actually, the "space" is an important organelle called the **central vacuole** (from the Latin word for "empty"). This organelle contains water and various storage products, has a single membrane, and can occupy from 5 to 95 percent of the total cell volume.

Far from empty, the central vacuole's main role is to take up space. The vacuole fills up a plant cell with what amounts

to a bag of water that presses a small amount of cytoplasm and all the cell's organelles into a thin layer just below the cell membrane (see Figure 3.7b). This makes the surface-to-volume ratio quite favorable. When plenty of water is available, a plant's many vacuoles store it and keep the cells plump, giving firm shape to the leaves, stems, and other plant parts. The next time you neglect a houseplant and it wilts, you'll know that its vacuoles need refilling.

Although the central vacuole contains mostly water, it can also store sugars, proteins, ions such as Na$^+$, pigments, and other materials. If you gently tear a section of an orange in

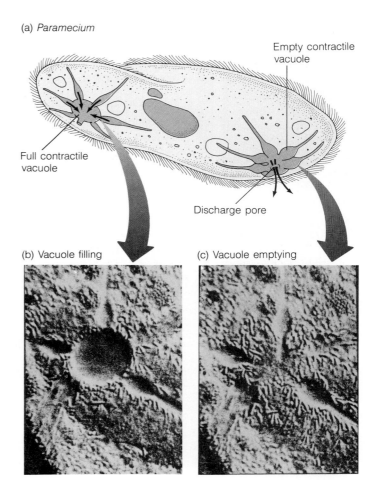

(a) *Paramecium*

Empty contractile vacuole

Full contractile vacuole

Discharge pore

(b) Vacuole filling

(c) Vacuole emptying

FIGURE 3.22 ■ The Contractile Vacuole: Wastewater Management. (a) *Euglena, Paramecium,* and many other large, single-celled organisms living in fresh water have contractile vacuoles, an organelle that collects and pumps out water. In *Paramecium,* the contractile vacuoles are near the cell's surface, expelling water one or more times per minute. Water then exits via the discharge pore. (b) Tiny ducts lead water into the nearly filled contractile vacuole of a paramecium. (c) When the vacuole contracts, the fluid is forced out.

two, you will observe small, delicate spindle-shaped bags containing orange juice. These bags are actually single cells, each having a huge central vacuole packed with the sweet pungent juice most of us like so much. Vacuoles in other plant cells can also harbor poisonous substances that protect plants from hungry animals, or they may contain specialized products such as rubber and opium.

Single-celled organisms that live in fresh water, such as the protists *Euglena* and *Paramecium,* have another type of vacuole that eliminates the water that tends to accumulate in these aquatic cells. This "bailing out" is done by **contractile vacuoles** that collect water and actively pump it out. *Paramecium* has a contractile vacuole at each end of the cell (Figure 3.22). Water collects quickly in the vacuole and is squeezed out through the discharge pore every 15 to 60 sec-

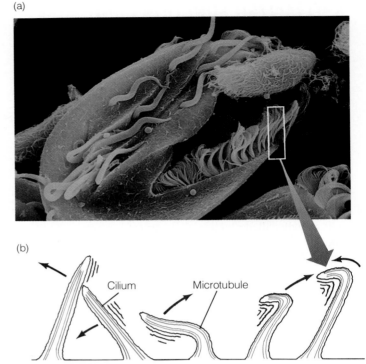

(a)

(b)

Cilium   Microtubule

FIGURE 3.24 ■ **Cilia: Hairlike Projections That Beat in Concert.** Cilia tend to occur in vast numbers, like short, movable hairs over the cell surface. (a) The slipper-shaped *Paramecium* cell in the upper right corner can swim quickly forward or backward by the wavelike motions that sweep continuously and sequentially across its thousands of surface cilia. The much larger *Euplotes* cell, with its own fringe of waving cilia, is attempting to consume the *Paramecium.* (b) As in flagella, microtubules span the interior length of each cilium and enable it to flex.

onds by contractions based on the activity of the cytoskeleton. In this way, the contractile vacuole prevents excess water from accumulating and, perhaps, eventually bursting and killing the cell.

Cellular Whips: Cilia and Flagella

For cells such as *Euglena* that pursue sunlight, or for a mammal's white blood cells that chase invading microorganisms, movement is an important cell function, and it is made possible for these and most other mobile cells by the contraction of certain cytoskeletal fibers (see Figure 3.16). Many cells possess either of two specialized organelles of movement: flagella or cilia, both of which are made from microtubules. Some free-living cells like *Euglena,* as well as the sperm cells of many animals and some plants, have a fine whiplike organelle called a **flagellum** that extends from the cell surface and undulates, pushing the cell forward through the water (Figure 3.23). Without this propulsion, a human sperm cannot bring about fertilization. Some Maori tribesmen from New Zealand, for example, are sterile owing to a genetic defect that renders the sperm's flagellum nonfunctional.

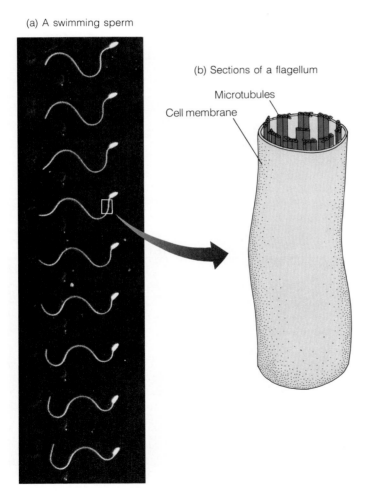

(a) A swimming sperm

(b) Sections of a flagellum

Microtubules

Cell membrane

FIGURE 3.23 ■ **How Flagella Move.** (a) A large wave moves down a sperm's flagellum, forcing the sex cell forward. The sperm in this photo are from a tunicate, a small, simple inhabitant of tide pools. (b) A flagellum has nine pairs of microtubules arranged in a circle, with two microtubules in the center. When the microtubules slide past each other, the flagellum bends. Both flagella and cilia develop from short cylindrical structures called centrioles (see Figure 3.7), which remain at the base as *basal bodies.*

Certain single-celled protists bear not one flagellum but thousands of shorter projections called **cilia** (singular, cilium) all over the cell surface (Figure 3.24). The cilia beat in concert like the oars of a medieval galley ship and allow the cell to swim quickly. In some protozoa and in the cells of certain multicellular organisms, cilia can serve a different function, sweeping fluid and particles across the stationary cell. In cells that line the human lungs, for example, cilia sweep dust particles out toward the air passages to eventually be expelled in mucus or swallowed. These cilia are often damaged in smokers and contribute to their cough and to their greatly increased chance of developing lung disease.

Cell movements of a different kind cause the devastating migration of cells when a cancer begins to spread, or *metastasize.* These creeping movements of animal cells rely on the cytoskeleton, just as a running squirrel depends on its skeleton (Figure 3.25). Using magnification, one can see that the rear of the cell attaches to a surface and that the front edge of the cell protrudes forward in sheetlike ruffles that appear to flutter. The ruffling edge thrusts forward, and the lower cell surface sticks to the substrate. Eventually, the rear end snaps free of its surface attachment, and the cell moves forward.

Cell Coverings

The outer boundary of the cell cannot be said to fall strictly at the plasma membrane, because virtually all cells secrete coverings—either strong cell walls or fluffy coatings—that protect the delicate membrane and confer other advantages as well. In addition, while some cell types, like *Euglena,* are free-living and independent, many live within multicellular organisms and are surrounded by an **extracellular matrix,** a meshwork of secreted molecules that serves as scaffold and intercellular glue.

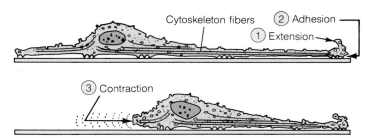

FIGURE 3.25 ■ **Cell Movements: Creeping, Ruffling, and Inching.** The changeable shape of an animal cell, as well as the ability to creep and glide forward, is maintained by the dynamic cytoskeleton. As an animal cell in a Petri dish creeps forward, the front end flattens like a ruffle blowing in the wind as microfilaments rapidly extend (1) and contract. As the leading edge stretches forward, sticky areas (adhesion plaques) on the lower surface cling (adhere) to the substrate below (2). Eventually, the cell contracts (3), the trailing edge snaps forward in inchworm fashion, and the cell locomotes toward the ruffling edge.

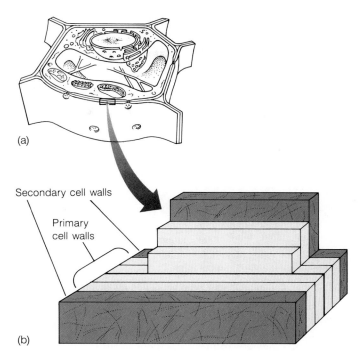

FIGURE 3.26 ■ **Plant Cell Walls.** **(a) Plant cells are surrounded by walls made up mostly of cellulose, the most abundant organic molecule on earth. (b) As a newly divided plant cell matures, it lays down a porous primary cell wall just outside the plasma membrane that remains flexible until a secondary cell wall, usually more rigid than the first, is deposited inside the primary wall, close to the plasma membrane.**

Cell walls made largely of cellulose surround plant cells on all sides (Figure 3.26). First the cell lays down a **primary cell wall** immediately outside the plasma membrane. The primary cell wall remains stretchy and flexible until the cell it surrounds has stopped growing and begun to mature. This wall is also porous, allowing water, gases, and solid materials to pass through to the plasma membrane. Many plant species lay down a **secondary cell wall** inside the first; it can be very rigid from stiffening materials such as lignin, the major component of woody tissue. If you are studying at a wood desk, you and your work are being supported by thickened, dead cell walls—all that remains of the once living tree.

Some fungi and protists have cellulose in their cell walls as well as a wide variety of other molecules. Bacteria and archaebacteria also secrete cell walls, but these are single-layered and made up of complex materials that include sugars, lipids, and amino acids rather than cellulose. Some biologists consider plant and bacterial cell walls to be a form of extracellular matrix, but that material is more dramatically illustrated in animals. Virtually all animal cells secrete and become surrounded by a meshwork of extracellular molecules, mostly fibrous proteins linked to polysaccharides (Figure 3.27). The most common of these fibrous proteins is **collagen,** which, as we saw in Chapter 2, has stiff, ropelike polypeptide chains wound around each other into fibrils. Col-

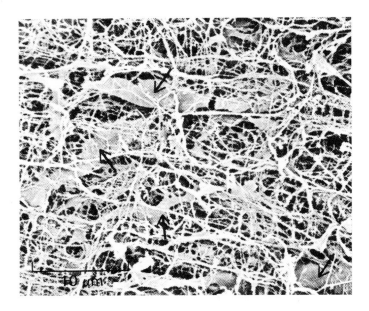

FIGURE 3.27 ■ **Extracellular Matrix: An Exterior Meshwork.** Most animal cells are surrounded by a meshwork of protein and polysaccharide molecules. This extracellular matrix cushions the cell, strengthens the tissue the cell is part of, and helps establish and maintain the cell's shape. The most common extracellular material in animals is collagen. This scanning electron micrograph reveals the collagen fibers surrounding the eye cells from a chick embryo.

cells and is a factor in their ability to spread throughout the body. This shows that proteins of the extracellular matrix can dramatically affect cell behavior as well as cell shape.

Links Between Cells

Cells in multicellular organisms are not only embedded in an extracellular matrix; they are also attached to neighboring cells by physical linkages. Like the matrix materials, these junctions help weld cells together into functional tissues and organs; but they also allow the cells to communicate freely, and they enable the activities of the various cell types to be coordinated. While some linkages allow materials to flow between cells, others, in organs such as the urinary bladder, prevent fluid leaks between cells. Other junctions, called adhering junctions, hold cells together. People with genetically defective adhering junctions often lose large patches of skin from the slightest scrape. Figures 3.28 and 3.29 describe tight junctions, gap junctions, and other specific types.

lagen molecules constitute 25 percent of all the protein in a typical mammal and most of the material in tendons, which act like pulleys and cables that enable muscles to move bones. One can usually feel the cords that run behind the knee by bending the leg and pulling the heel backward. Those cords— the tendons—are almost entirely made up of proteins in the extracellular matrix.

Another protein in the extracellular matrix (fibronectin) abundantly surrounds normal cells, but is absent from cancer

FIGURE 3.28 ■ **Junctions and Links Between Cells.** Numerous junctions allow linkage and communication between cells, as the figure lists. More specifically, a *tight junction* is a band around the cell that forms a tight seal and prevents molecules from passing between it and the next cell. A *belt desmosome* is a ring of actin filaments that encircles a cell and helps it adhere to adjacent cells. A *spot desmosome* is like a tiny spot weld or button that joins two cells and anchors intermediate filaments. A *gap junction* is a protein-lined pore that connects adjacent cells and allows communication.

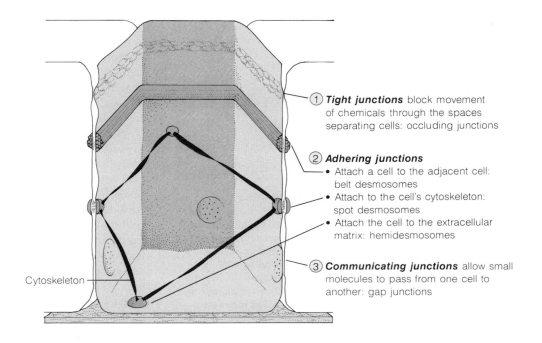

① *Tight junctions* block movement of chemicals through the spaces separating cells: occluding junctions

② *Adhering junctions*
- Attach a cell to the adjacent cell: belt desmosomes
- Attach to the cell's cytoskeleton: spot desmosomes
- Attach the cell to the extracellular matrix: hemidesmosomes

③ *Communicating junctions* allow small molecules to pass from one cell to another: gap junctions

Cytoskeleton

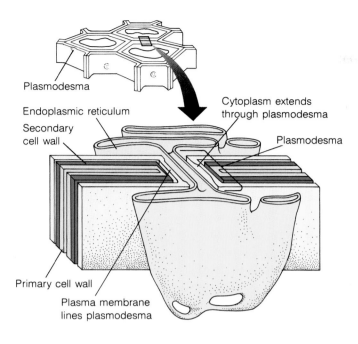

Plasmodesma

Endoplasmic reticulum

Secondary
cell wall

Cytoplasm extends
through plasmodesma

Plasmodesma

Primary cell wall

Plasma membrane
lines plasmodesma

FIGURE 3.29 ■ Communicating Links Between Plant Cells. In plants, encasement by the cell wall would isolate and block intercellular coordination if it weren't for special links. Called *plasmodesmata*, these plant cell connectors are delicate bridges of cytoplasm that pass through the walls of adjacent cells and link them, allowing the rapid exchange of materials and sometimes electrical signals.

CONNECTIONS

This chapter explored the roles of cells and their components. Cells share a set of basic "housekeeping" tasks that keep them alive, and eukaryotic cells have a set of complex organelles—plasma membrane, nucleus, cytoplasm, ribosomes, and internal membranes—associated with those tasks. In simpler prokaryotic cells, the plasma membrane, ribosomes, and a DNA-containing region carry out similar functions.

Cell structures are based on the same biological molecules discussed in Chapter 2. Lipids make up cell membranes; specific proteins in the membranes give the organelle its particular function; carbohydrates give strength or store energy; and nucleic acids direct the cell's activities.

We've also seen that in eukaryotes, which make up the vast majority of all cells, specialized organelles carry out specialized tasks; among them are the chloroplasts for photosynthesis, cilia and flagella for movement, and vacuoles for storage, space filling, and water excretion. Finally, we have seen that deficiencies in cell structure and function can lead to weak bones, undeveloped brains, sterility, and other serious conditions.

The focus on cells so early in this book provides a basis for understanding more complex living systems, for as the pioneers of cell biology discovered, all living things contain cells, and all the biochemical processes of life go on inside them. Thus, cells are truly "life modules," and to understand their activities is to understand much about life itself.

Highlights in Review

1. All cells share certain basic functions responsible for keeping them alive, and they also share certain physical structures—the machinery of life.

2. Each cell's specialization depends on its shape, size, and the particular proteins and organelles it contains.

3. The parts of a cell do not in themselves make up a living unit because life is an emergent property based on the physical organization and integrated activities of all cell parts.

4. Two centuries after the discovery of cells, German scientists developed the cell theory: (1) All living things are composed of one or more cells; (2) cells are the basic living units, and all the chemical reactions of life take place within cells; (3) all cells arise from preexisting cells.

5. There are two major kinds of cells. Prokaryotic cells have a plasma membrane and cytoplasm, but no membrane-bound nucleus or other membrane-bound organelles. Eukaryotes have a plasma membrane and cytoplasm, as well as a nucleus and several kinds of membrane-bound organelles. Members of the domains true Bacteria and Archaea are prokaryotes, and members of the domain Eucarya (plants, animals, fungi, and protists) are eukaryotes.

6. Most cells are extremely small: About five average animal cells can fit across the thickness of this page. Cell size is limited by the surface-to-volume ratio, because the larger a cell's volume, the greater are its needs for material exchange at the cell surface.

7. A cell has certain life-support tasks. It must take in raw materials and expel wastes; absorb and convert energy to usable form; build biological molecules for repair, growth, and duplication; and regulate and coordinate these tasks.

8. A plasma membrane surrounds the cell. It keeps harmful substances out, sequesters useful substances within, and regulates the traffic of raw materials and wastes into and out of the cell. The membrane is a semifluid bilayer of phospholipid molecules studded with protein molecules that accomplish the actual "traffic control."

9. Some materials enter or exit the cell via inpocketing and outpocketing of the cell membrane. This is how cholesterol is cleared from the blood.

10. Most of a cell's day-to-day activities are determined by information molecules issuing from DNA. In a eukaryotic cell, the nucleus is often the largest organelle. The nucleus is surrounded by a nuclear envelope, and it contains a dark-staining region, the nucleolus, that produces ribosomal RNA.

11. Besides the nucleus and plasma membrane, the rest of the cell is made up of the cytoplasm, which is mostly water containing dissolved proteins, small molecules, and suspended particles. The cytoplasm acts as a pool of raw materials and contains a latticework, the cytoskeleton, that suspends the organelles.

12. Ribosomes are small spheres of RNA and protein that exist either free in the cytoplasm or attached to the endoplasmic reticulum (ER). Ribosomes are the workbenches on which proteins are made.

13. The endoplasmic reticulum is a network of flattened, hollow tubules and sheets extending throughout the cell and forming interconnected channels. Rough ER is studded with ribosomes and is involved in the production and transport of proteins, such as the enzymes that digest a person's food. Smooth ER lacks ribosomes and produces, stores, and secretes steroid hormones and other lipids, including vitamin D necessary for strong bones.

14. Materials made in the ER can be stored in vesicles or enter the Golgi apparatus—a stack of saucer-shaped, baglike membranes that modify the molecules produced in rough and smooth ER and package them for export from the cell or for storage in the cytoplasm.

15. Lysosomes are spheres containing digestive enzymes that recycle worn-out cell parts. The lack of even one of these enzymes can cause lethal Tay-Sachs disease.

16. Mitochondria are small sausage-shaped bodies with a smooth outer membrane and a highly convoluted inner membrane. Mitochondria use oxygen to produce chemical fuel (ATP) for the cell's activities.

17. Plants and some protists like *Euglena* contain plastids, which are small bodies with double membranes. Chloroplasts are green plastids that carry out photosynthesis. Chromoplasts store yellow, orange, and red pigments in flower petals and other structures, while leucoplasts store starch granules in a potato, for example.

18. A central vacuole filled with water and some stored molecules takes up most of the space in plant cells. A wilted plant's vacuoles are not filled with water. Contractile vacuoles pump excess water from cells that live in fresh water.

19. Organelles of movement, the flagella and cilia have a similar array of microtubules and can move entire cells, for example, sperm, or can move substances past cells, for example, clearing the lungs and airways of debris.

20. Plant cells have outer walls made mostly of cellulose. The primary cell walls remain flexible until the cell's growth slows; a secondary cell wall is often deposited inside the first, adding rigidity. The paper in this book is made of plant cell walls. Animal cells have an extracellular matrix that surrounds them in a supportive meshwork and that acts like intercellular glue. This meshwork is the main component of tendons.

21. Cells often have important physical links to their neighbors, including spot desmosomes, delicate bridges of cytoplasm (gap junctions and plasmodesmata) that allow passage of materials and signals between cells, and "rubber gaskets" between cells (tight junctions) that prevent leakage from one side of a cell sheet to another.

Key Terms

aerobic respiration, 72	lysosome, 71
cell membrane, 61	microbody, 71
cell theory, 54	microfilament, 68
cell wall, 75	microtubule, 68
central vacuole, 73	mitochondrion, 71
chlorophyll, 73	nuclear envelope, 66
chloroplast, 73	nuclear pore, 66
chromoplast, 73	nucleoid, 55
chromosome, 67	nucleolus, 67
cilium, 75	nucleus, 65
collagen, 75	outer membrane, 72
contractile vacuole, 74	plasma membrane, 61
crista, 72	plasmodesmata, 77
cytoplasm, 67	plastid, 73
cytoskeleton, 67	primary cell wall, 75
desmosomes, 76	prokaryotic cell, 55
electron microscope (EM), 62	resolving power, 62
endocytosis, 65	ribosome, 67
endoplasmic reticulum, 67	rough ER, 70
eukaryotic cell, 55	scanning electron microscope (SEM), 62
exocytosis, 65	
extracellular matrix, 75	secondary cell wall, 75
flagellum, 74	selectively permeable membrane, 64
fluid mosaic model, 63	
gap junction, 76	smooth ER, 67
Golgi apparatus, 70	surface-to-volume ratio, 60
inner membrane, 72	tight junction, 76
intermediate filament, 68	transmission electron microscope (TEM), 62
leucoplast, 73	
light microscope, 62	vesicle, 70
lipid bilayer, 63	

Study Questions

Review What You Have Learned

1. How did Robert Hooke and Anton van Leeuwenhoek add to our knowledge of cells?

2. State the different parts of the cell theory.

3. How are prokaryotic cells structurally different from eukaryotic cells?

4. What are the functions of a cell's plasma membrane, nucleus, and cytoplasm?

5. Although a cell is tiny, it contains a number of organelles, including ribosomes, endoplasmic reticulum, Golgi apparatus, lysosomes, and mitochondria. What role does each of these organelles play in the life of the cell?

6. True or false: Since bacteria do not contain a nucleus, they do not possess DNA. Explain your answer.

7. Discuss how, in cellular terms, a magnolia tree and a spiny anteater show a division of labor.

8. How are plant and animal cells similar? How are they different?

9. Describe three types of cell movement.

10. Compare the coverings around plant cells and animal cells.

Apply What You Have Learned

1. What prevents water from passing through the lipid bilayer of a plasma membrane?

2. A science fiction writer describes an amoeba growing to the size of an elephant, then devouring horses and cows. From your knowledge of surface-to-volume ratio, would you agree with the science in this fiction? Explain.

3. A scientist discovers and carefully describes a new species of single-celled organism. What different aspects of the cell would she investigate by using a phase-contrast microscope, a transmission electron microscope, or a scanning electron microscope?

For Further Reading

ALBERTS, B., D. BRAY, J. LEWIS, M. RAFF, K. ROBERTS, and J. WATSON. *Molecular Biology of the Cell.* 2d ed. New York: Garland, 1989.

BRETSCHER, M. S. "How Animal Cells Move." *Scientific American,* December 1987, pp. 72–80.

DARNELL, J., H. LODISH, and D. BALTIMORE. *Molecular Cell Biology.* New York: Scientific American, 1986.

PRESCOTT, D. M. *Cells.* Boston: Jones and Bartlett, 1988.

WEISS, R. "Cystic Fibrosis Treatments Promising." *Science News* 139 (1991): 132.

The Dynamic Cell

Red Blood Cells: Activity and Order

Find the pulse in your wrist or neck, and feel the blood surge against your fingers for a few seconds. With each pulse of blood, more than 5 million red blood cells pass beneath your fingers, and these are but a small fraction of the 25 trillion that course through your blood vessels, picking up life-sustaining oxygen as they race through your lungs and delivering it, within seconds, close to each cell of your brain, muscles, and every other part of your body. Red blood cells are bags of red-pigmented protein molecules that transport oxygen, but they are much more too. Like all other cells in the body, they are dynamic living units that carry out mechanical, chemical, and transportation tasks that keep them—and us—alive.

Every second, the marrow inside your bones produces nearly 2 million new red blood cells (Figures 4.1 and 4.2a). Each new red blood cell has a nucleus, ribosomes, mitochondria, endoplasmic reticulum, and many of the other intricate organelles we encountered in Chapter 3. All of these operate at an incredibly rapid rate to churn out millions of protein molecules, especially the red-colored oxygen carrier hemoglobin and various kinds of enzymes, for the mature red cell. After a few days, the immature blood cell performs an intriguing movement: It enters the bloodstream by squeezing through a blood vessel wall into one of the many fine vessels that crisscross the bone marrow (Figure 4.2b and c). As the cell squeezes through, its rigid nucleus pops out, but the cytoplasm remains. Once the cell enters the blood vessel's hollow center, it is swept away by the moving bloodstream. The mature cell is now bright red and disk-shaped with a dent in the center. It lacks a nucleus and has the flexibility of soft rubber, allowing it to pass through narrow vessels in the fingers, toes, nose, ears, and elsewhere.

Because the nucleus has been expelled and because the mitochondria and ribosomes disintegrate within a few days,

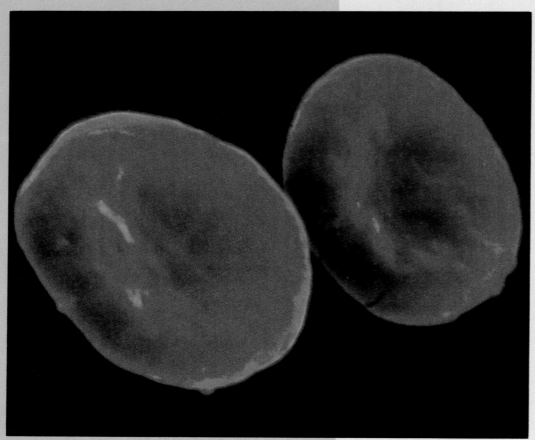

FIGURE 4.1 ■ **Red Blood Cells Look Like Life Savers—and They Are.** Red blood cells carry oxygen to other body cells and cart away carbon dioxide.

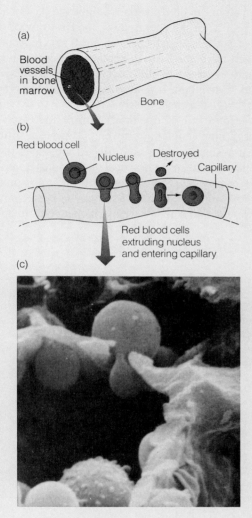

(a)

Blood
vessels
in bone
marrow

Bone

(b)

Red blood cell

Nucleus   Destroyed

Capillary

Red blood cells
extruding nucleus
and entering capillary

(c)

FIGURE 4.2 ■ Maturation of the Red Blood Cell. (a) Red blood cells arise in the bone marrow. (b) Each red blood cell enters the bloodstream by squeezing through the wall of a fine blood vessel called a capillary. During the squeezing process, the nucleus pops out of the cell. This loss limits the cell's life span, but gives it greater physical flexibility for bending and stretching as it passes through narrow vessels. (c) A scanning electron micrograph of a red blood cell squeezing into a capillary (magnification 8545 ×).

the mature red blood cell cannot divide (and so cannot reproduce), nor can it form new proteins. It is therefore one of the simpler cells in the body and provides a good case study for some of the more fundamental activities associated with living cells. A red blood cell can survive an average of about 120 days, and every second during that time, it carries out dozens of tasks simultaneously. The millions of hemoglobin molecules inside the cell operate like miniature traps, snapping open and shut, holding and then releasing oxygen. Enzyme

molecules within the cell work extremely quickly, cutting and joining other molecules, releasing energy, and generating cell parts. The cell's entire water content moves back and forth across the plasma membrane 100 times a second. And glucose is constantly burned to fuel these activities.

The red blood cell is undeniably dynamic, and it shares this state of constant, rapid, yet orderly activity with all other living cells. The goal of this chapter is to reconstruct the dynamic cell—to bring the cell parts described in Chapter 3 back together and back to life—and to examine the kinds of tasks cells perform.

All living cells represent a fascinating contradiction. The intricate structure of each cell is a model of orderliness and precision, and the coordinated activity of the parts goes on continually. With their elements acting in concert, cells carry out *chemical tasks,* such as breaking down food, *transport tasks,* such as moving materials in and out, and *mechanical tasks,* such as squeezing through pores or lashing flagella back and forth. Yet despite the organized structure of cells and the continual processing we call life, a fundamental law of matter in the universe says that all things tend toward *disorder.* We can easily observe this tendency in the nonliving world: Given enough time and neglect, a building's paint will peel, its roof will leak, its walls will fall down, and its materials will turn to dust or rust. With such universal tendencies, then, how can a living thing remain in a steady state of order and activity?

The answer is that living things, the cells they are made of, and the maintenance of those cells depend on a continual flow of energy from the environment. The energy is then processed and transformed into a different state for storage or use. The special agents of this energy flow are the protein catalysts called enzymes and the energy-carrying molecules called ATP. Cells also depend on a flow of materials—mostly water, ions, nutrients, and wastes—to help maintain their order and activity, and this chapter will describe how such materials pass into and out of the cell.

Our consideration of the dynamic cell will answer several questions:

■ How does a cell use energy from the environment as fuel for life-support activities?
■ What are the basic laws of energy flow in the universe, and how do they affect living cells?
■ How do chemical reactions occur within cells, and how do energy carriers and enzymes propel and speed them?
■ What is metabolism, and why is it so crucial to life?
■ How does the cell transport materials, thus maintaining the appropriate chemical contents within the cell?

CELLS AND THE BASIC ENERGY LAWS OF THE UNIVERSE

A dynamic cell—which means any living cell—is like a bustling metropolis that runs on energy. Materials stream in and out through the "city walls"—the plasma membrane. Instructions issue from the "city government"—the DNA housed in the nucleus or nucleoid region—for cell maintenance and the synthesis of materials. Ribosomes, like microscopic factories, churn out proteins, and in eukaryotic cells, the endoplasmic reticulum and Golgi apparatus can process and ship these materials to other "cities." Lysosomes break down wastes and recycle useful raw materials. The plastids of plant cells produce and store carbohydrates like miniature gardens and warehouses. And the subcellular power plants called mitochondria process a steady stream of usable energy "packets" that fuel the collective commerce of the cell.

Just like cities, automobiles, airplanes, and other entities that accomplish work, the dynamic cell needs a constant flow of energy to carry out its tasks. And significantly, the flow of that energy proceeds according to certain universal laws. We saw in Chapters 2 and 3 that cells are made up of organelles, that these cell parts are in turn constructed of atoms and molecules, and that these constituents act according to the chemical and physical laws of the universe. A great deal can therefore be understood about the behavior of cells by learning about the laws of energy.

The Laws of Thermodynamics

We use the term *energy* in many ways: We call an active person energetic, and we also consider energy a major political issue. But what *is* energy? In its most basic form, **energy** is the power to perform chemical, mechanical, electrical, or heat-related tasks.

There are two major states of energy in the universe: potential and kinetic. **Potential energy** is energy that is stored and ready to do work; a huge boulder poised at the edge of a cliff (Figure 4.3a), the water pent up behind Grand Coulee Dam, and the chemical bonds inside a piece of firewood all contain stored energy, capable of being released and accomplishing work, such as smashing a hole in the road at the bottom of the cliff, turning the wheels of a hydroelectric generator, or heating a frying pan full of bacon. **Kinetic energy** is the energy of motion—of a hurtling rock or of tumbling water, for example (Figure 4.3b).

Clearly, potential energy can be converted to kinetic energy. Moreover, energy can be converted from one *form* to any of several others in both the potential and kinetic states. We are all familiar with these forms of energy: Light, heat, electrical, chemical, and mechanical energy are detectable all around us. And we have all seen their interconversions: Electrical energy can be converted to light and heat in a light bulb; the energy of reacting chemicals inside a battery can be con-

verted to electricity to run a radio or flashlight; the heat from a coal fire can be converted to the mechanical energy in the wheels of a steam locomotive; and the atomic energy inside a hydrogen atom can be converted to light and heat in a nuclear bomb. The first of the energy laws, the **first law of thermodynamics,** addresses such conversions. It says that energy can be changed from one form to another, but during these conversions, it is *conserved;* that is, it is neither created anew nor destroyed. Thus, there is a constant amount of energy in the universe, but it can change form.

The **second law of thermodynamics** addresses the fact that energy conversions are never 100 percent efficient. It states that systems always tend toward greater states of disorder. What this means in practical terms is that given enough time, the energy in every system will undergo spontaneous conversions from one form to another, but these conversions are inefficient, and some energy will change form and inevi-

(a)

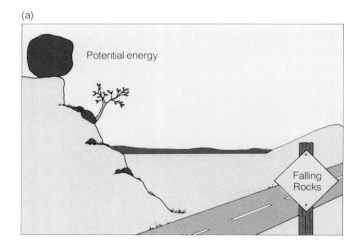

Potential energy

Falling Rocks

(b)

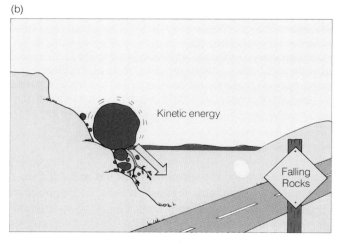

Kinetic energy

Falling Rocks

FIGURE 4.3 ■ Potential and Kinetic Energy. (a) A boulder poised at the top of a cliff has potential energy; it can do work, such as thundering down the hill, uprooting a tree, and smashing a hole in the pavement. (b) Potential energy is converted to kinetic energy as the rock rolls.

tably be lost to the surrounding environment as heat. **Heat** is nothing more than the random commotion of atoms and molecules. Since randomness is the opposite of order, the inefficient conversion of energy from one form to another (with some energy lost in a form that can no longer do work) results in increasing disorder in the system. Scientists use the term **entropy** as a measure of the disorder or randomness in a system. Entropy accounts for the energy that escapes the system during transformations and is no longer usable. Anyone who has ever tried to keep a room tidy is already quite familiar with the concept of entropy: Without a constant input of energy to clean up dust and dirt, and put clothes and books away, entropy tends to increase, and things go downhill fast.

Consider one chain of energy conversions and the steadily increasing entropy it represents. The nuclear reactions taking place in the sun release light but also massive amounts of heat, which are forever lost in space (Figure 4.4). Sunlight enters the earth's atmosphere and strikes the leaves of a bush. Most of the light is reflected as heat and once again lost, simply radiating away into the atmosphere in a low-quality, unusable form. (Gathering up such heat energy for use would require the expenditure of more energy than you could recover!) Some of the light is reflected as the green light we see as the plant's color. And a small amount is absorbed as light and is converted by the plant to chemical energy; but during this process, still more heat is lost. Finally, a deer grazes on the leaves, and the animal's cells release the stored chemical energy, which fuels ongoing life processes; but as always, the conversion is inefficient, and energy radiates into the air as body heat from the warm-blooded animal.

Only a tiny fraction of the original solar energy is saved in the chemical form that can be expended to "purchase" order—in this case, precise movements by the deer's muscles as it bounds across the forest floor. At the same time, a great deal of energy has been irreversibly lost as the random jostling of molecules accomplishing no work. Physicists predict that eventually, the energy from all the blazing stars, radioactive rocks, and other energy sources in the universe will have been converted again and again until all is lost as shimmering heat and nothing remains but a colossal cloud of randomly arrayed atoms. Fortunately, this "heat death of the universe" is billions of years off.

Cells and Entropy

The significant question for us here is, How do the energy laws apply to the cells of living organisms—islands of incredible order in a universe that is slowly winding down as a result of entropy? And the answer is that cells, too, are subject to the same tendency to wind down and become disordered. They only remain alive and healthy—orderly and active—at the high price of a constant flow of energy. This flow must provide enough energy to fuel all of the cell's synthesis and maintenance activities while still leaving enough leftover energy to satisfy the second law—the inevitable loss of heat

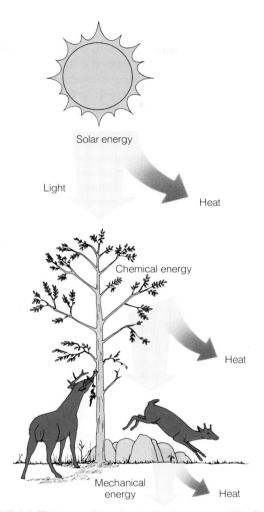

FIGURE 4.4 ■ Energy Flow in the Environment. The flow of energy through an ecosystem is predicted by the laws of thermodynamics. The first law of thermodynamics states that energy can be changed from one form to another but is neither created nor destroyed. Here, nuclear energy from atomic fusions taking place in the sun is converted to light, to chemical energy in the plant's tissues, and to mechanical energy in the animal's tissues. The second law of thermodynamics states that all such interconversions are inefficient to some degree. Thus, with each interconversion in this chain, some energy is lost as heat, diffuses away, and becomes uselessly disorganized. This explains why there are many plants, fewer deer, even fewer cougars, and no cougar-predators—animals specialized to attack and consume these cats.

to the surrounding environment. As the previous example illustrates, the original energy source for virtually all living things is the sun, and light energy is converted by plants to chemical energy and stored in the bonds of carbohydrate molecules. Plant cells can then use the energy from these molecular storehouses to fuel their activities, and animals, fungi, and many kinds of microorganisms can obtain their energy indirectly by consuming plant matter or other plant eaters as food.

There is no doubt that this cascade of energy conversions—sun→plants→organic molecules→life processes within cells—satisfies the second law, because it takes place spontaneously, and heat is lost at every step. But within the cell, there is a delicate balance between order-producing activities, such as maintenance, repair, and protein synthesis, and activities that mainly generate heat. A living cell, in a real sense, is a temporary repository of order purchased at the cost of a constant flow of energy. If that energy flow is impeded, order quickly fades, disorder reigns, and the cell dies. The impediment can be lack of food or an injury or random errors accumulating in the cellular constituents that maintain cell order (such as the nucleus and ribosomes).

The red blood cell is an interesting case study for energy flow, entropy, and death, because, as we saw earlier, most of the machinery for maintaining cellular order is discarded early in the cell's life cycle. The mature cell obtains its energy source—glucose molecules—through its plasma membrane from the yellowish liquid portion of blood, and enzymes in the cell break these sugars down and provide fuel for the cell's activities. However, over time, entropy takes its toll. The precise ordering of molecules in the cell membrane and internal enzymes begins to decline, and since there are no organelles for synthesis and repair, the balance is tipped further and further toward disorder. After about 120 days, the withered cell is removed from circulation and dismantled. In contrast, a typical human brain cell can live for 75 years or more by expending much of its energy in maintenance and repair

activities. Surprisingly, that long-lived brain cell is, in a sense, younger than the red blood cell. So long as its machinery for repair remains intact, the brain cell continually replaces aging organelles, and this constant refurbishing amounts to a cellular fountain of youth; at any given time, most of the cell parts will be less than one month old, and during the cell's lifetime, the entire contents (and the cell membrane, too) will be replaced hundreds of times.

Clearly, the nonstop organizational activity of a living cell bears a steep price tag in energy requirements. But exactly *how* cells use that energy to defer entropy is our next topic.

CHEMICAL REACTIONS AND ENERGY FLOW IN LIVING THINGS

We have seen that plants and other photosynthetic organisms convert light energy to chemical energy, which in turn fuels the activities of nearly all living things, directly or indirectly. Once inside a cell, this energy in chemical form fuels the maintenance of order through **chemical reactions:** transformations of sets of molecules into other kinds of molecules, with a concomitant shifting of energy content from the bonds of one set of molecules to the bonds of the other. Thus, just as energy must flow into the cell, it must also flow from one set of compounds to another within the cell to power cellular activities. Let's see how.

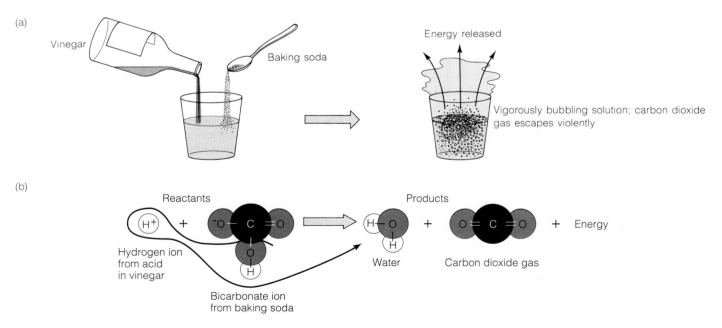

FIGURE 4.5 ■ **Energy-Releasing Reactions.** (a) **Vinegar and baking soda react with each other in an exergonic, or energy-yielding, reaction, manifested by vigorous bubbling. (b) During the reaction, the reactants are transformed into products. A hydrogen ion (H$^+$) from the acetic acid in the vinegar joins with an OH group from the baking soda's bicarbonate ion, and water (H$_2$O) forms. Taking the OH from bicarbonate leaves CO$_2$, the gas that bubbles away; then, energy is released, and the products are more disorganized than the reactants.**

Chemical Reactions: Molecular Transformations

During a chemical reaction, starting substances, the **reactants,** interact with each other to form new substances, the **products.** There are often two reactants and two products, but two reactants can be transformed into one product, and one reactant can become one or two products. This general relationship can be represented by the equation

$$A + B \rightarrow C + D$$

where A and B represent reactants and C and D stand for products.

As a child, you may have done some simple kitchen chemistry, mixing vinegar and baking soda and watching the vigorous bubbling that results (Figure 4.5a). The reactants in this case are hydrogen ions (H^+) from the acetic acid in the vinegar and bicarbonate ions (HCO_3^-) from the baking soda. As Figure 4.5b shows, an OH cleaved from a bicarbonate ion is combined with a hydrogen ion. As a result, H_2O forms, along with CO_2, or carbon dioxide gas, the cause of the bubbling. This reaction takes place spontaneously, as soon as the ingredients are combined. There is no need to add heat, electricity, or any other form of energy to make it "go." A reaction like this that releases energy is called **exergonic** (meaning "energy out"). Heat is released during the reaction between vinegar and baking soda because of the inefficiency of energy conversions. As new products form from the reactants, the energy-containing bonds between the carbon, hydrogen, and oxygen atoms become rearranged, and during this conversion, some energy is released as heat, and the entropy (disorder) of the system increases as the carbon dioxide molecules bubble off randomly into the air. This satisfies the second law of thermodynamics: The products contain less energy and are more disordered than the reactants.

Some exergonic reactions are like the rock poised at the top of a hill: They need some energy input before they will go. Nevertheless, even though a little bit of energy is needed to get the reaction started, the overall result is a release of energy.

A great number of cellular activities involve exergonic reactions and give off heat. A striking example is the contraction of muscle cells. The movement of muscle cell fibers is based on numerous exergonic reactions. So if you've ever wondered why running, aerobic dancing, weight lifting, or any other muscular activity makes you heat up and start to sweat, it's because of exergonic reactions!

There is another category of chemical reactions that do not proceed spontaneously and do not give off heat. These are called **endergonic** ("energy in") reactions. A familiar endergonic reaction takes place when you cook an egg (Figure 4.6); the added energy (heat) causes the egg white proteins to change configuration and solidify. Endergonic reactions are very important to living things because they include many of the underlying molecular transformations that bring about order in the cell, such as the building of proteins, the replace-

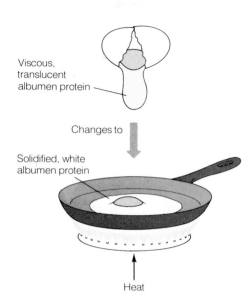

FIGURE 4.6 ■ Energy-Requiring Reactions. It's strange to think that frying an egg can induce an endergonic, or energy-requiring, reaction—but it does. The addition of heat energy alters the hydrogen bonds in the proteins in egg white. This changes the proteins' shape, and we can see the result: A clear, viscous solution of the proteins becomes opaque, white, and solid.

ment of worn sections of membrane, and the generation of new ribosomes. All endergonic reactions, including those that bring about construction and maintenance activities in a cell, fail to "go" unless energy is supplied in some form, because the energy of the chemical bonds in the reactants is *less* than the energy of the bonds in the products.

If the order and activity of the dynamic cell depend on an appropriate source of energy for each and every endergonic reaction, where does that energy come from? The answer demonstrates the beautiful economy of nature: The energy for the cell's endergonic reactions comes from the cell's exergonic reactions. The two kinds of reactions are energetically *coupled,* so that the leftover energy of one reaction powers the energy needs of the other, just as a spinning treadwheel could power a blender (Figure 4.7).

Significantly, the cell parts we studied in Chapter 3 are organized in ways that take advantage of coupled reactions. Ribosomes, for example, are constructed in such a way that the energy-producing reactions that provide power for building new proteins take place right next to the energy-requiring reactions that add new amino acids onto a growing protein chain.

Metabolism: Chains of Reactions

Sometimes the energy from one reaction passes to another reaction, and sometimes the chemical product of one reaction becomes a reactant in the next reaction. Regardless of what

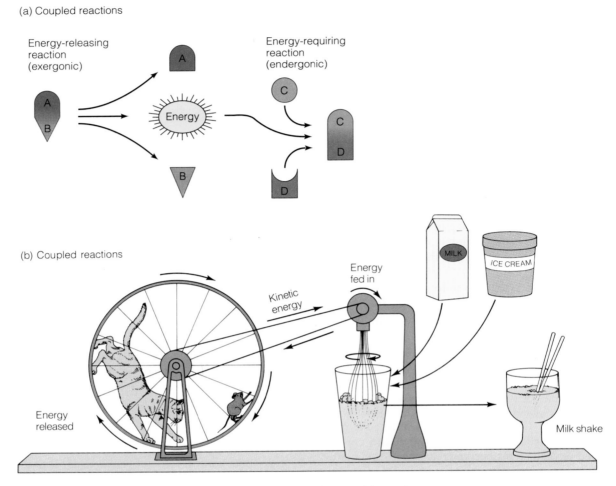

(a) Coupled reactions

Energy-releasing
reaction
(exergonic)

Energy-requiring
reaction
(endergonic)

Energy

(b) Coupled reactions

Kinetic
energy

Energy
fed in

MILK

ICE CREAM

Energy
released

Milk shake

FIGURE 4.7 ■ Energetic Coupling. In living things, energy-releasing (exergonic) reactions are coupled to energy-requiring (endergonic) reactions. (a) Energy is released from the splitting of compound AB, energy is required for the combining of C and D into CD, and the energy from the first reaction can drive the second reaction. (b) By whimsical comparison, energy from a spinning treadwheel (powered by the muscles of a hungry cat and a frightened mouse) can be mechanically coupled to a mixer that makes milk shakes out of ice cream and milk.

passes between the components, such interlinked chains of reactions within cells are called **metabolic pathways** (Figure 4.8). They allow the living cell to subdivide a big chemical change—one that might, for example, release a tremendous amount of heat—into a number of smaller steps, liberating energy in packets small enough for the cell to use efficiently.

Some metabolic pathways break down complex molecules to simpler products, with an accompanying release of energy. Such pathways are collectively called **catabolism,** and an individual breakdown and energy-releasing reaction, such as the one shown in Figure 4.8, is called a *catabolic reaction.* In this figure, a starting material is modified, joined to another substance, and then split in two. Other metabolic

pathways bring about the synthesis of complex molecules from simpler precursors and usually require an input of energy. One such reaction is called an *anabolic reaction,* and a series of them constitute **anabolism.** Together, catabolism and anabolism—that is, sets of reactions that break down molecules and build up molecules, that give off energy and require it—constitute **metabolism:** the energy and material changes within a living thing that arise from interrelated chemical reactions and allow a cell to utilize energy from its environment and survive. Metabolism is so important to life that we devote all of Chapters 5 and 6 to its key features. But here we must discuss the agents of cellular energy flow—ATP and enzymes—to understand how energy allows the cell to carry out transport, chemical, and mechanical work.

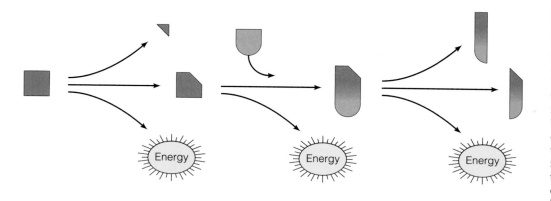

FIGURE 4.8 ■ A Metabolic Pathway. In living things, reactions usually occur in long series called metabolic pathways, where one of the products of a reaction becomes one of the reactants in the next reaction in the chain. In a metabolic pathway of the catabolic type, as shown here, a more complex molecule is broken down to simpler molecules, and energy is released. Metabolic pathways of the anabolic type require an energy input and bring about the building of more complicated molecules from simpler ones.

ATP: The Cell's Main Energy Carrier

If the energy from one type of reaction in a living thing can drive another such reaction, what form does that energy take? Light? Electricity? Heat? Nuclear power? The logical assumption might be heat, since we saw that exergonic reactions give off heat. But the heat given off in a cell dissipates, and there is no mechanism in cells for gathering it up and putting it to good use. Instead, most of the energy released during an exergonic reaction is quickly trapped in the bonds of an energy carrier that can transfer energy from one molecule to another. This crucial energy-carrying molecule is called **ATP** (Figure 4.9a). When energy is released from a

reaction, it can be used to alter a molecule called ADP (adenosine diphosphate), which has two phosphate groups and lower energy content, into ATP (adenosine triphosphate), which has three phosphate groups and higher energy content (Figure 4.9b). Likewise, when ATP is cleaved, it releases ADP and a small packet of energy (Figure 4.9c). The significance of ADP and ATP, as well as a few other similar energy carriers, is that they function as intermediates, or links, between energy exchanges in the cell.

Energy intermediates are important because for energy to flow from one chemical reaction to another, the reactions must have some chemical compound in common—that is, a compound that appears among the products of one reaction

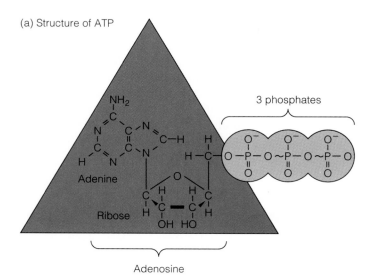

(a) Structure of ATP

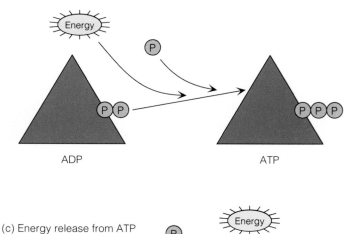

(b) ATP synthesis

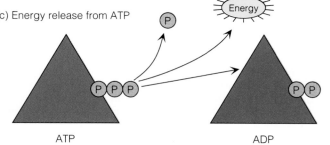

(c) Energy release from ATP

FIGURE 4.9 ■ The Structure and Function of ATP. ATP, or adenosine triphosphate, is the main energy carrier in living cells. (a) ATP consists of a portion called adenosine and, as the name *triphosphate* implies, three phosphate groups. (b) ATP is synthesized by the addition of a phosphate group and energy to ADP, adenosine diphosphate. (c) Energy is released from ATP when the terminal phosphate is cleaved.

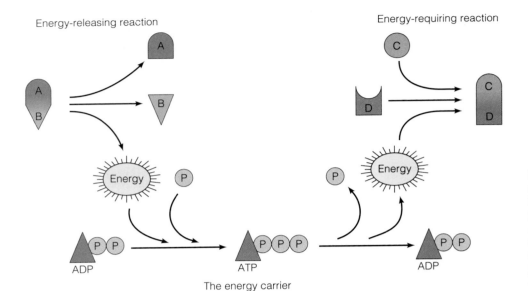

Energy-releasing reaction   Energy-requiring reaction

The energy carrier

FIGURE 4.10 ■ **How ATP Serves as an Energy Carrier.** Energy released from an exergonic reaction can make possible the synthesis of ATP from ADP and phosphate. Energy now trapped in ATP can be liberated when a phosphate is cleaved, and it can power an endergonic reaction.

and the reactants of the next. ATP often appears in the equations for both exergonic and endergonic reactions (Figure 4.10).

An analogy for ATP is money. When a farmer sells his or her crops and then uses the money to buy new tools, the intermediate for both exchanges is money. ATP is sometimes considered a form of currency in the cell—the chemical coin of the realm—that is "saved up" during energy-yielding exergonic reactions and "spent" during energy-costly endergonic reactions. The coupling of reactions by the common intermediate ATP shows the relative efficiency of cellular processes

FIGURE 4.11 ■ **ATP Powers Living Reactions.** The glowing of this bioluminescent fish, *Kryptophanaron alfredi,* in the inky recesses of the Caribbean Sea, is fueled by ATP. The fish consumes food, and this is broken down into small molecules. Bacteria within special pouches under the fish's eyes and along its sides extract chemical energy from the small molecules and store much of the energy in the form of ATP. Later, the cells break down the ATP to ADP, and the liberated energy helps to form a chemical that releases energy as light. Chemical energy is thus converted to light energy, and ATP is the intermediary.

("relative" because some energy is lost with each exchange). And the critical activities of both single cells and multicellular organisms—the storage and use of hereditary information, the building of cell parts, the transport of certain materials, movement, and even glowing in the dark—all involve ATP (Figure 4.11).

The ATP "budget" of a single cell reads like the annual financial ledger of a small nation. Consider the bacterium *Escherichia coli,* which resides (usually harmlessly) in the human intestine. An *E. coli* cell contains about 5 million ATP molecules, but it spends 2.5 million *per second* to fuel the work of building DNA, RNA, protein, and lipid molecules. Thus, its energy reserves can "buy" just 2 seconds worth of maintenance and growth. The cell must continually break down food molecules, store the energy in ATPs, and stockpile its currency in the cytoplasm to keep up with the staggering energy costs of resisting entropy and staying alive. The human body—with its trillions of cells—cycles an estimated 40 kg (about 88 lbs) of ATP molecules into ADP and back again each day, representing quadrillions of chemical reactions and energy exchanges daily.

Chemical Reactions in the Cell: Enzymes Make Them Go

Chemical reactions can take place in a test tube or inside a living cell. However, the reactions in cells—even those fueled by ATP—could not take place continually or quickly enough without the molecular agents called *enzymes.* Biologists and physicians know this because many human diseases are caused by a single defective enzyme, even though all the thousands of other enzymes in the body function perfectly. Some children, for example, are born without the ability to synthesize a particular enzyme that the body must have in order to change an inactive form of vitamin D to the active

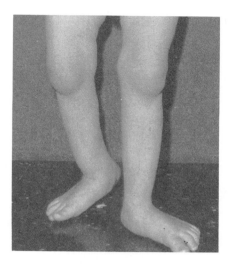

FIGURE 4.12 ■ **Rickets Reveal the Importance of Enzymes in the Body.** A specific enzyme with a complicated name (25-hydroxy-cholecalciferol-1-hydroxylase) is just one of thousands of human enzymes. If this single enzyme functions improperly, however, a child's body cannot carry out a certain chemical reaction that converts vitamin D to its active form. Lacking the vitamin, the child cannot build strong bones. More normal growth and development is possible if the child gets enough active vitamin D.

form. Without active vitamin D, the children cannot generate strong, healthy bones, and they develop the disease called rickets, with its characteristic curved legs among the symptoms (Figure 4.12). Fortunately, if these children get enough of the active form of vitamin D in their diets, they can bypass the need for the enzyme and develop and grow normally.

Why do human cells require a specific enzyme to convert vitamin D to an active form? Why can't that conversion just occur spontaneously? The answer is that vitamin D *does* change spontaneously from inactive to active form, but the conversion happens very slowly—so slowly that too little of the active form would be available to ensure strong bones and

normal growth. When a specific enzyme is present, however, the reaction occurs far faster than without the enzyme. How, then, do enzymes—which are simply one class of protein molecules—speed up chemical reactions? To understand the answer, we must consider a concept that biochemists call activation energy.

In a very real sense, there is a barrier, or hill, separating the reactants and products in any chemical reaction, whether in a cell or in a test tube. For vitamin D to be converted from inactive to active form or for a pair of imaginary reactants, which we'll call AB and C, to be converted to A and BC, the reactants must collide with each other. That collision cannot be a gentle bump, but instead must be a solid, energetic thump so that the chemical bonds in the reactants can be broken and the new bonds in the products formed. For a fleeting instant during the conversion, chemical bonds in the reactants are distorted like springs being stretched, and this fleeting intermediate state, A~B~C, cannot be reached without a very energetic collision between the molecules. The intermediate springlike state is the **transition state,** and because energy input is required to achieve it, it is often characterized as a barrier, or hill, separating reactants and products (Figure 4.13).

Now, most molecules jostling about and colliding randomly do not have enough energy of motion (kinetic energy) to achieve and overcome the "energy hill" of the transition state; they simply bounce off each other. (Similarly, cars colliding at a very low speed—say, 2 mph—usually bounce off and leave each other unchanged.) Some individual molecules, however, do jostle about with enough kinetic energy that they collide in a so-called **productive collision,** an impact that generates the springlike transition state A~B~C, and AB and C can be converted to A and BC. The energy of impact great enough to cause molecules to cross the "hill" is called the **activation energy** (see Figure 4.13).

In a living cell, most molecules do not on their own collide with sufficient energy to cross the barrier and react—at least not fast enough to keep pace with the cell's needs for energy and materials. In principle, one way the reactions in a

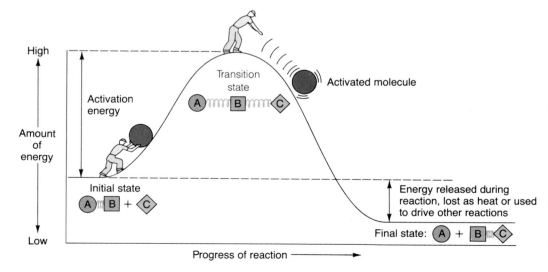

FIGURE 4.13 ■ **Activation Energy.** Reacting molecules need to collide with a certain minimum amount of energy before they can reach the springlike transition state. The amount of energy needed to reach the transition state, the activation energy, is much higher than the initial energy state, and that transition state resembles a hill, or barrier, when graphed. Once activated molecules reach and then pass the transition state, they proceed to the final state, the new products. As the graph shows, the overall energy content in the bonds of the products is lower than it was in the reactants. The excess energy has been lost as heat.

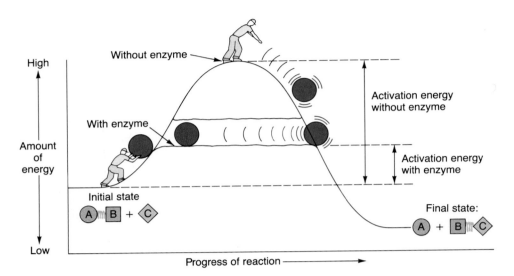

FIGURE 4.14 ■ Enzymes Lower the Energy of Activation. An enzyme can lower the activation energy— the barrier that must be crossed by reacting molecules—without itself being changed or used up. It's as if the enzyme "bores a tunnel" through the activation energy "hill." In this way, enzymes speed up biological reactions.

cell could be speeded up would be for the speed of colliding molecules to increase. But this increase of speed would require an increase of temperature, and high temperatures would speed all the reactions in a cell, not just those needing a boost. This overall increase would cause the cell to become disorganized, and it would die. Instead, the molecules called **enzymes** play by far the most important role in speeding reactions in living cells. Enzymes are **catalysts,** agents that speed other reactions without themselves being used up or changed. Enzymes act by lowering the activation energy needed for biochemical reactions to take place but not becoming altered themselves (Figure 4.14). The action of an enzyme, in effect, bores a tunnel through the energy barrier hill, allowing the reactants to react and become products without the need to overcome the high activation energy barrier.

Enzymes not only allow reactions to proceed at the relatively low temperatures compatible with life processes, but they also allow only specific reactions to take place. As we have seen, one specific enzyme speeds up the conversion of vitamin D to its active form. Another enzyme, called *carbonic anhydrase* (enzyme names often end in the suffix *ase*), catalyzes, or speeds up, the reaction shown in Figure 4.5b and helps red blood cells transport carbon dioxide from tissues such as active muscles to lungs, where the CO_2 can be exhaled. Each single molecule of carbonic anhydrase facilitates this transport at the amazing speed of 600,000 times per second! This enables the conversion to proceed nearly a million times faster than it would without a catalyst.

Carbonic anhydrase and most of the thousands of other specific enzymes act by lowering the activation energy and hence speeding the *rate* of a chemical reaction. Let's take a closer look, now, at *how* enzymes carry out their work.

Enzyme Form Facilitates Biological Reactions

Enzymes are like virtually all other biological molecules in one important way: Their functioning is directly related to their form. Certain RNA molecules called *ribozymes* have recently been shown to catalyze some reactions among other RNAs. However, protein enzymes are by far the most versatile biological catalysts.

Like other proteins, each enzyme has its own unique three-dimensional shape but most share an important feature: a deep groove, or pocket, on their surface called the **active site** (Figure 4.15a). The shape of the enzyme's active site fits a specific reactant or set of reactants. These specific reactants, or **substrates,** fit into the groove rather like a key in a lock, and once locked in, they are acted on by the enzyme.

The enzyme holds, or *binds,* the substrates in place, forming an **enzyme-substrate complex** (Figure 4.15b). Biochemists used to think that the shape of the active site was rigid and unvarying, just like the inside of a lock. However, they now believe that the site can change shape as substrates bind to it, improving the "fit." This in turn orients the substrates near to each other and then subtly changes the shape of the substrates, straining their bonds and helping them to reach the transition state (Figure 4.15c). By means of this activity, the enzyme effectively lowers the activation energy, making it easier for substrates to react and form products (Figure 4.15d). Figure 4.16 shows a three-dimensional computer graphic of an actual enzyme, lysozyme, the enzyme in egg white that kills any bacteria that might infect and kill the incubating embryo. The deep groove shown here is the active site, the portion of the enzyme that actually dismantles bacterial cell walls.

(a) Substrates bind to the enzyme's active site.

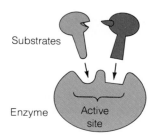

Substrates

Enzyme   Active
site

(b) The enzyme's active site changes shape, improving fit
and orienting the substrates.

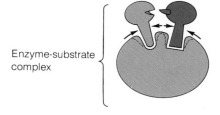

Enzyme-substrate
complex

(c) The enzyme strains the shape of the substrates, facilitating
a productive collision.

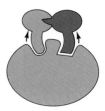

(d) The reaction finishes, and the enzyme returns to normal.

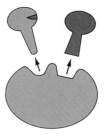

FIGURE 4.15 ■ Enzymes: Structure and Function. This figure
shows how an enzyme, represented by the purple globular shape,
facilitates the reaction between two substrates, one without a protruding "nose" and the other with the "nose." (a) The substrates
bind to the enzyme's active site. (b) This enzyme-substrate complex brings the substrates into close proximity and orients them
appropriately. (c) These changes strain the bonds of the substrates
and help them react. (d) When the reaction is complete, the
enzyme regains its initial state and is ready to catalyze a new
reaction.

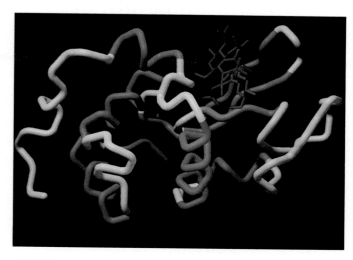

FIGURE 4.16 ■ Lysozyme, the Major Enzyme in Egg White.
Lysozyme protects a chick embryo from bacterial infection. The
enzyme binds substrate (shown in red)—the bacterial cell wall—
in its central V-shaped cleft, which is visible near the center of
this three-dimensional computer-generated model of lysozyme.

Because reactions catalyzed by enzymes underlie the
dynamic cell's simultaneous, nonstop activity, enzymes can
be seen as the effectors of the life process, enabling the cell to
carry out transport, chemical, and mechanical survival tasks.
Together with ATP, enzymes help ensure the flow of energy
that enables the cell to resist entropy—the tendency to
become disordered, fall apart, and die.

METABOLISM: THE DYNAMIC CELL'S CHEMICAL TASKS

The living cell, bustling with activity, carries out an important
set of integrated chemical activities that transforms energy.
Called metabolism, this set of activities involves thousands of
simultaneous chemical reactions catalyzed by enzymes.
Metabolism accomplishes two things: It harvests the energy
needed to fuel work by breaking down complex molecules
(catabolism); and it enables the building of cell parts for
growth, maintenance, and cell division (anabolism). Chapters
5 and 6 detail the steps involved in harvesting chemical
energy from food molecules and in converting the energy
from sunlight into those food molecules in the first place.
Metabolism can be seen as the "engine" that powers all other
cellular work: The harvesting of energy from the sun and food
molecules produces a great abundance of ATP and other cellular energy carriers that fuel the activities of enzymes and
allow cells to transport materials, build cell parts, and move
about.

The construction of new cellular parts in a cell such as *E.
coli* proceeds at a spectacular rate and keeps pace with the
cell's ability to grow and divide into two cells every 20 min-

utes or so. A bacterial cell can generate 1400 protein molecules per second at a cost of more than 2 million ATP molecules! In that same second, the cell can also generate part of a DNA molecule, 12.5 RNAs, 12,500 lipid molecules, and 32.5 polysaccharide molecules at an additional ATP cost of almost 300,000 ATPs. And an *E. coli* is small and simple compared to a typical plant or animal cell. Even in a slow-growing eukaryotic cell—say, a rat liver cell that divides only once every two or three months—there is a continuous turnover of cellular components. For example, in just three weeks' time, all the proteins in a rat liver cell are replaced, and every seven or eight days, all the phospholipids in the membrane are replaced and the old ones dismantled into raw materials.

While the chemical tasks of a cell are prodigious, they could not take place unless the transport tasks brought raw materials into the cell and carted off the cellular wastes.

TRANSPORT TASKS IN THE DYNAMIC CELL

The cell carries out chemical tasks simultaneously with a set of transport tasks. Like any bustling metropolis, the living cell keeps enough of the right materials on hand to meet the constant demands of construction, maintenance, export, and other activities. To do this, the cell expends energy in building and maintaining its outer barrier, the plasma membrane, and it also ferries and pumps certain materials in and out at an additional high energy cost. Clearly, the flow of energy is closely tied to the flow of materials, and both are requisites for life.

Cells are minute pools of liquid surrounded by lipid bilayer membranes (see Chapter 3). The dynamic cell is like an aquatic metropolis—more like Venice than like Rome. Most of a cell's internal activities take place surrounded by water molecules, and the outside of the plasma membrane must remain wet. Evidence suggests that life arose in the sea, and living things have retained an "inner sea" ever since. The pool inside each cell is a bit like seawater; it is a rich solution of ions and dissolved gases, but it also contains sugars, amino acids, and other raw materials, usually with a distinctly different composition than the fluid outside the cell.

The cell's inner pool is called the **intracellular fluid,** and the materials for maintaining that pool are taken up from the fluid outside the cell (Figure 4.17). A single-celled organism is almost always surrounded by fresh water or seawater or by the body fluids of another organism that it inhabits. In a multicellular organism, the fluid outside the cell is usually **extracellular fluid,** which fills all the spaces around and between cells.

Simple animals usually have a single extracellular fluid, while complex animals have two types: the clear extracellular tissue fluid in the spaces between cells and the blood that flows through the animal's vessels. Materials can move back and forth between the intracellular fluid, extracellular fluid, and blood across the membrane barriers that separate them.

Such movements are governed in part by **gradients,** differences in the concentration or pressure of materials in one part of a contained area compared to another.

Materials—ions, water, amino acids, and so on—tend to move *down* concentration gradients, that is, from regions of higher concentration to regions of lower concentration. When a substance enters or leaves a cell by moving down such a gradient, the process is called **passive transport:** The material moves without energy expenditure by the cell. Cells, however, also transport certain materials in or out *against* gradients; this means that there is already more of a certain substance inside the cell than outside, for instance, but the cell needs even more. Movement against gradients, like rolling a rock uphill, requires energy expenditure by the cell and is called **active transport.**

Passive Transport, Diffusion, and the Second Law of Thermodynamics

The cell needs a constant supply of materials so that maintenance activities can take place and new cell parts can be built. What's more, the cell must get rid of excess ions, organic

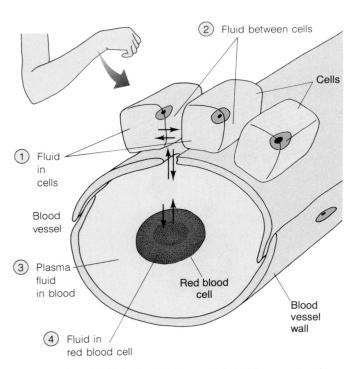

FIGURE 4.17 ■ Biological Fluids and Their Movements. A small blood vessel from a person's arm illustrates the fluid compartments in and around cells. There is intracellular fluid inside the cells that surround the blood vessel (1) and also inside the red blood cells (4). There is extracellular fluid between the cells (2). And there is fluid called plasma in the blood (3). Water and materials from these various fluids can move back and forth across cell membranes.

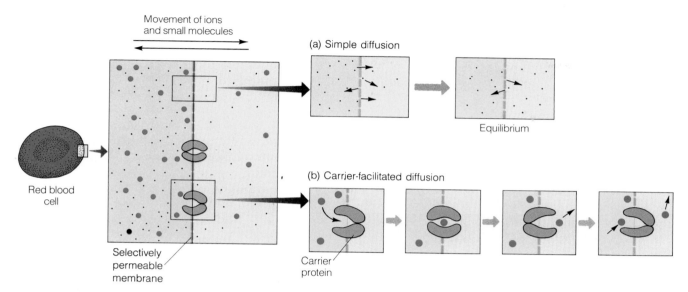

Movement of ions
and small molecules

(a) Simple diffusion

Equilibrium

(b) Carrier-facilitated diffusion

Red blood
cell

Selectively
permeable
membrane

Carrier
protein

FIGURE 4.18 ■ **Diffusion in the Cell.** (a) The cell membrane is selectively permeable; it allows the passage of nonpolar and small polar molecules by *simple diffusion.* (b) Other somewhat larger molecules, however, such as glucose or the nitrogenous waste product urea, cannot diffuse through the membrane without the activity of special carriers in a process called *carrier-facilitated diffusion.* Specific carrier proteins span the membrane and change shape to allow the solute to cross the bilayer, then spontaneously change back again, ready for the next incoming molecule. Because the solute always moves down a concentration gradient, this is true diffusion, with no energy expenditure by the cell. But carrier-facilitated diffusion can allow substances to diffuse across a membrane that would otherwise block their passage, thus enabling the cell to take in or get rid of substances quickly enough to keep pace with metabolism. In a red blood cell, both simple and facilitated diffusion take place simultaneously across the plasma membrane.

wastes, and other materials. Many of the substances that enter or leave the cell via passive transport follow gradients in or out. The tendency of materials to move from areas of high concentration to areas of low concentration is called **diffusion,** and you can witness the results of diffusion if you pour cream in your coffee. Billions of collisions send water, cream, and coffee molecules bouncing off each other in random directions like billiard balls, until eventually, the cream molecules have spread evenly throughout the cup.

A popular classroom experiment is to build a diffusion system that acts something like the membrane of a red blood cell or other cell. This usually involves a container separated in half by a cellophane sheet punctured with tiny holes through which water and other molecules can move. Both the dissolved molecules and water will migrate back and forth passively, with the net migration always in the direction of higher concentration to lower. Although molecules move in both directions across the sheet, more move from the high-concentration side to the low. When the concentrations on both sides are equal, equilibrium is reached, and the same number of molecules move in both directions at the same rate, and so no further change occurs. Most materials enter or leave cells via this process of *simple diffusion* (Figure 4.18a).

The lipid bilayer surrounding the cell is permeable to many small molecules and lipid-soluble molecules. It does not, however, allow large or electrically charged molecules, including ions, to pass: It is a *selectively permeable membrane.* How then, do larger or charged molecules pass in and out? The selectively permeable membrane contains *pores* that will allow charged molecules such as ions, water, and certain other small polar molecules to pass freely. Most large and charged molecules still cannot penetrate *either* the membrane itself *or* the pores and thus cannot diffuse in or out of the cell passively. For some large or electrically charged molecules, this problem is overcome by means of a special type of diffusion called **carrier-facilitated diffusion,** which is still passive but involves carrier molecules that allow certain substances like glucose and the waste product urea to pass through a membrane while many other molecules cannot move across (Figure 4.18b).

Passive Transport and the Movement of Water

Since a cell contains mostly water, and since its delicate outer membrane must be immersed in fluid, water itself is probably the most important substance to enter or leave the cell, and it does so by simple diffusion. The movement of water through a selectively permeable membrane (such as a cell's plasma

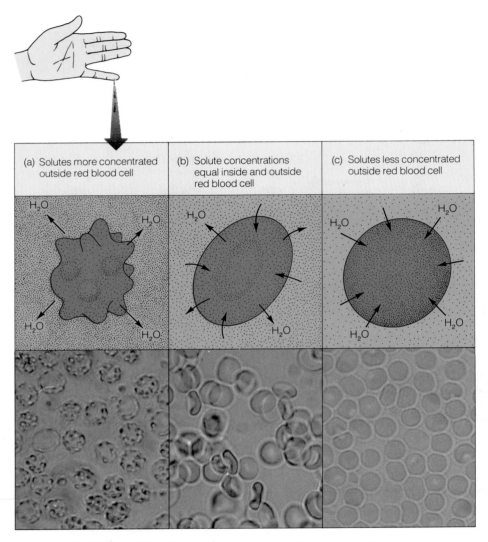

FIGURE 4.19 ■ Osmosis and the Red Blood Cell. Water moves across a selectively permeable membrane such as the plasma membrane of a red blood cell by osmosis, that is, passively from regions where water is most concentrated (i.e., where the solute concentration is low) to regions where water is least concentrated (where the solute concentration is high). (a) When a cell is placed in a solution that is highly concentrated in ions or other solutes (a hypertonic solution), water moves out by osmosis, leaving the red blood cell shriveled. (b) In a solution with solute concentrations similar to those of the cell interior (an isotonic solution), water moves in and out at equal rates, and the cell maintains its normal shape. (c) When a cell is placed in pure water, a solution lacking solutes entirely (a hypotonic solution), water rushes into the cell, and the cell expands rapidly, sometimes bursting. The red blood cells in these photographs are magnified about 400 times.

membrane) in response to gradients of concentration or pressure has its own name: **osmosis.** Nevertheless, the principles involved in osmosis are similar to those governing the diffusion of all materials, and in fact, the flow of water molecules is directly related to the concentration of other materials.

Water moves from a region where it is present in high concentration or pressure to one where it is present in low concentration or pressure. In this case, high concentration refers to the relative number of water molecules, not to sugar or salt molecules or some other solute in the solution. Therefore, pure distilled water has a higher concentration of H_2O molecules than does water containing a solute such as salt.

The red blood cell is a good example of the importance of diffusion and osmosis to living cells. As it races through the blood vessels, the red blood cell is bathed in a yellowish fluid called *blood plasma,* and significantly, the total concentration of salt ions, sugars, and other solute molecules in this plasma is very similar to that in the intracellular fluid pool inside the cell. As a result, the extracellular plasma is said to be **isotonic** relative to the fluid in the cell. (*Iso* means "same," and *tonic* refers to osmotic properties.) Thus, an isotonic solution outside the cell has the same osmotic properties as the fluid in the cell's interior.

Surprisingly, 100 times the total volume of water in the red blood cell moves rapidly back and forth across the cell's plasma membrane every second. Yet there is no *net* water movement in or out because the concentration of water molecules is similar on both sides; hence, the cell retains its normal disk shape (Figure 4.19b). But what would happen if you placed red blood cells in a beaker of pure distilled water? In pure water, the fluid outside the cells would be **hypotonic** (*hypo* means "low") compared to the fluid inside each cell, since distilled water has a vastly lower concentration of ions, proteins, and other solutes than the cell's cytoplasm. At the same time, however, the concentration of water molecules in the beaker is far greater than in the cells, and so water will tend to move into the cells by osmosis, causing the cells to swell or even burst from the rapid influx (Figure 4.19c). (The concave shape of the red blood cell is an evolutionary "safety device" that allows maximum expansion before bursting.) One could alternatively prepare a **hypertonic** solution—one with a higher concentration of salts, sugars, proteins, and other solutes but a lower concentration of water than the red blood cell's intracellular fluid. Since the concentration of water molecules would be lower outside the cell than in, water would move outward by osmosis, leaving the cell seriously shriveled (Figure 4.19a).

You may have noticed that the day after you eat a big helping of a very salty or sweet food—say, a large bag of potato chips or a quart of ice cream—your weight goes up a few pounds. The reason is that the solute concentration of body fluids increases, and so you retain more of the water you drink as it moves into the fluids and offsets the extra solutes. Near the end of a woman's menstrual cycle, hormones can also cause the solute concentration of cells to rise or fall, often bringing about the water retention and the "boggy" feeling some women experience at that time.

The solute concentration inside the cells of most plants is usually higher than that in the soil water around its roots, and therefore water tends to enter the plant and its cells. The cells do not burst, however, because they are contained by their strong cellulose walls. The water in the now-plump vacuoles exerts an outward force called **turgor pressure** that plumps up the cells. In an interesting tropical plant called the sensitive plant, cells with large plump vacuoles at the base of the leaf stalk hold the leaves erect (Figure 4.20a). When the plant is touched, however, certain ions are pumped out of each such leaf base cell, and then out of the vacuole. Water then leaves the vacuoles by osmosis, the vacuoles and whole cells shrink in size, and the leaves droop (Figure 4.20b).

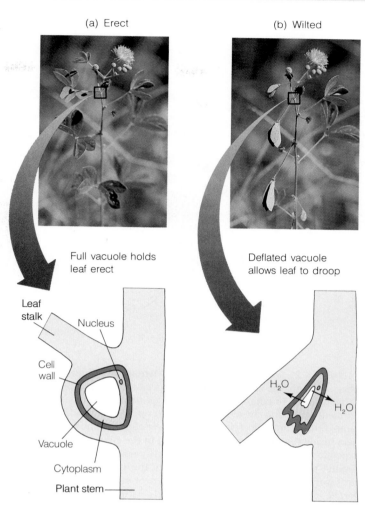

FIGURE 4.20 ■ Plant Cells and Osmosis. Water in the central vacuoles of plant cells exerts an outward force called turgor pressure that keeps the plant tissues firm. (a) In the sensitive plant, the appearance of the entire leaf depends on the turgor pressure of a single cell in the junction between the leaf's stalk and the plant's main stem. When turgor pressure in that cell is high, the leaf stands erect. (b) If you touch the plant, this cell pumps ions out, and water follows by osmosis. This lowers the turgor pressure, and the leaf looks wilted. Plants also exhibit another characteristic of water movement: *bulk flow.* Bulk flow refers to the overall movement of a fluid down a pressure gradient and to the fact that all the molecules of the fluid move together and in the same direction, like the water in a fast-flowing river. Concentrated sugar solutions, such as the sap in maple trees, move from leaves and roots to other plant parts by means of bulk flow.

Active Transport: Energy-Assisted Passage

Ironically, the principle of osmosis costs the cell dearly. Cells contain such an array of raw materials (relative to the extracellular fluid) that for the majority of cells, there is a continuous tendency for water to enter the cell and cause swelling.

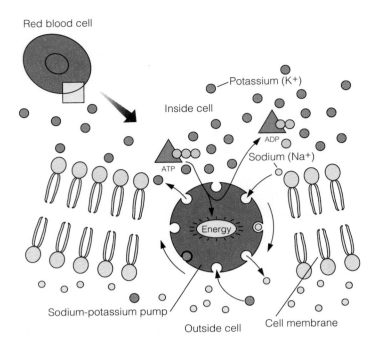

Red blood cell

Potassium (K+)

Inside cell

ADP

ATP

Sodium (Na+)

Energy

Sodium-potassium pump

Outside cell

Cell membrane

FIGURE 4.21 ■ Active Transport by Membrane Pumps. In the plasma membrane of a red blood cell, special membrane proteins actively pump in potassium ions (K+) against their concentration gradient (more ions inside the cell, fewer outside) and actively pump out sodium ions (Na+) against their concentration gradient (less inside the cell, more outside). Most of the energy being expended when a person is resting quietly goes to drive the sodium-potassium ion pump. Sometimes, the active transport carriers work only when a sodium, hydrogen, or other ion is *cotransported,* or ferried in or out simultaneously, with the passenger molecule. For example, milk sugar enters a bacterial cell by cotransport: For every hydrogen ion transported down its concentration gradient, a milk sugar molecule is transported up its chemical gradient. All membrane pumps and active transport carriers are highly specific for particular molecular passengers; moreover, they all expend ATP energy, and they all move substances against concentration, pressure, or other gradients.

At the same time, cells must stockpile certain materials for building, maintenance, and energy needs that are too large to move in by passive diffusion. Thus the dilemma: To prevent swelling, the cell must move certain materials *out* (with water following osmotically), and yet to continue functioning, the cell must move certain materials *in* against their concentration gradient. Moving a substance against such a gradient involves the energy-costly process of active transport. In many cells, active transport is accomplished by a set of proteins in the plasma membrane that act like pumps, pushing or pulling ions into and out of the cell.

One type of pump, a **sodium-potassium pump,** maintains osmotic balance in the cell but at a high energy cost—30 to 70 percent of the cell's entire energy budget. A red blood cell, for example, contains a high concentration of K+ but a low concentration of Na+, while the blood plasma outside the cell contains the reverse. The special membrane proteins in the pump actively move K+ in and Na+ out against their concentration gradients by expending energy in the form of ATP (Figure 4.21). This pumping results in a proper osmotic balance, since water "follows" the Na+ out. Thus, so long as pumping continues, the cell does not shrink or burst.

The red blood cell also needs a constant supply of glucose to fuel metabolism, but glucose molecules cannot enter the cell via simple diffusion. Most cells, in fact, need glucose as well as amino acids, lipids, and other raw materials for building cell parts, but these commodities are often too big to pass through a fixed membrane pore or cannot diffuse in because the concentration gradient is unfavorable. Each cell employs *membrane transport proteins* that actively bring in the needed molecules, enabling the cell to stockpile these commodities as fast as it uses them. The raw material floats up and binds to a transport protein that spans the membrane. The membrane protein then changes shape (for some compounds, at a cost of ATP energy), and in so doing, acts as a carrier that transports the "passenger," releasing it on the other side.

The inpocketing and outpocketing of cell membranes (endocytosis and exocytosis) also help many cells take in or expel particles of food, water, or wastes and thus accomplish their transport work to keep up with the staggering need for raw materials (review Figure 3.13). Once inside the cell, however, the materials from all routes of entry are used in the cell's equally important chemical tasks and to a certain extent in the mechanical tasks that generate movement.

THE DYNAMIC CELL'S MECHANICAL TASKS

As the cell carries out its chemical and transport tasks, it must also accomplish a set of mechanical tasks, often reflected by movements of the entire organism or its parts. Movement is one of the most striking characteristics of life, and it is the easiest kind of cellular activity to observe. Cells are capable of a tremendous range of movements. Muscle cells contract, sometimes hundreds or thousands of times per second, allowing an animal to swim, fly, or run (Figure 4.22). Amoebas creep along solid surfaces. Cells with flagella lash them about, enabling them to glide forward or backward. Ciliated cells can sweep particles across their surfaces or dart through the water as if propelled by thousands of tiny oars. And immature red blood cells pass through the narrow pores in the blood vessel walls and squeeze out most of their internal organelles. Moreover, all eukaryotic cells, whether obviously mobile and active or apparently stationary, have an array of internal movements.

Underlying mechanical actions in the cell are the rapid assembly or disassembly of cytoskeletal filaments or the sliding of filaments past each other (review Figure 3.16). These actions are accomplished by enzymes and fueled by ATP, and they represent a unique form of energy conversion. The chemical energy stored in ATP is released and converted directly to the mechanical energy underlying the rapid, orderly movements of the cytoskeletal components. This can be compared to the workings of a steam engine, where the chemical energy of coal is first converted to heat and then into the mechanical movement of the wheels. The conversion of chemical energy to mechanical energy in muscles is discussed in more detail in Chapter 29. For now it is simply worth noting that mechanical work is one of the cell's crucial activities, enabling it (or the organism the cell is part of) to pursue food, to divide and reproduce, and to carry out chemical and transport work. This is just one more reminder that all the work of the dynamic cell is inextricably intertwined into one ongoing process—life.

CONNECTIONS

In the time it takes you to blink, sneeze, yawn, or swallow, nearly every cell in your body will have performed thousands of individual activities. Membrane pumps will have transported billions of ions in and out of your cells with a concomitant flux of water molecules, back and forth, over and over. Enzymes will have sped up hundreds of metabolic reactions—with interconnected metabolic pathways like the strands of a web—that break down food and build up

thousands of proteins, lipids, and other molecules. Some cells will have rippled their cilia. Others will have squeezed out their nuclei. Still others will have divided in half. And inside them all, organelles will have shifted—vesicles moved, cytoskeletal elements lengthened or shortened, and carrier proteins drifted through the plasma membrane like icebergs.

It is practically impossible to describe or envision all the simultaneous activities in a cell, and yet they go on, second after second, making up the life process. Every swaying kelp plant, every bracket fungus, every gray whale breaching the ocean's surface is an isolated population of cells, all pumping ions, metabolizing compounds, moving organelles about, and resisting the universal tendency toward disorder—at least for a while.

The energy costs for cellular activity are enormous, and yet, collectively, life processes are exergonic. Cells are temporary repositories of sunlight in chemical form; they convert light into chemical packets that fuel their internal organization, then give off heat that disorders the environment ever so slightly by causing random jostling motions of the air or water around them. Efficient energy collection and use are so critical to survival that they have been a prime cutting edge for evolutionary processes. Natural selection rewards those organisms that can grab the most energy and use it efficiently, while the less efficient life forms tend to die out.

In the next two chapters, we will trace the energy flow through cells. As we do, try to envision the bustling metropolis—the dynamic cell—that surrounds and is empowered by that flowing energy.

FIGURE 4.22 ■ Dynamic Cells and Dynamic Animals. The little owl (*Athena noctue*) takes flight. This feat of locomotion is based on the contractile cytoskeleton inside muscle cells and requires a great deal of energy. Owls, although often large birds, have noiseless flight based on the soft outer edges of flight feathers, which you can see fanned out here.

Highlights in Review

1. The red blood cell, with its complex life history, shows just how dynamic even a simple cell can be as it carries out transport, chemical, and mechanical work that enables it to maintain a steady state of order and activity.

2. The ability to carry out organized biological work depends on a continual flow of energy from the environment—a flow that is mediated by enzymes and the energy carrier ATP and accompanied by a continual flux of materials into and out of the cell.

3. Energy, the capacity to do work, exists in two states: potential energy (stored energy) and kinetic energy (the energy of motion). These states, as well as different forms of energy, can be interconverted, but according to the first law of thermodynamics, energy is neither created nor destroyed in the process. Some energy, however, is lost and released as heat during such interconversions. The release, according to the second law of thermodynamics, adds to the entropy—the increasing randomness and disorder—of the system.

4. Cells, too, are subject to entropy, but the chain—sun → plants → organic molecules → life processes—makes energy available in chemical form, which the cell can then expend to "purchase" order through the construction and maintenance of cell parts and processes.

5. Chemical reactions, during which reactants are converted to products, underlie virtually all cellular energy conversions. Exergonic reactions proceed spontaneously and give off heat, while endergonic reactions will not "go" without energy input.

6. In cells, these two kinds of reactions are coupled so that the energy for endergonic reactions is generated during exergonic reactions. The energy is usually trapped in a carrier called ATP (although there are other such carriers as well), which serves as a chemical intermediate, linking all the energy exchange reactions collectively called metabolism.

7. Before reacting molecules can be transformed into products, they must be activated; that is, they must have sufficient energy to undergo a productive collision and reach a very brief intermediate transition state. The minimum amount of energy they need for a reaction, the activation energy, forms a barrier between reactants and products and would prevent most reactions in cells from occurring quickly enough if there weren't catalysts called enzymes to facilitate and speed up reactions.

8. Enzymes lower the activation energy needed for biochemical reactions to occur by binding reactants in the groove called the active site and forming enzyme-substrate complexes; orienting the substrates in such a way that appropriate regions of the molecules will collide; and changing the shapes of the substrates and helping them reach the transition state. Enzymes can speed up reactions a millionfold.

9. A major task of the dynamic cell is to maintain the appropriate mix of ions, molecules, and gases in the intracellular fluid so that materials are always on hand for metabolism—especially for the building of new cell parts. Cells accomplish this by passive and active transport.

10. Materials naturally move down gradients from areas of high concentration to areas of low concentration, and this simple process called diffusion accounts for most of the passage of materials through the cell membrane.

11. Water also moves down gradients from areas of higher water concentration (purer water with fewer solutes) to areas of lower water concentration (solutions with more solutes) in a process called osmosis. If the extracellular fluid is isotonic, there will be no net movement of water due to osmosis. However, if the extracellular fluid has a lower concentration of dissolved materials than the cell (a hypotonic solution), the cell will imbibe water and swell or burst; and if it has a higher concentration of dissolved materials (a hypertonic solution), the cell will shrink from water loss.

12. Cells expend a great deal of energy pumping ions in and out, helping maintain the correct intracellular concentration of each solute.

13. Cells carry out additional active transport work: Carrier proteins embedded in the plasma membrane ferry large molecules across, and the entire membrane can inpocket or outpocket, thus taking in or expelling water or materials.

14. The cell's mechanical tasks involve movements of the entire cell, of external organelles such as cilia and flagella, or of internal parts such as the chromosomes during cell division. The rapid assembly or disassembly of the cytoskeleton underlies all movement, and once again, enzymes facilitate the reactions that lead to these structural changes, and ATP is "spent" providing chemical energy.

Key Terms

activation energy, 89
active site, 90
active transport, 92
anabolism, 86
ATP, 87
carrier-facilitated diffusion, 93
catabolism, 86
catalyst, 90
chemical reaction, 84
diffusion, 93
endergonic, 85
energy, 82
entropy, 83
enzyme, 90
enzyme-substrate complex, 90
exergonic, 85
extracellular fluid, 92
first law of thermodynamics, 82
gradient, 92
heat, 83

hypertonic, 95
hypotonic, 95
intracellular fluid, 92
isotonic, 95
kinetic energy, 82
metabolic pathway, 86
metabolism, 86
osmosis, 94
passive transport, 92
potential energy, 82
product, 85
productive collision, 89
reactant, 85
second law of thermo-
    dynamics, 82
sodium-potassium pump, 96
substrate, 90
transition state, 89
turgor pressure, 95

Study Questions

Review What You Have Learned

1. Distinguish between potential and kinetic energy.
2. Discuss how living cells counteract the tendency toward entropy.
3. Classify the following activities as exergonic or endergonic, and explain why in each case: running; building of proteins; weight lifting; formation of new ribosomes.

4. Explain why biologists often describe ATP in a cell as energy currency.

5. What is activation energy? How do enzymes affect it?

6. Describe the structure of an enzyme, and explain why enzyme action is highly specific.

7. Compare active and passive transport.

8. True or false: During osmosis, water molecules diffuse through a membrane from a region of high concentration to low, until there is an equal concentration on both sides of the membrane; then the flow of water molecules through the membrane stops. Explain your answer.

9. Describe the differences between isotonic, hypotonic, and hypertonic solutions.

10. How does the sodium-potassium pump help maintain osmotic balance in the cell?

Apply What You Have Learned

1. Your grandfather might say, "For the most part, I'm less than a year old." In what way would he be right?

2. A student attempting to measure the optimal temperature for the activity of the enzyme pepsin in digesting egg white protein tried 17°C, 37°C, and 57°C. At which temperature would the enzyme's activity be greatest? Why?

3. People used to preserve codfish by adding large quantities of salt to the fish. How might this disrupt bacterial cells that might otherwise grow on the fish?

For Further Reading

ALBERTS, B., D. BRAY, J. LEWIS, M. RAFF, K. ROBERTS, and J. WATSON. *Molecular Biology of the Cell.* 2d ed. New York: Garland, 1989.

DARNELL, J., H. LODISH, and D. BALTIMORE. *Molecular Cell Biology.* New York: Scientific American, 1986.

HIRSCHHORN, N., and W. B. GREENOUGH III. "Progress in Oral Rehydration Therapy." *Scientific American,* May 1991, pp. 50–56.

PRESCOTT, D. M. *Cells.* Boston: Jones and Bartlett, 1988.

STRYER, L. *Biochemistry.* 3d ed. New York: W. H. Freeman, 1988.

How Living Things Harvest Energy from Nutrient Molecules

A Mountain Hike: Splitting Sugar for Muscle Power

Suppose you and a friend decide one day to hike up a steep trail that winds alongside a tumbling stream—a trail, let's say, like the many crisscrossing the Green Mountains of Vermont. Fortified by a hearty breakfast of pancakes and maple syrup, you start up the trail in the brisk morning air (Figure 5.1). As you ascend into the fragrant pine forest, the steepness of the terrain begins to tax you, and your heart starts to pound, your chest to heave, and your leg muscles to burn. Eventually, adopting a slower, steadier pace, you reach the summit and take in a panoramic view of blue-green forested ridges that gently fade to gray in the distance.

To carry out this simple weekend activity, the human body must rely on many fundamental life processes. Two of the most basic life processes are the subject of discussion here and in the next chapter. First, the food you eat for breakfast contains stored chemical energy. Plants such as maple trees and wheat soak up sunlight and capture light energy in the bonds of glucose molecules, which form the sugar in maple syrup and the flour in breakfast pancakes. Second, the muscles in the legs and elsewhere in the body harvest that stored chemical energy, which in turn fuels walking, breathing, and all the activities that carry two hikers up a mountain trail.

This chapter focuses on how living things strip energy from molecules such as sugars, and Chapter 6 describes how plants trap the sun's energy in those sugars by means of the process called photosynthesis. Together, these two chapters explain a central theme in biology: Plants, as well as certain bacteria and protists capable of photosynthesis, capture the sun's energy in sugars. The harvesting of energy from these sugar molecules then supports the life processes in virtually all living things.

The energy-harvesting processes that take place in muscle cells are also at work in plants, animals, fungi, protists, and

FIGURE 5.1 ■ **Energy for Hiking: From the Sun and Metabolism.** Maple trees in New England and fields of wheat in the Midwest trap solar energy and build sugars. These hikers are using the energy stored in maple syrup and the wheat flour of pancakes as they move steadily up the trail.

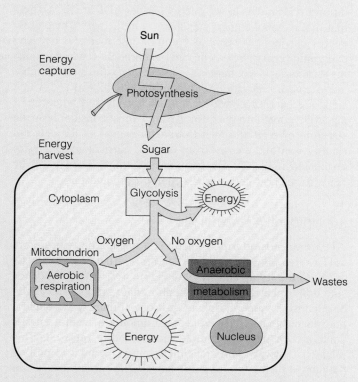

FIGURE 5.2 ■ **Overview of Energy Harvesting Pathways in the Cell.** Plants trap energy from sunlight in the simple sugar glucose through the process called photosynthesis. Virtually all cells break down glucose through a metabolic pathway called glycolysis. If oxygen is present, many cells can net a large additional energy harvest through the aerobic respiration pathway. If oxygen is absent, some cells can carry out fermentation (anaerobic respiration) with no further gain of ATP but a recycling of some waste products from glycolysis into necessary raw materials.

bacteria. In fact, this leads to a second theme: The same metabolic pathways fuel all life activities. The similarity of the pathways in all organisms is a powerful reminder of life's unity, its evolutionary descent from common ancestors.

Hikers wandering up a forest trail could come to a fork where one path leads, let's say, to the mountain summit where the air is thin, and the other path loses elevation, dropping down through the forest where the air is richer in oxygen. The pathway of energy metabolism is a bit like a branching trail. Leading up to the fork is the metabolic pathway called *glycolysis*, which is common to virtually all cells. Energy production can then proceed along a pathway of reactions called *fermentation* or along another pathway called *aerobic respiration;* the direction depends on the cell type and on environmental conditions. Thus, in most types of cells, energy metabolism is accomplished either by glycolysis and fermentation or by glycolysis and aerobic respiration (Figure 5.2). Muscle

cells, yeast cells, and certain other kinds of cells, however, are somewhat special in that they can follow the first pathway if oxygen is lacking and the second pathway if oxygen is abundant.

Glycolysis and fermentation take place in the cytoplasm of a cell. Although they enable cells to survive **anaerobic** conditions (when oxygen is scarce or absent), they are inefficient and incomplete: The harvest of ATP, the cell's main energy carrier, is low, and the excretion products still contain considerable energy. When conditions are **aerobic,** that is, when oxygen levels are high, muscle cells, yeast cells, and others break down sugars by means of glycolysis and **aerobic respiration.** Both pathways are continuous sets of chemical reactions that break and rearrange the bonds in organic molecules. While aerobic respiration begins in the cytoplasm with glycolysis, the next phases take place in the mitochondria. Together, glycolysis and the aerobic processes in the mitochondria break sugar down completely to water and carbon dioxide—the gas a person exhales with every breath (and with particular force during a difficult uphill hike)—and thus, the aerobic pathway results in a far greater harvest of ATP molecules from the starting substance glucose than does the anaerobic pathway.

We will encounter a third theme in our discussion of energy harvest: metabolic pathways are *gradual*; that is, they occur in steps. The energy stored in a nutrient molecule is not released all at once in an incinerating burst of heat; instead, molecular bonds are broken and rearranged in numerous small steps, and the sequentially released energy is stored for future use. Since the transfer of energy is always inefficient, some heat is lost forever at each step. Despite this inevitable waste, however, living cells harvest enough energy in the form of ATP to fuel their growth, maintenance, and reproduction.

This chapter will answer several questions about how living things harvest the energy they need to sustain life:

- How do cells build and use energy carriers such as ATP?
- What are the energy-harvesting reactions in a cell?
- How does glycolysis begin the energy harvest?
- What is the role of fermentation during metabolism in the absence of oxygen?
- How does aerobic respiration allow cells to harvest huge quantities of energy when oxygen is present?
- How does the cell control energy metabolism?
- What three systems supply energy to an exercising person?

ATP AND THE TRANSFER OF ENERGY FROM NUTRIENT MOLECULES

Like the bacterial cells considered in Chapter 4, muscle cells contain millions of ATP molecules and expend nearly half of them every second on life-sustaining cellular activities. Obviously, the supply of this energy currency must be regenerated speedily and continuously—and so it is. This regeneration takes place through the pathways of glycolysis, fermentation, and aerobic respiration—pathways that break down energy-storing compounds.

For all living cells, whether prokaryotic or eukaryotic, and whether capable of surviving under anaerobic or aerobic conditions, ATP and its lower-energy partner ADP are the carriers that link metabolic energy exchanges in the cell. Thus, these energy carriers can serve as common intermediates between exergonic (energy-releasing) and endergonic (energy-requiring) reactions (see Chapter 4). As we study the specific reaction steps of energy-releasing pathways, we will see which steps are exergonic, which are endergonic, and where ATP and ADP come into play. But first, we must consider how the structure of ATP allows it to transfer energy in the cell and exactly how the energy from organic nutrient molecules is channeled into ATP for temporary storage. Then we'll be ready to see how cells harvest energy from nutrients.

ATP Structure: A Powerful Tail

ATP, or adenosine triphosphate, is a nucleotide with a long tail of three phosphate groups, and this structure explains its utility to the living cell (review Figure 4.9). ADP, or adenosine diphosphate, contains two phosphate groups, and AMP, or adenosine monophosphate, just one (Figure 5.3). When a phosphate bond is broken, a large amount of useful energy contained in the whole molecule is readily released. This release of energy from a chemical bond is similar to pulling the plug at the bottom of a rain barrel. The energy released in the outflow was stored in the total water contents, not simply in the rubber plug.

The cleaving of one phosphate group from ATP yields 8 kilocalories of energy per mole* of ATP and leaves ADP plus the inorganic phosphate ion (designated P_i, see Figure 5.3). The cleaving of the next phosphate group from ADP yields another 8 kcal/mol and leaves behind AMP + P_i. Cleaving the final phosphate group from AMP yields only 2 kcal/mol. In most energy transfer reactions, ATP serves as the high-energy, or "charged," form of the molecule, and ADP as the low-energy, or "discharged," form.

The structure of ATP has three major consequences: (1) Energy is released when phosphate groups are cleaved; (2) energy is required to put phosphate groups back on ADP or AMP; and (3) phosphate groups can be transferred from ATP to other kinds of molecules, thereby energizing them and allowing them to participate in reactions that could not take place otherwise. This transfer process is called **phosphorylation,** whether the phosphate is transferred to an unrelated molecule such as glucose or to a molecule of ADP. Figure 5.4 shows the phosphorylation of glucose to glucose-6-phosphate, which is the mandatory first step in glycolysis. The transferred phosphate group in effect acts like a little battery pack, providing enough power so that the sugar molecule can readily take part in later reactions that produce energy. The phosphorylation of glucose costs the cell some energy, since ATP must be cleaved to provide the phosphate group and form the new bond. However, the expenditure is like priming

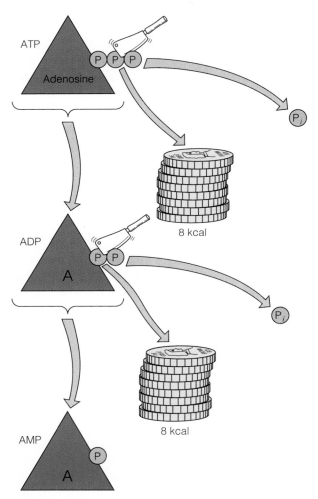

FIGURE 5.3 ■ Energy Currency for the Cell: ATP, ADP, and AMP Compared.   ATP, or adenosine triphosphate, has a long, three-phosphate tail, and the molecule contains a substantial amount of energy. ADP has a shorter tail with just two phosphate groups and contains less energy. AMP has just a one-phosphate tail and considerably less energy than either ADP or ATP. Cleaving one phosphate group from ATP yields 8 kcal of energy, one inorganic phosphate ion, and ADP. Cleaving another phosphate group yields an additional 8 kcal of energy, another inorganic phosphate ion, and AMP.

---

*kcal/mol; see the abbreviations chart on the inside back cover.

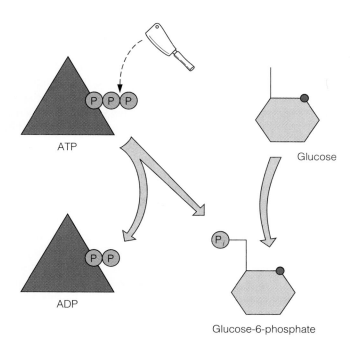

FIGURE 5.4 ■ Phosphorylation of Glucose. Glucose can be "supercharged" by the cleaving of ATP and the addition of the phosphate group to the glucose molecule. This glucose-6-phosphate can then participate in reactions in which a regular glucose molecule cannot take part.

a pump or spending money to make money; as glycolysis proceeds, that original "cost" is paid back with interest—the harvesting of additional ATPs.

"Electrical Currents" in Energy Metabolism

The structure of ATP may explain its importance to energy flow in the cell, but ATPs are not the only energy carriers to take part in glycolysis, fermentation, and aerobic respiration. The transfer of energy—particularly the phosphorylation of ADP molecules to build ATPs—involves a flow of electrons, much like an electric current, that is handled by the first three carriers listed in Table 5.1 on page 104.

The transfer of electrons from one molecule to another is called an **oxidation-reduction reaction.** Oxidation is the removal of electrons (which *increases* the net charge), while reduction is the addition of electrons (which *reduces* the net charge; Figure 5.5). For example, if a molecule with no charge gains an electron with its charge of -1, the result is a molecule with a reduced charge of -1. But notice that the process is reversible (see Figure 5.5); the molecule with a -1 charge can be oxidized by losing an electron, regenerating the uncharged form. Oxidation-reduction reactions always take place in pairs, with one molecule becoming oxidized (losing electrons) as the other is being simultaneously reduced (gaining electrons).

In biological systems, special energy carriers transfer a pair of electrons, along with a hydrogen ion (proton), from one molecule to another. One such carrier, called NADH (nicotinamide adenine dinucleotide), can give up two electrons and a hydrogen ion and become oxidized to NAD$^+$, while one of a large number of other compounds can trap the electrons and hydrogen ion and become reduced (see Figure 5.5). Likewise, the carrier FADH$_2$ can donate two hydrogen ions and two electrons to any of a large number of different molecules and become oxidized to FAD as the other compound is simultaneously reduced. Both NAD$^+$ and FAD are *coenzymes;* their presence is necessary before key metabolic enzymes can function.

Have you ever wondered what *vitamins* are and why nutritionists advise people to consume certain minimum levels in their daily diet? The answer is that the organic molecules we call vitamins are precursors to NAD$^+$ and FAD and other coenzymes with their important roles in aerobic respiration. We'll talk much more about vitamins in Chapter 24.

The overall message in this discussion of oxidation-reduction reactions is that as metabolic pathways proceed in the cell, coenzymes and other energy carriers pick up electrons and hydrogen ions and later pass them along to other carriers.

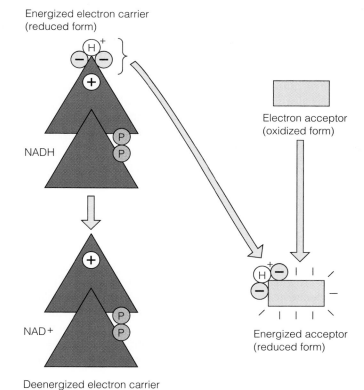

FIGURE 5.5 ■ Oxidation-Reduction Reactions and Electron Carriers. An oxidation is the removal of electrons; a reduction is the addition of electrons; and the two transfers are always coupled. Energy pathways such as glycolysis and aerobic respiration run on coupled oxidation-reduction reactions, often linked by energy carriers such as NADH. As NADH is oxidized, two electrons and one hydrogen ion (proton) are released, and these are available to reduce an electron acceptor.

TABLE 5.1   Common Electron Carriers

Electron Acceptor / Electron Donor	Full Name	Role
$\dfrac{NAD^+}{NADH}$	Nicotinamide adenine dinucleotide	Electron transfer during breakdown of fuels
$\dfrac{FAD}{FADH_2}$	Flavin adenine dinucleotide	Electron transfer during breakdown of fuels
$\dfrac{NADP}{NADP^+}$	Nicotinamide adenine dinucleotide phosphate	Electron transfer in manufacture of new biological molecules
$\dfrac{Cytochrome\ Fe^{3+}}{Cytochrome\ Fe^{2+}}$	Cytochrome proteins (Fe = iron)	Electron transport
$\dfrac{O_2}{H_2O}$	Oxygen and water	Final electron acceptor in electron transport chain in mitochondria

The flow of electrons through the carriers, specifically through their alternating oxidation and reduction, creates a current that is channeled in special ways into ATP formation. Thus, during the overall breakdown of a nutrient molecule such as glucose in a muscle or yeast cell, the sugar becomes oxidized while an end product is reduced, and in the process, energy is channeled into the production of ATPs.

Now let's begin to look at that overall breakdown.

AN OVERVIEW OF GLYCOLYSIS, FERMENTATION, AND AEROBIC RESPIRATION

All cells need energy to live, whether they are budding yeast cells growing in a tank of grape juice, the billions of cells that cooperate as an owl flaps silently or as you walk to class, or the cells in a willow leaf fluttering in the wind. The immediate *source* of that energy may differ: Yeasts absorb nutrients directly from their environment; humans chew and digest foods made mostly from plant and animal materials; and willow leaves generate their own carbohydrates via photosynthesis.

Organisms that must take in preformed nutrient molecules from the environment are classified as **heterotrophs** (literally, "other-feeders"), and they include the yeasts and all other fungi, most kinds of protists, bacteria, and humans and all other animals. Plants and the photosynthetic microbes that can generate their own nutrient molecules are considered **autotrophs** ("self-feeders"). When it comes to energy harvest, however, it doesn't matter whether an organism is a het-

erotroph or an autotroph, because all organisms share the same biochemical pathways for releasing energy stored in the molecular bonds of nutrient molecules.

As Figure 5.6 shows, energy harvest in all cells begins with the catabolism (breakdown) of glucose in the sequence of reaction steps called **glycolysis** (literally "glucose splitting"). The energy stored in the chemical bonds of glucose molecules originally came from the sun, and it was trapped in molecular form by the photosynthesis taking place in green plants, algae, and certain kinds of bacteria. The hikers we discussed earlier got the energy for walking up a steep hill secondhand by consuming pancakes and maple syrup, abundant sources of glucose molecules. The pathway of glycolysis breaks down the six-carbon sugar glucose into two molecules of the three-carbon compound pyruvate. The steps of glycolysis take place in the cytoplasm; they proceed in the presence or absence of oxygen, and they are mediated by a specific set of enzymes every cell possesses. Two ATPs are formed during glycolysis for each glucose split; this represents a harvest of less than 5 percent of the energy that could be released by burning the sugar molecule in oxygen via aerobic respiration. Biologists believe that glycolysis evolved in the earliest cells, perhaps 3 billion years ago, and was passed down to all other organisms. This explains the near-universal occurrence of glycolysis in living cells—even in plant cells that capture the energy in sunlight—and underscores the evolutionary relatedness of all life.

The pathways of fermentation and aerobic respiration were evolutionary add-ons to the biochemical antique of glycolysis (see Figure 5.6). In certain yeasts and bacteria, and in some animal muscle cells, fermentation proceeds when oxygen is absent. As in glycolysis, the reaction steps of fermentation take place via enzymes in the cytoplasm, but they go beyond glycolysis and convert the pyruvate formed during that pathway into ethanol and carbon dioxide, into the compound lactic acid, or into other end products, depending on the species. Fermentation does not result in additional ATPs, but it does regenerate the energy carrier NAD^+, making it available for its continuous role in glycolysis.

Aerobic respiration occurs only in the cells of aerobic organisms (life forms that employ oxygen for metabolism) (see Figure 5.6). There, the respiration pathway acts on the pyruvate generated during glycolysis to yield carbon dioxide and water plus an additional 34 ATP molecules per initial glucose molecule. The two phases of aerobic respiration, the *Krebs cycle* and the *electron transport chain,* operate only under aerobic conditions, and the various enzymes that catalyze the reaction steps are located in the mitochondria, accounting for that organelle's reputation as a cellular powerhouse. During the Krebs cycle, two ATPs form per glucose, and numerous electrons are passed to energy carriers. These electrons are removed sequentially in the electron transport chain by enzymes embedded in mitochondrial membranes, and the energy current (the flow of electrons) thus generated is used to produce a whopping harvest of 32 more ATPs per initial glucose molecule. The total ATP yield from the entire

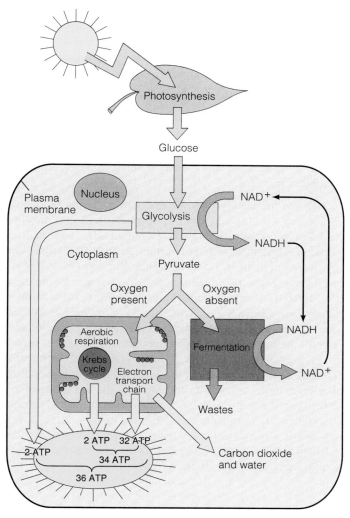

FIGURE 5.6 ■ **Energy Harvest: An Overview.** This diagram shows the energy tally for the aerobic and anaerobic pathways of glucose breakdown. The left side of the diagram shows the aerobic pathways: Glycolysis yields 2 ATP molecules per glucose molecule, and aerobic respiration (with its two parts, the Krebs cycle and the electron transport chain) yields an additional 34 ATPs for a total of 36 ATPs per initial glucose molecule. The right side of the diagram shows the anaerobic pathways: Glycolysis yields 2 ATPs as well as the reduced energy carrier molecule NADH; fermentation yields no additional ATPs, but regenerates the oxidized NAD⁺ needed for the next round of glycolysis to take place.

sequence, beginning with glycolysis and ending with electron transport in an aerobic cell, is glycolysis 2, Krebs cycle 2, electron transport chain 32, or a total of 36 ATP molecules per initial glucose molecule.

One can summarize energy-harvesting pathways as follows: In glycolysis, cells split glucose and produce a small amount of usable energy. When oxygen is absent, cells ferment the products of glycolysis, which yields no additional energy but does regenerate intermediates needed for continued glycolysis. When oxygen is present, most cells carry out

aerobic respiration on the products of glycolysis and produce large amounts of ATP. Studying the specific steps of each catabolic pathway—our next task—is as essential to appreciating life as studying symphony movements and play acts is to appreciating music and drama.

GLYCOLYSIS: THE UNIVERSAL PRELUDE

At this moment, practically every cell on earth is burning the six-carbon sugar glucose via the nine sequential reaction steps called glycolysis ("glucose splitting"). Glycolysis is the basis for energy metabolism in all living things, since it serves as a biochemical prelude to the pathways of fermentation and aerobic respiration. As this pathway proceeds in the cell's cytoplasm, each glucose molecule is split into two molecules of the three-carbon compound pyruvate, and two ATPs are harvested. The overall formula for glycolysis,

$$C_6H_{12}O_6 + 2\ ADP + 2\ P_i + 2\ NAD^+ \rightarrow$$
glucose
$$2\ C_3H_4O_3 + 2\ ATP + 2\ NADH + 2\ H^+ + 2\ H_2O$$
pyruvate

reveals that during the splitting of each six-carbon sugar (with its 12 hydrogen atoms) into 2 three-carbon pyruvates (each with 4 hydrogen atoms), the additional 4 hydrogen atoms are removed; 2 are transferred to the energy carrier NAD⁺, and the other 2 are released as hydrogen ions (H⁺). It is this transfer of electrons and hydrogen ions from the intermediate compounds in the pathway that creates an energy currency, which is then stored in ATP, releasing H_2O.

Each of the nine steps in the glycolysis pathway is diagrammed and described in detail in Figure 5.8. The significant outcome of each stage, or group of steps, can be illustrated as in Figure 5.7 and summarized this way:

Stage One: Energy Investment. Glucose is phosphorylated, "charged" with two high-energy phosphate groups at the cost of two ATP molecules. The reactions that bring this about represent the pump priming mentioned earlier; so far, energy has been spent, not harvested, producing a sugar with two phosphates. (This corresponds to Stage One in Figure 5.7 and to the more detailed steps 1 to 3 in Figure 5.8.)

Stage Two: Sugar Splitting. The high-energy sugar formed in Stage One is split into two molecules of the compound G-3-P, each of which has three carbons and one phosphate group. (See Stage Two in Figure 5.7 and step 4 in Figure 5.8 for details.) Each of the remaining steps must take place twice, once for each of the three-carbon molecules formed.

Stage Three: Oxidation-Reduction. The phosphorylated three-carbon G-3-P is oxidized, and energy is released; energy and hydrogens are transferred to NAD⁺, reducing it to NADH. In addition, inorganic phosphate groups are added to the three-carbon G-3-P, leaving it with two phosphates and forming a new compound called DPGA. (See Stage Three in Figure 5.7 and step 5 in Figure 5.8 for details.)

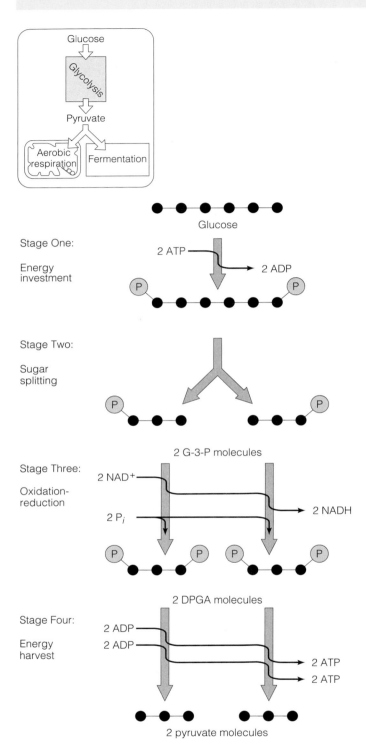

Stage Four: Energy Harvest.   The DPGA, with its three carbons and two phosphates, is next shaped and reshaped into a series of related compounds (as shown in Stage Four in Figure 5.7 and steps 6 to 9 in Figure 5.8). The phosphates are removed one at a time and transferred to ADP, generating a total of four new ATPs for each molecule of glucose dismantled via glycolysis. The importance of these transformations from one compound to another lies in the fact that energy is released and ATP is harvested. The final end products of glycolysis are two pyruvate molecules, each with three carbons. These pyruvates are the starting point for fermentation and aerobic respiration.

To summarize glycolysis, then, each molecule of glucose is boosted with the addition of phosphates and split into two G-3-Ps (which are three-carbon compounds), and each of these is rearranged and split further so that ATP forms and pyruvate is left over. For each initial glucose, two ATPs are used up in Stage One (see Figure 5.7) to prime the process, two NADHs form in Stage Three, and four ATPs form in Stage Four. (Since Stages Three and Four are repeated twice, once for each three-carbon compound, four ATPs are made.) Two of the four ATPs pay back the initial investment, leaving a net gain of two ATPs for each glucose molecule burned. This is less than 5 percent of the energy stored in the bonds of that sugar molecule, and it can be merely the beginning of the much larger energy harvest that takes place during aerobic respiration in the mitochondria. But the energy harvest from glycolysis alone is enough to sustain life in organisms that ferment, such as yeasts, or enough to power a hiker's short but explosive jump over a fallen tree in the trail.

FERMENTATION: "LIFE WITHOUT AIR"

Louis Pasteur, the famed nineteenth-century microbiologist who discovered fermentation, called it *"vie sans air"* ("life without air"), and indeed, **fermentation** is a major metabolic pathway for anaerobes—cells that can live where oxygen levels are very low or where oxygen is altogether absent. The specific biochemical reactions we call fermentation rest on the foundation of glycolysis and do not directly yield ATP. Fermentation does, however, enable certain kinds of cells to survive on the limited energy proceeds of glycolysis by recycling the intermediate compound NAD^+, needed at Stage Three for glycolysis to continue generating ATPs. Even for aerobic organisms like people, microbial fermentation has important implications. Fermenting yeasts and bacteria generate carbon dioxide and alcohol or lactic acid, which in turn help create many of our favorite drinks and foods, including wine, beer, bread, and cheeses. And human muscle cells, such as those in a hiker's legs can ferment for short periods of time when oxygen levels are low. (Such low levels of oxygen result when the muscle cells work so hard that they consume more oxygen than the person's breathing can supply.) This muscle cell fermentation can produce a quick burst of speed that can save a person's life—but it can also cause painful cramps.

FIGURE 5.7 ■ A Summary of Glycolysis. In Stage One, two ATP molecules are spent, and two released phosphates are added to glucose. In Stage Two, this charged glucose is split into 2 three-carbon molecules, each with one phosphate (G-3-P). In Stage Three, an oxidation-reduction reaction and the addition of phosphates result in 2 three-carbon molecules, each with two phosphates. In Stage Four, those phosphates are removed and energy is captured in four ATPs, two of which "pay back" the original investment. The net yield, then, is two ATP, two NADH, and two pyruvate molecules. A more complete explanation of glycolysis is presented in Figure 5.8.

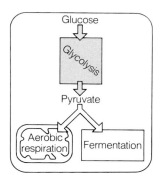

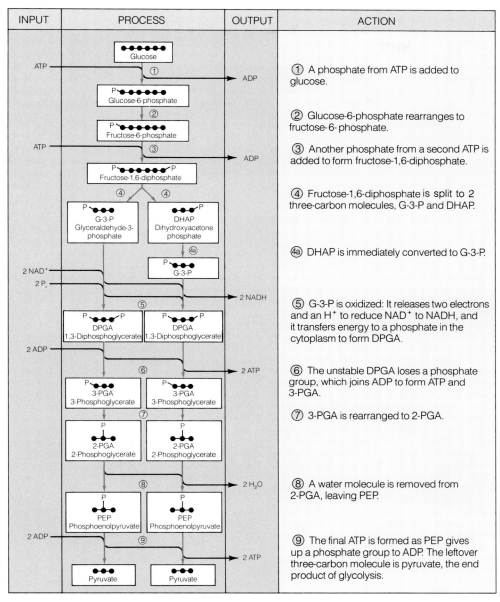

INPUT	PROCESS	OUTPUT	ACTION

① A phosphate from ATP is added to glucose.

② Glucose-6-phosphate rearranges to fructose-6- phosphate.

③ Another phosphate from a second ATP is added to form fructose-1,6-diphosphate.

④ Fructose-1,6-diphosphate is split to 2 three-carbon molecules, G-3-P and DHAP.

④a DHAP is immediately converted to G-3-P.

⑤ G-3-P is oxidized: It releases two electrons and an H$^+$ to reduce NAD$^+$ to NADH, and it transfers energy to a phosphate in the cytoplasm to form DPGA.

⑥ The unstable DPGA loses a phosphate group, which joins ADP to form ATP and 3-PGA.

⑦ 3-PGA is rearranged to 2-PGA.

⑧ A water molecule is removed from 2-PGA, leaving PEP.

⑨ The final ATP is formed as PEP gives up a phosphate group to ADP. The leftover three-carbon molecule is pyruvate, the end product of glycolysis.

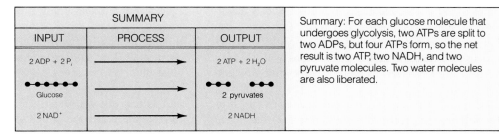

SUMMARY			Summary: For each glucose molecule that undergoes glycolysis, two ATPs are split to two ADPs, but four ATPs form, so the net result is two ATP, two NADH, and two pyruvate molecules. Two water molecules are also liberated.
INPUT	PROCESS	OUTPUT	
2 ADP + 2 P$_i$	→	2 ATP + 2 H$_2$O	
Glucose	→	2 pyruvates	
2 NAD$^+$	→	2 NADH	

FIGURE 5.8 ■ **Glycolysis in Detail: The Splitting of Glucose.** This diagram shows the main reactions of glycolysis, the first phase of energy metabolism in living things, and is a more detailed explanation of Figure 5.7. The first column shows the raw materials needed for the pathway, including the energy carriers ATP, ADP, and NAD$^+$ and inorganic phosphate (P$_i$); the second column shows how the initial six-carbon glucose molecule is split to 2 three-carbon molecules; the third column lists the products of the reactions; and the fourth column explains what happens in each step. The pathway is summarized at the bottom. The splitting of each glucose yields two ATP, two NADH, and two pyruvate molecules; in the process, two additional ATPs are first cleaved to ADP and then rephosphorylated.

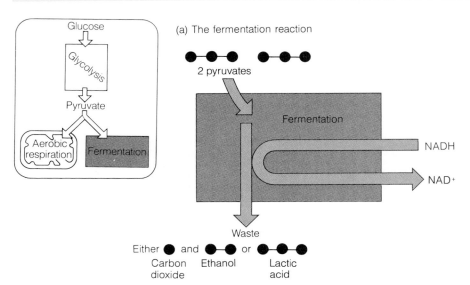

(a) The fermentation reaction

2 pyruvates

Fermentation

NADH

NAD⁺

Waste

Either ● and ●—● or ●—●—●
Carbon　　Ethanol　　Lactic
dioxide　　　　　　　acid

Glucose

Glycolysis

Pyruvate

Aerobic respiration

Fermentation

FIGURE 5.9 ■ Fermentation and Gourmet Food. (a) Fermentation recycles NADH, a product of glycolysis, to NAD⁺, a substance needed for glycolysis to continue. Fermentation also alters another product of glycolysis, pyruvate, to waste products. The waste molecules are different in different organisms. (b) A yeast cell growing under anaerobic conditions will carry out alcoholic fermentation. During this form of fermentation, enzymes split pyruvate molecules from the glycolysis pathway into ethanol and carbon dioxide. The carbon dioxide can then cause bread dough to rise. (c) In the absence of oxygen, certain bacteria and other microorganisms, as well as certain types of muscle cells, break down pyruvate from glycolysis into the three-carbon compound lactic acid. In a cheese factory, lactic-acid-generating bacteria help produce cheese.

(b) Baking and alcoholic fermentation

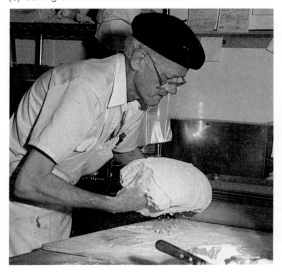

(c) Cheese making and lactic acid fermentation

Alcoholic Fermentation

In a few types of cells—the baker's and brewer's yeast called *Saccharomyces cerevisiae* being the most familiar—the three-carbon pyruvate produced during glycolysis is further metabolized in two reaction steps. These steps yield NAD⁺ (the oxidized form of NADH) as well as carbon dioxide and the two-carbon compound ethyl alcohol (also called ethanol), but no ATP (Figure 5.9a). NAD⁺ is necessary for energy production (Stage Three in glycolysis; see Figure 5.7) and hence for the continued life of the yeast cell. We capitalize on the carbon dioxide and alcohol that make possible our baking, brewing, and wine-making industries. Yeasts, by the way, are remarkable organisms, as Box 5.1 on page 116 describes.

For the yeast cell, the carbon dioxide and ethyl alcohol are simply metabolic waste products to be excreted. The significance of fermentation for the cell's continued survival lies in the regeneration of NAD⁺ from NADH. Glycolysis would

grind to a halt if NAD⁺ were not continuously available for Stage Three. Thus, the additional steps of alcoholic fermentation guarantee a supply of the critical energy carrier. For people, however, the waste products of yeast's alcoholic fermentation are highly useful. Bakers add *Saccharomyces cerevisiae* to flour and other ingredients, and once inside the anaerobic environment of a big wad of dough, the yeast begins to break down the amylose starch into glucose units and to ferment the glucose. The bubbles of excreted carbon dioxide gas cause the dough to expand, and the small amount of alcohol also released into the dough evaporates, filling the air with the appetizing aroma of baking (Figure 5.9b).

Yeasts occur naturally on grape skins, but vintners usually add special strains of *S. cerevisiae* to grape juice, seal the "must" (as the combination is called) into airless tanks for weeks, and allow the yeast to ferment the juice's natural "grape sugar," or glucose. The alcohol level eventually reaches 10 to 12 percent of the wine's volume—a level so high that most of the yeast is killed off. Brewers of beer or sake also employ *S. cerevisiae,* but they must carry out the preliminary steps of breaking down the starches in barley and rice to glucose, since the yeast cannot use these complex carbohydrates directly.

Producers of sauerkraut and pickled cucumbers, beets, tomatoes, and other such foods employ several kinds of fermenting yeasts and bacteria whose populations rise and fall in the pickling crocks as the fermentation process proceeds. Many of these organisms, however, produce lactic acid rather than alcohol as a by-product.

Lactic Acid Fermentation

Certain mammalian muscle cells, as well as the microorganisms that create yogurt, cheeses, and soy sauce, carry out lactic acid fermentation under anaerobic conditions. In these cells, the pyruvate generated during glycolysis is converted in a single reaction to lactic acid (see Figure 5.9a). (The term *lactic acid* refers to the chemical form excreted from the cell with an organic acid (COOH) group; see Chapter 2. In solution inside the cell, the form of the compound is usually lactate, with a COO^- group.) During the reaction, NADH is oxidized to NAD^+, and the pyruvate is reduced; the pyruvate accepts two electrons and a hydrogen ion from the carrier and becomes lactic acid. As in alcoholic fermentation, there is no additional ATP harvest, but the regeneration of NAD^+ allows glycolysis to proceed with its net energy gain. Like alcohol and carbon dioxide, lactic acid is a useless waste product to the anaerobic cell, but is highly useful to the dairy industry, imparting the sour taste of yogurt or the sharp taste of cheese to cultured milk or cream.

Probably the earliest fermented foods were sour milk products like yogurt, kefir, buttermilk, and sour cream, which resulted from the natural growth in milk and cream of fermenting bacteria such as *Lactobacillus bulgaricus* and *Streptococcus lactis.* Today, these sour foods, as well as most kinds of hard and soft cheeses, are produced through the deliberate culturing of lactic-acid-producing bacteria in milk or cream (Figure 5.9c). Soy sauce—the salt and pepper of oriental cooking—is made by inoculating soybeans with a fungus, then fermenting them with three kinds of lactic acid yeasts and bacteria for nearly a year.

Ironically, lactic acid can be a bane as well as a boon, because it sometimes forms inside animal muscle cells and leads to cramping. A vigorously exercising mammal, such as a hiker suddenly rushing up a steep slope to avoid a falling tree limb, may develop a leg cramp because the muscle cells are not getting sufficient oxygen. With oxygen in short supply, the pyruvate formed during glycolysis cannot be further broken down via aerobic respiration, fermentation begins in the muscle cells, and lactic acid is formed. If this waste product builds up, it can cause a painful leg cramp or a "stitch in the side" that dissipates only when the exercise lessens and the flow of oxygen to the tissues is once again sufficient to sustain aerobic respiration.

In a sense, lactic acid is a warning signal to slow down. But it can also remind us of the body's marvelous capacity to function even under adverse conditions. If a tired hiker, after avoiding the tree limb, were suddenly confronted by a rattlesnake in the middle of the trail, he or she could probably back off, then run to safety on the metabolic proceeds of glycolysis alone. One reason that exercise enthusiasts now emphasize *aerobic* exercise—exercise at a mild enough pace so that oxygen delivery outstrips oxygen use—is that it prevents muscle cramping. However, there are more important reasons we'll encounter later.

Anaerobic Metabolism: A Final Tally

The summary in Figure 5.10 shows that glycolysis releases two ATPs per molecule of glucose broken down, as well as the energy carrier NADH and two molecules of pyruvate.

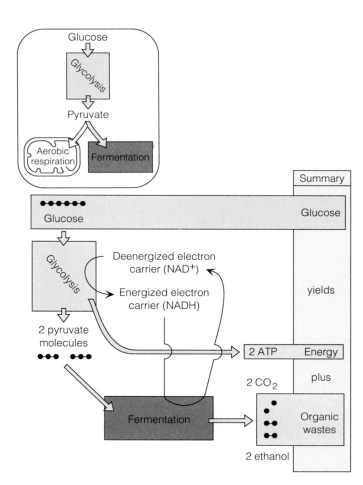

FIGURE 5.10 ■ Anaerobic Metabolism: A Final Energy Tally. Some cells can survive when oxygen is scarce. In those cells, the pyruvate molecules produced by glycolysis are shunted through the fermentation pathway, where they are broken down to organic wastes; in the process, the energy carrier NADH is recycled, but no new ATP forms. Thus, in anaerobic metabolism, fermentation enables glycolysis to continue its harvest of two ATP molecules per molecule of glucose.

FIGURE 5.11 ■ **Environmental Carbon Cycling: Anaerobic Respiration Plays a Key Role.** Hikers pausing for a rest by a clear mountain lake see only the sparkling upper layers of water, where photosynthesis in the tissues of algae and green plants fixes carbon and releases oxygen into the water. When water plants and oxygen-breathing fish die, their bodies sink to the lake bottom. These deep layers are colder and soon become depleted of oxygen, as oxygen-utilizing microorganisms begin to decompose the debris. Once the oxygen is fully depleted, the microorganisms switch to anaerobic metabolism and complete the decomposition of the organic matter to carbon dioxide and often methane gas, which hikers can sometimes see bubbling up to the surface. Without this anaerobic recycling, most of the carbon would be tied up in dead plants and animals.

During fermentation in muscle cells, yeast cells, and certain other kinds of microorganisms, the pyruvate molecules are degraded to wastes such as alcohol, carbon dioxide, and lactic acid, while the NADH is recycled to NAD^+. The energy yields of these two metabolic pathways may seem meager—just two ATPs per glucose. However, anaerobic metabolism (energy breakdown in the absence of oxygen) is crucial to the global recycling of carbon and the stability of the environment.

Organic matter from dead leaves, dead microorganisms, and other sources often sinks into an environment devoid of oxygen, such as the soft layers at the bottom of lakes or oceans (Figure 5.11). If it weren't for anaerobic decomposers—organisms capable of breaking down organic matter via anaerobic metabolism—most of the world's carbon would eventually be locked up in undecomposed organic material in these oxygen-poor environments, and deeper and deeper layers would build up. As a result, there would be too little carbon dioxide available as a raw material for photosynthesis, plants would be unable to generate new glucose molecules, and neither plants nor animals would survive.

Although simple and microscopic, anaerobic decompos-

ers clearly play an enormous role in the balance of nature. The vast majority of life forms, however, expire fairly quickly without oxygen. But their dependence on oxygen is based on a form of metabolism that provides an energy harvest 18 times greater than that provided by anaerobic metabolism.

AEROBIC RESPIRATION: THE BIG ENERGY HARVEST

While muscle cells as well as yeasts and other decomposers have the ability to metabolize sugars even when oxygen is unavailable (for a while, at least), the vast majority of living things, from euglenas to elephants, roses to redwoods, are made up of cells that require oxygen for their major metabolic pathway. Known as aerobic respiration, this pathway shunts the products of glycolysis through the Krebs cycle and harvests energy in the electron transport chain. Being 18 times more productive than glycolysis alone, this extended pathway can provide the huge amounts of ATP an active cell needs, and its superiority is clearly demonstrated: One out of every 500 yeast cells has a mutation, or change, in its DNA that prevents it from carrying out aerobic respiration even when oxygen is present. These so-called *petite* yeasts get by solely on glycolysis and fermentation, but they grow two to three times more slowly than their normal counterparts.

The reason for the greater efficiency of aerobic respiration is revealed by the overall equation for aerobic respiration:

$$C_6H_{12}O_6 + 6\,O_2 + 36\,ADP + 36\,P_i \rightarrow$$
$$6\,CO_2 + 6\,H_2O + 36\,ATP$$

The initial glucose molecule is broken down completely to inorganic waste products that are not energy-rich; thus, most of the energy residing in the molecular bonds of the sugar is released and a large proportion of it stored as ATP—36 ATPs per glucose, to be precise.

Aerobic respiration has two phases that use the products of glycolysis. The first phase is known as the Krebs cycle, named after the scientist who worked out its reactions, Sir Hans Krebs. The **Krebs cycle** of reactions cleaves the carbons from pyruvate, releases them as carbon dioxide, and stores energy in reduced carrier molecules. The second phase, the **electron transport chain,** then strips the electrons from the reduced carriers (Figure 5.12). The flow of electrons down this transport chain creates a current that is then used to build the cellular currency ATP.

The Krebs Cycle: Metabolic Clearinghouse

As in fermentation, the two pyruvate molecules produced from each glucose during glycolysis are the raw material for the Krebs cycle. Unlike fermentation, however, the important events of the Krebs cycle take place not in the cytoplasm but in the mitochondria of eukaryotic cells (or on the plasma membranes of bacteria). Recall from Chapter 3 that the mito-

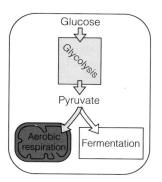

FIGURE 5.12 ■ **Aerobic Respiration: An Overview and Energy Tally.** Aerobic respiration breaks down the products of glycolysis via the Krebs cycle and the electron transport chain (a series of electron carriers embedded in the mitochondrial membrane). For every glucose molecule broken down through aerobic respiration, two ATP molecules are harvested during the glycolysis phase, and two more ATPs are gained during the Krebs cycle. The Krebs cycle also produces a set of reduced carrier molecules ($FADH_2$ and NADH), and these are oxidized in the electron transport chain with the accumulation of 32 more ATP molecules. (In some cells, notably those of the liver and heart, the ATP yield can be slightly higher—34 ATPs per glucose.) Altogether, during aerobic respiration, 36 ATPs are harvested per glucose molecule, 6 carbon dioxide and 6 water molecules are released as inorganic waste products (12 water molecules are produced, but 6 of these are used up for a net release of 6), and 6 oxygen molecules are used up.

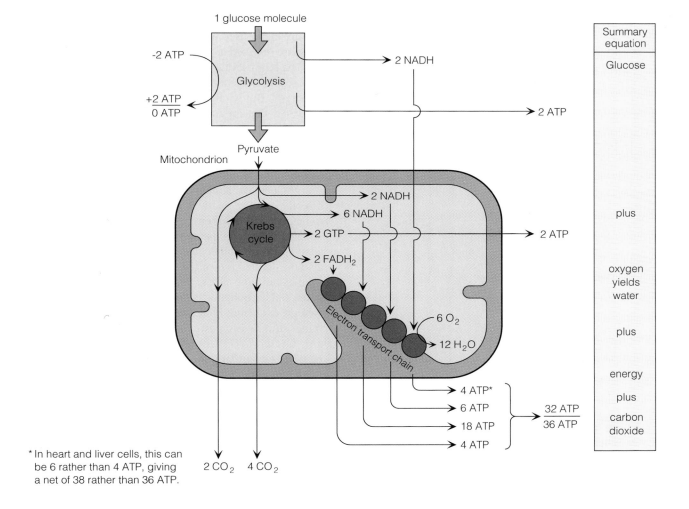

* In heart and liver cells, this can be 6 rather than 4 ATP, giving a net of 38 rather than 36 ATP.

chondrion has an outer membrane surrounding the outer compartment, as well as a convoluted inner membrane surrounding the inner matrix (Figure 5.13). The first, or preparatory, stage leading to the actual cycle of reactions takes place in the mitochondrial matrix, after the pyruvate molecules produced by glycolysis have been transported into the mitochondrion from the cytoplasm. Of the three carbons in each pyruvate, one is oxidized to carbon dioxide, and in the process, energy is stored in one molecule of NADH (Figure 5.13, Stage One).

The two leftover carbons are then shuttled into the Krebs cycle by a molecule called coenzyme A (Figure 5.14). These two carbons are next added to a four-carbon molecule (Stage Two), and from this six-carbon molecule, two carbons are stripped off one at a time and released as carbon dioxide molecules (see Figure 5.13, Stage Three, and for detail, Figure 5.14). As the reshuffling of carbons proceeds, electrons and hydrogen ions are transferred to the carriers NAD^+ and FAD, and some ATP eventually forms. Finally, the original four-

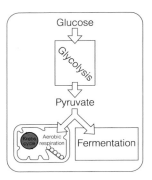

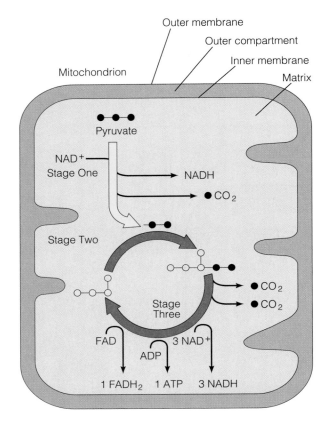

FIGURE 5.13 ■ The Source of Exhaled Carbon Dioxide: The Krebs Cycle Simplified. When an animal exhales, it is breathing out the waste gas carbon dioxide. That gas arises from the complete dismantling of the six-carbon skeleton of glucose. While the six carbons of glucose are split into 2 three-carbon pyruvate molecules during glycolysis in the cytoplasm, aerobic respiration takes place in the mitochondrion. A preliminary reaction in the mitochondrion removes one carbon (Stage One) and the reactions in the Krebs cycle itself remove two more carbons (Stage Two), giving three CO_2 molecules (each containing one carbon) for each of the two pyruvates. The total, then, is six CO_2 molecules per initial glucose molecule. In the Krebs cycle, enzymes add two carbons to a four-carbon compound and then reorganize the resulting six-carbon molecule, removing two carbons one at a time. This regenerates the four-carbon molecule—hence the cyclic nature of the Krebs cycle (Stage Three). In completing the cycle, five reduced electron carriers and one ATP are produced for each molecule of pyruvate. A more detailed version of the Krebs cycle is shown in Figure 5.14.

carbon compound is regenerated as the cycle completes one turn. In total, for each pair of pyruvate molecules and each two turns of the cycle, two ATP molecules and ten charged carriers form (eight of NADH and two of $FADH_2$) and four carbon dioxide molecules (representing the rest of the carbon atoms in the original glucose) are released.

The ATP harvest from the initial glucose molecule is therefore no greater during the Krebs cycle than during glycolysis. However, the ten reduced electron carriers represent a treasure trove of stored energy that will be released, bit by bit, by the components of the electron transport chain. You can follow the fate of the carbons during the Krebs cycle in Figure 5.14. The six carbon dioxide molecules (from two turns of the cycle) eventually diffuse out of the mitochondrion and are either used by the cell for photosynthesis (see Chapter 6) or are released as a waste product.

The Krebs cycle is an important intermediate phase in aerobic respiration for the dismantling of carbons and the storing of energy. But what makes it a metabolic clearinghouse? The answer is that organic nutrients other than glucose can enter the aerobic breakdown pathway at the Krebs cycle (Figure 5.15). If the supply of sugar falls, an animal or plant cell must begin to break lipids or proteins down into component parts. Some of these subunits can then be converted into pyruvate, acetyl-CoA, or Krebs cycle intermediates and be shunted directly into the pathway at the appropriate stage, eventually to be dismantled for energy harvest.

The clearinghouse concept extends still further, because acetyl-CoA and other raw materials from the Krebs cycle can also be shunted *out* of the pathway, to be converted and then used as building blocks for the biosynthesis of new fats, proteins, and carbohydrates. The Krebs cycle is, in a sense, like a foreign exchange bank: Just as one can change dollars into francs, francs into pesos, or pesos into pounds, so can the cell turn fats into ATP energy or turn the components of sugar metabolism into amino acids by means of the Krebs cycle.

An important exception to this clearinghouse principle is the human brain cell. For complex reasons, it can use only glucose as fuel, a fact that has major implications. First, it explains why sugary foods are mood elevators. Soon after a person consumes a candy bar or soft drink, glucose subunits are cleaved from the sucrose and enter the bloodstream and brain. Infused with their favorite fuel, the brain cells can function at peak efficiency, leading one to feel happier, smarter, and livelier—at least for a time. Second, this quirk of the brain cell explains why dieters are warned to consume at the very least 500 calories per day and to avoid liquid or powdered protein diets: The brain alone needs at least that many calories of glucose for normal functioning, and without it, a person can grow faint and even lapse into unconsciousness.

The Electron Transport Chain: An Energy Bucket Brigade

The final stage of aerobic respiration is really the crux: Here is where most of the energy of glucose is turned into ATP; where oxygen comes into play, making this respiration *aero-*

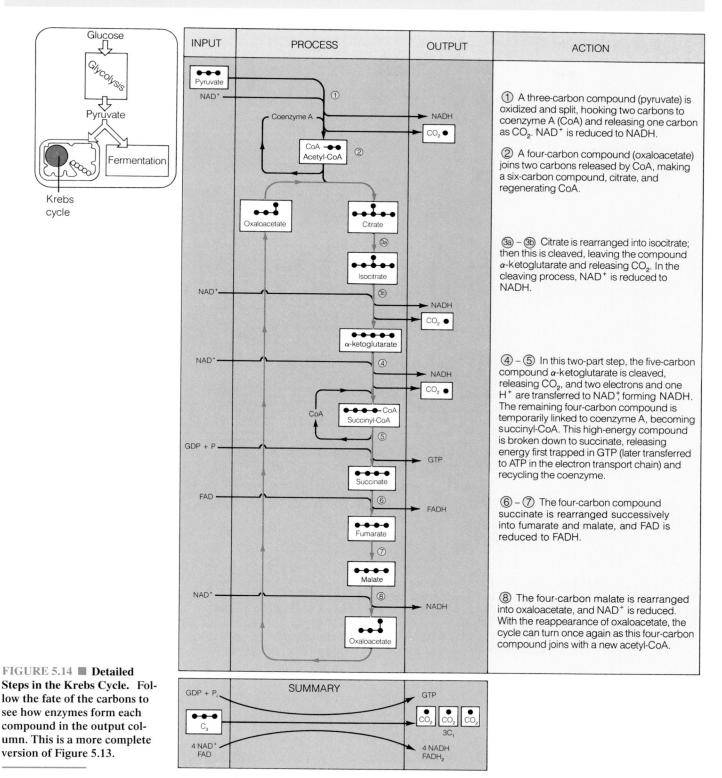

FIGURE 5.14 ■ **Detailed Steps in the Krebs Cycle.** Follow the fate of the carbons to see how enzymes form each compound in the output column. This is a more complete version of Figure 5.13.

bic; and where mitochondrial structure is so crucial to the survival of eukaryotes, which constitute the vast majority of living species.

The electron transport chain is both a series of structures and a series of events that are intertwined. To follow these structures and events, we must return to the detailed architecture of the mitochondrion. Recall that the inner mitochondrial membrane separates two fluid-filled compartments within the organelle (the inner compartment, or matrix, and the intermembrane space or outer compartment) and that the Krebs

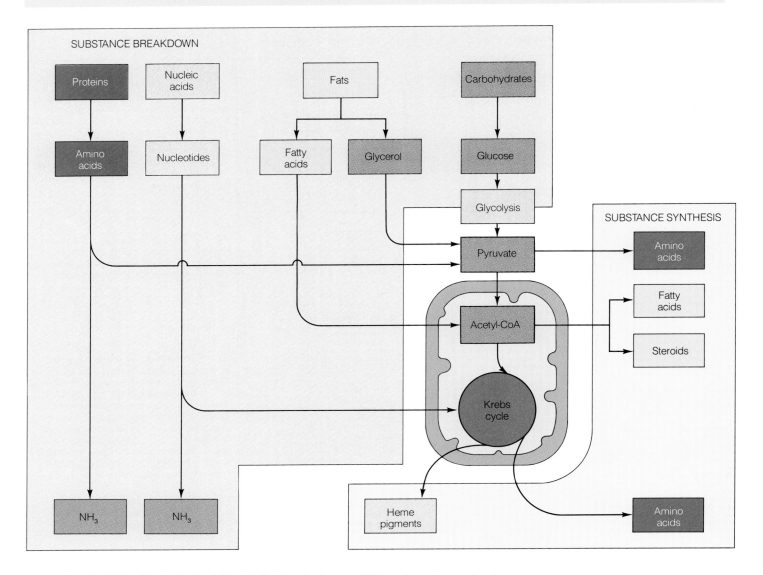

FIGURE 5.15 ■ The Krebs Cycle: A Metabolic Clearinghouse. While only certain molecules can feed directly into the Krebs cycle, biological polymers—proteins, nucleic acids, fats, and polysaccharides—can themselves be broken down into constituent parts, and these can be modified into intermediates that feed into the cycle. In this way, an organism can harvest energy not just from glucose but from any of the biological polymers. In addition, Krebs cycle intermediates can be removed from the cycle and modified into new materials for the cell, including amino acids, fatty acids, steroids, and iron-containing heme pigments.

cycle occurs within the inner compartment (Figure 5.16). Embedded within the inner membrane, however, are sets of complex proteins that protrude from and sometimes span the entire membrane. These proteins are capable of transporting the electrons that were loaded onto the carriers NADH and FADH₂ during the Krebs cycle.

The important thing about these proteins is their ability to oxidize or reduce other molecules. Electrons stored in NADH or FADH₂ during the Krebs cycle can thus be passed from one electron transport protein to the next in a sort of downward energy flow—an electron bucket brigade—during which

small amounts of energy are released with each downward step. This energy is then trapped as ATP.

The events of the electron transport chain represent a series of oxidation-reduction reactions in which the reduced carriers NADH and FADH₂ become oxidized; that is, they lose electrons and protons (hydrogen ions). At the same time, oxygen atoms gain the electrons and hydrogen ions and become reduced to water (H_2O). Oxygen thus serves as the final acceptor of electrons and hydrogen ions for the entire aerobic respiration pathway, explaining why it is called "aerobic." If oxygen cannot accept the electrons and protons once

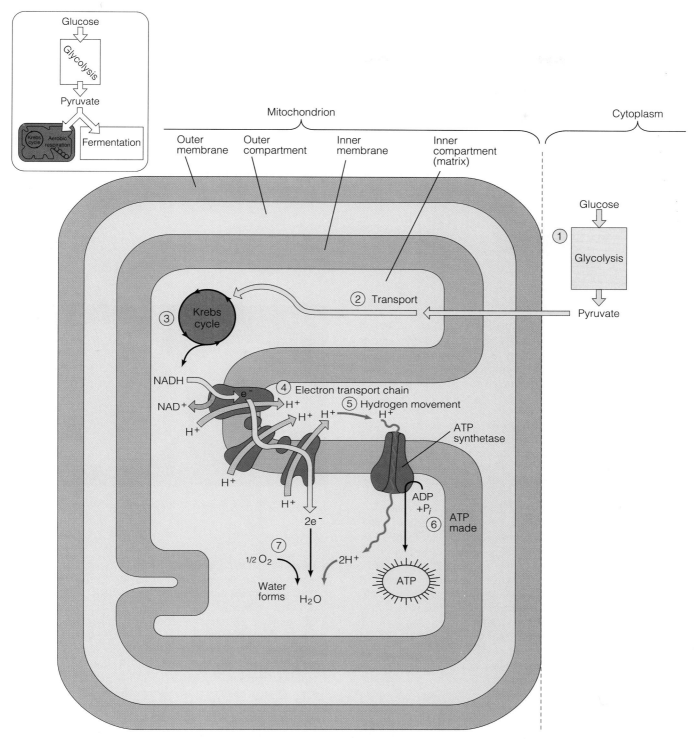

FIGURE 5.16 ■ Mitochondrion: Site of the Krebs Cycle and Electron Transport Chain. This diagram shows where the phases of aerobic respiration take place in the mitochondrion. Pyruvate molecules (1) generated during glycolysis in the cytoplasm are transported through the outer and inner mitochondrial membranes to the inner compartment (matrix) (2), where the Krebs cycle takes place (3). Then electrons from energy carriers are passed down the electron transport chain (4), and proteins use this released energy to pump hydrogen ions across the inner membrane (5). Finally, these hydrogen ions flow through the protein ATP synthetase back across the inner mitochondrial membrane, and the protein traps energy in ATP (6) much as a turbine captures the energy of water flowing over a dam and converts it to electricity. The final acceptor of the electrons is oxygen (7), which joins hydrogen ions and electrons and forms water.

they have passed down the electron transport chain, the entire process quickly stops, and the organism or cell, if strictly aerobic, dies. That's how cyanide kills living things. The C≡N chemical group (carbon bonded to nitrogen) in sodium cyanide binds tightly to a pigment molecule called cytochrome a_3 in the electron transport chain and prevents it from accepting the final electron transfer. The result is that ATP formation halts, and the cell quickly starves for energy and dies.

We can follow the transport of electrons down the chain of proteins embedded in the mitochondrial membrane as well as the link of this transport to the formation of ATP. First, recall that the raw material of the Krebs cycle, pyruvate, is generated during glycolysis, which splits glucose into pyruvate in the cell's cytoplasm (Figure 5.16, step 1). The pyruvate diffuses into the mitochondrion (step 2) and is dismantled during the Krebs cycle, with hydrogens and high-energy electrons being transferred to electron carriers (step 3). These high-energy NADH or $FADH_2$ molecules from the Krebs cycle move randomly about in the mitochondrial inner compartment and eventually bump into and react with one of the first two transport proteins (step 4). A series of oxidation-reduction reactions takes place, so that with each consecutive

---

BOX 5.1   H O W   D O   W E   K N O W ?

A YEASTY SUBJECT FOR RESEARCH

Grocery stores sell familiar-looking red and yellow packets in the baking section that contain a remarkable living species called *Saccharomyces cerevisiae*—a cell so small 2000 could line up end to end across your thumbnail (Figure 1). The organism is, in fact, the ingredient known as dry yeast or baker's yeast, and it is responsible for our bread, wine, and beer. But it is also much more than that. Yeasts have a marvelous metabolic flexibility that allows them to break down organic molecules via fermentation if oxygen is absent and via aerobic respiration if oxygen is present. Moreover, they are one of the hottest research subjects in modern biology and are helping biologists understand topics as different as cell shape, Down syndrome, and cancer.

Baker's yeast is a member of the kingdom Fungi, and like many other yeast species, it grows best on the surfaces of fruits, vegetables, and grains. It will, however, thrive in liquids, soil, food materials, rotting vegetation, and anywhere else sugars and carbohydrates can be broken down for energy and materials. When oxygen is absent, yeasts ferment the energy sources and release CO_2 gas and ethyl alcohol as by-products. The gas bubbles carbonate beer and champagne and cause bread to rise, and the alcohol makes beverages intoxicating.

Because yeasts are eukaryotic cells that are easy to grow and manipulate in the lab, they have become a common subject for molecular biologists. In fact, research interest in yeast has grown so exponentially in recent years that if the upsurge continued at this pace, not just scientists but everybody alive would be studying yeasts by 1995!

In the years since researchers invented genetic engineering techniques, yeast geneticists have succeeded in modifying the cells in instructive ways.

■ They have learned to add specific genes in specific places that will give the cells new traits. This technique, for example, has already enabled a California biotechnology firm to create a synthetic vaccine for the serious liver disease hepatitis B.

■ By exploring yeast's actin and tubulin proteins (see Chapter 3) and the genes that encode them, scientists have revealed interesting details about the cytoskeleton of eukaryotic cells and much about cell shape and cell division.

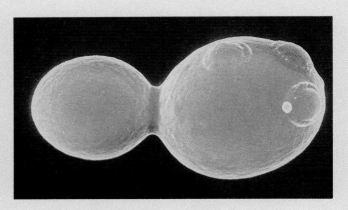

FIGURE 1   ■   **The Yeast *Saccharomyces cerevisiae*.**

■ Researchers have learned to create artificial yeast chromosomes, and by observing their behavior, learned more about the chromosomal mistakes that cause genetic conditions such as Down syndrome.

■ Certain yeast genes are switched on at some times and off at others. Studies of this switching are helping biologists to understand how specialized cells in higher organisms—muscle cells, brain cells, and nerve cells, for instance—can contain the same set of genes in each nucleus, yet still express very different characteristics, depending on cell type.

■ Finally, cancer researchers have found a gene called *ras* in yeast that is remarkably similar to the human *ras* gene implicated in lung, colon, breast, and bladder tumors. Basic studies have already revealed certain activities of the yeast gene that may help scientists learn how to turn off the *ras* gene in a human tumor.

The yeast cells we buy in the supermarket today are probably very much like the bumpy, budding organisms that ancient Egyptians tossed into their vats of grape juice (Figure 1). Modern techniques for probing and modifying them, however, are certain to generate new "designer yeasts" as well as a new understanding of these ubiquitous eukaryotes.

transfer of electrons from one protein to the next, a bit more energy is released.

Membrane proteins use the energy released by this flow of electrons to pump hydrogen ions from the inner compartment to the outer compartment of the mitochondrion (step 5). The accumulation of hydrogen ions in the mitochondrion's outer compartment represents potential energy, like water stored behind a dam: As the hydrogen ions flow back across the membrane into the inner compartment through special channel proteins, they give up some of their energy. This is the key event, because the energy released by the flow of ions down this *hydrogen ion concentration gradient* drives the phosphorylation of ADP to ATP, much as the flow of water over a dam turns a turbine that collects energy in the form of electricity (step 6). This phenomenon has been called **chemiosmotic coupling,** the coupling of energy formation to chemical and osmotic events, but biologists do not yet understand the exact mechanisms that bring it off. They do, however, know the bottom line for energy formation in most aerobic cells: The electron transport chain takes electrons from the energy carriers generated in glycolysis and the Krebs cycle, uses the electrons' energy to produce ATP molecules, and finally transfers the electrons and hydrogen ions to oxygen, thus producing water (step 7). For each set of energy carriers (NADH and FADH$_2$) that comes from a single glucose molecule, 32 ATPs can form during the electron transport phase— a whopping harvest of nearly 90 percent of the ATP currency "earned" during the breakdown of a glucose molecule via aerobic respiration.

The final tally for the entire three-part aerobic pathway is 36 ATPs per glucose molecule, as itemized in Figure 5.12. Despite the uncertainties that remain, biologists are quite sure that the unique internal structure of the mitochondrion makes possible the high efficiency of aerobic respiration in eukary-

otes. One bit of evidence for this is that aerobic bacteria lacking mitochondria can still carry out aerobic respiration via transport proteins embedded in their plasma membranes; however, their ATP harvest is generally less than half that of eukaryotic cells. We active multicellular organisms need huge amounts of ATP energy, and our billions of mitochondria provide it. Any impairment of mitochondrial function can have severe consequences for health and vitality, as Box 5.2 on page 119 explains.

THE CONTROL OF METABOLISM

We have seen that cells possess marvelous metabolic mechanisms—biochemical pathways that accomplish three major processes in the cell: breaking down organic nutrients; harvesting ATP energy; and building needed proteins, lipids, carbohydrates, or other biological molecules by shunting appropriate biochemical subunits into and out of the Krebs cycle clearinghouse. But the intricacies of metabolism raise new questions: What makes a cell begin burning its stores of lipids or proteins when glucose runs out? When the supplies of glucose return, what then stops the cell from burning all its lipids or proteins—literally eating itself up from the inside and thereby destroying its cellular structures? Finally, what triggers a cell to build molecules only when they are needed (see Figure 5.17)? Clearly, there must be a great deal of internal coordination and control over the cell's harvesting of energy and building of needed materials. But what form does it take?

We can create a simple analogy for the way that glycolysis and the breakdown of glucose is controlled in the cell. Imagine a shoe factory in which shoes pile up so high that they topple over and stop the assembly line; it is only after the

(a) End product accumulates

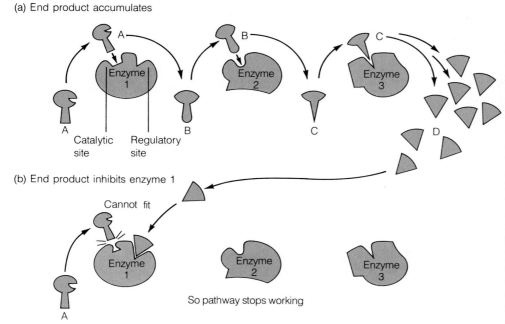

(b) End product inhibits enzyme 1

So pathway stops working

FIGURE 5.17 ■ How Cells Regulate Energy Metabolism. The accumulation of the end product of energy metabolism—ATP—can feed back and inhibit the activity of an enzyme early in the pathway of glycolysis and hence inhibit the entire pathway. (a) The pathway's end product (triangular molecules representing ATP) accumulates and (b) fills the regulatory site of enzyme 1. This alters the shape of the enzyme, thus inactivating it and turning off production of the pathway leading to more ATP. With glycolysis temporarily halted, no pyruvate is made, there is thus no fuel for the Krebs cycle, and aerobic respiration comes to a halt. When the excess ATP is used up, the regulatory site on enzyme 1 opens up, the enzyme is once more active, and the metabolic energy harvest can once again commence.

shoes are removed and shipped out that the line is free to start up again and make more shoes. Likewise, such high levels of ATP can build up in a cell that the cell requires no more of the molecular fuel. When this occurs, the ATP binds to a special regulatory site on an enzyme called phosphofructokinase (PFK) that catalyzes part of step 1 in glycolysis (review Figure 5.8). This binding shuts down enzyme activity. This, in turn, switches off the entire glycolytic pathway: If step 1 does

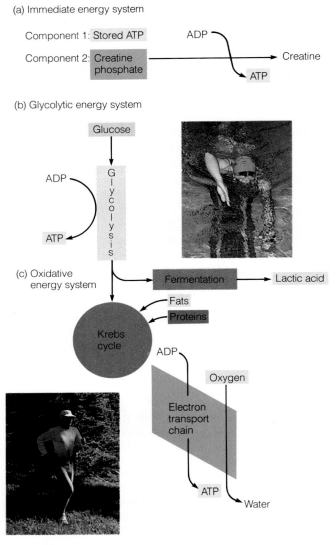

FIGURE 5.18 ■ Three Energy Systems for Fueling Muscular Motion. (a) The immediate energy system, based on ATP and creatine phosphate stored in muscle cells, can fuel a brief explosive activity. (b) The glycolytic energy system, based on glycolysis (anaerobic breakdown) of glucose in the muscle cells, can fuel short-duration, high-intensity activities, such as a 200 m swim. (c) The oxidative energy system, based on cellular respiration (aerobic breakdown) of glucose, glycogen, fatty acids, or amino acids stored elsewhere in the body and carried to the muscles through the bloodstream, can fuel long, sustained activities, such as a 50 km run in the mountains.

not function, there is no substrate for step 2, and so on. The regulatory advantages of this system are obvious: When the cell already has high levels of ATP, that very molecule can act as a control agent to turn off its own production. Then, when ATP levels drop, the excess ATP is pulled from the enzyme "machinery," and the glycolysis "assembly line" can once more resume. The buildup of a metabolic product and its inhibitory effect on a special enzyme such as PFK is called **feedback inhibition,** since the accumulation of the product in effect feeds back and turns off its own production (Figure 5.17). Feedback inhibition works because the special enzyme has two binding sites, the catalytic (or catalytically active) site and the regulatory (or *allosteric*) site. When a molecule—usually the end product of the pathway the enzyme is part of—binds to the enzyme's regulatory site, the enzyme changes shape, and its catalytic site stops binding the normal substrate; in other words, the enzyme's activity is turned off.

Through feedback inhibition and other forms of control, the cell's metabolic enzymes are turned on or off so that the cell burns glucose when it is available and burns lipids or proteins when glucose is lacking, and builds the appropriate biological raw materials just when they are needed for growth or maintenance activities. The control mechanisms within cells are truly elegant—efficient yet highly conservative—and occur in the smallest bacteria, the large muscle cells in a hiker's leg, and even the baker's yeast sealed in packets on the grocer's shelf. And the similarity of control mechanisms reminds us that all types of cells are related in a fundamental way by their metabolism. Such identical forms of control and regulation simply ensure that order rather than metabolic chaos reigns within the organism.

THE ENERGY SOURCE FOR EXERCISE

Like any other aerobic organism, an exercising person relies on the energy-harvesting pathways we discussed in this chapter. This is true whether the energy is needed for the powerful lunge of a center guard in football, for one leg of a relay event, which a swimmer will quickly churn through, or for a hiker's long steady climb up a mountain trail. Energy for different forms of exercise, however, comes from different parts of the metabolic pathway.

Exercise physiologists have determined that there are three energy systems—the *immediate system,* the *glycolytic system,* and the *oxidative system*—that supply energy to a person's muscles during exercise. The duration of physical activity dictates which system the body uses. The immediate energy system is instantly available for a brief explosive action, such as one football tackle, one bench press, or one put of the shot, and the system has two components (Figure 5.18a). One component is the small amount of ATP stored in muscle cells, immediately useful like the few coins you carry around in your purse or pocket. This stored ATP, however,

BOX 5.2

PERSONAL IMPACT

SARAH'S MITOCHONDRIA

One day in 1974, while walking in Manhattan with her mother, eight-year-old Sarah T. (not real name) suddenly felt like her legs were being pricked by tiny needles. The symptoms continued and worsened, and over the next few months and years, Sarah could no longer walk a city block without her muscles burning and her heart racing. At age 16, she was attending her high school classes in Washington state in a motorized wheelchair. Sarah's intensifying fatigue and burning muscles baffled her doctors; the only clue to the problem was a high level of acidity in her blood. After intense study, a team of medical geneticists, biochemists, and physicians working at universities in Oregon and Pennsylvania finally recognized that Sarah's exhaustion was not based on problems with her lungs or heart, but rather with the power supply to her muscles.

Like everyone else's muscle cells (Figure 1a), Sarah's generated lactic acid during physical activity. Hers, however (Figure 1b), made as much while strolling slowly or taking a shower as other people's do during strenuous exercise. Researchers deduced that Sarah's problem lay in her system of energy production—either in glycolysis or fermentation, or in the Krebs cycle and the electron transport system. But which of these was altered? And how could the researchers help her?

Experiments ruled out one respiratory pathway after another until biochemists traced Sarah's problem to her mitochondria. Researchers discovered that because of an altered protein in the electron transport chain (embedded in the mitochondrial membranes), her muscle cells could not transfer electrons to oxygen. Without this electron transfer, the electron transport chain—the muscles' major energy provider—could not make ATP. Instead, Sarah's muscles generated huge numbers of mitochondria, but still had to rely solely on the small ATP harvest provided by glycolysis and the fermentation that accompanies it. As we saw, however, fermentation in muscle cells creates lactic acid, and this was causing Sarah's blood acidity levels to climb so high that she occasionally blacked out.

Having identified the cause of her tremendous fatigue, researchers could look for a solution, a substitute acceptor that would allow electrons to cross the gap in the electron transport chain created by the abnormal protein in her mitochondrial membranes. That substitute electron acceptor turned out to be a simple mixture of vitamin C and a derivative of vitamin K.

Sarah remembers her first injection of the vitamins. "I was feeling dead all the time, then after taking the vitamins I felt so good I said, 'Hey, Mom, let's jog.' I was that much better, instantly." By age 20, she no longer used her wheelchair and had entered college. Sarah is not cured of her metabolic defect—the abnormal electron acceptor still fails to work in her mitochondria. But because a team of researchers understood cellular respiration, they could determine her problem and supply the appropriate substitute acceptor, so she can now generate much less lactic acid and enough ATP to lead a normal life. [Quote from *Old Oregon*, Fall 1986; article by Lisa Cohn.]

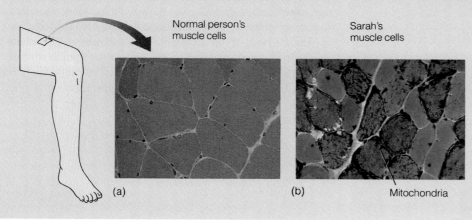

Normal person's muscle cells

Sarah's muscle cells

(a)

(b)

Mitochondria

FIGURE 1 ■ **(a) Normal muscle fibers. (b) Sarah's muscle fibers with masses of aberrant mitochondria (red) circling the fibers.**

runs out after only half a second—barely enough time to heave a shot or return a tennis serve, let alone run to catch a bus. The second component of the immediate energy system is a high-energy compound called *creatine phosphate,* which muscle cells store in larger amounts than ATP. Creatine phosphate is more like a few dollar bills than like the coins you carry. When the ATP in a muscle cell is depleted, creatine phosphate transfers an energetic phosphate to ADP; this regenerates ATP, and ATP can then fuel the muscle cell to contract and move the body. Even the cell's store of creatine phosphate, however, becomes depleted after only about a minute of strenuous work; thus, muscles must rely on more robust systems to power longer-term activities.

The glycolytic energy system, which depends on the glucose splitting of glycolysis in the muscles, fuels activities lasting from about 1 to 3 minutes, such as an 800 m run or a 200 m swim (Figure 5.18b). This storage form is more like an account at the bank than like dollars in your wallet, and it "purchases" more activity. Glycolysis in the muscle cell cytoplasm can cleave glucose in the absence of oxygen and generate a few ATPs. Fermentation takes the product of glycolysis a step further and makes the waste product lactic acid. Recall that glycolysis produces only two ATP molecules for each molecule of glucose. This is why the glycolytic energy system can sustain heavy exercise for only about 3 minutes and why lactic acid begins to build up in muscles by then.

Activities lasting longer than about 3 minutes—a jog around the neighborhood, an aerobic dance session, or a long uphill hike—require considerably more oxygen and employ the oxidative energy system (Figure 5.18c). This system can supply energy for activity of moderate intensity and long duration and is like money invested in stocks and bonds that give long-term steady income. The oxidative energy system is based on cellular respiration and includes the Krebs cycle and oxidative phosphorylation via the electron transport chain. It uses oxygen as the final electron acceptor and generates many ATP molecules per glucose molecule burned. Once an exercise session has ended, ATPs start to build up again. The form of metabolic regulation called feedback inhibition takes over, slowing glycolysis and with it the rest of the aerobic respiration pathway.

While the glycolytic energy system can make ATP only from glucose or glycogen (cleaved to release glucose), the oxidative energy system can produce energy by breaking down carbohydrates, fatty acids, and amino acids mobilized from other parts of the body and transported to the muscle cells by way of the bloodstream (review Figure 5.15). Clearly, anyone interested in melting away body fat should engage in aerobic (oxygen-utilizing) activities like strenuous walking, jogging, swimming, or bicycling, which primarily rely on the oxidative energy supply system and its ability to use fats as fuel.

CONNECTIONS

Energy metabolism lies at the heart of biology, and there are several lessons in its detailed consideration. Whether or not you remember the equations and reactions of glycolysis, fermentation, and aerobic respiration, you can easily recall that the pathways are complex and that they proceed in a stepwise fashion. Each step is mediated by a specific, often controllable enzyme, and the release of energy is gradual, rather than occurring in a single exergonic burst that would incinerate the cell. These stepwise reactions go on every second in every cell and break down the orderly arrangement of carbons in nutrient molecules and release them in the less-ordered form of carbon dioxide, thus increasing entropy. In the process, however, some of the released energy is stored as ATP that will buy the cell a temporary respite from its own tendency toward disorder and increased entropy by allowing for maintenance, growth, and reproduction.

We can also see why we owe such a debt of gratitude to mitochondria. Without them and their unique inner structure, large multicellular organisms like people and apple trees, with their vast energy needs, could not survive and would never have evolved in the first place.

Finally, the concepts of energy metabolism help us understand some of the other subjects we study in biology, like cell biology, genetics, and physiology. We learned in Chapter 3 that most cells must take in oxygen, give off carbon dioxide, and use up organic nutrients, and the fundamental reasons are now apparent. Oxygen is the electron acceptor for the electron transport chain, making possible the harvest of almost 90 percent of the aerobic cell's ATP. Without it, metabolism quickly ceases and the aerobic cell has only seconds or minutes to live on its ATP reserves.

What's more, our lungs act like bellows to draw in oxygen and expel carbon dioxide, and the stomach and intestines exist to extract glucose, fatty acids, and proteins from foods, making these raw materials available for the Krebs cycle and other metabolic sequences. These facts reflect the simple truth that we human beings are massive assemblages of individual cells, each with nearly identical metabolic requirements and activities, serviced by physiological systems that keep those metabolic fires burning.

Highlights in Review

1. A few kinds of cells, including baker's yeast and the muscles of a human hiker, can use the major pathways for catabolizing (breaking down) organic nutrients, called fermentation and aerobic respiration. Each of these sequences of reactions in turn rests on the fundamental pathway called glycolysis.

2. ATP links metabolic energy exchanges in the cell because of the energy stored in the bonds of its three phosphate groups. The cleaving of one or two phosphates releases a considerable amount of the energy stored in the whole ATP molecule.

3. ATP can transfer phosphates to other molecules, where they act like tiny battery packs that enable the newly charged molecule to participate in otherwise energetically unfavorable reactions.

4. Energy transfer in the cell involves a flow of electrons like an electric current. This transfer of electrons requires a simultaneous set of oxidation-reduction reactions in which electrons are moved from one substance (which thereby becomes oxidized) while they are gained by another substance (which becomes reduced). The high-energy electron carriers NADH and $FADH_2$ transfer electrons and store and release energy in this way.

5. Glycolysis means "glucose splitting," and during this pathway, a six-carbon glucose molecule is split into 2 three-carbon pyruvate molecules with the concurrent formation of two ATP molecules. Pyruvate can be further processed via fermentation if oxygen is absent or aerobic respiration if oxygen is present.

6. Glycolysis has nine reaction steps, during which glucose is first phosphorylated twice, then broken into two molecules of the charged three-carbon compound G-3-P. Enzymes then re-

arrange G-3-P repeatedly, and as energy is released, one NADH and a net of two ATPs form per initial glucose molecule. The three-carbon end product of glycolysis is pyruvate.

7. During alcoholic fermentation, which takes places under anaerobic conditions, pyruvate is further catabolized into the waste products ethyl alcohol and carbon dioxide. While no additional ATP forms, NADH generated during glycolysis is oxidized to NAD^+, which can then be reused. The production of bread, wine, and beer all depend on fermentation by yeast and on the waste products given off during the process.

8. During lactic acid fermentation, cells convert pyruvate to the organic waste product lactic acid. Again, no ATP is harvested, but NADH is oxidized to NAD^+, which is recycled in glycolysis. Sour milk products, most cheeses, and soy sauce are all produced by lactic acid yeasts and bacteria. Muscle cells can also ferment glucose to lactic acid for short time periods if deprived of oxygen, but the buildup of lactic acid leads to cramping and pain.

9. During aerobic respiration, each glucose molecule is fully oxidized to the inorganic waste products carbon dioxide and water, and a large harvest of ATP takes place.

10. The Krebs cycle is a set of reaction steps that act on pyruvate, stripping away the carbons and releasing them as carbon dioxide. In the process, two ATP molecules form, and a considerable amount of released energy is stored as the high-energy carriers NADH and $FADH_2$. The Krebs cycle is a metabolic clearinghouse: Breakdown products of proteins and lipids can enter aerobic respiration at this stage, and such raw materials can also be shifted out from the Krebs cycle for synthesis of needed biological molecules.

11. The electron transport chain consists of electron transport proteins embedded in the inner mitochondrial membrane. Some of these proteins strip electrons from the carrier molecules NADH and $FADH_2$ and pass the electrons along to other proteins in a series of oxidation-reduction reactions, releasing energy bit by bit. This energy fuels the pumping of hydrogen ions from the mitochondrion's inner compartment to its outer compartment. When the hydrogen ions flow back across the inner mitochondrial membrane to the inner compartment through a special protein, they drive the synthesis of ATP by chemiosmotic coupling.

12. The final tally for aerobic respiration in most eukaryotic cells is 36 ATPs per initial glucose molecule: 2 ATPs generated during glycolysis, 2 ATPs during the Krebs cycle, and 32 during the electron transport chain.

13. The appropriate use of organic nutrients and the timely production of biological molecules by the cell requires metabolic controls. The feedback inhibition of allosteric enzymes such as phosphofructokinase is one such control mechanism.

14. The immediate system, the glycolytic system, and the oxidative system provide energy to muscles for short-term (1 second), intermediate (1 to 3 minutes), or long-term (several minutes to hours) muscular activities.

Key Terms

aerobic, 101
aerobic respiration, 101
anaerobic, 101
autotroph, 104
chemiosmotic coupling, 117
electron transport chain, 110
feedback inhibition, 118

fermentation, 106
glycolysis, 104
heterotroph, 104
Krebs cycle, 110
oxidation-reduction reaction, 103
phosphorylation, 102

Study Questions

Review What You Have Learned

1. Prepare a table that compares fermentation with aerobic respiration. Use these headings: presence or absence of oxygen; amount of ATP produced; waste products.
2. State the basic difference between a heterotroph and an autotroph, and give an example of each.
3. What are two results of glycolysis?
4. How are alcoholic fermentation and lactic acid fermentation similar? How are they different?
5. Why is aerobic respiration much more efficient than fermentation?
6. Describe what happens to a pyruvate molecule during the Krebs cycle.
7. How is the Krebs cycle a metabolic clearinghouse?
8. The electron transport chain provides the final step in aerobic respiration. How does the electron transport chain operate?
9. What role does each of the following play in the electron transport chain? a mitochondrion; transport proteins; oxygen; NADH; FADH.
10. Explain feedback inhibition.

Apply What You Have Learned

1. Why do vintners seal the "must" in airless tanks even though yeasts are capable of aerobic respiration? What would happen if they were to bubble air through the must?
2. Explain how daily swimming helps some people lose weight.
3. Explain why a marathon runner may develop leg cramps.

For Further Reading

ALBERTS, B., D. BRAY, J. LEWIS, M. RAFF, K. ROBERTS, and J. WATSON. *Molecular Biology of the Cell.* 2d ed. New York: Garland, 1989.

DARNELL, J., H. LODISH, and D. BALTIMORE. *Molecular Cell Biology.* New York: Scientific American, 1986.

NEWSHOLME, E., and T. LEECH. "Fatigue Stops Play." *New Scientist,* September 22, 1988, pp. 39–43.

PRESCOTT, D. M. *Cells.* Boston: Jones and Bartlett, 1988.

STRYER, L. *Biochemistry.* 3d ed. New York: W. H. Freeman, 1988.

Photosynthesis: Trapping Sunlight to Build Nutrients

The Photosynthetic Champion

If you've ever traveled through rural Missouri, Kansas, Indiana, Illinois, Iowa, or Nebraska during the summer, then you know why the Midwest is called the Cornbelt. You can drive for hour after hour through a gently rolling sea of corn plants, the stalks and leaves looking a luminous green and rustling in the hot, humid air (Figure 6.1). The inland sea grows monotonous to the passerby, but it is of singular importance: Corn is by far our nation's largest crop. American farmers harvest 7 billion bushels each year from 70 million planted acres—a collective cornfield the size of Arizona. Biologists are still puzzling over the plant's ancestry and origins, but they are certain of its metabolic virtues: Corn is the most efficient of all the major grain crops at converting carbon dioxide and sunlight into organic nutrients—the kind of nutrients that support virtually all life on earth. Corn is a photosynthetic champion, and photosynthesis is the subject at hand.

Some 7000 years ago, Mexican or Central American Indians began to domesticate corn, and by the time Columbus arrived, Native Americans were cultivating nearly 200 varieties from Chile to southern Canada. Many now consider the domestication of corn to be humankind's greatest agricultural achievement, because no other plant—wild or domesticated—bears anything like the massive cob of starchy kernels we now harvest from corn.

Biologists are still debating over the ancestry of corn, but the closest living relative appears to be teosinte, a small, spiky plant that produces small, single rows of hard kernels. In comparison, not only is the modern ear of corn gigantic, but its seeds (kernels) are so tightly packed on the cob that corn plants cannot reproduce without human assistance. Despite this, the corn plant has an underlying trait that has made its modification worthwhile: a highly efficient mode of photosynthesis.

FIGURE 6.1 ▪ A Photosynthetic Champion.
Corn provides one-quarter of all the calories the world's people and farm animals consume.

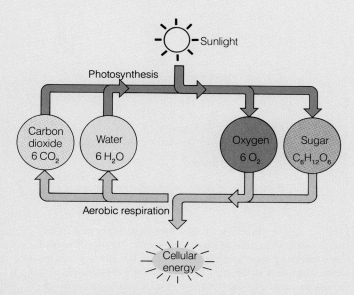

FIGURE 6.2 ■ Photosynthesis: The Reciprocal of Aerobic Respiration. Photosynthesis uses CO_2 and H_2O, generates O_2, and traps and stores solar energy in the chemical bonds of sugar molecules. Aerobic respiration uses O_2, expels CO_2 and H_2O, and liberates energy from the bonds of organic molecules.

Photosynthesis is the metabolic process by which solar energy is trapped, converted to chemical energy, and stored in the bonds of organic nutrient molecules such as glucose and other carbohydrates. Chapter 5 discussed the metabolic pathways that break down glucose and release energy, and here we find out where the glucose comes from in the first place.

Nearly all types of plants and algae, as well as some protists and bacteria, are capable of photosynthesis. In wheat, rice, and most other cereal plants, photosynthesis stops when the weather is very hot and dry. In corn, however, there are special structures and activities in the leaves and stems that allow photosynthesis to proceed even on a hot, dry day, and that helps give corn the title of photosynthetic champion. Even with the corn plant's special efficiency, however, the photosynthesis taking place within the individual cells is virtually identical to that in an oak leaf or kelp frond.

Photosynthesis begins with the trapping of light energy by special light-absorbing molecules, or *pigments*. Pigments can be red, orange, yellow, purple, or blue, but in most autotrophs ("self-feeders"; see Chapter 5), the green pigment *chlorophyll* predominates. Chlorophyll is both widespread and vitally important to life. The huge portions of our planet's surface covered by planted fields, pastureland, grassy plains, shorelines, and towering forests appear green because of the trillions and trillions of chlorophyll molecules produced by autotrophs. These light-trapping pigments are arguably the most important biological molecules, because they are the keys to photosynthesis. And photosynthesis may be the most important metabolic pathway to familiar life forms, since it generates the nutrients that virtually all cells break down via fermentation and aerobic respiration. Once photosynthetic pigments trap light, autotrophs use the energy to generate sugars and other organic nutrients, then break them down again for their own cellular energy needs. Heterotrophs (animals, fungi, and most microbes) must, of course, use autotrophs or other heterotrophs as food. But whether a heterotroph consumes a fallen log, a tomato leaf, or a corn-fed steer, the primary source of the nutrients is photosynthesis, acting on simple raw materials and powered by sunlight.

Photosynthesis is more or less the chemical opposite of aerobic respiration in that photosynthesis builds up what respiration breaks down, and photosynthesis uses up carbon dioxide and generates oxygen as a waste product (Figure 6.2). Nevertheless, photosynthesis resembles the other metabolic pathways in several fundamental ways. It involves a series of reaction steps, each catalyzed by a specific enzyme, as well as electron transfers and oxidation-reduction reactions. One phase of photosynthesis takes place in the membrane of the chloroplast (a specialized organelle similar to the mitochondrion) and involves a gradient of hydrogen ions and phosphorylation. A second phase takes place in another part of the chloroplast and is marked by a reaction cycle akin to the Krebs cycle. Thus, all the metabolic pathways that capture and release energy are chemically and functionally related, and many of the concepts in this chapter will already seem familiar.

Our study of photosynthesis investigates how living cells trap the energy of sunlight into the bonds of sugar molecules. In the process, we will answer a series of questions:

■ What are the major chemical events of photosynthesis, and where do they occur?

■ What is light energy, and how do pigments trap it?

■ What are the specific steps during which light energy is trapped and stored in energy carriers?

■ What are the steps by which cells build sugars from the energy in energy carriers?

■ What process gives plants like corn particularly efficient photosynthesis?

■ How do carbon atoms cycle from autotrophs to heterotrophs in a massive global exchange?

AN OVERVIEW OF PHOTOSYNTHESIS

There is a beautiful symmetry to the metabolic processes of respiration and photosynthesis that is revealed by their nearly opposite overall equations. In Chapter 5, we saw that when oxygen is present, aerobic respiration in mitochondria allows cells to break down glucose into carbon dioxide and water and to release chemical energy:

$$C_6H_{12}O_6 + 6\,O_2 \rightarrow 6\,CO_2 + 6\,H_2O + \text{chemical energy}$$

glucose   oxygen   carbon   water
                       dioxide

In photosynthesis, nearly the reverse takes place in the chloroplast. Light energy is trapped, transformed, and then used to convert carbon dioxide and water into glucose and oxygen:

$$6\,CO_2 + 6\,H_2O + \text{light energy} \rightarrow C_6H_{12}O_6 + 6\,O_2$$

carbon   water   glucose   oxygen
dioxide

In the first equation, chemical energy is released from glucose, and the products have less energy stored in their chemical bonds than do the reactants. In the second equation, solar energy is stored in the chemical bonds of glucose, and so the products contain more energy than the reactants. Clearly, living things must have both a source of energy as well as a means of releasing it, and for green plants and most other autotrophs, the direct energy source is sunlight.

During photosynthesis, light energy is trapped and used to remove electrons in a high-energy state from water, and these are then added to carbon dioxide; in the process, their energy is captured in the bonds of a glucose molecule. This all sounds simple enough, but what does it really mean? When sunlight strikes green chlorophyll or other colored pigments in the chloroplast of a corn leaf, let's say, some of the solar energy becomes trapped as it boosts electrons in the pigment molecules to higher energy levels. Then, before the electrons drop back to their original energy levels, they pass down an electron transport chain much like the one in the mitochondrial membrane, and the trapped solar energy is converted to chemical energy (see Chapter 5). As energy is released from the light-boosted electrons bit by bit, it is stored in the chemical bonds of the high-energy carriers ATP and NADPH. (NADPH is similar to NADH in function but slightly different in structure.) These events make up the first phase, or so-called **light-dependent reactions,** of photosynthesis. The reactions are driven by light energy and can take place only when light is available (Figure 6.3).

The newly created high-energy carriers ATP and NADPH then supply the energy needed for the second phase of photosynthesis, the **light-independent (dark) reactions,** during which the bond energy in the carriers is released and stored in the bonds of glucose molecules. The light-independent reactions can take place in darkness, but they do not *require* dark-

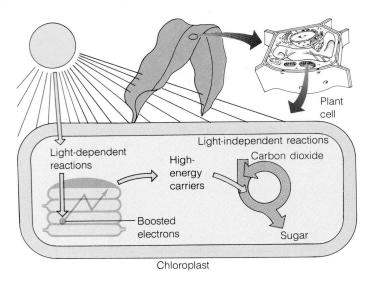

FIGURE 6.3 ■ Overview of the Light-Dependent and Light-Independent Reactions of Photosynthesis. Photosynthesis takes place within cells containing chloroplasts, usually in green leaves, stems, and other structures. The process includes two phases, the light-dependent and the light-independent reactions, both of which occur in the chloroplast. During the light-dependent reactions, light energy is trapped; it then boosts electrons to higher-than-initial energy levels, and as these electrons pass from one electron acceptor to another, the energy is stored in chemical form in high-energy carriers such as NADPH and ATP. During the light-independent reactions, the high-energy carriers supply energy to a cycle of reactions that fix carbon into biological molecules like glucose.

ness; they simply do not require the presence of light. During the light-independent reactions, energy from the carrier molecules converts carbon dioxide molecules to compounds containing carbon and hydrogen, such as the sugar glucose, $C_6H_{12}O_6$. These compounds can then diffuse away from the chloroplast and be used in glycolysis within the same plant cell, or they can be used to build cellulose or starch. In summary, the major consequences of photosynthesis are twofold: During the light reactions, solar energy is converted to and stored as chemical energy in the bonds of ATP and NADPH molecules; and during the dark reactions, this chemical bond energy is released and stored in a more stable form—the bonds of sugars and other nutrients.

The Chloroplast: Solar Cell and Sugar Factory

In a typical plant such as corn, a green multicellular organism made up of eukaryotic cells, both the light and dark reactions of photosynthesis occur in the dark green **chloroplasts** (see Figure 3.21). Each corn leaf cell may contain 40 to 50 chloroplasts, and each square millimeter of leaf surface more than half a million of the organelles. Chloroplasts are analogous to mitochondria in several ways: Both are elongated organelles about the size of bacteria; both carry out energy-related tasks in the cell; both have their own DNA in circular chromo-somes; and both have complex, membranous structures. However, while mitochondria are "powerhouses" that gen-erate ATP, chloroplasts are more a combination of solar cell and sugar factory that captures sunlight and generates glucose and other carbohydrates. Box 6.1 on page 126 describes the possible evolution of chloroplasts.

Recall that mitochondria have an outer membrane and a deeply folded inner membrane in which electron transport chain proteins are embedded. Chloroplasts have smooth outer and inner membranes lying side by side that collectively enclose a space filled with watery solution, the *stroma* (Figure 6.4). A third membrane lies within the inner membrane and forms a complicated system of stacked, disklike sacs, the **thylakoids**, interconnected by flattened channels. Each stack of thylakoids is called a *granum* (plural, grana), and each indi-vidual disc has an internal space, or compartment, within the third or *thylakoid membrane*. Chlorophyll and other colored pigments are embedded in this membrane, along with electron transport proteins.

The light-dependent reactions of photosynthesis take place in the thylakoid membrane and inside the thylakoid disc and provide the chloroplast's solar cell activity. The light-independent reactions begin in the stroma and continue in the cytoplasm, and these reactions collectively act as the sugar factory. After sugars are produced through photosynthesis, they leave the chloroplast and are either (1) linked into starch molecules and stored in plastids, (2) used as subunits in the cellulose of the plant cell wall, or (3) broken down in the plant cell's own cytoplasm and mitochondria (via glycolysis and aerobic respiration), where they power cellular activity. Ulti-mately, light energy is the power source for the photosynthe-sis that builds the sugars that fuels the cell. So before proceed-ing further, let's take a look at the nature of light and solar energy.

COLORED PIGMENTS IN LIVING CELLS TRAP LIGHT

Our earth is constantly bathed in light radiating from the sun—light with profound and nearly immediate effects on many organisms. Each minute, the solar energy striking a square meter of the earth's surface could raise the temperature of half a liter (about a pint) of water by 25°C. Plants, algae, and other autotrophs use less than 1 percent of this energy in photosynthesis, but they are so numerous on our planet that they collectively produce between 150 and 500 billion tons of carbohydrates every year—enough to build a giant mountain of cellulose and starch. Light travels so quickly and the pho-tosynthetic process takes place so rapidly that you can practi-cally eat sunlight in a fresh-picked leaf: If you pluck and immediately chew a growing lettuce leaf, you are consuming energy that left the sun just 8 minutes earlier and was con-

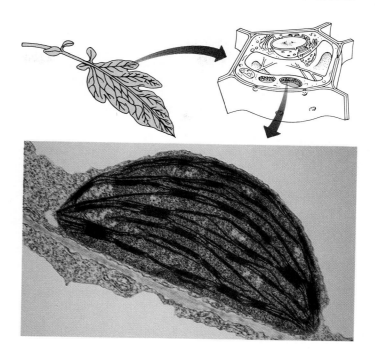

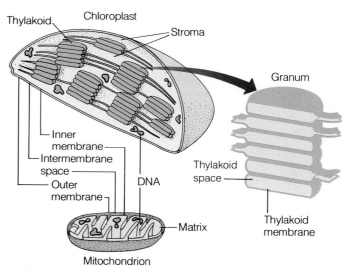

FIGURE 6.4 ■ **Chloroplast Structure.** Chloroplasts are mem-brane-bound organelles with outer and inner membranes and an intermembrane space between them, as well as a third set of mem-branes forming disklike sacs called thylakoids. Each thylakoid has its own membrane and internal space, and several sacs are often stacked together in a granum. A stroma, or matrix, surrounds the thylakoid stacks. Shown here is a chloroplast from a tomato leaf (magnification 25,000 ×). A mitochondrion is shown for size and shape comparison.

verted to the chemical energy in carbohydrate molecules almost instantly upon striking the plant. Photosynthesis is not only rapid, but also amazingly efficient. Box 6.2 on page 129 describes a plant that gathers light for photosynthesis in the inky recesses of the deep ocean.

BOX 6.1　　FOCUS ON THE ENVIRONMENT

WHEN YOU SEE GREEN, THINK OF A GREEN SEA

Look out across the emerald green grass in a park or the deep verdant hues in a pine forest, and you are seeing a piece of evolutionary history. Those oceans of color come from chlorophyll pigments bound up in chloroplasts. Many plant biologists think the banana-shaped organelles probably arose after ancestral photosynthetic bacteria took up residence in larger cells. Algae may have evolved from these ancient cellular communes and in turn given rise to land plants like grass and trees. While this endosymbiotic ("coming-to-live-inside-in-harmony") theory makes good sense, skeptics want evidence, and recently, marine biologists stumbled onto an interesting bit of it.

In 1988, Sallie W. Chisholm and a team of colleagues from Harvard University and Woods Hole Oceanographic Institution reported their discovery of a strange new kind of photosynthetic cell in six different ocean areas from the tropics to the North Atlantic. The cells are prokaryotic (have no nucleus) and are so tiny that the team could see them only with an electron microscope. They were also incredibly numerous, with up to 100,000 cells in every teaspoonful of seawater sampled. The cells have a kind of chlorophyll *a* never seen before, as well as chlorophyll *b*, alpha- (rather than beta-) carotene, and traces of other pigments.

The team places the new cells in the phylum Prochlorophyta, which contains species generally thought to be living relatives of the very early cells that gave rise to chloroplasts. One type of prochlorophyte called *Prochloron*, discovered in 1976, grows inside the bodies of sea squirts—and other little marine animals (Figure 1)—trading photosynthetic products for room and board. Some biologists consider *Prochloron*'s symbiosis, its ability to coexist peacefully, to be significant, since some similar photosynthetic bacterium probably entered into communal living with the ancestors of plant cells.

In 1986, researchers discovered a second type of prochloro-

FIGURE 1 ■ *Prochloron* **and a Sea Cucumber. The green tinge on this animal, a type of sea cucumber, comes from the symbiotic alga** *Prochloron.*

phyte—this one free-living—in shallow lakes in the Netherlands. The new type discovered by the Chisholm team in 1988 is also free-living, but is perhaps more significant for three reasons: (1) It contains both chlorophylls *a* and *b*, the same pigments in the chloroplasts of virtually all land plants, a striking evolutionary coincidence; (2) the internal structure of the new species is quite similar to the *Prochloron* cells that live within sea squirts and may, in fact, be a free-living counterpart resembling the ancient marine cells that existed independently before coming to inhabit plant cells; and (3) the newly discovered green cells are so abundant that they must have great ecological significance as fixers of carbon and producers of oxygen in the open oceans.

It is always exciting when biologists discover something so potentially important right beneath our collective noses. In this case, the discovery may also help explain the sea of green right outside our windows.

The photosynthetic conversion of light energy to chemical energy supplies the crucial energy source for most living things and thus helps ensure their survival. But other aspects of life can also depend on such conversions and can also involve the excitement of certain absorbing molecules (pigments). Our sense of vision, for example, depends on light-sensitive pigment molecules in the retina, a special layer inside the eye (see Chapter 28). And various plant species produce their flowers at appropriate times of the year (e.g., crocuses in early spring, marigolds in late summer). This seasonality is a result, in part, of the activities of other light-absorbing pigments (see Chapter 31). Clearly, light and pigments are vital to important biological processes.

Light, Chlorophyll, and Other Pigments

Visible light—the white sunlight we can separate into a rainbow of colors through a prism—is just a small part of the **electromagnetic spectrum,** which is the full range of electromagnetic radiation in the universe, from highly energetic

gamma rays to very low-energy radio waves (Figure 6.5). All such radiation travels through space behaving both as vibrating particles (called **photons**) and as waves. The less energetically a photon vibrates, the longer its wavelength; thus, radio waves can be many kilometers long, while gamma rays have wavelengths below 1 nm (one-billionth of a meter).

The photons of visible light have wavelengths in the narrow range of 380 nm (violet light) to 750 nm (red light). It is no coincidence that living things can absorb and use light within this restricted wavelength range. Gamma rays, with their shorter wavelengths, are so energetic that they disrupt and destroy the biological molecules they strike, while radio waves are so low in energy that they cannot excite biological molecules at all. We can separate and perceive the colors in sunlight because the pigments in our retinas are excited by different vibrational frequencies within the visible range, and nerve impulses fired off to the brain are interpreted as red, blue, violet, and other colors.

Objects like red apples, black panthers, or green leaves look colored because they absorb and reflect light. An apple looks red because pigment molecules in the skin cells absorb

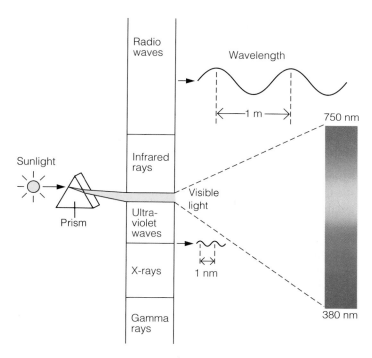

FIGURE 6.5 ■ **Visible Light and the Electromagnetic Spectrum.** We are constantly bathed by electromagnetic energy, from radio waves to infrared rays (heat), X-rays, and gamma rays. Each category has an energy range measured in oscillating waves of specific lengths (wavelengths). Only the small portion of the entire electromagnetic spectrum with wavelengths in the range of 380 to 750 nm is visible to us as white light that can be broken by a prism into colored light, each color with different wavelengths, as in a rainbow.

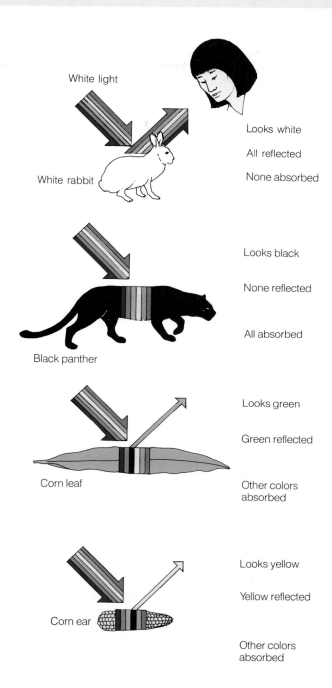

FIGURE 6.6 ■ **Pigments and Absorption Spectra.** The colors we see in a given object depend on which wavelengths of light the pigments in that object absorb and which are reflected back to our eyes. A white rabbit looks white because all wavelengths are reflected. Black pigment in a panther's coat absorbs light throughout the whole color spectrum and reflects nothing; thus, it looks black to us. Chlorophyll in a corn leaf absorbs many wavelengths of light, but green light is most strongly reflected, and thus the leaf looks green. Corn ears contain carotenoids, which absorb wavelengths in the blue range and reflect back red, yellow, and orange light. Because a leaf often contains both chlorophyll and carotenoids, its pigments can absorb and use for photosynthesis most of the wavelengths of visible light that strike it.

light from various parts of the visible spectrum and reflect only red light. A panther looks black because carbon-containing pigments in the animal's coat absorb all the wavelengths of light that strike them, reflecting none. A corn leaf looks green because its green chlorophyll pigments absorb violet, blue, yellow, and red light and reflect only green. This range of absorbed light is called the **absorption spectrum,** and each colored pigment displays its own unique spectrum (Figure 6.6). There is an essential difference, however, between chlorophyll molecules and the pigments in a panther's coat: Chlorophyll participates in photosynthesis as well as giving a green leaf its color.

Two main types of chlorophyll take part in photosynthesis: **Chlorophyll *a*** occurs in all photosynthetic autotrophs, and **chlorophyll *b*,** with its slightly different molecular structure, is found mostly in land plants and green algae. In all autotrophs, chlorophyll is accompanied by **carotenoid pigments,** which absorb green, blue, and violet wavelengths and reflect red, yellow, and orange light. Carotenoids are generally masked by chlorophyll and thus tend to be unnoticed in green leaves. However, they give bright and obvious color to many nonphotosynthetic plant structures, such as roots (carrots), flowers (daffodils), fruits (tomatoes), and seeds (corn kernels) (see Figure 6.6). And carotenoids are behind the glorious colors of autumn. As summer ends and the nights grow cool, chlorophyll begins to break down, allowing the gold and red carotenoids to show through and emblazon maple, oak, sumac, and other deciduous trees and vines (Figure 6.7).

In both the chlorophylls and carotenoids, molecular structure accounts for the light-absorbing properties. The chemical bonds in these and other pigment molecules are in constant motion, and the atoms share electrons; this creates an overall dynamic state and allows the pigments to absorb photons of light energy. Chlorophyll absorbs all but the green wavelengths, and carotenoids all but the red and yellow wavelengths. Functioning together in pigment complexes, chlorophylls and carotenoids can absorb most vibrational frequencies in the visible range and thus most of the available energy in visible light. The importance of these pigments, of course, is not that they absorb light, but what becomes of that captured energy.

THE LIGHT-DEPENDENT REACTIONS OF PHOTOSYNTHESIS

When sunlight shines on a corn leaf, the light-dependent reactions of photosynthesis proceed almost instantaneously, converting solar energy to chemical energy. These reactions take place in the thylakoid membrane and in the compartment inside the thylakoid disc, and they involve three main stages: First, light excites electrons to higher energy levels; second, that energy is stored in stable energy carriers; and third, water molecules are split, replacing the excited electrons with low-energy electrons and releasing oxygen. Let's consider these three stages in more detail.

Light Excites Electrons

Strange as it may sound, leaves have antennae: Embedded in the thylakoid membranes of each chloroplast in every photosynthetic cell are so-called **antenna complexes** (Figure 6.8). These are clusters of 200 to 300 chlorophyll, carotenoid, and other pigment molecules that act in concert during the harvest of light energy. The pigment molecules are arranged around a central chlorophyll *a,* the **reaction center.** That central chlorophyll is associated with a special set of proteins and other molecules that donate electrons to the center or accept electrons from it. This kind of pigment-protein complex is called a **photosystem,** and there are two different photosystems, called *photosystem I* and *photosystem II,* each differing slightly in the wavelength of light it absorbs.

The reaction center chlorophyll in a photosystem absorbs slightly lower-energy light than the other pigments; thus, when the other pigment molecules are struck by photons of light, they pass the energy they absorb (in the form of electrons) from one molecule to the next until the electrons are

FIGURE 6.7 ■ Autumn Glory.   When chlorophyll breaks down in autumn, carotenoids can shine through, painting the forest with orange, gold, and scarlet hues. This shot was taken in northern Pennsylvania.

BOX 6.2   FOCUS ON THE ENVIRONMENT

RED ALGAE THAT GROW IN THE DARK

Botanists from the Smithsonian Institution made a startling discovery not long ago in the western Atlantic Ocean off the Bahamas. They were exploring the uncharted seafloor near San Salvador Island in a helicopter-like research submarine capable of withstanding great water depths when they came upon an underwater plateau—a seamount—never before mapped. Far more surprising than the discovery of this new terrain, however, was the plant species they found there—a type of red alga growing in virtual darkness on the slopes of the seamount some 268 m (about 880 ft) below the ocean surface.

These algae are a purplish color and grow in hard, flat colonies. The cells (shown in Figure 1, magnified 1600 times and viewed from the top) are arranged in vertical columns. The side cell walls are lined with calcium carbonate (limestone) and are thus opaque and rigid. The cell tops and bottoms, however, are clear and allow the exceedingly dim light at this ocean depth to penetrate the columns of cells vertically.

The discovery of this odd plant is causing botanists to revise their long-held belief about photosynthetic efficiency: that a plant cannot grow in any environment where the light levels are less than 1 percent of the full sunlight striking the ocean surface at midday. Since 1 percent penetration stops below 215 m (about 705 ft), plants were—in theory—unable to grow at greater depths. Nature is no respecter of scientific theory, how-

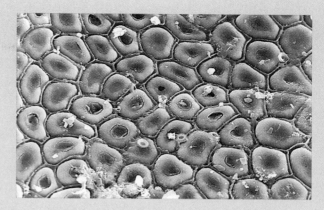

FIGURE 1 ■ **Cells of Deep-Dwelling Red Algae.**

ever, and the newly found red algae grow quite well with only 0.0005 percent of the peak surface sunlight. What's more, the plants are about 100 times more efficient at capturing sunlight and converting it to carbohydrates than are red algae growing near the ocean's surface. Like the corn plants that do so well on land, these strange new photosynthetic champions will help scientists probe the solar-powered metabolic pathway that is crucial to life on earth. And perhaps someday genetic engineers can transfer their highly efficient photosynthesis to the land plants we depend on for food.

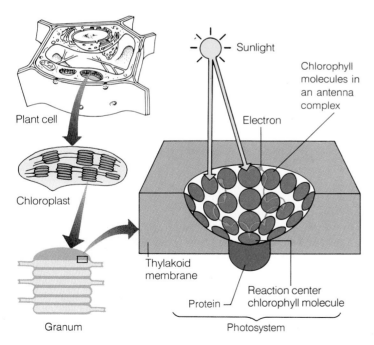

FIGURE 6.8 ■ **The Antenna Complex: A Cluster of Pigment Molecules.** Antenna complexes—clusters of light-absorbing pigment molecules—channel the energy from impinging photons of light to a central chlorophyll molecule, the reaction center. This chlorophyll is embedded in protein and associated with electron donors and acceptors into a complex called a photosystem. Light energy boosts the energy in a pair of electrons in the reaction center chlorophyll, and they are accepted by an electron acceptor in the photosystem.

finally transferred to the reaction center, which actually participates in photosynthesis (see Figure 6.8). Somewhat like Ping-Pong balls bouncing off a hard surface, the electrons bounce from pigment to pigment within the antenna complex at high speeds; each electron transfer takes only 10^{-12} second. Some biologists compare the passage of electrons in the antenna complex to the movement of energy between metal atoms in a radio wave antenna or between silicon and metal atoms in a solar cell or computer chip. But the antenna complexes are biological molecules, and the funneling of light energy among them results in an excited chlorophyll a molecule. This molecule possesses—albeit fleetingly—a pair of highly energized electrons (Figure 6.9, step 1).

The energized electrons of chlorophyll in themselves are not of much use to the cell. Their energy, however, is soon trapped and stored in more stable energy carrier molecules.

Energy Is Stored in Energy Carriers

The light-excited chlorophyll has a very short lifetime. It quickly passes its energized electrons to a series of electron acceptors, each of which passes them to the next acceptor in the series (Figure 6.9, step 2). The result of this electron "bucket brigade"—an electron transport chain—is the storing of energy in two kinds of molecules that can accomplish useful work in the cell: the energy carriers ATP and NADPH.

As the electrons pass from one acceptor to the next, they

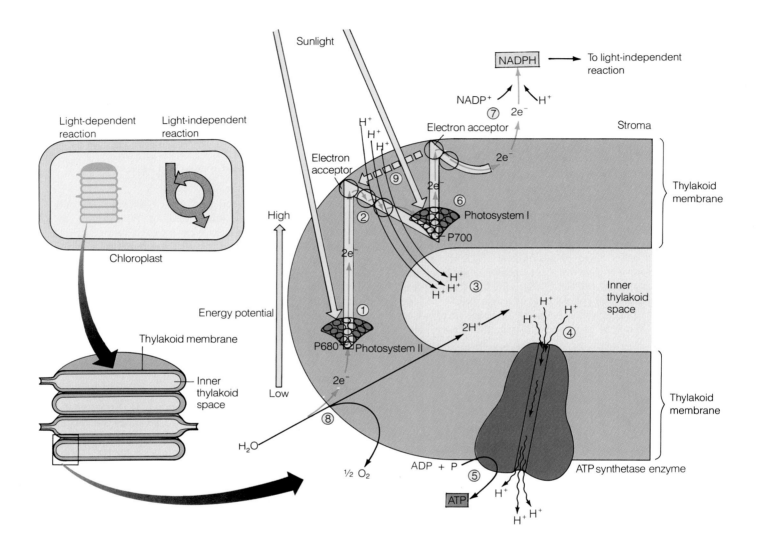

FIGURE 6.9 ■ The Light-Dependent Reactions of Photosynthesis. *First stage: Light excites electrons.* Energy from light striking an antenna complex is channeled to the reaction center chlorophyll in *photosystem II* (step 1). This chlorophyll absorbs light at 680 nm and is called *P680.* Light energy boosts the energy in a pair of electrons in this chlorophyll, and the electrons are received by an electron acceptor. *Second stage: Energy is stored.* The electrons are next passed from one electron acceptor to the next (step 2), and small amounts of energy are released in this electron transport chain, allowing the pumping of hydrogen ions across the thylakoid membrane (step 3). As the hydrogen ions flow out of the inner thylakoid space (step 4), ATP is synthesized, as it was in mitochondria (step 5). At the end of the electron transport chain, the electrons are accepted by a second reaction center chlorophyll (called *P700*) embedded in *photosystem I.* Absorption of another photon of light boosts more electrons (step 6), and the final electron acceptor is $NADP^+$, which becomes NADPH (step 7). *Third stage: Water is split.* In the final stage, water is split, yielding a pair of electrons (step 8), which replace those electrons lost from the reaction center chlorophyll. Water-splitting also yields two hydrogen ions, which enter the thylakoid space, and an oxygen atom, which combines with another oxygen, making O_2 gas. Steps 1, 2, 6, and 7 are called the Z scheme of photosynthesis (or noncyclic photophosphorylation), since the solid yellow line showing the path of energy flow follows a zigzag shape. In contrast, the more ancient cyclic photophosphorylation involves steps 6, 9, and 2, as the electrons follow a roughly circular path, and less ATP energy is generated.

lose a bit of their energy, and that becomes stored in ATP molecules by a process similar to the one taking place in mitochondria—the flow of hydrogen ions through a membrane protein that synthesizes ATP (Figure 6.9, steps 3 to 5; also review Figure 5.16, steps 5 and 6).

At the end of the electron transport chain, the electrons join a second photosystem and become boosted by light once more (Figure 6.9, step 6). This boost-drop-boost pathway of energy gain and release can be plotted as a zigzag path on a graph of energy content and hence is called the *Z scheme* of photosynthesis (Figure 6.9, steps 1, 2, 6, and 7, highlighted in yellow). Finally, the rebooted electrons are transferred one last time, and NADPH forms (Figure 6.9, step 7). This electron carrier stores seven times more energy than an ATP molecule and can move around in the chloroplast and the cell, where the energy is released once again and can perform useful work. (In an ancient type of photosynthesis involving only photosystem I, electrons travel the circular path shown in Figure 6.9, steps 6, 9, and 2. Called **cyclic photophosphorylation,** this pathway provides only a small amount of ATP. In contrast, the zigzag path we have just described is called **noncyclic photophosphorylation.)**

With the production of the stable energy carriers ATP and NADPH, the light-dependent reaction has trapped light energy into useful chemical energy. But it has removed an electron pair from the reaction center chlorophyll in photosystem II, leaving a hole that must be filled with another electron pair. That pair comes from the splitting of a water molecule.

Water Is Split and Oxygen Produced

When a person holds his or her breath, that person soon begins to feel anxious, because the body continues to use up existing oxygen from the blood as respiration goes on in the mitochondria of the cells, producing ATP at the cost of glucose and oxygen. In a matter of seconds, powerless to prevent it, the person gasps—then air enters the lungs, oxygen moves into the blood, and the mitochondria are once more supplied with the gas needed during ATP production. Clearly, a continual source of oxygen is vital for human life (and, in fact, all animal life), and the photosynthesis taking place in grass, trees, algae, and other autotrophs releases that oxygen as a waste product. The eco-political slogan "Have you thanked a plant today?" is apt indeed.

The photosynthetic pathway inside plant cells strips electrons from water molecules (Figure 6.9, step 8). These electrons replace the electrons that were boosted from chlorophyll by incoming light. The oxygen left over after water (H_2O) is split is then discarded as a metabolic waste product and is available for cellular respiration in all aerobic cells, whether prokaryote, plant, animal, protist, or fungus. The hydrogens left over when water is split are also used in the plant cell's chloroplasts during the production of ATP, much as hydrogen is used in mitochondria (review Figure 5.16).

Large aerobic organisms like ourselves would never have evolved if photosynthesis had not emerged long before and if oxygen had not been released into the atmosphere and oceans. But photosynthesis did appear, and by the middle of the twentieth century, biologists had probed the biochemical details of the light-dependent reactions and grasped their profound importance in absorbing light, converting it into chemical energy, and splitting water with the release of oxygen. Researchers have been trying ever since to build a simplified photosynthetic system—a synthetic leaf in a test tube—that can split water and store energy. The results, however, have been mixed. Such efforts remind us just how marvelous biological structures like chlorophyll molecules, thylakoid membranes, and electron transport chains truly are, and how simply and effectively they convert light energy into the chemical bond energy that fuels life.

THE LIGHT-INDEPENDENT REACTIONS OF PHOTOSYNTHESIS

A corn leaf is a far better solar collector than any mechanical solar device yet built. All such mechanical devices trap sunlight and convert it to chemical, electrical, or heat energy. But solar devices usually require expensive silicon crystals and rare metals, moving parts, and constant upkeep. A leaf, on the other hand, can track the sun's path without wheels and gears; it is mostly organic and biodegradable; it can sprout from a small seed and grow to large size in just a few weeks; and it uses only water and sunlight as it builds the energy carriers ATP and NADPH during the light reactions of photosynthesis. Moreover, the corn leaf does something no solar device can do but that all autotrophs do quite naturally: It builds organic nutrients such as glucose. To do this, a leaf requires only carbon dioxide from the atmosphere, the high-energy carriers generated during the light reactions, and a few enzymes and intermediates that function in a cyclic series of reactions. This cycle of reactions, called the light-independent reactions of photosynthesis, or the **Calvin-Benson cycle** (after the biologists who figured out its steps), takes place on the outer surface of the thylakoid membrane, facing the stroma. The light-independent reactions are a cycle with three main stages: First, carbon is fixed to a biological molecule; second, carbohydrates are manufactured as stable energy storage molecules; and third, the starting compound for the cycle is regenerated (Figure 6.10).

Fixing Carbon

In the first stage, called **carbon fixation,** carbon dioxide from the air becomes attached to a biological molecule, a process that makes carbon available for the fabrication of glucose and more complex carbohydrates (Figure 6.10, Stage One). The acceptor of carbon dioxide is a five-carbon molecule called

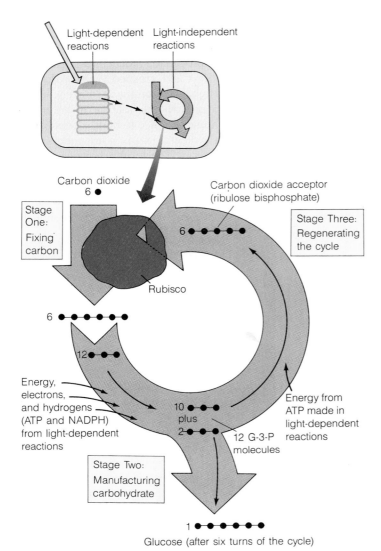

FIGURE 6.10 ■ **The Light-Independent Reactions of Photosynthesis.** This cyclic series of reactions takes carbon molecules from the air and fixes them into biological molecules. *Stage One: Carbon is fixed.* Six CO_2 molecules from the atmosphere are fixed, or added to six molecules of ribulose bisphosphate via the activity of the enzyme nicknamed rubisco. This forms six molecules of an unstable six-carbon compound, which breaks down rapidly to 12 molecules of a three-carbon compound; through this mechanism, the carbon has been fixed. *Stage Two: Carbohydrate is manufactured.* Enzymes act on energy, electrons, and hydrogens from the NADPH and ATP made in the light-dependent reactions, altering the fixed carbon into 12 molecules of the three-carbon compound G-3-P (glyceraldehyde-3-phosphate). Two of these G-3-P molecules can be siphoned from the cycle and joined, generating glucose or other compounds. The other 10 G-3-P molecules are left for the final step. *Stage Three: The cycle is regenerated.* Energy from ATP generated in the light-dependent reactions modifies the 10 three-carbon G-3-P molecules into 6 five-carbon ribulose bisphosphate molecules. With this starting compound regenerated, the cycle can continue, and more carbon (CO_2) can be taken from the air and incorporated into biological molecules.

ribulose bisphosphate. When the one-carbon molecule carbon dioxide is joined to this five-carbon molecule, the result is an unstable six-carbon molecule that breaks down to 2 three-carbon molecules. The addition of carbon to a biological molecule is accomplished by an enzyme nicknamed **rubisco** (for ribulose bisphosphate carboxylase). Unlike many enzymes, this one works fairly slowly, but because it is so central to the light-independent reactions and the building of carbohydrates, chloroplasts must build many rubisco molecules; in fact, nearly half the protein molecules in most chloroplasts are copies of this enzyme. Many biologists believe that this redundancy makes rubisco nature's most common protein—a fact that graphically demonstrates the importance of photosynthesis to life on earth.

The net result of carbon fixation, the first step in the Calvin-Benson cycle, is the linkage of inorganic carbon dioxide into a biological molecule within the plant cell. This fixed carbon is the molecular starting place from which the plant will build proteins, DNA, and cellulose (cell wall material). These biological molecules will in turn be consumed and end up as animal or fungal tissue, or even the wood in your house or desk.

Manufacturing Stable Storage Molecules

While the first stage of the Calvin-Benson cycle of light-independent reactions traps carbon into biological molecules, the second stage traps energy into stable storage molecules (Figure 6.10, Stage Two). Specifically, energy, electrons, and hydrogens generated during the light-dependent reactions are added to the newly fixed carbon, now in the form of three-carbon molecules. Roughly speaking, the carbon goes from CO_2 (carbon dioxide gas) to CH_2O (carbohydrate) through the addition of hydrogens and electrons. One of the main carbohydrates formed is glucose, the sugar that directly fuels human activity and the collective cellular tasks of most other organisms as well. Since glucose has six carbons ($C_6H_{12}O_6$), six carbon dioxides must be fixed, one at a time, in six turns of the cycle, before one glucose molecule can be generated.

Regenerating the Starting Compound

So far, the first two stages of the light-independent reaction have fixed both carbon and energy into useful storage forms such as glucose. All that remains, therefore, is for the five-carbon compound that began the cycle to be regenerated, a process requiring some ATP generated during the light-dependent reactions (Figure 6.10, Stage Three). Once this carbon dioxide acceptor is rebuilt, the enzyme rubisco can initiate a new cycle and the formation of more glucose.

What Makes Glucose a Better Energy Store Than ATP?

You may wonder why a corn cell doesn't simply use the ATP and NADPH from the light reactions directly to fuel cellular activity instead of converting it in a second reaction series to

(a) Corn: A C$_4$ plant

(b) Ice plant: A CAM plant

(c) The carbon dioxide pump

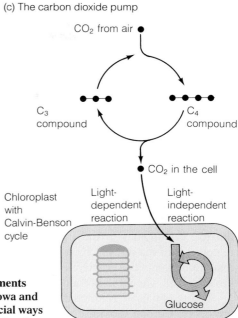

FIGURE 6.11 ■ **Photosynthesis in All Seasons: Some Plants Adapted to Hot, Dry Environments Have Special Methods for Fixing Carbon.**   Corn (a), which thrives in the hot summers of Iowa and Kansas, and ice plant (b), which flourishes along dry berms of California freeways, have special ways of fixing carbon that other plants lack. Recall from Figure 6.10 that the first stable product of carbon fixation in most plants has three carbon atoms. These are called C$_3$ plants. In contrast (c), corn and ice plant initially fix carbon into a compound with four carbons by what can be called a carbon dioxide pump. Hence, corn is called a C$_4$ plant, and ice plant is called a CAM plant based on the special type of carbon compound its carbon fixation involves. In C$_4$ plants, the carbon dioxide pump is in one type of cell, and the light-independent reactions of the Calvin-Benson cycle are in another type of cell. In CAM plants, the carbon-dioxide pump and the light-independent reactions are in the same cells, but they occur at different times during the day.

yet another form like glucose. The answer is that carbohydrates are far more compact and practical forms for storing chemical energy than are the high-energy carriers. The glucose molecule, for instance, is smaller than a single ATP, yet stores in its bonds the energy equivalent of 36 molecules of ATP. Glucose is also less reactive than ATP and NADPH in cells—in fact, you can store it for years in a jar on the shelf in your kitchen without its breaking down—and it can travel easily from place to place in an organism. Carbohydrates can be easily converted to other biological molecules (see Figure 5.14), and glucose and other sugars can be easily linked into huge inert storage molecules such as starch or glycogen (to be tapped only when needed) or into structural molecules like the cellulose in plant cell walls (see Chapter 3).

Despite the great practicality of storing energy in carbohydrates, the building of even a single glucose molecule is a relatively slow and costly process, requiring 18 ATPs, 12 NADPHs, and six turns of the Calvin-Benson cycle for each glucose formed. Happily for us and the rest of the living world, the sun provides an almost limitless energy source. Thus, day after day, this virtually inexhaustible energy stream can excite electrons, and the light-dependent reactions can store energy and split water. As we saw, this provides oxygen for the aerobic respiration that goes on in the cells of most of the world's organisms, as well as energy, electrons, and

hydrogen atoms used in the light-independent reactions. That latter cycle then fixes carbon into biological molecules and manufactures carbohydrates, and the cycle's starting materials are regenerated. The glucose molecule provides the plant, alga, or other autotroph with raw materials for repair and growth, as well as supplying compact carbohydrate fuel (glucose) for the metabolic release of energy via aerobic respiration. Heterotrophs, such as animals or protists, which consume autotrophs or other heterotrophs, reap the benefits of this carbon fixation even though they cannot perform the biochemical trick themselves. Thus, one set of reactions (the light-independent reactions), powered by the other set (the light-dependent reactions), lies at the very center of the web of life, fixing the carbon compounds that virtually all living things must have.

A SPECIAL TYPE OF CARBON FIXATION

Each summer, as the hot sun dries up people's unwatered lawns, annoyingly luxuriant tufts of green crabgrass flourish among an otherwise brown sea. Midwestern farmers often note that during a hot, dry spell, a field of corn can be green next to a dying field of beans (Figure 6.11a). And motorists may also notice that while wildflowers shrivel and die back along the roadside in the summer, ice plant remains plump and healthy (Figure 6.11b). Interestingly, many plants that evolved in hot, dry climates display a small variation on the principles of photosynthesis we just learned—a variation which enables them to continue to photosynthesize at times when other plants cannot.

When the weather is hot and dry, the tiny ports (or stomata) on leaf surfaces close, preventing water from escaping. This closure, however, also shuts out carbon dioxide and shuts in oxygen. Thus, in hot, dry weather, oxygen from the light-dependent reactions builds up in the leaf, and carbon dioxide decreases. This, in turn, inhibits the rubisco enzyme and leads to a process called **photorespiration,** which in normal plants causes carbon to be lost instead of fixed. The solution to this problem evolved in plants such as corn and ice plant: Carbon dioxide is supplied to the light-independent reactions not directly from the air, but by a special *carbon dioxide pump* that circumvents the problem of photorespiration and allows growth, even when it is hot and dry.

The essentials of a carbon dioxide pump are shown in Figure 6.11c. Carbon dioxide from the air is added to a three-carbon compound, making a four-carbon compound. When the four-carbon compound breaks down, it delivers carbon dioxide to the light-independent reactions directly and regenerates the original three-carbon compound of the pump. With the concentrated delivery of carbon dioxide directly to the light-independent reactions, photosynthesis can continue, and the plant will remain green and healthy.

There are two different types of carbon dioxide pumps in different groups of plants. Corn, crabgrass, sugarcane, and many other tropical plants are called **C₄ plants** because the first stable compound after carbon fixation has four carbons. In contrast, as we saw in Figure 6.10, Stage One, most plants are **C₃ plants,** with a three-carbon compound being the first stable product after carbon fixation. Succulents such as ice plant, jade plant, and other plants with water stored in thick leaves, as well as some cacti, have a slightly different carbon dioxide pump and are called **CAM plants** (CAM stands for *crassulacean acid metabolism*). Because of their carbon dioxide pumps, C₄ and CAM plants have a significant advantage over the more common C₃ plants—and this advantage accounts for the exasperating luxuriance of crabgrass.

PHOTOSYNTHESIS AND THE GLOBAL ENVIRONMENT

It is sometimes hard to relate metabolic processes like photosynthesis and aerobic respiration—which take place within microscopic cells and organelles—to ourselves and our daily lives. But consider these facts: Every cell in your body contains millions of carbon atoms and oxygen molecules that were once physically present in the tissues of the plants that live around you; and the plants in your house or backyard contain carbons and oxygens that were once present in animals,

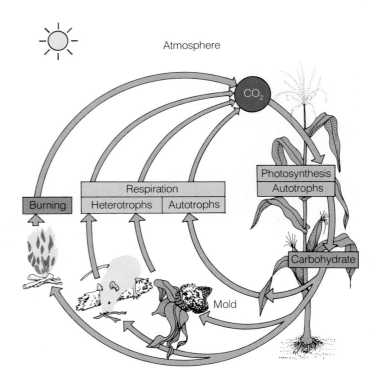

FIGURE 6.12 ■ The Global Carbon Cycle. Vast amounts of carbon cycle through the air, soil, and water as autotrophs fix CO_2 into organic compounds, and heterotrophs (along with nonliving combustion processes such as burning) break down those compounds and once again release CO_2.

fungi, or microorganisms. Aerobic respiration and photosynthesis are not simply chemical processes with reciprocal equations and opposite raw materials and products. They are metabolic engines that drive a global **carbon cycle,** a flow of organic compounds (with their carbon, hydrogen, and oxygen atoms) from autotrophs to heterotrophs to the atmosphere, soil, and water, and back to autotrophs once again (Figure 6.12). Most autotrophs fix carbon dioxide into organic compounds and release oxygen. Most heterotrophs break down those compounds and in the process consume oxygen and release carbon dioxide. Autotrophs fix that released carbon dioxide once again, release new oxygen, and the cycle continues.

These events ultimately involve all of the trillions of tons of carbon on our planet, whether it is currently tied up in wood, leaves, animal tissues, microorganisms, or rotting soil particles, and whether it is floating in the atmosphere as carbon dioxide gas, dissolved in seawater, or tied up as carbonates in rocks.

Biologists estimate that the earth has a 300-year supply of fixed carbon and a 2000-year supply of free (gaseous) oxygen; that is, if all autotrophs suddenly stopped functioning today, all organic compounds would be metabolized to carbon dioxide within 300 years, and no free oxygen would remain in the oceans and atmosphere after just 2000 years. This is an astonishingly short time, considering it took many millions of years for oxygen to accumulate to its current levels in air.

Figure 6.12 shows the global carbon cycle and the relationships between autotrophs and heterotrophs and between photosynthesis and aerobic respiration. Notice that the burning of fossil fuels (a form of fixed carbon) makes a significant contribution to the pool of carbon dioxide available for fixation by autotrophs. The burning of fossil fuels and of our rain forests may be causing a buildup of atmospheric carbon dioxide gas and leading to a phenomenon called the **greenhouse effect,** in which the gas traps heat from sunlight beneath the atmospheric layers, just the way the glass windows trap heat inside a greenhouse. This trapped heat may in turn be leading to increased global temperatures, changing climate patterns, additional photosynthesis by autotrophs, and a slight melting of the polar ice caps.

An extreme scenario for disrupting the carbon cycle invokes the burning and fires that would be generated during a nuclear war. Some scientists have calculated that such intense burning would not only raise carbon dioxide levels, but would also throw into the atmosphere massive clouds of smoke and ash that would block the penetration of sunlight. This, they speculate, would plunge the earth into a long, dark, cold **nuclear winter** and cause photosynthesis to diminish drastically and harm many or most living things. We'll return to these sobering topics in Chapter 36. The point to remember here is this: Just as the metabolic processes in individual cells have a massive collective effect—the global carbon cycle—worldwide disturbances of that cycle caused by human activity will have their own pervasive effects. Those effects, if we allow them to occur, could ultimately extend down to the basic biochemical levels and the mechanisms of energy exchange that fuel life processes on our planet, disrupting them to the point of exterminating complex organisms.

CONNECTIONS

Photosynthesis is far from universal. Of the 3 to 5 million living species cataloged so far, heterotrophs outnumber autotrophs by a wide margin. While nearly all living things break down organic nutrients for energy, only about half a million species can build those nutrients through solar-powered carbon fixation. Nevertheless, those half million support the millions of other living species. If photosynthesis stopped today, both heterotrophs and autotrophs alike would die from lack of energy in a surprisingly short time. Life depends largely on light energy channeled through the sequential reactions of photosynthesis.

Now that we have considered the details of photosynthesis, we can see how photosynthetic efficiency underlies the bounty of our nation's largest crop—corn, a plant with a special kind of carbon fixation that allows it to grow even under hot, dry conditions. We can also see that humankind's best efforts to reproduce the simplest photosynthetic system in the laboratory are eclipsed by the simplest green alga in the ocean. We can see why leaves have antennae and turn bright colors in fall. And finally, we can see that tampering with the environment could lead to disastrous changes in the fundamental interdependency of autotrophs and heterotrophs, aerobic respiration and photosynthesis.

This chapter ends the unit on cells and cellular chemistry. We have come a long way—from atoms and molecules to cell structure, mechanisms of water and solute transport, and finally to the flow of energy from the sun and other sources through the dynamic systems of living cells. Our next chapter deals with another of life's characteristics: reproduction. You will see, however, that cell structure, chemistry, and energetics are encountered again in this new topic, as well as throughout the rest of the book. The cell is the fundamental living unit, and its properties are closely reflected in all larger and more complex living things.

Highlights in Review

1. Photosynthesis is the metabolic process by which solar energy is trapped, converted to chemical energy, and stored in the bonds of carbohydrates. Nearly all living organisms depend on photosynthesis, either directly, by photosynthesizing themselves, or indirectly, by consuming photosynthesizers or their products.

2. The reactants and products of photosynthesis and aerobic respiration are reciprocal. The photosynthetic process uses carbon dioxide and water while capturing energy and producing sugars and oxygen. The respiration process burns sugars and oxygen while harvesting energy and releases carbon dioxide and water.

3. During photosynthesis, electrons and hydrogen ions are stripped from water molecules and added to carbon dioxide molecules in two sets of reactions. In the light-dependent reactions, solar energy is captured in high-energy electrons. These electrons and their energy are stored in ATP and NADPH. During the light-independent reactions, chemical energy from ATP is used to fix carbon dioxide from the air, and the energy is trapped in sugar molecules.

4. Chloroplasts are the sites of photosynthesis in all eukaryotic autotrophs. Chlorophyll molecules are embedded in the membrane of the thylakoid disc.

5. Visible light is just a small part of the electromagnetic spectrum, which also includes radio waves and X-rays.

6. Chlorophyll molecules absorb violet, blue, yellow, and red

wavelengths of light, reflecting only green wavelengths. Chlorophyll *a* occurs in all autotrophs, while chlorophyll *b* occurs mostly in land plants and green algae.

7. Carotenoids are red, yellow, and orange pigments that function as accessories to chlorophyll in the leaf and give many fruits, seeds, and roots their colors. In autumn, the chlorophyll breaks down, and the carotenoids show through in the brilliance of fall leaves.

8. Chlorophyll, carotenoids, and other pigments occur in antenna complexes—clusters of 200 to 300 molecules that trap light energy and pass it on to the reaction center chlorophyll at the center of photosystem I or photosystem II.

9. In the light-dependent reactions, electrons can follow a Z-shaped path. Incoming photons of light cause electrons to be sequentially boosted from photosystem II. Each boosted electron is passed to an acceptor and then moves down an electron transport chain in the thylakoid membrane, where ATP forms. The boosted electrons are replaced with electrons donated by H_2O, which is split, releasing O_2. At the end of the chain, the electrons are rebooted by light striking photosystem I. The boosted electrons then move down another electron transport chain and finally reduce $NADP^+$ to NADPH.

10. In the light-independent reactions (Calvin-Benson cycle), the enzyme nicknamed rubisco adds a CO_2 molecule to the five-carbon compound ribulose bisphosphate, generating a short-lived six-carbon unit which rapidly splits into 2 three-carbon molecules; carbon fixation has taken place.

11. Autotrophs store chemical energy for long periods in carbohydrates, which are far more compact and inert than ATP and NADPH and can be converted to other biological molecules. Next, energy, electrons, and hydrogens from ATP and NADPH are added to the three-carbon compounds, and carbohydrates form. In the final stage of the Calvin-Benson cycle, ribulose bisphosphate is regenerated.

12. If the weather is hot and dry, leaf stomata close and photorespiration occurs in leaves of C_3 plants, causing a loss of carbon. In a C_4 plant such as corn, or a CAM plant such as ice plant, a carbon dioxide pumping mechanism allows photosynthesis to continue even in hot, dry weather.

13. The reciprocal metabolic processes of photosynthesis and aerobic respiration drive a global carbon cycle—a flow of organic compounds, CO_2, and O_2 between autotrophs and heterotrophs.

Key Terms

absorption spectrum, 127
antenna complex, 128
C_3 plant, 134
C_4 plant, 134
Calvin-Benson cycle, 131
CAM plant, 134
carbon cycle, 134
carbon fixation, 131
carotenoid pigment, 127
chlorophyll *a,* 127
chlorophyll *b,* 127
chloroplast, 124

cyclic photophosphorylation, 131
electromagnetic spectrum, 126
greenhouse effect, 135
light-dependent reactions, 124
light-independent (dark) reactions, 124
noncyclic photophosphorylation, 131
nuclear winter, 135
photon, 126
photorespiration, 134

photosynthesis, 123
photosystem, 128
reaction center, 128

rubisco, 132
thylakoid, 125

Study Questions

Review What You Have Learned

1. Create a chart that compares aerobic respiration and photosynthesis, and use these headings: location in cell; starting materials; end products; energy released or stored.
2. What are the energy sources in the light-dependent reactions and light-independent reactions of photosynthesis?
3. Discuss the ways plant cells use the sugars produced during photosynthesis.
4. List the various types of photosynthetic pigments. Where is each found?
5. Describe how chlorophyll becomes activated by light energy.
6. Differentiate noncyclic from cyclic photophosphorylation.
7. What are the advantages when glucose serves as the end product of photosynthesis?
8. Trace the reaction steps that lead to the formation of glucose in the Calvin-Benson cycle.
9. How is a C_4 plant such as corn adapted to carry on photosynthesis even when most of its stomata are closed?
10. Describe the carbon cycle.

Apply What You Have Learned

1. What is the evidence that almost all life on earth is powered by solar energy?
2. Scientists trying to duplicate photosynthesis in the laboratory face several basic problems with respect to water and energy. What are they?
3. A water plant, *Elodea,* carries on photosynthesis in the light and releases bubbles of O_2, which an experimenter can collect in a test tube. Suggest one way of increasing the rate of O_2 production.

For Further Reading

ALBERTS, B., D. BRAY, J. LEWIS, M. RAFF, K. ROBERTS, and J. WATSON. *Molecular Biology of the Cell.* 2d ed. New York: Garland, 1989.

CAVALIER-SMITH, T. "Making a Real Discovery." *Nature* 342 (1989): 870.

LEWIN, R. A., and L. CHENG. *Prochloron.* New York: Chapman and Hall, 1989.

PRESCOTT, D. M. *Cells.* Boston: Jones and Bartlett, 1988.

STRYER, L. *Biochemistry.* 3d ed. New York: W. H. Freeman, 1988.

WILLIAMSON, P., and J. GRIBBIN. "How Plankton Change the Climate." *New Scientist,* March 16, 1991, pp. 48–52.

YOUVAN, D. C., and B. L. MARRS. "Molecular Mechanisms of Photosynthesis." *Scientific American,* June 1987, pp. 42–50.

PERPETUATION OF LIFE

Cell Cycles and Life Cycles

A Week in the Life of a Small But Famous Wound

The American presidency has proved, for too many of its esteemed office holders, to be not just a challenging job but a dangerous one. Former president Ronald Reagan, for example, was shot in 1981 and spent several weeks recovering from the serious wound. He also underwent three additional surgeries during his eight-year term in office—for an inflamed prostate, for colon polyps, and for a suspicious skin patch—all of these probably having as much to do with his age and life-style as with his demanding job. The last surgery took place in 1985, and it will no doubt be remembered by many because of the lighthearted newsphotos of Mr. Reagan standing on the White House lawn and pointing to a newly bandaged nose.

As the president described the routine surgery he had just undergone to remove a small irritated bump that may have been cancerous, he was unintentionally drawing the nation's attention to the remarkable biological phenomenon of *cell division* and the underlying *cell cycle* that brings it about. Before physicians discovered the small suspicious growth (Figure 7.1), skin cells at the site were proliferating faster than they were dying, thus accumulating in a crusty bump. The cells were failing to heed the normal signals that regulate the rate of cell division, and left unchecked, might have eventually spread to other areas. Doctors cut away the abnormally growing cells, leaving a small wound that would heal by the mechanisms of cell division and be protected under an inconspicuous patch of scar tissue.

During the surgical procedure, some of Mr. Reagan's normal skin cells were unavoidably injured, and in the week following surgery, these injured cells sent out signals that triggered nearby uninjured cells to begin dividing. These cell divisions allowed the wound gap to be filled and the sides of the wound to be knitted back together. Other signals then slowed the repair process, restoring the normal rate of division (just fast enough to replace sloughed-off cells).

Just one day after surgery, cells were already dividing in the deep layers of the skin and in the linings of tiny nearby blood vessels. Between days 2 and 5, spindle-shaped cells were stretching out along protein fibers in the blood clot that had formed at the wound's surface. By day 7, a new layer of cells covered the wound,

FIGURE 7.1 ▪ **A President and a Skin Lesion.** The red dot on Ronald Reagan's nose, clearly visible during this press conference, was later removed, leaving a small wound that healed and was covered by a tiny white scar.

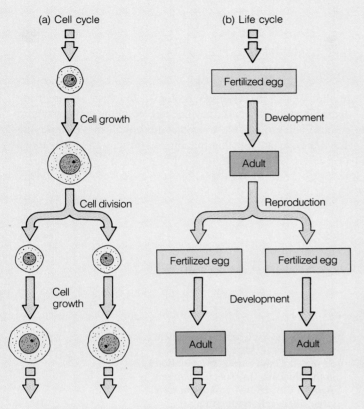

(a) Cell cycle

Cell growth

Cell division

Cell growth

(b) Life cycle

Fertilized egg

Development

Adult

Reproduction

Fertilized egg

Fertilized egg

Development

Adult

Adult

FIGURE 7.2 ■ The Cell Cycle and the Life Cycle Compared.
(a) During the cell cycle, a cell grows, then divides in two; the
progeny grow, then each divides in two; and so on. (b) During the
life cycle of a multicellular organism, a fertilized egg develops
into an adult organism that produces eggs and/or sperm. These
unite with other eggs and sperm and form new fertilized eggs,
which can grow to new adults, which can in turn produce new fer-
tilized eggs, and so on. Eventually, each adult dies, but its progeny
live on.

and skin cells were rapidly filling in below it. Protein fibers secreted by these new cells intermingled with the fibers in the blood clot, forming a scar. This scar looked pink at first because of the many new blood vessels supplying the area, but over time, it would turn whitish as the mass of protein fibers contracted and squeezed out some of the blood supply.

Cell division—whether the kind that replaces normal, everyday cell loss; the kind that fills in wounds; or the cancerous kind that eludes control—brings us to a new subject: *reproduction*. This, as we saw in Chapter 1, is a basic characteristic of life and the process by which organisms produce new individuals similar or identical to themselves. At the most fundamental level, the perpetuation of living things requires **cell division,** the splitting of one cell into two. For a single-celled bacterium or protist, cell division and reproduc-

tion are usually one and the same; the single-celled organism simply divides in half, forming two new daughter cells. For a multicellular organism, however, reproduction is usually more complex, involving specific organs that produce single cells, often called eggs in the female and sperm in the male. Egg and sperm unite to form a fertilized egg, still just one single cell, that proceeds to divide and then to develop into a new multicelled organism.

Before most cells divide, they make a copy of the genetic instructions. One copy is then apportioned to each of two daughter cells during division. The events that take place inside the cell between one division and the next are collectively called the **cell cycle,** and the net result of these events is cellular reproduction—one cell becoming two (Figure 7.2a). Each organism also has a **life cycle,** the events that take place between its initiation and its death, or sometimes between the reproduction of one generation and the reproduction of the next (Figure 7.2b)

Cell division is usually precisely controlled, and the importance of this regulation is a theme that will be repeated throughout this chapter. It helps explain, among other things, why a president's nose remains virtually the same shape year after year, why a wound on that nose starts healing and stops healing at the appropriate times, and why the loss of normal cell regulation, leading to cancer, can be so potentially dangerous.

Two additional themes will emerge as well. Once the genetic information has been copied, it must be apportioned correctly to reproducing cells and organisms. Cellular reproduction ensures that each daughter cell will receive clear instructions that will allow it to function normally. Another unifying theme is that both cell cycle and life cycle generally involve periods of growth. Without periods of enlargement, the cell's daughters and granddaughters would get smaller and smaller, and a newly reproduced plant or animal would never reach the size or maturity necessary to reproduce itself.

Our study of cellular reproduction answers several questions:

■ Where does genetic information reside in a cell?

■ What are the phases of the cell cycle?

■ How does mitosis distribute chromosomes during the cell cycle?

■ How is the cell cycle regulated?

■ How do the life cycles of multicellular organisms proceed from fertilized egg to embryo, immature stage, mature stage, and reproduction?

■ What is the special role of meiosis in life cycles?

■ What are the roles of meiosis and mitosis in evolution?

CHROMOSOMES: REPOSITORIES OF INFORMATION THAT DIRECT CELL GROWTH AND REPRODUCTION

The skin cells on a person's nose share a general strategy of cell reproduction with all other living cells: They take in nutrients, increase in size, and then divide in two. But what information directs the way a skin cell uses nutrients, grows, or divides? What prevents a skin cell from becoming two earthworm cells upon division, or one of the daughters of an oak cell from becoming a clover cell? The answer is that the cell nucleus contains the genetic information needed to govern what a cell looks like and how it performs.

Information for Directing Cell Growth: Stored in the Nucleus

In the 1930s, the German biologist Joachim Hämmerling determined that the cell nucleus rather than the cell cytoplasm governs cell growth. He arrived at this conclusion by studying *Acetabularia,* an alga that inhabits warm seas and looks like a

(a) *Acetabularia mediterranea*

FIGURE 7.3 ■ Classic Experiment with the Green Alga *Acetabularia* Revealed That Genetic Information Lies in the Nucleus. (a) Species of green algae in the genus *Acetabularia* have large single cells with either an umbrella-shaped cap or a daisy-shaped cap that will regenerate if removed. (b) Transplanting the stalk of the umbrella species to the foot of the daisy species eventually results in the regrowth of a daisy-shaped cap, proving that genetic instructions for cap shape lie in the cell nucleus.

(b) The transplant experiment
A. mediterranea

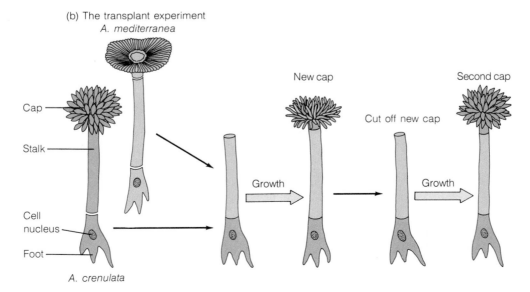

Cap

Stalk

New cap

Cut off new cap

Second cap

Growth

Cell nucleus

Foot

Growth

A. crenulata

(a) Fertilization

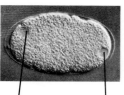

Sperm
nucleus

Egg
nucleus

(b) Nuclear fusion

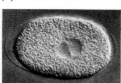

(c) Fertilized egg (zygote)

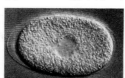

(d) Cell division by mitosis

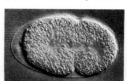

(e) 2-cell stage

FIGURE 7.4 ■ How Do We Know That the Nucleus Contains Hereditary Information? While studying a type of worm called a nematode, nineteenth-century biologists observed that during fertilization of the egg cell, the nucleus is the only component of the sperm that actually enters the egg. Since the offspring show hereditary traits from the father, they reasoned that the nucleus contains hereditary information. These photographs, taken through a modern microscope, show the sperm and egg nuclei at fertilization (a), during fusion (b), and forming a zygote, or fertilized egg (c). When the zygote divides by mitosis (d), it produces a two-cell embryo (e), which continues division. Eventually, many cells result and form the adult worm.

slender green toadstool. Each individual of *Acetabularia* is a huge single cell about 6 cm (2¼ in.) long, easily visible to the naked eye (Figure 7.3a). The cell has three main parts: a *foot,* which attaches the cell to a rock for support and contains the nucleus; a *stalk,* which grows upward from the foot; and a photosynthetic *cap,* which grows on the stalk and can resemble an inverted umbrella in some species and a daisy in others (Figure 7.3b). If the cap is cut off, the cell will regenerate another one identical to the first. But where do the directions for regenerating the new cap come from? From the cytoplasm in the stalk from which the new cap grows? Or from the nucleus, which is located in the foot?

To find out, Hämmerling took the stalk of an umbrella-cap cell (lacking both foot and cap) and transplanted it to the foot of a daisy cell (lacking both stalk and cap). This composite cell generated a new cap intermediate in shape between umbrella and daisy, suggesting that some information came from the stalk and some from the foot. When the researcher removed this regenerated cap, another grew—this time, daisy-shaped (see Figure 7.3b). Clearly, some material in the foot of the composite cell (originally a daisy-cap species) must have directed the construction of the new cap, and not some material in the stalk (originally from an umbrella-cap species).

From this and other experiments, biologists concluded that the nucleus of *Acetabularia* cells, and, in fact, the nucleus of all eukaryotic cells, is the central repository of information for constructing new parts of each new cell. But the role of the nucleus was discovered to be broader still, for it contains information used in constructing the entire new individual. The tiny roundworm *Caenorhabditis elegans,* a common worm-shaped inhabitant of the soil, is a good example. During fertilization, only the nucleus of the sperm cell actually enters the egg cell; it then moves through the egg cytoplasm and fuses with the egg nucleus. This is visible in Figure 7.4a

and b. This penetration and fusion initiate the life of a new individual, with hereditary information donated in equal amounts by the mother and father. The development of the whole organism then proceeds using both sets of information (Figure 7.4c–e). From this, we can conclude that the nucleus directs events not only in one cell and its cell cycle, but in the whole organism and its life cycle as well.

Genetic Information: Stored in the Chromosomes

What substance within the nucleus carries the information to build a cell or an entire organism? The answer to this came in 1883 from experiments with another nematode worm. Researchers found that each egg and sperm cell contained two bodies that took up stain and looked bright when viewed through a microscope. These two stained structures were named *chromosomes* ("colored bodies").

By watching fertilization under a microscope, biologists could see that chromosomes are the only structures that egg and sperm cells contribute in equal amounts to the fertilized egg and the new individual that grows from it. Thus, nineteenth-century scientists deduced that hereditary information lies in the chromosomes. Biologists know today that all organisms contain chromosomes, although the size, shape, and number of these structures differ from species to species. They also know that chromosomes contain DNA (see Chapter 2). Prokaryotes have a single long DNA molecule; the ends are joined into a circle, and the circle lies tangled in the center of the cell. Eukaryotes usually have several chromosomes, each containing just one long DNA molecule, but the molecule is linear, not circular. Within the chromosome, this long DNA molecule winds around proteins like string wrapped around a spool, and this packaging helps to decrease tangling. Electron micrographs reveal great loops of the DNA-protein

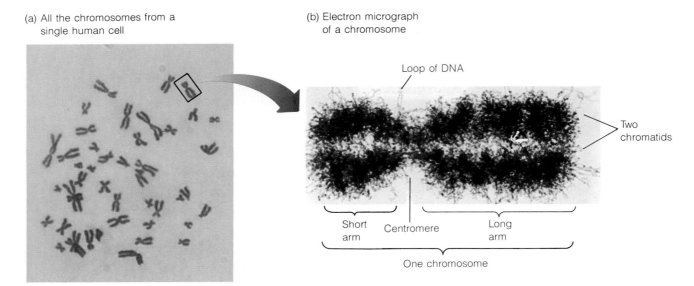

(a) All the chromosomes from a
    single human cell

(b) Electron micrograph
    of a chromosome

Loop of DNA

Two
chromatids

Short   Centromere   Long
arm   arm

One chromosome

FIGURE 7.5 ■ **Chromosomes: "Colored Bodies" in the Nucleus.** (a) Just before cell division, dark-staining structures called chromosomes become visible in the cell. A person has 46 chromosomes in each body cell nucleus. (b) In an electron microscope, a chromosome during cell division appears as two fuzzy rods connected at a point. Biologists call the entire structure a chromosome, and each of the two rods a chromatid. A single very long DNA molecule wrapped around little balls of protein makes up each chromatid. You can see these DNA-protein fibers flaring out from the edges of the chromosome.

thread extending out from the axis of a condensed chromosome (Figure 7.5b). Eukaryotes have at least two chromosomes in the nucleus: The nucleus of the nematode mentioned earlier contains 4; a giant sequoia nucleus, 22; a goldfish nucleus, 104; and the nucleus in a human skin cell, 46 (see Figure 7.5a).

The fact that chromosomes carry DNA, the genetic information needed to build a new cell, is important to our understanding of the cell cycle with its alternating phases of growth and cell division. Because each new daughter cell must receive its own set of chromosomes containing a copy of the hereditary material, the duplication and distribution of the chromosomes are central activities in the cell cycle.

THE CELL CYCLE

Just as it takes two very different sets of blueprints to build a suburban ranch house and an urban highrise, each living species has its own unique genetic blueprint. This is reflected in the numbers and sizes of chromosomes, which differ from species to species. Within a given organism, however, each body cell has the same hereditary material and a set of chromosomes of the same number, size, and shape. Thus, every skin cell on a person's nose has 46 chromosomes in the nucleus; and every cell in one type of roundworm has 12 chromosomes. When a given cell doubles in size and then divides, the chromosomes must also double in number and then be separated into two new daughter cells. The cell cycle of growth and duplication is thus echoed in the chromosomal behavior of duplication and separation. This is apparent in both single-celled organisms like bacteria and in multicellular eukaryotes like people.

The Cell Cycle in Prokaryotes: One Cell Becomes Two Through Binary Fission

The cell cycle in the intestinal bacterium *Escherichia coli* is an alternation of cell growth and cell division. The rod-shaped *E. coli* cell doubles in length and then separates in two as a partition forms in the middle. This cell division process is called **binary fission** (literally, "splitting in two"). The two new daughter cells increase in length, and under good conditions, can divide again in about 30 minutes.

During binary fission, each daughter cell receives an identical set of genetic information. Recall that a prokaryote's chromosome, its circular DNA molecule, is found in the dense-looking center of the cell rather than in a true membrane-bound nucleus (see Chapter 3 and Figure 7.6). The circular bacterial chromosome is attached to the cell membrane at one point, and as the cell doubles in size, this chromosome

(a) Binary fission in a bacterium

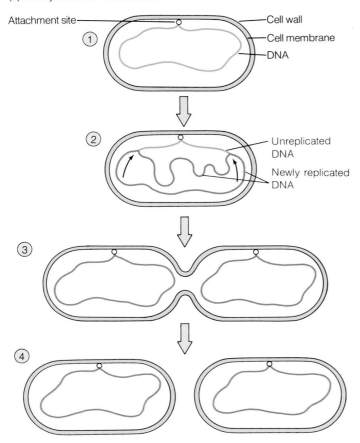

(b) Micrograph of a dividing bacterium

FIGURE 7.6 ■ **Cell Division in Prokaryotes.** **(a)** Bacteria reproduce by binary fission; that is, DNA doubles, and each cell elongates and divides in two. **(b)** Electron micrograph of a dividing bacterium, a species of *Shigella,* responsible for dysentery. A similar division process occurs in the intestinal bacterium *Escherichia coli.*

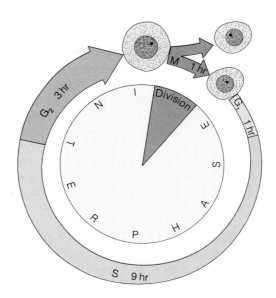

FIGURE 7.7 ■ **Four Phases of the Cell Cycle: M, G₁, S, G₂.** In a eukaryotic cell, growth (interphase) and division take place in four phases: an active growth phase (G₁) of varying length (here, about 1 hour); a long phase of DNA synthesis (9 hours or so), called S; a preparation for cell division (3 hours or so), called G₂; and then the rapid division phase (about 1 hour), called M for mitosis.

Cell Cycle in Eukaryotes: Four Phases of Growth and Division

The cell cycle in eukaryotes differs from that of prokaryotes in several ways, and the division of cells in a wound provides a good model. Such divisions are part of the four-phase cell cycle of the eukaryotic cell, with its large size (relative to a prokaryote) and its membrane-bound nucleus. Three of these four phases bring about cell growth, and only one results in cell division (Figure 7.7).

The growth period of the eukaryotic cell cycle is called **interphase,** because it intervenes between two division periods (see Figure 7.7). The three parts of interphase are labeled G₁ (for "gap 1"), S (for "DNA synthesis"), and G₂ (for "gap 2"). The fourth phase, the division period, is labeled M (for "mitosis").

■ **G₁ Initiates Interphase and Is an Active Growth Phase of Variable Length**   Early experimenters, studying the cell cycle through simple microscopes, observed two gaps, which they labeled *gap 1* (G₁) and *gap 2* (G₂), coming between the synthesis (S) and

doubles, or *replicates* (Figure 7.6a). The result is two side-by-side circles, each with its own point of attachment to the membrane (Figure 7.6a, step 3). The central partition then forms, cleaving the original cell into two daughters, each with a complete chromosome. Ribosomes and other cell structures, which have doubled in number during the growth phase, are parceled out equally during division. The two identical cells that result from binary fission then grow and repeat the cell cycle.

mitotic (M) phases. Biologists know today that G_1 and G_2 aren't gaps at all, but are important growth phases during which the cell generates new components in preparation for division. Without such growth, the cell would be halved in size and contents with each mitosis until division was no longer possible. During G_1, new proteins, ribosomes, mitochondria, endoplasmic reticulum, and other cell constituents are built in preparation for DNA synthesis and eventual cell division.

The length of the G_1 phase determines the length of the entire cell cycle. The G_1 phase can be quite short or very long, depending on the type of cell, its role in the organism, and conditions in the environment. For example, cells dividing the first few times in a newly fertilized egg have very short G_1 phases and hence divide very rapidly. Other cells, like the skin cells on a person's nose, normally have a long cell cycle and a long G_1 (a few days in length). But that can change: If a doctor removes some of those cells, the G_1 phase may shorten in the remaining skin cells at the edge of the wound, speeding up growth and division and enabling the wound to heal rapidly. The bark-forming cells of an aspen tree can also enter a shortened G_1 if bark is nibbled away by an elk or a deer, and can rapidly produce bark and thus protect the damaged area.

Whether a cell has a short or long G_1 depends on the time it takes to undergo an internal change that commits it to complete a cell cycle, beginning with the next phase, the S phase.

■ S Is a Period of DNA Synthesis

When cells leave the G_1 phase, they enter the S (synthesis) phase, during which the double-stranded DNA molecule in each chromosome is duplicated. During S, the proteins that form the little rods that DNA wraps around (the *histone* proteins) are also synthesized. When S ends, each chromosome is made up of two identical and parallel DNA molecules. By some poorly understood "bookkeeping" mechanism, every stretch of DNA in a chromosome is copied once and only once during each S phase. After this copying is complete, the cell enters G_2.

■ During G_2, the Cell Prepares to Divide

The cell continues to make many proteins during G_2, the second "gap" separating DNA synthesis and mitotic division. When a researcher artificially blocks this manufacture, the cell fails to enter the M phase. Clearly, some protein is needed to promote mitosis, and this is made during G_2. Only after that protein is produced can the cell leave the last growth phase and begin to divide, via the events of mitosis.

■ M Is the Fourth Phase and the Period of Cell Division

The M phase consists of two main events: **mitosis,** the division of the nuclear material, and **cytokinesis,** the division of the cytoplasm (Figure 7.8). During mitosis, the two sets of threadlike chromosomes are distributed to opposite ends of the cell, and two new daughter nuclei form. During cytokinesis, the daughter nuclei are separated into new daughter cells. Other cell constituents, such as mitochondria, chloroplasts, and ribosomes, duplicated during G_1, S, and G_2, are now apportioned to the two daughter cells. The result is two identical cells, each of which can now enter the G_1 stage of interphase. With this, the cycle is complete and ready to begin again—perhaps with the long G_1 of a normal skin cell or the short G_1 of a cell in a healing wound. In either case, the apportionment and separation of genetic material and other cell constituents are so crucial to cellular reproduction that we devote the next section to discussing these events fully.

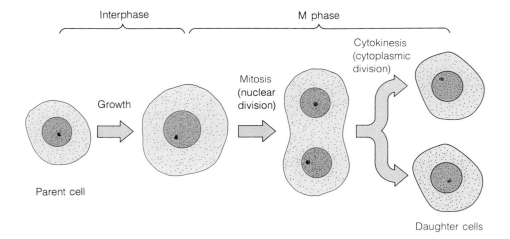

FIGURE 7.8 ■ Events of the M Phase: Division of the Nucleus and Division of the Cytoplasm. A cell grows during interphase and then enters the M phase. In the M phase, the nucleus divides during mitosis, and the cytoplasm divides during cytokinesis. Division results in two daughter cells identical to the original parent cell.

MITOSIS AND CYTOKINESIS: ONE CELL BECOMES TWO

Through a microscope, the G_1, G_2, and S phases of the cell cycle do not seem very active. By contrast, the two main events of cell division are spectacular: Mitosis, which is actually a division of the cell nucleus, is accompanied by a dramatic "dance of the chromosomes," while cytokinesis, the division of the cytoplasm, entails a rapid pinching off, or partitioning, of the cell that creates two new individuals. In the next few pages, we describe and illustrate the dynamic beauty of these cell movements.

Mitosis: Chromosome Choreography

The cell watcher equipped with a light microscope will notice two conditions that change radically once the cell enters mitosis (M): The cell, which doubles in size during interphase, is eventually halved during M; and the chromosomes, indistinguishable during interphase, become quite visible during M. Throughout most of the cell cycle, each chromosome's long DNA molecule is unraveled and tangled in the nucleus like wadded-up string. During M, however, the DNA becomes tightly wound up, and distinct chromosomes suddenly appear under the light microscope (review Figure 7.5), allowing the observer to witness the chromosome choreography as the phase proceeds.

When they first appear in the M phase, the chromosomes have already doubled (Figure 7.9 a and b), and each now consists of two identical rods, called sister **chromatids,** held together at a single point, the **centromere** (Figure 7.9b). Thus, a chromosome at the beginning of the M phase consists of two identical DNA molecules, each coiled up in a separate chromatid. The "dance of the chromosomes" involves a series of movements that cause the chromosomes to line up and the sister chromatids to separate to opposite ends, or **poles,** of the parent cell and then be partitioned into new daughter cells (Figure 7.9 b–d). Each of these daughters has one copy of each chromosome and hence is genetically identical to the other daughter and to the original parent cell. The dance is made up of five parts, summarized here and described in detail in Figure 7.10:

1. In **prophase,** the chromosomes condense, the nucleolus disappears, and a **mitotic spindle** forms—a weblike structure of microtubules that suspends and moves the chromosomes.
2. In **prometaphase,** the nuclear envelope disappears, the spindle enters the nuclear region, and the chromosomes attach to the spindle. Individual chromosomes jostle back and forth, as if objects of a tug-of-war between the two poles.
3. In **metaphase,** the spindle microtubules align the chromosomes in the middle of the spindle.
4. In **anaphase,** the centromeres divide and the spindle microtubules, in an organized manner, pull the chromatids

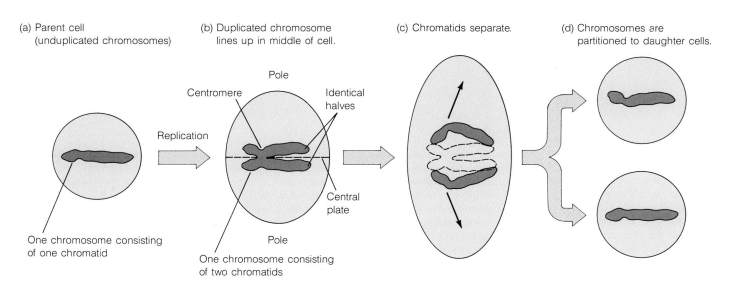

(a) Parent cell (unduplicated chromosomes)

(b) Duplicated chromosome lines up in middle of cell.

(c) Chromatids separate.

(d) Chromosomes are partitioned to daughter cells.

Pole

Centromere

Identical halves

Replication

One chromosome consisting of one chromatid

Central plate

Pole

One chromosome consisting of two chromatids

FIGURE 7.9 ■ Dance of the Chromosomes During Cell Division. **(a)** Each chromosome in the parent cell originally contains a single chromatid. **(b)** During interphase, each chromosome duplicates and comes to contain two chromatids. During the rapid M phase, each chromosome carries out a series of dancelike movements. First, the duplicated chromosome lines up along the cell's midline. **(c)** Then the chromatids separate toward opposite poles of the cell. **(d)** Finally, the chromatids—now individual chromosomes—become partitioned into separate daughter cell nuclei.

End of interphase

Mitosis =

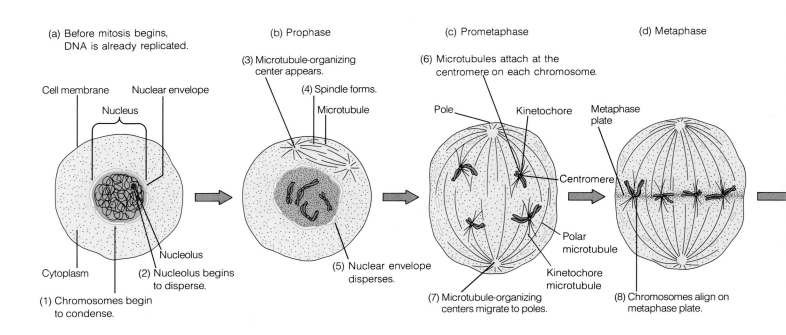

(a) Before mitosis begins, DNA is already replicated.

Cell membrane

Nuclear envelope

Nucleus

Nucleolus

Cytoplasm

(2) Nucleolus begins to disperse.

(1) Chromosomes begin to condense.

(b) Prophase

(3) Microtubule-organizing center appears.

(4) Spindle forms.

Microtubule

(5) Nuclear envelope disperses.

(c) Prometaphase

(6) Microtubules attach at the centromere on each chromosome.

Pole

Kinetochore

Centromere

Polar microtubule

Kinetochore microtubule

(7) Microtubule-organizing centers migrate to poles.

(d) Metaphase

Metaphase plate

(8) Chromosomes align on metaphase plate.

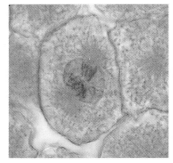

Early prophase

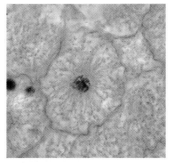

Prophase

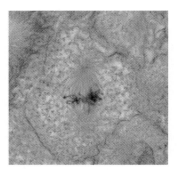

Prometaphase

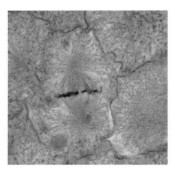

Metaphase

FIGURE 7.10 ■ The Phases of Mitosis. The cell's DNA has already replicated during interphase. As the cell enters the first part of mitosis, called prophase, the DNA reverts from its diffuse, tangled state during interphase and becomes more tightly packaged (1). Also, the nucleolus disperses (2).

The microtubule-organizing centers duplicate and spin out microtubules (3). Animal cells have centrioles, or short cylinders of microtubules, at the centers of their microtubule-organizing centers, but plant cells lack centrioles. The microtubule-organizing centers separate toward opposite ends, or poles, of the cell, spinning out the mitotic spindle, a gossamer structure that suspends and moves the chromosomes (4).

At the beginning of prometaphase, the spindle invades the nuclear region, and the nuclear envelope disperses (5). Microtubules attach to chromosomes by kinetochores, special sites located at the centromere of each chromosome (6). The chromosomes then jostle back and forth as the polar microtubules interact with the kinetochore microtubules, and the microtubule-organizing centers complete their migration to the poles (7). (Also see Figure 7.11.)

During metaphase, the chromosomes become aligned in a plane on the metaphase plate (8), a plane lying halfway between each pole, like the plane a knife makes when cutting a cantaloupe in half.

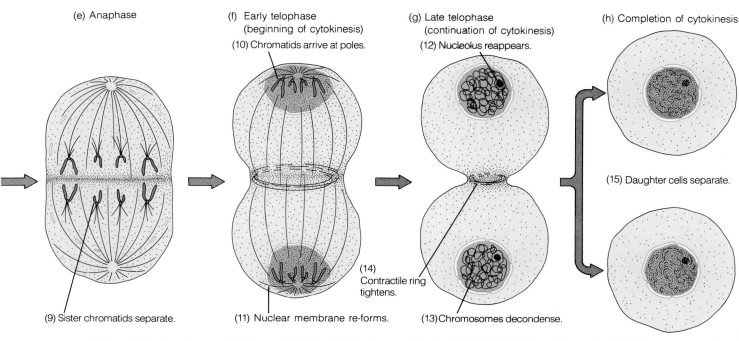

Cell division = M phase

Beginning of interphase

Division of nucleus

Cytokinesis = Division of cytoplasm

(e) Anaphase

(f) Early telophase
(beginning of cytokinesis)
(10) Chromatids arrive at poles.

(g) Late telophase
(continuation of cytokinesis)
(12) Nucleolus reappears.

(h) Completion of cytokinesis

(15) Daughter cells separate.

(14)
Contractile ring
tightens.

(9) Sister chromatids separate.

(11) Nuclear membrane re-forms.

(13) Chromosomes decondense.

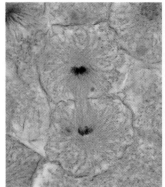

Anaphase

Next, during anaphase, the kinetochores separate, and microtubules pull sister chromatids apart, toward opposite poles (9).

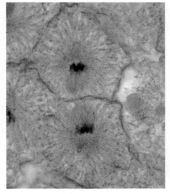

Early telophase

Early telophase marks the beginning of cytokinesis. The daughter chromatids (now independent chromosomes) arrive at each pole (10), after which the nuclear membrane re-forms around the chromosomes (11).

Late telophase

Then the nucleolus reappears (12), the spindle dissolves, and the chromosomes reel out again into a tangled mass of DNA and protein (13). In addition, in late telophase in animal cells, a contractile ring tightens around the cell's midline where the metaphase plate had been, creating a furrow (14).

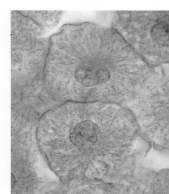

Completion of cytokinesis

Separation of the daughter cells completes cytokinesis (15). The micrographs show cells of the whitefish, magnified 450 times.

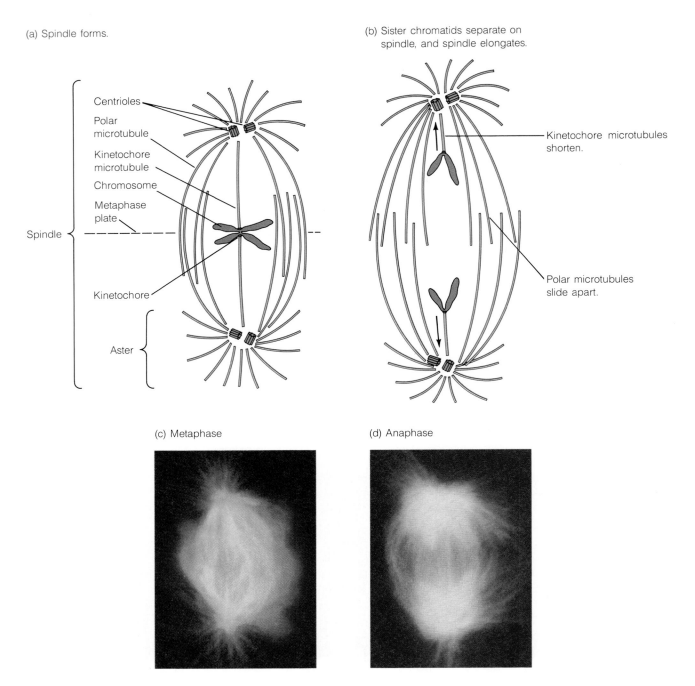

(a) Spindle forms.

(b) Sister chromatids separate on spindle, and spindle elongates.

Centrioles
Polar microtubule
Kinetochore microtubule
Chromosome
Metaphase plate
Spindle
Kinetochore
Aster

Kinetochore microtubules shorten.

Polar microtubules slide apart.

(c) Metaphase

(d) Anaphase

FIGURE 7.11 ■ What Moves the Chromosomes: Structure and Function of the Spindle. (a) As a spindle forms in animal cells, microtubules radiate around the centrioles, the cell's microtubule-organizing center, like a star, or *aster*. Longer *polar microtubules* extend from each pole toward the other and overlap at the metaphase plate in the middle of the cell. Another set of microtubules, the *kinetochore microtubules,* lead from the centrioles at the poles to each chromatid, attaching at the kinetochore, a special group of proteins at the chromosome's centromere. (b) During anaphase, sister chromatids move toward the poles, and the cell lengthens. The chromosome's centromere splits, and the sister chromatids separate, moving toward opposite poles along the kinetochore microtubules. A tiny protein-based motor in the kinetochore probably runs along the microtubule like a locomotive along a train track. The microtubule then dissolves after the "locomotive" passes by. The overlapping polar microtubules slide apart, generating the pushing force that lengthens the cell in preparation for division into two cells. (c) This photograph of a cell in metaphase shows the spindle microtubules stained yellow and the chromosomes stained red and lined up at the metaphase plate. (d) A cell in anaphase is longer and has chromosomes much nearer the poles.

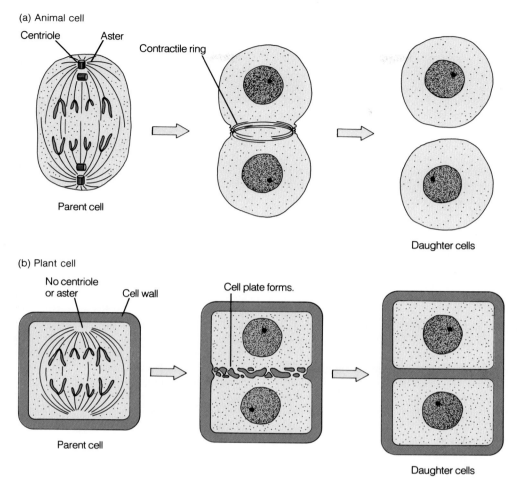

FIGURE 7.12 ■ Cytokinesis in Animal and Plant Cells. **(a) Animal cells are pinched in two from near the cell surface by a contractile ring that creates a furrow and eventually separates the daughter cells. (b) A cell plate partitions a plant cell from the inside out, forming two daughters.**

(now called chromosomes) apart and toward opposite poles. Figure 7.11 discusses the role of the spindle in separating the chromatids during anaphase.

5. In **telophase,** the chromosomes arrive at opposite poles, and the preparatory events reverse (the nuclear envelope reappears, the spindle dissolves, and so on).

Once telophase is over, the division of the cell nucleus (mitosis) is complete, and the cell now has two nuclei carrying identical sets of chromosomes. The M phase continues, however, as the cytoplasm must still divide and separate the new cells. Review each step of Figure 7.10 carefully. Box 7.1 on page 150 describes the effects of radiation on mitosis.

Cytokinesis: The Cytoplasm Divides

Toward the end of mitosis, the cytoplasm of most animal and plant cells begins to divide by means of cytokinesis (literally, "cell movement"). In both animal and plant cells, new cell

membranes form at or near the midline once occupied by the chromosomes at metaphase (Figure 7.12) and separate the two nuclei into the new cells. The details of cytokinesis vary, however, because animal cells have a pliable outer surface, while plant cells have a rigid cell wall.

Animal cells divide from the outside in as a circle of microfilaments called a **contractile ring** pinches each cell in two. During anaphase, the constriction of the ring, which contains actin, creates a dent, or **furrow,** in the cell surface, much as a purse string tightens around the neck of a purse (Figure 7.12a). This furrow deepens and eventually squeezes the cell in two.

Plant cells, with their rigid cell walls, retain their shape throughout the cell cycle and divide from the inside out (Figure 7.12b). During telophase, vesicles filled with cell wall precursors pinch off from the Golgi apparatus (called dictyosomes in plants; see Chapter 3) and collect in the cell's center, where the chromosomes align at metaphase. The many separate vesicles gradually fuse, forming a central partition, or **cell plate,** made of cell membrane and cell wall material. This

BOX 7.1   FOCUS ON THE ENVIRONMENT

RADIATION SICKNESS: A DISEASE OF MITOSIS

In the event of a nuclear war, vast amounts of radiation would be released into the environment. Many plants and animals not killed immediately by the blast would eventually die from exposure to environmental radiation. Most of us know these somber facts, but do not know *how* radiation sickness kills organisms.

To begin with, radiation breaks chromosomes. If a cell with a broken chromosome never divides, there will usually be no adverse effect on the cell's functioning. The cell will still be able to carry on in a normal fashion for the simple reason that diploid cells have two copies of each chromosome. Damage to one at a single point will usually be compensated for by the intact copy.

The real problem begins when an irradiated cell starts to divide. Since chromosomes are drawn to the poles of a dividing cell by spindle fibers attached only at the centromere, broken ends of the chromosome (not physically connected to the centromere) will be left behind during mitosis and will not be distributed normally to the daughter cells. Chromosomal information will be unbalanced, perhaps enough to cause the daughter cells to die (Figure 1). The consequence of irradiation is therefore the death of dividing cells, and the effects on the whole functioning organism can be devastating.

In an irradiated plant, growth stops, halting development of new leaves, bark, roots, and flowers. The plant continues to live, but only so long as the preexisting leaves, roots, and bark can sustain the plant. When those organs eventually die from age, predation, cold, or wind damage, they cannot be replaced, and the entire plant dies.

Animals suffer a similar fate. After radiation exposure, a victim's nerves, muscles, kidneys, and many other organs work nearly normally for quite some time. But if the individual is wounded, the wound does not heal, because skin lost from a scrape or burn does not regrow. As millions of red blood cells die of old age each day, they are not replaced, and the person becomes anemic. Nor can the body produce new white blood cells to fight infections, and the person might succumb to an otherwise relatively harmless disease. Infections are, in fact, the earliest cause of death for radiation victims. As cells lining the intestines die but are not replaced, small holes eventually develop in the gut. Bacteria seep into the body cavity, cause a massive infection that the body cannot ward off, and lead to the individual's death.

Even when radiation exposure is low, there can be long-term consequences. A very recent study of people who survived the nuclear detonations at Hiroshima and Nagasaki in 1945 found that their cancer rates have been three to four times higher than previously suspected and several times higher than nonirradiated Japanese alive then.

These grisly descriptions do two things: They underscore the unthinkable horror of nuclear war and the devastating consequences it would have for most life forms on earth and for the physical environment. The discussion also highlights the role of mitosis in our daily lives. Without the restorative powers of mitosis and its orderly distribution of hereditary information from one cell generation to the next, we would continue to live but a short and miserable time.

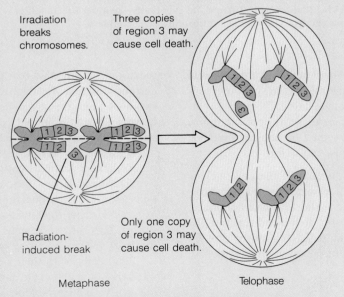

Irradiation breaks chromosomes.

Three copies of region 3 may cause cell death.

Radiation-induced break

Only one copy of region 3 may cause cell death.

Metaphase

Telophase

FIGURE 1 ■ **Mitosis in an Irradiated Cell with a Broken Chromosome Can Lead to Genetic Imbalance.**

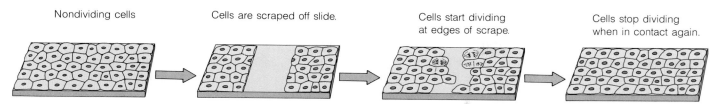

Nondividing cells   Cells are scraped off slide.   Cells start dividing   Cells stop dividing
                                                                 at edges of scrape.   when in contact again.

FIGURE 7.13 ■ **Contact Inhibition: Cells Stop Growing When They Contact Other Cells.** Cells will grow and divide only until they cover a slide or dish with a layer one cell thick. The contact of other cells at their edges inhibits further division. If a swath is cut through a layer of nondividing cells (as during a wound), cells along the edge will grow and divide until they are once more in contact with their neighbors.

fusion completes the central partition and divides the plant cell into two identical daughter cells, each with its own nucleus, ready to begin interphase.

Viewed collectively, the events of the cell cycle ensure that the cell grows, that the chromosomes are copied, that the chromosomes and other cell constituents are apportioned equally, and that the cell divides in two—in just that order. Cells, however, must sometimes cycle rapidly, sometimes slowly, and sometimes not at all. Obviously, the timing of such events must be controlled. But how?

REGULATING THE CELL CYCLE

An organism's survival depends, in part, on when and where cell division occurs. When a person gets a wound on the nose, skin cells divide, repairing the damaged tissue; but once the wound is healed, the skin cells must stop dividing, lest they give rise to a mass of skin tissue. This starting and stopping of cell division is controlled by external factors, such as cell contact and growth factor molecules that can diffuse between and within cells. Cell division is also controlled by internal factors, including specific regulatory proteins within the cell.

External Factors Regulating the Cell Cycle

■ **Cell Contact and Cell Division**   Mammalian cells growing in a Petri dish can serve as a good model for the physical factors that control cell division during wound healing. Mammalian cells in a dish behave like people loading into an elevator. Just as the people form a layer one person high and do not climb onto each other's shoulders, cells grow and divide only until they cover the dish with a layer that is one cell thick, and then remain in an extended G_1. If you remove a swath of cells from the dish, the cells at the edge of the empty path enter S phase and continue dividing until the space is once again filled; then they stop (Figure 7.13). The same thing happens when you cut your finger;

the cells at the edge of the wound divide rapidly, grow inward from all directions, close the gap, and stop dividing when they meet. In both instances, the cells are exhibiting **contact inhibition** of cell division and cell migration, which means that the cells stop growing when they contact other cells. Biologists do not know exactly how physical contact with other cells inhibits cell division, but they do know that chemical factors influence cell division rates.

■ **Growth Factors and Cell Division**   Just as the external physical factor of cell contact can control when and if a cell divides, external chemical factors can also affect cell division. Proteins called **growth factors** can enhance the growth and division of specific cell types. At the site of a cut or other wound, growth factors are released from broken cell parts, stimulating division in the cells that line damaged blood vessels and other nearby cells. There the growth factors bind to special receptors embedded in the cell membrane and trigger cellular events that culminate in the division of those cells, so that the tear in the skin becomes filled in by new cells. Whenever you cut your finger or scrape your skin, chemical growth factors are released, and together with the physical factor of cell contact, help heal the injury.

Although growth factors are normally released in very small amounts, genetic engineering techniques have made it possible for researchers to produce larger quantities that doctors can use to treat injured patients. In Oklahoma, for example, ammonia splashed into the eye of a man repairing industrial refrigeration systems. Even after three weeks of conventional therapy, his cornea (the protective covering of the eye) had failed to heal, and he could still see only blurs and shadows with the injured eye. Clinical molecular biologists at the University of Oklahoma treated his eye with drops containing epidermal growth factor made by genetic engineering, and in just four days, the man's cornea had healed. Within a few weeks, his vision had returned to nearly normal.

While cell contacts and growth factors are external elements that influence cell division, other regulatory factors work from within the cell.

Internal Factors Regulating the Cell Cycle

The external regulatory factors we have just seen must trigger some event inside the responding cell in order to influence cell division. In late 1989, a wave of excitement spread through the community of cell cycle researchers when they realized that the same two types of regulatory proteins, called *pp34* (for polypeptide 34) and *cyclin,* appear to be regulators of the cell cycle in eukaryotes as different as yeasts, fruit flies, and frog embryos.

The researchers realized that if they could understand how pp34 and cyclins operate to regulate the cell cycle, they could perhaps design new therapies to help heal patients' wounds. Numerous experiments revealed that the amount of pp34 does not change during a cell cycle, but the amount of the cyclin protein does go up and down. During G_1, cyclin and pp34 cause the cell to enter the S phase and continue through the cell cycle. This G_1 cyclin then disappears until the next cell cycle. During G_2, a second cyclin appears and binds to pp34, forming an active complex called *M-phase-promoting factor* or *MPF.* With MPF present, the cell enters mitosis. These recent studies showed researchers that the cell cycle is regulated by proteins whose levels rise and fall in rhythm with that cycle; that each step in the cycle probably causes the next to occur; and, remarkably, that the same proteins act in the cells of many kinds of organisms.

While the details are still fuzzy, it is evident that external regulatory factors—cell contact and growth factors—act on the cell's internal mechanisms to cause the levels of cyclins to rise or fall. Once the proper cyclin has appeared, the cell completes a cell cycle, and the new cells may help heal a wound. This critical control sometimes goes awry, however, and the result can be cancer.

Cancer: Cell Cycle Regulation Gone Awry

Since World War II, the life-threatening group of diseases called **cancer** has been on the rise, and today, cancer causes about one out of every three deaths each year in the United States. The biological basis of cancer is a loss of the normal cell cycle regulation we have been discussing. Cancer cells don't grow and divide more quickly than normal cells; the problem is they do not stop growing or moving after contact with other cells (Figure 7.14). Hence, cancer cells crawl over other cells, invade healthy tissues, and multiply into masses called **tumors.**

Researchers are fervently searching for the mechanisms that underlie this unusual cell behavior in the hopes of saving the thousands of cancer patients who die each year. Recent evidence indicates that some cancers are related to changes in a cell's DNA that can affect either growth factors or receptors for growth factors at the cell surface. A cancer cell, unlike a normal cell, might produce both the growth factor and its surface receptor site, so that it continuously stimulates itself to divide. Recent exciting research indicates that some cancer-

causing genes (so-called oncogenes) may exert their effects by interacting with the cyclin/pp34 proteins that regulate the cell cycle. There are many such speculations (described in Chapter 13) and still much uncertainty about cancer. But it is clear that the secrets of cancer will be found in the regulation of the cell cycle.

LIFE CYCLES: ONE GENERATION TO THE NEXT IN MULTICELLULAR ORGANISMS

For single-celled organisms, like bacteria or *Euglena,* the cell cycle just described is the same as the life cycle: Each cell division reproduces the entire organism and produces the next generation. Most multicellular organisms, however, like worms and humans, are too complex to simply split down the middle and somehow reorganize the two groups of cells into half-size versions of the original. Instead, new multicellular organisms must usually start from one or a few cells that divide repeatedly. The cells in the dividing cluster begin to take on various specialized roles, then finally develop into a new individual. At each stage, individual cells pass through the cell cycle again and again. The life cycle can thus be summarized as the process by which the immature stage undergoes development, forming the mature stage, which then carries out reproduction and forms new immature individuals.

Life cycles are clearly more complicated than cell cycles, and the last stage, reproduction, can actually follow one of two very different strategies: asexual reproduction or sexual reproduction.

Asexual Reproduction: One Individual Produces Identical Offspring

For many types of plants and a few kinds of animals, reproduction can be **asexual,** where a new but genetically identical offspring grows directly from a few body cells of a single parent. The group of cells divide mitotically and usually remain attached to the parent for a set length of time, depend-

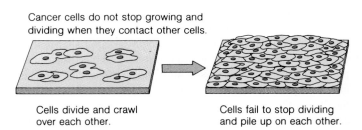

Cancer cells do not stop growing and dividing when they contact other cells.

Cells divide and crawl over each other.

Cells fail to stop dividing and pile up on each other.

FIGURE 7.14 ■ Cancer Cells Do Not Obey the Rules of Contact Inhibition. In cancer cells, the contact of neighboring cells does not inhibit growth; hence, cancer cells can form masses (tumors) that can invade healthy tissue.

(a) Cutting treated with hormone (b) Planting the cutting

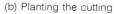

FIGURE 7.15 ■ **Asexual Reproduction in Plants: Artificial and Natural.** To make a cutting from a prize fuchsia, a gardener cuts off the new growth at the end of a young branch, dips the cut stem in a plant hormone that encourages rapid mitosis of cells at the cut surface (a), and sticks the stem into potting soil (b). In a few weeks, roots develop from the stem and establish the new plant. Since the cutting is genetically identical to the original plant, the eventual bush will have flowers the same showy size and color as the favorite original fuchsia.

ing on species. Eventually, the cell grouping can detach from the parent and begin to function as an independent organism.

You have observed asexual reproduction if you've ever left a potato in the pantry too long. Cells in the potato's "eyes" (buds of tan tissue on the surface) begin to increase in number by mitosis, draw energy from the potato starch, and develop stems and leaves. If cut from the potato and transferred to soil, these new individuals begin to photosynthesize, grow, and function as plants genetically identical to the one that produced the neglected kitchen potato. Farmers, in fact, usually plant sprouted eyes instead of potato seeds.

Farmers and gardeners also induce asexual reproduction artificially to duplicate a prize-winning specimen such as a fuchsia with stunning flowers. Using the technique described in Figure 7.15, a gardener can make a cutting that will grow into a new plant with showy flowers identical to the original's. More intricate ultramodern methods enable biologists to grow forests of identical miniature trees in their laboratories (see Chapter 31).

Asexual reproduction (also called vegetative reproduction) is not just a tool of humankind, however: It is a common survival strategy in nature. Plants such as strawberries send out runners, special arching stems that reach the ground about a foot from the parent plant, take root, and develop stems and leaves. In Southeast Asia, the bamboo plants in an entire forest are genetically identical, all derived from a single founder plant by asexual reproduction. In most cases of asexual reproduction, the new individual sprouts near the parent plant. This is a sensible strategy, since an area that is suitable for the parent should also be a benevolent environment for the identical offspring.

Certain animals also display asexual reproduction. The hydra, a relative of the jellyfish, can reproduce by **budding:** Cells in special regions undergo rapid mitosis and become organized into new hydras. Another means of asexual reproduction in animals is **regeneration:** If cut in half, a sea star, such as the orange one in Figure 7.16, can regenerate the missing portion. So can many other simple creatures. Obviously, though, a tail docked from a boxer puppy cannot regen-

erate the legs, torso, and head of a new dog. And even a roundworm cut in two will die. These and most other animals rely on sexual reproduction. And many animals and plants that usually reproduce asexually can switch to sexual modes when conditions dictate.

Sexual Reproduction: Gametes Fuse and Give Rise to New Individuals

Although asexual reproduction can lead to new individuals, the life cycle of most species involves **sexual reproduction.** Parents (usually two, but sometimes one) generate specialized cells called **gametes,** and when the gametes from opposite mating types (usually male and female) fuse, the life of a new individual begins.

Gametes are usually immobile cells called eggs or small cells called sperm that can move or be carried to the egg. In many organisms, such as people and ginkgo trees, each indi-

FIGURE 7.16 ■ **Asexual Reproduction in Animals.** The large orange arm of this sea star regenerates an entire new individual, including the six smaller orange arms.

vidual produces just one kind of gamete—egg or sperm. However, in pear trees, earthworms, and many other plant and animal species, each adult individual can produce both types of gametes. Sperm and eggs come from a specialized line of cells called the **germ line**. As we have seen, the body cells (also called **somatic cells**) of a multicellular organism eventually die. If before the organism dies, however, it reproduces sexually and its germ cells unite with those of another individual, then some portion of the organism's genetic heritage lives on in the offspring. Thus, the germ cells provide a form of immortality for multicellular organisms.

The fusion of egg and sperm is called **fertilization,** and the result is a single cell, the **zygote,** in which hereditary information from both parents unites and creates a new combination that is genetically unique (Figure 7.17a and b). The single-celled zygote may then undergo development, a period of rapid cell division and cell specialization, during which an immature form emerges, continues to grow, and changes into a mature adult capable of finishing the life cycle by undergoing meiosis and producing new gametes for a new generation (Figure 7.17c–e). Box 7.2 on page 155 describes an animal whose entire history of cell division has been studied, cell by cell.

Early in the chapter, we said that two kinds of cell division, mitosis and meiosis, characterize the life cycle of a multicellular organism. Mitosis and meiosis differ in that chromosomes in the daughter cells from a mitotic division are identical to those of the parents, while the daughter cells from a meiotic division have only half as many chromosomes as the parent cell. **Meiosis** (literally, "to make smaller") is a spe-

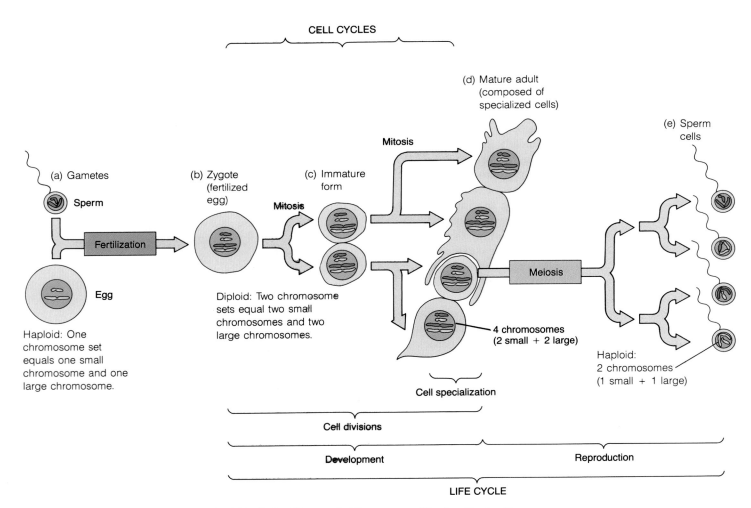

FIGURE 7.17 ■ **Chromosomes and the Life Cycle: Changes in Chromosome Number During Fertilization and Meiosis.** (a) Gametes (egg and sperm) are haploid, with just a single set of chromosomes, two in this example. (b) A zygote, formed at fertilization by the union of egg and sperm, represents a new genetic combination with the normal diploid (2n) chromosome number. (c) Cell division by mitosis gives rise to an immature form. (d) Further mitotic divisions and cell specializations lead to the mature adult—here, a fanciful, imaginary organism with just four cells. (e) Meiosis takes place in gamete-forming organs and results in the generation of eggs or sperm with the haploid (1n) chromosome number.

BOX 7.2　　　　　　　H O W　D O　W E　K N O W ?

THE WHOLE WORM CATALOG

Dig a shovelful of soil, and it will teem with transparent worms called nematodes that taper at both ends and move by wriggling and lashing back and forth. Nematodes are amazingly common and abundant: More of them inhabit the dirt in a garden planter box than there are people on earth, and enormous numbers of many nematode species live in bodies of water or in the bodies of plants and animals.

After 30 years of intensive study, biologists have amassed what amounts to a "Whole Worm Catalog," describing the nematode called *Caenorhabditis elegans* (Figure 1) in more detail than any other animal ever studied. While such a compendium may sound like boring reading compared to a spy novel or even a college textbook, this collected knowledge is going to deepen our understanding of animal life remarkably.

In the early 1960s, British biologist Sydney Brenner set out to learn everything there was to know about the embryonic development, reproduction, and day-to-day functioning of *C. elegans*. On top of that, he intended to do it quickly. In retrospect, the short time frame was overly optimistic, and three decades later, Brenner—along with researchers all over the world—are still trying to fully understand the simple animal just half the size of an *l* on this page. Nevertheless, their labors have produced a foundation from which to pose and answer some of the biggest questions in animal research.

They found, for example, that like virtually all animals, a *C. elegans* nematode starts as a single fertilized egg that divides and redivides until an embryo with major organs forms, a young worm containing just a few dozen cells hatches, and within three days, an adult has formed with precisely 959 cells. By tracing the fate of each dividing cell all the way from fertilized egg to adult, biologists learned the derivation of each of those 959 cells, and this gives them a unique window on some of the major puzzles in animal development: How do particular organs arise? What causes cell division to start and then stop when

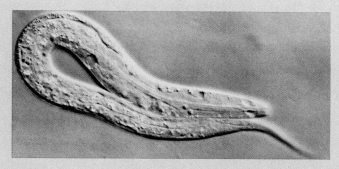

FIGURE 1 ■ **The Nematode *Caenorhabditis elegans*.**

organs are the right size? What happens to the embryo or adult if one of the very early cells is damaged or destroyed?

Another major section of the "Whole Worm Catalog" concerns the nematode's nervous system. Biologists have discovered that an adult *C. elegans* has exactly 302 neurons (nerve cells), and they've learned exactly how they connect and interact with each other. This "wiring diagram" fills a 340-page document and serves as a reference manual from which to study reactivity, movement, and other behaviors of the whole organism at the molecular level.

In what may be the most significant work of all, geneticists have completed a map of known genes in the *C. elegans* chromosomes, and they are hoping by the year 2000 to unravel the sequence of all 100 million nucleotide bases (see Chapter 2) in these collective genes. With such a detailed road map superimposed on the well-charted territories of the nervous system and development, biologists will be able to explore the terrain of animal function—integrated from the level of genes all the way to the level of observable structure and behavior—as never before.

---

cial type of cell division that produces gametes or spores, specialized cells whose chromosome number is half that of other body cells (see Figure 7.17). If this chromosome reduction did not take place to counteract fertilization, a doubling of the chromosome number would occur with each new generation.

Fortunately, meiosis does take place in sexually reproducing species, from single-celled yeasts to people and other multicellular organisms. Thus, eggs and sperm of the fanciful species shown in Figure 7.17 contain two unpaired chromosomes (one long and one short; Figure 7.17a), while each body cell has four chromosomes (two long and two short; Figure 7.17d). When the gametes of this species fuse, the zygote receives one long and one short chromosome from each gamete, and the normal chromosome number (four) is restored (Figure 7.17b). Mitotic divisions of the zygote then take place, and the growing individual continues to have four chromosomes per body cell.

Biologists use special terms to describe a cell's chromo-some count. A cell that is **haploid** (literally, "single vessel") has just one chromosome set in its nucleus, while a **diploid** ("double vessel") cell has two sets (Figure 7.17a and b). Thus, a gamete from our hypothetical organism with two unpaired chromosomes (one long and one short) is haploid (Figure 7.17a), whereas body cells, with two chromosome pairs, are diploid (Figure 7.17b–d). Likewise, a human egg or sperm is haploid and has 23 chromosomes, while a human body cell is diploid and contains 46. If a person is conceived without exactly 46 chromosomes, the results can be profound. Box 7.3 on page 157 describes one such consequence: Down syndrome.

Fertilization and meiosis play essential but opposite roles in a life cycle involving sexual reproduction: Fertilization doubles the chromosome number, while meiosis divides it in half. As in mitosis, a dramatic dance of the chromosomes during meiosis helps ensure that each daughter cell will receive the appropriate chromosome number.

(a)  (b)

FIGURE 7.18 ■ **Meiosis Occurs in Specialized Organs of Plants and Animals.** Mitosis takes place throughout the body, but meiosis occurs only in specialized organs, such as (a) the anthers (here, yellow ovals) or ovaries of flowers such as the columbine and (b) the sex organs of a male ibex.

MEIOSIS: A RESHUFFLING AND REDUCTION OF CHROMOSOMES

No earthworm, no horseshoe crab, no human child is ever genetically identical to its parent, and neither is any plant derived through sexual reproduction. We sometimes hear the phrase "like begets like," and there is a resemblance between generations that is, in fact, an important stabilizing factor in evolution. To evolve, however, means to change, and some mechanism must also allow for differences to arise between one generation and the next. Meiosis provides one such mechanism. While mitosis takes place throughout the body, meiosis occurs in just the few cells within reproductive organs that produce gametes (Figure 7.18). And unlike mitosis, meiosis reduces the chromosome number in gametes by half, so that the fusion that takes place during fertilization restores the diploid chromosome number characteristic for each species.

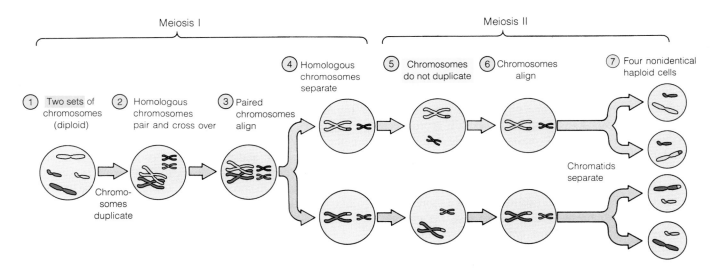

FIGURE 7.19 ■ **Summary of Major Events in Meiosis.** During meiosis I, chromosomes replicate, homologous chromosomes pair and cross over, and the homologues separate (1–4). During meiosis II, a chromosome's two chromatids separate, and each gamete (germ cell) receives a haploid set (5–7).

BOX 7.3   *P E R S O N A L   I M P A C T*

DOWN SYNDROME: HOW MISTAKES IN MEIOSIS ALTER DEVELOPMENT

Like 1 in every 1000 children born in the United States, the boy in Figure 1 has **Down syndrome.** Like most other people with this condition, he is mentally impaired; his heart is malformed; he has a special type of eye fold; he is short; his hands and feet are stubby; and his palm prints are abnormal. In searching for the cause of Down syndrome, researchers observed that a woman over 45 years old is 100 times more likely to have a baby with Down syndrome than a woman of 19.

When researchers closely examined cells from children with Down syndrome, they found 47 chromosomes in the nucleus of each cell rather than the normal human complement of 46. The cells had three copies of chromosome 21; for this reason, the syndrome is sometimes called *trisomy 21* (literally, "three copies of chromosome 21"). Geneticists traced this extra chromosome to a mistake in meiosis that could be linked to the mother's age.

Long before a woman reaches maturity—in fact, at the time of her own birth—she possesses reproductive organs (ovaries) that contain all the egg cell precursors (oocytes) she will ever have. Each precursor is arrested in prophase I, the first phase of meiosis, and can remain that way for many decades until a hormone signal causes the egg to resume meiosis in preparation for fertilization. Somehow, during that long stage of arrested meiosis, an egg is occasionally damaged in a way that prevents meiosis from occurring properly when it resumes. The older a woman gets, the more likely this damage becomes.

One mistake in meiosis that can result from the damage is **nondisjunction,** during which the homologues fail to separate or disjoin. The egg could then contain two copies of chromosome 21. When fertilized by a normal sperm with one copy of

FIGURE 1 ■ A Child with Down Syndrome.

chromosome 21, the zygote could then have three copies of chromosome 21 (trisomy 21). The embryo with trisomy 21 has a gene imbalance in each cell, and this alters normal development and leads to the characteristic birth defects seen in Down syndrome.

Unraveling the cause of Down syndrome was an important scientific achievement, although it cannot ameliorate the hardships faced by victims and their families. However, once researchers revealed the age connection, women could at least decrease their likelihood of producing a Down syndrome child in two ways: by planning to complete their families before age 40 and by seeking specific tests during pregnancy that reveal if a fetus has chromosomal defects (see Chapter 12).

The Stages of Meiosis

Meiosis changes diploid cells into haploid cells through a two-part sequence of events called meiosis I and meiosis II. In meiosis I, a diploid parent cell divides, forming two haploid daughters, and in meiosis II, those two daughters divide again, resulting in four haploid cells. Meiosis I is somewhat similar to the events of mitosis, except that sets of chromosomes act as pairs rather than acting independently as in mitosis. Meiosis II is even more similar to the events of mitosis than meiosis I, except that no DNA synthesis occurs during the interphase between meiosis I and meiosis II, and for this reason, the chromosome number remains haploid in the meiotic division of telophase II. Figure 7.19 shows a summary of the major events of meiosis I and II. Figure 7.20 illustrates and explains each of the five phases in meiosis I and II. The cells in both figures represent an organism whose body cells contain just two pairs of chromosomes: a long and a short one donated by the father's gamete and a long and a short one donated by the mother's gamete. Maternal and paternal copies of the same chromosome bearing very similar but usually not identical genetic information are called **homologous chro-**

mosomes. (Appendix A.2 compares mitosis and meiosis, side by side.)

Genetic Variation Arises During Meiosis

Mitosis leads to genetically identical daughter cells, but the cells produced during meiosis can only be regarded as *similar,* not identical. During meiosis, a reshuffling of the maternal and paternal chromosomes leading to a brand new chromosome combination takes place (Figure 7.21a). This reshuffling is called **genetic recombination,** and it occurs in two ways: through crossing over and independent assortment.

■ **Crossing Over: An Exchange of Chromosome Parts** The first meiotic mechanism that ensures variety in the gametes is chromosomal **crossing over,** a process during which corresponding parts of homologous maternal and paternal chromosomes are exchanged (Figure 7.22). During prophase of meiosis I, homologous chromosomes pair. While the homologues are joined, enzymes break the DNA molecules in each homologue, switch corresponding regions, and then rejoin them. Crossing over during meiosis results in a chromosome of mixed origin, with portions derived from both maternal and paternal chromosomes.

Meiosis I

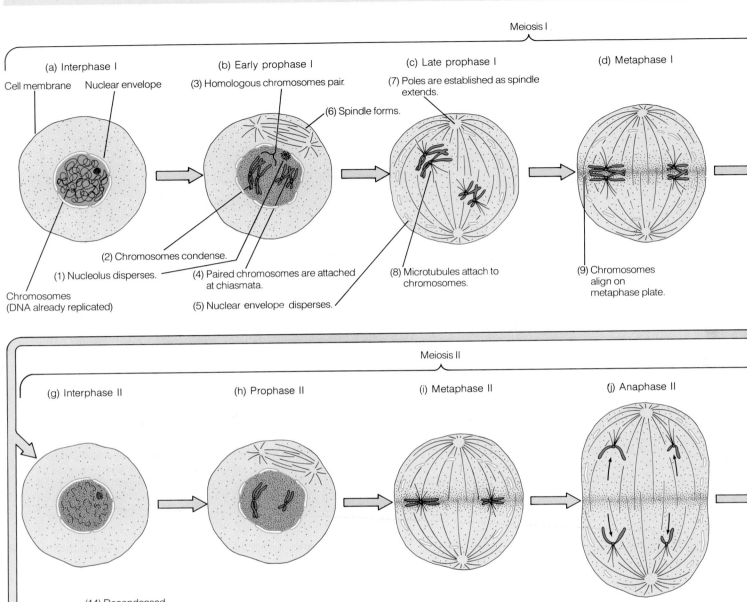

(a) Interphase I

Cell membrane   Nuclear envelope

Chromosomes
(DNA already replicated)

(1) Nucleolus disperses.

(2) Chromosomes condense.

(b) Early prophase I

(3) Homologous chromosomes pair.

(6) Spindle forms.

(4) Paired chromosomes are attached
    at chiasmata.

(5) Nuclear envelope disperses.

(c) Late prophase I

(7) Poles are established as spindle
    extends.

(8) Microtubules attach to
    chromosomes.

(d) Metaphase I

(9) Chromosomes
    align on
    metaphase plate.

Meiosis II

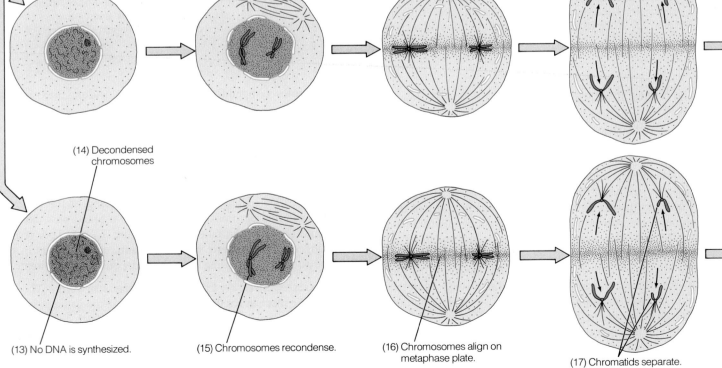

(g) Interphase II

(h) Prophase II

(i) Metaphase II

(j) Anaphase II

(14) Decondensed
     chromosomes

(13) No DNA is synthesized.

(15) Chromosomes recondense.

(16) Chromosomes align on
     metaphase plate.

(17) Chromatids separate.

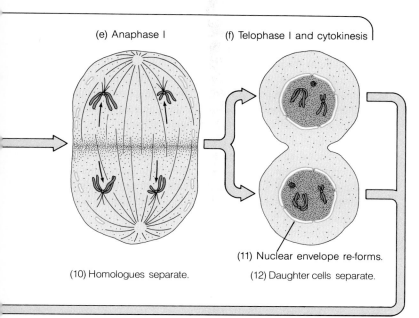

(e) Anaphase I

(f) Telophase I and cytokinesis

(11) Nuclear envelope re-forms.

(10) Homologues separate.

(12) Daughter cells separate.

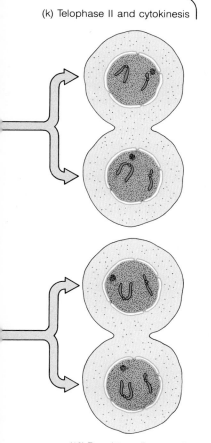

(k) Telophase II and cytokinesis

(18) Daughter cells separate.

FIGURE 7.20 ■ **The Phases of Meiosis.** During the S phase of interphase I, DNA is synthesized, as in a normal diploid cell at the corresponding phase of a conventional mitotic cell division.

1. In early prophase I, just as in mitosis, the nucleolus disperses.
2. The duplicated chromosomes condense and become visible through the light microscope.
3. Homologous chromosomes line up together. In the cells shown here, the two long chromosomes pair with each other, and the two short chromosomes pair side by side. The pairing of homologous chromosomes is the most significant event of prophase I and does not occur in mitosis.
4. Toward the end of early prophase I, homologous chromosomes separate slightly, but remain attached to each other at one or two places called the *chiasmata*. In these regions, one chromosome appears to cross over another.
5. In late prophase I, the nuclear envelope breaks down.
6. The spindle forms.
7. The spindle poles are established.
8. Chromosomes are attached to kinetochore microtubules.
9. During metaphase I, paired homologous chromosomes align at the metaphase plate. Both chromatids of each homologue orient toward the same pole, while the chromatids of the other homologue orient toward the opposite pole. (In mitosis, since homologous chromosomes do not pair, the two chromatids of each chromosome orient toward opposite poles instead of toward the same pole.)
10. During anaphase I, the homologues separate from each other, but sister chromatids remain linked at their centromeres and do not separate from each other as they do in mitosis.
11. Nuclear membranes re-form during telophase I, and the cytokinesis that follows produces two daughter cells.
12. Each of the daughters has two chromosomes consisting of two sister chromatids attached at their centromeres, with the haploid number of chromosomes.
13. During interphase of meiosis II, no DNA synthesis occurs, in contrast to mitosis and meiosis I.
14. The chromosomes decondense into stringlike masses. Since there is no DNA synthesis, the two daughter cells that have passed through meiosis I continue to have the haploid number of chromosomes but the diploid amount of DNA as they enter prophase II.
15. During prophase II, individual chromosomes become visible in the light microscope, the nuclear membrane breaks down, and the spindle forms.
16. During metaphase II, the chromosomes line up on the metaphase plate, and this time the two chromatids orient to opposite poles, as in mitosis.
17. During anaphase II, the two sister chromatids separate and become individual chromosomes.
18. After the nuclear membrane re-forms during telophase II, cytokinesis takes place and four cells result. Each of the four daughter cells has the haploid number of chromosomes and the haploid amount of DNA. The events of meiosis result, overall, in a single diploid cell giving rise to four haploid cells.

(a) Puppies with different phenotypes

(b) One possible chromosome arrangement

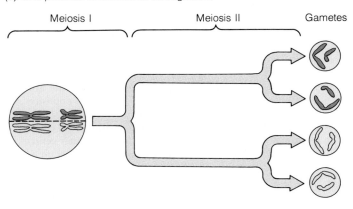

(c) Another possible arrangement

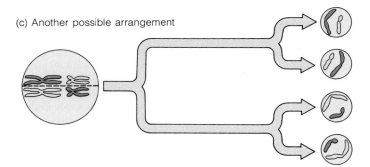

FIGURE 7.21 ■ Meiosis, Genetic Recombination, and Genetic Variation. (a) These puppy siblings of different sizes, colors, and fur patterns are visible evidence of the genetic variation produced by meiosis. (b, c) Independent assortment during meiosis can help provide the genetic variability evident in the puppies. In one of the mother dog's egg-forming cells, two chromosomes (dark blue) inherited from her father (the maternal grandfather of the puppies) lie on the same side of the metaphase plate during meiosis I (b), while in a different egg-forming cell, two chromosomes inherited from her father lie on opposite sides of the metaphase plate (c). These two arrangements give different combinations of chromosomes in the gametes, the eggs, as the diagram shows on the right. Since different puppies inherit different combinations of chromosomes from their maternal grandfather and grandmother, they can be genetically quite variable.

■ **Independent Assortment: Chromosomes Randomly Distributed to Gametes**   The second means of genetic recombination depends on the way chromosomes pair. The chromosomes in Figure 7.21b are arranged with the paternal chromosomes above and the maternal chromosomes below. Figure 7.21c shows another possible arrangement, with one maternal and one paternal chromosome on each side of the spindle. Because of this arrangement, the chromosome distribution in the gametes in part (b) differs from the distribution shown in part (c). Each gamete pictured has a haploid set of chromosomes (one short and one long), but a total of four combinations are possible based on parental origin. These combinations differ from each other and from the parent cells. Clearly, new genetic combinations can arise during meiosis, since maternal and paternal chromosomes are genetically different.

The assortment of parental chromosomes into gametes is entirely random, and thus all four combinations occur with equal frequency. Geneticists call the random distribution process **independent assortment** and have traced its origin to the way the homologous chromosomes line up along the metaphase plate during meiosis I.

The independent assortment of chromosomes during meiosis can be compared to choosing from the menu of a French country restaurant, where there are two choices for each course. The two choices—duck or veal, Brie or Camembert, apple tart or crème caramel, and so on—are like the two homologous chromosomes. Just as the potential number of unique meals depends on the number of courses, the potential number of unique genetic combinations in the gametes depends on the number of chromosomes. From a menu with six courses, two choices per course, one could make up 2^6 (or $2 \times 2 \times 2 \times 2 \times 2 \times 2$), or 64, different meals. For gametes, that number is 2^n, where n is the haploid chromosome number. For the cells diagrammed in Figure 7.21b and c, $n = 2$, so the species could form 2^2, or 4, different gamete types, as we saw. For a human, that number would be 2^{23}, or a potential of 8 million different chromosome combinations in the

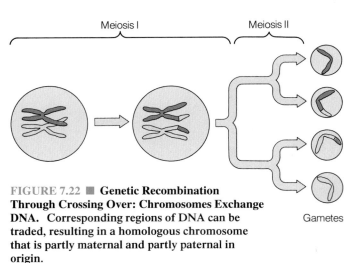

FIGURE 7.22 ■ Genetic Recombination Through Crossing Over: Chromosomes Exchange DNA.   Corresponding regions of DNA can be traded, resulting in a homologous chromosome that is partly maternal and partly paternal in origin.

nuclei of eggs or sperm, and the combining of maternal and paternal chromosomes in the zygote at fertilization increases the number of genetic combinations still further. This incredible genetic diversity based on independent assortment helps explain why an organism with several chromosomes is unlikely to produce two genetically identical gametes, and why, in turn, cucumbers, prize bulls, people, and all other organisms resemble but are never exactly like either parent.

Together, crossing over and independent assortment ensure that offspring produced through sexual reproduction are not genetically identical to their parents. When you see a litter of puppies, all with different fur color, pattern, and body size, you are seeing evidence that genetic recombination took place during meiotic divisions in the gamete-producing reproductive organs of the parent dogs. The puppies resemble their parents in many ways, but differ in at least a few, and some of

these differences may render the young more fit to survive. And therein lies the evolutionary significance of genetic recombination during meiosis.

MEIOSIS, MITOSIS, SEXUAL REPRODUCTION, AND EVOLUTION

Mitosis and meiosis play very different but equally important roles in the life cycles of sexually reproducing eukaryotic organisms (Figure 7.23). In preparation for fertilization and sexual reproduction, meiosis reduces the chromosome number in the gametes and brings about a critical genetic reshuffling. Once fertilization takes place, mitotic divisions of the zygote produce the growing individual, say, the 3 billion cells

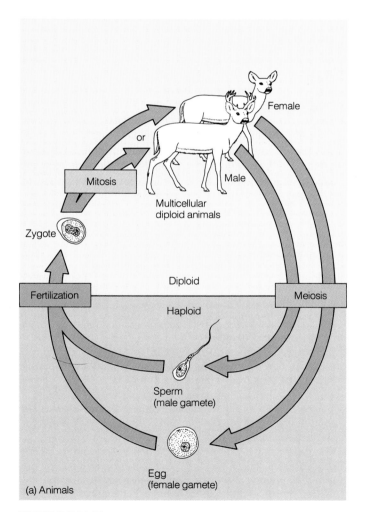

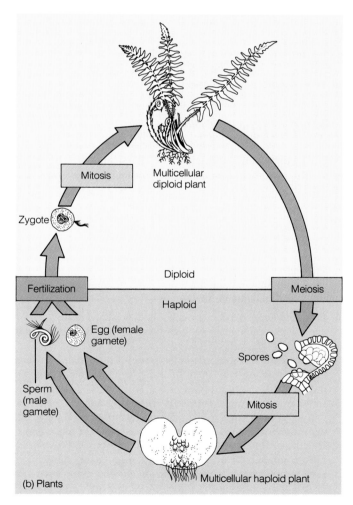

FIGURE 7.23 ■ **The Roles of Mitosis and Meiosis in the Life Cycles of Animals and Plants.** (a) In most animals, each adult produces either egg or sperm via meiosis, and the union of gametes at fertilization yields a zygote that divides mitotically to form a new adult. (b) In many plants, meiosis produces haploid spores that grow mitotically into multicellular haploid plants, sometimes free-living, as in this fern, sometimes dependent on the adult plant, as in pines or tulips. Each haploid plant then produces gametes through mitotic divisions, and the union of haploid gametes creates a zygote, restoring the diploid chromosome number in the future adult.

of a newborn human baby. In asexually reproducing organisms, such as potatoes and hydras, mitosis allows a few parent cells to generate identical offspring that will grow and once again reproduce, asexually or sexually.

It may seem odd that nature would invent two means of reproduction. But there are distinct advantages to each mode, depending on conditions at the time of reproduction. In an unchanging environment—say, the cold, dark mud at the bottom of the deepest oceans—asexual reproduction may be superior to the sexual mode. If a parent has a favorable genetic combination that allows it to survive these hostile conditions, then reproducing asexually would preserve the parent's time-tested genetic combination and automatically result in offspring suited to the dark, the cold, and the mud. In a changing environment, however, such as salt flats near the ocean shore, which are subject to periodic drying, flooding, and changing salt levels, one genetic combination might be advantageous under dry conditions but a disadvantage under wet conditions. Here, asexual reproduction might doom the offspring to the parent's losing gene combination, whereas sexual reproduction—with its reshuffling of genes during meiosis—might deal a few of the offspring a winning hand in surviving unpredictable conditions.

CONNECTIONS

Carefully timed and regulated cell divisions underlie many basic life processes, including reproduction, development, growth, wound healing, and the routine replacement of aging cells, tissues, and organs. We also saw that cell division is a key feature in the simple cell cycles of bacteria as well as the more complex life cycles of plants and animals.

One principle of the cell theory presented in Chapter 3 can be fully understood only in the context of cell division. We said there that new cells arise only from preexisting cells, and in this chapter, we saw that only through an alternation of enlargement and division can cells give rise to new generations. We can now extend the notion of life arising from preexisting life to include multicellular organisms: New plants, animals, and fungi arise only from preexisting members of their own species via asexual or sexual reproduction and mitotic and meiotic divisions.

Finally, we encountered a graceful "dance of the chromosomes" during both mitosis and meiosis that distributes hereditary information, either identical or reshuffled. We saw how mistakes or loss of normal control can have consequences as serious as Down syndrome and cancer. These mistakes help explain why radiation exposure can be so deadly for plants and animals. And the study of an example like Mr. Reagan's skin lesion helps us understand the fundamentals of cell division, development, and survival.

The chapters that follow build on the principles of cell division, delving into the nature of the hereditary information and the way molecular geneticists manipulate it to create new gene combinations, as well as covering such topics as sexual reproduction and the development of the embryo. All these concepts, however, rely on the gene. So in the next chapter, we explore the gene and the historic experiments that led to its discovery.

Highlights in Review

1. President Reagan's skin growth—both normal and abnormal—and wound healing is a good case study for understanding cell cycles and life cycles.
2. Studies with a single-celled alga revealed that the information for constructing each part of each new cell resides in the nucleus.
3. In multicellular organisms, only the sperm nucleus enters and merges with the egg nucleus during fertilization. The nucleus therefore directs the life cycle as well as the cell cycle.
4. Hereditary information lies in the chromosomes. Prokaryotes have a single circular chromosome, while eukaryotes have at least two separate linear chromosomes. The duplication and distribution of the chromosomes are central to the cell cycle.
5. The cell cycle is an alternation of growth and division.
6. Prokaryotic cells increase in size and then undergo binary fission. During this process, the circular chromosome is duplicated, each copy is localized at one end of the double-sized cell, and a partition forms, separating the two cell halves.
7. Eukaryotic cells have a four-phase cell cycle, with three phases devoted to growth and DNA duplication, and one to cell divi-

sion. The three-phase growth period is called interphase, and the division period, mitosis.
8. Interphase has three phases: G_1 (gap 1), during which the cell produces new proteins and cell parts; S (synthesis), during which DNA molecules are copied; and G_2 (gap 2), during which the cell makes more protein and prepares to divide.
9. During mitosis, or the M phase, duplicated chromosomes are apportioned equally to new daughter cells, and cytokinesis creates two new cells.
10. Mitotic division has five parts. In prophase, the chromosomes condense, the nucleolus disappears, and the mitotic spindle forms. In prometaphase, the nuclear envelope disappears, and the chromosomes attach to the spindle. In metaphase, the chromosomes become aligned in the middle of the cell. In anaphase, the chromatids are pulled to opposite poles. In telophase, the events of prophase and prometaphase reverse in preparation for cytokinesis. The cell now has two identical nuclei.
11. During cytokinesis in animal cells, a contractile ring of microfilaments makes a furrow in the pliable cell surface and even-

tually squeezes the cell into two daughters, each with an identical nucleus formed during mitosis. In plant cells, a cell plate made of cell wall and cell membrane material forms down the cell's midline and partitions the cell into two identical daughters.

12. Contact inhibition of cell division is an external control over the cell cycle. Growth factors also affect the timing of cell division.

13. Cancer cells do not stop growing or migrating upon contact with other cells and hence invade healthy tissue and form tumors.

14. While the cell cycle is an alternation between cell division and cell growth, the life cycle involves formation of the zygote, development of the immature stage, continued development to a mature stage, and reproduction.

15. Reproduction can be asexual and involve mitotic divisions alone, or it can be sexual and involve both meiosis and mitosis.

16. In asexual reproduction, a few body cells of a single parent divide mitotically, producing an offspring genetically identical to the parent.

17. During sexual reproduction, gametes form after meiotic divisions. Meiotic divisions halve the chromosome number, making the gamete haploid; fertilization then restores the diploid chromosome number in the zygote.

18. Meiosis has two parts, meiosis I and meiosis II. In meiosis I, homologous chromosomes pair and then separate to opposite poles, halving the number of chromosomes in the two daughter cells. In these cells, meiosis II proceeds without DNA synthesis, and chromatids separate to opposite poles, yielding four haploid cells.

19. The four daughter cells produced through meiotic divisions are similar but not identical to the parent cell. Independent assortment causes a random distribution of parental homologous chromosomes in the gametes, and crossing over results in a switching of parts of homologous chromosomes.

Key Terms

anaphase, 145	growth factor, 151
asexual reproduction, 152	haploid, 155
binary fission, 142	homologous
budding, 153	chromosomes, 157
cancer, 152	independent assortment, 160
cell cycle, 139	interphase, 143
cell division, 139	life cycle, 139
cell plate, 149	meiosis, 154
centromere, 145	metaphase, 145
chromatid, 145	mitosis, 144
contact inhibition, 151	mitotic spindle, 145
contractile ring, 149	nondisjunction, 157
crossing over, 157	pole, 145
cytokinesis, 144	prometaphase, 145
diploid, 155	prophase, 145
Down syndrome, 157	regeneration, 153
fertilization, 154	sexual reproduction, 153
furrow, 149	somatic cell, 154
gamete, 153	telophase, 149
genetic recombination, 157	tumor, 152
germ line, 154	zygote, 154

Study Questions

Review What You Have Learned

1. Describe how binary fission takes place in a prokaryotic cell.
2. Name and describe the four phases of cell division in a eukaryotic cell.
3. How do mitosis and cytokinesis differ?
4. Distinguish between a chromosome and a chromatid.
5. Arrange these phases of mitosis in their correct order: prometaphase; prophase; anaphase; telophase; metaphase.
6. Compare cytokinesis in animal and plant cells.
7. What might allow cancer cells to continue growing indefinitely?
8. How do mitosis and meiosis differ?
9. Explain how new genetic combinations can arise during meiosis.
10. Which is more advantageous in a changing environment, sexual reproduction or asexual reproduction? Why?

Apply What You Have Learned

1. A mutation causes a cell to produce cyclins continuously. How would this affect the cell, and what might be the result?
2. A plant breeder who has developed an attractive-looking pink geranium flower wishes to obtain many more flowers of the same variety. Should the breeder employ sexual or asexual methods of reproduction? Why?
3. Using your knowledge of human genetics, explain why a baby born to a 40-year-old woman runs a greater risk of being born with Down syndrome than a baby born to a 25-year-old woman.

For Further Reading

ALBERTS, B., D. BRAY, J. LEWIS, M. RAFF, K. ROBERTS, and J. D. WATSON. *Molecular Biology of the Cell.* 2d ed. New York: Garland, 1989.

CHANDLEY, A. C. "Meiosis in Man." *Trends in Genetics* 4 (1988): 79–83.

COWEN, R. "Speeding Up Wound Healing the EGF Way." *Science News* 136 (1989): 39.

KENYON, C. "The Nematode *Caenorhabditis elegans.*" *Science* 240 (1988): 1448–1453.

McINTOSH, J. R., and K. L. McDONALD. "The Mitotic Spindle." *Scientific American,* October 1989, pp. 26–34.

MURRAY, A. W., and M. W. KIRSCHNER. "Dominoes and Clocks: The Union of Two Views of the Cell Cycle." *Science* 246 (1989): 614–621.

MURRAY, A. W., and M. W. KIRSCHNER. "What Controls the Cell Cycle." *Scientific American,* March 1991, pp. 56–63.

ROBERTS, L. "The Worm Project." *Science* 248 (1990): 1310–1313.

VAUGHAN, C. "Hiroshima Study Shows Higher Risks of Low-Level Radiation." *New Scientist,* January 6, 1990, p. 28.

WEINBERG, R. A. "Finding the Anti-Oncogene." *Scientific American,* September 1988, pp. 44–51.

Mendelian Genetics

White Tigers and Family Pedigrees

In the dry season of 1951, hunters found four tiger cubs playing in the hunting preserve of the Maharajah of Rewa, in central India. Although the hunters were looking for bigger game, one of the cubs caused great excitement: Instead of the normal orange-and-black-striped coat, it had a pure white pelt! Lured into a cage by a bowl of water, the young tiger was brought to the maharajah's palace and named Mohan, meaning "Enchanter." Mohan grew into a magnificent creature, larger than most tigers, strong, and healthy. But he lacked the usual rich orange and jet black pigments of normal tigers; his coat was nearly pure white, with ashen gray stripes. What's more, his nose and paw pads were pink instead of black, and his eyes were ice blue with a tendency to cross (Figure 8.1).

Was Mohan just a fluke of nature, a unique occurrence? Or were his special traits *hereditary,* that is, capable of being passed to offspring and through them to future generations?

Mohan was mated with Begum, a normal tiger, but the results were disappointing: None of the cubs had Mohan's sparkling white coat, and it seemed that Mohan was a once-only occurrence, never to be repeated. But Mohan was kept in a cage with one of his daughters, Radha, and you can imagine the surprise of his keepers when they saw the offspring from an incestuous mating of the two, represented in the family tree in Figure 8.2. Some of the cubs in the second generation had beautiful white coats. To the maharajah's joy, the trait *was* hereditary. Although it appeared to have skipped Radha's generation, it had remained intact (though hidden) and had reappeared in the next generation. What could cause a hereditary trait to behave in this hit-and-skip fashion? And what was the relationship of the white pelt to the crossed blue eyes?

FIGURE 8.1 ▪ **A White Tiger with Blue Eyes That Tend to Cross.**

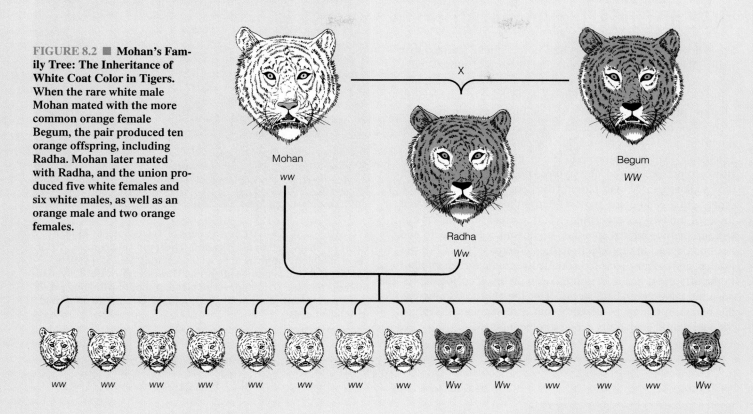

FIGURE 8.2 ■ Mohan's Family Tree: The Inheritance of White Coat Color in Tigers. When the rare white male Mohan mated with the more common orange female Begum, the pair produced ten orange offspring, including Radha. Mohan later mated with Radha, and the union produced five white females and six white males, as well as an orange male and two orange females.

Mohan
ww

Radha
Ww

Begum
WW

ww ww ww ww ww ww ww ww Ww Ww ww ww ww Ww

We will explore the answers to these questions as we investigate **heredity,** the transmission of physical, biochemical, and behavioral traits from parent to offspring. Observable traits, such as fur and eye color, are controlled by units called **genes,** which are specific, discrete portions of the DNA molecule in a chromosome; and the rules of heredity describe how genes and the traits they determine are passed from generation to generation. Simple and straightforward, the rules follow directly from the process of meiosis, the special cell division that occurs during the formation of eggs and sperm; and they hold in principle for all sexually reproducing organisms, from the single-celled *Paramecium* to peas and tigers.

Knowledge of genetics also helps reveal the workings of evolution by natural selection. The segments of DNA we call genes are passed along intact from parent to offspring. Genes that underlie an unfavorable trait (such as crossed eyes in a predator like the tiger, which relies on split-second coordination to fell a swift-moving deer) will occur less frequently over the course of several generations, since fewer individuals carrying that trait will survive and reproduce. By studying genes and the ways they interact with one another, we can begin to understand how individuals come to have different anatomical and physiological traits that in turn may affect their ability to survive and reproduce.

However, a key to understanding heredity is the *systematic study* of inheritance. Casual observation of genetic events can be deceiving, a point the maharajah discovered through Mohan and his descendants. But as a nineteenth-century monk and the pioneering geneticists who followed him revealed, genetic traits that seem to disappear or blend in an indistinguishable mix of characteristics *can* be sorted out *if* one studies them systematically. A clear hypothesis, a thoughtful experimental protocol, and careful quantitative analysis of results permit the detection of biological entities that cannot be directly perceived or seen.

This chapter explores the rules of heredity and the role of genes as traits are passed from parents to offspring. It answers several major questions along the way:

■ How did the monk Gregor Mendel discover and analyze genes in his quiet abbey garden?
■ What rules govern the inheritance of two genes controlling two different traits?
■ How are genes organized on chromosomes, and how do independent assortment and crossing over take place?
■ How do genes interact with one another and with the environment to govern the appearance, function, and behavior of organisms?

GENETICS IN THE ABBEY: HOW GENES WERE DISCOVERED AND ANALYZED

Since ancient times, gardeners and farmers have realized that many characteristics of domesticated plants and animals pass from parent to offspring. But exactly how the traits are transmitted was unclear. Individuals seem to have a nonspecific mixture of their parents' traits. For instance, two hibiscus plants can have flowers that differ in color, orientation and length of petals, and orientation of sepals (little green leaflike extensions below the petals). When plants with these two flower types are mated, the hybrid offspring's flowers have the petal length of one parent, the sepal orientation of the other, and a petal color and orientation intermediate between those of the two parents (Figure 8.3). Casual observations of similar mating results in other organisms gave rise to the notion that during reproduction, the hereditary "stuff" of the mother and father *blend* to produce the characteristics found in the offspring, just as cream mixes with black coffee to produce the tan-colored café au lait. This idea became known as the blending theory of heredity, but a monk named Gregor Mendel would eventually prove it to be incorrect.

One parent   Offspring   The other parent

FIGURE 8.3 ■ Casual Observations of Traits Such as Flower Color or Shape Led to the Blending Theory of Inheritance. Petal color in hibiscus appears to be a blended trait: Red and yellow parents give rise to orange offspring. However, most traits, even in hibiscus flowers, do not appear to blend. A parent with long petals and a parent with short petals, for example, give rise to offspring with long petals.

(a) Pea plant

(b) Cross-fertilization

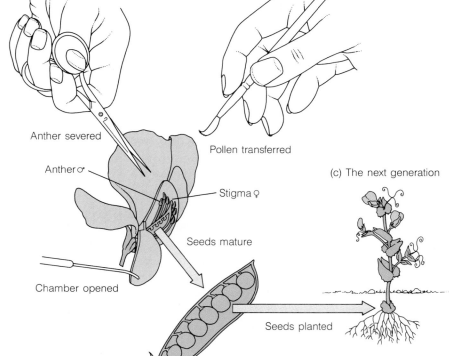

Anther severed

Anther ♂

Chamber opened

Pollen transferred

Stigma ♀

Seeds mature

(c) The next generation

Seeds planted

FIGURE 8.4 ■ The Advantages of Experimenting on Pea Plants. (a) Pea plants not only have attractive flowers, but their petal arrangement protects one pea flower from fertilization by the pollen of another pea flower. As a result, each flower fertilizes itself in nature. (b) An experimenter can cross-fertilize pea flowers precisely and deliberately and thus decide which will be the male parent and which the female parent of a new generation. Mendel was able to gently tease back the petals from the flower he chose to be the female parent, reach in with a pair of scissors, and sever the anthers, the source of pollen. Thus, he could eliminate the male gametes from that flower. He could then dust the stigma (the pollen-receiving structure of the female flower) with pollen from another pea plant he chose to be the male parent. (c) Finally, Mendel could collect seeds from the mated plant, grow them, and observe the characteristics of the offspring's flowers.

(a)

	Stem length	Flower color	Seed shape
Dominant characteristic (dominant allele)	Long	Purple	Round
Recessive characteristic (recessive allele)	Short	White	Wrinkled

(b)

FIGURE 8.5 ■ Mendel Studied Several Pairs of Traits in Pea Plants. (a) Each of the traits Mendel examined (stem length, flower color, seed shape, and so on) can appear in two forms: a dominant form and a recessive form. For the three traits mentioned here, the forms are long versus short stems, purple versus white flowers, and round versus wrinkled seeds. (b) In 1990, researchers discovered that the normal dominant (*W*) allele for the trait of seed shape causes starch to be stored in the seed, and this makes it plump and round. Seeds homozygous for the recessive (*w*) allele do not make or store this starch, and so they look shriveled and wrinkled.

New ideas about the nature of matter were helping scientists formulate new theories about inheritance. Gregor Mendel, an Augustinian monk who had studied for two years at the University of Vienna, had learned there that each physical object is made up of discrete atoms and molecules (see Box 8.1 on page 168). In the 1850s, he wondered if heredity could also be governed by similarly discrete particles that retain their identity from generation to generation, even though some may be hidden in a *hybrid* (the offspring of two individuals with differing forms of a given trait). Mendel put this new *particulate theory of heredity* to the test in a carefully devised long-term study that cataloged the number and types of pea plants in successive generations.

The Critical Test: A Repeatable Experiment with Peas

The particulate theory predicts that each hereditary factor will remain unchanged in a hybrid, whereas the blending theory predicts that each factor will be permanently diluted in the hybrid. Mendel realized that he could disprove one of these two predictions by checking the offspring of a hybrid generation: If the original parental forms reappeared in the second generation, this would show that the hereditary factors had passed through the hybrids unchanged and particulate; but if the original forms failed to reappear in the hybrid's offspring, then the factors would have been permanently blended.

Mendel chose the garden pea as his test subject because it

had several distinct advantages over other organisms (Figure 8.4a). First, he could purchase pea strains that demonstrated clear alternatives for single traits, such as short stem versus long stem or white flowers versus purple flowers. By selecting strains that differed in only one trait, he could study inheritance of one feature unconfused by all other variations. Second, Mendel could easily control which peas mated with which, because peas normally *self-fertilize*. Male and female sex organs occur within the same flower enclosed in a little chamber of petals, and this effectively ensures that the pollen—the source of male gametes—and the egg come from the same flower. A third advantage of pea plants was that Mendel could *cross-fertilize* plants simply by clipping off the pollen-producing anther before its maturation and dusting the egg-containing stigma with pollen from another selected plant (Figure 8.4b). From the mature seeds of this cross, Mendel could grow new pea plants (Figure 8.4c).

Mendel Disproves the Blending Theory

Mendel analyzed the inheritance of clear-cut alternatives for seven pea traits, including stem length (long versus short), flower color (purple versus white), and seed shape (round versus wrinkled) (Figure 8.5a and b). He began by demonstrating that he had 14 strains of **pure-breeding** peas. When self-fertilized, the parental plants always produced offspring like themselves; short stems begat short stems, long begat long.

BOX 8.1   BIOLOGY: A HUMAN ENDEAVOR

MENDEL AND HIS MENTORS

Without the help of a supportive family and insightful teachers, Gregor Mendel (Figure 1) might never have left the family farm. The son of a prosperous farmer in what is now Czechoslovakia, Mendel spent many of his early years learning the finer methods of flower and fruit tree cultivation, as well as the art of beekeeping. Nature fascinated the child, as did science and mathematics, and a teacher at the local elementary school recognized Mendel's exceptional talent and urged his parents to continue his education. They consented, but ran into trouble midway through his secondary schooling, when his invalid father developed financial problems and could no longer manage the farm or Gregor's schooling. An older sister, Veronika, took over the farm, and a younger sister, Theresia, gave Gregor part of her dowry to continue his studies.

In the fall of 1843, Cyril Napp, a broad-minded abbot who required his postulants to pursue secular as well as religious education, accepted Mendel into the Augustinian order at a very special monastery in old Brno. There Mendel came into contact with many of the prominent philosophers and scholars of the time as he developed his own "interdisciplinary major," combining studies of agriculture and winegrowing with theology and philosophy. While substitute teaching mathematics, Latin, German, and Greek at a local high school as part of his parish responsibilities, Mendel was so successful that the school administrators sent him to the University of Vienna to take a qualifying exam to become a certified teacher of natural history and physics.

Mendel flunked the examination two times. But Abbot Napp continued to support him through two years of study at the university. Ill health took Mendel back to the monastery, where he continued his teaching and became known as a schoolmaster who was much more pleased with a pupil's lively interest in a subject than with a mind full of memorized facts.

Mendel's tenacity in challenging orthodox thought stood the

FIGURE 1 ■ **Gregor Mendel with His Beloved Pea Plants.**

monk in good stead once he began to link his loves of nature and mathematics through the careful study of the reproductive behavior of pea plants. Ten years of intensive research yielded his now-famous principles of heredity, but the world wasn't ready to listen. At two public lectures of the Natural Science Society in Brno in 1865, biologists of the day listened politely as Mendel outlined his results and theories, and then they quietly ignored the work for 35 years. The theories were essentially mathematical, and at that time mathematicians and plant breeders worked in completely different realms.

Disappointed, Mendel returned to the monastery and eventually became abbot, trading most of his scientific studies for the administrative tasks of running a busy abbey. But the study of biology would never be the same. The farmer's son who saw a connection between the worlds of mathematics and agriculture had introduced quantitative analysis to a new field, and geneticists today are still reaping the results.

Next, Mendel carried out **monohybrid crosses,** matings between individuals that differ in only one trait. In one such monohybrid cross, he planted long-stem and short-stem seeds early one spring and let them grow into the **parental (P) generation.** Later that spring, when the parental plants had flowered, Mendel cross-fertilized long-stem plants with pollen from the short-stem plants; and in the summer, when the pods became swollen with plump peas, he collected the seeds. These seeds would produce the next generation, called the **first filial (F_1) generation,** meaning the first generation in the line of descent. Planted in the spring of the second year, the F_1 seeds of the long-stem/short-stem cross all grew into long-stem plants (Figure 8.6b). The characteristic that appears in the F_1 hybrid, such as long stems in peas or orange color in tigers, is said to be **dominant,** while the one that does not appear (short stems or white fur) is referred to as **recessive.**

What happens to the recessive characteristic? Does it disappear completely? Does it blend with the dominant characteristic? Or does it remain intact but hidden in the F_1 genera-

tion? Although he would have to wait a year to find out, Mendel knew exactly how to learn the answers to these questions. He allowed the long-stem F_1 hybrid plants to self-fertilize, and the next spring he planted the seeds of the **second filial (F_2) generation.** When the second generation of pea plants grew up, most of them had long stems, but significantly, there were some plants with short stems. Again, no stems of intermediate length appeared (Figure 8.6c). The reappearance of pure short stems among the offspring of long-stem hybrids was dramatic disproof of the blending theory and concrete evidence consistent with the particulate theory of heredity.

Segregation Principle for Alleles of One Gene

Mendel was not satisfied with saying that "some" of the F_2 plants had short stems. He counted the number and found that 787 of the F_2 plants had long stems, while 277 had short stems. These numbers showed about a 3:1 ratio (2.84:1) of long-stem to short-stem plants in the F_2 generation. When Mendel examined monohybrid crosses for all seven traits, he obtained similar results. In each case, all the F_1 hybrid plants were identical, showing only the dominant form of two alternatives and no plants with intermediate features. After self-

fertilization of the F_1 generation, the recessive form of the trait in question reappeared in about one-quarter of the F_2 plants, while the other three-quarters showed the dominant form. Clearly, the unit of heredity that produced short-stem flowers in the parental generation had been passed along to the F_1 generation, although it remained hidden.

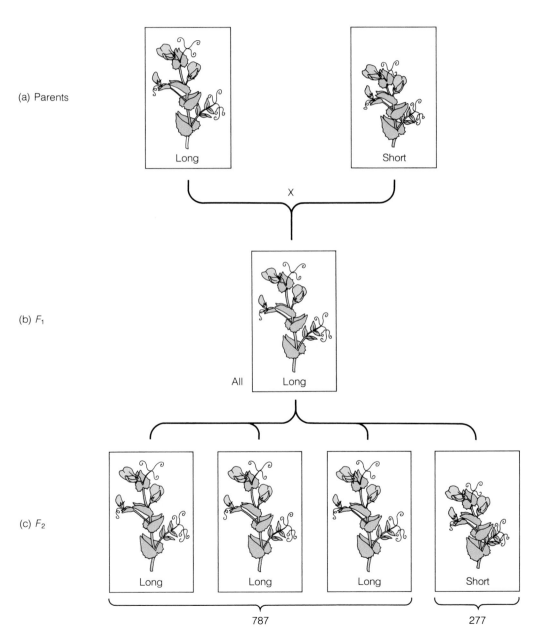

FIGURE 8.6 ■ Mendel's Evidence for the Particulate Theory of Heredity. When Mendel crossed long- and short-stem pea plants (a), he got all long-stem progeny in the F_1 generation (b), but in the F_2 generation (c), one-quarter of the progeny were short-stemmed and three-quarters were long-stemmed.

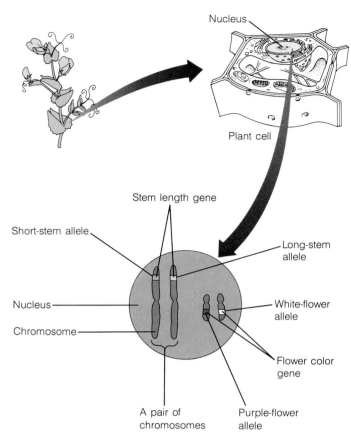

Nucleus

Plant cell

Stem length gene

Short-stem allele

Long-stem allele

Nucleus

White-flower allele

Chromosome

Flower color gene

A pair of chromosomes

Purple-flower allele

FIGURE 8.7 ■ Genes, Alleles, and Chromosomes. Although Mendel discovered the entities of inheritance we now call genes, he was not aware of the genetic bodies called chromosomes. Today we know that a cell's nucleus contains rod-shaped chromosomes organized in pairs. Specific genes, such as those for stem length or flower color, lie at specific locations on individual chromosomes. Each gene can occur in more than one alternative form, such as the allele for long stem versus the allele for short stem. The white-flower and purple-flower alleles are also shown here.

■ **Genes and Alleles**   The significance of these two findings was clear to Mendel. Since short stems reappeared in the F_2 plants, the hereditary factor that causes short stems had to be an individual unit, like a particle, and not like a liquid that could be mixed. This particulate factor of inheritance is now called a gene. A gene influences a specific trait in an organism, such as the length of a pea stem or the color of a tiger's coat. The gene is not the trait itself, but a factor that causes the organism to form the trait. Mendel also showed that each gene can have alternative forms we now call **alleles.** In pea plants, the gene for the stem length trait has two alleles, one causing long stems and one causing short stems. Likewise, the gene for coat color in tigers has two alleles, one for orange fur and one for white fur.

We now know that genes reside on *chromosomes,* threadlike bodies in the cell nucleus that contain DNA (Figure 8.7

and Table 8.1). Although an individual chromosome contains thousands of genes controlling hundreds of different traits, each chromosome will have just one allele (alternative form) for any individual gene.

■ **Dominant and Recessive Alleles**   Mendel realized that the reappearance of short-stem plants in the F_2 generation meant that the short-stem allele was present but invisible in the hybrid, F_1 plants. If the short-stem allele had not been present, it could not have passed on to the F_2 offspring. Then, because hybrids had long stems, he knew that the long-stem allele was also present in the hybrid generation, and so he concluded that a hybrid has one visible allele and one invisible allele.

The visible trait is said to be *dominant.* A tiger like Radha (see Figure 8.2), with one allele for an orange coat and one for a white coat, will always "look" just like a pure-breeding orange tiger with two alleles for orange coat. The allele that is "invisible," overshadowed each time it is paired with a dominant allele, is said to be *recessive.* The alleles for white coat in tigers and short stems in pea plants are recessive.

TABLE 8.1   The Rules of Heredity

1. A hereditary trait is governed by a sequence of DNA (or in some viruses, RNA) called a *gene.*

2. Genes reside on chromosomes.

3. The gene for each trait can exist in two or more alternative forms. Called *alleles,* these forms help determine an organism's external appearance, biochemical functioning, and behavior.

4. Most higher organisms have two copies of each gene in every body cell (they are *diploid*).

5. *Homologous chromosomes* are two chromosomes that are similar in size, shape, and genetic content.

6. A homozygote has two identical alleles of a gene; a *heterozygote* has two different alleles of a gene.

7. A heterozygote may have visible traits dictated by only one of the alleles, called the *dominant* allele. The hidden allele is called the *recessive* allele.

8. *Phenotype* is the way an organism looks and functions; *genotype* is an organism's genetic makeup.

9. Pairs of alleles separate, or *segregate,* during egg and sperm formation, and each gamete will be *haploid* (have one copy of each gene). At fertilization, sperm and egg combine, and the resulting zygote is diploid.

10. According to the principle of *independent assortment,* genes on different chromosomes assort into gametes independently of each other.

11. Linked genes lie on the same chromosome and tend to be packaged into gametes together.

12. Occasionally, two genes on the same chromosome may become separated owing to an exchange of alleles between that chromosome and its homologous chromosome, a process called *recombination* (*crossing over*).

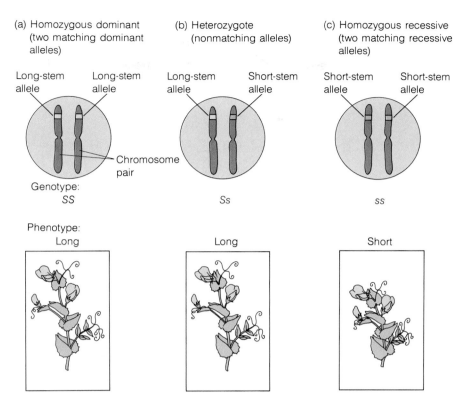

(a) Homozygous dominant (two matching dominant alleles)

(b) Heterozygote (nonmatching alleles)

(c) Homozygous recessive (two matching recessive alleles)

Long-stem allele   Long-stem allele

Long-stem allele   Short-stem allele

Short-stem allele   Short-stem allele

Chromosome pair

Genotype: *SS*

Ss

ss

Phenotype: Long

Long

Short

FIGURE 8.8 ■ Mendel's Synthesis. Mendel's genius was seeing an underlying pattern in the data he collected from dozens of experiments and thousands of pea plants. Couched in modern terms, Mendel's ideas suggest that underlying an organism's visible, physical makeup (its phenotype) is a set of two alleles (the genotype) that determine the phenotype. Because a dominant allele such as for long stems can hide a recessive allele such as for short stems, a homozygote (a) with two long-stem alleles will have the same phenotype as a heterozygote (b) with one copy of each allele, even though it has a different genotype. Only the homozygous recessive (c), with two matching short-stem alleles, will have the short-stem phenotype.

Modern biologists discovered that virtually all organisms possess two alleles for each trait, as Mendel predicted, because each cell has two copies of each chromosome (see Figure 8.7). One chromosome of each pair will have one allele (the short-stem allele of the stem length gene, let's say), while the other chromosome of the pair could have the other allele (the long-stem allele).

■ **Genotype Versus Phenotype** The existence of dominant and recessive alleles implies that two plants or animals with a different genetic makeup, or **genotype,** for a certain trait may have the same visible physical makeup, or **phenotype.** For example, a phenotypically long-stem plant may be genotypically either hybrid or pure-breeding for the long-stem trait (Figure 8.8a and b). By studying the hybrid plants, Mendel reached the important conclusion that each parent plant has two alleles for each gene. Hybrid organisms, with two different types of alleles for a given trait, are said to be **heterozygous** for that trait (Figure 8.8b), while pure-breeding organisms, with a pair of identical alleles, are **homozygous** for the targeted trait (Figure 8.8a and c). Thus, the pure-breeding long-stem and short-stem parents were homozygotes, while their hybrid offspring were heterozygotes.

■ **Segregation of Alleles in Gametes** Since each hybrid has one allele typical of each parent, Mendel suggested that each parent must donate one allele to each offspring. This implies that the two alleles in a parent separate from each other during gamete formation and produce eggs or sperm that have only one allele apiece. For example, if a parent's genotype includes two different alleles for the stem length gene—one long-stem allele and one short-stem allele, let's say—then the egg or sperm will bear either one allele or the other, but not both. A cell with just one allele of each gene is said to be *haploid* (*hap* means "one"), while a cell with two alleles of each gene is said to be *diploid* (*di* means "two"). Thus, egg and sperm cells are haploid, but parent pea plants are diploid. Generalizing from Mendel's pea experiments, we can define the **segregation principle:** Sexually reproducing diploid organisms have two copies of each gene, which segregate from each other during meiosis. When gametes form, they each contain only one copy of each gene.

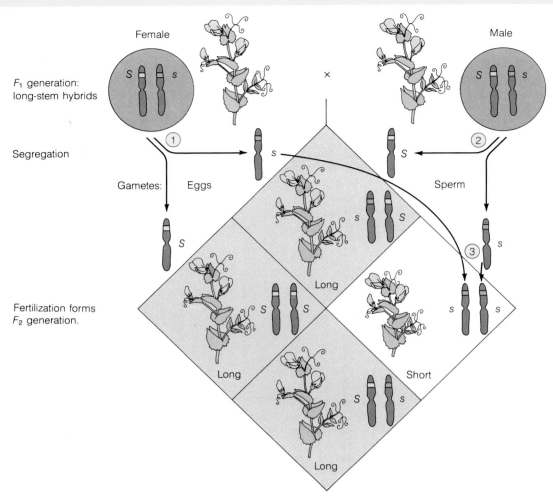

Genotypes and phenotypes of F_2 generation

FIGURE 8.9 ■ Predicting Genotypes and Phenotypes. Mendel correctly hypothesized that alleles separate during the formation of gametes (step 1 and 2) and randomly recombine at fertilization (step 3), so that offspring inherit one allele from each parent for a given trait (here, stem length). This is the segregation principle. By writing the two kinds of alleles from the female parent along one side of a square (step 1) and the two kinds of alleles from the male parent along the other side of a square (step 2), we can pair the alleles (step 3) and determine the genotypes that could result in the offspring. Here, parents heterozygous for stem length produce one homozygous long-stem plant (*SS*), two heterozygous long-stem plants (*Ss*), and one homozygous short-stem plant (*ss*).

Mendel's segregation principle can explain one of the mysteries concerning white-coated Mohan. Mohan's orange F_1 children, including Radha, were hybrids bearing one white-coat allele received from pure-breeding white Mohan and one orange-coat allele from their normal, pure-breeding orange mother. The fact that these F_1 heterozygotes had orange fur shows that the white-coat allele is recessive and the orange-coat allele is dominant.

■ Peas and Probabilities: Segregation Predicts Phenotypic Ratios Mendel's segregation principle can explain the *types* of hybrid and F_2 peas, but can it also account for the numbers of each type found? That is, can it explain why about one-quarter (instead of one-eighth or one-third or some other frac-

tion) of the F_2 generation is homozygous for the recessive allele? Mendel applied his analytical skills to show that it could.

Geneticists usually use letters to represent genes, with uppercase (capital) letters for dominant alleles and lowercase letters for recessive alleles. If we use *s* to designate the short-stem allele of the gene for stem length and *S* to designate the long-stem allele, then the hybrid F_1 generation is *Ss*. In the meiotic cell divisions that form gametes in the *Ss* hybrid, the alleles segregate, so that half the gametes end up with *S* and the other half with *s* (Figure 8.9, steps 1 and 2). Mendel showed that a mating between two F_1 plants will produce three long-stem for every short-stem plant if fertilization occurs randomly, that is, if the pollen bearing *S* unites equally often with eggs bearing *S* and *s*. An easy way to visualize the results of random fertilization is to draw a diagram called a **Punnett square** (after the British mathematician who first used it). To construct a Punnett square, write the pollen types along one side of a square (*S* or *s*) and the egg types down the other side (*S* or *s*) as shown in Figure 8.9. Then fill in the empty boxes with the genotypes of the offspring that would result from the fertilization of each egg type with each pollen type, as shown in Figure 8.9, step 3, for one of the four combinations.

Figure 8.9 shows four F_2 genotypes: *SS, Ss, sS,* and *ss.* But since the order of alleles is not important, *sS* and *Ss* are equivalent, and so there are really only three genotypes, found in the ratio 1*SS*: 2*Ss* :1*ss.* If we look at the physical characteristics of the plants themselves, however, we find that the 1: 2 :1 genotypic ratio produces a 3:1 phenotypic ratio. The reason is that the single *SS* genotype and both *Ss* genotypes have the same long-stem phenotype because *S* is dominant over *s*. So the phenotypic ratio expected from the segregation principle and random fertilization is (1 + 2) : 1, or three long-stem plants to one short-stem plant, a result close to what Mendel actually counted. With this in mind, look again at the pea pod shown in Figure 8.5b and count the numbers of wrinkled and round peas. Based on the rules we just discussed, can you see why such a ratio might be apparent in the seeds?

You can demonstrate the probability of obtaining the 3:1 relationship by tossing two different coins simultaneously. Let a dime represent a pollen grain, and let a penny represent an egg. Each coin has a head, representing the *S* allele, and a tail, representing the *s* allele. Thus, each coin has an equal number of *S* alleles and *s* alleles, just like a population of gametes from a heterozygote. Flip both coins at the same time (to represent fertilization) and record whether they land heads up or tails up. If both are heads, the genotype is *SS;* two tails reflect an *ss* genotype; and if one coin is heads and the other tails, the "offspring" will be heterozygous. Flip the coins 20 times, recording each outcome. Is the overall ratio close to the ratio derived from using the Punnett square? If not, a few more tosses may bring the figures closer.

Like the toss of a coin, the combination of pollen and egg is governed by the laws of chance. In a small number of trials, the results may differ substantially from those predicted for random tossing, but as the number of trials increases, we will obtain results closer to the mathematically predicted values. Mendel's genius was to deduce the existence of entities we now call genes from the ratios he found.

■ **A Testcross Can Distinguish Genotypes**   In addition to accounting for the ratios of phenotypes in each generation, Mendel developed the **testcross,** a valuable tool for determining the genotype of an individual organism. The testcross involves mating an individual of unknown genotype to a homozygous recessive, the purpose being to discover the unknown genotype. A pea plant with a long stem, for example, can be *SS* or *Ss.* If the long-stem parent is homozygous *SS,* then all the offspring will be *Ss* and have long stems. If, on the other hand, it is heterozygous *Ss,* then a mating with a homozygous recessive *ss* plant will generate offspring that are half long-stem (*Ss*) plants and half pure-breeding short-stem (*ss*) plants. Another instance in which the testcrosses proved useful to geneticists was the curly-eared cat discovered in 1981 (Figure 8.10).

Now think back to Mohan's mating with his hybrid daughter Radha (see Figure 8.2). We can consider it a testcross between the homozygous recessive white Mohan (whose genotype we could write as *ww*) and the orange

FIGURE 8.10 ■ Curly Cats. In 1981, a stray cat with oddly curled ears moved in with a family in Lakewood, California. Today, the animal's descendants are exciting feline fanciers all over the world. When that curly-eared cat (which resembled this black male) mated with a normal cat, about half the kittens had curly ears and the rest were normal. To explain these results, geneticists saw this as a testcross between the curly-eared cat, which had to be heterozygous for a dominant allele causing the curly ears, and a normal cat that was homozygous for the recessive normal allele.

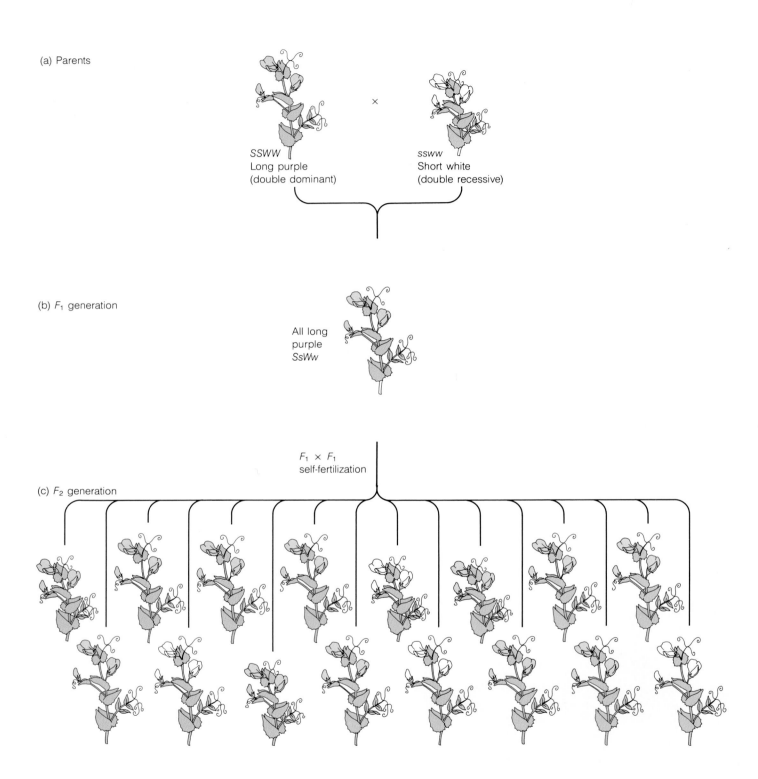

(a) Parents

SSWW
Long purple
(double dominant)

×

ssww
Short white
(double recessive)

(b) *F₁* generation

All long
purple
SsWw

F₁ × *F₁*
self-fertilization

(c) *F₂* generation

FIGURE 8.11 ■ **Results of a Dihybrid Cross.** (a, b) A double-dominant long-stemmed purple-flowered plant mated to a double-recessive short-stemmed white-flowered plant gives F_1 progeny that are all double heterozygous with long stems and purple flowers. (c) When the F_1 generation self-fertilizes, some of the F_2 generation represent new combinations: plants with long stems and white flowers and other plants with short stems and purple flowers. How can this be explained?

Radha, who clearly must have been heterozygous (genotype *Ww*). Viewed in this way, the genetic basis for the appearances of the second-generation cubs becomes clear: Some were homozygous white (*ww*) and others heterozygous orange (*Ww*).

INHERITANCE OF TWO INDEPENDENT TRAITS

The segregation principle gives a satisfying explanation for the inheritance of a single gene's pair of alleles. But what happens if we want to follow more than one gene pair at the same time? To answer this question, Mendel made a **dihybrid cross,** a mating between parents that differ in two characteristics. For his parental generation, he chose short-stem plants with white flowers (*ssww*) and long-stem plants with purple flowers (*SSWW*). (Purple is dominant over white, so we represent the purple allele by *W* and the white allele by *w*.)

In the F_1 generation produced by mating *SSWW* with *ssww*, all the offspring had purple flowers at the ends of long stems (Figure 8.11a and b); in other words, the dominant forms for both stem length and flower color appeared in the F_1 generation. Mendel then allowed the F_1 plants to self-fertilize and produce the F_2 generation. The outcome of this mating was not so easy to predict. Would purple flowers always be found with long stems, and white flowers always with short stems, as in the original parents? (Such offspring are referred to as **parental types.**) Or would some purple flowers appear on short stems, and some white flowers on long stems? (Such offspring are called **recombinant types.**)

Ratios Reveal Independent Assortment

When Mendel looked at his field of F_2 pea plants, he saw that indeed new combinations had formed (Figure 8.11c). Some of the plants were short purple and some long white. Evidently, the alleles had reassorted during the formation of gametes. Again, Mendel insisted on counting the number of each type of plant and found approximately the following ratio: $\frac{9}{16}$ long purple : $\frac{3}{16}$ long white : $\frac{3}{16}$ short purple : $\frac{1}{16}$ short white.

If, however, he looked at just flower color or just stem length, he found a phenotypic ratio of 3 : 1. For instance, there were $9 + 3 = 12$ purple-flower plants and $3 + 1 = 4$ white-flower plants ($12 : 4 = 3 : 1$). There were also 12 long stems for every 4 short stems. From these calculations, Mendel concluded that the 9 : 3 : 3 : 1 ratio could arise only if the gene for stem length and the gene for flower color assorted independently of each other so that four types of gametes formed. If *S* went into one gamete, then either *w* or *W* would also enter that gamete with equal probability, giving equal numbers of *Sw* and *SW* gametes. Likewise, *sw* and *sW* gametes would occur equally, making the four types in a 1 : 1 : 1 : 1 ratio.

Figure 8.12 uses the Punnett square to find what geno-

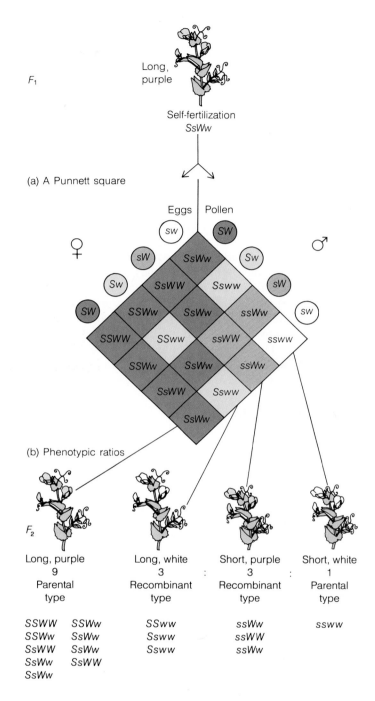

FIGURE 8.12 ■ Analyzing a Dihybrid Cross. (a) An experimenter can predict the genotypes and phenotypes of two traits simultaneously by constructing a double-sized Punnett square, listing the possible gene combinations in gametes from both parents, then filling in the boxes. (b) Some of the predicted offspring are parental types (long-stem plants with purple flowers and short-stem plants with white flowers), but some are recombinant types (long-stem plants with white flowers and short-stem plants with purple flowers). By comparing offspring's phenotypes and genotypes, one can predict the ratio of parental to recombinant types. Some letters in the genotypes listed here are inverted as in the Punnet square.

types we would expect from a random combination of these four types of gametes. Sixteen (4 × 4) combinations would be expected with a phenotypic ratio of 9 : 3 : 3 : 1.

Because the outcome of the dihybrid cross could be predicted by assuming that two gene pairs assort independently during the formation of eggs and sperm (see Chapter 7), Mendel proposed his second rule of heredity, the **principle of independent assortment:** Different hereditary factors segregate into gametes independently of each other.

Mendel's Results Ignored

Biologists of Mendel's time turned a deaf ear to the monk's analyses of inherited characteristics, and although he published his results in 1865 in a journal that was sent to about a hundred scientific libraries, the few plant breeders who cited them indicate by their remarks that they did not understand the significance of his three main concepts: (1) hereditary factors (or genes) are distinct units; (2) alleles separate during the formation of gametes and join together randomly at fertilization; and (3) two genes assort independently during sexual reproduction. Discouraged, Mendel continued breeding experiments with far less enthusiasm and soon became abbot of his monastery and gave up his scientific studies. For more than 30 years, his ideas lay dormant, until 1900, when botanists in Austria, Germany, and the Netherlands duplicated Mendel's experiments in other plants and recognized the importance of his work.

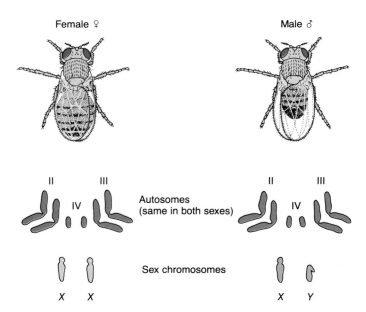

FIGURE 8.13 ■ Sex and Chromosomes. In fruit flies and many other animals, males and females have identical sets of autosomes but differently shaped sex chromosomes. Fruit flies have four chromosome pairs, three identical sets of autosomes and one pair of sex chromosomes. Males have one X and one Y sex chromosome, while females have two X chromosomes.

GENETICISTS LOCATE GENES ON CHROMOSOMES

While the report of Mendel's discovery of hereditary factors was gathering dust in European libraries, scientists were advancing rapidly on another, seemingly unrelated front—**cytology,** the science of cell structure. Particularly important were microscopic observations of how chromosomes move during mitosis and meiosis (see Chapter 7). Shortly after the rediscovery of Mendel's principles, and as a result of advances in understanding cell structure, biologists realized that there are several parallels between the inheritance of genes and the distribution of chromosomes during meiosis: (1) Two copies of each gene and two copies of each chromosome exist in each cell; (2) a pair of alleles and two homologous chromosomes both segregate into different gametes; and (3) different genes and separate chromosomes both assort independently when egg and sperm are formed. These correlations suggested the hypothesis that genes are physically linked to chromosomes. To test that suggestion, investigators would have to identify individual chromosomes and show that a specific trait is always transmitted along with a specific chromosome. They started by studying sex and eye color in the tiny fruit fly *Drosophila melanogaster.*

Different Sexes—Different Chromosomes

The two members of each chromosome pair are generally identical in shape. In males and females of the same species, this is true for all pairs of chromosomes except one—the sex chromosomes. Chromosome pairs with two identical members in both sexes are called **autosomes,** while chromosome pairs that have dissimilar members in males and females are called **sex chromosomes** (Figure 8.13). In fruit flies, tigers, and people, females have two identical sex chromosomes, called *X* **chromosomes,** while males have only one *X* chromosome and another unpaired, smaller chromosome called a *Y* **chromosome.** Although sex chromosomes are common in animals, they are rarely found in plants.

The distribution of sex chromosomes during meiosis can account for the appearance of equal numbers of males and females. An *XY* male is like the heterozygous parent, and an *XX* female is like the homozygous recessive in a testcross (Figure 8.14). If an *X* and a *Y* segregate in meiosis, then half of the sperm will contain a *Y* and the other half an *X* chromosome. If these sperm randomly fertilize a group of eggs, each egg containing an *X* chromosome, then half of the zygotes formed will be male and half female. Note that a male's single *X* chromosome has to come from his mother. Abraham Lincoln once said, "All I am and have I owe to my mother." This certainly was true for the characteristics related to his *X* chromosome.

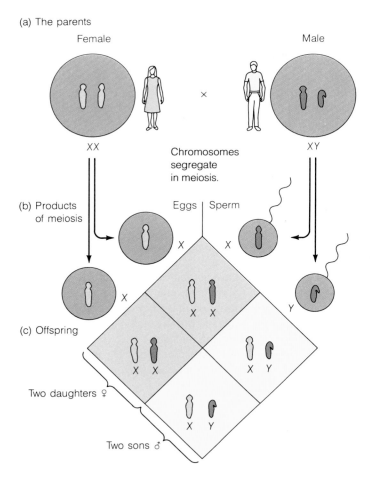

(a) The parents

Female   Male

×

XX   Chromosomes   XY
            segregate
            in meiosis.

(b) Products   Eggs │ Sperm
    of meiosis

                X   X

                X   Y

(c) Offspring

                    X X   X Y

Two daughters ♀

                X X   X Y

Two sons ♂

FIGURE 8.14 ■ Crisscross Inheritance of the *X* Chromosome.
**(a) Women have two *X* chromosomes, but men have one *X* and
one *Y* chromosome. (b) Sex chromosomes separate during the
meiotic divisions that produce eggs and sperm. (c) Daughters
inherit one of their *X* chromosomes from their father, while sons
inherit their only *X* chromosome from their mother. Sons inherit
their *Y* chromosome from their father.**

Sex and White Eyes

Since males and females have different chromosomes, we
know that at least one trait—gender—is regulated by chromosomes. But are there any others? Thomas Hunt Morgan
wanted to find out about genes and chromosomes but did not
have Mendel's monastic patience. Because he couldn't wait a
year between generations, he chose the fast-breeding fruit fly
Drosophila melanogaster as his experimental organism.
Although pests in the kitchen, fruit flies are beautiful under
the microscope, with brick red eyes, tan abdominal stripes,
and glistening black bristles. No longer than an "l" on this
page, fruit flies are also easy to raise and to breed, and they

develop quickly: In just 12 days, an egg will become a reproductive adult ready to produce hundreds of offspring.

One day, while looking in the microscope, Morgan saw an
intriguingly different kind of fly—a male with eyes of creamy
white instead of the usual brick red (Figure 8.15). A *mutation*—a permanent change in the genetic material—had
altered a gene for eye color from the normal red-eye allele (*W*)
to the white-eye allele (*w*).

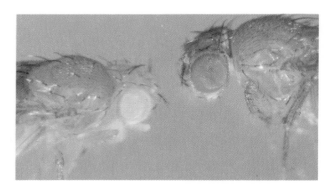

**FIGURE 8.15 ■ The White-Eye Mutation in *Drosophila*. The
fly on the right with red eyes is normal, or wild-type, in color,
while the fly on the left is a white-eyed mutant (magnification
60×).**

From a series of crosses, one of which is shown in Figure
8.16, Morgan realized that the gene for eye color is carried on
the *X* chromosome and is therefore a sex-linked, or ***X*-linked,**
gene. After discovering and experimenting with several *X*-
linked genes, Morgan and his students concluded that the *Y*
chromosome carries no allele of the gene for eye color or for
most other *X*-linked genes. Among the useful information to
come from their work on sex linkage was this generalization:
Genes are located on chromosomes.

Mutations Reveal That Each Chromosome
Carries Many Genes

Shortly after discovering the gene for white eyes, investigators found other mutations in fruit flies. One caused the flies
to have yellow rather than black bodies (it was called the yellow mutation); another produced purple eyes; while still
another gave rise to mutant flies that lacked a specific short
vein in the wing (they were called cross-veinless). As more
and more mutations were found, each one for a different gene,
it soon became apparent that flies have many more than four
genes. But fruit flies were known to have only four chromosome pairs. Hence, each chromosome must contain many different genes.

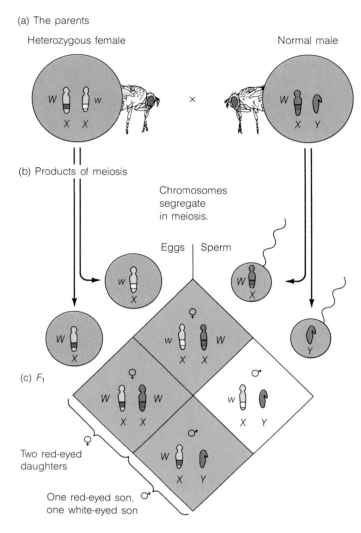

FIGURE 8.16 ■ One of Morgan's Experiments with the White-Eye Allele. (a) A female heterozygous for the white-eye allele has red eyes. (b) During meiosis, she will produce some eggs with a *W* allele and others with a *w* allele. Her normal mate will produce some sperm with an *X* chromosome bearing the *W* allele and others with a *Y* chromosome bearing no allele of the eye color gene. (c) In the F_1 generation, white eyes reappear in one-quarter of the offspring—but always in males. Half the males are white-eyed. A Punnett square for these crosses reveals that daughters receive an *X* chromosome from their father as well as from their mother, while sons receive an *X* chromosome only from the mother. Therefore, a son's phenotype will reflect whichever allele is carried on the *X* chromosome he got from his mother.

Linkage: Genes on the Same Chromosome Tend to Be Inherited Together

Morgan and his students then wondered how the many different genes located together on the same chromosome would be inherited. Would they assort independently as predicted by

Mendel's second principle? Or would they always be inherited together? Experiments like the one shown in Figure 8.16 revealed what is so often the case in the life sciences: Neither extreme hypothesis was true. Morgan's team found that two genes on the same chromosome were not always inherited together in the original parental combinations; usually, however, fewer new combinations were found than predicted by Mendel's law of independent assortment.

The pioneer geneticists now had to devise some new hypothesis, a new model for explaining this intermediate result. They suggested that gene pairs residing on the same chromosome are *linked*, like beads on a string, in a **linkage group:** a set of genes *usually* inherited together. Fruit flies have four linkage groups, the same as the number of chromosomes.

But if genes strung along the same chromosome tend to be inherited together, how could Morgan account for the infrequent but still significant number of crosses in which offspring received new combinations of alleles (Figure 8.17)? Morgan made an assumption that takes care of this quandary: An occasional exchange of chromosomal pieces, called crossing over, takes place during meiosis (review Figure 7.22).

Figure 8.18 shows how crossing over in meiosis can account for the recombination of linked genes. Crossing over

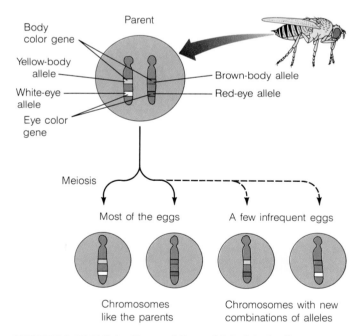

FIGURE 8.17 ■ Inheritance of Genes Linked to the Same Chromosome. A body color gene (with yellow-body versus brown-body alleles) and an eye color gene (with white-eye versus red-eye alleles) are both linked on the *X* chromosome. If one copy of the *X* carries both recessives (alleles for yellow body and white eyes) and the other *X* chromosome carries both dominants (alleles for brown body and red eyes), then most of the eggs will have the double-recessive or the double-dominant arrangement of alleles. New combinations (yellow body and red eyes, or brown body and white eyes) will be infrequent.

produces new combinations of alleles in each generation. Some of these new genetic combinations may provide an organism with novel physical structures or abilities and hence be favored by natural selection, thus contributing to evolution.

Crossing Over: Evidence That a Chromosome Is a Linear Array of Genes

Crossing over is significant in our understanding of genetics, not only because it allows for the reshuffling of genes on the same chromosome, but also because geneticists have exploited it to determine the order of genes along a chromosome through a process called **genetic mapping** (Figure 8.19). Genetic mapping is possible because genes that frequently recombine are farther apart on the chromosome than genes that almost never recombine. Comparing the recombination rates of different pairs of genes gives a measure of how

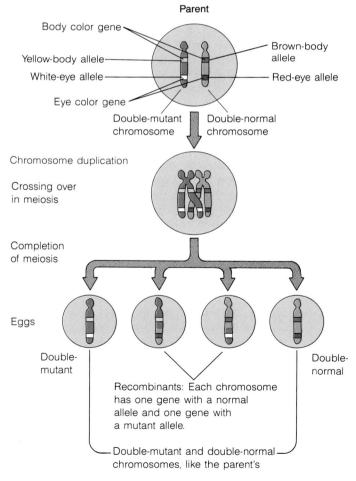

FIGURE 8.18 ■ **Genetic Recombination Is Due to Crossing Over of Chromosomes.** If crossing over occurs in a meiotic division, one observes 50 percent parental and 50 percent recombinant gametes.

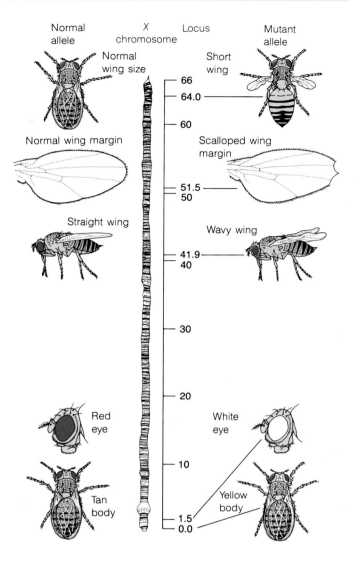

FIGURE 8.19 ■ **Map of the *Drosophila X* Chromosome. The distance between two genes is reflected in the frequency of recombination seen in crosses between flies with these two traits. Thus, one would expect to see more recombination between the yellow/ tan body color gene and the wavy/straight wing shape gene (separated by about 42 map units) than between the yellow/tan body color gene and the white/red eye color gene (separated by only 1.5 map units).**

far apart they are, just as timing a trip between two cities on a highway can give an idea of the distance between them on the road map.

From genetic maps prepared using this logic, geneticists drew two important conclusions: First, each gene is located on a chromosome at a specific place called a **locus.** In other words, the gene for white versus red eyes is in the same place on the same chromosome in any fruit fly, and likewise the gene for attached versus free earlobes is in the same chromosomal position in all people. Second, a chromosome is a linear

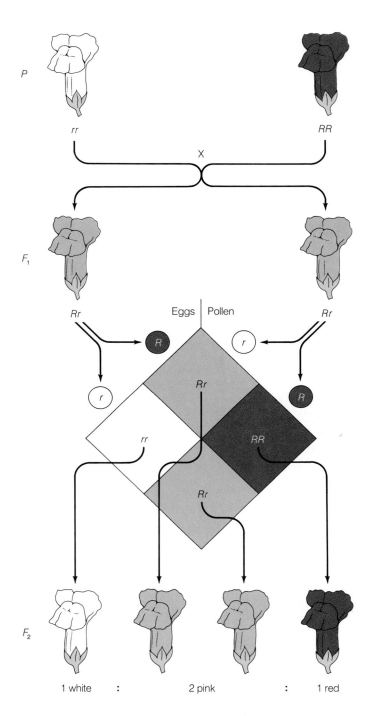

1 white   :   2 pink   :   1 red

FIGURE 8.20 ■ Flower Color in Snapdragons: Incomplete Dominance. A heterozygous plant with one red-flower and one white-flower allele (*Rr*) has pink flowers because a single red-flower allele in an *Rr* heterozygote makes less pigment than the two red-flower alleles in a homozygous *RR* plant. Since less pigment is made, the heterozygote appears pink.

array of genes, like beads on a string. In our discussion of human genetic diseases (see Chapter 12), we will see that the ability to identify a gene's location has enabled medical geneticists to isolate genes causing various human diseases (such as cystic fibrosis or muscular dystrophy or certain types of cancer), and the isolation of these genes may one day lead to effective therapies.

The series of experiments leading from Mendel to Morgan established fundamental principles about the gene, an entity so basic that its name reflects its role in *gene*rating our bodies, our cells, ourselves. The study of simple organisms, such as peas and flies, led to great concepts, namely, the location of genes at identified places on specific chromosomes and the rules governing the inheritance and recombination of chromosomes. As we continue, we will find that a gene does not perform its function in a vacuum, but interacts with other genes and the environment in specifying how an organism develops, operates, and lives.

GENE INTERACTIONS: EXCEPTIONS OBSCURE MENDELIAN PRINCIPLES

In biology, every rule is challenged by phenomena that seem to contradict it, and Mendel's principles of heredity are no exception. An early geneticist admonished his students, "Treasure your exceptions." An unexpected result, a fly that is not predicted by the "rules," makes a statement about how genes work or how an organism functions. As biologists studied more and more genes, they discovered that interactions between alleles of the same gene, between different genes, and between genes and the environment are often unexpected. On closer inspection, however, the unforeseen genetic behavior could usually be explained by slight extensions of Mendelian principles. The remaining sections describe such exceptions, and Box 8.2 discusses a fascinating new topic, *genomic imprinting,* or the different expression of genes in males and females.

Interactions Between Alleles

■ **Incomplete Dominance** According to Mendel, an allele is either dominant or recessive, but as later geneticists continued the study of inheritance, they found that some alleles fail to fall clearly into either category. Snapdragons, for example, can be white or red (Figure 8.20). When self-pollinated, white flowers breed true, and so do self-pollinated red flowers. But if pollen from a white flower fertilizes the eggs in a red flower, the heterozygous offspring have pink flowers, not white or red as Mendel's principle of dominance would have predicted. Since the phenotype of the heterozygote is intermediate between pure-breeding red flowers and pure-breeding white flowers, this is called **incomplete dominance.**

BOX 8.2 PERSONAL IMPACT

HINNIES, MULES, AND MENDEL'S RULES

It's a good thing Gregor Mendel didn't ride a mule; the principles of inheritance might never have been the same.

When Mendel crossbred plants that produced only round peas with others producing only wrinkled peas, all the offspring had round seeds. This was true whether the male gametes came from the round-seeded plants and the female gametes came from the wrinkled-seeded plants or vice versa. The gene for seed shape, in other words, acts the same in the offspring regardless of which parent donates it. From such simple observations, Mendel derived his straightforward laws of heredity.

Imagine, though, how scientific history might have been changed had Mendel trotted about Brno on a big brown mule, coarse-woven robes flapping in the wind. Had he fancied mules, Mendel would certainly have known that breeders produce the hybrids by crossing a male donkey with a female horse, and that while a mule can be as tall as its mother, it has the ears, legs, hooves, tail, and stubbly mane of the donkey father (Figure 1a). Mendel would also surely have been aware that the reciprocal cross of a female donkey and a male horse produces not a mule but a hinny—a hybrid that looks basically like its equine father and resembles its donkey mother much less (Figure 1b).

Mules and hinnies are obvious exceptions to the kind of genetic equivalence Mendel discovered in pea plants. And who knows? Enough examples like this, and the good monk might have gotten discouraged and gone back to mathematics!

Modern geneticists have studied the curious differences between mules and hinnies, along with a collection of similar genetic oddities that result from what they call "genomic imprinting." This phrase refers to a hypothetical imprinting process that takes place during gamete formation and causes a gene to be expressed one way in the offspring if the male parent donates it, and another way if the female parent donates it.

Several human diseases show evidence of genomic imprinting, including Huntington's disease and the Prader-Willi and Angelman syndromes. Huntington's is a severe degenerative condition that is inherited as a dominant trait and tends to onset in middle age (details in Chapter 12). About 10 percent of the time, however, the disease onsets in children, and of those young victims, 9 out of 10 inherit the disease from their father. A child with Prader-Willi syndrome is usually short, obese, and mentally retarded and tends to have inherited both copies of chromosome 15 from the mother. A child with Angelman syndrome is usually retarded, laughs excessively, and moves in a jerky, puppetlike manner. Research shows that many such children inherit a complete chromosome 15 from the father, but lack pieces of the maternal chromosome 15.

Geneticists are not certain how genomic imprinting occurs, although some think it may involve DNA methylation—the attachment of methyl groups to parts of the DNA molecule—and with that, the turning on or off of certain genes. They do know, however, that genomic imprinting represents an important exception to Mendel's basic laws of equivalent inheritance. And many suspect that imprinting is playing unknown but significant roles in normal development and genetic diseases—not to mention in phenotypic patterns as dramatic as the hinny and the mule.

(a) Mule: Offspring of a female horse and a male donkey

(b) Hinny: Offspring of a female donkey and a male horse

FIGURE 1 ■ A Mule (a) and a Hinny (b).

We can explain incomplete dominance if we assume that two doses of the red-flower allele make enough red-colored pigment in snapdragon petals to produce a red color, and white-flower alleles make no pigment at all, leaving the petals snowy white. Thus, a heterozygous plant with one red-flower allele and one white-flower allele would make an intermediate amount of pigment, and hence the petals would look pink. This intermediate phenotype still does not indicate a blending inheritance, because when two heterozygous pink flowers are mated, the offspring have the expected Mendelian ratio of one red to two pink to one white. The alleles emerge from the hybrid unchanged.

Now look back at Figure 8.3, and see if you can explain the results with the hibiscus flowers in terms of the particulate theory and dominant, recessive, and incompletely dominant genes.

■ **Codominance**   In hybrids that exhibit **codominance,** two alternative alleles are fully apparent in a hybrid; thus, both phenotypes show up in the organism. A familiar example is the blood type gene called *ABO* (Figure 8.21). Allele *A* causes a marker—a certain sugar called type A—to appear on the surface of red blood cells, and the person is said to have blood type A. Allele *B* causes a different marker—a sugar called type B—to be on the blood cell surface, giving blood type B. Someone who has two *A* alleles has only the A sugar, and a person who has two *B* alleles has only the B sugar. But a heterozygote with one *A* allele and one *B* allele will have both A and B sugars on the surface of the red blood cells and hence will have blood type AB, a codominant phenotype.

■ **Many Alleles of One Gene**   Although Mendel studied only two alleles for each of his seven pea plant genes, some genes have more than two alleles. Human blood groups illustrate this concept. In addition to the two codominant alleles *A* and *B*, this blood group gene has a third allele that is fully recessive to both *A* and *B*. This recessive allele is called *O*, and a person with two doses of *O* has neither the A nor the B sugar marker and has blood type O. Since *O* is recessive, an *AO* heterozygote will have blood type A, and a *BO* heterozygote will have blood type B. Although there are three alleles of the

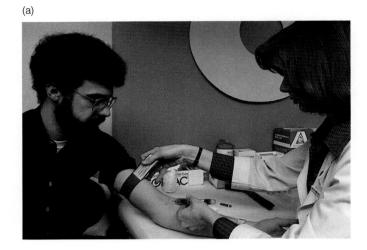

(a)

(b) Blood types

Phenotype		Genotype
Blood type	Cell surface molecule	
A	Red blood cell	AA or AO
B		BB or BO
AB		AB
O	(Neither A nor B)	OO

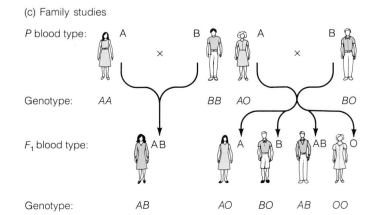

(c) Family studies

P blood type: A   B   A   B

Genotype: *AA*   *BB*   *AO*   *BO*

F₁ blood type: AB   A   B   AB   O

Genotype: *AB*   *AO* *BO* *AB* *OO*

FIGURE 8.21 ■ Blood Types Illustrate Codominance and Multiple Alleles.   (a) One way of investigating a person's genotype is to learn what types of genetic markers are on the surface of the person's cells. Researchers often study blood cells as a source of human cells. (b) Alleles for blood type (which particular cell surface marker occurs on a person's red blood cells) are multiple: They can be either *A, B,* or *O* and they occur in the combinations listed here in the genotype column. Neither *A* nor *B* is dominant—they are codominant: If both are present, both are expressed fully in the phenotype, leading to the AB blood type. (c) Matings between a person of blood type A and a person of blood type B can give very different progeny, depending on the genotype of each parent.

(a) Black   (b) Chocolate   (c) Yellow

BBEE   bbEE   BBee
BBEe   bbEe   Bbee
BbEE   bbee
BbEe

FIGURE 8.22 ■ Masking Genes: Coat Color in Labrador Retrievers. Labrador retrievers can be (a) black, (b) chocolate, or (c) yellow, and their coloring is controlled by the black-coat gene (*B* or *b*) and the extension gene (*E* or *e*). When a dog receives two doses of the *e* allele, it will be yellow; *ee* masks the effects of the black-coat gene. When the dog inherits *EE* or *Ee*, however, the black-coat gene can be expressed in a black (*BB* or *Bb*) or chocolate (*bb*) coat color, giving a 9 : 3 : 4 ratio instead of the expected 9 : 3 : 3 : 1 in a cross between two parents heterozygous for both genes (*BbEe*).

ABO gene in the human population, a single individual can never have all three at once, because as Chapter 7 explained, each child gets only one allele of a gene from each parent.

The *ABO* gene has three alleles, but some genes have even more than that. Such sets are called *multiple allelic series,* and the medical effects of one kind of multiple allelic series are quite well known. The genes controlling tissue rejection in heart, liver, or kidney transplantations all have many alleles. As part of the *major histocompatibility complex,* such genes encode certain proteins that appear on a cell's surface; these substances serve as identification markers on each individual's tissues and organs and help our bodies distinguish our own cells from organisms like bacteria, viruses, or parasites that might invade and cause disease.

Since there are several genes in the major histocompatibility complex, each with multiple alleles, it is very unlikely that two unrelated (or even related) persons will have precisely the same constellation of alleles at all loci. That is why a person with kidney or liver disease must often wait a long time before becoming matched to a suitable donor. If there are too many allelic differences between the tissues of donor and recipient, the cells of the recipient will kill the cells of the donated organ. Even with drugs (called immunosuppressants) that quell these reactions, multiple alleles for tissue types constitute a major obstacle to life-saving transplantations for a large number of people.

Interactions Between Genes

■ Masking Genes: How Can You Tell a Yellow Dog from a Chocolate-Colored One?

Some genes mask the presence of others and thereby alter telltale phenotypic ratios. The coat color of Labrador retrievers provides a familiar example. Dogs homozygous for the dominant *B* allele of the black-coat gene have a pure black color, while dogs homozygous for the recessive *b* allele have a soft, chocolate-colored coat (Figure 8.22). *Bb* heterozygous animals are also black, indicating complete dominance of *B* over *b*.

But some Labrador retrievers are yellow. How does that come about? Yellow versus dark (black or chocolate) is controlled by another gene called the extension gene, with alleles *E* and *e*. Homozygous recessive *ee* dogs are yellow, no matter which alleles have at the *B* locus; an *ee* genotype, in other words, masks the *B* locus. Conversely, the *E* allele allows the genotype at the *B* locus to show through, and so *BBEE*, *BBEe*, *BbEE*, and *BbEe* animals will be black, and *bbEE* and *bbEe* dogs will be chocolate. Geneticists call a situation like this, in which one gene masks the effect of another gene, **epistasis.** Because the *ee* genotype masks the expression of any allelic form of the black-coat gene, you can't tell by looking at a yellow Labrador what its genetic constitution is at the *B* locus.

■ Traits Controlled by Many Genes

Until now, we have talked about genes whose alleles determine clear alternatives: white versus orange fur in tigers, and purple versus white flowers in peas. Some traits, however, are not qualitatively distinct like these, but vary quantitatively over a range of values. The phenotypes of **quantitative traits** can be measured: the length of a tobacco flower in millimeters, the amount of milk produced by a cow per day in liters, the height of a person in meters. Unfortunately, quantitative traits are complicated for the geneticist because they are generally **polygenic;** that is, they are controlled by several interacting genes rather than by a single pair of alleles at a single locus.

A case in point is skin pigmentation in people (Figure 8.23a). Blacks of African descent and Whites of Northern European descent have similar sets of genes, differing only in a few, such as interacting genes for skin color.

Matings between blacks and whites produce children with intermediate skin color, and matings between F_1 individuals produce children with a wide range of skin pigmentations—a few as light as the white grandparent and a few as dark as the black grandparent, but most in a middle range following a bell-shaped curve. This result can be explained if four genes inherited in a normal fashion act together to control the difference in skin color between African blacks and European whites. If each African allele adds color and each European allele takes away color, then the observed distribution makes sense.

Understanding the inheritance of quantitative traits is especially important for agricultural geneticists, who through crossbreeding and selecting recombinants, try to develop new

(a)

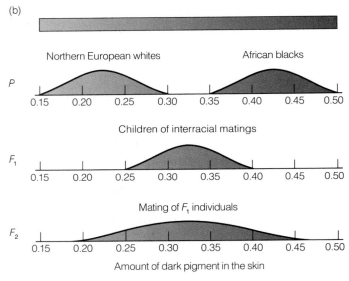

(b)

Northern European whites   African blacks

P

0.15 0.20 0.25 0.30 0.35 0.40 0.45 0.50

Children of interracial matings

F_1

0.15 0.20 0.25 0.30 0.35 0.40 0.45 0.50

Mating of F_1 individuals

F_2

0.15 0.20 0.25 0.30 0.35 0.40 0.45 0.50

Amount of dark pigment in the skin

FIGURE 8.23 ■ Human Skin Color Is a Result of Polygenic Inheritance. (a) Skin color differences between Africans and Europeans are probably due to differences in only four genes. (b) Matings between blacks and whites usually give intermediate skin colors in the children, and matings between F_1 individuals show skin color variations that approach a standard bell curve in the F_2 generation, with some black individuals and some white individuals resulting, but mostly brown-skinned people of many shades from dark to light.

genotypes that will, for instance, grow more bushels of wheat per acre or more pounds of pig per pound of feed. However, because many polygenes interact with each other and the environment, it is necessary to use very careful experiments and sophisticated statistical procedures to distinguish the contribution of each factor. We should therefore be grateful for

the perseverance of quantitative geneticists, for without them, modern farmers would produce only a fraction of the corn, wheat, and pork they now provide for our tables.

Multiple Effects of Individual Genes

In the preceding section, we saw that a single phenotype, such as human skin color, can be affected by more than one gene. The reverse is also true. In a phenomenon known as **pleiotropy,** a single gene may determine several different phenotypes. The white tiger Mohan shows the action of a pleiotropic gene. The primary effect of the white-coat allele is an absence of *melanin,* the dark pigment in the hair, skin, feathers, and scales of vertebrates. However, the same mutant allele also seems to give rise to crossed eyes, not only in tigers, but in other species that have a similar mutation—for instance, in some albino people (Figure 8.24). We do not yet know why this one allele produces two apparently unrelated phenotypes. However, it is easy to see how the process of nat-

FIGURE 8.24 ■ Albinos Illustrate Pleiotropy. A mutant allele of a single gene for melanin production in humans can affect several phenotypic traits, including the pigmentation of eyes, hair, and skin and the occurrence of crossed eyes.

FIGURE 8.25 ■ Siamese Cats and Environmental Effects on Gene Expression. Siamese cats have dark ears, nose, paws, and tail because those extremities are cooler than the rest of the body, and the enzyme involved in producing dark coat pigment can only function at this lower temperature. Thus, a gene's environment can sometimes influence its expression.

ural selection might keep the frequency of the white-coat allele very low among tigers. Crossed eyes could make it more difficult for a tiger to hunt; thus, homozygous white tigers might have a somewhat harder time surviving to sexual maturity. Over the course of several generations, the white-coat allele would be selected against.

For some pleiotropic genes, researchers have been able to pin down a single molecular or cellular cause for the multiple phenotypes. A recessive gene found in the Maori, the aboriginal people of New Zealand, causes defects in a single protein necessary for the action of cilia and flagella. Cilia normally clear the respiratory tracts of dust and debris, so if cilia fail to work efficiently, because of either mutation or smoking, respiratory disease may ensue. And if flagella fail to propel a man's sperm, he will be sterile. Indeed, Maori men homozygous for the gene have frequent respiratory problems and are sterile.

Environmental Effects on Gene Expression

The expression of a gene can be altered not only by other genes, but also by the environment. One striking example is found in a small relative of the white tiger—the Siamese cat (Figure 8.25). Siamese house cats have little color except on their ears, face, tail, and feet. The mutant allele responsible for this hair color pattern in the cat is a different form of the same gene responsible for an orange coat in normal tigers and the white coat of Mohan. But unlike the other two alleles in this multiple allelic series, the expression of the special allele in Siamese cats is temperature-sensitive. The enzyme that catalyzes the production of dark pigment in these cats is unable to work at normal body temperature. But at the lower temperatures found in the cat's extremities, it is catalytically active and can produce melanin that darkens the ears, paws, and tail.

It is often difficult to distinguish the effects of environment from those of genotype on the final phenotype of an individual because the interactions are complex, and controlled experiments are hard to perform.

CONNECTIONS

The story of Mendel's experiments is a remarkable illustration of a scientist's ability to form concrete conclusions from indirect data. Mendel did not know about chromosomes or meiosis. Yet, from his careful analysis of stem length and flower color in peas, he was able to infer the existence and behavior of factors now called genes—tiny objects he would never see. It was left to microscopists to point out the correlations between genes and chromosome distribution in meiosis and to conclude that genes reside on chromosomes.

Mendel's insights also had tremendous theoretical impact once they were rediscovered. By convincing biologists to think of genes as distinct units that pass through a hybrid unchanged, Mendel replaced the blending theory with the particulate theory of inheritance. And by demystifying, through relatively simple computations, how alleles pass through a hybrid intact, he gave evolutionary theorists the kind of mechanism they needed to explain the origins and maintenance of biological diversity.

Ironically, one of the factors most responsible for Mendel's success—his choice of distinct gene alternatives—led to the conclusion that has been most revised over the years: that alleles are either strictly dominant or strictly recessive. Interactions between alleles, between different genes, and between genes and the environment largely qualify Mendel's rule of dominance.

The amazing array of gene interactions makes it hard to penetrate the complex overall relationship of genotype to phenotype. But research has shown that many of the observable traits we have discussed are caused by some chemical in the organism's body. Color in the hair or skin of tigers, dogs, or people is caused by the dark pigment melanin; the color of pea flowers is caused by a red pigment; and cell surface substances that thwart organ transplants differ because they are composed of different protein-carbohydrate complexes. Somehow, genes control these chemical differences. Exactly how they achieve that control will be the subject of the next chapter.

Highlights in Review

1. All inherited traits, from pelt or petal color to stem length and even certain behaviors, are controlled by one or more genes, discrete stretches of DNA found in every cell of a plant or animal.

2. Genes serve as the inherited units through which natural selection shapes each species throughout evolution. Over the course of generations, genes that emerge as adaptive for the species are more likely to be passed along to offspring than are genes that are maladaptive.

3. Genes reside on chromosomes, strung along end to end like beads on a string. The fact that genes are on chromosomes explains the close relationship between the rules of heredity and the principles of meiosis.

4. The genes for each trait can exist in two or more alternative forms called alleles. At conception, every individual receives one allele of each gene from each parent. If the two alleles are identical, the individual is said to be homozygous for that trait. If the two alleles are different, the individual is heterozygous for that trait.

5. The two alleles for a given gene in each cell are carried on two chromosomes that are similar in size, shape, and genetic content, which are called homologous chromosomes. Each chromosome carries many genes. Mendel's first law, the principle of segregation, states that when gametes form, homologous chromosomes—and hence the gene copies they contain—separate, giving each sperm or egg cell only one copy of each gene.

6. Mendel's second law, the principle of independent assortment, explains that the genes on different chromosomes are distributed into gametes independently of each other. However, genes that are on the same chromosome tend to be packaged into gametes together as linked genes.

7. Genes that reside on chromosomes that determine gender are said to be sex-linked. Genes on all other chromosomes are said to be autosomal. Occasionally, the alleles of two linked genes may become separated because they exchange with the corresponding alleles on the homologous chromosome; this process is called crossing over, or recombination.

8. A heterozygote may have visible traits dictated by only one of the alleles, called the dominant allele. The hidden allele is called the recessive allele. Visible traits determine an individual's phenotype, while the underlying genetic constitution that produces that phenotype is known as the genotype.

9. Sometimes alleles are neither clearly dominant nor clearly recessive. In heterozygotes showing incomplete dominance, the phenotype is unlike either parent but is intermediate between those of the homozygous parents. In heterozygotes exhibiting codominance for a specific trait (usually the characteristic of an enzyme or other protein), the characteristics of *both* phenotypes show up fully.

10. Interactions with the environment or with other genes can mask the presence of a particular allele. Many traits, including skin color, are polygenic, which means that they are controlled by several interacting genes rather than by a single pair of alleles. Conversely, a single gene may influence several different traits in a phenomenon known as pleiotropy.

Key Terms

allele, 170
autosome, 176
codominance, 182
cytology, 176
dihybrid cross, 175
dominant, 168
epistasis, 183
first filial (F_1) generation, 168
gene, 165
genetic mapping, 179
genotype, 171
heredity, 165
heterozygous, 171
homozygous, 171
incomplete dominance, 180
linkage group, 178
locus, 179
monohybrid cross, 168
parental (*P*) generation, 168

parental type, 175
phenotype, 171
pleiotropy, 184
polygenic, 183
principle of independent assortment, 176
Punnett square, 173
pure-breeding, 167
quantitative trait, 183
recessive, 168
recombinant type, 175
second filial (F_2) generation, 168
segregation principle, 171
sex chromosome, 176
testcross, 173
X chromosome, 176
X-linked, 177
Y chromosome, 176

Study Questions

Review What You Have Learned

1. Why did Mendel use garden peas in his experiments?

2. Trace the results of a cross between pure round-seeded pea plants and pure wrinkle-seeded pea plants through the F_1 generation; through the F_2 generation.

3. State Mendel's principle of segregation.

4. The F_2 generation of Mendel's pea plants showed a 1 : 2 : 1 genotypic ratio and a 3:1 phenotypic ratio. Explain what causes the genotypic ratio to differ from the phenotypic ratio.

5. Show the genotype of all possible gametes made by a hybrid pea plant with long stems and purple flowers, *LlPp*.

6. How are the sex chromosomes of sperm and egg alike? How are they different?

7. If two homologous chromosomes contain alleles *AD* and *ad*, respectively, what genotypes would occur after crossing over?

8. How is the inheritance of flower color in snapdragons an exception to Mendel's principle of dominance?

9. Three Labrador retrievers have the different gene combinations *BBEE*, *BBee*, and *bbEE*. What is the color of each dog?

10. True or false: Linkage is an exception to Mendel's principle of independent assortment. Explain your answer.

Apply What You Have Learned

1. To observe the laws of chance, toss a dime and a quarter simultaneously 10 times and tally the number of head-head, head-tail, and tail-tail combinations. How close are your results to a 1:2:1 ratio? Now toss the coins 90 times more, for a total of 100 trials. How close are your new results to a 1:2:1 ratio? Explain.

2. In guinea pigs, black fur is dominant over white fur. How could an animal breeder test whether a black guinea pig is homozygous or heterozygous?

3. In humans, red-green color blindness is a recessive X-linked trait. Use a Punnett square to show the possible genotypes of offspring of a color-blind man (X^-Y) and a normal homozygous woman (X^+X^+), where X^+ contains the normal allele of this gene for color vision, and X^- contains the allele for color blindness.

For Further Reading

BHATTACHARYYA, M. K., A. M. SMITH, T. H. ELLIS, C. HEDLEY, and C. MARTIN. "The Wrinkled Seed Character of Pea Described by Mendel Is Caused by a Transposon-like Insertion in a Gene Encoding Starch-Branching Enzymes." *Cell* 60 (1990): 115–122.

HALL, J. G. "Genomic Imprinting: Review and Relevance to Human Disease." *American Journal of Human Genetics* 46 (1990): 857–873.

LEYHAUSEN, P., and T. H. REED. "The White Tiger: Care and Breeding of a Genetic Freak." *Smithsonian,* April 1971, pp. 24–31.

OREL, V. *Mendel.* New York: Oxford University Press, 1984.

SAPIENZA, C. "Parental Imprinting of Genes." *Scientific American,* October 1990, pp. 52–60.

SAVORY, T. H. "The Mule." *Scientific American,* December 1970, pp. 102–108.

SUZUKI, D. T., A. J. F. GRIFFITHS, J. H. MILLER, and R. C. LEWONTIN. *Introduction to Genetic Analysis.* 4th ed. New York: W. H. Freeman, 1986.

DNA: The Thread of Life

A Sudden Epidemic and Unbeatable Bacteria

In 1955, a medical disaster struck a small community in Japan. A life-threatening form of dysentery raced through the population, and at the height of the epidemic, the four antibiotics being prescribed to treat the victims suddenly became useless. The bacterium responsible for the dysentery outbreak, *Shigella dysenteriae,* had somehow acquired resistance to the antibiotics—the mainstays of modern medicine—and people were dying in frightening numbers.

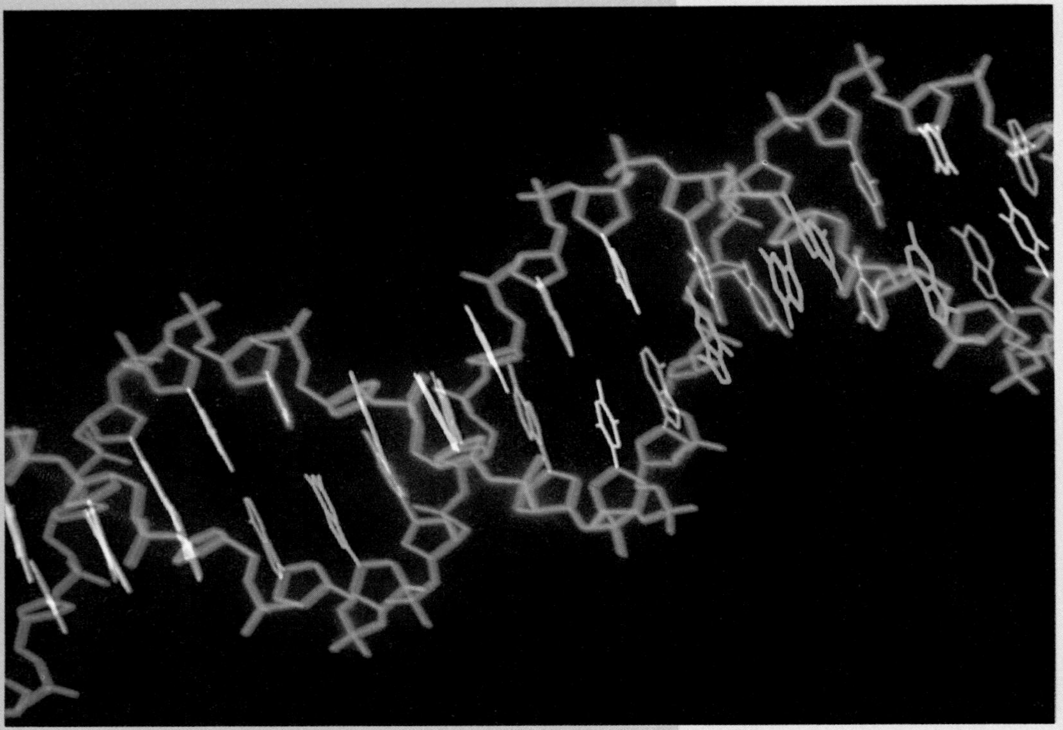

Biologists had never before witnessed such rapid evolution of *multiple drug resistance* in a bacterial species, and the occurrence left them confused and struggling to solve the mystery. In the past, mutant bacterial strains had developed resistance to one or another antibiotic, but never to several drugs all at once. Still more disturbing, the resistance spread rapidly from one kind of bacterium to another, ultimately appearing in the species that cause typhoid fever, pneumonia, and severe urinary tract infections. How could bacteria become resistant to penicillin and several other antibiotics all at once? And how could resistance to those drugs spread from one bacterial species to another?

Geneticists searched intensively and found that multiple drug resistance was caused by minute circles of DNA (Figure 9.1). Called **plasmids,** these circles can exist either inside or outside bacterial cells, but can only multiply inside. Thus, plasmids can be seen as the ultimate parasites—entities composed of pure genetic material with a few powerful genes, just enough to ensure their own survival and reproduction in a bacterial host. One DNA circle, or plasmid, called pBR322 (Figure 9.2), contains only two genes, each of which inactivates a different antibiotic. Another important DNA sequence is located at a special site along the circle of plasmid DNA; this sequence essentially says, "Start copying here." Copying, or replication, leads to the formation of new copies of the plasmids within the host bacterium. The plasmid DNA has no other genes and needs no others. Because each plasmid provides its host with the genetic information needed to protect

FIGURE 9.1 ■ **DNA: The Code of Life.** This computer-generated image of a DNA molecule shows the twisted ladder structure so central to its information-bearing capacity.

A plasmid is a circle of DNA.

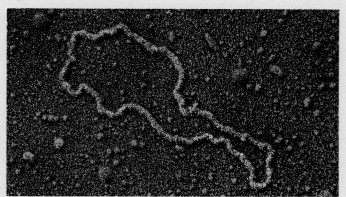

FIGURE 9.2 ■ **Plasmids: The Ultimate Parasites.** A plasmid is a circle of DNA that can survive inside an *E. coli* bacterium or other cell and make the cell resistant to antibiotics. This electron micrograph shows a plasmid enlarged 30,000 times. The plasmid, called pBR322, bears a gene for resistance to the antibiotic ampicillin, a gene for resistance to tetracycline, and a short sequence, the origin of replication, where copying of the plasmid DNA begins.

the cell from antibiotics, the bacterium survives to replicate itself, and the plasmid DNA is replicated as well. When the bacterium happens to release the replicated plasmids, they can spread to other bacteria, bringing to their new hosts the valuable ability to survive in the presence of antibiotics.

A plasmid represents the simplest hereditary system and displays two hallmarks of genes discussed in Chapters 7 and 8: It is a single molecule that can store *information,* and it can undergo *replication.* Since plasmids are nothing more than DNA and yet can bestow on their bacterial hosts the dramatic genetic trait of drug resistance, they make an ideal case study for the topics covered in this chapter: what DNA is, what genes are, how genes replicate, and how they function.

Biologists have learned a great deal about DNA and genes from the simple genetic systems of plasmids and bacteria. One of their most important discoveries is the unifying nature of genetic material: Both the structure of the genes and the way genetic information is encoded and replicated are fundamentally the same in all living organisms, from bacteria to whales. Thus, the DNA in your cells looks and acts essentially the same as the DNA in the bacteria on your skin and in your intestines. In fact, the structures of different kinds of genes are so similar that molecular geneticists can easily remove a piece of your DNA and splice it into the DNA of a plasmid. A bacterium carrying this altered plasmid then replicates your DNA along with the plasmid DNA circle. The result: a human gene multiplying in a bacterial cell!

By studying simple forms of DNA, biologists have also discovered a second unifying principle: The key to the crucial activity of DNA lies in the molecule's striking double helix structure. This is one more example of how form dictates function in living things. Just as a bat's large ears are the secret to its acute directional hearing, the three-dimensional structure of DNA is the secret to its means of storing information and replicating itself.

Finally, knowledge of how genes are built and how they operate has led molecular biologists to develop powerful new ways to manipulate DNA. (These manipulations are described more fully in Chapter 11.) Today's experiments in genetic engineering will shape the world for generations to come, allowing scientists to generate new kinds of proteins with important applications in medicine, agriculture, and industry. The decisions governing the development of gene-splicing technology will require the informed participation of all citizens and should be based on the best possible understanding of what DNA is and how it works.

We shall see in this chapter that genes are made of DNA. Plasmids are a simple demonstration of this fact, since they contain genes and consist only of DNA. The structure of that DNA is simple and elegant. DNA is like a ladder twisted along its long axis, the "ladder's" two uprights forming a pair of intertwined corkscrews; in other words, it is a **double helix.** This structure determines how DNA serves as the template for its own replication and how the information encoded in genes is translated into a protein. We will also see that genes work by specifying the amino acid sequence of the proteins. During the Japanese dysentery epidemic, the antibiotic resistance genes of the plasmid DNAs inside bacteria caused the antibiotics to break down into harmless compounds. The actual breakdown of the antibiotic molecules was facilitated by an enzyme whose sequence of amino acids was specified by a gene in the plasmid DNA.

As we retrace the steps that led from Gregor Mendel's abstract idea of a gene to the modern concrete reality of a self-replicating molecule with a well-known shape and activity, we will address several fundamental questions:

■ What is the proof that genes are made of DNA?
■ How are immensely long DNA molecules packaged?
■ How is DNA's structure the key to self-replication, and how does DNA function within cells?
■ How do genes determine the structure of proteins and consequently the function of cells?

WHAT ARE GENES?

Mendel's experiments showed that discrete hereditary factors determine a pea plant's physical appearance: the color of its flowers, the length of its stem, and the shape of its seeds. But Mendel himself had no idea what genes were or how they worked. How could the information in an invisible gene control something as visible as a flower's pink color? To answer such a question, we must first know the chemical nature of the gene, then ask how that chemical functions. The first step toward determining the chemical nature of the gene was the demonstration that genes in higher organisms are located on chromosomes (see Chapter 8); it thus followed that some substance in the chromosomes must constitute the genes. Since chromosomes in higher organisms contain mainly protein, RNA, and DNA, one of these substances was presumed to be the genetic material. But which one?

Geneticists at first thought that only proteins could carry the complex information in genes. Proteins have 20 different subunits (amino acids), while DNA has only 4—the bases adenine (A), thymine (T), guanine (G), and cytosine (C) (see Chapter 2). Clearly, 20 amino acid subunits can form many more combinations (and carry much more information) than can 4 bases, just as you can form more 5-letter words from an alphabet of 20 letters than from an alphabet of 4 letters.

Bacterial Transformation: Evidence for DNA

The notion that genes are made of protein was challenged in 1944. Researchers at that time found that they could transfer an inherited characteristic—the ability to cause pneumonia—from one strain of bacteria to another simply by exposing a harmless bacterial strain to DNA extracted from a pathogenic (disease-causing) strain. The pathogenic DNA taken up by the harmless bacteria caused the bacteria to make a protein that results in pneumonia (Figure 9.3). This process of transferring an inherited trait by an extract of DNA is called **transformation**.

Transformation is the process that can allow plasmids to spread so rapidly from one bacterium to another, as they did in the dysentery epidemic. Transformation proves that genes are made of DNA, but in 1944, few geneticists would accept the transformation experiments as evidence of that fact. Many were not yet ready to give up their cherished notion that genes are made of proteins.

Bacterial Invaders: Conclusive Proof from Viral DNA

The definitive experimental proof of the gene-DNA connection came from experiments with phages. Properly called **bacteriophages** (which means "bacteria eaters"), these viruses consist of nothing but a core of DNA surrounded by a protein shell (Figure 9.4a and b). They infect bacterial cells by attaching themselves to the cell surface and injecting their

FIGURE 9.3 ■ How Do We Know Genes Are Made of DNA? **(a)** The nonvirulent R (rough) strain of pneumococcus, on the right, grows on the same Petri dish with the virulent S (smooth) strain, on the left (magnification $1640\times$). The S strain secretes a capsule around its cell wall that protects it from a host animal's immune system. **(b)** When a researcher injects a mouse with live S strain bacteria, the rodent dies, whereas **(c)** a mouse injected with live R strain bacteria lives. This shows that the S strain is virulent. **(d)** When the S strain is killed by heat and then injected, the mouse lives. **(e)** However, when the heat-killed S strain is mixed with the live R strain and injected, the mouse dies. A blood sample from the dead mouse shows that mixing the heat-killed S strain and the R strain transforms the live R strain into an S strain. Subsequent experiments proved that the DNA of the heat-killed virulent strain was incorporated into the nonvirulent strain, rendering it virulent—and deadly.

genes into the cell (Figure 9.4c). Once inside the bacterium, the phage genes commandeer the cell's protein-making machinery so that the cell makes almost nothing but phage proteins. Finally, the cell bursts open and releases about a hundred new phages, each of which infects any bacterium it touches, repeating the cycle.

Since phages are much smaller and simpler than cells, they make ideal subjects for determining whether protein or DNA is the hereditary material. Microbial geneticists Alfred Hershey and Martha Chase infected *Escherichia coli* bacteria with phages whose DNA and protein were labeled by two separate radioactive compounds that could be distinguished from each other. They found that phage DNA entered the bacterial cells, but phage protein did not (see Figure 9.4c). Thus, it was the DNA that carried the information to make new phages.

But is a simple bacteriophage representative of more complex living organisms? The answer is yes—or mostly yes. Subsequent research has confirmed that DNA is the genetic material in all cells and many viruses. (In some viruses, such as the virus that causes acquired immune deficiency syndrome, or AIDS, the genes are made of RNA.) The experiments described here changed the gene from an abstraction, a vague something that determines heredity, into a tangible chemical that biologists could see and manipulate. Nowadays, molecular biologists break open cells, separate the DNA from the proteins, add ice-cold ethanol to the DNA, and put it in

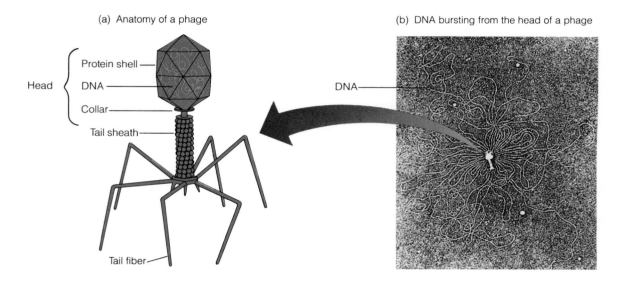

(a) Anatomy of a phage

Head
Protein shell
DNA
Collar
Tail sheath
Tail fiber

(b) DNA bursting from the head of a phage

DNA

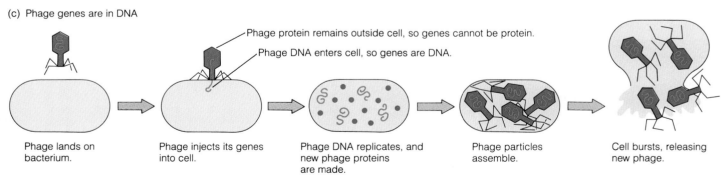

(c) Phage genes are in DNA

Phage protein remains outside cell, so genes cannot be protein.

Phage DNA enters cell, so genes are DNA.

Phage lands on bacterium.

Phage injects its genes into cell.

Phage DNA replicates, and new phage proteins are made.

Phage particles assemble.

Cell bursts, releasing new phage.

FIGURE 9.4 ■ **Structure and Life Cycle of a Bacteriophage.** (a) The bacteriophage T4 looks like a crystal on legs. The crystalline head of a phage consists of a protein shell surrounding DNA; a columnar tail and leglike tail fibers are also made of pure protein. A relatively large amount of DNA is packed into the head of a phage. (b) The DNA from a phage has spewed out of the head. If this DNA were stretched out, it could wind around the perimeter of the photo six times. (c) In the phage life cycle, the phage lands on a bacterial cell and injects its genes; this directs the cell to make new phage particles, and these burst out and infect additional cells. Since phage DNA enters the cell, but phage protein does not, Hershey and Chase concluded that genes are made of DNA.

FIGURE 9.5 ■ **DNA Is a Stringy White Material.** In a laboratory, researchers isolate DNA from the cells that carried it. Here, a biochemist is separating DNA from solution.

the freezer. The next morning, at the bottom of the test tube is a stringy white precipitate a bit like powdered sugar—but it is pure DNA, pure genes (Figure 9.5).

Just knowing that genes are made of DNA, however, still does not explain how a gene works. The way a gene stores information and replicates is revealed by the form of the DNA molecule itself.

DNA: THE TWISTED LADDER OF INHERITANCE

A delicate butterfly and the plant on which it rests both have genes encoded in molecules of DNA. DNA can replicate so perfectly that the offspring of two swallowtail butterflies look just like the parents, and yet the genetic material can still contain enough variation so that some swallowtails turn out black rather than yellow (Figure 9.6). The variability inherent in DNA is the basis of natural selection, the key to evolution, and the explanation for life's stunning diversity. What is it about DNA that makes this possible, and how does its structure account for both the unity and the diversity of life?

To see how the relatively simple DNA molecule perpetuates life, we will examine it from the smallest subunits to the entire chromosome in which the DNA is packaged.

DNA: A Linear Molecule

One fact underlies DNA's simplicity of structure and function: DNA is a linear molecule. Despite this simplicity, however, the linear DNA molecule contains almost infinite possibilities for instructions. The structure of a single strand of DNA can be imagined as a long chain with links of four different colors, each representing a different nucleotide. A nucleotide consists of three parts: a phosphate, a sugar, and a base (Figure 9.7). The four kinds of nucleotides are identical except for the bases they contain—either *adenine* (A), *cytosine* (C), *guanine* (G), or *thymine* (T). So in genetic shorthand, we often refer to DNA as a chain of bases, rather than as a chain of nucleotides. For example, a DNA chain, or sequence, might consist of 15 nucleotides, each with one of the four kinds of bases. But it would be described according to the order of its bases: ACCCGTCCGTGTTAG.

FIGURE 9.6 ■ **DNA: The Universal Hereditary Material.** The same kind of biological molecule can encode genetic information for these yellow and black swallowtail butterflies.

BOX 9.1

THE HUMAN GENOME PROJECT

In October of 1990, the journal *Science* included a large foldout wall chart printed with figures that resemble military decorations and candy-striped worms. The poster was, in fact, an up-to-date map of the human genome, placing genes of known function and location, regions where the nucleotide sequence has been worked out, and other genetic data in their correct positions on blown-up, diagrammatic versions of the human's 22 chromosomes plus *X* and *Y* (Figure 1). By looking at the bands of yellow, orange, blue, and magenta, one can interpret how complete—or more accurately, how incomplete—the mapping effort was at that time.

Although molecular geneticists had by then already spent over five years and $100 million of federal research funds in what will eventually be the largest scientific project ever undertaken, there are far more empty spaces on the human gene map than data to plug in. Some have labeled the Human Genome Project the ultimate measure of humankind—an effort that could revolutionize medicine, biology, and psychology. Noting that only a tiny fraction of the map is completed, however, critics wonder whether the benefits can possibly outweigh the tremendous costs.

First suggested in 1985, the Human Genome Project will cost more and take longer than building the atom bomb or landing people on the moon. Hundreds of researchers will labor simultaneously (1) to create a low-resolution physical map that places the 1000 or so currently known human genes in the proper positions on the 23 pairs of chromosomes; (2) to locate and identify the 50,000 to 100,000 additional genes (some 4400 of which occur on an average chromosome); and (3) to determine the nucleotide sequences of all 3 billion bases in the human DNA—even sequences that don't code for proteins or that simply repeat each other.

The project will cost at least $3 billion and won't be completed before the year 2005. The result will be a 500-volume encyclopedia of human genetics spelling out the detailed instructions for every protein in the body, and with them, a high-resolution view of how these proteins function normally day to day and perhaps participate in mental and physical illness. Collectively, humans have more than 4000 hereditary diseases, including sickle-cell anemia, cystic fibrosis, and Huntington's disease. These three and a few hundred others can be traced to single-gene mutations. Most diseases, however, including cancer, diabetes, Alzheimer's, and heart disease, probably have roots in multiple genes, and the interactions between them will be challenging to work out, even with a genome map.

To some critics—including 23 Harvard professors who published a dissenting view of the Human Genome Project in *Science* shortly before the 1990 map appeared—the promised high-resolution view is a mixed blessing. Knowing the complete nucleotide sequence, wrote Bernard Davis and colleagues, would be "like viewing a painting through a microscope," with researchers having to "plow through 1 to 2 million 'junk' bases

FIGURE 1 ■ **The Human Genome Map.**

before encountering an interesting sequence"* and then trying to discover its unknown function. Even senior statesman of molecular biology Sydney Brenner (see Box 7.2 on page 155) jokes that working out nucleotide sequences is so tedious that a penal colony could hand out sequencing projects as prison sentences. And everyone from the critics to Nobelist James Watson, who is heading those parts of the project overseen by the National Institutes of Health, agree that only 3 percent of the bases in the total sequence code for proteins of any kind.

Other major criticisms are financial and ethical. The last decade or so has seen major cuts in funding for biomedical research. The money allocated for Genome Project work in 1991 alone could provide sizable ($200,000) research grants to over 750 investigators. Critics also worry that a human gene map could lead to genetic discrimination, since doctors, insurance companies, and employers could "read" your inherited tendencies toward heart disease, cancer, or dementia, let's say, even though you have no symptoms now and may never have them.

Despite these legitimate concerns, many scientists are convinced that the benefits will far exceed the risks and point to the wealth of basic information already unlocked from DNA sequencing in viruses, bacteria, plants, and other animals. James Watson, probably the Genome Project's most eloquent spokesman, sees it this way: "When finally interpreted, the genetic message encoded within our DNA molecules will provide the ultimate answers to the chemical underpinnings of human existence."†

---

*Bernard D. Davis et al., "The Human Genome and Other Initiatives," *Science* 249 (1990): 342–343.

†James D. Watson, "The Human Genome Project: Past, Present, and Future," *Science* 248 (1990): 44–49.

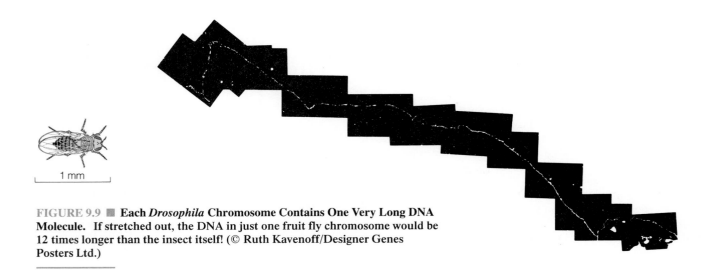

FIGURE 9.9 ■ **Each *Drosophila* Chromosome Contains One Very Long DNA Molecule.** If stretched out, the DNA in just one fruit fly chromosome would be 12 times longer than the insect itself! (© Ruth Kavenoff/Designer Genes Posters Ltd.)

Packaging DNA

Watson and Crick derived the twisted-ladder structure of DNA from studies of chemically purified DNA. But the DNA in living cells is more complex. First of all, it comes in two forms: circular and linear.

The DNA of many prokaryotic cells loops back on itself to form a circle. DNA in plasmids (review Figure 9.2), bacterial cells, and viruses that infect mammalian cells is circular in form, as is the DNA of the chloroplasts and mitochondria in eukaryotic cells.

In contrast, the nuclear DNA of eukaryotes and many viruses is linear. In these organisms, single DNA molecules are also extraordinarily long and are organized into chromosomes, each chromosome consisting of a single, long, tightly wound DNA molecule. The DNA molecule from the largest of the four chromosomes of the fruit fly *Drosophila*, for example, is about three times as long as the word *genetics* printed here (that's 12 times as long as the fly itself!) and carries a few thousand genes (Figure 9.9). In spite of the molecule's length, each of the fly's millions of cells contains two copies of this chromosome as well as pairs of the other three chromosomes. How do cells package such a huge molecule into a structure as small as a chromosome?

The enormous length of DNA in a eukaryotic cell cannot be wadded up haphazardly, because if it were, the separation of DNA molecules during cell division would be as difficult as untangling two kite strings. What happens instead is that DNA, like a proper kite string, is wound around spools, the spools being made of proteins called **histones** (Figure 9.10a and b). A single histone spool wrapped with two loops of DNA (140 base pairs long) is called a **nucleosome.** Adjacent

nucleosomes pack closely together to form a larger coil, somewhat like a coiled telephone cord. This coil, in turn, is looped and packaged with scaffolding proteins into **chromatin,** the substance of a chromosome (Figure 9.10c and d). It is this orderly packing of DNA that prevents massive tangles during cell division.

From base pair to double helix to spool-wrapped histone, the structure of DNA is a boundless biological resource. How that structure makes DNA work to create diversity out of unity is the true marvel of biology.

MYSTERIES OF HEREDITY UNVEILED IN DNA STRUCTURE

When Watson and Crick first publicly presented their model of the double helix at a meeting of geneticists and biochemists in Cold Spring Harbor, New York, in the spring of 1953, they knew it would generate excitement, for it suggested ways that DNA might replicate and store information. But even Watson and Crick were startled by the enthusiastic reception their peers gave the double helix model. In the eyes of a scientist, a hypothesis takes on power and beauty in proportion to how well it explains diverse facts simply and coherently; and the scientific community agreed that the double helix model was a powerful tool.

In the laboratory, geneticists quickly found that complementary base pairing and linear variation in base order do indeed account for a gene's ability to replicate itself and store information. This section explores how genes can do this.

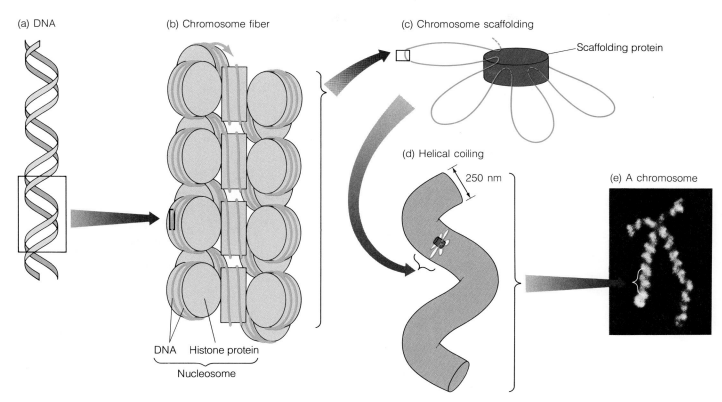

(a) DNA

(b) Chromosome fiber

(c) Chromosome scaffolding

Scaffolding protein

(d) Helical coiling

250 nm

(e) A chromosome

DNA   Histone protein

Nucleosome

FIGURE 9.10 ■ How DNA Is Packaged into Chromosomes. DNA is packaged in a precise way that allows its enormous length to be contained in a chromosome without tangling or loss of function. The DNA double helix (a) wraps around protein spools, forming nucleosomes (b). Nucleosomes interact with each other, probably forming a coil, or solenoid, in a chromosome fiber. The chromosome fiber loops in and out of a central core of scaffolding proteins (c) that also gently coil (d). This coiling is visible in an intact chromosome (e). The chromosome fibers shown in part (b) can also be seen in Figure 7.5, along with the coiling shown in this figure.

DNA: A Molecular Template for Its Own Replication

People have always recognized that like begets like. It is obvious at some family reunions, for instance, that red hair, a big nose, or short stature has been passed along for several generations. For traits to be conserved this way, genes must produce exact copies of themselves, and those copies must be transmitted from parent to child. Geneticists discovered that DNA is passed from generation to generation with extraordinary accuracy because the molecule itself is a **template,** or pattern, for its own replication (Figure 9.11).

The Three Stages of Replication

Because each kind of base will link up with only one other kind of base, the order of bases in the parent DNA molecule determines the order of bases in the daughter molecules.

Thus, DNA replication follows directly from the principle of complementary base pairing and can be divided into three steps:

1. Unwinding. For replication to occur, the two strands of the double helix must first separate from each other (Figure 9.11a). The unwinding of DNA is catalyzed by special enzymes that break the rungs of the ladder—the weak hydrogen bonds attaching the bases on one strand to the bases on the other strand. The bases on each side of the unwound duplex are thus left unpaired. The Y-shaped junction at the spot where the two strands are unwinding is called the *replication fork,* and it is at this fork that the DNA is copied.

2. Complementary base pairing. The unpaired bases generated by unwinding can then pair with (form hydrogen bonds with) any free nucleotides that happen to drift into the area. Yet the pairing is anything but random. An A base on one DNA strand will pair only with a free T base

(a)   Unwinding

Strands separate.

(b)   Pairing

Free nucleotides diffuse in
and pair up with bases on the
separated strands.

(c)   Joining

Each new row of bases is linked into a continuous strand.

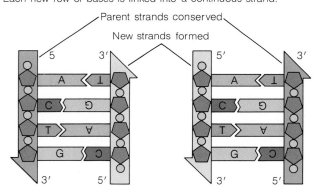

(d) The replication fork

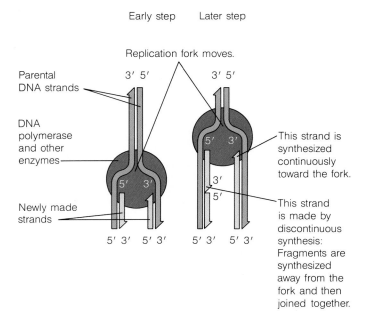

FIGURE 9.11 ■ DNA Replication. When a parent DNA strand
replicates, the strands separate (a) and previously unattached
nucleotides (that have accumulated in the nucleus) diffuse in and
pair up with appropriate unpaired bases (b). The new bases then
link together (c), forming new strands. The two new molecules
are identical to the parent DNA molecule, and in each, one strand
is inherited intact from the parent and one is newly formed. The
two newly forming DNA strands (light blue) behave somewhat
differently at the replication fork (d). The enzyme DNA polymerase moves only in one direction (5′ to 3′). Thus, at an early step in
replication, one of the newly synthesized strands is synthesized in
the direction of the fork, but the other is synthesized in the opposite direction (away from the fork). At a later step in replication,
when the fork has moved forward, the strand made in the direction of the fork can be synthesized continuously. The other newly
made strand, however, is made discontinuously in short fragments, starting near the fork and moving away from it. These
fragments are then joined together by enzymes, and a new intact
strand of DNA is generated.

that happens to float in, and an attached C can pair only
with a free G. Thus, the order of bases in the old strand
specifies the order in the new strand according to the rules
of complementary base pairing (Figure 9.11b).

3. Joining. After bases pair with their complementary bases,
    covalent bonds link them together. This joining, called
    polymerization, is catalyzed by an enzyme called *DNA
    polymerase* (Figure 9.11c). Since this enzyme can synthe-
    size DNA in only one direction, there are differences in

the way the two newly made DNA strands are synthesized. See Figure 9.11d. (A drug called AZT, which is
helping prolong the lives of AIDS sufferers, prevents the
normal joining of bases during DNA synthesis, as Box 9.2
on pages 200–201 explains.)

The continuous repetition of these three steps along the
length of the DNA molecule—unwinding, base pairing, and
joining—produces two DNA duplexes from one, and each

FIGURE 9.12 ■ Is DNA Replication Semiconservative? If DNA replication is semi-conservative, then one strand of each double helix will be derived intact from the parent double helix (brown in the sketch), and the other strand will be newly formed (yellow in the sketch). Follow the path of one original parental (brown) strand through two rounds of DNA synthesis. To see if the semiconservative path predicted by the hypothesis is true, researchers grew cells from a hamster ovary for two rounds of DNA synthesis—two cell cycles—in the presence of a compound that fluoresces bright yellow. All newly made strands will therefore show up bright yellow in the chromosomes. DNA molecules with one newly made strand and one parental strand will have half as much fluorescent material and will appear brown. The results of this experiment, as shown in the photograph, are consistent with the hypothesis. Each so-called "harlequin" chromosome comes from the ovary of a Chinese hamster. It is shown here immediately after the second period of DNA synthesis, so it consists of two chromatids, each made up of a single DNA molecule (review Chapter 7). The DNA molecule in the brown chromatid has one entire strand of the double helix from the original parent and one entire strand newly synthesized, while the DNA molecule in the bright yellow chromatid has both strands made of newly synthesized DNA. This is the result one would expect if DNA replication were semiconservative.

"Harlequin" chromosomes

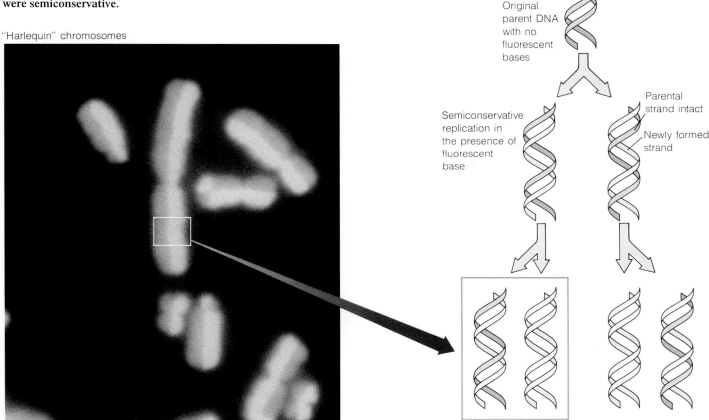

new duplex is identical in base sequence to the original. A quick look at Figure 9.11d shows that each of the two daughter DNA duplexes has one strand inherited intact from the original parent, while the other strand is completely new. Since only one of the two strands in the daughter molecule is inherited intact—or conserved—from the parent molecule, this type of replication is called **semiconservative replication.** Figure 9.12 depicts experiments that proved that DNA replication is semiconservative in higher organisms. As far as biologists have been able to detect, all living creatures share this mode of DNA replication. Occurring in the S phase of the cell cycle, before mitosis, DNA replication is fundamental to the phenomenon of "like begets like."

■ **The Amazing Accuracy of Replication** Considering the complexities of unwinding, base pairing, and polymerization required for semiconservative replication, it is a wonder that DNA molecules are ever copied correctly. Yet DNA synthesis is incredibly accurate: An error creeps in only once every 10^9

BOX 9.2   *PERSONAL IMPACT*

DNA SYNTHESIS, AZT, AND TREATING AIDS

Thanks to researchers who understood the basic biological mechanism of DNA synthesis, there is a drug that helps people suffering from AIDS. In 1985, medical researchers at Duke University showed that a drug called zidovudine (AZT) effectively slows the progression of acquired immune deficiency syndrome, a virus-caused illness that destroys the immune system and leaves the body susceptible to life-threatening infections and cancers (details in Chapter 22). These studies were preliminary, however, and AZT did not receive approval from the U.S. Food and Drug Administration for two more years. This delay triggered much anger and protest from AIDS patients, hundreds of whom died during the interim period. Why was the U.S. government initially hesitant about legalizing the drug? And how, in the first place, does AZT therapy work?

AIDS is caused by the HIV virus, which has about nine genes stored in an RNA molecule—a single-stranded molecule of heredity related to the double-stranded DNA. In order for the HIV virus to successfully infect a cell of the human immune system (a white blood cell), the virus's RNA genes must be copied into the matching DNA form. This RNA-to-DNA copying is carried out by a viral enzyme (called *reverse transcriptase*) that synthesizes DNA from the viral RNA template much as the normal DNA-synthesizing enzyme (*DNA polymerase*) makes DNA from a DNA template (review Figure 9.11).

In both DNA-to-DNA copying and RNA-to-DNA copying, synthesis proceeds in a single direction and accomplishes a hooking up of two nucleotides, with the phosphate of one nucleotide connected by an oxygen atom to the sugar of the previous nucleotide in the chain (Figure 1a, step 1). The Duke University biologists knew that DNA polymerase acts by bringing an OH group that hangs down from the sugar of one nucleotide together with another OH group that protrudes upward from the phosphate of a second nucleotide (Figure 1a, step 2). As a result, an H from one joins an OH from the other, making H_2O, or water (step 2). In the process, the enzyme leaves the remaining oxygen stuck as a bridge between the two nucleotides, and the newly forming DNA chain grows one nucleotide longer.

The AIDS researchers at Duke knew two other important things: (1) Both the cell's DNA polymerase and the virus's reverse transcriptase enzymes can pick up the drug AZT instead of the normal nucleotide containing adenine (A) and incorporate it into growing DNA; and (2) because of AZT's structure (missing the critical OH that forms the link to a second nucleotide), synthesis of a new DNA chain halts at the point where AZT is inserted. They reasoned that if AZT is incorporated in this way and prevents the viral RNA from being copied into DNA, the virus could not infect additional immune system cells, and the spread of AIDS through the body would be slowed.

In July of 1985, AZT was first administered to AIDS patients at Duke University Medical Center. During this initial study, AIDS sufferers improved considerably from the treatment. After reviewing this work, however, scientists and physicians from the FDA insisted that they could not approve and release the drug nationally until additional studies were carried out. This was their reason: It was clear that AZT blocks synthesis of DNA from the viral RNA template. It was not yet clear, though, how much the drug also blocks necessary DNA synthesis in the patient's own cells. If this blockage was too pronounced, it could lead to a condition similar to radiation sickness (see Box 7.1 on page 150) or to cancer and would be, in the long run, worse for patients than other AIDS therapies already available.

As these additional, longer-term tests were going on, many AIDS patients protested and demonstrated to gain faster access to AZT (Figure 1b). Fortunately, the follow-up studies confirmed that AZT therapy does block viral replication more effectively than it blocks DNA synthesis in the patient's own cells and that while there are significant side effects from this cellular blocking, such as headaches, mood swings, anemia, and digestive disturbances, in general AZT prolongs AIDS patients' survival, and improves their quality of life. In March of 1987, the FDA approved AZT as a prescription drug for AIDS patients, and today, it is helping to keep thousands of people alive as the search for better treatments and prevention goes on.

---

bases. To approach this level of accuracy, you would have to type this entire book a thousand times and make only one typing error!

For many organisms, survival requires extreme accuracy of DNA replication. For example, an animal **genome** (the total of all the genes in a single egg or sperm) contains about 3×10^9 base pairs. On average, each egg or sperm would have about three errors, given the actual error rate. If the rate were much higher, the DNA's information would be so altered that the organism wouldn't be able to function.

The very low rate of errors during replication depends in part on a process similar to the one used to catch errors in the typesetting of this book: proofreading. DNA polymerase removes any base from the end of the newly formed strand

that is not hydrogen-bonded to another base, then corrects any mistakes that occur during base pairing, rather like a correcting typewriter that lifts a wrong letter, then replaces it with a correct one.

The necessity for such an amazing level of accuracy derives from the two main functions of DNA: information storage and replication. If an error in replication causes the sequence of bases to change, then the genetic information will also change. Such changes in genetic information, called *mutations,* can cause an individual to grow or function abnormally. Human diseases like hemophilia (bleeder's disease) and muscular dystrophy are caused by mutations. Fortunately, the accuracy of complementary base pairing and the proofreading action of enzymes prevent nearly all such information

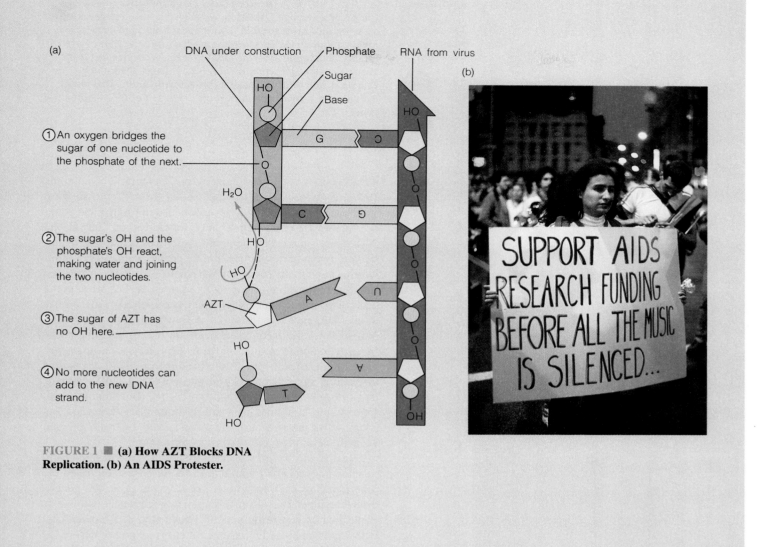

FIGURE 1 ■ **(a) How AZT Blocks DNA Replication. (b) An AIDS Protester.**

changes from taking place during DNA replication. While mutations like hemophilia prove that DNA's information content is important, biologists wanted to know exactly *how* DNA stores information. A view of what genes do, particularly the function of information storage, emerged well before Watson and Crick studied the structure of DNA and from a harmless but startling genetic defect in human babies.

How Does DNA Store Information?

■ **Wrong Genes, Wrong Enzymes** Much to the consternation of their parents, some otherwise **normal** babies produce urine that stains their diapers jet black. This rare condition, called *alkaptonuria,* occurs when a defective gene causes the babies to excrete a breakdown product of the amino acid tyrosine that turns black on contact with the oxygen in the air.

Around the time that Mendel's laws were rediscovered about 90 years ago, the English physician Archibald Garrod had an insight: He suggested that the substance these infants excreted is an ordinary product of metabolism that in most people is broken down by an enzyme (Figure 9.13a). But in people with alkaptonuria, he speculated, this enzyme does not work, so the product is never broken down, but instead accumulates to excess (Figure 9.13b). The excess substance, Garrod thought, must be excreted in the urine, where it turns black when exposed to air. He also observed the effects in dark spots on the outer ear (Figure 9.13c).

(a) Cells of normal infants   (b) Cells of abnormal infants

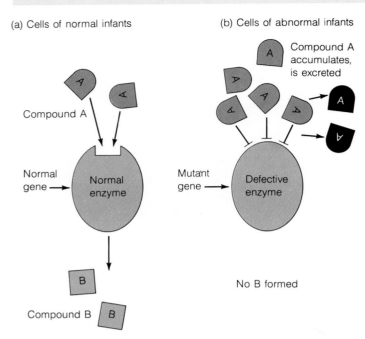

FIGURE 9.13 ■ **A Defective Gene and a Defective Enzyme: Causes of a Strange Condition.** A single genetic change causes some infants' urine to stain their diapers black, and causes dark pigmentation in the ears of some adults. **(a) In the cells of normal infants, compound A is converted to B by an enzyme. (b) If the enzyme is defective, compound A accumulates, the infants excrete it, and it turns black on contact with air. (c) This man's ear has a dark patch and dark spots as a result of the alkaptonuria gene.**

(c)

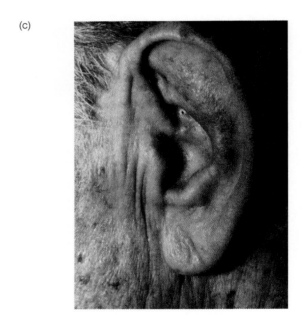

metabolism went unrecognized for 35 years. Then, in the 1940s, geneticists George Beadle and Edward Tatum devised sophisticated laboratory experiments to determine what genes do. Instead of relying on chance mutations, as Garrod had done, they planned to generate their own mutations. Naturally, they could not mutate human genes. Instead, they studied the pink mold *Neurospora crassa*. You may sometimes find this organism growing on a forgotten heel of bread in the back of your kitchen cabinet, and microbiologists often raise it on culture media in test tubes. Although structurally simple, *Neurospora crassa* can make nearly every compound it needs for growth and reproduction from sugar, salts, and just a single vitamin. Beadle and Tatum irradiated *Neurospora* spores to produce mutants that, like people, could not synthesize certain vitamins or amino acids (Figure 9.14). One mutant mold, for example, could not grow unless thiamine (vitamin B_1) was added to its diet, because the mold lacked a certain enzyme needed to complete the synthesis of thiamine.

Beadle and Tatum subjected their collection of mutants to genetic analysis and found that different genes controlled the inheritance of each nutritional requirement. The two geneticists concluded that (1) each step in a biochemical pathway is controlled by a specific gene; (2) each step in a biochemical pathway is catalyzed by a specific enzyme; and therefore (3) a gene causes a specific enzyme to be formed. They summarized their work with the phrase *one gene–one enzyme,* which implies that each gene regulates production of only one enzyme. According to this hypothesis, the plasmid involved in the Japanese dysentery epidemic must have carried a gene for an enzyme that breaks down certain antibiotics. Similarly, a person with brown eyes must have genes that control enzymes that catalyze the production of brown pigment in the eye. By inference, a person with blue eyes lacks these genes, the enzymes, and the brown pigment.

■ **Hypothesis Revised: One Gene–One Polypeptide**   Garrod, Beadle, and Tatum showed that genes somehow determine the presence of an enzyme, but a crucial question remained: In what way? By what mechanism does a gene cause an enzyme or other protein to be produced? The answer came from studies of young African Americans suffering from an inherited blood disease.

In 1905, a young black man suffering from pains in his joints and abdomen, chronic fatigue, and shortness of breath consulted a Chicago physician. A blood test showed that the

In the first application of Mendel's laws to humans, Garrod studied the families of affected children and noted that alkaptonuria is an "inborn error of metabolism," inherited as a recessive mutation like Mendel's short-stem pea plants. He also reasoned that if a mutant allele causes the *absence* of enzyme function, then the normal allele is responsible for the *presence* of enzyme function. In other words, a gene functions by allowing a specific enzyme reaction to occur.

■ **The One Gene–One Enzyme Hypothesis**   As with Mendel's ideas, the significance of Garrod's work on inborn errors of

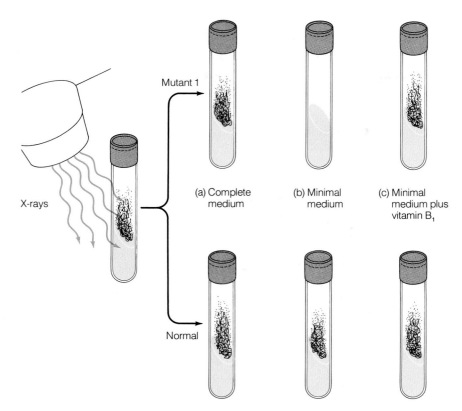

FIGURE 9.14 ■ **Analysis of Mutant *Neurospora* Strains Suggests the One Gene–One Enzyme Hypothesis.** Beadle and Tatum zapped *Neurospora* spores with X-rays; some spores remained unchanged, but in others, the DNA was altered and mutations occurred that blocked the mold's ability to make needed substances. To distinguish between the normal and mutant molds (which looked identical from the outside), Beadle and Tatum grew irradiated spores on different kinds of nutrient media. Complete medium provided all the vitamins, minerals, and nutrients *Neurospora* needs to grow, and both normal and mutant strains grew luxuriantly on it (a). Minimal medium provided only the raw materials with which a normal mold could synthesize its own complex nutrients. If a mutant could not synthesize the complex substances needed from the raw materials in the minimal medium, it could not grow on the deficient food (b). The mutant mold represented here can grow only if vitamin B_1 is added to the minimal medium (c). Apparently, the genetic inability to make vitamin B_1 itself blocks the mutant's growth on minimal medium. Spore irradiation experiments like this led Beadle and Tatum to the one gene–one enzyme hypothesis.

man's blood had too few red blood cells and that many of the cells were shaped like crescents, or sickles, instead of normal biconcave discs (see Chapter 4 and Figure 9.15). Later it was discovered that the sickle-shaped blood cells tended to clump together and lodge in very small blood vessels—especially in the joints and abdomen—blocking blood flow and causing pain. The sickle-shaped cells were also fragile and easily destroyed, causing a shortage of red blood cells, or anemia. Because red blood cells carry oxygen from the lungs to the rest of the body, a deficiency of these cells results in a chronically inadequate oxygen supply and a feeling of breathlessness and fatigue. The man had what is now called *sickle-cell anemia,* and this genetic condition now affects nearly 60,000 people in the United States.

FIGURE 9.15 ■ **Normal and Sickled Blood Cells.** Analysis of blood cells showed that a gene specifies the order of amino acids in a polypeptide. This scanning electron micrograph of blood from a child with sickle-cell anemia shows a sickled cell (left), then a normal cell, and then three more abnormally shaped red blood cells (right).

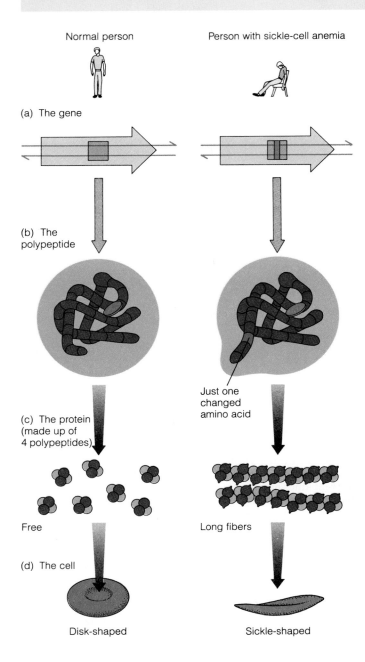

Normal person   Person with sickle-cell anemia

(a) The gene

(b) The
polypeptide

Just one
changed
amino acid

(c) The protein
(made up of
4 polypeptides)

Free   Long fibers

(d) The cell

Disk-shaped   Sickle-shaped

FIGURE 9.16 ■ **Mutation in a Single Gene Causes Sickle-Cell Anemia.** As the figure shows and the text describes, a mutant gene sequence leads to an altered hemoglobin protein, changed blood cell shape, and widespread effects on the body.

Scientists knew that red blood cells are packed with the protein hemoglobin, which carries oxygen from the lungs to the rest of the body. When they compared the molecular structure of hemoglobin taken from normal people with the hemoglobin of sickle-cell patients, they found a subtle difference. Recall that a hemoglobin molecule consists of two pairs of polypeptide chains: two α chains and two β chains (see Figure 2.28). Researchers found that the α chains of normal and sickle-cell hemoglobins are identical, but the sickle-cell β chains differ from normal β chains by just 1 amino acid out of the 146 present in each chain. In normal β chains, the sixth amino acid is glutamic acid, whereas in sickle-cell β chains, it is valine (Figure 9.16). This small change is enough to distort the shape of the molecule, and the distortion causes groups of the molecules to join together to form long fibers instead of remaining as individual globs. These fibers in turn alter the shape of the red blood cell and bring about the suffering and early death of sickle-cell anemia victims. The clear connection in sickle-cell anemia between a change in a single gene and the substitution of a single amino acid had two results: It enabled doctors to diagnose the disease more accurately, and it also showed geneticists that a gene specifies the order of amino acids in a single polypeptide chain.

Discovering the basis for sickle-cell anemia revealed how a simple genetic mutation can skew the functioning of an entire organism. A mutant gene alters the shape, and hence the function, of a certain polypeptide and protein (e.g., hemoglobin). Cells carrying the mutant protein cannot do their job (e.g., carry oxygen). And sickness at the level of the whole organism (fatigue, pain, and breathlessness) may result. This discovery also suggests how a simple circle of plasmid DNA might allow a bacterium to cause an epidemic: If the plasmid DNA bears a gene that encodes an antibiotic-destroying enzyme, then the bacteria will be able to grow and cause disease even when the antibiotic is present.

Geneticists were able to fathom a great deal about genes and their activities by discovering that they are made of DNA and that they specify the sequence of amino acids in polypeptides. However, this still left the important question of how information—the information for red hair or brown eyes or short stature—is stored in DNA. Innovative experiments soon gave the answer: The sequence of bases contains the genetic message. The next chapter reveals the details of how the information in a gene—a mere stretch of nucleotide bases—can be "expressed" as an enzyme that destroys antibiotics or builds eye pigments or helps create a flower petal.

Sickle-cell anemia has been found in about 1 out of every 50 West Africans and in about 1 out of every 400 American blacks. Within these populations, the disease is inherited as a recessive mutation in simple Mendelian fashion: Only a child who inherits a sickle-cell allele from both parents suffers from the anemia. In Chapter 16, we will see why the disease is found primarily among those of West African descent; here we focus on the molecular basis of sickle-cell anemia, for that mechanism sheds light on how genes specify proteins.

BOX 9.3 PERSONAL IMPACT

ANTIBIOTICS, CATTLE FEED, PLASMIDS, AND YOUR HEALTH

A few years ago, an outbreak of severe food poisoning in the northern United States killed one person and sent 17 others to the hospital. Most people infected by the bacteria *Salmonella newport* just feel out of sorts, but these hospitalized people had to be treated for severe diarrhea, cramps, nausea, and vomiting. Researchers found that the *Salmonella* bacteria causing the disease contained a plasmid bearing resistance to three different antibiotics: amoxicillin, penicillin, and tetracycline. Through some additional detective work by government epidemiologists, the plasmids were traced to a herd of beef cattle in North Dakota being fed large doses of antibiotics to stimulate weight gain. The feeding of antibiotics, a nearly universal practice among American livestock growers, causes animals to gain weight up to 10 percent faster than cattle not fed anitibiotics (Figure 1). Unfortunately, the antibiotics also selectively promote the growth and spread of bacteria containing plasmids with antibiotic resistance genes, and this poses a serious health hazard to anyone who eats the meat of cattle contaminated with these bacteria.

The routine application of antibiotics to cattle feed lies behind the dangerous increase of antibiotic-resistant bacteria in humans, but it's not the only cause: Many people also take antibiotic drugs unnecessarily. Twelve of the 18 victims in the *Salmonella* outbreak had been taking antibiotics—three without a doctor's prescription. If they had not been taking the drugs, they would probably not have become ill. The drugs had killed off the victims' own plasmid-free bacteria, so the resistant *Salmonella* encountered no competition from other bacteria and could multiply rapidly. To complicate matters, a recent study shows that 60 percent of physicians prescribe antibiotics for the common cold, even though it is caused by viruses, which remain

FIGURE 1 ■ **Antibiotics in the Feedlot. Cattle fed antibiotics gain weight faster, but their meat can spread genes for antibiotic resistance.**

untouched by antibiotics. This indiscriminate use of antibiotics by livestock farmers, patients, doctors, and others threatens human health.

It is ironic that despite the simplicity of plasmids and our excellent understanding of their structure and life cycle, their simple strategy for getting ahead—based on gene action and natural selection—continues to cause health problems today, more than 30 years after researchers first discovered them in Japan.

CONNECTIONS

The experiments discussed in this chapter revolutionized the geneticist's view of the gene, giving Mendel's abstract theory a physical reality—DNA, the stringy white matter at the bottom of a test tube or in the nucleus of a cell. As genes became tangible, geneticists learned to manipulate them, take them apart, and see how they work. Modern knowledge of gene structure and function in turn allowed geneticists to probe previously impenetrable biological mysteries. These included such subjects as how a single DNA base pair change results in a mutated gene, and how that mutation causes a single amino acid substitution in a polypeptide, which can in turn cause a change in hemoglobin protein structure and a resulting lethal case of sickle-cell anemia. Likewise, knowledge of DNA replication revealed how resistance to antibiotics could spread through several species of bacteria in a hospital plagued by a dysentery epidemic. And from that it was clear why the widespread and indiscriminate use of antibiotics can promote the increase of antibiotic resistance plasmids, so that drugs are no longer the reliable cure-alls they once were for

serious infections such as dysentery, pneumonia, and salmonella (see Box 9.3).

The conceptually simple DNA molecule—just a couple of strings of bases twisted around each other—unifies all life. It provides spinach leaves with the information required to become green and collect sunlight, and cardinals with the information to produce scarlet feathers and find a mate. DNA also diversifies life. Spinach leaves are green and cardinal feathers red because the DNA in the pigment-forming cells differs in base sequence. For the same reason, spinach leaves differ from oak leaves, cardinals from robins, and so on throughout the kingdoms of life—the plants, animals, fungi, and microorganisms.

Ultimately, this central molecule of life provides each species with a means of obtaining energy and reproducing—of harvesting energy directly from sunlight, or indirectly from the nutrients stored in food, and then of passing along exact copies of the genes to offspring via the remarkable fidelity of DNA's template mechanism during replication. How DNA blueprints are actually decoded to form the proteins that carry out cell function is the subject of Chapter 10.

Highlights in Review

1. The transfer of antibiotic resistance from one bacterium to another by plasmids illustrates the power and simplicity of gene function.
2. Genes have two major functions: They replicate themselves and they carry information that specifies amino acid sequences in the proteins of each organism.
3. Genes look and act fundamentally the same in all living things.
4. Genes are made of DNA, not protein.
5. The key to how genes function—both to specify polypeptides and to replicate—lies in the unique double helix structure of DNA.
6. Information is encoded in DNA in a linear fashion and can be read in only one direction.
7. DNA is composed of two intertwined chains of nucleotides. Each nucleotide is made of a sugar group, a phosphate group, and one of four bases. Thousands of nucleotides are connected in a chain—sugar to phosphate to sugar—with the bases hanging off each of the sugars.
8. The four types of bases can be in any order, so that each gene has a different—but specific—base sequence. From this variability comes DNA's remarkable capacity to store almost infinite amounts of information.
9. The two chains in each DNA molecule are oriented in opposite directions.
10. Each of the four bases tends to form weak hydrogen bonds with only one of the other three bases. A joins with T, and C joins with G. This phenomenon is called complementary base pairing.
11. Because of complementary base pairing, the bases on the two polynucleotide chains bond to each other, holding the two chains together, with the bases facing inward and the sugar-phosphate backbones facing outward.
12. The two bonded chains are twisted together into a double helix, like a ladder twisted along its long axis.
13. DNA in eukaryotes is extremely long and must be packaged neatly so that it doesn't tangle during cell division. It is wound around spools of proteins called histones and then coiled upon itself repeatedly.
14. DNA's double helix structure is the key to how it functions.
15. Genes work by specifying the structure of proteins, including the enzymes that control chemical reactions in the cell and the other proteins that make the structure of the cell.
16. In general, each gene specifies one polypeptide, that is, one chain of amino acids.
17. Because of base pairing, each chain acts as a template for the creation of a new copy of the other chain. Thus, two new strands are created, one from each half of the old chain in a process called semiconservative replication.
18. Replication occurs in three stages: Enzymes *unwind* the two chains of the double helix from each other; nucleotides float in and *pair* with their complements, forming a new chain; and an enzyme called DNA polymerase *joins* the nucleotides in the new chain together—sugar to phosphate to sugar.
19. DNA replication is incredibly accurate, with less than 1 error in every 10^9 bases. This accuracy is due to DNA polymerase, which removes incorrect (i.e., unpaired) bases from the end of the newly forming nucleotide chain.
20. DNA stores information in the order of its bases. The order of the bases specifies the order of amino acids in polypeptides.

Key Terms

bacteriophage, 190
chromatin, 196
complementary base
    pairing, 194
double helix, 189
genome, 200
histone, 196

nucleosome, 196
plasmid, 188
semiconservative
    replication, 199
template, 197
transformation, 190

Study Questions

Review What You Have Learned

1. What is the relationship of genes to enzymes?
2. Compare the structure of the hemoglobin molecules in normal and sickled red blood cells.
3. Explain the one gene–one polypeptide hypothesis.
4. How did research on transformation prove that genes are made of DNA?
5. List the steps by which a phage infects a bacterium.
6. How are the four kinds of nucleotides similar? How are they different?
7. Which of these base pairs is unlikely to be found in a DNA molecule, and why? AT; CG; TA; GA; GC.
8. Describe the three steps of DNA replication.
9. What mechanism helps prevent errors in DNA replication?
10. What would be a result of a change in DNA base sequence?

Apply What You Have Learned

1. Owing to a mutation in a single gene, albino people do not form melanin. Using the one gene–one polypeptide principle, propose a model to explain an albino's pale skin, hair, and eyes.
2. The virus that causes the human disease AIDS has about nine genes but consists only of RNA and protein. What do you think is the genetic material in the AIDS virus? How could you test the accuracy of your hypothesis?
3. A botanist discovers a previously undescribed plant with burgundy flowers in the Amazonian rain forest. This plant's DNA has 18 percent adenine. What percentage is thymine? Guanine? Cytosine?

For Further Reading

ANGIER, N. "The Gene Dream." *American Health Magazine,* March 1989, pp. 103–108.

DAVIS, B. D., et al. "The Human Genome and Other Initiatives." *Science* 249 (1990): 342–343.

FELSENFELD, G. "DNA." *Scientific American,* May 1985, pp. 58–67.

JOHNSON, J. "The Oldest DNA in the World." *New Scientist,* May 11, 1991, pp. 44–48.

LOUPE, D. E. "Breaking the Sickle Cycle." *Science News* 253 (1989): 360–362.

MITSUYA, H., R. YARCHOAN, and S. BRODER. "Molecular Targets for AIDS Therapy." *Science* 249 (1990): 1533–1544.

Novick, R. "Plasmids." *Scientific American,* December 1980, pp. 102–127.

Radman, M., and R. Wagner. "The High Fidelity of DNA Duplication." *Scientific American,* August 1988, pp. 40–47.

Schmitz, A. "Murder on Black Pad. The Rapist-Killer Haunting an English Village Left One Telltale Clue: A Piece of His DNA." *Hippocrates,* January/February 1988, pp. 48–58.

Watson, J. D. *The Double Helix.* New York: Atheneum, 1968.

Watson, J. D. "The Human Genome Project: Past, Present, and Future." *Science* 248 (1990): 44–49.

Watson, J. D., and F. H. C. Crick. "Molecular Structure of Nucleic Acids: A Structure for Deoxyribosenucleic Acid." *Nature* 171 (1953): 737–738.

Watson, J. D., N. H. Hopkins, J. W. Roberts, J. A. Steitz, and A. M. Weiner. *Molecular Biology of the Gene.* 4th ed. Menlo Park, CA: Benjamin/Cummings, 1987.

Zuckerman, S. "Antibiotics: Squandering a Medical Miracle." *Nutrition in Action,* January/February 1985, pp. 9–11.

How Genes Work: From DNA to RNA to Protein

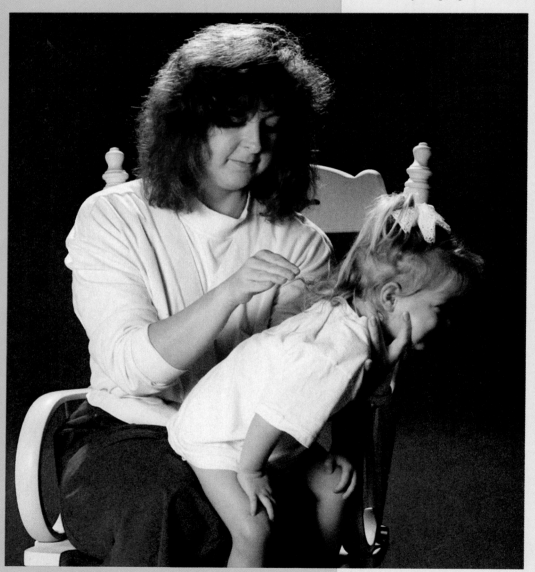

FIGURE 10.1 ■ **Manual Therapy for a Child with Cystic Fibrosis.** A "percussion session," during which a mother thumps her child's back vigorously, to loosen and help expel the sticky mucus clogging the victim's lungs. This therapy decreases the infections, scarring, and gradual loss of lung function that characterize cystic fibrosis. A mutated gene leads to this devastating condition.

Cystic Fibrosis: A Case Study in Gene Action

A tiny infant receives a parent's tender kiss on the cheek, and the mother or father receives a piece of disquieting information: The baby tastes so *salty*. Why? This exchange may be the first indication that the newborn has cystic fibrosis, the most common lethal genetic disease among Caucasians. One in every 20 people of northern European ancestry carries one copy of the recessive mutation which leads to the disease, but shows no symptoms. Only a child who inherits one copy from each parent (and is thus a homozygous recessive like a pea plant with a short stem or white flowers) will develop this life-threatening illness.

Cystic fibrosis is essentially a disease of clogged ducts; the recessive gene and the faulty protein it encodes (described shortly) lead to a buildup of sticky mucus in lung passages, pancreas ducts, sweat glands, and sperm ducts. As a result, the individual with cystic fibrosis tends to have difficulty breathing, as well as dangerous bacterial infections in the lungs, stomachaches due to poor absorption of essential fats in the diet, a salty secretion on the skin, and, in adult males, usually sterility.

Right now, doctors must treat the symptoms of cystic fibrosis one by one rather than correcting the genetic defect itself. As we will see later in the chapter, that basic genetic correction may be possible in the not-so-distant future and is one of the most exciting medical prospects for the early 1990s. The current treatments, however, include mainly "percussion sessions" during which a parent thumps the child firmly on the back to dislodge gummy mucus and allow easier breathing for a while (Figure 10.1); a special diet; powerful antibiotics to fight the lung infections; and pills containing certain digestive

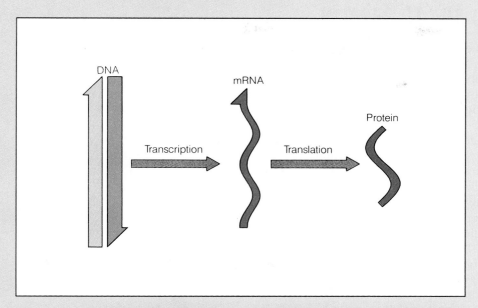

FIGURE 10.2 ■ Path of Information Flow in a Cell: DNA to RNA to Protein. Enzymes transcribe DNA into a messenger RNA molecule; ribosomes translate the mRNA into the polypeptide chain of a protein.

enzymes to replace those blocked up in the pancreas by clogged ducts. Despite these symptomatic treatments, nearly half of the affected individuals fail to survive to age 20. That's why the prospect of gene therapy for cystic fibrosis is so important.

How does a mutation in a single gene cause defects in so many systems (the lungs, pancreas, sweat glands, sperm ducts)? What makes one human gene out of approximately 100,000 so important? And how can physicians combat the effects of the deadly gene more effectively? The answers to these questions are crucial to the health of about 50,000 young people in the United States and are based on the principles of gene action discussed in this chapter.

We know that each gene specifies the synthesis of one polypeptide (see Chapter 9), so it is not surprising that a disease like cystic fibrosis—the result of a single defective gene—results from errors in the synthesis of a single protein. In cystic fibrosis, the protein is called CFTR (for cystic fibrosis transmembrane regulator). This protein is embedded in the cell's plasma membrane and probably transports chloride ions (Cl^-) out of the cell. Sufferers of cystic fibrosis make a faulty CFTR protein, and so chloride transport is abnormal. Our broad goal in this chapter is to discuss the links between genes and proteins, whether normal or abnormal, and the cause of devastating inherited diseases such as cystic fibrosis.

Three unifying themes will emerge as we follow the processes by which a mutation in a single gene results in a defec-

tive protein and ultimately in a range of defects in the individual organism. The first theme is that information flows from DNA to RNA to protein (Figure 10.2) and from proteins to the building of an organism's phenotype. The gene for the CFTR protein in a healthy person, for example, is normally transcribed into an RNA molecule, which is then translated into CFTR, and this, in turn, regulates normal ion passage through cell membranes, and the person does not suffer the symptoms of cystic fibrosis. A second unifying theme is that protein synthesis requires a large expenditure of energy, and cells have evolved ways that minimize that energy cost by regulating *gene expression*—the translation of genetic information into proteins. A final theme is that basic genetic mechanisms are essentially universal: All creatures, from bacteria to field mice to flowering dogwood, share the same approach to protein synthesis.

This chapter will answer several important questions about how genes function, controlling protein shape and activity and ultimately the activities of living things:

■ How does information flow from DNA to RNA to proteins?

■ How can even slight alterations in DNA lead to diseases like cystic fibrosis?

■ What determines when and where the information in a gene will be used in a cell?

THE PATH OF INFORMATION FLOW FROM DNA TO RNA TO PROTEIN

Genetic research in the mid-twentieth century revealed that a gene is made of DNA and that it acts by specifying the amino acid sequence in a polypeptide chain. The next question arising from this work was simple but profound: Exactly how does the information in DNA become decoded and translated into protein structure?

The answer unfolded over decades of research and can be summarized this way: Genetic information within each cell flows from DNA to RNA to protein (see Figure 10.2). Research further revealed that protein synthesis is a two-step process. In the first step, information in a portion of the DNA molecule is copied, or *transcribed,* into an RNA molecule. RNA is an information-storing molecule with a structure somewhat similar to DNA; we will describe RNA shortly. This first step of making RNA from DNA is called **transcription**—a word that implies that information in one dialect (the base sequence of DNA) is copied into a different dialect (the base sequence of RNA) in the language of nucleic acids. In the second step, the information in one type of RNA molecule is *translated* by two other types of RNA molecules into sequences of amino acids, or polypeptide chains. This step is called **translation** because genetic information is translated from the language of nucleic acids into the language of proteins.

The three kinds of RNA involved in the two-step process of protein synthesis are messenger RNA, ribosomal RNA, and transfer RNA. **Messenger RNA (mRNA)** carries from DNA the information, or message, for manufacturing proteins during the translation phase, while both **ribosomal RNA (rRNA)** and **transfer RNA (tRNA)** assist in the translation of mRNAs to protein. All three types of RNA have a similar chemical composition and are themselves transcribed from DNA.

Both transcription and translation are essential to every living cell. Without them, proteins—the building blocks of cells—could not be made.

Transcription: Information Flows from DNA to RNA

The production of new proteins—proteins that transport ions out of cells, proteins that trap sunlight in a leaf, or proteins that help replace skin cells that have peeled away after a sunburn—begins with transcription. The story of transcription, the synthesis of a molecule of RNA from a DNA template, begins with the structure of RNA.

■ **RNA: Usually a Single Strand** Like DNA, RNA consists of a long string of nucleotides linked by sugar-phosphate bonds. However, RNA differs from DNA in four respects. First, in each RNA molecule, the base uracil replaces the thymine found in DNA (Figure 10.3a). As in DNA, the sugar, phos-

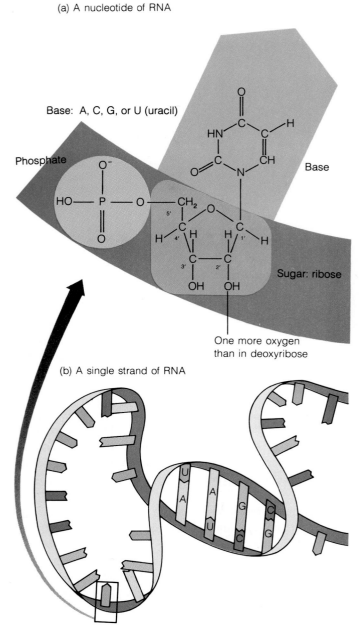

(a) A nucleotide of RNA

Base: A, C, G, or U (uracil)

Phosphate

Base

Sugar: ribose

One more oxygen than in deoxyribose

(b) A single strand of RNA

FIGURE 10.3 ■ The Structure of RNA. (a) A base (A, C, G, or U) plus a ribose sugar and a phosphate together form a single ribonucleotide. The base shown here is U (uracil). (b) Many RNA nucleotides join together to form an RNA molecule that can loop back and form base pairs with itself. A pairs with U, and C pairs with G.

phate, and base combine to form a nucleotide. But in this case—and here is the second difference—it is a **ribonucleotide,** because the sugar in RNA is *ribose* instead of the deoxyribose in DNA (see Figure 10.3a). Deoxyribose has one less oxygen atom than ribose. That is the only structural difference

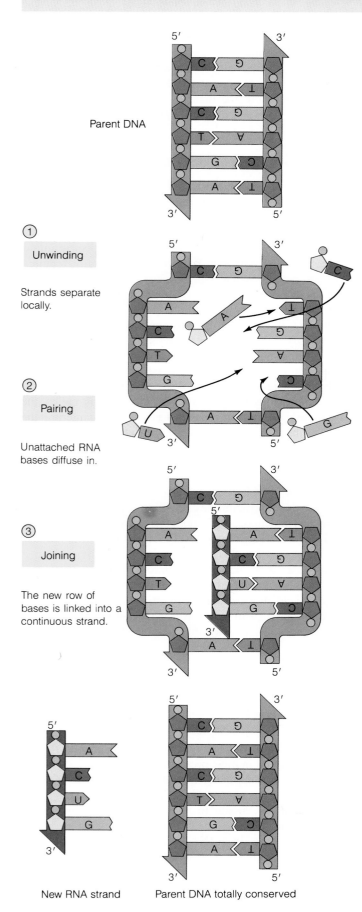

Parent DNA

① Unwinding

Strands separate locally.

② Pairing

Unattached RNA bases diffuse in.

③ Joining

The new row of bases is linked into a continuous strand.

New RNA strand   Parent DNA totally conserved

FIGURE 10.4 ■ How DNA Is Transcribed into RNA.
(1) *Unwinding.* **Just as in DNA replication, the two strands of the helix separate. This requires energy because the hydrogen bonds must be broken.** (2) *Complementary base pairing.* **The four types of ribonucleotides float in and pair with their complementary bases on only one strand of the DNA duplex.** (3) *Joining the ribonucleotide to the growing RNA.* **Once the first two nucleotide units are in place, the large enzyme RNA polymerase begins joining the sugar-phosphate backbone together. New nucleotides are added to the growing chain in only one direction (5′ to 3′).**

arise from this single alteration. Third, RNA usually consists of a single strand (Figure 10.3b), whereas DNA usually consists of two. In some kinds of RNA, however, a single RNA molecule will fold back on itself and form short double-stranded regions connected by complementary base pairs (see Figure 10.3b). In these folded regions, A pairs with U instead of with T as it does in DNA. (C pairs with G in both RNA and DNA.) Finally, RNA molecules are much shorter than the DNA molecules that make up chromosomes. Each DNA molecule carries many genes, but an RNA molecule usually contains information from only one gene. Each short mRNA molecule that carries information to make a protein is called a **message.** For each polypeptide, there is a message with a unique length and base sequence.

■ **Base Pairing and the Process of Transcription**   Replication and transcription are DNA's two most important functions. Both replication and transcription result in a new nucleic acid molecule with a base sequence specified by the old molecule, but in DNA replication, the new strand is DNA, while in DNA transcription, a strand of DNA is copied into a new strand of RNA. During transcription, as sections of the DNA duplex unwind (Figure 10.4, step 1), ribonucleotides (not deoxyribonucleotides) float in and pair with the deoxyribonucleotide bases on one DNA strand (Figure 10.4, step 2). The enzyme **RNA polymerase** then joins the ribonucleotides together, thus forming a single strand of RNA (Figure 10.4, step 3). During transcription, nucleotides are joined in only one direction (5′ to 3′), as in DNA replication. Finally, the newly made RNA strand diffuses away, and the parental DNA is wound back up again.

Although the replication and transcription processes are similar, there are important differences (Figure 10.5). First, while both strands of a DNA duplex are copied during DNA replication, only one strand is usually transcribed into RNA for any given region of DNA. This is an important concept, because modern geneticists define a gene as a length of DNA that is transcribed into an RNA molecule. The DNA segment depicted in Figure 10.5 would thus contain two genes.

A second difference between DNA replication and transcription is that one gene can be transcribed from one strand, while the next gene may be transcribed from the other strand.

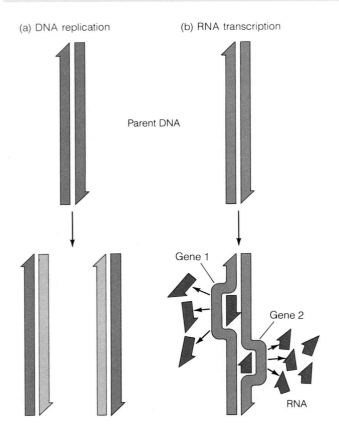

FIGURE 10.5 ■ DNA Replication and Transcription Compared.
(a) During DNA replication, both DNA strands of the parent molecule are copied exactly, but only once, and the new and old strands remain together. (b) During RNA transcription, only a small portion of a DNA molecule is copied, but many times, and the new RNA strand separates from the old DNA template strand.

Third, during replication, the whole DNA molecule is copied; but during transcription, a limited number of DNA bases are transcribed at a time. Most transcribed regions are between 100 and 10,000 nucleotides long, and a single DNA molecule can have several thousand transcribed regions. Fourth, during replication, only one copy of each gene is made, but in transcription, a single gene may be transcribed thousands of times. Thus, a cell can make thousands of copies of an individual RNA molecule, each bearing the encoded message from a single gene. Finally, after DNA replication, the new and old strands of DNA join together, but after transcription, the two original DNA strands rejoin, and the newly formed strand of RNA separates from the DNA and moves from the nucleus into the cytoplasm, where the RNA then performs its function, as we shall see shortly.

Accurate transcription is essential to life and good health, as Box 10.1 on page 224 (describing poisonous mushrooms and the emperor Claudius Caesar) illustrates. Despite its crucial role in living cells, however, the transcription of a gene

does not alone guarantee proper gene function; RNA must first be translated into a serviceable protein. The mutant cystic fibrosis gene, for instance, is transcribed into an mRNA that is translated to an abnormal CFTR protein, and that altered protein brings about the many health problems in children with cystic fibrosis. We'll return to this later.

Translation: Information Flows from RNA to Protein

While transcription of DNA to RNA occurs in the nucleus of a eukaryotic cell, translation into protein takes place in the cytoplasm after the newly made RNA moves out of the nucleus and into the cytoplasm. During translation, the information in an RNA message is converted into a specific amino acid sequence in a polypeptide chain. This translation requires some equipment. When a beginning language student sits down at a desk to translate a paragraph from French into English, a French/English dictionary acts as a sort of *code*. The student, the *translator*, interprets and applies this code; and the *desk* holds in place the passage in French and the blank writing paper ready to receive the translation into English. In a similar way, cells need a code for translating words from the language of nucleic acids (bases) into the language of proteins (amino acids). This code is called the **genetic code.** Moreover, cells need a translator to interpret the code. During protein synthesis, the translator is transfer RNA. Finally, cells need a "desk" on which the translation can take place, and that desk is ribosomal RNA.

■ **The Genetic Code: Dictionary of Life**   The genetic code determines which amino acids the cell will translate from each base sequence. The language of ribonucleic acids has an alphabet of only 4 letters (the 4 bases A, C, G, and U), while the language of proteins has 20 (the 20 amino acids). Obviously, one base alone cannot encode an amino acid; there just aren't enough bases. One might guess that two bases could encode an amino acid. For example, AU might encode one amino acid, CA another, and so on. But that wouldn't work either, for there are only 16 ways to combine the four bases into pairs:

AA	AC	AG	AU
CA	CC	CG	CU
GA	GC	GG	GU
UA	UC	UG	UU

Clearly, 16 is still not enough "words" to encode the 20 amino acids found in most proteins.

The next step up, using three bases to make a "word," provides 64 ways to combine the four bases into code words—more than enough to encode the 20 amino acids (Figure 10.6). And three is indeed the magic number. A set of three adjacent bases in mRNA is called a **codon,** and codons specify amino acids. Geneticists were able to decipher which codon corresponds to which amino acid by adding artificial messages to an extract of bacterial cells containing all factors

Codon			Amino acid		Codon			Amino acid		Codon			Amino acid		Codon			Amino acid
U	U	U	Phe		A	U	U	Ile		C	U	U	Leu		G	U	U	Val
U	U	C	Phe		A	U	C	Ile		C	U	C	Leu		G	U	C	Val
U	U	A	Leu		A	U	A	Ile		C	U	A	Leu		G	U	A	Val
U	U	G	Leu		A	U	G	Met (START)		C	U	G	Leu		G	U	G	Val
U	C	U	Ser		A	C	U	Thr		C	C	U	Pro		G	C	U	Ala
U	C	C	Ser		A	C	C	Thr		C	C	C	Pro		G	C	C	Ala
U	C	A	Ser		A	C	A	Thr		C	C	A	Pro		G	C	A	Ala
U	C	G	Ser		A	C	G	Thr		C	C	G	Pro		G	C	G	Ala
U	A	U	Tyr		A	A	U	Asn		C	A	U	His		G	A	U	Asp
U	A	C	Tyr		A	A	C	Asn		C	A	C	His		G	A	C	Asp
U	A	A	STOP		A	A	A	Lys		C	A	A	Gln		G	A	A	Glu
U	A	G	STOP		A	A	G	Lys		C	A	G	Gln		G	A	G	Glu
U	G	U	Cys		A	G	U	Ser		C	G	U	Arg		G	G	U	Gly
U	G	C	Cys		A	G	C	Ser		C	G	C	Arg		G	G	C	Gly
U	G	A	STOP		A	G	A	Arg		C	G	A	Arg		G	G	A	Gly
U	G	G	Trp		A	G	G	Arg		C	G	G	Arg		G	G	G	Gly

FIGURE 10.6 ■ **The Genetic Code.** The codon/amino acid dictionary shows how just four ribonucleotide bases (U, C, A, and G) can be combined into 64 possible arrangements of three bases in sets called codons, and how these can form 20 different amino acids plus three stop codons. For example, U in the first position, A in the second position, and C in the third position (UAC) equals Tyr, or tyrosine. Note that the code is redundant: UUU and UUC can both mean phenylalanine; CAU and CAC can both mean histidine; and so on. Translation begins at an AUG, or start, codon, and this codon also encodes the amino acid methionine.

necessary for translation and monitoring to see which amino acid was present in the resulting polypeptide. They found that the message UUUUUUUUUUUU, for example, translated into the polypeptide Phe-Phe-Phe-Phe, indicating that the codon UUU specifies the amino acid phenylalanine. Similar experiments revealed the meanings of all 64 codons. The codon dictionary is shown in Figure 10.6. Notice that several different codons can specify the same amino acid. For example, both AAA and AAG specify the amino acid lysine. In other words, some codons are synonymous: They mean the same thing. On the other hand, each individual codon specifies only one amino acid. The genetic code is redundant, but not ambiguous.

When researchers tested a complicated message such as AGCAGCAGCAGCAGC, they obtained an unexpected result. This message translated into not one but three different polypeptides, each consisting of only one kind of amino acid: either all serine (codon AGC), all alanine (codon GCA), or all glutamine (codon CAG). We can understand how this happens if we assume that translation can start at any place along the message; thus, we can write the message in any of the following ways, with imaginary breaks separating the codons:

AGC AGC AGC AGC AGC
A GCA GCA GCA GCA GC
AG CAG CAG CAG CAG C

This result shows that translation starts at one point and then reads off bases in groups of three. Codons do not overlap,

and no bases are skipped. The result also shows that the place where translation begins determines the meaning of the message. The mRNA message given in the previous paragraph can be divided into codons in three ways, or three different **reading frames.** Differences in the reading frame thus allow different meanings from the same stretch of messenger RNA. To appreciate the importance of the reading frame in a message, just consider for a moment howc hangesi nr eadingf ramec ana ltert hism essage.

In cells, there is a special codon that signifies where translation should start. This **start codon,** AUG, also encodes the amino acid methionine. As geneticists deciphered the 64 codons, they found that three of the codons—UAA, UAG, and UGA—encode no amino acid at all. Wherever these codons appear in a message, the translation stops. Hence, they are called **stop codons.** With the completion of the artificial message experiments, researchers were able to fully decipher the genetic code.

■ **Universality of the Genetic Code** The scientists who broke the genetic code were studying the protein-making apparatus of bacteria, but this does not mean that the code holds only for bacteria. Numerous experiments on other life forms have proved that nearly all organisms use the same genetic code. The only exceptions are certain mitochondria and some protists, in which a few codons differ from the norm. It is an amazing testament to the unity of all living things that human cells and bacterial cells speak exactly the same genetic lan-

(a) A tRNA molecule

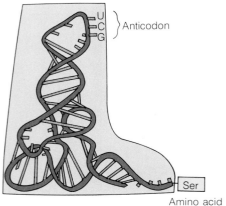

(b) The anticodon in tRNA aligns with the codon in mRNA.

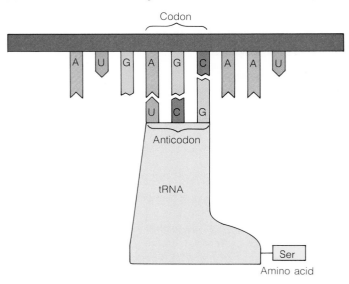

FIGURE 10.7 ■ The Translator: tRNA. **(a)** Each transfer RNA is a ribbon about 75 bases long that is looped back on itself and folded in space by complementary base pairing to form a bootlike shape. At one end of each tRNA is its specific anticodon, in this case, UCG; at the other end is attached the amino acid corresponding to this anticodon, here, Ser (serine). **(b)** The tRNA's anticodon forms base pairs with a codon in mRNA.

guage. A gene from a human cell translates into exactly the same amino acid sequence, whether the gene resides in the nucleus of that human cell or is artificially placed in a bacterial cell.

The near universality of the genetic code is strong evidence that all living things derive from ancestral cells with a singular or similar origin early in our planet's history (see Chapter 15). Developed at the dawn of life, this ancient code has apparently been inherited intact for billions of years.

The cracking of the genetic code was a big step toward understanding how cells make proteins. Besides showing how many bases a code word contains, it revealed that protein syn-

thesis starts and stops at special codons and that the reading frame is important in translation. However, breaking the code failed to show how cells translate codons into amino acids.

The information in an mRNA message is translated into polypeptides that will become part of a cell's proteins. To make this translation, the cell uses the genetic code plus tRNA as the "translator" and rRNA (in ribosomes) as the "desk" that holds things in place so that the translation can proceed smoothly and efficiently.

■ **The Translator Molecule: tRNA**   Transfer RNAs are single strands of RNA that translate mRNA messages into amino acids and then transfer those amino acids to the growing protein chain. About 75 bases long, tRNAs loop back on themselves, forming a characteristic shape rather like a lumpy ski boot (Figure 10.7). Each tRNA corresponds to a different amino acid and has a different base sequence. Nevertheless, the characteristic lumpy boot shape remains the same. The tRNAs work rather like human translators. To translate a message from French to English, a person must find the English word that matches each French one. Similarly, to translate a message from RNA into protein, tRNA molecules must match a nucleic acid word, or codon, with a protein word, or amino acid. How does tRNA accomplish this?

Each tRNA molecule has at one end a sequence of three bases, called the **anticodon,** that recognizes its complementary codon in an mRNA; at the other end, the tRNA carries one of the 20 amino acids, attached by covalent bonds (Figure 10.7a). Enzymes attach the right amino acid to the right set of RNAs. For example, one of the six tRNAs with the amino acid serine attached to one end has the anticodon UCG at the other end (see Figure 10.7a). The UCG anticodon on the tRNA pairs up with the AGC codon on an mRNA, and this connection makes AGC a codon for serine (Figure 10.7b).

The codon-anticodon connection is the key to how amino acids line up in proper sequence for polypeptide synthesis: The order of codons in mRNA specifies the order of anticodons—each bonded to a specific amino acid—in the tRNA lineup. To join the ordered amino acids to each other is then the job of the ribosome.

■ **Ribosomes: The Protein Production Desk**   *Ribosomes* support messenger and transfer RNAs and contain enzymes that link amino acids to each other on the growing polypeptide chain. In eukaryotes, each ribosome is a conglomerate of four types of ribosomal RNA molecules. Approximately 75 proteins are wrapped about these rRNAs, forming beadlike structures, the ribosomes. Each ribosome consists of one large and one small subunit separated by a groove (Figure 10.8a). When an mRNA strand has been "threaded" through the groove, the ribosome slides down the mRNA like a pulley on a rope. As it goes, the ribosome joins the ordered amino acids, one by one, to the growing polypeptide chain. Usually, a linear cluster of ribosomes (called a *polyribosome*) moves together in a train along each mRNA message, translating multiple copies of the same polypeptide (Figure 10.8b). Every cell contains

(a) A ribosome

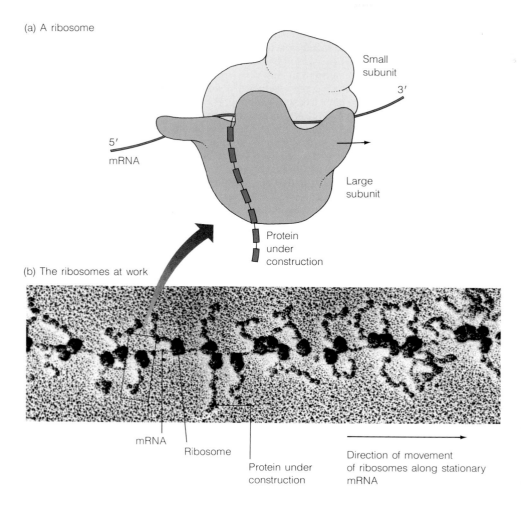

Small
subunit

3′

5′
mRNA

Large
subunit

Protein
under
construction

(b) The ribosomes at work

mRNA

Ribosome

Protein under
construction

Direction of movement
of ribosomes along stationary
mRNA

FIGURE 10.8 ■ Ribosomes: The Desks on Which Translation Occurs. (a) The large and small ribosomal subunits are each made up of RNAs of specific lengths as well as many kinds of associated proteins. The base sequences of these ribosomal RNAs are similar in all life forms, but are different enough in different species to show genetic relationships such as those shown in Figure 1.16. (b) The ribosomal subunits join to the mRNA, then slide along the mRNA like a train on a track or a pulley on a rope, and the order of bases in the mRNA can then be translated into a specific sequence of amino acids in a polypeptide. The electron micrograph shows an mRNA along which ribosomes are "marching." An ever-elongating polypeptide chain trails from each ribosome. Researchers were able to photograph this example of translation in action in cells from the salivary glands of *Chironomus,* a kind of gnat (magnification 17,500×).

tens of thousands of ribosomes, located wherever proteins are being synthesized, be it in the cytoplasm of a eukaryotic cell or in the DNA-containing region of a prokaryotic cell. Working together, then, the ribosomal "desk," the tRNA "translator," and the mRNA message bring about the life-sustaining sequence of events called **protein synthesis.**

Protein Synthesis: Translating Genetic Messages into the Stuff of Life

The synthesis of proteins is one of the cell's most important tasks, since the activities of proteins underlie nearly all cellular functions. For muscle cells to flap the wings of a honeybee, for seaweed to synthesize chlorophyll, or for oxygen to be transported in the blood of a kangaroo or child, a protein (often an enzyme) must be constructed that will then carry out a special role. Some cells make a tremendous amount of a single protein, and the result can be a spider's gossamer web, or a bear's long, sharp nails (Figure 10.9).

(a)   (b)

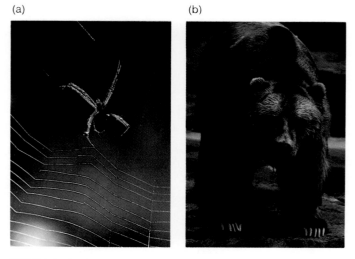

FIGURE 10.9 ■ Protein Synthesis Made Manifest. The results of abundant protein synthesis are visible in (a) a spider's silk web and (b) a bear's thick coat and curving claws.

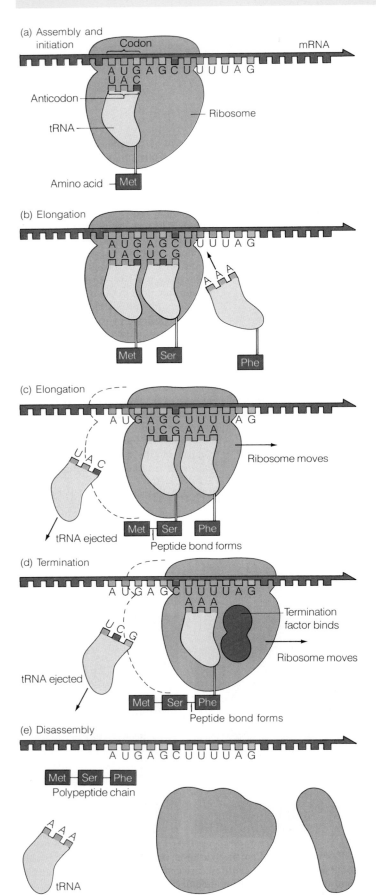

(a) Assembly and initiation

(b) Elongation

(c) Elongation

tRNA ejected
Peptide bond forms

(d) Termination

tRNA ejected
Peptide bond forms

(e) Disassembly

Met—Ser—Phe
Polypeptide chain

tRNA

FIGURE 10.10 ■ The Main Events of Protein Synthesis.
(a) *Assembly and Initiation.* For translation to begin, several components must be assembled. A special initiator tRNA molecule charged with the amino acid methionine binds to the small ribosomal subunit, and then the small and large subunits and the mRNA come together. The anticodon of the initiator tRNA binds to the AUG start codon on the mRNA. (b) *Elongation.* A second charged tRNA forms base pairs between its anticodon and the second codon in the mRNA, thus aligning the first and second amino acids, each still attached to its tRNA. (c) *Continued Elongation.* During peptide bond formation, the bond between the first tRNA and its amino acid is broken, and enzymes forge a peptide bond between the first and second amino acids. The first tRNA, now uncharged, is ejected from the message and recycled, and the ribosome moves down the message one codon. Now a dipeptide (Met-Ser in this case) is hooked to the second tRNA. To continue peptide elongation, the foregoing steps are simply reiterated many times until the complete polypeptide is formed. A charged tRNA binds to the third codon. (d) *Termination.* A peptide bond is formed between the dipeptide and the newly arrived amino acid forming a tripeptide, the tRNA from the second amino acid is ejected, and the ribosome moves once more. At the end of the message, the ribosome reaches a stop codon along the mRNA such as UAG, a codon for which there is no tRNA. At this point, a termination factor binds in place of tRNA and translation halts. (e) *Disassembly.* The two ribosomal subunits separate from each other and release the mRNA, and the polypeptide is cleaved from the last tRNA molecule. The polypeptide is then free to do some job in the cell, and the translation machinery can then reassemble and initiate the synthesis of another polypeptide molecule.

Protein synthesis, the assembly of amino acids into a polypeptide, has three main stages: initiation, elongation, and termination. These stages are pictured and described in detail in Figure 10.10. During the **initiation** stage, the first in a series of tRNAs associates with both a ribosome and an mRNA, forming a complex of many molecules. During the steps of **elongation,** the sequential enzymatic addition of amino acids builds the growing polypeptide chain. Finally, the **termination** stage brings the growth of a particular polypeptide chain to a halt when the ribosome has reached the stop codon.

As in other complex biological processes, a minor change in protein synthesis can have major effects in the organism. The damaging effects of cystic fibrosis stem from a simple change in the DNA of the CFTR gene—a change that alters a single amino acid in the CFTR protein and disrupts the protein's function. The next section explores precisely how such mutations can alter the function of genes.

GENE MUTATION: A CHANGE IN BASE SEQUENCE

A **mutation** is a change in the base sequence of an organism's DNA. This alteration at the level of genotype often has no effect on the organism's phenotype, but sometimes it can cause a massive—often deleterious—shift in the way an organism looks or functions. Geneticists recognize two general categories of mutations: **chromosomal mutations,** which affect large regions of chromosomes or even entire

(a) Original DNA

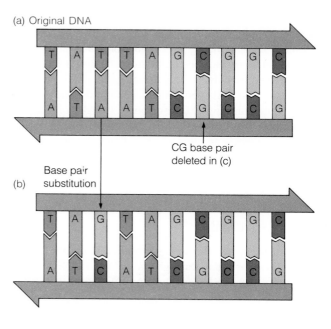

Base pair substitution

(b)

CG base pair deleted in (c)

(c)

Base pair deleted

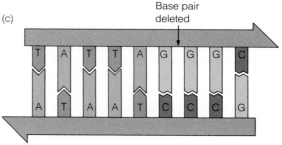

(d)

Base pair insertion

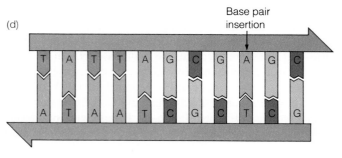

FIGURE 10.11 ■ Mutations in Individual Genes. Changes in the nucleotide base sequence of an original DNA (a) can occur in several ways. (b) In a base substitution, one base pair [a TA pair in part (a)] can be replaced by another (here GC). (c) In a base deletion, a base pair is removed from the DNA. (d) In a base insertion, a base pair is added to the DNA. In each case, the sequence is altered so that the gene is mutated and may function very differently, if at all. Sometimes several adjacent bases may be deleted or inserted at once. The most common mutation causing cystic fibrosis is the deletion of three adjacent base pairs.

chromosomes and hence the locations of many genes, and **gene mutations,** which alter individual genes. Chromosomal mutations include changes in chromosome number, as in Down syndrome (see Chapter 7), and changes in chromosome structure (detailed in Chapter 12). Here we focus on the simplest level: single-gene mutations, or changes in the base sequence of a single gene.

The simplest sort of mutation in a gene occurs when one base pair replaces another, for example, when a TA pair in the parental molecule is replaced by a GC pair in the mutated DNA (compare Figure 10.11a and b). Geneticists call this a *base substitution* mutation. In other mutations, a single base pair might be removed from the DNA entirely, as when a CG *base deletion* alters the DNA in Figure 10.11a to the shorter DNA in Figure 10.11c. Finally, a single base can be added to the parental DNA molecule causing a *base insertion* (compare Figure 10.11a and d). Such changes in one or a few bases are called *point mutations.*

These simple base changes can sometimes have profound effects on the structure of the protein encoded by the mutated gene. Consider, for example, the X-ray photograph in Figure 10.12 of the massively enlarged skull of a child. This child has *thalassemia,* an inherited disease common in Americans of Mediterranean descent. In children with thalassemia, red blood cells are rapidly destroyed by the spleen, and this organ becomes enlarged. The sick child's bone marrow, the site where new blood cells originate, increases greatly in size, generating new blood cells that replace the destroyed ones.

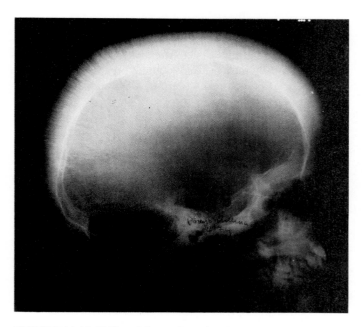

FIGURE 10.12 ■ How Mutant Proteins Cause Thalassemia. A child with the blood disease thalassemia has a defective gene for hemoglobin, the oxygen carrier in red blood cells. The spleen removes altered red blood cells, eventually causing severe anemia. Red blood cells are replaced by excessive bone marrow. The rapid growth of bone marrow in the skull causes an enlarged head and a peculiar "hair on end" appearance of the bone when X-rayed.

The peculiar "hair-on-end" appearance of the enlarged skull reflects that enlargement and rapid blood cell production.

Children with thalassemia have a mutation in a gene that encodes hemoglobin, the red-pigmented protein that transports oxygen in red blood cells. One such mutation is a single base substitution in DNA that changes a CAG codon into a UAG codon in mRNA. The original CAG codon specifies the amino acid Gln (glutamine) according to the genetic code in Figure 10.6, while UAG means stop. (A mutation to a stop codon, by the way, is called a *nonsense mutation*.) With the CAG to UAG substitution, translation of the protein is terminated early, and the polypeptide has only 39 instead of the normal 146 amino acids. This short, nonfunctional polypeptide causes the spleen to destroy the red blood cells quickly. The skull enlarges as more red cells are produced, but the child still has too few blood cells, and this leads to an early death. The mutation that causes this tragedy is a single base pair change among the 3 billion in the child's DNA. Imagine typing a term paper a thousand times as long as this book, only to get a failing grade because your paper has one misspelled word. That is very much like what has happened to this child with thalassemia.

Base insertions and deletions can have effects just as dramatic as base substitutions. A *frameshift mutation* results from the insertion or deletion of a base pair, either of which changes the reading frame, the starting point for the division of bases into codons, or sets of three. When the reading frame changes, every single amino acid from that point on also changes, producing an unrelated polypeptide. Some children with thalassemia have this sort of mutation.

Although base substitutions, deletions, or insertions are the same in all organisms, their effects are much more complex in a child than in a bacterium. The next section shows how mutations in intricate multicellular organisms precipitate cascades of effects in many seemingly unrelated physiological processes.

How a Gene Mutation Changes the Phenotype

Mutations can have diverse effects. The crossed eyes and white pelt of Mohan the tiger, for example, are the result of a single mutation (see Chapter 8). To understand how a single genetic mutation causes a debilitating condition, let's examine how the cystic fibrosis mutation alters the DNA, the mRNA, and the CFTR protein and how the altered protein might, in turn, cause a person to have chronic lung infections, poor digestion, salty sweat, sterility, and a shortened life.

The normal protein encoded by the cystic fibrosis gene has 1480 amino acids; amino acid number 508 in this long string is a Phe (phenylalanine). Most people who suffer from cystic fibrosis have a deletion of three base pairs from this region of their DNA. This DNA deletion causes the mRNA to be one full codon shorter, the CFTR protein to be one amino acid shorter, and the Phe that should be in position 508 to be missing. As mentioned earlier, CFTR protein is normally embedded in a cell's plasma membrane and pumps chloride ions out of the cell. Without a normal CFTR protein, ions cannot be pumped out of the cell properly, and this leads to an imbalance in fluid transport and osmosis in the ducts in the lungs, digestive tract, skin, and sex organs. Because of this, the mucus in the ducts becomes dry and sticky rather than smooth-flowing, clogging the ducts and making them prone to infection.

This new understanding of cystic fibrosis is significant because medical researchers can now design treatments aimed at the specific defect—the altered gene and protein and their consequences. This would have been impossible without a thorough understanding of transcription, translation, and the molecular basis of mutation. Most mutations, like the ones causing cystic fibrosis and thalassemia, have a major, usually negative impact. So where do mutations come from, and how do organisms protect themselves from such genetic changes?

The Origin of Mutations

Mutations arise either through spontaneous errors during DNA replication or later, through the effects of physical or chemical agents on the DNA. During replication, the wrong base can be inserted into a DNA molecule. Usually, this kind of replication error is edited out by the proofreading of DNA polymerase, so such mutations are very rare. More commonly, physical and chemical agents called *mutagens* change DNA structure. Mutagens are everywhere and include ultraviolet rays from the sun, chemicals in cigarette smoke, and even many natural substances from plants or fungi. For example, aflatoxin, a natural compound found in moldy peanuts, is an extremely potent mutagen.

Despite the ubiquity of mutagens in our environment, a permanent mutation is a relatively rare event in most people, and when mutations are common, the consequences are dire. People with the inherited skin disease called xeroderma pigmentosum, for example, accumulate mutations rapidly in their cells. The girl shown in Figure 10.13a suffers from this inherited condition and has dozens of skin cancers where ultraviolet light from the sun has mutated DNA in her skin cells. Medical geneticists have discovered that in normal people, *DNA repair enzymes* constantly patrol the DNA, moving up and down the long strands, detecting and fixing damaged or unpaired bases (Figure 10.13b–d). People with xeroderma pigmentosum, however, have defective DNA repair enzymes; thus, mutations from what would be a normal amount of exposure to the sun accumulate in their cells and cause numerous skin cancers.

Cancer-causing substances are known as *carcinogens*, and they act by generating mutations. A mutagen has the potential to cause cancer if it happens to affect a gene that regulates cell growth. Since the agents that cause mutations in bacteria are identical to those that damage human DNA, nearly anything that increases the rate of mutation in bacteria also increases the mutation rate—and potentially the cancer

(a) Skin cancer from unrepaired DNA damage

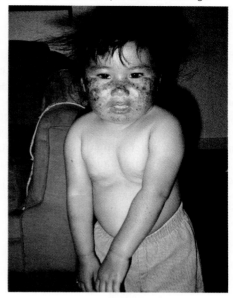

FIGURE 10.13 ■ How Enzymes Patrol DNA and Repair Mutations. (a) This Navajo child suffers from the skin disease xeroderma pigmentosum. She has developed hundreds of skin cancers, visible as small dark spots on the facial skin, which has been exposed to the sun, but not on her arms and body, which are protected by clothing. Deficient DNA repair leaves the skin cells vulnerable to the mutagenic effects of ultraviolet light. When ultraviolet light from the sun strikes and damages DNA in a normal person, (b) enzymes find the damage, (c) cut it out, and (d) make a new reciprocal copy of the old strand and ligate, or join, the two strands, leaving the DNA as good as new.

rate—in people. Owing to this fact, researchers can test the carcinogenicity (cancer-causing ability) of chemical and physical agents by exposing certain bacteria to suspected carcinogens and measuring mutation rates in the bacteria. With this test (called the Ames test after its originator), researchers have generated an extensive catalog of mutagenic substances that are all potentially carcinogenic. Individuals who expose themselves to known mutagens, such as excess sunlight or cigarette smoke, place a heavy burden on their DNA repair enzymes, needlessly increasing the likelihood of cancer.

An individual with a defective DNA repair enzyme will have increased susceptibility to cancer in all of his or her cells. In contrast, a person with the cystic fibrosis mutation has abnormalities in only certain cells, especially in the ducts of the lungs and pancreas. Why is it that some mutations affect many cells, while others only affect certain cells? The answer lies in the phenomenon of gene regulation.

(b) Damaged DNA

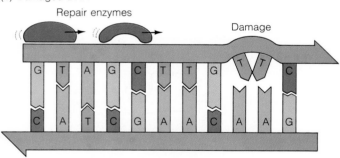

(c) Damaged DNA excised

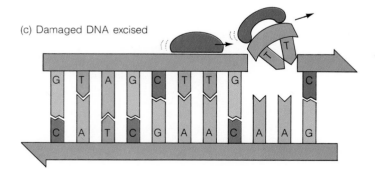

REGULATION OF GENE ACTIVITY

Different cells have different constellations of proteins. While all cell types produce the enzymes that repair DNA, only red cells make hemoglobin, only pancreas cells make the digestive enzyme trypsin, and so forth. Clearly, each cell regulates the processes of transcription and translation in such a way that it produces the right amounts of the right types of proteins as needed. The process that controls each cell's gene activity is called **gene regulation,** and it can occur at any one of five levels in the pathway of gene expression.

Levels of Control

Genes can be regulated at any of five levels (Figure 10.14).

■ **Transcriptional Regulation**   Cells usually regulate gene products by controlling the amount of mRNA transcribed from a gene. For example, developing red cells transcribe hemoglobin genes quickly and in great number, producing globin messenger RNA molecules and, from them, hemoglobin proteins. Every other cell in the body also contains the globin gene, but none transcribes it.

(d) Excised DNA filled in and repaired

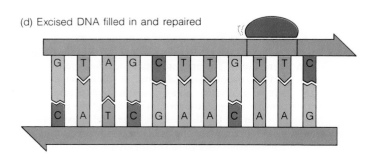

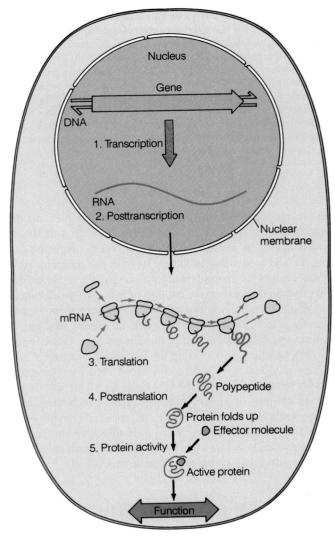

FIGURE 10.14 ■ **How Gene Expression Can Be Controlled at Five Different Levels.**

a mutation that blocks the normal posttranslational modification of the CFTR protein. This disease quite dramatically proves the importance of these posttranslational processes.

■ **Regulation of Protein Activity**   The presence of small molecules can alter a protein's function. For example, the end product of a biosynthetic pathway can decrease an enzyme's activity by binding to it (see Figure 5.17).

Both prokaryotes and eukaryotes use all five levels of control, but in different ways. The following discussion of prokaryotes emphasizes the transcriptional level of control because it is the most commonly used and best understood.

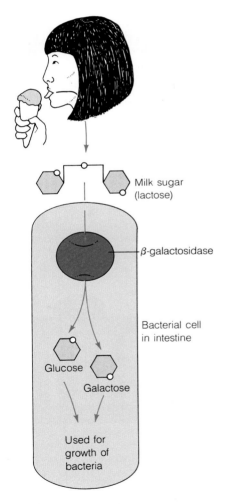

■ **Posttranscriptional Regulation**   Some mRNAs are chemically modified before they are transported from the nucleus and then translated in the cytoplasm. Without that modification, the mRNA may fail to reach the ribosomes and so lose its ability to be translated into a protein.

■ **Translational Regulation**   Cells can govern the rate at which mRNA is translated into protein by blocking access of the ribosomes to the messenger or by limiting the growth of the polypeptide.

■ **Posttranslational Regulation**   To function properly, some proteins must be modified after translation. Whole stretches of amino acids may have to be removed from a polypeptide; other chemical groups, such as carbohydrates and phosphates, may have to be added. Most people with cystic fibrosis have

FIGURE 10.15 ■ **An Enzyme Breaks Lactose into Simple Sugars in an *E. coli* Bacterium.**   The enzyme β-galactosidase cleaves lactose (milk sugar) into glucose and galactose. If you eat ice cream, *E. coli* bacteria in your intestine break down the lactose and use it for their own growth. A regulator protein can detect the presence of the lactose, turn on the β-galactosidase gene, and cause more enzyme to be made. Thus, when the milk sugar is present, the bacteria have the enzymes they need to quickly use it.

Gene Regulation in Prokaryotes

Even though bacterial cells have a simpler structure than eukaryotic cells, bacteria have streamlined and highly sophisticated mechanisms of gene control. Bacterial cells can change the kinds of proteins they make quickly and frequently. Consider, for example, the *Escherichia coli* bacteria growing in your intestine. Just after you have eaten a prime rib dinner and your digestive enzymes have reduced the protein in the meat to amino acids, the liberated amino acid tryptophan becomes abundantly available to the resident *E. coli*. It would be a waste of energy if *E. coli* cells, which synthesize the enzymes for making tryptophan, continued to do so once a big supply was abundantly available in the environment. Instead, they shut down production of tryptophan-making enzymes. Later, after a dessert of vanilla ice cream, the bacteria are floating in a sea of lactose, the main sugar in milk. The cells soon respond, making lactose-breaking enzymes.

A bacterial cell that makes unnecessary proteins expends energy and materials that could be used to prepare the cell for reproduction; thus, it grows and divides more slowly than a bacterium using its resources more efficiently. Bacteria have a gene regulation mechanism called an *operon,* and this prevents the production of unnecessary proteins.

The best-studied operon regulates lactose use. Scientists in the mid-1900s knew that the enzyme β-galactosidase breaks the milk sugar lactose into the two simple sugars glucose and galactose (Figure 10.15). *E. coli* cells growing in the absence of lactose contain none of this enzyme. However, if lactose is supplied, each cell produces thousands of β-galactosidase molecules in just 10 to 20 minutes. Somehow, lactose rapidly *induces* formation of the sugar-digesting enzyme.

In 1960, through some elegant experiments, French geneticists François Jacob and Jacques Monod figured out how *E. coli* bacteria regulate the synthesis of β-galactosidase in response to the presence or absence of lactose, the substrate the enzyme acts on. They found that a protein in the *E. coli* cell can sense the absence or presence of lactose. If lactose is absent, the protein signals the DNA not to make mRNA for the β-galactosidase enzyme (Figure 10.16a). Conversely, if lactose is present, the sensor protein allows the DNA to make mRNA for the enzyme (Figure 10.16b). Careful experiments showed that the sensor protein (which Jacob and Monod called the *repressor,* since it represses transcription) acts by binding to DNA. Specifically, it binds to a spot they called the

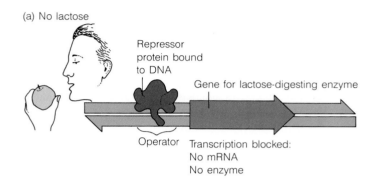

(a) No lactose

Repressor protein bound to DNA

Gene for lactose-digesting enzyme

Operator   Transcription blocked:
No mRNA
No enzyme

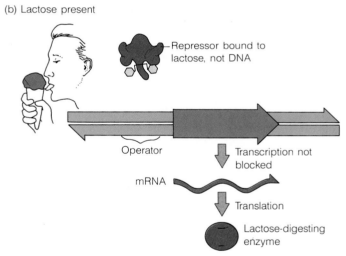

(b) Lactose present

Repressor bound to lactose, not DNA

Operator

mRNA

Transcription not blocked

Translation

Lactose-digesting enzyme

FIGURE 10.16 ■ How a Bacterial Cell Makes an Enzyme Only When It Is Needed.   A normal bacterial cell makes enzymes for digesting the milk sugar lactose only when that sugar is present in the cell's environment. This energy efficiency is due to the operon mechanism of gene regulation. (a) In the absence of lactose, a repressor protein binds to DNA at the operator, a region of DNA near the gene that encodes the digestive enzyme lactase. The presence of the repressor physically blocks the ability of RNA polymerase, the RNA-making enzyme, from transcribing the gene. When transcription is blocked, no mRNA is formed, it cannot be translated into protein, and hence no enzyme forms. (b) When lactose is present, it binds to the repressor, and the repressor is then unable to bind to DNA. Now RNA polymerase has access to the gene and transcribes it into mRNA. This message is translated, the enzyme is synthesized, and it can digest the sugar lactose, providing energy and materials for the bacterium. (c) A computer image shows how the repressor wraps an arm around one of the grooves in DNA and hence blocks transcription. DNA is shown in rust and gold, and protein in white, blue, red, and green.

(c) Repressor protein bound to DNA

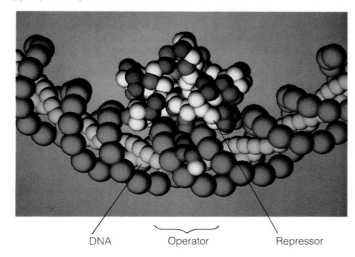

DNA   Operator   Repressor

operator, which is near the gene that encodes the lactose-digesting enzyme (Figure 10.16). When the repressor binds to the DNA, it blocks transcription of the gene for the enzyme; with no transcription, there can be no mRNA, and hence no lactose-digesting enzyme is formed. Jacob and Monod gave the name **operon** to the protein signaler and the group of genes that work together to regulate production of a given enzyme.

An important feature of gene regulation in bacteria is *transcription-level control,* whereby the cell regulates the amount of enzyme by regulating the transcription of mRNA. Another important feature is *simultaneous transcription and translation* (Figure 10.17a). Because bacteria lack a nuclear membrane, translation of an mRNA molecule can start at one end of the message even before transcription of DNA into mRNA has finished at the other. This simultaneous activity increases the speed with which bacteria can respond to changing levels of materials in their environment. The last important feature of bacterial gene regulation is *short-lived mRNAs.* The *E. coli* cell synthesizes an mRNA message in 4 minutes, but destroys it in 3. Thus, when the cell no longer needs to make a particular enzyme, it rapidly dismantles the machinery for making it. Again, this allows rapid response to changing metabolic needs.

Gene Regulation in Eukaryotes

While eukaryotes and prokaryotes share many mechanisms of gene regulation, such as controls at the transcriptional level and control by regulatory proteins, they also have divergent regulatory mechanisms based on their differences in size and complexity. Operons, common in bacteria, do not occur in eukaryotes. In nucleated cells, transcription occurs in the nucleus and translation in the cytoplasm (Figure 10.17b); consequently, eukaryotes cannot transcribe and translate simultaneously, and eukaryotic gene expression is slowed.

For example, the lactose that *E. coli* bacteria (prokaryotes) can process so readily because of the *lac operon* presents a problem to some mammals (eukaryotes). Baby mammals produce a lactose-digesting enzyme called lactase in their intestinal cells, and the enzyme allows them to digest milk sugar. But as the babies mature, they permanently lose the ability to make the lactase and along with it the means of digesting milk sugar. The mammalian regulation of lactase production—continuously on, then permanently off—seems crude compared to the environmentally sensitive bacterial lactose operon. But then, most mammals never drink milk as adults, so losing this digestive capacity is actually an efficient response.

Like other mammals, most of the world's people lose the ability to digest lactose. The only exceptions are Caucasians and certain dairying tribes of Africa. After early childhood, drinking milk gives most of the world's people a stomach-ache. Unfortunately, many adults without lactase have a taste for ice cream and other milk products. To circumvent this

problem, they may enlist the aid of *Lactobacillus,* the bacterium used to make yogurt. If a lactose-sensitive person takes a pill containing *Lactobacillus* just before eating a giant bowl of mocha almond fudge ice cream, the bacterial enzyme β-galactosidase digests the lactose before it can reach the intestine and cause discomfort.

It might seem odd that a person's enzyme production would be turned on or off. But there is an underlying reason: People, like most eukaryotes, develop from a single fertilized egg into a multicellular organism, and to make this transition, genes must be turned on and then permanently off at different stages of development.

Compare a colony of bacteria with a human embryo in the womb. Both consist of millions of cells, all derived from a single founder cell, a solitary bacterium or a fertilized egg. But all the cells in the bacterial colony are identical, while the cells in the embryo form a bewildering assortment of shapes and functions: Gut cells are cubes filled with opal droplets of

(a) Prokaryote: Simultaneous transcription and translation

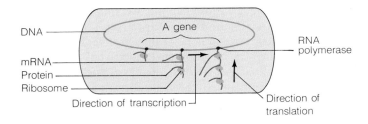

(b) Eukaryote: Separation of transcription and translation

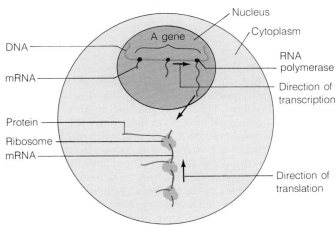

FIGURE 10.17 ■ Where Transcription and Translation Take Place in Prokaryotes and Eukaryotes. (a) In a bacterial cell, transcription and translation occur simultaneously along the naked, circular chromosome. (b) In a eukaryotic cell, transcription takes place in the nucleus, and the newly made RNA is transported to the cytoplasm, where translation then occurs.

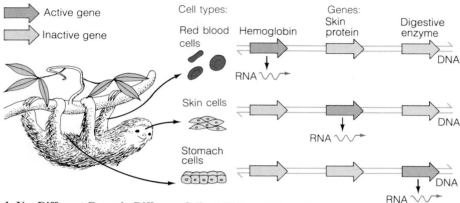

FIGURE 10.18 ■ **Mammals Use Different Genes in Different Cells at Different Times During Development.** In general, each cell in the body of a eukaryote has a complete set of genetic information. Thus, each cell has the gene for the oxygen-carrying protein hemoglobin normally found only in red blood cells; each cell has the gene for the fibrous protein keratin normally made only by skin cells and found in hair and claws; and each cell has the gene for digestive enzymes made only by cells of the digestive tract. In addition, each cell has 100,000 other genes for proteins made by other groups of cells. The reason that red blood cells accumulate hemoglobin but not keratin or digestive enzymes is that immature red blood cells transcribe the hemoglobin gene into RNA, but they do not transcribe the genes for keratin or digestive enzymes. Likewise, skin cells transcribe their keratin genes, but not hemoglobin or digestive enzyme genes, and digestive tract cells transcribe digestive enzyme genes but not genes for the other two proteins. This is called differential gene expression.

digestive enzymes; red blood cells are little scarlet disks packed with the oxygen carrier hemoglobin; and nerve cells have long branches that communicate with other cells.

The disparate methods of gene control in prokaryotes and eukaryotes explain why cells in a bacterial colony are all alike, while the millions of cells making up a plant or animal diverge into different forms. Bacteria turn on different genes under different conditions (e.g., they use the genes for lactose-digesting enzymes only when lactose is available). But mammalian cells express different genes in different cells (red blood cell precursors express hemoglobin genes, and gut cells express digestive enzyme genes) and at different times (the fertilized egg expresses neither hemoglobin nor digestive enzyme genes, but the red blood cell precursors and gut cells derived from the egg do; Figure 10.18). Bacteria regulate their genes conditionally, depending on what is present or absent in the environment, while multicellular eukaryotes regulate their genes spatially and temporally, depending on where the cell lies in the organism and where the organism is in its life cycle. An important goal of research biologists is to learn how the different cells in a multicellular organism express different genes during development.

■ **Gene Regulation and Development**   Although every cell in a multicellular eukaryotic individual has a full complement of genes, different cells require different proteins, and thus each cell expresses (transcribes and translates) different sets of genes and produces different sets of proteins. Evidence for the hypothesis that different cells express different genes comes from the giant chromosomes found in certain fruit flies (Figure 10.19). Within these special chromosomes, 1000 DNA

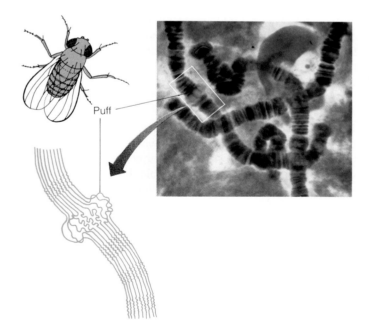

FIGURE 10.19 ■ **Unwinding Chromatin Makes a Section of DNA Accessible for Transcription.** The salivary glands of fly larvae, such as the midge *Chironomus* or the fruit fly *Drosophila*, have giant chromosomes with easily visible bands, or regions of tightly coiled chromatin (a combination of DNA and protein), as well as puffs, or regions that have ballooned out and where the DNA is being actively transcribed. Different puffs appear at different times in development and in different tissues. This corresponds to the prediction of the hypothesis that different cells are different because they express different genes.

BOX 10.1   *PERSONAL IMPACT*

CAESAR EXPERIMENTS WITH RNA SYNTHESIS

How long can a person live without transcribing DNA? The Roman emperor Claudius Caesar provided an answer in A.D. 54 with an experiment he unwittingly—and no doubt unwillingly—performed with his wife Agrippina. She mixed into Caesar's favorite dish of edible mushrooms (*Amanita caesarea*) a few of the poisonous species *A. phalloides* (Figure 1). These poisonous mushrooms contain a substance called *α-amanitin*, which inhibits transcription in eukaryotic cells by blocking the activity of RNA polymerase. Even very small amounts of α-amanitin prevent body cells from synthesizing messenger RNA.

For the first 10 hours after Caesar ate this delicacy, all seemed well. But as he digested the fungus, the α-amanitin entered his bloodstream and was absorbed by his liver and kidneys, where it began to block transcription. About 15 hours after his repast, with no new mRNA to make new proteins, Caesar's liver cells stopped functioning, and nausea, diarrhea, and delirium began to hit him. Two days later, he died of liver failure. It is highly doubtful that Caesar learned to appreciate the valuable role of RNA polymerase in DNA transcription. But perhaps, in a general way, Agrippina did.

FIGURE 1 ■ *Amanita caesarea.*

duplexes lie bundled together side by side. Regions that are actively undergoing transcription appear as *chromosome puffs*—areas where the DNA is spun out from the axis of the chromosome in giant loops and where RNA is being made. Different cell types have puffs in different regions of the chromosome, as would be expected by the differential gene expression hypothesis.

In people, both red blood cell precursors and nerve cells, for example, have genes for making hemoglobin and genes for making nerve communication molecules. But only developing red blood cells express the hemoglobin genes, while only nerve cells express the genes for communication molecules. These two cell types also contain the genes for making eye pigments, tooth enamel, toenail proteins, and the CFTR protein, but developing red blood cells and nerve cells never express these and many other genes. In fact, most mature cells use only 7 percent of their genes—a different, though overlapping, 7 percent for each cell type.

Knowing that different cell types express different genes, we can now understand the range of symptoms a person with cystic fibrosis experiences. The cystic fibrosis gene must be operating in some cells and necessary for their function, but unused and unnecessary in other cells. In fact, recent work shows that the mRNA from the cystic fibrosis gene is found in duct cells—air ducts, pancreatic ducts, sweat ducts, and sperm ducts—but is not found in brain cells and many other cell types, suggesting that the gene is transcribed only in certain cells. This fact makes it easier to understand the duct-related problems of affected people, since it is only in duct cells that the CFTR protein normally functions (and in cystic fibrosis sufferers functions abnormally). Evidently, gene regulatory mechanisms ensure that only duct cells transcribe the

gene leading to cystic fibrosis. In eukaryotes, these regulatory mechanisms probably act very similarly to the equivalent mechanisms already discussed for bacteria: Regulatory proteins bind special regions of DNA (in eukaryotes, these are called *enhancers*), and this binding alters the transcription rate of the gene in a way that is specific to each cell type.

Now, let's apply this idea of gene regulation to the development of an embryo. As a one-celled zygote develops into a larger, more complex organism, the multiplying cell groups begin to take on their various functions. Cells that are at first unspecialized soon become nerve cells, skin cells, blood cells, or duct cells as a result of the process of *cell differentiation*, and this specialization process is due to selective gene expression. In eukaryotes, then, gene regulation underlies differentiation, which in turn allows development to proceed, changing a fertilized egg into an embryonic mouse, plant, or person. Biologists don't yet know how different genes are selected for expression in different cells or how the selection, once made, is stabilized.

■ **Regulation, Development, and Genetic Disease**   Biologists have been working hard to learn how genes are regulated during embryonic development, since understanding this process will ultimately help physicians treat genetic diseases such as sickle-cell anemia and thalassemia. Recall that people with these conditions have defective genes for hemoglobin. Researchers discovered some time ago that an adult mammal expresses a different gene for hemoglobin than does a fetus enclosed in the mother's womb (Figure 10.20). People and other mammals, in fact, have a *multigene family* for hemoglobin, a group of genes related in structure and function that code for closely related blood pigment molecules.

In people with sickle-cell anemia and thalassemia, the gene for the adult form of hemoglobin is defective; but the gene for the fetal form of hemoglobin is normal. If biologists could fully understand gene regulation during development—precisely how DNA-binding proteins and enhancers work, for example—perhaps they could prevent the fetal hemoglobin gene from turning off during development. If that fetal gene were continually transcribed, even in the child and adult, the individual would have sufficient amounts of normal hemoglobin to prevent the symptoms of sickle-cell anemia or thalassemia, even though the mutant gene for the adult form of hemoglobin remained in the cells unexpressed. Many medical researchers are trying to learn how to manipulate the expression of these genes during development.

Recent experiments suggest that researchers may be able to develop a therapy for cystic fibrosis based on the principles of transcription and translation discussed in this chapter within a few years' time rather than the decades many thought it might require. When geneticists isolated the cystic fibrosis gene, they discovered that, like most other eukaryotic genes, it was broken into many different parts—in this case scattered along a portion of chromosome 7. In general, eukaryotic genes differ from prokaryotic genes in their possession of **introns,** stretches of DNA that *intr*ude into the gene but do not appear in the final mRNA (and hence are not translated into protein). The remaining portions of a gene, which *do appear* in the mRNA, are called **exons,** since they are *ex*pressed. The cystic fibrosis gene has 24 exons. When DNA is transcribed, the product of transcription is called the *primary transcript,* a precursor to mRNA that includes an RNA

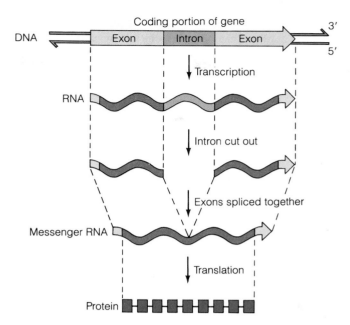

FIGURE 10.21 ■ Introns and Exons: Structure of a Eukaryotic Gene. In eukaryotes, only certain parts of a gene actively encode proteins. The whole coding portion of a gene is transcribed into a primary transcript, but introns, or intrusive stretches of DNA, are spliced out before the actual mRNA forms. This message then serves as the instruction for translating the information in a gene into a polypeptide. The figure shows one intron, but the cystic fibrosis gene has 24, and other genes can have as many as 50.

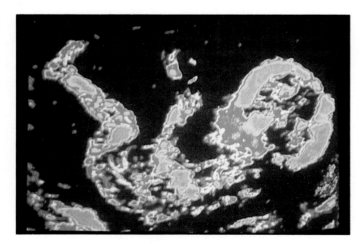

FIGURE 10.20 ■ Regulation of the Hemoglobin Gene Family: A Fetus Expresses a Special Hemoglobin Gene. The third-trimester human fetus shown in this sonogram is transcribing a different hemoglobin gene than in an adult. The fetal hemoglobin is not affected by the mutations that cause sickle-cell anemia and thalassemia. Therefore, physicians hope that they will be able to manipulate the regulation of this gene in their patients with these diseases, turning on the fetal gene, letting the fetal gene take over from the defective adult gene, and ameliorating the symptoms of disease.

copy of both the introns and exons (Figure 10.21). The primary transcript is then processed (or edited) by cleaving out the introns and splicing the remaining exons together. Surprisingly, some introns act as a nonprotein enzyme (or *ribozyme*) that splices itself out. The result is an mRNA ready to be exported from the nucleus, now bearing the actual exon sequences that will be translated into protein.

In 1990, researchers used genetic engineering techniques (discussed in the next chapter) to build a nonmutated copy of the cystic fibrosis gene possessing only exons (expressed regions) and no introns. They inserted a synthetic gene into airway cells and pancreas duct cells that they had removed from a person with cystic fibrosis and grown in the laboratory. Wonderfully, the inserted gene was transcribed into an mRNA, this messenger was translated into properly functioning CFTR protein, and this worked to pump chloride ions normally: The diseased cells had been cured! Officials of the Cystic Fibrosis Foundation now hope that researchers will be able to design a spray that can deliver the engineered gene to airway cells in a living person, and that through transcription and translation of the CFTR gene, their cells and respiratory system might function normally. (No genetic change in airway cells would be transmitted to a person's children, of course, since airway cells do not give rise to eggs and sperm.) Formidable obstacles remain in designing effective delivery systems for engineered genes, but before long, medical workers will be applying the basic biology of transcription, translation, and gene regulation described in this chapter to diseases like cystic fibrosis and sickle-cell anemia and helping to improve the quality of life for millions of sufferers around the world.

CONNECTIONS

In a very real sense, the possibility of helping people with cystic fibrosis and other genetic diseases stems directly from life's basic unity at the level of genes and molecules. Mendel initially discovered genes in plants; later experimenters learned that genes are made of DNA by studying viruses; research on fungi supported the one gene–one enzyme concept; and studies with bacteria elucidated the mechanisms of transcription, translation, and gene regulation. All these principles are being applied now to helping people with genetic diseases, and the future uses of this knowledge are potentially limitless. Few genetic applications would be possible, however, without our modern techniques for manipulating genes, which grew out of a remarkable episode in scientific history that began in the late 1970s. This ability led to a revolution in biological engineering that is still going on and is the subject of Chapter 11.

Highlights in Review

1. Genes work by specifying the base sequence of mRNA, which specifies the amino acid sequence of a polypeptide. Information in DNA is copied to RNA through a process called transcription, and information in RNA is transferred to protein through a process called translation.

2. Messenger RNA (mRNA), transfer RNA (tRNA), and ribosomal RNA (rRNA) differ from DNA in that they are made of ribonucleotide subunits rather than deoxyribonucleotide subunits. A single oxygen atom on the sugar ribose differentiates the two kinds of nucleotides. RNA consists of only a single strand of nucleotides, not a double strand, as in DNA; and the RNA base uracil replaces the DNA base thymine. RNA molecules are also generally much shorter than DNA molecules because they tend to code for only one or a few genes rather than for the whole genome.

3. Like replication, transcription is directed by base pairing and carried out by a polymerase enzyme. But in transcription, ribonucleotide bases pair with the deoxyribonucleotide bases on the DNA strand, and the enzyme is RNA polymerase rather than DNA polymerase. During transcription, a few genes may be copied thousands of times, while during replication, the whole genome is copied just once. The newly formed RNA strand separates from the DNA molecule in transcription, instead of joining with it as does a newly replicated DNA strand.

4. The genetic code, transfer RNA, and ribosomes together bring about translation. The genetic code is identical in nearly all organisms. A group of three bases is a codon. Except for the start codon and the three stop codons (which tell RNA polymerase where to begin and end translation), each codon specifies one amino acid. Several different codons may specify the same amino acid, but no codon specifies more than one amino acid.

5. The tRNAs translate the mRNA codons into an amino acid sequence. At one end of each tRNA molecule is an anticodon; at the other end is a specific amino acid. The anticodon pairs up with the mRNA codon, so that the amino acids are ordered according to the mRNA codon sequence.

6. Ribosomes hold mRNAs, tRNAs, and amino acids in place until the amino acids can be joined together into a polypeptide. Each mRNA may have several ribosomes running along it at once, all translating polypeptides off the same mRNA.

7. A change in the base sequence, a mutation, may consist of a base substitution, insertion, or deletion. Usually, DNA repair enzymes detect and fix altered DNA; if these enzymes fail, a permanent mutation occurs. A change in the sequence of DNA bases results in a change in the sequence of RNA bases, which can result in a change in the sequence of amino acids—an altered polypeptide.

8. Mutations in the sequence of base pairs in DNA can result in genetic defects or cancer.

9. Cells limit the production of unnecessary proteins by regulating gene activity. They can accomplish this by regulating transcription, modifying the mRNA after transcription, regulating translation, modifying the polypeptides after translation, and adjusting protein activity itself. The most efficient level at which to regulate—and the most common—is transcription.

10. A classic example of transcription-level control in prokaryotes is the lactose operon of *E. coli*. In the absence of lactose, a repressor protein binds to an operator sequence (a special site on the DNA) and so prevents transcription of genes for lactose-digesting enzymes. In the presence of lactose, the repressor protein binds to the lactose instead of to DNA. RNA polymerase then transcribes the structural genes that code for enzymes that digest lactose.

11. Prokaryotes change the kinds of proteins they make both often and rapidly, which accommodates changes in their environment. They achieve flexible gene regulation dependent on environmental conditions by using transcription-level control, by transcribing and translating simultaneously, and by destroying mRNAs almost as fast as they are made.

12. All cells in eukaryotic organisms have the same genes, but some cells use one set of genes while others use another set. These differences arise through the developmental process called cell differentiation. How eukaryotes achieve this cell differentiation is one of the great mysteries of biology.

13. Unlike prokaryotic genes, eukaryotic genes that are regulated simultaneously are not necessarily located near each other; they may even be on different chromosomes. Like prokaryotic genes, eukaryotic genes may be flanked by regulatory sequences of DNA that are responsive to gene regulators made of protein.

14. In eukaryotes, genes are clustered according to their historical relatedness in multigene families. The hemoglobin gene family in humans encodes several structurally similar globins, some of which are expressed at different times in development.

15. Our increasing understanding of genetics gives us hope that we may one day be able to help victims of genetic defects.

Key Terms

anticodon, 214
chromosomal mutation, 217
codon, 212
elongation, 216
exon, 225
gene mutation, 217
gene regulation, 219
genetic code, 212
initiation, 216
intron, 225
message, 211
messenger RNA (mRNA), 210

mutation, 216
operon, 222
protein synthesis, 215
reading frame, 213
ribonucleotide, 210
ribosomal RNA (rRNA), 210
RNA polymerase, 211
start codon, 213
stop codon, 213
termination, 216
transcription, 210
transfer RNA (tRNA), 210
translation, 210

Study Questions

Review What You Have Learned

1. What is the difference between transcription and translation?
2. How do RNA and DNA differ?
3. During transcription, what will the order of the bases in mRNA be if the base sequence in DNA is CTAGCT?
4. Refer to the codon dictionary in Figure 10.6 and give the possible codons for the following amino acids: lysine; phenylalanine; glycine.
5. Explain why the genetic code is identical in humans, bacteria, and other organisms.
6. Assume that there is a CAG anticodon on a tRNA. Which mRNA codon does the anticodon pair up with? Which amino acid is attached to this tRNA?

7. What are three causes of gene mutation?
8. Explain the functioning of a bacterial operon.
9. Explain the statement, "You can take the cell out of the liver, but you can't take the liver out of the cell."
10. Compare the basic features of gene regulation in prokaryotes and eukaryotes.

Apply What You Have Learned

1. How could a scientist use the Ames test to determine whether the chemicals in a new hair dye are carcinogenic?
2. Although Jacob and Monod received the Nobel Prize for their elucidation of the operon model, that model does not apply to a horse or a maple tree. Why not?
3. A mutation occurs in a human cell that changes the anticodon of a transfer RNA from AUG to AUU. The tRNA carries tyrosine. What will be the likely consequences?

For Further Reading

ALBERTS, B., D. BRAY, J. LEWIS, M. RAFF, K. ROBERTS, and J. D. WATSON. *Molecular Biology of the Cell.* 2d ed. New York: Garland, 1989.

DAVIES, K. "The Search for the Cystic Fibrosis Gene." *New Scientist,* October 21, 1989, pp. 54–58.

FLINT, J. A., V. S. HILL, D. K. BOWDEN, S. J. OPPENHEIMER, P. R. SILL, S. W. SERJEANTSON, J. BANA-KOIRI, K. BHATIA, M. P. ALPERS, A. J. BOYCE, D. J. WEATHERALL, and J. B. CLEGG. "High Frequencies of Alpha-Thalassaemia Are the Result of Natural Selection by Malaria." *Nature* 321 (1986): 744–750.

ROBERTS, L. "Cystic Fibrosis Corrected in Lab." *Science* 249 (1990): 1503–1504.

SCHULMAN, L. H., and J. ABELSON. "Recent Excitement in Understanding Transfer RNA Identity." *Science* 240 (1988): 1591–1592.

STEPHENS, J. C., M. L. CAVANAUGH, M. I. GRADLE, M. L. MADOR, and K. K. KIDD. "Mapping the Human Genome: Current Status." *Science* 250 (1990): 237–244.

WEISS, R. "Cystic Fibrosis Treatments Promising." *Science News* 139 (1991): 132.

WEISS, R. "Upping the Antisense Ante: Scientists Bet on Profits from Reverse Genetics." *Science News* 139 (1991): 108–109.

Genetic Recombination and Recombinant DNA Research

FIGURE 11.1 ■ **Supermouse.** Genetic engineers produced a giant mouse twice as big as its normal sibling by inserting a gene for human growth hormone into one of its chromosomes.

Building a Bigger Mouse

Could a human gene direct a mouse cell to produce a protein? Or are our genes completely foreign to the cells of all other organisms? The answer is clear-cut and came from an experiment that produced a giant mouse (Figure 11.1). This hefty rodent is just one example of the biologist's powerful new ability to manipulate genes through the laboratory procedures called **recombinant DNA technology**, this chapter's main subject.

In the late 1970s, molecular geneticists from Seattle, Philadelphia, and Stockholm isolated the gene for *human growth hormone*, which causes rapid growth in human muscles, bones, and connective tissue. They took this human gene, inserted it into a mouse chromosome, and found that mouse cells receiving the extra DNA transcribed and translated it correctly into human growth hormone. Once circulating in a mouse's bloodstream, the hormone caused the animal to grow rapidly and continuously. Some mice treated this way grew to about twice the size of normal mice. For the first time, a human gene had been expressed in another animal, and the results were phenomenal.

Molecular biologists were able to achieve these results by using recombinant DNA technology to isolate the growth hormone gene from human cells, insert the gene into bacterial cells, turn the bacterial cells into tiny copying machines that churned out hundreds of copies of the growth hormone gene, and finally insert these copies into the chromosomes of mouse embryos.

■ **Growth Hormone and Body Size.** "Spud" Webb and "Tree" Rollins illustrate the effects that differing amounts of growth hormone can have on human height.

Using recombinant DNA technology (also called *genetic engineering techniques*), scientists can break apart and rejoin two different DNA molecules. By analogy, if you took pieces of red and green string, cut each in two, and retied a green piece to a red one, you would have *recombined* the parts of the two strings. This type of recombination also occurs in nature during the crossover phase of meiosis. But whether it occurs naturally during meiosis or artificially in the laboratory, recombination provides new arrangements of old genes.

By employing special enzymes extracted from bacteria and viruses, molecular geneticists can now cut and paste DNA from all sorts of species. Furthermore, recombinant DNA researchers can, in theory, create any genes they desire and insert these test-tube "designer genes" into plant or animal cells. There, the designed genes are replicated along with the host's own DNA and expressed in the host as if they had occurred naturally. Besides creating giant mice, such manipulated genes may someday help researchers to improve crops and farm animals, mass-produce rare drugs, and treat myriad human diseases.

Genetic recombination has occurred spontaneously and naturally for millions of years in the sex organs of sexually reproducing organisms (see Chapter 7). This natural recombination provides much of the variation that, through natural selection, is the basis of evolution (see Chapter 1).

Three themes will recur throughout this chapter. First, genetic recombination—whether artificial or natural—depends on complementary base pairing. The elegant structure of DNA thus determines yet another aspect of gene function. Second, genetic recombination is a key feature in the evolution of life. During meiosis, recombination creates new genotypes that can be acted on by natural selection. Finally, the genetic recombination that geneticists now conduct in the test tube offers tremendous potential for understanding the basic mechanisms of cell function, for developing new agricultural products, and for achieving any number of advances in medicine and the drug and chemical industries. Despite this great potential, genetic engineering also poses environmental and ethical risks, as we will see.

This chapter will reveal answers to the following questions:

■ How do DNA molecules recombine in nature, and what is their evolutionary significance?
■ How do geneticists recombine DNA molecules in the laboratory and insert new genes into bacterial, plant, and animal cells?
■ What potential prospects and problems lie in using recombinant DNA technology (genetic engineering) for human benefit?

RECOMBINATION IN NATURE

Whether it occurs in the test tube or in the testes, genetic recombination creates a DNA molecule with a new arrangement of genes. Before discussing the ways molecular biologists recombine DNA in the laboratory, we must examine the tried and true recombination methods "invented" by nature long ago and understand their consequences for the history of living organisms.

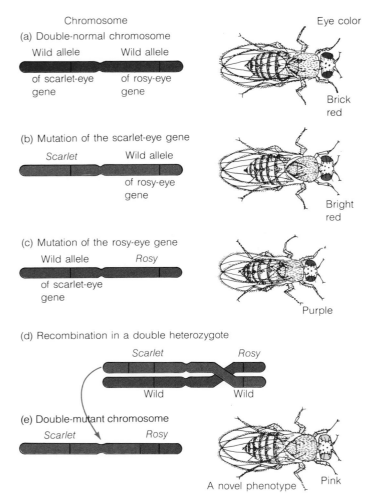

FIGURE 11.2 ■ **How Recombination Creates Novel Phenotypes. (a) Wild-type flies have brick red eyes owing to the action of the normal wild-type alleles of two genes widely separated on the same chromosome. (b) A mutant allele of one of the genes leads to bright red eyes. (c) A mutation in the other gene causes purple eyes. (d) Recombination (crossing over) during meiosis between a chromosome with a mutant scarlet-eye allele and its homologue with a mutant rosy-eye allele can yield a chromosome bearing both mutant alleles. (e) The double-mutant recombinant chromosome causes pink eyes—a novel phenotype. In some situations, such novel phenotypes might give the organism an evolutionary advantage. (See Figure 7.22 for a specific diagram of crossing over.)**

Recombination: Universal Source of Variation

Genetic recombination is a central feature of reproduction in all organisms. It occurs in bacteria, in the ovaries and testes of mice and humans, and in the ears and tassels of corn. Recall that during a meiotic cell division in a eukaryotic cell, chromosomes can cross over and produce new arrangements and combinations of genes (see Chapters 7 and 8). The phenotype of an organism carrying these recombined genes will differ from the phenotype of either parent—or, for that matter, from the phenotype of any ancestor at all.

Consider the novel offspring—unlike either parent—that can result from recombination in a diploid organism such as the fruit fly. Wild-type (normal) flies have brick red eyes (Figure 11.2a). Flies homozygous for the scarlet-eye allele of one gene have bright red eyes, however (Figure 11.2b), while flies homozygous for the rosy-eye allele of a different gene on the same chromosome have purple eyes (Figure 11.2c). Recombination can occur in a fly that is double heterozygous for scarlet and rosy in such a way that both mutant alleles now lie on a single chromosome (Figure 11.2d and e). The remarkable offspring homozygous for this recombinant chromosome will possess a strange and beautiful new phenotype—pale pink eyes (Figure 11.2e), different from either of the parents or the typical wild-type fly. This novel phenotype is due not to a new mutation, but rather to a new association of alleies resulting from recombination.

Contribution to Evolution

The preceding example shows that recombination produces new genetic combinations, with phenotypes sometimes unique and not predictable from the original mutations. Since the pink-eyed phenotype is different from either the rosy or scarlet parental type, flies containing the new genetic arrangement for pink eyes may survive and reproduce differently than their parents. If, for example, pink-eyed flies could see better under certain light conditions and therefore reproduce more successfully in a given environment, their new combination of genes would become more common. Thus, by creating new potentially successful variants, recombination plays an important role in evolution by natural selection.

Recombination occurs in sexually reproducing organisms—sometimes in surprising ways (see Box 11.1). The mechanisms by which recombination occurs are universal, and this is because DNA has the same basic structure and function throughout the kingdoms of life. Species characterized by recombination have greater genetic variation on which natural selection can act, and thus they have the edge in adapting to environmental change. For example, a fruit fly species that underwent recombination would express many more phenotypes—including new ones such as pink-eyed individuals. If the environment changed somehow and suddenly favored these new recombinants, then the species might still persist despite the environmental challenge.

BOX 11.1   BIOLOGY: A HUMAN ENDEAVOR

BARBARA McCLINTOCK AND JUMPING GENES

Geneticist Barbara McClintock made a startling discovery in the mid-1940s that explained why Indian corn can have brilliantly decorated kernels. The mechanisms were so complicated, however, and her studies so advanced for that era that her colleagues largely misunderstood and ignored the results for 25 years. That, however, scarcely fazed McClintock—nor was she noticeably fazed when her work won her a belated Nobel Prize in 1983 (Figure 1). In her life as well as her scientific career, McClintock has been an absolute original.

From earliest childhood, McClintock had what her biographer calls "a capacity to be alone"—an emotional self-containment that allowed her to pursue her interests without need for outside approval, encouragement, mentors, heroes, or followers. As a genetics student in the 1920s, for example, while the other coeds wore flowing skirts and long tresses, McClintock bobbed her hair and donned knickers. She wasn't a tomboy, she "just couldn't be bothered" with these confining symbols of femininity while working in the experimental cornfields. About that same time, an incident occurred that demonstrated her extreme degree of concentration. McClintock became so focused on a geology final exam that she forgot her own name and struggled for 20 minutes to sign the test booklet. After graduation, while working in isolation at the remote Cold Spring Harbor laboratory in New York, this same laser-focus enabled her to unravel the mysteries of corn genetics and almost single-handedly raise maize to the status of the fruit fly as an experimental subject.

McClintock knew, as hundreds of generations of farmers had known before her, that some corn strains grow *variegated* kernels, or pale seeds with bright irregular patches of purple. Such irregular coloring indicates that even within the same kernel, a gene can code for pigment production in some cells but fail to do so in neighboring cells.

McClintock tried to map the gene responsible for turning pigment production on and off and made the unprecedented observation that the gene in question jumped around the corn genome from one chromosome to another. This was surprising, because geneticists believed that each gene occupied a constant, precise location on a specific chromosome. The phenomenon was due to a specific class of genetic elements, the *jumping genes,* or **transposons,** which violate the normal rules of genetic recombination.

Transposons are double-stranded DNA sequences that undergo "illicit" recombination: They recombine with sequences to which they are not homologous, in contrast to normal DNA molecules that recombine with other sequences only if their base orders are similar.

Transposons not only leap about in the genome, but also alter the action of the genes they leap into. If a transposon skips out of its original position and enters the gene for purple pigment in a corn kernel, it will alter the DNA sequence of the pigment gene, disrupt its action, and prevent pigment production. If the transposon later hops out of that same gene, the normal gene structure may be restored, purple pigment can reappear once

FIGURE 1 ■ **Nobel-Prize-Winning Geneticist. Barbara McClintock in 1947, at work on illicit recombination in corn.**

more, and the offspring of these few cells make dark spots. Since the gene "jumps ship" in only a few cells, the kernel is mostly light with a few dark spots.

While McClintock first found transposons in corn, they were later discovered in bacteria, fruit flies, and even people. Some transposon-like viruses in mammals take this freeloading a step too far: They cause cancer and kill the host. Other times, when a transposon leaps, it breaks a chromosome in two or carries a part of the chromosome to a new location, and this genetic alteration can seriously affect cell functioning. In corn, flax, and other organisms, environmental stress can trigger the movement of many transposons at once, amplifying a small "tremor" into a genetic earthquake. While these kinds of massive gene rearrangements can be disruptive, they can also add to the genetic diversity that is the foundation of evolution by natural selection.

Although long in coming, the approbation McClintock finally received acknowledges the genius of an intensely focused maverick with the "capacity to be alone." It was a truly remarkable feat to ferret out some of the basic mechanisms of genetic flexibility long before other geneticists had even discovered the structure of DNA or how it operates in the living cell. McClintock's philosophy of research is characteristically straightforward but profound: When an experimenter sees something that doesn't fit, there's a reason for it, and he or she must find out what it is. People who see exceptions as "aberrations and contaminants" simply "miss what is going on."

Source: Evelyn Fox Keller, *A Feeling for the Organism: The Life and Work of Barbara McClintock* (San Francisco: W. H. Freeman, 1983).

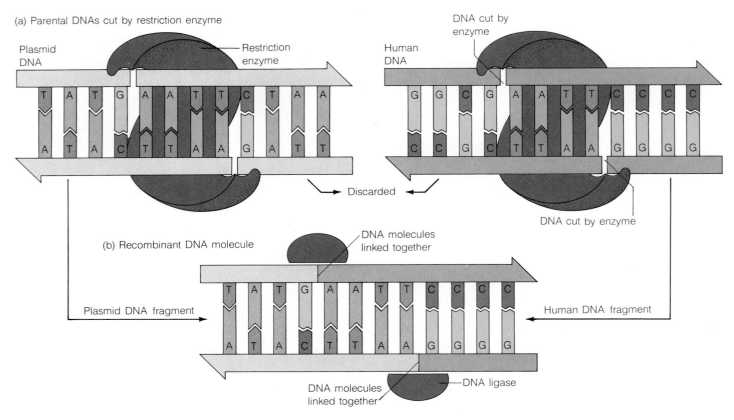

FIGURE 11.3 ■ How Geneticists Create Recombinant DNA Molecules. A biologist makes a recombinant DNA molecule by cutting two different DNAs and then gluing the ends of the two molecules together. (a) *Cutting DNA.* A restriction enzyme recognizes a specific sequence of nucleotides in DNA and cuts the DNA in or near that sequence. Here the enzyme *Eco* RI recognizes GAATTC and cuts after the G in both parental molecules, the human DNA (dark blue) and the plasmid DNA (pale blue). (b) *Gluing DNAs together.* The cut ends of the two parental DNAs are "sticky," since they can form complementary base pairs with any other DNA cut by the same restriction enzyme. Here the AATT from the sticky end on the human DNA pairs with the TTAA from the plasmid sticky end. A DNA repair enzyme called DNA ligase can then form covalent bonds in the DNA's sugar-phosphate backbone and glue the two foreign DNAs together, making a recombinant molecule.

Species in which genes recombine are more adaptable and more likely to persist and evolve than species that cannot undergo recombination. Their variation is nevertheless completely random—undirected and indeterminate. Since recombination in nature can and does occur anywhere along a chromosome, it is purely a matter of chance whether new gene combinations help, hinder, or make no difference to an individual organism.

Not so, however, with artificial recombination in the laboratory. Genetic engineers with their recombinant DNA technology have removed the element of chance. Using sophisticated instruments, specific enzymes, and other materials, they can now create new gene combinations tailored to serve specific human purposes. The rest of the chapter describes the techniques and applications of recombinant DNA.

THE POWER OF RECOMBINANT DNA RESEARCH

Organisms as different as lions and dandelions, pines and porcupines, *E. coli* and broccoli are the result of mutations and recombinations, followed by natural selection for those newly generated phenotypes that favor survival and reproduction. Because mutation and recombination in nature occur randomly, however, the production of new genes, proteins, and phenotypes is extremely slow.

Today, thanks to a revolution in techniques that began in the early 1970s, genetic engineers can design and construct brand-new genes in just a few weeks—genes that might have taken thousands of years to appear naturally in organisms inhabiting streams, oceans, fields, or forests. These revolutionary microtechniques include the capacity to cut DNA molecules in specific places and to paste selected DNA fragments together into a specially designed recombined DNA molecule. Box 11.2 on page 236 describes a standard laboratory technique, gel electrophoresis, and how it can be employed in the gene-splicing process. And Box 11.3 on pages 240–241 explains how a young researcher invented one of genetic engineering's most powerful tools.

How to Construct a Recombinant DNA Molecule

Experimenters use two pivotal tools to control recombination in the test tube, and neither tool is larger than a molecule. The first tool is a **restriction enzyme,** a protein that can cut DNA at specific, well-defined points along the ladder of base pairs. The second tool is **DNA ligase,** a protein that can join DNA strands together, creating a new, artificially recombined DNA molecule. (In nature, DNA ligase repairs damaged DNA; review Figure 10.13.)

■ **Enzymatic Scissors** Researchers obtain restriction enzymes, their scissors for cutting DNA, from bacteria. These microorganisms naturally produce restriction enzymes with which they cut certain viral DNAs into pieces, thus "restricting" the types of viruses that can infect them. A restriction enzyme recognizes a specific base sequence of a few base pairs wherever it might occur in an organism's DNA and cleaves the DNA at a consistent place in or near the same sequence. Some restriction enzymes, because of their shape, cut the DNA in a "staggered" fashion, with one end of the DNA protruding. Other restriction enzymes, however, cut the DNA strands straight across. One restriction enzyme called *Eco* RI, for example, recognizes the sequence GAATTC and cleaves DNA between the G and the A on both strands, as indicated in Figure 11.3a. When the weak hydrogen bonds between strands break, the two DNA fragments can separate, leaving single-stranded ends protruding.

■ **Molecular Glue** The unpaired sequences protruding from a staggered cut—biologists call them "sticky ends"—can easily re-form hydrogen bonds with complementary unpaired sequences on the ends of other DNA molecules. In fact, any two DNA molecules with complementary protruding sequences can join together, no matter how unrelated the rest of the two DNA molecules may be. This tendency for sticky ends to join together through complementary base pairing turns out to be a boon to genetic engineers: It allows them to recombine DNA molecules from organisms as different as people and bacteria. For instance, in Figure 11.3a, the DNA indicated in dark blue could represent the human growth hormone gene cut by *Eco* RI, while the DNA indicated in pale blue could represent a similarly cut bacterial plasmid, a circle of DNA that can replicate in a bacterial cell (see Chapter 9).

Hydrogen-bonded base pairing between the AATT of the human DNA and the TTAA of the plasmid DNA can hold the two fragments together (Figure 11.3b). Once this happens, DNA ligase can make strong covalent bonds between the opposing ends of the two molecules. The covalent bonds join the two fragments at the designated sites.

Restriction enzymes and DNA ligase allow researchers to design and construct recombinant DNA molecules in the test tube. But for the molecules to be useful, many identical copies must be made (a process called cloning), and that's a job for bacterial plasmids.

How to Clone a Human Gene

A bacterial plasmid is the molecular geneticist's beast of burden. These small circles of DNA carry antibiotic resistance genes and can ferry short strands of foreign DNA into bacteria, where they can be replicated. If, for example, a geneticist splices a DNA fragment—say, the gene for human growth hormone—into a plasmid and then inserts that plasmid into a bacterium, then as the bacterium replicates its own larger chromosome, it will also make copies of the plasmid containing the human gene. After harvesting the plasmids (each of which carries a copy of the gene for human growth hormone) from the multiplied bacterial cells, the geneticist isolates the growth hormone genes from the plasmids. (This process is a bit like scribbling a short note in the middle of a typed letter, then making 100 photocopies of the letter: Both the letter and the note get copied.)

Figure 11.4 traces the steps that researchers follow to **clone** a DNA fragment—that is, to select a desired DNA sequence and reproduce many copies of it. First, they cut the circular DNA of plasmid molecules with a restriction enzyme, opening them up and forming two sticky ends on each plasmid. They also cut DNA containing the gene they want to clone with the same restriction enzyme, so that it is cut into thousands or millions of fragments, each of which also has sticky ends (Figure 11.4, step 1). The sticky ends of a plasmid DNA molecule and one of the fragments of human DNA adhere by base pairing, and treatment with DNA ligase completes the circle (step 2). In step 3, the researchers transfer the plasmid—with the hitchhiking human DNA—into a bacterial cell; this transfer of a heritable trait by a piece of DNA is the process of transformation that caused the epidemic we encountered in Chapter 9. Then, in step 4, the researchers culture the bacterial cell (which now contains the plasmid bearing human DNA), and the cell divides and redivides into a clone of cells—a group of identical daughters all descended from the same founder cell. An antibiotic that the researchers add to the bacterial growth medium kills any bacterial cell that does not contain a plasmid, with its antibiotic resistance gene.

Now the researchers face their biggest problem. They started with millions of fragments of human DNA (only a few of which are pictured in step 1 of Figure 11.4). Each fragment

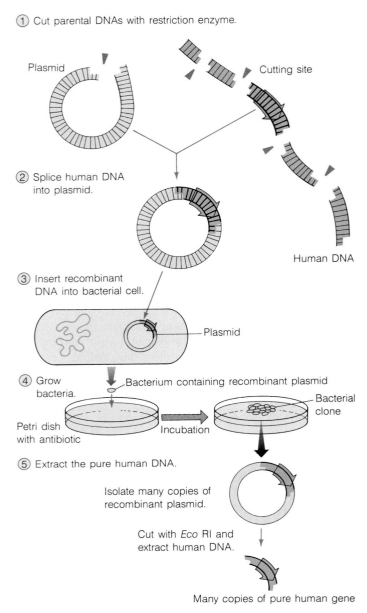

① Cut parental DNAs with restriction enzyme.

Plasmid

Cutting site

② Splice human DNA into plasmid.

Human DNA

③ Insert recombinant DNA into bacterial cell.

Plasmid

④ Grow bacteria.

Bacterium containing recombinant plasmid

Bacterial clone

Petri dish with antibiotic   Incubation

⑤ Extract the pure human DNA.

Isolate many copies of recombinant plasmid.

Cut with *Eco* RI and extract human DNA.

Many copies of pure human gene

FIGURE 11.4 ■ How to Clone a Human Gene. As the text describes, there are five main steps to cloning a human gene, beginning with human DNA, a plasmid, and a restriction enzyme and ending with many copies of the human gene.

was spliced into a different plasmid, and each plasmid entered a different bacterium. These bacterial cells then multiplied into millions of identical-looking colonies (one of which is shown in step 4). Here, then, is the problem: Which one of the colonies contains bacteria bearing the desired gene, and how can the researchers find this "needle in a haystack"?

Biologists identify the bacterial colony containing the gene they wish to isolate by using a *probe,* a marker molecule that binds specifically to the desired gene. A probe for the

human growth hormone gene, for example, might be a radioactive DNA copy of the messenger RNA that encodes growth hormone. Because this radioactive probe can form complementary base pairs only with the growth hormone gene, any bacterial colony that becomes radioactive after exposure to the probe must contain the human growth hormone gene.

After identifying the rare colony with its desired gene, the researchers can isolate and separate the plasmids—still carrying the human DNA—from the cloned bacterial cells and then extract the human gene from the plasmids (Figure 11.4, step 5). At this point, they have cloned a human gene. The steps shown in Figure 11.4 allowed researchers to mass-produce the human growth hormone gene through cloning.

To construct the giant mouse described in the chapter opener, genetic engineers combined the structural gene for human growth hormone with the regulatory region of a mouse gene to make a single fused gene, then inserted it into a plasmid, cloned the fused gene, and injected about a thousand copies of it into fertilized mouse eggs (Figure 11.5). The mouse eggs were allowed to grow in a laboratory dish to early embryo stages, and then researchers implanted them into a foster mother's womb. When the baby mice were born, some had incorporated the artificial gene into their genomes and expressed the gene by synthesizing human growth hormone and growing to double the weight of their littermates. Thus, a plasmid carried the human gene, and one of the mouse's own regulatory genes ensured that the gene was expressed.

Molecular biologists originally devised their techniques for manipulating DNA—cutting and pasting DNA and cloning—as a means of understanding gene structure and function. Researchers gleaned much of the knowledge about eukaryotic gene regulation (see Chapter 10) from such applications of recombinant DNA research. The next section describes how geneticists have employed their artificial genes for other purposes.

PROMISES AND PROBLEMS IN RECOMBINANT DNA RESEARCH

Genetic engineering holds tremendous promise for humankind. Imagine a drug that can kill the viruses causing the common cold, available in great supply and for a few dollars; or an enzyme that can arrest a heart attack already in progress; or bacteria that can make automobile fuel from discarded corn stalks; or other bacteria that can protect crops from late spring freezes; or corn that contains proteins with the nutritional value of beef proteins; or gene replacements to cure people and animals of crippling genetic diseases; or engineered mice that can be used to screen for and identify cancer-causing agents.

But now consider the potentially less favorable side to genetic engineering. While bacteria may someday be engineered to break down oil and then used to disperse oil slicks,

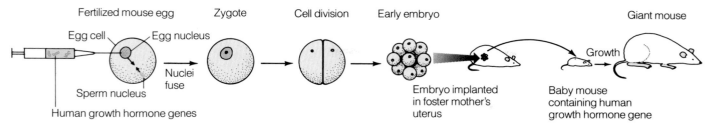

FIGURE 11.5 ■ **Genetically Engineering a Giant Mouse.** Researchers injected thousands of copies of human growth hormone genes into a fertilized mouse egg cell, which grew into an embryo in the laboratory. They then implanted this early-stage embryo into a foster mother and waited for giant mice to show up in the litter of young.

what if these hypothetical bacteria moved beyond oil slicks and began to devour the world's dwindling oil supplies? Or what if it became possible for people to genetically engineer children? Many feel that recombinant DNA research has the potential to make certain dreams come true, but others worry about the social and ecological impacts that might result. Let's look at the promise, the present performance, and the nagging questions that arise from recombinant DNA research.

Reshaping Life: The Promise of Genetic Engineering

People have three ways of exploiting "made-to-order" genes for their own purposes. The first way is to alter bacteria so they make a protein such as human insulin, which can be used to treat life-threatening diabetes. The second way is to alter an organism to make it more useful. For example, we could engineer corn or cattle to grow faster and resist diseases better. The third way is to alter the human genome itself to cure disease.

■ **Mass-Producing Proteins** People have been exploiting microorganisms for thousands of years: Cheese, yogurt, beer, and bread are all manufactured with the aid of microorganisms. Now, through recombinant DNA technology, we can use microorganisms to manufacture virtually any protein we want by inserting into them a gene for the desired protein. By 1990, the Food and Drug Administration (FDA) had approved 11 drugs made with genetic engineering techniques; more than 100 are now being tested or reviewed.

The ten-year-old girl in Figure 11.6 was born unable to produce growth hormone and seemed destined to live out her life as an abnormally small individual. But in just one year of treatment with human growth hormone—mass-produced by genetically engineered bacteria containing the human growth hormone gene—the girl grew about 12.6 cm (5 in.). Using standard cloning techniques, scientists at a biotechnology firm near San Francisco isolated the gene for human growth hormone and then inserted it into bacteria. Grown in an enormous vat, the bacteria then produced commercial quantities

of human growth hormone—enough for treating the thousands of children in the United States who make too little of the hormone to grow to normal height. The precious substance is now available in pharmacies (by prescription), and physicians hope someday to use this synthetic growth hormone to treat burns, slow-healing fractures, and bone loss diseases, in addition to abnormally short stature in children.

Biotechnologists have also succeeded in cloning the gene for growth hormone from cows, producing large amounts in bacterial cells and harvesting the protein for commercial use. When injected into dairy cows, this hormone can increase milk production by 10 to 20 percent, and the U.S. Department

FIGURE 11.6 ■ **A Growth Spurt Thanks to Genetic Engineering.** As the marks on the doorframe show, this child grew several inches during treatment with growth hormone made through genetic engineering techniques.

BOX 11.2   HOW DO WE KNOW?

MANIPULATING DNA MOLECULES BY ELECTROPHORESIS

How does a researcher handle a DNA molecule—an incredibly long, thin, fragile strand of base pairs thousands of times thinner than anything we can see or touch?

One popular technique that molecular geneticists use is **electrophoresis,** the movement of electrically charged molecules in an electric field. Geneticists generally set up the electric field in a gel, a jellylike medium that provides both support and water-filled spaces through which DNA molecules can move. DNA is negatively charged, owing to the many phosphate groups in its backbone. When placed in an electric field, DNA molecules will migrate toward the positive pole of the field, with molecules of different sizes migrating at different rates: Like football players, the small ones move fast, the really large ones move more slowly, and medium-sized ones travel at speeds in between. Researchers capitalize on this difference in migration rate to separate molecules of different sizes from each other, even though they cannot manipulate the molecules directly.

Starting with a solution of DNA molecules in a test tube, a geneticist can cut a long piece of DNA into several smaller fragments of different sizes by adding restriction enzymes (see Figure 11.3). The geneticist can then pour the mixture into a small well cut into a slab of gel and apply an electric field across the slab. After 2 or 3 hours, the different fragments will have migrated different distances, depending on their size, and so will have separated into distinct bands in the gel. The researcher can see the DNA bands by staining them with a material that shines a bright orange-pink in ultraviolet light (Figure 1). The genetic engineer can then carefully slice the desired band from

FIGURE 1 ■ **Electrophoresis. A molecular geneticist loading samples of DNA onto a gel for electrophoretic analysis of the molecules.**

the gel, extract its DNA with a series of chemical solutions, and eventually insert the DNA into a plasmid and then incorporate it into a cell from a mouse, a tomato, or a cow. If the DNA fragment isolated by electrophoresis contains a gene that makes the tomato more nutritious or the cow less fatty, then electrophoresis—with its capacity for revealing and manipulating DNA—will have materially enhanced the quality of our lives.

---

of Agriculture has certified that milk from treated cows is safe to drink. This advance might seem like a boon to the dairy industry, but using a product of recombinant DNA may have a significant economic downside. Agricultural economists have pointed out that farmers already produce a surplus of milk products. Given this surplus, competition from large milk-producing operations, which can more easily afford to use the expensive drug, may drive small family-owned dairies out of business.

Growth hormone is just one type of protein now mass-produced through engineered bacteria. Bacteria have also been fitted with genes for insulin hormone that is chemically identical to the insulin made in the human pancreas. Until recently, diabetics had to rely on insulin obtained from pigs or cows. This animal insulin differs slightly from human insulin and so is less effective at preventing the diabetic from suffering wide fluctuations in blood sugar, fluctuations that can lead to blindness, coma, and death.

Genetically altered cells are also producing a number of other medically useful proteins. These include an enzyme called *tissue plasminogen activator* (*TPA*), which can dissolve the blood clots that block arteries during a heart attack, and a protein called *gamma interferon,* which can protect laboratory-grown human cells from infection by hepatitis virus and

herpes virus. In laboratory tests, interferon can also stimulate the growth of tumor-killing cells. This partial list of early successes only hints at the many useful products that genetically engineered microbes will ultimately provide for us.

■ **Improving Plant and Animal Stocks** Recombinant DNA will accelerate a genetics project farmers began 10,000 years ago: to improve crops and livestock. In the past, breeders selected individuals with desirable traits—traits that arose by mutation and naturally occurring DNA recombination—to be parents of the next generation. Today, geneticists can select livestock such as cows, sheep, or pigs and insert into them specific genes that increase bone and tissue growth or milk production, just as researchers inserted the human growth hormone gene into the mouse. The greatest impact on human lives, however, will be made through advances in the genetic engineering of plants.

Today, only 30 plant species make up nearly 93 percent of the human diet. And our dependence on these 30 species will only increase in the future. Global human population continues to explode, even as agricultural land gives way to encroaching cities and deserts and as energy for producing fertilizers and pesticides becomes increasingly scarce and expensive. For decades, botanists have been breeding better crop plants to help meet the growing worldwide demand for food. With the current capacity to insert genes directly into individual plants and animals, they can begin to circumvent these time-consuming breeding programs and use genetic engineering techniques to improve the efficiency of photosynthesis; decrease crop dependence on fertilizers; improve the nutritional quality of seeds, grains, and vegetables; and increase plant resistance to pests, salt, drought, and extreme temperatures.

For example, researchers have introduced a gene into tomato plants that renders them resistant to some plant viruses. The protective gene, which they took from one partic-ular virus, prevents this and certain other disease-causing viruses from infecting plant cells. Normal tomato plants become stunted when infected with disease viruses, but the *transgenic* plants (plants carrying a foreign gene in their chromosomes) resist viral infection and grow luxuriantly (Figure 11.7). Researchers are also using gene transfer technology to design new varieties of oranges, cassava, wheat, and rice that need far fewer chemicals to resist insects and disease-causing fungi and that use nutrients more efficiently. These genetic improvements should lower the farmer's costs in producing the food, as well as the environmental pollution due to pesticides and fertilizers. Note, however, that while increased food

FIGURE 11.7 ■ **Tomatoes Engineered for Disease Resistance. The tomato plants on the left are transgenic organisms. Molecular geneticists transferred a gene from a plant virus into the tomato's chromosomes. The plant cells now make a harmless viral protein, and if a disease tries to infect the plant cell, it is blocked by the preexisting viral protein. The plants on the right are normal garden-variety tomatoes. Researchers exposed both plant rows to a disease virus. The transgenic plants produced as many pounds of the luscious red fruit as uninfected tomato plants, while nonengineered controls that were infected produced only 30 percent of the normal yield.**

production may help to alleviate hunger in the short term, it does not solve the world's largest and longest-term problem—human overpopulation.

A natural question arises from achievements in genetics research: If beneficial genes can be inserted into plants, lab animals, and livestock, then why not into people?

■ **Human Gene Therapy** Gene therapy, altering a person's genes to combat disease, is a field still in its infancy. In theory, gene therapy could be applied in either of two ways: Genes could be inserted into the somatic (body) cells or into the germ cells (the cells that give rise to the sperm and eggs). The first procedure, *somatic cell gene therapy,* is the straightforward treatment of an individual's disease—a simple extension of current medical practice. But the second kind, *germ line gene therapy,* would also affect the genetic makeup of the treated individual's offspring. Each type of therapy has important implications.

In late 1990, physicians carried out the first instance of somatic gene therapy—the introduction of genetically engineered cells into a human being to cure disease. In this case, the patient was terminally ill with melanoma, a particularly rapidly growing skin cancer, and had failed to respond to conventional cancer therapies. Steven Rosenberg of the National Cancer Institute removed a portion of one patient's tumor and extracted from it a type of white blood cell called tumor-infiltrating lymphocytes, or TILs. These cells naturally home in on a tumor and crawl into the mass of tumor cells. Previous experiments had shown that specially treated forms of these cells can slow tumor growth (Figure 11.8).

Rosenberg inserted into the patient's own TIL cells a gene coding for tumor necrosis factor, a protein that kills cancer cells in mice. He then returned the genetically engineered white blood cells to the patient's body. Rosenberg hopes that the altered TIL cells will move to the tumor, invade it, and there produce the anticancer protein. If they make enough of

(a) Before treatment

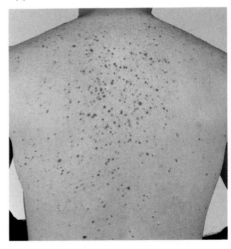

(b) After treatment

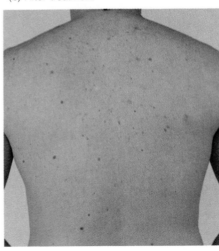

FIGURE 11.8 ■ **Using Gene Therapy to Fight Cancer.** Experiments using nonrecombinant DNA methods have implicated certain white blood cells in tumor regression. (a) A male patient with dozens of potentially lethal melanoma skin cancers on his back. (b) White blood cells called tumor-infiltrating lymphocytes (TILs) were isolated from the patient and treated so that they increased their production of a certain anticancer protein. When put back into the patient, the TILs caused many of the melanomas to regress. (c) In the first trial of gene therapy, physicians removed TIL cells from a different group of patients, each with an incurable skin cancer (step 1). Next, medical geneticists inserted into the TIL cells the gene for a potent tumor-killing protein called tumor necrosis factor (TNF) (step 2). When the engineered TIL cells were reintroduced into the patients, they homed in on the tumors and invaded them (step 3). The physicians hope that the production of TNF by the genetically engineered cells will destroy the skin cancer cells (step 4) and allow patients to survive.

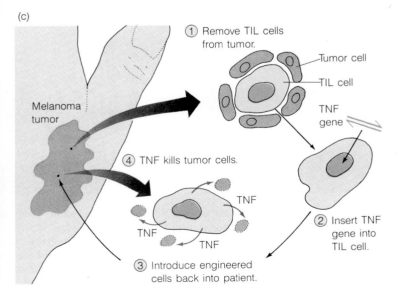

this tumor necrosis factor, it should destroy the mass and cure the patient. The results of the therapy were not known at the time this book went to press, but there is legitimate optimism about the prospects.

Since the gene therapy in this case altered white blood cells (which are somatic, or body, cells) but not sex cells, the altered genes would not be passed on to the patient's children. For most people, this sort of gene therapy presents no new ethical problems: As in conventional surgery or drug therapy, a single individual is treated, he or she benefits, and the defective genes remain to be perpetuated in the human population.

Gene therapy of germ line cells is another matter, since it constitutes a totally new approach to medicine and raises complex issues of safety and ethics. In this type of gene therapy, recombinant DNA would be inserted into human sex cells via techniques similar to those used to make the giant mouse (see Figure 11.1). Not only would the treated individual be affected, but so would all the individual's descendants. While researchers have had some success with techniques for this kind of gene transfer in mice and sheep, they must still solve many technical and ethical problems before treating human sex cells. One problem is that the success rate is low— only 6 successes out of 300 injected eggs in a typical experiment with mice. Another problem is that the inserted DNA sometimes causes a new mutation, possibly creating a new defect as serious as the disease the therapy was meant to cure. Finally, the inserted normal gene usually does not replace the native defective gene, but is simply added to the genome, so that the defective gene can still be inherited by some of the children of the treated individual.

At present, germ line therapy seems out of the question for purely technical reasons; however, it will someday be possible and will pose significant long-term ethical questions. Since its application would alter the genetic makeup of children not yet born, it could change the course of human evolution. Would a treatment that causes a heritable change infringe on the rights of future generations? Are the expected benefits of germ line gene therapy worth the unknown risks? Could a problem be solved in more traditional ways, whose risks are understood and whose benefits are clear? These and other questions have led scientists and lay observers to debate—at times heatedly—the potential misuse, accidental or deliberate, of recombinant DNA techniques.

Recombinant DNA and Environmental Risks

Some critics have suggested that introducing organisms altered by recombinant DNA methodologies into the environment poses significant risks. Professional ecologists (scientists who study the interactions of organisms with each other and with their physical surroundings) agree that some dangers do exist. A dangerous new weed could conceivably be created if, say, a new variety of rice engineered with genes for salt tolerance escaped from cultivated fields and invaded the brackish water in the mouths of rivers. Or an old pest might

FIGURE 11.9 ■ **Unexpected Environmental Consequences and the Introduction of Novel Organisms.** The kudzu vine (which is not genetically engineered) was brought to the American South to control erosion. But the hardy plant competed with native vegetation and is now destroying some forests. The Ecological Society wants regulatory agencies to carefully evaluate genetically engineered organisms to be certain their biological properties will inflict minimal harm on the environment.

become worse if, for example, weeds interbred with a genetically engineered domestic crop and gained a gene for herbicide resistance. One can imagine yet another scenario in which the genes encoding enzymes that can degrade lignin, a major constituent of wood, became more widespread, thus altering the cycling of substances in an ecosystem.

There have, of course, been many ecological disruptions with natural, unengineered organisms; a prime example is the devastating growth of kudzu, a fast-growing, large-leaved Japanese vine that was introduced into the southeastern states to control soil erosion but has ended up choking tens of thousands of acres of native and commercial forests (Figure 11.9). The Ecological Society of America, the professional

BOX 11.3 BIOLOGY: A HUMAN ENDEAVOR

KARY MULLIS AND COPIES OF COPIES OF COPIES

As Kary Mullis tells it, he invented one of the most powerful tools of modern molecular biology while driving down a dark country road one night in April of 1983. Mullis had been working for the Cetus Corporation in Emeryville, California, synthesizing nucleic acid probes with which to identify specific base sequences in DNA. If a researcher wants to study that sequence further, he or she usually needs to generate millions or billions more copies. In the early 1980s, it could take weeks of lab work to amass such a stockpile—still no more than a small test tube full of DNA.

Working with probes and the subsequent tortuous copying procedures was not only difficult but somewhat repetitious and tedious, and it left Mullis lots of time to brainstorm easier approaches to manipulating DNA. That spring night in 1983, he was musing about new ways to create, as it were, a molecular photocopy machine that could churn out unlimited copies of a desired piece of DNA. By the time Mullis reached his cabin near Mendocino, he had solved the problem in his head, and by the following Monday, had tested it in the lab. The result: polymerase chain reaction, or PCR.

The technique is so fundamentally simple, writes Mullis, that other DNA workers wonder, "Why didn't I think of that?" Starting with a long double-stranded DNA molecule containing a short sequence he wished to copy and study, he heated the long molecule to separate the strands. Next he added "primers," short pieces of DNA that can be designed to "bracket" the desired DNA sequence, attaching to either end of the piece on one strand as well as its complementary piece on the other separated strand (Figure 1). Mullis then added DNA polymerase, the enzyme needed to synthesize new chains of DNA bases, and waited for it to generate a copy of each bracketed DNA segment. As long as (1) heat is supplied at the right time to separate the DNA strands, (2) primers are present to bracket the

desired piece of DNA, and (3) the polymerase enzyme is available to copy the designated segment, cycle after cycle of replication can take place with the copied strands serving as templates for the next round of copying. In this chain reaction, 2 copies become 4, 4 become 8, then 16, 32, 64, 128, and so on, up to a million copies in just 20 relatively rapid cycles.

Mullis had to make some adjustments to the process, including using DNA polymerase from a hot springs bacterium that can withstand the near-boiling strand-separating heat jolt needed for each duplication cycle. Eventually, he was able to automate the entire process and enclose it in a machine housing. He had invented a DNA copying machine that has since become standard equipment in most DNA laboratories.

Like a paper copier, a PCR machine has innumerable uses, some of them ingenious and colorful. For example, by employing PCR, crime investigators can take DNA fingerprints (see Box 12.3 on page 260) from the cells in a tiny speck of dried blood or at the base of a single human hair and search for a criminal's identity. With conventional techniques, the sleuth needs 1000 times more blood (a sizable stain) or 1000 strands of hair (a yanked-out handful!) to collect enough DNA to study.

PCR is also allowing physicians to diagnose genetic defects in very early embryos by removing the DNA from a few sloughed-off or collected cells and amplifying it. They are also using PCR to detect the presence or absence of AIDS virus in the infants of mothers with the immune system disease. And finally, biologists can now study the tiny strands of DNA they retrieve from ancient specimens—Egyptian mummies, woolly mammoths, even human brains preserved in old bogs—by copying the genetic material with a PCR machine.

Kary Mullis's spring evening brainstorm not only made him a wealthy man, but also enriched the study of biology immeasurably.

---

society of scientists studying the environment, agrees that the modern methodologies of recombinant DNA and genetic engineering can produce genetically novel organisms that can benefit agriculture, waste management, and detoxification of chemicals. In light of this, while the society does recognize the risk in introducing any nonnative species, it lent its support in 1989 to the development of genetically modified organisms, as long as those products of biotechnology are carefully evaluated by regulatory agencies before being released into the environment. They contend that any organism proposed for environmental release should be evaluated according to its biological properties, and not simply according to whether developers used recombinant genetics or traditional breeding techniques.

While ecologists have been concerned with the environmental effects of genetically novel organisms, others have voiced concern about the consequences for human health, as well as the ethics of gene manipulation.

Recombinant DNA: Novel Problems of Safety and Ethics

Questions about both the safety and morality of recombinant DNA research have cast a shadow over its promise of a better future for the human race. What, for example, would happen if bacteria transformed with a cancer-causing gene escaped from a laboratory and infected people? Would an epidemic of cancer decimate the population? This possibility so horrified scientists in the mid-1970s that molecular biologists agreed among themselves to stop research on recombinant DNA for several months as a small group of researchers conducted a potentially dangerous experiment—a "worst-case scenario."

In a laboratory specially designed to contain dangerous microorganisms, they spliced cancer-causing genes into the DNA of bacterial plasmids and then exposed mice to the potentially cancer-causing bacteria. As it turned out, the mice developed no more tumors than did unexposed mice. Evidently, the bacteria's newly spliced cancer genes were unable

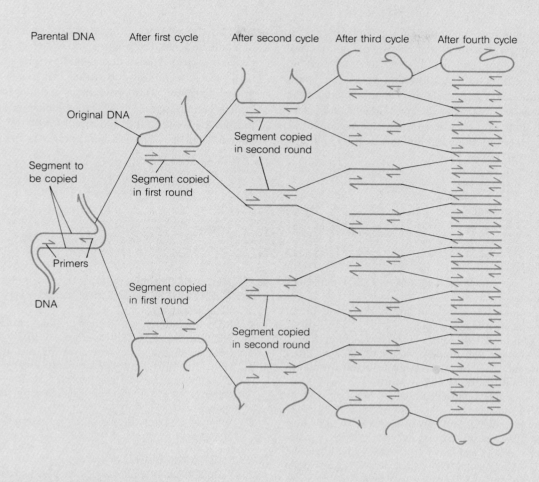

Parental DNA   After first cycle   After second cycle   After third cycle   After fourth cycle

Original DNA

Segment to be copied

Segment copied in first round

Segment copied in second round

Primers

DNA

Segment copied in first round

Segment copied in second round

FIGURE 1 ■ Polymerase Chain Reaction.

to infect the mouse cells and cause cancer. The public's worst fears were allayed, and the National Institutes of Health allowed recombinant DNA researchers to resume their work under strict guidelines designed to prevent the escape of recombinant organisms into the outside world.

Many other questions remain to be resolved. Most people would agree that inserting a normal gene into mutant somatic cells containing a faulty gene is merely an extension of current medical practices and ethics. But significant disagreements arise over using gene therapy to "improve" an already healthy child. For example, prospective parents might ask physicians to insert additional growth hormone genes into their newly conceived embryo so they could produce a taller basketball player or a beefier football player than might otherwise be produced. Other parents may want a child with greater intelligence or a more winning personality. From a scientific standpoint, such applications of recombinant DNA research now seem only a remote possibility. Untold numbers

of genes control each of these complex traits in unknown ways. Geneticists may never be able to discover and clone all the relevant genes and introduce them into a person with any degree of control. However, society should consider these issues before biotechnologists achieve such breakthroughs. And if history repeats itself, the potential for human gene therapy will arrive more quickly than we expect.

Only through free and open discussion will potential problems be solved and appropriate guidelines be devised to ensure that recombinant DNA technology can help produce higher-yield crops, better livestock, and healthier humans while minimizing risk. "Sunshine," as U.S. Supreme Court Justice William O. Douglas once said, "is the best disinfectant." With proper safeguards developed in an open forum, recombinant DNA technology may one day help alleviate some types of human suffering, as did anesthesiology and antibiotics, and improve agriculture to make food more available for our planet's burgeoning population.

CONNECTIONS

The last few chapters have taken us on a genetic journey. It started with the simple statement that genes exist and control the way organisms look and act. It continued through an exploration of gene structure and how that structure enables genes to determine an organism's traits. And it included the discovery that geneticists can isolate individual genes, alter them, and insert them into the chromosomes of organisms as different as bacteria, tomatoes, and humans, and by so doing change the organisms' traits.

This chapter concentrated on how genes recombine, both in nature and in the laboratory, and what role such recombined genes may play. It also described how biologists are already using DNA technology to make drugs such as insulin and growth hormone, to improve crops and livestock, and to cure human hereditary diseases. The next chapter ends the genetic journey by looking into the rapidly growing field of human genetics and the progress being made in detecting and preventing genetic defects.

Highlights in Review

1. During genetic recombination, two different DNA molecules may break and rejoin.
2. Genetic recombination is a natural process that occurs in all sexually reproducing organisms.
3. New combinations of alleles can create novel phenotypes.
4. Although mutation is the ultimate origin of all new DNA sequences, genetic recombination reshuffles those mutations, which maximizes the expression of variation.
5. Recombination occurs naturally only between homologous regions on DNA molecules.
6. Biologists can recombine the DNA of any species in the test tube by using enzymes. Restriction enzymes cut DNA molecules at specific places and leave protruding ends. When the ends of two DNA molecules have paired, DNA ligase enzyme can be used to link the two molecules together.
7. Foreign genes can be inserted into bacterial, plant, or animal genomes. Even in this strange new environment, the foreign gene can often be expressed to make its characteristic protein.
8. Genes designed and made by molecular geneticists can be used to cure human disease and to improve crop species.
9. Society must confront the questions of safety and ethics that recombinant DNA technology poses.

Key Terms

clone, 233
DNA ligase, 233
electrophoresis, 236
recombinant DNA
   technology, 228

restriction enzyme, 233
transposon, 231

Study Questions

Review What You Have Learned
1. Explain how geneticists used the human gene for growth hormone to produce a giant mouse.
2. If a fruit fly with scarlet eyes mates with a fruit fly with rosy

eyes, could they produce offspring with pink eyes? If so, how?
3. True or false: Recombination occurs in very few organisms. Explain your answer.
4. Both mutation and recombination produce new DNA sequences. How do the two processes differ?
5. Compare the actions of restriction enzymes and DNA ligases.
6. Give specific examples of the improvements biologists could bring about in animals and plants by inserting beneficial genes.
7. Describe how somatic cell gene therapy may help combat cancer.
8. Why is research on recombinant DNA subject to strict guidelines?

Apply What You Have Learned
1. An old superstition holds that each of us has a double somewhere in the world. How would you answer such a claim?
2. Tissue plasminogen activator (TPA) is an enzyme that dissolves blood clots. Physicians now use it to stop a heart attack caused by a clot blocking a blood vessel of the heart. From principles you learned in this chapter, outline how you would design and manufacture TPA using recombinant DNA technology.
3. A circus owner hires you as a consultant on a project to develop new animals for the circus. He proposes to use recombinant DNA technology to create a Pegasus (a winged horse) and a Minotaur (half man, half bull). Is his project likely to be successful? Why or why not?

For Further Reading

BASKIN, Y. "DNA Unlimited." *Discover,* July 1990, pp. 77–79.

FRIEDMANN, T. "Progress Toward Human Gene Therapy." *Science* 244 (1989): 1275–1281.

GASSER, C. S., and R. T. FRALEY. "Genetically Engineering Plants for Crop Improvement." *Science* 244 (1989): 1293–1299.

JOYCE, C. "U.S. Approves Trials with Gene Therapy." *New Scientist,* August 11, 1990, p. 19.

MULLIS, K. B. "The Unusual Origin of the Polymerase Chain Reaction." *Scientific American,* April 1990, pp. 56–65.

SCHMECK, H. M., JR. "New Test That Finds Hidden AIDS Virus Is a Sleuth with Value in Many Fields." *New York Times,* June 21, 1988, pp. B11–B12.

SCHNEIDERMAN, H. A., and W. D. CARPENTER. "Planetary Patriotism: Sustainable Agriculture for the Future." *Environmental Science and Technology* 24 (1990): 466–473.

STAHL, F. W. "Genetic Recombination." *Scientific American,* February 1987, pp. 90–96.

TIEDJE, J. M., R. K. COLWELL, Y. L. GROSSMAN, R. E. HODSON, R. E. LENSKI, R. N. MACK, and P. J. REGAL. "The Planned Introduction of Genetically Engineered Organisms: Ecological Considerations and Recommendations." *Ecology* 70 (1989): 298–315.

WEATHERALL, D. J. "Gene Therapy in Perspective." *Nature* 349 (1991): 275–276.

WICKELGREN, I. "Please Pass the Genes." *Science News* 136 (1989): 120–124.

Human Genetics

Diet Soft Drinks and Human Genetics

Have you noticed warning labels on cans of diet soda that read "Phenylketonurics: Contains phenylalanine" and wondered what they mean? The answer is that the amino acid phenylalanine can cause mental retardation in the tiny percentage of young children who are born with a certain genetic defect. Fortunately, advances in the field of human genetics—the subject of this chapter—allow physicians to detect this condition even before a child is born and prevent its devastating effects (Figure 12.1).

Most children (and adults) can digest the low-calorie sweetener called aspartame now used to flavor many diet soft drinks and other foods. The discovery of aspartame was hailed as a major advance because unlike some earlier sweeteners, it is nontoxic and quite safe in moderate doses; the substance occurs naturally (at very low levels) in grapefruits; aspartame is intensely sweet yet low in calories and thus can help prevent obesity; and it contains large amounts of the amino acid phenylalanine, a component of virtually all proteins. Most people have a dominant allele of a gene that encodes the enzyme phenylalanine hydroxylase (PAH), which breaks down phenylalanine by converting it to the amino acid tyrosine. Unfortunately, a small percentage of people lack the PAH enzyme. These people inherit two recessive mutant alleles of the PAH gene and as a result do not produce the enzyme. They are called *phenylketonurics,* and their disease is called *phenylketonuria,* or *PKU,* because any phenylalanine they ingest in foods builds up in their bodies and is transformed not to tyrosine but to compounds known as phenylketones (Figure 12.2). These mousy-smelling substances are excreted, giving urine a telltale odor.

Odd-smelling urine is of little medical importance in itself, but the production of phenylketones carries an additional, more serious consequence: Phenylketones cause nerve cells in the brain to develop abnormally. If a child born with the disease PKU drinks mother's milk, diet soda, or any other food containing phenylalanine, his or her brain development will be stunted, seizures will start about the age of six months, and mental age rarely advances beyond age two.

In the past, 1 out of every 100 mentally retarded children suffered from PKU. But today, thanks to the techniques of modern human genetics, the recessive alleles that lead to PKU

FIGURE 12.1 ■ Genetic Diseases Like PKU Can Be Controlled. In this brother-sister pair from the Midwest, the disease phenylketonuria is under control thanks to early diagnosis and carefully controlled diet.

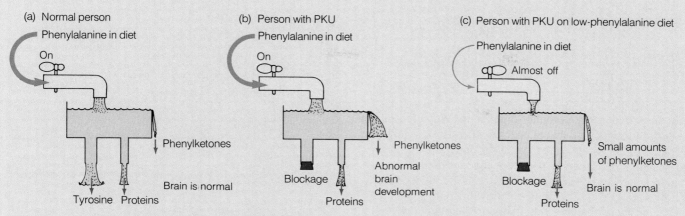

(a) Normal person

(b) Person with PKU

(c) Person with PKU on low-phenylalanine diet

FIGURE 12.2 ■ PKU and Metabolism. We can think about a person's phenylalanine metabolism as a flow-through process, during which the amino acid pours in from a dietary tap and flows out to other pathways. (a) In a healthy person, most of the phenylalanine in foods is converted to tyrosine by an enzyme, less is incorporated into proteins, and a tiny amount is converted to waste products called phenylketones. (b) A PKU patient lacks the enzyme that normally converts phenylalanine to tyrosine, so phenylalanine accumulates and is changed to large quantities of phenylketones, which build up and prevent normal brain development. (c) If a PKU patient severely reduces the dietary intake of phenylalanine, then most phenylalanine goes into proteins, so few phenylketones reach the brain, and the child can grow and live normally.

can be detected before the child is born; then, after birth, the parents can closely control the child's diet to exclude phenylalanine and thus prevent mental retardation.

PKU is just one of about 100 genetic conditions that physicians can now detect in the unborn fetus. And the production of the enzyme PAH is just one of 3000 human traits that geneticists have studied using both modern and traditional tools. People are like pea plants, fruit flies, *E. coli* bacteria, and all of a geneticist's other laboratory favorites in a fundamental way: The genes in our 23 pairs of chromosomes determine our physical traits, from hair, eye, and skin color to height and build and even to the shape of our earlobes, the width of our fingernails, and the length of our toes.

Despite the fact that genes determine physical traits, humans are uniquely difficult genetic subjects for several reasons: First, humans don't choose mates and produce offspring to satisfy a geneticist's curiosity; the investigator must search for existing subjects and matings that happen to express traits of interest. Second, there is almost never a true F_2 generation available for study because brothers and sisters rarely mate. Third, a given couple seldom produces more than ten children, and these days, usually fewer than three. Therefore, sample populations are small, making statistical analysis tricky. Finally, it could take more than half of a geneticist's career to follow the traits in even a single human generation, not to mention the many generations needed for substantive analysis. For these and other reasons, human geneticists have had to rely on traditional tools like family histories and karyotypes (photographs of chromosomes arranged in order of decreasing size) as well as to forge new tools like cell fusions and gene-mapping techniques to probe the human genome.

As we move through the chapter, we'll encounter the theme that people are unique genetic subjects again and again. And we'll also discover a second theme: that gene mapping and other modern techniques applicable to human genetics grew out of research on more classic genetic subjects, such as fruit flies and bacteria. Some of the same molecular-level research that brought about the recombinant DNA revolution is leading us into an era of dramatic changes and advances in human genetics. The field of human genetics can therefore be seen as an application of pure research done on other organisms. Finally, we'll see a third, rather sad theme emerging: The small genetic changes that alter a person's physiology, such as the recessive alleles that lead to PKU, usually have a negative impact on physical functioning. Our bodies are machines finely tuned by evolution; removing a random genetic "bolt" or loosening a "screw" is likely to impair, not improve, the machine's smooth functioning.

Our discussion of human genetics—how people inherit family resemblances, metabolic defects, and other characteristics—deals with a number of questions:

■ What standard genetic techniques do researchers use to study the inheritance of human traits?
■ How have special molecular techniques allowed scientists to map human genes?
■ How are scientists and physicians currently treating human genetic diseases?
■ How are geneticists helping to prevent new cases of genetic disease by identifying adult carriers, detecting defective genes in the unborn, and counseling prospective parents?

HOW TRADITIONAL TECHNIQUES REVEAL THOUSANDS OF HUMAN GENETIC CONDITIONS

People are wonderfully complex organisms: We come in a nearly infinite variety of shapes, sizes, and colors and have a huge range of individual traits. Many of our traits are *polygenic,* meaning that many genes work together to determine individual characteristics such as how well our livers function; how well we can see distances; how long our limb bones are; what our facial features look like; how intelligent we are; whether our skin is clear or dotted by acne; whether our feet are narrow or wide; and whether we are susceptible to various types of cancer, diabetes, allergies, heart attacks, or early senility (Alzheimer's disease).

(a) Attached earlobe

(b) Unattached earlobe

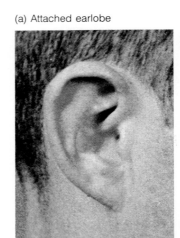

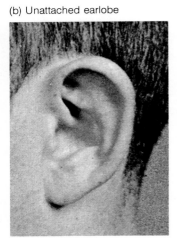

(c) Ear pit

(d) Hairy ear

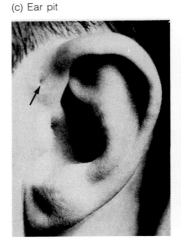

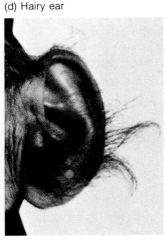

FIGURE 12.3 ■ All Ears: Some Human Genetic Traits Caused by Single Genes. The alleles of a single gene determine whether your earlobes are attached (a) or unattached (b), whether or not you will have ear pits (c), and whether or not tufts of hair will grow on your ears (d).

Pedigree of a family with PKU

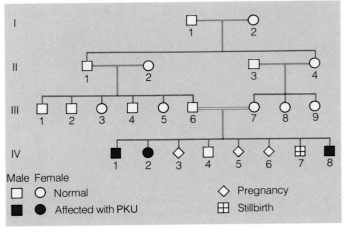

FIGURE 12.4 ■ Pedigree of a Family with PKU. In a typical pedigree, a single horizontal line connecting a male and a female represents a mating; a double line indicates a mating between relatives (*consanguinous* mating). A horizontal line above a series of symbols designates all the siblings of one family arranged in birth order from left to right. Vertical lines connect parents to siblings. Generations are labeled with roman numerals, and each individual in the pedigree is numbered. In this family, children with PKU (IV1, IV2, and IV8) had parents that were first cousins and phenotypically normal (III6 and III7). Both parents were heterozygous for a recessive PKU allele inherited from I1 or I2; this would explain why some offspring were homozygous for the disease.

At the same time, geneticists have discovered more than 3000 human conditions determined in simple Mendelian fashion by single genes. These include such traits as hair on the middle finger joint or elbow; attached or unattached earlobes (Figure 12.3a and b); widow's peak; ear pits (Figure 12.3c); extra fingers or toes (*polydactyly*); baldness; hemophilia; color blindness; counterclockwise cowlicks in the hair; dry, brittle ear wax; albino skin and hair coloring; thalassemia (see Chapter 10); and PKU. Most of these traits do not require medical treatment, and some, like baldness and color blindness, cannot be treated at this time. Others, however, like PKU, can be corrected if detected early enough. To study and treat a genetic condition, a geneticist must know whether it is caused by a single gene or by many genes or perhaps by an alteration in chromosome shape or number. To determine these things, human geneticists have borrowed two traditional tools from others in the field of genetics: the analysis of family histories and the study of chromosome variations. Let's look at each tool and the kinds of conditions it can uncover.

Human Pedigrees: Analysis of Family Genetic Histories

To follow the inheritance of a trait through all the members of an extended family, a geneticist gathers and studies medical records from all the family members he or she can locate.

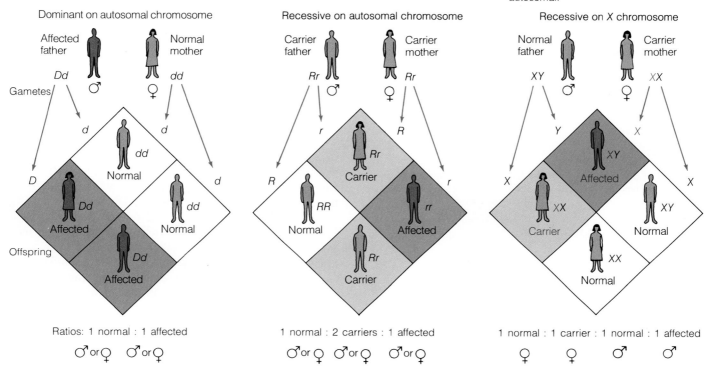

(a) *Is a trait dominant or recessive?*

1. If a child has a rare trait and one parent is affected, the trait is usually dominant.

2. If a child has a rare trait but neither parent is affected, the trait is usually recessive.

(b) *Is a trait X-linked or autosomal?*

3. If a son inherits the trait from his father, it cannot be X-linked and must be autosomal.

4. If all or nearly all the affected people are males, the trait is probably X-linked.

5. If both males and females are affected about equally, the trait is probably autosomal.

FIGURE 12.5 ■ Human Inheritance of Traits Determined by a Single Gene. (a) With dominant genes on an autosomal chromosome, all affected offspring have at least one affected parent. With one normal and one affected parent, half the offspring will be affected, half will be normal (1). For recessive genes on autosomal chromosomes, affected children usually have two unaffected heterozygous parents, and the ratio of offspring will be one normal to two carriers (normal phenotype) to one affected (2). (b) With recessive genes on an X chromosome, males are affected more often than females, and inheritance is from mother to son, not father to son. With a normal father and a carrier mother, half the daughters will be normal and half will be carriers, while half the sons will be normal and half will be affected.

From such records, the investigator then draws up **pedigrees,** orderly diagrams of a family's relevant genetic features. Figure 12.4 diagrams a family with PKU expressed in certain members. A pedigree shows the family relationships, sex, genotype, and phenotype of each member. With it, a geneticist can determine whether a given trait is dominant or recessive and whether a gene lies on a sex chromosome or on an autosome (a chromosome other than a sex chromosome). In human genetics, a trait on the X chromosome is referred to as *X-linked* (see Chapter 8). The human Y chromosome may contain only genes for sexual phenotype.

A pedigree can look rather formidable, with its maternal aunts, paternal grandparents, mothers, fathers, and siblings all marching through genetic history in tidy geometric formations (see Figure 12.4). Nevertheless, the rules for analyzing a pedigree are fairly simple, as Figure 12.5 shows.

Using Mendel's principles, geneticists can tell whether a trait is based on single-gene inheritance, whether it is due to dominant or recessive alleles, and whether the gene resides on autosomal or sex chromosomes. Table 12.1 on page 248 lists a number of human traits that are inherited in simple Mendelian fashion and that are discussed in this chapter. Here we discuss three of those traits showing different patterns of Mendelian inheritance.

TABLE 12.1   Some Human Genetic Conditions

Disease	Effect
Recessive Allele on Autosomal Chromosome	
Albinism (chromosome 11)	Missing enzyme; unpigmented skin, hair, and eyes
Cystic fibrosis (chromosome 7)	Defective membrane protein; excessive mucus production; digestive and respiratory failure
Phenylketonuria (PKU) (chromosome 12)	Missing enzyme; mental deficiency
Severe combined immunodeficiency (chromosome 20)	Missing enzyme; no immune response
Sickle-cell anemia (chromosome 11)	Abnormal hemoglobin; sickle-shaped red cells; anemia; blocked circulation
Tay-Sachs disease (chromosome 15)	Missing enzyme; buildup of fatty deposit in brain; no mental development
Thalassemia (chromosome 16 or 11)	Abnormal hemoglobin; anemia; bone and spleen enlargement
Recessive Allele on X Chromosome	
Duchenne muscular dystrophy	Missing membrane protein; muscle cells die
Green weakness (color blindness)	Abnormal green-sensitive light-absorbing pigment in eyes
Fragile X syndrome	Easily broken X chromosome; mental retardation
Hemophilia	Lack of a clotting factor; uncontrolled bleeding
Dominant Allele on Autosomal Chromosome	
Hypercholesterolemia (chromosome 5)	Missing protein that takes cholesterol from the blood; heart attack by age 50
Huntington's disease (chromosome 4)	Progressive mental and neurological damage
Conditions Due to Variations in Chromosome Number	
Down syndrome (an extra chromosome 21)	Mental retardation; heart abnormalities
Klinefelter syndrome (XXY)	Defect in sexual differentiation
Turner syndrome (XO)	Webbed neck; sterility

■ **PKU: A Recessive Allele on an Autosomal Chromosome** Most genetic defects are caused by *autosomal recessive* alleles, and PKU is a classic example. The pedigree in Figure 12.4 shows that two brothers and one sister in the same family inherited PKU but that neither parent showed the trait. Since neither parent showed the PKU trait, it must be recessive (see Figure 12.5). Also, since the affected girl's father is not himself affected, it cannot be X-linked and so must be on an autosomal chromosome. Finally, since the trait is recessive, the affected children must be homozygotes; and since the parents pass on the gene but do not show its effects, they must be heterozygotes, or **carriers,** with their dominant normal allele masking their recessive mutant one. The genetic condition called **albinism,** in which the homozygote's skin cells and hair follicles fail to make the dark pigment melanin, is another autosomal recessive condition.

■ **Duchenne Muscular Dystrophy: A Recessive Allele on the X Chromosome** The most common X-linked genetic disease is Duchenne muscular dystrophy (DMD), a degenerative muscle condition that strikes 1 out of every 3500 boys and results from inheriting a recessive allele. The DMD gene is the largest yet discovered, having over 2 million base pairs. This immense size probably explains why this gene mutates more frequently than perhaps any other human gene, with a new mutation occurring about once every 20,000 births. The muscle cells of an affected boy die slowly, beginning in the legs, and by age five, the child with DMD cannot stand up easily and he rises in a characteristic way (Figure 12.6a). At about 20 years of age, the muscles of his diaphragm degenerate, and he can no longer breathe.

Research based on recombinant DNA techniques has shown that the normal allele of the DMD gene encodes a protein that is located in the plasma membrane of muscle cells. In 1990, experimenters showed that this protein is somehow involved in regulating the entry of calcium ions into muscle cells. DMD patients lack the protein, and as a consequence, high levels of calcium ions enter muscle cells and cause them to degenerate. Based on this finding, physicians are planning to test drugs that might slow the entry of calcium into muscle cells and perhaps slow or prevent muscle degeneration.

On another front in the fight against DMD, physicians took immature muscle cells from normal donors and injected them into the feet of three boys with the degenerative disease. These normal muscle cells fused with the patient's own defective muscle fibers, and the strength in their feet muscles increased. Members of the Muscular Dystrophy Association hope that through drug therapy, muscle cell injections, and other future treatments will emerge for this condition.

Since Duchenne muscular dystrophy lies on the X chromosome, it is usually passed to boys by their mothers, who are heterozygous carriers (they transmit the trait but don't show it themselves). Another X-linked recessive allele causes hemophilia, or bleeder's disease, in 1 out of every 5000 male births. Box 12.1 on page 251 describes this disease and its inheritance in more detail.

(a) How a boy with DMD rises   (b) Pedigree of DMD, an *X*-linked gene

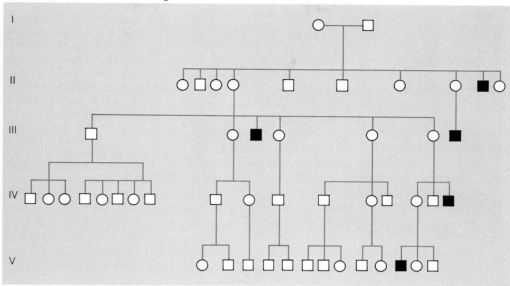

FIGURE 12.6 ■ Duchenne Muscular Dystrophy, Caused by a Recessive Allele on the *X* Chromosome. (a) To rise from the floor, boys with this disease must "climb up themselves," as this five-year-old is doing, because their lower limbs grow weak before their upper bodies do. (b) A pedigree of a family with Duchenne muscular dystrophy shows the characteristic inheritance of an *X*-linked gene: Nearly all the affected people are male, and a son never inherits the condition from his father. To simplify the pedigree, the geneticist has not drawn in unaffected spouses.

■ **Huntington's Disease: The Result of an Autosomal Dominant Allele**   In contrast to PKU and DMD, which appear in infants or young children, Huntington's disease does not manifest itself until ages 35 to 50. Researchers do not understand the physical basis for the disease or its timing, but they have chronicled a set of distressing symptoms in Huntington's patients that include progressive degeneration and death of the nerve cells, irregular and jerky movements, intellectual deterioration, and often severe depression. There is as yet no effective treatment for Huntington's disease, and the patient dies a sad and protracted death. Recent research, however, shows that nerve growth factor, a protein that stimulates cell division in nerve cells (see Chapter 7 as well as Figure 13.20), prevents the kind of brain cell deterioration usually seen in patients with Huntington's disease. Researchers are hoping that this nerve growth factor might be used effectively as a treatment for Huntington's.

Pedigrees have revealed that Huntington's disease is inherited as a dominant allele on an autosomal chromosome; thus, the allele passes to an afflicted child directly from an afflicted parent (Figure 12.7). Whether afflicted or not, all the victim's children suffer years of anxiety, because they have a 50/50 chance of inheriting the disease themselves and worry that they could be passing it on to their own children. Until recently, they had to wait until middle age—usually after they had reproduced—to learn their fate. In the mid-1980s, how-

ever, molecular geneticists developed a way to detect the gene earlier in the life of a potential victim (a person whose parent had Huntington's)—if, that is, they choose to know. Box 12.2 on page 258 explains their dilemma.

Chromosome Variations Reveal Other Genetic Conditions

While pedigrees provide the human geneticist's first tool, *karyotyping*—the preparation and study of chromosomes using special dyes and labeling techniques—enables geneticists to probe conditions caused not by single-gene mutations, but rather by changes in chromosome number or structure.

■ **Effects of Chromosome Number and Activation on Phenotype**   As we saw in Chapter 7, people normally have 46 chromosomes: 22 pairs of autosomes and 2 sex chromosomes, *XX* in females and *XY* in males (Figure 12.8). If a chromosome set deviates from that pattern, development is generally abnormal. Recall that three copies of autosomal chromosome 21 causes Down syndrome (see Box 7.3 on page 157). People can also inherit too many or too few sex chromosomes, with dramatic consequences. A person with one *X* and no *Y* chromosome (*XO*) is a sterile female with **Turner syndrome.** This is characterized by folds of skin along the neck, a low hairline at the nape of the neck, a shield-shaped chest, and failure to develop adult sexual characteristics at puberty.

(a) Woody Guthrie

(b) Arlo Guthrie

(c) Pedigree of a family with Huntington's disease

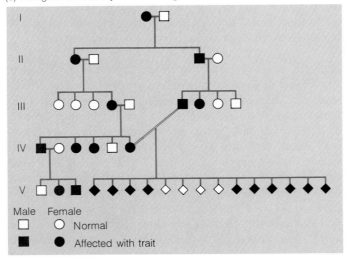

Male   Female
☐   ○   Normal

■   ●   Affected with trait

FIGURE 12.7 ■ Huntington's Disease, Caused by a Dominant Allele on an Autosomal Chromosome. (a) Woody Guthrie, singer and writer of "This Land Is Your Land," "So Long, It's Been Good to Know Ya," and other popular songs, died of Huntington's disease. (b) His son, Arlo Guthrie, a popular singer in the 1970s ("You Can Get Anything You Want at Alice's Restaurant" and "The Train They Call the City of New Orleans") had a 50/50 chance of inheriting the condition. (c) This pedigree of a family with Huntington's disease was collected by Nancy Wexler from a family in Venezuela in 1987 (see Box 12.2 on page 258). Note that a quick look at the pedigree shows that every person with the trait has at least one parent with the trait. The 14 children of the marriage between cousins are shown as diamonds, without an indication of sex or birth order for confidentiality.

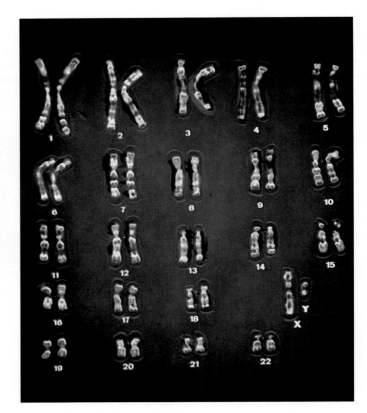

FIGURE 12.8 ■ The Human Genome: A Karyotype. To view human chromosomes, geneticists remove white blood cells, stain and photograph their nuclei, then cut the chromosomes from the photo with scissors and arrange them in pairs in decreasing size, as shown here. Geneticists have been able to localize many genetic abnormalities to specific chromosomes. Examples include Huntington's disease, chromosome 4; cystic fibrosis, chromosome 7; interferon deficiency, 9; albinism, sickle-cell anemia, β-thalassemia, 11; phenylketonuria, 12; Prader-Willi syndrome, Tay-Sachs disease, 15; α-thalassemia, 16; severe combined immunodeficiency, 20; Alzheimer's disease, 21; color blindness, Duchene muscular dystrophy, hemophilia, X; testis determining factor, Y.

BOX 12.1   P E R S O N A L   I M P A C T

THE ROYAL HEMOPHILIA

Viewed through the eyes of a geneticist, a single mutant allele in one *X* chromosome of a small Russian boy may have helped to change the history of the world.

The story begins with Queen Victoria of England, who lived from 1819 to 1901 and reigned for 63 years. A mutation in either Victoria's mother's egg or her father's sperm led to the creation of an allele for hemophilia, or "bleeder's disease," a disorder in which a person cannot produce a protein needed to bring about rapid blood clotting. Victoria inherited one mutant allele and one normal allele and was thus a heterozygous carrier for hemophilia who did not herself show the disease. She passed the mutant allele on to two of her five daughters and one of her four sons. Through intermarriage, the mutant hemophilia allele was subsequently spread throughout the nobility of Europe.

Victoria's granddaughter Alexandra (a carrier, like her grandmother) married Nicholas, later crowned czar of Russia (Figure 1). Their son and heir to the throne, Alexis, was a hemophiliac who suffered serious hemorrhaging from even small bruises that would go unnoticed in a normal child. One of Alexandra's ladies-in-waiting recalled a poignant time in 1914 when Alexis's nose began to bleed so relentlessly that the last hour of the unhappy child seemed to be at hand. Desperate, Alexandra turned to Rasputin, a self-styled religious mystic who seemed able to put her son Alexis into a trance. The lack of movement may have helped the hemorrhages to heal, and on this occasion, as on others, Rasputin seemed to help the boy recover. Alexandra was convinced that Rasputin had wrought a miracle, and through her, the mystic gained influence over Czar Nicholas. History shows that the czar's own actions during this period helped precipitate the Bolshevik revolution and ultimately the royal family's brutal murder.

FIGURE 1 ■ A Family Disease. Queen Victoria's granddaughter Alexandra, her husband Czar Nicholas of Russia, and their son Alexis, who suffered from hemophilia.

The man who became prime minister after the czar's overthrow stated, "If there had been no Rasputin, there would have been no Lenin." A geneticist might add, "Had there been no mutant allele in the royal families of Europe, beginning with Queen Victoria, there would probably have been no need to call in Rasputin in the first place." [R. K. Massie, *Nicholas and Alexandra* (New York: Dell, 1967), pp. 529–530.]

About one newborn male in a thousand has two *X* chromosomes and one *Y* chromosome (*XXY*), a condition called **Klinefelter syndrome.** These people develop as sterile males with small testes (sperm-producing organs), long legs and arms, and somewhat diminished verbal skills, although their IQ scores are normal. Most men with Klinefelter syndrome manage well socially and economically, and many are unaware of their chromosomal abnormality until they marry and are unable to father a child. If identified early, a Klinefelter male can be treated with male hormones in early adolescence to enhance his eventual sexual performance, prevent osteoporosis, and improve his mood and general sense of wellbeing.

Many boys who inherit one *X* and two *Y* chromosomes (*XYY*) grow to be over 6 feet tall and have below-average intelligence. Some geneticists and lawyers have suggested that an extra *Y* chromosome causes innate antisocial tendencies and point to the fact that 2 percent of men in prison have the pattern, compared with only 0.1 percent of men in the general population. This argument has been presented as a legal defense in a few cases. However, others point out that 96 percent of *XYY* males lead normal lives and that perhaps society treats tall but mentally slow boys in a way that promotes antisocial behavior.

Curiously, only changes in the number of sex chromosomes and in the smallest autosomal chromosome (chromosome 21) show up in relatively healthy human infants. Embryos that receive too many or too few of the other autosomal chromosomes have such disturbed development due to gene imbalance that they die in the womb or soon after birth. A child with trisomy 21 (Down syndrome) survives because presumably few genes on that autosome are sensitive to disruption by extra copies in the cell.

Extra *X* chromosomes are tolerated because a special mechanism makes only one *X* chromosome genetically active in each cell. Two weeks after fertilization, when a human female embryo consists of only about 500 to 1000 cells, one of the *X* chromosomes in each cell becomes genetically inactive. At this stage, the embryo consists of two groups of cells: One group eventually gives rise to the baby's body cells, while the other develops into parts of the placenta, the spongy organ through which nourishment passes from the mother to the developing young. In the placenta-forming cells of a female embryo, the father's X chromosome is alway inactivated, but in the baby-forming cells, either the mother's X chromosome or the father's X may become inactivated. Once an embryonic cell inactivates an X, however, all of the thousands or millions of daughter cells derived from it will have the same inactive X.

Since a female embryo inherits one X from her mother and one X from her father, some cells inactivate the mother's X and other cells inactivate the father's X. Thus, a female mammal is a mosaic of cells containing maternal or paternal active X chromosomes. You can see this mosaicism in the patches of black and orange fur in a calico cat (Figure 12.9a), as well as in women with a certain skin condition (Figure 12.9b).

Women have similar mosaic patterns for various traits in each of their organs. A woman whose father carries an X-linked allele for color blindness, for example, will have patches of eye cells that see color normally and patches that are color-blind, depending on which X chromosome is inactivated in each group of cells.

■ Exchanges of Chromosome Parts While extra chromosomes can cause conditions such as Down and Klinefelter syndromes, even a less drastic alteration of chromosome pattern can have serious consequences. Researchers have found that over 30 types of cancer are caused by broken chromosomes, or **chromosome translocations**—instances in which a part of one chromosome moves spontaneously to a new location on a different chromosome. In one such translocation common among Central Africans, the long arms of chromosomes 8 and 14 break in specific places, and the free ends exchange places (Figure 12.10). If the 8/14 translocation occurs in one type of white blood cell at any time during a person's life, it can lead to a cancer called Burkitt's lymphoma. This cancer can grow into an enormous jaw tumor within a matter of days. Fortunately, treatments with anticancer drugs can rapidly shrink the tumor and lead to long-term survival in about half the cases.

When a translocation occurs between chromosomes 8 and 14, a cancer gene, or **oncogene,** called *myc* is relocated. In its normal spot on chromosome 8, *myc* encodes normal amounts of a protein that regulates cell growth. But in its new position on chromosome 14, it is influenced by its new neighbor, a gene rapidly transcribed in normal white blood cells. This influence causes *myc* itself to become rapidly transcribed, somehow causing the white blood cell to grow and reproduce out of control. This proliferation then leads to cancer. Survivors of Burkitt's lymphoma can usually reproduce without fear of passing on the chromosome translocation to their offspring, since it normally occurs in a somatic (body) cell rather than a germ line (egg or sperm) cell.

■ Fragile X Chromosome and Mental Retardation In the mid-1980s, researchers discovered a chromosome alteration that is second only to Down syndrome (trisomy 21) as a genetic

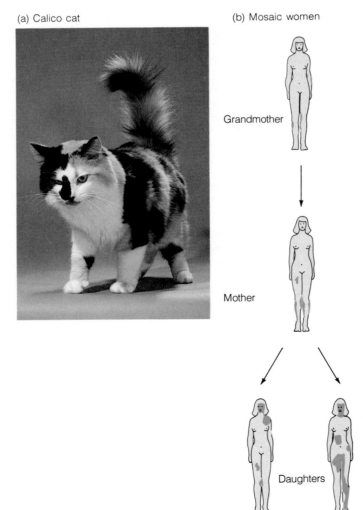

(a) Calico cat   (b) Mosaic women

Grandmother

Mother

Daughters

FIGURE 12.9 ■ Calico Cats and Mosaic Women. Mammals have evolved a mechanism that makes only one X chromosome genetically active in each cell. Since a female embryo inherits one X from her mother and one X from her father, some cells inactivate the mother's X and other cells inactivate the father's X. Thus, a female mammal is a mosaic of cells containing maternal or paternal active X chromosomes. You can see this mosaicism in a calico cat's tricolored coat, with its patches of black and orange fur (a). Women and girls are mosaics, too. This was revealed by studying women who are heterozygous for an X-linked gene called anhidrotic dysplasia. These women have a mutant allele that prevents the development of sweat glands; thus, heterozygous females have some patches of skin that lack sweat glands and other patches that produce sweat glands normally (b). Different skin patterns occur even in identical twins and depend on the random activation and inactivation of X chromosomes that took place in the embryo. While women with anhidrotic dysplasia reveal mosaic patterns from X chromosome inactivation in a very dramatic way, all women have similar mosaic patterns for various traits in each of their organs.

(a) Translocation between chromosomes 8 and 14

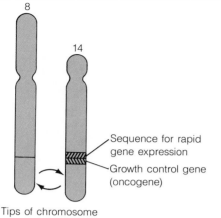

Sequence for rapid gene expression

Growth control gene (oncogene)

Tips of chromosome exchange

(b) A child with Burkitt's lymphoma

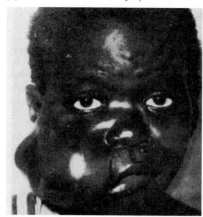

FIGURE 12.10 ■ Chromosome Transloca-tion Can Turn On a Cancer Gene. (a) In a reciprocal chromosome translocation, the tips of two chromosomes can exchange places. If the tip of chromosome 8 and the tip of chromosome 14 exchange in a white blood cell precursor, a growth control gene (an oncogene) from chromosome 8 comes to lie near a DNA sequence that specifies a high level of gene expression. This can cause the growth control gene to be expressed improperly, and the white blood cells divide without ceasing, causing a tumor called Burkitt's lymphoma. (b) A jaw tumor like this child's is a common symp-tom of Burkitt's, although such a tumor can also be triggered by a viral infection.

cause of mental retardation. The condition is called **fragile X syndrome,** and a person with this condition has X chromo-somes whose tips break off easily at a specific place (Figure 12.11). People with fragile X syndrome often have large, pro-truding ears and long faces, and as babies are hyperactive and make eye contact poorly. So far, researchers haven't deter-mined the link between a broken X chromosome tip and brain changes, but this syndrome underscores the importance of regular chromosome structure to normal embryonic development.

Translocations and broken chromosomes rearrange and delete genes. And significantly, a gene's specific location along a chromosome is crucial to its functioning, because adjacent DNA sequences often serve to turn neighboring genes on and off at appropriate times. Because a gene's loca-tion is so important, geneticists have tried to map as many human genes as possible and assign them to precise locations along the 23 human chromosomes. To do this, the researchers have applied new methods first developed on nonhuman sub-jects, and these methods are allowing them to detect and treat

(a) Men with fragile X syndrome.

(b) A broken X chromosome

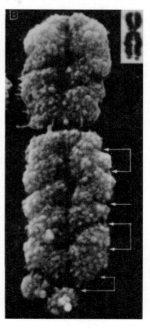

FIGURE 12.11 ■ Fragile X Chromosome and Mental Retardation. (a) Three brothers who have fragile X syndrome each display the elongated face, jutting chin, squarish forehead, and large ears that characterize the condition. (b) The lower tip has broken off the right chromatid of this fragile X chromosome from a mentally impaired male.

a much broader range of serious human genetic diseases than ever before.

THE NEW GENETICS: A REVOLUTION IN THE MAPPING OF HUMAN GENES

In 1911, geneticists noticed the striking form of color blindness called *green weakness,* a condition in which males inherit a visual defect that prevents them from seeing greens but allows them to see other colors normally (Figure 12.12). By studying pedigrees and finding that a man passes color blindness through his daughters to his grandsons, geneticists were able to assign green color blindness to the human *X* chromosome. The traditional tools of pedigrees and karyotyping, however, are so imprecise when applied to our 23 pairs of chromosomes that another half century elapsed before geneticists could use these techniques to assign the first gene to a specific autosomal chromosome.

Since 1970, thanks largely to two new techniques, geneticists have traced more than 700 human genes to specific places on sex and autosomal chromosomes. One of the meth-

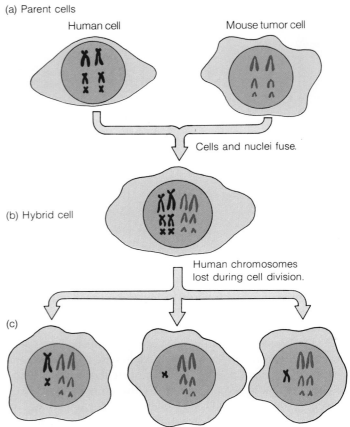

Each cell line has a different subset of human chromosomes.

FIGURE 12.13 ■ Mating Body Cells to Map Human Genes. (a) Geneticists bring together a human cell and a mouse tumor cell and cause them to fuse together using special treatments. (b) The fused hybrid cell contains one giant nucleus with complete sets of both human and mouse chromosomes. (c) Over several cell generations, the hybrids lose human, but not mouse, chromosomes. Since human chromosomes are lost by chance, different cell lines come to contain different human chromosomes. Any human trait that is retained by a hybrid cell must be contained on one of the human chromosomes that remain in the cell, allowing the trait to be mapped to a specific chromosome.

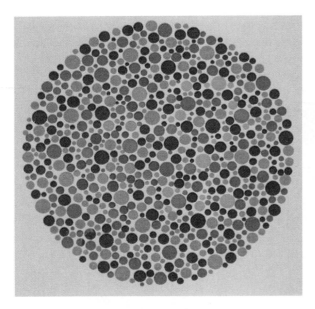

FIGURE 12.12 ■ Color Blindness Test: What Number Do You See? The first human gene that geneticists ever located was the gene for green weakness, which lies on the *X* chromosome and causes a partial color blindness. About 5 percent of males of northern European descent have this type of color blindness. The gene causes a decrease in the amount of a light-sensitive pigment in the eyes, and affected people see reds as reddish brown, bright greens as tan, and olive greens as brown. If you look at this plate and see only the number 4, you are green-blind. If you see only the number 2, you are red-blind. If you see the number 42, you have normal color vision. If the 2 is not very clear, then you are an intermediate called "green-weak." And if the 4 is not clear, then you are a "red-weak" intermediate.

ods uses fused somatic cells, and the other uses DNA fragments called RFLPs. Both would have amazed Gregor Mendel, and they have been called the "New Genetics." Let's look at these exciting new tools and see how the gene mapping they make possible is revolutionizing the study and treatment of genetic diseases.

Bypassing Sex: Somatic Cell Genetics

Geneticists can mate any two flies or pea plants they choose, but they do not conduct controlled mating of people. Nevertheless, researchers have learned how to "mate" normal human body cells in a way that bypasses sex and yet gives information for mapping genes. In the 1960s, researchers discovered a way to combine a human cell with a mouse tumor cell and treat them so that they fuse into a single cell with two nuclei (Figure 12.13). These nuclei subsequently merge (in a process analogous to fertilization) into a giant nucleus with 46 chromosomes from the human and 40 from the mouse. This fused cell with its huge nucleus is called a **somatic cell hybrid**.

As the hybrid cell divides, mistakes in mitosis apportion chromosomes into the new daughter cells incorrectly. Gradually, for obscure reasons, human but not mouse chromosomes are lost in a process genetically analogous to meiosis. Eventually, a researcher can establish a bank of cell lines, all derived from the original hybrid and each containing 40 mouse chromosomes but a different small subset of human chromosomes. A cell bank like this of somatic cell hybrids can be used to map human genes to a specific chromosome.

Let's say a researcher is interested in localizing the gene for PAH, the enzyme PKU victims cannot produce. If all hybrid cells that contain a complete chromosome 12 produce PAH, while other hybrids that lack chromosome 12 lack the ability to make PAH, then the PAH gene must lie on chromosome 12. Comparisons like this have revealed the locations of more than 700 genes since 1970, including those that encode the proteins collagen, insulin, growth hormone, and interferon and those whose mutant alleles lead to hemophilia and Tay-Sachs disease, a condition involving severe mental retardation, visual abnormalities, and death in early childhood. This number is expected to rise even more dramatically, however, with the advent of a new procedure for probing DNA: RFLPs.

Mapping Variations in DNA Structure

After a geneticist discovers a new mutation and determines whether it is dominant or recessive, autosomal or *X*-linked, the next thing he or she wants to know is where it occurs on a particular chromosome. Locating a gene on a map of the human genome (see Box 9.1 on page 195) is like locating yourself on a road map. When driving through New England with its many towns and villages, it is easy to find your position precisely on a road map; you just identify your location relative to the nearest town. It is much harder to do that in a state like Nevada, however, with so few towns and such great distances between them. Until the mid-1980s, human geneticists were "touring Nevada": They had located very few *genetic markers,* or identifiable sites such as genes for particular diseases or for individual blood types, as reference points for placing new genes on the map.

About that time, however, geneticists discovered something in human DNA that they had already observed in bacterial and *Drosophila* DNA: While the order of bases is extremely similar in DNA from corresponding places on homologous chromosomes—say, the identical region on a chromosome 13 from two different people—these sequences will occasionally differ in a single base pair. Since restriction enzymes recognize and cleave DNA molecules at specific base sequences (review Figure 11.3), a specific enzyme may

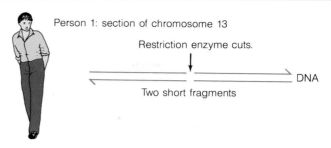

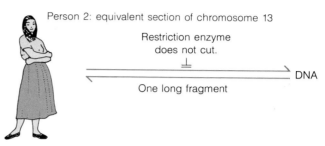

FIGURE 12.14 ■ The Principle of RFLPs. RFLPs (restriction fragment length polymorphisms) get their name in the following way: Enzymes called *restriction* enzymes cleave long strands of DNA into *fragments* by cutting at specific sites in the DNA's base sequence. Because of variations in DNA sequence, different people's fragments are of different sizes, and geneticists call these size differences *length polymorphisms*. RFLPs are due to mutations in the DNA sites that are cut by restriction enzymes. A restriction enzyme would cut DNA from person 1 into two fragments, but a mutation in the cutting site of the DNA from person 2 could cause the enzyme to no longer cut at that DNA position. Since a cutting site has been removed, the DNA remains joined in one very long fragment. These different fragment sizes can be used to help detect or predict diseases caused by mutated genes that contain cutting sites or that lie near cutting sites.

cut one person's DNA at this site, but not another person's (Figure 12.14). Cutting two corresponding (homologous) DNAs at different places is a bit like cutting two foot-long strings at different places: You get fragments of different lengths. These fragments are inherited according to Mendel's rules and can be used as genetic markers. Such DNA variations are called *restriction fragment length polymorphisms,* or **RFLPs** (pronounced "riflips"), and they are quite frequent in human populations. RFLPs greatly expanded the number of genetic markers for establishing reference points ("familiar towns") on the human gene map.

The more precise mapping made possible by using RFLPs has been significant in two ways. First, it allows a genetic disease like PKU to be located on the map near a particular RFLP. Since a researcher can identify a RFLP in the tiny bit of DNA obtained from just a few cells, he or she can tell with some certainty whether a particular family member is at risk for PKU. With this fragment system, even unborn fetuses can be identified as either carriers of the gene or potential victims of the disease. (A technique similar to RFLPs has also been used to take DNA "fingerprints" and to solve crimes and cases of disputed paternity; see Box 12.3 on page 260.)

A well-defined genetic map has a second significance: It helps a molecular geneticist to clone a desired gene—that is, to use recombinant DNA methods to isolate, purify, make copies of, and analyze a gene for a given disease, let's say. By knowing exactly where the gene lies on a particular chromosome, the researcher can isolate it more easily in a test tube.

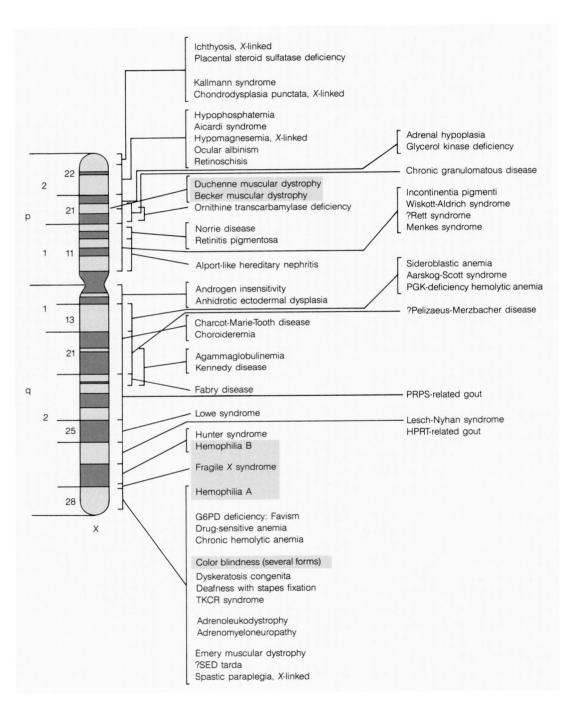

FIGURE 12.15 ■ Anatomy of the Human *X* Chromosome. This map of a human *X* chromosome shows the locations of several genes that code for genetic diseases. A combination of experimental techniques—somatic cell genetics, banding pattern, recombination mapping, and RFLP markers—has led to the precise location of genes coding for Duchenne muscular dystrophy, hemophilia, fragile *X* syndrome, and other conditions.

Once he or she analyzes the isolated disease-causing gene, medical researchers can devise new treatments to help victims of the disease. Recall that with muscular dystrophy, isolating and analyzing the gene allowed researchers to determine the structure and function of the protein (a membrane protein associated with calcium transport in muscle cells) and to design possible new therapies (drugs to correct the calcium transport and muscle cell injections).

The RFLP technique is extraordinarily powerful, and hopes are high that many disease-causing genes can now be mapped, probed, and predicted with far greater accuracy and speed. A map of the X chromosome pinpointing the location of various human diseases is shown in Figure 12.15. The RFLP technique has already been applied to PKU, hemophilia, Duchenne muscular dystrophy, dyslexia (confusion of left and right), cystic fibrosis, Lesch-Nyhan syndrome (a movement disorder that includes uncontrollable arm and leg spasms), a certain type of adult-onset diabetes, and other genetic conditions. By 1990, in fact, researchers had assigned 1800 functional genes and more than 2000 RFLPs to exact positions on the human gene map. And each week, they are adding about a dozen more. As Box 9.1 on page 195 explained, the location of all of our 50,000 to 100,000 human genes may be found by the year 2005.

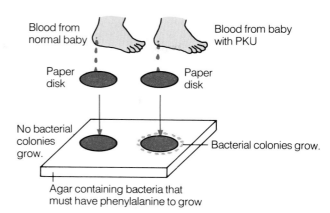

FIGURE 12.16 ■ Test for Detecting PKU in Newborns. A technician collects a tiny drop of blood from the newborn's heel, drops it onto a disk of filter paper, and places the disk on the surface of a block of agar containing bacteria that must have high levels of phenylalanine to grow. No colonies grow around a disk containing a normal baby's blood because of its low amount of phenylalanine; but because the blood of a baby with PKU has 20 times more phenylalanine than normal, bacterial colonies will grow in a halo around the disk.

HELPING PEOPLE WITH GENETIC DISEASES

Gene mapping and sophisticated tools of the New Genetics are allowing modern physicians to do something doctors only dreamed of a few decades ago: to understand the entire pathway of a genetic disease—from mutant genes to absent or defective proteins to physiological effects on the patient—and to intervene successfully so that the patient can lead a nearly normal life. To help, the doctor must first identify the victim of a genetic disease, diagnose the condition, and then, armed with detailed knowledge about the disease, treat it at one of the three levels at which a mutant gene causes its damage: physiological repercussions (the effect on the whole patient), mutant gene product (the protein that is absent or produced improperly), or the altered gene itself. Physicians treat different genetic diseases at different levels, depending on which is most effective or, in some cases, which is currently available to them.

Identifying Victims of Genetic Disease

Genetic diseases manifest themselves at various points in the life cycle, and time of onset is important, since detection must precede treatment. Several genetic conditions appear only in

adulthood: Huntington's disease of middle age is an obvious one, but the list also includes breast cancer and adult-onset diabetes based on inherited tendencies, as well as less harmful traits, such as acne or hairy ears, which show up only after puberty.

The majority of genetic traits, however, become noticeable in early childhood. As the body develops, the child begins to walk and talk, and physical or behavioral abnormalities begin to show. Duchenne muscular dystrophy is an example, as are certain types of mental retardation.

Finally, some genetic traits are detectable at birth. Albinism, for example, or extra fingers and toes (polydactyly) are easily observed. And certain less obvious conditions, such as PKU, can be detected in the hospital nursery. (If you were born after 1962, you were probably tested for PKU.) Blood from a baby with PKU, taken from the infant's heel, has 20 times the normal amount of phenylalanine and can be identified as described in Figure 12.16. Early detection is crucial in a condition like PKU, where brain damage accumulates with each ingestion of more phenylalanine. This simple blood test has spared thousands of children from mental retardation.

The earliest possible diagnosis for other conditions—whether their onset is at birth, during childhood, or in adulthood—is critical in enabling the physician to prevent as much damage to the body as possible from the effects of the defective gene. This is true whether the effects manifest themselves as cancer, muscle or nerve deterioration, uncontrolled bleeding, or diabetes. The next step, of course, is treatment.

THE RISKS OF GENETIC SCREENING

In the early 1970s, it became common practice for American blacks to undergo genetic screening tests for sickle-cell anemia. While this enabled physicians to detect and treat many previously overlooked cases of the blood disease (see Chapter 9), it also brought with it a subtle social cost: The tests revealed carriers (heterozygotes with no disease symptoms) as well as those with full-blown anemia. This gave patients new information for seeking treatment and making reproductive decisions, but on the basis of the test results, some employers and insurance companies also denied people jobs and cheap premiums.

The recent revolution in genetic techniques has given us a newfound ability to map human genes, to find markers for specific diseases, and to screen unborn fetuses, children, and adults for various genetic conditions. While there is great enthusiasm in many quarters over the potential these techniques present for studying the genetic roots of many diseases, there is also considerable caution.

People at risk for Huntington's disease, for example, can now be tested to learn whether they carry the dominant harmful allele (and will inevitably develop the degenerative condition). Yet many of them have little desire to know their fate in advance. For example, Nancy Wexler (Figure 1), a prominent Huntington's researcher, is herself the daughter of a woman who died from Huntington's and has a 50 percent chance of showing the disease. She prefers, however, to live life without knowing her genetic status so that, as she explains it, she doesn't have to worry that the disease is coming on "every time I knock over a cup of tea."

Drawing parallels with sickle-cell anemia, some bioethicists fear that insurance companies and employers will require in-depth genetic screening for new customers or job seekers and may reject anyone with a gene that merely raises the probability of someday developing heart disease, cancer, or other conditions. Activists like attorney Jeremy Rifkin worry further that industrial firms will select for (even someday engineer people for) the ability to withstand occupational exposures rather than cleaning up their factories. And some social and religious groups are concerned that detailed fetal screening will result in more abortions, since at the present, most genetic defects that can be detected cannot be corrected or, in some cases, even treated effectively.

Some observers of the revolution in genetic screening have proposed the creation of new panels, laws, and rules. These, they say, should govern and probably restrict the speed and extent to which insurance companies, employers, and individuals can make use of detailed genetic information. Some geneticists, however, feel that existing bodies, such as the Recombinant DNA Advisory Committee of the National Institutes of Health, are already sufficient to meet the ethical and legal concerns posed by the rapid expansion of genetic screening. People on all sides of the issue seem to agree that widespread public awareness of the issues is essential to prevent abuses.

FIGURE 1 ■ **Charting a Genetic Killer. Nancy Wexler studies the extensive pedigree of a family with Huntington's disease.**

Physiological Therapy

A mutant gene can adversely affect the patient's entire physical health, and thus, many treatments intervene at the level of physiology. Once a PKU sufferer is identified, for example, the doctor prescribes a diet that prevents the buildup of phenylketones and hence protects the infant's brain. The diet is a monotonous but nutritious mush of dried milk powder (treated to remove phenylalanine), cornstarch, fresh fruits and vegetables (except beans and peas, which are high in phenylalanine), and butter. The baby and its parents walk a dietary tightrope, because the infant's caloric intake must be constantly high enough to prevent its body from breaking down its own proteins and thereby releasing phenylalanine from within. This high-calorie diet must be initiated within the first two or three months of life if the baby with PKU is to develop nearly normal intelligence. If treatment is delayed past six months, it is ineffective at preventing brain damage. The child must avoid phenylalanine completely until age six and to a lesser degree throughout life; hence the warning notice on cans of diet soft drinks that contain phenylalanine.

Other physiological treatments for genetic disease include dark sunglasses and chemical sunscreens for albinos and the surgical removal of extra fingers and toes from polydactyl children.

Protein Therapy

Some genetic diseases can be treated by giving the patient the gene product—the protein—they cannot make. In hemophilia, for example, the gene mutation blocks the synthesis of a clotting protein; thus, the blood fails to clot normally. However, after receiving blood transfusions that contain this clotting factor, a hemophiliac is able to form blood clots and is relatively safe from bleeding to death. A pituitary dwarf remains short because he or she lacks the gene for the protein growth hormone. Injections of that hormone will stimulate growth and allow the abnormally short child to partially "catch up" with his or her classmates. Likewise, some diabetics lack the gene for normal insulin production but can respond to injections of insulin. Protein therapies such as these can work as long as the needed protein is carried by the blood, but unfortunately, that is true for only a few diseases. Most proteins must be positioned in a specific organ—for example, the PAH enzyme in cell membranes in the liver—and so injecting PAH into a PKU victim serves no purpose. Ultimately, gene therapy may be the answer.

Gene Therapy

Treating a genetic disease at the physiological or protein level may help the patient's symptoms, but it leaves the defective gene unchanged. With gene therapy, the physician could sub-stitute a normal gene for the defective one; this would encode the normal protein, thereby allowing the person's body to function in a healthy way.

Physicians carried out the first instance of gene therapy in late 1990 to cure a four-year-old girl of an inherited condition called severe combined immunodeficiency disease (SCID). Patients with SCID cannot make a specific enzyme called adenosine deaminase (ADA) needed by white blood cells to mount an effective immune response and thus protect the body from bacteria, viruses, and the infections they cause. Children who lack the ADA enzyme usually contract and die from pneumonia or influenza and can survive for extended periods only if raised in a sterile chamber, fed sterilized food, and protected from physical contact with everyone—even their parents.

To help the little girl with SCID, W. French Anderson and colleagues from the National Heart, Lung, and Blood Institute inserted a normal copy of the gene for ADA into a certain type of disease-fighting white blood cell. Just 4 hours after governmental regulatory agencies had approved their petition to perform the first gene therapy, Anderson and co-workers were transfusing the genetically engineered cells slowly into the girl's blood as she sat watching grayish liquid drain from a plastic bag through a tube in her left hand (Figure 12.17). In

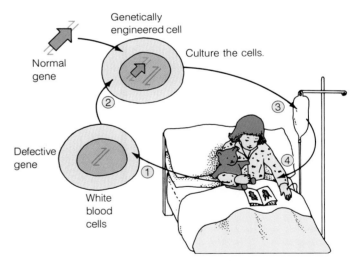

FIGURE 12.17 ■ **Immune Deficiency and the First Trial of Human Gene Therapy.** This sketch of a four-year-old little girl traces the steps of the first attempt to use gene therapy to treat an inherited disease. The child cannot fight off infections because her white blood cells lack a certain enzyme. Researchers at one of the National Institutes of Health in Bethesda, Maryland, took special white blood cells from the little girl (step 1) and inserted into them the normal gene that codes for the enzyme she lacks (step 2). Next they cultured the genetically engineered cells and raised billions of them (step 3). They then reinserted these billions of engineered cells into the little girl (step 4). It is hoped that the cells will make the missing enzyme and thus treat the disease.

DNA FINGERPRINTING CAN SOLVE CRIMES

Law enforcers—from Sherlock Holmes to the FBI—have collected and analyzed fingerprints as a way to identify crime suspects. To the crime fighter's consternation, the whorls and swirls on the fingertips can be altered surgically or covered with gloves, and in some violent crimes like rape or assault, fingerprints may not be left. Nevertheless, criminals usually leave behind unique and far more revealing bits of identity—their DNA—in drops of blood, in cells at the base of fallen hairs, or in semen. And geneticists are learning to read these clues.

In 1985, British molecular geneticists discovered segments of the human genome that are especially variable. Patterns of these segments are so utterly unique within each individual that they can serve as a DNA fingerprint. So-called *hypervariable regions* are short stretches of DNA, 20 to 60 bases long, repeated a variable number of times. The lengths of DNA fragments carrying the repeated sequence can reveal different alleles for the hypervariable region. And since there are several different hypervariable regions in a complete set of human chromosomes, the number of potential length combinations is enormous.

Family studies prove that hypervariable sequences are inherited in a normal Mendelian fashion: All fragments can be traced to one or the other parent and in turn to the grandparents. By studying 54 members of a family with four living generations, geneticists found that each individual had different fragment lengths—a unique genetic fingerprint. This pedigree potential has allowed geneticists to resolve cases of disputed paternity. In one case, British immigration authorities claimed that a boy was an illegal alien, but the technique revealed his true mother—a British citizen.

In crime cases where fingerprints are unavailable, DNA can be collected from the crime scene instead. For example, scientists have taken DNA from dried bloodstains on cloth that had remained in a normal room for four years, and from the cells at the base of a single hair. The DNA in sperm collected from old semen stains is also retrievable in sufficient quantities for unambiguous identification.

Not long ago, a dangerous rapist might have gone free if he wore a disguise and if the dazed victim was unable to pick out the assailant in a police lineup. Today, however, a rape case like this can be quickly solved if the assailant is among the suspects. In one such case, technicians took a vaginal swab from a woman six hours after sexual intercourse, and geneticists separated out cells from the woman's own reproductive tract from the male's sperm cells. Then they purified the sperm DNA, treated it with restriction enzymes, and compared the fingerprints (gel electrophoresis banding patterns) from the semen DNA, from DNA in the blood of the prime suspect, and from DNA in the victim's blood. The DNA fingerprints of the semen and man's blood sample matched perfectly. The probability of a match occurring by chance is less than 3 in 10^{11}. Since there are only 5×10^9 people alive on earth, the suspect had to be the rapist.

Prosecuting attorneys have already used DNA fingerprints in hundreds of trials, and 200 or more defendants have been convicted on the weight of this evidence as provided by expert witnesses (Figure 1). In a few widely publicized court cases in 1988 and 1989, however, the legal strategy of relying on DNA data backfired, and the judge threw out the "fingerprint" evidence. As it turns out, the procedures are so complicated that outside of a geneticist's lab—say, in an FBI crime lab or commercial forensics lab—the accuracy and dependability of the evidence can drop considerably. Numerous observers are calling for the passage of stringent federal standards for crime lab procedures, the use of DNA fingerprinting, and the use of molecular evidence in trials—before those techniques themselves get indicted.

FIGURE 1 ■ **Expert Witness Examines DNA Evidence in a Recent Murder Trial.**

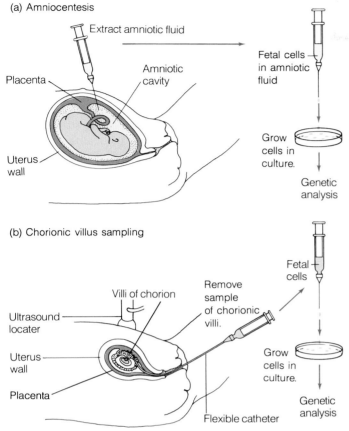

(a) Amniocentesis

Extract amniotic fluid

Placenta

Amniotic
cavity

Uterus
wall

Fetal cells
in amniotic
fluid

Grow
cells in
culture.

Genetic
analysis

(b) Chorionic villus sampling

Villi of chorion

Ultrasound
locater

Uterus
wall

Placenta

Remove
sample
of chorionic
villi.

Fetal
cells

Grow
cells in
culture.

Genetic
analysis

Flexible catheter

**FIGURE 12.18 ■ Prenatal Diagnosis of Defective Genes and
Chromosomes. As the text explains, the differences between
(a) amniocentesis and (b) chorionic villus sampling lie in when
and how the fetal cells are removed. In both techniques, the cells
are cultured and the chromosomes analyzed for genetic
abnormalities.**

less than a week, she was able to leave the hospital. The medical team hopes that the engineered cells will proliferate and allow her body to resist disease, but they will not know until many months after this book is printed. The families of children with genetic diseases, as well as the scientific community at large, anxiously await the outcome of this historic trial, since these early efforts may eventually lead to effective gene therapy to correct many more genetic disorders at a fundamental level.

PREVENTING NEW CASES OF GENETIC DISEASE

The recent revolution in genetic techniques has allowed scientists and physicians to carry detection and treatment one step further: Not only can they help those with existing genetic problems, but they can also help prevent the conception and birth of children doomed to a lifetime of suffering. They do this by detecting deleterious alleles in potential parents and in the unborn and by counseling the parents as to their options.

Prenatal Diagnosis

Tremendous progress has been made in the area of diagnosing genetic conditions in fetal cells before birth. In the mid-1960s, geneticists pioneered a method of collecting fetal cells called **amniocentesis,** and today, this procedure is used routinely for most pregnant women over 35 or for those with family histories of genetic disease (Figure 12.18a). The physician inserts a needle through the mother's abdominal and uterine walls and removes a few milliliters of the amniotic fluid that surrounds and cushions the fetus. This fluid contains sloughed-off fetal skin cells, which are collected and grown in the laboratory, then tested for defective chromosomes or genes. In these cells, certain defective genes can be detected by the absence of normal enzymes, such as the enzyme needed to prevent Tay-Sachs disease (see Table 12.1). Over 100 genetic abnormalities can be identified through amniocentesis, but the technique does have a drawback: It cannot be accomplished before about 14 to 20 weeks of pregnancy, and by that time, the mother may have already felt the fetus move and may find it emotionally hard to consider an abortion, even if test results reveal massive genetic defects.

In a more recent procedure called **chorionic villus sampling,** the physician removes fetal cells from the developing placenta, the organ that nourishes the fetus (Figure 12.18b). The fetal chromosomes can then be examined and the RFLPs studied. The mutation that causes sickle-cell disease, for example, eliminates a restriction enzyme cutting site present in normal DNA. Thus, the sickle-cell allele can be detected through RFLP analysis with 100 percent accuracy. This technique can also detect fetuses with PKU or Huntington's genes, but with somewhat less accuracy. Geneticists predict that soon, most common genetic diseases will be detectable in the unborn on the basis of RFLPs. Unfortunately, treating fetuses for these gene defects is a more distant prospect. It may be possible someday to replace defective fetal genes with normal DNA sequences and thus cure the baby's condition before birth. For now, however, the options are limited mainly to detecting carriers and providing genetic counseling.

Detecting Carriers

For decades, geneticists have been able to identify carriers of defective alleles through pedigrees: Children homozygous recessive for a trait like PKU have parents who are heterozygous carriers. This approach, however, only has predictive value for the additional children of those same specific sets of parents. Sensitive biochemical analyses can now reveal the presence of a defective enzyme in unaffected carriers. A heterozygote for PKU or Tay-Sachs disease, for example, has just one rather than two alleles making the normal enzyme. A heterozygote, therefore, has less enzyme than a person with two normal alleles. In this way, the person can be shown to be a carrier even if he or she hasn't yet produced children. These techniques won't work for some genetic ailments, particularly those whose defective protein has yet to be identified. However, RFLPs can sometimes be used instead; carriers for Duchenne muscular dystrophy and Huntington's, for example, are found this way.

The reason carriers of a genetic disease do not themselves exhibit disease symptoms is that their normal gene copy masks the defective copy. If during conception they could somehow donate only eggs or sperm containing the normal gene copy (not the defective one), they would not transmit the disease. Recent experiments suggest that this selectivity may one day be possible, beginning with female carriers. Recall from Chapter 7 that a newly ovulated egg has completed only the first meiotic division of meiosis (review Figure 7.19, step 4) and that the huge egg remains attached to its small sister cell (the polar body). Because homologous chromosomes separate during the first meiotic division, only one of the two cells—either the egg or the polar body—gets the homologue with the defective gene. Thus, if the polar body contains that defective copy, then the egg must have the normal copy. Researchers in Chicago devised a way to remove the polar body without injuring the egg and to test its DNA. They performed this test on eight egg/polar body pairs from a female carrier of a genetic disease and determined that two of her eggs were free of the defective gene. They combined these with her husband's sperm and reimplanted the eggs into her uterus; unfortunately, she did not become pregnant for reasons unrelated to the genetic test.

If this sort of test becomes more generally available to both men and women, it could allow the healthy carriers of genetic disease to produce disease-free children. For now, however, prospective parents who know they carry defective genes face hard choices. So do the parents of a fetus revealed as defective through prenatal diagnostic techniques. Genetic counselors try to help people with these difficult dilemmas.

Genetic Counseling

Genetic counselors provide information to families at risk for transmitting genetic defects. They acquaint the prospective parents with the manifestations of the disease they may pass along, with the potential suffering of an affected child, and with the emotional and financial impact on parents and siblings. Counselors can assess the couple's probability of having an afflicted child, help them consider the available options, and assist in their acceptance of the difficult reproductive decisions they must make.

■ **Decision Making: Who Shall Be Born?**   Decisions about genetics are momentous ones. A family may elect to forgo reproduction and adopt an unrelated child. Or they may try to beat the odds, and then, if they don't, resign themselves to the consequences of a genetically defective child. Or they may conceive, undergo prenatal testing, and then terminate any pregnancy involving a defective fetus. But each such decision entails tremendous emotional costs and pits possible objections to abortion and birth control—which are strong among several groups of Americans—against the realities of suffering children born with genetic defects, and their families.

As geneticists learn to detect more and more human genetic conditions, these difficult decisions are apt to become commonplace—and yet no less onerous. Someday, RFLP analysis will reveal a couple's risk of having a child who might develop cardiovascular disease, schizophrenia, diabetes, alcoholism, cancer, or Alzheimer's senility. Should the couple choose not to conceive or to abort a fetus found to carry a RFLP linked to one of these conditions? Or are these acceptable risks of being alive? How genetically perfect, in other words, would a child have to be if the parents had enough information to choose their offspring's physiological traits in advance? Society will face such choices in the not-too-distant future.

In addition to the reproductive issues regarding genetic screening, there are other issues as well, such as the use of the information by insurance companies and employers (see Box 12.2 on page 258). But no matter how much complexity the New Genetics may add to our reproductive lives, the exciting tools of this field will almost certainly solve more problems than they create, allowing society to begin truly eliminating the suffering by victims of genetic disease for the first time in history.

CONNECTIONS

Despite the unique challenges involved in studying a person's genes and chromosomes, understanding human genetics in general and a condition like PKU in particular requires that we apply the same genetic principles governing all other organisms. By simply recalling Mendel's laws of segregation, a physician can accurately predict the likelihood that two carriers will produce a child with PKU. From the one gene–one polypeptide concept, researchers can deduce that an enzyme is missing or defective in PKU victims. By knowing that a phenotype is the result of a genotype interacting with environmental influences, physicians can successfully protect PKU sufferers from mental retardation by altering their diet. And one could even have predicted the RFLP technique and its ability to mark defective genes by recalling the great variability in each person's DNA—the same variability that fuels evolution.

These principles, gleaned from research on bacteria, fruit flies, and other simple organisms, have modified the study of our own species and have opened a new era in genetics. They have allowed scientists to detect several genetic disorders early enough to prevent organic damage and to predict several others so that if parents wish, they can prevent the birth of an infant doomed to suffer and perhaps die prematurely. The ideal solution would be to eliminate gametes bearing defective genes by screening and removing them so that they can't participate in fertilization—an approach now in the experimental stage. Perhaps before such manipulation becomes possible, however, geneticists may learn how to alter the genotype of a defective fetus in the womb or to influence the way its phenotype unfolds so as to prevent the mutant allele from exerting deleterious effects. To understand how something like this could be done, we must, in the next two chapters, cross over to the realm of developmental biology. This area is the logical extension of genetics, and it explains how the genetic information in a single cell, the fertilized egg, directs that zygote to divide, develop, grow, and take on the shape and characteristics of a complex organism such as a human baby.

Highlights in Review

1. Human genetics is a challenging field because people are difficult genetic subjects; it is an important field because human genetic diseases such as PKU are a serious burden on individuals, families, and society.

2. Some of the same research techniques that enable researchers to splice genes and engineer new gene sequences have led us to a new era of genetics and the ability to map human genes, detect genetic diseases more easily, and treat them more effectively.

3. Human mutations are usually deleterious because the body is a smoothly functioning machine that has been "tuned" by evolution.

4. Human pedigrees, or diagrams of a family's genetic history, can reveal whether a trait is dominant or recessive and whether it lies on an autosomal or sex chromosome.

5. PKU and albinism are due to recessive alleles on autosomal chromosomes. Sufferers are homozygotes, and their parents are heterozygous carriers.

6. Duchenne muscular dystrophy and hemophilia are both caused by recessive alleles on the *X* chromosome. *X*-linked genes are passed to a son from his mother.

7. Huntington's disease is due to an autosomal dominant allele; even a heterozygote with just one dose of the mutant allele will show symptoms of nerve degeneration in middle age.

8. Changes in chromosome number and structure can lead to improper body functioning. Examples include Turner syndrome (an *XO* individual), Klinefelter syndrome (an *XXY* individual), and *XYY* males. An additional chromosome 21 results in Down syndrome. Translocations of chromosome parts can result in a form of cancer such as Burkitt's lymphoma. A fragile *X* chromosome, with a tip that tends to break off, can lead to mental retardation in male children.

9. Somatic cell genetics is a relatively new technique that fuses mouse and human cells to produce a somatic cell hybrid. The subsets of chromosomes in these cells can be more easily mapped, and researchers have located over 700 human genes using this technique.

10. RFLPs serve as genetic fingerprints—fragments of DNA produced by restriction enzymes that break the DNA at slightly different locations, depending on heritable variations in the order of base pairs in DNA. Some of these fragments bear specific genes and when radioactively labeled can serve as markers that help locate and map human genes. The Huntington's disease gene was mapped this way, and RFLPs are being applied to the study of PKU, hemophilia, and other diseases.

11. Some genetic diseases, like Huntington's, don't show up until adulthood. Others, like Duchenne muscular dystrophy, become apparent by early childhood. Still others, like PKU, can be detected at birth with a blood test.

12. Early detection allows the physician to treat a genetic disease by physiological means (e.g., prescribing a strict phenylalanine-free diet for PKU victims), by protein therapy (e.g., providing blood-clotting factor to hemophiliacs), or by attempting to repair defective genes in people with blood diseases.

13. Detection of a genetic anomaly in adult carriers or in an unborn fetus gives couples a choice of preventing the birth of a child with genetic problems. Genetic counselors can help a couple assess their options, which include adoption, birth control, abortion, and acceptance. In the future, perhaps defective fetuses can be repaired in the womb or gametes carrying genetic mutations can be screened out before fertilization.

Key Terms

albinism, 248
amniocentesis, 261
carrier, 248
chorionic villus sampling, 261
chromosome translocation, 252
fragile *X* syndrome, 253

Klinefelter syndrome, 251
oncogene, 252
pedigree, 247
RFLP, 255
somatic cell hybrid, 255
Turner syndrome, 249

Study Questions

Review What You Have Learned

1. What are the special challenges in studying human genetics?
2. Explain why boys inherit *X*-linked recessive traits from their mothers.
3. Give an example of a genetic defect caused by each of the following: change in chromosome number; chromosome translocation; a fragile *X* chromosome.
4. What is an oncogene and how does it operate?
5. How do researchers use somatic cell fusion in gene mapping?
6. How can geneticists use RFLPs to predict whether a person has a certain defective gene?
7. Compare amniocentesis with chorionic villus sampling.
8. What are the purposes of genetic counseling?
9. How is a woman a genetic mosaic?

Apply What You Have Learned

1. A couple's first child has the disease PKU. What is the likelihood that their next child will suffer from PKU?
2. A normal woman whose grandfather had hemophilia worries that she may produce a hemophiliac son. What is the likelihood of this if the woman is homozygous normal? If she is heterozygous normal?

3. Recently, a woman claimed that a certain businessman fathered her baby. How could a molecular geneticist confirm or disprove her claim?
4. A couple has a son with hemophilia. What is the probability that the next daughter will have the disease? What is the probability that the next son will have the disease?
5. Consider the three pedigrees shown in the figure below, and determine if the trait is inherited as an *X*-linked recessive, an autosomal recessive, or an autosomal dominant. Filled-in symbols indicate affected people.
6. While in medical school, you are presented with a case of a man who developed Huntington's disease at the age of 40. There had been no history of the disease in his family. One of your classmates argues that because neither parent showed the disease, the man inherited a recessive allele for the disease from both parents. Another student maintains that the disease is due to a new mutation in the germ cells of one of the man's parents. A third student says that the man inherited the allele for the disease from one parent and that the disease had not been observed in the family previously because all the carriers had died before showing the symptoms. Discuss each of these arguments, stating whether it is likely to be correct and describing what additional information you might need in making your evaluation.

(a)

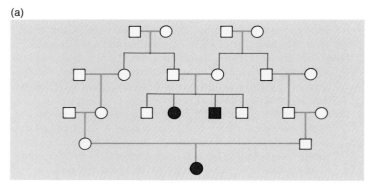

(b)

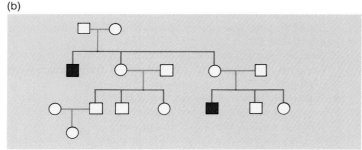

(c)

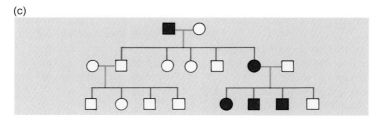

For Further Reading

BARNES, D. M. "'Fragile X' Syndrome and Its Puzzling Genetics." *Science* 243 (1989): 171–172.

CULLITON, B. J. "Mapping Terra Incognita (Humani Corporis)." *Science* 250 (1990): 210–212.

HALL, S. S. "James Watson and the Search for Biology's 'Holy Grail.'" *Smithsonian,* February 1990, pp. 41–49.

JAROFF, L. "Giant Step for Gene Therapy." *Time,* September 24, 1990, pp. 74–76.

JOYCE, C. "Your Genome in Their Hands." *New Scientist,* August 11, 1990, pp. 52–55.

KOLATA, G. "Genetic Screening Raises Questions for Employers and Insurers." *Science* 232 (1986): 317–319.

MANGE, A. P., and E. J. MANGE. *Genetics: Human Aspects.* 2d ed. Sunderland, MA: Sinauer, 1990.

McKUSICK, V. A. *Mendelian Inheritance in Man.* 9th ed. Baltimore, MD: Johns Hopkins University Press, 1990.

MOTULSKY, A. G. "Societal Problems in Human and Medical Genetics." *Genome* 31 (1989): 870–875.

NEUFELD, P. J., and N. COLMAN. "When Science Takes the Witness Stand." *Scientific American,* May 1990, pp. 46–53.

ROBERTS, L. "Ethical Questions Haunt New Genetic Technologies." *Science* 243 (1989): 1134–1136.

ROBERTS, L. "Huntington's Gene: So Near, Yet So Far." *Science* 247 (1990): 624–627.

ROBERTS, L. "To Test or Not to Test?" *Science* 247 (1990): 17–19.

VERMA, I. M. "Gene Therapy." *Scientific American,* November 1990, pp. 68–84.

WEISS, R. "Predisposition and Prejudice." *Science News* 135 (1989): 40–42.

Reproduction and Development: The Start of a New Generation

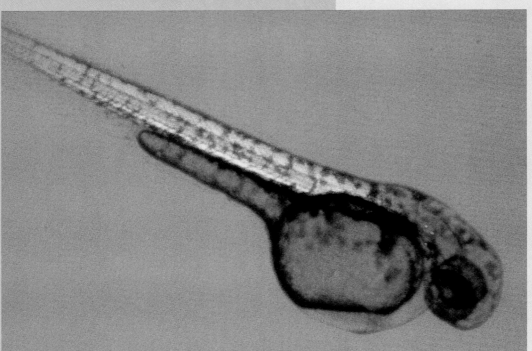

FIGURE 13.1 ■ **Embryonic Zebra Fish. Just three days old, this embryo was photographed with polarized light, revealing developing muscles (blue), eye (dark sphere), and yolk (central red sac).**

The Omnipotent Egg

As dawn breaks over India's Ganges River, life begins to stir. People come down to the river to wash and draw water, birds begin to call, and gavials—Indian crocodiles—haul themselves out on the muddy bank to soak up the sun. The most frenzied morning activity, however, goes on among the small, striped zebra fish that live in the slow-moving river.

As light penetrates the greenish water, the zebra fish converge into a mass of flashing silver bodies. Darting first one way then another, pairs of zebra fish skim the submerged rocks along the river bottom. Within each pair, the male excitedly chases the larger, rounder female and frequently rams its head into her swollen abdomen. When the male bumps the female, she lays a few tiny eggs the size of pinheads, then rushes away. As the eggs float slowly downward and settle between the rocks, the male releases a cloud of sperm around them and then lingers to swallow any eggs that do not drop to safety. As the morning light continues to brighten, the chasing and bumping subside, and the fish eventually rise to feed at the water's surface.

Meanwhile, lying safely among the rocks, the translucent, grayish eggs slowly begin to change their appearance. Fertilization by sperm has triggered a series of events within each egg. A thin film lifts from the egg surface and forms a clear, protective globe around the now darkened cell. Inside each egg, clear cytoplasm streams upward, forming a transparent cap above a dark granular yolk. Within 40 minutes, a partition forms and cleaves the egg in two. Over the next two days, hundreds of such cleavages will generate thousands of cells inside each translucent globe. A recognizable embryo will take form within this mass, followed by a tiny fish. The life of a new individual has begun (Figure 13.1).

The fusion of egg and sperm constitutes the process of *fertilization* and triggers an orderly increase in complexity called **development.** During development, the single-celled zygote divides by mitosis and produces thousands or millions

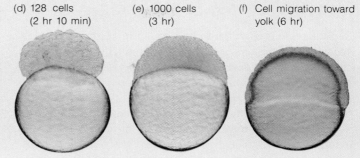

(a) 1 cell

(b) Clear cap and yolk mass

(c) 2 cells (40 min)

(d) 128 cells (2 hr 10 min)

(e) 1000 cells (3 hr)

(f) Cell migration toward yolk (6 hr)

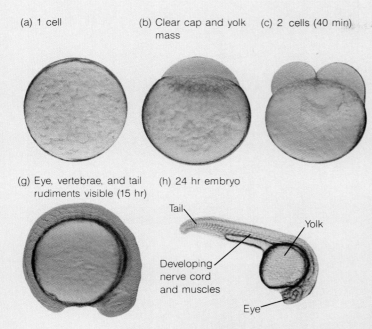

(g) Eye, vertebrae, and tail rudiments visible (15 hr)

(h) 24 hr embryo

Tail

Yolk

Developing nerve cord and muscles

Eye

FIGURE 13.2 ■ How a Fertilized Egg Develops into a Zebra Fish. A newly fertilized egg (a) rapidly separates into a clear cap of cytoplasm above a darker yolk mass (b). Within 40 minutes, the zygote divides into two cells (c). About two hours after fertilization, there are 128 cells in the cluster (d), and at three hours, the embryo consists of 1000 cells (e). After six hours, some of these cells have started to migrate downward, around the yolk (f), while others have begun to organize into the embryonic cell layers, visible as a thickened region at the top and sides of the yolk. At 15 hours (g), the thickened part of the embryo has developed an eye (here, lower right), the rudiments of vertebrae along the back (top of photo), and a tail (lower left). Finally, just 24 hours after fertilization (h), the fertilized egg has become a fish larva, almost fully developed and ready to hatch. Researchers removed the transparent globe of the egg shell in order to take these photographs. The fish larva is shown at a lower magnification than the earlier stages.

of cells (Figure 13.2), each in a prescribed place and with a defined shape and function. Initially, the developing organism, or **embryo,** grows inside a protected space—the jellylike globe of a zebra fish, the hard-shelled egg of a quail, the seed coat of a bean sprout, or the womb of a mare. But development continues long after hatching, germination, or birth as the new individual matures to adulthood and reproduces.

One of the most profound mysteries in biology is how a single cell, the fertilized egg, becomes an individual with limbs, heart, eyes, and other body structures. In the first half of this century, embryologists studied and carefully described the development of these body structures. They were able to decipher the patterns of cell division, how cell sheets fold and fuse as organs form, and which proteins accumulate as cells assume their adult functions. Today, there is a renewed sense of excitement among developmental biologists as they enter a second great revolution in embryology, going beyond what happens as an embryo forms and grows to what makes it happen—the actual molecular mechanisms of development. Their emerging realization is a main theme in this chapter: Remarkably similar molecules and mechanisms direct the development of all animals. For example, the same types of molecules direct the development of certain brain structures in a fly larva, a zebra fish, or a person. Plants and animals also share some features of development, but the details are sufficiently different between the two kingdoms that we devote this chapter only to animal development and Chapters 30 and 31 to aspects of plant development.

As we explore the marvels of animal development, two additional themes will recur. First, development is an ordered sequence of irreversible steps, each depending on previous events and causing future events. As the fertilized egg undergoes many rounds of mitosis, the resulting cells interact with each other and cause each to become progressively committed to performing tasks that are increasingly specialized. As the specialized cells interact, they set up new conditions that cause the next stage to develop. Finally, each of an embryo's cell types becomes committed to a different set of specialized tasks because each cell expresses a distinctive set of genes at specific times in development. The patterns of gene expression are controlled by regulatory substances originally contained in the egg or made by neighboring cells. Sometimes these gene expression patterns go awry, and the result can be a deadly cancerous tumor—development deranged.

This chapter on animal development will answer:

■ What events ensure that animals mate and that egg and sperm unite at fertilization?
■ How are fertilized egg cells transformed into embryos?
■ What mechanisms cause organs to form in an embryo?
■ How is cancer like development gone awry?
■ How do eggs and sperm form and launch a new generation?

MATING AND FERTILIZATION: GETTING EGG AND SPERM TOGETHER

Reproduction is essential to life: Each organism exists solely because its ancestors succeeded in producing progeny that could themselves develop, survive, and—if lucky and well adapted to the environment—reach reproductive age. In animals as different as zebra fish and zebras, the basic elements of reproduction are the same: A tiny mobile sperm cell fuses with (*fertilizes*) a huge, immobile ovum, or egg cell. The ovum contributes a set of chromosomes, a storehouse of nutrients, and crucial developmental information, while the sperm furnishes a second set of chromosomes and acts as a trigger for development. The meeting of egg and sperm requires events and processes at both the macrolevel and the microlevel. At the macrolevel, entire organisms must **mate,** that is, behave in a way particular to the species that will bring egg and sperm safely into proximity at appropriate times (Figure 13.3). At the microlevel, egg and sperm must make contact and fuse through a series of precise events that depend on the inherent structure and behavior of the gametes themselves. Let's first consider the range of common mating behaviors in animals, then turn to the events of fertilization.

Mating Strategies: Getting Organisms Together

Cooperation is the key to mating. Since in most animal species, individuals of each sex produce only one type of gamete, some degree of cooperation is necessary to ensure that when an egg from a female's body is ready to be fertilized, sperm from a male are available. Individuals often cooperate even in hermaphroditic species like earthworms and sea slugs (in which individuals make both eggs and sperm). *Hermaphrodites* usually pair up and reciprocally fertilize each other's eggs. Most animals are not hermaphrodites, however, and for mating to occur, individuals of both sexes in a species must first recognize each other and then synchronize the production and release of gametes.

■ **Mate Recognition** Animal mates recognize each other mainly by visual, auditory, and olfactory cues. Male birds are often splendidly colored, which visually attracts females: The male peacock's tail, the indigo bunting's vivid blue plumage, the male frigate bird's expandable red throat sac, and the widowbird's long, fluttering tail (see Figure 38.20) all stimulate the female's sexual interest at appropriate times of the year. Such visual cues abound throughout the animal kingdom: The lion's mane, the elephant seal's proboscis, a woman's breasts and buttocks, a man's muscular arms and chest, and the stickleback fish's red stripe are all examples.

Crickets respond to whirring sounds made by rubbing bristled back legs against each other, and these auditory cues are common noises on a summer's night. Olfactory cues

(a) External fertilization in frogs

(b) Internal fertilization in mantises

FIGURE 13.3 ■ **Bringing Egg and Sperm Together: The Marvels of Mating in the Animal Kingdom.** (a) *External fertilization.* The male golden toad (*Bufo periglenes*) attracts a mate partly by behavioral adaptations and partly by its eye-catching skin color. The male clasps the larger female, and the physical proximity causes both sexes to release gametes simultaneously. (b) *Internal fertilization.* Programmed mating behavior brings animals together, and the male releases sperm inside the female's reproductive tract. Praying mantises will continue to mate even if the female eats the male's head, which sometimes occurs.

include the odor signals (*pheromones*) that mammals and insects generate—chemicals given off by one individual that affect the behavior of others of the same species. Male pigs, for example, exude a strong-smelling chemical called androstenone (also known as "boar taint"). If a sow in the receptive phase of the sexual cycle catches even a faint whiff of boar taint, it will assume a sexual stance and wait to be mated.

■ **Synchronization of Gamete Production and Release**   Within a given species, eggs and sperm must reach maturity at the same time for fertilization to succeed. Some species mate only at specific times of year (often in spring), during which all adult members reach physical readiness and mate in a frenzy of activity—like the zebra fish in the Ganges. Environmental cues, such as gradual changes in day length, often trigger the synchronized maturation of gametes. Hens' eggs are more plentiful—and hence cheaper to buy—in springtime for this reason. The release of gametes is often synchronized as well and is usually stimulated by behavior. For example, when the male zebra fish rams the female's abdomen, both are stimulated to release gametes simultaneously.

This kind of synchronized release into the environment is a prelude to **external fertilization,** a process that takes place in many water-dwelling species and in which eggs and sperm are deposited directly into the surrounding water, meet by chance, and fuse. External fertilization can occur as in zebra fish or sea urchins, with little or no contact between the mating adults. In other species, such as frogs and salamanders, the males clasp the females firmly, their bodies touch for prolonged periods, and this touching stimulates the simultaneous release of gametes (see Figure 13.3a). Since unfertilized eggs survive only a short while, this synchronized release greatly increases the likelihood that an ovum will still be healthy when encountered by sperm.

Animals that inhabit land, including most mammals, birds, reptiles, insects, and snails, employ a second major mechanism to ensure gamete union: **internal fertilization,** wherein the male deposits sperm directly into the female's body and the gametes meet in a chamber or tube (see Figure 13.3b). Their synchronized meeting is based on internal chemical signals, or hormones, as well as on behavioral cues. This type of male deposition, also called **copulation,** has a distinct advantage: Sperm can be concentrated and protected within the female's body until eggs are available for internal fertilization, thus helping ensure that viable gametes meet at appropriate times.

Fertilization: The Actual Union of Egg and Sperm

While mating brings the eggs and sperm into close contact at appropriate times, fertilization—the actual fusion—depends on a series of events that stem from the structures and activities of the gametes themselves.

■ **Gametes: Cells Specialized for Reproduction**   Eggs and sperm are strikingly different—and *vive la différence!* An egg is usually the largest single rounded cell in an animal's body;

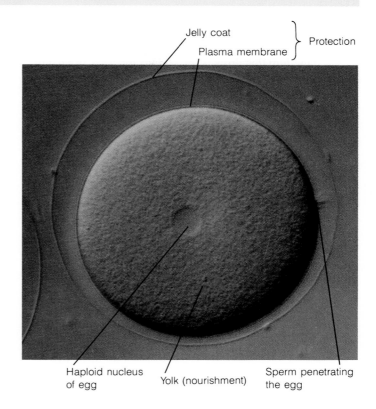

FIGURE 13.4 ■ **The Egg.** This single large egg cell of a sea urchin protects the new individual, instructs its development, and contributes half the chromosomes. Caught at the moment of fertilization, a sperm penetrates the egg from the right. Because this egg warehouses nourishment for the embryo, it is 1000 times the volume of the sperm, which just donates a haploid set of chromosomes.

the yolk of an ostrich egg, which is actually a baseball-sized **ovum,** plus a huge store of food for the future embryo, is the largest animal cell on earth. An ovum (the name for an egg cell after it leaves the ovary) is enormous and performs many jobs (Figure 13.4): It donates a haploid nucleus; it protects the developing embryo inside jellylike protein coatings, strong membranes, sacs of fluid, and sometimes hard or leathery shells; it nourishes the embryo with **yolk,** which contains rich stores of lipids, carbohydrates, and special proteins; it provides the machinery for protein synthesis in the form of tRNAs, ribosomes, and mRNAs, enabling the zygote to undergo very rapid cell division without slowing down for gene transcription; and finally, it directs development of the early embryo by special substances in its cytoplasm that control the expression of its genes. An ovum is like a computer loaded with a program, ready to play out a logical sequence of actions when the sperm's activity commands the program to RUN.

Unlike the egg, the sperm is one of the smallest cells in

the body, stripped down to just a few elements: a compact haploid nucleus; several mitochondria, which provide energy; a long flagellum that propels the sperm; and a sac of enzymes that digests a path through the egg's protective coatings (Figure 13.5). The sperm's streamlined size and shape reflect its narrow function: reaching the egg, penetrating its coatings, and delivering a haploid nucleus into the egg's cytoplasm. This penetration and delivery, together with the activation of the resting egg, make up the actual fertilization event.

■ **Fertilization: A Fusion That Triggers Development**   Fertilization unleashes dramatic and irreversible events that permanently transform the activities and appearance of the egg. As the lashing flagellum propels a sperm headlong into an egg's jelly coating, specific protein molecules on the sperm's surface bind to other proteins, called sperm receptors, on the egg's surface. Much the way a key fits into a lock, the binding of protein and receptor ensures that an egg will be fertilized by sperm from its own species and no other. Men whose sperm lack these recognition proteins are unable to father children (see Figure 13.5b and c).

After sperm/egg recognition takes place, the sperm's tip literally detonates like a microscopic bomb: In the so-called *acrosomal* reaction, the sac of digestive enzymes at its tip explodes, releasing the contents and forming a harpoonlike protein needle in some species that pierces through to the ovum's plasma membrane (Figure 13.6). The sperm's plasma membrane fuses with the ovum's, and the sperm nucleus plunges into the egg's cytoplasm. With this event, the egg, previously one of the body's most inactive cells, springs into a flurry of biochemical activity—a sleeping cell suddenly awakened by a sperm.

Within seconds, a wave of chemical reactions sweeps across the surface of the newly aroused egg, causing that surface to harden and present a barrier to the entry of any additional sperm. This is important, because fertilization by a second sperm would create a zygote with three sets of chromosomes—one set from each of the two sperm and another set from the egg. Such triploid embryos usually die.

Shortly after fertilization, the egg's oxygen consumption skyrockets, as does its rate of protein synthesis. Male and female nuclei converge and fuse, forming the zygote's single diploid nucleus. In some eggs, fertilization also triggers dramatic relocations of cytoplasm. For example, a clear cap of cytoplasm collects at the top of the zebra fish egg above the yolk (review Figure 13.2b). And a sperm's entry into a frog egg sparks a "shudder" of activity within the cytoplasm that sweeps around the egg just beneath the surface and generates a crescent of gray-colored cytoplasm at the egg's equator directly opposite the sperm's entry point.

In a few rare animal species, reproduction occurs without fertilization via the process called **parthenogenesis.** For example, in some lizards, there are no males, and embryos develop from unfertilized eggs that spontaneously become diploid by an aberrant meiotic process. In honeybees, unfer-

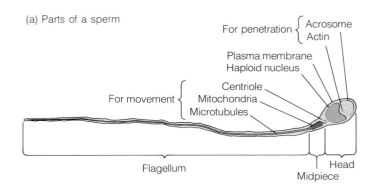

(a) Parts of a sperm

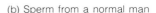

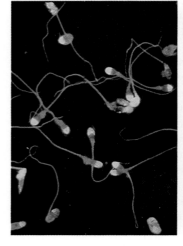 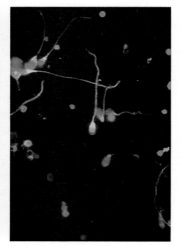

(b) Sperm from a normal man   (c) Sperm from a sterile man

FIGURE 13.5 ■ The Sperm.   This mobile cell contributes half the chromosomes during fertilization and triggers development in the zygote. (a) The streamlined sperm cell contains its payload, the haploid nucleus, plus factors that allow it to recognize and penetrate the egg, as well as organelles that power the lashing tail and allow it to swim. (b, c) Some men may be infertile because their sperm cells lack the normal form of a protein necessary for recognition and penetration of the egg. This protein (here stained bright yellow) is located around the front tip of a normal man's sperm. A sterile man's sperm has less of the protein, and it is distributed abnormally. This difference supports the hypothesis that the protein is crucial for fertilization.

tilized eggs develop parthenogenetically into haploid males, while fertilized eggs develop into diploid females. Regardless of whether development begins with parthenogenesis or with fertilization of egg by sperm, the new individual is still no more than a single cell. To become a young zebra fish, roundworm, or honeybee, the initial cell must subdivide again and again and grow into millions of cells. That subdivision begins in early embryonic development.

PATTERNS OF EARLY EMBRYONIC DEVELOPMENT

Just a dozen hours or so after the remarkable events of zebra fish fertilization, a multicelled embryo has begun to take visible shape, with distinct head and tail regions, definite right and left sides, and the first signs of a backbone along the dorsal surface. The sequence of events during early development is quite logical. First, the fertilized egg cell divides into many cells (cleavage); next, these cells become rearranged into three layers of cells (gastrulation); then these three layers interact with each other and develop into individual tissues and organs via the process of neurulation and organogenesis. Specifically, the inner cell layer forms the lining of the gut, lungs, liver, and pancreas; the outer layer forms the skin and nervous system; and the middle layer forms all the rest of the embryo's organs, including muscles, bones, and blood. Eventually, the fully developed embryo grows larger in size and matures sexually through the process of growth and the events surrounding gametogenesis. Table 13.1 summarizes these key events of development. Let's consider each event in turn.

The steps of fertilization

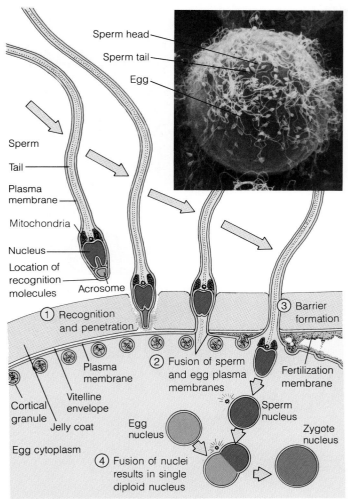

FIGURE 13.6 ■ Fertilization: Egg and Sperm Fuse, Forming a Genetically Unique Zygote. The scanning electron micrograph (insert) shows a flotilla of sperm cells surrounding a sea urchin egg. Several events take place at the cell surface. First, molecules on the sperm's surface recognize the egg, and the explosion of the needlelike acrosome allows the sperm to penetrate the coatings surrounding the egg's surface (1). Next, the plasma membranes of sperm and egg fuse (2), and then a sweeping wave of chemical reactions cause the breakdown of the granules below the ovum's plasma membrane and the establishment of the fertilization membrane, a barrier to penetration by any other sperm (3). Finally, the sperm and egg nuclei fuse, and a novel genetic combination is established within the zygote's newly formed diploid nucleus (4).

TABLE 13.1	Key Events of Development	
Event	What Happens	Consequences
Fertilization	Egg and sperm fuse	A unique genetic combination is created
Cleavage	Fertilized egg subdivides into a ball with many cells	Developmental determinants are partitioned into different cells
Gastrulation	Cells rearrange and produce an embryo with three cell layers	Precursors of internal organs migrate to inside of embryo; cells interact with new neighbors
Neurulation	Neural tube forms	Conditions are established that will generate other body organs
Organogenesis	Body organs form	Cells interact, differentiate, change shape, proliferate, die, and migrate
Growth	Organs increase in size	Adult body form is attained
Gametogenesis	Eggs and sperm develop	Reproduction becomes possible

Cleavage: Formation of an Embryo with Many Cells

The early developmental process called **cleavage** consists of a series of rapid, synchronous mitotic divisions that transform the zygote's single large cell first into a solid ball of cells called the **morula** (Latin for "mulberry"; visible in Figure 13.9) and then into a hollow ball of many small cells called the **blastula** (Figure 13.7). Cleavage divisions are not interrupted by growth periods, in contrast to cell division in later life (see Chapter 7); thus, as the egg is cleaved in half, then quarters, eighths, sixteenths, and so on, the cells get smaller and smaller. As cleavage continues, the embryo comes to have several hundred cells, or **blastomeres,** each about the size of a normal body cell. The ball, or blastula, however, is about the same size as the original ovum.

The business of cleavage is the production of many cells from one. While this requires the synthesis of new proteins, in frogs, sea urchins, and zebra fish, the source of the genetic information that encodes those proteins is the mother, not the embryo itself. How can that be? Recall that the egg is packed with messenger RNAs transcribed from the mother's genes. It is the translation of these mRNAs that produces new proteins during cleavage; the embryo's own genes are not transcribed into mRNA at this time. Biologists speculate that without needing to pause for transcription, the embryo's DNA can be replicated more rapidly, cells can divide more quickly, and the large number of cells needed for the next stage of development can be produced at an earlier time.

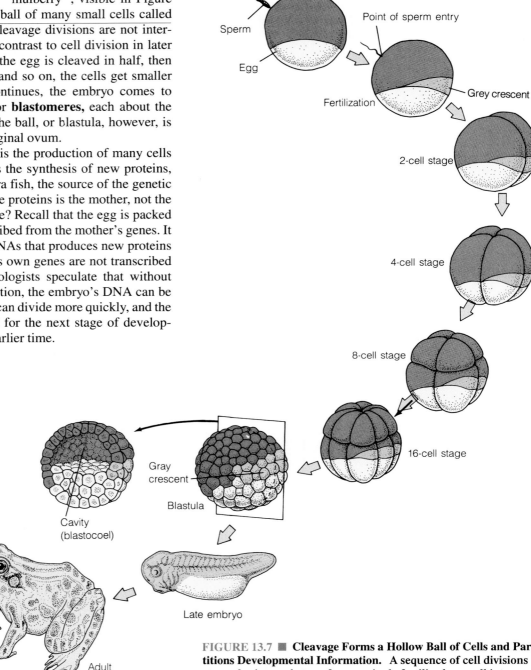

FIGURE 13.7 ■ Cleavage Forms a Hollow Ball of Cells and Partitions Developmental Information. A sequence of cell divisions at regular intervals transforms a single fertilized egg cell into a hollow ball of many smaller cells, called the blastula. Developmental information resides in a crescent of gray cytoplasm that forms in the fertilized egg opposite the point of sperm entry. As cleavage divisions proceed, the material is partitioned into only a few cells of the blastula and eventually becomes the embryo's dorsal region.

■ **Cleavage Patterns and Yolk Size**   The amount of yolk originally stuffed into the egg cell affects the pattern of cleavage divisions: Mammalian eggs, with little yolk, divide one way, while fish and bird eggs, with large amounts of yolk, divide another. After cleavage, developing mammals obtain nourishment from the mother's body via a spongy, blood-rich organ called the **placenta,** rather than from prepackaged yolk. For this reason, the eggs of mammals (including humans) are among the smallest in the animal kingdom, even though they are far larger than the other cell types in a mammal's body. Each cleavage division can pass completely through the tiny mammalian egg, creating equal-sized blastomeres.

Birds and fish undergo a pattern of cleavage very different from that of mammals. The huge egg cell's yolky mass is so large and dense that cleavage furrows cannot pass all the way through it (review Figure 13.2b). Therefore, in the blastula stage, a group of cells collect in a disc called the **blastodisc,** which looks like a lumpy beret sitting on top of the egg (review Figure 13.2c).

Regardless of whether cleavage extends completely through the embryo, as in mammals, or partitions only a small region of the cytoplasm, as in fish and birds, the blastomeres are no longer alike after cleavage: The subsets come to have different developmental capabilities.

■ **Cleavage Partitions Developmental Information**   We said earlier that the egg cell contains not just nutrients but also developmental information. But what does that mean? Egg cytoplasm contains chemicals called **developmental determinants** that will act as instructions for the developing embryo. These instructions become localized in particular regions of the egg, and after cleavage, certain instructions end up only in certain blastomeres. These determinants then activate the expression of specific genes in each blastomere, causing different cell groups to develop differently.

For example, a crescent of gray cytoplasm develops in a frog egg opposite the point of sperm entry (see Figure 13.7). As cleavage proceeds, this zone of gray cytoplasm, called the **gray crescent,** is partitioned into only certain cells. Cells containing this gray crescent material eventually become the embryo's dorsal (back) region. If a frog egg is treated to create two gray crescents, each on opposite sides of the egg, the result is a two-headed tadpole (Figure 13.8). Biologists concluded from this result that the gray crescent contains a developmental determinant that instructs cells to become dorsal parts of the embryo.

Recent experiments suggest that some developmental determinants may be mRNA molecules. For example, a certain mRNA must occur in the anterior of a fruit fly's egg for the embryo's head to develop, and an mRNA localized in the lower yolky half of a frog egg encodes a growth factor protein similar to one that can cause muscle cells to develop. Interestingly, while such determinants may help direct some developmental events in frogs and insects, other events in these

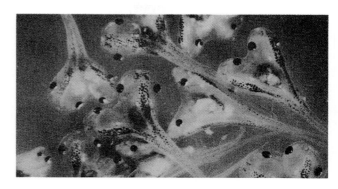

FIGURE 13.8 ■ **The Gray Crescent Determines the Embryo's Dorsal Region.**   These two-headed tadpoles formed after researchers manipulated eggs so that gray crescent material lay on two sides. The two resulting dorsal regions produced two heads and two spinal columns.

animals and most events in mammals rely on a different key to early development: the position of cells relative to one another.

■ **Cleavage Can Assign a Cell's Position—and Its Fate**   At the four- and eight-cell stages, all cells of a mammalian embryo are developmentally equivalent (Figure 13.9a). For example, if a researcher separates the four blastomeres of an early mouse embryo, the cells develop into complete mice instead of into parts of a single mouse. Likewise, if a human morula splits into two groups of cells (Figure 13.9b), each can grow into a complete human being. In fact, about a third of the cases of identical twins arise this way (Figure 13.9c), and this separation and subsequent development proves that the cells have not specialized at the time of splitting. (Fraternal twins arise from the fertilization of two separate eggs by two separate sperm, not the splitting of one morula.) With further development, however, some cells end up inside the ball of cells surrounded by other cells. The inside cells (called the *inner cell mass*) become the living embryo, while the outer cells become the *trophoblast,* a structure that helps nourish the early embryo (see Figure 13.9a). The trophoblast develops into part of the placenta, the blood-rich organ that supplies maternal nourishment to the mammalian embryo. (Identical twins can also form if the inner cell mass splits in two; a pair will develop sharing a placenta.) The placenta is sloughed as the *afterbirth.* A cell's fate—becoming either part of an embryo or part of a feeding structure—is cast simply by being at the right place at the right time. In some way, the inside position causes cells to express genes appropriate to one developmental pathway, while the outside position causes expression of other genes.

Clearly, a cell's early fate can be imposed on it by developmental determinants from the cytoplasm or by its position relative to other blastomeres. But regardless of which, the cleavage process accomplishes two things: First, it divides one large fertilized egg cell into many small cells; and second, it establishes differences between the small cells. These differences result in different genes being expressed in different cells of the embryo. As a result of this differential gene expression, embryonic cells begin a dramatic series of migrations—gastrulation—which establishes the animal's body plan.

Gastrulation: Establishing the Three-Layered Body Plan

The well-known embryologist Lewis Wolpert likes to say that the most crucial event in your life is not birth, marriage, or death, but gastrulation. While this process is neither celebrated nor even perceived by embryo or mother, **gastrulation** is the means by which a hollow ball of cells generated during cleavage is transformed into the rudiments of an animal, with front and back ends, right and left sides, a future skin outside, and a future gut inside (gastrulation means literally "gut formation").

The body plan of an adult animal helps us understand how such transformations occur and what they accomplish. Think of an animal's body as three concentric tubes or layers of cells: an inner tube forming the gut; a middle layer surrounding the gut and including muscle and blood; and an outer tube of skin with its sense organs (Figure 13.10). Gastrulation establishes the three cell layers, or **germ layers,** in the three

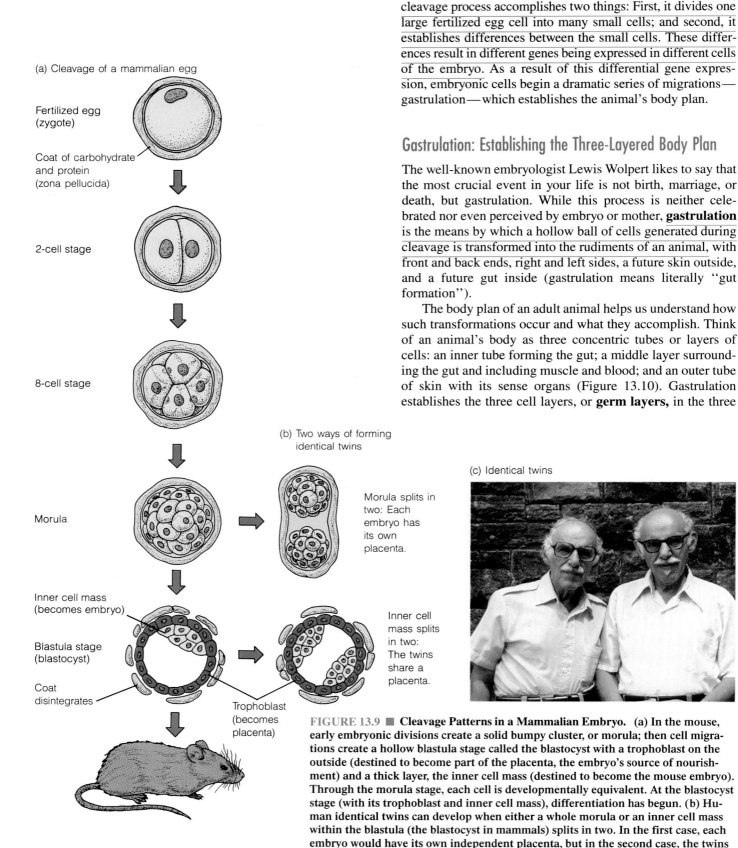

(a) Cleavage of a mammalian egg

Fertilized egg
(zygote)

Coat of carbohydrate
and protein
(zona pellucida)

2-cell stage

8-cell stage

(b) Two ways of forming
identical twins

Morula

Morula splits in
two: Each
embryo has
its own
placenta.

(c) Identical twins

Inner cell mass
(becomes embryo)

Blastula stage
(blastocyst)

Inner cell
mass splits
in two:
The twins
share a
placenta.

Coat
disintegrates

Trophoblast
(becomes
placenta)

FIGURE 13.9 ■ Cleavage Patterns in a Mammalian Embryo. (a) In the mouse, early embryonic divisions create a solid bumpy cluster, or morula; then cell migrations create a hollow blastula stage called the blastocyst with a trophoblast on the outside (destined to become part of the placenta, the embryo's source of nourishment) and a thick layer, the inner cell mass (destined to become the mouse embryo). Through the morula stage, each cell is developmentally equivalent. At the blastocyst stage (with its trophoblast and inner cell mass), differentiation has begun. (b) Human identical twins can develop when either a whole morula or an inner cell mass within the blastula (the blastocyst in mammals) splits in two. In the first case, each embryo would have its own independent placenta, but in the second case, the twins would share a single placenta. (c) Identical twins are evidence that the cells of an early embryo are developmentally equivalent.

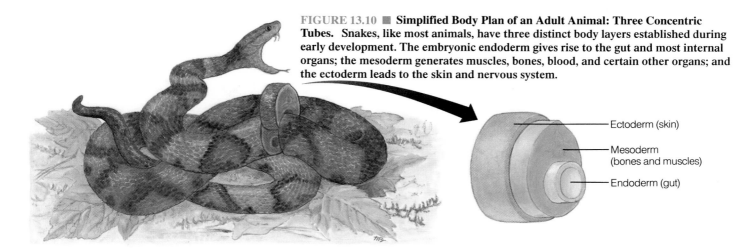

FIGURE 13.10 ■ **Simplified Body Plan of an Adult Animal: Three Concentric Tubes.** Snakes, like most animals, have three distinct body layers established during early development. The embryonic endoderm gives rise to the gut and most internal organs; the mesoderm generates muscles, bones, blood, and certain other organs; and the ectoderm leads to the skin and nervous system.

tubes. The embryo's inner layer, or **endoderm** ("inner skin"), gives rise to the inner linings of the gut and the organs that branch off from it, like the lungs, liver, pancreas, and salivary glands. The middle layer, or **mesoderm** ("middle skin"), gives rise to muscles, bones, connective tissue, blood, and reproductive and excretory organs. The outer layer, or **ectoderm** ("outer skin"), gives rise to the skin and nervous system. Figure 13.11 shows how an embryo's body parts and cells (except the gametes) can all be traced back to the three germ layers.

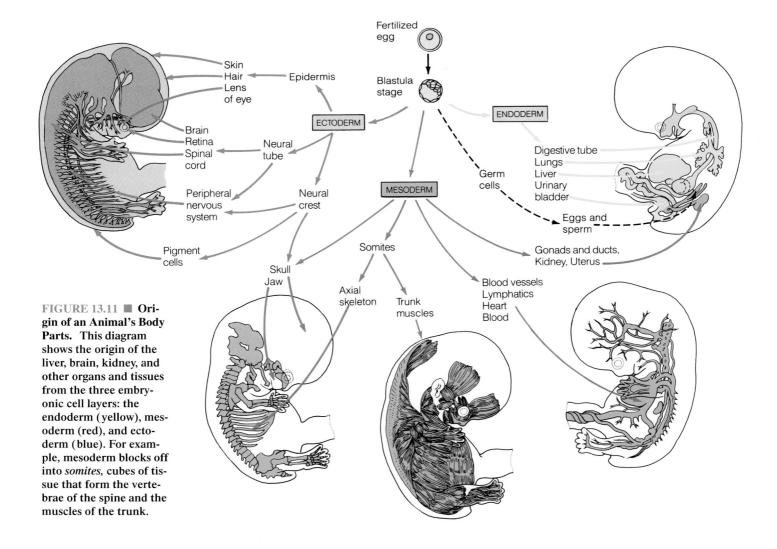

FIGURE 13.11 ■ **Origin of an Animal's Body Parts.** This diagram shows the origin of the liver, brain, kidney, and other organs and tissues from the three embryonic cell layers: the endoderm (yellow), mesoderm (red), and ectoderm (blue). For example, mesoderm blocks off into *somites,* cubes of tissue that form the vertebrae of the spine and the muscles of the trunk.

(a) Gastrulation of a sea urchin embryo

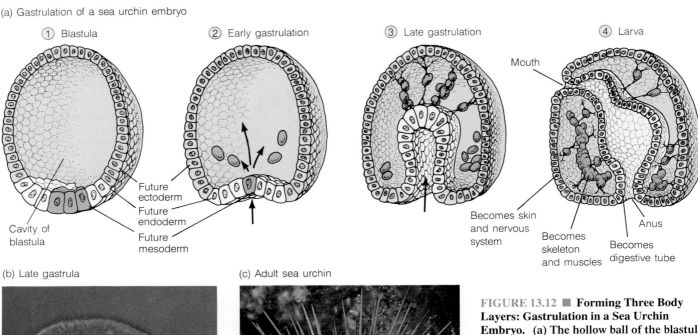

① Blastula   ② Early gastrulation   ③ Late gastrulation   ④ Larva

Mouth

Cavity of
blastula

Future
ectoderm

Future
endoderm

Future
mesoderm

Becomes skin
and nervous
system

Becomes
skeleton
and muscles

Becomes
digestive tube

Anus

(b) Late gastrula   (c) Adult sea urchin

FIGURE 13.12 ■ Forming Three Body Layers: Gastrulation in a Sea Urchin Embryo. (a) The hollow ball of the blastula (1) begins to indent at the bottom, and individual mesoderm cells (red) pop from the wall of the blastula into the cavity (2). The indentation of the endoderm (yellow) continues, pulled along by some mesoderm cells (3), until it reaches the ectoderm (blue), with which it fuses and forms a complete digestive tube through the embryo (4). The mesoderm forms the skeleton and muscles. (b) Photomicrograph of a gastrulating sea urchin embryo. (c) An adult sea urchin.

The gastrulation of a sea urchin embryo (Figure 13.12) demonstrates the main events of that developmental process in animals. First, mesoderm cells leave the surface of the blastula and enter the cavity of the hollow ball (Figure 13.12a, stages 1 and 2). Next, a dent appears in the lower part of the embryo and deepens, deforming the embryo like a finger pushing into a balloon (stages 2 and 3). This indentation produces the endoderm, and as it proceeds toward the top of the embryo, it eventually breaks through and forms the mouth (stage 4). The original opening becomes the anus, and the endoderm develops into the digestive tube. Cells remaining on the outside become the skin and nervous system. These two simple events—the movement of mesoderm and then endoderm to the inside of the blastula—thus generate a three-layered embryo.

In frogs, mesoderm cells begin to migrate into the embryo at the position of the gray crescent, and this migration establishes the embryo's dorsal side and posterior end (Figure 13.13). The opposite end becomes anterior, and so the right and left sides are also established. Thus, the gray crescent—

whose own position was determined by the point of sperm entry—plays a primary role in orienting the axes of the frog's body: head versus tail, left versus right side, dorsal versus ventral surface.

Neurulation: Forming the Nervous System

Gastrulation sets up the conditions for the next developmental stage, **neurulation,** or the development of the neural tube. The **neural tube** differentiates into brain and spinal cord.

As gastrulation ends in a vertebrate embryo, an arched "sandwich" of ectoderm, mesoderm, and endoderm lies above the yolk at what will be the embryo's back (Figure 13.14a and b). Along the embryo's midline, the mesoderm is organized into a rod, the **notochord,** which marks the future site of the backbone (the structure that distinguishes vertebrates from other animals). During neurulation, the cells of the notochord interact with the ectoderm cells lying just above them, and this interaction directs the sheet of ectoderm cells

FIGURE 13.13 ■ **Gastrulation in a Frog Embryo.** Since a frog egg has more yolk than a sea urchin egg, the mechanics of gastrulation are different, although the principles are the same: Mesoderm cells leave the surface of the blastula and enter the hollow ball's cavity, and endoderm follows. (a) The frog blastula in exterior view and (b) in cutaway view. (c) Invagination of mesoderm and endoderm begin. (d) The gut cavity enlarges, and the blastula cavity becomes smaller. (e) With gastrulation complete, a three-layered embryo has developed, and (f) an exterior view shows an embryo with a posterior (marked by the yolk plug, the site of the future anus), an anterior, a dorsal side (back), and a ventral side (belly).

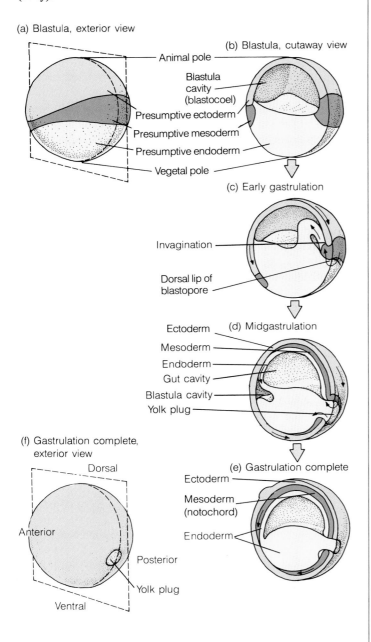

(a) Blastula, exterior view

(b) Blastula, cutaway view

Animal pole

Blastula cavity (blastocoel)

Presumptive ectoderm

Presumptive mesoderm

Presumptive endoderm

Vegetal pole

(c) Early gastrulation

Invagination

Dorsal lip of blastopore

(d) Midgastrulation

Ectoderm

Mesoderm

Endoderm

Gut cavity

Blastula cavity

Yolk plug

(f) Gastrulation complete, exterior view

Dorsal

Anterior

Posterior

Yolk plug

Ventral

(e) Gastrulation complete

Ectoderm

Mesoderm (notochord)

Endoderm

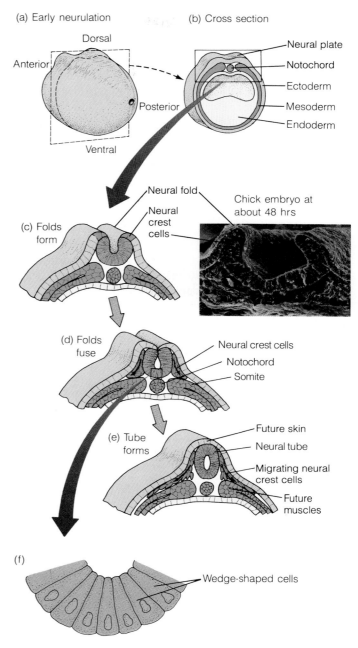

(a) Early neurulation

Dorsal

Anterior

Posterior

Ventral

(b) Cross section

Neural plate

Notochord

Ectoderm

Mesoderm

Endoderm

Neural fold

Neural crest cells

(c) Folds form

Chick embryo at about 48 hrs

(d) Folds fuse

Neural crest cells

Notochord

Somite

(e) Tube forms

Future skin

Neural tube

Migrating neural crest cells

Future muscles

(f)

Wedge-shaped cells

FIGURE 13.14 ■ **Neurulation: Formation of the Future Nervous System.** The events of neurulation, or neural tube formation, are easier to envision by considering only the upper surface of this frog embryo cut in half. (a, b) The neural plate, a region of tall ectoderm cells, takes shape on the embryo's dorsal (back) surface. (c) Folds form as cells in the neural plate change shape and cells from the side push in, causing a wrinkle in the sheet. The insert shows a chick embryo at this stage of development. (d) When the neural folds come together, the neural tube pinches off, the surface ectoderm (future skin) becomes a continuous sheet, and the neural crest (the cells at the crest of the neural fold) begin to disperse. (e) Changes in cell shape help the neural tube form. (f) Cells in the lower part of the neural tube become tall and wedge-shaped, helping the sheet to roll up.

to roll up into the neural tube. The surface layer of the embryo folds up, forced from the sides like two ripples in a sheet being pushed across a bed toward each other (Figure 13.14c). The two folds fuse and bud off the neural tube (Figure 13.14d and e). The cell shapes change from rounded to wedgelike, and this helps the tube roll up and close at each end (Figure 13.14f). This tube will eventually develop into the nervous system.

Unfortunately, sometimes mistakes occur during neurulation. Embryologists do not understand why, but in about 1 out of every 200 human births, the neural tube does not roll up completely and fails to close, producing the most common birth defect among American babies, so-called *spina bifida*. This anomaly leaves the spinal cord open to the outside, causing the infant pain and often paralysis. While it is not preventable at this time, the condition can be detected early in the pregnancy and improved by surgery after the baby is born.

So far we have discussed how the neural tube forms. But developmental biologists wanted to know what signals certain cells among the many thousands in the embryo to migrate in this way, fold up, and become a neural tube. To investigate this question, biologists Hans Spemann and Hilde Mangold in 1924 removed cells from the gray crescent area of a donor newt embryo and transplanted them into a host newt embryo

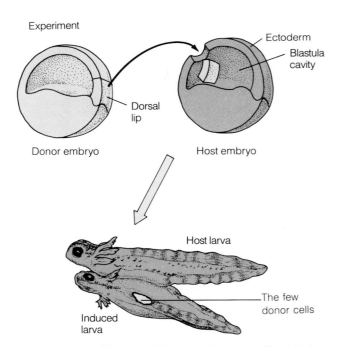

FIGURE 13.15 ■ **What Establishes the Embryo's Head-Tail Axis?** Experimenters transplanted cells from the gray crescent area (dorsal lip) of one newt embryo to the future belly of another newt embryo and watched as neurulation took place on both sides of the embryo and two animals emerged, joined belly to belly. Gray crescent cells induced the host's ectoderm layer to generate a new head-tail axis and eventually a new embryo.

in a region that would normally make belly skin (Figure 13.15). Remarkably, the implant caused a second neural tube to form on the belly of the host embryo. The host embryo became a bizarre set of Siamese twins attached belly to belly. The embryologists concluded that the transplanted cells containing gray crescent cytoplasm had liberated a signal that caused **primary embryonic induction;** that is, they had induced the host's ectoderm to form a neural tube, notochord, muscles, and eventually an entirely new embryo.

For almost 65 years, the identity of that inducing signal remained a mystery. But the recent discovery of a molecular candidate for that signal, a substance called *activin,* is exciting embryologists. Activin is a protein related in structure to another protein called *transforming growth factor β.* Ironically, when some cell types contact transforming growth factor β, they behave like cancerous cells; the related protein activin, however, promotes normal development, and without it, an embryo fails to develop a nervous system and dies. More work is needed to confirm that activin is Spemann and Mangold's organizing signal, but current results do suggest that activin triggers one of the most important functions in the embryo—the organization of the entire body plan.

The kinds of inductions we just saw (mesoderm inducing ectoderm to form a neural tube) and the developmental determinants (instructional chemicals) we discussed earlier both influence how cells develop. They trigger cells to become **committed** (*determined*) to a specific developmental pathway. Thus, cells become neural tube cells *or* skin cells, but not both. Once determined to become a particular type, the cells normally remain committed. In short, development is a one-way path. Somehow, the commitment process identifies those genes that each cell should express so that it can perform as a skin cell, for example, and those it should leave dormant so that it does not perform as any other cell type. Biologists are still not sure how this works, but a special piece of fruit fly DNA called a homeobox is giving them important clues (see Box 13.1).

With the normal completion of early development—cleavage, gastrulation, and neurulation—the embryo's major axes have been established. But the embryo is far from a finished animal. The next developmental priority is the establishment of the body organs that will allow the animal to move about, to search for and eat food, to digest, to escape predators, to find a mate, to reproduce—perchance to dream. That is the task of organogenesis.

DEVELOPMENT OF BODY ORGANS: ORGANOGENESIS

Let's say you were to travel to the banks of the Ganges, carefully collect some zebra fish embryos from the river bottom, then observe their continued development under a microscope

BOX 13.1   HOW DO WE KNOW?

A LEG GROWS FROM THE HEAD: THE HOMEOBOX

Truth is definitely stranger than fiction when it comes to a series of bizarre mutations in fruit flies: Some mutant flies have wings growing where the eyes should be; others have legs replacing the mouthparts; and still others have four wings growing where there should be two. Despite the eerie nature of these aberrations, they have great potential significance, because the genes that cause them may be keys to controlling development—in frogs, fish, and people, as well as in fruit flies.

These mutations cause one body part to be replaced by another, and they are called *homeotic mutations* (*homeotic* means "replacing similar parts"). Since they alter which organs grow in which spot, the mutations obviously affect fundamental switch points in development and thus are highly interesting to developmental geneticists. Researchers in this field found that many of the mutated genes share a DNA sequence (dubbed the *homeobox*) that encodes a stretch of about 60 amino acids. Several different homeotic genes appear to be involved in different developmental switches, but each contains a very similar homeobox. Scientists concluded that the homeobox might provide insights into how cells make developmental choices.

One homeotic mutation affects only the head and thorax ("chest" region), causing legs to replace antennae and the first pair of legs to resemble the second pair (Figure 1a and b). Researchers studying this mutation found something else that truly astonished them: The homeobox within that fly gene encodes a protein virtually identical to a portion of a human protein! They went on to find similar homeoboxes in earthworm, zebra fish, and mouse genes as well. For a particular sequence of nucleotide bases to be so carefully conserved during animal evolution, it must be doing something fundamental and perhaps similar in each kind of organism. Another discovery hinted at what that fundamental activity might be. The nerve cords of five- to ten-week-old human embryos express a *Hox* gene (as homeobox genes are called in vertebrates) very similar to one from a fruit fly. Figure 1c shows where this gene is expressed in the hindbrain of a developing mouse embryo, another vertebrate. Since key events in human development occur at this time and place, some biologists have suggested that the homeobox helps determine how a person's nervous system will grow.

A homeobox encodes proteins that bind to specific DNA sequences. Like the repressor in the *lac* operon (see Figure 10.16), these may regulate transcription of other genes and in this way control development. The occurrence of homeoboxes in flies, fish, and people, as well as their expression during embryonic development, reminds us of life's unity and how closely we share genetic and developmental mechanisms with other organisms.

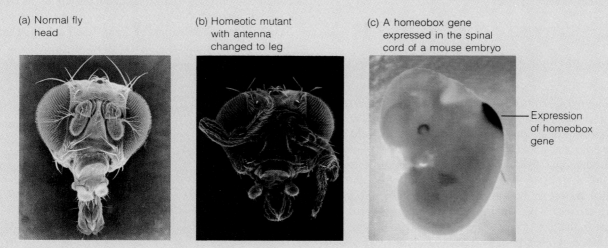

(a) Normal fly head

(b) Homeotic mutant with antenna changed to leg

(c) A homeobox gene expressed in the spinal cord of a mouse embryo

Expression of homeobox gene

FIGURE 1 ■ **Homeotic Mutants.** Sequences in genes responsible for these bizarre changes in fruit flies are nearly identical to sequences expressed during human development. (a) Head of a normal fly. (b) Antennae of a mutant fly become legs. (c) A homeobox gene in a mouse embryo is expressed in a specific part of the nervous system. Presumably the action of this gene is necessary to cause this part of the nervous system to develop correctly.

at 6-hour intervals for three days. As you watched each tiny fish develop still enclosed in its clear, jellylike globe, you would see various organs take shape in the translucent animal: eyes, tail fins, liver, kidneys, muscles, and others. This **organ-ogenesis** (literally, "origin of organs") involves changes in the shape of cells and cell groups (*morphogenesis*) and changes in the types of proteins individual cells make (*differentiation*).

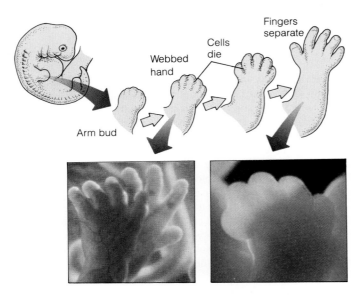

FIGURE 13.16 ■ **How Cell Death Generates the Shape of Your Hands.** Your hand begins as a paddle, roughly like the fins on a zebra fish or the webbed feet of a duck. The fingers "emerge" as four rows of cells, which radiate from the base of the paddle, die away. This pattern of cell death is genetically programmed according to species, because in ducks and fish, for example, most of the cells don't die, and webbing remains between the fin rays or toes. What's more, genetic mistakes can happen, leaving a duckling without its webs or a human baby with them.

Development of Organ Shape

The three cell layers created during gastrulation form the rudiments of the eyes, liver, muscles, and other organs. But the cells of those layers undergo a procession of changes during morphogenesis; some cells change shape, some divide rapidly, others die away, and still others move to new places as the organ develops. This cavalcade of proliferating, disappearing, and migrating cells goes on continuously, and the result can be endoderm bulging out from the gut and becoming lungs or liver; a pit sinking through the ectoderm and forming the future ear; or dispersed mesoderm cells aggregating into a block and becoming a muscle. Figure 13.16 shows how programmed cell death sculpts the developing hand, and Box 13.2 on pages 282 and 283 explains how the arm forms with correct axes.

■ **Cells Can Migrate Long Distances in the Embryo**   Recall that the neural tube arises as a sheet of ectodermal cells curls up and creates folds that fuse together (see Figure 13.14). Where these folds meet, a group of cells called the *neural crest* begins to migrate in defined pathways around the embryo like football fans leaving a stadium parking lot, giving rise to cells and organs in distant places. Some neural crest cells stay near the epithelium, migrate into the skin, and form pigment cells. If, as you look at your own arm, you can see brown moles or freckles or overall dark skin pigment, then you are observing

the results of embryonic pigment cells that crawled from the neural crest near your spinal cord across your back and down your arm.

Other neural crest cells migrate from the back of the embryonic head and around both sides of the developing mouth, forming the teeth, as well as some of the muscles, bones, and cartilage of the lower face (Figure 13.17). As the mass of migrating cells approach the front of the face, they proliferate and grow together at the "trough" between nose and upper lip. If the mass of cells fail to proliferate and grow together, they leave a space between the nose and mouth, or they leave a gap in the roof of the mouth. These birth defects, called *cleft lip* and *cleft palate,* respectively, can now be easily corrected by surgery.

While cell migration, cell death, and the folding of cell sheets can shape organs, another set of changes—this time at the protein level—occurs before the organ can carry out its specific functions.

Development of Organ Function: Differentiation

Within an animal's body, each cell type performs a specific service: Intestinal cells digest food; muscle cells create movement; eye cells detect light. A cell can carry out its particular functions in large part because of the distinctive proteins it contains—proteins that break chemical bonds in food molecules; or proteins that cause a cell to contract; or proteins that change form when struck by light. Each cell type that looks different and contains a different set of proteins is called a *differentiated cell* and has become specialized by the process of **differentiation.**

As we have seen, cell specialization begins when the commitment process identifies those genes that each cell will express as it differentiates. Early determination (commitment) events are caused by the action of specific molecules in the egg (cytoplasmic determinants), while later determination events are caused by interactions between developing cells

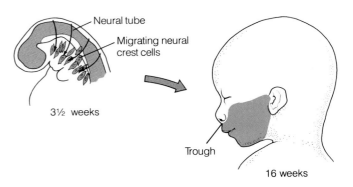

FIGURE 13.17 ■ **Migrating Neural Crest Cells Help Form the Face.** Neural cells move from the neural folds around the sides of the neural tube and embryonic head and help form the lower part of the face and jaw. The origin of the neural crest is shown in Figure 13.14d and e.

(embryonic induction). After determination occurs, cells are committed to a specific developmental fate, and regulatory proteins in determined cells select genes for use that are appropriate to the cell type—for example, genes for milk proteins in a mammary gland cell. After determination commits specific genes to future activity, the differentiation process causes these genes to produce their specialized proteins. During differentiation, mammary gland cells in the embryo actually synthesize milk proteins and acquire organelles for milk secretion owing to their ample exposure to pregnancy hormones.

Differentiation is the last in a series of five processes that enable a fertilized egg to divide, enlarge, and grow into an animal with fully functioning organs. Table 13.2 summarizes and reviews the five processes. Within an embryo, these five processes perform a developmental rondo: They are repeated over and over in different cell groups until the embryo has organs with proper shapes and specialized functions. The embryo is then ready to face the rigors of the world outside its transparent jelly globe, its eggshell, or its mother's uterus. But development is far from over at hatching or birth; it continues throughout the individual's lifetime.

DEVELOPMENT CONTINUES THROUGHOUT LIFE

In terms of day-to-day survival, an animal's most important activities are finding and consuming food and defending itself. But in terms of evolutionary survival, reproducing is equally crucial. The development that continues after hatching or birth helps accomplish both short- and long-term survival by allowing the organism to grow larger, to acquire adult characteristics, and to mature sexually.

Continual Growth and Change

If you've ever adopted a young pet, then you know that **growth** is the most apparent aspect of postembryonic development. The animal's weight doubles four or five times before stabilizing in adulthood. The same is true for a newborn child; the weight of a 7 lb infant can double to 14, 28, 56, and 112 lb or more as it develops throughout childhood and adolescence (Figure 13.18). The weight gain comes largely from an increase in cell number due to cell division throughout the body. Thus, a newborn with billions of cells will have trillions as an adult.

Different parts of the body, however, grow at different rates. Although their weights are similar, it is easy to tell an adult toy Manchester from a Doberman pinscher puppy because of differences in the proportion of the head to the body. There are often other obvious differences between young and adult animals: The adult may have a slimmer shape; bigger teeth or claws; stronger muscles; faster locomotion; coloration or textural differences in feathers, fur, or skin; the presence of antlers; or any of a thousand other characteristics that help the animal harvest food, find and attract mates, and defend itself and its territory. Moreover, one can often distinguish a young or middle-aged adult from an old adult by the gradual changes in outer body coverings, by declining energy levels and agility, by the accumulation of scars and injuries, and by size changes. Some animals—lobsters and sturgeon, for example—continue to grow throughout their lifetimes. Among many other animals, however, only particular organs constantly increase in size. An 80-year-old person usually has a large nose and large ears because the cartilage in these organs has never stopped growing.

TABLE 13.2 Processes That Bring About Development	
Process	What Happens
Formation and storage of cytoplasmic determinants	Chemical factors stockpiled in the egg and partitioned into different cells during cleavage (like gray crescent cytoplasm) cause cells to select and express a set of genes that initiate a developmental program, such as commencing gastrulation
Induction	One cell group causes another group to start developing into a different cell type; e.g., the notochord induces nearby ectoderm to form the neural tube
Commitment (determination)	Cells select one pathway of development; e.g., cells in a developing fly become a leg or antenna
Morphogenesis	Organs form via changes in cell shape, cell proliferation, cell death, and cell migration; e.g., neural tube rolls up; digits are carved; neural crest cells migrate
Differentiation	Specialized cell types develop; includes the production of specific proteins that allow a cell to carry out its particular role; e.g., skin cells make fibrous proteins; intestinal cells make digestive enzymes

BOX 13.2   H O W   D O   W E   K N O W ?

HOW DOES A CELL "KNOW" WHERE IT IS?

Sometimes, as a human embryo develops in the womb, one of its limbs "gets confused." Instead of creating a normal hand with a thumb on one side, a little finger on the other, and three digits in between (Figure 1a), it generates a hand lacking a thumb but possessing two little fingers that flank five middle digits (Figure 1b). A machinist who lived in Boston about 100 years ago had such a hand and reported that it not only helped him do his job, but also gave him "certain advantages in playing the piano."

An unusual appendage like this raises a special question: What makes a little finger form on one side of the hand but not on the other? Or, more generally: How does a cell in an embryo "know" where it is so it carries out its functions in the appropriate place? Over 20 years ago, Lewis Wolpert suggested that such cellular localizing was based on *positional information*—spatial cues that tell cells where they are in the embryo, perhaps via diffusible molecules. Experiments on chicken wings and other vertebrate limbs have revealed how positional signaling does in fact take place in embryos.

A chicken's wing develops much like a human arm (Figure 1c). It has a single large bone attached to the shoulder, two smaller bones in the lower limb, a wrist joint, and a series of digits, and it develops from a tiny bud. Within this bud, a shell of ectoderm overlies a core of mesoderm, and during development, the bud transforms from a rounded protuberance to an elongated limb with bones, muscles, and skin.

W. J. Saunders wondered whether one part of the limb bud was more important in establishing the "thumb-to-pinky" series of positions than any other. He found that if he transplanted a special region (called the polarizing zone) from the posterior portion of one wing bud to the anterior portion of another, the limb produced a wing similar to the hand of the

Boston machinist, with two mirror-image copies of digits 4 and 3 and a single central copy of digit 2 (Figure 1d). A century ago, when the Boston machinist was still a developing embryo, his limb bud must have been damaged and must have spontaneously undergone a similar transplantation, with the result being a hand that had one central copy of digit 2 and two mirror-image copies of fingers 5, 4, and 3.

It was later found that researchers can produce a similar mirror-image wing by applications of a substance called retinoic acid. (Retinoic acid is a derivative of vitamin A, which is itself essentially a molecule of carotene, the bright orange pigment in carrots.) In the late 1980s, it was shown that retinoic acid is distributed in a gradient across the limb, with a high concentration at the limb bud's posterior and diminishing concentrations toward the anterior edge. It was suggested that this concentration gradient directs the pattern of digit formation, allocating digits to final locations on the limb and specifying the type each will develop into (thumb, middle finger, pinky, and so on). Limb cells can "read" the amount of retinoic acid around them because it binds to a receptor on the cell surface for that particular substance. The binding of retinoic acid to the cell surface receptor causes new sets of genes to be expressed, and scientists think these genes, in turn, cause a group of cells to develop as a thumb or as a little finger (or equivalent digits).

In 1990, other researchers showed that a variety of retinoic acid receptors are present in other parts of developing embryos and that retinoic acid can turn on human *Hox* genes (see Box 13.1). Thus, the same rules of positional information that allow a hand to develop and carry out a wide range of coordinated movements may also broadly govern the development of other embryonic organs and parts.

(a) Growth

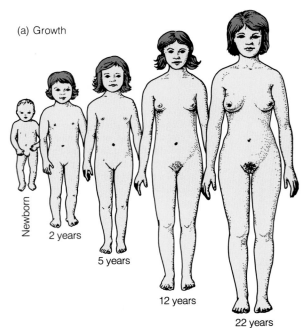

Newborn

2 years

5 years

12 years

22 years

(b) Maturation of the hand bones

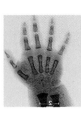 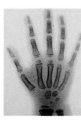   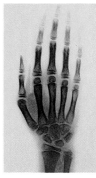 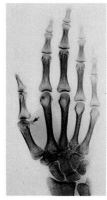

2 years   3 years

14 years   60 years

FIGURE 13.18 ■ **Growth: Weight Quadruples and Proportions Change Between Birth and Adulthood.** (a) As this girl grows into a woman, her weight multiplies from 7 lb to 14, 28, 56, and 112 lb. Also, her head grows proportionately smaller, her arms and legs longer, and her trunk less dominant. (b) X-rays of human hands reveal that the finger bones lengthen and enlarge and the knuckles become more and more prominent throughout life.

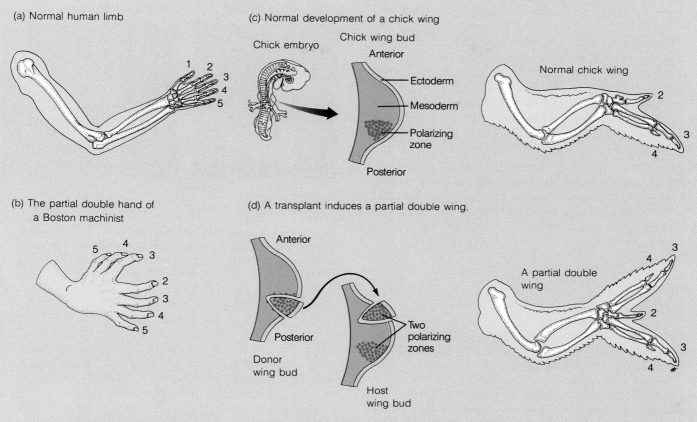

(a) Normal human limb

(b) The partial double hand of a Boston machinist

(c) Normal development of a chick wing

Chick embryo

Chick wing bud
Anterior

Ectoderm
Mesoderm
Polarizing zone

Posterior

Normal chick wing

(d) A transplant induces a partial double wing.

Anterior

Posterior

Donor wing bud

Two polarizing zones

Host wing bud

A partial double wing

FIGURE 1 ■ **Development of the Hand.** (a) A normal human hand contrasts with (b) that of a machinist from Boston. This man's hand had a double set of digits. (c) A normal wing bud in the embryonic chick leads to normal bones in the adult chicken wing. (d) When the posterior part of a limb bud is removed from a donor and transplanted to the anterior portion of a host wing bud, the result is a limb with two posterior axes and two mirror-image sets of digits.

Most organs, however, stop enlarging after an animal reaches its adult size, and development continues mainly through the generation of new cells that replace aged and dying cells. The orderly replacement of worn-out cells usually involves **stem cells,** generative cells that retain the ability to divide throughout life. Special stem cells in the skin generate new skin cells; others in the bone marrow generate new blood cells; some in the intestines make new intestinal cells; and stem cells in the gonads make new gametes. When a stem cell divides, one daughter differentiates into the specialized cell type (e.g., blood or skin cell) while the second daughter becomes another stem cell (Figure 13.19).

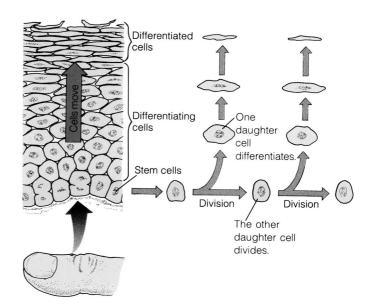

Differentiated cells

Differentiating cells

Cells move

Stem cells

One daughter cell differentiates.

Division

Division

The other daughter cell divides.

FIGURE 13.19 ■ **Stem Cells Divide and Renew Body Parts.** In the skin on your index finger, stem cells in the dividing layer continuously reproduce as time goes by. As each stem cell divides, one daughter remains an undifferentiated stem cell with the potential for future divisions, while the other daughter becomes differentiated into an epidermal cell and is pushed outward, becoming part of the finger's protective layer. The differentiated skin cell can no longer divide and eventually dies and flakes off as dry skin while younger cells push up from below.

Natural substances in the body called *growth factors* regulate the division and differentiation of cells so that growth and change occur only where and when needed. Growth factors can act like gas pedal or brakes to stimulate growth or inhibit it. Growth factors include *growth hormone,* which stimulates increased height; *platelet-derived growth factor,* which stimulates cells to divide and fill in wounds; *activin,* which may be Spemann and Mangold's embryonic organizer; and *nerve growth factor,* which helps nerve cells survive and grow (Figure 13.20). Together, stimulatory or inhibitory growth factors maintain the optimal number of cells for each body part, and under normal circumstances, those numbers will not change very much as the animal ages. Sometimes, however, something goes terribly wrong: Growth regulation fails, tissues grow unchecked, and the result can be a massive tumor or invasive cancer.

(a) Growth factor absent   (b) Growth factor present

Few nerve cell outgrowths

Many nerve cell outgrowths

FIGURE 13.20 ■ Growth Factors Regulate Cell Division and Differentiation. With growth factors absent, nerve growth is inhibited (a), but with the specific protein called nerve growth factor present, nerves grow luxuriantly (b).

Cancer: Development Running Amok

A tumor is, in a sense, an inappropriate continuation of development. A **benign** tumor is a clump of cells that continues to grow unchecked, but stays localized in a tightly packed group and thus can usually be removed surgically. A cancerous, or **malignant,** tumor, however, grows without ceasing and also spreads throughout the body (**metastasizes**), invading healthy organs and destroying them and often making surgical removal impossible (Figure 13.21a). In insidious ways, cancer cells mimic normal embryonic cells.

We saw that most fully differentiated cells seldom divide and instead arise from stem cells, which can divide but cannot perform as mature cells. Cancer cells appear to be stem cell derivatives that fail to completely differentiate and hence preserve their capacity to divide. Cancer cells (also called *transformed* cells) assume the rounded shape of embryonic cells (Figure 13.21b) and proliferate as if they were still part of a rapidly growing embryo. As we saw in Chapter 7 and earlier in this chapter, proteins called growth factors help regulate cell growth and differentiation (Figure 13.21c). Normally, these regulatory proteins are secreted by one cell, diffuse to another, and bind to a receptor on that recipient cell's outer surface. This binding stimulates signal proteins inside the cell, which in turn alert the nucleus to prepare for division.

In cancer cells, this normal process goes awry. Mutations occur in the genes that usually encode the growth factors, their receptors, or the signal proteins inside the cells. These mutations can then lead to cells proliferating out of control. The mutant forms of normal genes are called *oncogenes,* and different types of cancers may involve different oncogenes. For example, breast cancer cells produce many different growth factors, and scientists think that at least some of them may encourage the cancer cells to proliferate and invade new tissues. Significantly, women whose breast cancer cells have extra copies of a specific oncogene that encodes a growth hor-

mone receptor protein are much more likely to die than patients lacking the extra oncogene. In an exciting new approach to cancer therapy, researchers have found, for example, that they can block the growth of certain human breast cancer cells by giving the patient a specific antibody (a kind of protein formed by the immune system). This antibody binds to the mutant receptor protein and thereby blocks the attachment of the growth factor. This prevents the cell from receiving a signal to grow and thus inhibits the growth of the breast tumor.

We have just seen that unless prevented by competition from specific antibodies, oncogenes can transform normal cells into cancer cells. Researchers have recently discovered a different class of genes called *tumor suppressor genes,* whose protein products directly block tumor growth. Cancer biologists now think that before a full-blown cancer can begin, damage must accumulate in a number of tumor suppressor genes as well as in normal genes, rendering them oncogenes. Some of the accumulated mutations may affect cell growth, while others may affect cell movement. This is suggested by *metastasis,* the spreading of tumor cells and invasion of other healthy tissues, which is the hallmark of a malignant tumor. In a similar way, early embryonic cells invade the wall of the uterus and secure a nutrient supply, but that invasion is limited to one organ and ends before birth. Normal embryonic cells, such as neural crest cells, flatten out and then stick to and migrate along trails of a special protein (fibronectin) laid down in the embryo. Cancer cells, however, have very little of this protein, and researchers speculate that this may partially explain why the cells are round, not flat (review Figure 13.21b), and why they are "slippery"—they don't stick to their neighbors and so escape from them and crawl between nearby cells. If they escape into the blood or

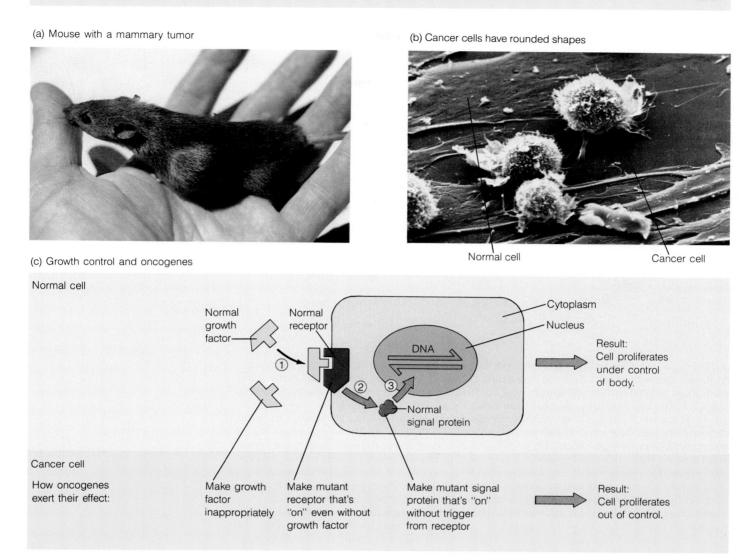

(a) Mouse with a mammary tumor

(b) Cancer cells have rounded shapes

Normal cell   Cancer cell

(c) Growth control and oncogenes

Normal cell

Normal growth factor

Normal receptor

Cytoplasm

Nucleus

DNA

Normal signal protein

① ② ③

Result: Cell proliferates under control of body.

Cancer cell

How oncogenes exert their effect:

Make growth factor inappropriately

Make mutant receptor that's "on" even without growth factor

Make mutant signal protein that's "on" without trigger from receptor

Result: Cell proliferates out of control.

FIGURE 13.21 ■ Cancer: Development Gone Awry. (a) A mammary tumor grows out of control in a mouse. (b) Cells in such a tumor have a different shape from normal differentiated cells. This scanning electron micrograph shows how the flattened growth characteristics of normal cells contrast with the rounded shape of cancer cells. (c) Growth in normal cells is controlled by growth factors, proteins made by one cell that can speed up the cell cycle in another cell (1). The growth factor diffuses to the target cell and binds to a receptor on the cell's surface (2). This binding triggers a change in a signal protein, which causes the nucleus to divide, and the cell to proliferate under control (3). In contrast to normal cells, cancer cells have mutations in genes that encode the growth factor, its receptor, or the signal protein. These mutations of the normal gene are called oncogenes, and they cause the cell to proliferate out of control.

lymph, these cells can spread far from the original tumor. If the cells are able to grow in their new locations and acquire an adequate blood supply, they may disrupt the structure of tissues enough to lead to the person's death. Once a large number of secondary colonies have formed, it is almost impossible to cure the cancer through surgical removal alone

of the primary and secondary sites. Radiation and/or chemotherapy are then needed as well.

With cancer, as with all diseases, an ounce of prevention is worth a pound of cure. Since we now know that cancers are due to mutations in genes that control normal growth and differentiation and that several mutations will usually accumu-

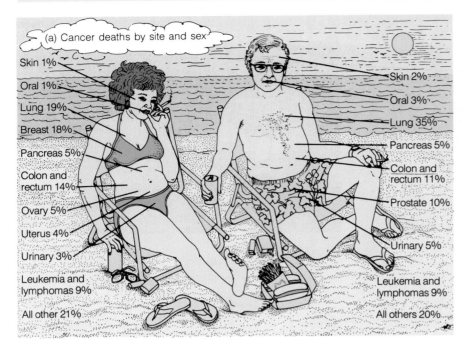

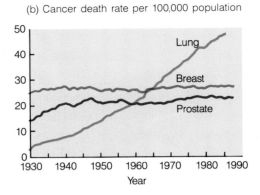

FIGURE 13.22 ■ **Cancer Risks and Life-Style Choices.** Each year during the 1980s, physicians diagnosed cancer in about 1.3 million Americans. Of those people, about 400,000 had skin cancer, while about 930,000 had cancer at other sites. (a) Today, one out of every five deaths in the United States is from cancer—a figure that has risen steadily since 1930 and is largely due, researchers believe, to the steep climb in lung cancers from cigarette smoking. (b) Since 1965, lung cancer has claimed more men than has any other type of cancer. From 1930 to 1986, breast cancer was the biggest cancer killer of women. Beginning in 1986, however, lung cancer began to claim more women's lives each year than breast cancer, probably because of increased smoking among women. You can help reduce your own risk of cancer. Cigarette smoking accounts for 30 percent of *all* cancer deaths (not just lung cancer) and should be avoided. Researchers believe that a high-fat, low-fiber diet increases the risk of breast, colon, and prostate cancers; that eating large amounts of salt-cured, nitrate-treated, or smoked foods can lead to esophageal and stomach cancers; and that the heavy use of alcohol is linked to cancers of the mouth, larynx, throat, esophagus, and liver. Finally, almost all of the 400,000 cases of skin cancer per year are due to overexposure to sunlight. A few changes in diet and life-style could add years to your life by reducing your cancer risk.

late before a cancer metastasizes, we must avoid repeated exposure to cancer-causing agents such as sunlight, cigarette smoke, and high-fat, low-fiber diets. A great deal of research data shows that people can markedly reduce their cancer risk in this way (Figure 13.22).

THE FORMATION OF GAMETES: THE DEVELOPMENTAL CYCLE BEGINS ANEW

We have seen that physical changes continue to accumulate after hatching or birth. As the young animal develops toward maturity, the body enlarges, the chest grows hair or swells to form breasts, the head sprouts antlers, and so on, depending on the individual's gender and species. Most of the changes

in physical form, physiology, and behavior are under the control of hormones (see Chapters 14 and 26), and many help prepare the organism for mating and reproduction—the start of a new developmental cycle. Whether the external changes are dramatic or subtle, sexual maturation always involves **gametogenesis:** the construction of *germ cells* (eggs and sperm), which are the living link between generations.

Special Egg Cytoplasm Leads to Germ Cells in the New Individual

In many species, the egg cell contains the seeds of its own rebirth in the form of **germ plasm,** a region of distinctive granular cytoplasm within the egg cell. As cleavage divides the fertilized egg, germ plasm is partitioned into only a few

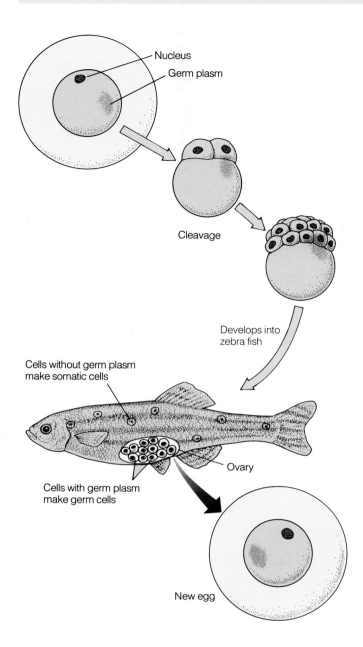

FIGURE 13.23 ■ Continuity of the Germ Plasm: Egg to Adult to Egg. Germ plasm, a special cytoplasm stored in the egg, becomes incorporated into only a few cells during cleavage. Cells containing the germ plasm become germ cells, with the ability to differentiate into sperm or eggs, which again contain germ plasm.

CONNECTIONS

The story of animal development carries us full circle from a single fertilized egg to an embryo, young animal, adult, and finally to the production of new eggs or sperm. Development helps show the unbroken continuity of living things, because none of an organism's physical structures and no event in its life cycle is made from scratch. Instead, each has a precursor in a previous part of the cycle, is constrained by prior structures and events, and develops as a result of chemical instructions or interactions among preexisting cells.

Embryology is a pivotal discipline in the biology of multicellular organisms. The actions of genes and proteins at the molecular level cause cells to become committed and to differentiate, divide, and specialize into the organism's body plan and organs. These embryonic organs become the functioning anatomical parts of organisms that we will study in later chapters of the book. The embryo is also the vehicle for evolutionary change. New species arise because new traits are selected that alter development. The gradual modification of a dinosaur's forelimb into the first bird's wing, for example, occurred because dinosaur embryos that had genes causing cells to change shape, divide, differentiate, migrate, or die away during the formation of an embryonic limb developed into adults that could glide through the air more effectively and hence survive and reproduce in greater number.

While this chapter has laid out the general principles of animal reproduction and development, there is a great deal more to say about how these generalities apply to our own species and how we progress through the life cycle from fertilized egg to fertilized egg. That is the subject of the next chapter.

blastula cells—the future germ cells (Figure 13.23). The germ plasm contains cytoplasmic determinants that in some unknown way instruct a cell to become a germ cell.

Once germ cells have been determined in an embryo, they migrate to a pocket in the mesoderm, which differentiates into a male gonad (a testis) or a female gonad (an ovary), depending on the embryo's sex. Inside the gonad, some germ cells mature into sperm or eggs. With the maturation of sperm and eggs, the process of development has gone full cycle, ready for another fertilization, more cleavage, gastrulation, and so on, as another individual's life begins.

Highlights in Review

1. Zebra fish are a good example for the study of animal development because through the clear protective globe surrounding the egg, we can watch the single-celled fertilized zygote dividing into thousands of cells and forming a recognizable embryo.

2. The egg cell is truly omnipotent; it gives rise to all other cell types, including other egg cells.

3. Three themes emerge from the study of development: Developmental processes are remarkably similar in all animals; development is an ordered sequence of irreversible steps; and embryonic cells express alternative genes at specific times.

4. Mating between animals usually begins with mate recognition, often based on visual or olfactory cues. Hormonal or behavioral signals then help to synchronize the production and release of gametes. The release can be external, with eggs or sperm deposited directly into a watery environment, or internal, with sperm deposited into the female's body.

5. Eggs are storehouses for nutrients and developmental instructions, while sperm are small, stripped-down cells specialized for rapid swimming and delivery of a haploid nucleus into the egg's cytoplasm.

6. During fertilization, the sperm's sac of digestive enzymes explodes, and the sperm penetrates the ovum's plasma membrane. The membranes of sperm and egg fuse, and the sperm nucleus plunges into the egg cytoplasm and fuses with the egg's nucleus, creating a diploid nucleus for what is now the zygote, the first cell of the new individual.

7. Cleavage is a series of rapid, synchronous mitotic divisions that in many species transforms the zygote into a hollow ball called the blastula.

8. Cleavage partitions instructional chemicals, or developmental determinants, into specific blastomeres. The gray crescent, a special cytoplasm that develops in a frog's egg opposite the point of sperm entry, is partitioned into a few blastomeres and induces an embryo's dorsal region to form. Cleavage can also assign a cell's position (e.g., outside or inside the cell mass of a mammal's blastula, or blastocyst) and with it the cell's developmental fate as part of the embryo or part of the trophoblast (future nutritive tissue).

9. Gastrulation is the process in which the blastula indents into a two-layered ball, and in which cells migrate between the two layers and form three layers—the ectoderm, mesoderm, and endoderm—the germ layers that give rise to all other cell types (except the germ cells).

10. Neurulation involves a folding of the ectoderm layer induced by underlying mesodermal notochord cells, forming the neural tube or future nervous system.

11. Cytoplasmic determinants and inductions work together and trigger determination within cells—the selection of a specific developmental pathway toward one cell type or another.

12. During organogenesis, the body's organs assume their form and function. The development of organ form, or morphogenesis, involves changes in cell shape, rapid cell proliferation, programmed cell death, and cell migration. The development of organ function, or differentiation, involves the selective expression of genes through the actions of regulatory proteins.

13. An animal's development continues after hatching or birth. The animal grows and takes on adult characteristics that enable it to collect food, defend itself and its territory, and find mates. In adulthood, cell division usually involves stem cells and is regulated by growth factors and other substances.

14. Benign or malignant tumors form as normal growth controls break down and cells proliferate unchecked; malignant tumors also spread, or metastasize, to other areas of the body.

15. The formation of gametes and subsequent fertilization complete the developmental cycle. Cytoplasmic determinants instruct the development of germ cells (eggs or sperm) in the embryo.

Key Terms

benign, 284
blastodisc, 273
blastomere, 272
blastula, 272
cleavage, 272
committed, 278
copulation, 269
development, 266
developmental
    determinant, 273
differentiation, 280
ectoderm, 275
embryo, 267
endoderm, 275
external fertilization, 269
gametogenesis, 286
gastrulation, 274
germ layer, 274
germ plasm, 286

gray crescent, 273
growth, 281
internal fertilization, 269
malignant, 284
mate, 268
mesoderm, 275
metastasize, 284
morula, 272
neural tube, 276
neurulation, 276
notochord, 276
organogenesis, 279
ovum, 269
parthenogenesis, 270
placenta, 273
primary embryonic
    induction, 278
stem cell, 283
yolk, 269

Study Questions

Review What You Have Learned

1. Various visual and olfactory cues lead to mating. What are some examples of each?
2. List the events of fertilization that follow the penetration of an egg by a sperm nucleus.
3. Is the egg of a whale (a mammal) larger than a chicken egg? Explain.
4. What role do developmental determinants play in the egg cytoplasm?
5. Which organs develop from each of the three germ layers?
6. Trace the steps of neurulation.
7. How do growth factors function?
8. How is a malignant tumor like a benign tumor? How is it different?
9. Compare the outcomes of egg fertilization and parthenogenesis.

Apply What You Have Learned

1. How can a person reduce his or her own risk of cancer?
2. An embryologist discovers a mutant zebra fish that lacks a notochord. What will happen to the development of the fish's nervous system?

3. An embryologist transplanted the posterior part of a fertile fruit fly egg to the anterior part of another egg. Why did the blastula-stage embryo develop germ cells at both ends?

4. Some drugs interfere with the formation of body parts during embryonic development. Why would it be more dangerous for a woman to take these drugs during the early part of a pregnancy than during the later part?

5. You are called as an expert witness in a lawsuit against a chemical company. The plaintiff is a woman who claims that her exposure to the company's chemicals when she was eight months pregnant caused her child to be born with a cleft palate and spina bifida. Would you side with the woman or with the chemical company? Defend your position.

 For Further Reading

BARINAGA, M. "Zebrafish: Swimming into the Development Mainstream." *Science* 250 (1990): 34–35.

CHERFAS, J. "Embryology Gets Down—To the Molecular Level." *Science* 250 (1990): 33–34.

DAY, S. "Genes That Control Genes." *New Scientist,* November 3, 1990, pp. 1–4.

DEROBERTIS, E. M., G. OLIVER, and C. V. E. WRIGHT. "Homeobox Genes and the Vertebrate Body Plan." *Scientific American,* July 1990, pp. 46–52.

GILBERT, S. *Developmental Biology.* Sunderland, MA: Sinauer, 1985.

HOFFMAN, M. "The Embryo Takes Its Vitamins." *Science* 250 (1990): 372–373.

KIMMEL, C., and R. WARGA. "Cell Lineage and Developmental Potential of Cells in the Zebrafish Embryo." *Trends in Genetics* 4 (1988): 68–74.

KLINE, D. "Activation of the Egg by the Sperm." *Bioscience* 41 (1991): 89–95.

MARX, J. "Oncogenes Evoke New Cancer Therapies." *Science* 249 (1990): 1376–1378.

NASH, J. M. "Cracking Cancer's Code." *Time,* November 12, 1990, p. 56.

RADETSKY, P. "The Roots of Cancer." *Discover,* May 1991, pp. 60–64.

SCHOENWOLF, G. C., and J. L. SMITH. "Mechanisms of Neurulation: Traditional Viewpoint and Recent Advances." *Development* 109 (1990): 243–270.

SUMMERBELL, D., and M. MADEN. "Retinoic Acid, a Developmental Signalling Molecule." *Trends in NeuroSciences* 13 (1990): 142–146.

VILE, D. "Cancer and Oncogenes." *New Scientist,* March 10, 1990, pp. 1–4.

WASSARMAN, P. M. "Profile of a Mammalian Sperm Receptor." *Development* 108 (1990): 1–17.

The Human Life Cycle

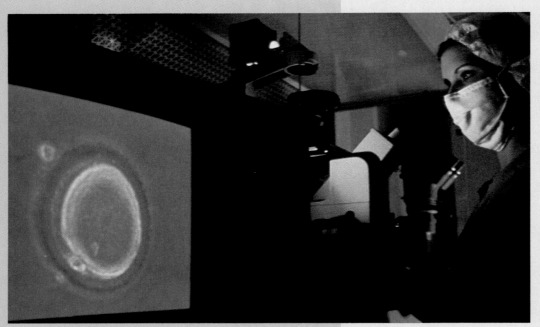

FIGURE 14.1 ▪ **Test-Tube Fertilization.** A technician examines the projected image of a ripe egg removed from a woman's ovary in preparation for in vitro fertilization—fertilization "in glass."

Fertilization in a Laboratory Dish

The birth of a baby girl in Oldham, England, in 1978 created a sensation around the world. Louise Joy Brown was a healthy and normal newborn in every respect save one: Her conception was revolutionary. She was the first baby in human history to be conceived in a laboratory dish. Biologists know more about the fertilization, embryonic development, growth, and maturation of humans than about those events in most other animal species, and this knowledge has enabled them to create "test-tube babies" like Louise Brown.

For nine years, Louise's parents, Lesley and John Brown, had failed to conceive a baby because a "roadblock" prevented Lesley's eggs from reaching her uterus, the organ that receives, protects, and nourishes the developing embryo. To circumvent this blockage, a team of pioneering physicians and researchers removed eggs from her ovary, mixed them with John's sperm, and reimplanted an early-stage embryo into her uterus. The procedure involved several steps. First, the medical team monitored the ripening of an egg cell still inside one of Lesley's ovaries; then they inserted a pencil-thin viewing tube (a laparoscope) near the cell, carefully sucked the ripe egg into a thin, hollow needle, and placed the egg in a laboratory culture dish (Figure 14.1). Meanwhile, they gathered a sperm sample from John, added the sperm to the culture dish containing the harvested egg, and waited for the sperm to fertilize the egg. Once the zygote cleaved to the eight-cell stage, the team drew the microscopic cluster into a flexible plastic tube and released the living cargo—now a viable embryo—into the mother's uterus. There it burrowed into the lining of the uterine wall and developed into a larger and larger organism. Nine months later, Louise was born, the first person conceived by **in vitro fertilization** (literally, "fertilization in glass") (Figure 14.2).

Such a technological feat is possible because biologists understand so many details of human fertilization, embryonic implantation, and fetal development. Much of the material we explore in this chapter will build on concepts of animal development from Chapter 13. Human embryos are amazingly similar to fish and chicken embryos at certain stages—right down to the tail and gill slits (Figure 14.3). Such similarities bespeak our close evolutionary ties to the other animals with backbones as well as the universal principles governing growth from a one-celled zygote to a multicellular plant or animal. Human life cycles resemble those of most other animals in another way: For a conception to occur and for a new generation to be perpetuated, males and females must reach sexual maturity, develop sexual characteristics, attract each other, and mate.

Once a person is conceived and born, development continues throughout childhood and adolescence. Like most animals, we cannot reproduce asexually, and so our bodies are mortal: They age, wear out, and eventually die after seven or eight decades of living. Nevertheless, our germ cells, produced in the gonads, or sex organs, give us the potential for genetic immortality (see Chapter 13).

As we discuss the physical structures and mechanisms that allow a couple to produce a baby and thus to begin a new human life cycle, several unifying themes will become apparent. First, the male and female reproductive systems are generally parallel in structure and function. They are two variations on a common motif: perpetuating the species. Second,

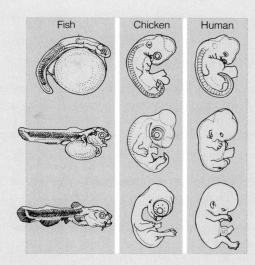

FIGURE 14.3 ■ **Vertebrate Embryos: Developmental Similarities Underscore Evolutionary Ties.** In their early stages, the embryos of fish, chicken, and human look amazingly similar (top row). In a middle stage, the fish embryo is elongating, but both chick and human still look alike and have curling tails. In a later stage, the differences are becoming more obvious, and the human embryo has lost its tail.

intricate networks of communication between body organs coordinate the manufacture and release of eggs and sperm, as well as an individual's development and maturation. These communications involve nerve signals as well as hormones, which are regulatory molecules produced in one part of an organism that cause cellular activity to change in other parts. Third, the pregnancy that commences once egg and sperm unite is a biological partnership between mother and young. While the mother's body provides the embryo nutrition and protection for nine months, the developing offspring chemically orchestrates the duration of pregnancy by signaling its own arrival in the uterus (and thereby ensuring safe lodging), stimulating birth, and initiating a milk supply. Finally, birth is the start of independent life, but it is not the end of development. People continue to grow and mature for about two decades before they reach peak physical performance and reproductive capacity and begin to age.

Our discussion of the human life cycle will answer the following questions:

- How do the human sexual organs develop in the embryo and child and function in the adult?
- How does human sexual activity unite egg and sperm, and how do birth control methods prevent reproduction?
- What remarkable steps transform a single-celled zygote into a multicellular human baby?
- How does growth continue after birth and throughout childhood, puberty, adulthood, and old age?

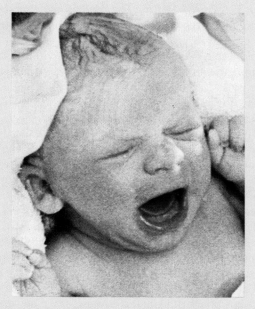

FIGURE 14.2 ■ **Louise Brown, the First "Test-Tube Baby."** After several years of trying to conceive without success, the Browns, with the help of laboratory techniques, produced a healthy baby girl.

MALE AND FEMALE REPRODUCTIVE SYSTEMS

Each human being has a complex system of primary and secondary sexual characteristics. *Primary sexual characteristics* are reproductive organs capable of passing along parts of a person's genome to the next generation via sexual reproduction. In contrast, *secondary sexual characteristics* are external features not directly involved in sexual intercourse, but nevertheless significant in reproductive behavior (Figure 14.4).

Male and female primary sexual characteristics are different but parallel in structure and function. Each sex has a set of *primary reproductive organs* consisting of the testes in males and the ovaries in females. The primary reproductive organs produce sperm or eggs, which arise through meiotic divisions, and they also produce sex hormones. Both sexes also have *accessory reproductive organs:* various ducts, chambers, and glands that act as organic plumbing to transport and store the eggs or sperm or, in the woman, to nurture the developing embryo. Let's see how human reproductive organs are built and regulated, starting with the simpler male system.

The Male Reproductive System

Male reproductive organs have just two jobs: making sperm by the millions and transferring them into the female's reproductive tract, where the male's gametes can encounter and fertilize eggs. The two jobs involve different subsets of organs and processes.

TABLE 14.1	Male Reproductive System
Structure	**Function**
Parts of Testis	
Seminiferous tubule	Produces and transports sperm
Interstitial cells	Produce testosterone
Accessory Ducts	
Epididymis	Stores and matures sperm
Vas deferens	Conducts sperm
Ejaculatory duct	Transports sperm
Urethra	Transports sperm and urine
Accessory Glands	
Seminal vesicle	Secretes fluids that aid sperm function
Prostate gland	Secretes fluids that aid sperm function
Bulbourethral gland	Secretes lubricants
External Genitals	
Scrotum	Contains testes
Penis	Organ of sexual intercourse; contains urethra

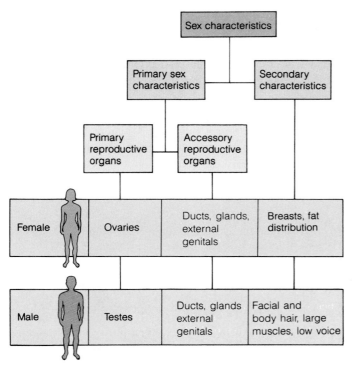

FIGURE 14.4 ■ Human Sex Characteristics.

■ **Organs That Produce Sperm** The primary male reproductive organs are the **testes** (singular, testis): smooth, oval structures about 4 cm (1.5 in.) long, held in a baglike **scrotum** (Figure 14.5a and Table 14.1). Each testis is subdivided into about 250 compartments (Figure 14.5b), and each compartment contains up to four highly coiled, hollow tubes called **seminiferous tubules,** which are about 70 cm (28 in.) long. If placed end to end, all the tubules in a man's testes would make a pipeline several hundred meters long but only about half the thickness of a sheet of paper. Sperm-forming cells (**spermatogenic cells**) in the tubule walls (Figure 14.5c) undergo meiosis and develop into sperm, while **supporting cells** (*Sertoli cells*) in those same walls surround and nourish the developing sperm. Connective tissue encases the seminiferous tubules, and embedded in it lie the **interstitial cells** (*Leydig cells*), which produce the male hormone testosterone.

To develop properly, sperm must have an environment several degrees cooler than the normal internal body temperature of about 37°C (98.6°F). In fact, the testes hang outside the body in a pouch of skin, the scrotum. If a man wears tight pants, exercises too hard, or sits in a hot tub, the temperature of the scrotum and testes can increase to a point where sperm development temporarily stops in some males. When he cools down, however, the spermatogenic cells can once again develop into sperm.

(a) Cross section of male pelvic area

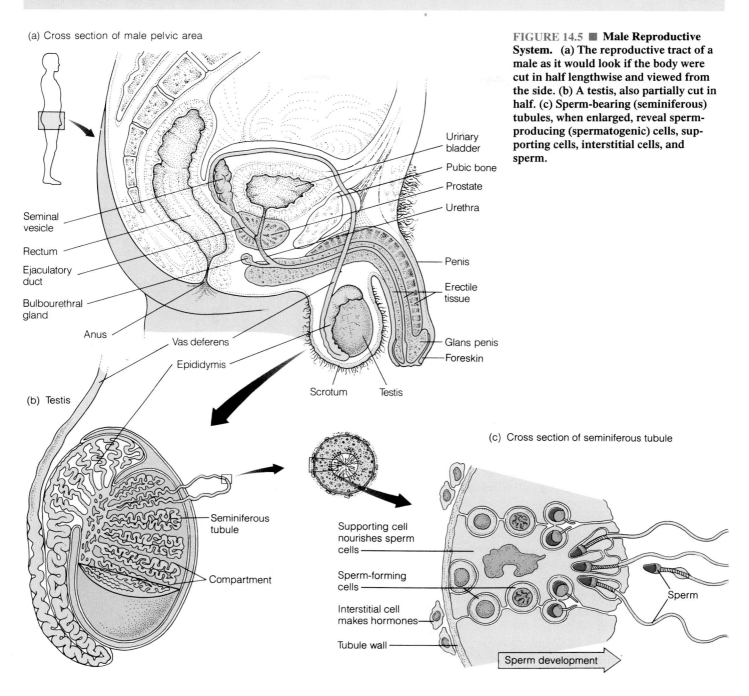

FIGURE 14.5 ■ **Male Reproductive System. (a) The reproductive tract of a male as it would look if the body were cut in half lengthwise and viewed from the side. (b) A testis, also partially cut in half. (c) Sperm-bearing (seminiferous) tubules, when enlarged, reveal sperm-producing (spermatogenic) cells, supporting cells, interstitial cells, and sperm.**

Seminal vesicle

Rectum

Ejaculatory duct

Bulbourethral gland

Anus

Vas deferens

Epididymis

(b) Testis

Urinary bladder

Pubic bone

Prostate

Urethra

Penis

Erectile tissue

Glans penis

Foreskin

Scrotum   Testis

Seminiferous tubule

Compartment

(c) Cross section of seminiferous tubule

Supporting cell nourishes sperm cells

Sperm-forming cells

Interstitial cell makes hormones

Tubule wall

Sperm

Sperm development

After attaining their streamlined shape (see Figure 13.5), sperm pass down the cavity of the seminiferous tubule and collect in the **epididymis,** a tightly coiled 7 m long tube on top of the testes. Here, final maturation takes place and the sperm are stored, ready for transport.

■ **Organs for Sperm Transport**   When a male is sexually stimulated during intercourse or masturbation or even unconsciously during sleep, sperm are rapidly transported and forcefully released from the body. The journey begins when smooth muscle cells in the walls of the epididymis contract

repeatedly and propel the sperm into a 45 cm (18 in.) long **vas deferens** (also called *ductus deferens* or sperm duct), a connecting tube that also has muscular walls that continue to propel the sperm (see Figure 14.5a and b). The vas deferens leads back into the body from the scrotal area; near the urinary bladder, it merges with the **ejaculatory duct,** a duct leading from the **seminal vesicle.** Secretions from this small gland bathe the sperm in a fluid that contains the sugar fructose and other nutrients as well as substances that regulate pH and stimulate muscular contractions in the female reproductive tract. The combination of sperm and fluid empties into the ejaculatory duct, which passes down the middle of a very important secretory organ, the **prostate gland.** This chestnut-shaped structure secretes a milky alkaline fluid that mingles with fluid of the seminal vesicle. The added secretion enlarges the volume of sperm and fluid and lowers the mixture's pH to about 7.5. This pH will help neutralize the natural acidity of the female reproductive tract, an acidity that protects her delicate tissues from microorganisms but also tends to inhibit sperm motility.

Once surrounded by milky fluid, the sperm passes from the ejaculatory duct to the **urethra.** This is a dual-purpose tube that can also carry urine from the bladder out through the **penis,** a cylindrical organ that transfers sperm to the female. A gland at the base of the penis, the **bulbourethral gland,** secretes a mucuslike lubricating substance, while spongy **erectile tissue** within the penis fills with blood and stiffens the entire organ, thereby facilitating entry of the penis into the female's vagina for intercourse.

At the peak of sexual excitement, **ejaculation** takes place: Strong muscular contractions of the urethral walls and other muscles forcibly expel **semen** from the penis. (Semen consists of sperm plus surrounding **seminal fluid,** the secretions of the seminal vesicle, prostate gland, and bulbourethral gland.) While only a small volume of semen is ejaculated (2 to 3 milliliters,* about a teaspoonful), it normally contains about 400 million sperm. So many sperm hurtling toward a single egg maximizes the chance for fertilization, and hormones help guarantee a continual supply of viable sperm.

■ **Hormonal Control of Sperm Production**   Hormones from the brain and testes work together in an interlocking feedback loop that governs the timing of sperm production. The basic scheme is simple: If sperm production falls below a specific level, hormones are released from the brain and transported in the bloodstream to the testes, causing these organs to increase production of both sperm and **testosterone,** a male steroid hormone chemically related to cholesterol. When the level of testosterone rises above a certain level, it acts on the brain, blocking further release of the brain hormones. With a lower concentration of brain hormones, the testes' production of sperm and testosterone falls. When sperm is released during sexual intercourse, the hormone levels fall below the set point, and the whole cycle starts once more.

The thin arrows in Figure 14.6 indicate, in a more specific way, how hormones regulate sperm production. The loop starts in the *hypothalamus,* a group of nerve cells located in the floor of the brain. When the level of testosterone is low, the hypothalamus secretes a *releasing hormone* (called *gonadotropin-releasing hormone,* or *GnRH*) into the blood (step 1). This peptide hormone flows in the blood directly to the *pituitary,* a pea-sized organ hanging from the base of the brain, where it stimulates the release of two peptide hormones called *LH* (*luteinizing hormone*) and *FSH* (*follicle-stimulating hormone*) (step 2). These hormones (called gonadotropins, since they stimulate the gonads) then move through the

bloodstream and activate cells in the testes. LH triggers the interstitial cells to produce and secrete testosterone (step 3), while FSH causes supporting cells to enhance formation of

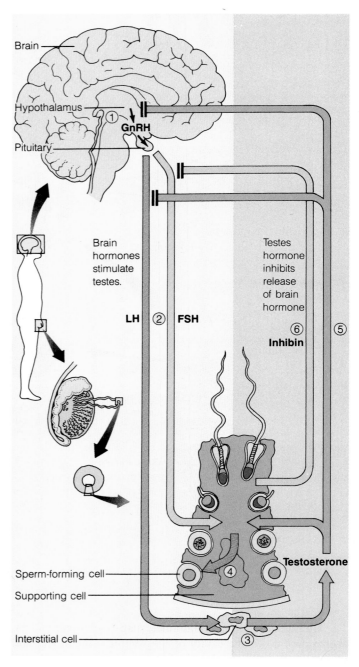

FIGURE 14.6 ■ How Hormones Control Sperm Production. Several hormones operating in interlocking feedback loops control the timing of sperm production. In the overview, brain hormones (steps 1 and 2) stimulate the testes to produce sperm and testosterone (steps 3 and 4). Testosterone also stimulates the testes to release inhibin. When the levels of both hormones have risen sufficiently, they inhibit the production of the brain hormones in a negative feedback loop (steps 5 and 6).

_____
*ml; see the abbreviations chart on the inside back cover.

sperm (step 4). Soon the sperm count rises. Meanwhile, testosterone circulates in the bloodstream at higher levels, and the interconnected loop feeds back on itself: High testosterone levels signal the hypothalamus to produce less releasing hormone (step 5). This, in turn, suppresses the release of LH and FSH, and without them, less testosterone and fewer sperm are manufactured. In addition, testosterone causes supporting cells in the testes to release the peptide hormone *inhibin,* which helps inhibit FSH production (step 6). When testosterone levels drop too low again, the hypothalamus is once more activated, and the whole cycle starts again: releasing hormone→LH and FSH→testosterone and sperm.

Testosterone has another effect besides triggering the release of brain hormones; it induces the development of secondary sexual characteristics at puberty. These include growth of mature genitals, male facial hair, and muscles; the stimulation of sperm production by sperm-producing cells; an increased interest in sex; and in many male animals and often in humans, the expression of more aggressive behavior. The hormonal feedback loop keeps the hormone levels and sperm production fairly constant and ever ready to reach and fertilize a human egg.

The Female Reproductive System

The female's reproductive organs not only produce and transport gametes, but also receive and nourish developing embryos and deliver babies. Like males, females have gonads that produce gametes and tubes that transport them, but they also have specialized modifications of the tube wall and more elaborate hormonal cycles.

■ **Production and Pathway of the Egg**   Women produce female gametes, or eggs, within two solid, almond-shaped organs called **ovaries,** which lie inside the body cavity just below the waistline (Figure 14.7a and b; Table 14.2). Each ovary is made up of cells called **oocytes,** which develop into eggs, and cells that surround and support the oocytes, called **follicular cells** (Figure 14.7c and d). An oocyte surrounded by follicular cells constitutes a unit called a **follicle,** and each ovary contains about 2 million follicles at birth.

Every 28 days or so, a process called **ovulation** takes place: A single follicle in one of the ovaries enlarges, its oocyte matures, the follicle ruptures, and a mature egg, now called an *ovum,* is released into the body cavity (see Figure 14.7c). The follicular cells left behind in the ovary change to the corpus luteum (described shortly). British researchers took the egg that became Louise Brown from her mother's ovary just at ovulation (see Figure 14.1). An ovum, the largest human cell, is about the size of the dot on this *i.* It moves toward the fringed opening of one of two 10 cm (4 in.) long tubes called *oviducts* (or Fallopian tubes) (see Figure 14.7b and c). Both right and left oviducts are lined with millions of cilia that act like paddles that sweep the ovum along. Sperm encounter the egg and fertilize it—usually as it moves down

the oviduct. The oviducts are contiguous with the thick-walled, pear-shaped **uterus** (see Figure 14.7a and b). Eventually, the ovum is swept into the narrow chamber within the uterus, where, if already fertilized, it can be nourished and develop into an embryo and fetus.

Most months, the ovum is not fertilized and instead degenerates or exits the uterus through the **cervix,** a muscle-lined opening with walls that secrete mucus. Depending on its consistency (which is regulated by hormones), this mucus can plug the opening during pregnancy or aid the movement of sperm. Outside the cervix lies the **vagina,** a hollow, muscular tube that receives the penis during intercourse, conveys uterine secretions to the outside, and can stretch into the birth canal that allows passage of the fetus. External reproductive organs surround the vaginal opening; these include protective tissues (labia minor and labia major) and tissues sensitive to sexual stimulation, such as the **clitoris,** and the lubricating **Bartholin's glands.** The functioning of these internal and external organs requires a hormonal control system that is parallel to, but more complex than, the male's.

■ **Hormonal Control of Egg Production and Uterine Preparation**
Once a human egg is ovulated and fertilized, it lodges in the uterus and there develops into an embryo. The uterine wall

TABLE 14.2 Female Reproductive System	
Structure	Function
Parts of Ovary	
Oocytes	Develop into eggs
Follicle cells	Support oocytes; produce estrogen and progesterone
Accessory Ducts	
Oviduct	Transports ovum; site of fertilization
Uterus	Sustains embryo throughout pregnancy
Cervix	Secretes mucus; reduces bacterial contamination
Vagina	Organ of sexual intercourse; acts as birth canal
Accessory Glands	
Bartholin's gland	Secretes lubricant
External Genitals	
Labium major	Protects other reproductive organs
Labium minor	Protects other reproductive organs
Clitoris	Sensitive to sexual stimulation

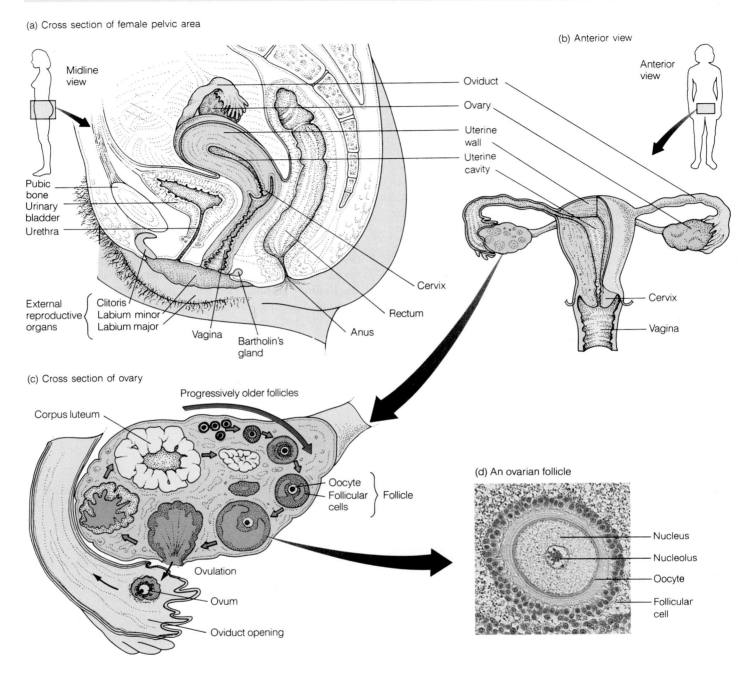

(a) Cross section of female pelvic area

Midline view

Pubic bone
Urinary bladder
Urethra

External reproductive organs
{ Clitoris
Labium minor
Labium major }

Vagina
Bartholin's gland

Oviduct
Ovary
Uterine wall
Uterine cavity

Cervix
Rectum
Anus

(b) Anterior view

Anterior view

Cervix
Vagina

(c) Cross section of ovary

Progressively older follicles

Corpus luteum

Oocyte
Follicular cells } Follicle

Ovulation
Ovum
Oviduct opening

(d) An ovarian follicle

Nucleus
Nucleolus
Oocyte
Follicular cell

FIGURE 14.7 ■ Female Reproductive System. (a) The female reproductive system as it would look if it were cut in half lengthwise and viewed from the side. (b) The reproductive system viewed from the front looks a bit like a pear with arms. (c) An enlarged view of one ovary and fringed oviduct opening. (d) A magnified view of a human egg cell, surrounded by a jellylike coat and follicular cells.

lining, or **endometrium** (see Figure 14.8), goes through stages of buildup and preparation to receive and support an embryo and then goes through stages of breakdown and outflow if no embryo arrives. Most women have a roughly 28-day cycle coordinated by hormones and called the **menstrual cycle.** During each menstrual cycle, the endometrium builds up and prepares for pregnancy, and an egg is ovulated. If the ovulated egg is not fertilized, the lining sloughs off, and men-

strual bleeding occurs, beginning a new 28-day cycle. In the menstrual cycle, ovaries make the female hormones **estrogen** and **progesterone,** the pituitary makes LH (luteinizing hormone) and FSH (follicle-stimulating hormone), identical to the brain hormones made in the male, and these hormones interact in a feedback loop (see Figure 14.8).

The menstrual cycle is usually counted from day 1, when menstrual flow begins and when levels of progesterone and estrogen in the blood are at their lowest. As in males, the drop in gonadal hormones triggers the hypothalamus to secrete releasing hormone (GnRH; see Figure 14.8, step 1). A decline in the production of this small releasing hormone (just ten amino acids long) appears to cause the menstrual cycle to cease temporarily in some women who are very thin or who exercise strenuously, such as ballerinas, long-distance run-

ners, gymnasts, or body builders. When present, the releasing hormone stimulates the pituitary gland to produce FSH and LH (step 2), which travel in the bloodstream to the ovary. As the name suggests, FSH stimulates follicles to grow (step 3), but usually only one follicle with its oocyte matures each month. The follicle grows rapidly and secretes increasing amounts of estrogen (step 4). This hormone causes the uterine

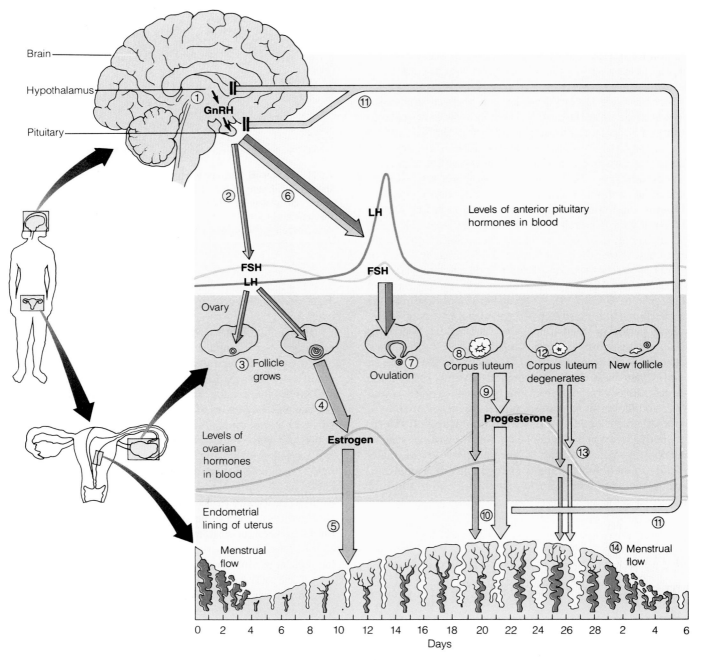

FIGURE 14.8 ■ How Hormones Control Egg Release: What Makes the Menstrual Cycle Cycle? As in the male, a welter of hormones operate in interlocking feedback loops, controlling the ripening and release of an egg every 28 days. (The text follows each step in detail.) The top two panels show the concentrations of hormones in the blood at various days in the menstrual cycle starting at day 1, the first day of menstrual flow. Included in the middle panel is a diagram of events in the ovary on different days, and the bottom panel indicates the thickness of the uterine lining in response to the cycling hormones.

lining to become thicker and more heavily supplied with blood (step 5). On about the fourteenth day of a 28-day cycle, the pituitary gland secretes a large pulse of LH and additional FSH (step 6), and these trigger the oocyte to complete the first meiotic division, which it began before birth. The developing follicle then ruptures and releases the egg (step 7).

The doctors who assisted in the in vitro fertilization of Louise Brown carefully monitored the levels of LH in Lesley Brown's body so that they could predict the exact time of ovulation and collect the ovum. This monitoring is relatively easy, since excess LH is secreted in the urine. Women who want to become pregnant can synchronize intercourse with ovulation by buying "dipsticks" coated with chemicals that turn bright blue when dipped in urine that contains peak levels of LH.

Once the ovum has left the ovary and begins its trek down the oviduct, the follicular cells left behind in the ovary enlarge and form a new gland, the **corpus luteum** (literally, "yellow body") (step 8). Corpus luteum cells continue to secrete estrogen, but they now begin producing large quantities of progesterone as well (step 9). Together, estrogen and progesterone promote the continual buildup of the uterine lining (step 10) and inhibit the hypothalamus from making releasing factors and the pituitary from releasing FSH and LH (step 11).

If the ovum does not encounter sperm on its downward journey and is therefore not fertilized, diminishing levels of LH and FSH allow the corpus luteum to degenerate on day 24 of the cycle (step 12) and thus to release less and less estrogen and progesterone (step 13). As these hormones diminish, the endometrium begins to slough off, and an approximately five-day-long period of menstrual flow starts again (step 14), marking the beginning of the next cycle.

The female's hormonal loops are similar to the male's in several ways. First, and most obviously, they involve the same releasing hormones from the brain (releasing LH and FSH). Second, hormone concentrations are self-regulating in that they feed back and turn their own production off. Third, reproductive hormones ensure the continuation of the species by making gametes available—continuously in the male, cyclically in the female. Finally, estrogen has developmental effects in the adolescent female that correspond to those of testosterone in the male.

SPERM MEETS EGG—OR DOESN'T: FERTILIZATION, BIRTH CONTROL, AND INFERTILITY

Once a female's hormones have done their job, an egg, released from the ovary, passes down the oviduct and is receptive to sperm (and thus fertile) for only a day or so. As in other animals, that gametic rendezvous depends on behavioral fac-

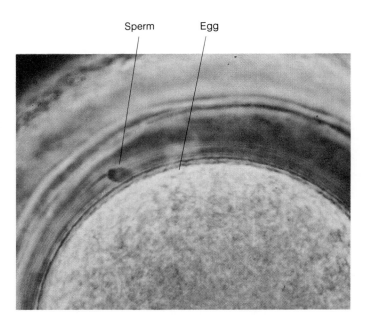

Sperm   Egg

FIGURE 14.9 ■ Human Fertilization. This photo documents the brief moment of fertilization, during which the speediest sperm enters the egg (here, near center of photo). The sperm nucleus will move into the egg cytoplasm, and the two nuclei will fuse, forming the unique genotype of a new individual.

tors: For fertilization, male and female must engage in **coitus,** or sexual intercourse, within a few days of the time of ovulation.

Human bodies are admirably suited to help sperm and egg converge. Nerve impulses generated during sexual stimulation increase blood flow to the male's penis, causing the organ's spongy erectile tissue to collect blood, swell, and stiffen. In the female, the clitoris, the tissues around the vagina, and the nipples of the breasts also swell and grow sensitive to stimulation. Lubricants flow from the male's bulbourethral glands and from the female's Bartholin's glands and vaginal wall, easing penetration and making the vagina more hospitable to sperm. Further, clitoral stimulation and pelvic thrusts build sexual excitation in both partners, and the excitement usually peaks with **orgasm,** involuntary muscular contractions of the vaginal and uterine walls or of the muscles lining the seminal vesicles, urethra, and other male structures. These contractions are highly pleasurable in both sexes and in the male cause the high-pressure ejaculation of sperm.

Of the 400 million living sperm simultaneously racing toward the ovum at 1 cm (0.39 in.) per minute, only a few tens of thousands navigate the cervix and enter the uterus, and fewer still encounter the ovum in the oviduct, where fertilization usually takes place. Nevertheless, over 100 sperm may collide with the ovum, the tip of each sperm detonating and releasing enzymes, and these enzymes will chew through the ovum's protective coats (review Figure 13.6b). The race ends when one sperm enters (Figure 14.9). This union triggers the ovum to complete meiosis II (LH triggered completion of meiosis I at ovulation) and to erect barriers to further sperm penetration. With this, the sperm nucleus moves toward the egg nucleus and fuses with it in the egg cytoplasm, thus forming a unique diploid genotype. The implantation of the developing embryo in the uterine wall marks the beginning of **pregnancy,** a series of developmental events that involve close cooperation between mother and embryo and transform a zygote into a baby.

Fertility Management and the Planet's Life-Support Systems

The drama of human fertilization happens quite naturally—and, many think, all too often. Every minute, at least 230 babies are born in the world, but only about 90 people die, leaving a net increase of 140 people per minute, or 1.4 million new people every week. Many eminent ecologists, including Paul and Anne Ehrlich, believe that human fertility is the single most serious problem in the world. Because of the exploding human population, food supplies are inadequate and trees and fossil fuels are burned to excess, causing global warming, the extinction of species, and the ruin of entire ecosystems. And the Ehrlichs point out that even in industrialized countries like the United States, where the birthrate is fairly low, population is still a serious problem because American parents use about 30 times more energy, food, and other resources raising a child than parents in poorer countries like Ecuador or Bangladesh. For personal reasons, as well as these environmental ones, many people around the world seek to avoid unwanted pregnancies.

Birth Control

Birth can be prevented either by blocking conception (**contraception**) or by terminating pregnancy. There are many good ways to block conception, including preventing an ovum from forming or preventing fertilization. Pregnancies can be terminated by inhibiting the fertilized ovum from implanting into the wall of the uterus or by preventing or interrupting embryonic growth after implantation. Table 14.3 on page 300 provides details for different types of intervention.

All techniques for blocking fertilization share the same goal—to prevent sperm from reaching the egg—but they vary widely in approach. Behavioral techniques include abstaining from intercourse totally, abstaining during a woman's fertile period (the so-called rhythm method, or natural family planning), and withdrawal of the penis before ejaculation. All three approaches are notoriously unsuccessful (see Table 14.3). The biggest problem with natural family planning is that the couple must remain abstinent for about 17 days per month. This is because it is currently difficult to predict the

day of ovulation (and thus the 24 hours or so when the egg can be fertilized) and because sperm can live in the female reproductive tract for up to three days, lying in wait, as it were, for the egg's release.

There is reason to hope that an accurate system of prediction can be made commercially available based on the timing of hormone peaks in a woman's blood. As you can see from Figure 14.8, the amount of estrogen in a woman's blood peaks just before she ovulates, and the amount of progesterone rises just after ovulation; these peaks, in effect, provide the red light and green light for sexual intercourse. Pharmaceutical companies have developed sensitive "dipstick" tests based on monoclonal antibodies (immune system molecules described in Chapter 22). One type now sold in drugstores can be dipped in a few drops of urine, and a color change indicates the increase in progesterone just after ovulation. A second type, still being developed, would work the same way but detect the increase in estrogen and hence predict ovulation, so that couples wishing to avoid pregnancy could abstain from intercourse between the estrogen peak and the progesterone peak. Despite the hopes for this better ovulation predictor, no new form of birth control has appeared in the United States since the introduction of the pill and IUD over 30 years ago. Box 14.1 on page 301 explores the reasons for this and the consequences for both the developed and developing world.

Chemical techniques for preventing conception, such as douching (rinsing the vagina) or applying spermicidal (sperm-killing) creams, foams, or jellies are only slightly more effective than behavioral techniques. Physical barriers such as a rubber condom (a sheath worn over the penis) or diaphragm (a dome covering the cervix) can be used much more successfully if combined with spermicides. Condoms can also help in preventing the spread of sexually transmitted diseases. Sterilization, a permanent barrier to conception, is achieved when a surgeon ties off a woman's oviduct (in a *tubal ligation*) or a man's vas deferens (in a *vasectomy*).

The pill, or oral contraceptive, is a hormonal strategy for blocking the production of eggs or sperm. The female pill is a mixture of synthetic hormones that block FSH and LH production and hence the maturation and ovulation of eggs. Pills have side effects some people find unacceptable; blood clots are a particular risk to women who are over 35 or who smoke. A recent contraceptive innovation called Norplant is implanted under the skin of a woman's upper arm. It slowly releases the same hormones used in birth control pills and can prevent conception for up to five years.

In the experimental realm, scientists have developed a birth control vaccine that causes the woman's immune system to destroy a pregnancy-sustaining hormone, thereby allowing any embryo that might have formed to be washed out with the menstrual flow. Even if fertilization occurs, the embryo's implantation into the uterine wall can be inhibited. An intrauterine device (IUD), a small plastic or metal insert worn in the uterus, can interfere mechanically with implantation of the embryo. While very effective, these devices are associated

TABLE 14.3 Birth Control Methods

Method	How It Works	Percent Accidental Pregnancies in First Year	Side Effects or Other Problems
Behavioral Techniques			
No method	Chance	Up to 80 or more	Unwanted pregnancies and risks associated with childbirth
Rhythm method	No intercourse during woman's fertile period (days 12–20)	20	Often ineffective because eggs can be released before or after this period
Withdrawal (coitus interruptus)	Penis is withdrawn before ejaculation	20	Depends on willpower and timing; small amounts of semen are often released before ejaculation
Chemical Techniques			
Spermicides (foams, creams, vaginal rinses)	Chemicals introduced to vagina before or after intercourse kill sperm	21	Applications often too little, too late, or not done at all, and sperm can survive
Physical Barriers			
Condom	Rubber sheath on erect penis catches ejaculated semen	12	Effectiveness depends on how carefully used; condom can slip and allow semen to leak; some people are allergic to condom material or coatings
Condom plus spermicidal foam	Same as above plus sperm-killing action of contraceptive foam	0–2	Effectiveness depends on how carefully method is used; helps protect against AIDS and other STDs
Diaphragm plus spermicidal jelly or cream	Round rubber dome with springy edge covers cervix and holds jelly or cream; diaphragm blocks sperm entry; chemicals kill sperm	2–10	Effectiveness depends on how carefully both are used; some people are allergic to diaphragm or chemicals
Contraceptive sponge	Disposable sponge containing spermicide blocks cervix, absorbs sperm, and kills them	18	Effectiveness depends on how carefully used; some people are allergic to the sponge or chemicals
Sterilization			
Tubal ligation	Doctor ties off a woman's Fallopian tubes, permanently blocking sperm passage	0.4	Slight chance operation will not completely block tubes; requires surgery
Vasectomy	Doctor severs and ties off a man's vasa deferentia, permanently blocking sperm release	0.15	Very slight chance procedure will not completely block vasa deferentia; simple procedure; can be performed in doctor's office; possible immune system reaction
Hormonal Strategies			
Birth control pills, implants	Synthetic estrogens and progesterones prevent normal menstrual cycle from occurring, so eggs are not released	3	Effectiveness depends on how conscientiously pills are taken; some women experience nausea, missed periods, other side effects; pills increase the risk of blood clots and strokes
Implantation Blockers			
Intrauterine devices (IUDs)	Prevent fertilized egg from implanting in uterine wall	6	Most no longer sold in U.S. because manufacturers fear lawsuits due to infections, scarring, and sterility risk

with increased risk of pelvic infections, potential scarring, and future infertility.

Abortion is the expulsion of a fetus, either spontaneously or deliberately induced. The Supreme Court ruled nearly two decades ago that the government cannot forbid deliberate abortions early in pregnancy. Many people, however, have strong moral objections to abortion, and most people would agree that safe and effective contraception is medically and

---

BOX 14.1   BIOLOGY: A HUMAN ENDEAVOR

CARL DJERASSI: POPULATION PESSIMISM, CONTRACEPTIVE CONCERNS

Carl Djerassi was born in Austria in 1923, and at the time, the world's human population was 1.9 billion. In the 1950s, about the time Djerassi won international acclaim for designing the first oral contraceptives, the population was 2.5 billion. By his sixtieth birthday in 1983, that number had swollen to 4.7 billion, and if he lives to celebrate his hundredth birthday in the year 2023, Djerassi will be sharing the planet with 8 billion other people. Never before has our species needed effective contraceptives so badly. But the father of the birth control pill is not optimistic about the prospects (Figure 1).

Djerassi has at least three main reasons for his pessimism. First, as he has stated in his articles and lectures, most of the population increase will come in developing countries, where cultural attitudes still tend to favor large families. In 1984, for example, Europe's population of 490 million was approximately equal to Africa's at 530 million. By 2023, however, Europe's population will have increased by only 4 percent to 510 million, while Africa's will have mushroomed nearly 300 percent to 1.4 billion—a pattern expected for much of Asia, India, and Latin America, as well.

Because of this demographic pattern, the populations of industrialized nations tend to be getting older, while Third World populations are getting younger, on average. This explains Djerassi's second cause for pessimism: Virtually all contraceptive research is done in industrialized countries (which Djerassi calls the "Geriatric World"), while it is the developing nations (his so-called "Pediatric World") that most need new birth control technologies. Pharmaceutical companies in the Geriatric World now spend most of their research dollars toward new drugs for the diseases of old age such as heart attacks, cancer, arthritis, and diabetes, not primarily for the health problems of young people, babies, or children.

Finally, Djerassi's third reason for pessimism is related to the second: In 1970, 13 companies (9 American, 4 European) were sponsoring contraceptive research. Today, only 4 companies remain in that field (3 European and just 1 American). This is partly the geriatric medicine phenomenon and partly, thinks Djerassi, a side effect of product liability lawsuits and antiabortion activism in the United States. These situations in the United States also help explain why European companies have introduced the only totally new contraceptives in the last 20 years.

FIGURE 1 ■ Carl Djerassi Lecturing on Birth Control.

If Djerassi could choose the new birth control methods of the future, he would pick a new spermicidal cream with antiviral properties to fight AIDS, herpes, and other venereal diseases; more kinds of once-a-month pills effective as menses inducers (like the French product RU-486); easily reversible forms of vasectomies (male sterilization); and reliable antifertility vaccines against which a woman wishing to become pregnant would take a short-acting reversal injection.

Because it takes nearly 20 years from the stage of product design through animal and human testing to the marketing of a new drug, Djerassi thinks it very unlikely that such new contraceptives will be available to help stem the mammoth population increase our world will surely face in the twentieth century. A pessimistic prospect, indeed.

emotionally superior to abortion. Still, there are several medically supervised abortion procedures currently available in the United States, and RU-486 is a menstruation-inducing pill developed in France and used widely in Europe (see Box 14.1 on page 301).

Overcoming Infertility

While limiting human fertility is an environmentally important goal, life without children is unthinkable for many people, and infertility can seem a personal tragedy. The parents of Louise Brown had a common and distressing problem: infertility. One in six couples cannot conceive without medical treatment, and that number is growing, partly because of the increase in venereal diseases (see Box 14.2 on page 305) and the use of IUDs and partly because of the common practice of deferring parenthood to the 30s, when fertility naturally declines. About half the time the woman is the infertile partner, and in most of these cases, her uterus or oviducts are blocked or scarred. When the male is the infertile partner, he usually has a low sperm count, or his sperm lack normal motility.

Today, physicians can successfully treat about 70 percent of infertile men and women through therapy with synthetic hormones, corrective surgery to unblock passages, or techniques like in vitro fertilization. Even with medical help, however, some women cannot produce eggs normally or carry a baby, and some men cannot make healthy sperm, and they are increasingly turning to gamete donors and surrogate mothers, although the legal issues surrounding this practice are complicated.

In addition, researchers have devised ways to freeze embryos. If the first attempt to artificially implant an embryo fails, physicians can retrieve another from cold storage and try again without having to repeat all the prior steps of in vitro fertilization. The first frozen-embryo baby was born in the United States in June 1986 and joins Louise Brown and more than a thousand other "test-tube babies" in a remarkable and rapidly growing club.

PREGNANCY, HUMAN DEVELOPMENT, AND BIRTH

Whether an embryo is conceived the old-fashioned way or with the help of laboratory technicians and Petri dishes, it follows the same course: It resides in the womb for approximately nine months and grows from a single fertilized egg to a bouncing bundle of 200 billion cells. The unique set of genes established at conception programs the shape, distribution, and functioning of the billions of developing cells.

Implantation and the Chorion: The Embryo Signals Its Presence

The zygote begins to cleave as cilia in the oviduct sweep it toward the uterus (Figure 14.10a), and on day 4 it reaches the morula stage (see Chapter 13). As this berrylike, solid ball increases to about 30 cells, a cavity forms in the middle, and the embryo is now called a blastocyst (see Figure 14.10a, day 5). The blastocyst forms two cell groups: the inner cell mass, which will become the embryo, and the trophoblast, which will form nutritive tissue. About six days after fertilization, when the blastocyst consists of 100 cells or so, it attaches to the uterine wall, secretes enzymes that break down a small portion of the lining, burrows in (or *implants* in the lining), and at day 7 establishes the first physical bond between mother and young.

Trophoblast cells contribute to the **chorion,** a fluid-filled sac that surrounds the embryo. The chorion is the embryo's three-way ticket to survival: It absorbs nutrients from the mother's blood and passes them on to the rapidly dividing embryo; it develops into the larger placenta that sustains the embryo throughout the nine months of gestation; and it produces a hormone called **human chorionic gonadotropin (hCG),** which prevents a new menstrual cycle, the onset of which would flush the embryo from the uterus (see Figure 14.10a). As hCG enters the mother's bloodstream (starting on day 10 after fertilization) and is carried to the ovaries, it mimics the action of LH and FSH. Specifically, hCG prevents the corpus luteum from degenerating as it normally would two weeks after ovulation. Instead, the corpus luteum continues to make estrogen and progesterone, and these hormones in turn maintain the uterine lining, prevent menstruation, and allow pregnancy to continue for the first two months (Figure 14.10b).

The production of hCG, the first major biochemical event of pregnancy, signals the embryo's presence and positively confirms pregnancy. Home pregnancy tests use a simple but ultrasensitive system for detecting hCG, and 97 percent of the time, they accurately reveal a pregnancy just days old.

The chorion, arising from the trophoblast, grows and enmeshes with maternal tissue, forming the dark red, spongy *placenta* (Figure 14.10b and c). The placenta enlarges as the pregnancy continues and serves as the vital link between mother and embryo. The placenta is an exchange site where a thick tangle of embryonic blood vessels encounters blood-filled spaces in the uterine lining. Embryonic and maternal bloods do not mingle, but nutrients and oxygen pass from the mother's blood across embryonic vessel walls and into the embryo's blood, and carbon dioxide and other wastes pass back in the reverse direction. After a few weeks, the placenta begins to make estrogen and progesterone, and these hormones maintain the uterine lining, block the production of LH and FSH, and thus prevent menstruation and maintain the pregnancy. The corpus luteum then slowly degenerates.

FIGURE 14.10 ■ Early Development and Implantation of the Human Embryo. (a) In the human, the early stages of development take place as the zygote travels down the oviduct toward the uterus. The egg starts to implant itself in the uterine wall about the sixth day after fertilization. By the tenth day, the chorion forms and gives off the hormone human chorionic gonadotropin (hCG), which maintains the corpus luteum's production of estrogen and progesterone and thus prevents the uterine lining from sloughing off in a new menstrual cycle. (b) After 60 days, the placenta is well established and produces the hormones that prevent menstruation for the remainder of the embryo's nine-month gestation. (c) Narrow fingers of tissue, or chorionic villi, project from the chorion, and each eventually houses a tiny blood vessel. The maternal blood fills the spaces around the villi, and exchange of gases and materials takes place across the delicate layer separating the maternal and fetal blood supplies.

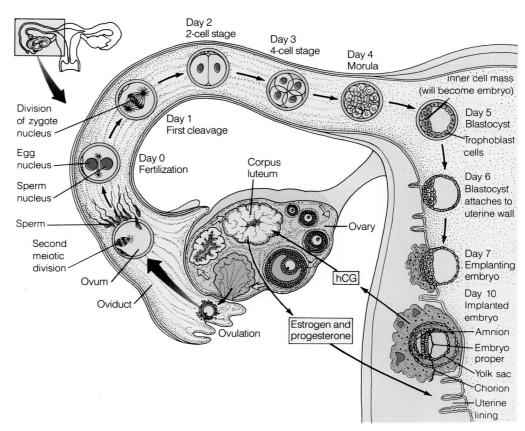

(a) Ovulation to implantation

(b) The pregnant uterus at 2 months

(c) A portion of the placenta

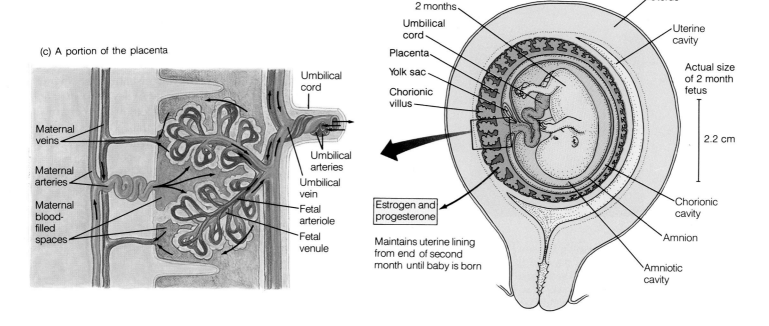

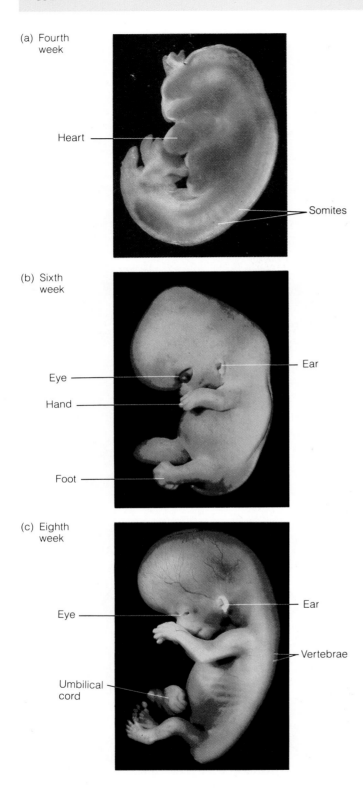

(a) Fourth
week

Heart

Somites

(b) Sixth
week

Eye

Hand

Foot

Ear

(c) Eighth
week

Eye

Umbilical
cord

Ear

Vertebrae

FIGURE 14.11 ■ The Developing Human Embryo. Photos show recognizable details in the stages of the developing human embryo. **(a)** By the fourth week, the embryo has a future back-bone and brain. **(b)** In the sixth week, eyes, ears, hands, and feet are beginning to form. **(c)** By the eighth week, the embryo has limbs, vertebrae, eyes, ears, and a jaw.

Developmental Stages in the Human Embryo

Even before a woman suspects she may be pregnant, the embryo has embarked on the early stages of development. Eight days after fertilization, the inner cell mass differentiates into ectoderm and endoderm layers, and the two layers form a disc within the trophoblast that resembles a pair of salad plates stacked inside a basketball. The cells in the upper plate grow upward and form a hollow ball. The space within this ball becomes the **amniotic cavity,** enclosing a watery, salty fluid that keeps the embryo moist and cushions it from blows (blue in Figure 14.10a, day 10, and 14.10b). The embryo sloughs off cells into this fluid; much later these cells can be collected and analyzed during amniocentesis (see Chapter 12). Cells in the lower plate grow downward and form another hollow ball (the *yolk sac*) by day 10 (golden color in Figure 14.10a, day 10). Parts of this sac eventually migrate into the placenta and form part of the **umbilical cord,** a lifeline connecting the offspring to the placenta and maternal blood supplies.

In the third week of pregnancy, when the embryo is smaller than the length of this *l,* gastrulation establishes the embryo's three-layered body plan. During the next eight weeks, neurulation and organogenesis proceed, transforming the tiny beanlike mass into an organism with characteristic human shape.

The neural tube rolls up midway through the third week of pregnancy, and a few days later, blocks of mesoderm called **somites** begin to pinch off on both sides of the neural tube, initiating the process of organogenesis (Figure 14.11a). Over 40 pairs of somites will form and differentiate into muscles, tissue layers beneath the skin, the backbone, and parts of the skeleton. A tube—the developing heart—begins pumping blood early in the fourth week and swells to form an anterior bulge, clearly visible in Figure 14.11a. As in developing frog embryos, pouches poke out from the primitive gut, then grow and branch, forming lungs, liver, and pancreas. By the end of the fourth week, the enlarging head bends toward the heart bulge, and rudiments of the arms, legs, ears, and eyes have begun to form but are not yet distinguishable (see Figure 14.11a). During the fifth week, nostrils develop, paddle-shaped hands are shaped from buds, and elbows, upper arms, and shoulders take form. By the start of the sixth week, the 9 mm (0.5 in.) long embryo is half head, with a vaguely alien face (Figure 14.11b). As the first eight weeks end (Figure 14.11c), the embryo, now called a fetus, has the rudiments of all its organs, including the start of its sex organs.

Sex Differentiation: Variations on One Developmental Theme

The primary sex organs have an interesting—if somewhat startling—course of development. The embryo produces gonads, sex ducts, and external genitals before the eight-week mark, but you can't tell a boy from a girl at this stage, because sexual structures are *indifferent*: They have yet to complete differentiation and establish the embryo's gender.

The fascinating thing about sex differentiation is that in the absence of a specific biochemical signal from the *Y* chromosome, these indifferent structures become female. Without

---

BOX 14.2   P E R S O N A L   I M P A C T

SEXUALLY TRANSMITTED DISEASE: A GROWING CONCERN

One consequence of America's sexual revolution during the 1970s and 1980s was a dramatic rise in the incidence of *venereal diseases*—diseases of the genital tract and reproductive organs caused by microorganisms. These diseases are part of the broader category of *sexually transmitted diseases* (STDs)—infectious diseases of any body region that can be passed to a partner through sexual contact. Regardless of terminology, however, the problem remains: A person with multiple sex partners has a substantial risk of contracting a sexually transmitted disease. Over 10 million cases are treated each year in the United States. And if undetected or untreated, such diseases can lead to severe complications.

The most common STD is *chlamydia,* an infection by the bacterium *Chlamydia trachomatis* picked up through sexual contact with an already infected person (Figure 1). A woman may have no symptoms of chlamydia at all, or she may experience pelvic pain, painful urination, vaginal discharge, fever, and swollen glands near the groin; a man may have a discharge from the penis or painful urination. The infection can be simply and effectively treated with antibiotics, but if undetected or untreated, it can lead to severe infection of the reproductive organs and even sterility. Because a person with chlamydia can be symptom-free, pregnant women can unknowingly pass the infection on to their newborns. This STD, in fact, is the most common infection in newborns, with 100,000 cases per year.

The second most common STD is *herpes genitalis,* caused by the herpes simplex virus type 2 (HSV type 1 causes cold sores and fever blisters). At least 20 million Americans have herpes, and 300,000 to 500,000 new cases arise annually. The virus causes watery blisters to form around the genitals; these break and form painful open sores that eventually heal. The virus can lie dormant for weeks, months, or years. Then, due to sunlight, emotional stress, or other triggers, the virus can break out once again and cause a new cycle of pustules and sores. Right now, there is no cure for herpes, and only partially effective antiviral drugs are available. If birth coincides with an active herpes phase in the mother, the newborn can suffer damage to the brain, liver, or other organs.

The fastest-spreading, and in many ways most worrisome, STD is *venereal warts,* caused by the papilloma virus. These small, painless, cauliflower-like bumps grow around the sex organs, rectum, or mouth and are passed through skin-to-skin contact. Getting rid of the warts is usually no problem; they can be burned or frozen off or surgically removed. More ominously, however, researchers are finding that the papilloma virus (which

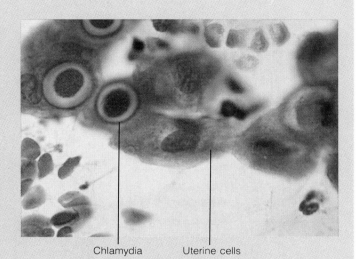

Chlamydia   Uterine cells

FIGURE 1 ■ *Chlamydia trachomatis* **Causes the Most Common Sexually Transmitted Disease.**

can remain in the body even after the warts are removed) has been present in 90 percent of the cervical cancer tissues studied so far.

Perhaps the best-known STDs are *gonorrhea* and *syphilis.* Both are caused by microorganisms; both are contracted through sexual activity with an infected person; both can have mild initial symptoms (a discharge or painless sore) or no symptoms at all; and both can be successfully treated with antibiotics. As in chlamydia, failure to find and treat gonorrhea at an early stage can lead to severe infection of reproductive organs or sterility. Failure to treat syphilis can also result in widespread damage to the heart, eyes, and brain and can result in severe damage or death to an unborn child.

Other STDs include pubic lice (crabs); scabies (parasites that burrow under the skin); certain types of vaginal yeast infections; trichomonas (flagellated protists that infect the vagina or penis); and acquired immune deficiency syndrome (AIDS; see details in Chapter 22).

Clearly, sexually transmitted diseases have become a serious threat. Planned Parenthood recommends that sexually active adults (particularly those with numerous sex partners) use condoms to prevent the passing of infections; be alert to sores, bumps, discharges, painful urination, or pelvic pain; and get tested regularly for STDs. This is especially important before or during pregnancy.

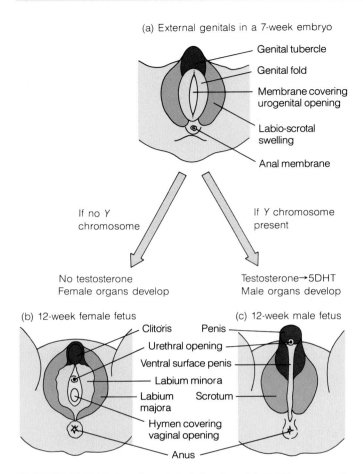

(a) External genitals in a 7-week embryo

Genital tubercle
Genital fold
Membrane covering urogenital opening
Labio-scrotal swelling
Anal membrane

If no Y chromosome

If Y chromosome present

No testosterone
Female organs develop

Testosterone→5DHT
Male organs develop

(b) 12-week female fetus

(c) 12-week male fetus

Clitoris   Penis
Urethral opening
Ventral surface penis
Labium minora
Labium majora   Scrotum
Hymen covering vaginal opening
Anus

FIGURE 14.12 ■ Development of Gender. (a) Until seven weeks of development, the external genitals of male and female embryos look alike, with a bulb (the genital tubercle), a fold, a swelling, and an opening. (b) In the absence of a *Y* chromosome, certain genes cause female genitals to develop. (c) In the presence of a *Y* chromosome, the testes develop and secrete testosterone, which is converted to a related hormone (5DHT, or 5-dihydrotestosterone) that induces the development of male genitals.

a *Y* chromosome, genes become active and cause gonads to become ovaries, the pre–vas deferens cells to die, and the sex ducts to develop into oviducts and uterus. And the external genitals follow suit (Figure 14.12). Nature, however, has ensured an alternative means of sexual differentiation by providing a set of signals that can turn indifferent gonads into male instead of female organs.

The signals begin with a specific gene on the *Y* chromosome called the testis-determining factor, or TDF. Researchers recently isolated a likely candidate for this gene and found that the gene product is probably a DNA-binding protein. This suggests that the TDF gene acts by regulating the activity of other genes, causing each of the two indifferent gonads to develop into a testis. Once differentiated, the two testes then begin to produce the steroid hormone testosterone and a pro-

tein called Müllerian inhibiting hormone. The testosterone maintains the male duct system, but Müllerian inhibiting hormone kills the duct cells that could develop into oviducts and uterus. Clearly, this substance made by male fetuses blocks the growth of new female reproductive system cells. Biotechnologists are hoping that they can one day manufacture enough Müllerian inhibiting hormone to treat women suffering from cancers of the reproductive organs by stopping these abnormal cells from growing and dividing. Testosterone from the embryonic testes has one additional effect: Some of it is converted to a second hormone, 5DHT (5-dihydrotestosterone), which causes the external genitals to become penis and scrotum (see Figure 14.12c).

Unfortunately, things don't always work so smoothly. Very infrequently, an *XY* male inherits a mutation that prevents his testosterone from being converted to 5DHT in normal amounts. At birth, the male baby's genitals look female, and all during childhood, he appears to be a girl. At puberty, however, a surge of testosterone is released from the boy's testes, the clitoris enlarges into a penis, the labia fuse to form a scrotum, the testes descend into it, and the individual begins to look and act like the male that he always was!

Fetal Life: A Time of Growth

The pace of fetal change slows as the weeks pass from the first **trimester** (three-month period), with its rapid-fire developmental activity, through the second and third trimesters. A few finishing touches are added, such as hair, eyelashes, nails, and more recognizable facial features. But the main activity is growth. The fetus becomes 600 times heavier between the eighth week and birth at about the thirty-sixth week. Figure 14.13 shows the enlargement during part of this period. The mother may feel flutters, rolls, kicks, and punches from within as the fetus shifts position and moves its limbs, starting at about three months. Subtler movements usually go unfelt. These include swallowing (four months), thumb sucking (five months), heartbeat detectable by a doctor's stethoscope (five months), handclasping (six months), and silent crying (eight months). The development of body and behavior are programmed in the offspring's genes and usually proceed in a normal and healthy way. But environmental factors can cause disruption.

Mother's Contribution to the Fetal Environment

The development of a human fetus requires a real partnership. Hormones from the embryo help establish and maintain the pregnancy; the mother's body changes continuously to accommodate the growing young; and the mutually produced placenta acts as a support system as well as a barrier to many harmful substances. Because the foods, drugs, and other chemicals the mother takes into her body profoundly affect the baby, she must be especially careful during pregnancy.

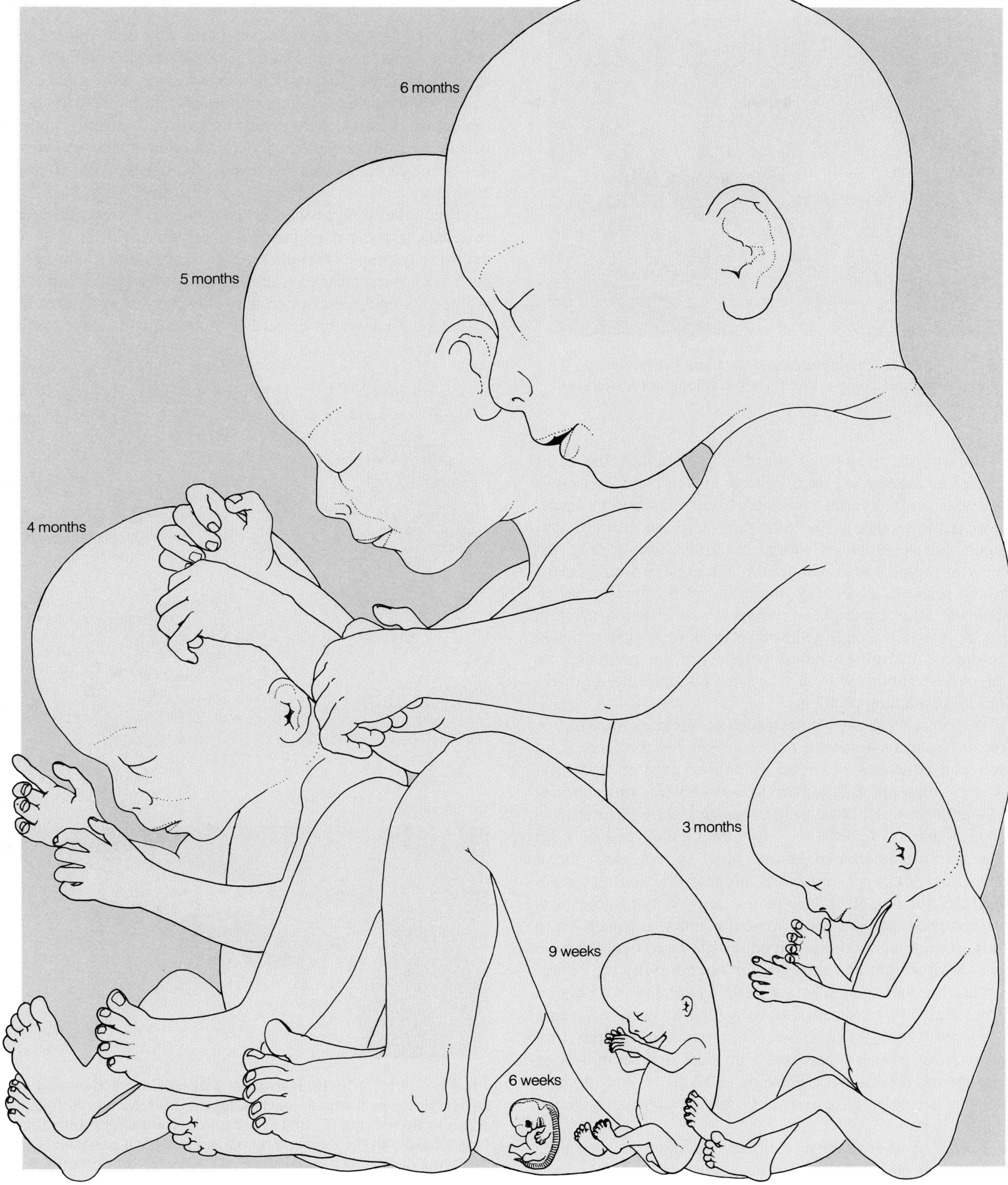

FIGURE 14.13 ■ The Fetus Grows.   Growth of the fetus, showing actual sizes between six weeks and six months.

FIGURE 14.14 ■ **Thalidomide and the Fetal Environment.** This child is learning to compensate for his developmental alteration.

For proper growth of muscles and bones, the fetus requires a constant supply of protein and calcium, as well as fatty acids, carbohydrates for energy, vitamins, and minerals. Because the mother's diet must provide these, obstetricians usually advise pregnant women to drink lots of milk, eat plenty of protein, and take vitamin and mineral supplements. Protein intake is especially important in the final trimester, when the fetus experiences the greatest expansion in brain size. A maternal weight gain of 25 lb or so is now believed appropriate for most women to help prevent premature or underweight infants with their greater susceptibility to infections and breathing problems.

A mother's life-style habits, such as smoking, drinking, or drug use, can have severe repercussions for the fetus. Two such consequences are fetal alcohol syndrome and fetal tobacco syndrome. Babies born to women who drink substantial amounts of alcohol during pregnancy show greater incidence of mental retardation, emotional abnormalities, cleft palates, underdeveloped hearts, slow growth, and facial anomalies. Many obstetricians suggest that their patients avoid alcohol altogether during pregnancy. What's more, miscarriage (premature expulsion of the fetus) is much more likely in smoking mothers, and babies are more likely to have a low birth weight and thus greater susceptibility to respiratory disease and sudden infant death syndrome (suffocation during sleep). Overall, infants with fetal tobacco syndrome have a death rate 10 to 100 percent higher than children from nonsmoking mothers. Mothers who take amphetamines or cocaine risk infants with neurological defects, and mothers who take heroin or other narcotics often give birth to addicted infants.

Some of the saddest chapters in modern medicine revealed that prescription drugs also carry a risk of fetal damage. In the fall of 1960, doctors in Europe and America witnessed the sudden appearance of a new birth defect: Affected infants were mentally and emotionally normal, but they had shortened, twisted legs, and their hands grew directly from the shoulders (Figure 14.14). After a two-year epidemic of the defects, researchers traced the cause to a new and popular sedative and antinausea drug called thalidomide. While this drug calms adult nerves, it alters the embryonic nerves that help direct proper limb development and growth. Ironically, thalidomide does not cause these effects in mice or rats, and so its devastating effects were not discovered during premarketing laboratory tests.

Premarket drug testing has become much more stringent in recent years, but pregnant women must still be wary of exposure to drugs, other chemicals, and certain viral diseases such as German measles (rubella). As late as 1988, about 100 women taking a prescription drug for severe acne ignored the manufacturer's warning and became pregnant while on the

(a) A positive feedback loop propels labor

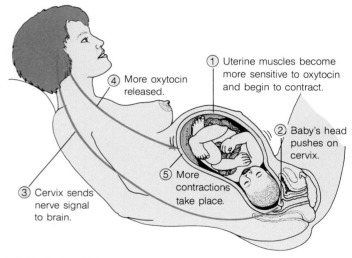

(b) Shortly after birth

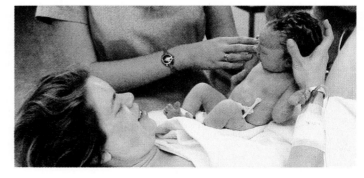

FIGURE 14.15 ■ **Birth: Hormones Trigger Contractions and Delivery.** (a) As birth approaches, uterine muscles become more sensitive to oxytocin (1), and they begin to contract, pushing the baby's head into the mother's cervix and causing it to widen (2). The stretched cervix sends a signal via nerves to the brain (3), which causes more oxytocin to be released (4). This increase in oxytocin causes more uterine contractions, and the cycle escalates (5). (b) The cycle continues until the baby is expelled through the birth canal. Finally, the mother and newborn can begin to recover from the arduous process.

FIGURE 14.16 ■ **Development Continues Long After Birth: The Many Faces of George Burns.** The actor at ages 37, 58, 77, and 90.

medication. As a result, they gave birth to mentally retarded babies with abnormally small jaws and very low-set ears. The drug, called Accutane, is a relative of retinoic acid, the developmental signaling molecule we discussed in Chapter 13. It's not surprising, therefore, that such a drug would affect a developing embryo or fetus. Pregnancy is clearly a time when a woman must take special care of her general health and nutrition—and, indirectly, her baby's.

Crossing the Threshold: The Magic of Birth

As a pregnant woman waits, the fetus prepares both itself and its mother for the impending labor and delivery. The fetal lungs grow, and particular valves develop in fetal blood vessels, readying them for the newborn's first breath. Special brown fat is stored around the neck and down the fetus's back, and this will produce heat after the baby is expelled from the warm uterus. Special reserves of carbohydrates are laid down in the heart and liver that tide the baby over until it can suckle milk, and the placenta makes and secretes a hormone that prepares the mother's breasts to produce milk. In the last couple of months before the birth, the uterine muscles become more excitable, contracting periodically (so-called "false labor") and building strength as the time of birth approaches.

During this last trimester, the uterus's sensitivity to the hormone **oxytocin** increases a hundredfold (Figure 14.15a). This substance is made in the brain's hypothalamus region and is released from the pituitary. When the sensitivity to oxytocin passes some threshold, labor begins. Nerves from the cervix signal the brain and pituitary to release more oxytocin, and this causes the uterine muscles to contract even more. Oxytocin also stimulates the uterine wall to release **prostaglandins,** hormones that enhance uterine contractions still further. This causes the release of even more oxytocin, and so on, in an ever-building *positive feedback loop.*

The cervix widens as the contractions grow longer, stronger, and more regular. Each contraction starts at the upper end of the uterus and moves toward the cervix, which the baby's head pushes further and further open as he or she

moves toward the vagina, or birth canal. The squeezing eventually causes the amniotic sac to burst (the "water breaks"), and as the cervical opening reaches a width of about 10 cm (4 in.), delivery is usually only minutes away. With considerable pushing of the abdominal muscles, the mother is able to force the baby's head past the pelvic opening, and then the shoulders and hips emerge. The baby, still attached to the uterus via the umbilical cord, takes its first gasp of air. Soon the blood vessels in the cord cease pulsing, and attendants sever it. Maternal blood vessels that supply the uterus clamp down, which prevents excessive blood loss, and with a few more uterine contractions, the placenta is expelled.

Recent studies show that the heavy stress of being born is actually beneficial in that it causes stress hormones (adrenaline and noradrenaline) to surge through the infant's body, clearing its lungs, promoting normal breathing, mobilizing stored energy, and sending extra blood to the heart and brain. Although the transition from being a semiaquatic fetus to an air-breathing land animal is difficult, most babies make it quite well. At the same time, the mother goes through a physical transition to nonpregnancy, and both she and the father must adjust psychologically to being parents.

GROWTH, MATURATION, AND AGING: DEVELOPMENT CONTINUES

While birth marks the end of fetal development, it is not the end of growth and change (Figure 14.16). People continue to traverse a series of more gradual thresholds as they move from infancy to childhood, puberty, adulthood, and old age.

Infancy and Childhood: Growing Up Fast

Within hours after a horse gives birth to a foal, the newborn animal is standing, walking, nuzzling, and following its mother. A newborn human infant, however, is helpless— unable to feed, clean, or defend itself, or sit up, talk, or walk about. **Infancy,** the period from birth to about age 2, and **childhood,** from age 2 to about 12, are times of profound physical and mental change and development during which the child can grow to 70 percent of his or her adult height and weight, quickly amass a large vocabulary of spoken and written words, and develop coordination, perception, strength, agility, personality, and social skills.

Growth in a human follows a regular pattern that is quite evident in the infant and child. Growth is fastest in the head and body center and slowest toward the periphery. At birth, the brain is the most developed organ and is 25 percent of its

final adult weight. By six months after birth, the brain has reached 50 percent of adult weight; by 30 months, 75 percent; and by age ten, 90 percent. Body weight lags far behind brain weight, advancing from 5 percent of adult weight at birth to only 50 percent at age ten. In keeping with the center-outward growth pattern, the child's trunk grows faster than its arms and legs, and it gains coordination over gross limb movements before finger and toe movements.

The details of childhood development are too numerous to consider here, but we summarize a few of the major milestones of physical and mental growth in Figure 14.17.

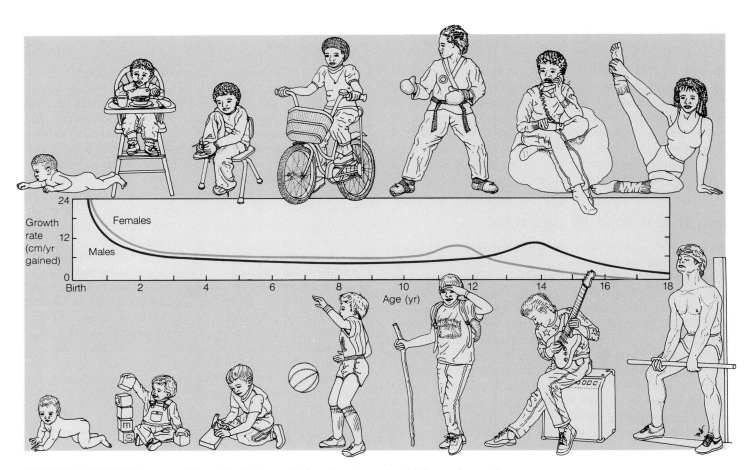

FIGURE 14.17 ■ Milestones of Growth. Between birth and age two, the child experiences the major stages of brain growth. Height increases 18 to 20 cm (7 to 8 in.) per year. The child learns to sit up, stand, walk, maintain eye contact, babble, and imitate. As powers of mental reasoning begin to emerge, the child learns the elements of grammar, acquires a small vocabulary of spoken words, develops some degree of eye-hand coordination, and starts to feed itself. Between ages two and five, height increases 7 to 10 cm (3 to 4 in.) per year, and the child reaches half its adult height. The child also attains control over elimination and develops many fine motor skills (painting, using scissors, building puzzles) and gross motor skills (running, climbing, riding a tricycle, playing ball). The child completes the basic process of language acquisition at this time and has a vocabulary of several thousand words. The child also acquires more reasoning and problem-solving skills and becomes more social. Between ages 5 and 12, the child's height increases 2.5 cm (1 in.) per year, the limbs elongate, the muscles grow, and coordination improves, allowing the child to roller-skate, swim, dance, play baseball, and so on. The child adds to its vocabulary, reasoning skills, and problem-solving ability; a sense of causality and morality begins to emerge; and by age 12, this individual lies at the threshold of adolescence, sexual maturity, and attainment of adult size.

Puberty: Sexual Maturity

Adolescence is marked by the most dramatic physical changes since fetal life: **puberty,** the maturation of the reproductive system and the development of secondary sexual characteristics. In a teenage girl between about 11 and 13 years of age, estrogen produced in the ovaries enters the bloodstream and causes target tissues (around the nipples and hips) to differentiate into their sexually mature state. The visible results are the deposition of fat in hips and other areas (creating body contours) and the enlargement of the breasts. Normally accompanying these changes are a growth spurt that brings the girl close to her adult height, the onset of the menstrual cycle, and a heightened interest in sex.

In the male between ages 13 and 15, testosterone produced in the testes enters the bloodstream, reaches target tissues in the genitals, face, and other parts of the body, and stimulates the development of body and facial hair, enlarged muscles and genitals, and a lower voice. This is usually also accompanied by a growth spurt and an interest in sex.

Adulthood: The Longest Stage in the Life Cycle

With modern nutrition, sanitation, and health care, physical maturity can begin at the end of adolescence and last for 50 to 75 years or more. Some additional development of adult features—more shifting of fat, further enlargement of sex organs, and continued growth of body hair—extend into the early 20s. But after that, all the body's organ systems are set and perform at their peak efficiency for about a decade. Sometime after age 30, the process of **aging** begins, a progressive decline in the maximum functional level of individual cells and whole organs. Very gradually—almost imperceptibly from year to year—the muscles lose strength, breathing and circulation become less efficient, hair follicles lose their ability to make pigment, the skin becomes less elastic, and reproductive functions decline. About age 50, females experience **menopause,** a cessation of the menstrual cycle, while males may lose some *potency,* or ability to maintain an erection. Sometime after age 60, people reach **senescence,** or old age, when the decline in cell and organ function is less gradual and more profound. Many researchers have studied aging and proposed intriguing theories about its causes.

What Causes Aging?

Researchers group most hypotheses about aging into two opposing categories: genetic clock hypotheses and wear-and-tear hypotheses. Many biologists support the idea of a genetic clock, arguing that there is a timetable for aging and death specified by the genome. Just as the genes regulate the timing of organ formation, cell death in the formation of fingers and toes, and sexual maturation, they also regulate when our various organs and systems will cease functioning. Figure 14.18 illustrates an experiment on aged cells that uncovered what may be an "aging substance," perhaps coded for by "aging

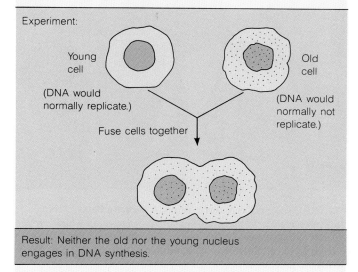

Question: Is cell aging due to a diffusible substance?

Hypothesis: A diffusible substance present in an old cell blocks DNA synthesis.

Experiment:

Young cell
(DNA would normally replicate.)

Old cell
(DNA would normally not replicate.)

Fuse cells together

Result: Neither the old nor the young nucleus engages in DNA synthesis.

Conclusion: A substance in the old cell diffuses to the young cell's nucleus and blocks its division

FIGURE 14.18 ■ **Is There an Aging Substance?** Biologists wondered whether a young cell contains some substance that allows it to synthesize DNA repeatedly and divide many times, or whether an old cell contains something that causes it to cease dividing. To test these possibilities, they fused a young cell to an old cell and observed that neither nucleus in the hybrid cell synthesizes DNA or divides. Their conclusion: Some substance in old cells blocks the continued DNA synthesis and cell division that is characteristic of young cells. Control experiments showed that a young cell fused to another young cell continues to enter the DNA synthesis phase of the cell cycle, while an old cell fused to another old cell does not begin to synthesize DNA.

genes." Candidates for such genes include the tumor suppressor genes (like the retinoblastoma gene) that put a brake on cell division.

There is tantalizing evidence for these genetic clock hypotheses. First, cells seem to have preset limits to the number of times they can divide. Skin cells from a human infant will divide about 50 times and then stop, while similar cells from a 90-year-old will divide only a few times.

Second, organs seem to age in a preprogrammed way. Each year, starting at about age 30, we experience a 1 percent decline in the maximum function of various organs. This includes lung capacity, the amount of blood the heart pumps, the strength and coordination of muscles, the cleansing power of the kidneys, and the highest sounds we can hear.

Third, certain genetic conditions can bring about symptoms of aging. Children with *progeria* (early aging) begin to lose their hair, show wrinkles, and experience arthritis and

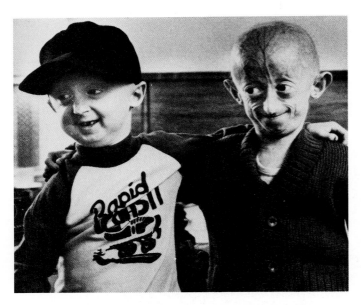

FIGURE 14.19 ■ **Premature Aging: A Genetic Basis?** Some children have a genetic condition called progeria, in which certain signs of aging develop by age five or six. The boy on the left (age 9) from Texas and his new acquaintance (age 8) from South Africa are just about to tour Disneyland.

heart attacks by age five or six (Figure 14.19). These children, however, do not suffer the whole spectrum of age-related illnesses, which includes cataracts, diabetes, and cancer. Certain other people inherit a dominant mutation that appears to cause Alzheimer's disease, a kind of senility. The brains of Alzheimer's patients shrink and contain far greater numbers of tangles and knots, called senile plaques, than do the brains of normal older people. While the complete genetic basis for diseases like progeria and senility remains unsolved, the conditions do seem to support the genetic clock concept.

Other biologists believe that our preprogrammed genes are less likely to cause aging than are wear and tear: the accumulation of random errors in DNA replication, transcription, or translation that lead to the disarray of information systems, or the accumulation of metabolic by-products that disable enzymes, other proteins, and lipids. Informational errors may be due to environmental insults such as sunlight, radiation, or chemicals or may themselves result from a buildup of internal metabolic by-products. One class of culprits may be the toxic, chemically active, oxygen-containing molecules called free radicals that arise naturally during energy metabolism (see Chapter 2). The relentless piling up of metabolic garbage probably interferes with normal cell function, including protein synthesis, and perhaps contributes to the 1 percent per year decline in organ function we all experience.

Aging Gracefully

Whatever the major causes of aging may be, there are several good reasons to think that most of us can and will live long, healthy lives despite the inevitabilities of decline and mortality.

First, compared to most animals, humans have very long **life spans,** or maximum potential ages. The oldest verified age is just over 120 years, as of 1990.

Second, human **life expectancy,** the maximum probable age a person will reach, has never been higher (Figure 14.20). For most of human history, most people died between ages 20 and 40. "Life," as English philosopher Thomas Hobbs wrote, was indeed "solitary, poor, nasty, brutish, and short." But thanks to advances in agriculture, emergency medicine, obstetrics, and public sanitation and immunization programs, average life expectancy in North America has increased steadily to its current levels of 78 for women and 71 for men.

Third, studies show that good habits lead to longer life. A man who eats regular meals (including breakfast), exercises regularly, sleeps an adequate amount, maintains his ideal weight, does not smoke, and limits alcohol consumption can live an average 11 years longer than a man who follows three or fewer of those practices. A woman who observes all six healthy practices can add seven years to her already longer life expectancy. As an added bonus, those observing these six positive life-style habits remain healthier during their extra years than cohorts with fewer good habits.

Fourth, the only known environmental parameter that invariably increases the life span of laboratory animals is a diet restricted in calories. Rats given a calorie-restricted diet *with adequate nutrition* live about 30 percent longer than rats allowed to feed freely. There is some collateral evidence that caloric restriction can slow human aging, perhaps, biologists speculate, by slowing the genetic clock, by decreasing metabolic rate and hence the generation of free radicals, or both.

Finally, the dread many people feel over the prospects of growing old may be the result of misconceptions and negative stereotyping. A recent study revealed that 95 percent of the elderly live independently, not in institutions; most are in reg-

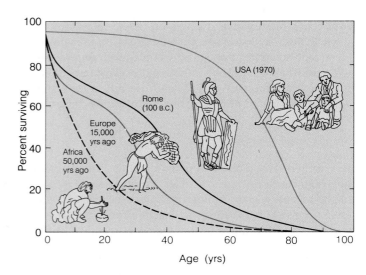

FIGURE 14.20 ■ **Expect a Long Life.** Life expectancy in the United States is high for both females and males. This is a relatively new development; historically, human life expectancy trailed off rapidly after 40.

FIGURE 14.21 ■ Aging Gracefully. Good habits, good health, and good times can help a person live a long and happy life. Old age need not be a time of dependency and illness.

ular contact with their families, not isolated or lonely; most are vigorous and active, not frail and sedentary; and most are financially secure, well educated, and integrated into their communities. Clearly, the last decades of life can hold great satisfactions rather than dependence and illness (Figure 14.21). A satisfying old age is enhanced by good health and social support systems one establishes much earlier in life.

CONNECTIONS

In this chapter, we followed the human life cycle from sperm and egg production in the male and female reproductive systems, through mating, fertilization, fetal development, childhood, sexual maturation, and adulthood. We saw that the developing embryo undergoes remarkable changes during the first few weeks of life, including the establishment of all major body organs and the differentiation of gender. The reproductive plumbing, the hormones that regulate gamete production, and the precursors of the external genitals are all variations on a central theme, modified in slightly different directions in the two sexes.

Reproduction solves a central problem for the human species. Our bodies are mortal—they age, wear out, and die. The maturation of sex organs and the blossoming of secondary sex characteristics in adolescence make it possible for male and female to start new life cycles through their children. Reproduction is our only means of continued existence as well as our anchor to parents, grandparents, great-grandparents, ancestors, and progenitors. How far back does the unbroken chain extend? And where did it all start? We may never know the answers for certain, but the laws of nature, operating today as they did on earth billions of years ago, allow us to hypothesize how life could have started and how lineages could have begun. Chapter 15 bridges the gap from genes to the diversity of life forms by examining modern explanations for the origins of life.

Highlights in Review

1. The first "test-tube baby," Louise Joy Brown, symbolizes the sophisticated knowledge biologists have amassed on human reproduction and the human life cycle.

2. Four themes become apparent when discussing the human life cycle: Male and female reproductive systems are structurally and functionally parallel; intricate communication networks are necessary for eggs and sperm to grow, be released, and unite and for a baby to be born; pregnancy is a biological partnership between mother and young; and development continues for decades after birth.

3. Collectively, male reproductive structures make sperm and transfer them into the female's reproductive tract, where the sperm encounter and fertilize eggs.

4. Testosterone, LH and FSH, and releasing hormone function together in interlocking negative feedback loops that maintain a continual sperm supply.

5. Female reproductive structures make and transport eggs, prepare and nourish developing embryos, and deliver babies.

6. Estrogen, progesterone, LH and FSH, and releasing hormone function together in interlocking negative feedback loops that bring about the cyclic production and release of eggs (the menstrual cycle).

7. The corpus luteum produces estrogen and progesterone and promotes the buildup of the uterine lining (endometrium). If an egg is fertilized, this lining is maintained throughout pregnancy. If the egg is not fertilized, the corpus luteum degenerates, and the lining is sloughed off.

8. Sexual attraction and intercourse can bring about a convergence of egg and sperm in the female's oviduct, followed by fertilization, the fusion of one sperm nucleus with the egg nucleus.

9. Birth control is achieved by either contraception or termination of pregnancy. Contraception relies on techniques for blocking fertilization; pregnancy termination depends on preventing implantation or preventing embryonic growth.

10. To overcome infertility, men or women may need hormonelike drugs, corrective surgery, in vitro fertilization, insemination with donor gametes, or the help of a surrogate mother.

11. Pregnancy is a partnership between embryo and mother. After the embryo implants in the uterine lining, it develops a chorion, and this produces a hormone, hCG, that prevents the corpus luteum from degenerating and hence the lining from being sloughed off. Pregnancy tests detect hCG levels in the urine. The chorion grows and enmeshes with maternal tissue to form the placenta, which nourishes the embryo, removes wastes, and screens out bacteria and toxins.

12. The human embryo develops much as do other mammalian embryos. It produces an amniotic sac; undergoes gastrulation,

neurulation, and organogenesis; and grows rapidly during gestation.

13. Sexual differentiation in the fetus begins with an indifferent embryo that is neither male nor female. In the absence of a signal from the *Y* chromosome, the gonads, sex ducts, and external genitals express genes causing them to develop as female structures. If a *Y* chromosome is present, the indifferent gonads becomes testes, the organs produce testosterone and Müllerian inhibiting hormone, and male ducts and external genitals develop.

14. The mother's nutrition, life-style habits, and taking of prescription drugs can affect fetal development.

15. The fetus programs itself and its mother for delivery. Among other changes, it stores special fat and carbohydrate reserves, secretes hormones that prepare the mother's breasts for milk production, and cause her body to release the signals (prostaglandins and oxytocin) that initiate labor and delivery.

16. Infancy and childhood are periods of phenomenal physical growth and mental development, including the acquisition of language. Puberty is marked by an additional growth spurt and by the maturation of sexual organs and the development of secondary sexual characteristics. During adulthood, the body reaches its peak physical performance, then gradually declines until senescence sets in, when the body ages more rapidly.

17. Aging may be caused by genetic programming, by wear and tear, or by other factors. Regardless, it is possible to age gracefully, particularly by observing good health habits throughout life.

Key Terms

aging, 311	oocyte, 295
amniotic cavity, 304	orgasm, 298
Bartholin's gland, 295	ovary, 295
bulbourethral gland, 294	ovulation, 295
cervix, 295	oxytocin, 309
childhood, 310	penis, 294
chorion, 302	pregnancy, 299
clitoris, 295	progesterone, 296
coitus, 298	prostaglandin, 309
contraception, 299	prostate gland, 294
corpus luteum, 298	puberty, 311
ejaculation, 294	scrotum, 292
ejaculatory duct, 293	semen, 294
endometrium, 296	seminal fluid, 294
epididymis, 293	seminal vesicle, 293
erectile tissue, 294	seminiferous tubule, 292
estrogen, 296	senescence, 311
follicle, 295	somite, 304
follicular cell, 295	spermatogenic cell, 292
human chorionic gonadotropin (hCG), 302	supporting cell, 292
	testis, 292
infancy, 310	testosterone, 294
interstitial cell, 292	trimester, 306
in vitro fertilization, 290	umbilical cord, 304
life expectancy, 312	urethra, 294
life span, 312	uterus, 295
menopause, 311	vagina, 295
menstrual cycle, 296	vas deferens, 293

Study Questions

Review What You Have Learned

1. List the following structures in their proper order to show the route sperm travel from their place of origin to the outside world: epididymis, vas deferens, seminiferous tubules, prostate gland, ejaculatory duct, urethra.

2. Why might male bicycle racers have reduced fertility?

3. Outline the feedback loop that maintains the sperm supply.

4. Where is an egg produced? What pathway does the egg follow from site of production to the outside world?

5. Create a chart of female hormones using the following headings: Name of Hormone; Where Formed; Effect.

6. Of the 400 million human sperm released during coitus, how many will fertilize the egg? Explain.

7. What are some causes of infertility in women? In men?

8. Why is the chorion important to the developing embryo?

9. Does an embryo's blood mix with its mother's in the placenta? Explain.

10. How does the *Y* chromosome stimulate male differentiation?

Apply What You Have Learned

1. Why might a pregnant woman suffer a miscarriage if her body does not produce enough progesterone?

2. Researchers have isolated a gene on chromosome 21 that may be responsible for Alzheimer's disease. What is the practical value of this finding?

3. Human cells seem to divide no more than 50 times. What might be the evolutionary advantage of this limit?

4. To increase their strength and muscle mass, female athletes may take anabolic steroids, which are similar to testosterone. What effect might these drugs have?

5. An anabolic steroid was found in the urine sample of a sprinter who won a gold medal in the Olympic Games. The sprinter claimed that someone must have added the banned substance to his sample. The laboratory technicians did not believe this, however, because the amount of the natural male steroid testosterone in his urine was much lower than in a normal male. From what you know of the feedback loops that control the production of testosterone, would you support the arguments of the sprinter or the technicians? Explain your choice.

For Further Reading

ARAL, S. O., and K. K. HOLMES. "Sexually Transmitted Diseases in the AIDS Era." *Scientific American,* February 1991, pp. 62–69.

GIBBONS, A. "Gerontology Research Comes of Age." *Science* 250 (1990): 622–625.

MCLAREN, A. "What Makes a Man a Man?" *Nature* 346 (1990): 216–217.

MOORE, K. L. *Essentials of Human Embryology.* Toronto: Decker, 1988.

OHLENDORF-MOFFAT, P. "Surgery Before Birth." *Discover,* February 1991, pp. 59–65.

ULMANN, A., G. TEUTSCH, and D. PHILBERT. "RU486." *Scientific American,* June 1990, pp. 18–24.

WEINDRUCH, R., and R. L. WALFORD. *The Retardation of Aging and Disease by Dietary Restriction.* Springfield, IL: Thomas, 1989.

PART THREE

LIFE'S
VARIETY

Life's Origins and Diversity on Our Planet

FIGURE 15.1 ▪ **Swamp Mud and Methane Microbes.** Prokaryotes called methanogens inhabit the oxygen-depleted ooze below the still, shallow waters in cypress swamps such as this one at Greenfield Lake, North Carolina. The cells release odorous molecules that collect in pockets some people call swamp gas.

Methane Generators: Relatives of the Earliest Microbes?

What do sewage sludge, swamp mud, rotting manure, and a cow's breath have in common? The most obvious shared characteristic is an awful smell. And the reason for it is that a particular kind of bacterium lives in the cow, the mud, and the sludge and gives off methane gas (CH_4) and hydrogen sulfide (H_2S), with its unpleasant odor of rotten eggs and decay (Figure 15.1). The methane-producing bacteria are of special interest to us in this chapter because they may be the closest living relatives to some of the very first life forms on earth. Studying the characteristics of this ancient prokaryotic group, the conditions in which they thrive, and how they differ from other organisms provides a good background for understanding how life originated and diversified on our planet.

More than a dozen species of prokaryotes produce methane as a by-product of their metabolism, and biologists refer to this group as **methanogens** ("methane-generating organisms"). Depending on the species, the cells can be shaped like rods, spheres, spirals, or clusters (Figure 15.2), and they grow and thrive where there is little or no free oxygen (see Chapter 5). Sewage sludge, swamp mud, stagnant water, lake and ocean bottom sediments, rotting manure, and the digestive tracts of cows and certain other animals all lack free oxygen and hence can harbor methanogens. There the cells obtain energy from the hydrogen and carbon dioxide in their surroundings and produce the metabolic waste product methane. Without the activity of methanogens, people who operate sewage treatment plants could not process sewage as efficiently or collect methane gas to burn for heat and electricity. What's more, the organic matter in thousands of acres of wetland sediments would decompose much more slowly. Nevertheless, the methanogens are a comparatively small and restricted group today because oxygen-free environments are now relatively rare on our planet.

Methanogens weren't always an odd minority, though. Early in the earth's history, the entire planet was devoid of free oxygen, and many biologists believe that methane prokaryotes and similar anaerobic species were the dominant—

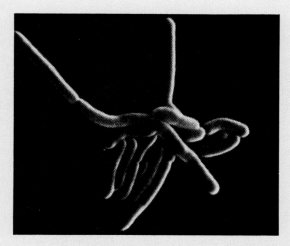

FIGURE 15.2 ■ Methane-Producing Microbes: Ancient Life Forms. Cells of the prokaryote *Methanobacterium ruminantium* (magnification 30,000×) inhabit a cow's intestines, live anaerobically, and produce methane gas.

and probably the only—life forms to exist. Methanogens have several traits that set them apart from all other living cells and that suggest a very early origin:

■ The hydrogen and carbon dioxide they metabolize were probably common in the earth's first atmosphere.

■ Methanogens cannot tolerate free oxygen, and the first atmosphere had virtually none.

■ Their cell walls, cell membranes, and electron transport chains contain molecules different from those of the true bacteria.

■ They fix carbon (see Chapter 6) via a unique pathway unused by other cells.

■ The RNA in their ribosomes is different in structure from that of true bacteria, and certain other cell parts and processes differ as well.

There are many suggestions for how the methanogens may have emerged on the primitive earth, and we describe the most widely accepted hypotheses in this chapter. Methanogens were probably not the direct progenitors to most other living things. However, the ancestors that did lead to life's magnificent diversity of species were very likely other simple bacteria that probably emerged shortly before the methanogens, and in similar ways.

Three themes will emerge as we consider the origins of life and how it diversified into the millions of species that have lived and died throughout the earth's history. First, life arose as a direct result of the physical conditions on our planet (Figure 15.3), the molecules that were present, and the ways those molecules interacted. Biologists do not consider life some magical force that animated lifeless materials, but rather an emergent property based on the behavior of the materials that make up living things. Second, we may never know exactly how life originated, because the conditions and materials present 3.5 billion years ago have long since changed. Scientists can only reconstruct those conditions, simulate the processes, gather fossil and chemical evidence, and speculate. Third, life and earth coevolved; their histories are intertwined. The biological activities of the early organisms caused massive and permanent changes in the earth's atmosphere, its land, and its cycles. In turn, geological activities that shaped and reshaped the earth strongly affected the evolution of living things. The result of this coevolution was a multiplicity of habitats on our planet and the diversification of living organisms to more than 10 million species. Biologists are cataloging those species today, and the science of assigning species to categories, called **taxonomy,** helps reveal the evolutionary relationships between various groups of organisms.

Our discussion of life's origin and diversification will answer the following questions:

■ How did our planet form, and what made it conducive to life's origin?

■ What steps might have preceded the emergence of living cells from nonliving precursors on the early earth?

■ How did living things influence the earth, and how did changes in the earth affect the course of life's history?

■ How do scientists categorize the millions of living species into related groups and use these classifications to further understand evolutionary history?

FIGURE 15.3 ■ Earthrise. The blue planet emerges above the moon's horizon. The appearance of life about 4 billion years ago was just one stage in the evolution of our water-covered planet—an evolution that began with the Big Bang some 18 billion years ago and included the formation of our planet about 4.6 billion years ago.

EARTH AS A STAGE FOR LIFE

One of the twentieth century's most dramatic achievements, the exploration of space, has provided tremendous new knowledge about our solar system's central star, the sun, and the nine planets that orbit it. The more we explore and learn, however, the lonelier we feel, for many astronomers believe that we earthly organisms inhabit the solar system alone. Why did living things arise, flourish, and survive only here?

How the Earth Formed: From Big Bang to Big Rock

Most astronomers believe that the universe began with the so-called Big Bang, an event about 18 billion years ago during which an enormous explosion created all the matter in the universe, 99 percent of which is hydrogen and helium atoms. Radio astronomers find that all matter—in comets, meteors, planets, stars, or scattered particles—is still rapidly expanding outward as a result of the Big Bang.

Trillions of stars formed (and new ones continue forming) as gravitational attraction pulled the clouds of hydrogen and helium atoms together. Within a forming star, the gravitational pull becomes so strong and the compression so great that internal temperatures soar and the sphere of gases ignites. All of the heavier elements form by means of nuclear fusion reactions inside the stars, and clouds of these elements often come to orbit young stars. In our own solar system, such clouds orbiting the sun cooled and combined into the planets, moons, and countless smaller bodies. Geologists think the earth formed 4.6 billion years ago (Figure 15.4).

A Black and Blue World

After its formation, the earth is thought to have been a huge barren ball of ice and rock, devoid of oceans and atmosphere. Deep within the planet, however, radioactive elements (see Chapter 2) released energy that combined with the crush of gravity to heat the rock slowly into a molten mass that did not cool for several hundred million years (Figure 15.5a). As the cooling took place, a dense core of iron, nickel, and other heavy elements formed, surrounded by a liquid outer core, a thick mantle of hot rock, and at the surface a hardened crust of cool black rock. Molten rock frequently erupted through this surface, building cinder cones that punctuated the flat, black landscape, spewing lava and issuing clouds of gases from the planet's interior. The earth's first atmosphere began to form from these vapor clouds, and gravity held a blanket of carbon dioxide, water vapor, carbon monoxide (CO), nitrogen (N_2), hydrogen sulfide (H_2S), plus small traces of hydrochloric acid (HCl), ammonia (NH_3), and methane (CH_4) as the first air. Moisture in this air fell to the earth's steaming surface as rain, evaporated, and fell again, eventually cooling the surface and creating a shallow ocean that covered the entire globe and was broken only by the sharply rising black cones spewing forth more lava and gases.

Because of the volcanic ash in the air, the sky would have looked pale blue; it would also have been dotted with clouds and frequently streaked by flaming meteors and asteroids—small hunks of primordial rock hurtling in from space. This planetary study in black and blue, with its oxygen-free air, was the stage on which life made its first appearance (Figure 15.5b).

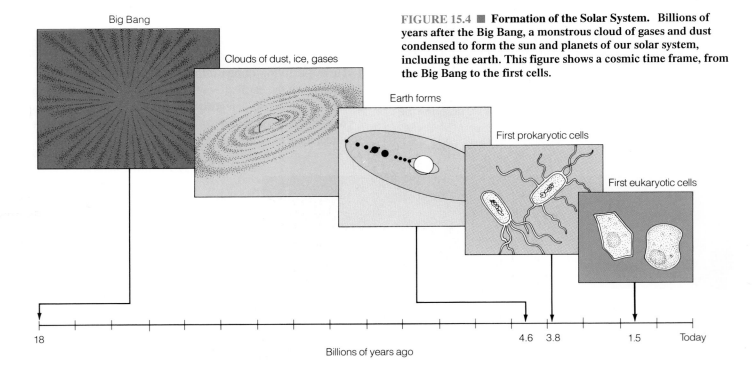

FIGURE 15.4 ■ **Formation of the Solar System.** Billions of years after the Big Bang, a monstrous cloud of gases and dust condensed to form the sun and planets of our solar system, including the earth. This figure shows a cosmic time frame, from the Big Bang to the first cells.

Big Bang

Clouds of dust, ice, gases

Earth forms

First prokaryotic cells

First eukaryotic cells

18   4.6   3.8   1.5   Today

Billions of years ago

(a) Earth: 4 billion years ago

(b) Earth: 3.5 billion years ago

FIGURE 15.5 ■ The Primordial Earth. (a) A molten scene about 4 billion years ago as the forces of gravity and radioactive decay melted the initial ball of rock and ice and created a red inferno of lava that boiled and smoked for millions of years. (b) By about 3.5 billion years ago, the planet had cooled, and the landscape was nothing but somber shades of black and blue. (Artist's conceptions.)

Earth's Advantageous Place in the Sun

Life as we know it—based on liquid water and carbon compounds—could have arisen only on a planet with sufficient amounts of both. Mercury and Venus, the sun's closest satellites, are blisteringly hot, with daytime surface temperatures higher than a self-cleaning oven's. Under such conditions, all water occurs as vapor and all carbon compounds as inorganic gases. The earth orbits between Venus and Mars—a planet that may once have had flowing water and living things, but is now barren, with all of its water frozen and most of its carbon trapped in rocks. Of all the planets in our solar system, only earth had a winning combination of composition, geological activity, size, and distance from the sun. Because of our planet's molten core and volcanic eruptions, carbon dioxide and water vapor were (and are) thrown into the atmosphere. Because of our planet's size and distance from the sun, gravity held the gases like an encircling blanket, which in turn trapped enough of the sun's heat energy to keep the surface temperate and most of the water liquid.

Ancient Earth and the Raw Materials of Life

One of the biggest puzzles in the study of life's origins concerns the availability of organic building blocks for the earliest cells. Did such building blocks, including simple sugars, amino acids, fatty acids, and nucleotides, exist in the black and blue world we have described? And if so, where did they come from?

Evidence suggests that organic compounds did exist on the early earth and could have accumulated in various ways. First of all, astronomers have recently discovered enormous clouds of organic molecules in the spiral arms of our galaxy and hypothesize that such clouds take part in the coalescence of planets (Figure 15.6a). Thus, some organic compounds

(a) Organic molecules exist in a nebula.

(b) Simple organic molecules can join, forming subunits of biological molecules.

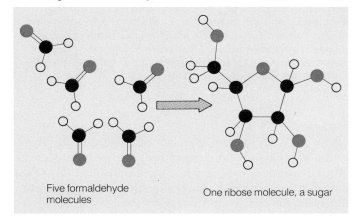

Five formaldehyde molecules

One ribose molecule, a sugar

FIGURE 15.6 ■ Organic Raw Materials of Life. (a) Clouds of organic molecules aglow in the light of Orion nebula M42 take part in a planet's formation. Such organic molecules from space may have contributed organic materials for early life, although they were not of biological origin. (b) Simple organic monomers (here, five formaldehyde molecules) link to form one ribose molecule, a simple sugar found in RNA.

FIGURE 15.7 ■ The Urey-Miller Experiments: Simulations of Earth's Primordial Conditions Produce Biological Subunits. Stanley L. Miller (pictured) and Harold C. Urey, working at the University of Chicago in 1955, designed a simple system to re-create the conditions on the early earth. They filled a flask with gases they thought might have been present in the atmosphere 4 billion years ago (including methane, CH_4; ammonia, NH_3; water vapor; and hydrogen gas), then shot bolts of electric current through the gaseous mixture to simulate primordial lightning storms. They continued the experiment for a week and watched as the water in the "shallow sea" (a connected flask below the gas chamber) grew pink and then dark red with organic material generated by the sparks. They analyzed the contents of this liquid and found a high concentration of amino acids and sugars. This experiment, and dozens of others like it altering only the composition of the "atmosphere" and the source of energy (ultraviolet light, heat, radioactivity, and so on), led to the "primordial soup" theory. This is the idea that life arose in a rich broth of biological monomers in some warm, shallow ocean basin on the primitive planet. As the text explains, this is just one of the many theories for the origin of organic molecules.

were very likely present from the earth's beginning. Radio astronomy data reveal compounds such as formaldehyde (H_2CO), hydrogen cyanide (HCN), and ammonia (NH_3) in the cosmic clouds. And laboratory experiments simulating conditions on the early earth show that formaldehyde molecules can join in various ways to form sugars (Figure 15.6b), and cyanide molecules can form amino acids.

A second source of organic raw materials might have been a class of meteorites called carbonaceous chondrites. These nondescript dark gray stones formed when the planets arose,

and they continue to hurtle through space, occasionally falling to earth even now. Chemical analysis of a single carbonaceous chondrite has revealed all five nucleotide bases: adenine, guanine, cytosine, thymine, and uracil. Some scientists hypothesize that millions of tons of such meteorites struck the earth during its early history, perhaps providing prebiotic ("before life") organic compounds.

Finally, many scientists point out that natural energy sources—lightning, the ultraviolet component of sunlight, heat from volcanic activity—could have provided energy for reactions that turned atmospheric gases into *monomers* (the units of organic polymers; see Chapter 2). Figure 15.7 shows and describes a historic experiment in which chemists re-created their concept of the early atmosphere and oceans and generated monomers with simulated lightning. Modern variations on this experiment, substituting different combinations of atmospheric gases and energy sources, continue to show that organic compounds form easily under conditions probably close to those found on earth more than 4 billion years ago. Many biologists conclude that the raw materials of living things did indeed exist on earth and could have combined as a result of the inherent properties of those molecules and the energy sources and physical conditions present at the time. The stage was then ready for the drama of life to begin.

THE UNSEEN DRAMA: FROM MOLECULES TO CELLS

Many scientists believe that a record of the entire sequence of chemical events leading to life's emergence may well have been laid down in sediments that solidified into rocks starting about 4 billion years ago. However, we'll almost certainly never find that fossilized sequence, because geologists have excellent evidence to prove that the earth was heavily bombarded by giant meteors for the first 800 million years of its history. Every square kilometer of the earth's initial crust probably melted during these impact events, and so the oldest earth rocks that can ever be discovered are probably no more than 3.8 billion years old.

Now here is the fascinating—and frustrating—thing: Fossils of the earliest cells, similar in many ways to the methanogens mentioned earlier, already existed in rocks nearly 3.5 billion years old, and cellular traces have been found in rocks 3.8 billion years old. Clearly, life emerged sometime before that period, but the evidence melted eons ago! Biologists, therefore, must be content to simulate, experiment, and speculate on the steps in that emergence with little hope of ever finding irrefutable proof of how it actually happened. One plausible scenario involves five steps, each of which obeys the laws of chemistry and physics: Small molecules join into long chains, these chains make copies of themselves, chains of different types interact, cell-like compartments take shape, and coordinated cell-like activities emerge (Figure 15.8).

Molecules Join into Long Chains

Biological polymers could have formed in various ways. Experiments show that when solutions of amino acids are frozen solid or heated to dryness, small polypeptides form spontaneously. Perhaps "primordial soup"—seawater containing organic precursors—collected in ancient tide pools, became concentrated via solar evaporation, and finally dried in the hot sun or froze at night, with polypeptides resulting. Other experiments show that under certain conditions, clay can trap light or heat energy and transfer it to organic molecules. Thus, inorganic clays, plentiful on the early earth, could have acted both as substrates and as primitive catalysts in the polymerization of polypeptides, polynucleotides, polysaccharides, and perhaps lipids.

Chains Copy Themselves

Most biologists agree that the first polynucleotides would have included simple, single-stranded RNA molecules, which, as we know from modern RNAs, fold up into three-dimensional shapes (see Chapters 9 and 10). Such shapes might have been more stable in water than unfolded polynucleotides, and hence the tighter, more stable shapes would have been selected for by chemical evolution (i.e., would have tended to remain intact while natural chemical processes broke apart other polynucleotides). Experiments also show that RNA strands can self-replicate: If RNA is added to a solution containing bases (A, U, G, and C), new complementary RNAs form. What's more, certain RNAs (referred to as **ribozymes**) can splice out pieces of themselves that in turn act as primitive enzymes that catalyze the breaking and rejoining of other RNA molecules. In the quiet tide pools or soggy clays of this so-called "RNA world," RNA molecules would have had both a genetic, informational role based on their structure and a catalytic role in making and breaking chemical bonds based on their function.

Molecular Interactions Take Place

We know today that genetic information flows from genes to proteins. So how did primitive RNAs come to contain meaningful genetic information, and then how did primitive polypeptides become associated with the RNAs? The change from random RNA structure to information-containing RNA structure to association with proteins is a very big one, and biologists do not really know how it happened. It is, in fact, *the* crucial change, the coevolution of chicken and egg, as it were. So far, no one has a detailed hypothesis. Researchers have observed, however, that amino acids and nucleotides in watery solutions do tend to associate with each other. People have even searched deep-sea vents for evidence of such associations taking place today (see Box 15.1 on page 325).

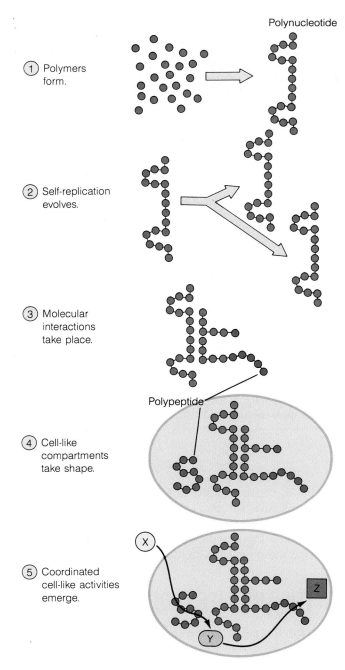

1. Polymers form.

2. Self-replication evolves.

3. Molecular interactions take place.

4. Cell-like compartments take shape.

5. Coordinated cell-like activities emerge.

Polynucleotide

Polypeptide

X   Y   Z

FIGURE 15.8 ■ Five Possible Steps in the Emergence of Living Cells. No one knows exactly how living cells arose, but biologists propose that polymers such as polypeptides and polynucleotides must have formed from monomers (1) and that the polynucleotides (primitive RNAs) began to self-replicate (2). Interactions between these RNAs and polypeptides (simple enzymes) took place, so that the information encoded in the base sequences of the RNA molecules came to specify an amino acid sequence in the polypeptide chains (3). Eventually, cell-like compartments formed around the interacting molecules (4), and finally, coordinated cell-like activities emerged inside the "protocells," including simple metabolic pathways that would allow the compartments to take up and use energy from the environment (5).

Over the course of millions of years of such associations, certain nucleotide sequences must have come to signify specific amino acid sequences, and genes must have come to code for proteins. The two functions of the early RNAs—storing genetic information and catalyzing reactions—were later taken over by other molecules. The more stable nucleic acid DNA became the genetic material, and the more versatile proteins became enzymes, the biological catalysts. RNA molecules, however, retained the central role of translating the information stored in DNA into the amino acid sequence of proteins (see Chapter 10).

Cell-Like Compartments Take Shape

There is very clear evidence that polypeptides and phospholipids spontaneously form tiny spheres under certain conditions of heat, dryness, and pH (Figure 15.9). Some biologists have suggested that such spheres may have enclosed self-replicating associations of genes and proteins. Some of these compartmentalized systems, or "protocells," would then have been selected on the basis of their more efficient absorption of raw materials and use of energy and the fidelity of their replication. One group of these protocells (which some biologists are now calling the *progenotes*) were the earliest common ancestors of all living things that survive today. There are several kinds of strong evidence that modern organisms—from single-celled methanogens to morel mushrooms to humankind—all arose from this one population of progenotes. These forms of evidence include the basic similarities in the structures of RNA molecules, the kinds of chemical reactions cells perform, and the common genetic code.

Coordinated Cell-Like Activities Emerge

Somehow, the protocells came to have metabolic pathways. Perhaps a competition between compartments for limited raw materials from the surrounding environment led to selection for series of enzymes that could modify more abundant materials into less available ones. In this way, metabolic pathways would have evolved with the ability to harvest energy from chemical bonds and synthesize new materials. At some point, the coordinated activities—replication, protein synthesis, metabolism, and repair—became so closely integrated and interdependent that the "compartments" were indistinguishable from true biological units—living cells—with all the fundamental characteristics outlined in Chapter 1. These early cells may have been like the methanogens of today, extracting energy from inorganic compounds such as hydrogen and carbon dioxide and building their own organic molecules, such as glucose and methane.

The pathway from the cosmic formation of organic molecules to the first cells probably occurred in a long series of small steps. Each of these small steps had a high probability of occurring, given the laws of chemistry and physics and the

FIGURE 15.9 ■ **Cell-Like Lipid-Coated Droplets Can Form Spontaneously in the Laboratory.** Under specific conditions of temperature, moisture, and acidity, tiny spheres will form in solutions that contain amino acids and phospholipids. These lipid-coated droplets (magnification about $10,000\times$) form as the solution of phospholipids cools. Perhaps the earliest cell membranes formed spontaneously under similar circumstances.

conditions existing on the earth 4 billion years ago. In fact, many biologists feel that once these rare conditions existed, life was bound to emerge. Nobel-Prize-winning cell biologist Christian de Duve expressed the sentiment this way: "The universe was pregnant with life."

EARTH AND LIFE EVOLVE TOGETHER

The earth has probably been inhabited for more than 80 percent of its history, or about 3.8 billion years. Fossil evidence suggests, however, that *multicellular* organisms—animals, plants, and some fungi—first appeared only about 700 million years ago. Biologists conclude that single-celled organisms ruled the earth for the vast majority of its history, modifying the planet in dramatic and permanent ways.

Early Life Forms Evolve and Change the Earth

The Age of Microbes lasted nearly 3 billion years, and fossils reveal that at least five major developments occurred during this immensely long period, followed by three more evolutionary milestones involving multicellular organisms. Table 15.1 summarizes these eight events.

■ The Earliest Life Geologists have discovered ancient rocks in southern Greenland that contain deposits of organic carbon—deposits that provide some of the earliest evidence of life. These rocks formed 3.8 billion years ago in ocean sediments shortly after meteorites stopped bombarding the earth and obliterating all traces of earlier events. Like today's bio-logical molecules, these carbon deposits are enriched in the lighter isotope of carbon (^{12}C) instead of the heavier carbon isotope (^{13}C) characteristic of inorganic deposits. This isotope ratio makes sense if the deposits had a biological origin. Although biologists do not know what kinds of cells generated the carbon compounds, they must have been anaerobic cells, since the primordial atmosphere lacked oxygen. These earliest cells were probably heterotrophs ("other-feeders") that obtained their raw materials from the carbon-containing organic molecules that fell to earth as meteors or were generated when lightning shot through the primitive atmosphere. In addition, the cells probably obtained their energy by fermentation.

■ Cells Synthesizing Their Own Organic Molecules Fossilized remains of structures called stromatolites have been found near the desolate Australian town of North Pole and dated at 3.5 billion years old. Modern stromatolites are pillow-shaped mounds built layer upon layer by massive colonies of anaerobic blue-green algae (cyanobacteria) (Figure 15.10). The Australian fossils contain tiny spheres and filaments considered to be bacteria closely resembling modern blue-green algae. Although *conclusive* evidence is lacking, these fossils suggest that ancient cells capable of making their own carbon-containing molecules by photosynthesis *may have been* among the earliest cells. These earliest photosynthesizing cells probably made use of sulfur compounds rather than water as electron donors and hence did not produce oxygen as a waste product (review Figure 6.9). Other fossils from this period reveal a trail of biochemical evidence left by cells that were probably similar to modern methanogens (prokaryotes), which use carbon dioxide, hydrogen, and hydrogen sulfide as substrates for energy metabolism.

TABLE 15.1   Steps in the Evolution of life

Event	Billions of Years Ago	Consequences
Full diversity of life forms present	0.4	Complete ozone screen; atmosphere same as today; 20% oxygen; large, active fishes, land plants
Shelled animals and early land plants appear	0.55	Diversity evident in fossil record; 10% oxygen in atmosphere
Many-celled organisms appear	0.67	Fossils and tracks made; oxygen and ozone accumulate in atmosphere to about 7%
First eukaryotic cells appear	1.5	Mitosis, meiosis, genetic recombination, and aerobic respiration occur; 2% oxygen in atmosphere
Oxygen-tolerating blue-green algae appear	2.0	Ozone screen begins to form; iron deposits appear on earth's surface
Strong evidence of photosynthetic organisms	2.8	Oxygen is given off into atmosphere, but still is less than 1%
Autotrophs, methane-generating bacteria, and sulfur bacteria appear; suggestive evidence of photosynthesis	3.5	Little change in atmosphere
The origin of life	3.8	Primordial atmosphere lacks oxygen

FIGURE 15.10 ■ **Modern Stromatolites: Evidence of Photosynthetic Cells.** These pillow-shaped mounds were built by large colonies of photosynthetic cells (blue-green algae) in Hamlin Pool, along Australia's west coast. Fossils of similar structures built 3.5 billion years ago may be evidence of some of the earliest cells on earth.

■ **Photosynthesis and the Release of Oxygen Gas** Australian rocks dating from 2.8 billion years ago show *definite* filaments that strongly resemble today's blue-green algae. From this evidence, biologists have concluded that photosynthesizing cells capable of releasing oxygen gas into the environment *had evolved* by 2.8 billion years ago. Life had started to modify the planet.

■ **Oxygen Tolerance and Aerobic Cells** For over a billion years after oxygen-producing photosynthetic cells evolved, there was little or no accumulation of oxygen in the atmosphere. Instead, iron compounds in seawater reacted with the oxygen from photosynthetic cells, forming iron oxides, which fell to the ocean floor by the ton. In other words, the oceans rusted! Geologists have found iron deposits dating from this time lying in horizontal bands in deep rock strata at sites all over the earth. After most of the iron in seawater reacted this way, oxygen probably began to build up in the atmosphere. Ironically, since oxygen is poisonous to anaerobic cells, this initial buildup would have been harmful to both the early heterotrophs and the photosynthesizers.

About 2 billion years ago, as oxygen levels continued to build, cells became fossilized in the Lake Superior area—filaments and spheres and occasionally thick-walled cells that were probably resistant to the harmful effects of oxygen. Such fossils of thick-walled cells suggest that oxygen given off by blue-green algae all over the earth had begun to accumulate in the atmosphere—perhaps reaching as much as 1 percent of today's oxygen concentration—and that cells began to evolve mechanisms for avoiding oxygen poisoning.

The accumulation of oxygen would have had two other consequences. First, it would have allowed the evolution of cells that obtain their energy by aerobic respiration, with oxygen as the final electron acceptor. These cells might have been similar to the common bacterium *E. coli.* Second, the increase in oxygen would have had a significant impact on the global environment with the formation of an ozone screen. Sunlight striking gaseous oxygen (O_2) creates ozone (O_3), which in turn absorbs ultraviolet light. The accumulating ozone would have protected living things from some of the damage that ultraviolet light inflicts on DNA. Because of this newly developing ozone shield, cells could begin to live closer to the planet's surface.

■ **Eukaryotic Cells** Fossils from 1.5 billion years ago show clear evidence of large cells with a membrane-bound nucleus: eukaryotic cells. The nucleus most likely originated in a prokaryotic cell related to methanogens, but lacking a nuclear envelope (Figure 15.11). This forerunner of eukaryotic cells probably obtained nourishment by phagocytosis, that is, engulfing food particles in pockets of its plasma membrane (review Figure 3.13). These invaginations of the plasma membrane probably came to surround and protect the host cell's naked DNA (Figure 15.11a and b). A similar mechanism probably explains how eukaryotic cells got their mitochondria, chloroplasts, flagella, and nuclei. Put forth by Lynn Margulis and others, the **endosymbiont hypothesis** suggests

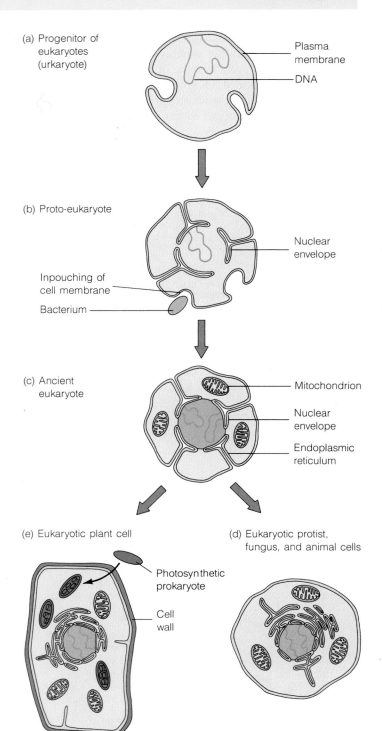

FIGURE 15.11 ■ **The Endosymbiont Hypothesis: How Eukaryotes May Have Evolved.** Many biologists believe that a prokaryotic host cell, called a *urkaryote*, lacking a cell wall and specific internal organelles but capable of inpouching its cell membrane (a) may have merged with symbiotic bacteria that could perform respiration in the presence of oxygen (b). These bacteria then began to survive in the host cell's cytoplasm. The inpouching may have given rise to the endoplasmic reticulum and the nuclear membrane in this ancient eukaryote (c). The nonphotosynthetic symbionts could have evolved into mitochondria in the ancient eukaryote (d). And the photosynthetic "invaders" could have given rise to chloroplasts (e).

BOX 15.1　　　　　HOW DO WE KNOW?

THE SEARCH CONTINUES

One reason scientists continue to explore the solar system and the remote corners of our planet is to search for new life forms and for clues about life's origins. Recent investigations of Martian soil and of hot-water vents far below the ocean surface are cases in point.

In 1976, the United States sponsored the *Viking* mission to Mars (Figure 1). Each landing vehicle bore robot arms and electronically controlled minilaboratories that together could grab and test the red, barren soil for organic molecules and evidence of microbial metabolic activities. While the tests revealed no trace of life as we know it, scientists concluded from photographs that Mars at one time had flowing rivers, lakes, and seas and that life may have evolved there but died out. Traces of that evolution may well remain in Martian rocks, and a return mission could perhaps answer some of the big puzzles surrounding the emergence of life from organic building blocks.

In a 1990 report, the U.S. National Academy of Sciences urged President Bush and his administration to fund another search of Martian soil and rocks for organic precursors to life. The report also recommended that a high priority of the U.S. space program be an expanded search for organic precursors in comets, in asteroids, and on the moon.

Three years after the *Viking* mission, a tiny submarine called *Alvin* descended 2500 m (more than 1.5 mi) into the deep ocean abyss off the southern tip of Baja California. There the two oceanographers aboard made a spectacular discovery: a field of tall chimneys spewing great plumes of hot, mineral-laden water. Return trips to these *hydrothermal vents* and to others elsewhere in the Pacific revealed that the chimneys lie above rock beds crisscrossed by tiny cracks and subtended by zones of hot magma. Seawater percolates down through the cracks, is heated, then seeps back upward through the chimney bearing hydrogen, methane, ammonia, hydrogen sulfide, carbon dioxide, carbon monoxide, and cyanide—compounds that can serve as precursors to the organic building blocks of life. What's more, explorers found strange communities of marine animals surrounding the vents, including clams, mussels, and large tube worms whose bodies were packed with chemoautotrophic bac-

FIGURE 1 ■ **Experimental Probe on the Surface of Mars.**

teria like the methanogens we have been discussing. Such bacteria obtain their carbon from CO_2 and obtain their energy from the bonds of inorganic compounds such as H_2S.

One researcher has even hypothesized that submarine hot springs were the site of life's origins and that the bacteria found there recently might be descendants of the earliest cells. Furthermore, he suggests that a sequence of chemical events leading from inorganic precursors to living cells could still be going on today in the dark, hot recesses of the abyssal vents and that ribozymes or other self-replicating RNA molecules might even be found there.

Recent support for the theory of a deep-sea vent origin of life has come from geologists and planetary scientists studying the earth's heavy bombardment period. Comets and meteors would have generated more than just craters and tidal waves: They would have flash-boiled the oceans and pressure-cooked the atmosphere. After each impact event, scientists believe, thousands of years would have passed before rain once again began to fall, replacing the boiled-off upper layers of ocean water hundreds of meters deep. Given such devastating surface activities, deep-sea vents begin to look like mild, stable "seed beds" for the origins of life. Perhaps the answer awaits some future deep-sea mission.

---

that mitochondria, chloroplasts, and flagella originated as free-living bacterial cells that were engulfed by or became attached to a generalized ancestral cell type (Figure 15.11b–e). Together, the host and guest organisms formed cellular communes, viable single organisms with each member adapted to the group arrangement and deriving benefit from it. The term *endosymbiotic* literally means "mutual benefit within."

Mitochondria, for example, are remarkably similar in size and shape to modern aerobic bacteria, and significantly, these ATP-generating organelles have their own DNA, which replicates independently of the eukaryotic cell's genetic material. Chloroplasts are similar in size and shape to certain photosynthetic prokaryotes (review Box 6.1 on page 126), and like mitochondria, they have separate, self-replicating DNA with bacteria-like genes. The inclusion of energy-processing organelles such as these would have been metabolically

advantageous for ancestral cells large enough to accommodate them and tolerant of oxygen.

■ **Multicellular Organisms**　The Age of Microbes ended about the time that multicelled animals (**metazoans**) and plants appeared 670 million years ago. Fossils from five continents show impressions in sediment left by soft-bodied marine animals that crawled in the sand or mud. Scientists estimate that for such active organisms to appear, the atmosphere must have contained about 7 percent oxygen—about one-third of today's 20 percent.

■ **Hard-Shelled Animals**　Within another 120 million years or so, animals with hard, protective outer shells (such as clams and horseshoe crabs) left fossil imprints. The size and activity of these organisms were increasing, as was the atmospheric oxygen (to probably 10 percent, or half of today's levels). Besides oxygenating the air, living things had begun affecting the earth in other dramatic ways by then. Methane from methanogenic bacteria and carbon dioxide from aerobic organisms were accumulating in the atmosphere. Global cycles that involved living and nonliving processes were developing for hydrogen, oxygen, carbon, nitrogen, phosphorus, and other elements (see Chapter 36). The remains of living things were creating massive deposits of silicon, manganese, and sulfur, in addition to iron, and building thick layers of soil and organic sediments.

■ **Large Plants and Animals**   By 400 million years ago, the atmosphere was essentially like today's, with 20 percent oxygen content. The ozone screen was in place, and very large, complex life forms were appearing in profusion. Large fishes swam in the ancient seas, and the primitive land plants grew along moist shores. Within another 200 million years, amphibians, reptiles, birds, and mammals would move about the continents, and large stands of conifer trees and flowering plants would grow abundantly.

Scientists in the eighteenth and nineteenth centuries sometimes spoke with puzzlement about the "sudden appearance" of hard-shelled life forms in the fossil record. The appearance of shelled animals, however, was preceded first by soft-bodied ancestors and before them by an immensely long history of single-celled species. Only because the collective activities of these microbes over billions of years altered the earth in permanent ways were larger, later life forms—ourselves included—able to evolve.

The Earth Evolves and Alters Life

Just as life's evolution affected the earth, the earth, too, has changed geologically and influenced the evolution and diversification of living things. That change, however, was slow to start. Geologists are convinced by various types of evidence that for hundreds of millions of years after the earth formed, the face of the planet remained nearly the same: Black volcanic islands jutted upward through blue ocean, and as the lava from nearby cones accumulated, some areas of high, flat terrain—the first continents—emerged from the seas.

The early landmasses were probably flat areas of lighter rock upon huge *plates* of heavier basaltic rock. Today, the earth has eight major rock plates (Figure 15.12a). Most of the plate surfaces are submerged beneath deep ocean; only about one-eighth of their surface area bears the higher, drier continental masses. Crustal plates rest on molten mantle material and drift slowly as convection currents in the mantle cause upwelling and the formation of new crustal material at the junctures, or **rifts,** between the great plates. As new crust forms in the rifts beneath the ocean, it pushes the plates a few centimeters apart each year and causes their leading edges to collide with those of other existing plates. With nowhere to go, the colliding edge of one such plate is slowly driven down beneath the edge of the other at zones called deep-ocean

trenches (Figure 15.12b). Stresses far below the ocean or land surfaces, created by the slow, inexorable movements of plates, cause most of the geological activity that molds the face of the earth, including the earthquakes that rock the Middle East and Pacific Rim nations, the volcanoes of the Pacific Northwest and many of the world's mountainous regions, and the uplifting of the Rocky Mountains and other ranges.

Geologists believe that the building and movement of crustal plates, a process called **plate tectonics,** began more than 4 billion years ago, but because crustal movement was very slow, stresses did not build up quickly, and the face of the earth did not change much for the first 3 billion years or so. This time span, from 570 million to 4.6 billion years ago, is usually called the *Proterozoic* ("early life") era and is one of four time spans, or **geological eras,** into which scientists divide the earth's history (Figure 15.13). Each era is further

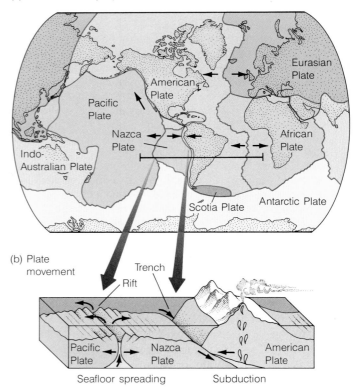

FIGURE 15.12 ■ **Plate Tectonics: Continents Gliding and Colliding. (a) The continents and oceans ride at the surface of massive crustal plates. (b) At the junctions (*rifts*) between certain plates,** new crustal material oozes up from the earth's core and pushes the plates apart. Geologists call this *seafloor spreading.* Here, it is occurring between the Pacific and Nazca plates. Where the leading edges of plates collide, one is generally pushed below the other, causing a deep trench to form (a process called *subduction*). Here, the Nazca plate is being subducted beneath the American plate. Subduction causes most of the violent geological activity that takes place in countries rimming the Pacific Ocean, the Mediterranean, and elsewhere.

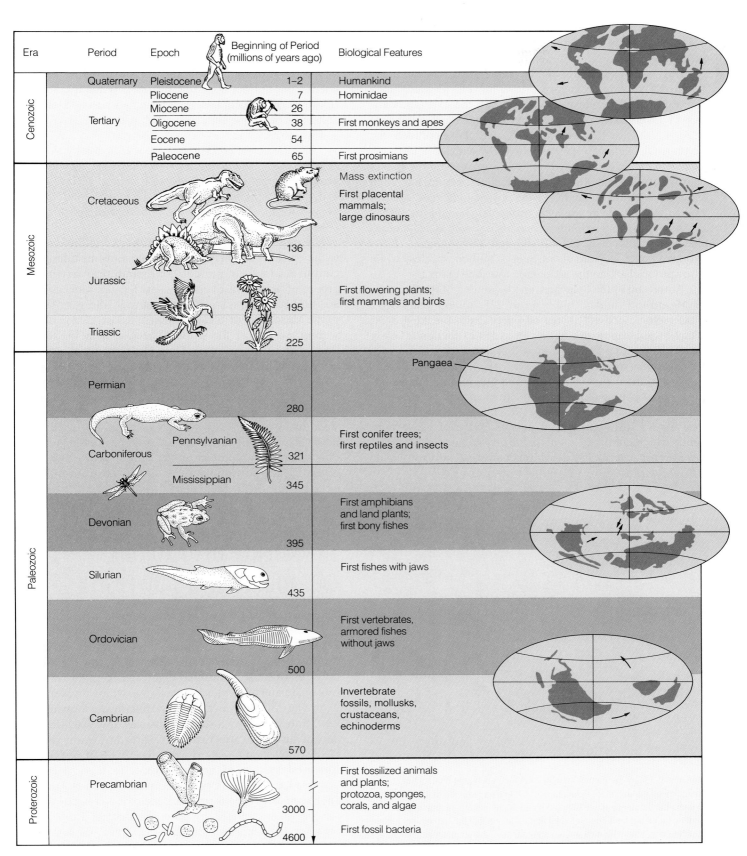

Era	Period	Epoch	Beginning of Period (millions of years ago)	Biological Features
Cenozoic	Quaternary	Pleistocene	1–2	Humankind
		Pliocene	7	Hominidae
	Tertiary	Miocene	26	
		Oligocene	38	First monkeys and apes
		Eocene	54	
		Paleocene	65	First prosimians
Mesozoic	Cretaceous			Mass extinction; First placental mammals; large dinosaurs
	Jurassic		136	First flowering plants; first mammals and birds
	Triassic		195	
			225	
Paleozoic	Permian		280	
	Carboniferous	Pennsylvanian	321	First conifer trees; first reptiles and insects
		Mississippian	345	
	Devonian		395	First amphibians and land plants; first bony fishes
	Silurian		435	First fishes with jaws
	Ordovician		500	First vertebrates, armored fishes without jaws
	Cambrian		570	Invertebrate fossils, mollusks, crustaceans, echinoderms
Proterozoic	Precambrian		3000	First fossilized animals and plants; protozoa, sponges, corals, and algae
			4600	First fossil bacteria

Pangaea

FIGURE 15.13 ■ Earth's History: Geological Eras and Periods. This chart relates the emergence of life forms to changes in the earth's landmasses, starting with the Paleozoic era.

subdivided into **geological periods.** The Proterozoic had just one period, the Precambrian, and as we saw, the Precambrian was dominated by the evolution of bacteria and single-celled eukaryotes.

During the next major era, the *Paleozoic* ("ancient life"; 225 to 570 million years ago), crustal movements were much more pronounced, and plates bearing all the existing continents converged into a single super landmass called Pangaea ("All Earth"; see Figure 15.13). As the plates collided, the continental landmasses on top of them buckled and tilted, huge mountain ranges were uplifted, warm, shallow inland seas were filled and drained, and ocean basins were altered in number, size, and position. These physical changes, accompanied by changes in climate, created new habitats or altered existing ones. Some groups of organisms evolved adaptations that allowed them to exploit the new environments; these groups diversified into more than 1 million species, while other groups lacking the adaptations became extinct.

At the end of the Paleozoic and beginning of the third era, or *Mesozoic* ("middle life"; 65 to 225 million years ago), plate tectonic activity increased once again, breaking apart Pangaea and leading to a massive wave of extinctions of early aquatic and land organisms, as well as to the formation of yet more habitats for new species.

The island continent of Australia provides a good example of how plate tectonics and continental drift affected the evolution of plants and animals. At the time Pangaea existed (see Figure 15.13), mammals had not yet evolved the placenta, which nourishes young in the uterus. Instead, offspring were born at a very early developmental stage and grew inside

a pouch or marsupium, nourished by milk (Figure 15.14). These small primitive mammals called *marsupials* thrived all across Pangaea, as well as on the giant landmass of Australia after it broke apart and drifted away from Pangaea. Thousands of species of mammals with placentas evolved on other continents, but did not arise on Australia. Without competition from placental mammals, the marsupials radiated into a rich collection of unique animals, including the kangaroos, koalas, and wombats.

Along with the breakup of Pangaea, the Mesozoic saw the coming of the giant reptiles, the dinosaurs, as well as the growth of great swampy forests of tropical plants and trees and the beginning of the birds, mammals, and flowering plants. As the Mesozoic ended, the current geological era began.

During the *Cenozoic* ("recent life"; 65 million years ago to the present), extensive crustal plate movement brought the continents to their present configuration, carrying many groups of organisms to entirely new locations and conditions and further encouraging the divergence of lineages. At the end of the Mesozoic era, a catastrophic mass extinction occurred, wiping out half of the world's species, including the dinosaurs. After this mass extinction, birds, mammals, and flowering plants diverged into large and varied groups of species.

In the next four chapters, we will consider each major group—the single-celled organisms, the fungi, the plants, and the animals—in detail. Here it is important simply to understand the overall effects of geological activity on life's evolution: The continually changing earth put pressures on living things that led to the appearance and disappearance of millions of species over time. Earth and life truly coevolved.

FIGURE 15.14 ■ **How Continental Drift Affects Evolution.** The kangaroo is a marsupial—a mammal with a pouch. At the time Australia began to drift toward its present position, mammals with placentas had not yet evolved, and while they arose on other continents, they did not arise in Australia. Australian marsupials thus evolved into a wide array of species without competition from placental mammals, thanks largely to continental drift.

THE SCIENCE OF TAXONOMY: CATALOGING LIFE'S DIVERSITY

Since before the time of Aristotle, naturalists have delighted in searching out, describing, naming, and grouping the myriad kinds of living things. Long ago, however, they realized that they needed a system to categorize, or *classify,* the life forms they discovered. In the mid-eighteenth century, a Swedish biologist named Carolus Linnaeus solved the problem of categorizing life's diversity by assigning every organism then known to science to a series of increasingly specific groups, depending on the structural traits shared by group members. He also invented a way to name each type of organism: the **binomial system of nomenclature.** In this system, each organism is assigned a two-word name: a **genus** name and a **species** name. A genus (plural, genera) is a group of very similar organisms related by common descent from a recent ancestor and sharing similar physical traits. We humans, for example, are the only modern-day members of the genus *Homo.* A species, as we saw in Chapter 1, is a unique group

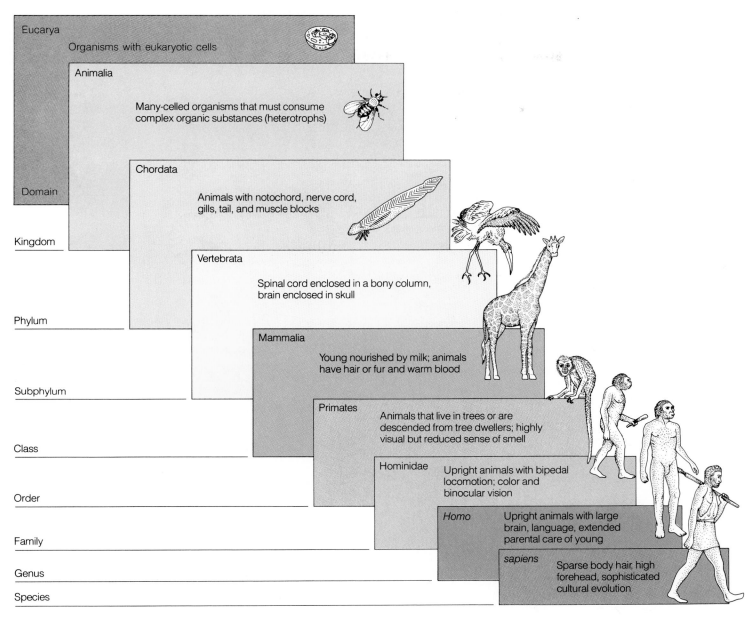

FIGURE 15.15 ■ Taxonomic Classification of Human Beings. Each higher taxonomic group encompasses a wider variety of organisms.

within a genus, whose members share the same set of structural traits and can successfully interbreed only with members of the same species. Our species name is *sapiens* ("wise"). Note that the genus name begins with a capital letter, and both names are written in italics.

Linnaeus placed each organism, with its genus and species name, in a series of categories, or **taxonomic groups,** based on shared characteristics. Biologists recognize several major levels beyond species and genus: Members of similar genera belong to the same **family,** members of similar families to the same **order,** similar orders to the same **class,** simi-

lar classes to the same **phylum,** similar phyla to the same **kingdom,** and now, similar kingdoms to the same **domain.** Figure 15.15 shows our own complete classification from domain to species.

You can see that we are members of the kingdom Animalia, and during Linnaeus's time, all organisms were considered either animals or plants (kingdom Plantae). The more biologists learned, however, the more species and structural traits they discovered that were neither plantlike nor animal-like. Since the late 1950s, biologists have generally recognized five kingdoms: the **Monera,** or single-celled prokary-

(a)

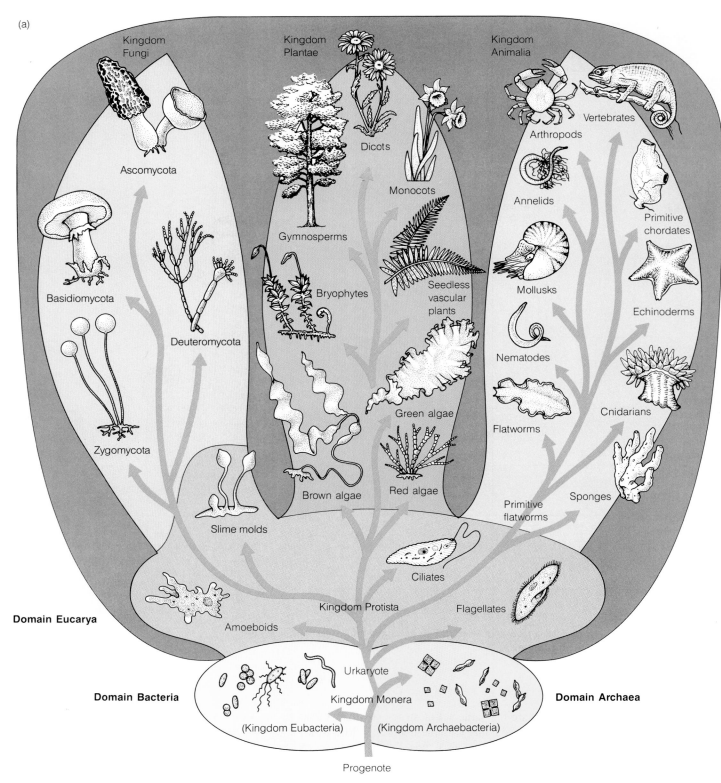

FIGURE 15.16 ■ **The Five Kingdoms and the Three Domains.** Living things can take extremely
different forms. (a) The millions of species in the five kingdoms (monerans, protists, fungi, plants,
and animals) all evolved from earlier organisms, and their oldest ancestors can be traced back to the
first prokaryotic cells on the ancient earth. Many evolutionary biologists believe that some common
ancestor, the progenote, gave rise to all the prokaryotes; that these simple cells gave rise to the more
complex single-celled eukaryotes, the protists; and that branches of protist-like cells then gave rise to
the fungi, plants, and animals.

(b)

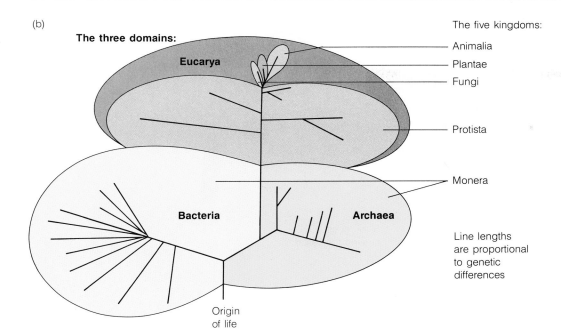

FIGURE 15.16 ■ (b) Taxonomists emphasizing the genetic similarity of RNA and DNA molecules now superimpose three domains on the five-kingdom scheme. These are the domains Eucarya (in gray), which includes the protists, fungi, plants, and animals; the domain Bacteria (in pale yellow); and the domain Archaea (darker yellow). The lengths of the black lines are proportional to the genetic differences between related organisms.

otes, including familiar bacteria like the ones that cause tooth decay and the less familiar ones that live in harsh environments, like the methanogens; the **Protista** (also called Protoctista), or single-celled eukaryotes, such as *Euglena;* the **Fungi,** or multicellular heterotrophs, such as mushrooms or molds that decompose other biological tissues; the **Plantae,** or multicellular autotrophs, such as algae, mosses, ferns, and flowering plants; and the **Animalia,** or multicellular heterotrophs, which lack cellulose and usually exhibit movement and include insects, clams, birds, and reptiles (Figure 15.16a).

While the five-kingdom system categorizes organisms based on their cell structure, genetic comparisons of different organisms suggest that life falls into just three main groupings, called domains (Figure 15.16b). As we saw in Chapter 1, the domain called Eucarya includes four of the five kingdoms—animals, plants, protists, and fungi (see Figure 15.16b). A second domain, called Bacteria, includes most disease-causing prokaryotes and some other members of the kingdom Monera. A third domain, Archaea (formerly called Archaebacteria), includes the prokaryotes that inhabit harsh environments, such as methanogens living at the bottom of swamps or the cells inhabiting acidic hot springs. Interestingly, members of Archaea are more similar genetically to members of Eucarya than they are to members of Bacteria. In other words, the readers and writers of this book are more closely related genetically to a methanogen growing in a swamp than the methanogen is to the tooth decay bacterium growing on our teeth! Microbiologists believe that the earliest ancestor of the eukaryotes (see Figure 15.11a) was closely related to the progenitor of methanogens and other Archaea cells, whereas the Bacteria arose from a more distantly related ancestor.

While the three-domain system, first proposed by Carl Woese in 1990, is logical and almost certainly correct, too little time has passed for the community of biologists to fully debate, digest, accept, or reject the concept. By 1991, at least one objection had surfaced: While Woese emphasizes evolutionary relationships based on a few genes from the nuclear genome, researcher Lynn Margulis argues that any taxonomic scheme should also include genomes in chloroplasts and mitochondria, which probably originated as endosymbionts (see Figure 15.11). This book generally follows the five-kingdom approach, but bear in mind the fundamental molecular distinctions that separate the prokaryotes into two of the three domains of life.

Domains and kingdoms and other taxa do more than simply divide up and list living things; they help show evolutionary relationships between species. Figure 15.16 depicts and explains the relationships between the kingdoms and shows the branching evolutionary pathways that led from common ancestors toward specialized groups. No species alive today, of course, was the progenitor of any other species now existing. Some modern organisms, such as the methane bacteria, may have changed less over evolutionary time than the lineages leading to crabs, say, or to daisies, but nevertheless, all lines have experienced 3.8 billion years of evolutionary change. All organisms now alive are at the tips of their branches in the great tree of life.

Taxonomists have placed nearly 2 million species on this tree of life. They haven't always agreed, however, on how to group the organisms into hierarchies. Over the years, in fact, taxonomists have devised three competing approaches to categorizing living things: the phenetic, cladistic, and evolutionary schools. Taking an example like the birds in Figure 15.17,

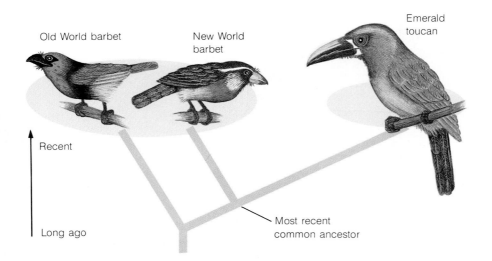

FIGURE 15.17 ■ Schools of Taxonomy. Taxonomists using a phenetic approach categorize bird species by physical similarities, as in the large and small ovals. Those using a cladistic approach would perch the birds on the evolutionary branches shown here.

taxonomists in the phenetic school might use the physical similarities (phenotypes) of the small red and yellow Old World barbet and New World barbet to place these species in the same taxonomic group. At the same time, they would assign the larger, brighter emerald toucan to a separate category represented by the smaller oval in Figure 15.17. Taxonomists in the cladistic school emphasize genealogy. They would rely on DNA studies that revealed the New World barbet to be more similar genetically to the emerald toucan than to the Old World barbet and would place them, instead, on the family tree shown in Figure 15.17, in which the New World barbet and the emerald toucan share a more recent common ancestor.

Taxonomists in the evolutionary school take genealogy into account but also consider the history of each branch and place organisms on longer or shorter branches, depending on the degree to which their biological characteristics have changed from the original ancestors in their lineage. But they still have a leviathan task ahead, because as many as 8 million as yet undiscovered species are believed to inhabit the oceans, rain forests, and other hard-to-explore regions.

Regardless of general system, describing living species is a science with a deadline. As Box 15.2 explains, much of life's diversity—perhaps 2, 4, or even 5 million yet undiscovered and undescribed species—exist in the tropical rain forests. As humans continue to destroy these forests, we could perpetuate a mass extinction far worse than the one that killed the dinosaurs 65 million years ago and in the process cut off many branch tips—even whole limbs—from the tree of life that coevolved with our planet during 4 billion years of history.

CONNECTIONS

Understanding the origins of life is a profound scientific challenge because the drama unfolded on a stage that no longer exists and under conditions that changed forever once photosynthetic organisms began to thrive and give off oxygen. Nevertheless, by knowing both the characteristic behavior of matter—particularly organic molecules—and the attributes of all living things, biologists have reconstructed a plausible sequence of the prebiotic events that led to the emergence of the first cells. While there are no cell-like fossils from 4 billion years ago to suggest the form of the universal ancestors, or progenotes, we can study living fossils like the methanogens, which probably evolved directly from the earliest cells. From methanogens we can infer the kinds of physical challenges that all early organisms must have faced, as well as the biological solutions that arose.

Perhaps someday we will find, near some abyssal deep-sea vent or on a distant planet, more specific proof of how life evolved. In the meantime, though, the biochemical history laid down in living methanogens and other cells is a helpful substitute. The bad smells that come from sewage treatment plants and swamp mud can serve as sensory reminders of life's enormously long history and its primitive beginnings on an equally primitive planet. We cannot fully appreciate living things or understand their evolutionary relationships, however, without first comprehending the intertwined pasts of life and the planet on which life arose by natural processes.

The remaining chapters in this section explore life's splendid diversity, kingdom by kingdom.

BOX 15.2   FOCUS ON THE ENVIRONMENT

HOW MANY SPECIES INHABIT THE EARTH?

Biologists know how a cell makes protein, how brain hormones regulate the menstrual cycle, and how to transfer genes from one organism to another. They don't, however, know how many other species share this planet with us. Some think the figure could be 3 million, some think upwards of 50 million. But two things are certain: Virtually all of those species will turn out to be insects, and we're killing them off at a frightening rate.

Over the past 230 years, biologists have discovered, named, and recorded 1.5 million species of living organisms. This has been basically a hit or miss prospect, based on where naturalists have gone, what they have been interested in collecting, and how much time (and funding) they have had for studying the unknown organisms and determining whether they can in fact claim and name them as brand-new species.

Of the 1.5 million, about 5 percent are single-celled prokaryotes and eukaryotes; about 22 percent are fungi and plants; and about 70 percent are animals. Most of the animals are invertebrates (they lack backbones). The sea harbors the most varied groups of invertebrates, with 90 percent of all the higher taxa (phyla and classes). But these thousands and thousands of taxonomic categories contain only 20 percent of the individual animal species. By far the largest number of species are animals that inhabit land. And of these, nearly a million species so far named are insects. Ours is a planet of terrestrial insects (Figure 1). And there are probably millions of species still unnamed.

For many years, biologists estimated that an accurate number of total species would be about 3 to 5 million. They based this estimate on the fact that two-thirds of the currently known species live in temperate zones (where, coincidentally, almost all the scientists have lived). Since a long-accepted formula held that perhaps twice as many species inhabited tropical regions as temperate ones, it seemed likely that there were at least another 1.5 million species to be found and studied.

This estimate lost favor, however, when in the 1970s and 1980s biologists went out to conquer the "last biotic frontiers"—the deep-sea bed and the canopy (treetop habitat) of tropical rain forests. It has been a common occurrence for researchers to collect mud from a previously untested spot of ocean bottom or to gather the plants and animals clinging to the rarified world at the top of an individual tropical treetop and to find that 90 percent of the organisms collected were unknown to science!

For example, Terry Erwin, a researcher working in the Panamanian rain forest, set off insecticidal fog bombs in individual jungle trees and collected all kinds of beetles that died and dropped off. On average, 163 species of beetles fell from each tree species. He reasoned that since there are at least 50,000 tropical tree species, since beetles represent just 40 percent of animals in the jungle treetops (almost all the others are also insect species), and since half again as many species live on the ground as up in the forest trees, there could easily be 30 million additional species of insects and their relatives to describe in the rain forests alone. Adding in tropical plants, sea bed organisms, and all other undiscovered species in other climatic zones, the figure of 50 million new species—despite its enormity—begins to look possible.

We clearly have a very long way to go to collect and catalog them all, but the prospects of this are dismal, because people are cutting down rain forests far faster than scientists can knock the bugs out of the treetops and study them. Depending on how one estimates the total number of species on earth, we could be losing forever somewhere between 20,000 and 150,000 species a year. Without the tropics and all its terrestrial invertebrates, we would cease to be the planet of the insects. Long before that, however, we would cease to exist as a species that categorizes insects *or* cuts tropical forests—as later chapters will explain.

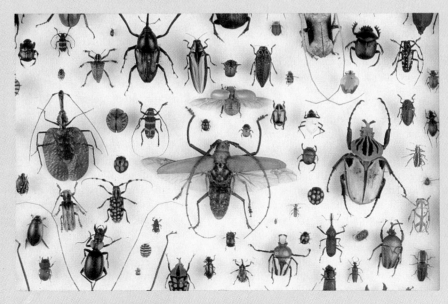

FIGURE 1 ■ **Earth: A Planet of Terrestrial Insects.**

Highlights in Review

1. Modern methane-producing bacteria live in restricted, oxygen-free environments, such as sewage sludge, swamp mud, and animals' intestines, and may be more similar to the earliest cells than any other organisms now living.

2. Three themes emerge from the story of life's origins and diversity: Life arose as a direct result of the physical conditions on our planet; because those conditions changed, we may never know exactly how life originated; and the earth and life co-evolved, each influencing the other's evolution.

3. After the Big Bang 18 billion years ago, the sun and its planets coalesced from clouds of hydrogen, helium, and heavier elements. By 4 billion years ago, the basic structure of the earth was set: a core of heavy metals surrounded by a liquid mantle of hot rock, a hard crust of cool black rock, an oxygen-free atmosphere, and a covering of blue oceans.

4. The earth had an advantageous combination of composition, geological activity, size, and distance from the sun, and thus life was able to evolve and prosper here but not on neighboring planets.

5. The organic building blocks of life probably existed on the early earth, owing to clouds of such molecules in space, carbonaceous chondrites, or natural reactions involving atmospheric gases.

6. Several stages in the origin of life could have resulted from the characteristic behavior of molecules. The organic building blocks could have formed biological polymers. The early polynucleotides, including RNAs, could have replicated themselves. RNAs and primitive polypeptides could have interacted, giving rise to the two-way association between genes and proteins. Cell-like compartments could easily have formed from polypeptides or phospholipids and enclosed self-replicating gene-protein systems in solution. The compartments could have developed metabolic pathways. And eventually, living cells could—and did—emerge.

7. Milestones in life's early history are recorded in sedimentary rocks from around the world. The earliest anaerobic cells probably existed by 3.8 billion years ago. By 3.5 billion years ago, primitive blue-green algae had evolved. By 2.8 billion years ago, blue-green algae definitely existed and were giving off oxygen into the seas and air. By 2 billion years ago, cells had begun to tolerate oxygen, causing a layer of rusted iron to appear at the ocean's bottom and a layer of ozone to form in the atmosphere. By 1.5 billion years ago, eukaryotic cells had emerged. And by 670 million years ago, multicellular aquatic animals appeared. Within another 400 million years or so, a profusion of species radiated into aquatic and land habitats.

8. The earth's plate tectonic activity caused the continents to tilt and change positions, the oceans and seas to fill and drain, and the mountain ranges to form. These physical changes altered survival pressures and created new habitats that affected the evolution and divergence of living species.

9. Scientists have begun to classify all living things into three domains—Bacteria, Archaea, and Eucarya—plus five kingdoms—Monera, Protista, Fungi, Plantae, and Animalia—and assign each organism to a series of lesser taxonomic subdivisions including phylum, class, order, family, genus, and species. These categories help reveal evolutionary relationships and show how a common ancestral group may have given rise to over 10 million living and extinct species.

Key Terms

Animalia, 331
binomial system of
   nomenclature, 328
class, 329
domain, 329
endosymbiont hypothesis, 324
family, 329
Fungi, 331
genus, 328
geological era, 326
geological period, 328
kingdom, 329
metazoan, 325

methanogen, 316
Monera, 329
order, 329
phylum, 329
Plantae, 331
plate tectonics, 326
Protista, 331
ribozyme, 321
rift, 326
species, 328
taxonomic group, 329
taxonomy, 317

Study Questions

Review What You Have Learned

1. How do methanogens differ from most other living cells?
2. What was the Big Bang?
3. Cite the possible sources of organic compounds on the early earth.
4. Rearrange the following theoretical steps to reflect the probable sequence of events as living cells arose: self-replication, molecular interactions, coordinated cell-like activities, formation of polymers, development of cell-like compartments.
5. What important events in the earth's history may have occurred in the following time periods? About 4.6 billion years ago. About 2.8 billion years ago. About 1.5 billion years ago.
6. What is the evidence that mitochondria were originally free-living organisms?
7. What causes most of the geological activity that alters the earth's surface?
8. What did Linnaeus contribute to taxonomy?
9. Define species.
10. List the following taxonomic groups in ascending order, from least inclusive to most inclusive: family, phylum, species, class, kingdom, order, genus, domain.

Apply What You Have Learned

1. The *Viking* mission to Mars in 1976 found no evidence of life on that planet. Why, then, is a return mission planned?
2. Scientists have discovered autotrophic bacteria thriving in deep-ocean hydrothermal vents. Why might—or might not—a future expedition discover photosynthetic prokaryotes in the same area?
3. You are engaged in a debate with someone who argues as follows: "You scientists claim that life evolved spontaneously on earth some 4 billion years ago, but you say that all life on the planet is related to a common ancestor. If life arose out of lifeless matter billions of years ago, it should do so today, and there should be many unrelated lineages of life." How should you answer this argument?

For Further Reading

CHYBA, C. D., P. J. THOMAS, L. BROOKSHAW, and C. SAGAN. "Cometary Delivery of Organic Molecules to the Early Earth." *Science* 249 (1990): 366–373.

DE DUVE, C. *Blueprint for a Cell: The Nature and Origin of Life.* Burlington, NC: Patterson, 1990.

DICKERSON, R. E. "Chemical Evolution and the Origin of Life." *Scientific American,* September 1978, pp. 70–86.

GRAY, M. W. "The Evolutionary Origins of Organelles." *Trends in Genetics* 5 (1989): 294–299.

HORGAN, J. "In the Beginning." *Scientific American,* February 1991, pp. 117–125.

KUNZIG, R. "Stardust Memories: Kiss of Life." *Discover,* March 1988, pp. 68–76.

MARGULIS, L. *Early Life.* Boston: Jones and Bartlett, 1982.

MARGULIS, L., and R. GUERRERO. "Kingdoms in Turmoil." *New Scientist,* March 23, 1991, pp. 46–50.

STORK, N., and K. GASTON. "Counting Species One by One." *New Scientist,* August 11, 1990, pp. 43–47.

WHITTAKER, R. H. "New Concepts of Kingdoms of Organisms." *Science* 163 (1969): 150–160.

WOESE, C. R., O. KANDLER, and M. L. WHEELIS. "Towards a Natural System of Organisms: Proposal for the Domains Archaea, Bacteria, and Eucarya." *Proceedings of the National Academy of Sciences USA* 87 (1990): 4576–4579.

Life As a Single Cell

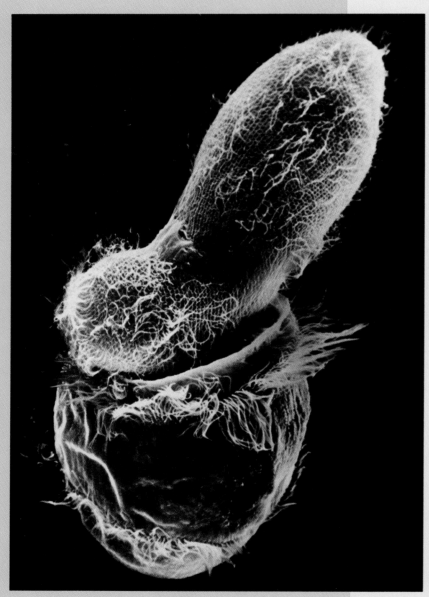

FIGURE 16.1 ■ **A Single-Celled Hunter and Its Prey.** The juglike *Didinium* (left) swallowing a slipper-shaped *Paramecium* cell.

The Complex Life of a Microbial Hunter

Watching through a microscope as the single-celled hunter *Didinium nasutum* stalks its prey is a bit like viewing a silent horror film. *Didinium* is globular in shape, and two rows of lashing cilia propel it through the water in a zigzag course in search of food. The food is usually slipper-shaped *Paramecium* cells, which are themselves gliding around in search of small food particles. But *Didinium* has a deadly weapon: a cone-shaped region that protrudes from one end of the cell and is ringed by poison darts. With this "snout," *Didinium* can stab a *Paramecium* cell, impaling it long enough to fire the darts and paralyze the hapless victim. Then, as the tip of its cone stretches open wider and wider, the hunter swallows alive the prey cell larger than itself (Figure 16.1). Digestive enzymes immediately begin breaking down the meal, and within 20 minutes, *Didinium* is ready to hunt again.

Didinium and *Paramecium* are both protists—that is, members of the kingdom Protista, along with 35,000 other species. As we discussed in Chapters 1, 3, and 15, the protists are single-celled eukaryotes. They are generally larger and more complex than prokaryotes and contain a true nucleus and other membrane-bound organelles, such as mitochondria, the endoplasmic reticulum, and chloroplasts.

In contrast to the protists, the other group of single-celled organisms, the more than 2500 species of prokaryotes, are among the simplest living cells. In the five-kingdom system of classification, all prokaryotes are members of the kingdom Monera. But modern molecular genetic data suggest that all life belongs to three domains (review Figure 15.16b). In this three-domain classification system, the prokaryotic kingdom Monera is split into two domains, Bacteria (also called Eubacteria) and Archaea (also called Archaebacteria), while the third domain, Eucarya, contains all eukaryotes, including *Didinium* and its fellow protists, plus animals, plants, and fungi in their kingdoms. Species in the domain Bacteria are the prokaryotic cells most often encountered in soil, water, and the body. Species in the domain Archaea are prokaryotic species living in harsh environments that are very hot, acidic, salty, or anaerobic (Figure 16.2). Prokaryotes in both domains lack a membrane-bound nucleus and other cell organelles and, like protists, exhibit many life-styles.

No matter what the environments or life habits of prokaryotic cells and protists like *Didinium,* they all share a fundamental characteristic: the ability to carry on all life processes

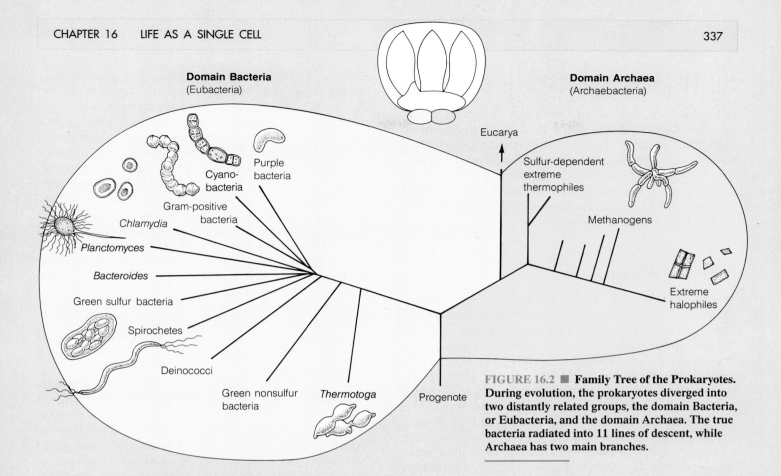

Domain Bacteria
(Eubacteria)

Domain Archaea
(Archaebacteria)

Eucarya

Purple
bacteria

Cyano-
bacteria

Gram-positive
bacteria

Chlamydia

Planctomyces

Bacteroides

Green sulfur bacteria

Spirochetes

Deinococci

Green nonsulfur
bacteria

Thermotoga

Progenote

Sulfur-dependent
extreme
thermophiles

Methanogens

Extreme
halophiles

**FIGURE 16.2 ■ Family Tree of the Prokaryotes.
During evolution, the prokaryotes diverged into
two distantly related groups, the domain Bacteria,
or Eubacteria, and the domain Archaea. The true
bacteria radiated into 11 lines of descent, while
Archaea has two main branches.**

as an independent single cell. No individual body cell
(somatic cell) of a multicellular organism such as an apple
tree or ladybug can live independently under normal conditions, nor do such cells ever approach the complexity of most
protists. *Didinium* can swim, jab, swallow, excrete, digest,
shoot darts, detect the presence of prey, and reproduce sexually, among other abilities. Most of the behaviors we see in
larger, more complicated beings appeared first in the single-
celled kingdoms.

While protists are often very complex single cells and prokaryotes are much simpler ones, there are still simpler entities,
the viruses, that exhibit some but not all the properties associated with life (see Chapter 1). Viruses are not cells, but particles composed of proteins, nucleic acid, and sometimes lipids. Unlike the monerans and protists, a virus cannot carry on
all life processes independently, and it can reproduce only
after entering a living cell. As we will see, viruses can cause
diseases and are invaluable to biological research.

Two more themes underlie our discussion of monerans
and protists in this chapter. Single-celled organisms are ubiquitous on our planet because of their success obtaining nutrients and reproducing in many different environments. Every
pinch of soil, glob of mud, droplet of water, and square centimeter of skin, hair, and leaf surface on this planet literally
teem with microbes. A small pinch of soil can contain a billion organisms and from 20 to 100 or more separate species.
Microbes are so numerous on our skin and in our intestines

that some biologists estimate that our bodies have nine bacterial cells to every one cell we make with our own genes. As
a group, microbes are ubiquitous because they reproduce rapidly and can metabolize virtually any source of organic matter
(as well as some inorganic sources) and because they can
withstand extremes of heat, cold, dryness, acidity, salinity,
pressure, and radiation that would kill other organisms.

The fact that microbes are so abundant in so many habitats
leads logically to another theme: Single-celled organisms
have tremendous ecological, medical, economic, and scientific importance. Every year, microbes fix millions of tons of
carbon, nitrogen, and other elements into biologically useful
compounds, release massive quantities of oxygen, and
decompose vast amounts of organic matter. At the same time,
thousands of microbial species live in close physical association with plants, animals, and other organisms, sometimes
benefiting their hosts, and sometimes harming them. Finally,
single-celled organisms are invaluable as subjects in scientific
research and as tools of biological engineering.

This chapter investigates the following questions regarding single-celled organisms and viruses:

■ How do prokaryotic cells obtain energy and materials,
reproduce, and interact with the environment?
■ Are viruses alive?
■ How do protists "make a living," and how do they cause
diseases?

THE LIVES OF PROKARYOTIC CELLS

Prokaryotic cells have an amazing range of habitats, living anywhere from arctic snowdrifts to deep-sea hydrothermal vents. The thousands of species in the domains Archaea and Bacteria are related in a great family tree; Figure 16.2 may serve as a helpful visual reference for sorting out their relationships as you read this chapter. Members of Bacteria and Archaea do differ in some fundamental ways, as we shall see, but let's first consider their common traits.

Prokaryotic Cell Structure

Prokaryotic cells have an outer cell wall that surrounds the plasma membrane, which in turn surrounds a noncompartmentalized cytoplasm dotted with ribosomes; they generally lack membrane-enclosed organelles (review Figure 3.4). A circular strand of DNA, usually coiled into one region of the cell, serves as the single chromosome, and metabolic activities, such as electron transport and photosynthesis, take place on the plasma membrane, which sometimes folds inward into the cell's interior. The outer cell wall is a strong but flexible covering made primarily of sugar-protein complexes called *peptidoglycans*. In so-called *gram-positive cells,* these occur in a single broad layer, while in *gram-negative cells,* the peptidoglycan layer is covered by an outer layer containing proteins and lipopolysaccharides (fat-sugar complexes). The dif-

ferences in cell wall characteristics cause gram-positive bacteria to stain purple and gram-negative bacteria to stain red (Figure 16.3). This staining technique, the Gram test, is one of the first procedures a microbiologist will perform on a newly found or unfamiliar bacterial species; it is important because it suggests the types of antibiotics that might be effective in fighting a **pathogenic** (disease-causing) bacterium. (An antibiotic is a chemical made by one microorganism that can slow the growth of or kill another microorganism.) In the summer of 1976, for example, many delegates at an American Legion convention in Philadelphia developed fever, chills, and pneumonia. Biologists isolated a new bacterium from the sick legionnaires and found it to be gram-negative. Streptomycin blocks protein synthesis and hence the growth of such gram-negative bacteria, while penicillin is more likely to block the growth of gram-positive bacteria by interfering with the building of the bacterial cell wall. So-called broad-spectrum antibiotics can attack both groups, and one of these—erythromycin—is especially effective against Legionnaires' disease.

Most prokaryotes are between 1 and 10 μm long, only about one-tenth the size of a eukaryotic cell. The range of sizes is a result of the diverse shapes, which include **cocci** (spheres), **bacilli** (rods), **spirilla** (spirals), and **vibrios** (curved rods) (Figure 16.4). The names of prokaryotes often reflect their shape. For example, *Bacillus anthracis,* the cause of anthrax, is rod-shaped, while *Streptococcus pyogenes,* the pathogen that causes strep throat, is spherical.

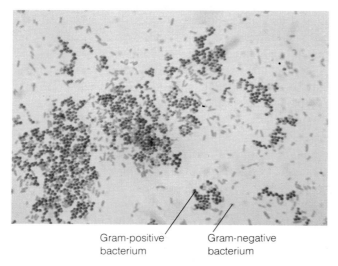

Gram-positive bacterium   Gram-negative bacterium

FIGURE 16.3 ■ Gram Staining Depends On Bacterial Cell Wall Structures. Danish bacteriologist Christian Gram developed a set of staining procedures that differentiate bacteria into two classes, based on the construction of their cell walls. Gram-positive cells, such as the *Staphyloccus aureus* shown here, stain purple; gram-negative cells, such as the *E. coli* shown here, stain red. A structural difference in their cell walls accounts for both the differential staining and the ability of many gram-negative bacteria to resist antibiotics and cause disease.

Nutrition in Prokaryotes

Prokaryotes can be heterotrophs, which must consume organic nutrients, or autotrophs, which can make their own organic nutrients. Together, these two modes of nutrition enable prokaryotes to survive on an enormous range of energy sources and hence to be ubiquitous in air, soil, water, and other life forms.

Most prokaryotes are heterotrophs, and collectively they consume an enormous range of compounds, from biological materials and organic substances like methane to inorganic substances like hydrogen sulfide, hydraulic fluids, toxic herbicides, and cancer-causing industrial wastes such as vinyl chloride and PCBs (polychlorinated biphenyls). Scientists have been developing ways to encourage the growth of prokaryotes naturally present in the environment to break down pollutants like these. After the massive Alaskan oil spill in 1989, cleanup crews seeded the oily shoreline with a "fertilizer" containing nitrogen and phosphorus (Figure 16.5). These nutrients, in combination with the carbon-containing organic compounds in the shoreline sludge, promoted the growth of local prokaryotes with a natural appetite for greasy hydrocarbons, and within two weeks, the sprayed areas had dramatically improved.

Many prokaryotes are **saprobes;** that is, they live on dead or dying organisms. Without the activity of these microbial

(a) Bacillus

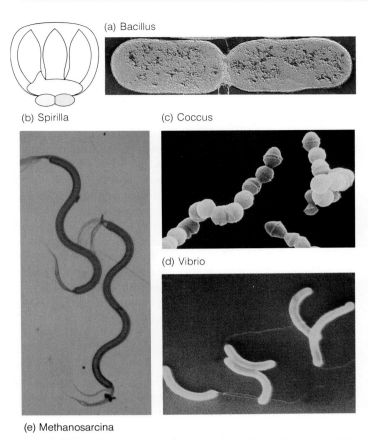

(b) Spirilla   (c) Coccus

(d) Vibrio

(e) Methanosarcina

FIGURE 16.4 ■ Prokaryotic Zoo: A Diversity of Shapes Among the Microscopic. Bacteria come in a wide range of shapes and sizes, including (a) rod-shaped bacilli, such as *Shigella,* (b) spiral-shaped spirilla like the spirochetes shown here, (c) round cocci, such as streptococci, and (d) comma-shaped vibrios, such as *Vibrio cholerae.* Note the whiplike flagella at the ends of the cells in (b) and (d). (e) A cluster of archaebacteria (*Methanosarcina mazei*), each cell with characteristic segmentation resembling the top of a cloverleaf roll.

"Fertilizing" bacterial growth

FIGURE 16.5 ■ Bacteria Help Clean the Environment. In 1989, millions of barrels of oil spilled from the tanker *Exxon Valdez,* despoiling hundreds of miles of Alaskan coastline. Cleanup crews sprayed nitrogen and phosphorus fertilizer on the fouled beaches, which encouraged the growth of naturally occurring oil-consuming bacteria. The microbes destroyed oil in the fertilized plots much more rapidly than they did in the unsprayed plots.

decomposers (as well as the multicellular saprobes, the fungi), the earth would quickly accumulate a thick layer of fallen leaves, dead animals, and other organic matter that would choke out living things. Other heterotrophic prokaryotes live on or inside other living organisms as harmful **parasites** or as neutral or beneficial **symbionts** (literally, "organisms that live together"). Most disease-causing bacteria are parasites, while the beneficial symbionts include intestinal bacteria that generate vitamins or help digest cellulose in cows and other herbivores.

A few prokaryotes are autotrophic; these self-feeding species include green and purple photosynthetic bacteria, cyanobacteria (blue-green algae), and chemoautotrophs such as sulfur bacteria and methane bacteria, which can obtain the carbon they need from carbon dioxide, and energy from the bonds of inorganic compounds such as hydrogen gas, hydrogen sulfide, sulfur, iron, or certain nitrogen compounds.

Prokaryotic Reproduction

Prokaryotes usually reproduce into identical daughter cells by binary fission or simple cell division (see Figure 7.6). If the dividing cells remain connected, small clusters or long filaments can form (see Figure 16.4c, e). Simple cell division can be incredibly fast and efficient, with fission occurring as often as every 20 minutes. If a single *Escherichia coli* bacterium and all its progeny continued dividing every 20 minutes, after just 48 hours there would be 2.2×10^{43} cells, or a mass of bacteria 4000 times the weight of the earth. This doesn't happen, of course, because the microbes have neither an unlimited source of nutrients nor ideal conditions of temperature and humidity. Nevertheless, the capacity for rapid reproduction combined with a high mutation rate (about one mutant cell per million dividing cells) means that prokaryotes can adapt quickly to changes in the environment. Prokaryotes also have several means of genetic recombination, including *conjugation*, the direct exchange of DNA through a strand of cytoplasm that bridges two cells (the bridges are called sex pili), and indirect exchanges, such as *transduction* and *transformation*. Transduction is the transfer of genes from one bacterium to another via a virus, while transformation is the transfer of genes by the uptake of DNA directly from the surrounding medium (see Figures 9.3 and 9.4).

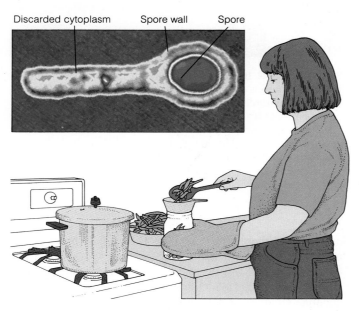

Discarded cytoplasm   **Spore wall**   **Spore**

FIGURE 16.6 ■ A Bacterial Endospore. Certain species of bacteria can withstand extreme environmental conditions (freezing, boiling, high pressure, radiation, and so on) by forming tough-walled endospores, or resting cells. An endospore (colorized blue) has formed in the *Clostridium botulinum* cell shown here. The endospore can survive adverse conditions, and when conditions improve, it may germinate and release a new individual. Special care must be taken when canning food to use sufficient heat and pressure to kill all *Clostridium botulinum* endospores.

Many prokaryotes have yet another reproductive adaptation that enables them to survive unfavorable conditions and then divide quickly when conditions improve. Some species form small, tough-walled resting cells called **endospores** (or simply *spores;* Figure 16.6). Spore formation is triggered by worsening environmental conditions. Spores can withstand extremes of heat, cold, drought, and even radiation for long periods. When conditions improve, the spores grow into new bacterial cells. Spores are the reason why surgical instruments must be sterilized with high heat and pressure, and they are also the reason why home-canned foods must be processed so carefully. The anaerobic bacterium that causes botulism poisoning (*Clostridium botulinum*) forms spores that can withstand several hours of boiling at 100°C (212°F). If these spores aren't destroyed during canning, they can later germinate into cells that produce one of the most poisonous substances known. The toxin blocks the transfer of nerve signals to muscles, and so the person becomes limp. Botulism poisoning is almost always fatal unless the victim is treated quickly with an antitoxin and placed on an artificial respirator.

Prokaryotic Behavior

Some prokaryotes are nonmotile—unable to move under their own power—and hence are subject to the currents and movements of their surroundings, whether that be water, blood, plant sap, or some other fluid. Many prokaryotes, however, are able to move along gradients of attractants (such as nutrients and light) or repellents (such as noxious chemicals) and thus find and exploit appropriate food sources or avoid predation. Some bacteria, like the ones in Figure 16.4b and d, have flagella (structurally different from eukaryotic flagella) that rotate and move the cell. Some photosynthetic bacteria move by a poorly understood gliding mechanism that allows them to remain in the light. Tests show that an *E. coli* cell can detect a change in nutrient concentration of just 1 part per 10,000. That's the equivalent of a person distinguishing between two jars, one containing 10,000 marbles and the other 10,001. The flagellum then propels the cell toward the higher concentration of the nutrient. Some bacteria, like the one in Figure 16.7, contain small crystals of iron oxide (magnetite) that act like little compasses and enable the bacteria to detect up from down and to move downward into nutrient-rich sediments.

Types of Prokaryotes

In recent years, the techniques of molecular biology have revolutionized how biologists classify prokaryotic cells. Prokaryotes used to be classified on appearance, as well as on the biochemical tasks they perform, but similar-looking bacteria were often very different biochemically, and this made classification difficult. Today, microbiologists can categorize bacteria according to the similarity of their ribosomal RNA

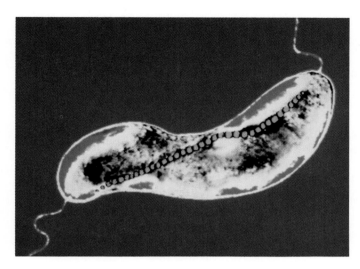

FIGURE 16.7 ■ A Single-Celled Prokaryote with a Magnetic Personality. This soil bacterium contains a chain of 36 magnetic metal spheres that show up orange in this false-color electron micrograph.

(rRNA). Recall that ribosomes, the "desk" on which proteins are translated (see Figure 10.8), contain RNA molecules. If two species have rRNAs with very similar nucleotide sequences, then they are probably closely related; otherwise, they are probably distant relatives. With such rRNA sequence data from many species, taxonomists can now draw family trees showing evolutionary relationships (see Figure 16.2), including support for membership in particular kingdoms and domains.

■ **Archaebacteria** Chapter 15 explained the evolution of the archaebacteria (members of the domain Archaea) and their common characteristics. Table 16.1 reviews some of those distinguishing traits. The 20 or so archaebacterial species fall into two major evolutionary lines. One group includes methane producers (methanogens), such as the *Methanobacterium ruminantium* that inhabits a cow's intestines, and extreme **halophiles** (salt lovers) like *Halobacterium halobium*, which tolerate extreme salt concentrations in salt flats and places like the Great Salt Lake and the Dead Sea (Figure 16.8). The second group contains the sulfur-dependent extreme thermophiles (heat lovers). These bacteria use sulfur in their energy metabolism and survive in hot sulfur springs and volcanic mud pots (review Figure 3.5). One of these remarkable archaebacteria grows best at water temperatures several degrees above boiling!

■ **True Bacteria** Recent analysis of RNA molecules suggests that there are 11 main evolutionary lineages among the true bacteria (see Figure 16.2). For various reasons, however, microbiologists continue to use the phenotypic classification based on bacterial form, physiology, and ecology as presented in *Bergey's Manual of Systematic Bacteriology* and other standard reference books. Our discussion of bacteria follows the evolutionary approach, moving through the domain Bacteria in Figure 16.2 in a counterclockwise direction, and Table 16.1 on page 342 gives the classification based on phenotypes. We discuss only the most important groups.

The largest and most diverse group of true bacteria are the *purple bacteria* such as *Rhodospirillum*. Many of these bacteria are purple and carry out photosynthesis with chlorophyll pigments that differ from those in plant cells. Other members of the group are heterotrophs, including the common intestinal bacterium *E. coli* (see Figure 3.9); the bacteria in Figures 16.4a and d; the *Rhizobium* species that fix nitrogen in peas and beans (see Figure 32.8); the bacterium that causes Legionnaires' disease; and many others. This group also includes *rickettsias,* tiny rod-shaped parasitic bacteria that absorb nutrients from a host cell (usually in a mammal). Rickettsias are transmitted by ticks, fleas, and lice and cause typhus fever, Rocky Mountain spotted fever, and other serious diseases. Some ancient member of the purple phototrophic bacteria probably gave rise to mitochondria in eukaryotes (review Figure 15.11).

An environmentally important group of true bacteria is the **cyanobacteria** (also called blue-green algae). Cyanobacteria, like plants, contain chlorophyll *a,* which allows photosynthesis and the generation of carbohydrates. Many cyanobacteria can also fix nitrogen, that is, reduce the N_2 from the air to organic nitrogen compounds (NH_3 and NH_4^+) for their own protein synthesis. Therefore, many cyanobacteria require only water, sunlight, a few inorganic nutrients, and the carbon dioxide and nitrogen gases readily available in air. As a result, cyanobacteria can live almost anywhere: in fresh or salt water; on damp rocks, soil, and tree trunks; and in hot, cold, or dry climates. Because cyanobacteria are so widespread,

FIGURE 16.8 ■ Halophiles: Salt Tolerators. These salt pans in the southern part of San Francisco Bay look ruddy because of the trillions of individual halophilic prokaryotes in the domain Archaea that thrive under such highly saline conditions.

TABLE 16.1 The Prokaryotes

Domain and Evolutionary Group	Groups Based on Structure and Physiology	Characteristics and Significance
Archaea (Archaebacteria)		
Methanogens and relatives	Methane producers	Grow in oxygen-free environments and produce methane
	Extreme halophiles	Inhabit highly salty environments, like salt lakes and solar evaporation ponds
Sulfur-dependent extreme thermophiles	Sulfur-dependent, heat-loving archaebacteria	Inhabitants of sulfur hot springs
Bacteria (Eubacteria)		
Purple bacteria and relatives	Purple photosynthetic bacteria	Some can form a purple layer in anaerobic zone of lakes; mitochondria arose from ancient purple bacteria similar to *Rhodospirillum*
	Pseudomonads	Versatile heterotrophs in the soil; useful in cleaning up industrial pollutants
	Enteric bacteria	Inhabit animal intestinal tracts, aiding digestion (*Escherischia coli*) or causing disease (*Salmonella*)
	Myxobacteria	Gliding bacteria whose colonies make complex multicellular spore-forming stalks
	Nitrogen fixers	Free in soil or associated with plants (*Rhizobium*); fix nitrogen in biological form
	Rickettsias	Live only inside other cells; cause typhus fever and Rocky Mountain spotted fever
Cyanobacteria	Cyanobacteria (Blue-green algae)	Photosynthetic system like a plant, and also fix nitrogen, thus most self-sufficient organisms; earliest organisms to release oxygen; precursors of chloroplasts
Gram-positive bacteria	Endospore formers	Heat-resistant spores can cause botulism or tetanus (*Clostridium* and *Bacillus*)
	Lactic acid bacteria	*Lactobacillus* makes yogurt; *Streptococcus* causes strep throat
	Gram-positive cocci	*Staphylococcus* causes toxic shock syndrome and other infections
	Actinomycetes	Cause leprosy and tuberculosis; produce antibiotics
	Mycoplasmas	Smallest of all cells; no cell wall; one cause of pneumonia
Chlamydiae	*Chlamydia*	Leading cause of blindness in humans; cause of most common sexually transmitted disease; cells live inside other cells; can't make ATP
Planctomyces and relatives	Appendaged bacteria	Cell wall consists of protein
Bacteroides and relatives	A branch of the gram-negative rods	The major anaerobic cells in the large intestine
Green sulfur bacteria	A branch of the phototrophic bacteria	Live on the warm fringes of sulfur hot springs
Spirochetes	Spirochetes	Corkscrew shape; cause syphilis and Lyme disease
Deinococcus and relatives	A branch of gram-positive cocci	Highly resistant to radiation; found near atomic reactors
Green nonsulfur bacteria	A branch of phototrophic bacteria	Grow in mats at the fringes of hot springs
Thermotoga	Anaerobic fermenters	Live in oceanic vents at highest temperature of any true bacteria

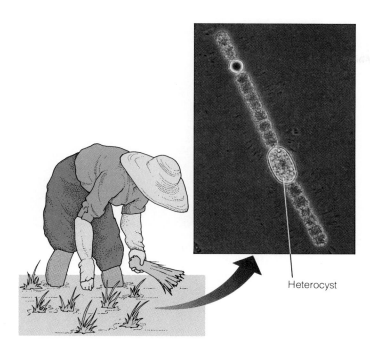

Heterocyst

FIGURE 16.9 ■ Cyanobacteria: Hardy Prokaryotes in Many Environments. Cyanobacteria (blue-green algae) often grow as filaments, with individual cells joined end to end, interspersed by an occasional thick-walled heterocyst—a cell that shuts out oxygen and allows nitrogen fixation to take place. Asian rice farmers fertilize their crops naturally with cyanobacteria that grow symbiotically with a water fern. When this plant dies, nitrogen is returned to the fields.

True bacteria include many medically important species that stain gram-positive. *Streptococcus mutans* (Figure 16.10) causes tooth decay. *Streptococcus pyogenes* causes strep throat. And *Staphylococcus aureus* can cause "staph" infections such as *toxic shock syndrome,* a sometimes fatal vaginal infection that occurs when the bacteria multiply in and around a tampon during menstruation. Fortunately, certain other gram-positive bacteria—the *actinomycetes*—produce more than 500 different natural substances that fend off competing microorganisms. Pharmacologists use those deterrent chemicals to create antibiotics such as streptomycin, tetracycline, and neomycin. Other gram-positive cells include the *mycoplasmas,* simplified parasitic bacteria that lack cell walls and live inside animals, plants, and sometimes other single-celled organisms. Only 0.2 to 0.3 μm long, mycoplasmas are the smallest free-living cells, and they can cause a dangerous form of pneumonia as well as urinary tract and other infections.

Bacterial species of the genus *Chlamydia* cause the most frequent sexually transmitted disease in North America (see Box 14.2 on page 305). These bacteria are even smaller than some viruses and have no way of making their own ATP; instead they must obtain energy from the cell they infect.

Another type, the *spirochetes,* have an undulating spiral shape and include the agents that cause *Lyme disease* and syphilis (see Box 14.2 on page 305). Spirochetes introduced into the body by the bite of a deer tick can cause the circular rash characteristic of Lyme disease, often followed by heart or nervous system disorders and, up to two years later, arthritis, especially in the knees.

their ecological role is tremendous; they contribute millions of tons of oxygen and fixed nitrogen and carbon to the environments and organisms around them. One nitrogen-fixing species of cyanobacteria (an *Anabaena* species) grows symbiotically with a water fern (Figure 16.9). Before planting rice, a farmer will allow dense blooms of the water fern to grow in the paddy. When the rice plants grow large, they crowd out the water fern as well as its cyanobacterial ally. Nitrogen released from the dead ferns acts as a fertilizer for the developing rice crop and promotes high yields without the need for potentially toxic chemical fertilizers.

In most nitrogen-fixing cyanobacteria, nitrogen fixation occurs only in special thick-walled cells called *heterocysts* (see Figure 16.9). Heterocysts represent a rare instance of division of labor among prokaryotes and the simplest instance in any living kingdom. The thick walls help protect nitrogen-fixing enzymes from oxygen generated during photosynthesis in adjacent cells.

Since cyanobacteria, like plants, contain chlorophyll *a,* many biologists believe that an ancient cyanobacterium may have entered a eukaryotic cell hundreds of millions of years ago and become the first chloroplast (review Figure 15.11). Later eukaryotes containing both this photosynthetic organelle and other membrane-bound organelles may, in turn, have given rise to the first generations of plants (see Chapter 17).

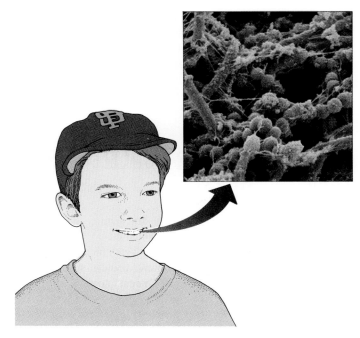

FIGURE 16.10 ■ Our Teeth Teem with Bacteria That Cause Decay. Here we see groups of rod-shaped *Streptococcus mutans,* which, if not brushed away, can cause cavities.

The Importance of Prokaryotic Cells

The ubiquitous prokaryotes have a tremendous influence on their environment and on other living organisms. Their greatest impact is ecological. Together they give off oxygen and cycle carbon, nitrogen, and other elements, breaking down enormous quantities of dead animals, fungi, and plants as well as human and animal wastes, pesticides, and pollutants that would otherwise poison the environment.

At the same time, prokaryotes have great medical impact, both negative and positive. Bacteria cause hundreds of human diseases, including "staph" and "strep" infections, blood poisoning, tetanus, venereal diseases, and dental caries, as well as thousands of diseases in other animals and plants. Bacteria do confer some medical benefits, however. *E. coli* and other inhabitants of the human gut, for example, produce vitamins K and B$_{12}$, riboflavin, biotin, and other cofactors that we probably absorb and use. Usually harmless residents like *E. coli* may also blanket the intestinal walls so heavily that harmful pathogens cannot gain access and pass through into the circulating blood. And many herbivorous (plant-eating) mammals, including cattle, sheep, and rabbits, would be unable to digest grasses and leaves without the cellulose-decomposing bacteria in their intestines.

Manufacturers harness prokaryotes to produce foods—including cheese, yogurt, pickles, soy sauce, and chocolate—and to generate chemical reagents, such as butanol, fructose, and lysine. People use other kinds of bacteria to clean up environmental poisons (see Figure 16.5) or to extract gold from low-grade ores. Researchers are also studying cells that inhabit hot springs in the hopes of constructing enzymes that function at high temperatures and could speed up industrial chemical reactions.

Finally, prokaryotes have had an immeasurable scientific impact. Much of what we know about biochemistry and molecular biology comes from the study of bacteria. And the techniques of genetic engineering are largely based on bacterial chromosomes and plasmids (see Chapter 11). Without prokaryotic research subjects, biology would be a century behind its current state.

VIRUSES AND OTHER NONCELLULAR AGENTS OF DISEASE

Earlier, we discussed the sexually transmitted prokaryote *Chlamydia,* a minute organism that must enter a host cell to obtain energy, survive, and reproduce. Nevertheless, it is still a cell, with its own plasma membrane, protein-synthesizing machinery, and metabolism. In contrast, some biological agents are not organized as cells, but depend on living cells for continued existence—and probably even evolved from them.

Best known are the **viruses,** geometric packages of genes that are 1000 to 10,000 times smaller than most prokaryotic cells. A virus particle is, in effect, a minute package of DNA

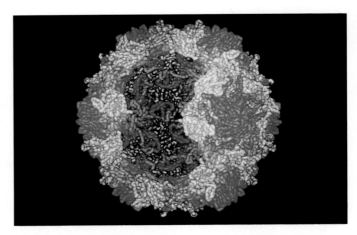

FIGURE 16.11 ■ **The Polio Agent: An RNA Virus.** Like all viruses, the poliovirus has a protein coat surrounding genetic material; it is RNA, not DNA. Since the virus attacks nerve cells that enable muscles to contract, muscles that power arms and legs may stop functioning and hence atrophy.

or RNA surrounded by a protein coat, or **capsid,** and occasionally other materials (Figure 16.11). Viruses infect cells by attaching part of the capsid to the exterior cell wall or membrane and then either entering the cell or simply injecting their DNA or RNA into the cell, leaving the capsid outside (review Figure 9.4). Once inside, the viral genes commandeer the cell's protein-synthesizing machinery, sometimes stopping production of cellular proteins entirely and preempting the machinery for the production of new virus particles complete with capsids. Eventually, the cell bursts and dies, liberating thousands of viruses that may then infect additional cells. Many biologists do not consider viruses to be alive because these agents lack the machinery for self-reproduction or metabolism.

There are hundreds of kinds of viruses, many of which cause plant and animal diseases such as those listed in Table 16.2. Some, like the rhinoviruses that cause colds and the influenza viruses that cause flu, are transitory parasites: We encounter them in sneeze droplets or mucus sprayed into the air by cold or flu sufferers, or we pick them up from contaminated hands or surfaces, and they begin to multiply in our body cells. Eventually, our immune system destroys them (see Chapter 22). Other types of viruses, however, like herpes simplex I and II, which cause cold sores and genital herpes, respectively, insert their DNA or RNA into the genomes inside nerves or other body cells and take up permanent residence. There they lie dormant, erupting and causing symptoms only occasionally when triggered by fever, sunlight, or other environmental stimuli. Since viruses are not cells, they are not killed by the antibiotics physicians use to block the growth of disease-causing prokaryotes. One of the biggest challenges in modern medicine is to develop drugs that can fight viruses. Box 16.1 on page 347 describes recent efforts to fight the common cold, genital herpes, and the flu, as well as explaining the environmental facts behind "new" disease viruses like the AIDS-causing virus HIV.

The **viroids** are a group of intracellular parasites that lack a protein coat, consist only of small RNA molecules, and seem to affect mainly plants. Some diseases of potatoes, cucumbers, citrus trees, and artichokes can be traced to viroids (Figure 16.12).

Finally, the **prions** are the smallest and strangest disease-causing agents that can be transmitted from one animal to another. Prions appear to lack genetic material entirely and to consist of nothing but proteins, yet they are implicated in serious nerve and brain diseases, including scrapie in sheep and goats; mad cow disease (see Box 16.2, on page 351); Creutzfeldt-Jakob disease and kuru in humans; and possibly Alzheimer's disease, the most common form of senility and a leading cause of death. Biologists are not sure how prions reproduce or cause disease, but the disease-causing agent is an abnormal form of a protein encoded by the infected animal's genes. Like viruses and viroids, prions may be evolutionary remnants of parasitic prokaryotes that left behind all their lifelike characteristics and retained only the structures necessary to infect host cells and be reproduced by them.

FIGURE 16.12 ■ **Viroids: Parasites That Infect Crop Plants.** Here, viroids have infected artichoke leaves and damaged or killed rings of tissue in numerous places.

THE PROTISTS: SINGLE-CELLED EUKARYOTES

Our discussion turns now from the single-celled prokaryotes and the noncellular viruses to the 35,000 species of single-celled eukaryotes, the protists (Figure 16.13), including the

TABLE 16.2	Noncellular Agents of Disease		
Agent	Constituents	Example	Disease
Viruses	DNA plus protein	Parvovirus	Gastroenteritis
		Herpes simplex I, II	Herpes
		Epstein-Barr virus	Mononucleosis, Burkitt's lymphoma
		Smallpox virus	Smallpox
	RNA plus protein	Paramyxovirus	Measles
		Togavirus	Rubella (German measles), yellow fever
		Rhinoviruses	Common cold
		Myxovirus	Influenza
		Poliovirus	Poliomyelitis
		Paramyxovirus	Mumps
		Rhabdovirus	Rabies
		Retroviruses	Cancer (some forms)
			AIDS
Viroids	RNA only	PST viroid	Potato spindle tuber disease
		Exocortis viroid	Citrus exocortis disease
Prions	Protein only	Various prions	Kuru (a brain disease in Borneo and elsewhere); Creutzfeldt-Jakob disease (a brain disease); scrapie (a disease of some farm animals)

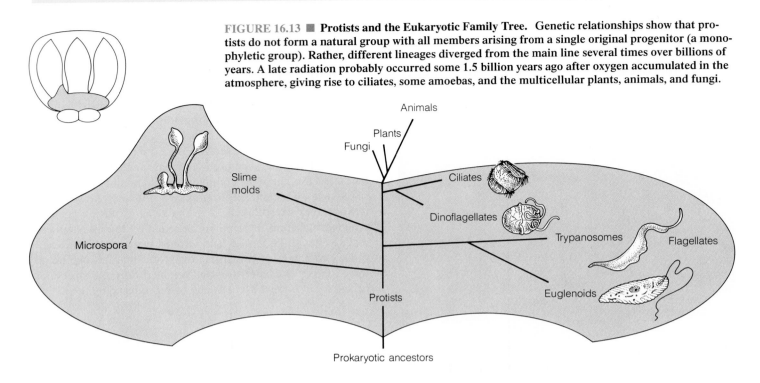

FIGURE 16.13 ■ Protists and the Eukaryotic Family Tree. Genetic relationships show that protists do not form a natural group with all members arising from a single original progenitor (a monophyletic group). Rather, different lineages diverged from the main line several times over billions of years. A late radiation probably occurred some 1.5 billion years ago after oxygen accumulated in the atmosphere, giving rise to ciliates, some amoebas, and the multicellular plants, animals, and fungi.

gluttonous *Didinium,* described at the beginning of the chapter. Together, members of the kingdom Protista (also called Protoctista) have a tremendous impact on health and the environment. We will discuss one protist that kills more people each year than any other disease, as well as other species that pump massive quantities of oxygen into the air and help support the vast numbers of fish and other aquatic animals.

General Characteristics

As a eukaryotic cell, the average protist is 10 times longer than a prokaryote and has 1000 times the volume. This larger size is made possible by the cell's compartmentalization and by the extensive membranous surface area of the endoplasmic reticulum, nuclear membrane, and invaginated mitochondrial membranes. The expanded membranous surface area enables the cell to carry out sufficient levels of metabolism, protein synthesis, and other cell functions needed to support a large cell volume (see Figure 3.7 and Table 3.1 for a review of eukaryotic cell organelles). The evolution of cell organelles allowed several different types of protists to emerge (Table 16.3). Some protists became animal-like hunters, like *Didinium.* Others became plantlike producers of carbohydrates. Still others became funguslike decomposers. And a few evolved with strange hybrids of the different life-styles.

Protozoa: The Animal-Like Protists

The hunters of the microbial world are the **protozoa** (literally, "first animals"), cells that usually stalk and consume other

cells or food particles. Each of four major protozoan phyla is distinguished by its means of locomotion: by flagella, by amoeboid movement, by no means of movement, or by cilia (as in *Didinium*).

■ **Flagellates** Protists in the phylum *Mastigophora* (the so-called flagellates) have only an outer plasma membrane, with no protective cell wall. In contrast to bacterial flagella, their flagella have the same internal structure and contain the identical protein (tubulin) as in the tail of a man's sperm. This evidence suggests that eukaryotic flagella have a very ancient origin. Most flagellates live as parasites or harmless symbionts within other organisms. A common human parasite, now a scourge to campers, is the flagellate *Giardia lamblia* (Figure 16.14a). Two other important flagellates are *Trichonympha,* a genus of cellulose digesters that lives in the guts of termites and enables the termites to utilize wood as food, and *Trypanosoma brucei gambiense,* which can cause African sleeping sickness (Figure 16.14b). Trypanosomes are dangerous pathogens because they can evade the human immune system by changing their surface coats. Trypanosomes are spread through the bite of the tsetse fly and can inflict their lethal damage on domesticated animals as well as on people. Trypanosomes and the tsetse flies that spread them prevent most types of dairy and beef cattle from thriving in an infested zone of Africa, which is nearly 4 million square miles in area—a region the size of the United States.

■ **Amoebas and Related Protists** Members of *Sarcodina* range from creeping, bloblike amoebas that live in moist terrestrial habitats (see Figure 3.13) to delicate glassy-shelled ocean

TERMINATING THE TERMINATOR

In the movie *The Terminator,* Arnold Schwarzenegger plays the robot-that-can't-be-killed—a fiendish nonliving antagonist that keeps on getting up and pursuing his terrified victim, despite being dropped and shattered, shot full of holes, electrocuted, and blown to bits.

Viruses are vaguely reminiscent of this character. They keep arising, assuming new, deadly forms, and attacking people, and they're not alive, so they're dreadfully hard to defeat. Where, then, do new dangerous viruses like the AIDS agent come from, and how will we fight these molecular terminators?

Researchers have determined that many new human viruses began as animal pathogens. The AIDS virus, for example, probably attacked only African monkeys until the continual advance of our species into formerly untouched or little-used wilderness areas brought about increased contact. The frequency of long-distance air travel was a secondary factor; once the virus mutated and became pathogenic to humans, it was dispersed fairly quickly by infected people flying from Africa to other parts of the world.

Here are some other examples of this "old new virus" phenomenon:

- The Marburg virus moved from Ugandan monkeys into German research scientists in the late 1960s.
- Rift Valley fever once infected only sheep and cattle, but began to infect people in South Africa around 1975, was carried to Egypt, and there infected millions and killed thousands.
- Rodent-borne Seoul virus, which causes hemorrhagic fever (with its high fever, internal bleeding, and kidney damage) was carried from Asia, appeared in rat populations in Baltimore, and has caused blood reactions (the formation of antibodies) in over 1000 Maryland residents so far.
- In Argentina, farmers clearing the pampas to plant maize have had increased contact with a mouse that bears the virus for Argentine hemorrhagic fever, which is now infecting field workers there.
- Brazilian cocoa farmers have picked up Oropuche fever from an insect that lives in old cacao bean husks and bears the virus.
- Asian influenza A virus killed 20 to 30 million people around the world in 1918. The virus lives in ducks and other birds and apparently hopped from birds to barnyard pigs and from pigs to people. If a deadly mutation arose again in this virus, researchers fear that another influenza pandemic could appear and once again kill millions.

Based on the frequency with which viruses mutate into human pathogens and get dispersed by global travel—not to mention the hundreds of human viruses already causing diseases—researchers have, for 30 years, been working hard to develop weapons against these nonliving enemies.

Antibiotics that kill living prokaryotes and protists leave virus particles untouched. Until recently, therefore, most drugs prescribed for colds, flu, herpes, and other viral infections have done little more than treat uncomfortable symptoms and control secondary infections, while the body's immune system gears up to destroy the virus on its own. Luckily, the research efforts are paying off, and antiviral drugs are beginning to arrive.

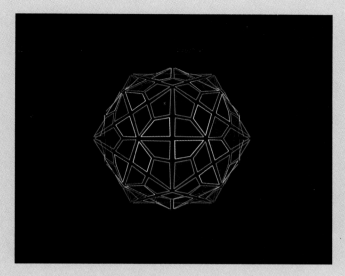

FIGURE 1 ■ **Virus Structure. This computer graphic of a picornavirus reveals an icosahedron—a soccer ball with facets.**

Scientists have announced two drugs that in different ways fight picornaviruses—a group that includes the agents of polio and the common cold (see Figure 16.11 and Figure 1). Recall that a virus must shed its capsid, or outer protein coat, in order to enter a cell and be replicated. One drug, called WIN 51,711, prevents picornaviruses from shedding their coats. Mice infected with poliovirus and then given WIN 51,711 remained unparalyzed.

Researchers have also used computers to design other drugs that will bind to specific capsid proteins in picornaviruses, including the one pictured in Figure 1. One drug locks into the cavity inside one barrel-shaped capsid protein called VPI and acts like a girder, preventing the viral protein from unfolding. A virus particle "locked" shut like this should be unable to release its RNA, even if it gains access to the interior of a cell. Thus, it can't replicate and spread to new cells. This drug has not yet been tested.

A drug with a much longer history, called interferon, is proving effective against the human cold. Interferon is a natural substance produced by the immune system in response to viral infections. Interferon released from one cell attaches to a neighboring cell and somehow prevents it from replicating viruses. With the advent of gene splicing and other techniques of bioengineering, a few companies have been able to produce large quantities of interferon, which drug researchers have incorporated into a very effective nasal spray. Clinical trials are now under way to get the necessary government approval to market both of these drugs.

Yet another drug, called acyclovir, is proving effective against herpes infections. Acyclovir blocks the activity of certain herpes viral enzymes and thus can help control the herpes infections that cause shingles, cold sores, and genital herpes, a sexually transmitted disease affecting about 20 million Americans.

Researchers think that, within a decade, physicians will have many such drugs with which to control viruses—to "terminate the terminators."

(a) *Giardia*   Flagellum

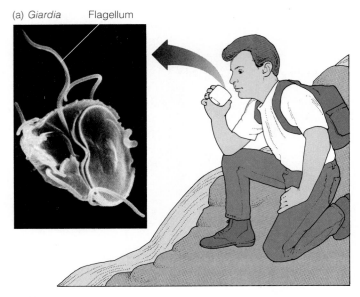

(b) Trypanosomes   Flagellum

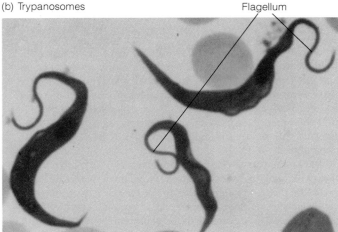

FIGURE 16.14 ■ **Some Flagellates Are Agents of Disease.** (a) A drink from a clear-looking but contaminated mountain stream can lead to infection by *Giardia*. This flagellated protist lines the inner surface of the intestines, causing severe abdominal gas, cramping, and diarrhea. (b) An eel-like trypanosome (magnification 5000×) wriggles among a person's lozenge-shaped red blood cells. The parasite enters the bloodstream with the bite of a tsetse fly, then invades the brain and spinal cord.

their shells contribute to the formation of thick limestone and other rock deposits. The chalky White Cliffs of Dover formed this way.

■ **Spore-Forming Protists**   A third group of protozoa (*Sporozoa*) are internal parasites residing in a wide range of animals, and they do not move under their own power like the other groups of protozoa. The sporozoa in the subgroup *Apicomplexa* have a complex of rings and tubules at the cell's apex and a sporelike stage in which the cells lack any means of locomotion. The best-known member of this group is *Plasmodium vivax,* the cause of malaria (Figure 16.16). This protist is spread by tropical mosquitoes. Despite the ubiquity of bacteria and bacterial diseases, the protozoan disease of malaria is actually humankind's most prevalent infectious disease. At any one time, 100 million people are ill with malaria, and about 1 million people die of malaria each year—most of them very young African children. Vigorous attempts at eradication of mosquitoes with DDT and treatment of the protozoan with drugs like chloroquine have been met with a growing evolution of resistance to these chemicals by the organisms, the result being the natural selection of mosquitoes and plasmodial parasites that are harder to kill. Scientists are working hard to develop vaccines against this disease.

Another sporozoan subgroup is *Microspora,* parasites that live inside cells in nearly every group of animals. Microspora were one of the very earliest branches off the main line of eukaryotic evolution (review Figure 16.13). Interestingly, microspora have no mitochondria, so they resemble the postulated cell which became the host for the bacteria that

species. However, all members of this group display **pseudopodia,** limblike cellular extensions that serve in locomotion and feeding. Pseudopodia (literally, "false feet") can project outward in any direction, pulling the cell along and immobilizing prey so that the sarcodine can engulf it by phagocytosis (see Chapter 3). Several marine sarcodine species form beautiful shells around their soft cell bodies. *Foraminiferan* cells secrete whitish calcium-based shells that can look like spiral seashells or chambered nautiluses, while *radiolarians* produce silicon-based shells that look like miniature glass ornaments (Figure 16.15). These protozoa are so plentiful that

(a)   (b)

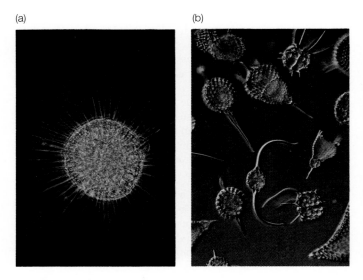

FIGURE 16.15 ■ **Sarcodina: Pseudopodia and Glassy Shells.** Sarcodines extend narrow ribbons of cytoplasm, or pseudopodia, to capture prey or pull themselves forward. (a) *Actinosphaerium* resembles a pincushion with hundreds of glass needles. (b) Radiolarians, such as this mixed group of species, have intricately patterned glassy shells.

invaded eukaryotic cells and led to the first mitochondria. More work on this little-known group of organisms could provide exciting new data for our understanding of the evolution of organisms with true nuclei.

■ **Ciliates**   The fourth protozoan phylum includes the aquatic **ciliates,** one of which is the hungry stalker *Didinium.* Ciliates are characterized by rows of cilia that bend with coordinated oarlike motions, propelling the cell or helping to sweep food particles into its mouth organelle. This "mouth" can be an oral groove, such as that possessed by *Paramecium,* or an expandable pore, such as *Didinium*'s "snout" (see Figure 16.1).

Ciliates have several organelles with functions analogous to those of animal organs: anal pores that discharge wastes; microfilaments and microtubules that support and contract like bones and muscles; tiny toxic darts (*trichocysts*) used in hunting prey; and food vacuoles that are filled with enzymes and function like digestive organs. Each ciliate also contains a giant nucleus with many sets of chromosomes (a polyploid **macronucleus**) that directs cell activities, as well as one or

TABLE 16.3 The Kingdom Protista		
Section, Phylum or Division, and Subgroup	Common Members	Characteristics and Significance
Animal-like Protists (Protozoa)		
Flagellates (Mastigophora)	Trypanosomes	Each moves by means of a flagellum; cause sleeping sickness
Amoebas (Sarcodina)	Amoebas	Move by pseudopodia; cause amoebic dysentery
	Foraminiferans, radiolarians	Shells contribute to limestone deposits
Sporozoa	*Plasmodium vivax*	Cause of malaria and many human diseases each year
Ciliates (Ciliophora)	*Paramecium*	Feeds on bacteria in ponds
	Didinium	Hunts other protists
Plantlike Protists		
Euglenophyta	*Euglena*	Moves by flagellum; contains chloroplasts
Pyrrhophyta	Dinoflagellates	Flagella in grooves; cause of red tides
Chrysophyta	Golden-brown algae, diatoms	Photosynthetic; glassy shells that serve as abrasives
Funguslike Protists		
True slime molds (Myxomycota)	*Physarum*	Many nuclei in one large cell; break down organic matter on forest floor
Cellular slime molds (Acrasiomycota)	*Dictyostelium*	Single cells converge and form slug; break down organic matter on forest floor
Water molds (Oomycota)	*Phytophthora infestans*	Cause late potato blight; thrive in dampness; damage plants

(a)

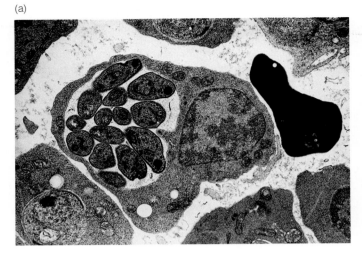

(b)

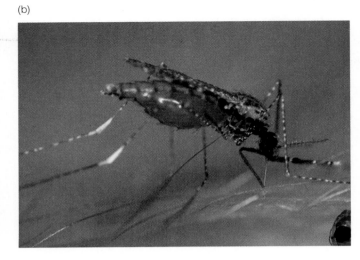

FIGURE 16.16 ■ **A Tropical Mosquito Spreads the Malarial Parasite.** *Plasmodium vivax* is the cause of malaria, humankind's most prevalent infectious disease. Once the protist infects cells, as shown in (a), symptoms develop, including severe headaches, fever, and general discomfort. Malaria is spread by tropical mosquitoes (b) and afflicts 100 million people each year, killing 1 million of them. Vigorous eradication of mosquitoes with DDT and the prescription of quinine drugs has fostered a growing natural resistance among the organisms, resulting in mosquitoes and parasites that are harder to kill. Researchers are now working on a vaccine to help protect against malaria.

more small diploid nuclei (**micronuclei**) that undergo meiosis and are exchanged during sexual reproduction (conjugation; Figure 16.17).

Carbohydrate Producers: The Plantlike Protists

Three divisions of organisms within the kingdom Protista contain chlorophyll and carry out photosynthesis: the *euglenoids,* the *dinoflagellates,* and the *golden-brown algae* and *diatoms.* These protists are part of the mass of cells called **phytoplankton** ("floating plants") that grow near the surface of fresh and marine waterways. This mass of cells has tremendous ecological significance in that it releases oxygen and forms the base of the aquatic food chain: Larger organisms graze on the phytoplankton, still larger creatures eat the grazers, and so on. Many plantlike protists swim via flagella; thus, they have both plantlike and animal-like characteristics.

■ **Euglenoids**   The word *euglena* comes from the Greek for "good eye," and indeed, each of the many green, spindle-shaped euglenoids has an eyespot, or **stigma.** This structure shades a light receptor and allows the aquatic cell to swim in the direction that maximizes its photosynthetic rate. It is tempting to speculate that since euglenoids have chloroplasts, like plant cells, and a flagellum, like certain animal cells, their ancestors may have given rise to both types of multicellular organisms. Molecular analysis, however, shows that euglenoids are related to trypanosomes and other flagellates and diverged from the main eukaryotic line eons ago (see Figure 16.13). Euglenoids may even contain chloroplasts with an origin independent from the chloroplasts in green plants.

■ **Dinoflagellates**   The dinoflagellates are a distinctive group with occasional massive economic impact. *Dinoflagellate* means "spinning cell with flagella," and indeed, these plantlike marine protists have two flagella that cause them to spin as they swim. One flagellum winds about the middle of the cell like a belt in a groove and causes the spinning motion, while another projects backward in a second groove and propels the cell (Figure 16.18a). Some dinoflagellates wear a coat

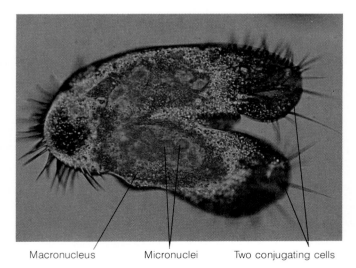

Macronucleus   Micronuclei   Two conjugating cells

FIGURE 16.17 ■ **Conjugation in Ciliates.** Ciliates like the *Stylonychia* shown here exchange micronuclei, which carry chromosomes as well as material for forming a new macronucleus, the organelle that governs the cell's activities.

BOX 16.2 PERSONAL IMPACT

MAD OVER MYSTERIOUS FIBRILS

The year 1990 was a good one for British vegetarians; they could go on eating their regular healthy diets while all around them, beef eaters went crazy over "mad cow disease." Cattle were dropping by the thousands in British pastures, their brains turned to sponge. School officials took beef off school menus. France banned the sale of one of Britain's biggest agricultural exports. The merrymaking British tabloids had a field day. And the culprit behind it all was a mysterious nonliving disease agent far smaller and simpler than a virus.

Mad cow disease (formally called bovine spongiform encephalitis, or BSE) is a close cousin to (perhaps even the same disease as) scrapie in sheep and goats. This neurological condition usually starts with jerky, uncoordinated movements, progresses to a degenerated brain with the perforated look of a bath sponge, and inevitably ends in death. Ranch mink and captive deer and elk can contract similar diseases, and so can people: the degenerative condition called kuru has been seen in cannibalistic tribes who eat human brains (presumably brains already infected with the kuru agent). And the very similar Creutzfeldt-Jakob disease (CJD) is sometimes contracted from hospital instruments during eye or brain surgery.

It is tempting to think that a person might contract CJD from eating the muscle tissue (meat) of a scrapie-infected sheep or "mad" cow. And in fact, millions of British drove themselves to distraction with just this worry. Research, however, suggests that if anything, only the consumption of infected brains and eyes might convey the scrapie agent. Unfortunately, some sausages, hot dogs, and luncheon meats do contain these ground-up organ meats, so many British swore off their usual "bangers" at meals.

British agricultural officials were able to trace the outbreak of mad cow disease to a change in feeding practices in the late 1970s and early 1980s. At that time, financially strapped farmers started giving their herds low-quality feeds with protein coming not from the preferred sources of soybean or fish meal, but from the ground-up bones, heads, and offal of other grazing animals—including, as it turned out, sheep infected with scrapie. The key to the problem is that the scrapie agent can lie dormant and undetected for many years before causing neurological symptoms. Thus, it was well over a decade before the consequences of the cheap cattle feed would be known.

Because of this long time lag, biologists used to call the BSE/scrapie agent a "slow virus." While it is certainly slow, it is apparently not a virus at all: For years, researchers have tried without success to identify the scrapie infectious agent and have centered on two theories: that it is a strange protein particle or prion that needs no nucleic acid to replicate itself, or that it is a virino, a tiny bit of DNA (perhaps mitochondrial DNA) that commandeers the host's protein-making machinery to generate a protective coat around itself. The same proteinaceous fibrils are found in the spongy brains of animals with scrapie or BSE.

As research on these mysterious diseases continues, the furor has died down over the mad cow outbreak of 1990. But many British meat eaters have been left with a permanent distaste for animal brains, eyes, and the processed meats that contain them.

(a) A dinoflagellate

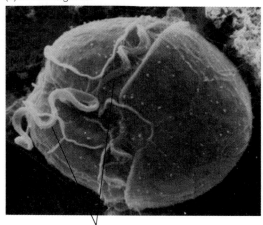

Flagella

(b) A red tide

FIGURE 16.18 ■ Dinoflagellates: Armored Protists That Cause Red Tides. (a) The microscopic phytoplankton *Protogonyaulax catenella* has the dual flagella in grooves characteristic of so many dinoflagellates and wears a coat of cellulose armor. (b) Some dinoflagellates cause dangerous red tides—dense blooms of certain species that can tint the water blood red and produce deadly toxins, as happened during this red tide along the coast of Ethiopia. The toxins act as nerve poisons, killing people who eat fish and shellfish that have consumed the dangerous dinoflagellates.

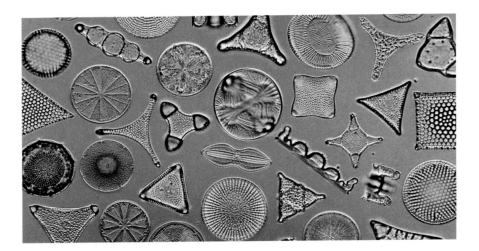

FIGURE 16.19 ■ **Diatoms: Golden-Brown "Pillboxes."** Diatoms have a golden-brown pigment, as well as jewellike shells. Diatoms have silicon instead of cellulose in their cell wall and store oils rather than starches. Diatom shells fall to the ocean floor and accumulate in crumbly white sediments called diatomaceous earth, which people use in toothpaste and swimming pool filters. Diatoms and golden-brown algae contain chlorophyll and are so abundant that biologists estimate they contribute more oxygen to the atmosphere than all land plants combined.

of armor—an elaborately embossed cellulose cell wall—while others have only a cell membrane. Dinoflagellates contain chlorophylls *a* and *c* as well as carotenoids. A few contain enzymes that cause the organisms to glow in the dark and the ocean waves to burn with a bluish glow at night, outlining the splash of a midnight swimmer with eerie sparkles. Despite their beauty, dinoflagellates can cause **red tides,** dense blooms that can tint the water blood red and produce deadly toxins that act as nerve poisons (Figure 16.18b). These can build up in fish and shellfish, killing them outright or poisoning people who gather and eat them. This is why coastal areas often ban the collection and consumption of shellfish during May through August, when red tides can occur.

■ **Golden-Brown Algae and Diatoms**   These protists are the most common and perhaps the most exquisitely beautiful of the phytoplankton species. Most golden-brown algae and diatoms have a golden color as a result of a carotenoid pigment as well as their chlorophyll pigments. The cells have silica, the main component of window glass, instead of cellulose in their cell walls, and they store oils rather than starchy compounds. Beyond these similarities, they are quite diverse. Some are amoeba-like, some have one or two flagella, and some are nonmotile. Golden-brown algae are smaller than diatoms but extremely numerous in phytoplankton layers and thus major contributors to marine food chains. Diatoms get most of the popular attention, however, because of their jewellike, glassy shells (Figure 16.19). The intricate patterns of tiny holes and channels in these rigid shells function as pores for material transport to and from the cell. The walls are actually two half shells, or *valves,* that fit together like a lid and container of a pillbox. These shells fall to the ocean floor and accumulate in crumbly white sediments called diatomaceous earth, which people use in toothpaste, swimming pool filters, and other applications. So abundant and ubiquitous are the golden-brown algae and diatoms that some biologists estimate that they may contribute more oxygen to the atmosphere than all land plants combined.

The Funguslike Protists

Among the most bizarre members of the kingdom Protista are the **slime molds.** Like mushrooms and yeasts, these glistening organisms derive food and energy by secreting digestive enzymes that break down organic matter and then absorbing the digested material back into the cell. They also form multicellular reproductive structures. There are three types of funguslike protists: the true slime molds, the cellular slime molds, and the water molds. A **true slime mold** exists as a large, flat, often fan-shaped mass that moves about slowly like a giant amoeba in damp soil, leaf litter, and downed wood. This mass is called a *plasmodium* and consists of a continuous cytoplasm surrounded by one plasma membrane, but containing many diploid cell nuclei. **Cellular slime molds** exist as single-celled, amoeba-like organisms that glide along in search of bacteria. When food is scarce, both types of slime molds send up brightly colored **fruiting bodies,** reproductive structures that look like golf balls on tees. Spores produced in the balls are scattered by wind or rain and germinate in new locations, dispersing the species (Figure 16.20).

The third group, the **water molds,** inhabit soil and water. Some are single-celled and others form fuzzy, branching filaments called *hyphae.* Like the true slime molds, water molds have several nuclei within a common cytoplasm and form large, immobile egg cells (their formal name, in fact, is Oomycota). After fertilization, they form spores that disperse by swimming. Thus, oomycotes thrive only when it is damp (hence the name "water molds"). One parasitic water mold did just that during the unusually cool, wet weather that struck Ireland in 1845–1847. This species causes *potato blight,* a disease that rapidly rots and kills growing potato vines. The Irish depended so heavily on potatoes for their daily diet that more than a million people starved, and 1.5 million more emigrated, mostly to North America (Figure 16.21).

The funguslike protists show a clear trend toward the kind of cellular cooperation one sees in multicellular organisms. The existence of funguslike, animal-like, and plantlike pro-

(a) Life cycle of cellular slime mold

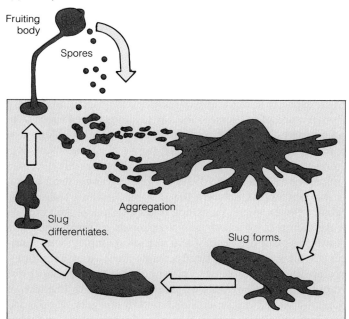

(b) Fruiting bodies of slime mold

FIGURE 16.20 ■ **Life Cycle of the Cellular Slime Mold.** (a) Cellular slime molds normally move about like amoebas—as single cells in search of bacteria. If environmental conditions worsen, however, they can aggregate, form a slug, and slither to some likely location. There they stop and differentiate into a fruiting body with a stalk and a spore chamber that eventually releases new, genetically recombined spores. These, in turn, germinate into new amoeba-like cells that individually crawl off once again in search of bacteria to consume. (b) Close-up of slime mold fruiting bodies on a tree trunk.

FIGURE 16.21 ■ **Potato Blight and Ireland's Food Staple.** This Irish potato farmer may have had relatives who lived through the great potato famine of the mid-nineteenth century, caused by a water mold.

tists suggested to biologists earlier in this century that each protist group may have given rise to the corresponding multicellular group—the fungi, animals, or plants. Recent evidence from molecular genetics, however, modifies our understanding of protistan evolution.

Evolutionary Relationships Among the Protists

The survey of protists we just completed was based on phenotypes, and recent molecular data show how the groups are related genetically. The protists, as we have seen, are very diverse in appearance and habits, and taxonomists now doubt that a common progenitor gave rise to all the protists. Instead, many separate ancestral organisms probably led to the various types of protists. (Taxonomists call the protists a *polyphyletic* group—one with many different ancestors.)

One of the oldest lines was surely *Microspora,* parasites of many animals and even other protists (see Figure 16.13). Microspora are evolutionary relics; they contain a membrane-bound nucleus that undergoes a unique mitotic division, and they lack internal organelles such as mitochondria or chloroplasts. The microspora probably split from the main line while the earth's atmosphere was still anaerobic (about 2.5 billion years ago), and they probably resemble the host cells to the first mitochondria (see Figure 15.11b and c).

Somewhat later, the flagellates diverged, and then the slime molds, and then, over a more recent and fairly short period of evolutionary time, a rapid radiation led to ciliates, some amoebas, fungi, plants, and animals. Some biologists suggest that this radiation was an evolutionary response to the accumulation of oxygen in the earth's atmosphere about 1.5 billion years ago.

Finally, the evidence summarized in Figure 16.13 indicates that the large organisms we see in our everyday lives—toadstools, grass, robins—may have arisen from a common ancestor rather late in evolutionary history. (Taxonomists call them a *monophyletic* group.) Odd as it may seem, the fungi, plants, and animals represent just one recent sprig on the great tree of life, and we devote the next three chapters to their characteristics and to how that sprig may have appeared and diverged.

CONNECTIONS

Ironically, the most numerous and in many ways the most important organisms on earth are also the smallest and the oldest. All around us, in the soil, air, and water, unseen trillions of single-celled organisms take in nutrients, metabolize them, and produce new generations in as little as 20 minutes. This capacity to reproduce quickly and to absorb and utilize a broad range of nutrients is the key to single-celled success. In some ways, our own survival depends on the ubiquity of prokaryotes and protists: Golden-brown algae and cyanobacteria together produce most of the oxygen we breathe; soil microbes "clean up the dirt" by breaking down dead organisms; phytoplankton form the first link in the oceanic food chain. Nevertheless, parasitic prokaryotes and eukaryotes do take their toll. To a pathogenic organism, a human body or a towering plant is an ecosystem in itself, complete with nutrients, fluids, relatively steady temperatures and humidities, competition for resources, places to live, and a chance for rapid reproduction. Tens of generations—entire lineages of cells—live and die on our teeth and in our guts. People and other multicellular organisms are available habitats, and these cells are merely opportunists that move in and multiply. In the next chapter, we begin to look at those multicellular "habitats" to see how they arose during evolution.

Highlights in Review

1. *Didinium nasutum* exemplifies the quintessential characteristic of prokaryotes and protists: the ability to carry out all life processes as an independent, free-living single cell.

2. Single-celled organisms are ubiquitous owing to their spectrum of life-styles and efficient reproduction, and they are important ecologically, medically, economically, and scientifically.

3. Most prokaryotes are surrounded by a cell wall, lack membrane-bound organelles, and have a circular strand of DNA functioning as a single chromosome. Many are pathogenic (they cause diseases), and they can be of several shapes.

4. Most prokaryotes are heterotrophs—either saprobes or parasites—but some are autotrophs, obtaining energy from light (photosynthesis) or from inorganic chemicals (chemoautotrophy).

5. Most prokaryotes reproduce via binary fission; some also exhibit conjugation or other forms of genetic exchange, and some form spores.

6. Prokaryotes can be motile or nonmotile; motile species can follow gradients of light, chemicals, or gravitational fields.

7. Genetic evidence suggests that the kingdom Monera should be broken into two domains: Archaea, or Archaebacteria, which evolved very early and whose members tend to have unusual forms of metabolism, and Bacteria, or Eubacteria, whose members are more common and include the gram-negative and

gram-positive species, the actinomycetes, the rickettsias, the mycoplasmas (the smallest living things), and the cyanobacteria, which contain chlorophyll, have simple nutritional requirements, and show division of labor between cells.

8. Viruses are nonliving agents of disease. Each virus particle is made up of a protein capsid surrounding some DNA or RNA. The particle injects its genetic material into a living cell, which then generates new virus particles. Viroids are small RNA molecules devoid of a protein coat, while prions are protein particles devoid of genetic material but capable of causing serious brain and nerve disorders in animals.

9. All members of the kingdom Protista are eukaryotic cells, complete with membrane-bound organelles. Many are very complex cells with specialized organelles for hunting, photosynthesis, or decomposing organic matter.

10. Protozoa are animal-like protists and include the flagellated mastigophorans, the sarcodines with their amoeboid movement, the ciliates with their numerous cilia, and the sporozoa with their sporelike nonmotile stage.

11. The plantlike protists contain chlorophyll and carry out photosynthesis. They include the euglenoids, the dinoflagellates, and the golden-brown algae and diatoms.

12. The slime molds are funguslike protists that carry out extracellular digestion of organic matter.

Key Terms

bacillus, 338
capsid, 344
cellular slime mold, 352
ciliate, 349
coccus, 338
cyanobacterium, 344
endospore, 340
fruiting body, 352
halophile, 341
macronucleus, 349
micronucleus, 350
parasite, 339
pathogenic, 338
phytoplankton, 350

prion, 345
protozoan, 346
pseudopod, 348
red tide, 352
saprobe, 338
slime mold, 352
spirillum, 338
stigma, 350
symbiont, 339
true slime mold, 352
vibrio, 338
viroid, 345
virus, 344
water mold, 352

Study Questions

Review What You Have Learned

1. How do protists and prokaryotes differ?
2. Why are microbes important organisms?
3. Describe the cell structure of a typical prokaryote.
4. How do gram-positive and gram-negative bacteria differ?
5. Describe the shape of the following bacterial species: *Bacillus subtilis, Staphylococcus aureus, Spirillum volutans,* and *Vibrio cholerae.*
6. How do a saprobe, a parasite, and a symbiont differ?
7. True or false: Only sexually reproducing eukaryotes are capable of genetic recombination. Explain.
8. Identify three habitats where archaebacteria might be found.
9. What are heterocysts? Why do biologists credit them with an important role in the early evolution of cyanobacteria?
10. Create a chart using the following headings: Type of Bacteria; Characteristics.
11. How are some types of bacteria useful to humans?
12. Trace the steps by which a virus infects a cell.
13. Define viroid and prion.
14. Name the four major groups of protozoa, and give the chief characteristics of each.
15. What is phytoplankton? Why is it important?
16. What characteristics distinguish golden-brown algae and diatoms from other types of phytoplankton?
17. How do true slime molds and cellular slime molds differ?

Apply What You Have Learned

1. A newly hired laboratory worker believes that he can thoroughly clean surgical instruments using hot, soapy water. Is he correct? Explain.
2. A science fiction novelist writes a story about genetically engineered bacteria that escape from a laboratory and infect all land plants, causing them to stop producing oxygen. As a result, all animal life perishes. Is this scenario plausible?
3. Campers in the wilderness decide that it is safe to drink from a cool, clear, backcountry stream. What illness may result from this decision? Why?
4. A student argues that because viruses have such simple structures, they must have been the first organisms to arise on earth. Is his argument likely to be correct? Why or why not?

For Further Reading

AIKEN, J. M., and R. F. MARSH. "The Search for Scrapie Agent Nucleic Acid." *Microbiological Reviews* 54 (1990): 242–246.

BEGLEY, S., and T. WALDROP. "Microbes to the Rescue!" *Newsweek,* June 19, 1989.

BROCK, T. D. *The Biology of Microorganisms.* 5th ed. Englewood Cliffs, NJ: Prentice Hall, 1988.

CHERFAS, J. "Malaria Research—What Next?" *Science* 247 (1990): 399–403.

CORLISS, J. "The Kingdom Protista and Its 45 Phyla." *Biosystems* 17 (1984): 87–126.

DANIELS, T. J., and R. C. FALCO. "The Lyme Disease Invasion." *Natural History,* July 1989, pp. 4–10.

MAHMOUD, A. A. F. "Parasitic Protozoa and Helminths: Biological and Immunological Challenges." *Science* 246 (1989): 1015–1021.

MARGULIS, L., J. O. CORLISS, M. MELKONIAN, and D. J. CHAPMAN. *Handbook of Protoctista.* Boston: Jones and Bartlett, 1990.

OLDSTONE, M. B. A. "Viral Alteration of Cell Function." *Scientific American,* August 1989, pp. 42–48.

PAIN, S. "BSE: What Madness Is This?" *New Scientist,* June 9, 1990, pp. 32–34.

POSTGATE, J. "The Malleable Microbe." *New Scientist,* February 16, 1991, pp. 38–41.

POSTGATE, J. "Microbial Happy Families." *New Scientist,* January 21, 1989, pp. 40–44.

SCOTT, A. "Viruses Work to Improve Their Image." *New Scientist,* May 19, 1988.

SLEIGH, M. *Protozoa and Other Protists.* New York: Edward Arnold, 1989.

STANIER, R. Y., J. L. INGRAHAM, M. L. WHEELIS, and P. R. PAINTER. *The Microbial World.* 5th ed. Englewood Cliffs, NJ: Prentice Hall, 1986.

Plants and Fungi: Decomposers and Producers

FIGURE 17.1 ■ **A Giant Sequoia. The majesty General Sherman Tree is the largest living organism on earth, and towers over neighboring trees and visitors in Sequoia National Park.**

Redwoods and Root Fungi

The most massive organisms ever to exist on our planet are still alive today: the giant sequoia trees growing on the western slope of the Sierra Nevada in central California. These plants can reach heights of 84 m (275 ft) and live for 3000 years. California has still taller (but slimmer) trees—the coast redwoods, which can exceed 110 m (360 ft). And the state has still older but smaller trees—the bristlecone pines, which can survive for more than 5000 years. Still, *Sequoiadendron giganteum* has a kind of majesty based on sheer bulk that is unsurpassed in life's history (Figure 17.1).

The biggest sequoia on record, the General Sherman Tree, stands in a quiet grove in Sequoia National Park and has a total weight estimated at 6150 metric tons (13,567,400 lb). That's a mass equivalent to 1200 adult African elephants or 50 full-sized blue whales! The trunk of this gargantuan plant is 11.1 m (36.5 ft) in diameter—about the size of a large classroom. Midway up the trunk sprouts a limb that is 2.1 m (7 ft) thick and 38 m (125 ft) long: This single branch is larger than the typical tree of any species east of the Mississippi. Sequoias also have the thickest bark on earth, a woody cinnamon-red layer up to half a meter thick and resistant enough to protect the living tissue inside for 30 centuries against fire, frost, and drought. The plant reaches reproductive age at 10 to 14 years, but can continue producing cones and seeds for another 3000. And the wood from a single tree could fence an 8000-acre ranch and shingle the roofs of 80 houses.

Sequoias do, however, have one seeming anatomical deficiency: Their roots push outward only 15.2 m (50 ft) or so from the trunk and extend downward less than 1 m (39.4 in.) into the soil. By contrast, a chestnut tree less than one-sixth the sequoia's size can have a wider root system, and a corn plant can have roots twice as deep! How can such a shallow and relatively small root system anchor such a massive tree and take in enough water and nutrients to support its growth? The answer is that the roots are infected with a minute fungus that sends billions of tiny hairlike extensions into the soil around the sequoia roots. These extensions, in turn, absorb far more water and other nutrients than the plant's roots could on their own. Because of them, the largest living organisms can survive and expand to monumental girth on poor, dry soils that could scarcely support green beans, pumpkins, or rose plants.

(a) Kingdoms and domains of life

(b) A fungus/plant interaction

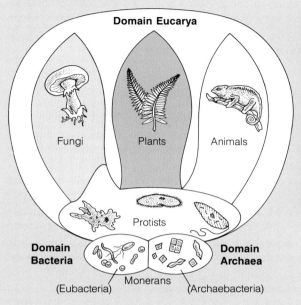

FIGURE 17.2 ■ Decomposers and the Interdependence of Two Kingdoms. (a) Plants and fungi are entirely separate kingdoms of life that arose from protistan ancestors, but in nature, the survival of one often depends on the growth of the other. (b) Here, a *Pycnoporus* fungus grows on a decaying log, recycling its nutrients.

With this chapter, we move from the single-celled kingdoms to the multicellular realms of the fungi and plants, including the organisms with the greatest numbers of cells, the sequoias, and the fungal inhabitants of their roots, so crucial to the big trees' survival. Biologists believe that both kingdoms arose from single-celled ancestors, the fungi from decomposers as much as 900 million years ago and the plants from photosynthetic precursors some 400 million years ago (Figure 17.2a). Because fungi are much older and are for the most part smaller and simpler organisms, we will consider them first.

We will see that fungi are the great decomposers: They break down organic matter and in the process make nutrients available to other life forms. This role gives fungi vast ecological and economic importance. We will also see that fungi live in a variety of ways, but virtually all fungi have free-living haploid and diploid phases (some single-celled, some multicellular) as well as reproductive structures called spores. Finally, we will see that fungi and plants are highly interdependent (Figure 17.2b). Most fungal species survive by breaking down plant matter, while some 90 percent of all plants that live on land have the kind of root-fungi combinations that the giant sequoias depend on. Many biologists believe that land plants could never have evolved without the previous presence of fungi on land.

In our discussion of the plant kingdom, some additional themes will emerge. First, plants are the great producers; each year, the earth's plants collectively produce 150 billion tons

of carbohydrates, an amount that could fill the boxcars of 100 trains, each long enough to stretch from the earth to the moon! As we saw in Chapter 6, many monerans and protists, most fungi, and virtually all animals live directly or indirectly on the carbohydrates that plants produce.

Second, plants have a distinctive life cycle based on an alternation between two multicellular phases: a diploid generation that forms spores and a haploid generation that forms gametes (egg and sperm cells). The key to understanding plant life cycles is this central alternation between multicellular haploid and diploid generations.

Throughout our discussion of plants, we will find one final theme: Over geological time, plants moved from the oceans to the dry continents, and as plants evolved, a series of physical systems and physiological trends emerged that helped them meet the challenges of life on land. New structures for internal transport, vertical support, reproduction, dispersal of the young, and so on, enabled complex land plants to diverge and greatly outnumber the kinds of simpler plants that remained in bodies of water or at their damp fringes.

This chapter answers several questions:

- How do fungi live, grow, and reproduce?
- How do the major groups of fungi influence the environment and the human economy?
- How do plants live, grow, and reproduce, and what might have been their evolutionary pathways?

AN OVERVIEW OF THE KINGDOM FUNGI: THE GREAT DECOMPOSERS

Nature works through balanced ecological cycles. Plants and photosynthetic microbes, for example, remove carbon from the air and trap it in carbohydrates. If these were the earth's only organisms, soon all the carbon would be locked up in their cells, and, lacking a carbon source, they could no longer grow. This doesn't happen, however, because other organisms eventually decompose living or dead tissue, unlocking compounds containing carbon, nitrogen, phosphorus, and other elements required for life and returning those elements to the soil and air. The earth's major decomposers are the 100,000 or more species of fungi (including the familiar mushrooms, molds, and puffballs) along with the various decomposing bacteria and protists discussed in Chapter 16.

Most fungi are *saprobes:* They decompose nonliving organic matter, such as fallen wood and leaves and dead animals. During the long evolution of the fungi, however, some species changed from decomposers of nonliving tissue to parasites that attack living things. Today, parasitic fungi are the main cause of plant diseases, with 5000 different fungal species attacking crops in fields, gardens, and orchards. Other types of parasitic fungi attack people and domestic animals, causing athlete's foot and vaginal yeast infections. And the saprobes have an indiscriminate "appetite" that gives them an economic impact second only to the fungal parasites. Fungi will consume everything from leather and cloth to paper, wood, paint, and other materials, slowly reducing old buildings, books, and shoes to crumbled ruin.

The Fungal Body: A Matter of Mycelia

Members of the kingdom Fungi are distinguishable by the way they obtain nutrients and by their unique physical structure. Fungi are heterotrophs; that is, they cannot build their own organic nutrients and so must consume them. To do this, they secrete enzymes that break down organic matter; then they absorb the released nutrients through their cell membranes. Because the actual digestive process takes place outside the organism's body, it is called *extracellular digestion.* An equivalent type of food breakdown takes place in an animal's hollow gut.

The body of a typical mushroom or bracket fungus looks solid, but is actually made up of cells joined into filaments called **hyphae** (singular, hypha) packed tightly together into a mat called a **mycelium**, rather like steel wool (Figure 17.3). The mushroom body is the aboveground portion, while an extensive, loose mycelium can penetrate the soil beneath it for many square meters. Each hypha has tubular side walls, made mostly of chitin, surrounding the cell's plasma membrane and cytoplasm, as well as incomplete, or perforated, cross walls that separate the cells lying end to end. Through these perforations, the cytoplasm of the interconnected cells can flow freely.

Hyphae can grow very quickly, and just one day's new hyphal growth in a tangled mycelium can easily exceed 1 km (0.62 mi). This explains how mushrooms can literally spring up overnight on damp logs or soil. This meshwork construction gives a fungus a large surface-to-volume ratio, which enables even a very large mushroom to digest and absorb sufficient amounts of nutrients to grow rapidly.

Fungal Reproduction: The General Case

In Chapter 7, we saw that many multicellular organisms can reproduce by either asexual or sexual means, depending on external conditions. In the fungi, asexual reproduction is the most common mode. Pieces can break off the hyphal meshwork and grow into new individuals; hyphal cells can divide or bud; or the fungus can produce asexual **spores,** cells that are dispersed and that divide mitotically into new, genetically identical fungi. The hyphae and the asexual spores are usually haploid, and the haploid phase dominates the fungal life cycle. Spores are adaptations for survival, able to withstand extreme conditions of dryness or cold, then produce a new fungus when conditions improve.

Fungi can also reproduce sexually (Figure 17.4). While fungi are neither male nor female, each haploid individual is one of two mating types: a *plus* or *minus* mating type (Figure 17.4). The haploid cells of either mating type can grow into hyphae and reproduce asexually. Eventually, a haploid cell of the plus mating type can fuse with a haploid cell of the minus

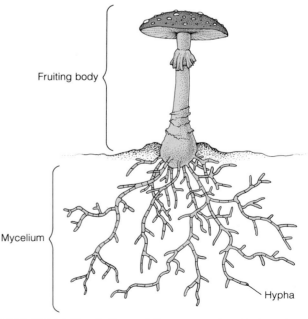

FIGURE 17.3 ■ Fungi: General Structures. A typical mushroom is just a small portion of the total organism. The rest, a tangled mat of mycelium, stretches below the ground surface. The mushroom, or fruiting body, produces spores that germinate and grow into new mycelial mats.

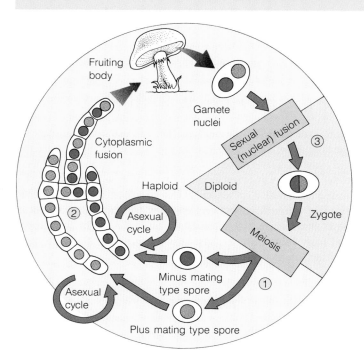

FIGURE 17.4 ■ Generalized Life Cycle of Fungi. The haploid phase dominates the life cycles of fungi. Spores can be of the minus or plus mating type. They can be produced either sexually or asexually, and they grow into hyphae. The cytoplasm of two cells from hyphae of opposite mating types can fuse and produce a cell with two nuclei (a dikaryon). This cell can give rise to more hyphae, which may form a fruiting body. Within the fruiting body, the two nuclei of opposite mating types may fuse, giving a diploid zygote. The zygote undergoes meiosis and once more produces haploid spores.

mating type, forming a cell with two haploid nuclei (a *dikaryon*). These double-nucleated cells can divide, hyphae can form, and these threadlike structures can join and give rise to a *fruiting body*. A mushroom is actually a fungal fruiting body. Inside the fruiting body, the haploid nuclei can become gametes, and the nuclei of two gametes can fuse, forming a diploid zygote (a process called sexual fusion). Meiosis in this diploid zygote produces haploid cells of the two mating types, and the cycle is complete. There are many variations of the fungal life cycle, but the essential feature is that the haploid stage dominates, and the diploid phase is often relegated to a single cell, the zygote.

Fungal Interactions with Plants

Despite the damage caused each year by saprobes and parasitic species, fungi as a group are certainly more beneficial than harmful. It is easy to imagine the earth becoming buried by a deep layer of dead wood and plants were it not for the activity of fungi, since they are the main decomposers of cellulose and lignin, the hard substances in wood and the two most abundant organic compounds on earth. But out of sight, below the soil surface, fungi and plant roots—like those of

the giant sequoias—interact symbiotically in important ways. Symbiotic associations between roots and fungi are called **mycorrhizae,** literally meaning "fungus roots."

Biologists have discovered and named several hundred species of mycorrhizal fungi associated with the roots of perhaps 90 percent of all land plants. In many woody plants, mycorrhizae form hairy mantles outside and between the cells of the plant's roots and expand the surface area for uptake of water and minerals (Figure 17.5a). In the majority of land plants, however, different types of mycorrhizal fungi grow

(a) Uninfected and infected roots

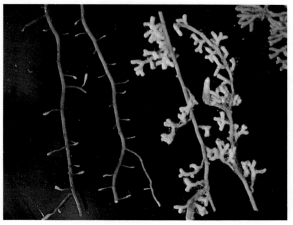

Without mycorrhizae   With mycorrhizae

(b) Some mycorrhizae grow inside plant cells.

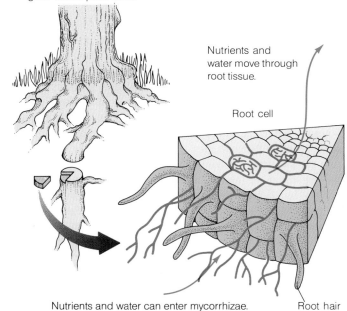

Nutrients and water can enter mycorrhizae.   Root hair

FIGURE 17.5 ■ Mycorrhizae: Fungus-Plant Associations. (a) Mycorrhizal fungi can grow between and outside the cells of plant roots. Shown here are bare pine roots (left) and pine roots associated with white mycorrhizal fungi (right). (b) Some species of mycorrhizae can grow inside plant root cells. Their microscopic hyphae (red) branch into the soil amid the root hairs. The fungus absorbs water and nutrients, and the host plant benefits.

FIGURE 17.6 ■ **Lichens: Alga-Fungus Associations.** Lichens can be colorful or camouflaged, but in each type, a fungus and photosynthetic algal cells grow in such close association that they appear to be a single organism. Several types of crustose lichens grow on this boulder in Colorado.

inside the root cells, send out microscopic hyphae, and also take up from the soil large amounts of nutrients and water that benefit the plant (Figure 17.5b). Sequoias have this type of fungus, and such infected plants can take in ten times more phosphorus and other nutrients than they could without the fungal symbionts. In exchange for augmenting the plant's nutrition, the fungus gets a home and probably other benefits not yet understood.

A second important type of fungus-plant interaction can be found in the 18,000 species of **lichens:** gray, orange, or greenish encrustations that grow on bark, soil, or rocks (Figure 17.6). A lichen is an association between a fungus and an alga, with the fungus forming a dense hyphal mat around the algal cells. The fungus lives off the alga's photosynthetic products, while the alga receives protection and a supply of water, among other benefits. Lichens take almost no nutrients from the substrate on which they live, and they are often the first organisms to inhabit lava flows or other newly exposed rocks. While lichens can survive in hostile environments, they are particularly sensitive to the industrial pollutant sulfur dioxide. Therefore, they have recently been used as early indicators of polluted environments.

While fossilized cell clusters that look like fungal hyphae occur in 900-million-year-old rocks, the oldest undisputed fungal fossils reveal that fungi grew on land 450 to 500 million years ago—long before plants left the oceans. What's more, fossilized mycorrhizae have been identified among the roots of the oldest fossilized land plants. This evidence suggests that "fungus roots" helped plants make the transition to land. Without fungus-plant interactions, plants would have evolved in very different ways, and life's giants, the sequoias, might never have appeared.

A SURVEY OF SAPROBES AND PARASITES: THE MAJOR FUNGAL GROUPS

Fungi are immobile organisms that are usually classified by the shapes of their spore-producing structures. Spores are the only means a fungus has of dispersing to new areas, and thus they are of prime importance. As we saw, spores can be formed asexually or sexually. Let's look at four major divisions of fungi, focusing on how they form spores, as well as other significant characteristics (Table 17.1) (In the fungal and plant kingdoms, a division is the equivalent of a phylum in other kingdoms.)

Zygomycota: Bread Molds and Other Zygote Fungi

The typical members of this group are filamentous and grow in cottony masses. All produce dark, thick-walled spore-forming structures in which fertilization generates diploid zygotes; the name of the group, in fact, comes from these structures. *Rhizopus,* the common fuzzy whitish or grayish mold that grows on bread, is a zygomycote. In *Rhizopus,* a mat of hyphae penetrates the bread and then sends up erect stalks, each bearing an asexual spore-producing structure that can split open and spread haploid spores. Spores can lie dormant and withstand extremes of drought and cold, then germinate later and produce a new hyphal mat. Most of the ecologically important mycorrhizal associations between fungi and plant roots involve zygomycote species.

Ascomycota: Mildews and Other Sac Fungi

This largest division of fungi includes mostly saprobes but some important plant parasites as well. Many grow in dense fuzzy mats; they have hyphae with perforated cross walls; and they can reproduce both asexually and sexually. During sexual reproduction, ascomycotes produce spores in a little sac called an **ascus** (Figure 17.7). In many species, rows of these sacs are borne in a cup-shaped structure called an **ascocarp.** George Beadle and Edward Tatum used a bread mold for their classic genetic studies, and it was an ascomycote (see Figure 9.14).

Surprisingly, the single-celled yeasts are ascomycotes. Although yeasts usually bud, that is, reproduce asexually by pinching off new smaller cells, they can also reproduce sexually by forming ascospores. Yeasts, which we employ in the brewing and baking industries, are probably the most economically useful fungi. Another yeast, *Candida albicans,* causes vaginal yeast infections.

Some ascomycotes are used directly as food, including the morels and truffles so favored by gourmet cooks. Other asco-

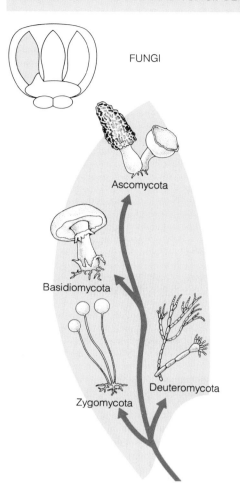

FUNGI

Ascomycota

Basidiomycota

Zygomycota

Deuteromycota

TABLE 17.1	Fungi: The Great Decomposers	
Division	Examples	Characteristics and Significance
Zygomycota (bread molds; 600 species)	Common black bread mold	Produce diploid spores and a cottony mat of hyphae on breads, grains, or other foods and organic materials
Ascomycota (sac fungi; 30,000 species)	Pink bread mold, brewer's yeast, morels, truffles	Produce spores in an ascus, or sac, borne in a cup-shaped body; because they include the yeasts, they are the most economically useful fungal group; powdery mildews harm fruit trees and grain crops
Basidiomycota (club fungi; 25,000 species)	Common field mushrooms, giant puffballs, bracket fungi, toadstools, smuts	Produce spores in club-shaped basidia; the fruiting body is the familiar mushroom or toadstool, which can be extremely poisonous
Deuteromycota ("imperfect" fungi; 25,000 species)	Species that produce penicillin	Lack sexual reproduction and instead produce asexual spores; various species are used in making drugs, cheeses, and soy sauce

(a) A cup fungus

(b) Reproduction in a cup fungus

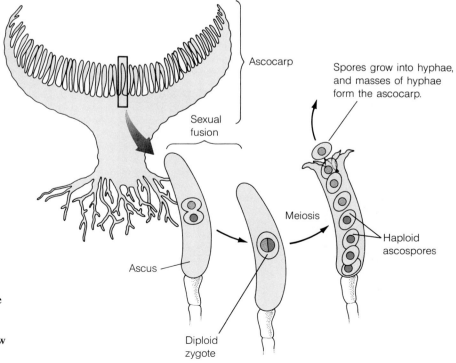

Ascocarp

Sexual fusion

Spores grow into hyphae, and masses of hyphae form the ascocarp.

Meiosis

Ascus

Haploid ascospores

Diploid zygote

FIGURE 17.7 ■ Sexual Reproduction in Cup Fungi. Cup fungi (a) have a visible structure, the ascocarp (b), which houses a series of little sacs, or asci; meiosis within each ascus yields ascospores, and these blow about, germinate, and grow into new cup fungi.

mycotes, however, attack our food crops, and powdery mildews parasitize apple and cherry trees as well as grain crops. Bread baked with rye infected by the ascomycote *Claviceps purpurea* causes ergotism, or ergot poisoning, characterized by hallucinations, convulsions, premature labor, and gangrene of the arms and legs. Some historians now believe that the "possessed" witches of Salem suffered from ergotism. Still other ascomycotes cause Dutch elm disease and chestnut blight, diseases that have robbed us of millions of our most beautiful hardwood trees.

Basidiomycota: Mushrooms and Other Club Fungi

The most familiar fungi—including edible mushrooms, bracket fungi, puffballs, toadstools, and smuts—are all basidiomycotes (Figure 17.8). Each produces **basidiospores** in club-shaped structures called **basidia** (singular, basidium). The spores are blown or carried by animals to new locations, where they germinate into new hyphae. The spore-producing abilities of the club fungi are truly prodigious. An average-sized grocery store mushroom can produce about 16 billion spores, and a giant puffball can produce 7 trillion. If all these spores germinated and developed into adult puffballs, they would collectively weigh 800 times the weight of the earth! Obviously, since we are not buried by puffballs, their spores rarely germinate and reach maturity.

The life cycle of basidiomycotes follows the general pattern we discussed earlier. The mushroom we see is just the fruiting body, or **basidiocarp,** the reproductive portion of the individual fungus made up of densely packed hyphae. Below the soil, sometimes extending for many meters, is a large mat of loosely tangled hyphae that sends up mushrooms at times appropriate for spore production and dispersal. Figure 17.9 shows the life cycle of a common field mushroom. Notice that

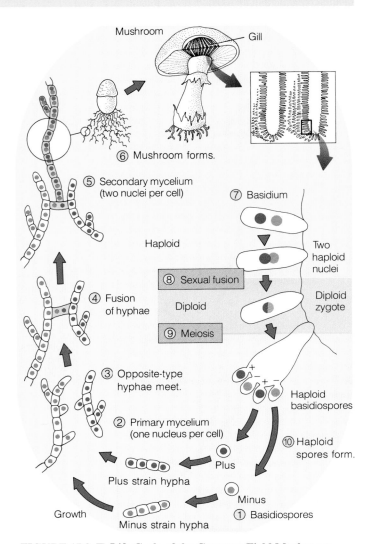

FIGURE 17.9 ■ Life Cycle of the Common Field Mushroom. See the text for a step-by-step explanation of the basidiomycote life cycle, from basidiospore to mushroom to new basidiospore.

(a) *Amanita* mushroom

(b) A giant puffball

FIGURE 17.8 ■ Common Forest Fungi. Most familiar forest fungi are basidiomycotes. (a) The deadly poisonous *Amanita muscaria* grows in a forest in New York State. (b) A giant puffball, *Calvatia gigantica*, dwarfs this boy's hand and arms.

haploid basidiospores are produced in the gills underneath the cap and that each spore has a nucleus of the plus or minus mating type (Figure 17.9, step 1). Once the spores have been dispersed, they grow into new threadlike *primary mycelia* (step 2). If two primary hyphae of opposite mating types encounter each other (step 3), a cell from each hypha fuses to form a single cell with two nuclei (step 4), and this divides into a *secondary mycelium,* or *dikaryon,* in which each cell has two nuclei, one plus and one minus (step 5); the nuclei of the dikaryon do not fuse until just prior to meiosis in the basidium. The secondary mycelium spreads underground and produces new mushrooms (basidiocarps) aboveground (step 6). Protruding from the gill surfaces of the new mushrooms are the special club-shaped cells, the basidia (step 7). Each basidium has two haploid nuclei, which fuse and form a short-lived diploid zygote, the only diploid cell in the life cycle (step 8). Meiosis quickly follows, and each basidium produces a new generation of haploid basidiospores (step 9), which are then released (step 10), ready to begin a new cycle.

One relative of the common field mushroom and with a similar life cycle is white rot fungus. This organism produces a unique and powerful enzyme that scientists are trying to harvest to break down dangerous environmental pollutants (see Box 17.1 on page 367).

Deuteromycota, or Fungi Imperfecti

Nearly 25,000 species of fungi seem to lack sexual reproduction (hence the name "imperfect" fungi). Most appear to be ascomycotes that reproduce only with asexual spores called *conidia* that are pinched off from the tips of spore-forming hyphae called *conidiophores*. Each fungus in this group is classified by how it produces conidia. We use many kinds of Fungi Imperfecti, including *Penicillium* and *Aspergillus* species for making penicillin and other drugs, Roquefort and other blue cheeses, and citric acid, soy sauce, and sake (Japanese rice wine). *Aspergillus flavus,* a mold that grows on stored peanuts and various grains, produces a compound called *aflatoxin,* one of the most potent cancer-causing substances ever identified.

We now shift our focus to a second group of organisms, the plants, with which fungi share so many interactions and which produce the carbohydrates that directly or indirectly support most other life forms.

THE PLANT KINGDOM: AN OVERVIEW

Plants can be gigantic like the sequoia, minute like the duckweed on a pond's surface, brilliantly colored, or drably camouflaged. Regardless of differences, however, plants have certain common physical characteristics as well as a fascinating evolutionary history.

Plant Life Cycles

Plants are multicellular organisms that generate their own organic nutrients through photosynthesis. Most plants have leaves (or equivalent structures) that act as solar collectors, stems (or the equivalent) that support these collectors, and roots (or similar structures) that anchor the plant and absorb water and nutrients. Plants also have a wide range of reproductive structures and processes, and they alternate between haploid and diploid forms, as do the fungi. In plants, however, both haploid and diploid phases are multicellular. In a sizable minority of species, each phase is a free-living green plant. The haploid phase is the **gametophyte** ("gamete-making plant"), and as the name implies, each gametophyte can produce male and/or female gametes. The diploid phase is the **sporophyte** ("spore-making plant"), and meiosis results in the production of spores, not gametes. Figure 17.10 shows

these alternating phases in the life cycle of a green alga. Notice that meiosis takes place in the sporophyte, producing haploid spores. These spores develop into haploid gametophytes, which make gametes. Sexual fusion then occurs between the haploid gametes, restoring the diploid chromosome number and yielding the sporophyte generation.

In algae (simple aquatic plants), the haploid phase dominates the life cycle and is the conspicuous plant we see. In a few algae and in many simple land plants, such as mosses, the diploid and haploid phases are each conspicuous green plants, as in the life cycle shown in Figure 17.10. Finally, in more complex land plants, like coconut palms and hibiscus, the diploid phase dominates and is the conspicuous plant we recognize. The majority of plants fit this last description.

Trends in Plant Evolution: A Changing Planet and the Challenges of Life on Land

Evidence suggests that plant life originated in water and that plants were probably derived from photosynthetic protists. Primitive green algae may have been the ancestors to all land plants, with the first transitional species surviving on the damp fringes of the ancient oceans some 400 million years ago (see Figure 15.13).

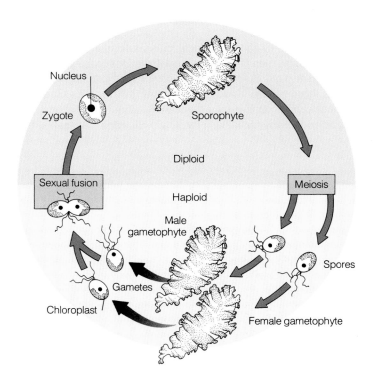

FIGURE 17.10 ■ **The Plant Life Cycle: Alternation of Generations.** Plants have an alternation between a multicellular haploid generation (a gametophyte, or gamete-making plant) and a multicellular diploid generation (a sporophyte, or spore-making plant). One phase is usually larger and longer-lived than the other, but both are present at some time in the plant's life cycle.

Plate tectonic activity during the early Paleozoic (400 to 570 million years ago) caused the margins of the continental landmasses slowly to buckle and relax dozens of times (see Chapter 15). This raising and lowering of the land allowed vast, shallow inland seas to accumulate and drain again and again. This climatic pendulum would have stressed aquatic plants, submerging them at some times and leaving them high and dry at others. Clearly, such environmental pressures would have favored organisms able to withstand desiccation, as well as to procure energy, reproduce, and absorb water and minerals on dry land. Biologists hypothesize that plants with exactly these abilities—derived from mutations and random genetic combinations—became the earliest surviving land plants. As we survey the plant kingdom, we will see evolutionary trends emerge, including the appearance of a vascular system that transports materials, the dominance of the diploid generation (the sporophyte), and the production of seeds, which helped improve the dispersal of young and the survival of new generations (Figure 17.11). Let's begin now with the algal ancestors of the land plants and trace the appearance of land plants step by step.

ALGAE: ANCESTRAL PLANTS THAT REMAINED AQUATIC

If one thinks of the geological eras as four acts in a long-running play, then the plant stars of the Proterozoic were the algae. But what exactly are algae? We have already encountered blue-green algae (cyanobacteria) in the kingdom Monera and golden-brown algae in the kingdom Protista. Are algae monerans, protists, plants, or all three? Actually, the term *algae,* from the Latin word for "seaweed," has been used to describe simple aquatic plants and has come to include aquatic photosynthetic organisms in all three kingdoms. The algae of the plant kingdom are mostly multicellular aquatic organisms with simple reproductive structures. They are classified in three divisions: the red algae, Rhodophyta; the brown algae, Phaeophyta; and the green algae, Chlorophyta. Table 17.2 describes these and the other divisions of the plant kingdom.

Algae probably arose about 600 million years ago from photosynthetic protists and were the only plants "on stage" that early. They have never exited the evolutionary play, however, and since their early stardom have diversified into 12,500 species that vary from tiny threads to giant kelps and tangled seaweeds. Algae live in virtually all bodies of water and sometimes on damp soils, rocks, trees, and even snow banks. Although they are generally simpler than the complex plants now dominating land habitats, modern algae continue to serve a critical ecological role: Since aquatic habitats are so widespread (covering nearly 75 percent of the earth's surface) and algae are so abundant, biologists believe that collectively the algae capture 90 percent of all the solar energy trapped by plants, notwithstanding the vast forests and prairies on land.

The aquatic habitat is a relatively benign and unchanging place, and its properties helped shape the organisms that live there. Because water supports the algal plant body, algae lack rigidity and usually undulate gently with waves and water currents. Since water surrounds the plant on all sides, individual algal cells absorb moisture and minerals directly from the surrounding water. Plant shape also reflects this direct contact with water: Most algae are quite flattened and translucent, with maximum surface area and cell exposure for absorbing water, minerals, and sunlight. Finally, reproduction can be asexual, involving the fragmenting of cells or body parts, or it can be sexual, with the production of eggs and sperm. The

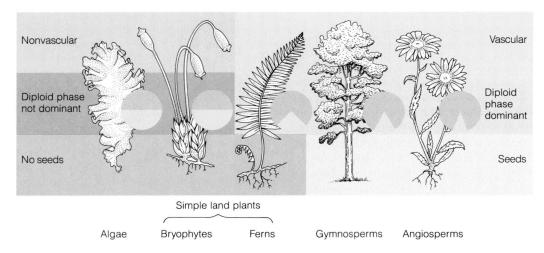

FIGURE 17.11 ■ **Trends in Plant Evolution.** In the evolution of the major plant groups, three main trends emerged: an increase in the importance of the diploid (sporophyte) phase of the life cycle, as shown by the green sectors in the life cycle circles; the evolution of a vascular system (shown in red in Figure 17.18a); and the evolution of seeds.

TABLE 17.2 Characteristics of the Major Plant Divisions

Division		Examples	Evolutionary Trends and Significance
Algae*			
	Red algae (Rhodophyta; 4000 species)	Many fanlike or filamentous types	Single-celled, colonial, or multicellular; reproduction asexual or sexual; haploid generation dominates; light-absorbing red or purple pigments can function at great water depths; used commercially for extracting thickening agents
	Brown algae (Phaeophyta; 1500 species)	Giant kelps and *Sargassum*	The largest algae; have tubelike conducting cells, but not true vascular tissue; diploid generation dominant in many species; kelp harvested for chemicals and cattle feed
	Green algae (Chlorophyta; 7000 species)	*Ulva*, *Chlamydomonas*	Produce carotene, chlorophyll, like the land plants; many have conspicuous haploid and diploid generations; may have been ancestors to the land plants; found in fresh water
Nonvascular plants			
	Bryophytes (Bryophyta; 20,000 species)	Mosses, liverworts, hornworts	Waterproof coatings, rigid tissues for upright growth on land, rootlike rhizoids; haploid generation (gametophyte) is dominant; often the first plants to colonize an area
Vascular plants (300,000 species)			Vascular tissue for transport and vertical support on land; most have roots, stem, and leaves; diploid sporophyte usually dominant; most successful and diverse group of plants
	Seedless vascular plants (Sphenophyta, Pterophyta, and Lycophyta; 12,000 species)	Ferns, horsetails, club mosses	Vascular pipelines; rhizomes, stems, and fronds; gametophyte can be tiny, independent plant or grows from sporophyte; grow on shady forest floors in low-lying damp areas
Seed plants			Produce seeds; gametophyte reduced to a few cells, dependent on sporophyte; do not depend on water for reproduction
	Cycadophyta (100 species)	Cycads	The three divisions of seed plants are cone-bearing remnants of ancient groups
	Ginkgophyta (1 species)	Ginkgos	
	Gnetophyta (70 species)	*Welwitschia*	
	Coniferophyta (Pinophyta) (600 species)	Pines, sequoias, firs	Naked seeds produced in cones; usually have needle-like leaves or scales; produce pollen; well-developed vascular system; true roots, stems, and leaves; sporophyte is dominant and supports gametophyte; conifers harvested in great numbers for wood products
	Anthophyta (Magnoliophyta)	Flowering plants	Produce flowers, seeds, and fruits; economically useful for food, drugs, landscaping
	Monocots (50,000 species)	Lilies, corn, onions, palms, daffodils	Leaves with parallel veins; seedlings have just one "seed leaf" or cotyledon; flower parts usually occur in multiples of three; seed stores much endosperm
	Dicots (225,000 species)	Roses, apples, beans, daisies	Leaves with netlike veins; seedlings have two cotyledons; flower parts in multiples of four or five; seed stores little endosperm

*Some biologists consider red, brown, and green algae to be protists because certain of their members are single-celled.

main secret to the algae's success is a range of photosynthetic pigments that can absorb the light of the different wavelengths that penetrate to varying water depths. Botanists use these same pigments to distinguish between red, brown, and green algae.

Red Algae: Deepest-Dwelling Plants

Most red algae are small, delicate plants that occur as thin filaments or flat sheets with an ornate, fanlike appearance; they generally live in shallow tropical ocean waters. Some,

FIGURE 17.12 ■ **Red Algae: Deep Dwellers.** Red algae can live in cold, inky waters to depths of 268 m (884 ft) below the ocean surface or in shallow tropical seas.

however, are single-celled or colonial, and a few species survive at depths of up to 268 m (879 ft; see Chapter 6 and Figure 17.12). These deep-sea denizens are 100 times more efficient at capturing sunlight than are the red algae found in shallow waters, and they rely on reddish accessory pigments called *phycoerythrins* that absorb light in the blue-green range—the only wavelengths that can penetrate to great depths. Red algae have chlorophyll *a* and the bluish pigment phycocyanin in addition to phycoerythrins, but it is mainly the reddish compounds that give the plants their stunning range of colors— from pink to purple to reddish black. At great depths, phycoerythrins absorb light energy and pass some of it to chlorophyll.

The conspicuous body of a red alga is the haploid gametophyte, which produces a gel-like protein that can be extracted commercially to make the laboratory growth medium agar; these algae also produce the starchy substance carrageenan, used as a stabilizing agent in ice cream, puddings, cosmetics, and paint.

Brown Algae: Giants of the Algal World

Most of the 1500 species of brown algae inhabit cool, offshore waters and occur as small multicellular plants. However, there are notable exceptions: The largest members of the algal world are the **kelps,** brown algae that can grow 100 m (more than 325 ft) long and float vertically like tall trees (Figure 17.13). Huge floating masses of the brown alga *Sargassum* thrive in the Sargasso Sea, north of the Caribbean, sometimes entangling hapless divers and ships alike.

Brown algae range from golden brown to dark brown to black. While they also contain chlorophylls *a* and *c,* it is the golden-brown carotenoid pigment called *fucoxanthin* that colors the plants and enables them to collect the blue and violet wavelengths of light that penetrate medium-deep water. These deep-water pigments explain why kelps can grow so tall—the sequoias of the sea.

Many brown algae possess complex structures analogous to parts of land plants: Leaflike **fronds** collect sunlight and produce sugars; the stemlike **stipe** supports the plant vertically; and the rootlike **holdfast** anchors the plant to submerged rocks. Special tubelike conducting cells carry sugars produced in the fronds to the deeper plant parts. These tubes function like the more specialized internal transport systems of land plants, but are not related tissues; both types reflect similar evolutionary solutions to the need for internal transport in a large multicellular organism.

In kelps, the diploid sporophyte dominates the life cycle, but in most smaller brown algae, the gametophyte and sporophyte generations are both free-living plants that resemble each other. The gametophyte produces gametes, while the sporophyte produces mobile spores called zoospores, whose flagella propel them to new locations where they develop into adult gametophytes.

Green Algae: Ancestors of the Land Plants

Most species of green algae live in shallow, freshwater environments or on moist rocks, trees, and soil, although a few inhabit shallow ocean waters. Green algae usually occur as single cells or as multicellular threadlike filaments, hollow balls, or wide, flat sheets. An intriguing exception, the "siphonous" algae, are *coenocytic;* that is, they have many nuclei in a single large cytoplasm surrounded by a cell wall, similar to the hyphae of most fungi.

Green algae are most notable for producing carotenoids and chlorophylls *a* and *b;* together these pigments absorb the sunlight penetrating shallow water or air with maximum efficiency. Green algae share this pigment combination only with the land plants. This is a main reason why botanists consider

FIGURE 17.13 ■ **Kelps: Underwater Giants.** As a diver looks up at tall, fluttering kelp through the sunlight that slants below the surface, the plants tower like forest trees and can reach 100 m (more than 300 ft) in height.

BOX 17.1 FOCUS ON THE ENVIRONMENT

AN ANTIPOLLUTION FUNGUS

White rot fungus, a basidiomycote relative of the edible field mushroom, has an appetite that would stun a teenager: This organism can digest anything from dead wood to DDT, and scientists are after its secrets in hopes of harnessing a new antipollution "device" for cleaning up the environment.

Phanerochaete chrysosporium (Figure 1) grows as small, white mycelial mats in the soil and inside rotten logs. It generates an enzyme called *ligninase,* which can degrade lignin, a molecule that resists breakdown by nearly everything else and is the hardening agent in wood. The presence of lignin can help protect a woody plant from organisms that would gorge on its simpler-to-digest cellulose. Nature is full of exceptions, however, and the white rot fungus is one of them. This living thing evolved with an enzyme that makes quick work of lignin. The white rot fungus secretes ligninase and then, after a time, simply soaks up the subunits of the degraded wood. It's lucky for living trees that this fungus is a saprobe instead of a parasite. And it may be lucky for the rest of us that scientists are studying *P. chrysosporium,* because the fungus has important implications for cleaning up pollution.

First of all, lignin itself is a nuisance to the paper industry, since it gives paper bags their brown color and turns newsprint yellow when exposed to air and sunlight. To combat these effects in higher-quality paper, papermakers add strong paper

FIGURE 1 ■ **White Rot Fungus (*Phanerochaete chrysosporium*).**

bleaches during production, but these, when present in industrial effluents, cause severe pollution of lakes and rivers. Researchers are looking for ways to avoid using such bleaches by employing white rot fungus to digest lignin during papermaking. Second, chemists have discovered that the carbon skeletons of DDT, dioxin, and many other highly toxic and carcinogenic pollutants are similar to those of lignin molecules, and the ligninase will degrade them just as handily. Scientists are working on ways to feed these pollutants to white rot fungus and thereby clean up additional industrial processes. Strange as it seems, this voracious resident of fallen logs and forest soils may someday be a priceless tool for industrial society.

---

green algae the direct ancestors of land plants. Green algae are also notable for a range of reproductive styles. In most species, the haploid phase dominates; the unicellular *Chlamydomonas* is an example of this haploid dominance.

Chlamydomonas is an oval cell that typically lives in freshwater pools and moist soils and is propelled by two flagella (Figure 17.14). The most prominent organelles are an eyespot that orients the cell toward light and a large, cup-shaped chloroplast that nearly fills the cell. The single adult

cell bears a striking resemblance to a spore from the multicellular marine green alga *Ulva,* or sea lettuce. Some biologists place the single-celled *Chlamydomonas* in the kingdom Protista. We use a more traditional taxonomy, however, and consider all the red, brown, and green algae, whether single-celled or multicellular, to be plants.

Ulva, an inhabitant of tide pools, grows a delicate, leaflike body, or **thallus,** which resembles sheets of green cellophane and is just two cells thick (Figure 17.15). *Ulva* displays an

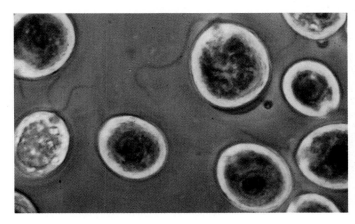

FIGURE 17.14 ■ **Green Algae: Plant Ancestors.** Green algae can occur as multicellular filaments or balls, or as single cells, such as *Chlamydomonas* (shown here), a common, single-celled resident of ponds and ditches. These cells have flagella, eyespots, and a cup-shaped chloroplast.

FIGURE 17.15 ■ *Ulva:* **A Sheetlike Green Alga with Nearly Identical Alternating Haploid and Diploid Phases.** An observer of *Ulva,* a green alga undulating in a shallow tidepool, would find it impossible to tell male or female gametophyte from sporophyte.

alternation of conspicuous haploid and diploid phases, which are multicellular and (to the human observer) virtually identical (review Figure 17.10). The haploid thallus, the gametophyte, produces gametes, while the identical diploid thallus, the sporophyte, produces spores.

The aquatic green algae probably gave rise to the simplest land plants, which still rely on standing water for reproduction. The next section recounts physical trends that emerged and allowed life to inhabit the land.

SIMPLE LAND PLANTS: STILL TIED TO WATER

In the long-running play of plant evolution, it wasn't until midway into the second act—the Devonian period of the Paleozoic era—that plants made the transition to land. There were two groups of players in this act: the **bryophytes,** or mosses and relatives, which grow low to the ground and require standing water for reproduction; and the **seedless vascular plants,** including ferns, horsetails, and club mosses, which grow taller than the bryophytes and have an internal transport system, but still require standing water to reproduce. The seedless vascular group played the starring role in this time period, but both were later upstaged by the seed-forming vascular plants, the gymnosperms and angiosperms, with their seeds and freedom from moisture-dependent reproduction. Botanists believe that an algal progenitor made the transition to land and then later diversified into the bryophytes and seedless vascular plants, the latter in turn giving rise to the seed-forming vascular plants (Figure 17.16).

The challenge of life on land remains today and includes (1) a way to prevent desiccation—in particular, a way to seal in water and minimize evaporation while still allowing for the exchange of oxygen and carbon dioxide; (2) a way to support the plant body in the absence of buoyancy from surrounding water; (3) a way to absorb water and minerals from the new substrate—the soil, mud, or sand—and transport them through the plant body; and (4) a way to reproduce sexually without shedding gametes directly into an ocean, lake, or pond. As we shall see, the relative success of each group of land plants depends, in part, on how its physical structures evolved to meet these challenges.

The Bryophytes: Earliest Plant Pioneers

The modern bryophytes are by far the most diverse lower land plants, and they include mosses, liverworts ("lobed plants"), and hornworts ("horn-shaped plants"). Second only in number of species to the flowering plants, bryophytes are small organisms that either lie flat or stand less than 2 cm (¾ in.) tall (see Figure 17.17a and Table 17.2). They are often the first plants to colonize a new area, and in most of the species, the sporophyte grows like a miniature street lamp from the leafy

gametophyte. These small, simple plants have successfully met two of the main challenges of terrestrial surroundings. They have waterproof coatings that prevent desiccation, as well as tiny "portholes" in their leaves that allow gas exchange through the coatings. Bryophytes also have tissues that are rigid enough to help keep the low plants upright, but they lack the kind of internal vascular system that absorbs or transports water and gives greater rigidity to other land plants.

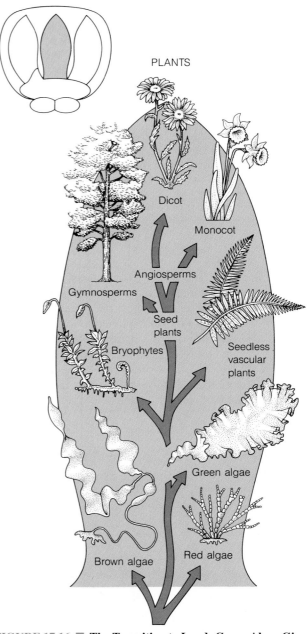

FIGURE 17.16 ■ The Transition to Land: Green Algae Give Rise to the Land Plants. Within the plants—one of the three kingdoms of multicellular organisms—green algae led to the simple land plants, such as mosses and liverworts. Such early land plants, in turn, gave rise to the larger, cone-bearing gymnosperms and to the flowering plants.

(a) A moss

(b) The life cycle of a moss

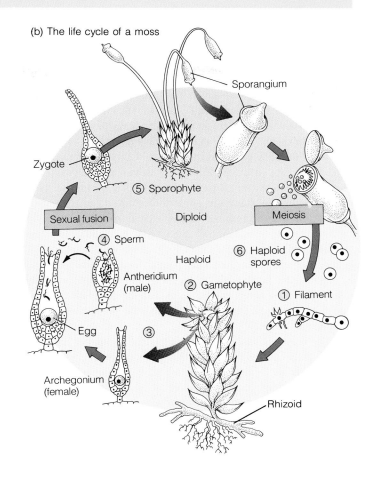

FIGURE 17.17 ■ **Mosses: The Most Familiar Bryophytes.** (a) Although mosses grow in thick, velvety carpets on fallen trees or patches of soil, each individual moss is a leafy gametophyte (haploid) that periodically bears a slender, green-brown sporophyte (diploid). Hairlike rhizoids anchor the plant to its substrate. (b) As the text explains, the moss life cycle alternates between the two conspicuous phases. A spore chamber, or sporangium, produces spores, which germinate and grow into gametophytes (1, 2); and these haploid plants form egg- and sperm-producing organs (3) that make gametes (4), which in turn fuse and grow into new sporophytes (5) that once again produce spores (6).

Instead of true roots, bryophytes have hairlike **rhizoids** that act only as anchors, absorbing neither water nor minerals. Water reaches the individual plant cells by slowly diffusing through the entire organism, as in a simple wick system. Because of this dependence on direct diffusion, the plants are limited in size and are restricted to growing in shady, moist places. What's more, bryophytes must live where it is at least seasonally wet because they have retained swimming sperm that can reach and fertilize an egg only when the plant is drenched.

Most moss species have life cycles typical of the bryophytes and reveal the kind of alternation between equally conspicuous gametophytes and sporophytes that we first saw in algae. The moss gametophyte begins as a tiny threadlike filament that very closely resembles certain green algae (Figure 17.17b, step 1). This filament grows into a haploid adult gametophyte—usually one of many in a densely packed velvety mat (step 2)—and develops both male and female reproductive structures. The sperm-producing organ is the **antheridium**, and the egg-producing organ is the **archegonium** (step 3). Flagellated sperm cells released from the antheridium (step 4) swim to the eggs, which remain in the protective archegonium. The diploid zygote that results from fertiliza-

tion divides mitotically and forms a **sporangium,** a little street-lamp-shaped structure that remains attached to and dependent on the gametophyte and that represents the entire diploid sporophyte generation (step 5). Following meiosis within the sporangium, haploid spores are produced and released (step 6), and these germinate into new threadlike gametophytes.

While many bryophyte species have survived in moist habitats, this pioneering group did not itself give rise to other more complex land plants, and the bryophyte life-style is considered an evolutionary dead end.

Horsetails and Ferns: The Seedless Vascular Plants

The real stars of the Paleozoic, the **vascular plants,** evolved at least one more of the solutions to life on land. Vascular plants have specialized cells that grow end to end in long internal pipelines extending from the tips of rootlike structures in the ground, up through the stem, to the leafy photosynthetic surfaces (Figure 17.18a). These vascular pipelines can transport water, minerals, and the products of photosynthesis throughout the plant and also lend vertical support that allows the plants to grow taller than the bryophytes—in fact to grow as tall as the General Sherman sequoia tree. Together with a layer of waterproofing on the outer cells of the leaves and stems, the vascular system enabled this new group to occupy drier habitats than bryophytes and thus move farther inland into unoccupied areas. The group includes the horsetails (Division *Sphenophyta*, Figure 17.18b); the lycopods, or club mosses and relatives (Division *Lycophyta*); and the familiar ferns and tree ferns (Division *Pterophyta*; see Figure 17.18c and Table 17.2). In the tropical climates of the Paleozoic, these first vascular plants grew in vast primordial forests and reached sizes much larger than today's tree ferns. However, the climate gradually grew colder and drier as the Mesozoic dawned, and these giants exited the stage, leaving behind the more diminutive horsetails, lycopods, and ferns (usually only 4 cm to 1 m in height, or about 1½ in. to 1 yd) that we see today on the shady forest floor. In a different sort of legacy, the bodies of the fallen giants were incompletely decomposed by fungi and bacteria, and the remaining organic compounds were carbonized into the enormous coal deposits people now mine.

The seedless vascular plants lack true roots, but have horizontal stems called **rhizomes** that grow on or just beneath the ground surface and survive from year to year. Growing up from each rhizome are many erect, leafy stalks. In horsetails and lycopods, the erect parts are true **stems;** in ferns, they are the central stalks of large leaves called fronds. At the end of each growing season, the stems and leaves die and are replaced by new aerial parts from new places on the slowly expanding rhizomes. This replacement is the way seedless vascular plants reproduce asexually. The prevalence of **vegetative** (asexual) **reproduction**—reproduction without genetic variation—is probably one reason these plants have changed so little over the last 400 million years.

In ferns, the undersides of the lacy fronds of the familiar fern adult—the diploid sporophyte—are dotted with **sori** (singular, sorus), or spore-bearing structures. Each spore develops into a tiny heart-shaped gametophyte. As Figure 17.19 shows, this green gametophyte (step 1) can produce either eggs or swimming sperm or both (step 2). Fertilization requires standing water, and after its occurrence (step 3), the zygote develops into a new adult sporophyte (step 4). The sori on the leaves of adult sporophytes contain globes called sporangia in which meiotic divisions produce haploid spores (step 5). When the sporangium breaks open, the spores are catapulted into the air; they then land and germinate into new gametophytes (step 6).

Horsetails, lycopods, and ferns show a continuation of the trend toward dominance by the diploid generation (review Figure 17.11): In these seedless vascular plants, the sporophyte is the conspicuous adult, and the gametophyte is a small free-living green plant. These plants still retain swimming sperm, like the bryophytes, and thus require standing water for sexual reproduction. In a sense, the seedless vascular plants are the botanical equivalents of the amphibians, animals such as frogs and salamanders, which can live on dry land but must return to water to lay their eggs. Interestingly, giant amphibians were the dominant land animals during the warm, swampy Paleozoic. Along with the giant lycopods, horsetails, and ferns of that era, the amphibians were replaced in the third act, the drier Mesozoic, by other stars that did not require standing water for reproduction: the seed plants and the reptiles.

(a) Vascular pipelines

(b) Horsetails

(c) Tree ferns

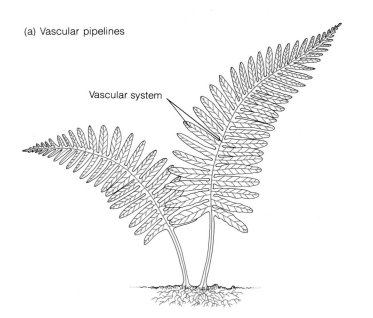

Vascular system

FIGURE 17.18 ■ **The Seedless Vascular Plants.** With their vascular pipelines for support and transport of water and nutrients (a), the horsetails, ferns, and tree ferns can grow larger and taller than the mosses and relatives. (b) A horsetail, *Equisetum.* (c) A tree fern, *Dicksonia antarctica,* in a grove of eucalyptus trees.

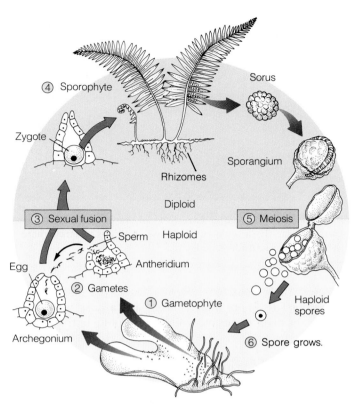

FIGURE 17.19 ■ Fern Life Cycle. As the text explains, the spore chamber in the sorus produces spores, which grow into green, heart-shaped haploid gametophytes. These make eggs and/ or sperm, which can fuse and grow into a new large, leafy diploid sporophyte.

GYMNOSPERMS: CONQUERORS OF DRY LAND

As the Mesozoic era began about 230 million years ago, Pangaea was breaking apart, the continental seas had retreated, and new masses of high, dry land became open to plant and animal colonization (see Figure 15.13). While the early land plants were tied to swampy lowlands, riverbanks, and coastal lagoons, new lines of plants appeared with reproductive modifications that broke the final barriers to complete colonization of land. These organisms were the **gymnosperms** (''naked-seed'' plants), and they included the now extinct seed ferns as well as the cycads, ginkgos, and conifers, such as pines, firs, sequoias, and redwoods (see Table 17.2).

The reproductive innovations of the gymnosperms include the pollen grain and the seed, as well as a further shifting toward dominance by the diploid generation. Pollen grains are immature male gametophytes composed of just two nonmotile cells plus a dry outer coat. Pollen grains are pro-

duced in huge numbers by male cones borne on the conspicuous seed plant, the sporophyte, and are disseminated by the wind. In effect, the pollen grain ''airlifts'' the sperm cells to the egg cells harbored inside special chambers in female cones, also borne on a mature sporophyte. Both male and female gametophytes are reduced to small nonphotosynthetic structures housed entirely by the sporophyte generation; thus, they embody the trend toward a dominant diploid phase. One could say that the water needed for fertilization is provided by the moist tissues of the parent plants themselves, rather than by splashing raindrops or standing water, and transportation of sperm is usually by ''flying'' rather than swimming. Another seed plant distinction is that gametophytes develop not from identical *homospores,* as in ferns, but from different-shaped *heterospores,* either smaller **microspores,** which eventually give rise to sperm, or larger **megaspores,** which ultimately give rise to eggs.

Once the pollen ''airlift'' and subsequent fertilization have taken place, the gymnosperm embryo develops inside an *ovule* that sits exposed on the surface of a sporophyll, or leaf of the female cone. (The term *naked seed* refers to the absence of a vessel surrounding the ovule.) Eventually, the ovule and enclosed embryo develop into a **seed,** which is encased in a tough coat housing both the young plant and a supply of food for the early stages of its growth. Seeds are the most successful dispersal devices to have evolved in the plant kingdom, and they greatly facilitated the further colonization of land. Reptiles—the dominant land animals during the Mesozoic, when gymnosperms were the ruling land plants—evolved parallel structures: eggs with leathery shells that could be laid and hatched on land. These freed them from the dependence on standing water displayed by amphibians. Hundreds of gymnosperm species still survive today, including the modern cycads, ginkgo trees, and conifers.

Cycads: Cone-Bearing Relics

A small group of tropical plant species often mistaken for palm trees, the **cycads** (Division *Cycadophyta*) are cone-bearing remnants of a group that flourished at the time of the dinosaurs, over 200 million years ago. Like the seeds of all gymnosperms, cycad seeds develop ''nakedly'' on open reproductive surfaces in cones. The most familiar American cycad is *Zamia pumila,* which grows as a native plant in the sandy-soiled forests of Florida, with leaves resembling palm fronds (Figure 17.20a). People often use these and other cycads as ornamental plants in warm regions, but sometimes they gather and eat cycad seeds and roots. Scientists recently found that a natural chemical in cycad seeds can cause weak muscles, tremors, and malaise. Residents of Guam suffered an epidemic of such symptoms when they had to rely on cycad seeds for food during the Japanese occupation of their island in World War II. This cycad compound may help researchers learn more about neurological disorders like Parkinson's disease and Alzheimer's disease.

Ginkgos and Gnetophytes: Odd Oldsters

Just a few species remain today of two formerly widespread plant divisions. *Ginkgo biloba*, the maidenhair tree, is the only surviving ginkgo species (Division *Ginkgophyta*). However, this stately tree, with its beautiful golden autumn foliage, is still planted along many urban streets because its fan-shaped leaves are particularly resistant to smoke and air pollutants. Ginkgos produce round, fleshy cones that resemble fruits (Figure 17.20b). In late summer, these cones, found only on "female" trees, begin to overripen and smell strongly of rancid butter. For this reason, gardeners prefer the pollen-producing "male" trees for ornamental plantings. Interestingly, the leaves of the ginkgo fall each year; it is **deciduous,** like many flowering trees.

The gnetophytes (Division *Gnetophyta*) include one of the most unusual vascular plants on earth: *Welwitschia,* a low-growing, cone-producing native of southwestern Africa that, when dug up, looks like an overgrown turnip. It has a disk-

(a) A cycad

(b) A ginkgo

FIGURE 17.20 ■ **Cycads and Ginkgos: Two Cone-Bearing Relics.** (a) Cycads such as this native of Florida, *Zamia pumila,* resemble palms but are ancient cone bearers. (b) The graceful fan-shaped leaves of the ginkgo remind some people of a maiden's flowing hair. In autumn, the fleshy yellowish cones ripen and fall, and the tree gives off a strong rancid smell.

FIGURE 17.21 ■ **Conifer Forest.** Cool, fragrant forests of pine, spruce, cedar, fir, or other conifers cover millions of acres where the altitude or latitude is high.

shaped woody stem, long, twisting leaves that stretch across the ground for more than a meter, and a deep taproot that allows the plant to survive without rain for up to five years.

Conifers: The Familiar Evergreens

The cone-bearing conifers are the most familiar and largest remaining group of gymnosperms, and they include many ecologically and economically important species. Millions of acres in mountainous regions are dominated by pine, spruce, fir, and cedar, much of which is harvested for wood, paper, and resins (Figure 17.21). Smaller conifers, such as yew, hemlock, juniper, and larch, are often used for the graceful landscaping of buildings and parks. Finally, the largest living things, the giant sequoias, are conifers as well as living reminders of the Mesozoic era, an age of giants.

Conifers have two distinctive characteristics: their narrow leaves and familiar woody cones. Their needle-shaped leaves are covered by a waterproof **cuticle,** or waxy layer. The needle shape resists desiccation, having little surface area relative to volume. In spring, at the start of the conifer life cycle, a large pine tree sprouts thousands of soft male cones and the more familiar large, hard female cones, each with a central axis and whorls of radiating woody scales (Figure 17.22, step 1). In spore-bearing structures (sporangia) of the soft male cones, diploid cells (microspore mother cells) undergo meiosis (step 2). The haploid products (the microspores) each develop into a *pollen grain,* the tiny male gametophyte plant. When the pollen grains are released, they shower the area around the tree with golden dust (step 3). Borne on spring breezes, these grains occasionally encounter female cones on other trees and sift down into the open scales, becoming fixed to a sticky fluid secreted by the **ovule,** a structure that houses the sporangium. When the fluid dries, it pulls the pollen inside the integument and thereby completes pollination.

An entire year will pass before actual fertilization takes place. During this time, the pollen grows a **pollen tube,** a pathway through the wall of the female sporangium. Eventu-

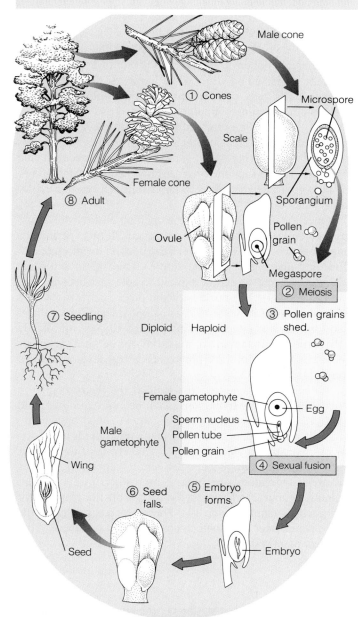

FIGURE 17.22 ■ Life Cycle of the Pine. The text gives a step-by-step account of the life cycle of a pine tree, a typical conifer, from the production of cones to the scattering of seeds and the sprouting of tender new seedlings.

ally, two sperm nuclei develop inside the pollen tube and advance toward the developing egg. Meanwhile, during the year-long period of pollen tube growth, a diploid cell (the megaspore mother cell) inside the ovule undergoes meiosis and forms four haploid spores. Three of these spores degenerate, and one divides mitotically into a multicellular female gametophyte containing haploid eggs. A single sperm nucleus fuses with a single egg, completing the process of fertilization (step 4). After fertilization, the embryo begins to form while still inside the ovule's protective coat (step 5). At the same time, this protective coat develops into the seed coat, and the

rest of the female gametophyte tissue divides and forms the food supply for the expanding embryo. By late summer of the second year, the scales of the woody female cone open, and mature seeds fall to the ground (step 6). Those that aren't consumed by hungry animals germinate the following spring into tiny pine seedlings (step 7) that produce the characteristic evergreen needlelike leaves and, later, the reproductive cones of an adult (step 8).

Conifers were able to conquer higher and drier reaches of the continents during the Mesozoic because of their numerous evolutionary advances: drought-resistant leaves; protective seed coats; greatly reduced male gametophytes that depend on the sporophyte and give rise to airborne pollen grains; equally reduced female gametophytes that give rise to eggs protected inside the ovule; a well-developed vascular system that produces wood and stiffens the trunk, branches, and roots; and, as we saw with the sequoias, the tendency to form mycorrhizal associations and the ability to survive for many centuries. Because of these modifications for life on land, conifers are still a successful group, with more than 600 modern species. Nevertheless, a final set of evolutionary "inventions" gave the flowering plants still greater success.

FLOWERING PLANTS: A MODERN SUCCESS STORY

With the advent of further continental uplift, drier and colder conditions became more commonplace on the earth's land-masses. The scene was set for the latest act in the evolutionary play—the modern, or Cenozoic, era—during which the mammals and the flowering plants (formerly called *angiosperms* and now commonly referred to as either Anthophyta or Magnoliophyta) would become the dominant land animals and plants. Both groups of organisms evolved with reproductive structures that protect and nourish developing embryos. Mammals evolved the placenta and the womb, while the flowering plants evolved a new reproductive structure, the ovary, which houses and protects the ovules; they also evolved fruits and flowers, which promote efficient pollination and seed dispersal. These innovations occurred along with other adaptations for life on land, such as broad leaves for efficient solar collection, and enabled the flowering plants to radiate into more species than the combined numbers of all other plant groups.

Today, flowering plants are the most common and conspicuous species in the earth's tropical and temperate regions, and from this monumentally successful group come virtually all of our crop plants (wheat, rice, corn, soybeans, fruits, and vegetables) and beverages (coffee, tea, colas, and fermented drinks), as well as spices, cloth, medicines, hardwoods, ornamental plantings, and, of course, flowers—symbols of beauty, affection, and renewal throughout human history. Table 17.2 shows the two major groups of angiosperms, the **monocots** and the **dicots,** and their main characteristics.

Flowering Plant Reproduction: A Key to Their Success

As in gymnosperms, the flowering plant life cycle displays the complete dominance of the diploid (sporophyte) generation and the complete dependence of the haploid (gametophyte) generation on the conspicuous adult plant. The older term *angiosperm* means "seed in a vessel," and indeed, seed development takes place within a structure at the base of the flower, called the *ovary,* which gives the ovules much more protection than the gymnosperm cone scale. The wall of the ovary eventually matures into a fruit that surrounds the seeds and aids in dispersal. In some cases, animals attracted to the fruit's colorful skin or luscious flavor eat the fruit and then excrete the seeds along with feces in some new location. Plants have myriad dispersal mechanisms, including "parachutes" that loft the seeds in the wind, hooks for hitchhiking on passersby, and tasty fruits (Figure 17.23).

The ovary is just part of the female reproductive structure (*carpel*) within the flower (Figure 17.24). A sticky top, or **stigma,** and a slender neck, or **style,** lead to the ovary, and inside the ovary are one or more ovules containing megaspore mother cells. The male flower part is the *stamen,* made up of the **anther,** a club-shaped structure that houses microspore mother cells, and the *filament,* a slender stalk. The microspore mother cells give rise via meiosis to immature male gametophytes, or pollen grains. A ring of colorful petals surrounded by a ring of often small green sepals completes the flower.

In spring, when the flowers of a given species open, each megaspore mother cell inside the ovary gives rise, via meiosis, to four haploid megaspores (Figure 17.24, step 1). Normally, three of these degenerate (step 2), and one divides into the female gametophyte, including one egg cell (step 3). Simultaneously, the anthers develop. The microspore mother cells inside the anthers undergo meiosis and form micro-spores (step 4), and these develop into immature male gametophytes, or pollen grains (step 5). When the anthers split open, mature pollen grains are released. In some species, these are dispersed by wind, while in others, they are carried by bees, birds, bats, or other animals, called **pollinators.** By chance, pollen grains will fall or be carried to a sticky stigma (step 6), and a pollen tube will begin to grow rapidly down through the style (step 7). Recall that in gymnosperms, this growth is slow, taking up to a year. In flowering plants, a tube 900 times longer than the original grain can form in just 5 to 10 hours—a fact that ensures rapid fertilization and evolutionary success.

Once the tube has reached the opening of the ovule housing the female gametophyte and its egg cell, two sperm cells migrate down, and a **double fertilization,** an event unique to flowering plants, takes place (step 8). One sperm cell fuses with the egg and forms the diploid zygote that becomes the embryo (step 9). The nucleus of the other sperm fuses with two *polar nuclei* in the female gametophyte and forms a triploid ($3n$) nucleus (step 10). This nucleus divides mitotically and forms **endosperm,** a nutritive tissue that becomes enclosed, along with the embryo, inside the developing seed (step 11).

As summer progresses, the wall of the ovary begins to enlarge into a fleshy **fruit** (step 12) or becomes modified into a winged structure such as a maple "squirt," the "parachute" of a milkweed seed, or one of the many other kinds of dispersal mechanisms (review Figure 17.23). We tend to think of fruits as round, sweet objects, but certain vegetables, such as cucumbers, tomatoes, and squash, are really fruits, and so are walnuts, peanuts, and pecans. Fruits fall off the plant or are harvested by animals (step 13). These processes disperse the seeds, which later germinate in new locations and there, grow into new plants (step 14).

(a)   (b)   (c)

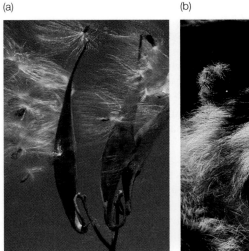  

FIGURE 17.23 ■ How Plants Disperse Seeds. (a) Butterfly weed seeds, *Asclepias tuberosa,* blowing in a summer breeze. (b) Prickly seeds of the common burdock, *Arctium minus,* clinging to a terrier's face. (c) Cedar waxwing feeding berries to its young.

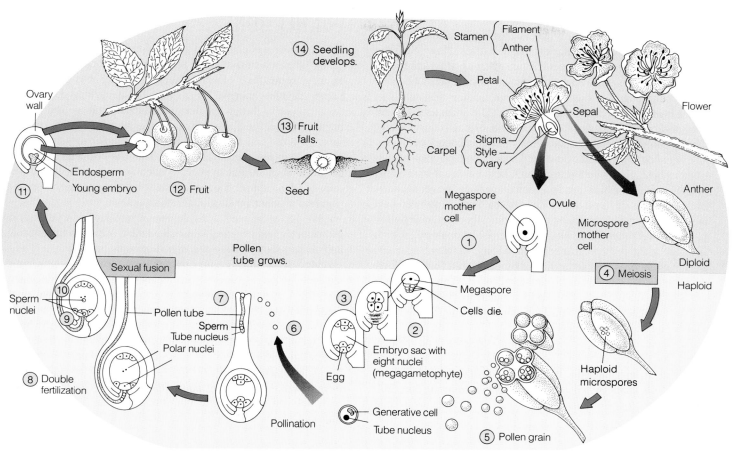

FIGURE 17.24 ■ **Life Cycle of a Flowering Plant.** As you follow the description given in the text, you can see the many steps in the anthophyte life cycle. The ovary produces eggs (1–3), the anthers make microspores that develop into pollen (4–6), and their double fertilization gives rise to embryo and endosperm inside a seed (7–10). Seeds usually form inside fruits, fleshy or dry, and when the seeds are released, they germinate into seedlings—the new generation of free-living sporophytes (11–14).

Flowering Plants and Pollinators: A Coevolution

The success of the flowering plants can be only partly explained by the protective enclosures missing in the naked-seed gymnosperms. Flowers, fruits, and seeds also help ensure pollination and seed dispersal by attracting animals. Most flowers attract insect, bird, or mammal pollinators that inadvertently carry loads of pollen from one plant to another as they forage for sweet nectar or for the protein-rich pollen grains. A flying or walking animal dusted with pollen and scouting for more of its favorite food will substantially increase the chances of pollination (Figure 17.25). And once fruits have developed on a plant, animals are attracted either by sight or smell to harvest and eat the fruits. The tough, highly resistant coats of many plant seeds allow them to pass through an animal's digestive tract intact. The seeds are then deposited in new locations, and this helps the plant spread.

The mutual dependence of flowering plants on pollinators and seed dispersers and of these animal species on the same

FIGURE 17.25 ■ **Sipping Beauty: A Study in Coevolution.** An Anna's hummingbird drinking from a red and magenta fuchsia blossom. Both organisms evolved with compatible anatomy.

plants for nutrition is no coincidence: It is the result of **coevolution,** a natural selection for increasing interdependence based on selective advantages for both parties. Plants that had genes for flowers with sweet nectar, a strong fragrance, or a bright color might have survived in greater numbers because more animals would have visited them in their constant foraging for food and carried away pollen. The animals that were attracted to the new food sources probably also survived in greater numbers, thus perpetuating the interdependence.

This coevolution has produced specialized physical structures in both plants and animals that help ensure the trade-off of nutrients for pollination and seed dispersal. The mouthparts of many types of bees, butterflies, moths, birds, and even bats are precisely the right shapes for tapping the nectar or pollen of the flowers they visit. Hummingbirds, for example, which see best in the red spectrum and have long, narrow beaks and a poor sense of smell, are attracted by bright red fuchsia flowers with their long, slender shape but little fragrance (see Figure 17.25). Conversely, flowers pollinated by bees tend to be yellow or blue and very fragrant, corresponding to the bees' vision and ability to perceive odors.

It requires energy, of course, for a flower to produce pigment molecules, nectar, and special parts not directly involved in reproduction. But plants pollinated by the wind, such as oak trees or sequoias, must produce more than 200 times as much pollen as a plant with its own army of animal pollinators.

CONNECTIONS

This chapter has characterized the fungi and plants in terms of the three major themes of this book: how organisms obtain energy, reproduce effectively, and evolve. As producers, plants capture energy, fix it, and build carbohydrates. As decomposers, fungi dismantle carbohydrates and other organic molecules, use some of the nutrients, and release more into the soil and air, where other life forms—including plants—can use them once again. It is the collective activity of the tons of fungal hyphae in each acre of topsoil and the tons of plant matter rooted in that soil that gives decomposition and production so global an impact.

We have devoted considerable space to life cycles and reproductive modes, and while the details are often unfamiliar and even strange, they are significant in three ways: They show what is unique about a group; they show the trends from simpler to more complex processes, revealing evolutionary directions; and they show how specific kinds of organisms recreate themselves and leave progeny.

As we consider the animal kingdom in the next two chapters, we will see an extension of a theme introduced here: the interrelatedness of the kingdoms. Just as plant roots need fungi for maximum absorption and nutrient recycling, and fungi need plant and animal matter for energy, so do the animals rely on all the other kingdoms in their role as mobile multicellular heterotrophs—the great consumers.

Highlights in Review

1. Giant sequoia trees, the most massive organisms ever to live, are members of the plant kingdom. These trees depend on tiny root fungi for their survival.

2. Fungi are the earth's major decomposers of organic matter, and they are either saprobes, which break down dead tissue, or parasites, which attack and consume living tissues.

3. Fungi are multicellular heterotrophs that carry out extracellular digestion by secreting enzymes outside the fungal body and then reabsorbing digested materials. The fungal body is made up of filamentous hyphae packed into a mycelial mass.

4. The fungal life cycle is dominated by the haploid phase. Fungi can reproduce asexually by fragmenting, budding, or producing asexual spores. They can also reproduce sexually: Hyphae of opposite mating types produce gametes that meet when the hyphae encounter each other. The gametes then fuse and form a zygote, the zygote undergoes meiosis and generates spores, and the spores then grow into new hyphae.

5. Mycorrhizae enable plants to absorb far more soil nutrients than if the fungi were absent from the plant's roots. Lichens, associations between fungi and algae, need few nutrients not available from the surrounding air and so are often pioneer species that inhabit bare rock.

6. Bread molds like *Rhizopus* are zygomycotes, and they form and release spores. Most mycorrhizal fungi, which associate with plant roots, are zygomycotes.

7. Ascomycotes produce ascospores in an ascus. Morels and truffles are examples.

8. Edible mushrooms, toadstools, and puffballs are basidiomycotes, and they produce basidiospores on club-shaped structures called basidia. The visible mushroom is the aboveground fruiting body; a tangled mycelial mass lies below ground.

9. The deuteromycotes (Fungi Imperfecti) include *Penicillium* and *Aspergillus* species and appear to be ascomycotes that reproduce only (or mainly) asexually with conidia.

10. Plants have a characteristic alternation of multicellular haploid and diploid generations, the haploid generation being the gametophyte (the gamete-producing plant) and the diploid generation being the sporophyte (the spore-producing plant).

11. Physical trends during plant evolution include the appearance of a vascular system, the growing dominance of the diploid generation, and the production of seeds in higher land plants.

12. Algae probably arose from photosynthetic protists some 600 million years ago. The algal plant body is usually filamentous or flat and thin. The haploid phase dominates the life cycle, and reproduction involves swimming eggs and sperm.

13. Red algae have red pigments that allow some species to gather light and survive far below the ocean surface.

14. Brown algae contain a brown pigment that also allows photosynthesis in deep water. Many species have plant parts called fronds, stipes, and holdfasts, analogous to leaves, stems, and roots. Some kelps grow to 100 m tall (more than 300 ft).

15. Green algae contain both chlorophylls *a* and *b*, most efficient in shallow water and in air; land plants also contain these chlorophylls. Green algae were probably the precursors to the first land plants.

16. Simple land plants include the bryophytes (mosses, liverworts, and hornworts) and the seedless plants (horsetails, lycopods, and ferns). Bryophytes lack a vascular system or true leaves, stems, or roots, and they have swimming sperm.

17. The seedless vascular plants have internal pipelines for transport and vertical support, as well as horizontal stems called rhizomes. Fertilization still requires standing water.

18. Gymnosperms include the cone-producing cycads, ginkgos, and conifers. Male cones produce pollen grains, the male gametophytes, and female cones produce megaspores, ovules, and later, seeds.

19. Conifers include the familiar evergreens, such as pine, fir, cedar, redwood, and sequoia. Conifers have needlelike leaves and woody female cones.

20. The pine life cycle involves airborne pollen, a slow-growing pollen tube, fertilization within the ovule's protective coat, and the maturation and dissemination of seeds.

21. Flowering plants evolved the plant ovary, flowers, and fruits, which protect the ovules and aid in pollination and seed dispersal. Flowering plants are the most diverse plant group, with more than 250,000 species, many economically important.

22. Sexual reproduction in flowering plants involves the production of pollen grains in male flower parts; transport by gravity, wind, or animal pollinators to female flower parts; the rapid growth of a pollen tube; a unique double fertilization leading to an embryo and to endosperm, or nutritive tissue; and the formation of a fruit surrounding the seeds.

23. Flowering plants and animal pollinators have a long and mutually beneficial evolutionary history.

Key Terms

anther, 374
antheridium, 369
archegonium, 369
ascocarp, 360
ascus, 360
basidiocarp, 362
basidiospore, 362
basidium, 362
bryophyte, 368
coevolution, 376
cuticle, 372
cycad, 371
deciduous, 372
dicot, 373
double fertilization, 374
endosperm, 374
frond, 366
fruit, 374
gametophyte, 363
gymnosperm, 371
holdfast, 366
hypha, 358
kelp, 366
lichen, 360

megaspore, 371
microspore, 371
monocot, 373
mycelium, 358
mycorrhiza, 359
ovule, 372
pollen tube, 372
pollinator, 374
rhizoid, 369
rhizome, 370
seed, 371
seedless vascular plant, 368
sorus, 370
sporangium, 369
spore, 358
sporophyte, 363
stem, 370
stigma, 374
stipe, 366
style, 374
thallus, 367
vascular plant, 369
vegetative reproduction, 370

Study Questions

Review What You Have Learned

1. Describe the extracellular digestion that takes place in a fungus.
2. What are mycorrhizae? How do they affect a plant's nutrition?
3. What are the four major divisions of fungi? In each division, which structure produces spores?
4. Outline the life cycle of a field mushroom.
5. Name the stages in a plant's alternation of generations.
6. How did changes in the earth's surface affect the evolution of land plants?
7. What are the three divisions of algae? How do they differ?
8. Is it true that algae are too simple to have ecological importance? Explain.
9. What adaptations enable a bryophyte to live on land?
10. Compare the gametophyte and sporophyte stages in a bryophyte, a seedless vascular plant, and a seed plant.
11. Compare the life cycles of a fern and a pine.
12. Name the female and male reproductive structures of a flowering plant.
13. In flowering plants, fertilization involves two sperm nuclei. How do the roles of the two nuclei differ?

Apply What You Have Learned

1. A farmer growing broccoli sprays insecticide on his crop. The insecticides drift over his neighbor's apple trees, which are in full bloom. How might he be ruining his neighbor's crop?
2. Biologists sometimes differ in their approaches to classification. Why might the unicellular *Chlamydomonas* be considered a protist by some and a plant by others?
3. Believing she can increase the productivity of her fruit trees by killing soil fungi, a homeowner sprays fungicides on the soil around the trees. Is this procedure likely to improve fruit yield? Why or why not?
4. Employees of a timber company planted fir seedlings on logged-over land in an area subject to frequent drought. All the seedlings died. Three more times they planted seedlings, with the same result each time. The chief forester then suggested that they hire a mycologist (a scientist who studies fungi) before trying to plant again. Why did he make this suggestion?

For Further Reading

CHAPMAN, A. R. O. *Functional Diversity of Plants in the Sea and on Land.* Boston: Jones and Bartlett, 1987.

GENSEL, P. G., and H. N. ANDREWS. "The Evolution of Early Land Plants." *American Scientist* 75 (1987): 478–489.

NIKLAS, R. J. "Aerodynamics of Wind Pollination." *Scientific American,* July 1987, pp. 90–95.

ROSS, I. K. *Biology of the Fungi: Their Development, Regulation, and Associations.* New York: McGraw-Hill, 1979.

WARING, P., and A. MÜLLER. "Fungal Warfare in the Medicine Chest." *New Scientist,* October 27, 1990, pp. 41–44.

Invertebrate Animals: The Quiet Majority

FIGURE 18.1 ▪ *Nasutitermes* **Termite. This insect protects the communal termite nest by squirting intruders with a sticky substance from its nozzle-shaped head.**

Self-Protecting Termites

Termite mounds are a common sight in the world's tropical regions. These hard earthen structures look like obelisks, pyramids, and furrowed monuments, and they can reach 6.7 m (22 ft) in height. The mounds are built and inhabited by millions of jostling white termites living in complex societies. Sequestered in their elaborate nests, the termites represent a concentrated source of food for larger animals. Only a few kinds of animals, however, including certain types of anteaters, make use of this rich food source. The reason is that termites have an amazing arsenal of self-defenses that keeps most potential enemies at bay. For example, more than 500 termite species in the genus *Nasutitermes* have a caste of fierce soldiers in their societies that protects the workers, queen, and others in the colony. At the first sign of an intruder, each soldier rushes forward and squirts an irritating gluey substance from its conical head (Figure 18.1). The anteater or other hungry intruder is just as likely to retreat with a painful tongue as it is to collect a meal and will often seek out a nest of less defended ants or termites as an alternative.

The members of *Nasutitermes* are **animals,** multicellular heterotrophs that move about under their own power at some point in their life cycle. Such organisms are members of the fifth and largest subdivision of all living things, the animal kingdom. This kingdom includes well over a million species—more than three times the number of all other living species combined. It will take us two chapters to survey this unparalleled diversity.

When people think of animals, they usually envision **vertebrates:** animals with backbones, such as goldfish, leopards, horses, and eagles. However, 95 percent of all animal species are **invertebrates:** animals without backbones, like termites, earthworms, and oysters. Because we humans, as well as most of our domesticated animals, are vertebrates, it is sometimes hard to acknowledge that we are vastly outnumbered and outranked in ecological importance by the simpler invertebrates. In this chapter, we will trace the history of the invertebrates. And as we do, we will encounter a host of colorful, strange, and sometimes fantastic organisms that will expand our notion of what animals are and how they live. The invertebrates are so numerous and diverse that no single species can begin to typify the entire group. *Nasutitermes,* however, reflects the major evolutionary trends that occur among the invertebrates, as well as distinct specializations for communication and defense.

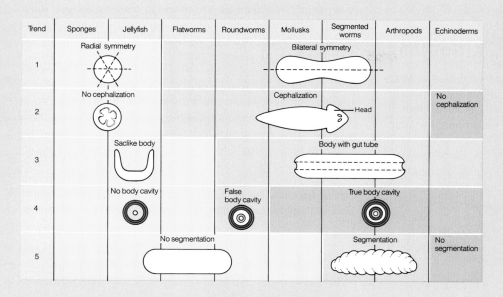

Trend	Sponges	Jellyfish	Flatworms	Roundworms	Mollusks	Segmented worms	Arthropods	Echinoderms
1	Radial symmetry				Bilateral symmetry			
2	No cephalization				Cephalization — Head			No cephalization
3	Saclike body				Body with gut tube			
4	No body cavity			False body cavity	True body cavity			
5	No segmentation					Segmentation		No segmentation

FIGURE 18.2 ■ **Evolutionary Trends in the Invertebrates.** *Trend 1:* Away from radial symmetry and toward bilateral symmetry. *Trend 2:* Development of a head (cephalization). *Trend 3:* Away from a saclike body and toward a body with a gut tube. *Trend 4:* Toward a true body cavity. *Trend 5:* Toward segmentation of the body.

As we survey the major phyla of invertebrates, we can trace the origin of the basic physical features most animals share, including a head, mouth, nervous system, heart, stomach, and appendages. There were five major anatomical and physiological trends leading to this "standard equipment"; they spanned hundreds of millions of years of evolution, and they reflected successful evolutionary solutions to the problems of survival (Figure 18.2). The first was a trend away from a circular body plan (so-called radial symmetry) and toward a body with symmetrical right and left halves (bilateral symmetry). Along with this came a second trend: cephalization, or the development of a head, with its centralized sensory apparatus that detects environmental stimuli such as touch, light, sound, and chemicals and its concentrations of nerve cells that integrate and interpret stimuli and control physical responses to them. Bilateral symmetry and cephalization provided many advantages, including various ways to move through the water or soil and more ways of interacting with other organisms and the physical surroundings. Further specializations unique to each species, such as the termite's bazooka-like head, conferred still more advantages.

In the third trend, invertebrate animals evolved away from a simple, saclike body with a single opening at one end toward a more complex, elongated body containing a food-digesting tube, the gut, with openings at both ends. Among other benefits, this trend in the evolution of body structures led to a more complete breakdown and use of food, making more energy available for rapid running, swimming, slithering, or flying.

A fourth trend in these animals with a tube-within-a-tube body plan was away from enclosure of the tube in solid tissue and toward suspension of the tube in a fluid-filled space called a coelom. This cushioned the gut, helped support the whole body from within, and allowed other internal organs to develop more complex forms.

A fifth trend was toward segmentation, the development of a series of body units, each containing similar sets of muscles, blood vessels, nerves, and other structures. Segmentation allowed animals to develop specialized body parts, such as legs, wings, and antennae. In some animal groups, these appendages became modified still further into pincers, fangs, paddles, wing covers, and other attachments that perform very specific tasks.

Although many species of organisms with all five trends emerged, thousands of types evolved and survive today without one or more of these features. Clearly, while these trends confer certain advantages, bilateral symmetry, cephalization, a one-way gut, the coelom, and segmentation are not prerequisites for survival.

This chapter will answer several questions:

■ Which characteristics do animals share, and how do invertebrates differ from vertebrates like ourselves?

■ What are the sponges, the simplest invertebrates, and how do they survive with simple saclike bodies?

■ What are the radial animals, the jellyfish and relatives?

■ What are the flatworms and roundworms, with their heads and bilateral symmetry?

■ What are the mollusks, including clams, snails, and squid, the most intelligent invertebrates on earth?

■ What are the segmented worms, with their advances in digestion and circulation?

■ What are the arthropods, the lobsters, spiders, insects, and relatives?

■ What are the echinoderms, or sea stars and relatives, with the first internal skeletons—but not the last?

ANIMALS AND EVOLUTION: AN OVERVIEW OF THE ANIMAL KINGDOM

Animals are quite distinct from members of the other king-doms. Unlike prokaryotes and protists, animals are *multicellular*. In fact, some of the largest animals have trillions of cells. And unlike plants, animals are *heterotrophs;* they can-

not manufacture their own food and instead must ingest it and break it down metabolically for its energy content. Unlike most fungi, which are also multicellular heterotrophs, animals can move about at some point in their life cycle—usually throughout the entire cycle—to search for food and mates and to avoid danger. (Fungi and plants often produce mobile spores, but these do not generally move about under their own power.) Finally, animals are diploid, and their primary mode of reproduction is sexual; because each individual grows and changes from a single fertilized egg to a multicellular organism, it passes through various distinct stages of development.

As a group, animals have a long and interesting evolutionary history, and because they have left a more complete fossil record than most members of the other kingdoms, that history

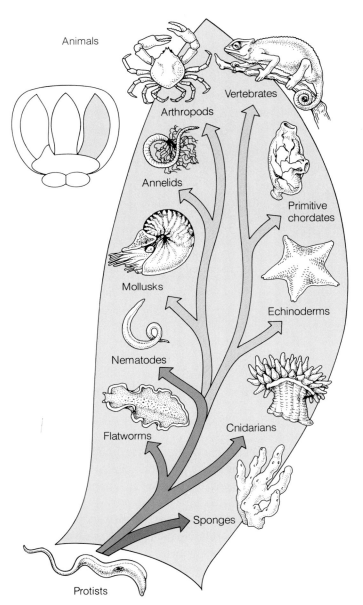

FIGURE 18.3 ■ **Evolutionary Relationships of the Major Animal Phyla.** Animals arose from protistan ancestors. The sponges were an early branch off the main line and may have had an independent origin. The cnidarians (jellyfish) were also an early branch off. The other branches, however, have diverged into large groups, including the mollusks, arthropods, and vertebrates. Two main lines of descent characterize later animal evolution.

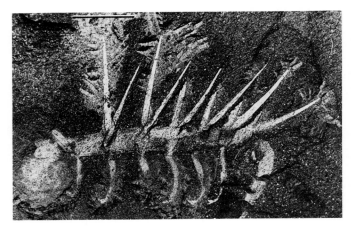

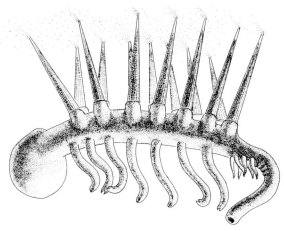

FIGURE 18.4 ■ *Hallucigenia:* **A Vision from the Mists of Invertebrate Evolution.** A small, strange animal with a bulbous head and seven sets of rigid pointed spines projecting from a tubelike body—*Hallucigenia*—lived more than 570 million years ago. It may have perched stiffly on unjointed legs, or perhaps the spines projected from its back and it walked on the opposite tubelike appendages. Fossil impressions of this living nightmare were discovered in the Burgess Shale deposits high in the Canadian Rockies. Here, archaeologists found dozens of now-extinct invertebrates (mostly soft-bodied and many as peculiar as *Hallucigenia*) that lived at the dawn of animal evolution. *Hallucigenia* may be an ancestor of *Onychophora* (see page 396).

is better understood. The earliest fossils of soft-bodied animals are burrows, trails, and impressions found in rocks from the Edicara region of southern Australia and are dated at about 700 million years old. Many zoologists (biologists who study animals) believe that animals arose from protistan ancestors, with the sponges and cnidarians (including the jellyfish) probably diverging in two independent lines very early from the main line leading toward bilateral animal groups. The branching diagram in Figure 18.3 shows the probable relationship of the sponges and other animals to the single-celled protists that preceded them. Soft-bodied animals existed for at least 120 million years before other species started generating hard parts (shells, scales, teeth, and bones); the earliest fossils of these hard structures were found in 580-million-year-old rocks (Figure 18.4).

The fossil record shows that the overall result of geological change and continued evolution was a great radiation of the animal kingdom—the emergence of large numbers of animal species with a variety of forms adapted for various environments. All the animals that ever existed—including invertebrates such as the first spiderlike animals to crawl on land and the giant dragonflies of primordial forests, and vertebrates such as the leather-winged pterosaurs and the lumbering mastodons and ground sloths of the last ice age—were products of the great radiation. At least 35 to 37 phyla evolved (animals in a phylum share the same general body plan), most of which still have living members and include more than a million individual species. We will discuss only the most important groups, beginning with the sponges, the simplest animals (Table 18.1).

TABLE 18.1   Key Invertebrate Animals

Phylum	Examples	Number of Species	Notable Features
Porifera	Sponges	9000	Asymmetrical saclike bodies; central opening; body wall perforations; spicules; no evolutionary descendants; no true tissues
Cnidaria	Jellyfish, hydras, corals, sea anemones	9000	Radial symmetry; grow as polyps or medusae; two tissue layers; gastrovascular cavity; nerve cells, nematocysts, planula larvae; no body cavity
Platyhelminthes	Flatworms, including tapeworms, flukes	20,000	Bilaterally symmetrical with a head; three tissue layers; true organs and organ systems; life cycles often complex and include two or more hosts; no body cavity
Nematoda	Roundworms	12,000	Bilaterally symmetrical with a head and gut tube; three tissue layers with unlined (false) body cavity; hydroskeleton; extremely common in soils and as parasites on other animals and plants
Mollusca	Snails, clams, octopuses, squid, slugs	120,000	Bilaterally symmetrical with a head and gut tube; three tissue layers with a lined body cavity; gills, mantle; open circulatory system; very common marine and freshwater organisms
Annelida	Segmented worms, including earthworms, polychaete worms, leeches	15,000	Bilaterally symmetrical with a head and gut tube; lined body cavity; segmentation; hydroskeleton; move with bristles pushing against ground; crop, gizzard; closed circulatory system; earthworms occur widely and help aerate soils
Arthropoda	Spiders, mites, ticks, scorpions, millipedes, centipedes, insects, lobsters, shrimp	1,000,000	Bilaterally symmetrical with a head and gut tube; lined body cavity; segmentation; exoskeleton for support and protection; specialized segments and appendages; jointed legs; tracheae and gills; acute senses; most diverse phylum in living world
Echinodermata	Sea stars, sea urchins, sea cucumbers	7000	Gut tube and lined body cavity; a head in some larvae and adults; no segmentation; first endoskeleton; unique water vascular system for locomotion; separate evolutionary line from Mollusca, Annelida, and Arthropoda

SPONGES: THE SIMPLEST ANIMALS

More than 9000 animal species—mostly marine creatures, but including some freshwater inhabitants—make up the phylum Porifera, the **sponges.** For centuries, people considered sponges to be plants, since the adults are **sessile,** or stationary, and are permanently attached to rocks, pilings, sticks, plants, or other animals. Sponges, however, are not plants, but multicellular animals that filter and consume fine food particles, bacteria, or protozoa from the water.

The Sponge's Body Plan

The simplest sponges resemble vases or irregular-shaped clusters of tubes (Figure 18.5a). Whether shaped like a branching vase or a cluster of tubes, each "container" has a **central cavity,** a large opening, and hundreds of tiny perforations, or **pores,** through the body wall. If you place red dye in water that surrounds a living sponge, you can see the colored water being drawn into the pores along with food particles. Special cells lining the cavity remove and digest the suspended food particles; and the water and waste matter exit through the central opening. Sponges range from the size of a pinhead to the size of a wine barrel. This large size is possible because canals permeate the body and provide avenues for the transport of food and wastes to and from the deepest parts.

Most sponges are asymmetrical and lack specialized tissues. The sponge's body wall is a thin "sandwich" with a layer of flattened cells outside; an inner lining containing *"collar cells"* (choanocytes), each with a long flagellum surrounded by a delicate collar of microvilli (see Figure 18.5b);

and a gelatinous filling containing several different kinds of wandering *amoeboid cells.* The outer cell layer, which is simpler but equivalent to the epidermis of more complex animals, protects the sponge. The constant beating of the collar cells' flagella draws water in through the pores and pushes it out through the central opening. The embedded amoeboid cells help digest food particles and differentiate into gametes.

The sponge's cells are supported by a "scaffolding." In some sponges, this consists of fibers of a protein related to hair protein, which forms a network surrounding the various cell types. Many sponge cells also contain tiny pointed **spicules** made of silica or calcium carbonate; these act as scaffolding and also protect the sponge from predators by presenting them a spiky mouthful (see Figure 18.5b). The irregular, tan bath sponges sometimes sold commercially are the tough fibrous skeletons that remain after people harvest the animals, trample them, leave them to decay, then process their "scaffolding."

In certain sponge species, if the sponge is broken into pieces and strained through fine mesh, the cells can live independently for days. Eventually, however, they regroup into a single functioning organism and once again develop pores, channels, and an inner scaffolding.

Sponge Activities and Reproduction

Although a sponge looks inert from the outside, its cells are continuously active. An average sponge moves about 166 L (45 gal) of water a day through its canals and pores. Food particles in this water collect at the base of the collar cells, and the collar cells and neighboring amoeboid cells take in

(a) Red vase sponge

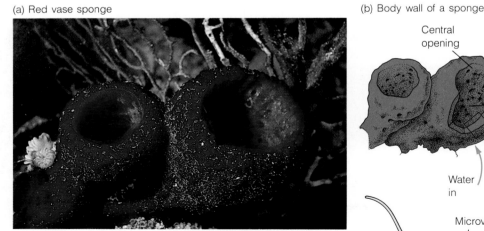

(b) Body wall of a sponge

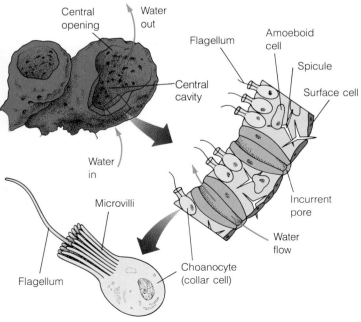

FIGURE 18.5 ■ The Sponges: The Simplest Animals. (a) This red vase sponge (*Mycele* species) is a common denizen of coral reefs. The simple vaselike body has a central opening and a central cavity. (b) The body wall contains choanocytes and amoeboid cells and is traversed by incurrent pores and channels, through which water flows, drawn by the flagella. Food particles collect on the microvilli of the choanocytes.

the particles via phagocytosis and digest them. The amoeboid cells then help distribute materials to all body cells. Sponges excrete wastes and gases through openings in the body wall or directly through the cells of the thin body wall itself; sponges have no gills, kidneys, or other organs. Sponges can reproduce asexually by budding off new individuals, small cell clusters called *gemmules.* Sexual reproduction involves eggs and sperm.

Zoologists believe that sponges evolved from protozoa called *choanoflagellates,* which have collars and flagella, live in colonies, and produce glassy spicules. The sponges, with their primitive cell layers rather than individual cells, represent a definite advance beyond colonial protozoa in that the sponge cells function together as a single individual, rather than as a cluster of independent, single-celled organisms, as in a protozoan colony. Within such a protozoan colony, each cell retains the ability to reproduce; in sponges, however, sexual reproduction requires a cellular division of labor. Sponges are not directly related to any other animal groups, nor did they give rise to more complex organisms.

CNIDARIANS: THE RADIAL ANIMALS

Some of the most ephemeral and beautiful invertebrates are members of the phylum Cnidaria (pronounced "ni-DAIR-e-ah," from the Greek word for "nettle"). This phylum includes the translucent hydras, the gossamer jellyfish, the sea anemones, and the colorful corals, and it has at least 9000 modern species (Figure 18.6). Most live in the oceans, but a few, such as the hydras, inhabit fresh water. **Cnidarians** have true tissues, cells of a similar form organized in a way that allows them to perform a specific function. Cnidarians also have a radial body plan; that is, they are circular with a central axis and structures that radiate outward like the spokes of a wheel. They also possess fearsome weapons: tentacles armed with stinging devices that assist the animals in self-defense and in capturing food.

"Vases" and "Umbrellas": Cnidarian Body Plans

Cnidarians grow in a range of sizes, colors, and strange appearances, but on close inspection, all have one of two basic radial forms: the **polyp,** a hollow, vaselike body that stands erect on a base and has a whorl of tentacles surrounding a mouth near the top (Figure 18.7a), or the **medusa** (plural, medusae), an inverted umbrella-shaped version of the polyp, with tentacles and mouth pointing downward (Figure 18.7b). Sea anemones, corals, and most hydras are polyps, while jellyfish are medusae. While sea anemones, hydras, and jellyfish usually live as independent individuals, thousands of coral polyps live together in huge colonies like the one shown

(a) A hydra   (b) A sea anemone

(c) A coral colony   (d) A jellyfish

FIGURE 18.6 ■ **The Cnidarians: Ephemeral Invertebrates.** Cnidarians are radial animals, often translucent and billowing, but protected by stinging weapons called nematocysts. (a) A brown hydra. (b) *Condylactis gigantea,* a pink-tipped sea anemone. (c) *Adelogorgia phyllosclera,* a coral. (d) *Aequorea victoria,* a jellyfish.

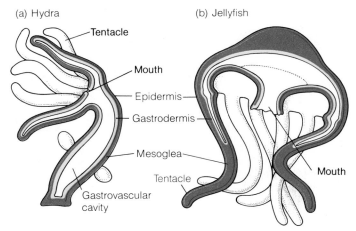

(a) Hydra   (b) Jellyfish

Tentacle
Mouth
Epidermis
Gastrodermis
Mesoglea
Tentacle
Mouth
Gastrovascular cavity

FIGURE 18.7 ■ **The Cnidarians: Two Tissue Layers.** A polyp, such as this hydra (a), and a medusa, such as this jellyfish (b), both have just two layers of cells. The blue layer is the epidermis (outer skin), and the yellow layer is the gastrodermis (stomach skin). Sandwiched between the two tissue layers is a noncellular mass of proteins and carbohydrates, the mesoglea ("middle glue"), indicated in red.

in Figure 18.6c. Each individual polyp in the colony is encased in its own cup-shaped limestone skeleton. Together, the millions of tiny coral cups (some inhabited but most empty) can form giant reefs and atolls. These include the only biological structure visible from space, the 1900 km long (1200 mile) Great Barrier Reef off Australia's eastern coast, as well as islands such as Bermuda, the Bahamas, and Fiji. These reef-building corals get their spectacular colors and obtain much of their energy from microscopic photosynthetic algae that live symbiotically inside the cells of the coral. Unfortunately, changes in the ocean are causing an epidemic of coral "bleaching"—the expulsion of the algae from the coral cells. Bleaching often leads to death of the coral (Figure 18.8). Researchers fear that global warming will accelerate this trend and perhaps destroy the largest and most impressive structures ever made by living organisms.

Each cnidarian has a three-layered body wall with an outer skin (*epidermis*), an inner "stomach skin" (*gastrodermis*), and a jellylike substance in between called **mesoglea** (literally, "middle glue"; see Figure 18.7). In polyps, the mesoglea layer is thin, but in medusae, it can be thick; in fact,

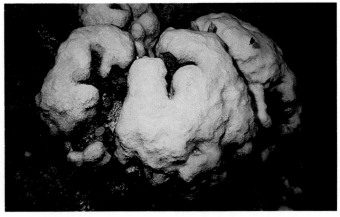

FIGURE 18.8 ■ **Coral Bleaching.** **Pollution and temperature changes in the oceans are causing coral colonies to expel their photosynthetic algae. This "bleaching" starves the coral polyps, and the coral colony dies. Here are "before" and "after" photos of the same coral reef.**

jellyfish resemble jelly because of the extreme thickness of the mesoglea. In one monstrous North Atlantic jellyfish that is nearly 3 m (10 ft) across and weighs a ton, virtually all the mass is mesoglea.

Cnidarians have three important adaptations relating to the capture and digestion of food. Embedded in the epidermis of the tentacles are remarkable organelles called **nematocysts,** or stinging capsules. When triggered by the approach or contact of an enemy or potential prey, the tube everts like a sock turned inside out. But this "sock" has a sharply pointed end that can strike the prey and release paralytic toxin. Several human deaths have been attributed to nematocysts from *Chironex,* a large tropical jellyfish. The better-known Portuguese man-of-war can inflict nasty stings, but rarely kills people.

Once a cnidarian has immobilized a prey or simply chanced to find a food morsel, it swallows the food and digests it within the body's central cavity, the *coelenteron,* or **gastrovascular cavity,** the second important adaptation (review Figure 18.7). Some gastrodermis cells produce enzymes that begin to break down the food; this enzymatic food breakdown within a cavity is called *extracellular digestion,* and cnidarians and all of the more complex animals make use of it. Extracellular digestion allows animals to digest larger food pieces and thus expand their range of food sources, compared to the intracellular digestion used by sponges and most protists.

A third adaptation helps cnidarians like jellyfish and anemones detect prey, coordinate body movements, and capture and swallow the victim. **Nerve cells** arranged in a loose network permeate the animal's tissues and enable it to detect stimuli and activate cells in response. Cnidarians and all more complex animals have nerve cells and nervous systems (see Chapter 27).

One of the most amazing adaptations is the ability of certain cnidarians to regenerate lost parts or even a complete body. If a freshwater hydra, for example, is cut into pieces or split almost in half, each piece or half will grow into an entire animal. In fact, a piece the size of a pinhead can regenerate a new animal as large as the original. This ability is quite different from the tendency of existing but disrupted sponge cells to re-form an organism. With cnidarians (as well as certain kinds of sponges), hundreds of thousands of new cells are regenerated from a tiny set. Regeneration is advantageous because if a predator rips off a tentacle or other body segment, the cnidarian is not permanently disabled.

Cnidarians Have Complex Reproductive Cycles

Sea anemones and corals reproduce asexually by budding or sexually by releasing eggs and sperm into the water. Adult jellyfish live as medusae and develop either egg- or sperm-producing organs (gonads), which release gametes into the water (Figure 18.9, step 1). When fertilized, the egg develops into a *planula,* a small, flat, solid larva propelled by cilia (step 2). The planula attaches to a rock or other object, develops

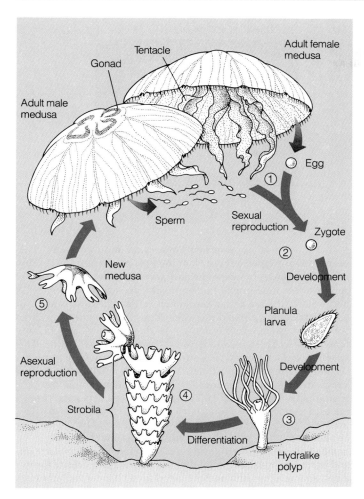

FIGURE 18.9 ■ Life Cycle of a Jellyfish. The text describes the jellyfish life cycle, from adult medusa to planula larva, sessile polyp, and new medusa.

into a stationary polyp (step 3), and then differentiates into a colony of polyps, or *strobila,* (step 4). The strobila resembles a stack of saucers, and in a type of asexual reproduction, each saucer in the stack develops tentacles and swims away as a new medusa (step 5). At maturity, the medusa develops either male or female gonads, and the cycle continues.

FLATWORMS: THE BEGINNING OF BILATERAL SYMMETRY AND CEPHALIZATION

The simplest invertebrates to display bilateral symmetry and "headedness"—traits seen in most animal species—are the **flatworms** (Figure 18.10). The phylum Platyhelminthes (from the Greek words for "flat worms") includes 20,000 or more species of flattened worms. Common planarians are free-living flatworms that inhabit freshwater lakes, rivers, or

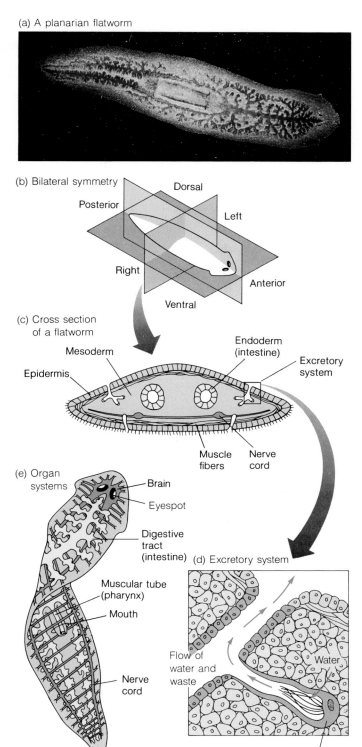

FIGURE 18.10 ■ Flatworms: Simple Animals with Primitive Organ Systems. (a) A planarian (red digestive tract). (b) Its elongated body shows bilateral symmetry; the right and left sides are mirror images of each other. The animal is also cephalized; the head is anterior, the tail posterior. (c) The planarian is characterized by three distinct tissue layers. The intestines and excretory and nervous systems play important roles. (d) The flame cells of the excretory system remove excess water from the body. (e) Flatworm organ systems.

Life cycle of a tapeworm

The "head," or scolex, of a tapeworm

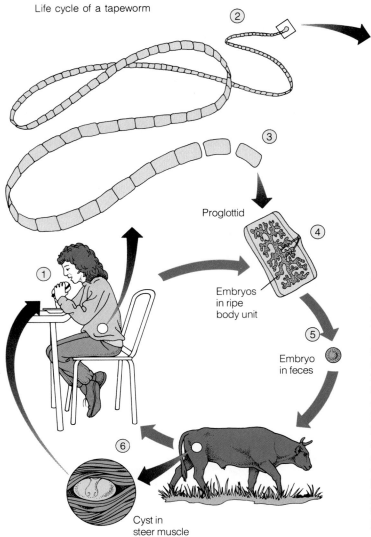

Proglottid

Embryos
in ripe
body unit

Embryo
in feces

Cyst in
steer muscle

FIGURE 18.11 ■ **Tapeworms: Appearance and Life Cycle.** The beef tapeworm infects about 60 million people worldwide. It can reach 19 m (60 ft) in length inside a person's intestines and is spread by the ingestion of raw or undercooked beef. The life cycle begins when the larvae are eaten (step 1). Each larva attaches itself to the host's gut and there develops into an adult (step 2). In the adult stage, tapeworms may cause the host to lose weight, suffer chronic indigestion, and have persistent diarrhea. The head of an adult beef tapeworm has a structure that attaches the parasite securely to the host's intestinal wall (see inset). Behind the head is a long string of many sections (proglottids), each containing reproductive organs (step 3). The many sections detach and are released in the person's feces (step 4); there the segments expel their eggs (step 5). If a cow eats food contaminated by egg-bearing human feces, the egg hatches in the new host, and the newly hatched organism penetrates the gut wall, enters the bloodstream, and bores into muscles (step 6). In the muscle, the parasite forms a cyst and develops into another larval form. If a person later eats undercooked beef containing the larva, it can once again infect a human host. If a person is accidentally infected directly with the parasite, steps 4 and 6 can take place in his or her tissues and lead to serious medical consequences if the larvae invade vital organs like the brain or eyes.

bodies of salt water (Figure 18.10a). Other flatworms, such as flukes and tapeworms, are parasites that exist on the host's body tissues or food reserves (Figure 18.11).

Bilateral Symmetry and Cephalization

Flatworms have **bilateral symmetry;** the right and left halves are mirror images, but the *anterior* and *posterior* (front and back) ends are different, and so are the *dorsal* and *ventral* (top and bottom) surfaces (Figure 18.10b). In addition, the flatworm displays **cephalization** (from the Greek word for "head"); one end functions as a head containing both a nerve mass that serves as a brain and specialized regions that can sense light, chemicals, and pressure. Since flatworms are bilaterally symmetrical instead of radial, when they move forward, the head region—with its brain and senses—encoun-

ters a new region first. Depending on the data collected by the head region, an animal can continue forward or back up and try a different direction. This evolutionary adaptation is so successful that heads occur in almost all animals more advanced than flatworms.

Distinct Tissues, Organs, and Organ Systems

Flatworms display two additional advances seen in all higher animals. They have three distinct tissue layers, and they develop true organs and organ systems. Recall that sponges and cnidarians have three-layered body walls, but the middle layer is a gelatinous material containing scattered cells. In flatworms, the middle layer, or mesoderm, is made up of living cells (not jelly) and lies between an outer cell layer, or epidermis, and an inner cell layer, the endoderm (Figure

18.10c). The mesoderm is important because it gives rise to muscles and other organs. **Organs** are structures made up of two or more tissues that function together, and flatworms have a number of organs that carry out digestion, movement, nervous responses, excretion, and reproduction.

Organs usually occur in **organ systems,** sets of organs with related functions, and flatworms have five such systems. The *digestive system* of the common planarian has a *pharynx,* or muscular tube, with a "mouth" at the end that can thrust out of the body during feeding, as well as a branched **intestine** lined with cells that absorb nutrients (Figure 18.10e). The intestine, with only one opening, is a blind tube; food enters and wastes exit through the same opening, the mouth. The *excretory system* rids the body of excess water and includes a network of water-collecting tubules adjacent to **flame cells** (Figure 18.10d). These saclike cells contain clusters of cilia that appear to flicker like flames; as the cilia move, they drive water into the tubules and out through pores in the body wall.

Planarians move by means of cilia on their external surfaces and by layers of contractile muscle cells that lie below the epidermis in a *muscular system.* Muscles receive signals from the *nervous system,* which includes the light-detecting cells of the eye, the brain, longitudinal "trunk lines," or **nerve cords,** and a network of lateral nerves (see Figure 18.10c and e). Finally, there is a *reproductive system.* Most flatworms have both testes and ovaries (thus, they are *hermaphroditic*), and they pair up and exchange both sperm and eggs with another individual.

A key to the flatworm's success is its flatness. Because it is so thin, oxygen and carbon dioxide diffuse straight through the thin layers to every cell, nutrients diffuse outward from the branching intestine to all body cells, and the animal needs no extra internal or external support.

Flatworm Life-Styles

Free-living flatworms, such as planarians, must hunt and avoid danger to survive, and their nervous systems and senses are especially crucial. Planarians have tiny **eyespots** that sense light and movement, as well as regions on the head that detect food (see Figure 18.10e). By contrast, parasites like flukes and tapeworms live a more sheltered life, with most of their needs provided for by a host. Parasitic flukes display prodigious reproduction. A larva of the Chinese liver fluke, for example, can infect a snail and reproduce annually, releasing a quarter of a million larvae. These, in turn, can infect a fish, and if the fish is insufficiently cooked, parasitize a person.

The blood fluke of Southeast Asia is responsible each year for 200 million cases of a human disease called **schistosomiasis,** which is second only to malaria in claiming human victims. The flukes live in aquatic snails at one stage of their life cycle and in humans at another. They penetrate human skin, migrate to the lungs, intestines, and other tissue, and cause fever, diarrhea, and pain in the stomach, back, groin, or legs.

Parasitic tapeworms are usually quite flat, with heads (each called a *scolex*) that are little more than knobs with hooks or adhesive suckers around the mouth that attach to host tissues (see Figure 18.11). Their bodies are made up of hundreds of units, and their reproductive organs are extremely prominent and active. Since shedding eggs and larvae is the only means that flukes and tapeworms have to reach new hosts and continue their life cycle, they must produce thousands; this helps ensure that a few will enter new hosts and survive. The tapeworms that infect cows' intestines have 800 to 900 body units filled with reproductive organs, and they shed embryos that bore into the cow's muscles and form protective cysts (see Figure 18.11). If a person consumes undercooked beef bearing these cysts, the cysts can grow into new tapeworms in the person's intestines.

In the earth's tropical zones, parasitic flatworms, with their tremendous capacity for reproduction, are major sources of debilitating diseases for people and their livestock. Their free-living cousins have little direct impact on people, but the group, overall, has had great evolutionary importance. Biologists have reason to believe that a small, bilaterally symmetrical creature similar to a flatworm probably gave rise to all the more complex animal groups.

ROUNDWORMS: ADVANCES IN DIGESTION

In sheer numbers of individuals, the most abundant animals on earth are **roundworms,** members of the phylum Nematoda. Roundworms, as the name implies, are round in cross section rather than flat, and most are very small; many, in fact, are microscopic (Figure 18.12). A cubic meter of rich soil can contain 3 billion nematodes, and biologist A. M. Cobb once

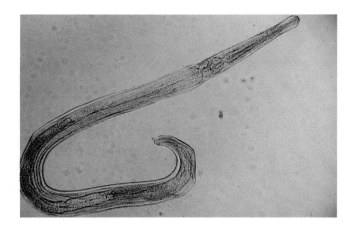

FIGURE 18.12 ■ The Roundworm: Phylum Nematoda. Nematodes thrive in a wide variety of damp habitats and were the first animals to have a body cavity. Here is the nearly transparent pinworm *Enterobius vermicularis,* which infects humans.

said that if all our planet's lands and seas were swept away but the nematodes somehow stayed in place, a clear outline of the earth and its geological features would remain. Nematodes thrive in mud, sand, soil, running or standing water— even in acidic fruit juices—and have been found in habitats from boiling hot springs to the Antarctic Ocean and the top of Pike's Peak. When the environment grows harsh, nematodes can curl up, dry out, and shut down metabolically for up to 30 years. In this "latent" state, they can withstand extremes of heat, cold, radiation, and chemicals that would quickly kill most other animals. Then, when conditions improve and water is once again present, the animals rehydrate and revive: instant nematode! While many roundworms are free-living, hundreds of the 12,000 or more nematode species are plant and animal parasites that damage crops and cause diseases.

Roundworm Advances: A Fluid-Filled Body Cavity and a Two-Ended Gut

The roundworm's success is often attributed to two new characteristics. While they share bilateral symmetry and cephalization with the flatworms, roundworms also possess an unlined body cavity called a **pseudocoelom** ("false body cavity"), as well as a one-way tubular gut, or digestive tract, with openings at both ends—the mouth and anus.

Sponges, cnidarians, and flatworms have no spaces between their two or three tissue layers; the layers are solidly packed together and surround the central digestive cavity like a single envelope (Figure 18.13). With the unlined body cavity (pseudocoelom), however, the roundworms show the start of a trend toward a true fully lined body cavity, or **coelom,** seen in more complex organisms. Roundworms have three tissue layers, but a space lies between the innermost layer and the outer two. All animals more complex than roundworms have a fluid-filled space not between two body layers but within the middle layer, the mesoderm. This space, lined on all sides by mesoderm, is a true coelom. The advantages of a coelom (or pseudocoelom, which works basically the same way) are fourfold. First, because of this space within the body cavity, the reproductive and digestive organs can evolve more complex shapes and functions. Second, in a fluid-filled chamber, the gut tube and other organs are cushioned and thus better protected. Third, since a liquid cannot be compressed, the pseudocoelom (or true coelom, when present) can act as a **hydroskeleton** (literally, "water skeleton"), providing support and rigidity for the soft animal. Finally, the activities of a suspended gut can take place undisturbed by the activity or inactivity of the animal's outer body wall.

One can easily see certain benefits of this architecture in the roundworms. The roundworm's gut has two separate openings, the anterior mouth and posterior anus. Food enters the mouth and moves in just one direction through the gut tube, with wastes exiting the anus. Regions of specialized function developed along the gut for grinding food into small pieces, breaking it down with enzymes, absorbing nutrients and water, and expelling wastes. Over time, roundworms

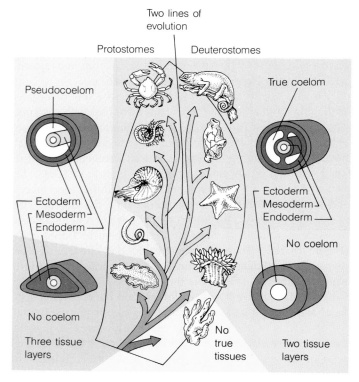

FIGURE 18.13 ■ Evolution of the Coelom. More complex phyla evolved three tissue layers with a split in the middle, the true coelom (upper right). This mesoderm-lined, fluid-filled space cushions internal organs. The figure traces the evolution of the coelom. Sponges have no coelom, nor do cnidarians like sea anemones and jellyfish, although cnidarians do have two tissue layers (lower right). Flatworms also lack a coelom, but they do have a complete set of three tissue layers (lower left). The roundworms form a false coelom (pseudocoelom), a cavity not completely lined with mesoderm (upper left). The more complex phyla (annelids, mollusks, arthropods, echinoderms, and chordates) have a true coelom (upper right). Note how the path of animal evolution splits into two major lines of descent, protostomes (left trunk) and deuterostomes (right trunk).

became highly efficient digesters, capable of consuming a wide variety of foods and therefore of inhabiting environments all over the world.

Impact of Roundworms on the Environment

Many types of free-living roundworms help consume rotting plant and animal matter and thus are ecologically important decomposers, like the bacteria and fungi. However, the parasitic roundworms get most of the fame—or infamy. At least 1000 nematode species parasitize plants, and some observers estimate that they consume fully 10 percent of all crops annually. Nearly 50 species parasitize people, entering in food or contaminated water or through bare skin. They cause a list of diseases, including trichinosis (from eating undercooked worm-infested pork), ascariasis (a common disease in tropical regions, characterized by lung infections and intestinal blockages due to masses of worms), hookworm (an infestation in the internal organs, also common in the tropics), and elephantiasis (Figure 18.14). Clearly, much of the nematode's evolutionary success comes at the expense of other organisms, ourselves included.

TWO EVOLUTIONARY LINES OF ANIMALS

Our discussion so far has seemed roughly linear—a history of increasing physical complexity in successive animal groups, as summarized in Table 18.1 on page 381. It may seem logical, therefore, to assume that evolution is also linear, with each group giving rise to the next more complex set of animals. Zoologists, however, think that evolution was not linear. They believe that an ancient progenitor similar to a free-living flatworm gave rise both to the roundworms and to all the other bilateral phyla. Zoologists also think that a branching off took place very soon after bilateral organisms arose, leading to two great animal lineages (see Figure 18.13). The first branch contains only invertebrates, including the next three phyla we will consider, the mollusks, annelids, and arthropods. The second branch includes the echinoderms and our own phylum, the chordates (see Chapter 19). The two branches are distinguished by very different patterns of early embryonic development.

Recall that during animal development, the embryo forms a hollow ball (the blastula). This ball then indents, with the infolding cells becoming the digestive tract (see Figure 13.12). In the evolutionary branch that contains only invertebrates, the initial indentation becomes the mouth—hence this evolutionary line is called the **protostome** ("first mouth")

line—and the second indentation becomes the anus. In the other evolutionary line—the one that yielded the vertebrates—the initial indentation becomes the anus, and a second opening becomes the mouth; biologists call organisms with this pattern the **deuterostome** ("second mouth") line. The two lines of descent also show differences in the way the eggs cleave and the way the body cavity (coelom) forms. Such differences provide the kind of evidence biologists use to understand the pathways of animal evolution.

Let's look at each of the diverging evolutionary roads, beginning with the protostomes.

MOLLUSKS: SOFT-BODIED ANIMALS OF WATER AND LAND

The phylum Mollusca, or **mollusks,** is a collection of more than 120,000 species of soft-bodied animals, including snails, slugs, clams, oysters, squid, and octopuses. All but the land snails and slugs are aquatic. Although mollusks are soft, some members secrete a hard shell; therefore, unlike the sketchy histories left by the groups we've covered so far, a fairly good fossil record of the mollusks remains. The phylum probably arose from wormlike, bilaterally symmetrical ancestors sometime before the early Cambrian (review Figure 15.13).

Molluscan Body Plan

You may have the impression from eating raw oysters that mollusks are slimy blobs with no distinct shape. However, there is an identifiable body plan to all mollusks, no matter how different or soft they look inside their shells. As Figure 18.15 shows, each mollusk has a **foot,** a muscular organ used

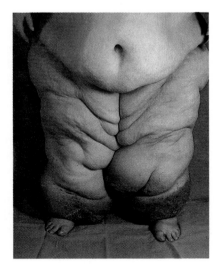

FIGURE 18.14 ■ **Elephantiasis. Grotesque limb enlargement can occur when roundworms block lymphatic vessels (see Chapter 21) and the tissues accumulate fluid. This woman with elephantiasis has hugely swollen legs but normal feet.**

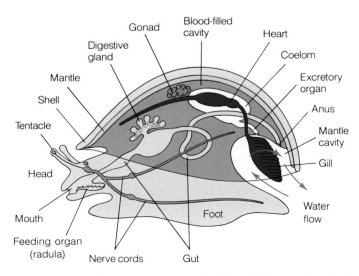

FIGURE 18.15 ■ **The Molluscan Body Plan. This generalized creature has the foot, head, mantle, and other major molluscan organs.**

for gripping or creeping over surfaces; a **head** housing the mouth, brain, and sense organs; a visceral mass containing internal organs (heart, gut, sex organs, and excretory and respiratory apparatus); and a **mantle,** a thick fold of tissue that covers the visceral mass and in some mollusks secretes the calcium carbonate that makes up the hard shell. The space between the mantle and the visceral mass is called the **mantle cavity,** and suspended in this space are **gills,** special surfaces for exchanging oxygen and carbon dioxide dissolved in water.

Molluscan Advances

The mollusk's respiratory gills function in tandem with a **circulatory system** that includes a **heart** with chambers for pumping blood; **vessels,** or blood-carrying tubes, that pass through the gills; and a blood-filled cavity that bathes internal organs and tissues (see Figure 18.15). The circulatory system is considered an *open circulatory system* because the blood flows through the open spaces part of the time and through vessels the rest of the time. As blood passes through the gills, it releases carbon dioxide and picks up oxygen; the blood then carries its oxygen cargo away from the gills and toward all internal cells. Together, the respiratory and circulatory systems provide each cell in the body with oxygen and nutrients in an efficient manner, and they have allowed many mollusks to reach large sizes; some, as we will see, are veritable monsters of the deep.

Although many are simpler, some mollusks have well-developed nervous systems with large brains and acute senses. Most members of the phylum have a special feeding organ called the **radula** (see Figure 18.15). This strap-shaped structure bears rows of tiny teeth and works something like a cheese grater, rasping off successive layers of food quickly. Finally, most mollusks produce highly mobile fringed larvae (*trochophores*); their similarity to certain larvae of the next animal group, the segmented worms, suggests that as different as they look, mollusks and annelids are related historically.

Some Classes of Mollusks

■ **Gastropods**   Snails, garden slugs, and sea slugs, or **nudibranchs,** are all **gastropods,** members of the class Gastropoda (meaning "belly foot"). Most snails are identifiable by their coiled shells, as well as by *torsion*—an internal coiling or twisting of the body mass during embryonic development. Torsion causes the anus and gills to lie directly over the head, but other modifications during development prevent self-pollution. Land snails and slugs are the only terrestrial mollusks, and they move slowly on their ventral feet along glistening **slime trails** (Figure 18.16a). While most aquatic snails have gills, land snails have air-breathing lungs, and they can close off their shell opening to retain moisture during dry weather. Some snails and most other gastropods have two antennae and two eyes. When a shelled gastropod senses danger, it can rap-

(a) A land snail

(b) A nudibranch

FIGURE 18.16 ■ **Gastropods: Snails, Slugs, and Nudibranchs.** **(a) Gastropods can be drab and familiar garden pests like this Hawaiian land snail, sliding across its slime trail. (b) The soft, shell-like mollusks called nudibranchs are often dazzlingly colored and patterned, such as this animal, the Elegant eolid (*Elabellinopsis iodinea*) of California's Monterey Bay.**

idly withdraw into the space, or mantle cavity, inside the shell. Many slugs and nudibranchs—fantastic-looking, vividly colored sea creatures—lack a shell completely (Figure 18.16b). Slugs protect themselves by remaining hidden beneath leaf litter or in other damp spots. Nudibranchs often produce sour or bitter substances that repel attackers. Like the strawberry frogs described in Chapter 1, their bright colors serve as a warning that a bad meal lies ahead for the predator.

■ **Bivalves**   The word **bivalve** means "two valves" and refers to the two half shells that enclose each member of this class: the oysters, clams, mussels, scallops, and relatives. Strong muscles close the valves and stretch an elastic ligament at the hinge. When the muscles relax, the shell snaps open. Rapid closing of the shell protects the animals from predators and in scallops also produces a self-propelling water jet. When the half shells are open, particles can enter and irritate the soft tissue. In oysters, a region of the mantle responds to injury or foreign particles by secreting layers of nacre—mother-of-pearl—around the offending object. Pearls are the result of the reddish, whitish, or black nacre deposits (Figure 18.17).

Bivalves are **filter feeders;** they have gills that, in addition to collecting oxygen and releasing carbon dioxide, can strain out and collect tiny food particles suspended in water. Beating cilia draw water across the gills, and a mucous layer traps the particles, which are then passed into the mouth. One biologist has estimated that each week, all the water in San Francisco Bay is drawn through the bivalves that lie half-buried in the silt and sand at the bottom of the bay!

■ **Cephalopods**   The squid, the octopus, the chambered nautilus, and other **cephalopods** (literally, "head feet") are the most complex mollusks and evolved as fast-swimming predators of the deep sea (Figure 18.18). In these creatures, the foot is modified: It bears a circle of eight or ten arms, each studded with suckers, and it terminates in a funnel, or **siphon.** Thus, a single organ, the foot, has become specialized for land travel in the gastropods and for hunting, swimming, and feeding in the cephalopods. The cephalopod mantle is also modified into a muscular enclosure; this can expand and draw water into the mantle cavity or contract and force it out of the siphon, jet-propelling the mollusk backward. These explosive bursts can carry the animal to safety or bring its suckered tentacles within reach of prey. Once captured, the prey is moved toward the mouth, which contains a pincerlike beak and a razor-sharp radula.

The coordination for hunting and feeding in cephalopods depends on acute senses, a large brain, and the most complex nervous system among the invertebrates. The largest cephalopods, the giant deep-sea squid, can weigh 450 kg (1000 lb) and reach nearly 18 m (60 ft) in length. Giant squid have the largest eyes in the animal kingdom—highly sensitive organs that can form images like our own but grow larger than a car's

(a) An octopus

(b) A nautilus

FIGURE 18.18 ■ **Cephalopods: Predators of the Deep.** Two of the most complex mollusks are (a) the tentacled lesser octopus (*Elgolone cirrhoser*), which lives in dark crevices in the Pacific Ocean floor, and (b) the chambered nautilus, with its coiled shell coated with nacre and divided internally by cross walls. Oliver Wendell Holmes called the nautilus the "ship of pearl."

FIGURE 18.17 ■ **Oyster and Pearl.** Some mollusks line their shells with nacre—calcium carbonate plus a gluelike protein called conchiolin. Nacre is what we call mother-of-pearl, and when an oyster secretes the material around a small irritant inside its shell, the result can be a lustrous jewel.

headlights. The octopods, with their radiating arms, may look sinister to some and dim-witted to others, but they actually have massive brains containing more than 170 million nerve cells. They can be trained to accomplish a number of tasks and are the most intelligent creatures without backbones. While members of the next phylum are far from intelligent, they and their descendants, the insects, are highly successful and show many innovations.

ANNELID WORMS: SEGMENTATION AND A CLOSED CIRCULATORY SYSTEM

Marine sandworms, common earthworms, and leeches are all members of the phylum Annelida, the **segmented worms,** whose 15,000 species belong to three classes. Most are members of the class Polychaeta (meaning "many bristled")—often colorful marine worms that burrow in the mud or sand and bear common names such as fireworms, clam worms, and feather dusters (Figure 18.19a). The familiar reddish earthworms are in the class Oligochaeta ("few bristles"). These ubiquitous inhabitants of moist soils, often numbering 50,000 or more per acre, literally eat their way through dense, compacted earth, excreting the displaced material in small, dark piles. Each year, earthworms carry as much as 18 tons of rich soil per acre to the surface.

The leeches are in a separate class, Hirudinea, and live mostly in fresh water. Many prey upon worms and mollusks, but the most infamous types parasitize large animals and suck their blood. A leech can consume three times its weight in blood and go for as long as nine months between meals (Figure 18.19b).

Annelid reproduction is generally sexual. Many marine annelids shed sperm and eggs into seawater, where the gametes unite and develop into trochophore larvae (which are similar to mollusk larvae). Earthworms and leeches are hermaphrodites; pairs reciprocally fertilize each other.

Annelids Show All Five Evolutionary Trends

The term *annelid* means "tiny rings" and refers to the external segments visible on members of this phylum. Annelids are the first group possessing all five physical traits we have been tracing: bilateral symmetry, cephalization, a tubular gut, a coelom, and segmentation. Although more than a million animal species evolved after the annelids, no new trends as basic as the emergence of these features occurred; all later animals, including humans, show variations on the anatomical themes first "stated" together in the annelids.

Earthworms—typical annelids—have 100 or more body segments (Figure 18.20a). Each segment is separated from the next by an internal partition, or **septum** (plural, septa), and each segment contains a set of internal structures. Most segments contain two excretory units called **nephridia** (Figure 18.20b). Each nephridium removes excess water and wastes from the body fluids by means of a ciliated funnel (reminiscent of the flatworm's flame cell) and excretes them through a pore in the segment wall. Each segment also contains a fluid-filled compartment of the coelom and is surrounded by circular and longitudinal muscles in the body wall. As the circular muscles squeeze against the incompressible fluids in the coelom and gut, force is transmitted to adjacent segments, creating a hydroskeleton, an internal skeleton made of fluid. Contractions in sequential segments produce waves of force that propel the animal forward. **Setae,** which are not true legs but pairs of bristles attached to each segment, push against the ground with each contraction and help the animal move. In addition to nephridia, muscles, and coelom compartments, each segment contains clusters of nerve cells connected to the brain by nerve cords.

(a) A polychaete annelid

(b) A leech

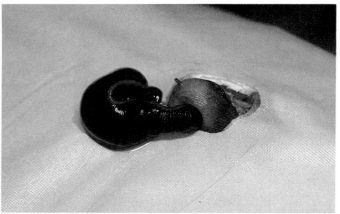

FIGURE 18.19 ■ Polychaetes and Leeches. Segmented worms range from the delicate to the despised, including (a) this feathery Christmas tree worm (*Spirobranchus giganteus*) and (b) this larger black leech. When surgeons reattach severed human fingers, they occasionally use laboratory-raised leeches during the patient's recovery. The leeches suck blood from the tissues around the reattachment sites, release anesthetics and anticoagulants, and thereby relieve pressure and decrease pain. This photo shows the tip of a reattached finger, projecting through a sterile plastic surgical drape.

More Annelid Characteristics

Organ systems for digestion and circulation run the length of the annelid worm and represent further evolutionary advances. The gut tube has two specialized regions not seen in roundworms: a **crop** for storing food and a **gizzard** for

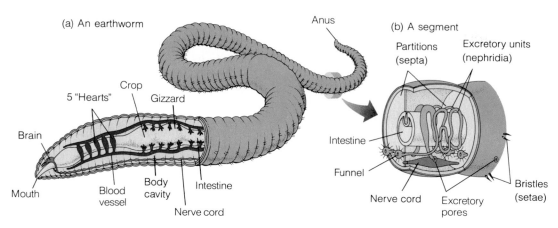

FIGURE 18.20 ■ **Earthworms: A Study in Segments.** **(a)** Anatomy of an earthworm, the most familiar annelid. **(b)** An enlarged view of a segment.

grinding it. After the gizzard has ground some food, the region called the intestine absorbs nutrients released from the food by digestive enzymes (see Figure 18.20a).

Annelids also have a *closed circulatory system:* blood carried entirely in tubes, or vessels, rather than partially in open sinuses. The contraction of several hearts, or muscular vessels, pumps the blood continuously through the closed circuit, and tiny vessel branches carry blood close to each cell in the body. The digestive and circulatory systems are interlinked, and blood passing through the gut picks up nutrients and transports them to all tissues. In turn, the blood picks up metabolic wastes from the tissue cells and transports them to the body cavity (coelom), from which they can be eliminated by the nephridia.

Advances in segmentation, digestion, and circulation allowed the annelids to grow longer and thicker than previous worms. Segmentation also had a special evolutionary significance: The descendants of the segmented worms, the arthropods, became the immensely successful group they are largely because of their segments and the appendages for chewing, sucking, flying, and running that developed from them.

ARTHROPODS: THE JOINT-LEGGED MAJORITY

The *Nasutitermes* termites introduced at the beginning of this chapter are members of the largest animal group on earth: the phylum Arthropoda. Arthropods include fossil trilobites, spiders, mites, ticks, scorpions, centipedes, millipedes, lobsters, crabs, and insects. Hard as it is to believe, the insects, numbering at least 1 million species, make up the great majority of all animal species (Figure 18.21). In fact, if you took a book that listed every animal species and pointed to any name at random, it would probably be an insect. And as Box 15.2 on page 335 explains, there may be many, many more insect spe-

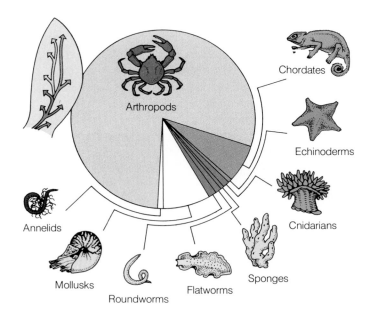

FIGURE 18.21 ■ **Arthropods: The Vast Majority of All Animal Species.** Compared to the insects and other arthropods, all other species make up a small slice of the animal kingdom. If the projections of 10 to 30 million total tropical insect species are correct, the rest of the animal kingdom would, in fact, be a microscopic slice! Our era could be called the Age of the Arthropods.

cies as yet undiscovered—perhaps 10 to 30 million more! The insects' astounding diversity and success are based on several characteristics shared by all arthropods: an external skeleton; modified, specialized segments; rapid movement and metabolism due to special respiratory structures; and acute sensory systems.

The Arthropod Exoskeleton

As animals—particularly land animals—grew larger and thicker, they needed support for the body. The key to the success of insects and other arthropods is the **exoskeleton,** or external skeleton, that completely surrounds the animal and provides strong support as well as rigid levers that muscles can pull against. In arthropods, the epidermis secretes a thick, hard cuticle that contains the polysaccharide chitin (see Chapter 2) in addition to sugar-protein complexes, waxes, and lipids that make the body covering waterproof. This mixture of substances makes a tough but somewhat flexible outer barrier for most arthropods that protects and supports the animal's soft internal organs. In crustaceans, such as crabs, lobsters, and shrimp, the exoskeleton also contains calcium carbonate crystals, which make it a hard, inflexible armor. Besides providing shieldlike protection from enemies and resistance to general wear and tear, the exoskeleton also prevents internal tissues from drying out. This is extremely important, since most arthropods live on land.

In all arthropods, the exoskeleton remains thin and flexible at the **joints,** the hingelike areas of the legs and body. The term *arthropod,* in fact, means "jointed foot." The presence of jointed appendages allows arthropods to move quickly and efficiently above the ground or seafloor instead of dragging the body directly along the ground on stubby legs or bristles. Muscles attached to the inside of the exoskeleton on either side of the joints provide the leverage needed to move each appendage. Exoskeletons do, however, have a major disadvantage: The animal cannot grow larger unless it **molts,** periodically shedding its constricting armor and producing a larger exoskeleton.

Specialized Arthropod Segments

While annelid worms have the same simple segments repeated throughout the body, arthropods have evolved different and highly modified segments that give them a far greater repertoire of activities. Their segments, first of all, are usually fused into a few major body regions; insects, for example, have three regions—the head, **thorax,** and **abdomen**—whereas in spiders and crustaceans, there is an abdomen plus one region with fused head and thorax (Figure 18.22a). During evolution, specific appendages for walking, swimming, and flying arose, with each species having its own modifications. Some arthropods also developed pincers or palps (feelers), which facilitate hunting and feeding. And from the head region grew other appendages: **mouthparts** that allow chewing and sucking and **antennae** that sense odors and vibrations. The brightly colored shrimp in Figure 18.22a, for example, has appendages modified for eating (mouthparts), grasping (pincers), walking (legs), mating (swimmerets), and fast swimming (tail). The appendages themselves became further modified during evolution. Just compare the flapping wings of the butterfly (Figure 18.22b) and the small buzzing wings of the housefly. The many modifications enabled the arthropods to exploit a wide variety of environments.

(a) A shrimp

Fused head and thorax   Abdomen

(b) An insect

Head   Thorax   Abdomen

FIGURE 18.22 ■ Arthropod Appendages: Modified for Different Tasks. (a) This brilliantly colored scarlet cleaner shrimp (*Hippolysmata grabbhami*) of the tropical western Atlantic has a fused head and thorax and a separate abdomen, as well as appendages modified for grasping, walking, mating, and swimming. (b) The giant swallowtail (*Papilio cresphontes*).

(a) Insect's tracheae

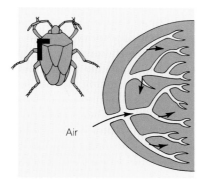

Air

(b) Lobster's gills

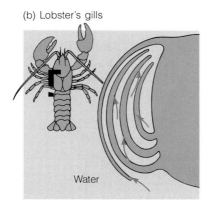

Water

FIGURE 18.23 ■ Arthropod Respiratory Organs. (a) An insect's tracheae are branching hollow tubes that carry air deep into the body tissues. (b) The gills of a lobster, an aquatic crustacean, also have a large surface area, but project outward to contact the water or moist air.

Arthropod Respiratory Advances

Special respiratory structures allow arthropods to metabolize and move rapidly, and in fact, some insects do both at the highest rates in the animal kingdom. Some tiny flies can generate enough metabolic energy to beat their wings 1000 times per second. High metabolic rates require rapid oxygen delivery, and the arthropods' efficient respiratory organs provide a large surface area for collecting oxygen and releasing carbon dioxide quickly: Insects have **tracheae,** branching networks of hollow air passages (Figure 18.23a); spiders and relatives have **book lungs,** chambers with leaflike plates for exchanging gases; and finally, aquatic arthropods have gills, flat tissue plates that act as gas exchange surfaces (Figure 18.23b). High metabolic rates and rapid movements allow arthropods to fly, run, and swim faster than any previous animal group and hence to escape predators with agility and to disperse over wider ranges.

Acute Senses

Most arthropods have antennae capable of detecting movement, sound, or chemicals with great sensitivity. The antennae of several male moth species, for instance, can detect the *pheromones,* or odor signals, given off by adult female moths at distances of over 11 km (about 7 mi). Arthropods also have various types of separate organs on the head and body that detect sound and taste, and many have special compound eyes made up of 2500 or more six-sided segments called **facets** (Figure 18.24). Many compound eyes are capable of color vision and can detect the slightest movements of prey, mates, or predators.

Important Arthropod Classes

The successful and highly divergent phylum Arthropoda is divided into a number of taxonomic classes, most of which will sound quite familiar.

■ **Centipedes and Millipedes**   The 3000 centipede species, members of the class Chilopoda, look a lot like annelid worms and have a series of flattened body segments. However, each segment bears a pair of true jointed legs (not bristles) that move the centipede swiftly in search of insects, worms, small mollusks, or other prey. Centipedes kill their prey with **poison claws,** modified legs on the first body segment (Figure

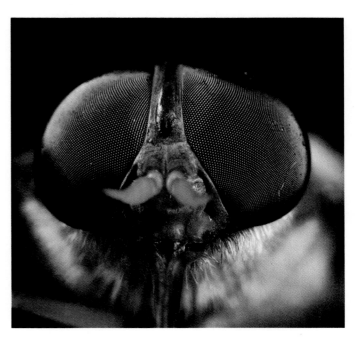

FIGURE 18.24 ■ The Many Facets of an Arthropod's Eye. The compound eye of this horsefly has hundreds of geometric facets. Each facet detects light and movement, and the insect sees a composite image created by the many individual segments.

(a) Centipede

(b) *Peripatus*

FIGURE 18.25 ■ Marchers of the Desert and the Forest Floor: A Centipede and *Peripatus*. Each of these animals has many segments and marches along on an army of small legs. (a) Centipedes have poison claws, while (b) this velvety *Peripatus* from Costa Rica ensnares its prey with gluey strands. *Peripatus* belongs to a phylum that shares traits with two other phyla, the arthropods and the annelids.

18.25). Some huge tropical centipedes are dangerous to humans, but the common varieties that lurk in damp basements are harmless and consume insects.

Millipedes, in the class Diplopoda, are slow-moving counterparts to the centipedes. They have two pairs of much smaller legs per segment; a round, not flattened, body; and a preference for decaying vegetable matter instead of live prey. Both millipedes and centipedes bear some resemblance to the members of a fascinating but obscure phylum, *Onychophora*. The few living members of this group, including the velvety *Peripatus* of tropical forests, are true missing links: They exhibit the unjointed legs of the segmented worms and the tracheae and molting displayed by arthropods.

■ **Crustaceans**   The 31,000 species of crabs, shrimp, lobsters, barnacles, crayfish, and sowbugs in the class Crustacea are so different from each other that only two generalizations apply: Almost all have an exoskeleton hardened with calcium salts

that covers most of the animal as a protective shell, or **carapace,** and all have two pairs of antennae.

The lobster, a familiar crustacean, has two major body regions: the anterior **cephalothorax,** which is covered by the carapace, and the posterior abdomen (Figure 18.26). Walking legs and pincers sprout from the cephalothorax, while the abdomen bears feathery swimming appendages.

Crustaceans are so numerous and so diverse that most bodies of water contain them. There are even a few terrestrial species, like the sowbugs. Many small aquatic species, such as fairy shrimp, brine shrimp, and copepods, serve as the primary food for various fish, whales, and other species. Barnacles, an aquatic species with a contrasting life-style, cling tenaciously to tidal-zone rocks; the natural cement they secrete is strong enough to keep the animals from washing away in pounding ocean surf—a feat comparable to a person standing upright in winds of 480 to 640 km per hour (300 to 400 mph). Some biologists are studying barnacle glue in hopes of developing medical cements for fastening dentures and rejoining broken bones.

■ **Spiders and Relatives**   Class Arachnida includes 55,000 species of **spiders,** ticks, mites, and scorpions. Arachnids belong to the subphylum Chelicerata, which also includes the extinct trilobites, the sea spiders, and the primitive-looking horseshoe crabs often seen on Atlantic coastal beaches. Arachnids lack antennae, have a cephalothorax and abdomen region, each divided into segments, and usually have six pairs of appendages (Figure 18.27). The first pair of appendages is modified into **chelicerae,** or poison fangs, used for killing prey or self-defense. The second pair holds the prey while the spider injects poison or enzymes. The other four pairs are walking legs. Spiders also have silk-spinning organs called

FIGURE 18.26 ■ The Blue Lobster: A Vividly Representative Crustacean. The lobster *Homarus americanus,* a denizen of the Atlantic Ocean and shown here in rare blue phase, can grow to 1 m (39.4 in.) in length and weigh 18 kg (40 lb). Like the shrimp (see Figure 18.22a), it has a fused head and thorax, as well as several sets of specialized appendages. Its mighty pincers help in food gathering and defense.

FIGURE 18.27 ■ Arachnids: Spiders and Relatives. This banded garden spider is fearsome but harmless to humans; it weaves a broad web and preys on small insects. Here, one has ensnared a grasshopper.

spinnerets at the rear of the abdomen; these reel out threads that are, size for size, stronger than steel. Despite their sinister reputation, most spiders are only capable of poisoning small animals. With their potent poisons, however, the black widow and brown recluse can be dangerous to people. Present-day scorpions have a body form very similar to the huge water scorpions that lived 500 million years ago and that probably gave rise to the land scorpions and spiders. Mites and ticks are, for the most part, merely irritating biters, but some ticks carry serious diseases, including Lyme disease and Rocky Mountain spotted fever, with its fever, rash, and joint pain.

■ **Insects** It is difficult, in a few paragraphs, to do justice to **insects,** the largest—and, by that measure, most successful—class of animals on earth. Most insects are terrestrial animals living in habitats from the tropics to the poles, but a few are aquatic. Many zoologists attribute insect success to the general arthropod traits we've already considered, but also to the insect's small size. Smallness enables the insects to exploit a vast array of microhabitats—the bark of a tree, the planar landscapes on the backs of leaves, the dense thickets of an animal's fur, or the universe of midair.

Specific body parts also help account for the insects' success. Many insects have organs for smelling, touch reception, tasting, seeing, and hearing in various parts of the body. An insect's head bears one pair of antennae; its thorax bears three pairs of legs and usually one or two pairs of wings; and its abdomen is usually free of appendages. A series of modified segments, the mouthparts, enables the insect to feed efficiently. Some, like mosquitoes, have superb pointed stylets for piercing and sucking. Others, like locusts and grasshoppers, have chewing mouthparts that can quickly decimate foli-

age (Figure 18.28). In 1985, a long, dark column of grasshoppers descended upon Idaho's Magic Valley, eating a swath about 3 km (2 mi or so) wide through the bean, beet, and potato fields. In one day, seven grasshoppers can eat as much as a grazing cow; and during a plague, they can occur 1800 to the square meter.

Insects have evolved various ways to grow, despite the confining exoskeleton, and various ways to thrive, despite the changing seasons. In some insects, like grasshoppers and cockroaches, the embryo becomes a miniature version of the adult without wings or mature reproductive organs. This organism feeds, grows, and molts five or six times until it reaches adult size, then does not molt again. In most insects, however, including butterflies, flies, and beetles, the embryo develops into an immature form, or *larva,* eats voraciously, then forms a transitional stage, or **pupa,** sometimes inside a cocoon. A complete change, a **metamorphosis,** takes place in the body within the pupal exoskeleton. Finally, a nonmolting, reproductively mature adult emerges (see Figure 26.1). In insects that metamorphose, the larvae and adults can be adapted to very different foods and environmental conditions. This successful evolutionary solution allows larvae to specialize in feeding and obtaining energy, and the adult to specialize in reproduction.

Perhaps the most fascinating of the arthropods are the **social insects:** the termites, ants, wasps, and bees. Most species of these insects live in large colonies with labor divided among **castes,** or subgroups, that differ in appearance and behavior. Termites, such as the *Nasutitermes* species, have the most complex social life of all the insects. There is a sexual caste of kings and queens, a worker caste with sterile males and females, and a soldier caste. Some soldiers have huge, biting mouthparts, and others have the conical, bazooka-shaped heads that squirt defensive chemicals on intruders (see Figure 18.1). Such insect colonies are highly successful. Ants, for example, may make up one-third the weight of all the ani-

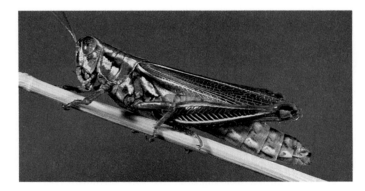

FIGURE 18.28 ■ Insects: Mostly Terrestrial and Highly Diverse. Grasshoppers like this common member of the genus *Melanoplus,* with its hearty appetite and chewing mouthparts, can denude a farmer's field in minutes.

mals on earth, with 200,000 ants for every person now alive! Insect colonies are highly evolved, functioning, in a sense, as a single well-coordinated organism with the capacity to simultaneously build homes and cities, defend their own, harvest food, and reproduce. When you consider that the roots of the animal kingdom lie in Precambrian creatures no more complex than sponges, cnidarians, and flatworms, social insects like ants or *Nasutitermes* termites demonstrate vividly the great evolutionary distance the invertebrates traveled.

ECHINODERMS: THE FIRST ENDOSKELETONS

An expert on invertebrates once called the echinoderms "a noble group especially designed to puzzle the zoologist." The 7000 members of the phylum Echinodermata, including the sea stars, brittle stars, sea urchins, sea cucumbers, and sea lilies, have an odd mixture of traits, some of which are innovations appearing in all higher animals and some of which are apparent regressions toward simpler forms. Besides their bizarre and interesting life-styles and appearances (Figure 18.29; also see Figure 13.12c), the **echinoderms** display two new features: an internal skeleton and a hydraulic pressure system that facilitates locomotion and is unique in the animal kingdom. Echinoderms are also deuterostomes with a developmental pattern that sets them off from the protostomes and other invertebrates and suggests an ancestry common to our own phylum, Chordata.

The word *echinoderm* means "spiny-skinned," and virtually all members have spines, bumps, spikes, or unappetizing projections that help to protect these slow-moving marine creatures from predators. Echinoderm larvae are bilaterally symmetrical, but adults of many species are radial, headless, and brainless. Nerve trunks run along each of the adult's arms (usually numbering five or multiples of five) and unite in a ring structure around the mouth (Figure 18.30a). This simple system allows coordinated—but very slow—movement of the limbs; a sea star travels only 15 cm (6 in.) per minute.

When echinoderms reproduce sexually, they shed sperm and eggs directly into the ocean, and the resulting fertilized eggs develop into larvae. But their most remarkable means of reproduction is by regeneration. A single sea star arm with a small piece of the central hub still attached can regenerate an entire body. A sea cucumber, when irritated, will sometimes violently expel its internal organs, leaving intact only small fragments, and then regenerate the lost parts.

The radial body plan is not an advanced trait, nor is the absence of a brain. Echinoderms also lack excretory and respiratory systems, relying mainly on diffusion for all exchanges with the outside. However, they do show one important evolutionary development: an **endoskeleton,** or true internal support system. Just below the outer epidermis

(a) Sea lily

(b) Sea cucumber

FIGURE 18.29 ■ Echinoderms: Spiny-Skinned Ocean-Goers. (a) Echinoderms such as this sea lily (*Cenometra bella*) from Palau in the South Pacific can resemble bright flowers blooming on coral reefs. (b) Or their camouflages can allow them to blend in like this sea cucumber living in the warm shallow waters off the coast of Fiji.

lie calcium-based plates that protect the internal organs and give rise to the outer spines and bumps. (We will consider other types of endoskeletons in Chapter 29.)

Echinoderms have a large coelom, and in many species, it is divided to form the unique **water vascular system,** a set of canals filled with modified seawater (Figure 18.30b and c). Hundreds of short branches off the canals become the tube feet that dot the ventral surface of a sea star, for example, and enable the animal to move across the seafloor. A bulblike portion of each tube foot can contract and force water from the canals into the foot. This pressure causes the foot to extend and a terminal sucker to grip the rock or sand below. This extension, adhesion, then release of countless tube feet moves the animal along. A sea star can also force open a tightly shut

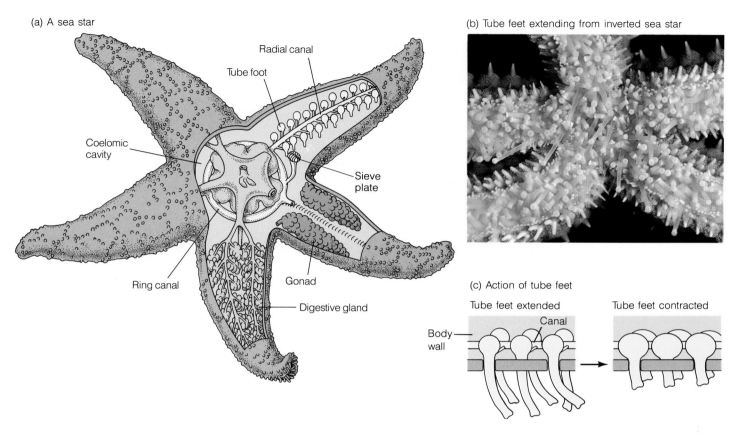

(a) A sea star

Radial canal

Tube foot

Coelomic cavity

Sieve plate

Ring canal

Gonad

Digestive gland

(b) Tube feet extending from inverted sea star

(c) Action of tube feet

Tube feet extended

Tube feet contracted

Body wall

Canal

FIGURE 18.30 ■ Sea Stars and Hydraulic Tube Feet.   (a) Sea stars have radial bodies with 5 to 25 arms or more, internally branching digestive pouches, and well-developed gonads. Most remarkable are the hundreds of tube feet on the ventral surface. (b, c) The tube feet can jut outward to an extended position under the hydraulic force of fluid from the internal water vascular system, or they can contract to rounded nubbins when the pressure diminishes. By sequentially jutting and contracting these tube feet in waves, the animal can slowly move along its substrate.

clam shell with its arms and tube feet, then evert its stomach through its mouth into the open shell and digest the prey. Capable of eating more than ten clams or oysters in a day this way, sea stars can be serious predators on commercial beds.

CONNECTIONS

The student of invertebrates enters a strange realm of sacs and vases, worms and parasites, swarming insects and monsters of the deep. Many of these creatures are unfamiliar and live in habitats we can explore only with the help of scuba gear, diving bells, backhoes, or microscopes. Nevertheless, the ani-

mals without backbones are the quiet majority, totaling 95 percent of all animal species. We tend to think of vertebrates as physically more complex than invertebrates, and many are. But the invertebrate world is full of astonishing structures like headlamp eyes, suckered tentacles, poison fangs, gossamer wings, and glue-spurting bazookas. Complicated processes like metamorphosis, caste systems, and intricate parasitic life cycles belie any attempt to dismiss all invertebrates as ''simple'' or ''primitive.'' In fact, the invertebrates, during an immense history that began more than 700 million years ago, ''tested'' a number of evolutionary ''ideas'' and came up with most of the physiological solutions employed throughout the animal kingdom in invertebrates and vertebrates alike. The next chapter examines some of these modifications in animals with backbones.

Highlights in Review

1. Animals are multicellular heterotrophs that can move about at some point in the life cycle, reproduce sexually, and pass through various stages of development.
2. Five important anatomical trends evolved in the invertebrates: bilateral symmetry; the development of a head with centralized sensory apparatus; a body plan with a two-ended gut; a coelomic cavity; and segmentation of the body.
3. Sponges are simple, sessile, saclike aquatic animals that show none of the five trends, nor complex tissues or organs.
4. Cnidarians (hydras, jellyfish, corals, and sea anemones) are radially symmetrical animals with stinging tentacles used for self-defense and feeding. They can be vaselike polyps or umbrella-like medusae.
5. Flatworms (planarians, flukes, and tapeworms) are bilaterally symmetrical and cephalized. Flatworms have three distinct tissue layers and develop true organs that accomplish digestion, nervous responses, body movement, excretion, and reproduction.
6. Roundworms (nematodes) are tiny, abundant invertebrates that parasitize animals and plants and decompose rotting materials. They were the first to have a two-ended gut with mouth and anus, which is cushioned in a fluid-filled chamber (the pseudocoelom) surrounding the endoderm, or inner tissue layer.
7. In one major animal lineage, the protostomes (mollusks, annelids, and arthropods), the indentation, or blastopore, of the developing embryo becomes the mouth, and a second opening becomes the anus. In the other major line, deuterostomes (echinoderms and chordates), the indentation becomes the anus, and a second opening becomes the mouth.
8. Mollusks (snails, clams, oysters, squid, and octopuses) have major body parts not found in most other invertebrates: the foot, visceral mass, mantle, mantle cavity, and gills.
9. Annelids (marine polychaetes, earthworms, and leeches) are segmented and display all five major anatomical advances. Body fluids act as a hydroskeleton. The gut has specialized regions for grinding and storing food, and the circulatory system is closed.
10. Arthropods (insects, crustaceans, spiders, ticks, mites, scorpions, millipedes, and centipedes) are the largest animal phylum. Each type has an exoskeleton; modified, specialized segments; a relatively high metabolic rate and rapid movement compared to other invertebrates; and acute senses.
11. Echinoderms (sea stars, brittle stars, sea urchins, sea cucumbers, and sea lilies) were the first to have endoskeletons. Most of the slow-moving, radially symmetrical marine animals have a water vascular system that allows locomotion.

Key Terms

abdomen, 394
animal, 378
antenna, 394
bilateral symmetry, 386
bivalve, 390
book lung, 395
carapace, 396
caste, 397

central cavity, 382
cephalization, 386
cephalopod, 391
cephalothorax, 396
chelicera, 396
circulatory system, 390
cnidarian, 383
coelom, 388

crop, 392
deuterostome, 389
echinoderm, 398
endoskeleton, 398
exoskeleton, 394
eyespot, 387
facet, 395
filter feeder, 391
flame cell, 387
flatworm, 385
foot, 389
gastropod, 390
gastrovascular cavity, 384
gill, 390
gizzard, 392
head, 390
heart, 390
hydroskeleton, 388
insect, 397
intestine, 387
invertebrate, 378
joint, 394
mantle, 390
mantle cavity, 390
medusa, 383
mesoglea, 384
metamorphosis, 397
mollusk, 389
molt, 394
mouthpart, 394
nematocyst, 384

nephridium, 392
nerve cell, 384
nerve cord, 387
nudibranch, 390
organ, 387
organ system, 387
poison claw, 395
polyp, 383
pore, 382
protostome, 389
pseudocoelom, 388
pupa, 397
radula, 390
roundworm, 387
schistosomiasis, 387
segmented worm, 391
septum, 392
sessile, 382
seta, 392
siphon, 391
slime trail, 390
social insect, 397
spicule, 382
spider, 396
spinneret, 397
sponge, 382
thorax, 394
trachea, 395
vertebrate, 378
vessel, 390
water vascular system, 398

Study Questions

Review What You Have Learned

1. List five trends in invertebrate evolution.
2. How does a sponge obtain nutrients?
3. True or false: A biologist might rightfully say that a sponge is basically a cluster of independent single cells similar to a protozoan colony. Explain.
4. What features would you see in a typical cnidarian?
5. Outline the life history of a jellyfish.
6. List five organ systems in the planarian, and name a specialized structure in each system.
7. What two new features arose in roundworm evolution?
8. Describe some important evolutionary adaptations of mollusks.
9. List the classes of mollusks, and name one member of each class.
10. List the classes of annelids, and name one member of each class.
11. Annelids show numerous evolutionary adaptations. List five.
12. Describe the structure of an earthworm segment.
13. What characteristics distinguish the arthropods as a group?
14. List the arthropod classes, and name at least one member of each class.
15. What are the main characteristics of echinoderms?
16. How does a sea star move?

Apply What You Have Learned

1. A biology student refuses to eat mackerel sushi in a Japanese restaurant, fearing that the uncooked mackerel contains fish tapeworm larvae. Are his fears well founded? Explain.

2. The laws of Islam and Judaism prohibit followers from eating pork. From a biological perspective, what might have been the benefit of this prohibition throughout much of history?

3. Commercial scallop farmers often kill sea stars found in their catch because the echinoderms feed on scallops. The farmers cut each sea star into several pieces and toss the bits overboard, but this strategy usually backfires. Why?

4. While on an oceanographic expedition, you find a deep-water organism previously unknown to science. Its body resembles a cup of branched tentacles, and the cup sits on top of a stalk attached to a rock. When you observe the animal under a microscope, you find that it has mucus-coated tube feet on its arms. The feet capture food particles in the water and pass them to cilia-lined grooves that sweep them into the central mouth. It has an anus but no excretory system and no brain. The order of events during its larval development shows it to be a deuterostome. What phylum would you place this animal in? Explain.

For Further Reading

BRUSCA, R. C., and G. J. BRUSCA. *Invertebrates.* Sunderland, MA: Sinauer, 1990.

GOULD, S. J. *Wonderful Life: The Burgess Shale and the Nature of History.* New York: Norton, 1989.

McMENAMIN, M. A. S. "The Emergence of Animals."*Scientific American,* April 1987, pp. 94–102.

MILLER, D., and C. VERON. "Biochemistry of a Special Relationship: Coral Reefs." *New Scientist,* June 2, 1990, pp. 44–49.

MORRIS, S. C. "Palaeontology's Hidden Agenda." *New Scientist,* August 11, 1990, pp. 38–42.

REAVEY, D., and K. GASTON. "Caterpillars in Wonderland." *New Scientist,* February 23, 1991, pp. 52–55.

STOLZENBURG, W. "When Life Got Hard." *Science News* 138 (1990): 120–123

WOOTTON, R. J. "The Mechanical Design of Insect Wings." *Scientific American,* November 1990, pp. 114–120.

The Chordates: Vertebrates and Their Relatives

FIGURE 19.1 ▪ Bats: Prehistoric-Looking Vertebrates. With their membranous wings and facial flaps for detecting sound waves, bats can look a fright, but most are beneficial insect eaters and plant pollinators.

Bats: Abundant Vertebrates with a Bad Reputation

Curiously, there are more species of bats than of all the more familiar mammals combined, such as bears, apes, seals, deer, and lions. Nevertheless, bats have a terrible reputation. Many people think they are vicious, filthy, stealthy, and dangerous. In reality, bats are overwhelmingly beneficial animals, and they serve well to introduce the phylum at hand, the **chordates:** animals with a stiff rod running down their back at some time during their life cycle.

Bats do fly silently about at night, but they are not the dirty, rabid vampires many people imagine. Rabies is rare in bats, and the animals clean themselves meticulously, like cats. Only three species rely on blood meals, flapping noiselessly up to cattle or other victims, making quick, painless incisions with razor-sharp teeth, and lapping up the flowing blood. The remaining 900 species eat fruit, pollen, insects, spiders, or other small animals. Together, these bat species consume enormous numbers of harmful insects, serve as the chief pollinators for over 130 species of flowering plants, and help reforest denuded lands by excreting millions of undigested seeds.

Bats are unique and fascinating animals—bad reputation and all. Their most obvious feature is a set of prehistoric-looking wings that in many species is made up of membranes stretching between the four limbs and tail (Figure 19.1). Bats are also known for the bizarre facial appendages that cause them to resemble the gargoyles on European cathedrals. The bat's facial appendages—both the ear flaps in front of the large ears and the nose leaves that sprout on the snout—help the animal receive sonar signals. As it flies, the bat generates ultrahigh-frequency sound waves through its open mouth or its nose; these sound waves bounce off objects in the dark, vibrate the thin ear flaps and nose leaves, trigger signals to the brain, and help the bat navigate without light.

Bats are the only mammals that truly fly (rather than glide), but like the birds, bats come in a wide range of sizes. The largest bats, the flying foxes, have a wingspan of 1.3 m (51 in.), while the smallest, Kitti's hog-nosed bats, are smaller than a bumblebee and are, in fact, the tiniest mammals on earth. In addition, bats are highly social: Thousands of individuals hang upside down together in caves or trees, and within these colonies, they assume specialized roles, such as guarding the cave entrance, scouting for food sources, or sounding an alarm when danger approaches. Even vampire bats can be social species: When solicited by a less successful roostmate, a bat who has just had a blood meal will regurgitate some of the blood into the hungry bat's throat.

With the bats and their relatives, we move away from the vast majority of animal species—the invertebrates—and focus on the remaining 5 percent—members of the phylum Chordata. All chordates have a central spinal or nerve cord and either a flexible rod (a notochord) or a series of interlocking bones called the backbone that protects that nerve cord and provides internal support for the body. Those chordates with a backbone (the great majority) are called *vertebrates* after the sets of interlocking bones, or **vertebrae,** that make up the bony column. Although there are well over a million invertebrate species compared to about 50,000 species of vertebrates, the vertebrates include the earth's fastest runners, highest fliers, deepest divers, most agile climbers, best problem solvers, and largest and most familiar animals, in the form of fishes, amphibians, reptiles, birds, and mammals. The abundant bats and the primates, the group that includes our own species, ***Homo sapiens,*** are in the class Mammalia of the subphylum Vertebrata of the phylum Chordata.

As we study the characteristics and evolution of the vertebrates and other chordates, we will see four unifying themes emerge. First, chordate evolution is a history of innovations that built upon the major invertebrate traits. Chordates display the basic traits that evolved first in the invertebrates: bilateral symmetry, cephalization, segmentation, coelom, and gut tube with mouth and anus. But the chordates developed another set of innovations in addition to those, including the spinal cord, a stiff rod (the notochord), gill slits, a tail, and blocks of muscles. These enabled chordates to grow larger than invertebrates, on average, and to successfully exploit their aquatic or land environments in new ways.

Second, chordate evolution is marked by physical and behavioral specializations. Looking just at the mammals, for example, and just at a single structure, the forelimb, one can see wide structural variation, with forelimbs of different types

specialized by natural selection for specific tasks: The bat's wing is especially adept at flying, the horse's foreleg and hoof at running, the human's agile hand at grasping, and the seal's flipper at powerful swimming. Looking just at the bats, the leathery wings, use of sonar, and complex social behavior are all specialized traits that allowed these animals to develop into successful nocturnal hunters and fruit feeders and to survive in caves and trees, safe from the reach of some of their enemies. Likewise, the success of the human species rests with a number of traits, including upright posture, large brain, agile hands, and cooperative social groups.

Third, as in many groups of organisms, evolutionary innovations and specializations led to **adaptive radiations** in the chordates: the development of a variety of forms from a single ancestral group. The earliest bats, for example, were probably clumsy fliers that ate only insects. These, however, radiated into 900 species that fly smoothly, eat a range of foods, and inhabit every type of environment except arctic tundras and extremely hot deserts. The fishes, amphibians, reptiles, birds, and mammals have all experienced great radiations during their group histories. (One great animal radiation, for example, resulted in a million species of insects.)

Finally, as we trace the evolutionary path that led to *Homo sapiens,* we will see that humans emerged with some unique physical traits and capabilities, but evolutionary principles common to all other organisms guided our emergence. We are a branch of the primate order, tested and preserved through natural selection like other organisms, even though our brain is the most complex structure in the known universe.

This chapter will consider the chordates in detail:

- What are the characteristics of the chordates, and how did they evolve?
- What chordates lack a vertebral column?
- What are the characteristics of the fishes, the earliest class of vertebrates and most successful aquatic animals?
- What are the characteristics of amphibians, the first vertebrates to live on land?
- What are the characteristics of reptiles, land animals freed from dependence on water for reproduction?
- What are the traits of feathered birds, the mostly airborne vertebrates?
- What are the mammals, the warm-blooded fur bearers, and how did they evolve?
- What are the primates, the mammalian order that gave rise to our own species, *Homo sapiens,* and what is our probable evolutionary history?

THE CHORDATES, INCLUDING THE SIMPLE TUNICATES AND LANCELETS

The diverse vertebrates—including many graceful, powerful, and fleet animals of land, sea, and air—evolved along one of the two great pathways of animal evolution, the deuterostome line. This path includes echinoderms (sea stars and their relatives) and acorn worms (phylum Hemichordata), marine invertebrates that share some traits with sea stars and some with chordates (Figure 19.2). The most primitive chordates include the tunicates (or sea squirts) and fishlike lancelets, small, inconspicuous, bottom-dwelling sea creatures without

vertebrae. These and all other chordates display bilateral symmetry, cephalization, a coelom, a gut tube with mouth and anus, and body segmentation—traits that first emerged in the invertebrates, and indeed, the chordates almost certainly evolved from ancient aquatic invertebrates. But chordates also share a number of new structures.

One of these new structures is the *notochord,* a stiff but flexible rod that provides internal support and runs the length of the animal (Figure 19.3). The notochord is present in tadpolelike tunicate larvae, but disappears in the adults. In lancelets, the notochord remains throughout life. In vertebrates, the notochord is present only in the embryo (see Figures 13.13 and 13.14); the vertebral column then develops in its place. A second new structure is the hollow *nerve cord,* or **spinal cord,**

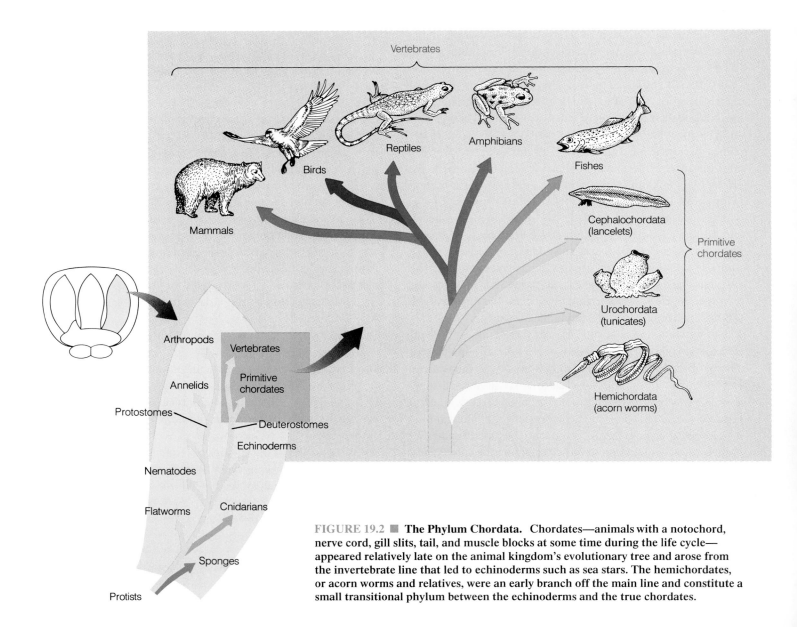

FIGURE 19.2 ■ **The Phylum Chordata.** Chordates—animals with a notochord, nerve cord, gill slits, tail, and muscle blocks at some time during the life cycle—appeared relatively late on the animal kingdom's evolutionary tree and arose from the invertebrate line that led to echinoderms such as sea stars. The hemichordates, or acorn worms and relatives, were an early branch off the main line and constitute a small transitional phylum between the echinoderms and the true chordates.

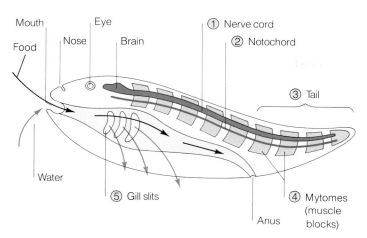

FIGURE 19.3 ■ Generalized Chordate Body Plan. Chordates show the major animal characteristics that emerged in the invertebrates, but five additional traits as well: the notochord, nerve cord, gill slits, myotomes, or muscle blocks, and tail. In complex chordates, such as our own species, some of the five chordate innovations occur only briefly during early embryonic development, then disappear.

which is a tube of nerve tissue that also runs the length of the animal, just above (dorsal to) the notochord. The nerve cord helps to integrate the body's movements and sensations. The nerve cord is present in chordates throughout embryonic and adult life.

Chordates also have **gill slits,** pairs of openings through the *pharynx,* the anterior region of the gut. If you still had a gill slit, it would be a tunnel from the back of your mouth to the outside of your neck. Tunicates and lancelets employ gill slits for filtering food. Young fish and certain amphibians breathe through gill slits, but later these tissues develop into true gills. In most reptiles, birds, and mammals, gill slits are vestiges occurring only in the embryo or developing into other structures. For example, one vestige of the gill slit is your eustachian tube, which reaches from the back of your throat to your middle ear, and from there to your outer ear canal. This is the tube that allows you to equalize pressure in your ears when in an airplane. Surrounding the notochord and nerve cord is a fourth chordate trait, muscle blocks called *myotomes.* Myotomes are actually modified body segments and the major signs of segmentation in chordates. When you eat fish, you can easily see these stacked muscle layers in the "flaky" white or pink meat. The notochord, nerve cord, and myotomes extend into the **tail,** the final chordate characteristic, a structure that protrudes beyond the anus at some point in each individual's development. Most chordates have a tail throughout life, but in some chordates, including humans, the tail appears only briefly in the embryo.

The chordate innovations—notochord, spinal cord, gill slits, tail, and muscle blocks—had dramatic evolutionary implications (Figure 19.4). The physical support of a notochord, and later of a vertebral column, allowed chordates to

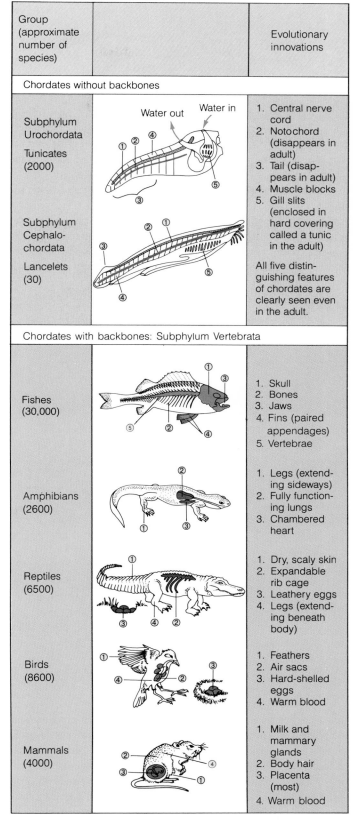

FIGURE 19.4 ■ Major Innovations in the Chordates. Each group of chordates shows a slightly different set of innovations, each modifying the adaptations of progenitor species.

Group (approximate number of species)		Evolutionary innovations
Chordates without backbones		
Subphylum Urochordata Tunicates (2000) Subphylum Cephalochordata Lancelets (30)		1. Central nerve cord 2. Notochord (disappears in adult) 3. Tail (disappears in adult) 4. Muscle blocks 5. Gill slits (enclosed in hard covering called a tunic in the adult) All five distinguishing features of chordates are clearly seen even in the adult.
Chordates with backbones: Subphylum Vertebrata		
Fishes (30,000)		1. Skull 2. Bones 3. Jaws 4. Fins (paired appendages) 5. Vertebrae
Amphibians (2600)		1. Legs (extending sideways) 2. Fully functioning lungs 3. Chambered heart
Reptiles (6500)		1. Dry, scaly skin 2. Expandable rib cage 3. Leathery eggs 4. Legs (extending beneath body)
Birds (8600)		1. Feathers 2. Air sacs 3. Hard-shelled eggs 4. Warm blood
Mammals (4000)		1. Milk and mammary glands 2. Body hair 3. Placenta (most) 4. Warm blood

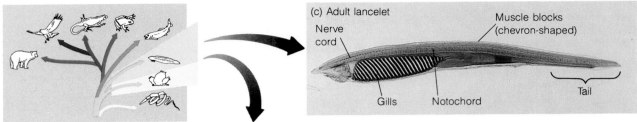

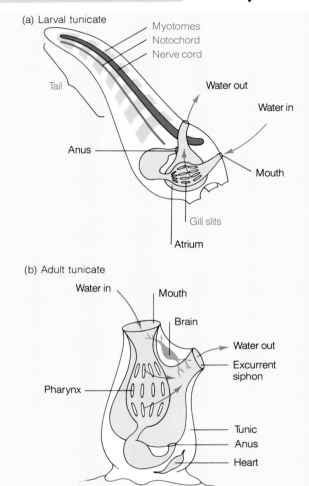

(c) Adult lancelet

(a) Larval tunicate

(b) Adult tunicate

FIGURE 19.5 ■ Chordates Without Backbones: Tunicates and Lancelets. (a) Tunicate larvae, with their streamlined body shape, resemble the generalized chordate (see Figure 19.3), but (b) adult tunicates are stationary filter feeders enclosed in a baglike tunic. (c) Lancelets, such as *Amphioxus,* are vaguely fishlike, but they rest vertically with transparent bodies planted tail first in the sand with only the anterior end exposed for capturing tiny food morsels.

Tunicates

Tunicates, or sea squirts, have all five chordate traits in their larvae: notochord, nerve cord, gill slits, muscle blocks, and tail. Only 1 cm (⅓ in.) long, tunicate larvae look like miniature tadpoles and move about by thrashing their long tails (Figure 19.5a). They eventually settle down—head first—and undergo a dramatic metamorphosis. The tail, with its notochord and nerve cord, disappears, and a stationary adult develops and remains attached to a submerged rock or harbor piling (Figure 19.5b). The adult has an outer envelope, or **tunic,** enclosing a large basket-shaped structure, the **pharynx,** perforated by hundreds of gill slits. Tunicates are filter feeders; beating cilia that line the gill slits draw water in through the mouth (or incurrent siphon) and sweep it out again through the *excurrent siphon.* Food particles suspended in the feeding current stick to a mucous sheet inside the pharynx and are passed to the stomach. If disturbed, a tunicate will often squirt water forcibly from its siphon. Hence the common name "sea squirt."

Lancelets

Lancelets are small, streamlined marine animals about 5 cm (2 in.) long that live half-buried—tail first—in the sandy bottoms of shallow saltwater bays and inlets (Figure 19.5c). All the major chordate traits are visible throughout embryonic, larval, and adult life—a contrast with tunicate development. Like tunicates, however, lancelets are filter feeders with cilia on their gill slits and a sticky, food-trapping mucous region of the pharynx; but if food becomes scarce, lancelets can pull up anchor and move to more fertile grounds. Some biologists believe that the ability to move as adults gives lancelets a dis-

grow larger and larger—as large, in fact, as dinosaurs and whales! The spinal cord improved centralized nerve coordination and control and allowed the evolution of acute senses, a wide range of behaviors, and greater intelligence than in most invertebrates. The gill slits, as we have seen, could be used for feeding or breathing. And the tail in its many manifestations became a major organ of locomotion (especially in fish), balance (in hopping animals), and even communication (in many birds and mammals).

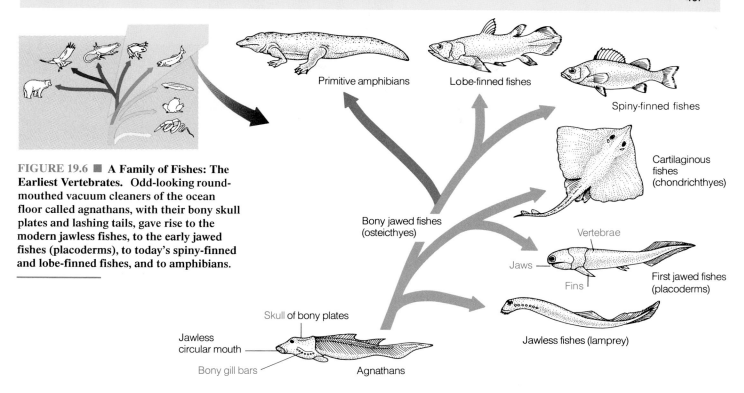

FIGURE 19.6 ■ A Family of Fishes: The Earliest Vertebrates. Odd-looking round-mouthed vacuum cleaners of the ocean floor called agnathans, with their bony skull plates and lashing tails, gave rise to the modern jawless fishes, to the early jawed fishes (placoderms), to today's spiny-finned and lobe-finned fishes, and to amphibians.

tinct feeding advantage over the stationary tunicates, and for this reason, they think that lancelets more closely resemble the mobile ancestor that led to the vertebrates.

FISHES: THE EARLIEST VERTEBRATES

Emerging 550 million years ago, at the dawn of the Paleozoic era the first fishes were streamlined filter feeders, about 30 cm (1 ft) long, that lived in the muddy bottoms of ancient seas. They had fixed circular mouths that lacked jaws and that could suck up sediments like organic vacuum cleaners; hence, these fishes are called **agnathans,** meaning "jawless" (Figure 19.6). While jawless fishes retained the chordate notochord for internal support and the gill slits for feeding, a **skull** protected the brain, and muscles rather than tiny cilia powerfully drew water and suspended food into the mouth. Muscle-powered gill slits allowed the agnathans to consume greater quantities of food than their earlier cousins and to grow 6 to 30 times larger. In addition, the gill openings were supported by gill bars made of *bone,* a protein meshwork hardened by calcium, phosphate, and other minerals. Bony plates beneath the skin of these early fishes served as armor that probably protected them against dangerous invertebrates, such as giant sea scorpions.

Modern jawless fishes have lost the bony plates and skulls of their ancestors and instead have very flexible internal skeletons of cartilage. The jawless lamprey is a parasite that at-

taches to its prey by suction, rasps through the victim's body wall with a sharp tongue, and then drinks its blood (Figure 19.7). At times, lampreys have been serious pests of commercially fished species in Lake Michigan.

FIGURE 19.7 ■ Lampreys Parasitizing Fish. Modern relatives of the ancient jawless fishes, lampreys live by rasping off the flesh and sucking the blood of other aquatic animals. Three lampreys (*Lampetra fluviatilis*) parasitize a large carp.

Advances During Fish Evolution

The ancient agnathans gave rise to a second class of fishes, the *placoderms* (early jawed fishes), about 425 million years ago (Figure 19.8). These descendants had three basic innovations that were so useful they appeared in nearly all the vertebrates that followed. First, the placoderms had **hinged jaws,** derived from agnathans' gill support bones (gill bars). Jaws allowed the placoderms to consume large chunks of food—kelp fronds, clams, other fish—instead of filtering small particles. As a result, some became huge; a person can stand upright in the fossilized jaws of one placoderm. Second, these jawed fishes produced the first vertebrae: They had a series of separate bones that fused with the notochord and arched over the spinal cord. The resulting vertebral column provided a site of attachment for muscles in the body wall, anchoring them as the body bent to and fro and allowing the muscles to grow larger and propel the fish more powerfully. The backbone also encased and protected the delicate spinal cord. Third, placoderms had the first **fins,** paired appendages that provided more control over swimming direction, speed, and depth (review Figure 19.6). The bony sockets in which the fin bones fit gave rise to the hips and shoulder joints of later land animals.

Modern Fishes

Although the placoderms were heavily armored with bony plates beneath the skin, their modern descendants, the cartilaginous fishes (**Chondrichthyes**), have skulls, vertebrae, and the rest of the skeleton made entirely of cartilage, a matrix of fibrous proteins without the minerals that harden bones. This class includes sharks, skates, and rays. Cartilage is lighter than bone and affords these predatory fish the speed and agility they need to catch prey. In skates and rays, large fins provide lift, much like graceful underwater wings, while

(a) A manta ray

(b) Jaws of a great white shark

FIGURE 19.9 ■ Cartilaginous Fishes: Hunters of the Sea. (a) A skeleton of cartilage is lightweight and flexible and gives this spotted eagle manta ray of Baja California a slight resemblance to a predatory bird flapping beneath the waves. **(b)** The fearsome jaws of a great white shark. Sharks are earth's most numerous vertebrate predators, and their jaws, inherited from placoderm ancestors, serve them admirably.

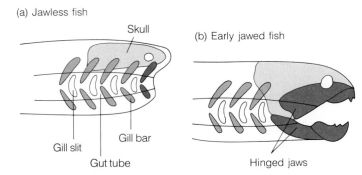

FIGURE 19.8 ■ Evolution of the Jaw. How the gill support bones (gill bars) of earlier jawless fishes may have evolved into the powerful jaws of the first jawed fishes, the placoderms. Jaws allowed placoderms to function as efficient grazers and hunters that preyed on kelp, fishes, and invertebrate neighbors. Their descendants include the sharks and the modern bony fishes.

in sharks, stiff fins knife through the water, helping these largest and most common oceanic hunters to maneuver easily toward their victims (Figure 19.9).

The other group derived from the ancient jawless fishes is the bony fishes (**Osteichthyes**), which include bass, bluegill, minnows, trout, tuna, and virtually all of today's familiar fresh- and saltwater fishes. There are two major groups of bony fishes. The oldest group are the **lobe-finned fishes,** which arose about 400 million years ago. These fishes had large, muscular, lobed fins that allowed them to "walk" across the bottoms of shallow bays, and they had **lungs,** or air sacs, for breathing air. When water levels fell, their innovations allowed them to survive by breathing air and migrating

(a)

(c)

(b)

FIGURE 19.10 ■ Bony, Jawed Fishes: The Most Diverse Vertebrates. The bony, jawed fishes, all descendants of the first jawed fishes, the placoderms, have radiated into the largest vertebrate phylum, with 30,000 living species and thousands of additional species that are now extinct. The oldest descendants, the lobe-finned fishes, had muscular appendages supporting their fins, as well as lungs that allowed air breathing during migration across land from pool to pool. Coelacanths (a) are living links to the early lobe-finned fishes, and they inhabit the deep waters off the Comoro Islands east of India. A recent census shows that there may be fewer than 300 of these ancient fish left, and that fishermen kill about three more each year. The first teleosts, or spiny-finned fishes, had delicate bones in their fins, as well as a swim bladder. The modern teleosts include brilliantly colored species, such as the blueface angelfish (b), and vertical swimmers, such as the bizarrely lobed weedy sea dragon (c), a sea horse, which seems to disappear when grazing in a clump of seaweed.

over land to other pools or bays. Present-day lobe-finned fishes include the **lungfishes** and the **coelacanths** (Figure 19.10a). An extinct group of lobe-finned fishes, the **rhipidistians,** were probably the first vertebrates to crawl out on land and live for extended periods.

A second group of bony, jawed fishes are the **spiny-finned fishes,** including **teleosts** such as trout and zebra fish. These lost their ancestors' fleshy fins, but gained much more versatile spiny fins with webs of skin over delicate rays of bone. They also lost the lungs, but developed instead a swim bladder, an internal balloon below the backbone that can change volume and allow the animal to adjust its swimming depth. These fishes radiated, early on, into a huge group. In

fact, the Devonian period (about 345 to 395 million years ago) is called the Age of Fishes because the teleosts and their relatives so dominated the earth's lakes and seas.

Even today, with 30,000 recorded species, teleosts are the largest group of vertebrates, and they owe much of their success to specializations that evolved as the group radiated. Some bony fishes have an antifreeze substance in the blood, allowing them to live below the polar ice. Some are brilliantly colored, an aid to courtship and a warning to predators (Figure 19.10b). Some have "flashlights" (luminescent organs) that enable them to find food or mates in the cold, black reaches of the ocean abyss. Many can flex muscular, streamlined bodies and live as fast-swimming predators, while the sea horses, with less flexible bodies, swim vertically, propelled by intricate undulations of the bony fins (Figure 19.10c). However, except for a very few air breathers, fish are relegated by their anatomy to life in water. The great landmasses were to be dominated by other vertebrates—descendants of the early fishes. Nevertheless, important evolutionary innovations within the fishes included the skull, bone, hinged jaws, paired fins, and vertebrae (review Figure 19.4).

AMPHIBIANS: FIRST VERTEBRATES TO LIVE ON LAND

Fossils show that during the Devonian, about 345 to 395 million years ago, the vertebrate lineage that included air-breathing, lobe-finned fishes produced the **amphibians:** vertebrates that can live both on land and in water (*amphi* means "both"; *bios* means "live"). Modern amphibians include the 3000 species of frogs, toads, salamanders, and wormlike apodes—animals whose ancestors overcame the formidable problems of life on land (including more efficient means of walking, breathing, and staying moist) beginning around 375 million years ago. Early amphibians had front and hind **legs** containing strong bones and powerful muscles. Extending sideways from the body, the limbs could support the animal's weight far better than lobed fins; and with two pairs of legs, the animal could move about on land—albeit slowly and clumsily—for greater distances, even without water's buoyant support.

Laborious walking, however, would have required a great deal of energy from food, and large quantities of oxygen would have been needed for aerobic respiration. Fossil evidence reveals that early amphibians had fully functioning lungs, or air sacs, that provided a site for gas exchange. Air was probably pumped in by swallowing movements, much as it is in modern frogs and toads (Figure 19.11a). Moreover, specialized chambers that arose in the heart and other modifications of the circulatory system tended to separate oxygenated blood en route to the body tissues from unoxygenated blood bound for the lungs—a separation that became complete in the mammals and birds (see Chapter 21). Finally, an amphibian's smooth, moist skin could absorb about half of the oxygen and release most of the carbon dioxide required by these animals.

These innovations provided sufficient oxygen but did not solve the problem of desiccation—drying out. Since the amphibian skin had to remain moist for gas exchange (to supplement the lungs), the animals were restricted to life near the water's edge. In addition, amphibians lay eggs with a clear, jellylike coating that must also stay moist, lest the embryos die before the fishlike tadpoles emerge. Because their eggs and larvae require water, amphibians must return to the water to reproduce, and in this sense, they are analogous to the first land plants whose sperm required standing water to swim from one plant organ to another.

Today, frogs, toads, salamanders, and a few relatives are the only remaining members of the class Amphibia and continue to display the evolutionary innovations of the amphibian lineage: legs, fully functioning lungs, and partial separation of oxygenated and deoxygenated blood in a chambered heart. These generally small vertebrates are very common in freshwater environments and show a range of interesting specializations. In some, the moist skin has become brightly colored, attracting potential mates and warning would-be predators of

(a) Play-actor frogs

(b) Salamander

FIGURE 19.11 ■ Amphibians: The Transition to Land. The brilliant gold and red pigments in the skin of (a) the tropical play-actor frog (*Dendrobates histrionicus*) and (b) the Chinese salamander (*Tylototriton verrucosus*) warn would-be predators of the amphibians' poison skin glands.

FIGURE 19.12 ■ Reptiles and Their Descendants, the Birds and Mammals. Cotylosaurs, the earliest reptiles, had several adaptations that allowed them to live and reproduce on dry land, including dry skin, large lungs and an expandable rib cage, advances in the circulatory system, and shelled eggs. Fossil evidence reveals that birds arose from thecodonts and mammals from therapsids, both direct descendants of the cotylosaurs that walked the earth some 300 million years ago. Thecodonts also gave rise to the dinosaurs, while the other groups of soaring, running, and swimming reptiles stemmed directly from the earlier cotylosaurs.

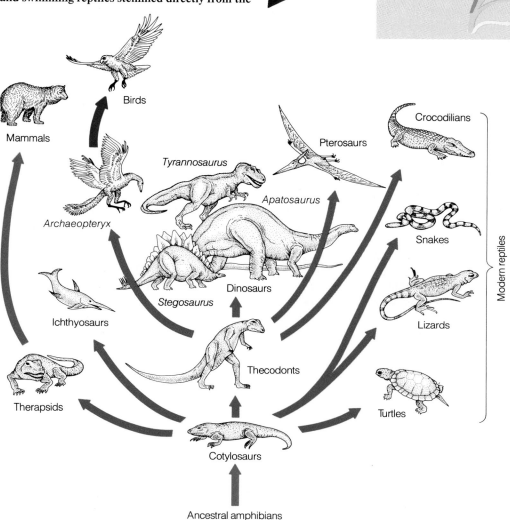

their poisonous glands (see Figure 19.11). And in others, the skin is dry and adapted to life in dry habitats. The nervous system of many amphibians, protected by the vertebral column and bony skull, has become well developed and accompanied by keen senses of sight and hearing. In addition, rapid-fire nervous reactions enable these animals to catch flies by flipping out their long tongues. And amphibian limbs are often specialized with webbed feet that allow efficient swimming and with thick muscles that propel hopping or running on land.

Unfortunately, amphibian populations all over the world began to decline about 20 years ago. Some scientists think that acid rain or global warming may be contributing to the animals' slow demise, because amphibians have porous skin that is especially susceptible to the quality and quantity of available water.

REPTILES: CONQUERORS OF THE CONTINENTS

As amphibians—which evolved far earlier—crawled about in profusion in the Carboniferous swamps, other, quite different, four-legged land animals called cotylosaurs—the earliest **reptiles**—were walking about among them (Figure 19.12).

Like the early seed plants, these reptilian creatures evolved important innovations for life on land that freed them from dependence on moist environments.

Cotylosaurs resembled crocodiles and had four innovations for terrestrial life (review Figure 19.4). First, these ancestral reptiles had dry, scaly skin that provided a barrier to evaporation and sealed in body moisture. This eliminated the skin surface as a major site for gas exchange, but a second reptilian innovation made up for this: respiratory modifications consisting of lungs with a larger surface area for gas exchange than in amphibians, an expandable rib cage that could draw in large quantities of air like a bellows, and a heart and circulatory system that separated oxygenated and deoxygenated blood more fully than the amphibian's two-cham-

bered heart. This allowed more oxygen to reach body tissues. Third, early reptiles displayed changes that eliminated reliance on open water for reproduction. Males had copulatory organs that could deliver sperm directly into the female's body rather than into the surrounding water, and females produced so-called **amniote** eggs. These eggs essentially encased the developing embryo in a pool of water (the amniotic sac), provided a source of food (the yolk), and surrounded both embryo and food with membranes and a leathery shell that prevented crushing and desiccation in open air. Finally, the cotylosaur's legs extended directly beneath the body rather than out to the side (as amphibians' legs did then and continue to do today); this extension provided better support and made walking and running easier.

(b) Galápagos tortoise

(a) Crocodile

(c) Lizard

(d) Burmese python

FIGURE 19.13 ■ Modern Reptiles. (a) Modern crocodiles, such as this saltwater crocodile from Australia, bear a striking resemblance to those ancient conquerors of land. (b) This giant land tortoise lives on Isla Isabela in the Galápagos archipelago, and like all turtles, hauls about a protective carapace. (c) The Komodo dragon of Indonesia can be up to 3 m (10 ft) long and weigh 115 kg (255 lb). The lizard has such a huge appetite that it will readily kill and devour other Komodo dragons and consume the dead bodies of nearly every other animal. (d) A Burmese python, also of the Indo-Malayan region, continues to grow for 20 years before reaching 8 m (27 ft) in length, and it can survive for upwards of 70 years.

The reptiles that followed the cotylosaurs built on the legacy of these innovations, and in fact, the cotylosaurs were pivotal to all subsequent vertebrate history (see Figure 19.12). They gave rise not only to the modern reptiles—the crocodiles, turtles, lizards, tortoises, snakes, and the tuatara—but indirectly to the birds and mammals as well. Descendants of the cotylosaurs became the thecodonts (small lizards that ran on two legs and gave rise to the giant reptiles we call collectively the **dinosaurs**) and the therapsids (small, heavyset, fiercely toothed animals that led to the mammals).

The warm Cretaceous was the heyday for dinosaurs ("terrible lizards") like the 15 m long (50 ft), meat-eating *Tyrannosaurus* and the 24 m long (80 ft), plant-eating *Apatosaurus* (formerly *Brontosaurus*). Contemporary to the dinosaurs were still other descendants of the crocodile-like cotylosaurs, including ichthyosaurs, which looked like long-necked sea monsters, and pterosaurs, the soaring reptiles with wings up to 8 m (26 ft) across.

The grand radiation of reptiles into forms that swam, flew, and lumbered across the earth spanned nearly 150 million years and inspired the common name for the entire Mesozoic era: the Age of Reptiles. Most of the great and diverse reptiles died out, however, in a massive extinction event at the end of the Cretaceous. Box 19.1 on page 414 considers the great mystery of the dinosaurs' demise.

A few small reptiles weighing less than 20 kg (44 lb) survived the extinctions and radiated once again during the Cenozoic, the current geological era beginning about 65 million years ago. Today, there are about 6500 species in the reptile class—animals that continue to show the innovations of dry scaly skin, expandable rib cage, fully separated oxygenated and unoxygenated blood supplies, copulatory organs, leathery amniote eggs, and legs extended directly beneath the body. Alligators, caimans, and crocodiles are all streamlined carnivores that inhabit warm climates (Figure 19.13a). The tortoises and turtles have a tough and unique protective structure, the shell, or carapace (Figure 19.13b). The lizards, snakes, and iguanas are elongated reptiles that inhabit wet, dry, or hot environments and sometimes reach great size. The heaviest lizards today are the Komodo dragons of Indonesia, which can weigh up to 115 kg (255 lb) (Figure 19.13c); and the longest snakes are the pythons of the same region, which can grow to 8 m (27 ft) (Figure 19.13d).

BIRDS: AIRBORNE VERTEBRATES

Except for domesticated animals, the most common chordates we see in our daily lives are *birds*. Curiously, some biologists have recently suggested that, from a zoological perspective, birds are actually dinosaurs! The first winged vertebrates—the giant, soaring pterosaurs—might seem like logical ances-

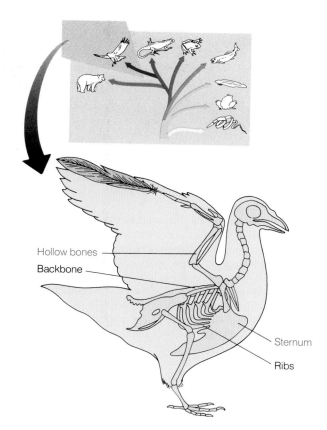

Hollow bones

Backbone

Sternum

Ribs

FIGURE 19.14 ■ **Feathers and Lightweight Bones: Innovations for Flight.** A generalized bird skeleton, showing the bladelike sternum and the reduced bones of legs and feet. Feathers have millions of individual, curving barbules that form the interlocking flight surface.

tors to the birds, but pterosaurs died out long before birds evolved. Instead, small, two-legged, lizardlike thecodonts are believed to be the real forerunners of the birds, and important fossil evidence supports this possibility. Six skeletons of the oldest bird, the crow-sized *Archaeopteryx,* have been found at different sites in rocks dated back to the Upper Jurassic period (150 million years ago). The fossil imprints suggest that this animal was a true intermediate: It had scaly skin; curving claws; a long, jointed tail; and sharp teeth like a reptile, but it had feathered forelimbs and tail like a bird.

Most avian evolutionary adaptations prepared the animals for efficient flight (see Figure 19.4). **Feathers** are exceedingly lightweight structures made of dead cells containing the protein keratin (see Chapter 2). Flight feathers, numbering 25,000 or more in a large bird, have a million tiny hooklike *barbules* that interlock and make the feather an aerodynamically sound "fabric." Fluffy down feathers, growing beneath the flight feathers, act as insulators.

BOX 19.1   FOCUS ON THE ENVIRONMENT

THE MYSTERY OF THE DISAPPEARING DINOSAURS

Dinosaurs once lumbered about on the very land where our houses, schools, and roads now stand. It is fascinating to ponder why the giant reptiles disappeared after flourishing for about 150 million years (Figure 1). Scientists have devised dozens of explanations based on fossils and mineral clues left in rock strata around the world, but the matter is far from settled.

Based on physical evidence, most paleontologists agree that the dinosaurs and probably 70 percent of the other plant and animal species on earth died out sometime around the so-called Cretaceous-Tertiary (KT) boundary, a time period about 65 million years ago.

One widely discussed theory holds that a massive comet or meteorite perhaps 10 km (6 mi) wide struck the earth about 65 million years ago, generating, upon impact, a crater some 150 km (90 mi) wide, and with it, an explosion 10,000 times bigger than one involving all the nuclear weapons now stockpiled. The resulting mushroom cloud would have spread a thick blanket of dust and smoke around the world, blocking out sunlight and preventing photosynthesis. What's more, the atmosphere's nitrogen gas would have literally burned up, combining rapidly into the kind of reddish brown smog one now sees over major cities (only thousands of times worse), choking off air breathers and leading to an acid rain with the pH of battery acid. These caustic rains would have leached toxic trace metals from the rocks, poisoning the waters. A long, dark, frigid winter would have set in until the dust settled out and the pollution rained down, and any organisms that survived this nasty scenario might have succumbed to a rebound period of intense greenhouse warming brought about by solar energy trapped under a layer of water vapor and CO_2 also thrown up during the impact.

Geologists have found indirect forms of evidence that a giant object struck the earth at the KT boundary, and some now think that the immense crater may be in the Caribbean Sea.

A competing theory holds that the eruption of giant volcanoes in what is now western India created a rock formation called the Deccan traps and led to a similarly gruesome environment at the time of the KT boundary. If such dire conditions were created by *either* an impact event *or* massive volcanism, the question becomes not how 70 percent of the animals and plants disappeared, but how 30 percent of them could have survived.

In a handful of other hypotheses, the dinosaurs died out because mammals ate their eggs; because the reptiles could not adapt to environmental changes; because their skeletons were too small to effectively support their massive bodies; and because they ate too many poisonous plants.

Many scientists maintain that no single event or condition could have caused the disappearance of numerous species over a geologically short period of a few thousand or few million years, and they suggest that at least four or more factors were involved: decreases in global temperature; dropping sea levels; broad shifts in seasonal climates leading to a decrease in the number of plant (and in turn, animal) species; and competition from smaller warm-blooded mammals.

What do you think?

FIGURE 1 ■ **Skeleton of *Albertosaurus*, with Young Onlookers at Chicago's Field Museum.**

(a) Flamingos

(b) Cassowary

FIGURE 19.15 ■ **Modern Birds.** (a) Vivid pink East African flamingos cluster in a courtship display. (b) The double-wattled cassowary (*Casuarius*) is the most colorful flightless bird, with its naked blue head skin, vivid red wattles, and jet black feathers.

A second set of avian innovations involves the skeleton and muscles. Birds have lightweight, hollow bones and a breastbone, or **sternum,** enlarged into a blade-shaped anchor for the powerful pectoral muscles that raise and lower the wings (see Figure 19.14). The legs are reduced to skin, bone, and tendons and can be folded up like an airplane's landing gear, reducing drag during flight.

Flight is a strenuous activity that requires plenty of oxygen for aerobic respiration of muscle and other tissue, and birds have a third set of modifications that ensures an adequate oxygen supply. First, they are *warm-blooded,* or **homeothermic.** They maintain a constant internal temperature despite environmental changes, and this steady temperature helps to maintain a steady production of ATP energy during cellular respiration, which in turn fuels the activities of flight and leg muscles. In fact, the earliest feathers may have been more useful as insulation than as aids to flight. *Cold-blooded,* or **poikilothermic,** animals, such as fish, amphibians, and reptiles, have body temperatures that vary, and hence grow active in warm environments but sluggish in cold ones. (See Chapter 20 for further discussion of warm- and cold-bloodedness.) Second, birds have a series of connected lungs and **air sacs** that exchange oxygen and carbon dioxide in an efficient one-way flow. And third, birds have a four-chambered heart that completely separates oxygenated and deoxygenated blood, so only the former reaches body tissues. (See Figures 21.6 and 23.7.)

Finally, in addition to the evolutionary innovations of feathers, warm blood, and air sacs, birds have hard-shelled eggs, which free them from dependence on water for reproduction.

So successful were the avian innovations that in the last few million years of the Cenozoic, birds radiated into a highly diverse class with more than 8600 species specialized for distinct modes of life. Flamingos, wading in large flocks, look like tall-stemmed pink flowers (Figure 19.15a). Condors with wingspans of 3 m (10 ft) soar on mountain updrafts, seldom flapping. Cassowaries, meanwhile, have powerful legs that enable them to run rapidly and deal deadly blows with their feet (Figure 19.15b). And champion migrators like the arctic tern have an innate sense of direction and the physical stamina to travel more than 40,000 km (25,000 mi) each year.

MAMMALS: RULERS OF THE CENOZOIC

This chapter began with the only true flying mammals—the bats—and now we arrive at the group that includes them—and our own species as well. While mammals arose from the fierce, heavyset reptiles called therapsids, they developed their own distinct body traits at least 180 million years ago. The very early mammals resembled shrews and probably scurried around the great dinosaurs' feet until the end of the

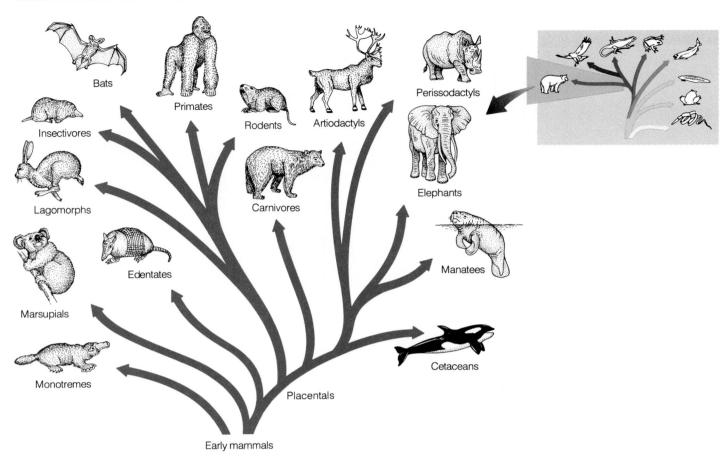

FIGURE 19.16 ■ Mammals: Diverse Land Animals of the Cenozoic Era. The current geological era has been called the Age of Mammals because this diverse group of complex, often large animals is so ubiquitous and successful.

Cretaceous. The mammals survived the mass extinctions of that time and radiated into 5000 modern species that give the current geological era, the Cenozoic, the appellation Age of Mammals (Figure 19.16).

Like birds, **mammals** are warm-blooded and have a four-chambered heart, but these two traits probably evolved independently in the two groups. Mammals have two unique innovations as well: **milk** and **body hair** or fur (see Figure 19.4). A fluid rich in fats and proteins, milk is produced in special glands called **mammary glands** and is used to nourish newborns. Many mammals have dense carpets of hair called **fur** covering the entire body, while others, such as certain monkeys and humans, have sparse body hair, and a few, including the whales and porpoises, have virtually no hair at all and rely instead on thick layers of fat called **blubber** for insulation.

The majority of mammals also have a special reproductive structure, the *placenta* (see Figure 14.10), that supports the growth of the embryo to a fairly complete stage of development before birth. These placental mammals descended from one branch of the earliest mammals. Two other branches led to nonplacental mammals with different ways of harboring embryos. The **monotremes,** which include only the duck-billed platypuses and spiny anteaters of Australia and New Guinea, lay leathery eggs and warm them until the young hatch. The females then nourish the young with milk. The **marsupials,** including kangaroos, opossums, koala bears, wombats, and dozens of other animals from Australia and the Americas, give birth to immature live young no bigger than a kidney bean. When the newborn emerges from the birth canal, it crawls upward into an elastic pouch of skin on the mother's ventral surface (the marsupium), attaches to a teat, starts to consume milk, and continues to develop inside the pouch.

With their body hair and successful reproductive strategies based on their evolutionary innovations of milk and the placenta, mammals can be compared to the flowering plants, which, with the evolution of flowers as reproductive organs, underwent a parallel radiation during the Cenozoic. Mammals radiated into a large and diverse class that can live in more environments and with more life-styles than any other class of animals except perhaps the birds (Figure 19.17).

Various specializations of structure and behavior underlie mammalian success.

Temperature Regulation

Mammals can keep a constant body temperature that is warmer than the ambient air on cool days and cooler than the air on extremely hot days. This is partly because of their homeothermic metabolism, partly because of their hair or blubber, and partly because they have evolved special behaviors such as hibernating and migrating. Bats, for example, gobble food insatiably when awake, then return to caves or similar hiding places with constant mid-range temperatures. They curl up tight and sleep in clusters, which retains warmth, and in winter, when food is in short supply, they hibernate, spending weeks in a dormant state with lowered body temperature.

Specialized Limbs and Teeth

Mammals display a range of limb types, with bones elongated, shortened, or broadened, depending on the animal's particular locomotor or food-gathering needs. In bats, the forelimb bones are light and strong, and the "finger" bones are greatly elongated and widely spread and support the flight membranes (see Figure 19.1). In moles and other digging mammals, the forelimbs are short and powerful, with oversized claws, while antelope forelimbs have strong, slender bones that allow swift running. Mammalian teeth are modified into chisels for gnawing wood (as in rats and beavers); flat molars for grinding grain (as in cows); and sharp points for tearing flesh (as in some bats and other meat eaters).

Parental Care

Mammals not only suckle their young for days or months, depending on the species, but usually guard the new generation fiercely and teach them survival skills. The extended care of young is especially pronounced in *primates* (monkeys, apes, and humans), as we shall see.

Highly Developed Nervous Systems and Senses

The keen senses of smell and hearing in most mammals, and of sight in a few, are instrumental in helping the animals find food, interact with mates and young, and avoid danger, just as these traits help birds, reptiles, and other animals but to differing degrees. The sonar of bats, which gives them the ability to navigate and locate objects in complete darkness, is but one example. Such sensory information is integrated and often stored in memory, enabling the animals to learn from experience and to minimize the effort they expend to survive. The brains of primates and cetaceans (dolphins, porpoises, and whales) represent the highest level of nervous system development among all animals.

(a) Duck-billed platypus

FIGURE 19.17 ■ A Modest Collection of Modern Mammals. Mammals produce body hair and make milk for feeding the young, but mammals are remarkably diverse in body size and appearance. (a) The duck-billed platypus is an egg-laying monotreme that nourishes its young. It protects itself with a venom-spurting spur, a rare poisonous weapon for a mammal. (b) The wombat, a marsupial from Australia, resembles a very large woodchuck and can live in deep burrows up to 100 ft long. (c) A North American cougar chases a snowshoe hare. (d) The Florida manatee is a mild-mannered, herbivorous "sea cow" that can weigh 1 ton.

(b) Wombat

(c) Cougar

(d) Manatee

EVOLUTION OF THE PRIMATES: OUR OWN TAXONOMIC ORDER

One of the most intriguing questions in modern biology is, How did human beings evolve? We have all the requisite traits for membership in the chordate phylum and the vertebrate subphylum, and with our warm blood, mammary glands, gestational mode, and body hair, we are obviously members of the mammalian class and the placental subclass. But our unique combination of behavioral abilities—including spoken and written language, agriculture, and extensive tool use—have allowed us to dominate the environment like no other animals before us. Despite this, there is ample evidence today—both fossil and genetic—that humans are simply one branch of the primate order and that our branch separated no more than 6 million years ago from the lineage leading to chimpanzees (Figure 19.18). Ethologist Desmond Morris called humans "naked apes," and so, it seems, we are.

The Primate Family Tree

Taxonomists recognize several dozen species in the order Primates and divide them into two suborders: Prosimii, the **prosimians** (which means "before monkeys"), and Anthropoidea, or **anthropoids** (literally meaning "humanlike").

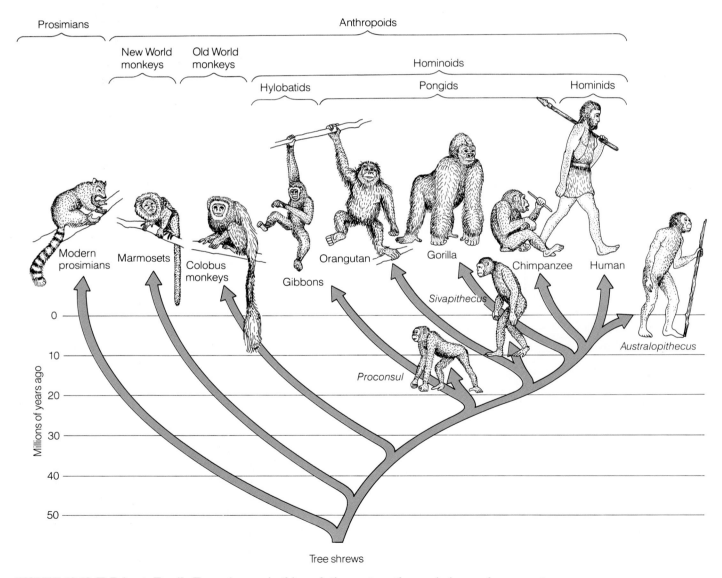

FIGURE 19.18 ■ Primate Family Tree. As seen in this evolutionary tree, the prosimians and marmosets (as well as other New World monkeys not shown here) branched off quite early from the tree shrews that gave rise to the primates. The Old World primates are much more recent lineages, with chimpanzees and humans diverging and evolving less than 10 million years ago. *Proconsul* and *Sivapithecus* are extinct primates, and some *Australopithecus* species evolved before the emergence of true humans.

Prosimians include small, tree-dwelling animals, such as the lemurs and tarsiers, while anthropoids include the Old and New World monkeys, the apes, and humans.

■ **Prosimians** Many primate taxonomists classify the present-day tree shrews of southern Asia as the most primitive primates. These squirrellike insectivores live in trees and have two hallmarks of all later primates: an **opposable thumb,** or a first digit that can touch the ends of all the other digits, and an acute sense of sight. Lemurs are fuzzy, tree-dwelling animals with foxlike muzzles, racoonlike tails, and long front and back limbs (see Figure 19.18); they survive today only on the island of Madagascar. Tarsiers are tiny inhabitants of forests in the South Pacific; they have agile hands and feet, flattened faces, and huge forward-directed eyes (Figure 19.19).

■ **Anthropoids** The anthropoids include the monkeys and apes, as well as our own species, *Homo sapiens*. The nonhuman primates that inhabit the forests of southern Mexico and Central and South America are the New World monkeys, characterized by a flat nose and grasping, or prehensile, tails, as in capuchins, marmosets, tamarins, and howler monkeys (Figure 19.20).

Old World monkeys live in tropical forests throughout the Eastern Hemisphere from Africa to India and Southeast Asia. Old World monkeys are characterized by their closely set, downward-pointing nostrils. These animals usually lack prehensile tails, often have brightly colored buttock calluses involved in sexual attraction, and have fairly large brains. The rhesus monkeys used in medical research are Old World primates (Figure 19.21), as are the langurs of India and Southeast Asia and the colobus monkeys, baboons, and mandrills of Africa.

FIGURE 19.20 ■ **A New World Monkey.** This is the mantled howler monkey (*Alouatta villosa*) of Guatemala. A related primate, the black howler monkey (*Alouatta caraya*) inhabits forests from Ecuador to Paraguay and produces a loud, sharp call that can be heard for many miles. The animal has enlarged, cavernous bones in the throat that act, in the words of one zoologist, as a sort of "bony trumpet."

FIGURE 19.19 ■ **Tarsiers: Primitive Primates.** Tarsiers are modern prosimians, with their huge eyes and ears and their knobby, adhesive toes; they live in the tropical forests of the Philippines, Sumatra, and Borneo, where they harvest fruit and hunt insects, lizards, and birds by night, leaping through the tree limbs like large frogs.

FIGURE 19.21 ■ **An Old World Monkey.** The rhesus monkey (*Macaca mulatta*), a favorite laboratory species, is widespread in the forests of Southeast Asia and northern India and lives in close-knit, gregarious social groups. The adult on the right is grooming its young.

(a) Orangutan

(b) Gorilla

(c) Chimpanzee

FIGURE 19.22 ■ The Apes, or Hominoids. The apes are anatomically and genetically our closest relatives. (a) The orangutans (*Pongo pygmaeus*) live in remote forests of Sumatra and Borneo. The name means "man of the woods," and indeed, the orangutan's face is capable of varied, sensitive expression. Their arms are longer than a chimpanzee's or gorilla's, and they move about rather clumsily in search of fruit and tender shoots. These animals are intelligent and somewhat social, and they construct sleeping shelters in the trees. (b) Gorillas (*Gorilla gorilla*) are the largest, strongest primates, and they inhabit the mist forests of western and central Africa. Male gorillas can be nearly 2 m (6½ ft) tall and weigh over 200 kg (450 lb), and while they will occasionally climb trees, they are mainly animals of the forest floor. The animals knucklewalk, and the males beat their breasts and roar loudly when aroused. Gorillas live as mixed-sex and mixed-age communities and migrate through the forests in search of abundant sources of fruit. Language experiments have revealed that gorillas are quite intelligent, have a good sense of humor, and can comprehend a large vocabulary of words and phrases. (c) Chimpanzees are our closest relatives and share 99 percent of our genes. Their brains are highly developed and resemble our own in external appearance and internal anatomy. They live in the forests of western and central Africa in family groups with a few males, several females, and numerous offspring. Chimpanzees forage during the day and huddle together at night on platforms that they construct from branches and leaves. They use tools (sticks and grass) in nature, and in captivity can be trained to carry out complex tasks, including how to communicate in American sign language. They express affection by hugging and kissing, and happiness by laughing and jumping up and down; and when unhappy, they cry or sob loudly.

The apes—the gibbons and orangutans of Asia, and the gorillas and chimpanzees of Africa—are the largest primates and our closest relatives. They are often called *hominoids* ("resembling or related to humans") after the name for the superfamily Hominoidae that contains the ape family (Pongidae) and the human family (Hominidae). Apes are large, tailless animals with long arms, large brains, and complex social behavior. Figure 19.22 describes the hominoids in more detail.

Primate Characteristics

Primates evolved several specializations for living in trees (and later, on grassy savannas) that have added to their success as a group and that form the background for human physical features.

■ **Vision**   An arboreal habitat is rich with visual information, such as fluttering leaves, moving spots of sunlight and shade, and tangled tree limbs. Thus, it is not surprising that primates evolved **stereoscopic vision:** overlapping visual fields with good depth perception that allow the animal to discriminate distances well. Stereoscopic vision is important for maintaining balance up in the trees and for catching small, mobile prey such as insects, a favorite prosimian food. Bony sockets protect the forward-facing eyes in the later monkeys and apes, and together, the eyes and brain create a realistic three-dimensional picture of the environment, which helps the animals to identify foods, mates, and predators.

■ **Brain**   The evolution of life in the trees and stereoscopic vision was accompanied by the development of powerful arms that allowed climbing and swinging movements, as well as dexterous hands with opposable thumbs for grasping tree limbs. The primates also evolved complex social systems, and these as well as vision and dexterity depended on large brains. Prosimians have fairly large brains for animals of their size and body weight; monkeys have proportionately larger brains than prosimians; and apes and humans have the most complex brains of any mammals—both in terms of size (in relation to the rest of the body) and number of nerve cells and in terms of the intricacy of internal connections between brain regions (see Chapters 27 and 28).

■ **Infant Care**   Along with their mobile life in the trees, primates evolved new reproductive strategies. Compared with most other mammals, they give birth to smaller litters of young, with most species producing just one young at a time. The infants tend to be helpless (especially those of apes and humans) and depend on their parent(s) for complete physical care for a long period after birth. The higher a species' intelligence, the more the parent must educate the young. This great investment of parental care pays off in the high survival rates of the young.

■ **Upright Gait**   Although humans are the only fully upright primates, most monkeys and apes spend more time vertical than horizontal, clinging upright to tree trunks, sitting upright on the ground, running on two legs with the tail providing balance, and leaping with body held vertically. Even though gorillas and chimpanzees walk on all fours by touching the ground with their feet and the knuckles of their hands, their arms are so long that the body is still fairly erect (Figure 19.23). Upright posture improves visibility and leaves the hands free for other activities. As we will see, this was one of the major factors in human evolution.

■ **Teeth**   Primates have teeth modified for an omnivorous diet of both plant and animal matter. The earliest mammals had 44 pointed (carnivorous) teeth, adapted for slicing up the bodies of insects. Early primates had similar teeth, but as later primates changed to diets of leaves or fruit, their back teeth, through natural selection, began to broaden and flatten, and they were able to consume a wider variety of foods.

Each of these primate characteristics—stereoscopic vision, large brain, extended infant care, upright gait, and omnivorous teeth—played an important role in the success of the primate lineages.

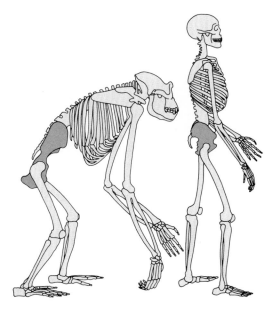

FIGURE 19.23 ■ Human and Gorilla Compared: A Matter of Skeletal Proportions.   The bones of the gorilla and the human are somewhat similar in shape and function, but the proportions are very different. The gorilla pelvis, for example, here shown in orange, tilts and cants the large rib cage and heavy neck and head forward, while the human pelvis is vertical and helps hold the entire skeleton upright.

Primate Evolution

How did early tree shrews lead to the modern groups of pro-simians, monkeys, apes, and humans? Genetic analyses and fossilized bones and footprints suggest that monkeys arose from prosimian ancestors about 50 million years ago and diverged into the New and Old World families (see Figure 19.18).

By 20 million years ago, a radiation of these hominoids was under way in African forests and grasslands. One of the hominoids was *Proconsul africanus,* a primitive tree-dwelling, fruit-eating ape that left a trail of fossils in East Africa. A later arrival, *Sivapithecus,* had heavy jaws, large

molars, and thick enamel, suggesting a diet of hard foods, such as seeds and nuts (Figure 19.18).

By comparing the similarity of DNA from pairs of living hominoids, geneticists can quantify the genetic differences between them. They have found that the genetic distance between people and chimpanzees is less than the genetic distance between chimps and gorillas. These numbers mean that a chimp is genetically closer to *you* than it is to a gorilla! Using other techniques, biologists have interpreted the genetic distances to mean that gorillas probably branched off the evolutionary line leading to humans about 10 million years ago, while chimpanzees branched off about 6 million years ago (see Figure 19.18).

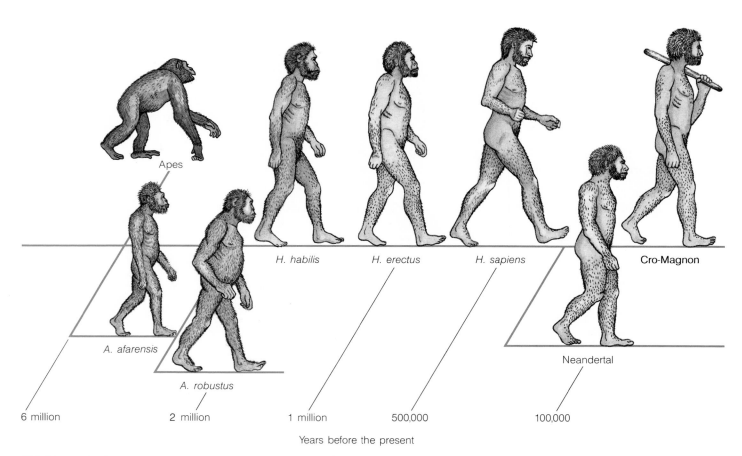

Apes

H. habilis   H. erectus   H. sapiens   Cro-Magnon

A. afarensis

A. robustus   Neandertal

6 million   2 million   1 million   500,000   100,000

Years before the present

FIGURE 19.24 ■ Family Album: Our Relatives. A family album of our hominid relations. The apelike *Australopithecus afarensis* (*Australopithecus* means "southern ape") lived about 3.75 million years ago and was the first fully upright hominid. The large-boned *A. robustus* lived about 2 million years ago, and recent evidence shows that its hand was adapted for precision grasping and it may have used tools. *Homo habilis* was an active tool user who lived about 1.5 million years ago and preceded *H. erectus.* Then *H. sapiens* emerged from *H. erectus* about 500,000 years ago. These species did not lead to humans in an advancing ladder, but occupied ends of branches in a family tree whose precise arrangement is still under investigation.

All of the final evolutionary branching of early primate species leading to *Homo sapiens* took place in less than 6 million years—an extremely short span in geological history. Using a 24-hour analogy, the earth formed at midnight, life appeared at 4:10 A.M., the first vertebrates appeared at 9:35 P.M., and the first humans arose only 38 seconds before the clock struck midnight. Let's examine that last "38 seconds."

THE RISE OF *HOMO SAPIENS*

Most anthropologists agree that, at present, the primitive tree-dwelling ape *Proconsul* represents the earliest identifiable genus on the path that eventually led to humans. Exciting fossil finds in Tanzania and Ethiopia in the mid-1970s, however, revealed a species that virtually everyone agrees was the earliest true *hominid* (early human) yet discovered: *Australopithecus afarensis* (Figures 19.18 and 19.24). *A. afarensis* had a very apelike skull and teeth, a brain only slightly larger than a chimpanzee's (450 ml), and long upper but short lower limbs. At the same time, however, the head sat on top of the backbone like ours does rather than projecting forward like an ape's, and the hands had humanlike bones (although chimpanzee-like joints). Most importantly, footprints show that the legs, feet, and pelvis were modified for fully upright walking on two legs (*bipedalism*). *A. afarensis* was small, standing only 1.1 to 1.4 m (3.5 to 4.5 ft) tall and weighing just 18 to 22.7 kg (40 to 50 lb). But its discovery proves fairly conclusively that an upright, two-legged stance was the first major human trait to evolve, long before an enlarged brain or evidence of tool use. It also suggests that these animals had hands free for carrying food or young and could see long distances in search of predator or prey.

Fossils show that australopithecines ("southern apes") emerged in Africa by 3 million years ago and that several species resulting from a radiation of the australopithecines lived simultaneously or in overlapping time periods in a multi-branched human family tree. All were upright animals with large faces, small brains, heavy jaws, and large, crushing molars up to 2½ times the size of our own (see Figure 19.24). The australopithecines were probably all vegetarians who used their massive teeth to process coarse, abrasive foods, such as tubers, seeds, and hard nuts. No conclusive evidence has been found to suggest that these early hominids made tools or hunted; thus, their behavior was probably much simpler than that of later humans.

Homo habilis

By 2 million years ago, one of the early australopithecines had given rise to the first human (or member of the genus *Homo*): *Homo habilis* (the "handy man"; see Figure 19.24). *H. habilis* individuals resembled their forebears, having large faces, big teeth, and trunks and limbs fully adapted for walking upright. At 1.5 m (5 ft) tall, however, they were larger than australopithecines, while their brains, at 700 ml, were half again as large as those of their predecessors. Even though this was still only half the size of a modern human's brain, it suggests a real increase in intelligence and probably a heightened degree of finger control.

Significantly, the first evidence of toolmaking and the butchering of animals appears in the fossil record about the time of *H. habilis*. Anthropologists speculate that the "handy man" hunted small animals and scavenged larger animals that lions or hyenas had killed. *H. habilis* individuals probably made crude stone tools by cracking rocks and flaking off chips, and they probably butchered their game. *H. habilis* and *Australopithecine robustus* probably lived at the same time and in the same areas, but the "handy man" survived and the "robust southern ape" became extinct. Why? Anthropologists speculate that the crude tools and larger brain of *H. habilis* may have enabled that species to compete better for food, or perhaps the "handy man" even killed *A. robustus* for meat. Meat eating set *H. habilis* apart from all earlier hominids and hominoids, and a way of life based on their newly evolved use of tools dominated the rest of human evolution.

Homo erectus

Whereas *H. habilis* had been restricted to the savannas and woodlands stretching from northeastern to southern Africa, a new species, *Homo erectus* ("erect man"), arose sometime before 1.5 million years ago and spread throughout northern Africa, southern Asia into Indonesia, and probably into southern Europe. A nearly complete skeleton of an *H. erectus* male was discovered in 1985 in northern Kenya and determined to be 1.6 million years old (Figure 19.25). The stage of bone growth showed that the male was a 12-year-old boy who would have been 6 ft tall had he lived to adulthood. The brain volume (800 ml) was larger than that of *H. habilis*, the teeth and skull had changed shape, and scientists speculate that the parts of the brain that govern abstract thought and reasoning may have undergone some internal reorganization. *H. erectus* made a new kind of stone tool that required finer skill: a sharp hand ax.

Both the geographical range and the new tools suggest a greater ability to deal with varied food resources and extremes of climate. Fossil evidence indicates that *H. erectus* hunted elephants, bears, antelope, and other large game in cooperative bands and transported, slaughtered, and cooked the meat over camp fires (which also provided warmth for long winter nights). *H. erectus* probably first used fire about 500,000 years ago as the species spread to the temperate zones of Europe and Asia. Cooperative hunting, cooking, and sharing were probably developed and maintained through *cultural transmission;* that is, they were taught to the next generation.

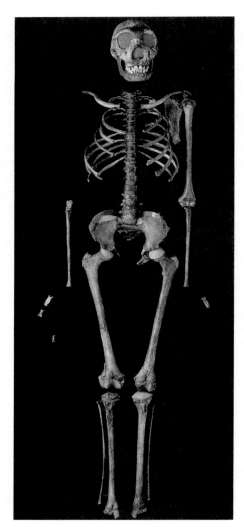

FIGURE 19.25 ■ *Homo erectus:* **A 1.6-Million-Year-Old Skeleton.   This nearly complete skeleton is that of a 12-year-old boy who would have been 6 ft tall as an adult.**

The discovery of this fairly complete *H. erectus* skeleton did reveal some very important clues about human evolution. The head of the 12-year-old boy was rather large, the pelvis was fairly narrow, and the boy had yet to reach adult size and sexual maturity. These factors suggest that the same pattern of human development we see today may have already emerged 1.6 million years ago. Modern human babies are born in a physically helpless state. The large head (and brain) could not pass through a woman's pelvis at birth if gestation were any longer and physical development in the womb any more complete. And our postbirth growth and maturation process continues for upward of 12 to 15 years. Initial helplessness and long growth phase ensure that our large brains will be able to store enormous amounts of survival information, while the long period of physical dependence during childhood will keep us in contact with potential teachers—parents, grandparents, siblings, and neighbors.

Homo sapiens

During the million years that early humans inhabited the Old World, their faces and teeth slowly decreased in size, and their brains enlarged. These gradual changes led imperceptibly from *H. erectus* to *Homo sapiens* by about 500,000 years ago.

Fossil evidence shows that *H. erectus* established populations in Africa, Europe, and Asia and were supplanted by *H. sapiens,* but anthropologists disagree about how that replacement occurred. Some suggest that our species arose in Africa and then spread throughout the Old World. Two groups of humans have occupied the earth in the millennia since that time: archaic *H. sapiens* (including Neandertals and others) and modern *H. sapiens* (including the Cro-Magnons and all the current races of people). Archaic humans dominated the Old World from about 400,000 years ago until they were superseded by modern humans between 30,000 and 100,000 years ago.

Neandertal fossils, first discovered in Germany's Neander Valley in 1856, were the first hominid remains to be studied scientifically. Neandertals have been much maligned as brutish, ignorant savages, but the accumulated facts speak quite differently. Living in Europe and the Middle East, Neandertals were short, stocky, powerfully built people with large protruding faces and characteristic projecting brow ridges. Their brains were similar in organization but *larger* (1400 ml) on average than a modern human's. The tools and other artifacts they left behind suggest an ability to deal with the environment through learned cultural behavior rather than brute physical force. They made spears and spearheads for hunting large game and scrapers for cleaning animal hides. They routinely built shelters, and their large front teeth were often worn, perhaps from chewing hides, as the Eskimo did traditionally, to make clothing. The skeletons of old and crippled Neandertals showed that the strong cared for the elderly and infirm, and they buried their dead ceremoniously with fine stone tools, servings of game meat, and even flowers—evidence of belief in an afterlife.

Modern humans did not descend from Neandertals but rather arose elsewhere (fossil evidence suggests Africa), migrated northward, and eventually outcompeted the archaics on their own turf. The two types may have coexisted for thousands of years, but by 34,000 years ago, the Neandertals had died out.

The Cro-Magnons looked distinctly different from their archaic cohorts. Their faces were smaller, flatter, and less projecting than the Neandertals', the heavy brow ridges had all but disappeared, and their skulls were higher and rounder. Their limbs were more slender (gracile, but still stoutly athletic compared to our own), their teeth were smaller, and most important, their culture was vastly more complex. Their tool kits contained knives, chisels, scrapers, spearheads, axes, and tools for shaping other tools of rock, bone, and ivory. They left dozens of cave paintings, engravings, and sculptures, suggesting a major development of symbolic forms of communication, probably accompanied by increased language abili-

(a)   (b)

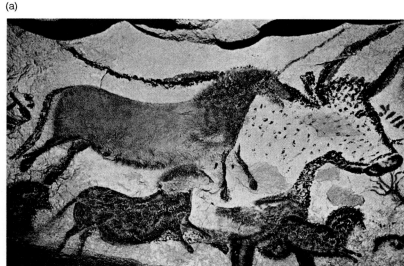

FIGURE 19.26 ■ The Origins of Art. Cro-Magnon peoples left numerous cave paintings throughout Europe and other continents beginning about 29,000 years ago. (a) This famous cave painting from Lascaux Cave in southern France depicts a bull and running horses. (b) The famous Venus of Lespugue, a statue carved of mammoth ivory, found in Lespugue, France, and dated at about 25,000 years ago, is one of humankind's first fertility symbols. It depicts a voluptuous, reproductive-age female with head bowed, probably as part of a fertility ritual. The appearance of art in human evolution indicates a major increase in the quality of information used by social groups and the need to encode it for future retrieval.

ties (Figure 19.26). Their living sites and shelters became larger and more complex, and human burials became more common and elaborate, indicating the establishment of religion.

The success of these cultural developments spurred a population expansion. Modern humans followed the herds of mammoths, woolly rhinoceroses, reindeer, and other game into the arctic regions of Eurasia and across the Bering land bridge into the Americas. They also built boats to carry them across uncharted waters to New Guinea and Australia. By 15,000 to 20,000 years ago, people had occupied virtually all the inhabitable regions of the earth.

Anthropologists have debated whether *H. sapiens* arose at a single site (probably Africa) and later spread around the world, or whether *H. erectus,* which was already widely distributed, evolved into *H. sapiens* at the same time in different places. Both fossil and genetic evidence supports the single-origin model. Sensational data come from analyzing mitochondrial DNA with restriction enzymes (see Chapter 11). Since animals inherit their mitochondria only from their mother's egg cell, the data suggest to some that all living people can trace their ancestry through their mothers, grandmothers, and great-grandmothers back to a single female who lived about 200,000 years ago, most likely in Africa. The media have dubbed this woman "Mitochondrial Eve." Males of many other families surely contributed genes in cell nuclei,

many other women were present in the human breeding population, and the time may have been earlier. Nevertheless, this one African woman was, in a sense, mother of us all. It was her descendants who left Africa and, in a series of major migrations, eventually populated the globe.

The accumulation of skills and knowledge, passed down through thousands of human generations in a *cultural evolution,* has allowed humans to invent agriculture and control food supplies, eliminate predators and some diseases, and blunt the effects of natural selection on themselves and their domesticated species. While we *Homo sapiens* arose via the same mechanisms and faced the same survival pressures as the other primates, mammals, and vertebrates before us, cultural evolution has changed our relationship to the planet. Today, we are the only creatures to control most aspects of our own existence, and with that control lies the future of all species.

CONNECTIONS

The evolutionary distance is great from a tunicate to a tiger or a Thailand farmer, yet each is a member of the same phylum, Chordata, and each arose from the same marine ancestors that lived more than 500 million years ago. Innovations, radiations, and specializations within the vertebrates led to the emergence of amphibians, reptiles, birds, and finally mammals, including the biggest, fastest-running, and most intelligent animals ever to live. But that should not overshadow the continued coexistence of millions of invertebrates, which may not have all of the mammals' complex physiological systems, but which have their own successful adaptations to life in modern environments.

It is interesting that only one mammal, *Homo sapiens,* is in a position to study the emergences, extinctions, and adaptations of fellow living things. Our species' enormous brain capacity and concomitant cultural development were the nat-

ural outcomes of primate evolution. That evolution went from primitive, shrewlike ancestors to students of biological science in just 65 million years—far less time than elapsed between the origin of the lobe-finned fishes and their first labored steps on land. We alone have the capacity to trace our historical roots, and that makes us unique. But the tortuous evolutionary pathway leading to the human brain was no different in principle from the roads that began at the origin of life and ended in the modern prokaryotes, protists, fungi, plants, and other animals. The following parts of the book will explore how animals and plants, once evolved, could overcome the day-to-day challenges of survival.

Highlights in Review

1. Members of the phylum Chordata possess a notochord, a hollow nerve cord, gill slits, and a tail with muscle blocks at some point in their life cycle.
2. Tunicates, or sea squirts, are stationary marine animals with a tadpolelike larva displaying the five chordate traits.
3. In the small, fishlike lancelets, adults show all chordate traits.
4. The fishes, arising some 550 million years ago, were the earliest vertebrates, and their innovations include a skull, bones, cartilage, vertebrae, jaws, and paired appendages (fins).
5. Early jawless fishes, or agnathans, gave rise to modern jawless fishes, such as the lamprey; to early jawed fishes, or placoderms, now extinct; to cartilaginous fishes, such as sharks, skates, and rays; and to modern bony fishes, or teleosts, of which there are 30,000 species.
6. The amphibians were the first vertebrates to live on land. Amphibian innovations include front and hind legs; fully functioning lungs and a moist, scaleless skin capable of gas exchange; and a chambered heart.
7. The amphibians radiated into a large, diverse group but now include only frogs, toads, salamanders, and wormlike apodes.
8. Reptilian descendants of the amphibians have dry, scaly skin; an expandable rib cage; a male copulatory organ; eggs containing the watery amniotic sac surrounded by a leathery shell; and legs directly beneath the body.
9. Early reptiles gave rise to modern reptiles, birds, and mammals. About 6500 reptile species survive today, including snakes, turtles, lizards, and crocodilians.
10. Bird innovations include feathers, homeothermy (warm-bloodedness), one-way air flow through the lungs, and hard-shelled eggs.
11. The mammals are warm-blooded and have a four-chambered heart like birds but show two additional innovations: body hair and milk production. A placenta also develops in most female

mammals during pregnancy. The monotremes, however, lay leathery eggs, and marsupials nurture their immature embryolike young in a pouch.
12. Mammals show four major types of specializations: constant warm body temperature (homeothermy); varied shapes of limbs and teeth; prolonged parental care; and highly developed nervous systems, senses, and behaviors.
13. Humans belong to the order Primates, along with the prosimians, the monkeys, and the apes. Primates, as a group, have evolved a keen sense of sight; powerful arms for climbing and swinging; dexterous hands with opposable thumbs; cooperative social systems; the most complex brains of any mammals; extended infant care and education; and a partially or fully upright gait, which frees the hands for carrying food or young and for using tools.
14. The earliest true hominid on which most agree is *Australopithecus afarensis,* a small but fully upright African protohuman dated at more than 3 million years ago. Other australopithecines arose over the next 2 to 3 million years, all vegetarians with massive faces and teeth.
15. *Homo habilis* arose from the australopithecines but had a larger brain and made crude stone tools. *Homo erectus* had a still larger brain, more sophisticated tool use, cooperative hunting, and perhaps language. Archaic *Homo sapiens,* including Neandertals, were stocky, powerful people with protruding faces, brow ridges, and very large brains. They made several kinds of stone tools, built shelters, cared for their elderly and sick, and buried their dead.
16. Modern *H. sapiens,* including Cro-Magnons, coexisted with the Neandertals and eventually took over their territories in Europe. They had higher, rounder skulls, more slender limbs, far more complete tool kits, and other examples of culture.

Key Terms

Study Questions

Review What You Have Learned

1. Which life stage of the tunicate, the larva or the adult, bears more resemblance to other chordates? Explain.
2. What characteristic structures do all chordates possess at some time in their life cycle?
3. How do members of Chondrichthyes and Osteichthyes differ? How are they similar?
4. Describe the lobe-finned fishes, and name a living example.
5. List three typical characteristics of amphibians.
6. List typical characteristics of reptiles.
7. Why do researchers consider *Archaeopteryx* proof that birds arose from early reptiles?
8. Give the main characteristics of birds. Of mammals.
9. Biologists group mammals according to mode of reproduction. List the three groups, and name at least two members of each.
10. List five characteristics of primates.
11. Summarize the evolutionary and genetic relationships between gorillas, chimpanzees, and humans.
12. How did *Homo habilis* and *Homo erectus* differ?
13. When did Neandertals live? What were the major features of Neandertal body structure and culture?
14. What evidence indicates that most lineages of *Homo erectus* became extinct and that *Homo sapiens* evolved in a limited geographical location?

Apply What You Have Learned

1. Cartoons frequently show dinosaurs and humans together. Is this picture scientifically accurate? Explain.
2. Many illustrations place *Homo sapiens* at the top of the evolutionary tree. Is this an appropriate representation of the evolutionary process? Why or why not?
3. Fishermen off the coast of Madagascar recently discovered some living coelacanths. Some biologists refer to these fish as "living fossils." Why?
4. An organism may evolve with adaptations that are beneficial early in its evolutionary history, but maladaptive as environmental conditions change later in evolutionary history. Describe a trait that was beneficial when humans existed in scattered bands, but is maladaptive when 5 billion people inhabit the globe.

For Further Reading

ALEXANDER, R. M. *The Chordates.* Cambridge, England: Cambridge University Press, 1975.

BENTON, M. J. "The Relationships of the Major Groups of Mammals: New Approaches." *Trends in Ecology and Evolution* 3 (1990): 40–45.

BURGIN, T., O. RIEPPEL, P. M. SANDER, and K. TSCHANZ. "The Fossils of Monte San Giorgio." *Scientific American,* June 1989, pp. 74–81.

CREE, A., and C. DAUGHERTY. "Tuatara Sheds its Fossil Image." *New Scientist,* October 20, 1990, pp. 30–34.

DIAMOND, J. "Dawn of the Human Race: The Great Leap Forward." *Discover,* May 1989, pp. 50–60.

GIBBONS, A. "Our Chimp Cousins Get That Much Closer." *Science* 250 (1990): 376.

GRIFFITHS, M. "The Platypus." *Scientific American,* May 1988, pp. 84–92.

KLEIN, R. G. *The Human Career: Human Biological and Cultural Origins.* Chicago: University of Chicago Press, 1989.

LUCAS, S. "The Rise of the Dinosaur Dynasty." *New Scientist,* October 6, 1990, pp. 44–46.

PIETSCH, T. W., and D. B. GROBECKER. "Frogfishes." *Scientific American,* June 1990, pp. 56–63.

SIMONS, E. L. "Human Origins." *Science* 245 (1989): 1343–1357.

STRINGER, C. "The Asian Connection." *New Scientist,* November 17, 1990, pp. 33–36.

STRINGER, C. B., and P. ANDREWS. "Genetic and Fossil Evidence for the Origin of Modern Humans." *Science* 239 (1988): 1263–1268.

WELLNHOFER, P. "Archaeopteryx." *Scientific American,* May 1990, pp. 42–49.

YOUNG, J. Z. *The Life of Vertebrates.* 3d ed. New York: Oxford University Press, 1981.

PART FOUR

HOW
ANIMALS
SURVIVE

CHAPTER 20

An Introduction to How Animals Function

A Whale of a Survival Problem

An intrepid visitor to the perpetually frozen Antarctic could stand at the coastline, raise binoculars, and witness a dramatic sight just a few hundred meters offshore: a spout as tall and straight as a telephone pole fountaining upward from the blowhole of a blue whale (*Balaenoptera musculus*), then condensing into a massive cloud of water vapor in the frigid air. The gigantic animal beneath the water jet would be expelling stale air from its 1-ton lungs after a dive in search of food. Then, resting at the surface only long enough to take four deep breaths of fresh air, the streamlined animal would raise its broad tail, thrust mightily, and plunge into the ocean again (Figure 20.1). The observer on shore might see such a sequence only twice per hour, since the blue whale can hold its breath for 30 minutes as it glides along like a submarine, swallowing trillions of tiny shrimplike animals called krill.

It is difficult to comprehend the immense proportions of the blue whale, the largest animal ever to inhabit our planet. At 25 to 30 m (80 to 100 ft) in length, this marine mammal is longer than three railroad boxcars and bigger than any dinosaur that ever lumbered on land (Figure 20.2). It weighs more than 25 elephants or 1600 fans at a basketball game. Its heart is the size of a beetle—a Volkswagen beetle. And that organ pumps 7200 kg (8 tons) of blood through nearly 2 million kilometers (1.25 million miles) of blood vessels, the largest of which could accommodate an adult person crawling on hands and knees. The animal has a tongue the size of a grown elephant. It has 45,500 kg (50 tons) of muscles to move its 54,500 kg (60 tons) of skin, bones, and organs. And this living mountain can still swim at speeds up to 48 km (30 mi) per hour!

Leviathan proportions aside, it is difficult to grasp the enormous problems that so large an organism must overcome simply to stay alive. For starters, a blue whale is a warm-

FIGURE 20.1 ■ **Blue Whale Plunging for Krill.** Earth's largest animal, the blue whale, represents a classic problem of homeostatis: how constant internal conditions can be maintained despite wide variations in environmental conditions.

Apatosaurus
(80 ft long)

Elephant
(12 ft tall)

Person
(6 ft tall)

Blue whale
(80-100 ft long)

FIGURE 20.2 ■ **A Whale to Scale.** A blue whale is longer and far heavier than an elephant or even an *Apatosaurus* (formerly *Brontosaurus*), the longest land animal that ever lived.

blooded animal with a relatively high metabolic rate; to stay warm and active in an icy ocean environment, it must consume and burn 1 million kilocalories a day. This it does by straining 3600 kg (8000 lb) of krill from the ocean water each day on special food-gathering sieve plates. In addition, each of the trillions of cells in the whale's organs must exchange oxygen and carbon dioxide, take in nutrients, and rid itself of organic wastes, just as a single-celled protozoan living freely in seawater must do. Yet a given whale cell—a liver cell, let's say—can lie deep in the body, separated from the environment by nearly 2 m (6 ft) of blubber, muscle, bone, and other tissues. For this reason, the whale needs elaborate transport systems to deliver oxygen and nutrients and to carry away carbon dioxide and other wastes. Finally, the galaxy of living cells inside a whale must be coordinated and controlled by a brain, a nervous system, and chemical regulators (hormones) so that the organism can function as a single unit.

Although blue whales are the largest animals that have ever lived, they share with all other animals the same fundamental physical problems of day-to-day survival: how to extract energy from the environment; how to exchange nutrients, wastes, and gases; how to distribute materials to all the cells in the body; how to maintain a constant internal environment despite fluctuations in the external environment; how to support the body; and how to protect it from attackers or from damaging environmental conditions. Blue whales have evolved with unique adaptations of form and function that meet such challenges and leave the animals suited to their way of life.

While the blue whale's solutions are specific to huge size, they represent general approaches that we will encounter

again and again in this part of the book. Our purpose here is to investigate these general shared approaches of animal anatomy and physiology. **Anatomy** is the science of biological structure; **physiology** is the study of how such structures work. While animals and plants share many basic elements of anatomy and physiology, there are also important differences; and for this reason, we separate the two subjects, covering animals in Part Four and plants in Part Five.

Our task in the present chapter is to give an overview of animal anatomy and physiology; we introduce the general principles here, and in Chapters 21 through 29 we delve into specifics. This chapter introduces the concept of **homeostasis,** the maintenance of constant internal conditions despite fluctuations in the external environment. (The word *homeostasis* comes from the Greek words for "same standing.") This process keeps an organism functioning on an even keel and maintains within a fairly narrow range of values the water and salt levels in body fluids; the levels of oxygen and carbon dioxide in the blood; the body temperature; and other aspects of physiology.

In this chapter, we will answer several questions about the general survival strategies of animals:

■ How do an animal's needs change along with its size?
■ What anatomical and physiological adaptations fit an animal to its environment?
■ How do feedback mechanisms help maintain a constant internal environment?

STAYING ALIVE: PROBLEMS, SOLUTIONS, AND THE ROLE OF PHYSICAL SIZE

Life arose nearly 4 billion years ago in a hostile environment—the black and blue world of churning volcanoes, searing ultraviolet light, and shallow seas we discussed in Chapter 15. Our planet, of course, is not so hostile now: By releasing oxygen, photosynthetic organisms helped form a protective ozone shield that screens out much of the damaging radiation; and the long, slow passage of time has changed the earth's geological features, weather, and climate considerably. Nevertheless, living organisms are subjected to wide variations in temperature, light, acidity, salinity, wind speed, and availability of water, minerals, and nutrients. These and other environmental factors create a shifting external setting to which organisms must adjust or die (Figure 20.3). At the same time, each living thing carries out certain activities basic to the life process (see Chapter 1), such as gathering energy, reproducing, and maintaining body integrity, and these activities (or the potential for them) must continue in a smooth, steady, orderly manner, or again, the organism will die. This steady state usually requires a continuous supply of energy and materials from the environment, as well as the dumping of wastes into it. The central problem for a living thing, therefore, is to maintain a steady state internally in the face of an often harsh and fluctuating external environment.

Solutions to the Central Problem: Homeostasis

Because cells are the basic units of life, essential life processes must be maintained in single cells as well as in larger organisms composed of many cells. The left-hand column of Table 20.1 lists these essential processes.

As we saw in Chapters 3 and 16, single cells are equipped with a number of organelles whose functioning allows the life processes to proceed in a constant and orderly way. We have also seen that diffusion across the plasma membrane is the means by which single cells exchange materials with their environment (see Chapter 4). A single-celled *Euglena* gliding about in a pond, for example, receives the gases, ions, and nutrients it requires by passive diffusion or active transport across its surface, then rids itself of waste materials in a similar way. The same is true for a multicellular organism like a flatworm, whose dimensions are thin and flat enough to bring each cell within diffusion distance of the outside world.

Cells and organisms must exchange materials with the environment, and diffusion is the means of transport at the cellular level. However, things become much more complicated for multicellular organisms too large and thick for each of their cells to be "serviced" by direct diffusion with the environment. As we saw in the case of the blue whale, a cell can be separated from the outside world by a distance of 2 m

FIGURE 20.3 ■ **Staying Alive: Maintaining a Steady State in the Face of a Changing Environment.** Despite the cold, howling wind, this Canadian musk ox has a constant warm internal temperature as a result of homeostatic mechanisms.

TABLE 20.1 Essential Life Processes	
Process	**Responsible Organ Systems**
Obtaining materials from the environment	Digestive system
Obtaining energy from the environment	Digestive system
Exchanging gases (oxygen and carbon dioxide) with the environment	Respiratory system
Regulating the composition of body fluids	Excretory system
Distributing materials within the body	Circulatory system
Defending the body from attack	Skin, immune system
Protecting the body from temperature extremes	Skin, circulatory system
Maintaining body shape	Skeleton, muscular system
Growing and reproducing	Reproductive system
Coordinating body activities	Nervous system, endocrine system

(6 ft) or more. Evolution has produced a number of special systems that overcome such distances by servicing the cells that lie deep inside multicellular organisms (see Table 20.1).

Figure 20.4 shows three of the systems—the digestive system, the respiratory system, and the excretory system— and the common and quite logical strategy that each employs. First, each has a supply or disposal tube leading from the outside environment deep into the animal's body. Second, each tube has special regions where substances can be exchanged between the contents of the tube and nearby body fluids bathing the tube. This exchange occurs by diffusion over short distances. Third, body fluids can be circulated to every cell in the body by means of bulk flow, a moving stream. And fourth, through diffusion, substances can move into each cell from the circulating fluids or be released by the cell and carried away. Consider the respiratory system, for example (see Figure 20.4): Oxygen enters the body in a tube and diffuses into the bloodstream in a special region, the lungs; bulk flow then carries the blood with its dissolved oxygen throughout the body; and finally, the oxygen can pass out of the blood and into individual body cells by diffusion.

Chapter 21 describes the process of gas exchange in more detail, along with the two or three different types of fluid most animals contain, each separated in its own "compartment": the fluid inside the cells (*intracellular fluid*); the fluid bathing the cells (*extracellular fluid,* or *tissue fluid*); and often a third fluid, *blood,* circulating partially or fully within a system of vessels. As materials pass from the fluid in one compartment to the fluid in another, the organism's homeostatic mechanisms maintain the proper composition of each type of fluid with only minor variations. This is more complicated than the lot of the one-celled organism, which maintains only a single kind of fluid inside the cell. This means that a multicellular

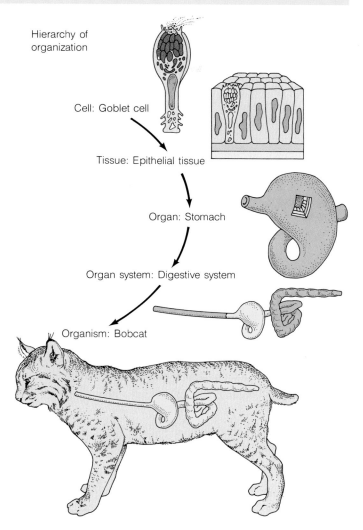

FIGURE 20.5 ■ Hierarchy of Organization in the Animal Body. The billions of cells in an animal's body are neatly organized into tissues, organs, and organ systems that carry out body functions efficiently. Organ systems working together in a coordinated fashion help an organism, such as an individual animal, stay alive.

organism has some control over the various aspects (e.g., temperature, salt content) of the immediate environment surrounding most of its cells. Still, each animal directly contacts the physical environment with its outermost surface and thus as a whole remains subject to environmental fluctuations.

Body Organization: A Hierarchy of Tissues, Organs, and Organ Systems

A multicellular organism, especially one as enormous as a blue whale, could not carry out its many survival tasks if its cells were crammed haphazardly into the body. Instead, the cells are organized into tissues, organs, and organ systems (Figure 20.5). A **tissue** is a group of cells of the same kind

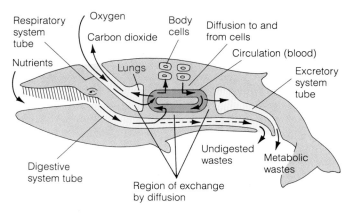

FIGURE 20.4 ■ Tubes and Diffusion: The Common Strategy for Exchanging Materials with the Environment. Animals obtain nutrients, exchange gases, and rid themselves of organic wastes by tubes that penetrate deep into the body. The tubes exchange substances with the blood by diffusion. Finally, substances diffuse to and from the blood and the individual cells. The digestive, respiratory, and excretory systems all work by means of a similar strategy involving tubes and diffusion.

(a) Type of epithelial tissue

(b) Functions of epithelial tissue

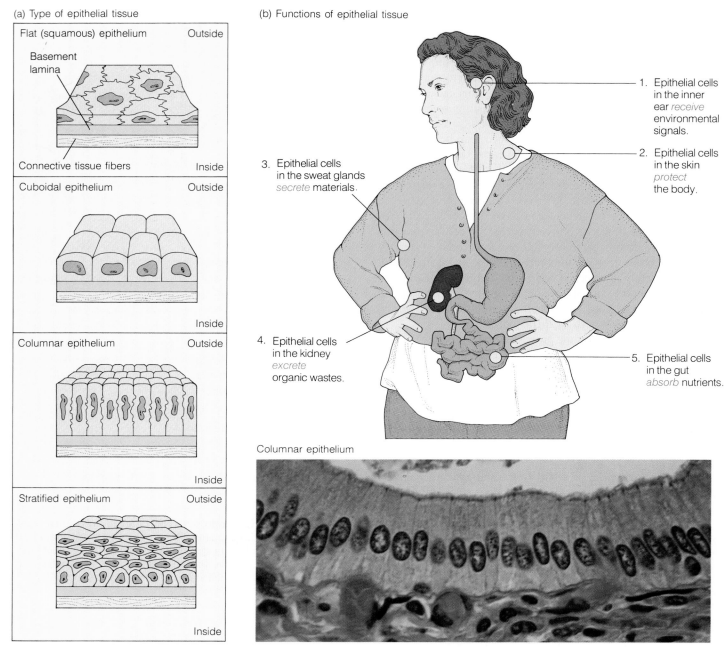

Flat (squamous) epithelium   Outside

Basement lamina

Connective tissue fibers   Inside

Cuboidal epithelium   Outside

Inside

Columnar epithelium   Outside

Inside

Stratified epithelium   Outside

Inside

1. Epithelial cells in the inner ear *receive* environmental signals.

2. Epithelial cells in the skin *protect* the body.

3. Epithelial cells in the sweat glands *secrete* materials.

4. Epithelial cells in the kidney *excrete* organic wastes.

5. Epithelial cells in the gut *absorb* nutrients.

Columnar epithelium

FIGURE 20.6 ■ Types of Epithelial Tissue and Their Roles in the Body. Epithelial tissue (or epithelium) consists of sheets of cells that cover body surfaces. Epithelium lines the outer layer of the skin, the inner surface of body cavities, and the chambers of the lungs and other organs. Closely packed epithelial cells have one surface free to face either the environment or a body cavity, while the other surface is pressed to a dense, tangled mat of proteins and carbohydrates called the basement membrane. Continuous cell division is common in epithelium, as cells are often worn away by abrasion both inside and outside the body. (a) Epithelial cells can be flat (squamous), cuboidal (cube-shaped), columnar, or stratified (arranged in layers several cells deep). The photo shows columnar epithelium lining a mammal's pancreatic duct. Regardless of shape or layering, epithelium does for animals what the cell membrane does for the single cell—act as thresholds through which substances must pass to enter or exit the body. (b) Epithelium functions in (1) reception of environmental signals; (2) protection of the body from invaders; (3) secretion of sweat, milk, wax, or other materials; (4) excretion of wastes; and (5) absorption of nutrients, drugs, or other substances. Damaged epithelia can greatly interfere with normal body functioning. For example, when cigarette smoke damages the cilium-covered epithelium which lines the breathing tubes, that epithelial lining can no longer remove harmful debris from the lungs and passageways as efficiently.

performing the same function within the body. For example, cells that line the outer surface of a whale's skin make up a tissue called the epithelium. An **organ** is a unit composed of two or more tissues that together perform a certain function. Your stomach, for example, is an organ containing four tissue types cooperating in the function of food storage and digestion. Several organs can form an **organ system,** which is defined as two or more interrelated organs that work together, serving a common function. The various systems listed in Table 20.1 are organ systems; the digestive system, for example, contains the salivary glands, liver, stomach, intestines, and pancreas functioning together in food processing. Collectively, all the organ systems make up an **organism,** in this case an individual animal.

The Four Types of Tissues

In the body of a mammal such as a blue whale or a human, there are over 200 types of cells but only 4 types of tissue: epithelial, connective, muscle, and nervous tissues.

Epithelial tissue consists of sheets of cells that cover or line body surfaces, like the outer layers of the skin or the inner layer of the intestine (Figure 20.6). Epithelial tissue also makes up the *exocrine glands,* which secrete substances like sweat onto the skin and digestive juices into the gut, and some of the *endocrine glands,* which secrete internal chemical messengers called hormones within the body. Epithelial sheets typically have one surface facing a space, like the air in the case of the skin or the urine-storing cavity of the bladder. The cells facing the space link tightly to each other by means of cell junctions (see Figure 3.28). These junctions bind adjacent cells together and stop the leakage of fluids from one side of the epithelium to the other. If these junctions did not exist, urine would leak out of the bladder's central storage cavity into the body. Junctions also link the inner surface of the epithelium to a thick meshwork of protein and polysaccharide fibers, the *basement membrane* or *lamina.* If it weren't for this firm bond, the skin would peel right off your body. Just beneath the basement lamina is the second tissue type, connective tissue.

Connective tissue binds other tissues together and supports flexible body parts. Reach up and gently squeeze the upper part of your outer ear—the stiff material you feel is a type of connective tissue called *cartilage* (Figure 20.7). Connective tissue includes not only cartilage, but bone, blood, and fat cells as well. Connective tissue generally lays down a network of protein fibers that holds other tissues together. One of these proteins is collagen, the most abundant protein in the body. Collagen is as strong as steel, weight for weight. A mutation in a collagen gene can cause the hereditary disease osteogenesis imperfecta, in which bones are so weak that the rib cage collapses at birth when the baby draws its first breath.

Muscle tissue contracts, or shortens, and allows an animal to move (Figure 20.8a; see also Chapter 29). Grab your upper right arm with the left hand and raise your right hand up to your chin. The hardening bulk you feel is muscle tissue as it shortens to move the arm. Muscle cells are generally quite long and spindle-shaped and filled with protein fibers. The fibers slide past each other like two parts of a telescope and cause the cell to shorten. There are three types of muscle: *smooth muscle,* which operates glands, blood vessels, and internal organs such as the intestine; *cardiac muscle,* which makes the heart beat; and *skeletal muscle,* which moves the bones, like those in your arms and legs. The contraction of muscle cells can be controlled by the fourth tissue type, nervous tissue.

Nervous tissue transmits electrochemical impulses; for example, you feel a pin prick on your finger tip because the pin deforms the nervous tissue and this triggers a nerve impulse (Figure 20.8b; see also Chapter 27). The essential feature of nervous tissue is its ability to transmit signals and thereby communicate. Nerve cells sense changes in their environment, process that information, and command muscles or secretory glands to act and thus adjust for the initial change. Nerve cells have very long, thin processes that act like telephone wires carrying messages from one place to another in the body. The longest cells of the body are nerve cells, or neurons. The activity of nervous tissue is so interesting that we devote two chapters to it—Chapters 27 and 28.

Most organs contain all four tissue types. In the small intestine, for example, the interior is lined with epithelial tissue. Connective tissue binds this epithelium to a layer of muscle tissue surrounding the organ. Finally, signals from nervous tissue can stimulate muscle contraction and speed the passage of materials through the small intestine.

Large Animals Face Additional Problems

We have seen that tissues organized into organ systems can overcome the problem of servicing billions of individual cells by transporting materials through tubes and close to the surface of each cell. But large animals have still other special needs not found in smaller, simpler animals: Large animals require strong structural support and sophisticated avenues of coordination and communication to integrate far-flung body parts.

Small aquatic organisms such as jellyfish and flatworms do not need extensive solid support because their tissue layers are thin, and seawater surrounds and supports them. Bulkier animals on land, however, need a scaffolding for muscle attachment and for keeping the organs from collapsing together into a quivering mass. The blue whale is a dramatic example. Even with its multiton skeletal system, it cannot live for more than a few minutes without the additional buoyant support of seawater. On land, the weight of its body quickly crushes its lungs and causes the animal to suffocate.

Large animals, like most others, require structures and processes that coordinate and integrate the actions of the various organ systems so that the organism functions efficiently as a single entity. The electrical signals of the nervous system

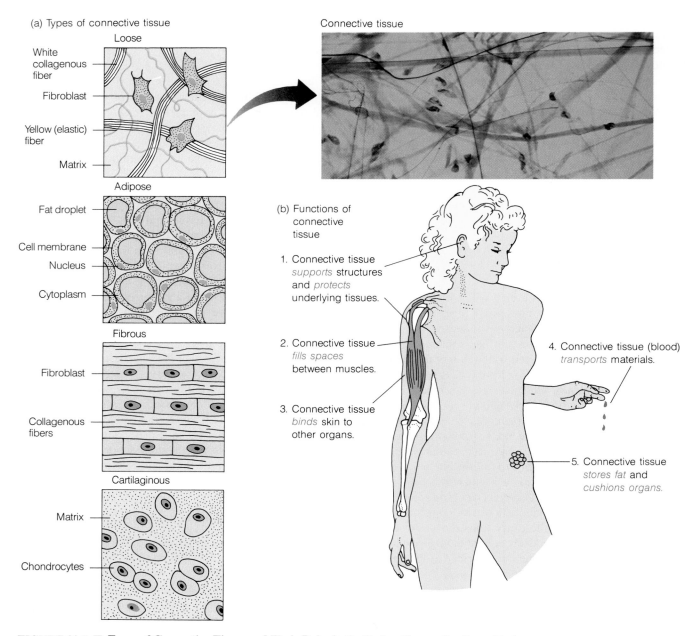

(a) Types of connective tissue

Loose

- White collagenous fiber
- Fibroblast
- Yellow (elastic) fiber
- Matrix

Adipose

- Fat droplet
- Cell membrane
- Nucleus
- Cytoplasm

Fibrous

- Fibroblast
- Collagenous fibers

Cartilaginous

- Matrix
- Chondrocytes

Connective tissue

(b) Functions of connective tissue

1. Connective tissue *supports* structures and *protects* underlying tissues.

2. Connective tissue *fills spaces* between muscles.

3. Connective tissue *binds* skin to other organs.

4. Connective tissue (blood) *transports* materials.

5. Connective tissue *stores fat* and *cushions organs*.

FIGURE 20.7 ■ Types of Connective Tissue and Their Roles in the Body. Connective tissue binds other tissues together and supports flexible body parts. It lies just under the basement membrane to which epithelium is attached, and it produces extracellular material that forms a matrix, or framework, for other structures. This matrix includes fibers of the protein collagen and elastic proteins, and it spaces connective tissue cells much farther apart than the densely packed epithelial cells. Connective tissue is so ubiquitous that the matrix material collagen is the body's most abundant protein, and weight for weight, collagen is as strong as steel. (a) The main types of connective tissue are loose connective tissue, which fills spaces between muscles and forms delicate membranes that connect epithelial layers to underlying organs; adipose tissue, or fat, which stores fat droplets and acts as a buffer around the kidneys, joints, and other areas; fibrous connective tissue, which contains a few cells but mostly collagen and elastic fibers, and which makes up tendons, ligaments, and similar tough, flexible structures; and cartilaginous tissue, or cartilage, which is made up of cells called chondrocytes, which secrete fibers in a gel-like matrix that is rigid enough to provide the framework of the nose and outer ears, attachment sites for muscles, and similar structures. The photo shows loose connective tissue from a mammal's breast. The small, brown ovals are cell nuclei. Some scientists also consider blood a connective tissue, as well as bone, which is made up of a hardened matrix of proteins and minerals plus living cells (see Chapter 29). (b) Connective tissue functions in a number of ways: (1) It supports and protects; (2) it fills spaces; (3) it binds skin to underlying organs; (4) it transports gases and nutrients; and (5) it stores fat and acts as a cushion.

(a) Muscle tissue

(b) Nervous tissue

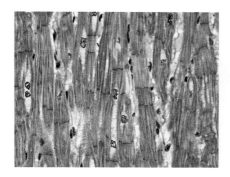
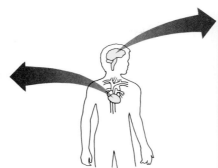
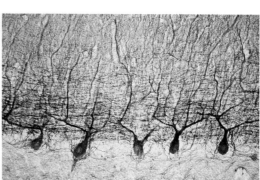

FIGURE 20.8 ■ **Muscle and Nervous Tissue.** (a) Muscle tissue, here cardiac muscle in a mammal's heart, can contract and shorten. The contraction of cardiac muscle can pump blood through the heart and blood vessels, while the contractions of smooth muscle can squeeze food and fluids through the stomach and intestines. The shortening of skeletal muscles can move body parts. (b) Nervous tissue, like this section from the human cerebellum, is the part of the brain that helps us maintain balance, among other basic functions.

rapidly integrate and regulate body functions, while the slower blood-borne chemical signals of the endocrine system cause slower, longer-lasting body reactions and changes. As we will see later in this chapter and throughout Part Four, both of these systems are necessary for an organism to maintain homeostasis—the internal constancy it needs to survive.

ANIMAL ADAPTATIONS: FORM AND FUNCTION SUIT THE ENVIRONMENT

Animals collect their organic nutrients from the environment, and the blue whale is the biggest heterotroph of all, with its 3600 kg (8000 lb) daily food requirement. It may seem ironic that the world's largest animal eats minute krill, but the blue whale's mouth is perfectly suited to harvesting this food source—the most abundant in cold ocean waters. The blue whale's sieve plates, or *baleen plates,* are evolutionary extensions of the ridges you can feel with your tongue on the roof of your mouth. But the whale has several hundred such ridges, each about 1 m (39.4 in.) long, that function in filter feeding like a giant sieve to collect bushels of krill, which the whale devours with one mighty swallow. In general, specialized anatomical adaptations allow different organisms to exploit their environments in different ways, and the adaptation's physical form is appropriate to its function.

The blue whale has numerous other anatomical and physiological adaptations that are fine-tuned to the animal's lifestyle and environment. Like all animals that swim rapidly, the blue whale has a highly streamlined spindle shape with hydrodynamic properties that allow it to move through the water with a minimum of drag. The ears have no external flaps, and the male genitals are retracted into a pouch. Besides streamlining the animal, this dearth of flaps and appendages also

helps to conserve heat by allowing a smaller body surface area for heat loss to the cold seawater. In strong contrast, a mammal like the desert hare in Figure 20.9 needs to dispose of body heat and has very large appendages—huge ears relative to its body size—that can act like organic radiators.

The whale's baleen plates and body shape are clearly fine-tuned to environmental constraints. But how do anatomical and physiological adaptations arise? The answer is evolution by natural selection. The traits we have been describing are

FIGURE 20.9 ■ **Tall Ears: Adaptions for Ridding the Body of Heat.** Unlike a whale, which conserves body heat when swimming in icy waters, an antelope jackrabbit (*Lepus alleni*) radiates excess internal heat during hot desert days. It does so through huge, thin ears with a large amount of surface area—organic radiators.

controlled by genes, and some of the blue whale's ancestors must have, for example, possessed genes for smaller ears. Presumably, whales with smaller ears could move through the water and obtain food more efficiently and thus could reproduce more efficiently than whales lacking these genes. Gradually, through natural selection, genes for smaller ear flaps became more frequent within the whale population. In general, the source of anatomical and physiological adaptations is natural selection, whether those adaptations are in a beetle, a sponge, a jackrabbit, or a whale.

HOMEOSTASIS: KEEPING THE CELLULAR ENVIRONMENT CONSTANT

During a regular checkup, your doctor can get a fairly accurate reading of how healthy you are by taking your temperature and blood pressure, listening to your heart and lungs, analyzing the contents of your blood and urine, and so on. Box 20.1 on page 440 describes the case of a young woman whose normal blood chemistry became badly disrupted because of hepatitis but was corrected through proper medical care. Since medical researchers have established a normal range of values for each measurement, chances are good that if yours fall within the correct ranges, your body is maintaining homeostasis, that is, keeping the internal environment at a constant healthy level. The take-home message is this: Your physiological systems continuously adjust to aspects of the environment in and around your cells, maintaining the extracellular environment within narrow limits suitable to the cells' efficient functioning.

To really understand how the body achieves homeostasis will require a reading of all the chapters on animal physiology, since the system covered in each chapter merely contributes to the overall stability of the internal milieu. Nevertheless, we can begin by looking at general strategies and then focus on one case history: how the blue whale stays warm in near-freezing polar waters but cool in tropical seas.

Strategies for Homeostasis: Feedback Loops

A highway driver keeping a vehicle within a narrow lane is a type of homeostatic system. This person must *sense* the situation at any given point ("Where am I on the road?"), *evaluate* it ("I'm too close to the shoulder"), then *act* (steer left). This requires three separate elements of the homeostatic system (Figure 20.10): a *receptor* to sense the environmental conditions (in this case, the eyes); an *integrator* to evaluate the situation and make decisions (here, the brain); and an *effector* to execute the commands (the muscles and bones of the hands and arms). Again and again in our discussion of physiological systems and their homeostatic functioning, we will see these common elements: receptors, integrators, and effectors. And we will see that the interplay of each set of

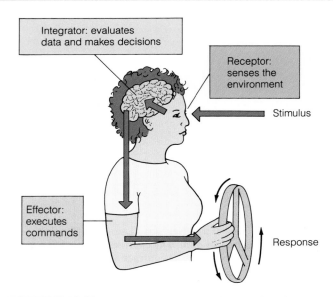

FIGURE 20.10 ■ Common Elements of Physiological Systems: Receptors, Integrators, and Effectors. These elements work together in feedback loops that help maintain homeostasis.

elements constitutes a **feedback loop,** a circuit of sensing, evaluating, and reacting that can resist change in the internal environment and maintain homeostasis.

Negative feedback loops resist change by sensing a stimulus (a change from a baseline condition, or set-point value), then activating mechanisms that oppose the trend away from the baseline. For example, a negative feedback system keeps your house warm in winter. As heat escapes through the walls, the inside temperature drops and the thermostat senses this change. It signals the furnace to turn on, and this raises the air temperature. Once the house is warm again, the thermostat stops sending the "turn-on" signal, and the furnace shuts off. This is a *feedback loop* because it starts with a temperature change (a decrease) and eventually feeds back to a temperature change (an increase). It is a *negative* loop because the response opposes the stimulus: The turning on of the furnace negates decreasing air temperature. Likewise, the whale's steady body temperature depends on an organic thermostat, and so does a person's.

Within the brain of a person or whale, a small region lying above the roof of the mouth, called the *hypothalamus,* acts as a biological thermostat. The hypothalamus has a "set point," a genetically determined ideal temperature for body cell activity (about 37°C, or 98.6°F in a person). Specialized hypothalamic nerves measure the temperature of blood flowing through the brain, while other nerves analyze information arriving from temperature sensors in the skin (Figure 20.11a). The hypothalamus then triggers mechanisms for increasing or dissipating heat. When the temperature of the skin and blood drops below a certain point, the hypothalamus signals effectors, the muscles, to begin to contract. At first these contractions are small and occur at different times in different muscle cells, but as the signal increases, muscle cells begin to fire in greater synchrony, and the individual begins to shiver. These contractions generate heat and thus help counteract the initial cold stimulus. If the environment grows too cold, however, and the body cannot keep up with heat loss, *hypothermia* may set in—lowered internal temperature, sensory confusion, and potential death. If, on the other hand, environmental temperatures are too high, the individual can experience *hyperther-*

mia—heightened internal body temperatures—which can also be damaging or even fatal.

While negative feedback loops resist internal change, positive feedback loops bring about rapid change. In a positive feedback loop, an initial change in one direction becomes pushed further and further in the same direction. The terrible screech you sometimes hear from a public-address system results from positive feedback. A microphone picking up sound from the speakers sends the sound back to be further

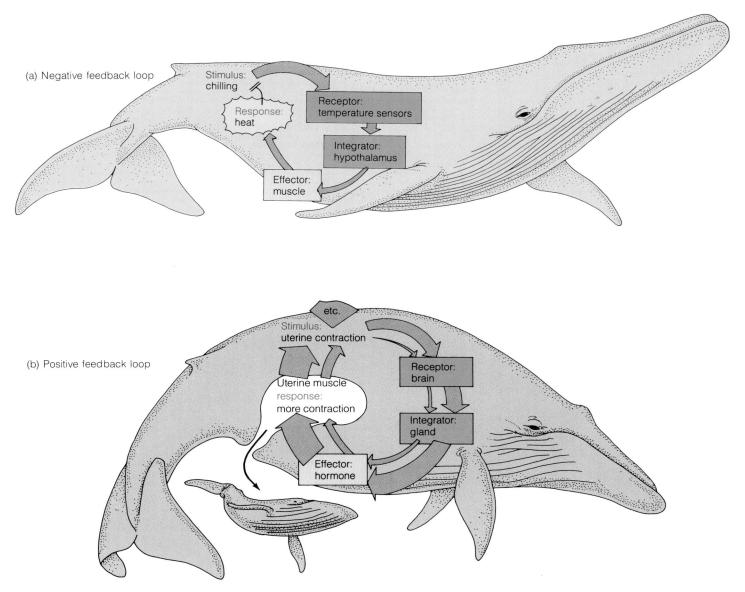

FIGURE 20.11 ■ Feedback Loops Prevent and Provoke Change: Negative and Positive Loops.
(a) When a whale is chilled, a negative feedback loop involving sensors in the whale's skin and hypothalamus region of the brain stimulates muscles to flex and warm the massive animal. Thus, the response of the control loop negates the original stimulus. (b) In a positive feedback loop, the response to the stimulus triggers *more* stimulus and *more* response in an increasing spiral. The hormonal loop that leads, in mammals, to stronger and stronger uterine contractions and eventually to birth is a positive feedback loop. Positive feedback loops generate changes, while negative feedback loops tend to resist them.

BOX 20.1   *PERSONAL IMPACT*

THE CASE OF MARTHA'S LIVER: A BLOOD TEST FOR HOMEOSTASIS

As a floor nurse at a major hospital, Martha depended on her boundless supply of energy. When a strange fatigue set in, however, and the whites of her eyes took on a yellowish cast, her upper abdomen began to hurt, and she grew nauseated, she was certain that she had contracted something serious and went to see her own doctor.

To the physician, the *jaundice* (yellowish pigmentation) suggested viral hepatitis, a disease in which a virus attacks liver cells. When aged red blood cells break down, they produce a dark yellow pigment called *bilirubin*. Now, a healthy liver can pick up the bilirubin and secrete it into the intestines for elimination from the body. In a diseased liver, however, pigment breakdown is disrupted, and the yellow substance builds up in the skin and eyes.

To test her suspected diagnosis of viral hepatitis, Martha's doctor drew some of her patient's blood (as the nurse in Figure 1 is doing with a different patient) and sent it to a medical laboratory for a *chem screen*. This is an analysis of 22 kinds of ions, small biological molecules, and enzymes found in every person's blood (Table 1, column A). If the person is healthy, his or her homeostatic mechanisms keep the quantity of each substance within a definable range of values (column B). But if an illness interferes with homeostatic feedback loops, one or more of the substances will appear in excessive or scanty quantities (column C).

For example, a healthy person has 0.1 to 1.2 mg of bilirubin per 100 ml of blood, but Martha had 3.6 mg, or three times more than the normal upper limit (column C). She also had up to 70 times the normal levels of five specific enzymes, abbreviated ALK PHOS, SGPT (ALT), GGT, SGOT (AST), and LDH. These enzymes appeared in her blood because the liver cells (which normally store the enzymes) were being broken down.

This particular pattern on the chem screen indeed indicates liver disease, and in combination with the other clues—jaundice, tender abdomen, fatigue, nausea, and Martha's exposure to patients—the doctor was able to confirm a diagnosis of viral hepatitis. Modern medicine as yet has no effective drugs for this condition, but Martha's physician prescribed bed rest and fluids, and within a few weeks, Martha's blood chemistry—and her old energy level—returned to normal.

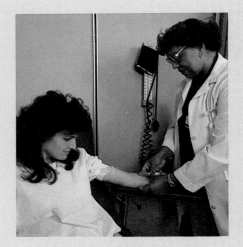

FIGURE 1 ■ **Taking a Blood Sample.**

TABLE 1   Readout of Patients' Blood Chemistry, with Hepatitis and Without

(A)	(B) Normal Range	(C) Hepatitis Patient
CHEM SCREEN		
SODIUM:	135–145 mEq/L*	140
POTASSIUM:	3.5–5.2 mEq/L	4.5
CHLORIDE:	95–109 mEq/L	104
CO2:	24–32 mEq/L	24
GLUCOSE:	65–110 mg/dl	93
BLOOD UREA NITROGEN:	9–20 mg/dl	6
CREATININE:	0.7–1.4 mg/dl	0.9
PROTEIN:	6.0–8.0 g/dl	7.1
ALBUMIN:	3.5–5.0 g/dl	4.2
GLOBULIN:	2.3–3.2 g/dl	2.9
CALCIUM:	8.5–10.5 mg/dl	9.4
PHOSPHORUS:	2.5–4.5 mg/dl	3.6
URIC ACID:	2.5–6.5 mg/dl	4.8
BILIRUBIN:	0.1–1.2 mg/dl	3.6
ALK PHOS:	30–110 IU/L	157
SGPT(ALT):	0–45 IU/L	2970
GGT:	3–60 IU/L	173
SGOT(AST):	7–40 IU/L	1990
LDH:	100–225 IU/L	500
IRON:	35–200 ug/dl	146
CHOLESTEROL:	130–240 mg/dl	142
TRIGLYCERIDES:	10–150 mg/dl	114

*mEq = milliequivalent weight; dl = deciliter (100 ml); IU = International Unit; μg = microgram. These units are commonly used in medical laboratories for measuring substances in the blood.

amplified. The microphone picks up that louder rebroadcast sound, the speakers reamplify it once more, and soon the sound mounts to an ear-piercing screech. In animals, positive feedback loops usually disrupt homeostasis for a specific purpose. An example is the positive feedback loop that drives labor and delivery in the birth of a person or whale (Figure 20.11b). The hormone oxytocin causes uterine contractions, and these strong muscular actions cause more oxytocin to be released, more contractions to occur, and so on. The feedback produces stronger and longer contractions spaced closer and closer together until, at the climax of the cycle, a human baby is born, or an infant blue whale (a whopping 7.5 m, or 25 ft, long!) is expelled into the sea. This explosive action stops the loop, and homeostasis—in this case, a noncontracting uterus—is restored.

Temperature Regulation: Homeostasis and Evolution

The maintenance of a fairly constant body temperature illustrates three general physiological mechanisms in addition to the action of negative feedback loops. The first two mechanisms govern fluid flow in the body, and the third involves an animal's behavioral adaptations.

The flow of body fluids, especially of blood, can help maintain body conditions such as temperature at constant levels. A blue whale has an enormous temperature problem, because it feeds in icy antarctic waters where it must resist chilling, and yet it must resist overheating during the months it spends in warm tropical waters to deliver its young. Whales have a layer of fat called *blubber* that is 30 cm (1 ft) thick and that completely encases the animal, much like a wet suit on a scuba diver. Blubber is a wonderful thermal insulator, capable of blocking the flow of heat from the body core outward into the ocean water. The blubber layer is crisscrossed by blood vessels, and when the animal is overheated—say, from exercising in warm water—blood can be shunted into these vessels. This conveys heat toward the body surface, where the surrounding lower-temperature water can absorb it (see Figure 20.12a). Your own skin may feel flushed and look red when you exercise heavily owing to a similar shunting mech-

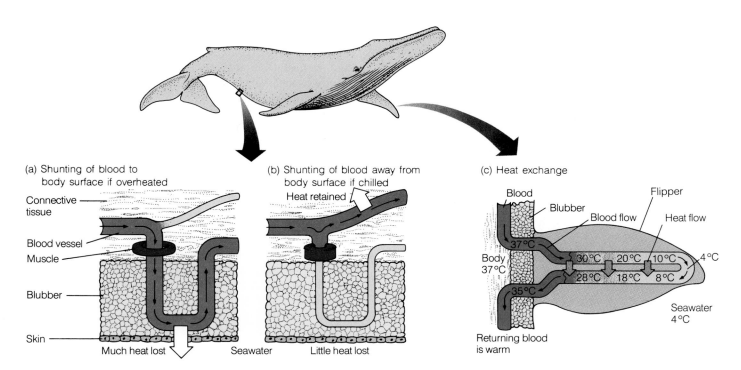

FIGURE 20.12 ■ **Temperature Regulation: Homeostasis and Evolution.** (a) A homeostatic mechanism involving tiny circular muscles that open blood vessels near the body surface allows an animal to get rid of excess body heat when it is surrounded by cooler seawater. (b) Closing the circular muscles that surround body surface blood vessels causes heat to be retained if the animal is chilled. (c) The blood vessels in a whale's flipper are arranged in a *rete mirabile* or "wonderful net." The vessels lie side by side in such a way that the current of hot blood flowing away from the body core moves counter to the current of warm blood flowing back toward the body, and heat flows directly from outgoing blood to incoming blood. With such a countercurrent flow arrangement, precious body heat is conserved rather than lost to the environment.

FIGURE 20.13 ■ **Gaping 'Gator: A Mechanism for Regulating Brain Temperature.** Alligator gaping, first observed by ancient Egyptians, helps reduce heat buildup in the animal's brain. Modern experiments prove that a basking 'gator can keep its body temperature high but its brain temperature lower and beneath the point of brain damage by locking its jaws open as it rests.

anism. When the whale is swimming in arctic waters, tiny muscles will divert blood from those same vessels, so that the animal's body heat stays in the core instead of circulating in the periphery (Figure 20.12b). Similar mechanisms that shunt blood to the intestines or muscles help keep other body conditions constant. The second fluid-related mechanism is called *countercurrent flow,* and it involves a special anatomical specialization that reduces heat loss through the whale's uninsulated flippers: a network of blood vessels that acts as a heat exchanger, passing heat from warm blood in the animal's core to the cool blood in the flippers (Figure 20.12c). (As we will see in Chapter 25, countercurrent flow also helps an animal's kidneys rid the body of wastes.)

Warm-blooded animals, such as mammals and birds (also called *homeothermic* or constant-temperature animals), employ negative feedback loops and fluid-shunting mechanisms in the maintenance of body temperature. An elevated body temperature, or fever, can help protect the animal from infection (see Box 20.2). The so-called cold-blooded animals (or *poikilotherms*), animals that obtain most of their heat from the environment, have a different set of evolutionary adaptations, many of them behavioral. A lizard or alligator, for example, will instinctively move in or out of the sun, do heat-generating push-ups, or bask with its jaws gaping open—all strategies that keep its body temperature between 35°C and 40°C (Figure 20.13). Similarly, an Indian python will rhythmically flex its muscles and warm incubating eggs; a honeybee will contract its wing muscles prior to flying in cool weather; and a box turtle will pant, thereby dissipating heat. Poikilotherms have physiological adaptations, too. Some freshwater fishes have two sets of cellular enzymes, one that functions best at cool temperatures and another that functions best at warm temperatures. And in some oceangoing fishes

such as tunas and sharks, hardworking red-colored muscles enable the animals to swim rapidly, and these muscles generate enough excess heat to release into the body core.

Whether the animal is endotherm or ectotherm, body temperature is just one feature kept in check by homeostasis. As we will see in later chapters, it takes many systems working simultaneously and under elaborate coordination and control to maintain an animal's total internal milieu despite the vagaries of the external world.

CONNECTIONS

The subject of animal anatomy and physiology flows naturally from the subjects covered in the first three parts of the book: the structures and activities of cells; how genes dictate those forms and functions; and how the five kingdoms of organisms evolved in all their diversity. Now we will see how members of the largest group, the animal kingdom, function internally and how this allows them to survive in and exploit their environments.

Throughout Part Four, we will focus primarily on the human. Not only do biologists know more about people than they do about any other organism, but understanding human physiology encourages us to take care of our bodies in intelligent ways. We will still, however, consider how other animals solve physiological problems in extreme environments—the dryness of the deserts, the intense pressures of

BOX 20.2
PERSONAL IMPACT

FEVER AND THE BODY'S NATURAL THERMOSTAT

Being feverish feels terrible: Your skin sends out heat like a radiator, and yet even under thick blankets you shiver and your teeth chatter; what's more, you may feel dizzy, thirsty, and tired and have a splitting headache on top of it all. Fever is the most obvious signal that the normal homeostasis of the body is disturbed, since one's temperature is no longer being maintained at a steady 37°C (98.6°F). The stoked-up internal furnace, however, is usually helping the body "burn out" foreign invaders, and the uncomfortable feelings of hot and cold are usually only temporary.

A fever can start after virus particles or bacterial cells invade the body and begin multiplying. The infectious agents can turn up the brain's natural thermostat—the hypothalamus—in an indirect way. As white blood cells attack the invaders, they leak out proteins called *pyrogens* (from the Greek words for "fire" and "origin"). Pyrogens then cause certain cells to release *prostaglandins* (see Chapter 26), and these hormonelike molecules act on the hypothalamus to turn up its thermostatic setting.

Just as turning up the thermostat in your home signals the furnace to fire up, the resetting of the hypothalamus to a higher value (say, to 43.3°C, or 110°F) triggers the heat-generating mechanisms of the body to step up their operation. The muscles start contracting, the body starts shivering, and this makes the temperature climb higher and higher. The sick person can still feel very cold, however, because the thermostat is set at 43.3°C (110°F), while the body temperature may be only 39.4°C (103°F). Elevated temperature helps to stop or slow bacterial growth, however, and soon the fever breaks: The thermostat is reset to 37°C, but since the body temperature is much higher—39.4°C—sweating begins, and the evaporative cooling provided by perspiration quickly reduces the body temperature.

This resetting is crucial, because a person can suffer brain damage if the body temperature exceeds 41°C (106°F). Care givers must quickly sponge the patient with ice water or alcohol to save his or her life. Aspirin can help reduce fever because it interferes with prostaglandin synthesis, and the thermostat shifts to a lower setting. But you should never exceed recommended dosage, because two much aspirin can be quite harmful.

In some cold-blooded animals, like lizards, a sick individual will actually induce its own artificial fever. Recall that a lizard will crawl into or out of the sunshine to keep its body temperature fairly warm and steady. A lizard with a bacterial infection will bake itself in the sun (Figure 1) until its body temperature soars and the bacterial growth slows. When researchers prevented sick lizards from baking out their infections, the animals died more frequently than those allowed to bask. Clearly, homeostatic mechanisms are crucial for survival.

FIGURE 1 ■ **Rainbow Lizard Basking on a Rock in Tanzania.**

deep oceans, the oxygen deficiency of high mountains, or the bitter cold of the earth's poles—so we can appreciate the full spectrum of solutions in the living world.

Although we separate individual organ systems into different chapters to make learning easier, keep in mind that this "dissection" is largely artificial. Each organ system depends on the others: The brain cannot function unless the blood delivers to it a constant supply of oxygen from the lungs and glucose from the digestive system; the blood cannot flow without the heart's muscular pumping action; the heart cannot pump without its own supply of oxygen and nutrients; the lungs cannot work without commands from the brain; the gut cannot operate efficiently without hormones; and so on. To give a sense of this dynamic interaction among organ systems, we will end Part Four by showing in detail how the human body responds to the physical stress of heavy exercise and yet continues to maintain the homeostasis it needs to survive.

Highlights in Review

1. The blue whale, because it is the largest organism ever to inhabit the earth, has extreme forms of the same survival problems it shares with all other organisms: maintaining homeostasis (a constant internal environment) despite fluctuations in the outside environment.

2. Single cells and very small or flat multicellular organisms exchange materials directly with the environment via diffusion.

3. Animals have special physiological systems to bring about exchanges between all cells and the outside environment. Each system is based on a tube leading from the outside; regions of the tube where exchange can take place; a circulation of body fluids; and diffusion of materials between circulating fluids and individual cells.

4. Homeostatic mechanisms keep the contents of body fluids stable.

5. In multicellular organisms, body cells are organized into tissues, organs, and organ systems. The four main tissues are epithelial, connective, muscle, and nervous tissue.

6. Large animals need strong structural support as well as a nervous system and an endocrine, or hormonal, system that coordinate and integrate the body's activities.

7. Specialized adaptations allow an organism to exploit its environment successfully. The forms of such adaptations are appropriate to their functions, and each arose by natural selection.

8. An animal's physiological systems work together, bringing about homeostasis in and around the individual cells.

9. Each individual homeostatic system has a receptor that senses environmental conditions, an integrator that evaluates those conditions, and an effector that executes commands.

10. Negative feedback loops sense movement away from a baseline condition, then bring about action that negates this alteration. Negative feedback systems help keep the internal temperatures of warm-blooded animals steady.

11. In a positive feedback loop, an initial condition triggers change, and the change triggers more change, until an explosive event occurs and stops the loop.

12. In the blue whale, anatomical, physiological, and behavioral mechanisms work together, helping keep the animal warm in icy polar waters and cool in warm, tropical seas.

Key Terms

anatomy, 431
connective tissue, 435
epithelial tissue, 435
feedback loop, 438
homeostasis, 431
muscle tissue, 435

nervous tissue, 435
organ, 435
organism, 435
organ system, 435
physiology, 431
tissue, 433

Study Questions

Review What You Have Learned

1. Define anatomy and physiology.

2. What does homeostasis mean? Give some examples of homeostatic mechanisms.

3. The digestive, respiratory, and excretory systems follow a common strategy in servicing the body's cells. What is it?

4. How are a tissue and an organ alike? How are they different? Use specific examples in your answer.

5. Does a given organ system function independently of other organ systems? Explain.

6. Name four special adaptations that suit a blue whale to its life in the sea.

7. How might the blue whale's adaptations have arisen?

8. How does negative feedback help the blue whale maintain a constant body temperature in cold water?

9. Give an example of a positive feedback loop.

10. How does an ectotherm regulate its body temperature?

11. When you develop a fever, what sequence of events is responsible?

Apply What You Have Learned

1. A circus promoter attempted to keep a whale alive on land by spraying salt water on its skin. The whale died anyway. Why?

2. True or false: Bacteria do not need to maintain homeostasis. Explain.

3. In Box 20.1, a healthy person's blood chemistry is compared with that of a hepatitis patient (see Table 1). What differences do you observe? What do these differences imply about the liver's role in maintaining homeostasis?

4. A science fiction movie features a monstrous flatworm as large as a blue whale. A character in the movie claims that the animal is a mutant with tissues and organs like a normal flatworm's but with cells thousands of times larger than normal. Is this scientifically believable? Why or why not?

 **For Further Reading**

KIESTER, E. "A Little Fever Is Good for You." *Science* 84 (1984): 168–173.

MINASIAN, S. M., K. C. BALCOMB III, and L. FOSTER. *The World's Whales: The Complete Illustrated Guide.* Washington, DC: Smithsonian Books, 1979.

SCHMIDT-NIELSEN, K. *Animal Physiology: Adaptation and Environment.* 3d ed. New York: Cambridge University Press, 1983.

SHERWOOD, L. *Human Physiology: From Cells to Systems.* St. Paul: West, 1989.

WURSIG, B. "The Behavior of Baleen Whales." *Scientific American,* April 1988, pp. 102–107.

Circulation: Transporting Gases and Materials

The Improbable Giraffe

Giraffes have always aroused irresistible curiosity in people. These strange African mammals are the third heaviest land animals after the elephant and rhinoceros, and they are by far the tallest: An adult male stands 5.5 m (18 ft) high, and a female 4.5 m (14¾ ft). Three basketball players standing feet to shoulders could scarcely look a male giraffe in the eye, and even a young giraffe just a few months old is tall enough to browse tender leaves from the high branches of an acacia tree (Figure 21.1).

As physiologists learned more about how animals function, they realized that a giraffe's body is even more enigmatic than it looks. The giraffe has, for example, a massive, muscular heart and thousands of miles of blood vessels. Yet compared to the way these same vital organs operate in smaller, shorter animals, the giraffe should be on its last legs: It has very high blood pressure. With its heart 2.5 m (8 ft) above the ground yet 2 to 3 m (6½ to 10 ft) below the head, the animal's legs and feet ought to swell with fluid, and upon standing up, the beast should faint from lack of blood to the brain. When a giraffe lowers its head 4 or 5 m (13 or more feet) to drink from a water hole, blood should come flooding down and blow out delicate vessels in the eyes and brain. Yet none of these things happen. A giraffe can rise (albeit slowly) from a prone position, gallop away at 48 km (30 mi) per hour, then stop to drink water with no ill effects whatsoever. The giraffe is indeed a physiological marvel and makes a good case study for the functioning of the **circulatory system**—the heart and blood vessels.

FIGURE 21.1 ▪ **Giraffes: Lofty Leaf Eaters.** The giraffe's long legs and neck enable it to browse from tree tops, but these same extremities pose special problems for breathing and blood circulation.

Physiologists have found that giraffes show numerous evolutionary adaptations of the circulatory system that help them survive on land despite their height and massive weight. High blood pressure helps distribute blood with its cargo of oxygen, nutrients, and other materials throughout the lanky body. High-pressure fluid outside the thick-walled blood vessels counteracts the pressure inside and prevents leakage and swelling in the extremities. And in the head, large blood vessels branch out into a network of tiny channels before reaching the brain and eyes, and this *rete mirabile,* or "wonderful net," lowers and regulates the blood pressure reaching those delicate organs, avoiding "blowouts."

While the giraffe's unusual physiological problems are met in unique ways, its basic needs are the same as all other animals': As a heterotroph, an animal needs a source of energy (sugars and other organic molecules), a supply of oxygen (for aerobic respiration), and a means of dumping carbon dioxide and other wastes. Since an animal is multicellular but the exchange of materials takes place by diffusion or by active transport across cell surfaces, each cell must lie either very close to the outer environment or very close to an internal supply route. During evolution, this reliance on diffusion has constrained the overall shapes that animals can take. Thus, they are either very small like a nematode worm (review Box 7.2 and Figure 18.12), very flat like a tapeworm, or, if large and thick, possessing a complex internal transport system capable of delivering life-giving supplies to within diffusion distance of every body cell.

A large animal like a giraffe has similar strategies for exchanging all types of materials with the environment, just as Chapter 20 explained for the blue whale: "Tubes" (actually, entire physiological systems) convey the different commodities to or from the outside, and the circulatory system distributes them internally. In most mammals, the digestive tube conveys nutrients, the respiratory tube carries oxygen, and the renal tube transports organic wastes (review Figure 20.4), and all of these tubes have close physical contact with parts of the circulatory system. Branching vessels of this system lie close to each body cell, and a fluid tissue, the blood, courses through the vessels, carrying the ions, nutrients, gases, and organic wastes that the cells must import or export to continue healthy functioning. The circulatory system also transports products of the hormonal system (involved in internal communication and control), as well as cells and molecules of the immune system (involved in the body's defense). Because the body's other physiological systems are so closely tied to the circulatory system, and because the delivery of their separate commodities requires a dependable means of internal transport, circulation is a central activity in maintaining homeostasis, the steady state conducive to life (see Chapter 20). Malfunctions of the circulatory system, especially heart attacks, kill half a million people every year. This is more than any other single condition in North America.

Three unifying themes will emerge from our discussion of animal circulation. First, the physical principles of diffusion and bulk flow underlie material transport. Diffusion allows the conveyance of materials between cells and their surroundings, and bulk flow allows the conveyance of large amounts of blood throughout the body. Second, the unique characteristics of blood allow many of an animal's homeostatic activities to take place. Blood is a collection of solids, liquids, organic molecules, and ions. Blood has the capacity to transport and deliver materials efficiently, to help balance pH and maintain body temperature, and even to seal leaks in the circulatory system. Third, adaptations for material transport reflect the demands and constraints of an animal's environment and way of living. Without its high blood pressure, its specialized network of blood vessels in the brain, and other adaptations of its circulatory system, the giraffe, for example, could not survive in the dry African savanna by browsing from tall trees and stooping to drink from seasonal water holes. While this chapter focuses mainly on people, our study of circulatory adaptations in other animals will help illuminate the general principles of circulation and material transport.

In our coverage of the circulatory system, we answer several important questions:

- What properties of blood contribute to its transport function?
- What are the main organs of the circulatory system, and how did they evolve?
- How does the circulatory system function in humans and other animals?
- How does the lymphatic system, the second network of transport vessels, aid in fluid homeostasis?

BLOOD: A MULTIPURPOSE LIQUID TISSUE FOR INTERNAL TRANSPORT

As a large mammal, you have about 5 L (more than a gallon) of blood in your body, coursing through an amazing 80,000 km (49,600 mi) of vessels. This moving stream of liquid tissue operates like a rapid courier service, circulating oxygen from the lungs, nutrients from the digestive tract, and regulatory substances such as hormones from one region of the body to another. Blood helps control the pH of body fluids, conveys immune cells specialized to defend the body, and distributes heat and assists in cooling (see Chapter 20). These and other functions are all based on the individual components of blood and the roles they perform.

Blood: Liquid and Solids

By whirling a sample of blood in a laboratory centrifuge, a technician can separate the components of this life-giving liquid. The most obvious separation will be into an upper fluid layer and a lower layer of cells and cell fragments (Figure 21.2). The top 55 percent is the pale yellow liquid called **plasma,** which is more than nine-tenths water. Plasma's main function is to transport blood cells and other particles, including salts, sugars, and fats from the foods we eat. Plasma also contains a storehouse of important dissolved proteins. One group of plasma proteins, the *globulins,* includes *antibodies,* defensive molecules that attack invaders (see Chapter 22). The *albumins,* another group, bind to toxic substances in the blood, among other activities. A third plasma protein, *fibrinogen* (described shortly), is essential for blood clotting.

Beneath the plasma layer in centrifuged blood is the remaining 45 percent, composed of cells and cell fragments divided into two layers: The first is a thin gray band representing less than 1 percent of blood volume. It is made up of white blood cells (important in the body's defense) and cell fragments known as **platelets** (important in blood clotting). The second is a much wider band containing only red blood cells. A physician can use the width of this band of red blood cells as a diagnostic tool. If it is less than 45 percent of total blood volume, the person has too few red blood cells and is probably *anemic,* perhaps suffering fatigue and shortness of breath.

Red Blood Cells and Oxygen Transport

The human body contains 25 trillion red blood cells—fully one-third of all its 75 trillion cells. Three features help red blood cells, or **erythrocytes** (from the Greek for "red" and "cell"), perform their vital function of transporting oxygen to tissues. The first two features concern the presence of proteins: hemoglobin, which transports oxygen, and the carbonic anhydrase enzyme, with its role in carbon dioxide transport. Third is the cell's biconcave disk shape that is a bit like a doughnut without a hole (Figure 21.3). Among other benefits, this disk shape allows all of the red blood cell's hemoglobin molecules to lie near the outer membrane, which oxygen must diffuse across. If red cells were spheres like the white blood cells in Figure 21.3, some hemoglobin would be "buried" deep in the cell and would have little access to incoming oxygen.

Recall from Chapter 4 that human red blood cells lose their nucleus, mitochondria, and most other organelles as the cells mature. Biologists believe that this evolutionary adaptation may make room for more hemoglobin and increase the cell's oxygen-carrying capacity. Regardless of why it happens, it dooms the erythrocyte to die after about 120 days. Replacement red blood cells develop from stem cells in the bone marrow (see Chapters 4 and 13). (Other stem cells generate white blood cells and the cells that give rise to platelets.)

A negative feedback loop controls red blood cell replacement when the red cell count decreases because of either normal red cell death or sudden blood loss from a serious wound (Figure 21.4). When there are fewer red cells in the blood, the liver and kidney receive less oxygen and begin to produce the hormone *erythropoietin.* This hormone travels through the bloodstream to the bone marrow, where it stimulates stem cells to divide more quickly; as more red cells are produced, they carry more oxygen throughout the body, including the liver and kidney. As a result, these organs slow their production of erythropoietin, and division of bone marrow stem cells

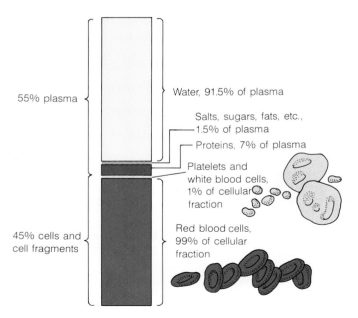

FIGURE 21.2 ■ The Fractions of Human Blood. If centrifuged, blood will separate into yellowish plasma and denser formed elements—red and white blood cells and cell fragments called platelets, which aid in blood clotting. Clearly, most of the blood is water and red blood cells.

Water, 91.5% of plasma

Salts, sugars, fats, etc., 1.5% of plasma

Proteins, 7% of plasma

55% plasma

Platelets and white blood cells, 1% of cellular fraction

45% cells and cell fragments

Red blood cells, 99% of cellular fraction

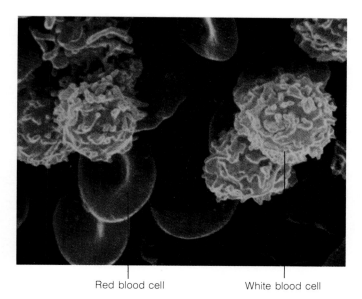

Red blood cell   White blood cell

FIGURE 21.3 ■ Blood Cells Transport and Defend. In this colorized scanning electron micrograph, red blood cells (which transport oxygen) are clearly distinguishable as dark red biconcave disks, while white blood cells (which help defend the body) look like fuzzy pink balls (magnification 6000×).

they retain their nucleus when mature and have a changeable, amoeba-like shape that allows them to squeeze through the walls of capillaries and patrol the fluid-filled spaces between cells. These characteristics help leukocytes to defend the body against invasions by microorganisms and other foreign materials.

There are five classes of white blood cells, and together they travel in the blood like a mobile militia. Two groups, the **neutrophils** and **monocytes,** are specialists at phagocytosis ("cell eating"): They are attracted to damaged or infected tissues and consume bacteria, viruses, and other cell debris. **Eosinophils** carry enzymes that break down foreign proteins and break up blood clots. **Basophils** release an anticlotting agent at the site of an injury and produce a chemical substance called *histamine* that delays the spread of invading microorganisms. Histamine also produces hives and swells up an allergy sufferer's nasal membranes (hence the use of *antihistamine* drugs). A final leukocyte group, the **lymphocytes,** are active in the immune responses that protect the body against infectious agents and foreign intruders (such as transplanted organs). There is so much to say about white blood cells and the body's immune defense that Chapter 22 is devoted entirely to the subject.

slows. This same cycle of events can be triggered by changing altitude or exercising vigorously. A mountain climber who travels from Milwaukee to the Himalayas, for example, will encounter air with a much lower concentration of oxygen molecules; thus, the blood will transport less oxygen to the liver and kidneys, and these organs will make more erythropoietin. The hormone will stimulate the bone marrow and thereby augment the climber's red blood cell supply. The increased number of red blood cells will then carry more oxygen throughout the body. Likewise, a runner in training will use massive amounts of oxygen, thus decreasing the amounts reaching the liver and kidneys during each training session. This also triggers a higher output of erythropoietin and a rise in red blood cell production.

Researchers have made progress recently in manufacturing hormones that stimulate the growth of red and white blood cells. This promises to decrease the need for blood transfusions, simplify bone marrow transplants, and bolster the fight against disease-causing microorganisms, AIDS, and some cancers.

White Blood Cells: Defense of the Body

For every 1000 red blood cells, there are only 2 white blood cells, or **leukocytes** (literally, "white cells"; see Figure 21.3). These white cells, however, are larger than the red cells, and

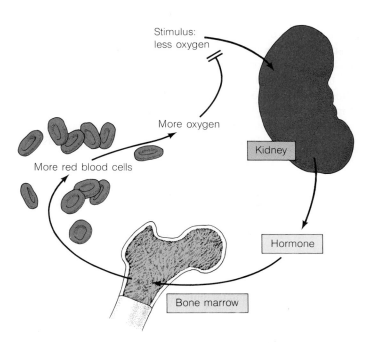

Stimulus: less oxygen

More oxygen

More red blood cells

Kidney

Hormone

Bone marrow

FIGURE 21.4 ■ Replacing Lost Blood Cells: A Negative Feedback Loop. When the number of red blood cells drops and the kidney receives less oxygen, it releases a hormone (erythropoetin) that stimulates the bone marrow to make more red blood cells. Increased numbers of red blood cells carry more oxygen to the kidneys (and other organs), and the kidneys stop releasing the hormone. The initial stimulus—lowered oxygen levels—is negated by the final result—more red blood cells. These additional cells thus transport more oxygen.

Platelets: Plugging Leaks in the System

Streaming along with the red and white blood cells are millions of cell fragments called platelets, or **thrombocytes.** These irregularly lobed bits and pieces have broken off from large specialized cells in the bone marrow. Platelets are crucial blood components, because like certain chemicals we can add to the water in a car's radiator, platelets help to plug small leaks in the circulatory system. As we will see later in the chapter, were it not for the blood-clotting action of platelets, an animal's blood might literally drain away through even a minor wound.

Blood, with all its separate components, is capable of delivering oxygen and nutrients, carting off waste gases, and patrolling and defending the body. But static, stagnant pools of blood could not accomplish these tasks. The key to carrying them out is constant movement: The blood must be pumped on a continuous circuit through the body, and that is the function of the circulatory system.

CIRCULATORY SYSTEMS: STRATEGIES FOR MATERIAL TRANSPORT

Just as delivery trucks would be useless in a town without roads, blood would be of little value to an animal unless it could circulate its cargo of gases, nutrients, and other substances throughout the body. A constant flow of blood with its storehouse of materials passes within 1 mm (0.04 in.) of each body cell, and this short distance allows diffusion of those materials to keep pace with the cells' consumption. Similarly, wastes that would accumulate and poison the cell can quickly diffuse into a nearby vessel and be swept away in the bloodstream. While the circulatory systems of various animals carry out both these functions, different environments place different demands and restrictions on the functioning. To fully appreciate the human circulatory system, we must compare it with parallel systems in other organisms. A major evolutionary trend emerges from such a comparison: An increase in the efficiency of the circulatory system coincides with an increase in the animal's activity level—activities that require adequate supplies of blood-borne oxygen and nutrients. This trend is reflected in the two main types of circulatory systems, open and closed.

Open Circulatory Systems

In animals with open circulatory systems, including most arthropods and mollusks, a blood equivalent called **hemolymph** circulates in the body. But hemolymph is contained in blood vessels, or circulatory pipelines, only part of the time. Short blood vessels pipe hemolymph into and out of the heart, but in the rest of the body, the fluid sloshes freely around and through the animal's tissues.

The spider in Figure 21.5a has a simple heart that is little more than an elongated, pulsating tube. The heart's pumping action sends hemolymph in the direction of the animal's brain, where it leaves the tube and percolates toward the back of the animal through the open body cavity, bathing the tissues in the animal's gut and other internal organs. Finally, the hemolymph returns to the heart tube through special pores. Open circulation is not a very efficient transport system, but spiders and insects can hop, fly, and lead active lives anyway because their respiratory apparatus, the branching, hollow book lungs, or tracheae, operate independently of the heart, bringing oxygen close to each cell.

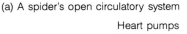

(a) A spider's open circulatory system

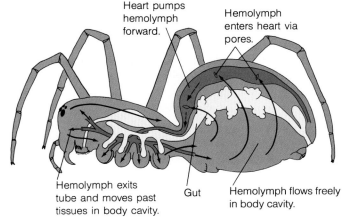

Heart pumps hemolymph forward.

Hemolymph enters heart via pores.

Hemolymph exits tube and moves past tissues in body cavity.

Gut

Hemolymph flows freely in body cavity.

(b) A fish's closed circulatory system

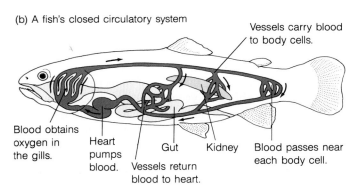

Vessels carry blood to body cells.

Blood obtains oxygen in the gills.

Heart pumps blood.

Vessels return blood to heart.

Gut

Kidney

Blood passes near each body cell.

FIGURE 21.5 ■ Open and Closed Circulatory Systems. (a) In the open circulatory system of an arthropod (here, a spider) or a mollusk, the blood (called hemolymph) moves in a vessel, but is then dumped into a sinus (an open area) in which it freely bathes the internal tissues before reentering the vessel. (b) In the closed circulatory system of a vertebrate (here, a sea bass), the blood remains in a system of interconnected vessels as it circulates through the body. Here, red signifies oxygenated blood leaving the gills and heart and traveling to the body tissues, and blue signifies deoxygenated blood returning from the tissues in vessels leading to the heart and gills.

Closed Circulatory Systems

Segmented worms and more complex animals, such as vertebrates, have closed circulatory systems, in which the circulating blood is completely contained within a system of vessels. In a representative vertebrate, such as the sea bass, the heart pumps blood through tiny vessels in the gills, an interface with environmental oxygen (Figure 21.5b). Once the blood has picked up oxygen, it courses through smaller and smaller vessels that spread throughout the fish's tissues, delivering oxygen to every body cell. On its circuit, the blood passes by the gut, where it picks up nutrients, and areas such as the kidneys, where organic wastes are removed from circulation and excreted from the body (see Chapter 25 for details). Eventually, the continued pumping of the heart pushes the blood around once more to the gills, where it dumps carbon dioxide and picks up new oxygen.

A closed circulatory system has several advantages over an open one. First, fluid contained within a network of closed tubes can be shunted to specific areas where it is needed, much as a farmer can dispatch the water in irrigation pipes to different fields. Second, since the fluid is completely contained within pipelines, more pressure can be exerted on it;

thus, the fluid can be forcefully distributed to areas distant from the pump (like a giraffe's head or feet). You can demonstrate the results of pressurized fluid by placing your thumb over the opening of a garden hose: You can squirt water on a tomato plant—or on an unsuspecting friend—much farther away than you could by dribbling the water out unobstructed. These features allow closed circulatory systems to deliver blood throughout larger, more complex, and more active organisms in a much more efficient way than an open system could.

Blood Circulation

In a fish, blood flows in a simple loop from the heart, to the gills (where it is oxygenated), to body tissues, and back to the heart (Figure 21.6a). A fish's heart has two chambers, a less muscular one called the **atrium** (plural, atria), which receives oxygen-depleted blood from veins, and a more muscular cavity, the **ventricle.** The ventricle receives blood from the atrium and pumps it through an artery to the small capillaries in the gills, where oxygen is absorbed into the blood and carbon dioxide released to the environment. Oxygenated blood

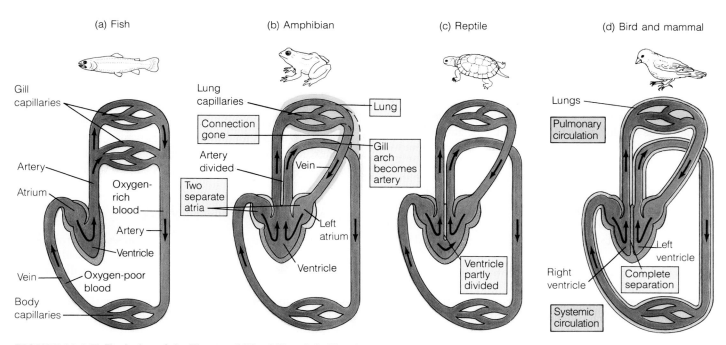

FIGURE 21.6 ■ Evolution of the Heart and Blood Vessels in Vertebrates. (a) A fish has a two-chambered heart (one atrium, one ventricle) and a single loop of blood vessels leading through the gills and to the body tissues. (b) An amphibian has a three-chambered heart (two atria, one ventricle) that allows some mixing of oxygenated and deoxygenated blood, plus a separate loop of blood vessels to the lungs (the pulmonary circulation). The innovations are highlighted in yellow. (c) In reptiles, the heart has three chambers (two atria and a partially partitioned ventricle) that further decrease the mixing of oxygenated and deoxygenated blood. (d) In birds and mammals, the heart has four chambers (two atria and two ventricles) that together prevent blood mixing, and the vessels are arranged in pulmonary (lung) and systemic (body) circulatory loops.

then leaves the gills and collects in large vessels that pipe fresh blood throughout the fish's body. After the oxygen and other materials are delivered to tissue cells, deoxygenated blood returns to the atrium through vessels called veins, and the cycle is repeated.

A significant feature of this single-loop system is the way it separates oxygenated from deoxygenated blood and delivers only highly oxygenated blood to muscles and other organs. The single-loop circulatory system of a fish, however, cannot deliver blood to most organs at a high pressure. Blood leaving the heart is under high pressure, but once it fans out through the dense network of fine capillaries in the gills, pressure diminishes and remains low as the blood moves through the body. While this low pressure is adequate to meet the metabolic demands of a "cold-blooded" (poikilothermic) animal that is supported and buoyed up by water (see Chapter 20), it could not push blood quickly enough through the body of a terrestrial vertebrate, whose active cells require high levels of oxygen and nutrients. Land vertebrates, instead, have a two-loop, dual circulatory system that meets the needs of both high pressure and the separation of oxygenated from deoxygenated blood: Blood is first pumped to the lungs, where it takes on oxygen, after which it flows back to the heart, where it is pumped out a second time—at high pressure—and then circulates to body tissues.

■ **Blood Circulation in Land Vertebrates** Although air-breathing lungfishes were probably the first vertebrates to crawl on land, adult amphibians and reptiles were the first vertebrates capable of living their entire lives on land. All such creatures have a heart-lung circulatory loop that increases blood pressure, as well as another evolutionary feature: a third heart chamber that enhances the heart's ability to keep oxygenated and deoxygenated blood separate. Amphibians have two atria instead of the single atrium found in fish: One atrium receives oxygenated blood arriving from the lungs, and one receives oxygen-depleted blood from body tissues (Figure 21.6b). Both atria empty into a single ventricle, and in it, there is some mixing of oxygenated and deoxygenated blood. Nevertheless, because of the position of the large vessels leading out of the lone ventricle, most of the oxygen-rich blood is shunted to the tissues, and most of the oxygen-depleted blood travels to the lungs.

During the evolution of the reptile lineage, the three-chambered heart underwent a significant change. In particular, in some reptiles, such as the crocodile and turtle, a thin membranous partition (septum) partially divides the single ventricle (Figure 21.6c). This modification further decreases the mixing of oxygenated and deoxygenated blood in the heart and thereby provides another measure of increased efficiency in oxygen delivery. In the birds and mammals that descended from reptilian ancestors, the septum enlarged until it eventually divided the ventricle into two fully separated cavities. All birds and mammals alive today have a four-chambered heart with an atrium and ventricle on the left side

and an atrium and ventricle on the right side. In a sense, these complex animals have two hearts that pump blood in two completely separate circulatory loops (Figure 21.6d). In the "lung loop," or **pulmonary circulation,** blood arrives from the tissues, enters the right side of the heart, and is pumped directly to the lungs, where it picks up oxygen. The oxygenated blood then enters the left side of the heart and is pumped out into a "body loop," the **systemic circulation.** This double loop means that the blood flowing to the muscles, brain, and extremities will have the highest possible oxygen content and that "used," oxygen-poor blood can be shunted from an isolated right heart to the nearby lungs without ever mixing with oxygen-rich blood. It also means that the left heart chamber can send blood out through the body loop at high enough pressure to reach all the tissues quickly.

To appreciate how the four-chambered heart and separate circulatory loops operate in birds and mammals and how blood pressure and blood flow are controlled, let's examine the mammalian circulatory system in detail, using the human system as our model.

CIRCULATION IN HUMANS AND OTHER MAMMALS: THE LIFE-SUSTAINING DOUBLE LOOP

It takes an immense network of circulatory tubing to deliver blood close to each of your cells (Figure 21.7a). If all of the blood vessels in your body could be removed and placed end to end, the single long pipeline would encircle the earth twice. To keep the blood flowing through all these vessels, your heart must beat with a regular rhythm for seven to nine decades, and the flow must be maintained at a high enough pressure to force blood into your brain, nose, toes, and all the tissues in between. Both the anatomy and the activity of the blood vessels and heart make these tasks possible.

Blood Vessels: The Vascular Network

Blood moves away from the heart in **arteries,** which are large, hoselike vessels with thick, multilayered, muscular walls (Figure 21.7b and c). The contraction of these wall muscles, along with the beating of the heart, helps keep blood under pressure—the force we call blood pressure. The easily felt vessels through which blood pulses in your wrists and neck are arteries. Arteries branch and form smaller vessels called **arterioles.** These vessels are too small to be seen with the unaided eye and have thinner, less muscular walls. Arterioles branch again into the dense, weblike network of delicate **capillaries;** these minute vessels permeate the fingertips, earlobes, and all the tissues of the body. Capillaries are microscopic vessels only about 8 μm (0.0003 in.) in diameter and rarely more than about 1 mm (0.04 in.) from any body cell. A

(a) The human circulatory system

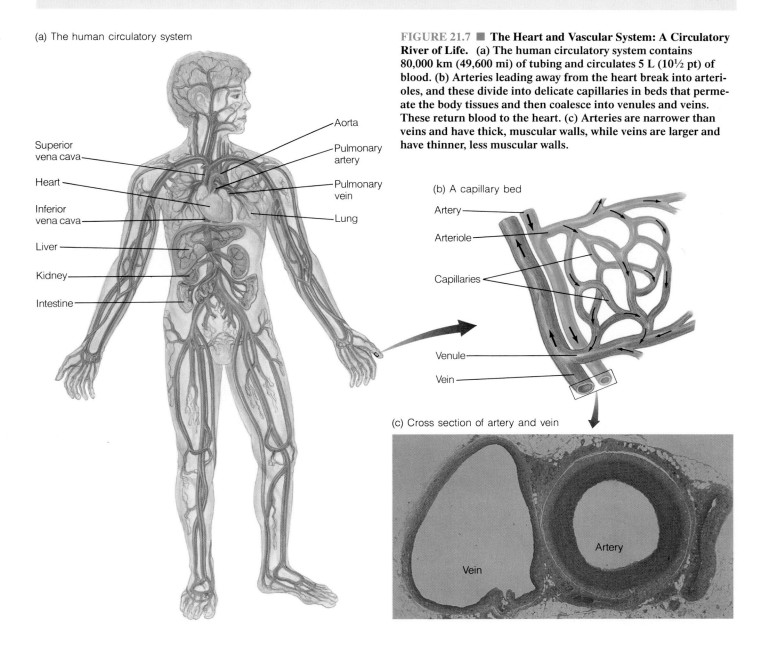

FIGURE 21.7 ■ **The Heart and Vascular System: A Circulatory River of Life.** (a) **The human circulatory system contains 80,000 km (49,600 mi) of tubing and circulates 5 L (10½ pt) of blood. (b) Arteries leading away from the heart break into arterioles, and these divide into delicate capillaries in beds that permeate the body tissues and then coalesce into venules and veins. These return blood to the heart. (c) Arteries are narrower than veins and have thick, muscular walls, while veins are larger and have thinner, less muscular walls.**

Superior vena cava

Heart

Inferior vena cava

Liver

Kidney

Intestine

Aorta

Pulmonary artery

Pulmonary vein

Lung

(b) A capillary bed

Artery

Arteriole

Capillaries

Venule

Vein

(c) Cross section of artery and vein

Vein

Artery

capillary's diameter is so small that red blood cells can pass through in single file only, and so delicate that their walls are just a single cell thick. The vessels are so numerous, however, that they make up almost all of the 80,000 km (49,600 mi) of a person's blood vessels, and if all were open at the same time, they could contain the entire 5 L (10½ pt) of human blood. The capillaries' ultrathin structure is one key to the efficiency of the circulatory system: Materials can readily diffuse outward or inward through the single cell layer.

Capillaries are interwoven into other tissues in networks called **capillary beds** that link the arterial and venous blood vessels (see Figure 21.7b). The arterial side of each bed con-

veys fresh blood to the capillaries, and oxygen, nutrients, carbon dioxide, and metabolic wastes move across the capillary walls into and out of the extracellular fluid and nearby tissue cells. The depleted blood then continues to move through the bed to the venous side; there the capillaries leading away from the tissues coalesce into larger vessels known as **venules,** which in turn merge and become **veins.** These carry blood toward the heart, where the cycle of circulation begins again.

Because venous blood has already traveled some distance from the heart and been slowed by the tight passage through the narrow capillaries, and because veins have a larger diameter and thinner, less muscular walls than arteries (see Figure

(a) Valve open

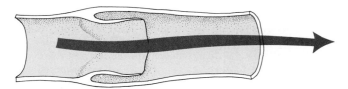

(b) Valve closed

(c) Varicose vein

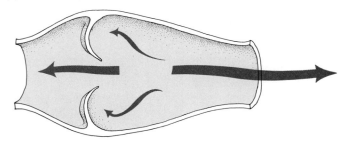

(d)

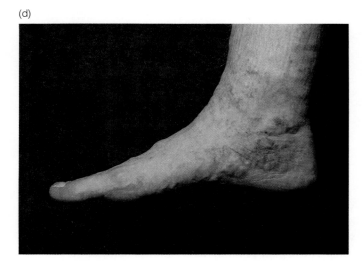

FIGURE 21.8 ■ **Valves Maintain a One-Way Blood Flow in Veins.** (a) When blood moves in one direction, it forces the valve open, and the fluid flows unimpeded. (b) If blood begins to flow in the opposite direction, the backward force pushes the valve flaps together and closes the valve. Thus, the valve allows blood to flow in only one direction—toward the heart. (c) In a varicose vein, the valve is damaged, and blood flows backward, causing the vein to distend and look very large and blue in a limb, usually the lower leg and foot, as shown in (d).

21.7c), venous blood is under lower pressure. This low-pressure fluid flows in the proper direction (toward the heart's right side) and does not backtrack or pool up in the extremities owing to a system of **valves**—tonguelike flaps that extend into the internal space, or **lumen,** of the vein (Figure 21.8). Like a one-way door, when a valve is pushed from one direction, it opens and allows blood to pass, but when pressure is exerted from the other direction, the valve stays locked tight. The continuous flow of venous blood against gravity, toward the heart, often depends on contractions of the large body muscles surrounding the veins. The constant flow of blood from heart to arteries to capillary beds to veins and back to heart depends partially on the structure of arteries and veins, as we have just seen, but is perpetuated by the action of the beating heart.

The Tireless Heart

At the physiological center of the circulatory system lies the heart (see Figure 21.7a). Roughly the size of a large lopsided apple, your heart pumps a teacupful of blood with every three beats, 5 L (10½ pt) of blood every minute, and upward of 7200 L (70 barrelfuls) of blood every day of your life. Over a lifetime, this amounts to 2.5 billion heartbeats and 18 million barrels of blood—enough to fill a small ocean. Although medical engineers have tried valiantly, replacing the living circulatory pump has proved exceedingly difficult.

This tireless organ, like the hearts of all mammals, has four chambers divided into two pairs: the right atrium and right ventricle, and the left atrium and left ventricle. Each atrium-ventricle pair plays a specialized role in circulating the blood. The right atrium receives deoxygenated blood from the body tissues via two large veins, the **superior vena cava** and the **inferior vena cava** (see Figures 21.7 and 21.9, step 1). These vessels collect blood from the upper and lower body, respectively. The deoxygenated blood is then pumped from the right atrium into the right ventricle through the petal-shaped *atrioventricular valve,* which prevents blood from flowing back when a ventricle contracts (Figure 21.9, step 2). The right ventricle in turn pumps blood past the half-moon-shaped *pulmonary semilunar valve* into and through the Y-shaped **pulmonary artery** (step 3) to the lungs, where gases are exchanged in the capillaries in the lungs.

In the next step of the circuit, freshly oxygenated blood collects in the pulmonary veins (the only veins in an adult person that carry oxygen-rich blood) and enters the heart's left atrium (step 4), which pumps the oxygenated blood through a second atrioventricular valve into the left ventricle (step 5). The thick walls of the muscular left ventricle contract around this blood in a wringing motion until enough pressure develops to push open the *aortic semilunar valve* and squirt the blood into the **aorta** (step 6)—the main artery leading to the systemic circulation. This body loop conveys blood to capillary beds in distant tissues, then returns it via venules and veins to the venae cavae and back to the right atrium again.

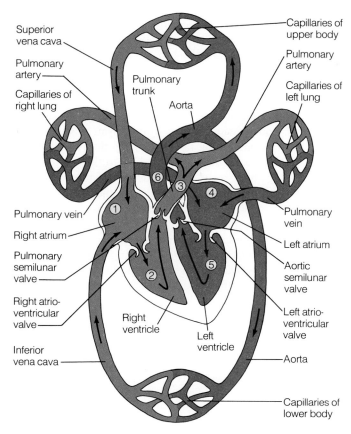

FIGURE 21.9 ■ **The Pumping Heart. Blood flows in a circuit from the venae cavae to the right atrium (1), right ventricle (2), pulmonary artery (3), lungs, pulmonary veins, left atrium (4), left ventricle (5), aorta (6), and body tissues.**

weakly (e.g., when you pick up a pencil) or more strongly (when you pick up a book as heavy as this one), and the strength of contraction depends on what percentage of the muscle fibers contract at any given time. By contrast, all the fibers in the cardiac (heart) muscle contract with each heartbeat; thus, the healthy heart is always pumping at its maximum strength. Blood flow through the heart is increased mainly by faster beating.

What accounts for this specialized characteristic of heart muscle? First, heart muscle cells are linked electrically by special regions, the *intercalated discs.* Gap junctions (see Figure 3.28) between cells in these disc regions allow electrical impulses that stimulate muscle contractions to spread instantaneously from muscle cell to muscle cell. As a result, neighboring sections of the heart wall contract and relax together in a superbly coordinated pumping action that keeps blood flowing smoothly through the system.

Second, cardiac muscle contracts automatically—that is, without stimulation from the nervous system. (Nerves do play a role in speeding or slowing the heart rate, however, as Chapter 27 describes.) Different subsets of heart muscle cells have different intrinsic rates of contraction. Some, called **pacemaker cells,** contract slightly earlier than others, and because each contraction spreads quickly throughout a region of the muscle, pacemaker cells ignite contractions in the entire heart and set the rate of the heartbeat.

Pacemaker cells are located near the upper right atrium in a region called the **sinoatrial (SA) node** (Figure 21.10). Immediately before a beat, an electrical impulse spreads from the SA node across the walls of both right and left atria, causing the two chambers to contract in unison. A second node, the **atrioventricular (AV) node,** is another area of modified

Together, the four valves in the heart prevent blood from reversing its normal flow; in a person with diseased valves, the heart's pumping strength is diminished by backwashing blood, which is sometimes heard as a heart murmur. Severely diseased valves can be replaced with either valves from a pig's heart or a mechanical device. The valves also make the "lub-dub" sounds you can hear through a doctor's stethoscope. As the ventricles begin to contract, the pressure of surging blood pushes the atrioventricular valves shut, and you hear a low-pitched "lub." Then, when the ventricles begin to relax at the end of systole, pressure in the aorta and pulmonary artery rapidly forces valves in both arteries to slam closed, producing the quicker, higher-pitched "dub."

How the Heart Beats

Like the muscles in your arm, each of the heart's four chambers is made up of specialized cells organized into contractile fibers. In your arm, however, your biceps muscle can contract

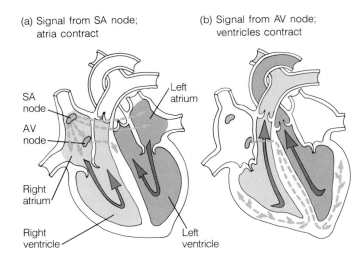

FIGURE 21.10 ■ **Electrical Impulses Drive the Heartbeat. (a) Pacemaker cells in the SA node send an impulse across the right and left atrial chambers, which respond by contracting. (b) Upon reaching the AV node, the signal is delayed for a split second, after which it reaches the ventricles and triggers their contraction, a squeeze that forces blood to the lungs and the rest of the body.**

cardiac muscle cells that is positioned at the junction where the atria and ventricles meet. When the electrical signal generated by the SA node reaches AV cells, there is a brief delay before the signal is passed to the ventricles. Upon reaching the ventricles, the signal triggers a contraction of the chambers that squirts blood into the pulmonary artery and the aorta, where it begins its circuit of the body. The brief delay in the transmission of the pacemaker impulse at the AV node is necessary for an efficiently beating heart: If the atria and ventricles were to contract simultaneously, blood in the atria might flow back into the veins instead of forward into the ventricles.

People sometimes have heart conditions in which the beat is irregular, and they can often be helped with an artificial pacemaker. This small electrical stimulator, powered by a battery or by the decay of a small amount of radioactive material, is implanted beneath the skin of the shoulder or abdomen. Then, by way of electrodes threaded through veins into the heart, the pacemaker sends rhythmic electrical impulses that stimulate the cardiac muscle to contract at an appropriate time.

Each time a heart's atria or ventricles contract, they are said to be in a **systole** phase. The opposite, or relaxed, phase is known as **diastole.** Together, these two phases make up the **cardiac cycle**—the contraction/relaxation sequence of atria and ventricles that makes up a single heartbeat. Recent evidence suggests that as the ventricles relax in diastole, they actively suck blood into their chambers.

Blood Pressure: The Force Behind Blood Flow

With each heartbeat, blood courses through your arteries under high pressure and at high speed (33 cm, or about 13 in., per second). If a person severs the aorta, blood spurts out more than 2 m (6½ ft) high! Blood pressure initiated by the heart's strong contractions serves an important purpose. It delivers blood quickly and continuously to cells far from the heart.

The "blood pressure" that a doctor evaluates with a tight cuff around an extremity is a measure of the force that blood exerts against artery walls as the heart alternately pumps (systole) and rests (diastole). The muscular, relatively inflexible walls of arteries help to maintain the blood pressure, but the farther blood travels from its central pump and the more it is dispersed through the miles and miles of arterioles and capillaries, the more the pressure falls. Blood flows from the areas of higher pressure to areas of lower pressure when the heart muscle contracts, just as toothpaste gushes from a tube when it is squeezed. The consumption of high-cholesterol foods can contribute to artery narrowing, to a consequent increase in blood pressure, and to a heightened risk of heart attacks and strokes, as Box 21.1 explains.

Blood pressure is measured in millimeters of mercury (mm Hg), that is, how high the pressure could lift a column of mercury. A normal reading for a person at rest is 120 mm of

mercury during systole and 80 mm of mercury during diastole, or 120/80. A giraffe's blood pressure, in comparison, averages 260/160—the highest in the animal kingdom. The giraffe is not sick, though; without such high pressures, blood could never move up the animal's long neck and reach its head.

The giraffe has two special behaviors that offset its naturally high blood pressure: It rises from a prone position fairly slowly, thereby preventing fainting during the few moments required for the heart to pump blood up to the brain; and it spreads its legs wide apart when it drinks (Figure 21.11), an adaptation that biologists believe moves the animal's heart closer to the ground and reduces the rush of blood toward the brain. Just as important, however, is the *rete mirabile,* or "wonderful net," of capillaries in the head that can open rapidly, diverting and slowing down rushing blood before the surging fluid can overtax vessels in the brain or eyes. The giraffe also has very thick-walled arteries that resist the pressurized blood inside and a thick hide that, applying pressure from the outside, acts a bit like the support hose on a person's legs.

Once blood has been forced through narrow capillaries, the pressure originally exerted by the heart is nearly spent. This low pressure allows veins to serve as the body's "blood reservoir," capable of holding as much as 80 percent of total

FIGURE 21.11 ■ **The Giraffe's Ungainly Drinking Posture Helps Protect Its Brain from a Rush of Blood.** With the heart closer to the ground, the downward rush of blood slows. Also, giraffes have a special adaptation that prevents blood from pooling in the limbs: a thick, tight hide that acts like "support hose."

BOX 21.1   P E R S O N A L   I M P A C T

CHOLESTEROL AND HEART DISEASE

The next time you're in a library or movie theater, notice the person sitting next to you: Current statistics indicate that either that person or you yourself will die of *atherosclerosis,* a disease sometimes referred to as "hardening of the arteries." What causes an artery to "harden"? Decades of research have focused on the role of cholesterol, a glistening white, fatty substance that is manufactured naturally by the liver. Cholesterol is an important component of cell membranes and is present in foods, especially eggs, meat, and dairy products. Scientific evidence suggests that cholesterol can build up in an artery and lead to the formation of *plaques,* thickened regions of the artery wall that occlude the passageway and prevent blood from flowing freely (Figure 1). As blood transit becomes increasingly impeded in a given artery, a blood clot may form and block the flow completely, leading to a heart attack or, if the affected artery is in the brain, a stroke. Each day, 1600 Americans suffer strokes, and 3400 suffer heart attacks. That's 1.2 million heart attacks annually. Together, cardiovascular diseases (ailments of the heart and blood vessels) are the leading cause of death in this country.

Numerous studies suggest that cholesterol is the main culprit. One long-term analysis of diet and heart attack risk among 362,000 middle-aged men showed clearly that the more cholesterol circulating in the blood, the higher the chances of a heart attack. The National Heart, Lung, and Blood Institute in 1990 urged everyone from children to seniors to eat diets low in saturated fats (animal fat, egg yolks, high-fat dairy products, hardened margarine), with fewer than 10 percent of total daily calories coming from those sources.

Physiologists now know that within a person's bloodstream, cholesterol travels attached to carrier molecules called lipoproteins; high-density lipoproteins (HDLs, or "good cholesterol") have very few cholesterol molecules attached, while low-density lipoproteins (LDLs) have great numbers of attached cholesterols. (Like oil on water, cholesterol floats; that's why LDLs are called "low density.") Evidence suggests that by exercising regularly, cutting back on saturated fats, eating more vegetable and fish oils, and avoiding smoking, heavy drinking, and obesity, a person will have higher levels of HDLs. And the HDLs, in turn, can help prevent plaques from building up in the blood vessels, thereby decreasing the risk of heart disease and strokes. People with high levels of LDLs, on the other hand, seem to be at greater risk of developing atherosclerosis, and researchers think the problem lies with the number of LDL receptors on the outer membrane of their cell surfaces.

If one has high numbers of LDL receptors on his or her cell surfaces, then cholesterol can be taken into cells, and little will remain circulating in the blood where it can build up in plaques. Unfortunately, many Americans have low levels of LDL receptors. As a result, their cells remove *less* cholesterol from the bloodstream, and the fatty material continues to circulate. This can contribute to plaque buildup, which can in turn predispose the person to atherosclerosis, heart attacks, and strokes.

Some individuals have low levels of functioning LDL receptors because of an inherited mutation in the LDL receptor gene,

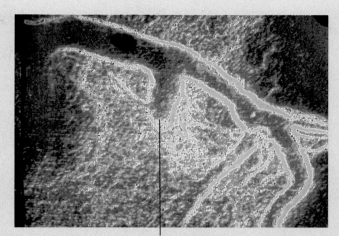

Flow blocked here.

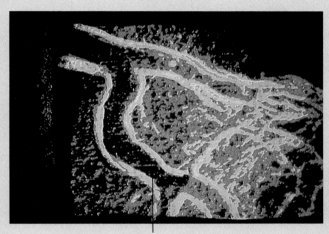

Blood flow resumes.

FIGURE 1 ■ **A Plaque-Clogged Artery. Before surgery (above) and after (below).**

and these people can suffer heart attacks in their 40s and 50s. In many other people, however, the characteristic may be related to a high-fat diet and a resulting chronic oversupply of cholesterol in the blood. In effect, an inundation with cholesterol is prevented when the metabolic machinery that manufactures a cell's LDL receptors is shut down, and thus the blood cannot be sufficiently cleansed of the circulating fatty sludge.

This understanding of the link between the body's handling of cholesterol and the development of atherosclerosis has generated a search for drugs that can stimulate cells to build new LDL receptors. Another approach is represented by the drug Lovastatin, which inhibits the action of a liver enzyme that makes cholesterol. In the meantime, the most prudent course is the one that health experts have been recommending all along: Eat fewer fatty foods, get more exercise, maintain ideal weight, don't smoke, and drink moderately if at all.

blood volume at any given time. It is also why venous blood tends to pool in inactive limbs and why physical activity helps maintain good blood circulation.

The movement of muscles in the arms, legs, and rib cage gently milks blood through individual veins, while the one-way valves in each vessel's central cavity prevent backflow and keep the blood moving toward the heart. If no muscular milking takes place over a long period of time, blood and other fluids accumulate in the extremities. In a rather macabre example of such fluid accumulation, some scholars believe that the lingering death of Jesus and other crucifixion victims in ancient times was due in large part to blood pooling in the legs.

Shunting Blood Where It Is Needed: Vasoconstriction and Vasodilation

From moment to moment, the distribution of blood varies in an animal's arteries, veins, and capillaries, depending on oxygen use. During an exercise session, for example, blood flow to your muscles increases; then later, if you eat a sandwich, blood flow to the intestines rises and nutrient molecules are absorbed from the gut into this increased blood supply (see Chapter 24). Variations in the amount of blood flowing through the vessels in a particular body region are generally controlled by the action of nerves or, as in the examples just cited, by the chemical messengers called hormones. If blood is needed in one region of the body, say, the arm and leg muscles, then contraction will take place in the walls of arteries or arterioles in other body regions. Contraction of the vessel walls is called **vasoconstriction** ("vessel constriction"; Figure 21.12a), and as a result of this process, the diameters of the vessels decrease, blood flow is reduced, and more blood can be shunted to other regions, such as the arms and legs, where it is needed to deliver oxygen to rapidly metabolizing muscle cells. In capillaries within the body regions that require a greater blood flow, a reverse process, **vasodilation** (Figure 21.12b), takes place; the vessels' internal spaces are enlarged, and more blood can flow through.

Vasoconstriction results when a small cuff of smooth muscle, the **precapillary sphincter** located near the capillary entrance, contracts. When we get extremely cold, precapillary sphincters may close down capillary beds near the skin sur-

(a) Vasoconstriction

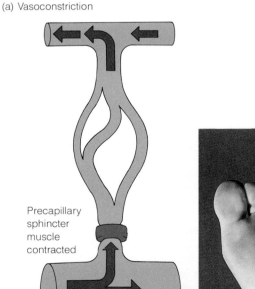

Precapillary sphincter muscle contracted

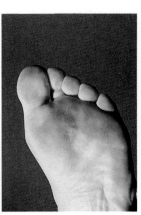

(b) Vasodilation

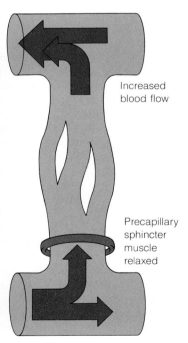

Increased blood flow

Precapillary sphincter muscle relaxed

FIGURE 21.12 ■ White Toes, Red Face: A Sign of Blood Flow Control. (a) In the foot of a person with Raynaud's syndrome, whitish, mottled regions are a sign of restricted blood flow. The mechanism at work is a set of tiny cuffs, or precapillary sphincter muscles, that constrict and close off blood flow to the capillaries. (b) The full flush of heavy exercise comes from relaxed sphincter muscles, dilated vessels, and increased blood flow to the skin, where excess heat can disperse to the environment.

face in the fingers or toes; the remaining blood in these tissues may become oxygen-depleted, and the skin can take on a yellowish or bluish cast (see Figure 21.12a). Likewise, eye drops that "get the red out" contain chemicals that constrict the precapillary sphincters in the tiny capillaries in the eyeball. When we are very warm, say, during active exercise, vasodilation allows the capillary beds to become engorged, and the skin can feel hot and look red (see Figure 21.12b).

How Bleeding Stops: The "Clotting Cascade"

Although precapillary sphincters can shunt blood away from a capillary bed, more drastic mechanisms are sometimes needed to stop blood from flowing to a particular spot, such as a finger wound made with a sharp paring knife (Figure 21.13a). In a cut vessel, several separate events take place that halt the flow of blood. First, smooth muscles in the walls of the damaged vessel contract (vasoconstrict) and partly close off the vessel, and circulating cell fragments called platelets react by releasing the hormone *serotonin,* which keeps the muscles contracting. Then platelets at the injured site begin sticking to each other and to rough surfaces at torn edges of the wound, forming a plug. The next stage is **coagulation,** the actual formation of a blood clot. The entire multistage process is subject to rigid controls, and for good reason; a clot that forms unnecessarily, such as in vessels of the heart or brain, can lead to a heart attack or stroke. The agents of such fine control are substances that circulate in the blood plasma, stimulating coagulation when a vessel is injured and inhibiting it most other times.

The formation of a blood clot (coagulation) has been dubbed the "clotting cascade" because it involves a series of steplike changes in blood proteins (Figure 21.13b). Ultimately, these transform **fibrinogen** (which, you may recall, normally circulates dissolved in the plasma) into tough, insoluble threads of a related protein, **fibrin.** Fibrin threads in turn become woven into a strong, wiry mesh that traps red blood cells, creating a blood clot (Figure 21.13c). What sets the fibrinogen-to-fibrin changeover in motion? The trigger is tissue damage: Injured cells release activators, which, in a cascade of reactions, stimulate the conversion of an inactive protein (prothrombin) into an enzyme (thrombin) that in turn changes the structure of fibrinogen. The altered fibrinogen units then line up together and form the fibers of fibrin.

People with "bleeder's disease," or hemophilia, have a mutant gene that codes for a defective protein in the clotting cascade (see Chapter 12). Unless a hemophiliac receives periodic intravenous injections of solutions containing the normal protein, even a minor injury can lead to major blood loss and even death.

Despite the ability of the circulatory system to plug leaks, fluid still tends to ooze from blood vessels. Complex animals have thus evolved a "second circulatory system" that drains away this leaked fluid.

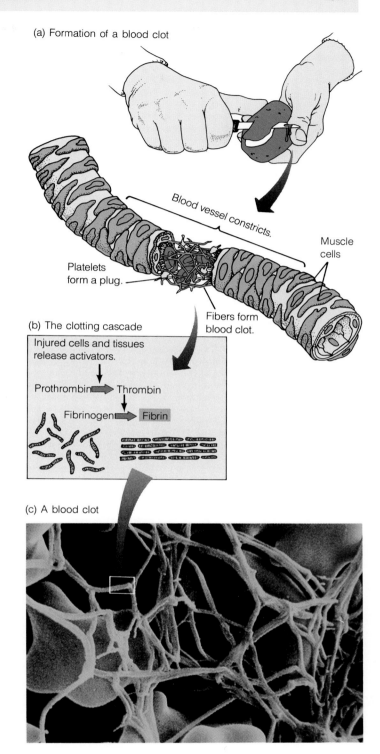

(a) Formation of a blood clot

Blood vessel constricts.

Muscle cells

Platelets form a plug.

Fibers form blood clot.

(b) The clotting cascade

Injured cells and tissues release activators.

Prothrombin ➡ Thrombin

Fibrinogen ➡ Fibrin

(c) A blood clot

FIGURE 21.13 ■ The Formation of a Blood Clot: A Life-Saving Cascade. (a) When a wound severs a tiny blood vessel, numerous processes are set in motion, including vasoconstriction of the vessel, the formation of a platelet plug, and then coagulation, or blood clotting. (b) A cascade of proteins acting on each other ultimately converts prothrombin to the enzyme thrombin. Thrombin removes a small piece from the fibrinogen protein, turning it into fibrin, and fibrin molecules line up and form a fiber. (c) Fibrin threads and red blood cells are clearly visible in this blood clot.

(a) The lymphatic system

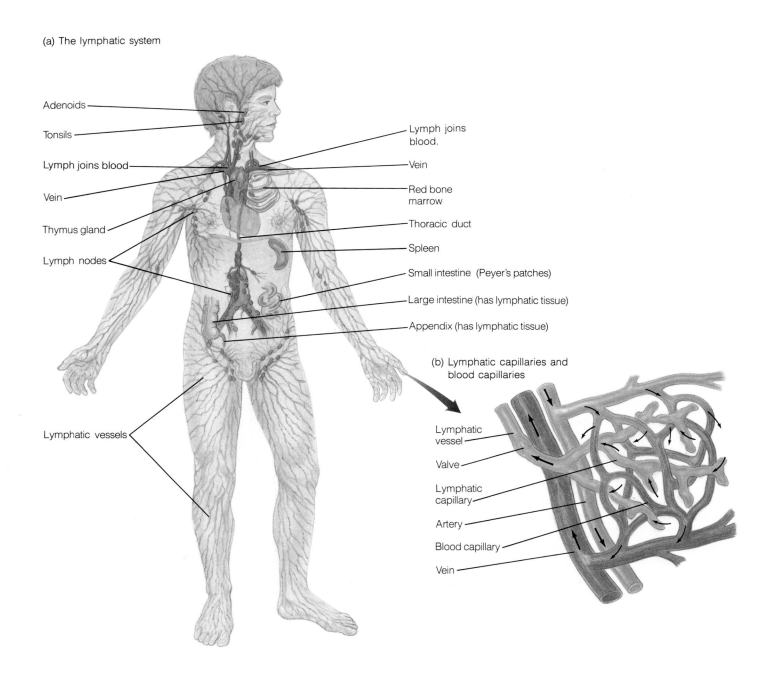

Adenoids

Tonsils

Lymph joins blood

Vein

Thymus gland

Lymph nodes

Lymph joins blood.

Vein

Red bone marrow

Thoracic duct

Spleen

Small intestine (Peyer's patches)

Large intestine (has lymphatic tissue)

Appendix (has lymphatic tissue)

(b) Lymphatic capillaries and blood capillaries

Lymphatic vessels

Lymphatic vessel

Valve

Lymphatic capillary

Artery

Blood capillary

Vein

FIGURE 21.14 ■ **The Human Lymphatic System.** (a) The vessels and organs of the lymphatic system. (b) How excess fluid, which diffuses out of the blood capillaries, collects in a lymphatic capillary and is returned to the blood circulation.

THE LYMPHATIC SYSTEM: THE SECOND FLUID HIGHWAY

One consequence of a highly pressurized blood system is that fluids can be forced through the walls of delicate vessels and can build up in the extracellular spaces. The second system of fluid-containing vessels, the **lymphatic system,** is a network of tubes that collects fluid that has been forced out of capillaries and returns it to the bloodstream (Figure 21.14a). As Chapter 22 describes, lymphatic vessels also play a major role in the functioning of the immune system.

Unlike the blood vascular system, the lymphatic system is made up of capillaries and larger vessels that do *not* form a circulatory loop. Instead, lymphatic capillaries are minute tubes with closed ends that permeate and drain the body tissues. Like creeks merging to form a river, these capillaries coalesce to form larger lymph vessels that move fluids toward the heart (Figure 21.14b). The lymphatic capillary walls are permeable to extracellular fluid and its contents. Excess fluid in the spaces between cells seeps into this tributary system and forms the **lymph,** which also contains proteins, dead cells, and sometimes invasive microorganisms, such as bacteria and viruses. Eventually, some of the smaller vessels drain their contents into two major lymphatic vessels that disgorge a steady stream of lymph into veins of the circulatory system near the heart. The lymph fluid then becomes part of the blood plasma once again.

To prevent debris in the lymph from mixing into the blood, there are several bean-shaped filtering glands called **lymph nodes** at intervals along the lymphatic vessels. As lymph moves along a vessel, it percolates through the nodes, like oil passing through an oil filter, and in the process, dead cells and other debris are filtered out. The nodes also harbor large numbers of infection-fighting lymphocytes (white blood cells) that can attack bacteria and other foreign materials. You have no doubt experienced painful swollen lymph glands in the neck, underarms, or groin area as these organs have increased their filtering activities during a bout of infection. Unfortunately, just as lymph vessels can carry dead cells away from an infection site, they can also transport migrating cancer cells to new sites (see Chapter 14). This is why doctors often remove lymph nodes and vessels along with cancerous tissue in patients with cancer of the breast or other sites.

Like veins, lymphatic vessels transport fluids under low pressure; valves keep the lymph flowing in one direction, and the squeezing force of contractions in muscle tissue surrounding the lymph vessels propels the fluid. A person with inactive muscles, say, an injured skier in a hospital bed, can suffer lymph accumulation and resultant tissue swelling called *edema.* In rare cases, the lymph vessels draining an extremity become blocked entirely, as by the infectious parasite that causes elephantiasis. The results can be swelling so enormous that the affected limb becomes barely recognizable (see Figure 18.14).

The lymphatic system also has two organs, the **thymus** and the **spleen,** that are somewhat separate from the network of vessels. The thymus is a soft, V-shaped organ located at the base of the neck in front of the aorta. The thymus plays a role in the body's immune system (see Chapter 22).

An adult's spleen, which is about the size and shape of a banana, lies just behind the stomach and acts as a spongelike holding area for a portion of the body's blood supply. Much of the spleen consists of filtering tissue that removes impurities from the blood (including old red blood cells), and the organ harbors dense masses of white blood cells as well. Surgeons can remove a person's injured spleen and cause no severe consequences to overall health. This suggests that the spleen performs no absolutely essential function.

CONNECTIONS

The circulatory system plays a central role in an animal's ability to cope with its environment, to maintain a steady internal state for every cell, and to survive. Different environments place different demands on the ways that oxygen and other necessary materials can be obtained and distributed to cells, and the evolutionary solutions to these problems have been varied. Some animal bodies, like sponges, are very porous, while others, like jellyfish or flatworms, are very thin, and materials can move in and out by direct diffusion. Most large animals, however, have complex circulatory systems that propel blood throughout the body. Regardless of approach, however, physical constraints, such as diffusion distances across cell membranes, the squeezing of red blood cells through narrow vessels, and the mechanics of bulk flow, limit the kinds of solutions that have evolved.

Although this chapter has focused on the circulatory system in isolation, we must keep in mind that an organism functions as an integrated whole. The heart muscle could not continue to contract without a steady supply of oxygen diffusing across gill or lung surfaces. Heart rate is regulated by the nervous and hormonal systems, and these, too, depend on a steady supply of blood, with its cargo of nutrients and oxygen. The interconnections go on and on, and they help explain why diseases of the heart have such catastrophic consequences. Only now are medical science and the population at large beginning to understand why diet, exercise, and other lifestyle factors are so important, why we cannot afford to ignore the warnings about cigarette smoking, cholesterol in foods, unmitigated stress, and too much alcohol: They can create internal conditions that make it impossible for the body's interconnected organ systems to continue to function.

In the next chapter, we will expand our view of the events that take place when the body is under attack. In particular, we will examine the roles of circulatory and lymphatic functions in the body's bastion of self-defense: the immune system.

Highlights in Review

1. As in all vertebrates and most other animals, the giraffe's circulatory system moves body fluids and helps maintain the animal's homeostasis by delivering oxygen, nutrients, and other materials to each body cell and removing metabolic wastes.
2. The blood has a fluid portion, the plasma, containing dissolved ions and membranes, and formed elements, the red and white blood cells and platelets.
3. The shape of a red blood cell and the millions of hemoglobin molecules contained in the cell allow it to transport oxygen efficiently. The replacement of dead and damaged red blood cells is controlled by a hormone acting in a negative feedback loop.
4. White blood cells squeeze through capillary walls and patrol the fluid-filled spaces between cells, defending the body against invasions by microorganisms and other foreign materials.
5. Arthropods and mollusks have open circulatory systems, with blood moving partly in vessels but mostly flowing freely around and through the animal's tissues. Segmented worms and vertebrates, however, have closed circulatory systems, with blood traveling in an elaborate system of closed vessels.
6. Fishes have a two-chambered heart with a single atrium and a single ventricle. Adult amphibians and reptiles have three-chambered hearts with two atria and one ventricle. Birds and mammals have four-chambered hearts with two atria and two ventricles. In birds and mammals, blood is pumped in a double loop—a pulmonary loop to and from the lungs and a separate "body loop," the systemic circulation.
7. A mammal's complex circulatory system is characterized by thick, muscular arteries that lead to narrower muscular arterioles and a network of fine capillaries with walls only one cell thick, allowing easy diffusion of substances into and out of the blood. Capillaries lead blood into venules, which coalesce into veins. Veins have valves that prevent the backflow of blood; thus, venous blood moves only toward the heart.
8. In the mammalian heart, cardiac muscle fibers are electrically linked by special cell junctions in the intercalated discs. The cardiac cycle consists of the contraction (systole phase) and relaxation (diastole phase) of the four heart chambers.
9. Blood pressure is a measure of the outward force that blood exerts on vessel walls as the heart alternately pumps and rests. Vasodilation and vasoconstriction help distribute blood to local body regions as needed.
10. The "clotting cascade" transforms the protein fibrinogen into a tough mesh of fibrin filaments that trap red blood cells.
11. The lymphatic system, including vessels, lymph nodes, the thymus, and the spleen, transports and filters lymph. The filtered lymph then flows back into the circulatory system, where it once again becomes part of the plasma.

Key Terms

Study Questions

Review What You Have Learned

1. What are the special adaptations of a giraffe's circulatory system?
2. List the constituents of blood plasma.
3. What are three special characteristics of red blood cells?
4. Describe the negative feedback loop that governs red blood cell replacement.
5. What route does blood follow in the single-loop circulatory system of a trout?
6. List the different types of vessels that transport blood on its circuit through the human body.
7. Place the following structures in their proper order to show the path of blood circulation in the human body: inferior and superior venae cavae, left atrium, left ventricle, pulmonary artery, right atrium, right ventricle, pulmonary vein, lungs, capillary beds, aorta.
8. What is the sequence of events during a heartbeat?
9. How do the systole and diastole phases of a heartbeat differ?
10. What keeps blood from flowing backward in an artery? In a vein?
11. Describe the sequence of events that produces a blood clot at a wound site.
12. Describe the function of lymph nodes.
13. Name two large lymphoid organs, and give the function of each.
14. Should a person try to rid his or her body of all cholesterol? Explain.

Apply What You Have Learned

1. Create a chart to record your pulse at different times of the day and during active and quiet periods. What do you conclude about the relationship of heart rate and time of day? Of heart rate and physical activity?
2. A nurse takes the blood pressure of a patient who has sat in the waiting room for half an hour and finds that the diastolic pressure is 90. Later, after leaving the medical office, the patient's

diastolic pressure drops to 75. What could account for the difference?

3. An out-of-shape office worker saves enough money to go trekking in Nepal. She knows that she will need to get in shape to accommodate the high-altitude hiking she will do. She could either lift weights or swim. Which should she choose, and why?

4. A scientist searching for fossilized dinosaur hearts claims that he can provide evidence on the question of whether the dinosaurs were warm-blooded or cold-blooded by determining whether they had three- or four-chambered hearts. How might this finding bear on the question?

For Further Reading

BROWN, M. S., and J. L. GOLDSTEIN. "How LDL Receptors Influence Cholesterol and Atherosclerosis." *Scientific American,* November 1984, pp. 58–67.

CAROLA, R., J. P. HARLEY, and C. R. NOBACK. *Human Anatomy and Physiology.* New York: McGraw-Hill, 1990.

GOLDE, D. W., and J. C. GASSON. "Hormones That Stimulate the Growth of Blood Cells." *Scientific American,* July 1988, pp. 62–70.

LILLYWHITE, H. B. "Snakes Under Pressure." *Natural History,* November 1987, pp. 59–67.

PALCA, J. "Getting to the Heart of the Cholesterol Debate." *Science* 247 (1990): 1170–1171.

PEDLEY, T. J. "How Giraffes Prevent Edema." *Nature* 329 (1987): 13–14.

ROBINSON, T. F., S. M. FACTOR, and E. H. SONNENBLICK. "The Heart As a Suction Pump." *Scientific American,* June 1986, pp. 84–91.

VINES, G. "Diet, Drugs and Heart Disease." *New Scientist,* February 25, 1989, pp. 44–49.

The Immune System and the Body's Defenses

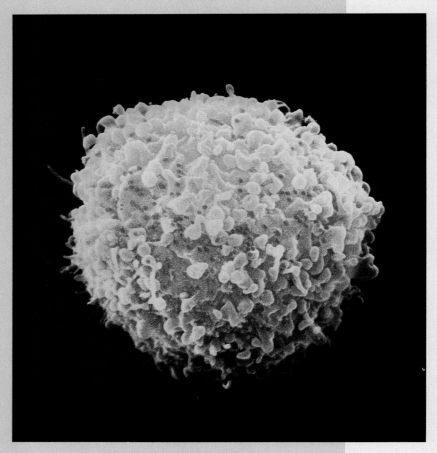

FIGURE 22.1 ▪ T Cell Infected by AIDS. A T cell, part of the body's protective armory of immune cells, normally helps fight off infections by viruses. But the AIDS virus (shown here as colored particles emerging from a stricken T cell) attaches and incapacitates those immune cells. When that happens, the whole immune system suffers a breakdown. People infected by the AIDS virus die of diseases such as pneumonia, skin rashes, and blood poisoning, which a healthy immune system could have resisted.

AIDS, The New Plague

Baby Anna was, to all appearances, a normal infant, and by the end of her first year of life, she could speak a few words and stand on her own. At about 18 months, however, she came down with a serious lung infection and for the next 8 months neither learned new words nor developed any new skills of perception and coordination. When she turned two, she again contracted a severe respiratory infection, and this time she regressed mentally and physically to the level of a seven-month-old. Although Anna recovered from the infection, she was prone to others and never regained her normal pattern of growth and development. Anna's mother, recognizing some of the symptoms that she herself was suffering, sadly realized that she had passed on to her child the disease known as AIDS—acquired immune deficiency syndrome.

AIDS brings into focus the central role of the **immune system**—the network of organs, cells, and molecules that defends the body from invaders of all kinds. This protective system is crippled in people with AIDS, leaving them unguarded from infections by viruses, bacteria, and parasites they encounter in the environment—agents that seldom harm people with an intact immune system. AIDS patients also tend to develop certain rare types of cancer. In fact, the pattern revealed by the lethal disease AIDS shows just how crucial the immune system is to our day-to-day survival.

AIDS is caused by a virus called HIV (human immunodeficiency virus; Figure 22.1) that is transmitted in blood, blood products, and semen. These modes of transmission have allowed the virus to spread easily in certain populations. People who use intravenous drugs, for example, are exposed to AIDS-tainted blood when they share needles with an infected person. Anna's mother, a heroin addict, became infected this way, and as with about 35 percent of the children born to AIDS-infected women, such as the child in Figure 22.2, the virus crossed the placenta and infected the baby. The passing of AIDS virus in semen makes it a devastating sexually transmitted disease that can only be prevented through the use of safe sex practices (see Table 22.1 on page 480 and Appendix A.3). By 1991, more than 170,000 Americans had been diagnosed with AIDS. Homosexual and bisexual men accounted for two-thirds of the cases, but epidemiologists expect the incidence of new cases in these groups and in intravenous drug users to level off by 1995, while the incidence among heterosexuals continues to rise.

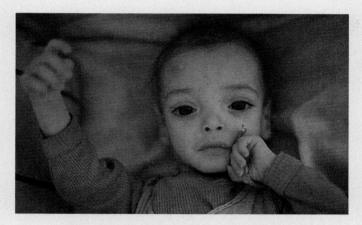

FIGURE 22.2 ■ AIDS-Infected Child. This child cannot fend off microorganisms from the environment or those that passed to her from her mother because her immune system was destroyed by the AIDS virus.

When active, the AIDS virus slowly and irreversibly incapacitates the body's natural network of defense, specifically by attacking white blood cells. The virus also infects brain cells, and their destruction can lead to impaired thinking, speaking, and performance. Infected brain cells led to baby Anna's mental deterioration, for example.

A healthy person's resistance to infection, called **immunity,** is the work of several trillion white blood cells. White blood cells attack foreign cells and molecules directly or make specially shaped protein molecules called *antibodies* that mark the intruders for destruction. A child with a normal, intact immune system successfully overcomes several bouts of cold and flu and gains resistance to reinfection as a result of a special "memory" function of the immune system. Tragically, Anna will probably die in childhood from a form of pneumonia or cancer that most people easily overcome.

AIDS is the most threatening epidemic of the twentieth century and a terrifying global problem. In some cities in Africa, AIDS is already the leading cause of death. Scientists who track epidemics think that by the turn of the century, the AIDS virus will have infected as many as 6 million people worldwide, including 1 million children. What's more, AIDS kills 85 percent of its victims within three years of diagnosis, and at present, there is no cure. Perhaps never before has the general public been made so aware of the effects of a healthy immune system and how it shields us daily from dangerous agents in the environment.

The interacting cells and molecules of the immune system circulate in the blood and lymph or lodge in organs such as the skin and spleen. As they circulate, they carry out three essential activities, which provide the recurring themes of this chapter (Figure 22.3). The first activity is recognition: The cells and molecules of the immune system recognize foreign invaders by identifying molecular patterns on the outer surface of cells and thus distinguishing all varieties of *nonself* (i.e., molecules and cells originating outside the body) from *self* (cells and molecules made by the body and belonging to it).

The second activity is communication between cells; for example, white blood cells that recognize invaders signal other white blood cells, while simultaneously, certain kinds of immune system cells communicate with hormone-producing cells and nerve cells.

A third response, elimination, defeats the attack: Elements of the immune system selectively eliminate foreign invaders by producing specific proteins that target viruses or bacteria for destruction by enzymes or by activating killer white blood cells that directly dispose of virus-infected cells.

In our brief tour of the immune system, we will see how variations on the themes of recognition, communication, and elimination make it all possible. This chapter will provide answers to the following questions:

- What major structures and activities account for a lifelong protection against diseases?
- How do the immune system's major players function in defense?
- What gives antibodies such enormous versatility in recognizing invaders and helping to destroy them?
- How can white blood cells produce millions of different antibody molecules shaped to recognize and attack specific invaders?
- How do immune system cells and molecules work together to dispose of invaders without harming the tissues of the organism's own body?
- How can physicians manipulate the immune system with vaccines and drugs to help protect patients?
- How can allergies be a type of "normal" immune response?

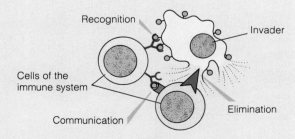

FIGURE 22.3 ■ Essential Activities of Immune System Cells. Immune system cells *recognize* invaders, *communicate* with other immune system cells, and *eliminate* invaders.

THE BODY'S DEFENSES: MAJOR PLAYERS, MAJOR ACTIVITIES

The immune system deploys both specific and nonspecific defenses. Specific responses involve individual defense molecules tailor-made for each kind of microbe or other invader. Nonspecific responses include physical barriers to infection as well as certain physiological reactions that the body produces in the same way for every intruder. These nonspecific mechanisms act faster than the specific responses that lead to immunity, but are not as effective at warding off disease.

Immediate Nonspecific Protection

A slip of your paring knife while peeling potatoes can open a chink in your skin—the body's first line of defense against unwanted outsiders. The skin forms a protective barrier that, along with mucous membranes of the nose, throat, and digestive passages, keeps out most minute organisms capable of multiplying within the body and causing infection (Figure 22.4). Thus, through a break in this smooth, elastic barrier, dirt and soil bacteria clinging to the knife's blade can enter the bloodstream.

Eventually, a blood clot will seal the gap and scar tissue will form, but even before that, the injured skin sets up a series of nonspecific physiological reactions: First, skin and blood cells release chemicals that raise the temperature; next, blood flow to the area increases; shortly thereafter, scavenging white blood cells known as *phagocytes* converge on the area in search of invaders (Figure 22.5). The rise in temperature slows bacterial multiplication; the increased flow of blood cleanses the wound and brings in more white blood cells; and the phagocytes engulf bacteria and cell scraps without disturbing the body's own intact cells. These physiologi-

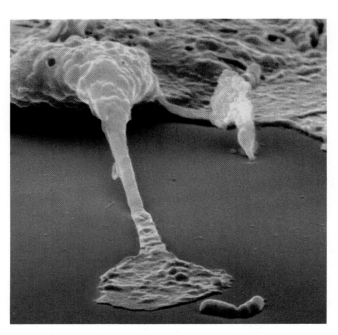

FIGURE 22.5 ■ **Phagocyte Stalking Bacteria: Nonspecific Internal Defenses Guard Against Invaders.** As this macrophage (white cell) approaches the small, rodlike bacteria (foreground, tinted green), a mobile cellular extension (pseudopodium) of the macrophage reaches out toward the invader (magnification 5000×).

cal reactions, which result in the redness, heat, and swelling we know as *inflammation,* clear the cut of debris and often keep the bacteria from spreading.

Within 4 hours of an invasion, a mysterious type of lymphocyte called a *natural killer cell* carries out a nonspecific counterattack along with the inflammation response. In an enigmatic way not yet understood by immunologists, these cells can recognize bacteria, certain kinds of cancer cells, and virus-infected cells and can kill them on first contact. Biologists know how crucial natural killer cells are to immunity because in 1989, a woman nearly died during the early phases of a herpesvirus infection. It turned out that her immune system was intact except that it lacked nonspecific natural killer cells. Fortunately, she survived with antibiotics and fought the infection, and later, her system mounted a normal specific immune response against the herpes.

Specific Immunity Against Specific Targets

Suppose that the bacteria that entered the wound in your finger begin to multiply faster than inflammation, scavenging phagocytes, and natural killer cells can destroy them. The body then falls back on its most powerful line of defense, an all-out **specific immune response.** During this response, certain white blood cells produce specially shaped proteins (antibodies) that either float in body fluids or attach to the surface

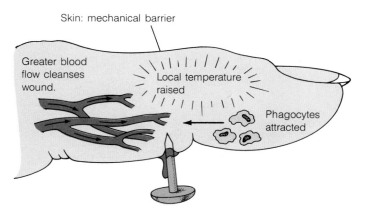

FIGURE 22.4 ■ **Nonspecific Defenses.** These external and internal defenses protect us mechanically and via the inflammatory response.

of white blood cells and attack the invader. Behind this specific response lies the entire arsenal of the immune system: its several kinds of white blood cells, the proteins they secrete, and the organs and fluids in which they multiply, differentiate, and eliminate invaders. Let's survey the principal players (Figure 22.6), then see how they interact in the defense of a cut.

■ **The Cells and Organs Behind the Immune Response**   Of the several types of white blood cells that cooperate in generating immunity, the lymphocytes are the driving force. *Lymphocyte* comes from the root words for "colorless cell," and indeed, these white blood cells are small, round, and relatively devoid of color. There are two main types of lymphocytes: **B cells (B lymphocytes),** which make and secrete the antibody proteins that coat invaders and mark them for destruction, and **T cells (T lymphocytes),** which kill foreign cells directly and help regulate the activities of the other lymphocytes. A third kind of white blood cell involved in specific immunity is the **macrophage** (literally, "big eater"; see Figure 22.6). Macrophages are large, specialized phagocytes that stimulate lymphocytes to attack invaders, then help clean up by consuming debris when the lymphocytes are through.

   Cells of the immune system—B cells, T cells, and macrophages—constantly circulate in the bloodstream and lymph; thus, like most other physiological systems, the immune system is tightly linked to the circulatory and lymphatic systems (see Chapter 21). Moreover, lymphocytes and macrophages arise from stem cells in the bone marrow, as do red blood cells. Unlike red blood cells, however, which mature into oxygen-carrying biconcave disks in the bone marrow, these white blood cells mature inside other organs: the *spleen, thymus,* and *lymph nodes* (see Figure 22.6). These and other *lymphoid organs*—so called because they are connected by the lymphatic system as well as by the bloodstream—trap foreign cells and molecules and harbor white blood cells.

■ **Molecules Involved in Defense**   No survey of the principal players in a specific immune response is complete without mention of antibodies and antigens. **Antibodies** are the protective proteins made by B cells and are secreted into the blood and lymph (see Figure 22.6). Different B cells make antibodies of slightly different shape. During an immune response, a specific part of the antibody molecule binds to a substance projecting from the surface of an infecting bacterium or other foreign agent. Substances that are bound by antibodies are often proteins but can be other kinds of mole-

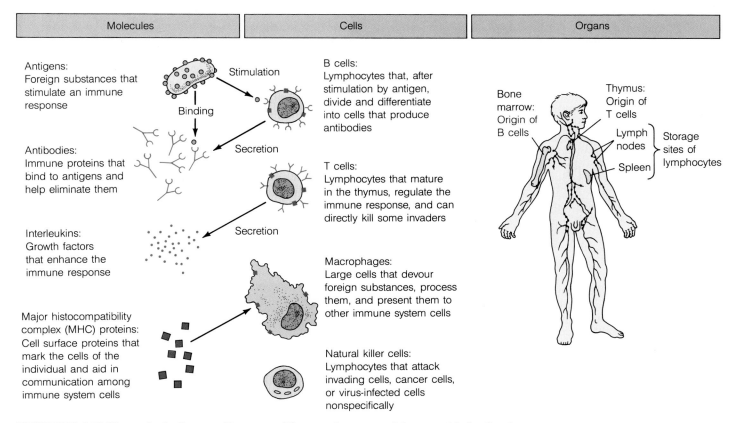

FIGURE 22.6 ■ **Players in the Immune Response.   These are just some of the many kinds of molecules, cells, and organs that play a part in the functioning immune system.**

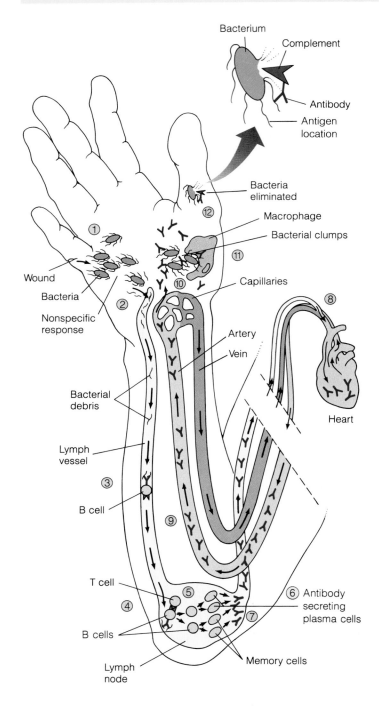

cules. They are examples of **antigens,** foreign substances that *gen*erate an immune response and trigger the production of antibodies that fight the foreign invaders. Invaders can have many antigens; the outer coats of most viruses, bacteria, fungal cells, and larger parasites contain many different proteins functioning as antigens. Because antibodies bind to antigens of reciprocal shape, they are often likened to locks and antigens to keys.

■ **A Specific Response: Immune Players in Action**   With this cast of characters in mind, let us return to our example of the cut finger and follow the steps involved in the specific immune response. As we saw earlier, once the skin is broken, the nonspecific defense responses battle the bacteria that have invaded the wound, and the battle creates debris (broken bacterial cells and cell parts; Figure 22.7, step 1). This debris begins to accumulate near the wound and enters the bloodstream and lymphatic system (step 2). There, the debris encounters B cells (step 3), and antibodies anchored in the membrane of some of these circulating B cells bind to the antigens. This binding leads to a specific immune response.

B cells carrying antibody-antigen complexes move to the lymph node nearest the wound—in this case, probably in the elbow—and settle in (step 4). Here, helper T cells (which have already received communication from macrophages about the bacterial protein) interact with the B cells and antigen-antibody complexes (step 5). The helper T lymphocytes secrete *interleukins,* or protein growth factors that stimulate the B cells to divide into 2, 4, 8, 16, and eventually hundreds of antibody-secreting factories known as daughter B cells, or plasma cells (step 6). Every second, each plasma cell secretes into the lymph fluid 1000 antibody molecules that can bind specifically to the antigens on the invading bacteria (step 7).

The secreted antibodies now leave the lymph node, and when the lymph drains back into the bloodstream near the heart (step 8), the antibodies enter the blood and circle back to capillaries near the cut (step 9). Here they pass into spaces between the cells and combine with their targets—the bacterial proteins (step 10). When millions of antibodies bind to bacterial proteins, they neutralize the bacteria by causing them to clump together. Such clumped bacteria complexed with antibodies can be eliminated from the body in two ways: The antibody coating can attract macrophages that devour, digest, and eliminate the antibody-bacteria complexes (step 11), or special molecules called *complement* proteins can bind to the antibody-bacteria complexes and perforate the bacterial membranes, killing the bacterial cells (step 12). There is a lag time between the start of an infection and a fully active immune response. This is because it takes four to ten days for a few B cells to divide and redivide into thousands or millions of plasma cell "factories" that can in turn secrete sufficient numbers of antibodies to stem the tide of infection. Once those antibodies are racing through the body, however, the bacteria that have grown in the cut are killed, and the wound is well on its way to healing.

FIGURE 22.7 ■ **How the Immune System Fights a Bacterial Infection: Events of a Specific Immune Response.** As the text explains, a finger wound touches off a multistep immune response, beginning with inflammation and other aspects of a nonspecific response (step 1) and proceeding through the recognition of antigens on the bacterial invader and formation of antibodies with reciprocal shape (steps 2 to 7), the clumping of antibodies and bacteria (steps 8 to 10), and the elimination of killed invaders (steps 11 and 12). In this figure, the bloodstream is pink, lymph vessels are yellow, bacteria are green, and antibodies are red.

ANTIBODIES: DEFENSE MOLECULES

Healthy newborn infants do not yet have a fully functioning immune system. They must rely for much of their protection against infection on antibodies they received from their mothers across the placenta or in the mother's milk. The antibodies provided by the mother, like those the newborns will later make themselves, are *immunoglobulins,* or globular proteins of the immune system. Each one consists of two identical *heavy chains* of amino acid and two identical *light chains.* Linked together in one unit, the four chains are arranged so that the molecule looks like the letter Y (Figure 22.8). "Heavy" and "light" refer to molecular weight, and each heavy chain is two or three times longer than a light chain.

The antibody's two basic functions are localized at opposite ends of the Y: The tips of the two branches each recognize and bind to an antigen of complementary shape the way an enzyme binds its substrate (see Chapter 4). Once the branches of an antibody bind to an antigen, the stem of the Y stimulates macrophages or complement proteins to attack and destroy the entire antibody-antigen complex (see Figure 22.7, steps 11 and 12).

In recent years, researchers have learned that the stem section of an antibody molecule always has the same or a very similar amino acid sequence and hence the same shape. The tips of the two branches, however, have a huge variety of different amino acid sequences and thus a huge range of sometimes radically different shapes. These shapes can bind in complementary lock-and-key fashion to an enormous array of antigens—each antibody to an antigen of a particular shape (see Figure 22.8b). Recently, molecular biologists have altered these branches so that the antibodies can catalyze chemical reactions. These so-called *abzymes* may allow researchers in the future to create specific catalysts tailored for any desired reaction.

Antigen-Binding Sites: Specificity and Diversity

Although any given antibody molecule binds to an antigen of a single shape, each person has more than 100 million distinct types of antibody molecules, each with a unique binding site. These recognize and combine with 100 million different antigens, including all the pollen from flowers, grasses, and trees as well as proteins in the outer coats of thousands of different viruses, bacteria, and molds. Because of recent advances (see Box 22.1 on page 470), this exquisite specificity can now be tapped to manufacture antibodies for various medical, diagnostic, and industrial purposes.

Amino acid variation gives antibodies the ability to recognize and respond to invaders, enabling the healthy immune

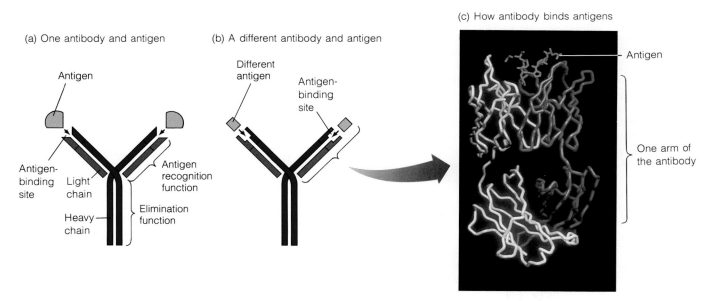

(c) How antibody binds antigens

(a) One antibody and antigen

(b) A different antibody and antigen

Antigen

Antigen-binding site

Light chain

Heavy chain

Antigen recognition function

Elimination function

Different antigen

Antigen-binding site

Antigen

One arm of the antibody

FIGURE 22.8 ■ Different Antibodies Bind Different Antigens. Antibodies are Y-shaped molecules with specific antigen-binding sites located at the tips of the Y's arms. Antigen molecules, wedge-shaped in (a), rectangular in (b), fit into the binding sites of antibody molecules like keys in locks. Light and heavy chains of the antibody join, making the antigen-binding site. The elimination of an antigen bound to an antibody depends on the stem of the Y made exclusively by heavy chains. (c) A computer graphic reveals how an antigen—in this case, a small peptide (top, in red)—binds to an antibody (here, just one arm of the Y is shown in dark blue for the heavy chain, light blue for the light chain). The side chains of the red antigen interact with the complementary surface of the antibody. These are the elements that fit together like lock and key.

system to promote resistance against measles, mumps, and recurring colds. The diversity of antibody shape and activity gives physicians a powerful diagnostic tool as well. An accurate test for specific antibodies to the AIDS virus is an example. The antibodies are produced as part of an early response to the virus and can combine with it and no other. Unfortunately, these antibodies do not protect against the virus, because the virus wreaks its havoc *inside* cells, where it is hidden from the antibodies. Long before symptoms emerge,

however, the test for those antibodies can show whether the virus is or ever was in a person's blood. And that information, although it doesn't indicate whether the person will go on to develop AIDS, can be used to identify blood from exposed people and prevent transmission of the AIDS virus in blood transfusions. Hopefully, one day it may also help doctors detect the presence of the virus early enough to forestall the progress of the disease into a full-blown case of AIDS—a tragic diagnosis for any patient.

BOX 22.1   H O W   D O   W E   K N O W ?

MONOCLONAL ANTIBODIES: TOOLS FOR MEDICINE AND RESEARCH

Once biologists and physicians began to understand how the human immune system works, they realized that they would have a powerful medical tool if only they could isolate a B cell able to make antibodies to a particular antigen—say, a marker on a gonococcal bacterium that causes venereal disease—and then gather up millions of those tailor-made antibodies and inject them into a patient to help fight his or her venereal infection. But how were they to isolate such a B cell and keep it and its daughters alive in a culture dish, and then how were they to gather enough of the desired antibody to use?

The answer came in 1975 when researchers Cesar Milstein and Georges Köhler, working together in England, developed a way to clone individual B cells and produce large quantities of antibodies. They began by injecting a mouse with a specific antigen, say, a gonococcal cell's surface protein, then removing the mouse's spleen with its store of B cells a few days later. Next they fused normal B cells from the spleen with cancerous B cells, or myelomas, whose forte was indefinite, rapid cell division. The result was a set of revved-up hybrid B cells called *hybridomas,* capable of churning out large numbers of specific antibodies of the desired type and of dividing and redividing into large clones of identical cells. The team's final steps were to choose the clone of hybridomas that made the specific desired antibody, grow huge numbers of the cells, then harvest the secreted monoclonal antibodies, antibodies from a single clone of plasma cells.

By now, commercial laboratories are selling more than 1 billion dollars worth of the cloned biologicals each year. Monoclonals are being used in basic research to help locate and identify specific proteins, in the diagnosis of medical conditions, and in the treatment of human diseases. For example, home test kits for detecting pregnancy (Figure 1) contain monoclonal antibodies that bind to human chorionic gonadotropin (see Chapter 14). Another laboratory test kit contains monoclonal antibodies to the enzyme prostatic acid phosphatase, a protein present at high levels in men with prostate cancer; use of the kit aids early detection of that dangerous disease. And medical laboratories are now using monoclonal test kits to detect surface proteins on gonococcus, chlamydia, and herpesvirus. With such tools, phy-

FIGURE 1 ■ **Home Pregnancy Tests Are Based on Monoclonal Antibodies. This woman is using such a test to determine whether a missed menstrual period is due to pregnancy. A special hormone is secreted when a woman is newly pregnant, and some of this hormone is excreted in her urine. When monoclonal antibodies bind to the hormone, they generate a color change in the test materials, which usually indicates pregnancy.**

sicians can cut three to six days off the time needed to confirm a patient's venereal disease and thus stop the rampant pathogens even quicker.

Potential treatments based on monoclonals include agents that can fight tumor cells; others that can block the rejection of newly transplanted organs; and still others that can fight certain autoimmune conditions such as lupus and rheumatoid arthritis. Monoclonal antibodies are true magic bullets, taking aim on specific molecules and leaving the rest of the body unharmed.

(a) IgG can cross placenta.

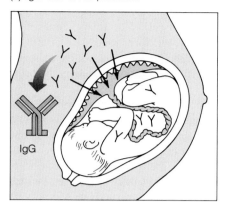

(b) IgA is secreted in body fluid.

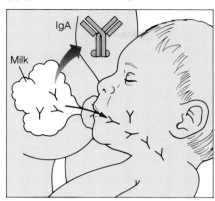

(c) IgE is involved in allergies.

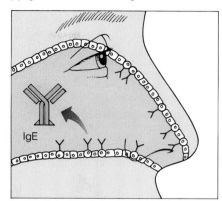

FIGURE 22.9 ■ Different Antibody Classes Have Different Stems and Different Destinations. (a) IgG can cross the placenta and confer immunity on the fetus. (b) IgA is secreted in saliva, tears, and milk and can help a breast-feeding infant resist infection. (c) IgE molecules bind to mast cells in the skin and mucous membranes lining the nose, throat, airways, and intestines and are involved in allergic responses.

How Antibodies Trigger Elimination

While the arms of a Y-shaped antibody bind to an antigen, the stem of the Y triggers the elimination of an invader. Slight variations in the amino acid sequence of the stem (and hence slight differences in stem shape) result in different elimination functions (Figure 22.9).

One stem class called *IgG,* or *immunoglobulin G,* circulates in the blood and efficiently eliminates viruses and bacteria. IgG is the only class of antibodies that can cross the placenta from a pregnant woman into her baby. The blood protein fraction called *gamma globulin* is rich in IgG and is often injected into a patient who has just had surgery or into a traveler bound for a country with contaminated food or water to provide for immediate protection. Another class of antibodies, IgA, has a different-shaped stem that allows these molecules to make their way into bodily secretions such as saliva, tears, milk, and the mucus of the intestinal lining. IgA molecules that pass to a nursing infant through the mother's milk can help protect the child from diseases to which the mother is immune. This is a strong rationale for breast-feeding a baby. A third antibody type, IgE, binds to specialized cells in the skin and intestinal lining. Besides helping to fight parasites, cells bound to IgE are responsible for the irritating condition we call allergy, as Box 22.2 on page 473 describes.

The key to the amazing effectiveness with which antibodies guard the body and destroy its enemies lies in their two-tiered construction, with the arms of the Y geared for recognizing invaders and the stem of the Y geared for eliminating them. To discover how the white blood cells can form a family of molecules with one basic shape but millions of uniquely shaped recognition sites, we must examine one particular class of white blood cells, the B lymphocytes, which form and secrete antibodies.

B CELLS: MOBILIZATION AND MEMORY

Every ready-to-go B cell is a small, colorless, rough-surfaced sphere (review Figure 21.3). Each B lymphocyte, however, carries in its membrane close to 100,000 copies of the kind of antibody it will secrete when stimulated by an antigen of reciprocal shape. Immunologists—scientists who study the immune system—have discovered that while all B cells carry antibodies on their membranes and can produce thousands of the molecules, each individual B cell carries and secretes antibodies with binding sites of only one shape. "One cell, one antibody" is the shorthand for this phenomenon.

The B Cell in Action

Although each B cell is studded with antibodies of a single type, our bodies contain millions of different B cells lodged in lymph nodes or racing through the blood and lymph fluids, and each B cell is able to respond to a different antigen. Taken together, our B cells can respond to nearly every antigen imaginable on a bacterium, fungal spore, pollen grain, or other invading entity (Figure 22.10, step 1). When an antigen enters the body, it *selects* from the entire population of millions of B cells only those few cells carrying the best-fitting antibody receptors, and it binds to those receptors (step 2).

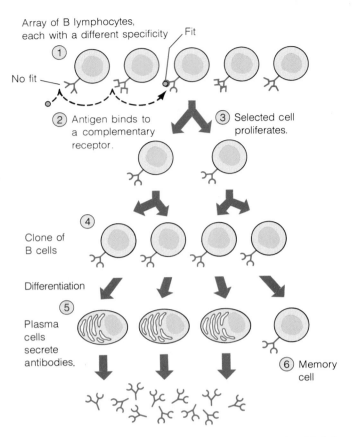

FIGURE 22.10 ■ **B Cells in Action: The Clonal Selection Model of Antibody Production and Immunological Memory.** Each of the millions of types of B cells can recognize an antigen of specific shape (1). Antigen binding selects a B cell (2) to proliferate (3) into a clone of identical B cells (4). These differentiate into plasma cells, which can secrete antibodies to the specific antigen (5), or memory cells, which can respond months or years later (6). Plasma cells are filled with endoplasmic reticulum (shown here in red), which enables the lymphocytes to churn out and export thousands of antibody proteins each minute.

cells (step 6). Like the original B cell, memory cells can divide and redivide when stimulated by the same type of antigen that triggered the original B cell. With this second round of division, the memory cells will form their own clones of both plasma and memory cells, all with the ability to make antibodies of the same specificity as those carried by the original B cell.

The first time a specific antigen stimulates an animal, only a small number of cells respond; this first response is known as a *primary response.* But the next time the animal encounters the same antigen, it is better prepared. It has a large pool of memory cells partway along the road to becoming differentiated plasma cells, and the net result is a stronger, swifter reaction to the invader (a *secondary response*). Memory B cells thus serve as remnants of former encounters with antigens and constitute a kind of **immunological memory** that can make the difference between resisting disease and succumbing to it.

By the time a healthy newborn grows to adolescence, he or she will have developed many sets of protective memory cells, providing immunity to many kinds of infections. The infant or child with AIDS, however, loses (or never develops) the ability to generate such immunological memory.

In people with fully functioning immune systems, memory cells can survive for several decades, and this explains why most of us rarely contract measles or chicken pox a second time. If a grandparent had measles as a second-grader and is then exposed to measles while babysitting a grandchild, a large pool of memory cells in the older person is stimulated to produce antibodies that help eliminate the viruses before they have a chance to cause disease.

The message about B cells is an important one for understanding the immune system: Each type of antigen stimulates specific individual B cells to form a clone of daughter cells. Some of these identical daughter cells will secrete antibodies that provide immediate protection, while others will lie in wait for a later exposure to the same antigen and give long-term protection.

The binding of an antigen to the receptor on each individual B cell stimulates the cell to divide rapidly and thus selects it to proliferate (step 3) into a clone, or group of daughter cells all descended from the same parent cell (step 4). In just ten days, a single B cell can generate a clone of 1000 daughter cells. Because many members of this clone secrete antibodies, the combined process of selection by antigen and proliferation into a clone is called the *clonal selection mechanism* of antibody formation.

Some of the daughter cells of each expanded clone stop dividing and differentiate into *plasma cells* (step 5), the antibody factories described earlier. Plasma cells focus all their energy on antibody production and survive only a few days. Other cells of the clone do not differentiate immediately into antibody-secreting factories. Instead, they become *memory*

Antibody Synthesis: Reshuffling a "Small" Deck of Genes

At this moment, your body contains B cells with antibody receptors shaped to match virtually any antigen you could ever encounter: a rare virus from Borneo, a pollen grain from a Mongolian steppe grass, and even molecules of an organic chemical just invented by a chemist in her laboratory! We know from earlier sections that each antigen is recognized by a differently shaped antibody. We also know that protein shape depends on amino acid sequence (see Chapter 2), which in turn depends on the information in genes (see Chapter 10). This whole situation leads to a tantalizing puzzle: How can your cells contain the huge number of genes necessary to code for millions of antibodies of different shapes, including mol-

BOX 22.2   PERSONAL IMPACT

ANATOMY OF AN ALLERGY

Does pollen from grass, trees, or ragweed make you sneeze? If so, you may be one of the 35 million Americans with allergies—extreme immune reactions to foreign substances or allergens. Allergens can range from pollen and house dust to pet dander, bee venom, milk, molds, and mites. Researchers are not completely certain how these substances can set off bouts of sneezing, wheezing, swelling, and itching, but they are studying the basic mechanisms of allergy in search of more successful treatments.

In the 1870s, German bacteriologist Paul Ehrlich studied the tissues of a person's physical barriers—the skin; mucous membranes of the eyes, nose, and throat; and linings of respiratory passages, lungs, and intestines—all of which have direct contact with elements of the environment. He noticed that each of these tissues has a type of cell packed with 1000 or more large, globular granules. Ehrlich named them mast cells after the German word for a fattening feed, because he assumed that the cells gobbled the granules. By the early 1960s, scientists had discovered that these mast cell granules account for the runny nose, watery eyes, and sneezing of hay fever; the diarrhea and stomach cramps of food allergies; and the wheezing of asthma. Allergens somehow trigger mast cells to disgorge histamines and other chemicals, and these agents then dilate the capillaries in the local area and cause them to leak fluid into surrounding tissues. This in turn can lead to swelling and redness and cause constricted airways, mucous secretion, lack of intestinal absorption, and other dysfunctions.

When an allergic individual first encounters an allergen—say, a noseful of ragweed pollen (Figure 1, step 1)—he or she will not experience a reaction, but plasmacytes will generate millions of IgE molecules for that antigen. These antibodies will move through the bloodstream, and between 100,000 and 500,000 of them will attach by their tail to the surface of each mast cell in the area of contact (step 2). The next time the allergic individual encounters allergen molecules, the allergens will form bridges between adjacent IgE molecules on the mast cell surface (step 3). Significantly, this bridging triggers a cascade of biochemical events that causes mast cells to "degranulate," or explosively release their granules and the histamines and other substances they store (step 4).

Physicians often treat allergy patients with a series of desensitization shots after determining the offending allergen—whether pollen from grass, pine, or goldenrod; house dust; milk, shellfish, or other food; or some combination of these. The object of desensitization is to deliver small quantities of allergen in a way that causes the body to make IgG but not IgE. Later, when the patient encounters naturally occurring allergen in the environment, the massive number of IgG molecules in the body then bind to the allergen molecules. This binding takes place before the allergen can link up with IgE on the surface of mast cells, cause an explosive release (degranulation), and lead to sneezing, watery eyes, swelling, or other symptoms.

To help the millions of allergy sufferers, researchers have been working intensively to understand the immunological mechanisms involved in allergy and to plan for improved drug treatments in the future. They have found, for example, that T cells make enhancer and suppressor molecules that turn the B

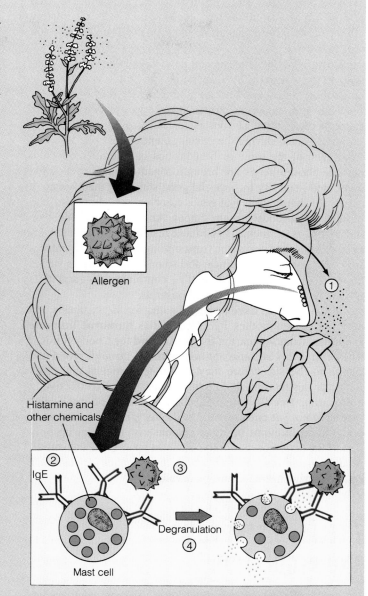

FIGURE 1 ■ Anatomy of an Allergy. The mast cell's role in allergy.

cell's production of IgE on and off. A drug that mimics the IgE suppressor might defuse the trigger on the mast cell "bombs"; thus, lab workers are working hard to create such suppressor drugs. Researchers have also learned why antihistamines—the standard drug approach today—do not work as well as some would like: While mast cell granules spew out histamines, they also release compounds called leukotrienes, which are 100 times more powerful in causing allergy symptoms. Antileukotriene drugs, also under development, may work alongside IgE suppressors to someday stop allergy miseries once and for all.

ecules you will probably never encounter and that haven't even been invented yet? The answer lies in the shuffling and reshuffling of a small deck of genes before and immediately after you were born, which created all the B cell types you will ever need.

How, then, does this special gene reshuffling work? Contrary to the normal rules of genetics, an amino acid chain in an antibody is encoded not by a single gene, but by three or four different genes separated on a specific chromosome. Each of these three or four genes is a single member of a group of genes. During the development of a B cell, one gene from each of the three or four groups is randomly selected, and the chosen genes are brought together, making a single combined gene. By this special gene shuffling mechanism, a relatively small number of genes can create an almost infinite number of antibody protein shapes. An analogy for this gene reshuffling is the menu at a Chinese restaurant, where customers can choose 1 appetizer from among 6 choices, 2 entrees from a list of 15 possibilities, and so on. If the menu is varied enough, the number of potential combinations—and thus unique dinners—can be in the thousands.

Immunologists have a special name for the synthesis and secretion of antibodies by B cells: the **humoral immune response,** from *humor,* the medieval word for "body fluids." They call it this because antibodies travel through blood and lymph to the site where they encounter antigens and mark them for destruction. Antibodies are particularly effective at eliminating free-floating microbes, viruses, and molecules that have not yet entered body cells. As effective as the antibody-based humoral immune response is, it makes up only part of the action of immunity. The other part is the **cell-mediated immune response,** in which the lymphocytes called T cells directly eliminate foreign invaders.

T CELLS: DIRECT COMBAT AND REGULATION

Unlike B cells, whose only function is to produce antibodies, T cells kill foreign cells directly and also have important regulatory roles in the immune system. T cells directly destroy body cells transformed by infection or cancer (Figure 22.11), and they attack cells transplanted from other animals. And by communicating with B cells and with each other, T cells can initiate, amplify, diminish, or stop immune responses.

How T Cells Work

In the light microscope, T cells look exactly like B cells: small, spherical, and colorless. But T cells pass through the thymus for processing and maturation and lodge in the lymph nodes, spleen, and skin, whereas B cells begin to mature in the bone marrow before eventually lodging in either the lymph nodes or the spleen. The thymus shrinks as a person

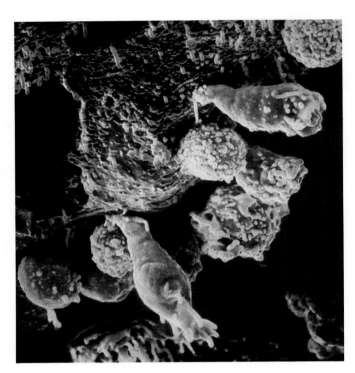

FIGURE 22.11 ■ T Lymphocytes Destroying a Cancer Cell. This scanning electron micrograph captures several rounded T cells (white) in the process of attacking a single larger cancer cell (brown). The constant vigilance of T cells prevents the many genetically transformed cells that arise in the body from multiplying into cancerous tumors.

matures, and the immune system functions less effectively. This explains why many senior citizens need flu shots at the start of each winter's flu season.

Like B cells, T cells have a receptor protein bound to their outer membrane. Ironically, it is this T cell receptor that binds to the AIDS-causing virus, allowing the virus to enter and infect the cell. Experimenters have recently engineered versions of this T cell receptor that can dissolve in fluids, bind to the AIDS virus, and keep the virus from spreading to noninfected cells—at least in the test tube. Researchers hope to start human trials of this modified T cell receptor to help AIDS victims. While the T cell receptor can bind to HIV viruses dissolved in body fluids, it can bind antigen only when a macrophage presents antigen to it along with self-marking membrane molecules called **histocompatibility proteins** (from the Greek *histos* for "tissue"). Histocompatibility molecules are a key to the two main functions of T cells: their ability to directly kill cells bearing antigens and their ability to regulate the immune response.

Each person's cells have a unique array of histocompatibility proteins serving as cellular fingerprints that distinguish "self" from "nonself." Histocompatibility proteins were first recognized for their role in determining whether an animal accepts or rejects a tissue graft, such as skin transplanted from a donor onto a burn victim. Rejection occurs when the histocompatibility "fingerprints" of the graft differ from those of the recipient. The T cells recognize that the grafted tissue is "nonself" instead of "self," and attack it.

■ **Effectors Are Eliminators**   Half of all T cells are effectors that eliminate cells with foreign antigens on their surface. *Cytotoxic* T cells (literally, "lethal to cells") poke holes in the membranes of foreign cells as well as in self-cells altered through infection by viruses or transformed by cancer (Figure 22.12a). When physicians transplant a heart or kidney, they try to find a donated organ with closely matched histocompatibility proteins. Unless the donor is an identical twin to the recipient, however, the physician must administer drugs that suppress immune responses and keep cytotoxic T cells from causing the body to reject the foreign organ. Surprisingly, some effector T cells are not cytotoxic; instead of killing cells

(a) A T cell kills a virus-infected cell.

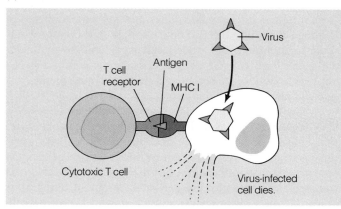

(b) A T cell can regulate other lymphocytes.

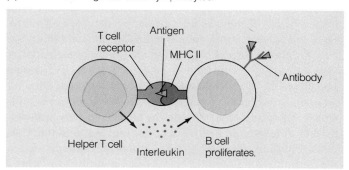

FIGURE 22.12 ■ **How Histocompatibility Proteins Help T Cells Kill and Regulate.** (a) A cytotoxic T cell does not respond to an isolated antigen; it recognizes the antigen only when it is combined with a major histocompatibility protein of class I (MHC I), a type carried by all body cells. When it sees an antigen thus combined, the T cell can kill the body cell infected by the virus. (b) Helper T cells regulate B cells and effector T cells by recognizing major histocompatibility proteins of class II (MHC II), which occur mainly on the surfaces of lymphocytes and macrophages. When an antigen and an MHC II are present, the helper T cell releases the protein interleukin, which acts on the nearby B cell, activating it to proliferate and form antibody-secreting plasmacytes.

directly, they release enzymes and hormones that wall off an infected area, cause inflammation, and remove the offending antigens. Before the first day of school, every child and teacher must have a routine test for tuberculosis that depends on a reaction by this type of effector T cell.

■ **Regulator T Cells Help or Suppress**   Half of all T cells control effector T cells and B cells rather than directly or indirectly disposing of antigens. *Helper T cells* signal B cells and effector T cells to spring into action (see Figure 22.12b). In contrast, *suppressor T cells* prevent helpers and effectors from taking action.

Both helper and suppressor T cells communicate with other cells by secreting molecules such as *interferon* and *interleukin* that alter the activity of specific targets. Significantly, cyclosporin—the main drug physicians use to block undesirable immune responses, such as the rejection of a transplanted organ—acts by inhibiting the production of interleukin by regulatory T cells. Researchers have also found that interleukins may help people infected with the AIDS-causing virus. They have injected people who harbor the AIDS virus with an interleukin called IL2, and it seems to activate cytotoxic T cells that kill AIDS-infected cells.

■ **The Biology of AIDS**   The central regulatory role of T cells becomes tragically evident in patients with AIDS, like baby Anna. The virus that causes AIDS infects and destroys a type of helper T cell required to activate effector T and B cells (Figure 22.13). Without activated cytotoxic T cells, cancers, like Kaposi's sarcoma, and protistan parasites, like *Pneumocystis carinii,* can grow unchecked. Without activated B cells, fewer antibodies are made and so bacterial infections become frequent. Virus particles also enter macrophages and accumulate in vesicles within those large cells. When such infected macrophages enter the brain, they can cause the neurological problems often seen in AIDS victims.

Since the AIDS epidemic first surfaced, researchers have wanted to find an *animal model* for the disease—that is, an animal whose immune system is destroyed by the same HIV virus that devastates our own. Chimpanzees, for example, can be infected with HIV, but don't sicken and die, while rhesus monkeys die but from a related virus, not HIV. In 1987, a young California researcher named Mike McCune had a brainstorm: Why not take a mouse born without a functioning immune system and transplant human immune system tissue into it? He used lymph nodes, part of a thymus, and some immature white blood cells from a human fetus in the hope that the white cells would grow, develop, and begin to confer immunity in the mouse just as in a human baby. Indeed, the chimeric animal, which the scientific press has since dubbed "the human mouse," did take on a functioning human immune system that can be devastated by the human HIV virus. By 1990, AIDS researchers all over the world were using the "human mouse," and hopes are high that it will lead to further progress in understanding and controlling the disease.

(a) How the AIDS virus causes harm

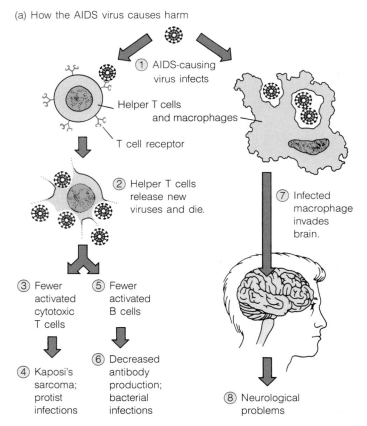

(b) An AIDS patient

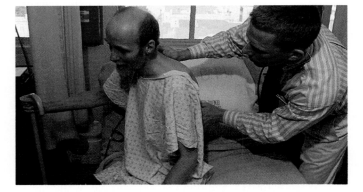

FIGURE 22.13 ■ How the AIDS Virus Wreaks Havoc in the Human Body. (a) The virus that causes AIDS infects helper T cells by binding to the T cell receptor (step 1). The virus proliferates inside the cell and kills the helper T cell (step 2). Helper T cells can no longer induce cytotoxic T cells to form (step 3), and thus these latter cells do not kill cancer cells, such as those that cause the multiple tumors of Kaposi's sarcoma, nor can they ward off infections of parasitic protozoa such as *Pneumocystis carinii,* which causes a rare form of pneumonia (step 4). Without helper T cells, B cells are not stimulated to produce antibodies (step 5), so the victim's body cannot combat viral and bacterial invaders (step 6). Macrophages take in the AIDS virus by phagocytosis (step 1), and the viruses accumulate in vesicles. Infected macrophages can enter the brain and cause neurological problems in some AIDS patients (steps 7 and 8). (b) In an AIDS patient, such as the one shown here, the AIDS-causing virus has destroyed the immune system, leading to infections and neurological problems.

To successfully block the life cycle of the AIDS virus, scientists must discover a great deal more about how it infects human cells, replicates, spreads to other cells, and destroys immune functioning. They do know, for example, that reverse transcription must take place from viral RNA to DNA (Figure 22.14). This knowledge led to the development of the AIDS drug AZT (see Box 9.2 on page 200), as well as to T cell receptor proteins that can be injected into the patient and there bind to envelope proteins on the AIDS-causing virus, which prevents them from spreading to new cells.

While AIDS patients have underactive immune systems, other people suffer from overactive systems or misdirected immunity.

Immune Regulation: A Balancing Act

The overriding task of the immune system is to protect us from foreign antigens without harming any part of ourselves. Without the ability to accomplish this task, the body would produce antibodies and killer T cells against its own proteins and hence destroy itself. So how does the powerful protective network tolerate self-components and yet not tolerate any invader that is marked nonself?

Many immunologists think that macrophages are a key to **self-tolerance**—the lack of an immune response to one's own molecules. The large, specialized phagocytes called macrophages engulf antigens and cellular debris, and these antigens become attached to the surface of the macrophage. The macrophage then presents the cell-surface-bound antigens to helper T cells. Macrophages do not, however, process self-molecules and present them to helper T cells. Because of this difference in macrophage behavior, self-substances do not trigger helper T cells. Moreover, during development in the fetus and newborn, immature B cells carrying antibodies that bind to self-substances may be killed. Thus, a combination of suppression by T cells and deletion of self-reactive B cells sets up a state of self-tolerance that is crucial to our survival. Loss of self-tolerance can lead to *autoimmune diseases*—attacks on certain cells or tissues by the person's own immune system. *Rheumatoid arthritis* is a form of autoimmunity that afflicts millions of people in the United States alone. In people with this condition, the joints swell painfully, the fingers can become gnarled and twisted, and everyday movements, like buttoning a shirt, can be painful or even impossible (Figure 22.15).

Pregnancy is an interesting exception to the immune system's recognition and elimination of foreign antigens. A pregnant mother tolerates foreign cells in the fetus (with its set of paternal genes and proteins). (See Box 22.3 on page 478.)

The take-home message about T cells is clear: While antibody-based humoral immunity protects us against small intruders like toxins, virus particles, and bacteria, T-cell-based cellular immunity attacks large invaders such as parasites, cancer cells, virus-infected cells, and tissue transplants.

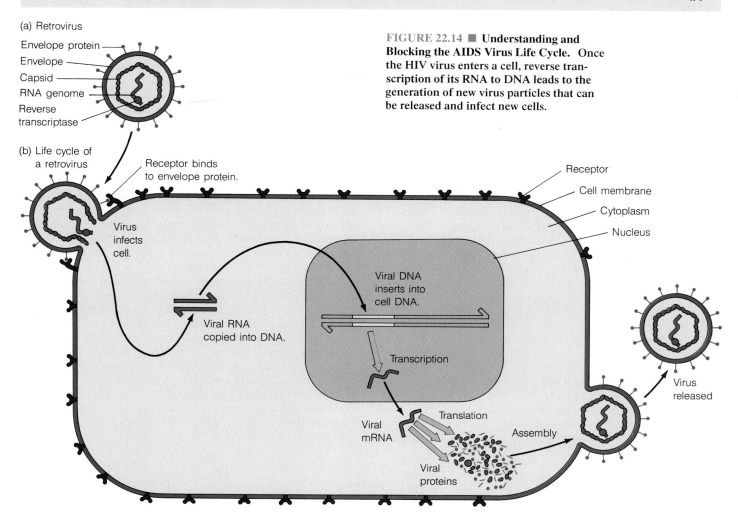

(a) Retrovirus

Envelope protein
Envelope
Capsid
RNA genome
Reverse transcriptase

FIGURE 22.14 ■ Understanding and Blocking the AIDS Virus Life Cycle. Once the HIV virus enters a cell, reverse transcription of its RNA to DNA leads to the generation of new virus particles that can be released and infect new cells.

(b) Life cycle of a retrovirus

Receptor binds to envelope protein.

Receptor
Cell membrane
Cytoplasm
Nucleus

Virus infects cell.

Viral DNA inserts into cell DNA.

Viral RNA copied into DNA.

Transcription

Virus released

Viral mRNA

Translation

Assembly

Viral proteins

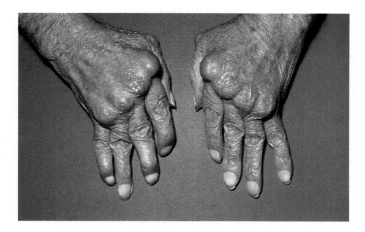

FIGURE 22.15 ■ Autoimmune Disease: A Breakdown of Self-Tolerance. In a person with rheumatoid arthritis, overly active immune cells spark inflammation and joint destruction. Some immunologists think that excessive, misdirected helper T cell activity is the major culprit. Anti-inflammatory drugs and the injection of gold salts provide some relief to many of the 7 million American victims of the disease, but researchers are still looking for more specific drugs with fewer side effects to fight autoimmune diseases.

MEDICAL MANIPULATIONS: SHORT- AND LONG-TERM PROTECTION

In the last 200 years, medical practitioners have learned to manipulate the immune system to promote health and prevent disease. They can quickly neutralize snake venom by injecting into a snakebite victim antibodies produced by and gathered from another animal. And doctors can prevent diseases for which there is no cure by activating a person's own immune responses with vaccines.

Passive Immunity: Short-Term Protection by Borrowed Antibodies

If you were bitten by a rattlesnake, there would be no time to spare. The venom contains neurotoxins, proteins that stop nerves from functioning, and you would need a quick-acting remedy to keep them from damaging your nervous system. The best treatment would be *passive immunization:* an injection of specific antibodies against the toxin—antibodies that

BOX 22.3

PERSONAL IMPACT

PREGNANCY, TOLERANCE, AND RH DISEASE

Many immunologists consider the human fetus to be a bit like a grafted organ: Although it contains some of the mother's own genes and proteins, it also contains the father's, including the same histocompatibility proteins that would lead the mother to reject one of his organs if transplanted. The mother's immune system obviously continues to function during pregnancy to protect her from foreign invaders. So why doesn't her body reject the half-foreign fetus?

Early studies showed that the uterus is a special immunological zone during pregnancy. At this time, a woman's body will still reject a skin graft from her husband anywhere on her exterior surface and will even reject material from the fetus if it is placed anywhere but the uterus. An embryo may be protected from its mother's immune system because it lacks histocompatibility markers. Protection may also come because the mother's IgG molecules block the offspring's antigens or because embryonic cells make protein signals that quiet cytotoxic T cells. Researchers are still studying the actual strategy of embryo protection, but it must be quite complicated, precise, and efficient to defend the tiny amount of tissues from the mother's powerful protective network.

This is not to suggest, however, that the fetus never comes under attack. In about one couple in 300, the woman's body rejects the fetus, perhaps because the man's tissue type is so similar to the woman's that the special protective strategies are not stimulated and the normal maternal defenses against foreign tissues go to work. And in about one couple in 15, a serious anemia called erythroblastosis fetalis, based on an attack by the mother's immune system, can occur in newborn infants.

Most people in North America are Rh-positive (they have Rh antigens on the membrane of their red blood cells), but some people are Rh-negative and lack the antigens. When an Rh-positive man and an Rh-negative woman produce a baby, it may be Rh-positive. If the baby's blood cells mingle with the mother's during delivery (Figure 1a) and its Rh antigens enter her bloodstream, her immune system may secrete anti-Rh antibodies. These don't affect the newly delivered baby, since a few days pass before the antibodies form in the mother's system. During a subsequent pregnancy, however, the antibodies are already in the mother's blood, and these can cross the placenta and attack the red blood cells of the new fetus (which will also be Rh-positive if the father is Rh-positive), leading to anemia, brain damage, or even death (Figure 1b).

A therapy has been found for couples where the male is Rh-positive and the female is Rh-negative. Since about 1970, it has been standard procedure to inject Rh-negative mothers with anti-Rh antibodies (called Rhogam) at the birth of the first child. These ready-made proteins bind to the Rh antigens on any fetal blood cells that enter the mother's circulation during delivery, cover them up, and prevent her own B cells from recognizing the antigens and making antibodies to them. Thus, maternal B cells do not gear up to produce anti-Rh antibodies, and since antibodies do not lie in wait for the next embryo, its tissues will be safe in the womb—the special immunological safe haven—until the fetus is born.

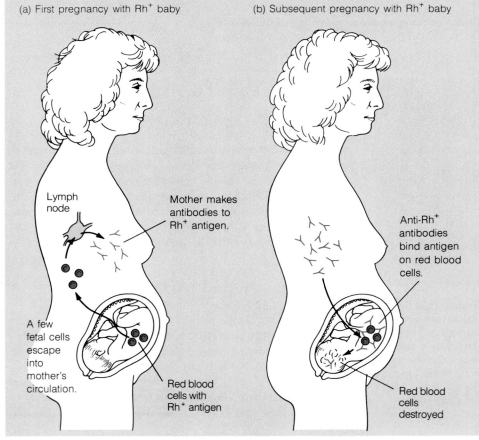

(a) First pregnancy with Rh⁺ baby

(b) Subsequent pregnancy with Rh⁺ baby

Lymph node

Mother makes antibodies to Rh⁺ antigen.

A few fetal cells escape into mother's circulation.

Red blood cells with Rh⁺ antigen

Anti-Rh⁺ antibodies bind antigen on red blood cells.

Red blood cells destroyed

FIGURE 1 ■ Pregnancy and Rh Disease. (a) A fetus's Rh-positive blood cells can escape into the Rh-negative mother and trigger an immune reaction. (b) In subsequent pregnancies, the mother's anti-Rh antibodies can enter the fetal bloodstream and attack fetal blood cells.

your own cells did not produce (Figure 22.16). Much earlier, technicians would have prepared a commercial product by injecting a horse or rabbit with inactivated snake venom and inducing the mammal to form antibodies. They would then have collected these antibodies, now called antivenin, which a doctor would then inject immediately into you, the snakebite victim. The antivenin would then circulate through your bloodstream and combine with (neutralize) the snake venom in a typical antigen-antibody fashion before the venom could reach your nerve cells.

Passive immunization has one great advantage: It works very fast. But it also has the disadvantage of acting for only a short time. Your immune system would soon recognize as for-

eign and eliminate the borrowed antibody molecules. With them would go the passive protection, leaving you vulnerable to snake venom if bitten again.

Active Immunity: Prevention by Altered Antigen

Unlike an antivenin that provides the short-term protection of passive immunity, vaccines against polio, diphtheria, or measles provide long-term protection by stimulating the body's own immune system to generate *active immunity*. Safer and more effective than most drugs, such vaccines provoke a specific response aimed at one microbe or toxin and nothing else in the body.

Edward Jenner, an eighteenth-century English country physician, developed the first vaccine almost 200 years ago. To help protect people from smallpox, a severely disfiguring disease that killed one out of every four people in Jenner's time, he injected people with a small amount of cowpox virus (the word *vaccination*, in fact, comes from the Latin word *vacca* for "cow"). Jenner based his technique on the observation that milkmaids who contracted cowpox from the cows they milked always recovered and almost never got smallpox. Indeed, the cowpox virus he injected caused mild sickness and discomfort, but it nonetheless served as effective protection against smallpox. By 1978, modern vaccines based on Jenner's original ones had successfully eradicated smallpox worldwide.

Modern vaccines are made of microbes and toxins that have been killed or otherwise modified in the laboratory so they cannot cause disease. The first shot with altered germs stimulates the production of antibody-producing cells and memory cells (review Figure 22.10). A booster shot, which is simply a second dose of the same vaccine, can then induce the memory cells to differentiate and form still more effector and memory cells. If a vaccinated person later comes in contact with live bacteria, virus, or toxin carrying those antigens, his or her body will already contain many antibodies and memory cells that can quickly eliminate the dangerous agents and prevent disease. Vaccination is a slow-acting process, requiring a booster shot and several weeks for the development of adequate protection. Nevertheless, the active immunity it stimulates lasts a long time, sometimes a lifetime.

New vaccines are the best hope for combating infectious diseases that cannot be treated or cured by other techniques of modern medicine. Unfortunately, many viruses, like different strains of the flu virus, as well as some other microbes, can change their surface properties so often that they evade active immunity. In the two centuries since Jenner's first vaccine, fewer than 20 safe, effective vaccines have been developed.

To conquer such microbial elusiveness, contemporary researchers are using recombinant DNA and other forms of biotechnology to fashion synthetic vaccines against influenza and other viruses that can foil the immunological memory. A vaccine against AIDS has been successful in chimpanzees, and trials are now under way to test such anti-AIDS vaccines in people. Since the AIDS virus attacks the very lymphocytes

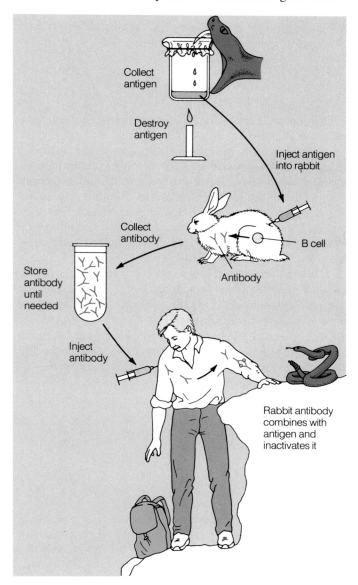

FIGURE 22.16 ■ Passive Immunization: Transferred Antibodies. As the text explains, technicians prepare commercial snakebite medicine (so-called antivenin) by collecting venom, inactivating it, and then injecting it into a rabbit. Later, antibodies are collected from the rabbit and injected into the snakebite victim, where they bind to and inactivate the toxic venom, rendering it harmless.

TABLE 22.1 Protect Yourself from AIDS

Risk Factors	Protection
■ Sex with an individual who has had a positive AIDS test, indicating exposure to the virus ■ Intravenous drug use ■ Injections of any kind with used hypodermic needles and syringes ■ Multiple sex partners ■ Sex with anyone who has had multiple sex partners ■ Transfusions with tainted blood ■ Sex with an intravenous drug user or with anyone engaging in other high-risk activities ■ Sex with any individual whose sexual history and exposure status (positive or negative) are unknown to you ■ Sex that involves contact with blood, such as sex during menstruation or anal intercourse ■ Other activities that involve contact with blood, including some types of work in medicine, dentistry, or undertaking	■ Find out your own exposure status and that of any potential sexual partner. ■ Avoid high-risk sexual activities. ■ Seek help for addiction, and use only sterile needles and syringes for any necessary injection. ■ Use condoms and, if possible, preparations that contain a virus-fighting compound such as nonoxynol 9. ■ Avoid unnecessary contact with blood, and accept only transfusions that test free of the AIDS virus. ■ Stay informed of new developments in the fight against AIDS. Seek additional information from your college or university health service, your local public health service, or your private physician.

that promote protection, since it readily changes the proteins in its outer coat, and since it can remain latent inside cells for many years, the development of a safe, effective AIDS vaccine may take many more years. In the meantime, observing general rules of good health as well as safe sex practices (see Table 22.1) are the best prevention techniques now available.

CONNECTIONS

Immune protection is a numbers game that helps each one of us adapt to living in our own particular environment of plants, animals, and microbes. In this game of chance, enormous diversity of specific immune responses is essential for health and survival. Specific immunity gives us the ability to resist the changing tides of microorganisms, chemical agents, and even tumor cells, including those never before encountered by us or our ancestors.

With each immune response, receptors on the surfaces of B and T cells recognize antigens, and this sets in motion a clonal selection mechanism that leads to the destruction and elimination of the invaders, as well as to future resistance to the same type of invader through immunological memory. A precise but dynamic balance of many elements produces this immunity to foreign substances while simultaneously allowing a tolerance to self-antigens.

The immune system is not self-contained. It relies on the circulation of blood and lymph (see Chapter 21) and on control and communication via the nervous and hormonal systems (see Chapters 26 through 28). This is one reason why undue stress or prolonged mental depression lowers our resistance to infection. In the next chapter, we will see that this same tight physiological interdependence underlies the smooth functioning of the respiratory system.

Highlights in Review

1. The vertebrate immune system protects an animal from disease by recognizing invading organisms or molecules, communicating information regarding the invader among different cells of the immune system, and eliminating the invader.

2. In a nonspecific immune response, blood flow and temperature increase locally around a wound, and white blood cells converge on the area and devour debris.

3. Macrophages are large white blood cells that devour debris in a wound, process the invader's molecules, and place portions of them on the surface of the macrophage in association with multiple histocompatibility proteins. Macrophages present these cell surface proteins to certain T cells.

4. Helper T cells help other lymphocytes carry out their immune function, while suppressor T cells suppress certain immune reactions, and cytotoxic T cells directly kill invading cells, cancer cells, or virus-infected cells.

5. B cells give rise to plasma cells, which produce antibodies, proteins that form a shape like the letter Y. The two arms of the Y can bind tightly to an antigen, such as molecules of an invader. The stem of an antibody molecule controls whether the antibody circulates in the blood, providing overall protection, or is secreted into a mother's milk, providing temporary protection to a newborn, or attaches to cells that line body surfaces, contributing to symptoms of allergy.

6. An antibody bound to an antigen marks the antigen for destruction (e.g., to be devoured by a macrophage).

7. Most lymphocytes probably arise in the bone marrow and then move to other organs, like the lymph nodes, spleen, or thymus, which give rise to T cells.

8. Each different B cell makes a different antibody that can bind a unique antigen. When a B cell binds an invading antigen and is stimulated by a helper T cell, it proliferates into a clone of cells. Some members of the clone become plasma cells, which secrete specific antibodies, while others become long-lived memory cells, which can respond like the original B cell when exposed to the same antigen again.

9. The cell-mediated immune response relies on T cells that directly eliminate foreign invaders.

10. Helper T cells can be invaded by HIV, the virus that causes AIDS (acquired immune deficiency syndrome). Without helper T cells, other lymphocytes are not adequately stimulated, and so few antibodies or memory cells or cytotoxic T cells are formed. Without the protection these provide, the AIDS victim dies from microorganisms or cancers that the intact immune system normally destroys.

11. In an autoimmune disease like arthritis, an individual's immune system inappropriately attacks molecules that arise in that individual's body.

12. In passive immunity, antibodies made by one individual are transferred to another individual. In active immunity, an individual's own immune system can be stimulated to eliminate a specific invader. A vaccine is an antigen altered so that it can no longer cause disease even though it can stimulate an immune response.

Key Terms

antibody, 467
antigen, 468
B cell (B lymphocyte), 467
cell-mediated immune response, 474
histocompatibility protein, 474
humoral immune response, 474

immune system, 464
immunity, 465
immunological memory, 472
macrophage, 467
self-tolerance, 476
specific immune response, 466
T cell (T lymphocyte), 467

Study Questions

Review What You Have Learned

1. When there is a break in the skin, what nonspecific immune responses occur?

2. What are the roles of B cells? T cells? Macrophages?

3. Why are lymphoid organs important to health?

4. Trace the sequence of events in the immune system's response to invading bacteria.

5. Name the classes of antibodies, and state the role of each class.

6. Is the Y-shaped molecular structure of antibodies the key to their effectiveness? Explain.

7. When the immune system encounters an antigen, how does its primary response differ from its secondary response?

8. How can a limited number of genes give rise to all the different types of B cell antibodies?

9. How does the humoral response differ from the cell-mediated immune response?

10. What is self-tolerance, and how does the immune system maintain it? Why is it a delicate balance?

11. How is the body protected by passive immunity? By active immunity?

Apply What You Have Learned

1. It is unlikely that a flu vaccine will confer lasting immunity against disease. Why?

2. A boy with measles receives a visit from his grandmother, who is sure she will not catch the disease. Why is she right?

3. Researchers believe that it will be many years before there is an effective vaccine against AIDS. What is their reasoning?

4. On a test you must pass to get your license to practice medicine, the examiner presents you with a case of a young patient who suffered from chronic infections. If the patient had a skin wound, for example, his skin would turn red, his tissues would swell and become warmer, and the redness and swelling would persist for weeks or months. His blood did not contain antibodies against the AIDS virus. The examiner gives you the following possible diagnoses: (1) The patient's immediate nonspecific immune response is hindered by his inability to produce histamines; (2) his helper T cells' production of interleukins is inadequate; (3) his immune system lacks the variety of antibodies necessary to defend his body against invaders; (4) his immune system has lost self-tolerance. Analyze each of these diagnoses and explain why they might or might not be likely explanations for the patient's symptoms.

For Further Reading

CLAYTON, J. "Confusion in the Joints." *New Scientist,* May 4, 1991, pp. 40–43.

COGHLAN, A. "A Second Chance for Antibodies." *New Scientist,* February 9, 1991, pp. 34–39.

COWLEY, G. "AIDS: The Next Ten Years." *Newsweek,* June 25, 1990, pp. 20–27.

GALLO, R. C., and L. MONTAGNIER. "AIDS." *Scientific American,* October 1988, pp. 47–48.

GAMLIN, L. "The Big Sneeze." *New Scientist,* June 2, 1990, pp. 37–41.

GREY, H. M., A. SETTE, and S. BUUS. "How T Cells See Antigen." *Scientific American,* November 1989, pp. 38–46.

HOOD, L. E., I. L. WEISSMAN, W. B. WOOD, and J. H. WILSON. *Immunology.* 2d ed. Menlo Park, CA: Benjamin/Cummings, 1984.

MONTGOMERY, G. "The Human Mouse." *Discover,* August 1989, pp. 49–55.

RENNIE, J. "The Body Against Itself." *Scientific American,* December 1990, pp. 106–115.

ROSENBERG, Z., and A. FAUCI. "Inside the AIDS Virus." *New Scientist,* February 10, 1990, pp. 51–54.

SMITH, K. A. "Interleukin-2." *Scientific American,* March 1990, pp. 50–57.

WECHSLER, R. "Hostile Womb." *Discover,* March 1988, pp. 83–87.

Respiration: Gas Exchange in Animals

Antarctic Deep Diver

By any serious measure, the Weddell seal has a difficult life. This large marine mammal lives on the windswept shores and coastal ice floes of Antarctica (Figure 23.1) and swims in waters that are near freezing year-round. Because the animal is so big—it can reach 3 m (10 ft) in length and weigh 350 to 450 kg (770 to 990 lb)—it has an enormous appetite. Yet its staple diet, the antarctic cod, lives just above the deep seafloor at depths of 600 m (2000 ft) or more. To catch its prey, the Weddell seal must routinely dive far deeper and hold its breath far longer than all but a handful of oceangoing mammals, and this deep-diving prowess has long interested biologists.

For more than 50 years, animal physiologists have been studying Weddell seals in the laboratory and seeking, through numerous experiments, to answer a central question: How can such a big organism stay under water for periods of more than an hour and yet still have sufficient oxygen for all its billions of cells? As in any aerobic organism, the cells of the Weddell seal must have a steady supply of oxygen and must rid themselves of carbon dioxide. How, then, can the heavy animal spend such a long time below the surface without risking death?

Laboratory studies have revealed some intriguing facts. First, the blood and muscles of the Weddell seal act as major oxygen reservoirs. The seal has twice as much blood as a person does per kilogram of body weight, and the seal's muscles have high levels of an oxygen-storing protein called myoglobin. Second, when submerged, the animal can experience a *dive reflex:* The heart rate slows, blood is

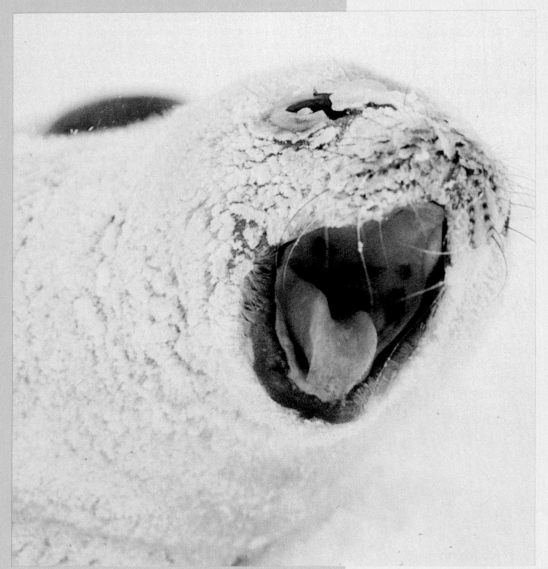

FIGURE 23.1 ▪ **The Weddell Seal: A Hard Life and a Special Talent for Deep Diving.** Encrusted with ice and snow after a howling storm, this Weddell seal (*Leptonychotes weddelli*) calls loudly. Weddell seals hunt for antarctic cod near the deep seafloor, remaining submerged for 20 to 70 minutes at a time.

FIGURE 23.2 ■ Skin Folds and Gas Exchange: A Matter of Surface Area. The Lake Titicaca frog has skin that falls into deep folds. These look strange but provide the animal with a large surface area through which oxygen can enter and carbon dioxide can escape. The surface area is so large, in fact, that this intriguing amphibian can survive without breathing through its lungs at all.

shunted away from most organs and toward those vital to navigation and controlling swimming movements (the brain, spinal cord, and retinas of the eyes), and the muscles may switch to anaerobic respiration (see Chapter 5), thereby requiring much less oxygen.

Recently, field biologists have developed a clever device for studying the animals as they dive for fish in antarctic waters: a small portable computer that can be glued to the animals and automatically take blood samples and other measurements as the animals swim. These field studies have shown that Weddell seals experience the dive reflex in nature much as they do in the laboratory, but only about 5 percent of the time, when they are exploring their environment or escaping predators and must remain submerged for an hour or more. When hunting for fish, the computers revealed, the seals usually dive for less than 20 minutes, their muscles remain aerobic (not anaerobic), and a blood storage organ, the spleen, releases great quantities of oxygen-bearing red blood cells, which help provide the gaseous element to all active cells and tissues for the dive's duration.

The Weddell seal, with its large volume of rich blood and other adaptations, is a good case study for our current subject, **respiration,** the process by which organisms exchange gases with the environment. While this seal has special physiological solutions to the problems of deep diving, it is like all other animals in the following way: It needs an organ of gas exchange large enough both to extract all the oxygen from the environment that the millions or billions of body cells require and to carry off the carbon dioxide that builds up in those cells as a result of cellular respiration (see Chapter 5). The seal's

lungs, the cod's gills, and the outer skin surfaces of some simple animals all play key roles in gas exchange, as do the air passages of many animals. Blood circulation is also closely tied to respiration, since, as Chapter 21 explained, gases are one of the major commodities carried in the blood.

Two unifying themes emerge in our discussion of respiration. First, the architecture of an animal's respiratory system is constrained by the principle of diffusion. Gases always enter and leave body fluids via diffusion, and this fact helps explain the shape and functioning of gills, lungs, and in some cases, skin (Figure 23.2). Second, an animal's activities are limited by the quantity of oxygen available; thus, the animal's respiratory architecture and its way of life are closely tied. A seal, with its highly complex lungs and rich blood supply, swims faster and is much more active than a flatworm, which exchanges gases directly across the body surface.

This chapter explores several questions about how animals exchange gases with the environment:

■ How does diffusion underlie all forms of respiration?
■ How do moist skin, gills, tracheae, and lungs compare as organs of gas exchange?
■ What organs together form the mammalian respiratory system, and how do they function?
■ How do physical mechanisms like gas pressure constrain gas exchange?

GAS EXCHANGE IN ANIMALS: LIFE-SUPPORTING OXYGEN FOR EVERY CELL

The subject of gas exchange in animals is filled with seeming contradictions: A person quickly airlifted from sea level to the top of Mt. Everest (an elevation of 8848 m, or 29,000 ft) would die from lack of oxygen in less than 5 minutes unless provided with an oxygen tank and breathing mask. Yet birds can fly over the same peak, flapping vigorously, using oxygen rapidly, and breathing the thin air with no ill effects. A large tuna, using its gills, can extract enough oxygen from seawater to meet the needs of its powerful swimming muscles, and yet a tiny mouse could not do this, even if its lungs could function in water. Some insects, such as stag beetles, are as large as mice and yet successfully extract oxygen from the environment using neither gills nor lungs. The explanation for these apparent puzzles lies in the wide variety of breathing structures and mechanisms that have evolved and that are capable of carrying out gas exchange in animals. These range from body shapes that allow simple diffusion, to gills for aquatic gas exchange, to the tracheae and lungs of land animals.

Diffusion: The Mechanism of Gas Exchange

Although animals meet their physiological needs for oxygen in different ways, all are aerobic organisms that must steadily take in oxygen if their cells are to burn carbohydrates and fats aerobically and expel the metabolic by-product carbon dioxide (review Figure 5.6). At the cellular level, oxygen and carbon dioxide enter and leave animal cells by diffusion, the spontaneous migration of a substance from a region of higher concentration to a region of lower concentration (see Figure

FIGURE 23.3 ■ **The Visible Worm.** The body of this bright yellow marine flatworm (*Pseudoceros* species) is sufficiently thin and flat that oxygen and carbon dioxide can diffuse into and out of each body cell from the surrounding seawater.

4.18). There are major differences, however, in how animals carry out *respiration,* which is the process of gas exchange between the whole organism and the external environment. (Recall from Figure 5.12 that the separate process of *cellular respiration* involves a series of oxygen-dependent chemical reactions in the mitochondria that supply the cell's internal energy needs.)

Two facts about diffusion in living creatures have shaped the evolution of respiratory structures. First, to enter and leave cells, gases must diffuse through a watery medium. Second, diffusion is much less rapid—an amazing 300,000 times slower—in water than in air. These facts have constrained the shape and activities of animal bodies. The porous nature of the sponge, for example, permeated as it is by channels, allows each body cell to lie only a fraction of a millimeter away from oxygen-bearing water (review Figure 18.5). And in flatworms, the thin, flat bodies are never more than 2 mm thick, so that oxygen and carbon dioxide can diffuse directly between cells and the animal's moist surroundings (Figure 23.3).

Most large animals employ a different, more complex strategy that relies not on direct diffusion from the environment, but on diffusion into and out of a circulating internal fluid (blood or its equivalent). This fluid picks up and transports oxygen across a moist surface—a physical interface with the environment—and distributes it throughout the body. Then it picks up carbon dioxide in the tissues and releases it back into the outside world, usually through the same moist interface. The specialized structures that have evolved as interfaces are gills, tracheae, and lungs.

Extracting Oxygen from Water: The Efficient Gill

Fish, tadpoles, and most other animals that extract oxygen from water have *gills:* organs specialized for gas exchange that develop as outgrowths of the body surface. Gills can be external (extending outside the body), or internal (housed entirely within), simple or complex, but each type provides a large surface area through which gases from the environment can diffuse and thus enter the organism's bloodstream.

Many amphibians, such as the axolotl, have frilly external gills that wave about in water currents (Figure 23.4). The elaborate "frills" increase the organ's surface area, and the outer membrane of the gill is only a single cell thick. This cell layer lies in contact with the equally thin walls of blood capillaries, which occur in dense networks (see Figure 21.7). As water moves past the waving gill surfaces, oxygen can diffuse in and carbon dioxide out through the gill membrane and capillary walls.

External gills like these lie dangerously exposed, making them vulnerable to predators and the elements. For this reason, most fishes have evolved more protected internal gills, which are nevertheless surrounded and supported by water, which prevents them from collapsing. Internal gills are often protected by stiff **opercular flaps** that are easily visible on

FIGURE 23.4 ■ The Oxygen-Gathering Gill: Getting Oxygen into the Blood. Like all axolotls (Mexican salamanders), this albino specimen sports elaborate external gills that have a large surface area for exchanging gases with the environment. The network of thin-walled blood capillaries gives the gills a reddish cast.

the side of a fish's head and that cover the openings, or gill slits, between the gill bars (Figure 23.5). Beneath each flap, the gill is subdivided into hundreds of flexible **gill filaments,** which are in turn composed of many thin, platelike structures. These delicate plates, or **lamellae,** hold the key to how a large, active, aquatic animal can respire efficiently in water, even though that medium holds ¹⁄₂₀ as much oxygen as air. Embedded within each lamella is a lacy meshwork of capillaries lying just one cell layer away from the water that passes through the gill filament. This proximity of blood to oxygen-bearing water means that oxygen readily diffuses across the cells into capillaries, while carbon dioxide diffuses outward just as easily. The pumping heart then circulates the oxygen-rich blood throughout the animal's body.

The steady opening and closing of a fish's mouth pumps a constant stream of water through the mouth to the gills, over the gills, then out through the opercular flaps. This water flow moves in the opposite direction to blood flow within the capillaries inside the gill's lamellae (Figure 23.5b–d). The result is a **countercurrent flow** (review Figure 20.12c), in which two fluids with different characteristics (here, different oxygen content) flow in opposite directions and can exchange materials along all their points of contact owing to diffusion from high to low concentrations. Thus, the animal can collect sufficient oxygen for aerobic cellular respiration and an active life-style, even though water holds so much less oxygen than air.

By contrast, land animals have access to the higher oxygen content of air, but they face a different threat: the drying out of their respiratory surfaces, which would bring gas exchange to a lethal halt. Part of the evolutionary solution to

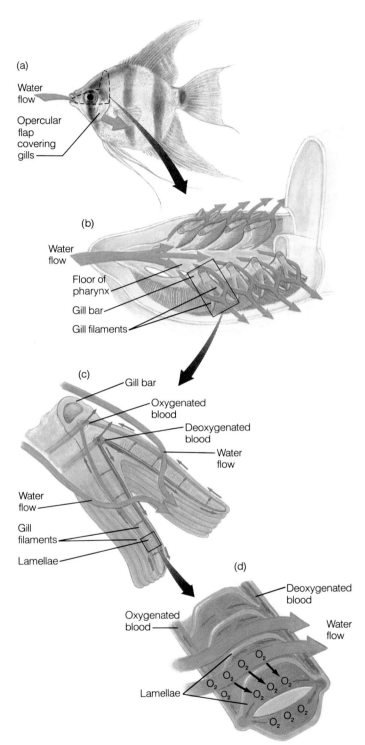

FIGURE 23.5 ■ Fish Gills: Flaps, Capillaries, and Countercurrent Flow Allow a Rich Harvest of Oxygen from Oxygen-Poor Water. Water moves into a fish's mouth (a) and passes across its gill filaments before exiting the gills (b). The water current runs in an opposite direction to the blood current flowing in tiny vessels inside the gill filaments, and by means of countercurrent exchange, a maximal amount of oxygen can diffuse into the blood vessels (c, d).

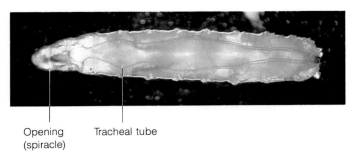

Opening
(spiracle)

Tracheal tube

FIGURE 23.6 ■ **Tracheal Tubes: Efficient Gas Exchange in Insects.** Two parallel tracheal tubes are clearly visible in this fruit fly larva. Air entering these tubes through the spiracles can fan out into the branching air capillaries, diffuse across fluid in the capillary tips, and reach internal tissue cells.

this problem has been the moist gas exchange surface of structures such as the branching tracheae or pouchlike lungs located deep within an animal's body.

Adaptations for Respiration in a Dry Environment: Tracheae and Lungs

The only animals that live surrounded entirely by dry air are terrestrial arthropods and mollusks, such as insects and snails, and land vertebrates, such as reptiles, birds, and mammals. Not coincidentally, the arthropods and vertebrates have evolved specialized internal respiratory channels: the *tracheae* and the *lungs,* respectively. These specialized structures help bring oxygen close to each body cell and carry away carbon dioxide.

■ **Tracheae: Tubes for Gas Exchange** Insects and certain other arthropods have a system of air tubes, the tracheae, that are physically linked to the outside world through holes known as **spiracles** located in the body wall (Figure 23.6). Inside the body, the tracheae branch into ever finer tubes (far narrower than a human hair) that end in **air capillaries,** which allow air from the environment to penetrate deep within the animal's tissues. Atmospheric oxygen entering the spiracles diffuses rapidly through the air in the tracheal tubes until it reaches a minute droplet of fluid pooled at the inner tip of each air capillary; here, the incoming oxygen dissolves in the fluid and diffuses across adjacent cell surfaces. In highly active tissues, such as the flight muscles of a honeybee, no cell is more than a few micrometers away from an air capillary. Oxygen reaches cells directly in insects, and they do not have a circulation of blood bearing gases.

■ **Lungs: Complex Air Bags** A few species of fish that live in the warm, stagnant, oxygen-poor water of swamps have evolved lungs, blind-ended internal pouches that connect to the outside by a hollow tube (Figure 23.7a). Lungs help sup-

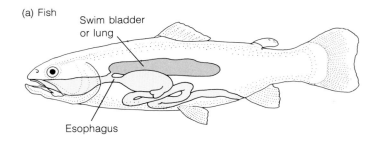

(a) Fish

Swim bladder
or lung

Esophagus

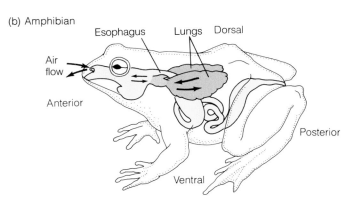

(b) Amphibian

Esophagus   Lungs   Dorsal

Air
flow

Anterior

Posterior

Ventral

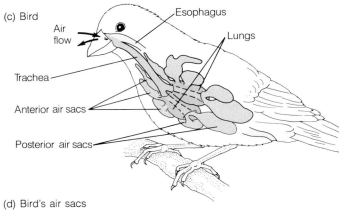

(c) Bird

Esophagus

Air
flow

Lungs

Trachea

Anterior air sacs

Posterior air sacs

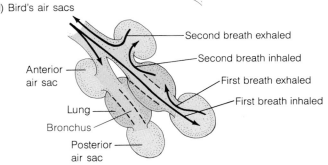

(d) Bird's air sacs

Second breath exhaled

Second breath inhaled

Anterior
air sac

First breath exhaled

First breath inhaled

Lung

Bronchus

Posterior
air sac

FIGURE 23.7 ■ **Pouches for Air Exchange.** (a) Since fish have gills, all but a few species lack lungs. Instead, many species have a swim bladder, an organ that holds gases and controls the animal's buoyancy in the water. (b) Amphibians have baglike lungs with relatively little interior surface area for gas exchange. The thin, moist skin—through which gases can readily diffuse—acts as a supplement to the lungs. (c) Birds have air sacs that interconnect with the lungs and lighten the animal. (d) The one-way flow of air through a bird's system of air sacs provides a high level of oxygen to the bird's actively contracting wing muscles and other organs.

plement the meager amount of oxygen that the gills supply in swamp fish. In many fishes, the lung pouch serves not as a true lung, but as a *swim bladder,* or gas bag, that enables the fish to maintain its position in the water without sinking or floating. In an amphibian, such as a frog (Figure 23.7b), the lungs are simple sacs with walls that are richly endowed with a dense lacework of blood capillaries. The wall of the lung is a thin, moist membrane through which oxygen can diffuse into the bloodstream and carbon dioxide can exit. Air flows in and out of the lungs via a single, two-way path. The simple saclike lungs of amphibians do not have as much surface area for diffusion of gases as the convoluted lungs of reptiles, birds, and mammals; but since amphibians also have moist skin through which gases can diffuse and supplement the lungs, they obtain enough oxygen to support their way of life.

In birds, the respiratory pathway is a bit different. Birds have air sacs and tubes called *bronchi* arranged in such a way that air flows in one direction through the lungs (Figure 23.7c and d). A bird must take two breaths to move air completely through the system of bronchi, lungs, and air sacs; the first draws air through the bronchi into the posterior air sacs and then into the lungs. The second breath draws in more air, and this pushes air from the lungs, into the anterior air sacs, and then out of the body. Since this one-way flow through the lungs prevents the mixing of fresh and "stale" air, birds can sustain extremely high levels of activity for much longer periods than we mammals can manage. And birds can flap actively at altitudes like those found at the top of Mt. Everest, where a normal human cannot even stand at rest without an oxygen tank.

A mammal's lungs do not have the benefit of a unidirectional current of air or water. Instead, like an amphibian's lungs, they operate by way of **tidal ventilation,** an in-and-out air flow rather like the ebb and flow of the tides. Air travels through an inverted tree of hollow tubes leading into the lungs, where gases are exchanged across a thin, moist membrane before the air moves out again through the same set of tubes. Usually, a quantity of air (about 500 ml, or a pint, in humans) is inhaled and exhaled in a regular rhythm. This tidal pattern means that fresh air enters the lungs only during half of the respiratory cycle and that a quantity of unexpelled, stale *dead air* filled with carbon dioxide remains in the lungs at all times, mixing with the fresh air that enters from outside. Special adaptations that increase the rate of gas exchange compensate for tidal ventilation in mammalian lungs.

RESPIRATION IN HUMANS AND OTHER MAMMALS

By the time you have finished reading this chapter—an hour, let's say—your efficiently operating respiratory system, superbly adapted to exchanging gases in a dry environment, will have drawn in oxygen and expelled carbon dioxide some 700 to 900 times. In contrast, a giraffe would have breathed

as many as 1200 times, which helps overcome the huge volume of dead air that must be cleared from the animal's very long windpipe before fresh air can enter the lungs. Despite such differences, giraffes, humans, Weddell seals, and all mammals share a basic set of respiratory structures and mechanisms.

Respiratory Plumbing: Passageways for Air Flow

When a mammal breathes in, air enters the respiratory system through the nose and sometimes through the mouth and is warmed and humidified by the moist mouth cavity or by the twin **nasal cavities,** chambers that open posteriorly into the *pharynx,* or throat (Figure 23.8a). The pharynx branches into a pair of tubes; one, the **esophagus,** leads to the stomach, while the other, the windpipe, or **trachea** (not the tracheae of insects), is the major airway leading into the lungs. At the anterior end of the trachea lies the **larynx,** or "voice box," housing the vocal cords. Just above the opening to the larynx is a flap of tissue called the **epiglottis,** which normally closes off the larynx during swallowing and thus prevents food from accidentally entering the lungs.

A few centimeters below the larynx in humans, the trachea branches into two hollow passageways called **bronchi** (singular, bronchus), each of which enters a lung. Finer and finer branchings of these tubes create an "inverted tree," with thousands of narrowed airways, or **bronchioles,** that eventually lead to millions of tiny, bubble-shaped sacs called **alveoli** (singular, alveolus). It is here, in the alveoli, that gas exchange takes place (Figure 23.8b). Each alveolus is surrounded by blood capillaries, and the inside of each tiny pouch is lined with a moist layer of epithelial cells. Human lungs contain roughly 300 million alveoli, and if the linings of all these delicate bubbles were stretched out simultaneously, they would occupy about 70 m^2—enough surface area to cover a badminton court, or 20 times the body's entire skin surface. At those places where the wall of a capillary lies near the outer wall of an alveolus, oxygen can easily diffuse out of the alveolus and into red blood cells squeezing down the center of the narrow capillary (Figure 23.8c). Conversely, carbon dioxide wastes can leave the blood, diffuse out of the capillary, enter the alveolus, and be expelled to the outside with the next exhalation.

Many of the cells that line the larger airways produce a sticky mucus ideally suited to capturing inhaled dirt particles or microorganisms. This mucus is continuously cleared from the bronchi by the beating of cilia, which sweep the mucus and any trapped debris up toward the pharynx, where they can be swallowed or expelled (Figure 23.9a).

Among Caucasians, the most common lethal genetic disease is *cystic fibrosis,* an inherited condition in which the lungs produce large quantities of a heavy, sticky mucus that interferes with gas exchange. Life expectancy is only 30 years among cystic fibrosis victims, but strides are being made in the detection and treatment of the disease.

(a) The human respiratory system

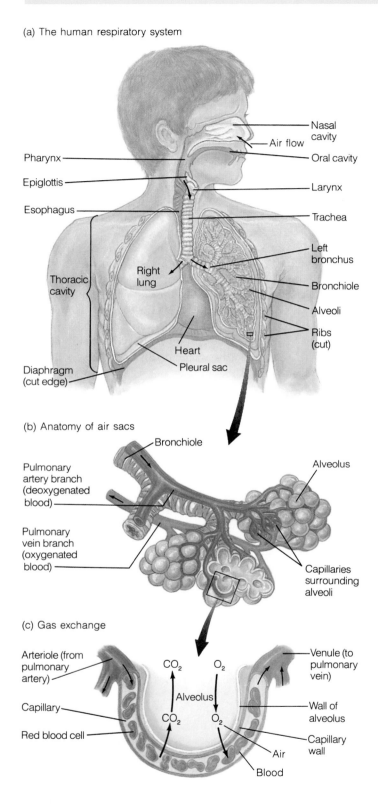

(b) Anatomy of air sacs

(c) Gas exchange

FIGURE 23.8 ■ **The Human Respiratory System: Air Passages in the Mouth, Throat, and Chest.** (a) Anatomy of the airways and lungs. (b) Close-up view of the branched passageways, or bronchioles, and the cluster of tiny air sacs, or alveoli. (c) The diffusion of oxygen and carbon dioxide between the alveoli and the blood passing in a nearby vessel.

(b) Diseased lungs of a smoker

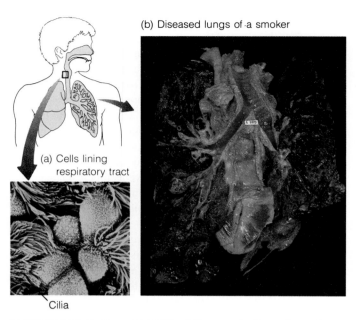

(a) Cells lining respiratory tract

Cilia

FIGURE 23.9 ■ **Bronchial Cilia: Microscopic Custodians of the Respiratory Tract.** (a) Hairlike cilia protrude from the cells that line the trachea and bronchioles and sweep out mucus and debris unless the cilia are paralyzed by cigarette smoke. When that happens, debris can reach and accumulate in the lungs. (b) Tar has blackened and clogged the delicate tissues of a smoker's lungs.

People who smoke cigarettes interfere with the delicate cells lining the lungs and bronchi. Inhaled tobacco smoke damages the cells lining the alveoli and paralyzes the cilia in the airways (Figure 23.9b). The smoke from a single cigarette can immobilize the cilia for hours and lead fairly quickly to a hacking "smoker's cough"—the respiratory system's attempt to rid itself of airborne garbage that accumulates because the cilia no longer sweep it out. Long-term abuse of the airways can lead to near total breakdown of the system; as cilia break down and other natural defenses are overwhelmed, it becomes increasingly likely that genetic changes in lung cells will go unrepaired and that the lungs will develop cancer or emphysema, a degenerative disease in which the alveoli steadily deteriorate. Also, the tars and gases in cigarette smoke damage blood vessel walls and increase one's chances of heart attack or stroke. Studies show that four out of every ten smokers—400,000 Americans each year—die as a direct result of their habit. This makes smoking by far the largest cause of preventable death each year.

Recent research shows that passive smoking—breathing the smoke from someone else's cigarettes—can have serious health consequences. In fact, passive smoking kills 53,000 Americans each year, making it the third leading cause of preventable death after direct smoking and alcohol abuse. Children raised around smokers are more likely to have respiratory and blood vessel problems. Adult nonsmokers who live

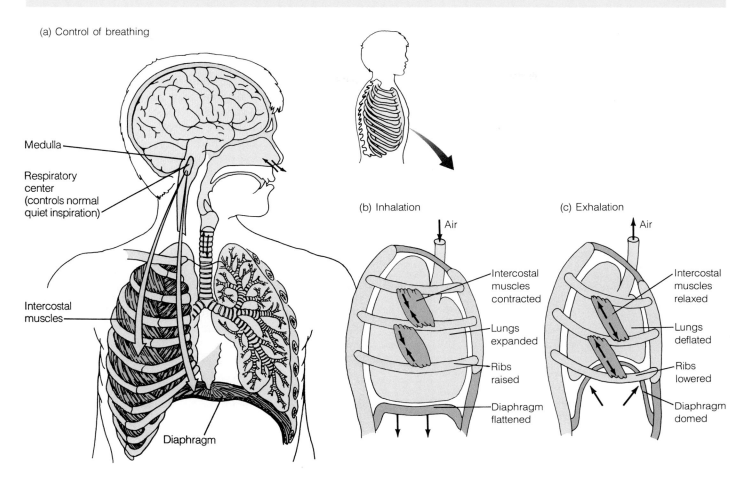

FIGURE 23.10 ■ Control of Breathing. (a) The medulla, a region in the brain stem, stimulates the rib muscles to contract. As a result, the diaphragm flattens out, and the lungs fill with air. (b) During inhalation, intercostal muscles contract and lift the ribs apart, and the diaphragm flattens. These two actions increase the size of the chest cavity and hence the lung capacity and draw air into the lungs. (c) During exhalation, the intercostal muscles relax, the diaphragm assumes a dome shape, the ribs move closer together, the chest capacity drops, and air is expelled.

with smokers have a 30 percent higher risk of death from heart attacks, and 15,000 passive smokers die of lung and other cancers in the United States each year, done in by another person's smoke.

Ventilation: Moving Air Into and Out of Healthy Lungs

Healthy lungs look like pink, spongy, deflated balloons. These soft, elastic sacs are suspended in the **thoracic cavity,** which is the region within the rib cage directly over the heart. In humans and other mammals, the lungs rest on a domed sheet of muscle, the **diaphragm,** but they don't simply hang there; each lung is enclosed in a fluid-filled **pleural sac** (review Figure 23.8). The fluid surrounding the lungs is under lower pressure than the air inside, and this pressure difference enables the lungs to remain slightly expanded even when no

air is being taken in. A "collapsed lung" results when a pleural sac is punctured and the fluid drains away. Breathing in and out, or **ventilation** of the lungs, is possible because the bellowslike activity of the diaphragm and expandable rib cage draws fresh air in and allows stale air to rush back out (Figure 23.10).

At the start of each inhalation, the *intercostal* muscles (which link adjacent ribs) shorten, lifting the ribs and pulling them apart slightly (Figure 23.10a). Simultaneously, the diaphragm muscle contracts, and the diaphragm moves downward toward the stomach; this causes its "dome" to flatten out (Figure 23.10b). As the chest cavity enlarges, the pressure of the fluid in the pleural sacs drops, and with it drops the air pressure in the lungs. As a result, air flows in from outside, moving from an area of higher pressure to one of lower pressure, through the airways to the alveoli, filling the lungs.

Exhalation, or the passive release of air from the lungs, results when these steps are reversed. The muscles that

FIGURE 23.11 ■ **Scaling Mt. Everest Without Supplementary Oxygen.** At the top of the world, where oxygen is thin, climbers such as Reinhold Messner trudge toward the summit, struggling to breathe in enough of that life-sustaining gas to keep the mind and muscles in gear.

expanded the chest cavity relax, the ribs are lowered, the diaphragm moves upward once again, and the pressure on the pleural sacs increases (Figure 23.10c). This pressure causes the lungs to deflate, squeezing air from the millions of alveoli out through the bronchioles, bronchi, trachea, and mouth or nose.

Control of Ventilation by the Brain

In 1980, Italian mountaineer Reinhold Messner became the first person to make a solo ascent of the highest peak in the world, Mt. Everest, without the aid of an oxygen tank (Figure 23.11). How was Messner able to achieve a feat that respiratory physiologists had calculated would be impossible? While recent measurements show that there is more oxygen at the top of Everest than previously thought, part of the answer lies in the incredible ability of Messner's body to extract oxygen from the air. Successful climbers like Messner have an enhanced hyperventilatory response: When placed in an oxygen-poor atmosphere, they tend to breathe much more deeply and rapidly than normal people. Let's look, now, at what controls breathing.

Respiratory control centers in the brain help determine when and how we breathe. The intercostal muscles that expand the chest cavity, for example, respond to nerve impulses generated by the medulla, a region in the brain stem that interconnects with the spinal cord (see Figure 23.10a). Breathing is involuntary; we cannot consciously prevent these nerves from firing or the chest from expanding (at least not indefinitely). Other neurons then act to inhibit respiratory muscle stimulation, causing intercostal muscles to relax and exhalation to begin. Taking an overload of chemical depressants, such as alcohol or barbiturates, can interfere with this normal process: These drugs inhibit nerve firing by the brain's respiratory centers and can cause breathing to become irregular—or to cease forever.

Does the human brain sense a rise in blood carbon dioxide or a decrease in blood oxygen? Three French physiologists answered this question in 1875 with a dramatic and disastrous experiment. The three ascended to a high altitude in a hot-air balloon equipped with bags of pure oxygen to be used as they perceived the need. Their balloons passed the 7500 m (24,600 ft) level (over 1000 m lower than the summit of Mt. Everest), and one of the scientists recorded—in an oddly scrawled handwriting—that they were feeling no ill effects. When the balloon finally descended, two of the three were dead. The survivor's sad conclusion: The human body is very poor at sensing its own need for oxygen.

In contrast, the brain readily senses an increase in carbon dioxide. If a person again and again rebreathes a small volume of air in a stuffy room, the concentration of carbon dioxide gradually rises, and this is detected by specific brain neurons. These in turn deepen and speed the person's ventilation and reestablish the blood's optimal oxygen–carbon dioxide balance.

One of the biggest mysteries still surrounding the respiratory prowess of the Weddell seal involves control by the brain. When the animal embarks on an hour-long underwater exploration or flees from a predator such as a killer whale, very early in a dive, the dive reflex sets in: The heart rate slows, and muscles change from aerobic to anaerobic metabolism. In a 20-minute feeding dive, however, the dive reflex is not triggered, either early or late. Somehow, soon after a dive begins, the seal's brain and body "decide" whether or not severe oxygen conservation measures are needed. At this point, researchers do not know how this "decision" is made. Researchers do, however, know this: Under normal circumstances, a mammal's brain automatically monitors various gas pressures in the blood and signals muscles in the chest to contract and cause breathing.

MECHANISMS OF GAS EXCHANGE

If a person is deprived of oxygen for as little as 10 minutes, brain damage or death is almost certain to occur. Clearly, ventilation must continue uninterrupted in the manner just described, and diffusion across respiratory membranes must occur without hesitation so that gases can be rapidly picked up and transported in the blood in sufficient quantity to service each cell throughout the body. Like ventilation, diffusion is based on gas pressure.

Partial Pressure and Diffusion

Scientists describe the concentration of a gaseous mixture in terms of **partial pressure** (P), the pressure exerted by each gas in a mixture. For example, air is about 21 percent oxygen, 0.03 percent carbon dioxide, and the rest mostly nitrogen. At sea level, the partial pressure of oxygen is 0.21 atmosphere, or $P_{O_2} = 0.21$.

If a leak developed in an airliner flying at 10,770 m (35,000 ft), the air in the cabin would still be 21 percent oxygen. But since the total air pressure at this altitude is only one-quarter of that at sea level, the partial pressure of oxygen would drop to 0.05 atmosphere (since $0.21 \times \frac{1}{4} = 0.05$)—not enough to sustain the passengers' lives. Gases always diffuse from places where partial pressure (i.e., the concentration) is high to places where partial pressure is low, and this characteristic is crucial to gas diffusion during an animal's respiration. When you inhale (Figure 23.12, step 1), the new air entering the alveoli (step 2) has a higher partial pressure of oxygen than the oxygen-depleted blood returning to the lungs from tissues in arms, legs, and elsewhere (step 3). Oxygen diffuses out of the tiny air sacs and into the bloodstream in nearby lung capillaries, where its concentration is low (step 4). It is then carried to distant body regions and there moves into tissue cells, where the oxygen concentration is lowest (step 5). Simultaneously, carbon dioxide moves out of the tissue cells, where its concentration is highest (step 6), moves across the capillary walls and into the bloodstream (where the concentration is lower), is carried back to the lungs (some as dissolved carbon dioxide gas and some as bicarbonate, HCO_3^-), and there moves into the alveoli, where the P_{CO_2} (carbon dioxide concentration) is lowest (step 7). From there, exhalation can carry the waste gas out of the body (step 8).

Although partial pressures drive gas diffusion at a rapid rate, this process alone is not efficient enough for the respiration of metabolically active animals. If your blood vessels were filled with tap water, that fluid simply could not take up enough oxygen to supply the needs of your active cells as you swam laps or jogged around a track. Evolution's answer to this problem is hemoglobin, an iron-containing blood component specialized to take on large amounts of oxygen quickly and to release it readily.

How Hemoglobin Transports Oxygen

In each cubic millimeter of blood, there are millions of red blood cells, each a biconcave disk filled with roughly 280 million molecules of the oxygen-binding pigment hemoglobin. Multiplying this number by the 25 trillion red cells in an average adult's bloodstream, one discovers that there are more

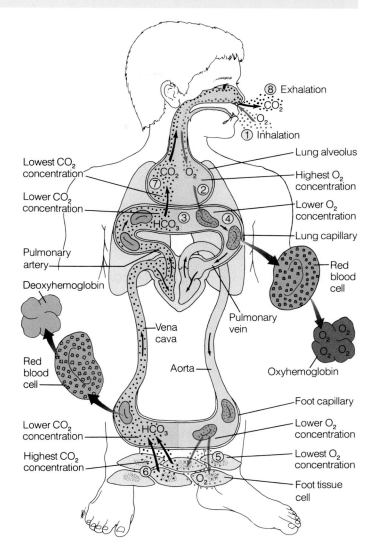

FIGURE 23.12 ■ **The Diffusion of Gases in the Body.** Oxygen entering the respiratory system diffuses through the delicate walls of alveoli and blood vessels and is carried in the bloodstream to distant body tissues (here, the feet). Carbon dioxide moves in the opposite direction. (See the text for step-by-step details.)

hemoglobin molecules in a person's body than there are stars in the Milky Way galaxy! As shown in Figure 2.28, a hemoglobin molecule consists of four protein chains, each attached to an iron-containing heme group. As blood cells move through capillaries in the lungs, these four iron atoms bind easily with the oxygen encountered there at high levels, and the binding produces oxygenated hemoglobin, or *oxyhemoglobin* (see Figure 23.12). This iron-oxygen combination is a brilliant red color and explains the bright color of the blood as it leaves the lungs. In contrast, blood returning to the lungs carrying hemoglobin depleted of oxygen (*deoxyhemoglobin*) is darker, with almost a bluish cast, causing vessels like those in the wrist to look blue.

FIGURE 23.13 ■ The Yak: Extra Red Blood Cells Help at High Elevation. *Bos grunniens,* the short-legged beast of burden in the Himalayas of Nepal and Tibet, has three or four times more red blood cells in each milliliter of blood than a person at sea level, but the yak's cells are small and contain less hemoglobin.

Hemoglobin's specialized structure enables it to act like an oxygen sponge and allows blood to soak up 70 times more oxygen than an equivalent quantity of water could absorb. When the blood reaches the muscles, fat, skin, and other body tissues, it encounters low oxygen concentrations, since the active tissue cells steadily use up the available oxygen. In such an environment, oxyhemoglobin molecules begin to give up their oxygen, and the liberated oxygen dissolves in the fluid portion of the blood and then diffuses into the tissue cells.

The oxygen-carrying ability of hemoglobin helps explain why Weddell seals have evolved with more blood and a greater concentration of red blood cells than people have. Curiously, the short-legged, long-haired cattle called yaks that live in the high Himalayas of Tibet and Nepal have lower hemoglobin concentration than a person's blood at sea level (Figure 23.13). But yaks have three or four times as many red blood cells per milliliter of blood as either humans or seals. Some observers suspect that yak hemoglobin itself may have a special affinity for oxygen, much as the hemoglobin of the human fetus can pick up the gas with particularly high efficiency.

A Special Mechanism That Unloads Oxygen to Active Cells

Highly active cells, like muscle cells in a climber straining to scale Mt. Everest or a seal pushing deep toward the ocean floor, produce considerable amounts of carbon dioxide. A small portion of the carbon dioxide dissolves in the plasma or combines with deoxygenated hemoglobin. Most of the carbon dioxide waste, however, reacts with water, forming carbonic acid (H_2CO_3), which in turn dissociates into bicarbonate ions (HCO_3^-) and hydrogen ions (H^+). This lowers the pH, making the local environment acidic. Additional acidity accumulates from the production of lactic acid by muscles working anaerobically, since the heart and lungs are unable to supply oxygen to the muscles as fast as they can use it up (review Figure 5.10). When muscles are working so hard that their cellular environment grows acidic, hemoglobin can give up its oxygen more easily to the oxygen-starved cells—a phenomenon called the **Bohr effect.**

During a Weddell seal's 1-hour dive, the anaerobic metabolism of the muscles creates a large quantity of the by-product lactic acid. This contributes to the Bohr effect in those muscles themselves, but not in other parts of the body, because the circulatory shunting mechanism keeps the blood in the muscles fairly separate from the general oxygenated blood supply, now going primarily to the brain, spinal cord, and retinas of the eyes. After the animal surfaces, its liver, lungs, and other organs clear out the lactic acid. But this clearing process can take nearly an hour, and during that time, the animal cannot dive again successfully, regardless of opportunities or threats in the environment. Despite this delay, the beauty of the general physiological mechanism is clear: Hemoglobin picks up oxygen in the lungs and releases it effectively under acidic conditions—the environment of working muscles.

CONNECTIONS

Respiration is a good example of how an animal's anatomical structures and functions are tuned to environmental demands and physical principles of nature. The respiratory exchange surface may be folds of skin, filamentous gills, finely branching tracheal tubes, lung sacs, or bubblelike alveoli. Regardless, in each case, the architecture allows gases to diffuse with maximum efficiency from the environment, through moist cell membranes, into the animal's cells, and back out again. The elaborate respiratory anatomy of a person or a Weddell seal is, at last analysis, merely a set of antechambers, transit pipes, and bellows that bring sufficient quantities of gas in and out of the multicellular organism. Always, though, the real action—the passage of oxygen and carbon dioxide based on gas pressure, diffusion distances, altitude, and other physical parameters—takes place at the delicate, moist cell surfaces of the respiratory system.

The interface of biological and physical principles will be equally clear when we discuss gas exchange in plants (see Chapter 32). In the meantime, we move on to two more physiological systems in animals—digestion and excretion—both central to the steady internal state we call homeostasis.

Highlights in Review

1. The Weddell seal, with its ability to dive to the deep sea bottom in search of food, is a good case study for respiration.
2. Respiration, the exchange of oxygen and carbon dioxide between an animal's body cells and its external environment, usually involves a moist exchange surface such as tracheae, gills, or lungs.
3. In the gills of aquatic animals, dense networks of capillaries lie just a few cells away from flowing water. A countercurrent flow between water and the blood in the capillaries enables fish gills to capture a large amount of the oxygen dissolved in water.
4. Air-breathing animals have internal respiratory channels or pouches with a moist surface through which gases diffuse into a fluid. Insects have branching tubes called tracheae, while vertebrates have lungs. In mammalian lungs, an in-and-out flow of air through an "inverted tree" of respiratory airways brings oxygen into contact with tiny sacs called alveoli.
5. Changes in breathing rate and depth are mostly involuntary and are regulated by nerves that detect changes in the level of carbon dioxide circulating in the blood.
6. The hemoglobin molecule is specialized to take on large amounts of oxygen and quickly release it to working tissues. Differences in the partial pressures of oxygen and carbon dioxide lead to their diffusion into or out of red blood cells and tissue cells.
7. Most of the carbon dioxide that enters the bloodstream is rapidly converted to carbonic acid and then to bicarbonate, a process that prevents the blood from becoming too acidic. The reaction is reversed in the lungs, and the carbon dioxide is exhaled.

4. What makes a bird's respiratory system highly efficient?
5. Place the following human respiratory structures in their proper order to show oxygen's path from the outside world to the blood: bronchioles, larynx, nasal cavities, bronchi, trachea, alveoli, capillaries.
6. List the effects of smoking on the body.
7. Do your lungs regulate your breathing rate? Explain.
8. How do oxyhemoglobin and deoxyhemoglobin differ?
9. How is the yak adapted to surviving in the "thin" air of the Himalayan mountains?
10. Trace the path of a carbon dioxide molecule from a blood capillary in your little toe to the air about to be exhaled from your lungs.
11. What is the Bohr effect?

Apply What You Have Learned

1. A heavy smoker tries to relieve her hacking cough by buying "extra-strength" cough drops at the pharmacy. Why will her strategy ultimately fail?
2. A condemned building collapsed and buried a worker in debris up to his chin. A co-worker struggled to help the victim breathe by keeping his mouth and nose clear, but the buried man suffocated anyway. Why?
3. As a research project, you study the breathing rates of students sitting in a crowded, unventilated classroom for an hour. What is your study likely to reveal, and why?
4. Skin divers who use snorkels instead of scuba tanks often hyperventilate before a deep dive. Repeated deep breaths maximize the oxygen levels and minimize the carbon dioxide levels in their blood. As a dangerous consequence, they may stay submerged for so long that they lose consciousness before the respiratory centers in their brains drive them to the surface for air. Considering both the levels of blood oxygen and carbon dioxide, what might cause this to happen?

Key Terms

air capillary, 486	larynx, 487
alveolus, 487	nasal cavity, 487
Bohr effect, 492	opercular flap, 484
bronchiole, 487	partial pressure, 491
bronchus, 487	pleural sac, 489
countercurrent flow, 485	respiration, 483
diaphragm, 489	spiracle, 486
epiglottis, 487	thoracic cavity, 489
esophagus, 487	tidal ventilation, 487
gill filament, 485	trachea, 487
lamella, 485	ventilation, 489

Study Questions

Review What You Have Learned

1. What are two adaptations that permit the Weddell seal to dive deep in the ocean?
2. How does countercurrent flow in gills maximize oxygen intake?
3. What path does oxygen follow from the outside environment to an insect's individual cells?

For Further Reading

ECKERT, R., D. RANDALL, and G. AUGUSTINE. *Animal Physiology; Mechanisms and Adaptations.* 3d ed. New York: Freeman, 1988.

FEDER, M. E., and W. W. BURGGREN. "Skin Breathing in Vertebrates." *Scientific American,* November 1985, pp. 126–143.

HILDEBRAND, M. *Analysis of Vertebrate Structure.* 3d ed. New York: Wiley, 1988.

MCRAE, M. "Altitude Ate My Brain." *Outside,* May 1990, p. 24.

READ, R., and C. READ. "Breathing Can Be Hazardous to Your Health." *New Scientist,* February 23, 1991, pp. 34–37.

WEST, J. B. "Physiological Responses to Severe Hypoxia in Man." *Canadian Journal of Physiological Pharmacology* 67 (1989): 173–178.

ZAPOL, W. M. "Diving Adaptations of the Weddell Seal." *Scientific American,* June 1987, pp. 100–107.

Animal Nutrition and Digestion: Energy and Materials for Every Cell

FIGURE 24.1 ■ **A Craving for Eucalyptus.** The koala (*Phascolarctos cinereus*, or "ash-gray pouched bear"), an Australian marsupial, survives on a diet of nothing but eucalyptus leaves. Specialized adaptations in its lengthy digestive tract allow it to extract the nutrients and water it needs from a diet that would poison or starve most mammals.

Up a Tree

What do you suppose would happen if, for some strange reason, you were forced to spend a week in a eucalyptus tree, sitting on its large, smooth limbs and relying on its aromatic, silvery green leaves for all your needs? Among other things, you would get very thirsty, hungry, and cold, and you might eventually try licking dew off the leaves, stuffing leaves into your clothing for warmth, and even eating a handful or two to stave off the gnawing hunger. You would find eucalyptus leaves to be leathery, however, and to have a bitter, nauseating, turpentine flavor.

Ironically, one mammal—the koala—is perfectly adapted to spending a lifetime in a eucalyptus tree (Figure 24.1). This cuddly-looking marsupial (mammal with a pouch; see Chapter 19) from Australia can meet all of its needs for shelter, food, and water in the very eucalyptus tree that would be so inhospitable to a stranded human. The koala's thick fur acts like a portable shelter and is effective at resisting cold wind. The koala has the equivalent of two thumbs on each paw, which helps it grip tree limbs tightly. Most remarkable, however, is the koala's ability to live on nothing but eucalyptus leaves. An exclusive diet of these pungent leaves— filled, as they are, with toxic oils and containing fairly low levels of nitrogen and other basic nutrients—would starve, poison, and desiccate other mammals. Yet a koala's digestive tract can detoxify the leaves and simultaneously extract all the nutrients and water this 9 kg (20 lb) animal needs.

The koala's specialized diet of eucalyptus leaves is an example of evolution in both trees and mammals. Most animals find eucalyptus leaves unpalatable because those leaves produce and store terpenes and other poisons. Millions of years ago, trees with the genes for making such protective substances suffered fewer attacks by herbivores (plant eaters) and thus reproduced more effectively than trees lacking those genes. At the same time, however, ancestral koalas with genes allowing them to exploit the virtually untouched resource of eucalyptus leaves were favored by natural selection and evolved a specialized digestive tract capable of utilizing compounds from the leaves—even toxic oils and gums—for complete nutrition.

The koala's ability—unique among mammals—to obtain energy and materials from eucalyptus leaves focuses our attention on animals' special adaptations for digestion and nutrition. Three unifying themes emerge from our discussion of these important physiological topics. First, animals must take in sufficient amounts of energy and materials in the form of food or they develop diseases and die. As heterotrophs, animals cannot manufacture their own energy-containing compounds from simple inorganic precursors the way plants and other autotrophs can. They must take in foods from the environment to provide themselves—and ultimately each of their millions or trillions of individual cells—with a supply of **nutrients,** substances that organisms extract from their surroundings that enable them to survive and reproduce. Animals, for example, need certain materials—amino acids, sugars, fatty acids, vitamins, and minerals—for aerobic respiration, replacement of worn-out parts, and growth. In the absence of specific nutrients, animals develop specific disease states. Lacking sufficient energy-containing compounds, animals starve.

Second, an animal's anatomy and physiology are tied closely to its diet and its nutrient needs. As Chapters 18 and 19 explained, each animal species has unique body coverings, appendages, muscles, internal organs, and physiological capabilities. Logically, these go hand in hand with different nutrient and energy requirements, and indeed, an animal instinctively consumes or learns to select a diet that will tend to fulfill its own requirements. Furthermore, its anatomical equipment for harvesting and digesting food will be precisely suited to its dietary needs. For example, a koala eating eucalyptus leaves and a butterfly sipping flower nectar are both herbivores eating only plant matter; but the koala has a mouth suited for crushing vegetation, the butterfly a mouth suited for sipping nectar, and each has a digestive system—shaped by natural selection—that can effectively use the food it con-

sumes. A Doberman pinscher and a crocodile are both meat-eating carnivores, with teeth that can easily tear flesh and a digestive tract well suited to utilizing large amounts of protein. And a person and a wild boar are both omnivores (*omni* means "all"), with teeth that can efficiently grind up both plant and animal matter and a digestive tract suited to handling both.

A third point will emerge as we discuss the ever-popular subject of food, eating, and **digestion**—the mechanical and chemical breakdown of food into small molecules that the organism can absorb and use. In most animals, digestion occurs in steps. The animal ingests relatively large pieces of food, breaks them down mechanically into small pieces, breaks these down chemically to their constituent nutrient molecules, absorbs the nutrients into the blood, transports them to cells, and eliminates unused residues. Different parts of the digestive tract carry out different specific steps in digestion, and the overall process is controlled by chemical signals (hormones) and nerve signals.

This chapter answers several questions about nutrition and digestion:

- What organic and inorganic nutrients must animals have in order to survive?
- How are different animals adapted to different kinds of foods?
- What structures and mechanisms allow people to take in food, break it down, and use the nutrients?
- How do nerve signals and hormones control digestion so that all the body's cells are continuously supplied with needed materials?

NUTRIENTS: ENERGY AND MATERIALS THAT SUSTAIN LIFE

All organisms require a source of energy and a dependable supply of carbon and nitrogen atoms. Plants get their energy directly from the sun, their carbon from the air, and their nitrogen from the soil. Animals, however, must take in all three as food molecules. The science of **nutrition** is concerned with the precise amounts of protein, carbohydrates, lipids, vitamins, and minerals and the total number of calories of food energy an animal must consume to stay alive and healthy. Without this knowledge, dairy farmers could not be sure their cows were producing the highest quality of milk, and parents might wonder if their children were growing and developing to their fullest mental and physical potential.

An animal can have a highly specialized diet: A koala will eat nothing but eucalyptus leaves. Likewise, there are moths that will consume nothing but the nectar of a few specific flowers, and other moths that exist only on the protein-rich tears of cattle. Most animals, however—ourselves included—need a more varied diet and cannot thrive without a mixture of foods from different sources. Because the human diet is both varied and familiar, we will use it as our main model in studying organic and inorganic nutrients, beginning with carbohydrates, then moving on to lipids, proteins, vitamins, minerals, and total calories.

Carbohydrates: Carbon and Energy from Sugars and Starches

The sugars in fruit and the starches in potatoes, rice, bread, and pasta are a rich source of energy and carbon atoms and provide the nutrients we call *carbohydrates.* As Chapter 2 explained, carbohydrates include monosaccharides, such as the glucose found in honey; disaccharides, such as sucrose, or table sugar; and polysaccharides, such as potato starch (see Figures 2.19 and 2.20).

Our digestive system can cleave glucose monomers from the sugars in fruit or from the starch in rice, wheat, or potatoes. After a meal, the glucose units pass into the bloodstream, and the circulatory system carries them to cells throughout the body, where the simple sugar serves as the main supplier of energy for glycolysis and aerobic respiration (Figure 24.2a). Cells in the brain and nerves are particularly sensitive to fluctuations in blood glucose levels. If starving, the body will break down first its fat stores, then its own muscle tissues, and convert the subunits to glucose, thus providing the sensitive nervous system cells with the levels they need to stay fully active.

People are often told to avoid table sugar and the desserts and processed foods containing it. Sugar is said to provide nutritionally "empty calories" and is blamed for causing, among other things, tooth decay. The fact is, sugar provides calories, our largest nutritional need. And while it doesn't

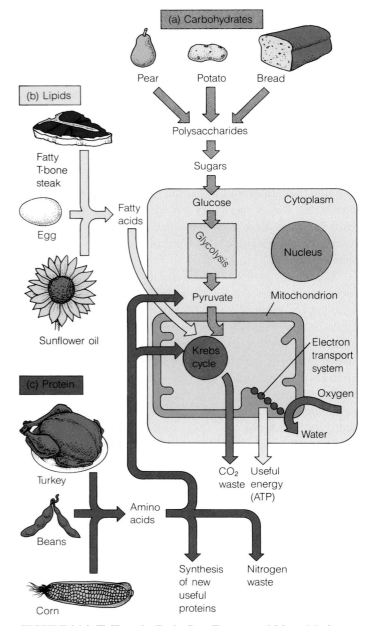

FIGURE 24.2 ■ How the Body Gets Energy and Materials from Foods. (a) Digestive processes cleave carbohydrate foods into their constituent polysaccharides and sugars. Ultimately, sugars are converted to glucose. Individual cells can then rearrange the atoms of this sugar by means of glycolysis and aerobic respiration. These processes create new kinds of carbon compounds that the body can use for materials and energy molecules such as ATP. (b) The digestive system breaks down lipids—the fats and oils in our daily diet—into fatty acids and other compounds. These can be modified and incorporated into the lipids the animal needs for storing energy or building cell membranes, or the steroids from which the animal's cells generate hormones. Or they can be converted to fatty acids, which can be cleaved to acetyl-CoA and enter the Krebs cycle and electron transport chain with its resulting harvest of the high-energy compound ATP. (c) Proteins in the diet are broken down into constituent amino acids, and these can be incorporated into new proteins. The carbon portion of amino acids can be modified and enter the Krebs cycle and the electron transport chain, where ATP and other high-energy compounds are formed. The "amino" part can be excreted as waste.

provide vitamins or minerals in significant amounts, neither do other sweeteners such as honey, brown sugar, raw sugar, corn syrup, or maple syrup. The issue is whether a person substitutes sugar-laden foods for more nutritious ones—a candy bar, say, instead of an apple. The causative role of sugar in tooth decay is a bit clearer: Sucrose readily promotes the growth of oral bacteria that produce acid and cause cavities. Whether you eat cake or a more complex carbohydrate, regular brushing is the answer to preventing tooth decay.

The cellulose fibers in plant cell walls, including the cells of leaf veins, the "strings" in celery stalks, the tissue surrounding grapefruit sections, and the woody core of a pineapple, are also complex carbohydrates, but ones that we humans cannot digest or utilize (review Figure 2.20). Herbivores, however, such as termites, koalas, and horses, have cellulose-digesting microbes in their digestive tracts that cleave glucose subunits from cellulose and provide the animals with a rich source of energy. The cellulose we consume in plant foods does provide fibrous "roughage," however, which helps to stimulate the mechanical movements that propel wastes through the large intestine.

Lipids: Highly Compact Energy Storage Nutrients

Most people have only to visualize the solid white fat of a bacon slice or the slippery golden oil in salad dressing to know what a lipid is. And many have only to reach down and "pinch an inch" to see that the body stores lipids as an energy reserve. About 95 percent of the lipids in the human body and in our normal diet are *triglycerides* (see Figure 2.21). The remainder are the *phospholipids,* which are crucial to membrane structure, and the *sterols,* including cholesterol and steroid hormones like estrogen and testosterone (review Figures 2.23 and 2.24).

Lipids are a compact energy source and the form most plants and animals use for long-term energy storage. Fats are oxidized during aerobic respiration and provide substantial amounts of usable energy (Figure 24.2b). After a meal of eggs, oily sunflower seeds, or fatty meat, an animal's digestive system breaks down lipids to fatty acids and glycerol. The fatty acids are broken down further, and the resulting products can then feed into the Krebs cycle.

Another reason animals often store fat rather than carbohydrates is that fats and water do not mix, whereas each gram of stored carbohydrate generally binds 2 grams (0.08 ounce)* of water. A bird storing carbohydrates and bound water rather than compact fat would be too bulky to fly. And one biologist estimates that if a person's body stored carbohydrates instead of fats, it would weigh twice as much. Without fat stores, walruses could not live in the Arctic (Figure 24.3), birds could not migrate long distances, carnivores would have more trouble surviving between irregular meals, and hibernators might not be able to survive the long winter without eating.

FIGURE 24.3 ■ Lipids: High-Energy Storage. Fats are a compact way to store calories. Active animals, such as walruses living near the Arctic Circle, have thick fat layers that act as superb insulation as well as a source of stored energy that the animal can use to generate heat and power activities.

Besides their storage function, fats also cushion internal organs, and if present in a thick layer, as in whales or walruses, can provide thermal insulation. Finally, certain essential nutrients, including vitamins A, D, E, and K, are fat-soluble, and lipids must be present for these vitamins to be absorbed and delivered to cells. The human body cannot generate one particular fatty acid—linoleic acid—from component parts; thus, our diets must contain this so-called **essential fatty acid** in order to build important components of the cell membrane involved in material absorption. Nevertheless, a high-fat diet increases one's risk of colon cancer and heart disease. In fact, according to a recent study of 88,000 nurses, a person who eats red meat (beef, pork, or lamb) daily has a risk of colon cancer three times higher than someone who ingests animal fat from those sources just two times a week or less.

Proteins: Basic to the Structure and Function of Cells

The body's structure and its vital activities depend on proteins. The body's most abundant protein, collagen, is a major constituent of skin, cartilage, tendons, and bone. Muscle tissue is largely protein, as are hair and the cornea of the eye. Enzymes, antibodies, hemoglobin, and some hormones are composed of protein molecules; without such proteins, most cellular activities would grind to a halt. What's more, there is a steady turnover of protein: Enzymes and cell constituents are continuously broken down and rebuilt, and newly formed cells replace dying cells. This constant turnover means that animals need a continuous supply of protein in food, from which their digestive systems can extract amino acids for the building of new protein as needed (Figure 24.2c).

A human is a large, active animal that requires about 1 g (0.04 oz) of protein per kilogram (2.2 lb) of body weight per day. As a rough measure, a college student of average height and weight needs approximately one-sixth of a pound of pure protein each day to replace losses. This is about what you would obtain from a cooked chicken breast. But knowing how much protein one needs does not answer the question of what *kind* of protein to eat. The body can synthesize many of the 20 amino acids it needs if the amine (nitrogen-containing) portion of the molecule is available (review Figure 2.25). The body cannot, however, manufacture eight so-called **essential amino acids:** lysine, leucine, phenylalanine, isoleucine, tryptophan, valine, threonine, and methionine (Figure 24.4). These must be obtained in the diet each day, since free amino acids are not stored. Children also need extra supplies of the amino acids histidine and arginine, since their bodies only make enough for maintenance, not growth.

If one or more of the essential amino acids are missing, the body's cells cannot synthesize the full spectrum of proteins necessary for replacing lost macromolecules or generating new cells. Why? First of all, the body cannot store amino acids; what it doesn't incorporate immediately into proteins, it excretes or converts to other biological molecules. Second, as Chapter 10 described, cells manufacture proteins by adding one amino acid at a time to a growing polypeptide chain. If even one of these amino acids is lacking, elongation of the chain stops. It's a bit like knitting a sweater with a repeating pattern of several colors of yarn: When you run out of yarn of one color, the project stops until you can get more. In the same way, if even one essential amino acid is absent, protein synthesis in the body comes to a screeching halt.

Most animal proteins in foods like meat, cheese, eggs, and milk contain the eight essential amino acids; however, many plant proteins do not. This is why strict vegetarians (those who avoid all animal products, including eggs and dairy products) must eat particular combinations of foods to ensure an adequate amino acid intake. One such combination is rice, which contains little lysine but an adequate amount of methionine, together with beans, which are deficient in methionine but contain a sufficient amount of lysine (see Figure 24.4). The incomplete nutrition provided by plant proteins also explains why some people in poor and underdeveloped nations show the swollen belly, patchy skin, and other symptoms of protein deprivation. Their diet, consisting mostly of cereal grains such as corn, wheat, or rice, all too often lacks one or more essential amino acids. Consequently, their bodies draw proteins from their own muscles and other tissues, dismantling them and using their essential amino acids for new protein synthesis. The rest of the amino acids present in

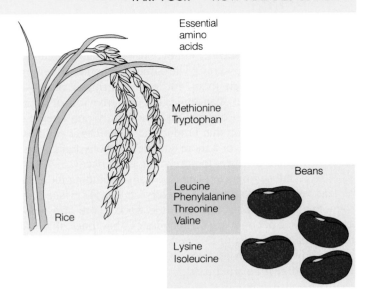

FIGURE 24.4 ■ **Essential Amino Acids.** The eight essential amino acids on this list must be included in the daily diet; if they aren't, the body begins to break down proteins in its own tissues and use the amino acids they contain for critical functions, such as maintaining the lungs, heart, and brain. Meat, fish, eggs, and other high-protein foods contain all eight essential amino acids. Rice contains only six, and beans a slightly different set of six, but a diet that includes both rice and beans in the same meal will provide all eight.

excess are simply excreted and wasted. As medical observers have noted sadly, when a human being starves, it is usually the lack of protein, rather than the lack of food energy, that leads to death.

Vitamins and Minerals: Nutrients of Great Importance

While the body requires relatively large amounts of protein and carbohydrate each day, it needs another set of nutritive substances, the vitamins and minerals, in only extremely small amounts. Even so, consumers are bombarded with persuasive advertisements for vitamin and mineral supplements. This phenomenon attests to their importance, but also to widespread misunderstandings about human nutrition.

Vitamins are organic compounds needed in small amounts for normal growth and metabolism, but which higher animals cannot themselves manufacture. Although vitamins are not a direct source of energy, most serve as precursors for coenzymes or activators for the enzymes that participate in cellular metabolism (see Chapter 5). Most animals, however, cannot make many of these cofactors themselves, and in most cases can store only small supplies in their cells. Thus, they must acquire a set of specific vitamins in their daily food.

Nutrition researchers have determined that humans require the 14 vitamins listed in Table 24.1. As noted earlier, vitamins A, D, E, and K are soluble in fat, while the B vitamins and vitamin C are water-soluble. Fat-soluble vitamins pass into the lymphatic capillaries in the digestive tract, from where they eventually reach the general blood circulation. Fat-soluble vitamins tend to be stored in the body's fat tissues; since accumulations can produce serious side effects, nutritionists warn against taking high doses of A, D, E, or K vitamins. Water-soluble vitamins, on the other hand, move directly into the bloodstream from the digestive system. Because amounts beyond the cells' immediate needs are excreted, taking large amounts of vitamin C, for example, is largely pointless; any excess beyond the normal daily requirement is filtered out by the kidneys and excreted in the urine soon after it enters the bloodstream.

Another prevalent myth is that vitamins from "organic" foods or natural sources are somehow healthier or more effective than vitamins from processed foods or vitamins that are manufactured directly. An "organic" tomato has no more nutritive value than one grown on a larger, more mechanized farm, although the latter may contain more pesticide residues. And since vitamin C, for instance, is the specific biochemical

TABLE 24.1   Vitamins

Vitamin	Sources	Functions in Body	Deficiency Symptoms
Water-Soluble Vitamins			
Choline	Egg yolks, liver, beans, peas, grains	Part of phospholipids; needed for nerve cell function	Not seen in humans
Vitamin C (ascorbic acid)	Dark green vegetables, citrus fruits, strawberries, brussels sprouts, other fruits and vegetables	Helps form collagen and bone; enzyme helper; blocks toxic effects of oxygen	Gum bleeding; hemorrhages under skin; rough skin; failure of wounds to heal; bone degeneration
Niacin	Milk, meats, cereals, and starchy vegetables	Part of enzyme helpers involved in electron exchange reactions	Sore skin; smooth tongue; diarrhea; mental confusion; irritability
Pantothenic acid	Widespread in foods	Central to energy metabolism	Rarely seen
Vitamin B_6 (pyridoxine)	Whole grain cereals, vegetables, meats	Enzyme helper involved in amino acid metabolism	Skin soreness; smooth tongue; abnormal brain activity
Vitamin B_2 (riboflavin)	Milk, meat, vegetables, whole grains	Helper of enzyme active in energy metabolism	Cracks at corner of mouth; sensitivity to light
Vitamin B_1 (thiamine)	Milk, dairy products, fruits, breads, vegetables	Helper of enzyme that removes carbon dioxide from nutrients	Beriberi (paralysis, swelling, heart failure); mental confusion
Folacin (folic acid)	Fruits, leafy and other vegetables, grains	Helper of enzyme involved in metabolism of amino acids and nucleic acids	Anemia; diarrhea; smooth tongue
Biotin	Widespread in foods	Constituent of many enzymes in metabolism	Not seen in humans
Vitamin B_{12}	Meat and dairy products	Helper of enzyme involved in nucleic acid metabolism	Anemia; nerve degeneration
Fat-Soluble Vitamins			
Vitamin E (tocopherol)	Vegetable oils, margarine, salad dressings	Counters toxic effects of oxygen	Breakage of red blood cells; anemia
Vitamin A (retinol)	Carrots, milk, vegetables, fruits	Part of visual pigment; helps maintain living tissues; promotes bone growth	Impaired night vision; dry eyes; diarrhea; lung infections; bone changes; tooth cracking and decay; impaired brain growth; anemia
Vitamin K	Cabbage, milk, green leafy vegetables	Essential for synthesis of certain proteins, including blood-clotting proteins	Unchecked bleeding
Vitamin D	Milk, sunshine, eggs, liver	Helps bones and teeth take up calcium for proper growth	Rickets (bowed legs, other bone deformities); tooth decay; blood changes; lax muscles

TABLE 24.2   Minerals

Mineral*	Sources	Functions in Body	Deficiency Symptoms
Major Minerals (more than 0.1 g/day)			
Calcium	Milk and dairy products, fish bones, collard greens, spinach, broccoli	Major part of bones and teeth; cell membrane integrity; helps collagen form; involved in nerve transmission	Fragility of bones (osteoporosis); stunted growth
Phosphorus	Widespread in foods	Major constituent of bones, blood plasma; part of DNA, RNA; needed for energy metabolism	Rare
Potassium	Bananas, orange juice, potatoes, tomatoes, other vegetables	Critical to normal heartbeat; principle positive ion in body cells	Mental confusion
Sulfur	Protein foods	Present in all body proteins; helps protein assume proper shape	Unknown
Sodium	Salt, common in foods	Constitutent of salt in body fluids; helps regulate fluid content of body	Rare; excess leads to water retention
Chlorine	Salt	Major negative ion in body fluids; part of HCl in stomach acid	Rare
Minor Minerals (less than 0.01 g/day)			
Magnesium	Nuts, legumes, cereals, dark green vegetables, seafood, chocolate	Constituent of bones; role in protein synthesis and energy metabolism	Tetany; prolonged muscle contraction; hallucinations
Iron	Oysters, liver, bran flakes, lean beef, spinach, greens, strawberries	Part of many major enzymes active in DNA synthesis and cellular respiration; hemoglobin and myoglobin	Anemia; exhaustion; headache; weakness
Iodine	Salt (iodized)	Part of thyroid hormone thyroxine; controls metabolic rate	Goiter (enlarged thyroid gland)
Zinc	Oysters, milk, egg yolks, meat, whole grains	Part of many enzymes; helps bone form; DNA and protein synthesis; wound healing	Retarded growth; small sex organs; loss of sense of taste
Selenium	Abundant	Part of many enzymes; antioxidant	Not seen in humans
Manganese	Widespread	Constituent of many enzymes in metabolism	Not seen in humans
Copper	Grains, shellfish, organ meat, legumes, dried fruit, fresh fruit, vegetables	Helps form hemoglobin, collagen, nerves	Rare; retarded growth, sluggish metabolism
Molybdenum	Many foods	Part of many enzymes	Not seen in humans
Fluorine	Fluoridated water	Normal formation of bones and teeth	Tooth decay
Chromium	Yeast, organ meats	Involved in carbohydrate metabolism	Stunted growth; adult-onset diabetes

*Arranged in order of decreasing amounts needed in daily human diet.

ascorbic acid and has a unique molecular structure, it will be the same whether it is extracted from rose hips or synthesized in a laboratory.

The body also needs small amounts of another type of nutrient: **minerals,** specific inorganic chemical elements (see Figure 2.3). *Major minerals* are those elements we need in amounts greater than 0.1 g each day; *minor minerals* are those we need in amounts less than 0.01 g daily (Table 24.2). An adult's body contains about 2 kg (4½ lb) of minerals, and of this, about three-quarters is the calcium and phosphorus in bones and teeth. The minerals that give tears, blood, and sweat their salty taste are sodium, potassium, and chlorine. Sulfur is found in many proteins, and magnesium is a constituent of enzymes that, along with calcium, is held in reserve in the bones.

The human body contains less than 1 teaspoon of minor minerals, but these elements still play critical roles. Perhaps most important is iron—an essential component of the oxygen-binding pigment hemoglobin. Iodine is a constituent of thyroid hormones, zinc is a common coenzyme (see Chapter 5), copper is needed for hemoglobin production, and fluorine is needed for healthy bones and teeth.

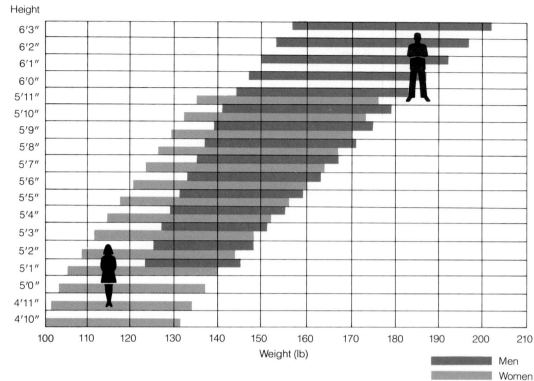

FIGURE 24.5 ■ Ideal-Weight Chart. Ideal weight depends on frame size and physical condition. At 6 ft and 195 lb, a heavy-boned, muscular football tackle would not be overweight, while a sedentary office worker with a medium frame probably would be. Height and weight are given for people not wearing shoes or other clothing.

A person's need for specific minerals can change over time. A woman, for example, needs more iron before menopause because she loses blood periodically, more calcium after menopause when hormonal changes alter the way the body uses that mineral, and more of both during pregnancy.

Because vitamins and minerals are required in such small quantities, most people in affluent countries get a more than adequate supply simply by eating a reasonably balanced diet. When poor diet leads to vitamin or mineral deficiencies, however, the consequences can be serious. The long-term absence of vitamin A, for instance, can lead to blindness, while the lack of B vitamins may lead to convulsions and other neurological disorders (see Table 24.1). As residents of a wealthy nation, we have a far greater problem with overeating and obesity than with malnutrition. An estimated 25 percent of teenagers and 50 percent of adults are either **overweight** (body weight more than 10 percent over ideal according to weight charts, such as the one in Figure 24.5) or **obese** (body weight more than 20 percent over ideal; Figure 24.6). Obesity has been shown to be a significant contributor to cardiovascular diseases, some forms of diabetes, joint problems, generally poor health, and an overall decrease in the quality of life, including reduced earning power based on early retirement for health reasons.

While obesity is common in North America, some people become so obsessed with body weight and thinness that it seriously threatens their health. In the disorder called *anorexia nervosa*, a person (typically a middle-class teenage girl) restricts her food intake severely, becomes cadaverously thin,

FIGURE 24.6 ■ Obesity: Overstorage of Physiological Fuel. Sumo wrestlers from Japan eat huge quantities of rice, fats, plus fish and other protein-rich food, to generate the body fat that gives them their formidable bulk. Although a Sumo wrestler, like this American champion, can easily weigh 300 lb, he is extremely strong, not simply fat.

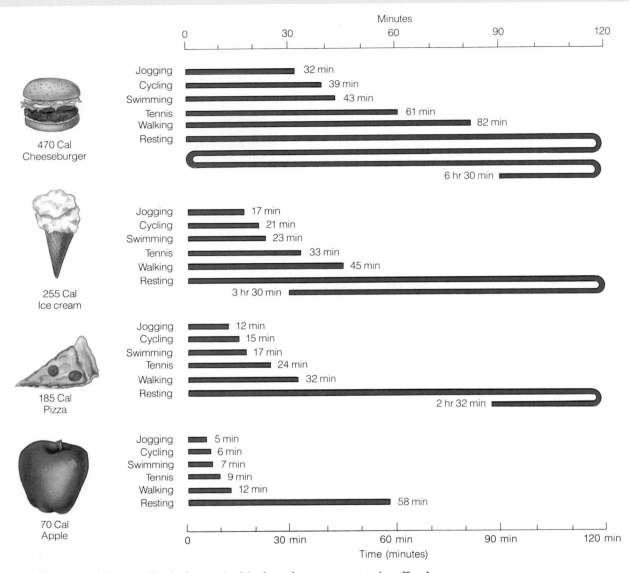

FIGURE 24.7 ■ Food and Exercise Equivalents. As this chart shows, you can swim off a cheese-burger in 43 minutes, but at rest, your body will take 6 hours and 30 minutes to burn those same 470 Calories!

and yet still sees herself as overweight. In a separate disorder called *bulimia,* a person (again, usually a young female) secretly binges on huge helpings of cake, ice cream, cookies, bread, or other foods, then purges herself with self-induced vomiting, laxatives, diuretics, fasting, or vigorous exercise. Both disorders have social, psychological, and probably neurochemical roots and can lead to organ damage and even death if untreated.

Food As Fuel: Calories Count

Like all animal cells, human cells require energy to carry out biochemical, mechanical, and transportation tasks. As Chapter 5 explained, cells derive energy from the chemical bonds in the fats, carbohydrates, and proteins an animal eats, digests, and absorbs; any energy an animal does not immediately use can be stored as glycogen in the liver or muscles or as fat in fat cells. Food energy is usually measured in kilocalories (kcal), also called Calories (Cal). Recall that a calorie is the amount of energy needed to raise the temperature of 1 ml of water 1°C; thus, 1000 cal, or 1 Cal, is the amount of energy it takes to raise the temperature of 1 L of water 1°C. A large apple contains about 100 Cal worth of energy-producing compounds; and jogging 1.6 km (1 mi) burns about 100 Cal of stored energy.

Each person has a minimum daily energy requirement that varies with age, sex, body size, activity level, and other factors. In general, a normally active female college student needs around 1800 to 2000 Cal a day to fuel her total meta-

FAT CELLS, SET POINT, AND WEIGHT LOSS

To quote the illustrious cat Garfield, for many overweight people, "Diet is die with a T." Despite their best efforts, fully 95 percent of all people who diet regain every bit of weight they lose. Why is it so difficult to lose weight and keep it off? Researchers have begun to investigate the possibility that like many animals, people have a fixed *set point*, a level of fat storage and body weight that is genetically determined and difficult—but not impossible—to alter.

The theory suggests that some people have naturally high set points (above ideal weights) while others have low ones (at or below ideal weights) and that the set point is based on the number and size of fat storage cells (Figure 1). Individuals vary in the number of fat cells their bodies contain, and fat cells in humans are smaller and much more numerous than in any other mammals save the hedgehog and the fin whale—both notoriously fat.

Once gained, fat cells appear never to be lost; they merely increase or decrease in size by storing more or less fat, depending on dietary excesses. Significantly, a person's fat cells tend to remain a given size and *to return to that original size* soon after a diet ends. Fat cells act as if they had a mind of their own, and in fact, they do appear to communicate with the brain. They seem to signal any drop in lipid stores, trigger increased appetite and eating behavior to compensate, and initiate a change in metabolic rate so that the body uses its calories more efficiently—all as if to defend the fat cell's genetically determined size.

In obese people, fat-shuttling enzymes may be overactive and may store fat molecules that would be burned for energy in naturally thin people. Research also suggests that "yo-yo" weight loss and gain from one failed diet after another may actually train the body to cling to every calorie, making it harder and harder to lose weight each time. Finally, it appears that the set point can be raised by the smell or taste of fatty food, an evolutionary adaptation, perhaps, to allow animals to take advantage of energy-rich resources when they find them.

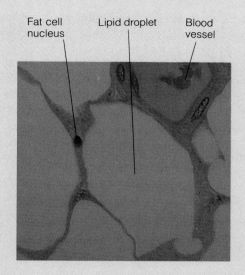

Fat cell nucleus   Lipid droplet   Blood vessel

FIGURE 1 ■ A Fat Cell with Its Large Stored Lipid Droplet.

Given this discouraging picture, how can one lower the set point and thus reduce the size of fat cells and with it the weight of the body? Evidence suggests that a moderate reduction in total calories (especially from lipids and sugars) is a beginning, but that dieting must be accompanied by a consistent increase in physical activity. Exercise seems to turn up the metabolic rate so that the body burns more calories—and not just during exercise sessions, but for all the hours at rest between bouts of exertion. Exercise decreases fat tissue and increases muscle mass; thus, the body looks and feels trimmer. Finally, moderate daily exercise reduces the appetite. Considering all these health benefits, regular exercise (four to five sessions per week) is probably the major reason why people who jog, swim, bicycle, or do another form of aerobics find it easier to control their appetites and to maintain their weights at lower levels—that is, to lower their set points.

---

bolic needs; a male college student needs about 2200 to 2500 Cal. Carbohydrates and protein each provide about 4 Cal/g, while fat provides more than twice as much, or 9 Cal/g.

Figure 24.7 provides a revealing look at the caloric values for several common snack foods and the amount of energy that a person must expend in various physical activities to work off that food. A person would have to jog for about 30 minutes, for example, to burn off the calories in a cheeseburger. When an animal's food intake exceeds its energy needs, the inevitable result is an increase in the amount of leftover energy stored as body fat. This basic biological fact means that the secret of weight control lies in taking in only as many kilocalories as the body needs for fuel. Sound weight-loss programs combine calorie reductions, primarily through diminished intake of sugar and fat, with increased physical activity. In fact, many physiologists believe that physical exercise can cause the body's basal metabolic rate to increase, that is, to burn calories at a faster rate both during

and after exercise. The result is less energy in and more energy out, with the differences made up from the body's fat reserves. Researchers have focused increasingly on the mechanics of weight gain and loss and have concluded that fat cells help determine our appetites as well as our waistlines (see Box 24.1). Some dieters find it is useful to focus on the biological function of eating: to take in the nutrients the body needs for energy, maintenance, and repair. As one wise saying goes: "Eat to live, don't live to eat" (see Appendix A.4).

While everyone needs carbohydrates, proteins, and lipids in their diets, there are health risks associated with many foods. Some scientists calculate that about a third of all cancer cases are due to cancer-causing substances naturally present in the food we eat. An example is the animal fat in red meat and the increased risk of colon cancer it confers, as mentioned earlier. What can consumers do to protect themselves? Nutritionists recommend eating a low-fat diet, thus reducing the risk of colon, breast, and prostate cancer; eating lots of fruits and vegetables, since the fiber in them moves all food through the bowel faster (thus decreasing the time that cancer-causing substances are in contact with tissues), and since the vitamins A and C they contain may inhibit the development of cancers; eating few salt-cured and smoked meats, which contain cancer-causing chemicals; and drinking alcohol in moderation or not at all, thereby decreasing the risk of cancers of the throat, liver, and urinary bladder. Finally, if recent studies on monkeys and other mammals can be generalized to people, a long-term reduction in total calories increases health and well-being as one ages.

DIGESTION: ANIMAL STRATEGIES FOR CONSUMING AND USING FOOD

Most animals, whether they eat eucalyptus leaves, raw meat, or a chef's salad, are consuming food in a form that cells cannot use directly. In all but the simplest animals, foods must

first be broken down into pieces, then into usable small molecules; these molecules must in turn be absorbed into the bloodstream and distributed to body cells, while unusable wastes are eliminated.

Recall from Chapter 18 that in the simplest animals—the sponges—some body cells take in tiny whole food particles directly from the water and break them down via enzymes, obtaining nutrients. This strategy is called **intracellular** ("within the cell") **digestion** (Figure 24.8a), and it circumvents the need for the mechanical breakdown of food or for a gut or other cavity in which to chemically digest foods. At the same time, however, intracellular digestion puts an upper limit on an animal's size and complexity and means that it can use only very tiny bits of material as food.

Animals could exploit the advantages of larger size after the evolution of **extracellular digestion**: the enzymatic breakdown of larger pieces of food into constituent molecules outside of cells, but usually within a special body organ or cavity (Figure 24.8b). Nutrients from the broken-down foods then pass into body cells lining the organ or cavity and can

(a) Intracellular digestion

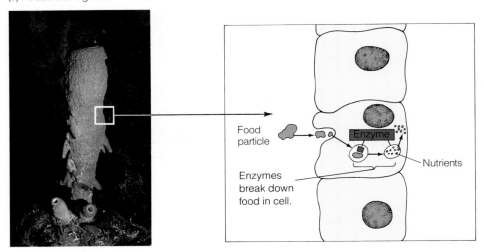

FIGURE 24.8 ■ Digestion: Intracellular and Extracellular. (a) A simple animal like this tube sponge has no gut or similar organ and instead carries out intracellular digestion: Tiny food particles are taken into the body wall cells via phagocytosis across the cell membrane (see Figure 3.19). Digestive enzymes within the cell then break the small particles into constituent molecules. (b) A hydra has a central digestive cavity that can take in and digest relatively large pieces of food. Cells lining the digestive cavity secrete enzymes into that central space. There, the enzymes break down food into constituent nutrients, and the nearby cells then absorb these nutrients.

(b) Extracellular digestion

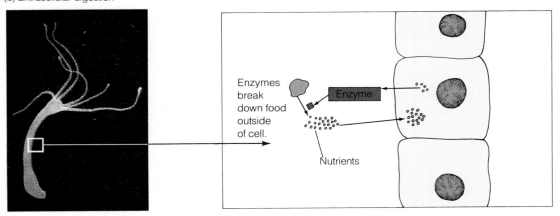

(a) The alimentary canal

(b) Steps of digestion

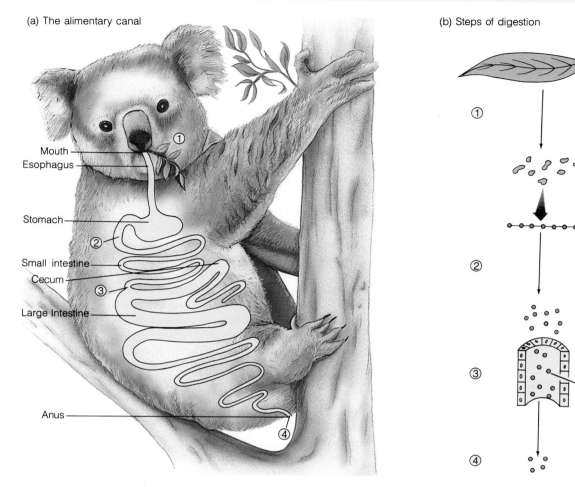

Mouth
Esophagus
Stomach
Small intestine
Cecum
Large Intestine
Anus

① Food ingested and broken into small pieces

② Large molecules (macromolecules) split into subunits (monomers)

③ Nutrients absorbed

④ Wastes eliminated

FIGURE 24.9 ■ **Four-Step Digestion Process: Ingestion, Digestion, Absorption, Elimination.** As the text explains in detail, the koala (a) ingests eucalyptus leaves, and its stomach enzymes break down the plant's macromolecules (mainly carbohydrates) into monomers (sugars). Cells in the gut wall then absorb these nutrients (b), and indigestible wastes eventually exit the gut tube.

then take part in energy metabolism or biosynthesis. Since the vast majority of animals rely on it, let's give extracellular digestion a closer look.

The Mechanisms of Extracellular Digestion

In a simple animal, such as a cnidarian or flatworm, extracellular digestion takes place in the *gastrovascular cavity,* an internal sac with a single opening through which whole particles of food enter and undigested wastes leave (review Figure 18.7). This system works fine for very flat or thin organisms, but has some important limitations: The animal does not mechanically break apart food pieces, which means that digestive enzymes in the saclike gut can work only on the surface of ingested food chunks, and no one area of the gut is specialized for a particular function.

In more complex animals, food enters one end of the digestive tract, the mouth, and moves in a single direction

through the gut tube, and wastes exit the other end, the anus; in between, a variety of specialized regions and structures perform special digestive roles. Since food lumps are often fairly large, animals need ways to break them into small pieces and thereby increase the surface-to-volume ratio of food particles so that the particles can be acted on more quickly and efficiently by digestive enzymes. Animals evolved two basic kinds of mechanical food-processing structures: teeth, located in the mouth, and gizzards near the gut's anterior entrance. Teeth can be flat for grinding, chisellike for clipping and gnawing, or sharp and conical for slashing and shearing. While birds lack teeth (hence the expression "scarce as hen's teeth"), they do have gizzards. A *gizzard* is a muscular sac often containing stones that grind against each other and pulverize food.

In virtually all vertebrates, the gut is divided into five main regions: mouth, esophagus, stomach, small intestine, and large intestine. Together, these form the digestive tract, or **alimentary canal** (Figure 24.9a). In addition, nearby **acces-**

sory organs, such as the salivary glands, liver, gallbladder, and pancreas, produce enzymes, bile, and other materials that are funneled into the tract at appropriate times and aid digestion. In general, then, digestion takes place within a long tube with specialized regions and nearby accessory organs. These organs add secretions that assist the complicated step-by-step process.

The Alimentary Canal: Food-Processing Pipeline

Within the alimentary canal, a four-step digestive process takes place that extracts nutrients from food (Figure 24.9b). First, the animal—a koala, let's say—ingests food—in this case, eucalyptus leaves. It breaks the leaves into small pieces with its grinding teeth (step 1), and enzymes from salivary glands are mixed with food as the animal chews and swallows. Then additional enzymes in the gut break down the large macromolecules of food—in this case, cellulose and plant juices from the leaves—into smaller molecules, or monomers (step 2). Next, the monomers are absorbed; they pass across the gut wall and into the animal's lymphatic system or bloodstream, which transports the small nutrient molecules to each cell in the body (step 3). Finally, most of the water in the food is absorbed into the circulation, and undigested residues are eventually eliminated (step 4).

For nutrients from food to move out of the alimentary canal and enter the bloodstream, they must pass through several cell layers. (We will trace the passage through these layers shortly.) The innermost layer facing the cavity of the digestive tract is a mucous membrane, or **mucosa** (Figure 24.10). Some mucosa cells produce digestive enzymes, while others secrete mucus, forming a thick film coating that protects the inner surface of the tract. This film keeps the stomach from digesting itself and lubricates food passing through. Just outside the mucosa is the **submucosa,** a connective tissue that is richly supplied with blood and lymph vessels and with nerves. Next is a composite layer made of muscle fibers, some encircling the gut and others running longitudinally. The contraction and relaxation of these fibers kneads the food, mixes it with digestive juices, and propels it along with rhythmic sequential contractions like toothpaste being squeezed through a tube. These rhythmic contractions are known as **peristalsis.** Finally, a thin outermost layer, the **serosa,** attaches the gut to the inner wall of the body cavity.

Evolution has resulted in a number of intriguing variations in extracellular digestive systems that allow animals to exploit very different foods. Carnivores, such as hyenas and wolves, have short intestinal tracts because meat is relatively easy to digest, whereas cows and rabbits have elaborate guts with several stomachlike pouches containing fermenting bacteria that help break down cellulose in the animals' vegetarian diets. Another variation on this theme is the koala's 2.5 m (8 ft) fermenting chamber, a blind pouch called the *cecum* lying coiled near the animal's stomach like a huge appendix (see Figure 24.9a). The largest such structure known in the animal kingdom, the koala's cecum is 30 times bigger than a person's appendix (which is only 8 cm, or 3 in., long) and contains bacteria that break down the cellulose in eucalyptus leaves and convert it to molecules the animal needs to survive

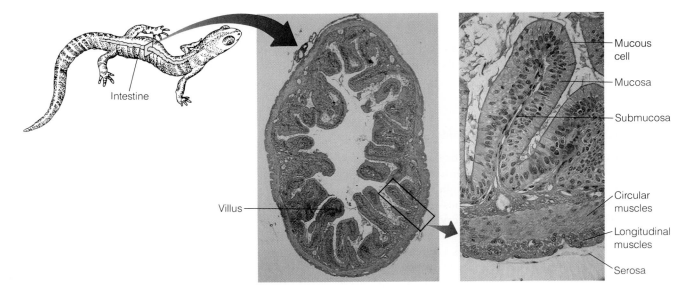

FIGURE 24.10 ■ **Anatomy of the Alimentary Canal: Four Concentric Layers Help Process Foods.**
A cross section of a salamander's intestine and an enlarged villus reveal mucosa, submucosa (red), two muscle layers (longitudinal and circular muscles), and a thin serosa.

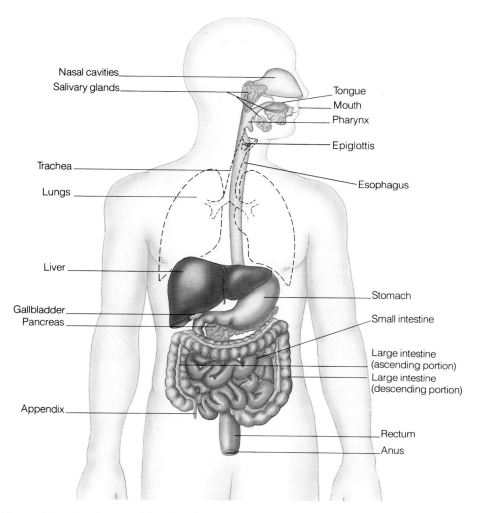

Nasal cavities
Salivary glands
Tongue
Mouth
Pharynx
Epiglottis
Trachea
Esophagus
Lungs
Liver
Stomach
Gallbladder
Pancreas
Small intestine
Large intestine
(ascending portion)
Large intestine
(descending portion)
Appendix
Rectum
Anus

FIGURE 24.11 ■ **The Human Digestive System.** The digestive system is essentially one long tube (which lines the mouth, curves around in fold upon fold, then exits at the anus) plus associated organs.

on its limited diet. A person's appendix has no known digestive function, although it does have some immune activity and may fight infections in the gut. The human appendix sometimes gets inflamed, leading to the medical emergency we call appendicitis. In the next section, we will see how the human digestive tract works on its normal omnivorous diet.

THE HUMAN DIGESTIVE SYSTEM

The human digestive tract (Figure 24.11), from mouth to anus, is roughly 8 m long, or about the height of a two-story house. All the sandwiches, brownies, apples, milk, and other foods a person eats pass through the central cavity, or lumen, of this tube and undergo one digestive process after another. To see in detail how the human alimentary canal liberates the

nutrients in foods and transports them across the gut tube layers, we will follow a meal—a turkey sandwich, let's say—throughout each step of its digestive journey.

The Mouth, Pharynx, and Esophagus: Mechanical Breakdown of Food

Your teeth are superbly adapted to the job of cutting, tearing, and grinding the plant and animal tissues you consume as an omnivorous mammal. (By contrast, a koala's teeth—especially its broad, flat molars—are well suited for grinding the fibrous, leathery leaves it eats all day and night.) Teeth can stand up to this regular wear and tear because the enamel covering them is the hardest substance in the body: The enamel on a person's permanent teeth will generate sparks if struck against steel. Working with the teeth is a muscular **tongue** (see Figure 24.11), the principal organ of taste, but also, in our species, an organ for forming the sounds of spoken language. When you take a bite of a sandwich, your tongue moves some of the food toward the molars for grinding and some to the incisors for cutting, shaping each small bite into a soft, moist lump, or **bolus,** that can be easily swallowed.

At the same time, each bite is also mixed with clear, watery **saliva,** which is secreted by three large pairs of **salivary glands** that lie in the tissues surrounding the oral cavity. Saliva contains primarily mucus and water, which moisten food particles and help them cling together in a bolus. Saliva also contains small amounts of *amylase,* a digestive enzyme that begins the breakdown of carbohydrates in the bread, lettuce, and tomato of the sandwich.

Swallowing often begins as a voluntary act once food reaches the back of the mouth. As the tongue pushes the bolus of food up and back against the roof of the mouth, however, sensitive touch receptors trigger the reflex action of the pharynx. When this happens, the **soft palate** rises, preventing food from entering the nasal cavities, and the flaplike epiglottis moves backward and downward, closing off the opening to the trachea, or windpipe (Figure 24.12). Simultaneously, the larynx (voice box) moves up, and muscle contractions in the throat push food or liquid into the **esophagus,** the "pipeline" to the stomach. By placing your fingers lightly on either side of your throat just below your chin as you swallow, you can feel the larynx move up and forward, opening the esophagus and receiving food.

To enter the stomach, the food bolus must pass through a **sphincter,** a ring of muscle located at the junction of the stomach and esophagus (Figure 24.13a). This muscular ring, known as the *cardiac sphincter,* usually remains tightly contracted, like the drawstring on a purse, and thus can prevent the stomach's contents from moving back up into the esophagus. The stomach's gastric juices are so acidic that they could severely damage the esophagus, which lacks the stomach's heavy mucous lining. People often experience "heartburn" when the stomach is very full; this burning sensation is due to small amounts of stomach acid seeping out past the sphincter into the unprotected esophagus.

The Stomach: Food Storage and the Start of Chemical Breakdown

Swallowing causes muscles in the esophagus to relax, enables the cardiac sphincter to open, and a bite of turkey sandwich to pass into the elastic J-shaped bag called the **stomach** (Figure 24.13b). The stomach not only stores food for later processing (an adaptation that enables large animals to feed less often), but also mixes and churns its contents with digestive juices through contractions of the stomach wall's muscle layers (Figure 24.13c). This churning helps expose as much of the food surface as possible to enzymes.

The average human stomach can comfortably hold about 1 L (a little over 1 qt), but the organ can stretch to accommo-

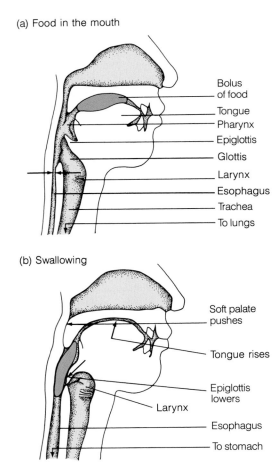

(a) Food in the mouth

Bolus of food
Tongue
Pharynx
Epiglottis
Glottis
Larynx
Esophagus
Trachea
To lungs

(b) Swallowing

Soft palate pushes
Tongue rises
Epiglottis lowers
Larynx
Esophagus
To stomach

FIGURE 24.12 ■ **Swallowing Reflex.** (a) Once the tongue shifts the bolus of moistened food to the back of the mouth, the pharynx is stimulated to a reflexive swallowing action. (b) The soft palate lifts, closing off access to the nasal cavities, the tongue rises, pushing the bolus backward, and the epiglottis lowers, closing off the opening to the windpipe (trachea). The larynx rises, and muscle contractions push the bolus down into the esophagus toward the stomach.

date much larger capacities. The notorious glutton Diamond Jim Brady reportedly ate *each day* several dozen clams and oysters, steak, roast beef, lobster, duck, several chickens, pheasant, two boxes of chocolates, several gallons of orange juice, pastries, ice cream, and more. At Brady's death, a coroner's report showed that his stomach was six times larger than that of a normal man!

When the stomach is full, waves of peristalsis in the stomach wall churn and mix the stomach contents with an acid bath of gastric juices secreted by glands in the stomach wall. Gastric juices are a mixture of water, hydrochloric acid (HCl), mucus, and **pepsinogen,** a precursor to the protein-cleaving enzyme called **pepsin.** The hydrochloric acid makes gastric juice acidic enough to kill off most bacteria or fungi contaminating foods. This acid contributes to the breakdown of food pieces into constituent protein fibers, fat globules, and so on, and it also allows pepsinogen to be converted to pepsin. Pepsin cleaves the peptide bonds that link amino acids in proteins. Therefore, during the time that protein-containing foods (such as the turkey in a sandwich) are in the stomach, they are partially digested to short polypeptide segments. Although digestion of the starch in the bread of the sandwich begins in the mouth, once food enters the stomach, very little additional digestion of starches and other carbohydrates (or of fats, such as those in mayonnaise) takes place. This must await passage of foods to the small intestine.

The result of the chemical activity and "mixing waves" in the stomach is a pasty, milky, and highly acidic soup called **chyme.** The mixing and churning of chyme gradually moves it toward the lower stomach, where another sphincter, the *pyloric sphincter,* controls the opening to the small intestine (Figure 24.13d). As the chyme is pushed against that opening, the sphincter relaxes just long enough for small, carefully regulated "doses"—on average, about a teaspoonful every 3 seconds after a meal—to be squirted into the small intestine so that further digestion and absorption can take place efficiently. It usually takes between 1 and 4 hours for a meal to be processed in the stomach and delivered, spoonful by spoonful, to the small intestine—less time for a high-carbohydrate meal, more time for a fatty meal.

The Small Intestine and Accessory Organs: Digestion Ends, Absorption Begins

Most of the chemical digestion of food and some of the absorption of nutrients take place in the small intestine, but this 6 m (19.7 ft) stretch of the gut tube does not accomplish these tasks alone; three accessory organs—the pancreas, liver, and gallbladder—dump in substances that aid digestion and absorption (see Table 24.3 on page 512).

■ **The Pancreas and Its Digestive Enzymes**   The **pancreas** is a narrow, lumpy organ situated near the junction of the stomach and small intestine (see Figure 24.11). It produces a host of digestive enzymes and secretes them into the small intestine, where food digestion continues after chyme leaves the stomach. These pancreatic enzymes include *proteases,* which break down proteins, *lipases,* which digest fats, and enzymes such as amylase, which break down carbohydrates. What's more, the pancreas also secretes bicarbonate ions (the main ingredient in indigestion remedies), which act like a buffer, neutralizing acid entering the small intestine from the stomach. This neutralization is vital, because unlike enzymes in the stomach, pancreatic enzymes cannot function in an acidic environment, and the acid could damage the small intestine. Table 24.3 lists the major pancreatic enzymes and their functions.

Some pancreatic cells secrete the hormones *insulin* and *glucagon* directly into the bloodstream. These hormones help keep steady levels of glucose in the blood (see Chapter 26).

■ **The Multipurpose Liver and the Gallbladder**   The smooth, irregularly lobed and roughly hemispherical **liver** is the largest gland in the human body (see Figure 24.11). It weighs about 2 kg (4½ lb) in an adult and performs biological tasks as diverse as destroying aging red blood cells, storing glycogen, and dispersing glucose to the bloodstream as circulating levels of the sugar drop. One of the liver's most important functions is to produce *bile salts*—molecules that are modified cholesterol and that act like detergents breaking up fat droplets in the small intestine. Bile salts travel through the bile duct to the **gallbladder** (Figure 24.14, step 1), a small pear-shaped sac on the underside of the liver, where they are stored as **bile,** a yellow-green liquid containing the bile salts, pigments from the breakdown of red blood cells, and other substances. The presence of fats in the small intestine stimulates the gallbladder to squeeze drops of bile down a duct into the small intestine (step 2).

Just as detergent mixed with cooking oil and water will cause the oil slick to disperse into thousands of tiny droplets, bile salts emulsify fat; that is, they disperse fat globules into

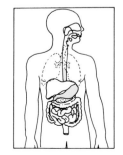 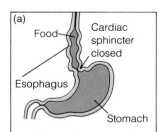 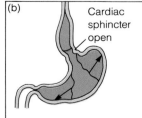 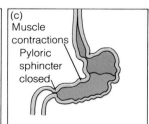 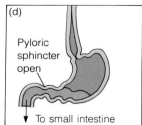

FIGURE 24.13 ■ **Function of the Stomach: Churning, Secretion, and Initial Food Breakdown.**   A food bolus (a) pushes on the cardiac sphincter, which opens, (b) allowing the food to enter the stomach. (c) Muscle contractions moving in peristaltic waves across the stomach mix the food with digestive enzymes and stomach acid, forming chyme. (d) The pyloric sphincter then opens in repeated brief bursts that allow small amounts of chyme to be squirted into the duodenum (upper part of the small intestine), where it is digested further.

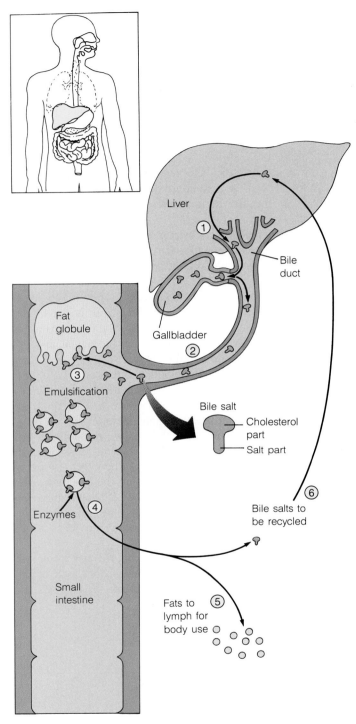

FIGURE 24.14 ■ Fat Digestion: Emulsification, Absorption, and Bile Salt Recycling. This diagram shows how a fat globule floating along in the small intestine after a meal is broken up into small droplets (emulsified) by bile salts. Bile salts are produced in the liver (1), stored in the gallbladder, and passed down the bile duct (2) to the small intestine. Bile salts help remove small groups of lipid molecules from fat globules (3). The lipid molecules are then acted on by enzymes and pass into cells lining the intestinal wall (4). The small group of lipid molecules become coated with protein and enter the lymph and eventually the blood circulation (5), and the bile salts are recycled and stored once again in the gallbladder (6).

tiny fat droplets (step 3), providing a larger surface area and thus better access by fat-digesting enzymes (lipases) and a more rapid breakdown of lipids into usable constituents.

Bile salts also aid absorption of fully digested fat molecules across the lining of the small intestine (step 4) and into the lymph (step 5). Bile salts are finally recycled back to the liver (step 6). Doctors sometimes give people who suffer from high levels of cholesterol in the blood a compound called cholestyramine, which binds to bile salts and passes down the intestinal tract without being absorbed. This prevents the recycling of cholesterol from bile salts and thus decreases the amount of cholesterol in the blood. Oat bran may lower cholesterol the same way.

Besides manufacturing the bile stored in the gallbladder, the liver picks up, stores, and sometimes synthesizes amino acids, glucose, and glycogen and stores vitamins and other compounds that cells need to function normally. Certain liver cells also contain special enzymes that can detoxify poisons. For example, the liver transforms molecules of ammonia—a toxic, nitrogen-containing waste created by the breakdown of amino acids—into urea, which is less toxic and is excreted in urine. The liver also detoxifies alcohol; but if the organ is overloaded with the drug year after year, it can become damaged and irreversibly scarred, producing a potentially fatal condition called *cirrhosis.*

■ **The Lengthy Small Intestine**   The small intestine is a remarkable coiled tube about 6 m (19.7 ft) long where carbohydrates and fats are largely digested and where protein digestion is completed to the stage where nutrients can be absorbed into the blood. The small intestine begins just below the stomach and has three main regions along its length: the upper section, or *duodenum,* the central *jejunum,* and the remainder, the *ileum.*

The small intestine is a marvel of compact biological engineering. If the surface of its inner lining were a smooth tube like a garden hose, there would be a relatively small surface area for digestion and absorption. Instead, however, that inner lining is so convoluted that it houses a huge absorptive surface. The intestinal lining is pleated into large numbers of folds (Figure 24.15a and b), and each fold is covered with fingerlike extensions known as **villi** (singular, villus), which project into the lumen and come into contact with chyme (Figure 24.15b). Further, the outer layer of each villus is carpeted with **microvilli,** microscopic brushlike projections of the cell membrane (Figure 24.15c). The combination of folds, villi, and microvilli creates a total surface area the size of a tennis court.

By the time a turkey sandwich is reduced to chyme and reaches the small intestine, it contains partially digested carbohydrates and proteins, as well as undigested fat. Here, enzymes secreted by the intestine and the pancreas combine with bicarbonate ions and bile salts and interact, gradually completing the digestion of the carbohydrates in bread to simple sugars, the proteins in turkey meat to amino acids, and the fats in mayonnaise to fatty acids and glycerol. These nutrient molecules are small enough to move across the plasma membranes of the microvilli and enter the intestinal cells. The amino acids and sugars then pass into blood capillaries along with most of the available vitamins and minerals from the tomato slice and lettuce, while lipids enter lymph capillaries (see Figure 24.15c). The remaining contents of the small intestine, including the small amount of indigestible "roughage" in the lettuce and tomato, pass into the large intestine.

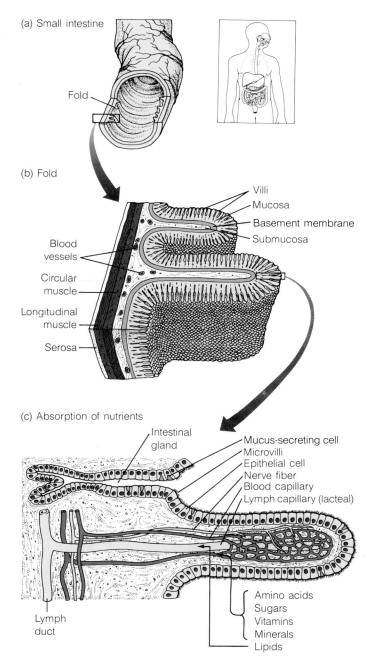

(a) Small intestine

Fold

(b) Fold

Villi
Mucosa
Basement membrane
Submucosa

Blood vessels

Circular muscle

Longitudinal muscle

Serosa

(c) Absorption of nutrients

Intestinal gland

Mucus-secreting cell
Microvilli
Epithelial cell
Nerve fiber
Blood capillary
Lymph capillary (lacteal)

Lymph duct

Amino acids
Sugars
Vitamins
Minerals
Lipids

FIGURE 24.15 ■ Highly Absorptive Lining of the Small Intestine: Folds Bearing "Fingers" Topped with "Brushes." Intestinal folds (a) are covered with villi (b). Each villus "finger" (c) encompasses blood and lymph capillaries that carry away nutrients absorbed through the brushlike carpet of microvilli, with their enormous combined surface area. Lipids enter the lymph, and other nutrients enter the blood.

The Large Intestine: Site of Water Absorption

The last 1.2 m (4 ft) of the alimentary canal is the large intestine, or **colon:** a stretch of the gut tube that ascends the right side of the body cavity, cuts across just below the stomach, then descends the left side and ends in a short tube called the **rectum** (see Figure 24.11 and Table 24.3). The colon has two main functions: to absorb water, ions, and vitamins from the chyme and to store the semisolid undigested wastes, or **feces,** until they are excreted.

The large intestine is twice as wide as the small intestine, but its walls have a much simpler structure: They lack the many folds and villi of the small intestine. The smoother surface has a lower surface area for absorption, but it also presents less resistance to the movement of chyme through the tube.

Each day, about 0.5 L (about 1 pt) of chyme (now minus the nutrients absorbed in the small intestine) reaches the colon along with 2 to 3 L (2 to 3 qt) of water, some from food and the rest secreted by the stomach and intestines themselves. The body cannot afford to lose this much water, however, and much of the water is reabsorbed as the chyme moves slowly through the colon over a period of 12 to 36 hours. This absorption gradually transforms the chyme from fluid to semisolid, and the wastes (including undigestible cellulose fiber from the bread, lettuce, and tomato of the turkey sandwich we started with) are stored until their pressure against the colon wall triggers a bowel movement, the muscular expulsion of wastes through the rectum and out the anus.

Digestion is never 100 percent efficient, and some nutrients invariably pass into the large intestine from the small. A variety of bacteria, including *Escherichia coli* and *Lactobacillus* and *Streptococcus* species, reside in the human intestine and live off these remaining nutrients; in the process, they produce a number of vitamins, including thiamine (vitamin B_1), vitamin B_{12}, riboflavin (vitamin B_2), and vitamin K. The colon absorbs these vitamins along with fluids. Many animals, including termites and herbivores, which ingest large amounts of cellulose, also benefit from the metabolic activities of "live-in" microorganisms. Without its cecum and the enclosed bacteria that release carbohydrates, vitamins, and minerals from eucalyptus leaves, the koala would starve. Likewise, without their own types of fermenting chambers and enclosed bacteria, cows and horses could never extract enough nutrients and calories from grass and hay to maintain life.

COORDINATION OF DIGESTION: A QUESTION OF TIMING

As we have seen, different parts of the digestive system carry out different, highly specialized tasks, and the result is an efficient use of food and nutrients. If these steps are tightly controlled and coordinated, the cells will get their steady input of

TABLE 24.3   Functions and Secretions of the Digestive Tube and Accessory Organs in Humans

Digestive Tube		Digestive System (stretched out)	Accessory Organ	
Function	Secretion		Function	Secretion

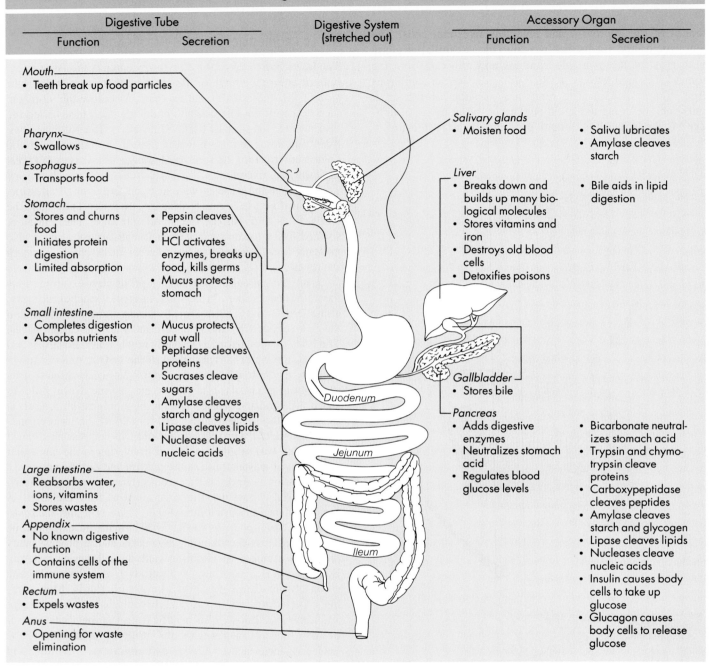

Mouth
- Teeth break up food particles

Pharynx
- Swallows

Esophagus
- Transports food

Stomach
- Stores and churns food
- Initiates protein digestion
- Limited absorption

Secretion (Stomach):
- Pepsin cleaves protein
- HCl activates enzymes, breaks up food, kills germs
- Mucus protects stomach

Small intestine
- Completes digestion
- Absorbs nutrients

Secretion (Small intestine):
- Mucus protects gut wall
- Peptidase cleaves proteins
- Sucrases cleave sugars
- Amylase cleaves starch and glycogen
- Lipase cleaves lipids
- Nuclease cleaves nucleic acids

Large intestine
- Reabsorbs water, ions, vitamins
- Stores wastes

Appendix
- No known digestive function
- Contains cells of the immune system

Rectum
- Expels wastes

Anus
- Opening for waste elimination

Duodenum

Jejunum

Ileum

Salivary glands
- Moisten food

Secretion (Salivary glands):
- Saliva lubricates
- Amylase cleaves starch

Liver
- Breaks down and builds up many biological molecules
- Stores vitamins and iron
- Destroys old blood cells
- Detoxifies poisons

Secretion (Liver):
- Bile aids in lipid digestion

Gallbladder
- Stores bile

Pancreas
- Adds digestive enzymes
- Neutralizes stomach acid
- Regulates blood glucose levels

Secretion (Pancreas):
- Bicarbonate neutralizes stomach acid
- Trypsin and chymotrypsin cleave proteins
- Carboxypeptidase cleaves peptides
- Amylase cleaves starch and glycogen
- Lipase cleaves lipids
- Nucleases cleave nucleic acids
- Insulin causes body cells to take up glucose
- Glucagon causes body cells to release glucose

---

energy and nutrients, even if meals are separated by long intervals, and the strong acid and powerful enzymes of the digestive system will not dissolve the gut itself.

The close control of digestion is achieved by the body's two great coordinators: the nervous system and the hormonal system. Nerves throughout the digestive tract allow communication between "downstream" and "upstream" regions so that the propulsion of food can be speeded or slowed appropriately. Nerve activity in the brain and spinal cord can also speed up or slow down digestive functions. For example, the sight, taste, smell, or even thought of food can cause signals from higher brain centers to travel via nerves to the salivary glands, causing them to secrete saliva, and to secretory glands in the stomach lining, inducing gastric juices—hydrochloric acid and pepsinogen—to begin flowing into the stomach (Figure 24.16). Food arriving in the stomach and pushing against the stomach wall can of course trigger the same response.

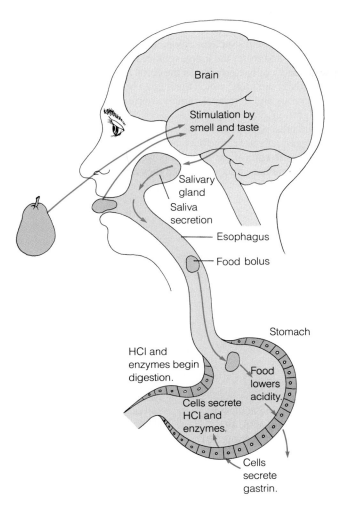

FIGURE 24.16 ■ Nerves and Hormones Coordinate Digestion. Your body must begin the digestive process when food is available—not when the stomach is empty. As the text explains, the control of digestion involves the senses, signals from the brain, and hormones secreted by the stomach and intestines.

Food in the stomach also has another effect: It lowers the overall acidity of the contents. This triggers cells in the stomach lining to secrete a hormone called **gastrin,** which stimulates nearby secretory cells to produce more hydrochloric acid (see Figure 24.16). This raises the acidity (lowers the pH) of gastric juice and at the same time helps speed the breakdown of the food. When the pH drops to about 2, the secretion of gastrin stops, and with it additional acid secretion. Thus, a negative feedback loop regulates stomach acidity, and food literally helps to stimulate its own digestion. People who are emotionally stressed for prolonged periods are susceptible to forming *ulcers,* craterlike sores in the mucosa of the stomach or small intestine. Apparently, ulcer victims have decreased secretions of the mucus protecting the stomach lining and/or

excessive gastric secretions—a combination that allows small portions of the stomach or intestinal lining to be literally digested away.

Several other hormones help coordinate the timing and amount of enzyme secretion with the presence of food. For example, **secretin** from the small intestine causes the pancreas to secrete bicarbonate that then enters the small intestine and neutralizes the stomach acid. Partially digested proteins cause the small intestine to release a second hormone, **cholecystokinin.** This hormone triggers the pancreas to release protein-digesting enzymes. Cholecystokinin also works on regulatory centers in the brain and produces the sensation of being "full." This is why nutritionists often advise dieters to eat the protein foods in a meal first so that they will feel satisfied sooner.

Working together, the body's rapid- and slow-acting control agents—nerves and hormones—fine-tune the secretion of digestive juices. As a result, enzymes and ions are instantly available to break food down into nutrients for the body's trillions of cells, and yet those powerful agents are present in the alimentary canal only when needed—when the animal has hunted or harvested food and consumed it.

CONNECTIONS

Digestion is the evolutionary solution to every animal's need for cell nourishment so that the organism can continue to live from day to day, but more importantly, so that it can reproduce and leave progeny. The relationship between nutrition and reproduction is displayed most dramatically by animals whose eating and reproductive stages are entirely separate. The silkworm caterpillar, for example, is a voracious eating machine, but once transformed into a moth, it has no mouth and spends the remainder of its short life surviving on nutrients stored by the caterpillar and generating new individuals.

Food is also crucial to the evolutionary processes in that an animal's physiology and anatomy are thoroughly tied to its diet. Intense competition for nutrients favors individuals who can exploit new environments; thus, through natural selection, mouthparts, digestive organs, and behaviors evolved that gave them an advantage in obtaining their own often highly specialized "slice of the pie." The koala with its eucalyptus leaves—a diet that would starve or poison a human—is a prime example of that process at work.

This chapter has focused mainly on the sequential steps by which the digestive system efficiently extracts nutrients from food and absorbs them for use by the body's vast collection of cells. In Chapter 25, we will consider a closely related system that is directly affected by the foods and liquids an animal consumes: the kidneys or equivalent organs, which regulate the balance of water and salts in the blood and body fluids.

Highlights in Review

1. Animals obtain from their environments nutrients that provide sources of energy and materials for body maintenance, growth, and reproduction. The animal's digestion process extracts carbohydrates, proteins, lipids, vitamins, and minerals from food.
2. People digest starches and many other carbohydrates into monosaccharides, but cannot digest cellulose fibers, which instead serve as roughage.
3. Most animals store energy as fats (lipids). Lipids in the human body and diet are primarily triglycerides, phospholipids, and sterols, including cholesterol.
4. In the human body, both the main structural molecules as well as the enzymes, which mediate cellular activity, are proteins. Proteins contain 20 different amino acids, of which 8, the essential amino acids, must be acquired in the diet.
5. The body requires but cannot synthesize small amounts of vitamins; many vitamins play essential roles in energy metabolism. Vitamin deficiencies can have severe consequences, but are rare in developed nations.
6. The body also requires small amounts of inorganic minerals such as calcium, phosphorus, sodium, and iron.
7. The energy in foods is usually measured in Calories. Fats contain the most Calories per gram because they can be more completely oxidized. When a person's energy intake exceeds expenditure, the body stores the excess as fat.
8. An animal's digestive system is closely tied anatomically and physiologically to its diet. All animals more complex than sponges carry out extracellular digestion of larger food pieces in a body cavity.
9. There are four basic steps in digestion: (1) the mechanical breakdown of food into small pieces by teeth or gizzards; (2) the chemical breakdown of large macromolecules (proteins, carbohydrates, fats, and nucleic acids) into their small subunits (amino acids, simple sugars, fatty acids, and nucleotides) by hydrochloric acid and enzymes in the stomach and by enzymes in the small intestine; (3) the absorption of the small subunits and other nutrients by the small intestine into the bloodstream; and (4) the absorption of water by the large intestine and the elimination of indigestible wastes.
10. The human digestive tract has four layers: the mucosa (the innermost layer), the submucosa, a layer of circular and longitudinal muscles, and the serosa. The muscle layer generates movement that mechanically mixes food with digestive juices and propels the solution along via rhythmic contractions known as peristalsis.
11. Amylase secreted by the salivary glands in the mouth begins the chemical digestion of carbohydrates. Gastric juices in the stomach containing hydrochloric acid and the enzyme pepsin begin the breakdown of proteins. The pancreas produces bicarbonate ions, which help neutralize the acidic chyme entering the small intestine, and enzymes that help complete chemical digestion.
12. The liver produces bile salts, which emulsify the fats in chyme for easier digestion. Bile is stored in the gallbladder. The liver also detoxifies various poisons found in the blood.
13. The lining of the small intestine has a huge surface area as a result of folds, villi, and microvilli. Nutrient molecules cross microvilli into intestinal cells, then enter blood or lymph capillaries.
14. The large intestine, or colon, slowly reabsorbs water from chyme and stores the indigestible residues for later elimination. A variety of bacteria live in the colon and there produce several kinds of vitamins, which the human body can absorb and use.
15. Nervous and hormonal signals coordinate the various steps of digestion.
16. The presence of food in the stomach or small intestine can also trigger cells to secrete hormones that stimulate digestion. Such feedback control ensures that powerful enzymes are present in the alimentary tract only when they are needed.

Key Terms

accessory organ, 505	nutrient, 495
alimentary canal, 505	nutrition, 496
bile, 509	obese, 501
bolus, 508	overweight, 501
cholecystokinin, 513	pancreas, 509
chyme, 509	pepsin, 508
colon, 511	pepsinogen, 508
digestion, 495	peristalsis, 506
esophagus, 508	rectum, 511
essential amino acid, 498	saliva, 508
essential fatty acid, 497	salivary gland, 508
extracellular digestion, 504	secretin, 513
feces, 511	serosa, 506
gallbladder, 509	soft palate, 508
gastrin, 513	sphincter, 508
intracellular digestion, 504	stomach, 508
liver, 509	submucosa, 506
microvillus, 510	tongue, 508
mineral, 500	villus, 510
mucosa, 506	vitamin, 498

Study Questions

Review What You Have Learned

1. How do herbivores, carnivores, and omnivores differ? Give an example of each.
2. List the categories of carbohydrates, and give an example of each.
3. What simple nutrients are produced during the digestion of carbohydrates, lipids, and proteins?
4. How do fats benefit the body?
5. How does the body use proteins?
6. What essential role do most vitamins fulfill?
7. Name six minerals that the body requires and explain why each is necessary.
8. What kinds of mechanical digestion occur in the mouth and stomach?

9. Give the details of chemical digestion in the stomach.
10. What is the role of the pancreas in digestion?
11. What is the role of the liver?
12. How do bile salts aid in the digestion and absorption of fats?
13. What digestive events take place in the small intestine?
14. List the functions of the colon.
15. Explain how nerves and hormones coordinate digestion.

Apply What You Have Learned

1. How can vegetarians ensure that they take in an adequate supply of essential amino acids?
2. Are rose hips from a health food store a better source of vitamin C than orange juice from the supermarket? Explain.
3. How does the drug cholestyramine help lower the level of blood cholesterol?
4. Your weight-lifting coach advises you to skip the vegetables and load up on meat at the dinner table, claiming that consumption of a high-protein diet will cause greater muscle development. Should you follow his advice? Why or why not?

For Further Reading

CAMPBELL-PLATT, G. "The Food We Eat." *New Scientist,* May 1988, pp. 1–4.

DAYTON, L. "Can Koalas Bear the Twentieth Century?" *New Scientist,* September 22, 1990, pp. 43–45.

DEGABRIELE, R. "The Physiology of the Koala." *Scientific American,* July 1980, pp. 110–118.

ECKERT, R., D. RANDALL, and G. AUGUSTINE. *Animal Physiology: Mechanisms and Adaptations.* 3d ed. New York: Freeman, 1988.

LEE, A., and R. MARTIN. "Life in the Slow Lane: For Australia's Koalas, Surviving on Eucalyptus Leaves Means Taking it Easy." *Natural History,* August 1990, pp. 34–43.

RATTO, T. "The New Science of Weight Control." *Medical Self-Care,* March–April 1987, pp. 25–30.

SHAPIRO, L. "Feeding Frenzy." *Newsweek,* May 27, 1991, pp. 46–53.

ULVNÄS-MOBERG, K. "The Gastrointestinal Tract in Growth and Reproduction." *Scientific American,* July 1990, pp. 78–83.

Excretion and the Balancing of Water and Salt

FIGURE 25.1 ▪ Kangaroo Rat: Remarkable Desert Survivor. Thanks to numerous behavioral and physiological adaptations, the kangaroo rat can survive in the sandy deserts of Arizona and New Mexico without drinking water.

Kangaroo Rats and Capable Kidneys

If, on some moonlit summer night, you happen to be driving along a country road in a desert area of southern Arizona or New Mexico, you may chance to see a small but remarkable mammal that is a living lesson on the wonders of the kidney. To catch a glimpse, you must stop the car, sit near a low earthen mound with five or six entrance holes, and wait patiently. If you are lucky, a diminutive grayish rodent resembling a cross between a hamster and a kangaroo will pop out and begin searching for dry seeds (Figure 25.1). This animal, the kangaroo rat, moves with long leaps of its large hind feet and drags its white-tipped tail along in the sand for balance. If it finds some seeds, usually from a mesquite bush or a small, tough plant called Mormon tea, it will quickly stuff the food into its cheek pouches with delicate paws, then hop off again in search of more.

The truly amazing thing about the kangaroo rat is that it can live in a sweltering, parched environment, eat nothing but dry seeds, and yet *never have to drink water.*

Biologists have long been intrigued by the seeming impossibility of a small desert animal that doesn't imbibe water. Through laboratory tests, they have pinpointed seven major behavioral and physiological adaptations that allow these rats to procure and conserve water in a supremely miserly fashion: (1) The rats come out of their burrows only at night when the air is cool and water evaporation is slow; (2) inside the burrow, they avoid heat-generating exercise; (3) each rat stores dry seeds in the burrow, and these absorb some of the moisture lost in the animal's exhaled breath; (4) kangaroo rats have air-cooled noses that trap and retain additional moisture that would have been exhaled; (5) their large intestines absorb virtually all the water from feces and eliminate only hard, dry pellets; (6) their kidneys produce a highly concentrated urine, which the rats excrete with little water loss; and (7) the animals gain so-called metabolic water—water molecules generated during aerobic respiration of the sugars in the seeds they eat. (Figure 5.16, step 7, shows how oxygen accepts hydrogen ions and forms metabolic water in the electron transport chain.) The result of all these adaptations is an

Water gain

Absorbed water (ml)

Metabolic water (ml)

60—
50—
40—
30—
20—
10—
0—

Water loss

60—
50—
40—
30—
20—
10—
0—

Feces

Urine (ml)

Evaporation (ml)

FIGURE 25.2 ■ Water Loss and Gain: A Balancing Act. Over a period of about four weeks, a kangaroo rat in the laboratory will eat 100 g (4 oz) of barley, will absorb about 6 ml (⅕ oz) of water from the air and gain 54 ml (1½ oz) more from oxidizing the barley. Balancing this gain of 60 ml of water is a loss of precisely 60 ml of water in its urine, feces, and exhaled breath.

animal in which water gain and water loss are both minimal but precisely balanced: Intake exactly equals evaporation plus excretion (Figure 25.2).

Living things are made up primarily of water. Most biochemical reactions take place in this universal solvent. Water helps stabilize an animal's body temperature within the range conducive to biological reactions. In most animals, blood or other watery body fluids transport gases and nutrients to every cell and carry away gaseous wastes. And the rapid movement of materials into and out of cells depends largely on the concentrations of both water and solutes (salts, organic materials, and so on) and on the resulting osmotic pressures between the intracellular and extracellular fluids.

In this chapter, we will see that a complex and marvelous organ, the kidney, regulates the balance of salt and water throughout the vertebrate body and also presides over the excretory system, a major physiological network. While an animal's digestive system rids the body of undigested solid wastes, the excretory system, with its kidneys and associated "pipes," cleanses the blood of organic waste molecules and carries them out of the body as urine or its equivalent.

Urine is a fluid that washes from the body uric acid or urea, the two kinds of nitrogen-containing waste products formed after protein and nucleic acid molecules are dismantled during cellular repair. The more concentrated these wastes become, the more dangerous they can be to the animal. The safest way for the body to remove them is to wash them away in a copious amount of dilute urine. But what if water must be strictly conserved, as in a desert animal like the kangaroo rat? How, in other words, does an animal's body rid itself of wastes safely while retaining the necessary concentrations of water and salt? In the vertebrates, the answer lies with the intricate functioning of the kidney. Regardless of external environment, the kidney maintains a constant and proper internal state by automatically removing nitrogen-containing wastes and salts in just the right amount of water so that the remaining fluid balances the body's solute content appropriately. Considering its role, one can easily see why the kidney is a major organ in the maintenance of homeostasis.

Three unifying themes occur in this chapter. First, while the excretory system maintains homeostasis in the blood, which flows near every cell, the critical issue for survival is the maintenance of constant conditions *inside* the trillions of individual body cells. If the pH, salt content, or waste concentration inside cells varies too much, individual cells will cease to function, and eventually the animal will die. As Figure 4.17 showed, the fluid inside cells (the intracellular fluid) undergoes a continual exchange of materials with the fluid in the spaces between cells (the extracellular fluid). Extracellular fluid, in turn, is in equilibrium with the watery fluid of the blood (the plasma). Thus, as the excretory system maintains the contents of the blood, it indirectly maintains the intracellular fluid within cells.

The second unifying theme holds that natural selection has varied the shape and function of the kidney and other excretory organs in ways that fit the animal to its environment. In the kangaroo rat, for example, fine tubules in the kidneys are thrown into extremely long loops that absorb water very thoroughly and leave only a low volume of highly concentrated urine to be excreted. Third, the kidney is the body's major blood-filtering organ, and its activities require a large expenditure of energy, since many molecules pass through the kidney's cell membranes via active transport (see Chapter 4). Without this continuous energy flow and without a healthy functioning kidney, poisonous wastes would build up and the body fluids would become too salty or too watery for life processes to proceed normally.

This chapter answers several intriguing questions about how an animal's body excretes wastes and balances salt and water:

■ What mechanisms remove nitrogenous wastes?
■ How do the kidney and other excretory organs function?
■ How do nerve signals and chemical messengers called hormones regulate the body's water content?
■ How do animals living in deserts, oceans, and ponds balance the body's salt and water content?

RIDDING THE BODY OF NITROGENOUS WASTES

Just as short ends of boards or scraps of cloth remain after you build a bookcase or sew a shirt, residues of nutrient molecules remain unused after cells process the nutrients and build from them new molecules and cell parts. **Excretion** is the process of removing the by-products of metabolism from the cells, the extracellular fluid, and the blood, as well as ridding the organism of excess water and salts (Figure 25.3; also review Figure 20.4). Excretion differs from elimination, the expulsion of undigested food as feces (see Chapter 24), in that undigested solid materials have never left the interior of the digestive tube nor entered the body or its cells. Let's look more closely, now, at the molecular by-products of metabolism and the excretion process that removes them.

Three Forms of Waste in Different Kinds of Animals

Cells can burn fats, carbohydrates, and proteins for energy, but the end products of their metabolism differ. After fats and carbohydrates are dismantled, water and carbon dioxide remain, and the gas is eventually expelled from the lungs or through the skin (review Figure 24.2). When cells metabolize proteins, the amino acid building blocks can be used directly to synthesize new proteins or can be broken down further to release energy. Only the carbon-containing (acid) portion of the amino acid molecule, however, is shunted through the energy-producing pathway (review Figures 24.2 and 5.15). The nitrogen-containing amino (NH_2) portion cannot be oxidized and becomes a waste product. The nitrogen-containing leftovers join with hydrogen atoms and become molecules of ammonia (NH_3), the strong-smelling alkaline ingredient often used in cleaning solutions.

Even at low concentrations, ammonia is highly toxic to cells, and all organisms must get rid of it quickly. Some organisms excrete ammonia directly, while others excrete the less toxic compounds **uric acid** and **urea** (Figure 25.4). Biologists divide organisms into three groups, depending on whether they excrete nitrogenous wastes as ammonia, uric acid, or urea.

■ **Ammonia As Waste: Dilution Is the Solution**   Removing nitrogenous wastes is not a serious problem for many aquatic organisms, because the ammonia can be dumped into the surrounding water, where it is diluted so much that it becomes nontoxic. Most bony fishes and aquatic invertebrates excrete most of their nitrogenous wastes as ammonia, often by secreting it across their gill surfaces into the vast quantities of surrounding water.

■ **Uric Acid: Nitrogen Wastes in Solid Form**   When a bird's cells produce ammonia from protein metabolism, the cells export the waste product, which then enters the blood, removing the danger to cells but creating the new problem of ammonia circulating in a limited volume of blood. A dangerous buildup is prevented because the animal's liver cells convert ammonia into the less toxic uric acid. This relatively insoluble substance continues to circulate suspended (not dissolved) in the blood, like sand washed along in a river, and is gradually removed by the kidneys. From here, the uric acid moves down narrow tubules called ureters that drain the kidneys, enters a single exit duct called the cloaca, and is expelled from the body as an insoluble white paste called guano. The white por-

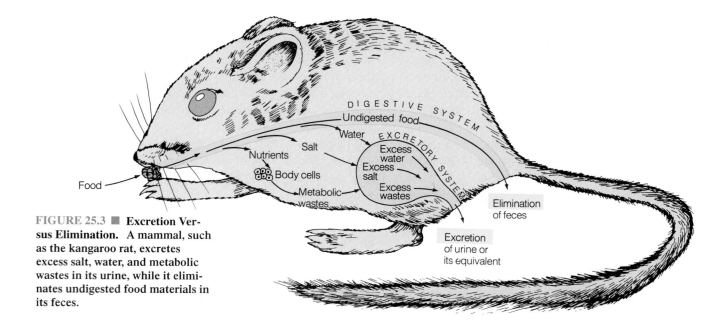

FIGURE 25.3 ■ **Excretion Versus Elimination.** A mammal, such as the kangaroo rat, excretes excess salt, water, and metabolic wastes in its urine, while it eliminates undigested food materials in its feces.

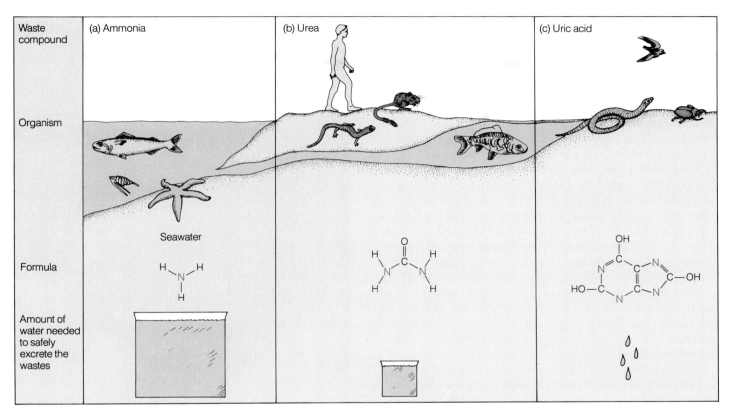

FIGURE 25.4 ■ **Three Kinds of Nitrogenous Waste Products.** (a) Some aquatic organisms, including protozoa, sea stars, and certain fishes, excrete nitrogenous wastes in the form of ammonia dissolved in large quantities of water. (b) Others, including amphibians, some fishes, and most mammals, excrete nitrogenous wastes as urea in a smaller amount of water. (c) Still others, including birds, reptiles, and insects, excrete the wastes in a paste of uric acid crystals containing just drops of water.

tion of the ubiquitous bird droppings that decorate window ledges, park statues, and the occasional human head or shoulder is expelled uric acid. In regions where large colonies of seabirds nest, whole islands can be whitened with a thick layer of guano (Figure 25.5).

Using uric acid as an excretory product makes it easy for birds to conserve water, since this relatively insoluble nitrogenous waste can pass out of the body as a paste containing little water. Uric acid produced by a chick developing within its shell can also be stored as a precipitate in almost solid form

FIGURE 25.5 ■ **Islands of Guano.** This island in the Pacific is black volcanic rock below and white guano above, the guano consisting of uric acid deposited by generations of cormorants, sea gulls, and other nesting seabirds. People sometimes harvest the nitrogen-rich guano from such islands and process it into fertilizer.

in a special part of the egg away from the embryo. Without an effective form of waste disposal like this, hard-shelled eggs might never have evolved. Reptiles, insects, and land snails also rely on uric acid excretion to rid themselves of nitrogenous wastes—a means of excretion that doubles as a water conservation strategy in both eggs and adults.

■ **Urea: Excreted by Mammals and Some Fishes**   A few kinds of fishes, as well as people, kangaroo rats, and other mammals, produce ammonia during protein metabolism, just as birds, reptiles, insects, and other animals do. Evolution, however, has provided them with a third solution to the problem of ammonia: the combining of ammonia and carbon dioxide to form urea. But just as dirty dishwater must go through sewer pipes to a central water treatment facility for final disposition, so must the less toxic, water-soluble urea be carried in the blood vessels to the kidneys for the formation and excretion of urine. A cell's nitrogenous wastes are treated in a multistep process: First, the ammonia is detoxified by conversion to urea in the liver; next, the urea is removed from the blood by the kidney; and finally, urea is concentrated in the urine and excreted from the body through a urinary plumbing system we will consider shortly.

Two exceptions to the mammal's normal urea pathway explain both human gout and a hibernating bear's lack of urination in winter. In humans, nitrogen from protein breakdown is carried off as urea, but nitrogen-containing waste products from nucleic acids are converted into uric acid. In some people, uric acid crystals precipitate out and form deposits in the big toe, elbow, or other joints and cause the excruciatingly painful condition known as **gout.** Medicines like colchicine can help relieve gout symptoms, and so can eating fewer meats that are high in nucleic acids, such as organ meats.

Figure 25.6 describes a special adaptation in American black bears that allows these hibernating animals to rid themselves of urea all winter long without awakening to urinate.

In bears, people, kangaroo rats, and all other vertebrates, the tasks we have been describing fall to the kidneys. Despite its complexity, however, the kidney achieves the same end as simpler excretory organs: Excretory organs rid animal bodies of nitrogenous wastes and maintain the salt and water levels within body fluids and thus within cells. Our next section describes how the kidney works as a biological jack-of-all-trades in waste excretion and salt and ion balance.

THE KIDNEY: MASTER ORGAN OF WASTE REMOVAL, WATER RECYCLING, AND SALT BALANCE

For many diners, eating tender, pale green shoots of asparagus is a pleasurable springtime event, but the gastronomic experience has a peculiar sequel: The next time they urinate, even

FIGURE 25.6 ■ **A Use for Nitrogenous Wastes.** The bladder (a urine-storing organ) of a hibernating bear absorbs urea, and the waste passes into the blood, where it is carried to the gut; there, bacteria break down the urea and use its nitrogen to make amino acids. The sleeping bear's body can then reabsorb and use the nitrogen in the amino acids to build new proteins and nucleic acids, thereby recycling the wastes into useful materials. This novel mechanism allows bears to hibernate all winter without a large waste buildup.

if just 20 minutes after eating, they notice the characteristic scent of asparagus. An inherited dominant allele causes most asparagus eaters to excrete an odorous sulfur-containing compound from the vegetable. A biochemical in the food seems to cross the gut, enter the bloodstream, be filtered out by the kidneys, and appear in the urine with amazing speed. Actually, the biochemical is acted on no faster than any other compound. The kidneys are simply marvels at processing body fluids and filtering out the urea, the sodium, potassium, and chloride ions, and the glucose, water, and other materials that need to be excreted to provide homeostasis. The key to a kidney's rapid functioning lies in its complicated internal structure and in the efficient plumbing system of which it is a part. The human kidney serves as a good model for kidney structure and function in vertebrates.

The Kidneys and Other Organs of the Human Excretory System

It is sometimes stated that micturition (the voiding of urine) is our most compelling bodily function. You will probably agree if you have ever been hungry, tired, and cold and had a full bladder simultaneously and can recall the order in which you addressed those needs. This urgency has to do with the necessary waste removal role of the excretory system and with its particular anatomy.

In humans, as in other vertebrates, the main organs of the excretory system are the **kidneys.** A person's two plump, dark red, crescent-shaped kidneys are each about the size of an adult's fist and are located in the abdominal cavity just below and behind the liver (Figure 25.7). For each kidney to carry out its vital blood-filtering activities, it must receive a large, steady flow of blood. This it does through the **renal arteries** (from the Latin *renes,* meaning "kidneys"), twin branches directly off the body's main blood vessel, or aorta. Cleansed blood then leaves each kidney in a large **renal vein** that drains into the inferior vena cava, the body's largest vein, which

leads directly to the heart. As the kidneys filter and remove excess water and waste substances from the blood, these materials collect as a concentrated urine in a central cavity in each kidney, then flow down a long tube called the **ureter.** The two ureters dump urine into a single storage sac, the **bladder,** which in an adult can hold about 500 ml (1 pt) of fluid. The bladder can distend to hold a bit more, but when it is full, it causes considerable pressure on nearby nerves and a feeling of urgency that can temporarily eclipse all other concerns. During **urination,** the urine eventually exits the body through the **urethra,** the single tube that drains the bladder.

A lamb kidney bought from the butcher and cut in half lengthwise looks a lot like a human kidney and reveals the anatomical features that enable the organ to filter water and solutes from the blood and form urine. A mammal's kidney has three distinct visible zones (Figure 25.8a). First, there is an outer renal **cortex** zone, a bit like a thick orange peel, where initial blood filtering takes place. Next there is a central zone, or renal **medulla,** divided into a number of fan-shaped regions (Figure 25.8b). The medulla helps conserve water and valuable solutes. The functional units of the kidney, twisted tubules called **nephrons,** reach from the cortex down into the

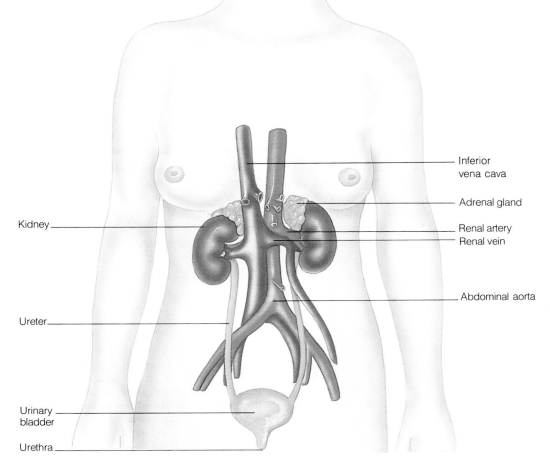

FIGURE 25.7 ■ **The Human Excretory System.** The kidney, bladder, drainage tubes, and major blood vessels make up the simple but critically important excretory system, with its roles in waste removal and control of the body's internal fluids.

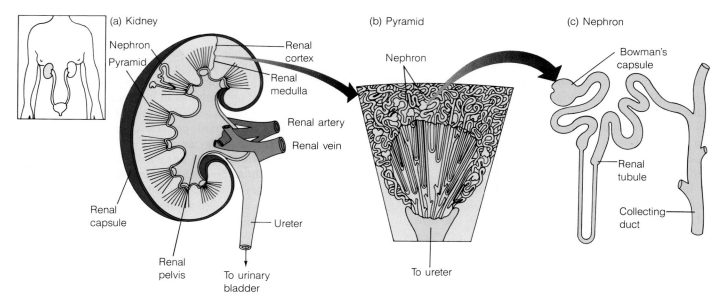

FIGURE 25.8 ■ Anatomy of the Human Kidney: A Blood-Cleansing Organ. (a) A cross section of the kidney reveals the basic anatomy. (b) A pyramid consists of many nephrons. (c) Each nephron is a twisted tubule that filters the blood.

medulla (Figure 25.8b and c). Finally, inside of these two zones lies the funnel-shaped, hollow inner compartment, or renal **pelvis,** where urine is stored before it passes into the ureter.

The Nephron: Working Unit of the Kidney

Have you ever cleaned out a desk drawer—sorting, discarding, saving certain items for future use, and then rechecking the drawer one last time? If so, you have approximated the kidney's general sorting and cleansing action and its three basic processes—filtration, reabsorption, and secretion—all of which involve the nephron (Figure 25.9). During **filtra-**

tion, one part of the nephron removes small molecules (glucose, amino acids, ions, water, urea, and so on) from the blood plasma, while leaving large protein molecules and blood cells behind in the blood. During **tubular reabsorption,** a different part of the nephron tubule sorts the small molecules, shunting the useful solutes back to the blood and retaining the wastes. Finally, during **tubular secretion,** regions of the nephron check the blood supply one last time and remove from circulation any ions, drugs, or other wastes that still remain and secrete them into the forming urine.

Each of the kidney's functional units, or nephrons, is a long, twisted, looping tubule. One end of the nephron is enlarged and cup-shaped like a punched-in basketball and lies

1. *Filtration:*
   Water and other small molecules filtered into nephron

2. *Tubular reabsorption:*
   Water, salts, and nutrients returned to blood

3. *Tubular secretion:*
   Some ions and drugs secreted into nephrons

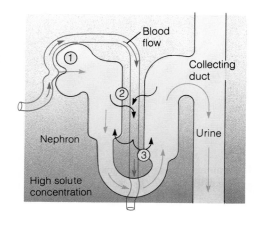

FIGURE 25.9 ■ Filtration, Reabsorption, Secretion: Basic Roles of the Nephron. Water and solutes like urea are filtered from the blood and enter the nephron at the Bowman's capsule (step 1). As the filtrate flows through the nephron's looped and twisted tubule, much of the water and valuable solutes are reabsorbed into the bloodstream (step 2). The gradient of salt concentration, here denoted by deepening shades of blue, is mainly responsible for the reabsorption of water. Simultaneously, regions of the nephron secrete various wastes, drugs, and other materials into the urinary fluid (step 3). Concentrated urine then passes through the collecting duct, renal pelvis, and ureters to the bladder.

in close contact with blood capillaries. The other end of the nephron drains away urine (Figure 25.10a). A human kidney contains roughly 1 million nephrons, each extending from the outer cortex through the medulla and draining into the renal pelvis. If all the nephrons in an adult's kidney were straightened out and placed end to end, they would form a microscopically slender tube about 80 km (49½ mi) long.

The nephron's specialized shape enables it to carry out filtration, reabsorption, and secretion in sequence and with admirable efficiency. The portion of the nephron that filters water and solutes from the blood is the enlarged, cuplike **Bowman's capsule.** Each Bowman's capsule is embedded in a matrix of tissue in the kidney's outer cortex, and it surrounds a tight bundle of capillaries called a **glomerulus,** much the way the fingers of your right hand can wrap around your left

fist (Figure 25.10a). This enclosing capsule of kidney tissue surrounding a tuft of blood capillaries forms the closest contact between the excretory and circulatory systems.

The fist-and-hand metaphor is almost literal, because the cells lining the Bowman's capsule, the so-called *podocytes,* have large central cell bodies with thousands of fingerlike projections that enclasp the capillaries completely (Figure 25.10b). Directly under the "fingers" of the podocyte is a basement membrane, a meshwork of protein fibers. Below that are the cells lining the blood capillary. These capillary wall cells are pierced with minute holes, and as blood moves along through the capillaries in the tuft, fluid from the plasma percolates through this delicate filter and into the Bowman's capsule. The fineness of these pores is ultimately responsible for the nephron's blood filtration activities: The Bowman's

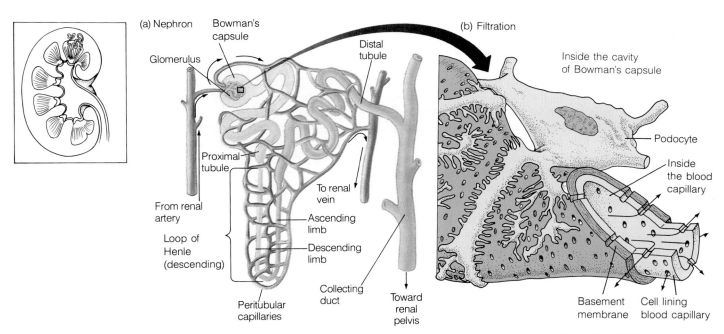

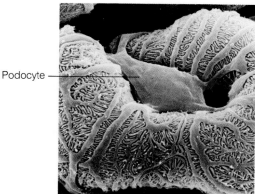

FIGURE 25.10 ■ **The Nephron: Twisted Tubule at the Center of the Kidney's Function.** (a) Each human kidney contains 1 million nephrons, and each nephron has a Bowman's capsule, a proximal tubule, an elongated loop of Henle, a distal tubule, and a collecting duct leading toward the renal pelvis. An arteriole enters the glomerulus (tuft of capillaries) inside the Bowman's capsule, and another arteriole leaves the glomerulus for the peritubular capillaries, which join to form the venule leaving the nephron. A portion of a capillary in the Bowman's capsule is enlarged in (b). The remarkable podocyte cells enclasp capillaries. The cells that line the walls of the capillaries are very thin and are punctured by tiny pores. Water, ions, and small organic molecules such as sugars filter through these pores, then percolate through the feltlike mesh of the basement membrane and into the kidney tubule. This scanning electron micrograph shows a podocyte cell enclasping a capillary (magnification about 2000 ×).

capsule allows wastes, water, and other small molecules to seep through these minute openings, while the crucial components of blood—red cells, antibodies, and other large proteins—remain behind in the capillaries. While the Bowman's capsule filters the blood, keep in mind that the blood is just a necessary intermediary between the important fluid inside the cells, the extracellular fluid, and the excretory system.

Follow the pathway of the blood in Figure 25.10a. The incoming arteriole branches into the glomerular tuft of capillaries. The glomerular capillaries do not merge directly into veins, but instead rejoin into an exiting arteriole that carries blood into a network of capillaries that twists about the outside of the nephron tubule's looped portion. This network is the **peritubular** ("around the tubule") **capillaries.** These capillaries finally merge into venous capillaries and then into a larger venule that connects to the renal vein; all these venous vessels carry blood away from the kidney.

While the Bowman's capsule is responsible for filtration, the looped part of the nephron, which has several sections, is responsible for reabsorption and secretion (described in detail later). Nearest the Bowman's capsule, the **proximal** ("near") **tubule** meanders through the kidney's cortex like a jumbled length of yarn (see Figure 25.10a). At the inner edge of the cortex, this tubule becomes thin-walled, straightens out, and dips sharply down into the medulla, then makes a hairpin curve and heads back up into the cortex. The U-shaped section of the nephron is known as the **loop of Henle** and is chiefly responsible for reabsorption of materials into the blood. The arm of the loop leading down into the medulla is the **descending limb,** while the portion leading back into the cortex is the **ascending limb.** The ascending limb leads to the part of the nephron farthest from the Bowman's capsule, the **distal** ("far") **tubule,** which is involved mainly in tubular secretion. Like the proximal tubule, the distal tubule winds about. Groups of distal tubules from various nephrons connect to large **collecting ducts,** which drain into the renal pelvis and carry away urine. Every 10 or 20 seconds, a small amount of urine collects in this hollow inner compartment, and peristaltic waves of the ureter sweep this urine away to the bladder for storage and eventual excretion through the urethra.

The Nephrons at Work

Nephrons are amazingly efficient at selectively removing wastes from the blood circulation while simultaneously conserving water, mineral ions, glucose, and other needed materials. In an adult human, 180 L (48½ gal) or more of blood—enough to fill a bathtub—pass through the 2 million nephrons in the two kidneys each day. At any given time, about one-fifth of the blood volume is passing through the blood vessels to and from the kidneys. We do not, of course, urinate 180 L a day, but instead the much more reasonable amount of about 1.5 L (3 pt) per day. Clearly, the nephrons accomplish a great deal of water conservation before they produce that smaller quantity of urine. Let us now look at each renal process in more detail to see how a nephron works.

■ **Filtration**   In this first stage of urine formation, blood pressure forces the plasma, or yellowish fluid portion of blood, out of the capillaries, through the tiny pores beneath the podocyte cells, and into the Bowman's capsule, just as high water pressure forces water out of a sprinkler. This transport of water, of sodium, potassium, and chloride ions, and of sugars, amino acids, and urea out of the capillaries and into the lumen of the Bowman's capsule is passive and does not require any special output of energy other than blood pressure. The fluid, or **filtrate,** in the capsule is still very much like blood plasma, except that it contains no large proteins (Figure 25.11).

■ **Tubular Reabsorption**   As soon as the filtrate enters the nephron's twisted proximal tubule, reabsorption begins and returns to the circulation most of the water, the sodium and chloride ions, the sugars, and the amino acids that were just filtered out of the blood in the Bowman's capsule (see Figure 25.11). These materials are moved out through the walls of the proximal tubule and pass back into the blood plasma of the peritubular capillaries entwining that proximal tube (review Figure 25.10). Instead of being driven passively by blood pressure, reabsorption of solutes in the proximal tubule depends on energy-costly active transport across the plasma membranes of nephron cells. As the ions are being reabsorbed in this way, 80 to 85 percent of the water in the original filtrate follows them passively back into the capillaries via the process of osmosis (see Chapter 4).

Next, the filtrate—minus most of its water but still containing urea and salts—passes down the descending arm and up the ascending arm of the loop of Henle. Here the cells of the ascending arm pump chloride ions (Cl^-) into the extracellular fluid surrounding the tissue cells of the kidney's medulla, and sodium ions (Na^+) follow passively. The result of this ion flux is that the innermost part of the medulla becomes very salty compared to the upper part. This means that the concentration of Na^+ and Cl^- is lowest in the cortex and grows increasingly higher toward the inner part of the medulla, the part containing the hairpin curve of the loop of Henle (see Figure 25.11). Since the tissue surrounding the loop is salty, water diffuses out of the loop's descending limb by osmosis and reenters the blood of the peritubular capillaries. This exodus of water ends, however, as the filtrate rounds the bend and moves up the ascending limb, because that part of the tube is not permeable to water (see Figure 25.11). From the ascending limb, the filtrate, still containing urea, some salt, and some water, passes toward the collecting duct. In that duct, once again, the walls are permeable to water; even though the filtrate passing through is already quite concentrated, the surrounding tissues are even saltier. Thus, more water diffuses out of the collecting ducts by osmosis and enters the blood of the peritubular capillaries.

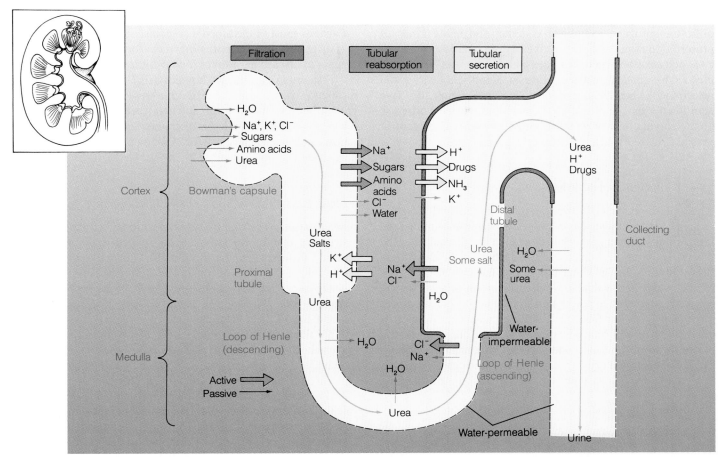

FIGURE 25.11 ■ How Materials Move Into and Out of the Nephron As Urine Forms. As a nephron extends through the kidney's cortex and medulla and dumps urine into the collecting duct, various substances enter and leave the filtrate. Broken lines represent segments of the nephron wall that are permeable to water, while solid lines represent wall segments impermeable to water. Narrow arrows represent passive diffusion of materials into or out of the nephron tubule, while wide arrows represent active transport against concentration gradients. Filtration activities are shown in blue, tubular reabsorption activities in green, and tubular secretion in yellow. Urine is shown as yellow. The text traces nephron function and material movements step by step.

By the time the filtrate (now called urine) has reached the part of the collecting tubule in the innermost (and saltiest) region of the medulla, much of the water has been reabsorbed. In fact, 99 percent of the water originally filtered from the blood in the Bowman's capsule has by now been returned to the body's circulation, and the urine has a high concentration of wastes relative to water content.

■ **Tubular Secretion** While the filtration and reabsorption activities of the kidneys' millions of nephrons are removing salt and urea from the blood and forming urine, another process—tubular secretion—goes on simultaneously in both the proximal and distal tubules (see Figure 25.11). Nephron cells remove a number of unwanted materials—including some hydrogen and potassium ions, ammonia, some drugs like pen-

icillin and phenobarbitol—from the blood in the peritubular capillaries surrounding the nephron and secrete them into the forming urine (see Figure 25.11). The walls of the collecting duct allow a certain amount of urea to pass back out rather than being excreted in the urine. This urea increases the brine-like concentration in the inner medulla and causes still more water to be removed and conserved. Besides cleansing the blood, tubular secretion can also help maintain an appropriate pH level in the blood (pH 7.3 to 7.4), since nephron cells secrete more hydrogen ions into the urine if the blood is too acidic and fewer hydrogen ions if the blood is too alkaline.

Tubular secretion is the physiological process that makes **drug testing** possible—checking a person's urine to see if he or she has taken drugs. Drug testing employs various laboratory techniques to detect even minute traces of the metabolic breakdown products of marijuana, cocaine, heroin, sleeping pills, tranquilizers, morphine, codeine, and many kinds of prescription drugs. If a person takes a drug overdose and loses consciousness and no one is sure which drug was consumed, physicians quickly test the urine to identify the drug and determine the best treatment for saving the patient's life. Two additional uses—testing athletes and employees for drug use on the playing field or on the job—are currently quite controversial (Figure 25.12).

Kidney Function: A Quick Review

Although the details of filtration, reabsorption, and secretion seem complex, these three processes can be summarized fairly simply. Water and solutes like urea are filtered from the blood capillaries and enter the kidney tubule. As the filtrate flows through the proximal tubule and loop of Henle, much of the water and valuable solutes are reabsorbed into the bloodstream owing to active transport of ions and the passive movement of water from the filtrate by osmosis. Simultaneously, the nephron removes various wastes, drugs, and other materials from the blood and secretes them into the urinary fluid. Concentrated urine then passes into the collecting duct, the renal pelvis, and then via the ureters to the bladder.

FIGURE 25.12 ■ **Kidney Function Makes Drug Testing Possible.**
The night before the 400 m freestyle event at the 1972 Olympics in Munich, Germany, swimmer Rick DeMont took a medication that his doctor prescribed containing ephedrine to counter DeMont's asthma. Although he won a gold medal for the swimming event, officials demanded that the athlete return his award because a test of his urine revealed the ephedrine, and the drug was on a list of banned substances for Olympic competitors. It is unlikely that the drug gave DeMont an advantage over other swimmers—except to take away his wheezing. He was, in effect, a victim of tubular secretion by his own kidneys. Here, DeMont (left) stands by the silver medalist; both represented the U.S. swim team.

Returning to our drawer-cleaning metaphor, the kidney removes most of the blood's contents, sorts and returns much of it to the blood circulation, then sorts the contents once again to discard harmful or redundant items.

An organ as central to homeostasis as the kidney is impossible to live without and difficult to replace. People with kidney diseases often spend many hours each week connected to a dialysis machine, a large instrument that mimics the filtration of a healthy kidney. Box 25.1 on page 529 describes this machine and how it works.

Kidneys Adapted to Their Environments

So close is the association between a kidney's function and its structure that biologists have discovered nephrons of distinctly different lengths in animals with markedly different life habits. A kangaroo rat in the desert needs to conserve water to an extreme degree, and each of its nephrons possesses a very long loop of Henle (Figure 25.13a). With this modification, its kidneys can reabsorb practically all the water as the filtrate moves through the long loop, and as a result, the animal can produce urine 25 times more concentrated than its own blood—the highest concentration of any mammal.

In contrast, an aquatic animal like a beaver takes in a great deal of water in its food and through its skin and must get rid of the fluid, not conserve it. In a beaver's kidney, the loops of Henle are short and reabsorb less water; as a result, the animal produces great quantities of urine with only twice the solute concentration of its own blood (Figure 25.13b). Humans have a combination of long and short loops of Henle and produce a urine of variable concentration, depending on water availability in the person's environment (Figure 25.13c).

REGULATING THE BODY'S WATER CONTENT: THIRST AND HORMONES

While the kangaroo rat and the beaver must deal with extremely dry or wet environments, other animals must cope with surroundings that fluctuate regularly—or irregularly—from wet to dry. An impala living on the plains of East Africa, for example, experiences distinct wet and dry seasons each year, while a person working out at a health club could inhabit a number of environments—an air-conditioned room; a warm, dry tennis court; a hot and humid steam room; a swimming pool. Although the environment outside an animal may change dramatically, the salt concentrations, pH, and waste concentrations inside the animal's cells cannot vary; otherwise, the proteins of the cell will cease to function. Therefore, the kidneys must be able to conserve water during the dry season or the tennis game but quickly remove it after a drink from a water hole or a glass of iced tea. Indeed, animals have two major mechanisms for regulating the activities of the kidneys and hence the water balance of the body: *thirst,* or the

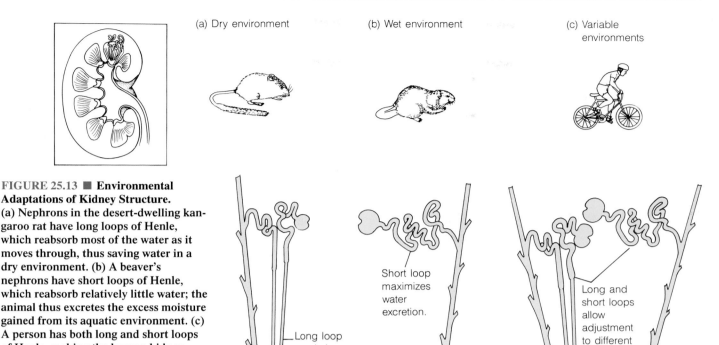

FIGURE 25.13 ■ Environmental Adaptations of Kidney Structure. (a) Nephrons in the desert-dwelling kangaroo rat have long loops of Henle, which reabsorb most of the water as it moves through, thus saving water in a dry environment. (b) A beaver's nephrons have short loops of Henle, which reabsorb relatively little water; the animal thus excretes the excess moisture gained from its aquatic environment. (c) A person has both long and short loops of Henle, making the human kidney capable of modulating water reabsorption and excretion, depending on the external environment.

desire to drink, which affects the amount of water taken in, and chemical messengers called **hormones** that alter the activities of the nephrons and hence the amount of water retained or excreted at any given time.

Thirst Regulates Drinking and Helps Maintain Water Balance

Although the kangaroo rat never needs to drink fresh water, most terrestrial organisms do. And even the loss of a small proportion of the body's fluid content—say, 1 to 2 percent from sweating or evaporation through the lungs during a tennis game on a hot day—can cause the concentration of solutes in the blood to rise. These rising levels trigger nerve cells in the brain's thirst center, located in the hypothalamus, and the individual becomes aware of thirst. When a person responds by drinking a large amount of water, the water distends the stomach walls and sets up nerve impulses that inhibit the thirst center even before the water is absorbed. Once the water enters the bloodstream from the stomach and small intestine, osmotic balance is restored and the concentration of solutes in the blood and inside body cells stabilizes until more water is consumed in foods or beverages or until sweating, exhaling, and forming urine remove fluids along with wastes.

Hormones Acting on Nephrons Control the Excretion of Water

Just as you can turn a water faucet on or off to increase or decrease water flow, your hormones can regulate the activity of nephrons so that they remove more water or return more to the blood. Recall that water reabsorption from the nephrons depends on (1) the passage of water across specific regions of the nephron tubule's walls, and (2) the salt concentration of the extracellular fluid in the kidney's medulla. Two different hormones act on these two factors to control water loss. A hormone called **antidiuretic hormone** (ADH, also called vasopressin) regulates the permeability of the distal tubule to water, while the hormone **aldosterone** regulates salt reabsorption.

■ **Antidiuretic Hormone**   Antidiuretic hormone is aptly named, because it prevents *diuresis,* the production of copious amounts of urine. ADH controls how much water the nephron reabsorbs from the filtrate and returns to the blood. ADH is produced in the brain region called the hypothalamus and is stored in a nearby region, the pituitary gland (Figure 25.14a). When the concentration of salt and other solutes rises—say, after a very salty snack of sardines and potato chips—the

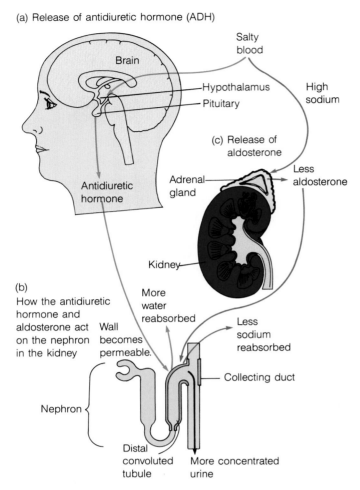

(a) Release of antidiuretic hormone (ADH)

Salty blood

Brain

Hypothalamus
Pituitary

High sodium

(c) Release of aldosterone

Antidiuretic hormone

Less aldosterone

Adrenal gland

Kidney

(b)
How the antidiuretic hormone and aldosterone act on the nephron in the kidney

More water reabsorbed

Less sodium reabsorbed

Wall becomes permeable.

Collecting duct

Nephron

Distal convoluted tubule

More concentrated urine

FIGURE 25.14 ■ How Hormones Regulate Salt and Water Balance. (a) Salty blood sensed by the brain's hypothalamus causes the release of antidiuretic hormone (ADH), which in turn (b) acts on the distal convoluted tubule and collecting duct, making them more permeable to water. Water then leaves the forming urine and reenters the blood. This conservation causes the release of more concentrated urine and helps prevent the blood from becoming saltier. When levels of salt in the blood decrease, ADH release is shut down, completing the negative feedback loop. (c) Salty blood also acts indirectly on the adrenal glands, decreasing the secretion of aldosterone and causing the distal tubule cells to reabsorb fewer sodium ions.

hypothalamus detects the changes and causes the pituitary gland to release some of its stored ADH into the blood. Upon reaching the kidney, ADH makes the walls of the distal tubule and collecting ducts temporarily more permeable to water, so more water is reabsorbed into the bloodstream, diluting the concentration of blood solutes. Since water is drawn from the filtrate into the blood, the urine becomes very concentrated and the blood becomes less salty. Finally, less salty blood causes the brain's secretion of ADH to decrease, thus completing a negative feedback loop.

As beer drinkers occasionally discover to their dismay, alcohol inhibits ADH secretion, leading to diuresis and the excretion of sometimes embarrassing quantities of urine. This causes dehydration and is a major reason for the "hangover" a person feels after drinking too much liquor; the body is dehydrated and must replace lost fluid.

■ **Aldosterone**   ADH affects reabsorption of water, and another hormone, aldosterone, affects absorption of salt. When the blood's concentration of sodium ions is high and the concentration of potassium ions is low, the adrenal glands (each of which sits on top of a kidney) decrease their secretion of aldosterone (Figure 25.14c). With less aldosterone, the distal tubules of the kidneys' nephrons reabsorb fewer sodium ions from the filtrate, and the urine becomes more salty. Aldosterone acts in an opposite way on potassium ions. Through this mechanism, sodium in the blood remains within the narrow range that is best for the activity of nerve cells and other crucial physiological activities.

The Kidneys' Role in Heart Disease

In recent years, physicians have warned people that diets high in sodium can elevate the blood pressure and strain a heart already weakened by atherosclerosis (see Chapter 21). The kidneys can counteract high salt content in the blood from a salty diet by excreting less water. This causes the total volume of blood in the circulatory system to increase and the blood pressure to go up, forcing the heart to work harder. The kidneys can also counteract falling blood pressure. When blood flow to certain cells near the glomerulus in the Bowman's capsule drops below a certain volume, the cells release the enzyme **renin,** which converts a blood protein into the hormone **angiotensin;** this hormone, in turn, causes blood vessels to constrict, thereby elevating blood pressure the way that squeezing on a water balloon increases pressure at the opposite end. Angiotensin also triggers the adrenal glands to secrete aldosterone, which results in sodium and water reabsorption and thus even higher blood volume and pressure.

Recently, physiologists have discovered yet another hormone—this one secreted by the heart itself—that helps keep blood pressure in check. When the blood volume and pressure increase, cells in the heart's right and left atria secrete *atrial natriuretic factor* (ANF), a hormone that causes sodium (natrium) to appear in the urine. Water follows the sodium by osmosis, thus increasing the quantity of urine, lowering the blood volume, and hence decreasing the blood pressure. Geneticists have recently isolated the human gene for ANF and hope to produce large quantities of it using recombinant DNA techniques (see Chapter 11) to treat people with chronic high blood pressure.

The message from this section is simple: Thirst and hormones regulate how much water we drink and how much salt and water we retain, and this, in turn, affects blood content, blood pressure, and other factors vital to homeostasis and health.

BOX 25.1 — HOW DO WE KNOW?

WHEN THE KIDNEY FAILS

It is easy to take the kidneys for granted: They operate silently, efficiently, and in most people continuously for a lifetime without a glitch. One only need learn about the consequences of kidney failure, however, or meet a victim of this condition, to understand how central the organs are to survival and just how difficult it is to imitate their natural functions.

The kidneys can lose their exquisite ability to cleanse the blood in several ways. Bacteria from feces can contaminate the urinary tract, attacking the kidneys. An autoimmune attack (see Chapter 22) on the kidneys by white blood cells can block and destroy glomeruli. Finally, poisoning by mercury, lead, or certain solvents can damage the kidney tissue. A person's nephrons are so numerous and so efficient that even if two-thirds of these tubules are destroyed, the individual can still live a fairly normal life. If the number drops to 10 or 20 percent, however, the person can suffer extreme tissue swelling as a result of salt and water retention, as well as a buildup of urea, hydrogen and potassium ions, and other metabolic by-products. These cause the blood to become very acidic and can lead to coma or—if the pH drops below 6.9—death.

Fortunately, biomedical researchers in the mid-twentieth century invented an artificial kidney, or *kidney dialysis machine,* which works via simple diffusion to take over some of the kidney's blood-cleansing functions (Figure 1). When the machine is turned on, waste-laden blood from the patient's artery is routed through a long, porous membrane bathed by a solution much like normal blood plasma. As it passes through, the wastes diffuse into the solution (from the region of higher waste concentration to the region of lower waste concentration). After circulating several times through many meters of tubing and back into the body, the patient's blood is sufficiently free from wastes to permit normal activity—at least for a while.

Unfortunately, the artificial kidney has some serious drawbacks: It carries out filtration but neither the reabsorption nor the secretion activities of a living kidney. Moreover, the patient's blood must be treated with an anticoagulant so that it

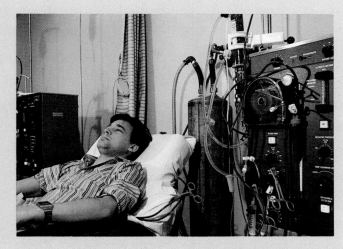

FIGURE 1 ■ **Kidney Dialysis Machine and Patient Having Blood Cleansed.**

does not clot as it passes through the machine, then treated again with a coagulant as it reenters the person's body so that he or she won't bleed too freely. Because of the drug treatments, the artificial kidney can only be used every two or three days, and because of the slowness of waste diffusion, each session may last from 5 to 10 hours. Also, the use of anticoagulants allows infections to occur with distressing frequency. People with kidney failure are literally captives of the dialyzer: It extends their life but diminishes the quality of it and gobbles much of their time.

Kidney transplants are an alternative to this captivity. Donors are limited to blood relatives or others with closely matching tissue types (see Chapter 22). And even with this careful screening, the patient may need long-term drug therapy to suppress immune rejection. Thousands of people are saved each year by dialysis machines and transplants. Clearly, though, healthy kidneys are nothing to take for granted.

STRATEGIES FOR SURVIVAL: HOW ANIMALS BALANCE SALT AND WATER

"Marooned at sea" is a favorite literary theme, and as Samuel Taylor Coleridge wrote in "The Rime of the Ancient Mariner," a sailor in this situation has "water, water everywhere, nor any drop to drink." Why can't a thirsty sailor simply drink seawater? The answer lies with the salt and water balance we have been discussing: Human kidneys cannot make urine as salty as seawater. For every 1 L (1 qt) of seawater humans consume, we produce 1.3 L (2¾ pt) of urine and quickly dehydrate. If a sailor had a pet kangaroo rat, the animal could easily drink from the ocean, since its kidneys make urine that is more concentrated than seawater. A shark, flounder, or sea gull can also drink seawater. Animals face different osmotic challenges, depending on their natural environments, and have evolved different kinds of physical adaptations for dealing with them. Let's look at some of the strategies that animals use on land, at sea, and in fresh water for **osmoregulation:** the regulation of osmotic balance.

Osmoregulation in Terrestrial Animals

On land, animals are submerged in an ocean of air and thus tend to lose water through evaporation across the body and respiratory surfaces, as well as through excretion of urine or its equivalent. The result, of course, is the need to drink water; the need to dump salt in the sweat and urine and sometimes through special salt glands; and finally, the need to conserve water through physical and behavioral mechanisms. An insect's waxy cuticle, for example, helps hold in water, as do

(a) A scaly skin retains moisture.

(b) White robes help slow evaporation.

FIGURE 25.15 ■ Land Animals: Body Coverings and Water Regulation. (a) This prehistoric-looking lizard of Costa Rican forests has a scaly hide that helps the animal retain much of its body moisture. (b) A Bedouin's long white robes and the camel's thick, light-colored fur block out heat and prevent excess evaporation as they rest in Egypt's Sinai desert.

FIGURE 25.16 ■ Brine Shrimp: Versatile Osmoregulators. (a) Brine shrimp in dilute seawater counter salt loss by actively uptaking salt across the gills and prevent waterlogging by expelling large quantities of water in the urine, owing to the action of the excretory gland. (b) The same shrimp can live comfortably in very salty water because the highly efficient gills act in reverse, pumping out excess salt, and the excretory gland now expels very little water.

a reptile's dry, scaly skin (Figure 25.15a), a bird's feathers, and a mammal's fur. The moist skin of an earthworm, snail, or amphibian, however, is subject to tremendous evaporative water loss, and so is the skin of a furless land mammal such as a human. Amphibians tend to remain in cool, wet places, and in many species, the bladder acts as a reservoir from which water can be removed to replenish body fluids when needed. Desert peoples tend to wear clothing that minimizes evaporation (Figure 25.15b), and they rest at midday, when the temperature is highest. Virtually all terrestrial animals choose home territories with access to dependable sources of water, and all varieties of animals have organs for balancing and excreting salt and water: the vertebrate's kidney, the insect's Malpighian tubules, the earthworm's nephridia, and so on (see Chapter 18). The kangaroo rat, with its seven adaptations for conserving water and excreting salt and urea (review page 516), is a prime example of osmoregulation on land.

Osmoregulation in Aquatic Animals

Aquatic animals are surrounded by water, not air, and for them, evaporation is usually not a major factor. They do, however, face a continual problem of osmoregulation stemming from the difference in salt concentration between their blood (and other body fluids) and the surrounding water. If the aquatic environment is hypertonic relative to (saltier than) their internal fluids (see Chapter 4), water tends to diffuse outward and leave them dehydrated. If the aquatic environment is hypotonic relative to (less salty than) their blood, water tends to diffuse in and "waterlog" them while solutes move out of their bodies toward regions of lower concentration. Various adaptations have evolved in different groups for excreting water, salt, or both.

■ **Aquatic Invertebrates** The easiest solution to osmoregulation is to have body fluids that are isotonic relative to (the same concentration as) the surrounding water, and that is exactly the situation with invertebrates such as oysters and sea stars. Biologists call these animals *osmoconformers;* their body fluids conform osmotically to their surroundings, whether salty or dilute. Other marine invertebrates, such as

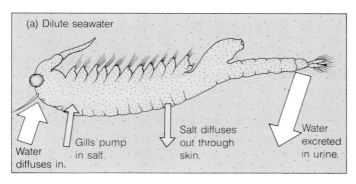

(a) Dilute seawater

Water diffuses in.   Gills pump in salt.   Salt diffuses out through skin.   Water excreted in urine.

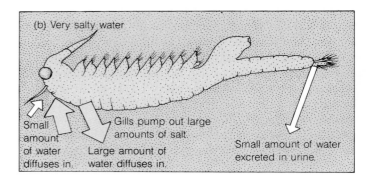

(b) Very salty water

Small amount of water diffuses in.   Gills pump out large amounts of salt.   Large amount of water diffuses in.   Small amount of water excreted in urine.

crabs and brine shrimp, are *osmoregulators;* they maintain body fluids at a steady solute concentration, regardless of environment. Brine shrimp can live in fresh water, normal seawater, and even much saltier places, like the Great Salt Lake in Utah or Mono Lake in eastern California, because the animals' exoskeletons impede water loss or salt movement. In fresh water or dilute seawater, where the animals tend to bloat with water but lose salt, a special *excretory gland* expels water while the gills actively transport salt back in (Figure 25.16). In seawater or briny lake water, the animal swallows salt water; then its gills work in reverse and pump out excess salt. Because of their highly efficient gills, brine shrimp can live in a wider range of osmotic habitats than most other organisms.

■ **Aquatic Vertebrates**   The body fluids of aquatic vertebrates can be more or less concentrated than their surroundings. In each case, however, they remain at steady concentrations through osmoregulation. In sharks, for example, and in crab-eating frogs (residents of coastal mangrove swamps), very little water flows into or out of the animal by osmosis because their body fluids have roughly the same osmotic condition as seawater (Figure 25.17a). Their blood accumulates high levels of the metabolic waste urea, and this raises the total solute

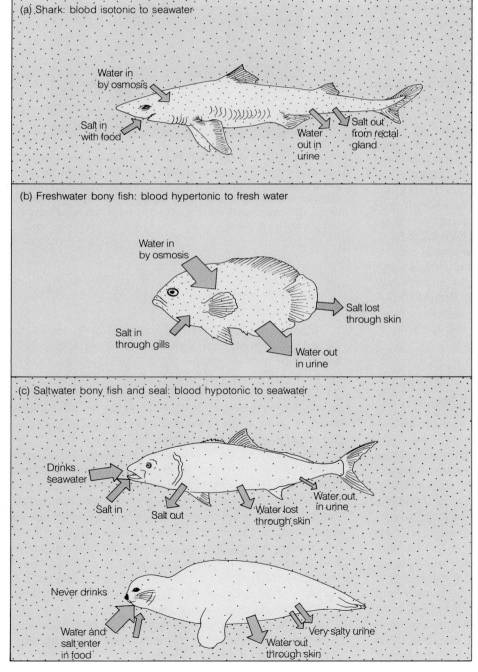

FIGURE 25.17 ■ **Solutions to Water Balance in Aquatic Vertebrates.** (a) A shark's blood is isotonic relative to seawater, owing to high levels of urea, not just salt. Therefore, salt still moves into the animal's fluids, and the shark's rectal glands excrete this excess salt. (b) A freshwater fish has hypertonic blood and thus absorbs water from, and loses salt to, the environment. Consequently, its gills actively pump in salt and its kidneys excrete large amounts of water in the urine. (c) Marine bony fishes and marine mammals have hypotonic blood and thus tend to lose water through the skin. To compensate, the saltwater fish drinks seawater, excretes little urine, and pumps out salt through the gills, whereas the marine mammal does not drink seawater and excretes very salty urine with little water content.

(a) Shark: blood isotonic to seawater

Water in by osmosis

Salt in with food

Water out in urine

Salt out from rectal gland

(b) Freshwater bony fish: blood hypertonic to fresh water

Water in by osmosis

Salt in through gills

Water out in urine

Salt lost through skin

(c) Saltwater bony fish and seal: blood hypotonic to seawater

Drinks seawater

Salt in

Salt out

Water lost through skin

Water out in urine

Never drinks

Water and salt enter in food

Water out through skin

Very salty urine

concentration of the blood to near that of the environment. Even though the total solute concentration is the same in shark blood and in seawater, salts do build up because the concentration of NaCl in the shark is lower than that in seawater. Sharks have a special *rectal gland,* however, which excretes excess salt.

In freshwater fishes and amphibians, the blood has a higher salt concentration than the environment. Since large quantities of water flow in across the skin, these animals do not need to drink water; instead, their kidneys produce large amounts of dilute urine that rids the body of excess water (Figure 25.17b). Unfortunately, salt also tends to leave the body and diffuse into the water, and so in freshwater fishes, the gills expend energy actively accumulating and absorbing salt from the lake or river water. In amphibians, the skin itself can accomplish this active salt transport.

Finally, in most marine bony fishes, reptiles, and mammals, the body fluids are hypotonic (less salty) relative to seawater, and thus, the animals tend to lose needed water and gain excess salt. As an evolutionary consequence, saltwater bony fishes drink seawater, retain most of the water, and produce very little urine, while their gills actively pump out excess salt (Figure 25.17c). Sea lions, whales, and other marine mammals are a bit like the kangaroo rat in that they never need to drink (taking in salt water only with food), and their very efficient kidneys conserve water strongly and excrete excess salt (see Figure 25.17c). Sea turtles, sea snakes, and other marine reptiles, as well as sea gulls and other oceangoing birds, drink seawater and pump out excess salt through special salt glands on the head (Figure 25.18).

It is clear from these descriptions that an animal's excretory system works continuously to offset undesirable losses or gains of water or salt from the environment, and this constant work must be fueled by food energy. Homeostasis—whether of body fluid content, oxygen levels in the tissues, heart rate, or body temperature—is costly but necessary to survival.

CONNECTIONS

Regardless of an animal's environment—be it desert, temperate forest, ocean, or freshwater pond—the organism faces problems in handling metabolic wastes and balancing salt and water levels in the body. The animal may have too little water or too much, an excess of salt or not enough; and inevitably, it will accumulate wastes, and these must be removed. Despite the variety of environments and problems, animals have surprisingly similar means of handling water, salt, and wastes. These include excretory tubes like Malpighian tubules or nephridia; kidneys containing scores of such tubules in the form of nephrons; and salt glands and gills that can absorb or get rid of salt as needed. Only occasionally do animals have novel solutions—such as the shark's urea-filled blood or the kangaroo rat's very long loops of Henle—to common excretory problems. The unity and diversity of life are reflected in these common and unique solutions to biological problems.

FIGURE 25.18 ■ Salt Glands Pump Out Excess Salt. Sea gulls, sea turtles, and many other marine vertebrates have special salt glands that secrete extremely salty droplets of water. Sea gulls can often be seen to shake their beaks rapidly from side to side; in so doing, they are usually shaking off salt droplets excreted from the salt gland.

Although waste removal and water and salt balance fall to the kidneys or equivalent organs, the excretory system interacts at many levels with other physiological systems. The kidney exchanges water and solutes with the circulatory system; it relies on the nervous system to initiate thirst and drinking or conserving behaviors; it depends on the digestive system to deliver water to the blood; and it is regulated by hormones that travel from the brain and other body parts.

The coordinated—and often energy-demanding— mechanisms that remove wastes and regulate water and salt balance in the blood exist because blood interacts with the extracellular fluid, which in turn exchanges material with intracellular fluid. If wastes do accumulate, or salts and pH levels are not maintained, the enzymes and other proteins inside cells will function improperly, and this imperils the entire organism. With a constant and appropriate internal environment, however, the animal can go about the more immediate tasks of feeding and defending itself and finding a mate, as well as the more immortal one of reproducing.

In the next chapter, we will continue our consideration of hormones and the roles they play in the chemical coordination of the body's physiological systems.

Highlights in Review

1. Excretory systems maintain constancy in an animal's cells by removing nitrogen-containing wastes generated by protein breakdown and maintaining the balance of water and salts in the animal's blood. Since materials are exchanged between blood and extracellular fluid, and in turn between extracellular and intracellular fluids, the excretory system indirectly helps to maintain the environment inside of cells.

2. Animals can be classified by their main form of nitrogenous waste: ammonia, uric acid, or urea.

3. The human kidney filters urea from the blood and excretes it into ureters, which lead to the saclike urinary bladder. Urine then leaves the body through the urethra.

4. The kidney's functional units are called nephrons, each consisting of a Bowman's capsule, a tangled proximal tubule, a hairpinlike loop of Henle, and a distal tubule that drains into a collecting duct.

5. Each nephron is the site of filtration, reabsorption, and secretion. The cuplike Bowman's capsule filters water, urea, and solutes from blood moving through the tuft of capillaries (glomerulus) that surround the capsule.

6. During tubular reabsorption, most of the water, glucose, amino acid molecules, and useful ions leave the nephron tubule and reenter the bloodstream via the nearby peritubular capillaries.

7. During tubular secretion, certain drugs and other organic molecules, as well as excess hydrogen, potassium, and other ions, are actively removed from the blood plasma and secreted into the urine. Water continues to leave osmotically, and the urine reaching the bladder contains only 1 percent of the water originally filtered from blood plasma.

8. In mammals, water balance is regulated by thirst, which is triggered by nerve activity in the hypothalamus of the brain, and hormones that control the functioning of the nephrons and hence the amount of water they reabsorb.

9. The kidney also helps regulate normal blood pressure and volume through the secretion of renin (which converts a blood protein to angiotensin, a hormone), which can increase blood volume and pressure.

10. Land animals are in constant danger of dehydration. Body coverings help conserve moisture, and so do the kidneys by forming concentrated urine. Other adaptations can be behavioral or physiological.

11. Aquatic invertebrates are osmoconformers or osmoregulators, while aquatic vertebrates are osmoregulators and have body fluids that are isotonic, hypertonic, or hypotonic relative to the surrounding water.

Key Terms

aldosterone, 527	excretion, 518
angiotensin, 528	filtrate, 524
antidiuretic hormone, 527	filtration, 522
ascending limb, 524	glomerulus, 523
bladder, 521	gout, 520
Bowman's capsule, 523	hormone, 527
collecting duct, 524	kidney, 521
cortex, 521	loop of Henle, 524
descending limb, 524	medulla, 521
distal tubule, 524	nephron, 521
drug testing, 525	osmoregulation, 529

pelvis, 522	tubular secretion, 522
peritubular capillary, 524	urea, 518
proximal tubule, 524	ureter, 521
renal artery, 521	urethra, 521
renal vein, 521	uric acid, 518
renin, 528	urination, 521
tubular reabsorption, 522	urine, 517

Study Questions

Review What You Have Learned

1. How do excretion and elimination differ? How are they alike?
2. List three types of nitrogenous wastes and organisms that excrete them.
3. Define glomerulus, podocytes, and peritubular capillaries.
4. True or false: The loop of Henle's main role is filtration. Explain.
5. Which constituents of the filtrate enter the Bowman's capsule?
6. What roles do active transport and osmosis play in the functioning of the proximal tubule?
7. By what mechanism does the sensation of thirst arise, and how does it help maintain the body's water balance?
8. Explain the connection between ADH, kidney function, and negative feedback.
9. How does aldosterone contribute to homeostasis?
10. Can the kidneys elevate blood pressure? Explain.
11. Explain the heart's role in regulating blood pressure.
12. How does osmoregulation differ in freshwater and saltwater fishes?
13. Explain how the excretory system interacts with other organ systems.

Apply What You Have Learned

1. Gout, caused by the accumulation of uric acid in joints, often plagued the nobility in previous centuries, but was seldom observed among poor people. What accounts for the difference in susceptibility?
2. Which kidney function enables sports officials to detect drug use by athletes? Explain.
3. If you drink too much alcohol at a party, you may feel "hung over" the next day. Why?
4. Over the course of a few months, a man's blood pressure rises substantially over its formerly normal levels. His doctor suspects cancer of an adrenal gland. Why did she make this diagnosis, and what blood test results might confirm it?

For Further Reading

CAROLA, R., J. HARLEY, and C. NOBACK. *Human Anatomy and Physiology.* New York: McGraw-Hill, 1990.

ECKERT, R., D. RANDALL, and G. AUGUSTINE. *Animal Physiology: Mechanisms and Adaptations.* 3d ed. New York: Freeman, 1988.

LECKIE, B. "Atrial Natriuretic Peptide: How the Heart Rules the Kidneys." *Nature* 326 (1987): 644–645.

SCHMIDT-NIELSON, K. *Desert Animals: Physiological Problems of Heat and Water.* New York: Oxford University Press, 1964, pp. 149–178.

Hormones and Other Molecular Messengers

The Worm Turns

On a wild cherry tree deep in a New England forest, a voracious eating machine is hard at work. A puffy, bluish green silk moth caterpillar, the size of your little finger, systematically devours one cherry leaf after another and grows fatter by the day (Figure 26.1a). Several times within a span of four weeks, the animal outgrows its confining outer cuticle, splits it, climbs out, and keeps on eating. Abruptly, however, the cecropia silkworm reaches a predetermined weight, stops eating, and begins to prepare itself for a remarkable transformation.

With glands similar to human salivary glands near its mouth, the larva spins a slender thread of pure silk. It attaches one end to a small branch, then winds the shiny strand around and around its body, forming a *cocoon,* a protective case within which a dramatic resculpturing soon begins (Figure 26.1b). Muscle cells that allowed the larva to move are destroyed, while cells stored in sacs within the caterpillar's body develop into the wings, legs, and antennae of the future adult moth. The animal, encased in a cocoon and partially transformed, is now a *pupa* and will spend the harsh New England winter in this form.

As spring approaches and the days grow longer, the cecropia pupa begins to stir. A hole appears in the cocoon, and over the next half hour, the animal—now a silk *moth* instead of a silk*worm*—extricates itself by pulling hard with spindly, newly formed adult legs (Figure 26.1c). Once it has fully escaped its confining case of silk, its abdomen begins to contract and pump blood into the damp, drooping wings. These then unfurl into a magnificent pair of red, black, and gold wings as wide as your hand and marked with white spots

(a) A caterpillar

(b) A cocoon

(c) The emerging moth

(d) The mature moth

FIGURE 26.1 ▪ **The Cecropia Silk Moth: A Voracious Eater Changes Form.** **(a)** A cecropia larva devouring cherry leaves. **(b)** A cecropia cocoon, an encasement of silken strands. **(c)** The cecropia moth emerges. **(d)** Its owlish wings unfurl.

resembling an owl's eyes (Figure 26.1d). Once dry, the wings can carry the silk moth off through the forest on its new mission: that of a flying reproduction machine. At this stage in the life cycle, the moth has huge gonads but no mouth. As it pursues (or is pursued by) a mate, it simply lives off the fat stored during its ravenous youth.

The remodeling of a caterpillar into a moth is a **metamorphosis,** literally, a "change in form," and the order and timing of the sequential changes that create the new form are closely regulated. A set of molecules called hormones do most of the regulating, and they and other molecular messengers—the broader category to which hormones belong—are the subject of this chapter.

Hormones are molecules produced by one set of cells (usually in a gland) and transported in body fluids to other parts of the body, near or far, where they bind to a target cell and cause a profound change in cell activity (Figure 26.2). In the silkworm, for example, a hormone called *ecdysone,* which is secreted by a gland in the thorax, diffuses through the larval body, and bathes all the cells. Cells sensitive to the hormone "bath" respond dramatically by differentiating first to their pupal form and then to their adult form. Chemicals like the hormone ecdysone, which leave cells, pass through one of several fluid or gaseous media, and regulate the activities of other cells, are called **molecular messengers.** Besides hormones, molecular messengers also include other important kinds of molecules that we will discuss.

Our goal in this chapter is to understand how molecular messengers in general and hormones in particular regulate physical changes in insects, people, and other organisms. We will also see how such molecules interact with each other and with the nervous system in the maintenance of homeostasis— the constant internal environment that is so crucial to an animal's survival (see Chapter 20).

Three themes will recur throughout this chapter. First, molecular messengers act in a similar way: They bind to a specific protein, called a receptor, on the outside surface or inside of a target cell, and this binding then triggers some activity. For example, the female silkworm's sex attractant leaves her body, wafts through the air, binds to receptors on the male's antennae, and triggers sexual excitation.

Second, hormones and other molecular messengers regulate physiological systems by preventing or provoking change. On the one hand, the body maintains homeostasis by preventing wide fluctuations in internal conditions, such as in

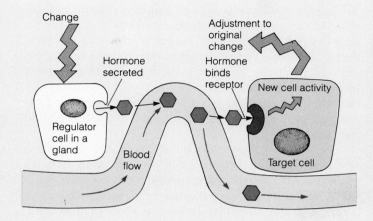

FIGURE 26.2 ■ A Model of Hormone Action. The action of hormones is simple but effective: A change in the body or in the environment triggers regulator cells, often in an endocrine gland, to secrete a hormone. The hormone travels from the regulator cell through fluid (usually the blood) to a target cell and induces a new activity in that cell. The new cellular activity adjusts the body to the initial change.

the concentration of blood sugar. On the other hand, for the body to undergo irreversible developmental events, such as metamorphosis in insects, it must undergo particular kinds of changes. Molecular messengers mediate both types of activities.

Third, very similar molecular messengers occur throughout the kingdoms of living things. A silkworm's metamorphosis and a child's puberty are both controlled by steroid hormones. This suggests an ancient evolutionary origin for the regulatory molecules and underscores the fundamental relatedness of all living organisms.

Our discussion will answer several questions about how hormones and other molecular messengers regulate body activities:

■ What general principles guide the action of molecular messengers?

■ What major hormones act in humans and other mammals, and how do they operate?

■ How do hormones prevent or provoke change?

■ How have molecular messengers changed during evolution?

MOLECULAR MESSENGERS: AN OVERVIEW

You are a composite reflection of your molecular messengers at work. These chemical regulators control the amount you eat each day and thus your body weight; how quickly you grew during childhood and adolescence; whether you are happy, depressed, or tense; how much salt and sugar flows in your blood; and a thousand other aspects of your physiology and behavior. The molecular messengers are clearly central to survival and normal functioning. But what are they, and how do they work?

The simplest answer is that all molecular messengers work by responding to a change or perturbation in the cell's environment and by triggering some new physiological activity. More specifically, the molecular messenger is secreted by a **regulator cell,** a cell that detects perturbations in the environment and emits the molecular messenger in response (see Figure 26.2). The messenger then diffuses through a fluid medium—either air, water, blood, or extracellular (tissue) fluid—and acts on one or more **target cells** that receive the molecular messenger and respond by carrying out the cellular activity that adjusts to the original perturbation.

Each target cell contains **receptors,** or proteins of specific shapes, each shape complementary to that of a particular molecular messenger. When a messenger comes along and binds to a receptor, like a key fitting into a lock, the receptor changes shape. This alteration heralds the messenger's arrival and triggers a change in cell activity. (Figure 26.3 describes an experiment with African pygmies that shows the importance of receptors to hormone action.) Some receptors lie in the cytoplasm. When the messenger enters the cell, it binds to a cytoplasmic receptor, and the messenger-receptor complex may in turn act on the DNA in the nucleus, triggering the formation of a new protein (and thus, a new activity for the cell). Some receptors, however, lie at the cell surface; when a messenger binds to one of these, the messenger-receptor complex triggers another molecule within the cytoplasm—a so-called **second messenger**—to act on a cellular protein, change its behavior, and in this way modify the cell's activity. Once a hormone or other molecular messenger binds its receptor, the hormone-receptor complex can therefore alter a cell's action by direct action on the DNA or by means of an intermediary, a second messenger inside the cell.

Some molecular messengers are secreted by isolated or individual cells within the animal's body, but most are produced by groups of cells organized into secretory organs called **glands.** A ductless gland that dumps molecular messengers directly into the extracellular fluid is called an **endocrine gland** (Figure 26.4a). The pituitary, adrenal, and thyroid glands are all endocrine glands, and most hormones are produced in these and about ten other endocrine glands. In contrast, a gland that dumps molecular messengers or other materials into ducts that generally lead out of the body is called an **exocrine gland** (Figure 26.4b). Exocrine glands make and

FIGURE 26.3 ■ Pygmies and Growth Hormone: The Importance of Receptors. The African pygmies grow to a height of only 4 ft 6 in. or so. Here, an adult male pygmy of the Efe tribe stands near an American researcher. Biologists know that growth hormone causes an increase in stature, so they wondered whether the pygmies' diminutive height was due to lack of growth hormone or to an absence of growth hormone receptors on their cell surfaces. To find out, early researchers injected pygmies with growth hormone and observed no additional growth, suggesting that they do not produce insufficient amounts of growth hormone. By next testing the pygmies' cells for growth hormone receptor, experimenters found that pygmies have only half as many receptors as do people of normal stature. They concluded that a pygmy's body may generate plenty of growth hormone but that the cells cannot detect it in normal amounts because of the scarcity of growth hormone receptors.

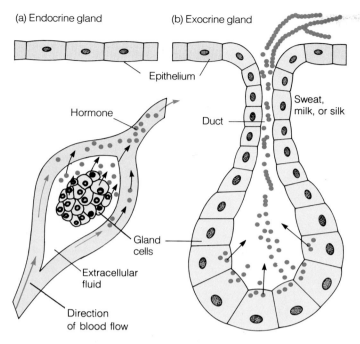

(a) Endocrine gland

(b) Exocrine gland

Epithelium

Hormone

Duct

Sweat, milk, or silk

Gland cells

Extracellular fluid

Direction of blood flow

FIGURE 26.4 ■ Glands With and Without Ducts. (a) An endocrine gland secretes hormones into the extracellular fluid. From there, the chemicals pass into blood vessels and travel to distant sites in the body. (b) An exocrine gland secretes milk, sweat, enzymes, or other materials into a duct, which leads the substance out of the body or into the digestive tract.

release body exudates such as milk, sweat, and the silken threads that silkworms and other arthropods spin into webs and cocoons.

Types of Molecular Messengers

Five kinds of molecular messengers leave individual cells or groups of cells (glands) and trigger activity in a different cell (or cells). **Paracrine hormones** are primitive hormones that act on cells immediately adjacent to the ones that secreted them. **Neurotransmitters** are nerve signal relay compounds. **Neurohormones** are hormones secreted by nerve cells. **True hormones** are nonneural hormones secreted by individual cells or by glands such as the pituitary or thyroid. And finally, **pheromones** are compounds secreted by one individual that affect the behavior of another individual of the same species. Each type of messenger can be distinguished by the organization of its secretory cells (as individual cells or parts of glands), its transport fluid (extracellular fluid, blood, air, or water), and the location of its target cells (adjacent to or at a distance from the secretory cells).

Paracrine hormones are secreted by individual regulator cells, diffuse through the extracellular space, and act only on adjacent target cells (Figure 26.5a). Regulator cells in our

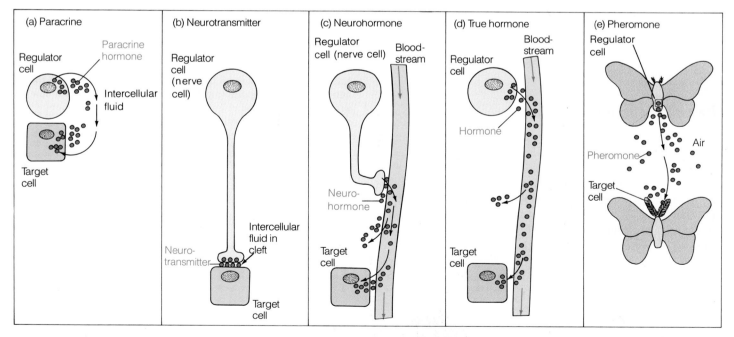

(a) Paracrine

Regulator cell

Paracrine hormone

Intercellular fluid

Target cell

(b) Neurotransmitter

Regulator cell (nerve cell)

Intercellular fluid in cleft

Neurotransmitter

Target cell

(c) Neurohormone

Regulator cell (nerve cell)

Bloodstream

Neurohormone

Target cell

(d) True hormone

Regulator cell

Bloodstream

Hormone

Target cell

(e) Pheromone

Regulator cell

Air

Pheromone

Target cell

FIGURE 26.5 ■ Molecular Messengers: Targets and Transport. (a) Paracrine hormones diffuse a short distance and act on an adjacent cell. (b) Some nerve cells can secrete neurotransmitters, and these diffuse across a narrow cleft to a target cell. (c) Some nerve cells can secrete neurohormones, but these pass into a blood vessel, travel some distance, then diffuse out again and reach a target cell. (d) True hormones are secreted by cells in endocrine glands and enter the bloodstream, travel some distance, diffuse out again, and act on a target cell. (e) Pheromones are secreted by regulator cells in exocrine glands, travel in a duct, leave the body, and stimulate target cells in another organism's body.

intestines, for example, detect protein molecules from food and secrete a paracrine hormone that causes adjacent target cells to secrete protein-digesting enzymes (review Figure 24.2 and Table 24.3).

Neurotransmitters are released by individual nerve cells and diffuse a very short distance through the extracellular fluid in a narrow space or cleft to a specific adjacent cell (Figure 26.5b). For example, a nerve cell running from your lower back down your legs releases a neurotransmitter that moves across a cleft to a muscle cell in the leg and causes the muscle to contract.

Neurohormones are also secreted by nerve cells, but travel through the bloodstream to target cells elsewhere in the body (Figure 26.5c). Nerve cell endings in the pituitary gland, for example, release the hormone oxytocin, which travels to the uterus and stimulates contractions during birth (see Chapters 14 and 20).

True hormones are usually secreted by endocrine glands, enter the bloodstream, and act on cells elsewhere in the body (Figure 26.5d). The silkworm hormone ecdysone, for instance, is secreted by a gland in the thorax and is carried in the blood to target cells throughout the body, where it stimulates adult organs to form.

Pheromones may be secreted by exocrine glands, leave the body via ducts, travel in air or water, and stimulate target cells in other organisms located kilometers away (Figure 26.5e). In a sexually mature female silk moth, for example, a scent gland in the abdomen secretes a pheromone that wafts on air currents for up to 11 km (7 mi) and randomly hits target cells in the male's antennae. This event can trigger wild sexual excitation that leads the male to fly long distances toward a potential mate.

Molecular Messengers and Receptors

Molecular messengers can activate target cells by binding to receptors inside a cell or at its surface. The major factor that determines whether a given hormone or other messenger will bind outside a cell or inside is whether that messenger is hydrophobic ("water fearing"; see Chapter 2) or hydrophilic ("water loving"). Messengers that are hydrophobic can easily pass into the cell through the hydrophobic cell membrane and bind to receptors within the cytoplasm or nucleus (Figure 26.6a, step 1). The messenger-receptor complex changes shape and then binds to a stretch of DNA along one of the chromosomes (step 2). This allows certain genes to be transcribed to mRNA (step 3) and thus new proteins to be made in the cell (step 4). This is the way ecdysone hormone from the silkworm brain causes the worm to form a pupa and then a new adult.

Hydrophilic messengers like proteins cannot pass through the hydrophobic (water-fearing) cell membrane of the target cell; their receptors are proteins partially embedded in the cell

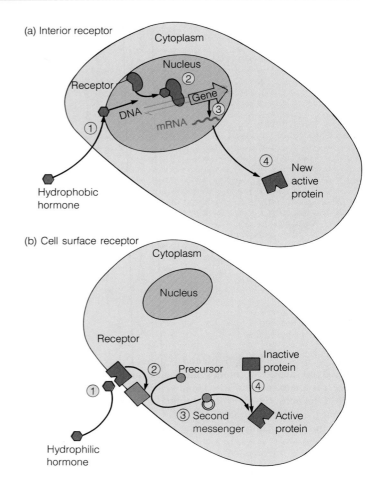

FIGURE 26.6 ■ How Hormones Effect Change: Gene Activation or Second Messengers. (a) Hydrophobic hormones, such as the steroid ecdysone, diffuse across a target cell's membrane (1) and bind to an interior receptor. The hormone-receptor complex then binds to specific places on the cell's DNA (2) and stimulates the activity of specific genes (3), leading to the production of a new protein that alters the cell's activity (4). (b) Hydrophilic hormones, such as epinephrine, cannot diffuse through the target cell's plasma membrane. Instead, hydrophilic hormones bind to a cell surface receptor (1), and the hormone-receptor complex touches off a chain of biochemical actions (2) that produces a second messenger (3), which in turn activates a protein and alters cell activity (4).

membrane (review Figure 3.12). The hormone-receptor complex causes a molecular relay team to trigger internal activity, and this is where second messengers enter the picture. When a hydrophilic messenger, such as the hormone epinephrine, binds to a surface receptor of a fat cell (Figure 26.6b, step 1), the receptor detonates a cascade (step 2) that changes the concentration of an intracellular second messenger (step 3). It is this secondary agent that causes an inactive protein inside the cell (the fat-splitting enzyme lipase in the case of fat cells) to suddenly become active (step 4) and thus alter the cell's activity, in this case leading to increased energy supply for the organism.

Hormones can alter two categories of second messengers: (1) cyclic nucleotides, such as the one called cyclic adenosine monophosphate (cAMP), and (2) two derivatives of a sugar-lipid component of the cell membrane plus calcium ions. Both types of second messengers are small molecules that appear inside a cell as the result of an external signal and act by changing the activity of specific enzymes or other proteins. The changed proteins, in turn, alter the activity of the entire cell, causing the organism to move its muscles, reproduce, burn fat, or function properly in a host of other ways that promote its survival.

It is significant that all molecular messengers in all animals work via receptors on or in target cells and that those with internal "relay teams" use just one of a few types of second-messenger compounds. Such facts attest to the evolutionary success of these few simple strategies for controlling physiological activity in a wide range of organisms.

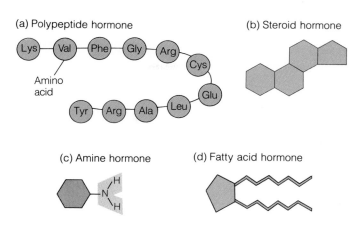

FIGURE 26.7 ■ **Generalized Structures of the Major Hormone Groups.**

HORMONES AND THE MAMMALIAN ENDOCRINE SYSTEM

Hormones are among the best-studied molecular messengers, and mammals—especially humans—are perhaps the best-studied sources. Biologists know in great detail, for example, how hormones govern the production and release of eggs, sperm, and milk and how the messengers stimulate the heart to speed or slow. Biologists know the chemical structures of the major hormones; they have mapped out most of the glands; and they understand how these secretory organs operate collectively and in concert with the nervous system, shepherding the various stages of reproduction and development and maintaining homeostasis within each system, organ, and cell.

Hormone Structure and the Endocrine System

Although mammalian hormones perform an incredible variety of tasks, most belong to just four molecular groups: the polypeptides, the steroids, the amines, and the fatty acid derivatives.

Polypeptide hormones are strings of from 3 to 200 amino acids and include oxytocin and luteinizing hormone (see Chapter 14 and Figure 26.7a). **Steroid hormones** contain four

joined rings of carbon atoms and are synthesized from cholesterol. Different steroid hormones have different atoms attached to the rings. Examples include estrogen and testosterone from the human ovaries and testes, respectively, and insect ecdysone from a gland in the thorax (Figure 26.7b). **Amine hormones** contain an amino group. Although they are derived from amino acids, the acidic functional group is often removed. This group of hormones includes thyroid hormones, which alter metabolic rates (Figure 26.7c). Finally, **fatty acid hormones,** such as *prostaglandins,* are derived from straight-chain fatty acids (Figure 26.7d). Prostaglandins were first discovered in semen and named after their supposed source, the prostate gland. They have since been discovered in most mammalian cells and tissues, however, and among other things cause muscles to contract and blood vessels to open or close (see Chapter 14). You can see the effect of prostaglandins on your own body. The next time you have a headache or menstrual cramps, consider that the pain is a result of the body's production of prostaglandins and their ability to cause blood vessels in the brain or muscle fibers in the uterus to contract. Drugs such as aspirin and ibuprofen block the production of prostaglandins and in this way relieve the pain for many people.

Mammals, as well as other vertebrates, have about a dozen major endocrine glands, similar in most cases to the major human endocrine glands shown in Figure 26.8. Together, these ductless glands make up the **endocrine system,** one of the body's major physiological networks. Much of the rest of this chapter is devoted to a gland-by-gland tour, beginning with the pituitary and hypothalamus, which together drive the activities of the ovaries, testes, thyroid, and adrenals.

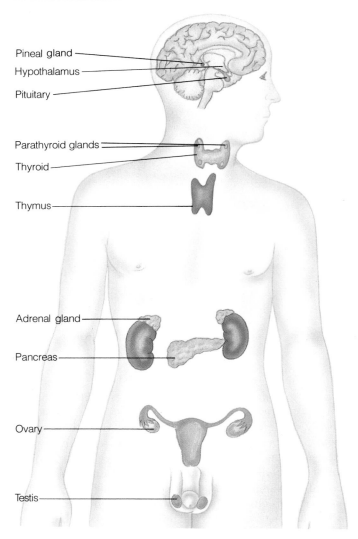

Pineal gland

Hypothalamus

Pituitary

Parathyroid glands

Thyroid

Thymus

Adrenal gland

Pancreas

Ovary

Testis

FIGURE 26.8 ■ Major Glands of the Human Endocrine System.
Several major endocrine glands secrete hormones that act on tar-
get cells nearby or at a distance in the body. Since males do not
have ovaries and females do not have testes, any individual has
only nine of the ten glands shown here.

Pituitary and Hypothalamus: Controlling the Controllers

Recall that together, the endocrine and nervous systems coordinate and integrate body activities, and the site of their major
interaction is the brain. At the base of the brain lies a collection of nerve cells called the **hypothalamus,** and hanging
from this region is a fingertip-sized bulb called the **pituitary**
gland (Figure 26.9). A bit like a light bulb in structure, the
mature pituitary has posterior and anterior lobes secreting
about ten kinds of peptide hormones, many of which help
control other glands throughout the endocrine system.

■ **Posterior Pituitary**   In the strictest sense, the posterior lobe
of the pituitary is more of a storage depot than an actual endocrine gland. This is because the two hormones it secretes,
oxytocin and antidiuretic hormone, or ADH (Table 26.1 on
pages 542 and 543), are actually made in the hypothalamus
and transported to the posterior pituitary in fine extensions of
the hypothalamic nerve cells called **axons** (Figure 26.9, step
1). The hormones are stored in the axon tips (step 2). When
signals from other parts of the brain stimulate the hypothalamus, it sends nerve impulses down to the axon tips in the posterior pituitary, causing the axon tips to release oxytocin or
ADH (step 3). After diffusing into the bloodstream, the hormone circulates and reaches target cells elsewhere in the
body.

A nursing mother demonstrates how the posterior pituitary works (Figures 26.9 and 26.10). As the baby begins to
suckle, sensitive nerves in the woman's nipples send signals
to the secretory cells in the hypothalamus, which in turn send
nerve impulses to their axon tips and cause the posterior pituitary to release the peptide hormone oxytocin into the bloodstream. When oxytocin reaches the musclelike target cells lining the milk ducts of the woman's mammary glands (which
are exocrine glands), it causes the target cells to contract, forcing milk into the ducts, out the nipple, and into the baby's
mouth. Figure 14.15 describes the role of oxytocin during
birth, and Figure 25.14 explains how antidiuretic hormone
(ADH) acts on the kidneys and helps maintain salt levels in
the blood.

■ **Anterior Pituitary**   The pituitary's anterior lobe makes and
secretes more than half a dozen of its own hormones (see
Table 26.1), but the hypothalamus controls the timing of their
secretion. As in the posterior lobe, nerve cell extensions, or
axons, run from the hypothalamus toward the anterior pituitary (Figure 26.9, step I). These axons relay impulses from
the hypothalamus based on how the hypothalamus monitors
conditions in the body and environment. The axon tips store
two kinds of small peptide neurohormones that stimulate or
inhibit hormone production by the anterior pituitary: **hypo-**
thalamic releasing hormones and **hypothalamic release-**
inhibiting hormones. When stimulated, nerve axon tips from
the hypothalamus liberate their inhibitory or releasing hormones into a special circulatory pathway that carries blood
directly to the anterior pituitary (step II). A specific releasing
hormone (step III) causes anterior pituitary cells to secrete
another specific hormone (step IV) into the bloodstream, and
eventually the hormone reaches target cells elsewhere. A
release-inhibiting hormone causes anterior pituitary cells to
stop secreting a specific hormone.

To see how this works, let's return to the nursing mother.
The anterior pituitary makes a hormone called prolactin
(PRL) that stimulates milk production and secretion. Most of
the time, however, a woman's hypothalamus makes a neurohormone called prolactin release-inhibiting factor, which prevents the anterior pituitary from secreting its prolactin. When

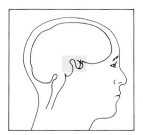

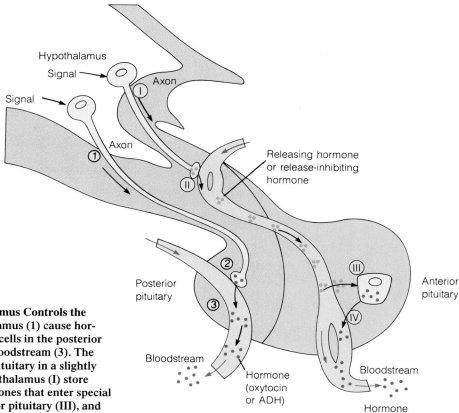

FIGURE 26.9 ■ How the Hypothalamus Controls the Pituitary. Signals from the hypothalamus (1) cause hormones stored in the endings of nerve cells in the posterior pituitary (2) to be released into the bloodstream (3). The hypothalamus controls the anterior pituitary in a slightly different way: Nerve cells in the hypothalamus (I) store releasing and release-inhibiting hormones that enter special blood vessels (II), travel to the anterior pituitary (III), and stimulate cells there to secrete hormones (IV), which in turn enter the bloodstream and travel to other body regions.

FIGURE 26.10 ■ The Pituitary Controls Milk Production and Release. When a baby nurses, the sucking action signals the mother's hypothalamus and pituitary to release the milk for the current feeding and to produce more milk for a future feeding. The mechanism is shown in Figure 26.9.

a baby suckles a woman's breast, the activity stimulates nerves that send signals to the hypothalamus, and these signals stop the release of the inhibiting factor. With the chemical inhibitor removed, the anterior pituitary can secrete prolactin, and the hormone can travel through the bloodstream to the mammary gland cells, which then produce and secrete milk into the ducts.

■ **Controlling the Other Endocrine Glands: Negative Feedback Loops**   In addition to their control of nursing, the hypothalamus and pituitary participate in several regulatory feedback loops that govern the activities of the testes, ovaries, thyroid, and adrenal glands. In each case, the hypothalamus produces a releasing or inhibiting factor that causes the anterior pituitary to secrete or not secrete a peptide hormone. If secreted, the peptide hormone then travels in the blood, eventually reaching the gonad or other gland, causing it to secrete a third hormone, generally a steroid or amine hormone. As the level of this third hormone builds, it eventually feeds back to the hypothalamus and/or pituitary and blocks additional hormone release. Figure 20.11 depicts the general strategy of feedback loops, and Figures 14.6 and 14.8 show how such negative feedback circuits affect the gonads. Here we will focus on the thyroid, parathyroid, and adrenals.

TABLE 26.1   Major Vertebrate Endocrine Tissues and Hormones

Tissue		Hormone	Target	Major Actions
Hypothalamus		Releasing and inhibiting hormones	Anterior pituitary	Stimulate or inhibit release of specific pituitary hormones
Anterior pituitary		Thyroid-stimulating hormone (TSH)	Thyroid	Stimulates synthesis and secretion of thyroxine
		Prolactin (PRL)	Mammary gland	Stimulates milk synthesis
		Adrenocorticotropic hormone (ACTH)	Adrenal cortex	Stimulates synthesis of sex steroids, mineralocorticoids, glucocorticoids
		Endorphins	Brain	Decrease pain
		Growth hormone (GH)	Many cells	Stimulates general body growth
		Luteinizing hormone (LH)	Ovary	Stimulates ovulation and synthesis of estrogen and progesterone
			Testis	Stimulates testosterone synthesis
		Follicle-stimulating hormone (FSH)	Ovary	Stimulates growth of ovarian follicle
			Testis	Stimulates sperm production
Posterior pituitary		Oxytocin	Mammary gland	Milk ejection
			Uterus	Uterine contraction
		Antidiuretic hormone (ADH)	Kidney	Increases water absorption
Thyroid		Thyroxine	Most cells	Increases metabolic rate and growth, causes metamorphosis in amphibians
		Calcitonin	Bones	Stimulates calcium uptake
Parathyroid		Parathyroid hormone	Bones	Stimulates calcium release into blood
			Digestive tract	Stimulates calcium uptake into blood
Adrenal medulla		Epinephrine	Circulatory system	Increases heart rate, blood pressure, and blood sugar
			Respiratory system	Increases breathing rate and clears airways
		Norepinephrine	Generally same as epinephrine	
Adrenal cortex		Sex steroids	Unknown	Stimulate growth of body hair in women
		Mineralocorticoids (e.g., aldosterone)	Kidneys	Increase sodium conservation
		Glucocorticoids (e.g., cortisol)	Many cells	Stimulate carbohydrate metabolism and decrease inflammation

TABLE 26.1 (continued)

Tissue		Hormone	Target	Major Actions
Pancreas		Insulin	Many cells	Stimulates glucose uptake from blood
		Glucagon	Many cells	Stimulates glucose release from cells into blood
Pineal		Melatonin	Hypothalamus	Blocks secretion of LH- and FSH-releasing factors
			Brain	Promotes sleep
Ovary		Estrogen	Many cells	Stimulates female development and behavior
		Progesterone	Uterus	Stimulates growth of uterine lining
Placenta		Chorionic gonadotropin (hCG)	Corpus luteum in ovary	Stimulates progesterone synthesis
		Placental lactogen	Mammary gland	Stimulates mammary gland development
Testis		Testosterone	Many cells	Stimulates male development and behavior
		Müllerian inhibiting hormone	Pre-oviduct cells	Kills pre-oviduct cells
Thymus		Thymosin	White blood cells	Stimulates differentiation
Gastrointestinal tract		Gastrin	Gut cells	Stimulates hydrochloric acid secretion
		Cholecystokinin (CCK)	Pancreas	Stimulates digestive enzyme secretion
Kidney		Erythropoietin	Blood cell precursors	Stimulates red blood cell production

The Thyroid and Parathyroid Glands: Regulators of Metabolism

The hypothalamus and pituitary control a small organ in the neck, the **thyroid gland,** which acts as the body's metabolic thermostat, regulating its use of energy as well as its growth. The most abundant thyroid hormone is **thyroxine,** an iodine-containing amine hormone that governs both metabolic and growth rates and stimulates nervous system function. A child with an underactive thyroid gland (a condition called *cretinism*) experiences low rates of protein synthesis and carbohydrate breakdown and thus low body temperature and sluggishness. The child's growth also becomes stunted, and mental development is retarded. An adult with an underactive thyroid gland and too little thyroxine secretion (a condition called *myxedema*) may gain weight easily and be slow mentally. Fortunately, doctors can treat these conditions by administering thyroxine.

A person can develop a **goiter,** a large lump on the neck caused by an enlarged thyroid (Figure 26.11). This enlargement can be understood in terms of the negative feedback loop that regulates thyroid activity. If the iodine concentration in the diet is low, and hence blood levels are low, the thyroid does not make much thyroxine, and the pituitary, sensing low

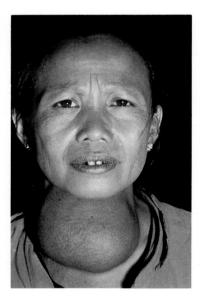

FIGURE 26.11 ■ **Goiter: The Result of a Disrupted Feedback Loop.** Normally, thyroid hormone, acting via the hypothalamus in a negative feedback loop, blocks the release of thyroid stimulating hormone from the pituitary. Without iodine, however—a key constituent of the thyroid hormone molecule—little thyroid hormone is made. With insufficient thyroid hormone to inhibit the hypothalamus, there is no longer a block to secretion of thyroid stimulating hormone by the pituitary, and so more of the stimulating hormone is produced. This thyroid stimulating hormone stimulates the thyroid gland to assume immense proportions, distorting the neck in a lump, or goiter.

levels of thyroxine, signals the thyroid to make more. The thyroid then responds by growing larger in a futile attempt to make more thyroxine. The iodine in iodized table salt intercepts the negative feedback loop and so can help prevent goiters from forming.

Alterations in the negative feedback loop that controls the thyroid can also cause this gland to become overactive. In May 1991, President George Bush began to complain of weakness and fatigue, and doctors found that uncoordinated quivering contractions in his heart's atrial chambers were causing it to race. The president's physicians diagnosed Graves' disease, a condition in which the body inappropriately generates antibodies that stimulate the thyroid to produce excess thyroxine. High levels of this hormone in turn lead to the sufferer's symptoms. Physicians treated President Bush with a radioactive form of iodine, an element contained in the thyroxine molecule. The president's overactive thyroid gland took up much of the iodine, and the radioactivity killed many thyroid cells. With fewer cells secreting thyroxine, the levels of the hormone fell, and physicians could then manage the president's symptoms with drugs, diet, and rest.

The thyroid gland and the associated **parathyroid glands,** a set of four small, dark patches of cells on top of the pale thyroid, work together and keep the level of calcium in the blood within very narrow limits. Without correct amounts of circulating calcium ions, nerves and muscles cannot function properly, and the person or other animal suffers nerve spasms, muscle contractions, and rapid death.

When blood levels of calcium (Ca^{2+}) start to rise—after you drink milk, for example—the thyroid releases the hormone **calcitonin,** which causes excess calcium to be deposited in bones. But if the calcium levels in the blood start to fall, the parathyroid glands release **parathyroid hormone,** and this causes bones to release more of the mineral and the body to pick up more of it from food. Note the counteracting relationship of these two calcium-regulating hormones and their interacting negative feedback loops. Regulating calcium levels is so crucial that the parathyroid hormone is one of only two hormones we absolutely need for survival. The other is aldosterone, a product of the adrenal glands.

The Adrenals: The Stress Glands

Sitting atop each of our kidneys is an **adrenal gland** (Figure 26.12a) that enables our bodies to react quickly to danger by fleeing or fighting. Each adrenal gland has a middle portion, or **medulla,** with an outer covering, or **cortex.**

Modified nerve cells in the adrenal medulla secrete two similar amine hormones: **epinephrine** (adrenaline) and **norepinephrine** (see Table 26.1). When an animal faces a sudden physical threat, such as a hungry lion or an angry boss, nerves trigger the secretion of the hormones (Figure 26.12b, step 1), which in turn act on target cells in the heart, lungs, intestines, and elsewhere, bringing about a so-called *stress response* (step 2). The hormones cause the heart to beat faster, blood

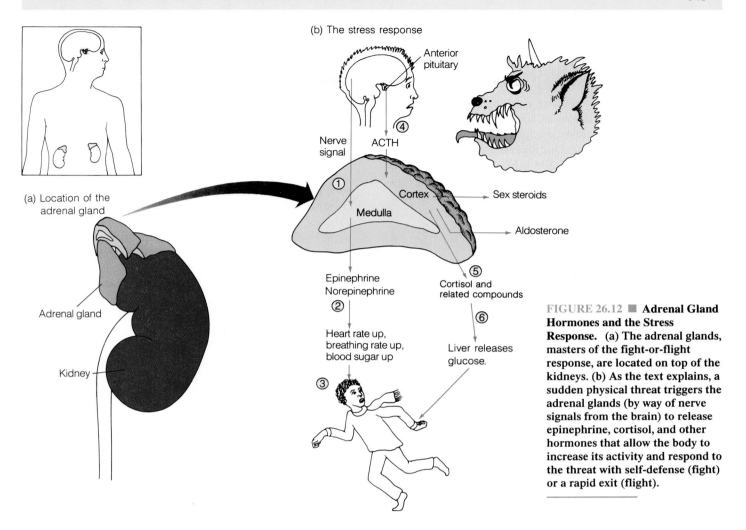

(a) Location of the adrenal gland

Adrenal gland

Kidney

(b) The stress response

Anterior pituitary

Nerve signal   ACTH

① Cortex   → Sex steroids

Medulla

→ Aldosterone

Epinephrine Norepinephrine ②   Cortisol and related compounds ⑤ ⑥

Heart rate up, breathing rate up, blood sugar up

Liver releases glucose.

③

FIGURE 26.12 ■ Adrenal Gland Hormones and the Stress Response. (a) The adrenal glands, masters of the fight-or-flight response, are located on top of the kidneys. (b) As the text explains, a sudden physical threat triggers the adrenal glands (by way of nerve signals from the brain) to release epinephrine, cortisol, and other hormones that allow the body to increase its activity and respond to the threat with self-defense (fight) or a rapid exit (flight).

sugar levels to increase, the breathing rate to speed up, and blood to be shunted away from organs such as the stomach and intestines. The result of all this is that the muscles receive more blood volume, oxygen, and sugar, and thus the animal is better able to defend itself or move away to safety (step 3).

The adrenal cortex, together with the medulla, secretes additional hormones that help an animal respond to stress. One, called **cortisol** (or hydrocortisone), is a member of the class of steroid hormones called *glucocorticoids.* As the name suggests, these hormones speed the metabolism of sugars as well as of proteins and fats. When an animal is stressed, perhaps by cold, strenuous activity, or pain, its hypothalamus sends a releasing hormone to its anterior pituitary, which secretes **ACTH** (adrenocorticotropic hormone) (Figure 26.12b, step 4). This triggers the adrenal cortex to produce cortisol and related compounds (step 5). Acting on target cells in the liver, fat tissue, and most other organs, cortisol causes stored proteins, lipids, and carbohydrates to be broken down and glucose to be rapidly generated (step 6), providing stressed cells with a source of quick energy. Inside anterior pituitary cells, ACTH can also break down, yielding β-**endorphin,** a hormone that acts on pain receptors in the brain—an action that decreases the amount of pain a stressed individual

senses. Recent studies show that when a dentist injects the hypothalamic releasing hormone for ACTH into a person before extracting his or her wisdom teeth, the patient releases more β-endorphin and experiences less pain.

Just as the body needs calcitonin and parathyroid hormones to maintain proper levels of calcium, it needs one final hormone from the adrenal cortex to help regulate mineral ions in body fluids. A steroid hormone and member of the *mineralocorticoid* class, *aldosterone* causes the kidney to conserve sodium and water and to excrete potassium (review Figure 25.14). Without this chemical messenger, the body would excrete too much sodium and water in the urine, the blood volume would drop precipitously, and death would come quickly.

The adrenal cortex also secretes steroid sex hormones. Box 26.1 on page 547 describes the strange story of adrenal and sex hormones in the spotted hyena of eastern and southern Africa.

By means of hormones and feedback loops, the adrenal cortex and the adrenal medulla act together to help an animal maintain homeostasis as well as overcome stress. What's at stake here is preventing or promoting change, and that is the major business of all glands and hormones.

HORMONES, HOMEOSTASIS, AND PHYSIOLOGICAL CHANGE

Minute by minute, as the sodium, calcium, or sugar levels in an animal's blood fluctuate up and down, hormones pull each one back into proper homeostatic range. Hormones also regulate daily or yearly cycles and irreversible developmental changes—for example, from child to adult or from caterpillar to moth. Because hormones, homeostasis, and change go hand in hand at all levels of physiological activity, their interactions warrant closer examination.

Hormones and Homeostasis:
Keeping Conditions Constant

One of the most painful reminders of our need to maintain a constant internal environment is **diabetes mellitus,** a condition in which blood glucose levels are not maintained within a normal range. Diabetics produce copious amounts of dilute, sugary urine and are nearly always thirsty. Without treatment, they risk *coma* (unconsciousness) due to severe dehydration or the accumulation of poisons in the blood. Diabetes is the most common hormonal disorder in humans, and it stems from defects in the *pancreas,* an elongated organ nestled near the intestines (Figure 26.13a). The normal pancreas has exocrine cells that secrete digestive juices into a duct leading to the central space within the small intestine, as well as islands of endocrine cells called the *islets of Langerhans,* which secrete the peptide hormones **glucagon** and **insulin.** A diabetic's islets cannot make enough insulin, and this creates homeostatic chaos.

In a person with a normal pancreas, glucagon and insulin work together by means of a feedback loop to make fine adjustments in blood sugar levels. After a breakfast of cereal and toast, nutrients enter the person's bloodstream, and blood sugar levels rise (Figure 26.13b, step 1). In response, one group of endocrine cells in the pancreas, the **beta cells,** secrete insulin (step 2), and this hormone causes target cells in the liver and other organs to remove glucose from the blood and store it as glycogen (step 3). Then, as daily activities cause blood sugar levels to fall (step 4), a second group of pancreatic endocrine cells, the **alpha cells,** detect the drop and respond by releasing glucagon (step 5). This causes liver cells to break down glycogen and release glucose, enabling blood levels to once again rise and fuel body action (step 6). Like a car's brake and accelerator, this dual control helps maintain precise blood sugar levels.

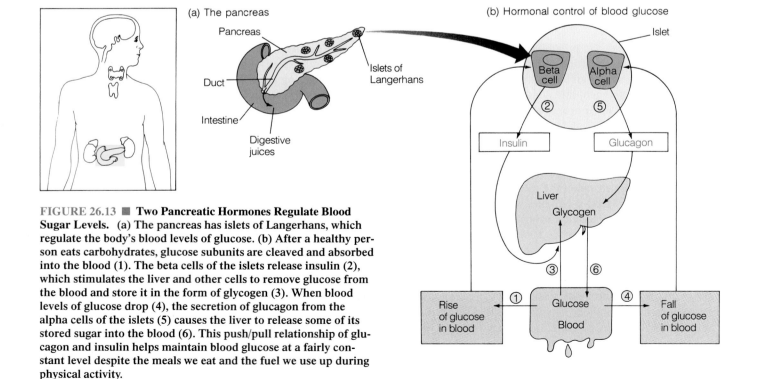

FIGURE 26.13 ■ Two Pancreatic Hormones Regulate Blood Sugar Levels. (a) The pancreas has islets of Langerhans, which regulate the body's blood levels of glucose. (b) After a healthy person eats carbohydrates, glucose subunits are cleaved and absorbed into the blood (1). The beta cells of the islets release insulin (2), which stimulates the liver and other cells to remove glucose from the blood and store it in the form of glycogen (3). When blood levels of glucose drop (4), the secretion of glucagon from the alpha cells of the islets (5) causes the liver to release some of its stored sugar into the blood (6). This push/pull relationship of glucagon and insulin helps maintain blood glucose at a fairly constant level despite the meals we eat and the fuel we use up during physical activity.

FEMALE HYENAS AND MALE HORMONES

Female spotted hyenas are one of nature's real oddities: A female hyena has genitals so masculine looking that human observers can rarely distinguish her from a male of the species. Specifically, she has a fully erectile pseudopenis and urogenital folds that are fused into scrotumlike pouches filled with fibrous tissue. What's more, she has an aggressive personality to match her masculine exterior: In every large clan of the fierce African predators, a single female monarch dominates the pecking order along with a court of other adult females who enforce dominion over all adult males through large body size and highly aggressive behavior. People have puzzled over the female hyena's peculiar anatomy and aggression since Aristotle's day, but only recently have biologists begun to solve the mystery. The answer lies, not surprisingly, with hormones—but hormones of an unexpected type.

In the early 1980s, a team of zoologists and animal behaviorists at the University of California at Berkeley captured 20 hyena cubs, then raised them to adulthood. They observed that starting in infancy, erection of the phallus is an important sign of excitement and social subordination for both male and female hyenas (Figure 1). Stephen Glickman and his team at Berkeley thought that the female's phallus might be a developmental by-product of male hormones shaping the female embryo's genitals and brain. These chemical messengers might, in turn, give her the powerful aggression, fierce behavior at a kill, and clear survival advantage that she and her cubs enjoy. If they were right, which hormone was causing the effects? The logical candidate was testosterone.

Tests showed that the dominant female does have testosterone levels that are higher than the average male's. The rest of the females, however, have less testosterone and other androgens than do males. But when the California team decided to test one particular weak androgen (male steroid) called androstenedione, they found that all female hyena cubs have eight times as much of the hormone as do males.

Androstenedione is just one compound in the metabolic pathway that begins with cholesterol and leads to active andro-

FIGURE 1 ■ **Phallic Erection in the Female Hyena.**

gens such as testosterone and dihydrotestosterone (see Chapter 14), as well as to active estrogens such as estradiol:

$$\text{Cholesterol} \rightarrow \text{androstenedione} \nearrow \text{estrogens} \searrow \text{androgens}$$

Research on rats, dogs, and monkeys has shown that androstenedione can have an unusual double influence on young animals. When experimenters injected androstenedione into infant male animals, the males grew up capable of the female's normal sexual posturing as well as their own gender's normal sexual mounting. And when researchers injected infant females, the adults had a male-type phallus and displayed mounting behavior, yet also showed normal female sexual functioning.

It is not yet clear how androstenedione works to engender both male and female attributes. It may have direct activity or it may be converted to other hormones with developmental and activational effects on target cells in organs such as the brain and gonads. Researchers are still studying such questions in the female hyena. But her unusual anatomy could lead to a new understanding of how hormones help spawn natural variations in sexuality and aggression throughout the animal kingdom.

In a person with diabetes, the pancreatic beta cells fail to generate insulin. Without sufficient insulin, body cells fail to remove glucose from the blood after a meal, so it builds up, and the kidneys remove the sugar along with lots of water from the blood. This results in copious, sweet urine, as well as in thirst and the danger of dehydration. What's more, without insulin, the liver, brain, and other body cells cannot remove glucose from the blood, and thus they "starve" in the midst of plenty. Although diabetics often crave carbohydrates, eating carbohydrate-rich foods does not help: Their glucose-starved cells burn lipids as fuel, and ketone by-products of lipid breakdown make the blood so acidic that it can lead to coma and death. The push-pull relationship of glucagon and insulin is clearly crucial to homeostasis and day-to-day health. Fortunately, diabetes can now be largely controlled

with insulin treatment. However, long-term damage to fingers and toes and to the retina due to poor circulation still occurs as some diabetics grow older.

Hormones and Cyclic Physiological Changes

An animal's life follows daily, monthly, and yearly cycles orchestrated by hormones. Evidence is mounting that such cycles are tied to daily, or **circadian,** light and dark periods and to the changing length of daylight during the four seasons. Additional evidence implicates a curious little brain structure, the **pineal gland** (review Figure 26.8 and Table 26.1), in the measurement of day length and the control of reproductive cycles, onset of puberty, and moods.

The pineal gland—so named for its resemblance to a pine cone—is a bumpy, thimble-sized knob lying deep in the brain. In lizards and certain other vertebrates, the pineal gland is called the third eye because it is structurally similar to the retina and actually perceives light directly. In humans and other mammals, the pineal gets secondhand information in the form of nerve stimulation from light receptors in the eyes. Regardless of species, however, the gland secretes an amine hormone called **melatonin** in response to darkness, and this hormone promotes sleep and inhibits the activity of the gonads via melatonin receptors in the hypothalamus.

While the evidence is not as clear-cut in people, melatonin may influence reproduction, puberty, and mood. In Finland, it is most common for couples to conceive twins and triplets (a sign of highly active ovaries) in July, when days are very long and melatonin levels are lowest, and it is least common in January, when nights are long and melatonin levels are highest. And just before puberty, melatonin levels in people suddenly drop by 75 percent, suggesting that the pineal hormone may have been inhibiting gonad development during childhood. Further research with a class of morphinelike brain chemicals called *endorphins* suggests that these peptides actually regulate the activities of the pineal gland, hypothalamus, pituitary, and gonads and thus may ultimately govern the onset of puberty. Melatonin may have yet another effect: Some people experience profound depression, oversleeping, weight gain, tiredness, and sadness when the days grow short in winter, then a rebound of better spirits in spring. Melatonin levels are blamed for this seasonal affective disorder syndrome (SADS), and sufferers have been successfully treated by exposure to bright lights for several hours per day in winter.

Hormones and Permanent Developmental Change

In addition to orchestrating cyclic change, hormones program an animal's progression through the life stages. As we just saw, a drop in melatonin precedes puberty in human teenagers, contributing to the ensuing surge of LH and FSH that directs the release of steroid hormones and the subsequent

FIGURE 26.14 ■ **A Tadpole Transformed.** The tadpole's thyroid gland makes thyroxine, a hormone that mediates the physical transition to froghood, including the resorption of the tail and the development of limbs. Review the other stages of frog metamorphosis in Figure 1.11.

enlargement of female breasts, deepening of the male voice, development of larger muscles, and other accompanying physical changes. (Box 26.2 explains how some athletes take illicit—and dangerous—steroids to artificially enlarge their muscles.) In tadpoles, thyroxine from the thyroid gland acts on target cells in the tail and flank, causing the animal to resorb its long tail and sprout frog legs (Figure 26.14). The most dramatic transformations, however, take place in insects like the silkworm and are mediated by two hormones.

Silkworms pass through five larval stages, during which, as we saw in the chapter introduction, the insect grows, produces a new exoskeleton, then sheds its old one, or *molts* (Figure 26.15). After a final growth spurt, it spins a cocoon and secretes a dark brown pupal exoskeleton instead of another bright blue larval one. In spring, as the days lengthen, the animal, still in its cocoon, secretes one last exoskeleton—an adult exterior resplendent with glistening red, black, and gold scales.

Two main hormones govern silkworm molting and metamorphosis from worm to moth. **Ecdysone,** a steroid, controls

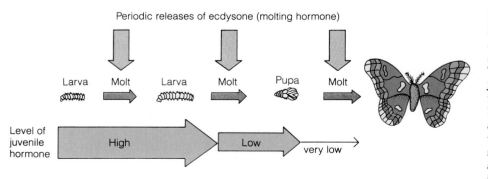

FIGURE 26.15 ■ **From Silkworm to Silk Moth: Hormones Provoke Metamorphosis.** As a tiny caterpillar grows, surges of ecdysone regulate the timing of each molt. Throughout early molt cycles, levels of juvenile hormone remain high and maintain the animal's larval form. Eventually, levels of juvenile hormone begin to drop, and with the next to last surge of ecdysone, the animal molts into a pupa. With the final surge of ecdysone, juvenile hormone levels are very low and the animal is transformed into an adult moth.

the *timing* of each molt. Biologists know this because if a researcher injects a larval insect with ecdysone, the animal begins to molt a short time later. The *type* of molt, however, is the province of **juvenile hormone,** a golden oil secreted by a gland attached to the brain. If juvenile hormone levels are high, the worm molts from one larval stage to another, but it remains a juvenile. If juvenile hormone levels are intermediate, the animal molts from larva to pupa. If concentrations are very low, the pupa molts to an adult. Biologists are still investigating how juvenile hormone seems to maintain youth almost magically in insects.

In both people and insects, steroid hormones provoke

changes in animal form. Such parallels reflect the retention of similar hormonal mechanisms throughout evolution.

HORMONES AND EVOLUTION: INVARIANT AGENTS OF CHANGE

Biologists have made exciting finds in recent years that reveal startling facts unsuspected just a decade ago: The simplest and the most complex organisms have similar molecular messengers. Scientists have found insulin-like peptides regulating

---

BOX 26.2   P E R S O N A L   I M P A C T

STEROID HORMONES AND THE INSTANT PHYSIQUE

There is an epidemic under way in North American locker rooms, gyms, and health clubs: the use of anabolic steroids to help build large muscles. The pressure to win is strong on the football field, the wrestling mat, the bodybuilding circuit, even the dating scene. That pressure has been so overpowering, in fact, that an estimated 7 to 8 percent of professional football players, 5 percent of college athletes, and 6.6 percent of high school boys are willing to take synthetic male hormones to become an overnight version of The Hulk. An instant physique can indeed help them win. But research shows that the gains are temporary, and the price is very high indeed.

Anabolic steroids such as stanozolol (the illegal use of which cost sprinter Ben Johnson his Olympic gold medal in 1988; Figure 1) are artificial forms of testosterone. Just as an upsurge of the real male hormone promotes rapid muscle development in adolescent boys, taking anabolic steroids can lead to bulking that gives an athlete an advantage. Unfortunately, the drugs also confer the other masculinizing effects of testosterone in exaggerated proportions: heavy hair growth, acne, and premature baldness. They also suppress the male's own androgens (male hormones), and this can lead to shrunken testes and enlarged breasts. And ominously, research shows that damage to the kidneys, liver, and heart are all too common in users of anabolic steroids, as are depression, anxiety, hallucinations, paranoia, and other psychological symptoms.

For many observers, the ethical issues are as serious as the psychological ones. Pumping the body up beyond its natural maximum size and strength is really a form of cheating—cheating one's competitors and ultimately oneself. Instant muscles, so easily obtained, are just as easily lost once the drug treatments stop. But the hidden damage may already be done. Consider Steve Courson, a former football player for the Pittsburgh Steelers and the Tampa Bay Buccaneers. Anabolic steroids damaged his heart so badly that he retired early, and doctors

FIGURE 1 ■ **Ben Johnson Lost an Olympic Medal for Taking Anabolic Steroids.**

were predicting he might live just five years longer. For too many healthy athletes, this shortcut to physical fitness can become a dead end.

sugar metabolism in both invertebrates (like mollusks and insects) and vertebrates (such as mammals and reptiles). They have found that a mammalian hypothalamic releasing factor will turn on some sexual activities in yeast, while a mating factor from the fungus will turn on some gonadal activities in rats. Clearly, molecular messengers must have a very ancient origin and must have been conserved for hundreds of millions of years.

Some biologists suggest that molecular messengers may initially have evolved in single-celled organisms as mechanisms that coordinate feeding or reproduction. As multicellularity arose, sophisticated organs evolved that govern the many individual coordination tasks, but these control centers relied on the same kinds of molecular messengers "invented" by the simpler organisms. Some of the messengers worked fairly slowly but had long-lasting action on distant cells; these became the modern hormones. Others worked more quickly, but influenced only adjacent cells for brief periods; these became the neurotransmitters. In the next chapter, as we see how nerve cells communicate, the familiar strategy of molecular messengers will once again emerge.

CONNECTIONS

In our consideration of molecular messengers thus far, we have seen that some cells detect perturbations in the environment and respond by secreting a regulatory substance into the surrounding fluid. That substance binds to a specific protein receptor in a target cell and triggers some new function in the target cell that often counters the initial perturbation. This pattern holds whether the organism is a person, hamster, silkworm, or yeast; whether the molecular messenger is a steroid, amine, polypeptide, or prostaglandin; and whether the activity is simple and direct or involves feedback loops, releasing hormones, and second messengers. The end result is an animal tuned to its environment, coping effectively with change, and surviving.

Our next subject, the regulators and receptors of the nervous system, have a kind of private intercellular communication that is like a rapid-fire tête-à-tête between neighboring cells. In contrast, the hormones, with their diffuse, long-distance activity, seem more like slowly whispered rumors that echo throughout the entire body.

Highlights in Review

1. Molecular messengers are secreted by regulator cells, are carried in a fluid medium, and trigger activity in target cells by binding to receptors inside the target cell or on its surface. In addition to the true hormones, molecular messengers include the neurohormones, neurotransmitters, paracrine hormones, and pheromones.

2. Endocrine glands secrete molecular messengers into the blood, while exocrine glands secrete molecular messengers or other substances such as milk or enzymes into ducts that can lead out of the body.

3. Hormones are molecular messengers that are secreted by endocrine cells into the blood; they alter the activity of distant cells.

4. Hydrophobic messengers can pass through a cell membrane and bind directly to receptors in the cytoplasm or nucleus. Hydrophilic messengers bind to cell surface receptors and trigger second messengers.

5. Mammalian hormones include polypeptides, steroids, amines, and fatty acid hormones. A mammal's set of 12 or so ductless glands is called the endocrine system.

6. The hypothalamus and the pituitary gland secrete many neurohormones and true hormones and control many other glands. The posterior pituitary secretes oxytocin and ADH. The anterior pituitary secretes prolactin and other hormones.

7. In general, hormones act via negative feedback loops when maintaining constant conditions. For example, two separate hormones act in opposing negative feedback loops, maintaining appropriate blood levels of calcium and glucose.

8. The thyroid gland secretes thyroxine, which regulates the rates of metabolism and growth, and calcitonin, which causes calcium uptake by bones. Parathyroid hormone from the parathyroid glands causes bones to release calcium, keeping blood levels above certain critical limits.

9. The adrenal glands secrete the stress hormones epinephrine and norepineprhine, which prepare an animal to fight or flee, and also make cortisol, aldosterone, and some sex steroids.

10. Hormones help maintain homeostasis. Diabetes, an inability to maintain correct levels of glucose in the blood, results from defects in the pancreas and insufficient insulin production.

11. The pineal gland helps govern daily and yearly cycles of physiological change by detecting day length and making melatonin.

12. Hormones also program an animal's irreversible developmental changes, such as puberty and metamorphosis.

13. The molecular messenger strategy (regulator cell→release of chemical agent→reception by target cell) has been conserved throughout evolution.

Key Terms

ACTH, 545
adrenal gland, 544
alpha cell, 546
amine hormone, 539
axon, 540
β-endorphin, 545
beta cell, 546
calcitonin, 544
cortex, 544
circadian, 547
cortisol, 545
diabetes mellitus, 546
ecdysone, 548
endocrine gland, 536
endocrine system, 539
epinephrine, 544
exocrine gland, 536
fatty acid hormone, 539
gland, 536
glucagon, 546
goiter, 544
hormone, 535
hypothalamic release-inhibiting
    hormone, 540
hypothalamic releasing hor-
    mone, 540

hypothalamus, 540
insulin, 546
juvenile hormone, 549
medulla, 544
melatonin, 548
metamorphosis, 535
molecular messenger, 535
neurohormone, 537
neurotransmitter, 537
norepinephrine, 544
paracrine hormone, 537
parathyroid gland, 544
parathyroid hormone, 544
pheromone, 537
pineal gland, 547
pituitary gland, 540
polypeptide hormone, 539
receptor, 536
regulator cell, 536
second messenger, 536
steroid hormone, 539
target cell, 536
thyroid gland, 544
thyroxine, 544
true hormone, 537

Study Questions

Review What You Have Learned

1. How does an endocrine gland differ from an exocrine gland?
2. List five kinds of molecular messengers and an example of each type.
3. How do hydrophobic molecular messengers produce their effects? Hydrophilic messengers?
4. Define second messengers, and explain how they function.
5. How do the hypothalamus and the pituitary interact?
6. Explain how an underactive thyroid affects a child. How does the same condition affect an adult?
7. What mechanism regulates the calcium level in blood? Explain.

8. What are the biological benefits of cortisol and epinephrine?
9. Describe the feedback loop that regulates the amount of glucose in your bloodstream.

Apply What You Have Learned

1. Will the metamorphosis of a silkworm caterpillar proceed normally if a researcher removes the glands that produce ecdysone? Explain.
2. In a nursing mother, does the infant's sucking stimulate milk production, or does the presence of milk in the nipple stimulate sucking? Explain.
3. A person with insulin-dependent diabetes will often eat a carbohydrate snack or meal before taking a dose of insulin. How does this snacking help the insulin do its job?
4. A health study of residents living near a nuclear weapons production facility reveals an unusually high incidence of thyroid cancer. Which of the following is a likely explanation for this cancer rate? Explain why you chose or rejected each explanation. (1) The cancers result from an inherited genetic weakness. (2) The cancers result from production facility emissions of radioactive iodine. (3) The cancers result from a lack of iodine in the water supply.

For Further Reading

ATKINSON, M., and N. MACLAREN. "What Causes Diabetes?" *Scientific American,* July 1990, pp. 62–71.

GILBERT, L. I., W. L. COMBEST, W. A. SMITH, V. H. MELLER, and D. B. ROUNTREE. "Neuropeptides, Second Messengers, and Insect Molting." *BioEssays* 8 (1988): 153–157.

GUILLEMIN, R., and R. BURGUS. "The Hormones of the Hypothalamus." *Scientific American,* November 1972, pp. 24–33.

RASMUSSEN, H. "The Cycling of Calcium As an Intracellular Messenger." *Scientific American,* October 1989, pp. 44–50.

ROSEN, O. M. "After Insulin Binds." *Science* 237 (1987): 1452–1458.

SNYDER, S. H. "The Molecular Basis of Communication Between Cells." *Scientific American,* October 1985, pp. 132–141.

How Nerve Cells Control Behavior

Cocaine: Highest Highs and Lowest Lows

A young couple from a midwestern state snorted cocaine from time to time to make themselves feel energized, more alert, and temporarily happier. The woman even continued this "recreation" into the early months of her pregnancy, but stopped after a while as the fetus grew.

Just before the baby was due, her husband gave her 5 g (0.2 oz) of cocaine as an anniversary gift and she snorted it to "celebrate." The drug, a nervous system stimulant, induced early labor and she delivered a baby boy. Tragically, the newborn was permanently paralyzed on his right side; the "hit" of cocaine the mother had taken just hours before passed through the placenta, entered the baby's bloodstream, and caused his blood pressure to soar. The high pressure burst blood vessels in his brain, and this damaged a large portion of that delicate organ, altering its control of the body's complex nervous system. The brain damage, in turn, impaired function in another part of the nervous system—specifically, the nerves that allowed movement of the baby's right arm and leg. Ironically, the mother took cocaine for its temporary stimulatory effects on her brain's reward center such as the one shown in Figure 27.1. However, the drug had a similar yet magnified impact on her son's nervous system, causing permanent and devastating damage.

Our subject in this chapter is the major integrative network in an animal's body: the **nervous system.** A heartbreaking event such as this baby's irreversible nerve damage points out one of the chapter's recurring themes: The nervous system controls and

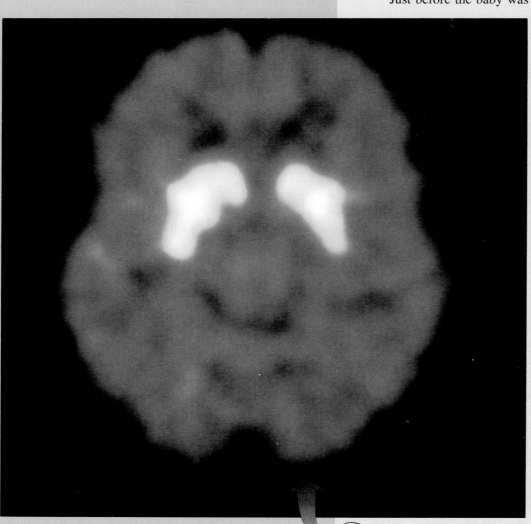

FIGURE 27.1 ■ **Cocaine High: A Change in Nerve Cell Function in the Brain.** Cocaine accumulates in the "reward center" of the brain (bright yellow) within just 15 minutes after a hit of cocaine, as shown in this PET scan, a special noninvasive view of a "slice" through a human brain.

coordinates the activities of an animal's body systems. This governing function works parallel to and at times in conjunction with the endocrine (hormonal) system (see Chapter 26), and it allows an animal to sense the environment, function as an integrated whole, and respond appropriately. Nervous systems do three things: They *receive* data about environmental conditions, they *integrate* the sensory data into a meaningful form, and they *bring about a change* in behavior, physiology, or information storage that helps the organism survive. The cocaine that the mother took altered the way her own nervous system integrated sensory data and regulated behavior (her mood and alertness) as well as the way the baby's nervous system helped orchestrate his circulation and blood pressure.

The governing function of the nervous system will be a recurring theme in this chapter, as will a second principle: The nervous and endocrine systems are able to work together coordinating and controlling the body because they share several important features. Both systems integrate body functions; both use molecular messengers (sometimes the identical molecules) for communication; and in both systems, the second-messenger strategy triggers the response inside certain target cells (review Figure 26.6). The nervous and endocrine systems do differ in substantial ways, however, including their speed of response (the nervous system responds more quickly) and the duration of their effects (hormonal effects often last much longer).

We will encounter a third recurrent theme: Although the nervous system plays several roles at different levels, the central task of the nervous system is *communication*. This communication can take place within an individual nerve cell; between neighboring nerve cells; between separate organs; and between different individuals in a community, allowing one animal to perceive the needs and internal states of another.

One final thread runs through this chapter: Communication within the body (and between individuals) is possible because nerve cells contact each other in special ways and are organized into highly elaborate networks. Nervous communication is a bit like telephone communication in that both depend on point-to-point contact—nerve cell to nerve cell for one, wire to wire for the other (Figure 27.2). (Hormonal information, in contrast, is more like radio communication, where a transmitter broadcasts a signal widely, but only special receivers—target cells or radios—tuned to the signal can pick up and interpret the message.) An unfortunate example of nervous communication occurred in the baby who was exposed to cocaine. Deep within the baby's developing brain, the stim-

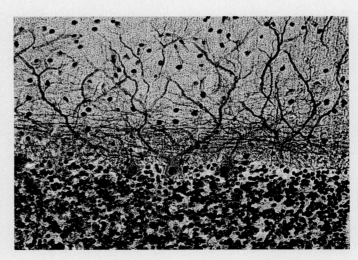

FIGURE 27.2 ■ **Nerve Cells Are Organized into Highly Elaborate Networks.** Neurons from a cat's cerebellum interconnect via delicate extensions, establishing a complex cellular network with point-to-point contact of nerve cell to nerve cell.

ulant altered the way one specific group of nerve cells communicated with another—those that control blood pressure—and the consequences were dramatic for the baby's behavior—especially the movement of his right arm and leg.

The behavior of humans and other animals depends on the way nerve cells are connected to each other and arrayed in precise patterns. To understand those connections and hence those behaviors, we must study the structure and activities of nerve cells, how those cells are organized into nervous systems, and how complex neural organs such as the brain can mediate thought, language, and sensation. In this chapter, we focus on individual nerve cells and their connections into nervous systems; then, in the next chapter, we concentrate on the brain and sense organs.

This chapter's discussion of nerve cells poses several questions:

■ How does a nerve cell's structure promote communication in the body?

■ How does a nerve cell convey an impulse or signal electrically, and how does that impulse carry information?

■ How do nerve cells communicate with other cells?

■ How are nerve cells organized into networks and simple nervous systems?

HOW NERVE CELL STRUCTURE FACILITATES COMMUNICATION

As you read this sentence, try wiggling the big toe on your left foot. Some of the nerve cells controlling that movement extend all the way from the lower part of your spinal cord to the muscles that move your toe—the longest individual cells in your body at nearly 1 m (39.4 in.) in length. A nerve cell running from a giraffe's spinal cord to its hoof can be three times as long, or nearly 3 m (10 ft)! Some nerve cells, in contrast, reach only a few millimeters, extending from one brain region to another. Although the human brain and nervous system contain more than 1000 different types of cells, all of them can be classified as either **neurons,** the nerve cells that accomplish the actual communication tasks, or **glial cells,** the support cells that surround, protect, and provide nutrients to neurons and may influence their function in ways biologists do not yet fully understand. Here we focus on the neurons.

Anatomy of a Nerve Cell

All neurons, including the ones that help wiggle your big toe, function by collecting information and relaying it to other cells in the body. Neurons are characterized by key structural features: the dendrites, soma, axon, and ends of the axon, or boutons (Figure 27.3).

Neurons gather information by a set of fine cell processes called **dendrites** (from the Greek for "little tree"). Dendrites pick up signals from the environment or from other nerve

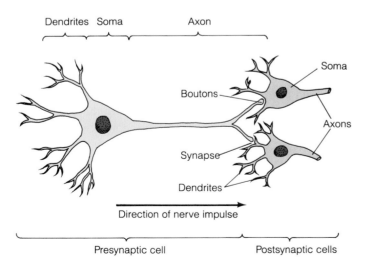

FIGURE 27.3 ■ **Neurons Transmit Information from One Cell to Another.** Shown here are the four parts of a neuron—dendrites, soma, axon, and axon terminals (boutons). At the synapse, or junction, between one cell's axon terminal and another cell's dendrite, information is passed from the presynaptic cell to the postsynaptic cell.

cells and pass these impulses to the soma. The **soma** is the neuron's main cell body. Housing the usual complement of intracellular organelles—nucleus, ribosomes, mitochondria, endoplasmic reticulum, and Golgi apparatus—the soma produces the proteins that make up the rest of the nerve cell, as well as the enzymes that determine its activity. The signals next pass through the soma to a long, tubular cell process called the **axon** (Greek for "axle"). The axon of some neurons extends from your spinal cord to your toe muscles, while the nerve cell's soma and dendrites are located in the spinal cord itself. Each neuron has many dendrites, but usually just one axon. A group of axons from many different neurons running in parallel bundles is called a **nerve.**

Eventually, the nerve signal reaches the often knoblike ends of an axon, the **boutons** (French for "buttons"), which terminate very close to another cell and serve as special regions of cell-to-cell communication. The place where a bouton approaches another cell is a junction known as a **synapse,** which often contains a microscopic cleft across which the message passes at lightning speed (see Figure 27.3). The synapse is the site where cocaine acts and alters communication between nerve cells. Boutons may form synapses with other axons, with the dendrites or soma of another neuron, with muscle cells, or with the secretory cells of endocrine or exocrine glands (see Chapter 26). The neuron that sends a message down its axon and across a synapse to another cell is the **presynaptic cell** (see Figure 27.3). The cell receiving the message that crosses a synapse is the **postsynaptic cell.** A given axon may branch and rebranch, forming synaptic junctions with up to 1000 other cells. Conversely, 1000 other neurons might form synapses with the dendrites and soma of a single neuron—rather like hundreds of hands reaching out to touch a central object simultaneously. An animal's communication network is literally a net, a lacework of interconnected neurons that allows the organism to carry out such complex behaviors as wiggling one big toe on one foot. Recall that the nervous system receives data, integrates it, and effects a change in physiology. Within a given neuron, the dendrites act as the receptors, the soma acts as the integrator, and the axon and boutons act as the effectors.

ELECTRIFYING ACTION: THE NERVE IMPULSE

Environmental signals can differ greatly and must be received and interpreted in different ways. A chili pepper, a chocolate bar, and an aspirin tablet all produce distinguishably different taste sensations. Listening to Brahms is nothing like hearing the screech of a braking car. And none of these inputs resembles the feeling of slamming your finger in a drawer. These sensations differ because their points of origin and destinations in the brain or spinal cord vary widely. At the level of a single neuron, however, any and every sensation of sufficient magnitude produces exactly the same kind of signal: a **nerve**

impulse, an electrochemical chain reaction, triggered by a series of abrupt, localized changes in the neuron's membrane, that allows specific ions to rush in and out of the cell.

A Nerve Cell at Rest

By focusing on a single patch of a resting neuron's plasma membrane, we can begin to see how a nervous impulse is generated. Neurons, like all cells, are bathed in an extracellular fluid that is a molecular soup with a high concentration of sodium ions (Na^+) and a relatively low concentration of potassium ions (K^+) (Figure 27.4a, step 1). In contrast, the fluid inside the neuron is relatively rich in potassium ions and low in sodium ions. Besides these differences, there are many more positively charged ions in a thin layer on the outside of the cell and many more negatively charged ions in a thin layer on the inside (Figure 27.4b).

The sodium ions outside the membrane are like the thousands of music fans pushing and straining to get into an amphitheater for a rock concert. And just as those rock fans represent potential energy waiting to be unleashed, the separation of positive and negative charges on either side of the cell membrane represents potential energy that, once unleashed, can activate a nervous impulse in that small region of the cell. A nerve impulse is generated in a neuron's membrane as ions (electrically charged atoms) flow into and out of the cell.

A Neuron's Resting Potential

Since there are more positive charges surrounding the neuron than occurring inside of it, the inside is less positively charged and hence is negative with respect to the outside of the cell (see Figure 27.4a, step 2). The charge difference equals an *electrical potential:* a measurable amount of potential energy. In a neuron at rest, measurements reveal that this *resting potential* is about 70 millivolts (mV). (A millivolt is one-thousandth of a volt. For comparison, a typical flashlight battery has a potential of 1500 mV.) Since the inside of the neuron is negative with respect to the outside, we say that the cell's resting potential is -70 mV (step 3).

Now, we know that a gate keeps rock fans outside a stadium. But what structure or force maintains a neuron's resting potential? In fact, three proteins embedded in the cell membrane function as two channels and a pump that keep the resting cell at -70 mV. The **potassium channel** is a protein-lined pore that allows potassium ions to pass through the membrane and leak down their concentration gradient from the inside of the cell to the outside (step 4). In contrast, a second protein-lined pore, the **sodium channel,** closes tightly and allows almost none of the sodium ions outside the cell to leak inward (step 5). The slow exodus of some potassium ions leaves behind a collection of negatively charged proteins too bulky to pass through the membrane, which gives the interior a slightly negative charge.

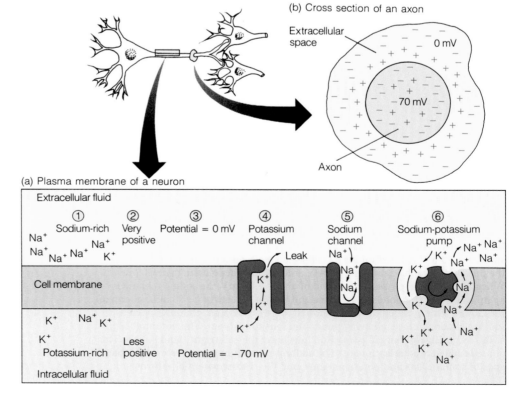

FIGURE 27.4 ■ Ions, Channels, and Pump Proteins in a Neuron's Membrane Establish the Cell's Resting Potential. (a) The extracellular fluid outside a neuron is rich in sodium ions but poor in potassium ions, while the intracellular fluid is rich in potassium ions, poor in sodium ions (1). The leakage of potassium ions out of the cell leaves the inside of the cell with fewer positive charges (2), and hence it is more negative than the outside of the cell. This results in an electrical potential of -70 mV inside the cell (3). Leaky potassium channels in the membrane allow potassium ions to leak out (4), but closed sodium channels prevent an influx of sodium ions (5). In addition, sodium-potassium pump proteins in the membrane (6) actively move two potassium ions in for every three sodium ions pumped out. (b) An excess of positive ions exists in a thin layer outside the neuron's membrane, while an excess of negative ions lies just inside the membrane.

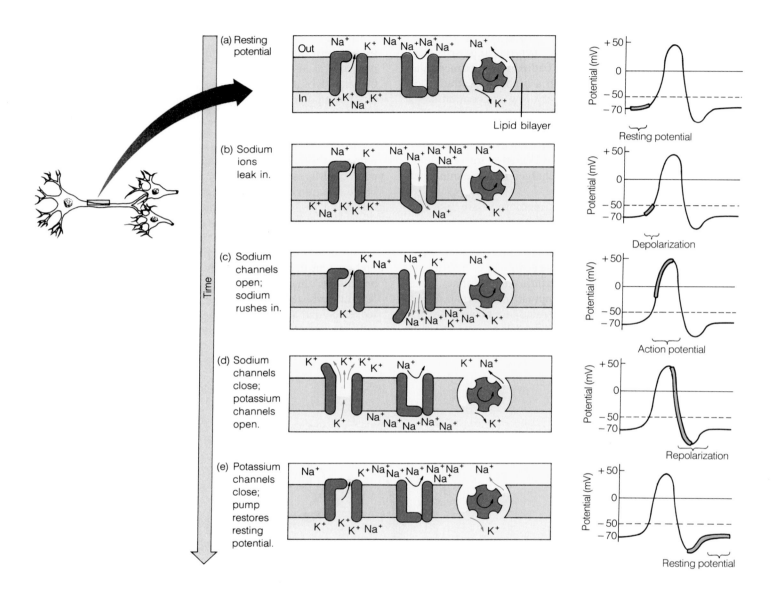

FIGURE 27.5 ■ How an Action Potential Is Generated in a Nerve Cell. A sudden change in the ion balance outside and inside of a neuron's plasma membrane leads to an action potential. (a) The neuron is quiet, and the resting potential stays at −70 mV. (b) A stimulus has allowed some sodium ions to leak in, raise the potential (graph, right-hand panel), and partially depolarize the cell. (c) When the potential passes a threshold of about −50 mV, the sodium channels open wide. Now the potential reverses and becomes positive (+50 mV, the top of the curve). This is called the action potential. (d) Potassium channels open as sodium channels close, allowing potassium ions to rush out. The potential begins to fall as positively charged potassium ions leave the cell (right-hand panel). This phase is called repolarization. (e) The sodium-potassium pump restores the balance of sodium ions and potassium ions, and the resting potential returns.

The third protein that helps maintain the neuron's resting potential is the *sodium-potassium pump* (step 6). The pump opposes the leakage of ions through the channels. For example, potassium diffuses down its concentration gradient out of the cell (step 4), but the pump brings potassium back into the cell. The reverse holds for sodium. Just as it would require energy for a child to keep a bunch of basketballs near the high end of a sloping playground, so the pump uses ATP to pass potassium and sodium ions against their path of diffusion. (Similar pumps work continuously in every animal cell and collectively use more ATP than any other body activity.) The actions of the three kinds of proteins in the neuronal membrane maintain a steady electrical potential of -70 mV between the cell's outside and its inside (see Figure 27.4b). In its resting state, the cell is said to be **polarized;** it has an imbalance of electrical charges.

The Action Potential: A Nerve Impulse

If a resting neuron already has an electrical charge—the resting potential—then what is a nerve impulse? A nerve impulse, or **action potential,** is the reversal of the charge on a resting cell. Whereas the resting nerve cell has a negative electrical charge, the action potential reverses the charge to positive. Let's examine how an action potential is generated in a nerve cell. When a specific neuron in the resting state (Figure 27.5a) is stimulated—perhaps by a chili pepper, a lark's song, or another nerve cell—some of its tightly closed sodium channels open, so that a few extra sodium ions leak into the cell (Figure 27.5b). As sodium ions continue to enter, the difference in charges between inside and outside decreases, and a gradual change takes place in the electrical potential across the membrane: -70 mV changes to -68 mV, to -62 mV, to -54 mV, and so on (see Figure 27.5b, right-hand panel). Since the electrical potential, or polarization, is decreasing, the cell is becoming **depolarized.**

When the difference in potential reaches a specific *threshold*—about -50 mV—something quite dramatic occurs: In that patch of membrane where the stimulus first arrives, most of the sodium ion channels open wide (Figure 27.5c). When sodium ions rush into the cell through the now-permeable membrane (like rock fans rushing in through unlocked turnstiles), the inside of the neuron becomes positively charged with respect to the outside. The electrical potential finally reaches a peak of about $+50$ mV, and this reversed polarity is the action potential.

Sodium channels remain open for only about 1 millisecond (one-thousandth of a second) and then close spontaneously. For a few milliseconds after closing, they cannot open again, and this state of temporary shutdown is called the **refractory period.** This property of the sodium channels limits the number of action potentials a neuron can experience (i.e., the number of times it can fire) each second.

About the time the sodium channels close, the potassium channels open fully, allowing potassium ions to rush out of the cell (Figure 27.5d). With this outpouring, the electrical potential falls to a level below the original resting potential of -70 mV. Eventually, the potassium channels close, and the continued "bailing" action of the sodium-potassium pump restores the neuron to its original resting potential (Figure 27.5e). The neuron is now ready for another environmental stimulus to cause sodium gates to open and trigger a new action potential.

The opening and closing of the gates in ion channels is the crucial step in the generation of a nerve signal, and you have probably experienced the effects of closed gates in your dentist's office. The novocaine that a dentist injects prevents the ion channels in nerve endings in the gums and tongue from opening their gates. Since ions cannot pass through the closed gates, there can be no action potential in the neurons that relay pain signals to your brain. With this lack of response, you feel little pain, even when the dentist's drill stimulates an exposed nerve.

Action potentials have a peculiar property: They are *all-or-none responses;* that is, either they do not occur at all, or when they do, they are always the same size for a given neuron, regardless of how powerful the stimulus may be. A jab to the ribs is more intense than a tender caress, not because the jab causes larger or faster action potentials, but because it causes more cells to fire more impulses.

Propagating the Nerve Impulse

Once a neuron receives a stimulus at a particular patch of its membrane, the resulting action potential generated in that local region is communicated along the length of the cell. To wiggle your toe, for example, nerve cells send (or "propagate") an action potential all the way down the axon from the spinal cord to the nerve cell boutons in the muscles that move your toe. That communication depends on the action potential passing from one patch of the membrane to an adjacent patch, then to the next, in a process called **propagation** of the nerve impulse.

The three frames of Figure 27.6 represent a portion of an axon's membrane at three different times. The figure shows how a wave of open sodium gates passes along the axon, a bit like football fans rising and cheering in a "wave" that circles the stadium. The open gates allow sodium ions to rush into the cell and generate a local action potential, which then moves along the axon.

Immediately outside a region experiencing an action potential, there are few sodium ions. Sodium ions nearby then diffuse into this space, moving from a region of high concentration to a region of low concentration (see Figure 27.6a). The diffusion of sodium ions away from the nearby patch causes this patch of membrane to lose its polarity (review Figure 27.5c). Since depolarization causes sodium channels to

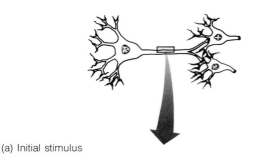

(a) Initial stimulus

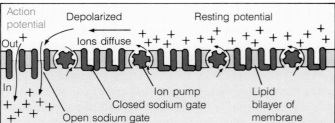

(b) A bit later

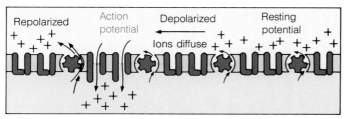

(c) Later still

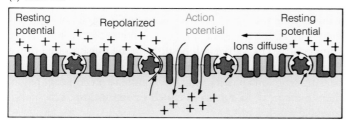

FIGURE 27.6 ■ How an Action Potential Propagates Along a Nerve Cell. As a neuron fires, an action potential moves along the plasma membrane for the entire length of the cell. For simplicity, we show only the movement of sodium ions (positive charges). (a) Depolarization occurs as sodium ions rush into the cell at the left of the figure. Outside the cell, sodium ions diffuse from an adjacent membrane patch on the right toward the first patch on the left, leaving the right patch with fewer sodium ions than normal. This decrease in sodium ions depolarizes the right patch. This leads (b) to an action potential in the patch to the right and a repolarization of the original membrane patch. In (c), the process is repeated. An impulse travels along the membrane in this way, with each adjacent patch depolarizing and reaching an action potential, until the wave of activity reaches the end of the entire nerve cell at the axon terminal.

fling wide open, now a full-fledged action potential is generated in this second patch (see Figure 27.6b). Meanwhile, the continuous pumping activity of the sodium-potassium pump repolarizes the first membrane patch. The result is that the action potential has moved along the neuron's axon, from left to right.

This series of three steps—ion diffusion, depolarization, and action potential—takes place sequentially again and again as the action potential travels the entire length of the axon, from your spinal cord, say, to your toe. Unlike the electrical current in a copper wire, which dies out over distance, a nerve impulse in a living neuron remains just as strong when it reaches the axon terminal in the tip of the toe as when propagation began back in the spinal cord.

Interestingly, the impulses in a given neuron travel in one direction only (from spine to toe, for example, but not from toe to spine), and the explanation for this is that the refractory period prevents reverse propagation. For a few milliseconds after any action potential, a membrane patch cannot experience another action potential; thus, in nature, the signal can be propagated forward to the next patch, but never moves backward toward the previous one.

In many invertebrates, a nerve impulse traveling down a neuron moves only about 2 m (6½ ft) per second. A tennis match would be a sluggish affair indeed if players had to wait 1 second for an impulse to travel from their brains to their toes. One way the propagation of nerve signals can be speeded up is for the diameter of the neuron to increase, since a larger axon has less resistance to the flow of ions, just as a wider pipe has less resistance to water flow than a narrower one. A few invertebrates, such as the fast-moving squid, have in fact evolved huge axons over 1 mm in diameter that carry impulses rapidly (Figure 27.7). A person, however, could never make do with giant axons: To contain the huge number of interlocking nerve cells we need to produce complex actions and thoughts, our heads would have to be ten times bigger. Instead, humans and other vertebrates have developed a means of rapid impulse propagation without giant neurons.

In vertebrates, fatty sheaths insulate neurons and speed impulse travel. *Schwann cells* and other special supportive glial cells extend along an axon like fatty sausages on a string and wrap their plasma membranes around the axon surface like sticky straps (Figure 27.8a). Together, the Schwann cells form the **myelin sheath,** a lipid-rich layer that insulates the axon from the extracellular fluid (see Figure 27.8). In the tiny gaps between these end-to-end Schwann cells lie bare regions known as *nodes of Ranvier,* and only in these uninsulated nodes can ions flow across the axon membrane. As a result of this cellular arrangement, the nerve impulse fairly leaps from one node to the next, bouncing along the axon 20 times faster than if the axon lacked the myelin sheath. Bounding propagation, or *saltatory conduction,* in a neuron with myelin insulation permits us the rapid reactions and movements we need to return a smashing tennis serve or perform a Rachmaninoff concerto on the piano. Even with saltatory conduction,

(a) A squid

(b) A giant axon from a squid

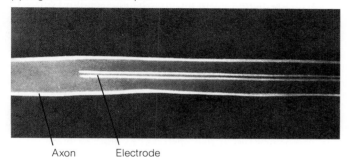

Axon   Electrode

FIGURE 27.7 ■ Giant Neurons of the Squid Propagate Nerve Impulses Rapidly. (a) Although this squid is only about 20 cm (8 in.) long, some squid can become true monsters of the deep; they can reach 15 m (49 ft) in length, their tentacles can be as large as good-sized tree trunks, and their suckers can be as wide as a person's fist. Nerve impulses can travel quickly in some axons of a squid because these axons are over 1 mm (0.039 in.) in diameter. (b) A neurophysiologist has inserted a fine glass electrode into a squid's giant axon to measure the cell's electrical potential.

however, nerve impulses travel at only about 85 m (279 ft) per second, 3000 times slower than the speed of electricity in a wire.

To see the value of myelin insulation, notice the mouse in Figure 27.8c that is shivering uncontrollably. This mutant mouse lacks myelin sheaths on its axons, and as a result, signals move erroneously between the axons from side to side as well as from end to end. Likewise, the human disease *multiple sclerosis* (MS), whose symptoms usually first appear in early adulthood, causes some nerve cells to lose their myelin sheath. Without the sheath, nerve impulses can no longer leap along the cell, and transmission slows, leading to double vision, weakness, and wobbly limbs.

In summary, then, a stimulus causes a neuron's membrane to depolarize, thereby generating an all-or-none action potential that propagates down the axon. If the axon is very wide or is wrapped in insulating myelin, then the impulse travels rapidly. But this gets the message only to the end of the first

neuron. For the message to be integrated in the brain or spinal cord and stimulate an appropriate reaction in the organism, it must cross from that initial cell to another cell and then another and another in a molecular relay. The next section explains how such internal relaying works.

HOW NEURONS COMMUNICATE ACROSS SYNAPSES

As we saw in the chapter's introduction, cocaine damaged a part of the baby's brain and paralyzed his right side. Nerve cells that normally carried signals to the right side of his body

(a) A neuron with myelin sheath

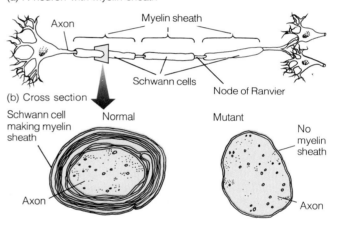

Axon   Myelin sheath

Schwann cells   Node of Ranvier

(b) Cross section

Schwann cell making myelin sheath   Normal   Mutant   No myelin sheath

Axon   Axon

(c) Normal and mutant mice

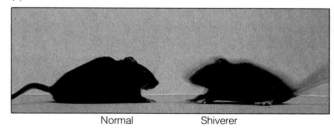

Normal   Shiverer

FIGURE 27.8 ■ Myelin Sheath: An Insulated Axon Propagates Impulses More Efficiently. (a) Vertebrate neurons may be wrapped in Schwann cells, which make up the insulating layer called the myelin sheath. Schwann cells are separated by bare regions, or nodes of Ranvier. The action potential can jump from one node to the next, allowing rapid propagation of the nerve impulse. (b) Cross section of neurons from a normal mouse (left) shows Schwann cells wrapped around axons. Cross section of neurons from a mutant *shiverer* mouse (right) reveals no encircling Schwann cells. (c) A normal mouse (left) and a mutant *shiverer* mouse (right). The shiverer is unable to control its body movements because without myelin sheaths, adjacent axons are not electrically insulated.

were destroyed; this interrupted communication, a primary function of the nervous system. This type of cell-to-cell signal transmission can involve a neuron passing information to another neuron or a neuron communicating with a muscle cell or a gland, but regardless, the signal must somehow cross the specialized junction, or *synapse,* between a neuron and a neighboring cell. A typical human neuron has so many branches that it forms between 1000 and 10,000 synapses with other cells. These junctions usually connect the axon of one cell to the dendrites or soma of another (see Figure 27.3), but synapses sometimes join axon to axon or dendrite to dendrite. In any case, synapses generally conduct signals in only one direction, from the sending (presynaptic) cell to the receiving (postsynaptic) cell.

Neurons have evolved two different ways of conveying messages to other cells. Some neurons are electrically coupled to others and communicate directly via this electrical connection, while uncoupled neurons are separated by a space, or cleft, and rely on chemicals that drift across the cleft through a fluid.

Electrical Junctions: Rapid Relayers

Most groups of animals, including coelenterates, arthropods, mollusks, and vertebrates (our own species included) have some neurons that connect to other cells through *electrical synapses.* In such cases, membranes of the sending (presynaptic) cell and the receiving (postsynaptic) cell are tightly fastened together by gap junctions (Figure 27.9). At such a junction, the ions that generate an action potential in the presynaptic cell can flow directly through pores into the postsynaptic cell and propagate the impulse in a direct electrical connection, somewhat like a phone line.

Message relay is especially rapid at these electrical junctions, and that speed is a valuable trait for certain kinds of neurons. In the giant motor neurons of the crayfish, for example, lightning-fast relay allows the animal to flip its tail instantly and escape a predator. Unfortunately, tight interconnections that allow such rapid transmission also leave little opportunity for the neuron to modify the ways it relays a signal. It is no surprise, therefore, that electrical junctions are vastly outnumbered in higher animals by chemical synapses, which function more slowly and require more steps, but also have more potential places to modify messages for specific conditions.

Chemical Synapses: Communication Across a Cleft

In vertebrates, most neuron-to-neuron and neuron-to-muscle cell communication takes place across junctions called *chemical synapses* (Figure 27.10). The axon ends at a knoblike bouton (see Figure 27.10a). Between this knob and the flat surface of the receiving cell lies a minute but distinct cleft

about 20 nm wide. The molecular messenger, or *neurotransmitter,* crosses this cleft and relays the nerve signal from sender to receiver (review Figure 26.5b).

■ **Crossing the Synapse** A neuron makes neurotransmitters in its cytoplasm, then loads the messenger molecules into small round packets called **synaptic vesicles,** which accumulate in the knoblike bouton (see Figure 27.10a). When an action potential reaches a bouton (see Figure 27.10b, step 1), it causes calcium ions (Ca^{2+}) to rush into the bouton. The increase in calcium ions stimulates the synaptic vesicles to fuse with the membrane of the presynaptic (sending) cell (step 2), liberating thousands of neurotransmitter molecules into the synaptic cleft (step 3). These molecules rapidly diffuse across the cleft and bind like keys to locklike receptor proteins embedded in the membrane of the postsynaptic (receiving) cell (step 4). This binding can cause the postsynaptic cell to generate an action potential (step 5). If the receiving cell is a muscle cell, it will contract; if it is another neuron, an action potential might be triggered in that receiver. The importance of receptor proteins is demonstrated by the condition myasthenia gravis. In people with this disease, the muscle cells have few receptors for neurotransmitter molecules; as a result, the muscle cells cannot respond normally to nerve signals, and the patients experience muscular weakness and fatigue.

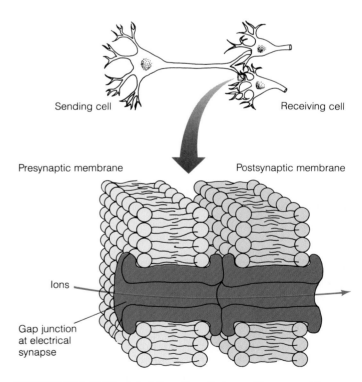

FIGURE 27.9 ■ **Electrical Junctions Between Neurons.** At an electrical synapse, the plasma membranes of two adjacent neurons are tightly joined at gap junctions that allow ions to flow from one cell to another and thus to propagate an action potential smoothly and immediately.

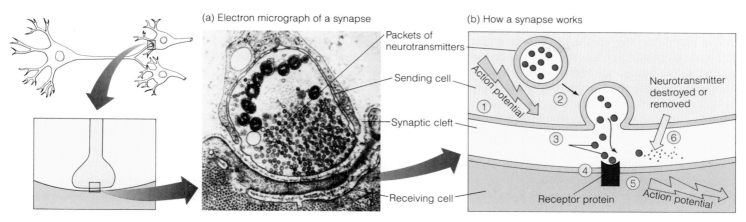

(a) Electron micrograph of a synapse

(b) How a synapse works

Packets of neurotransmitters

Sending cell

Synaptic cleft

Receiving cell

Action potential

Neurotransmitter destroyed or removed

Receptor protein

Action potential

FIGURE 27.10 ■ At a Chemical Synapse, Neurons Transmit Information Across a Cleft. (a) An electron micrograph of a synapse reveals that a cleft separates the sending (presynaptic) cell, with its bouton containing packets of neurotransmitter, from the receiving (postsynaptic) cell. (b) An action potential traveling down a sending cell and reaching the bouton (1) triggers the release of neurotransmitter (2); this messenger then diffuses across the synaptic cleft (3) and binds to a receptor protein (4), and this can trigger an action potential in the receiving cell (5). Cleanup activities then remove excess neurotransmitter (6), and as a result, the receiving cell fires a single time. This cleanup fails to occur normally in the "reward centers" of the brain in the presence of cocaine.

■ **Postcommunication Cleanup**   Once a neurotransmitter enters the cleft and triggers the receiving cell, it is enzymatically degraded (see Figure 27.10b, step 6) or reabsorbed into the bouton. Without this rapid "cleanup," the messenger molecules would remain in the cleft and would continuously stimulate the postsynaptic cell. This is what happens in a cocaine high. The drug blocks the cleanup of a specific neurotransmitter (dopamine) in cells of the brain's so-called "central reward system" (review Figure 27.1). Since the neurotransmitter acts for a longer time at these "pleasure" synapses, the person feels euphoria and excitement. But homeostatic mechanisms in the body soon begin to compensate by destroying the transmitter faster and more efficiently when cocaine is not present. Thus, without the drug, the reward centers of acclimated users are starved of stimulation, preventing them from feeling joy at normally pleasurable activities such as good food and good sex. Instead, the addict's body craves another "hit" of cocaine and thus begins a destructive cycle of drug-seeking behavior called **addiction.** In fact, cocaine addiction takes hold faster than addiction by any other drug, including heroin. Numerous animal species, from mice to monkeys, quickly learn to push a lever to obtain cocaine (Figure 27.11). Laboratory mice will experience their first "high," crave the drug, become addicted, and die from a self-delivered overdose, all within one week's time. Table 27.1 on pages 562 and 563 summarizes the known effects on the human body of many illicit, addictive drugs, as well as numerous legal prescription drugs.

FIGURE 27.11 ■ Animal Addict. If given unlimited access to cocaine and left alone, a mouse will learn to take doses of cocaine, become addicted, and eventually consume a lethal overdose of the drug—all in less than a week's time. The mouse shown here has learned to press the lever repeatedly with its paw, and this activates a spout to the left. When the mouse touches its tongue to the spout, a small amount of cocaine is delivered directly into its mouth. This continues until the animal dies from self-administering an overdose.

■ Types of Neurotransmitters Dopamine, the neurotransmitter involved in cocaine addiction, is just one of more than 50 different chemicals that can serve as neurotransmitters in the nervous systems of various animals. The best understood are *acetylcholine* and *norepinephrine*. Besides its role in transmitting nerve impulses within the brain, acetylcholine also relays messages from neurons to the skeletal muscles involved in maintaining posture, breathing, and the movement of limbs. Norepinephrine, found in synapses throughout the brain, seems central to keeping mood and behavior on an even keel. Researchers are not yet certain what the chemical does, but they believe that a deficiency of norepinephrine within the brain may underlie clinical depression, with its pessimistic moods, disturbed sleep, and low appetite and energy

levels. Conversely, they think that an excessive amount of the transmitter may help trigger mania, with its elevated or irritable moods, overactivity, and recklessness. In nerve cells outside the brain, norepinephrine works together with a second transmitter called epinephrine in controlling an animal's fight-or-flight response (review Figure 26.12).

Besides dopamine, acetylcholine, and norepinephrine, the growing list of neurotransmitters includes glycine and other amino acids, many peptides, and the specific substances serotonin and gamma-aminobutyric acid (GABA). Each of these plays specific roles in behavior, and their absence can lead to disease. Norepinephrine, as already mentioned, as well as serotonin, may be involved in clinical depression. And decreased quantities of dopamine are associated with the

TABLE 27.1   Some Common Drugs and How They Affect the Nervous System

Drug Class/Drug	Common Name or Source	Actions on Body	Effects on Mood	Dangers of Abuse
Opiates				
Morphine, heroin, codeine	Horse (heroin)	Causes drowsiness after binding to opiate receptors on neurons	Causes euphoria	Physical dependence, nausea, vomiting, constipation, death from respiratory failure in fatal doses
Stimulants				
Cocaine	Coke, crack	Raises heart rate and body temperature, dilates pupils	Causes euphoria, excitation	Convulsions, hallucination, cardiovascular damage
Amphetamine, methamphetamine	Dexedrine, crystal, meth, crank, cross tops	Increases heart rate, respiration, and blood pressure	Causes wakefulness, depresses appetite, increases alertness	Irregular heartbeat, chest pain, dizziness, anxiety, paranoia, hallucinations, convulsions, coma, cerebral hemorrhages in fatal doses
Nicotine	Tobacco	Causes vasoconstriction, racing heart, increased blood pressure	Acts as stimulant; causes euphoria	Dizziness, nausea, vomiting, withdrawal in addicted users
Caffeine, theophylline	Coffee, tea	Stimulates central nervous system and visceral muscles, increases heart rate and urine production	Causes some mood elevation, increases alertness	Anxiety, nervousness, insomnia, convulsions
Sedatives				
Alcohol	Beer, wine, liquor	Decreases respiration and body temperature, causes vasodilation, impairs vision, depresses central nervous system, numbs pain	Causes euphoria	Damage to liver, heart, brain, pancreas; respiratory failure in fatal doses

rigidity, weakness, and shaking of Parkinson's disease. These symptoms sometimes diminish when the patient takes oral doses of L-dopa, a chemical that brain cells can modify into the neurotransmitter dopamine.

Some compounds act as **neuromodulators** rather than neurotransmitters. Compared to the relatively quick, short-lived influence that a neurotransmitter has on the postsynaptic membrane, the neuromodulator may remain in the vicinity of the synapse for a longer time and make the receiving cell more or less responsive to neurotransmitters. The take-home message here is this: Neurons generally communicate with other cells by releasing a neurotransmitter at a synapse. The neurotransmitter diffuses toward a receptor on another cell and causes an alteration in the receiving cell's behavior.

■ **Excitatory and Inhibitory Synapses**   Chemical communication at synapses can either "encourage" or "discourage" the firing of the receiving (postsynaptic) cell. At an *excitatory synapse,* the molecular messenger brings the postsynaptic cell closer to firing a nerve impulse by causing a few channels for sodium ions to open and make the interior of the cell a bit more electrically positive. Thus, it takes fewer excitatory impulses to trigger an impulse in the postsynaptic cell.

In contrast, at an *inhibitory synapse,* the neurotransmitter decreases the likelihood that the postsynaptic cell will fire a nerve impulse by opening channels to potassium or chloride ions or both. By making the receiving cell's interior even more negative, it takes many more excitatory impulses than usual to trigger an impulse in the postsynaptic cell.

TABLE 27.1 (continued)

Drug Class/Drug	Common Name or Source	Actions on Body	Effects on Mood	Dangers of Abuse
Sedatives (continued)				
Methaqualone	Quaalude	Sedates, causes sleep, depresses central nervous system	Quieting	Nausea, vomiting, delirium, convulsions, coma, slow heart rate
Benzodiazepine	Valium, Librium	Relaxes skeletal muscles; decreases circulation, respiration, and blood pressure	Reduces anxiety, elevates mood	Depressed heart and lungs, decrease in muscular coordination, drowsiness; withdrawal can trigger seizures
Hallucinogens				
Cannabinoids	Marijuana, pot, grass, hash	Increases heart rate, causes vasodilation	Elevates mood, causes euphoria, sensory distortions	Lung damage from smoking; reduced sperm count, low testosterone in males
Lysergic acid diethylamide, psilocybin, mescaline	LSD, acid, mushrooms, peyote	Acts as a stimulant, raises heart rate and blood pressure, dilates pupils	Causes euphoria, hallucinations, sensory distortion	Irrational behavior
Dissociative Anesthetic				
Phencyclidine	PCP, angel dust	Kills pain, increases heart rate and blood pressure, causes swelling and fever	Elevates mood, causes perceptual disturbances	Coma, convulsions, psychosis, respiratory depression

■ **Summation** Since most neurons receive synaptic input from the axons of up to thousands of other cells, the ultimate activity in any given region of the cell results from the cumulative impact of all the molecular messages—some excitatory, some inhibitory—ferried across all the synaptic clefts of that cell. Whereas the firing of many impulses at an excitatory synapse might be enough to cause a postsynaptic neuron to depolarize, an inhibitory synapse nearby that receives an equal number of impulses could counteract the excitatory effect. In a process called **summation,** the postsynaptic cell sums up all the information impinging on it in a nearby region and over a short time interval. Eventually, if the "encouragements" outnumber the "discouragements," the threshold is reached within that cell, the chain reaction of electrochemical

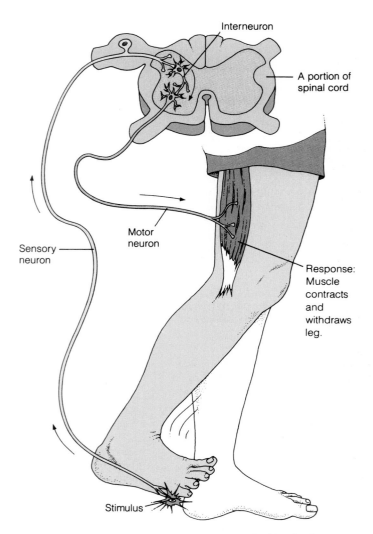

Interneuron

A portion of
spinal cord

Motor
neuron

Sensory
neuron

Response:
Muscle
contracts
and
withdraws
leg.

Stimulus

FIGURE 27.12 ■ Anatomy of a Reflex Arc. Jerking back your foot after stepping on a tack is based on a reflex arc that links the spinal cord and the muscles in the back of the thigh. A sensory neuron detects the tack and sends an action potential along its axon toward the spinal cord. The sensory neuron forms a synapse with an interneuron within the spinal cord. The interneuron links to a motor neuron that runs to certain leg muscles, causing the muscles to contract and lift the foot away from the tack.

events that constitutes an action potential is then automatically triggered, and the neuron fires an impulse in an all-or-none fashion.

The variety of neurotransmitters and neuromodulators at work, as well as the countless ways neurons link together, gives neuronal networks a rich repertoire of potential responses, despite the fact that the nerve impulse itself is exactly the same in each neuron and its neighbors. Our final section focuses on how networks of neurons—which are still relatively simple systems compared to the unbelievable intricacy of the brain and nervous system—interact to control behavior.

HOW NETWORKS OF NEURONS CONTROL BEHAVIOR

Although we have been discussing neurons as if they sit end to end like matchsticks on a table, they are actually arranged in delicate circuits—a spatial organization that allows the animal to receive, integrate, and act on information from its surroundings and thus to survive and reproduce. In many different animals, a basic type of nerve circuit called the **reflex arc** underlies simple behaviors. This simple neural loop links a stimulus to a response in a very direct way. If you step on a tack with your bare toe, for example, a reflex arc drives muscles in your leg to pull your toe away immediately, even before you consciously realize what has happened (Figure 27.12). Reflex arcs mediate behaviors that are generally rapid, involuntary, and nearly identical each time the stimulus is repeated and that require no conscious input from the brain.

Reflex arcs usually involve three kinds of neurons: sensory neurons, interneurons, and motor neurons (see Figure 27.12). A **sensory neuron** receives information from the external or internal environment—a tack stabbing a toe is this kind of information—and transmits it toward the spinal cord or brain, the nervous system's integrating centers. In those centers, **interneurons** relay messages between nerve cells and integrate and coordinate incoming and outgoing messages. While some reflex arcs do not have interneurons, complex behaviors rely on a network of many interneurons. Interneurons often connect with **motor neurons,** which send messages from the brain or spinal cord out to muscles or secretory glands. Motor neurons allow an organism to act, to respond to the information the sensory neuron brings in.

Learning Results from Changes at Synapses

Organisms constantly adjust to their environments. When conditions change, an animal's behavior changes, enabling the animal to meet new challenges or seize new opportunities. One phenomenon that allows an organism to adapt to a changing environment is **habituation:** a progressive decrease in the strength of a behavioral response to a constant weak stimulus when that stimulus is repeated over and over again with no

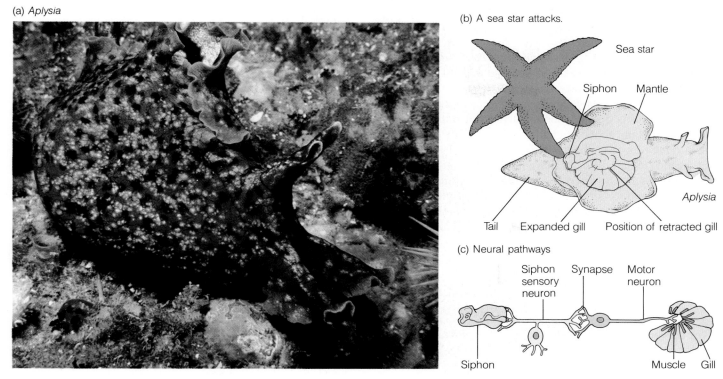

(a) *Aplysia*

(b) A sea star attacks.

(c) Neural pathways

FIGURE 27.13 ■ Simple Neural Circuits Control *Aplysia*'s Gill Retraction Response. (a) *Aplysia californica,* dweller of Pacific tide pools. (b) The touch of a predatory sea star causes *Aplysia* to retract its delicate siphon and gills. (c) The neural circuit that controls gill withdrawal is simple. Sensory neurons in the siphon detect the sea star's caress and send an action potential toward synapses. The release of neurotransmitter triggers motor neurons to cause muscles to contract and move the gill to safety. Habituation—in this case, a decreased contraction of the gill muscle after an experimenter repeatedly stimulates the siphon—occurs as a change in the strength of the synapse.

undesirable consequences. Habituation takes place when you get dressed in the morning. When you first put on your clothes, you can feel them against your skin, but within a few minutes, you become unaware of the constant light pressure. Biologists have investigated the cellular basis of this simple type of learning in the sea slug *Aplysia californica* (Figure 27.13a).

The sea slug has a tube called a siphon that draws water over the gills (Figure 27.13b). If a slug senses water currents from the movements of a nearby animal, say a predatory sea star, the slug retracts its siphon and gills. This response helps the animal to survive, since gills act like bait for hungry predators. The neural circuit that controls this reflex depends on two sets of neurons (Figure 27.13c). Branchlike dendrites of the sensory neurons in the siphon detect any distortion in the tissue around them and respond by initiating an action potential. The axons of these sensory neurons in the siphon form a synapse with motor neurons. The motor neurons, in turn, connect to muscle cells that can contract when directed by an incoming nerve impulse, withdrawing the gills to safety.

This gill retraction behavior is subject to habituation based on experience. The tenth blast of a brief jet of water from an experimenter's probe causes *Aplysia* to retract its gills only half as much as did the first blast. This shows that the animal has habituated and learned to ignore a stimulus with trivial consequences. But what events in the nerve pathway allow habituation? Researchers have found that after habituation, the sensory neuron releases less neurotransmitter. As a result, the motor neuron is less likely to fire, and muscles do not withdraw the gills as far.

What causes the sensory neuron to release less neurotransmitter? Although no one knows why, after habituation, fewer calcium channels open in the presynaptic cell, and so less neurotransmitter is released. Through their research, biologists were able to show that a simple learning process is due to a physical change in synaptic effectiveness.

After a sea slug has experienced an intense electric shock to its tail, it displays the opposite of habituation: **sensitization.** It withdraws its gills in response to the slightest touch. The explanation for sensitization also involves changes at the synapse: A neuromodulator released at the synapse causes more neurotransmitter to be released. With more neurotransmitter, more motor neurons fire more frequently, and the gill withdrawal reflex is stronger.

The simple learning responses of habituation and sensitization in *Aplysia* show that learning is due to changes in the effectiveness with which signals cross synapses. Can human learning be explained in the same way? Even simple behaviors, like scratching the end of your nose, must require the proper activation of a great many cells to ensure that the finger lands on the end of your nose and not in your eye and that the movement is a gentle scratch and not a stab. The more sophisticated the action—a ballet pirouette, say, or a reverse slam dunk of a basketball—the more cells and organization are needed to carry it out. Future researchers must determine whether straightforward rules that explain a simple behavior in *Aplysia* hold for all other animals as well. Perhaps the cellular events that control how a sea slug habituates to a jetting stream of water also control the way you come to understand principles for a biology quiz.

CONNECTIONS

The nervous system, along with the endocrine system, allows the animal body to work as an integrated whole. This chapter moved from the events within a single nerve cell to the communication between two neurons to neural networks and how they bring about and control meaningful behavior in animals. Underlying that integrated operation is the basic evolutionary principle that the nervous system, like the other physiological systems we have discussed, enhances an animal's chances of surviving and reproducing. The effect of cocaine in disrupting this integration—in causing, for example, the surge of blood pressure that destroyed part of a baby's brain—reminds us of the importance of the body's control systems.

The molecular messengers we encountered in this chapter, the neurotransmitters and neuromodulators, are similar in many ways to the hormones. Like hormones, neurotransmitters are released from one cell and diffuse through a fluid to a target cell, where they bind to a receptor protein and thereby change the target cell's activity. When evolution hits on a good idea, the pattern shows up again and again in slightly different forms.

Simple behaviors controlled by reflex arcs depend on just two or three interlinked neurons. More complex behaviors depend on millions of neurons organized into intricate nervous systems. The most intricate control center of all, the human brain, is also the most fantastically organized tissue that has yet evolved. Complex behaviors and their complex neural control organs are the subject of our next chapter.

Highlights in Review

1. The nervous and endocrine (hormonal) systems share several characteristics and work together in controlling and coordinating the body.
2. Nerve cells communicate in a point-to-point fashion, but the cells are arranged in complex networks that allow for rich behavioral repertoires.
3. Cells of the nervous system are classified as neurons or glial cells. Neurons communicate nerve impulses. They have a soma, cell processes called dendrites, an axon, and a bouton at the axon terminal. Glial cells provide nutrition, structural support, and insulation to neurons.
4. Animal responses are based on nerve impulses, chain reactions of ion flow that move along the neuron's membrane.
5. The extracellular fluid that surrounds a neuron is high in sodium ions but low in potassium ions, while the cell's interior is rich in potassium ions and poor in sodium ions. The cell's interior is electrically negative with respect to the exterior.
6. A sodium-potassium pump actively transports sodium out of and potassium into the cell against concentration gradients. Ion channels specific for either potassium ions or sodium ions open and shut, regulating ion flow across the cell membrane.
7. An action potential begins when sodium ions leak into a small region of the membrane and the cell becomes depolarized. Sodium channels in that membrane patch then open wide,

sodium ions rush in, and the inside of the neuron becomes positively charged. The sodium channels quickly close, a refractory period prevents them from reopening temporarily, and the potassium channels now open, allowing the balance of charges to become reestablished and the resting potential to return.

8. The nerve impulse propagates along the neural membrane from the patch that initially received the stimulus to adjacent patches. This occurs because the adjacent membrane patch loses some of its sodium ions and becomes depolarized, generating an action potential in the adjacent patch.
9. Propagation can take place more quickly in a giant axon or in one wrapped in a myelin sheath (formed by glial cells).
10. Neurons form junctions called synapses with other neurons and muscle cells. In electrical synapses, ions flow directly from one cell to the next. In chemical synapses, neurotransmitter molecules diffuse from one neuron, cross a gap, and trigger a nerve impulse in the next neuron.
11. Neurotransmitters include acetylcholine, involved in maintaining posture, breathing, and movement of limbs; dopamine, which is released in the pleasure centers of the brain; and serotonin, norepinephrine, and other molecules.
12. Neuromodulators remain near synapses and provide a longer, steadier influence on the receiving cell.
13. At an excitatory synapse, neurotransmitters "encourage" the

receiving cell to fire an action potential. At an inhibitory synapse, neurotransmitters "discourage" the receiving cell from firing.

14. Neurons can be linked into simple reflex arcs, composed of a sensory neuron that receives stimulation from the environment; an interneuron that integrates and coordinates incoming and outgoing messages; and one or more motor neurons that carry signals to the muscles or glands.

15. Habituation and sensitization are simple forms of learning. The physiological basis of these learned behaviors is a change in the effectiveness of signals crossing specific synapses.

Key Terms

action potential, 557	neuromodulator, 563
addiction, 561	polarized, 557
axon, 554	postsynaptic cell, 554
bouton, 554	potassium channel, 555
dendrite, 554	presynaptic cell, 554
depolarized, 557	propagation, 557
glial cell, 554	reflex arc, 564
habituation, 564	refractory period, 557
interneuron, 564	sensitization, 565
motor neuron, 564	sensory neuron, 564
myelin sheath, 558	sodium channel, 555
nerve, 554	soma, 554
nerve impulse, 555	summation, 564
nervous system, 552	synapse, 554
neuron, 554	synaptic vesicle, 560

Study Questions

Review What You Have Learned

1. List and describe the parts of a typical neuron.
2. What three kinds of protein complexes keep a neuron polarized? What role does each play?
3. Trace the events that lead to a nerve impulse.
4. True or false: A neuron's refractory period permits it to "fire" virtually without stopping. Explain your answer.
5. How is a nerve impulse propagated?
6. What is a myelin sheath? How does it affect transmission of an impulse along an axon?
7. How does an action potential cross a synapse?
8. Name two neurotransmitters, and state the function of each.
9. How do excitatory and inhibitory synapses differ?
10. Make a chart listing the three categories of neurons and the main function of each.

11. What is habituation? How does it produce a modified response?
12. What events lead to sensitization?

Apply What You Have Learned

1. Some poisonous snakes inject neurotoxins into their prey; others inject poisons that cause tissue damage. Which type of venom is likely to be better for capturing prey? Justify your answer.
2. A middle-aged woman with Parkinson's disease takes L-dopa. How does this substance relieve her symptoms?
3. Late one night, while driving down a lonely highway in the Nevada desert, a truck driver takes an amphetamine-laced "pep pill" to keep him alert. How does the pill function? What biological hazards does the driver risk in taking this pill?
4. While working as a neurobiologist, you discover a snake venom that paralyzes mammals. You suspect that the venom is disrupting or blocking one or more of the following features of neurons: (a) sodium channels, (b) sodium-potassium pumps, (c) synaptic vesicle release, (d) neurotransmitter receptors, (e) enzymes that degrade neurotransmitters. Your laboratory is equipped with microelectrodes that can stimulate and measure voltages in individual neurons, and analytical tools that can determine neurotransmitter concentrations at synapses. How would you determine the likelihood that the venom is affecting each of these neuronal features?

For Further Reading

BARINAGA, M. "The Tide of Memory, Turning." *Science* 248 (1990): 1603–1604.

BOZARTH, M. "New Perspectives on Cocaine Addiction: Recent Findings from Animal Research." *Canadian Journal of Physiological Pharmacology* 67 (1989): 1158–1167.

GOELET, P., V. F. CASTELLUCCI, S. SCHACHER, and E. R. KANDEL. "The Long and the Short of Long-Term Memory—A Molecular Framework." *Nature* 322 (1986): 419–422.

HOLDEN, C. "Street-Wise Crack Research." *Science* 246 (1989): 1376–1381.

HOLLOWAY, M. "Rx for Addiction." *Scientific American,* March 1991, pp. 94–103.

KANDEL, E., and J. SCHWARTZ. *Principles of Neuroscience.* 2d ed. New York: Elsevier, 1985.

KUFFLER, S. W., J. NICHOLLS, and A. MARTIN. *From Neuron to Brain.* 2d ed. Sunderland, MA: Sinauer, 1984.

CHAPTER 28

The Senses and the Brain

Keen Senses of a Night Hunter

Perched on an oak limb in the moonlight, a reddish brown owl rotates its head slowly. Like a ship with sonar, it scans the darkened English woods, listening for the slightest indication of a mouse scuttling across the forest floor at night. When it senses the rustle of leaves, the predatory bird swoops silently from its perch, deftly avoids a large branch looming out of the dim shadows, and pounces on the source of the sound with sharp talons outstretched. Grabbing a gray mouse with its feet and then snapping the rodent's skull with a powerful stroke of its hooked beak, the owl returns to its perch and shares the meal with its owlets.

This night hunter, the tawny owl (*Strix aluco;* Figure 28.1), is the most common owl in British forests and inhabits the woods throughout much of Europe, northern Africa, and western Asia. The tawny owl can fly silently because its outermost flight feathers end in soft fringes, not smooth, firm edges. And like other members of its avian order, the tawny owl has large, front-facing eyes. One might suspect that these piercing, ever-vigilant visual organs are chiefly responsible for the bird's hunting prowess, but in fact, a tawny owl's night vision is no better than that of some people who see particularly well at night. What's more, a simple experiment disproves the primacy of vision in the owl: If an experimenter ties a dry leaf to a mouse's tail and places the rodent in a dimly lit room with an owl, the rodent will scurry about in fright and the bird will pounce—not on the prey but on the rattling leaf!

FIGURE 28.1 ■ A Night Hunter Swoops to Catch Prey. The tawny owl can navigate in the dark by sounds and "mental maps" to catch a rodent and fly its prey back to the nest.

Clearly, a tawny owl's hearing must be very acute, and research shows that it is sufficiently sharp to allow the bird to capture prey in absolute darkness. Curiously, though, its ability to perceive sound is not superior to most people's. In both the tawny owl and the human being, the sense of hearing has reached its physical limits; both species can detect the smallest noise in a quiet environment. The owl, however, is superior at pinpointing the origin of a sound.

How does the tawny owl hunt so successfully in a darkened wood when it can see and hear only about as well as a very sensitive person? The bird's ability to *sense* slight perturbations in its environment is only part of the answer. The sensory information must also be *integrated*—that is, combined and processed—within the animal's brain. And the bird must then *react* in ways that allow it to capture its prey. The sensory phase, as we have seen, depends on keen sight and hearing. Integrating the sensory signals requires a precise mental map of where tree limbs project near the owl's perch, the location of rocks that could shelter small rodents, and which trails the mice usually traverse. This mental map serves as a guide the owl follows when sensory input is restricted, as it is at night. Similar mental maps of your own room allow you to walk to the bathroom in the dark without bumping into your desk. Such maps are based on complete familiarity with a room or corner of the woods, and the tawny owl lives its life in a restricted territory that it comes to know intimately. Finally, the bird's reaction involves coordinated muscular activity and continued sensing of body position and prey location.

Using the tawny owl as a central example, this chapter describes how sense organs work and how nervous systems integrate sensory information and thus coordinate behaviors that help animals survive and reproduce. The previous chapters in this part of the book have already demonstrated how sensing, integrating, and reacting are all vital to the maintenance of an animal's homeostasis in a changing environment. Chapter 27, in particular, showed the role of individual neurons in these activities. We will now see that nerve cells are organized into two higher-order networks: the central nervous system (CNS), which consists of the spinal cord and brain, and the peripheral nervous system, which contains all of the interacting nerve cells outside the CNS.

Four main themes will recur in our discussion of sensing and integrating. First, the nervous system's major role is to monitor internal needs and external conditions, process the information, and initiate appropriate behavioral responses.

Because of its nervous system, the tawny owl can sense hunger, detect a mouse scampering through fallen leaves, and respond by pouncing on the prey.

Second, sense organs contain excitable cells that are modified in ways that increase their sensitivity to one physical aspect of the environment. Thus, the owl's eye is most sensitive to light, its ear to sound, its nasal passages to smell, its mouth to taste, and its skin to pressure or temperature.

Third, nervous systems become more complex in more complicated animals, although they are built on similar kinds of nerve cells that act in the same ways. The nerve cells of a jellyfish resemble those of a sea slug and a tawny owl, but in higher animals, sense organs and nervous integrators become more centralized and contain many more cells.

Finally, different senses and different behaviors can be localized to specific regions or groups of regions in the brain. The human brain is the most intricately organized entity in the universe, and it is this structural organization that allows the brain to work. For example, as you read this page, a representation of the book going from left margin to right margin is mapped directly onto nerve cells in one region of your brain in a point-for-point spatial array.

Our survey of sense organs and nervous system function will answer several questions about how organisms sense the environment and integrate sensory information:

- How do animals hear and see?
- How did the brain and senses evolve?
- How do the main parts of a vertebrate's nervous system interconnect, and where in a person's brain are the seats of homeostatic control, motor coordination, sensation, emotion, facial recognition, and language?

WINDOW ON THE WORLD: SENSE ORGANS

An owl hears the rustling of a mouse in the fallen leaves. A cockroach detects light and scurries away from it. A person bites into a cherry and detects a sour taste. All such changes in the body and its surroundings are perceived by **sense organs:** groups of specialized cells that receive stimulus energy (such as wavelengths of light, sound waves, odor or flavor molecules, or pressure) and convert it into another kind of energy, the kind that can trigger a neural impulse. The nerve signals that sense organs relay to the brain answer two types of questions: What kind of stimulus did the sense organ receive (e.g., what is the pitch of the sound, the color of the light, or the nature of the chemical), and how strong is the stimulus (is it loud or soft, bright or dim, concentrated or dilute)? Our day-to-day survival depends on accurate answers to these questions.

A sense organ's receiving cells have bare nerve endings not covered by a myelin sheath (Figure 28.2). Pain receptors in the skin are examples of this. There are several different kinds of receptor cells, each a specific shape and able to make special proteins. Other affiliated cells screen out extraneous forces and amplify the effects of the specific stimulus before it reaches the receptor. For example, the eye's light receptor cells make light-sensitive proteins, while nearby support cells focus light on the receptors. In contrast, pressure-sensing cells in the skin detect physical bending, and the support cells mag-

nify the bending (see Figure 28.2b). While sense organs are generally "tuned" to specific sets of stimuli, any mechanical, chemical, or light (electromagnetic) stimulus of sufficient intensity can cause them to generate an action potential. A person bumped on the eye or head, for example, may "see stars" because cells in the visual pathway are sensitive to pressure as well as to light.

Just as the tawny owl depends mainly on its hearing and mental maps, different animals rely on different senses, depending on how they exploit their own particular environment. Bees are highly visual, for example, while bloodhounds rely heavily on the sense of smell. Some animals even utilize physical forces that we humans can detect only with highly sophisticated instruments. For example, rattlesnakes have a pair of sensory structures called pit organs beneath the eyes that enable the reptiles to "visualize" objects in the dark by detecting the heat patterns the objects give off (Figure 28.3); and certain fishes have sense organs that act like voltmeters that detect distortions in the electric field.

The two senses that people rely on most heavily in their day-to-day lives are hearing and vision—the same ones the tawny owl employs in its night hunting. For these reasons, we focus primarily on the mechanisms of hearing and vision in the following sections. Because smell and taste are so crucial to the survival of so many other vertebrates, however, we also describe those senses later in this section.

The Ear: The Body's Most Complex Mechanical Device

With over a million moving parts, the **cochlea** of the inner ear is the most complex mechanical apparatus in the human body and in the bodies of most other vertebrates. Cells in this coiled apparatus detect sound by first sensing subtle movements in

(a) Pain receptor   **(b) Pressure receptor**

Sensory neurons   Support cells

FIGURE 28.2 ■ **Pain and Pressure Receptors.** (a) The simplest sensory neurons, pain receptors, have nerve endings surrounded by a thin layer of support cells. Pressure on those support cells stimulates the receptor nerves directly. (b) Pressure receptors project into a capsule of support cells. The spatial arrangement of these cells increases the effects of pressure. The cells trigger an action potential in the pressure receptor neuron itself.

FIGURE 28.3 ■ **The Pit Viper's Pit Organ: A Heat Detector Senses Prey.** A rattlesnake—this one a resident of Venezuela— has a black cavity below and in front of each eye that works a bit like a night spotting scope. A thin membrane stretching across the back of the cavity detects heat and allows the snake to locate a rabbit, mouse, bird, or other warm prey in the dark.

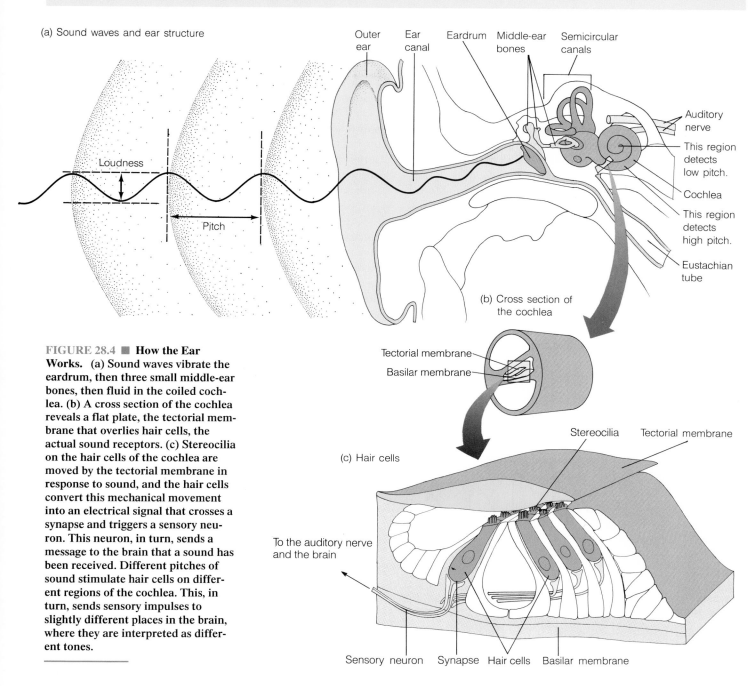

(a) Sound waves and ear structure

Loudness

Pitch

Outer ear

Ear canal

Eardrum

Middle-ear bones

Semicircular canals

Auditory nerve

This region detects low pitch.

Cochlea

This region detects high pitch.

Eustachian tube

(b) Cross section of the cochlea

Tectorial membrane

Basilar membrane

(c) Hair cells

Stereocilia

Tectorial membrane

To the auditory nerve and the brain

Sensory neuron   Synapse   Hair cells   Basilar membrane

FIGURE 28.4 ■ How the Ear Works. (a) Sound waves vibrate the eardrum, then three small middle-ear bones, then fluid in the coiled cochlea. (b) A cross section of the cochlea reveals a flat plate, the tectorial membrane that overlies hair cells, the actual sound receptors. (c) Stereocilia on the hair cells of the cochlea are moved by the tectorial membrane in response to sound, and the hair cells convert this mechanical movement into an electrical signal that crosses a synapse and triggers a sensory neuron. This neuron, in turn, sends a message to the brain that a sound has been received. Different pitches of sound stimulate hair cells on different regions of the cochlea. This, in turn, sends sensory impulses to slightly different places in the brain, where they are interpreted as different tones.

the fluid that surrounds them. The stimulated cells then send a nerve impulse to the brain. Finally, the brain interprets the sound of hands clapping, the shrill scolding of a Steller's jay, or the deep musical "hoo-hooo-hoooo" of the tawny owl.

A sound is really a set of compressed air molecules that pushes adjacent molecules, resulting in a wave of compressed air that travels along and eventually strikes the ear (Figure 28.4a). This compression could begin with two hands clapping together or with a mouse's foot scraping a leaf. An analogous compression begins when a stone hits the still surface of a pool of water and sets up concentric water waves. Sound

waves first strike structures such as the human's flexible, sculptured outer ear or the trough beneath the feathery ruff of an owl's face. The outer ear then funnels sound waves to the *eardrum,* a taut membrane that is stretched like a drum skin across the ear canal and that vibrates in time with the sound wave frequencies. A chain of three bones in the middle ear in turn transmits eardrum vibrations to the small, coiled cochlea, the sensory receptor that converts vibrations into nerve impulses. Overall, the cochlea has a snaillike shape, compact and coiled with elastic partitions that separate it into three fluid-filled chambers (Figure 28.4b). Attached to one partition, called the *basilar membrane*, are four rows of box-shaped *hair cells*, each cell crowned by an elegant bundle of threads called **stereocilia** that stand erect like a Mohawk haircut (Figures 28.4c and 28.5). Each hair cell bears about 100 stereocilia; in total, each human ear contains 1 million.

How do the inner-ear structures enable us to detect sound? When a sound wave vibrates the eardrum and middle-ear bones, causing one of those bones to compress the cochlear fluid, the basilar membrane deflects. This, in turn, causes the hair cells to move relative to the *tectorial membrane*, a flap that lies on top of them (see Figure 28.4c). This shearing force bends the stereocilia, and each hair cell converts the mechanical stimulation into an electrical signal. Finally, the hair cell transmits the signal across synapses to a sensory neuron that leads to the brain.

Receiving the signal, how does the brain know whether the incoming sound is the high-pitched whistle of a bull elk or the low growl of a grizzly bear? (After all, as Chapter 27 explained, at the level of the neuron, all action potentials are the same.) The answer involves the mechanical properties of the basilar membrane. Unlike a string on a plucked guitar,

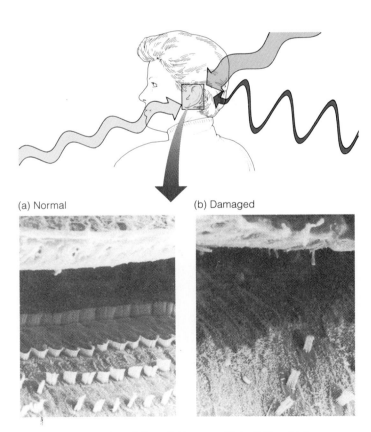

(a) Normal   (b) Damaged

FIGURE 28.5 ■ Loud Sounds Damage Hair Cells. (a) Intact hair cells from the human cochlea, highly magnified. **(b)** Violent vibrations from extremely loud sounds break off stereocilia from the hair cells, causing irreparable damage and permanent hearing loss.

which vibrates along its entire length when stimulated by the appropriate pitch, the basilar membrane vibrates in different places when stimulated by different pitches. High pitches excite motion near the wide part of the cochlea, while low pitches cause movement near the cochlea's narrow tip (see Figure 28.4a). Since each region of the membrane connects to a different part of the brain, different pitches stimulate different brain cells. In essence, a musical staff is inscribed in space across a certain part of the brain. A baby can distinguish the characteristic tones of its mother's voice because that voice stimulates hair cells at specific places in the cochlea, which in turn correspond to specific brain areas.

Animals can also distinguish loud sounds from soft ones, since loud sounds cause more neurons to fire action potentials and to fire more frequently. While a soothing voice causes the stereocilia on the hair cells to sway gently, loud sounds, such as those from a jackhammer or a heavy metal rock band, can break off the stereocilia and thus permanently damage the hair cells and along with them the sense of hearing (see Figure 28.5).

To fully sense its environment, an animal must know the direction of a sound as well as its tone qualities. Owls, for example, can localize sounds in the horizontal plane about as well as people, but the birds are much more accurate than people at locating sounds above or below them. This is because an owl's left ear is aimed higher than its right (Figure 28.6) and also because the owl has an internalized map in which each brain cell responds to sound from a different point in space.

■ **Detecting Balance and Motion**   While hair cells are crucial to hearing, they also figure prominently in detecting an animal's position in space. In a tawny owl, the **semicircular canals**—three curving tubes filled with fluid that function as organs of balance and motion detection (see Figure 28.4a)—register the bird's acceleration as its swoops down toward a scurrying mouse. Like the cochlea, the semicircular canals contain hair cells with attached stereocilia that project into the fluid filling the curving tubes. As the owl banks and dives, the detection of the motion is based on the swaying of these stereocilia on the hair cells, much as passengers standing on a bus sway when the vehicle rounds a corner. Likewise, hair cells in a trout's **lateral line organ**—a row of minute pores along the fish's side—detect slight changes in water pressure, which alert the animal to nearby prey or predators. In all cases, natural selection has utilized a simple mechanism, the hair cell, to convert a mechanical force into an electrical signal that can provide vital information and enhance the survival of different organisms in their own special environments.

The Eye: An Outpost of the Brain

The colors of the rainbow. A loved one's smile. Breathtaking mountain scenery. These delightful visual sensations are all possible because the eye converts electromagnetic radiation

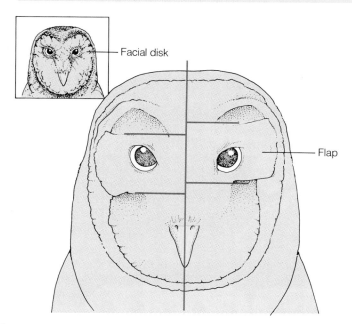

Facial disk

Flap

FIGURE 28.6 ■ How Owls Localize Sounds. Of all land animals, owls are the best at locating a moving target in three-dimensional space. While a person is as good as an owl at identifying the source of a sound in one plane—say, to the right or left while standing on the ground—owls are far better at localizing sounds that come from above or below—say, as the bird sits high in a tree. This superior ability is based on the asymmetrical positions of the owl's outer ears, revealed here with the owl's facial disk feathers trimmed away. A person can tell if a sound comes from the left, the right, or straight ahead because a sound from the left strikes the left ear first, and the brain interprets this as a direction. Owls can do the same, but can also localize sounds above or below their heads because the left ear and its sound-directing ear flap are much higher on the head than the right, as indicated by the parallel lines. Sounds from above will thus strike the left ear first, sounds from below, the right ear first. The brain compares the difference and interprets the source of the sound as above or below the owl.

in the form of light into neural energy, just as the ear changes the mechanical energy of sound into nerve impulses. Vision is our most important sense, and not surprisingly, the human eye is large—almost the size of a golf ball. For a tawny owl to see slightly better than the average person, its eyes must be even larger, even though its head is far smaller than a person's. If our eyes were proportionately as large, they would be the size of softballs!

■ **How the Eye Detects Light**   Light bouncing from a plant and striking the human eye first passes through the protective transparent outer layer, or **cornea** (Figure 28.7a). If gently place your finger over a closed eyelid and turn your eye from left to right, you can feel the bulge your cornea makes. After penetrating the clear cornea, light passes through a clear fluid and then enters the **pupil,** which is a black, circular, shutterlike opening in the **iris** (the eye's colored portion), then

traverses the **lens,** a circular, crystalline structure that focuses light through another fluid onto the **retina.** The retina is a multilayered sheet that lines the back of the eyeball and contains light-sensitive **photoreceptor cells,** or neurons; these begin converting light energy to electrical energy that can form brain patterns and create a visual image of an object. Because the lens is curved, it bends the light rays. As a result, the image of the object it projects on the retina is backward (with left and right reversed and upside down; see Figure 28.7a). Once the visual impulses reach the brain, however, they are integrated and sorted out so that the image conforms to reality.

Within the multilayered retina, the rear layer (closest to the back of the eye and just inside the eye's tough outer covering) consists of jet black pigment cells that protect the photoreceptors from extraneous light (Figure 28.7b). Nestled just in front of the pigment cells are two types of photoreceptor cells called **rods** and **cones** because of their distinctive shapes (see Figures 28.7b and 28.8). Rods are very sensitive to low levels of light, but cannot distinguish color, whereas cones need more light but can detect color. This is the reason you have trouble seeing colors in dim light. As one might expect, the retina of a tawny owl is jam-packed with rods, allowing it to pick up the dimmest light during night hunting; but since the retina contains few cones, the animal's color vision is probably very poor. In the retina's front layer, rods and cones synapse with sensory neurons that send axons toward the brain. Because these sensory neurons lie in front of the rods and cones, light must pass through the neurons before reaching the photoreceptors. The axons leading from individual retinal neurons collect into a bundle called the **optic nerve,** which exits the eye and leads to the brain (see Figure 28.7a and b).

■ **Rhodopsin, the Visual Pigment**   To understand how the eye works, let's concentrate on the nerve cells called rods and how they receive light. The front part of the rod contains the nucleus and forms synapses with other neurons, and the rear part detects light. The rear part contains stacks of membranous discs in which millions of molecules of the light-sensitive molecule **rhodopsin** are embedded (see Figure 28.7b). The protein part of each rhodopsin molecule snakes back and forth across the membrane and cradles the actual pigment portion, a small ring and chain structure called *retinal* (Figure 28.7c). (Eyes synthesize retinal from the compound that gives carrots their bright orange color; hence, there is wisdom to the old saying about eating your carrots!) The structure of retinal is really the key to vision, since it changes shape when light hits it: In the dark, retinal has a bent chain, but when it absorbs a small packet of light, the chain straightens out (see Figure 28.7c). This shape change unleashes a cascade of reactions that we eventually perceive as light. In a surprising link between the nervous and endocrine systems, this light-induced cascade in the rod cell depends on the same kind of intracellular second messengers that promote the action of polypeptide hormones.

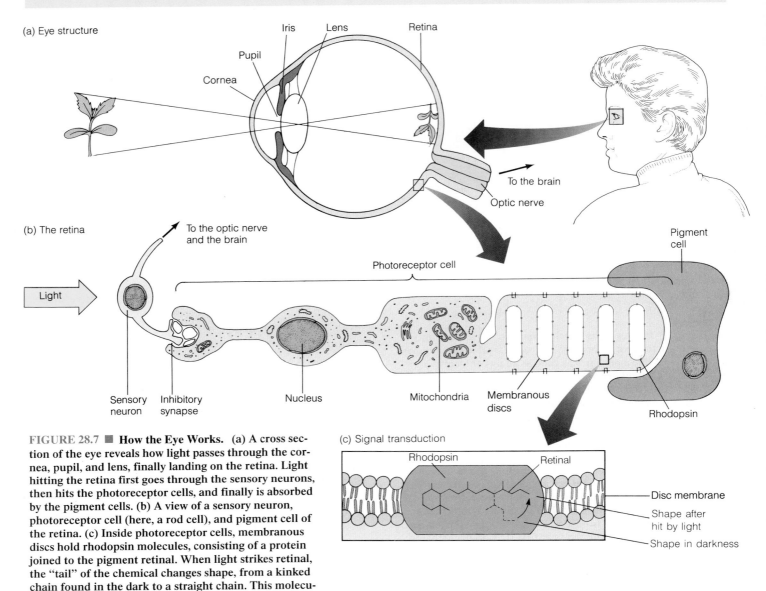

(a) Eye structure

Iris   Lens   Retina

Pupil

Cornea

To the brain

Optic nerve

(b) The retina

To the optic nerve
and the brain

Light

Photoreceptor cell

Pigment
cell

Sensory   Inhibitory   Nucleus   Mitochondria   Membranous
neuron   synapse   discs

Rhodopsin

(c) Signal transduction

Rhodopsin   Retinal

Disc membrane

Shape after
hit by light

Shape in darkness

FIGURE 28.7 ■ How the Eye Works. (a) A cross section of the eye reveals how light passes through the cornea, pupil, and lens, finally landing on the retina. Light hitting the retina first goes through the sensory neurons, then hits the photoreceptor cells, and finally is absorbed by the pigment cells. (b) A view of a sensory neuron, photoreceptor cell (here, a rod cell), and pigment cell of the retina. (c) Inside photoreceptor cells, membranous discs hold rhodopsin molecules, consisting of a protein joined to the pigment retinal. When light strikes retinal, the "tail" of the chemical changes shape, from a kinked chain found in the dark to a straight chain. This molecular change initiates a series of events that culminates in an action potential in a sensory neuron that leads to the visual centers of the brain. The brain interprets the action potential as a flash of light.

Recall that each kind of sensory neuron cell can tell the brain the strength of the signal and its type. In the case of neurons in the eye, brighter light stimulates more sensory neurons, which generate more action potentials in a shorter period of time; thus, the brain can perceive the increased intensity. Photoreceptors can also convey to the brain the position of the light source in the visual field. This is because photoreceptors in different parts of the retina receive light from different parts of the visual field and link up with different parts of the brain. For example, a horizontal line in the visual field stimulates a line of photoreceptors in the retina. This line links to sensory neurons in the visual center at the rear of the brain, and when the line of neurons there becomes excited, we see a horizontal line. We see color because cones are sensitive to light of different wavelengths.

Taste and Smell: Our Chemical Senses at Work

Although sight seems to be our dominant sense, taste and smell are our most direct and intimate links to the surrounding environment, and they play a far greater role in daily life than most people imagine. The tongue and nose receive and detect flavor and odor molecules—actual chemical tidbits of the environment. Thus, biologists consider taste and smell to be our *chemical senses.*

People often think of the tongue as a perceptual genius and the nose as a sensory dullard. However, precisely the opposite is true. The tongue is studded with small conical bumps, or papillae, which house the taste buds, and each bud consists of a pore leading to a nerve cell and surrounded by accessory cells arranged in an overlapping pattern that resem-

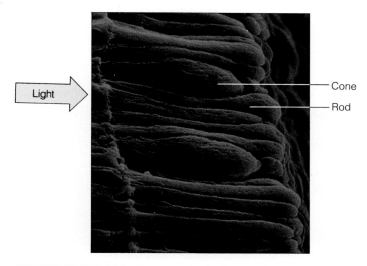

FIGURE 28.8 ■ Light Receptors in the Eye: Rods Can Detect Dim Light; Cones Register Color. Rods are tall and numerous, and this explains why our eyes can detect even extremely dim light. Cones are shorter and club-shaped, and each one can detect either blue, green, or red light.

hippocampus and hypothalamus) largely responsible for generating fear, rage, aggression, and pleasure and for regulating sex drives and reproductive cycles. A smell, therefore, stimulates the brain's centers of memory, emotion, and sexuality as well as the neocortex—the seat of conscious thoughts and learning—which surrounds the olfactory lobes.

Smells can also affect the endocrine system. Roommates in a women's dormitory, for example, often have synchronized menstrual cycles, and women with irregular periods often grow more regular when in a man's company on a routine basis. In both cases, the effects seem to be largely olfactory and due, perhaps, to pheromone-like molecules in the sweat, which act on the brain and its hormonal feedback loops with the body.

So far, our discussion has centered on the sensing portion of the sensing-integrating-reacting process, and the main message here is straightforward: An animal's brain can decipher the stimulus it receives as light, sound, taste, or smell because each sense organ is tuned to a different physical or chemical stimulus and sends a signal to a different part of the brain. The strength of the signal simply reflects the number of sensory neurons activated. To discuss the next step, neural integration, we need a more detailed knowledge of the nervous system and how it is organized.

bles an artichoke (Figure 28.9a). The nerve cells in taste buds have receptors capable of receiving flavor molecules, but they can distinguish only four general classes of flavors: sweet, salty, bitter, and sour. We can tell similar foods apart—beef from pork, beets from turnips, pickles from sauerkraut, honey from sugar—and sense the subtlety of their flavors because (1) these foods stimulate the four receptor types to different degrees, and (2) the volatile aroma molecules from the food travel into the nose or up the back of the throat and bind to receptors in the *olfactory epithelium.*

The olfactory epithelium consists of button-sized patches of yellowish skin high in the nasal passages. Neurons called olfactory cells lie embedded in these epithelial patches. One end of the nerve cell is a bit flower-shaped and binds odor molecules (Figure 28.9b). The other is a long, spindly axon that feeds into the olfactory nerve that carries signals directly to the olfactory bulb, or smell region, on the underside of the brain. Some scientists believe that olfactory receptors can distinguish a minimum of 32 primary odors, including sweaty, spermous, fishy, urinous, musky, minty, malty, and camphoraceous. If you take a bite of a sandwich and simultaneously pinch your nose shut, you can still perceive the four primary tastes, but not primary odors—and thus very little of the food's complex flavor.

People have long noted peculiarly intimate connections between smells, memories, and emotions. Who hasn't been temporarily overcome by a flood of memories, complete with appropriate emotions, when catching a whiff of a Christmas tree, or a puppy's fur, or a pipe like Grandfather smoked? The explanation for such odor déjà vu lies in the anatomy of the nose and brain. The olfactory lobes are closely connected to the limbic system, a series of small structures (including the

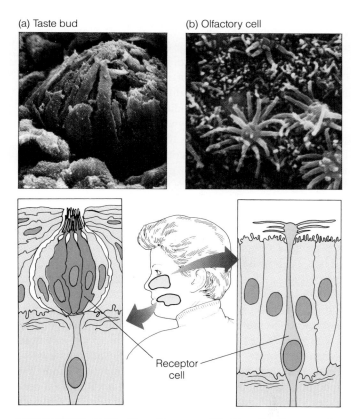

(a) Taste bud   (b) Olfactory cell

Receptor cell

FIGURE 28.9 ■ The Nose Knows and So Does the Tongue. A taste bud on the tongue (a) and odor-detecting cells in the nose (b) consist of receptor cells that generate a nerve impulse when stimulated by a specific chemical.

THE NERVOUS SYSTEM: FROM NERVE NET TO THE HUMAN BRAIN

Although an animal's eyes, ears, and other sense organs are crucial for detecting stimuli from within the body as well as from the environment, the animal's survival often depends on the ability to act—perhaps escape from a predator, pounce on prey, scout the area in search of food, or select a mate. Action requires both an integration of the sensory input and a coordination of different body parts. When an owl, for example, hears a rustling on the forest floor, it must be able to determine whether the sound comes from a mouse that the owl could eat or a prowling fox that could eat the owl. Those kinds of discriminations are made by neurons arranged in complex ways in the animal's central nervous system.

Trends in the Evolution of the Nervous System

Whereas the survival of a complicated animal depends on a functional nervous system, a simple animal like a sponge can get along with no specialized nerve cells at all. Each sponge cell communicates directly only with its immediate neighbors. Lacking overall neural integration, however, a sponge is little more than a reproducing, filter-feeding vase.

Among animals more complex than sponges, two general evolutionary trends regarding the nervous system become apparent: a concentration of neurons in the head and an increased number of interneurons.

Radial animals, such as hydras and jellyfish (see Chapter 18), have the simplest form of nervous organization in the animal world: the **nerve net,** a diffuse lacework of nerve cells and fibers that permits the conduction of impulses from one area to another. A stimulus anywhere on the body prompts a nerve impulse to spread across the network to other regions. The first evolutionary trend, a concentration of neurons in the head region, becomes apparent in animals more complicated than sponges, such as flatworms and roundworms (see Chapter 18), which move in a forward direction. These animals have sense organs concentrated in the body region that first encounters new environments. A flatworm's nerve net contains **ganglia** (singular, ganglion): distinct clumps of nerve cell bodies in the head region, which act like a primitive brain (review Figure 18.10). Distinct nerve trunks on either side of the body carry sensory information from the periphery to the head ganglia and carry motor commands from the head ganglia back to the muscles, allowing the animal to react.

In invertebrates such as mollusks, arthropods, and even segmented worms, the organization of the nervous system advances a step further. In these invertebrates, axons are joined into cables, or nerve trunks, and in addition to a centralized "brain," smaller peripheral ganglia help coordinate outlying regions of the animal's body (Figure 28.10). These ganglia represent the second evolutionary trend: The more complex an animal, the more interneurons it will have. Recall

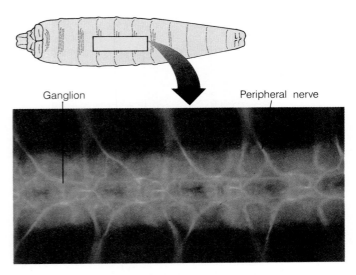

FIGURE 28.10 ■ **An Invertebrate Nervous System.** A fruit fly embryo illustrates some trends in the evolution of the nervous system. Sense organs are clustered at the head (here, to the right), along with a primitive brain. The glowing yellow stain reveals a central nerve cord punctuated by ganglia in each body segment. You can see how the ganglia are organized in a bilaterally symmetrical way. A lacework of axons from many interneurons connects the ganglia into one coordinated system with a brain at the anterior end.

that interneurons transmit signals from one neuron to another neuron (review Figure 27.3). Since interneurons in ganglia do much of the integrating that takes place in nervous systems, the more interneurons, the more complex behavior patterns an animal can display.

A consequence of the increasing numbers of interneurons was the evolution of the brain. The more complex the animal and the more complicated its behavior, the more neurons (especially interneurons) it will have concentrated into an anterior brain and bilaterally organized ganglia. Vertebrate brains, which we turn to next, are the culmination of this trend.

ORGANIZATION OF THE VERTEBRATE NERVOUS SYSTEM

Complex animal behavior—from the grouse's elaborate mating dance to the cobra's hypnotic weaving and the aerobic dancer's vigorous kicks and turns—depends on a highly organized nervous system. A vertebrate's nervous system is organized into two main units based on function and location: a **peripheral nervous system** and a **central nervous system (CNS)** (Figure 28.11). The peripheral nervous system is the "actor" or "doer"; it includes the sensory and motor neurons, and it connects the CNS with the sense organs, muscles, and glands of the body. The CNS is the nervous system's "thinker," or "information processor," and consists of the brain, which performs complex neural integration, and the spinal cord, which carries nerve impulses to and from the brain and participates in reflexes. The interplay between the central and peripheral systems allows an animal to sense environmental stimuli, integrate the information, respond appropriately, and in so doing, carry out the fascinating behaviors one sees in the animal kingdom (see also Chapter 38).

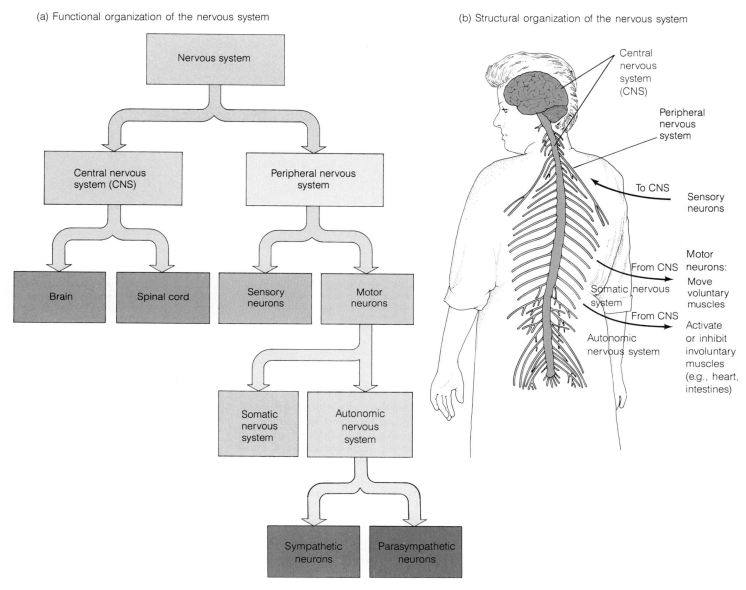

(a) Functional organization of the nervous system

(b) Structural organization of the nervous system

FIGURE 28.11 ■ Organization of the Vertebrate Nervous System. (a) Your nervous system has two main branches, the *central nervous system* (*CNS*), which controls the processing and integration of information, and the *peripheral nervous system,* which connects the CNS with sensory receptors, muscles, and glands. The CNS consists of the *brain,* which performs complex neural integration, and the *spinal cord,* which transports information to and from the brain and makes reflexes possible. The peripheral nervous system includes both *sensory neurons,* which carry information about the body and the environment to the CNS, and *motor neurons,* which convey instructions from the CNS to muscles and glands. Motor neurons are of two types: Those of the *somatic nervous system* control voluntary muscles, while those of the *autonomic nervous system* control internal organs, muscles, and glands. Finally, the autonomic nervous system itself has two components: the *sympathetic nervous system,* which controls energy expenditures (such as a rapid heart rate or an actively secreting adrenal gland), and the *parasympathetic nervous system,* which supplies or conserves energy (as in the stepped-up processing of food by the intestinal tract). (b) The information-sending and -receiving pathways for different parts of the body.

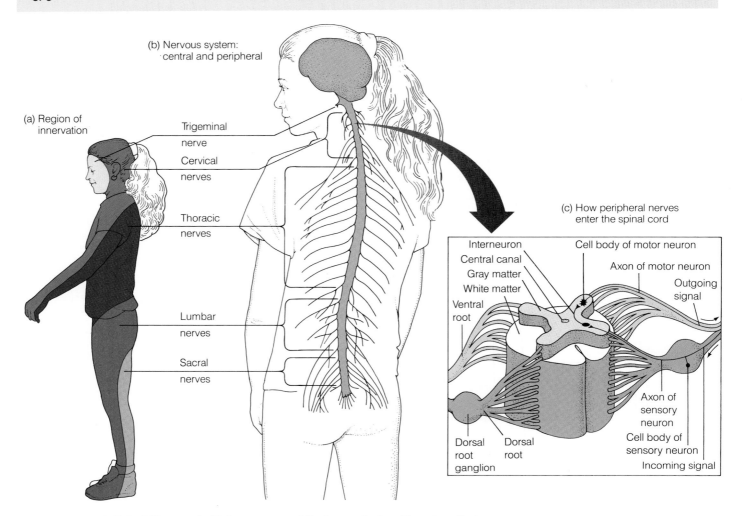

(b) Nervous system:
central and peripheral

(a) Region of
innervation

Trigeminal
nerve

Cervical
nerves

Thoracic
nerves

Lumbar
nerves

Sacral
nerves

(c) How peripheral nerves
enter the spinal cord

Interneuron
Central canal
Gray matter
White matter
Ventral
root

Cell body of motor neuron

Axon of motor neuron

Outgoing
signal

Dorsal
root
ganglion

Dorsal
root

Axon of
sensory
neuron

Cell body of
sensory neuron

Incoming signal

FIGURE 28.12 ■ Spinal Nerves. (a, b) A nerve map of the human body, with regions that correspond to particular stretches of the central nervous system and the nerves projecting from them.
Each part of the body is innervated by nerves emerging from a specific region of the spinal cord.
(c) Each major nerve has a dorsal root, which carries information from sense organs into the spinal
cord, and a ventral root, which carries impulses out from the spinal cord to the muscles and glands.
Gray matter consists largely of cell bodies and synapses. White matter consists of axons wrapped
with fatty myelin sheaths.

THE PERIPHERAL NERVOUS SYSTEM: THE NEURAL ACTORS

Peripheral neurons sense changes and cause actions, but do
little information processing. Peripheral neurons include *sensory neurons* like those in the eyes, ears, and skin, which carry
information about the environment to the CNS, and *motor
neurons,* which convey information from the CNS to muscles
and glands and trigger some activity.

Sensory Neurons

The peripheral nervous system is intimately connected with
the spinal cord, a bone-encased trunk of nerves carrying messages to and from the brain. Attached to the spinal cord are 31

pairs of spinal nerves, each a cable of thousands of axons
reaching out to different regions of the body (Figure 28.12).
To see how sensory and motor neurons are organized within
a spinal nerve, think about the reflex that causes you to pull
your finger from a flame. The dendrites of pain receptors in
your skin send action potentials to the cell body in the *dorsal
root ganglion,* a collection of sensory neuron cell bodies near
the spinal cord. The action potential continues along the sensory neuron's axon into the spinal cord, where it synapses
with interneurons. These interneurons are contained in the
CNS and carry a message to the brain, where it is interpreted
as pain. Other stimulated interneurons synapse in the spinal
cord with motor neurons that leave the CNS via the *ventral
root* of the spinal nerve and signal your muscles to quickly
pull your finger from the flame. (This is a reflex arc similar to
the one in Figure 27.12).

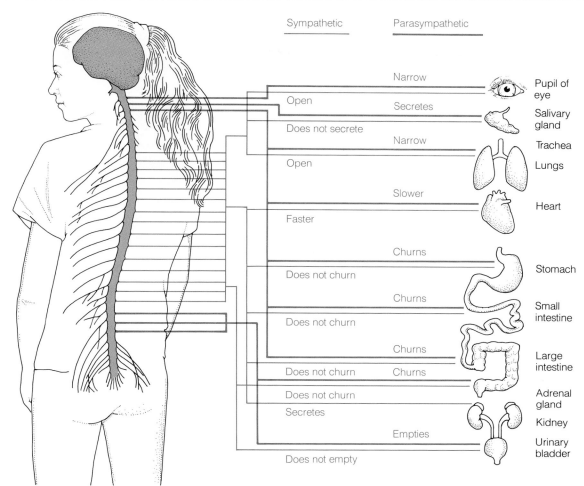

Sympathetic   **Parasympathetic**

Sympathetic	Parasympathetic	Organ
Open	Narrow	Pupil of eye
Does not secrete	Secretes	Salivary gland
Open	Narrow	Trachea / Lungs
Faster	Slower	Heart
Does not churn	Churns	Stomach
Does not churn	Churns	Small intestine
Does not churn	Churns	Large intestine
Does not churn	Churns	Adrenal gland / Kidney
Secretes		
Does not empty	Empties	Urinary bladder

FIGURE 28.13 ■ **Autonomic Branch of the Peripheral Nervous System.** The sympathetic and parasympathetic nerves of the autonomic nervous system run to many of the same organs but act in opposition, turning the organ's activity on or off, up or down.

Motor Neurons

The motor neurons in the ventral root of spinal nerves form two important systems: the *somatic nervous system,* which in general activates muscles under voluntary control (like those that allow you to ride a bicycle), and the *autonomic nervous system,* which regulates the body's internal environment by controlling glands, the heart muscle, and smooth muscles in the digestive and circulatory systems (Figure 28.13).

A person's autonomic nervous system operates primarily at the subconscious level, performing many of its duties through the spinal cord and lower centers of the brain. There are two sets of autonomic neurons that function in opposition, a bit like the accelerator and brakes of a car. *Parasympathetic nerves* leave the spinal cord in your neck and lower back, and *sympathetic nerves* leave your spinal cord in the central region of your back (see Figure 28.13). Sympathetic and parasympathetic neurons often innervate the same organ and help turn its activity up or down. For example, most sympathetic neurons secrete the neurotransmitter norepinephrine (noradrenaline), speeding and strengthening the heartbeat, while parasympathetic neurons secrete acetylcholine, slowing it.

In general, the parasympathetic nerves are a "housekeeping system" for the body, stimulating the stomach to churn, the bladder to empty, and the heart to beat at a slow and even pace during most daily and nightly activity. When an emergency arises, however, generating intense anger, fear, or excitement, the sympathetic nerves dominate, increasing heart rate, dilating the pupils of the eyes (which lets in more light), expanding the bronchioles of the lungs (which improves gas exchange), and slowing down nonessential digestive activities until the emergency is past. Many of us, when experiencing stressful changes that make us feel trapped and that we cannot respond to with satisfying physical activity (like fleeing a traffic jam or putting a pie in your boss's face) develop stress-related health problems, including high blood pressure. Some medications for high blood pressure (so-called *beta-blockers*) specifically inhibit the sympathetic synapses that stimulate the heart and blood vessels, thereby keeping the heart rate slower and steadier.

Both branches of the autonomic system are regulated primarily through reflex loops, such as the one that controls the emptying of the urinary bladder. When the bladder fills, the bladder wall stretches, the stretch receptors are activated, and they generate a signal. This signal enters the spinal cord via sensory neurons and stimulates motor neurons of the bladder muscle, causing that storage organ to contract and expel urine. In infants, this reflex loop is the only control over bladder function, so the organ empties whenever it fills. With toilet training, the child's brain gradually learns to exert control over this reflex loop.

Some cultures have explored conscious control over the autonomic nervous system—and with intriguing results. Some yogis in India, for example, claim that they can survive being sealed in an underground chamber for many days by lowering their metabolic rate and using far less oxygen than is normally required (Figure 28.14). Many such claims are surely exaggerations. But by measuring the yogis' breathing, pulse, and oxygen utilization with reliable instruments, scientists have concluded that some yogis can indeed consciously lower their metabolic rate below normal resting levels. The Western practice of biofeedback training, which uses electronic devices to indicate changes in blood pressure, blood flow to extremities, or other external states, is another example of the relationship between the conscious and autonomic nervous systems.

FIGURE 28.14 ■ **Mind over Matter: Some Yogis Can Consciously Lower Their Metabolic Rate.** This yogi from Madras, India, shown meditating, has learned to consciously control his autonomic nervous system.

THE CENTRAL NERVOUS SYSTEM: THE INFORMATION PROCESSOR

Although the peripheral nervous system collects data about the environment and directly controls the actions of the body's organs, all would be chaos if the central nervous system could not make sense of the data and coordinate the actions. The CNS can integrate sensing and reacting because of the intricate construction of the spinal cord and the brain and because of the millions of communication pathways within and between them.

The Spinal Cord: A Neural Highway

A cross section taken from a human spinal cord reveals a butterfly-shaped core of gray material surrounded by a bean-shaped field of white (see Figure 28.12c). The *gray matter* contains neuron cell bodies and unmyelinated nerve fibers; it is a zone of many synapses, where local neural traffic occurs (such as the reflexes discussed earlier). The *white matter,* on the other hand, is an interstate freeway; it consists mainly of thin axons that transport information long distances up and down the spinal cord to and from the brain. Since these long axons are insulated with a white, fatty myelin sheath (review Figure 27.8), the entire region has a whitish color. The spinal cord is responsible for the preliminary integration of signals from sensory neurons and interneurons. For example, when a flame touches your finger and you withdraw that digit reflexively, synapses between the sensory neuron, interneuron, and

motor neuron are taking place in the gray matter. Despite the preliminary role of the spinal cord in neural integration, most major data processing occurs in the brain.

The Brain: The Ultimate Processor

The human brain—a reddish brown organ that weighs about 1.4 kg (3 lb), contains 100 billion neurons, and has the consistency of vanilla custard—embodies our feelings and strivings, our knowledge and memories, our musical and verbal abilities, and our sense of the future. These complex behaviors and emotions can be localized to specific regions or groups of regions in the brain, as even minor damage to a given area can reveal. Thus, the anatomy of the brain provides a map for behavior, and certain behavioral disorders can be clues to the activities of specific brain regions.

The human brain has three interconnected parts (Figure 28.15): the **brain stem,** an extension of the spinal cord; the highly rippled **cerebellum,** attached to the brain stem; and the large, folded **cerebrum,** sitting atop the brain stem and spanning the inside of the head from just behind the eyes to the bony bump at the back of the skull.

The Brain Stem

The brain stem plays three crucial roles in sustaining life: It helps integrate sensory and motor systems, it regulates body homeostasis, and it controls arousal. The lowest part of the brain stem, the **medulla oblongata,** lies in the upper part of the neck at about the level of the mouth. The medulla helps keep body conditions constant by receiving data about activities in various physiological systems and by regulating numerous subconscious body activities, such as respiratory rate, heart rate, and blood pressure. Above the medulla lie the

pons and **midbrain,** regions that relay sensory information from our eyes and ears to other parts of the brain (see Figure 28.15).

The *hypothalamus,* still farther up the brain stem, regulates the pituitary gland and provides a link between the nervous and endocrine systems (see Chapter 26). The hypothalamus also has important roles in maintaining homeostasis, helping to regulate body temperature, water balance, hunger, and the digestive system. By electrically stimulating different regions of the hypothalamus, researchers can make an animal behave as if it feels alternately hot and cold, hungry and satisfied, angry and content. Some regions seem to be pleasure centers: A rat with electrodes permanently implanted in such areas will continue to press a foot pedal that turns on the current and stimulates the pleasure centers until the animal drops from exhaustion. Distinct regions of the hypothalamus are often quite small; moving an electrode only 0.5 mm (1/50 in.) may shift the induced sensation from extreme pleasure to extreme pain. Just above the hypothalamus is the **thalamus,** a relay area to the highly convoluted cerebrum that surrounds it (see Figure 28.15).

Extending throughout the brain stem from the thalamus down to the spinal cord is the **reticular formation,** a network of tracts that reaches into the cerebellum and cerebrum. When specific neurons of the reticular formation are actively firing, a person is awake, and when those neurons fall silent, the person sleeps or loses awareness. Intermediate levels of stimulation lead to dulling or sharpening of our senses. This selective shaping of neural input by the reticular formation and the thalamus helps a person to concentrate on a dinner partner's conversation in a noisy restaurant and plays a part in the groggy incoherence of an early-morning phone conversation.

The Cerebellum: Muscle Coordinator

Attached to the midbrain, or middle of the brain stem, is the cerebellum, a convoluted bulb that serves as a complex computer, comparing outgoing commands with incoming information about the status of muscles, tendons, joints, and the position of the body in space (see Figure 28.15). The result is a sculpting and refining of motor commands. A person whose cerebellum is damaged has jerky and exaggerated muscle movements. The cerebellum modifies motor commands initiated elsewhere in the brain until they are smooth and coordinated. Learning the finer points of a basketball pattern or a graceful ballet turn involves the cerebellum. Birds, including owls, rely on precise muscle coordination to avoid tree limbs and other obstacles and have large cerebellums relative to other parts of the brain (Figure 28.16).

Although muscle movement is refined in the cerebellum, it is initiated in the largest part of the brain: the cerebrum.

The Cerebrum: Seat of Perception, Thought, Humanness

The two side-by-side hemispheres of the brain's cerebrum fit like a cap over the brain stem. This mass of tissue embodies not only the attributes we consider human, such as speech, emotions, musical and artistic ability, and self-awareness, but also traits that we share with other vertebrates, such as sensory perception, motor output, and memory. In humans, the **cerebral cortex,** the cerebrum's highly convoluted surface layers, contains about 90 percent of the brain's cell bodies (Figure 28.17). During vertebrate evolution, this cap increased in size relative to other brain regions (see Figure 28.16). One of the great unsolved questions of modern biology is how these cells can control complicated traits. For now, biologists have many ideas but no single answer. One principle has been well established, however: All our basic cerebral functions, including sensations (like seeing or hearing), motor ability (movement), cognitive functions (like language and perception), affective traits (emotions), and character traits (like friendliness or shyness), can be traced to specific regions of the cerebrum. Clearly, the cerebrum is an enormously complex part of the brain—and one that demands a closer look. Let's first look at the sensory and motor functions of the cerebral cortex and then turn to more complex behaviors.

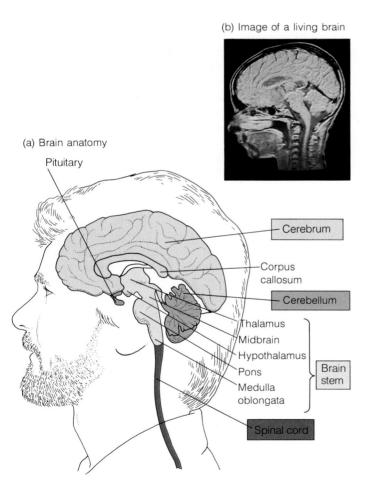

(b) Image of a living brain

(a) Brain anatomy

Pituitary

Cerebrum

Corpus callosum

Cerebellum

Thalamus
Midbrain
Hypothalamus
Pons
Medulla oblongata

Brain stem

Spinal cord

FIGURE 28.15 ■ Major Regions of the Brain. (a) The drawing reveals the brain stem, cerebellum, and cerebrum, as well as major parts of each section. (b) A photo of the brain taken with magnetic resonance imaging reveals the living tissues in remarkable detail (see Box 2.1 on page 28).

FIGURE 28.16 ■ **Brain Differences in Various Vertebrates.** While each of the vertebrates depicted here retains similar brain regions, the relative sizes are dramatically different. Compare the cerebrum of a fish, owl, cat, and monkey. The increasing size and surface convolutions correspond to increased intelligence and an enlarged behavioral repertoire.

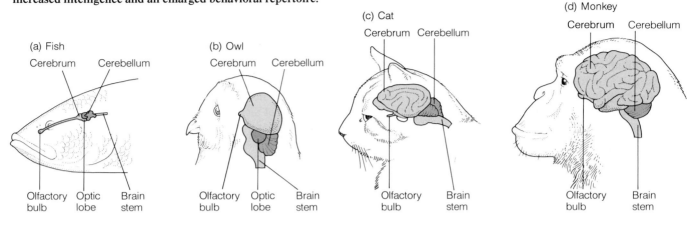

■ **Motor and Sensory Centers**   Brain surgery on patients with conditions such as epilepsy (a disorder that leads to seizures, or convulsions) has helped researchers map the cerebral cortex. Despite its billions of neurons, brain tissue has no pain receptors, so surgery can be carried out with only a local anesthetic for the scalp incision. By inserting extremely fine low-current electric wires into the exposed brain, neurosurgeons can stimulate specific regions of the brain and observe which muscles the patient moves or ask the alert patient to report the sensations he or she feels. With studies like these, scientists have mapped a projection of the body onto the surface of the cerebral cortex—that is, a map showing which brain regions control the sensations and motor functions of which parts of the rest of the body. If you run your right index finger from your left ear straight up to the top of your head, the arc traces the surface area devoted to the left **motor cortex,** the part of your brain that controls muscles on the right side of your body, including the ones that move your right hand. For example, the cortex region just above your left ear moves the muscles of your right jaw, while the part nearest the top of your head moves your right hip muscles (see Figure 28.17b). Likewise, the corresponding arc on the surface of the right side of your brain moves the left side of your body.

Just behind the motor cortex lies the **sensory cortex,** which registers and integrates sensations from body parts (see Figure 28.17c). (Note that the surface regions of the motor and sensory cortices are responsible for similar but not identical parts of the body, since motor nerves go to muscles, while sensory nerves go to sense organs such as touch receptors in the skin.) Again, the brain's left side receives sensations primarily from the body's right side, and vice versa.

On the maps of both motor and sensory cortices, the representation of body parts does not match their true proportions in the body: Lips, face, and fingers appear too large, and the trunk too small. This disproportionate mapping reflects the exceptionally large number of delicate touch receptors in face

and fingers and the large number of muscles necessary for speech and manual dexterity relative to the small numbers needed for sensation and movement of the trunk. Other specific brain regions are dedicated to vision, hearing, and smelling. Moreover, birds with particularly keen hearing, like owls, have an enlarged hearing region of the cerebral cortex, whereas birds that rely on the sense of touch and manipulation of the beak for feeding, like ducks, have a different region enlarged. Finally, the map can change with use, as revealed by recent experiments employing intense stimulation of a monkey's fingertip. This could explain the acquisition of skills such as learning Braille, the raised-dot writing system of the blind.

The sensory cortex is involved in a crucial evolutionary function: the perception of pain. While we adapt fairly quickly to certain sensations, such as the pressure of clothing on the skin, we don't adapt completely to pain, and this helps protect us by triggering behaviors that diminish the painful stimulus (excessive heat, pressure, chemicals, or electric jolts). The body's surface is richly supplied with specialized free nerve endings that perceive these kinds of stimuli and transmit nerve signals to an "electrical switching" region of the spinal cord, the dorsal horn (see Figure 28.12). From here the signals are relayed to the brain's thalamus and then to the sensory cortex, where they are integrated and interpreted and where a response can be fashioned—either an automatic one, like pulling your hand away from a hot pot lid, or a conscious one, like caring for a sore foot instead of going out dancing on it all night or skiing on it all day.

Nerve cells damaged by injury or infection release bradykinin, prostaglandins, and substance P, all of which render pain receptors more sensitive, heighten the sensation of pain, and lead to faster behavioral responses. When the injury is very dramatic, however, the body's nervous tissue can release natural opiates called *endorphins* along pain pathways. These opiates dull pain, and the suppression can allow a wounded person to drag him- or herself to safety before the pain sensations begin. All of these biochemicals help point out the evolutionary significance of pain.

■ **Higher Cerebral Function**   Many brain capacities in addition to sensory and motor functions can be mapped to specific regions, and once again, brain injuries have led researchers to

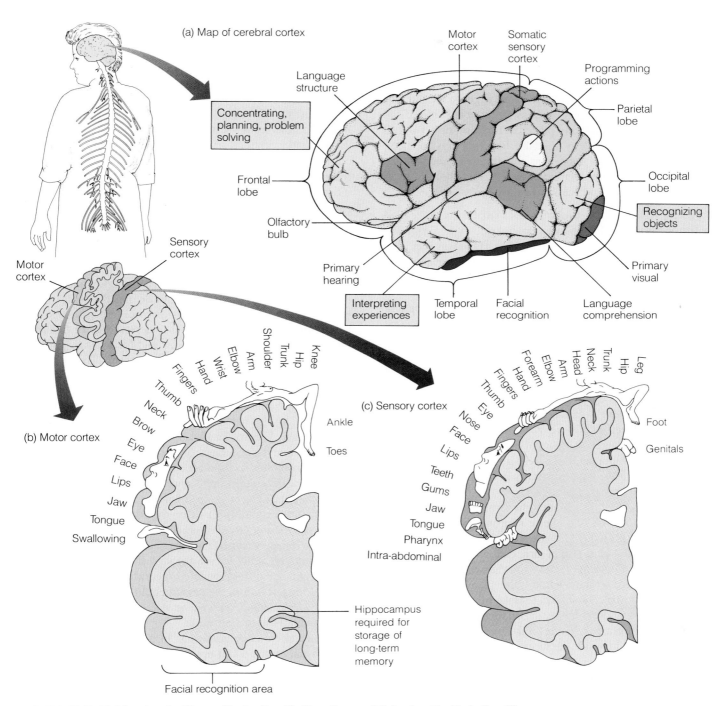

FIGURE 28.17 ■ Mapping the Human Brain: Specific Functions and Behaviors Reside in Specific Surface Regions of the Cerebral Cortex. (a) A map of the cerebral cortex reveals the mental functions associated with various regions of the brain's convoluted surface. The motor cortex can be traced to one "slice" and the sensory cortex to another. (Association cortex is shown in tan.) (b) If one were to lift the motor "slice" and map the regions that controlled the movements of corresponding body parts, the map would look like the strangely distorted human, or "homunculus," shown here. (c) If one did the same with the sensory "slice," the map of regions and corresponding body sensations would look somewhat different.

connect function and location in the cortex. One cortical area of the cerebrum, for example, is dedicated to recognizing faces (see Figure 28.17a), and a person who suffers an injury to this part of the brain can no longer recognize the faces of loved ones, although the injured person can still recognize their voices. Epileptics with disorders in specific areas of the cortex near the ears become moralistic and socially more aggressive and lose all interest in sex, whereas epileptics with disorders in other areas do not show these character traits. A person with damage to the right side of the parietal lobe just behind the sensory cortex may completely ignore the left half of the body, failing to wash or dress the affected side. If the patient's left arm is passively placed before him or her in full view, the person will deny that the limb belongs to them. Regions of the frontal lobe near the forehead seem to be involved in short-term memory, anxiety, and the ability to weigh consequences and act accordingly.

■ **Memory** Deep in the cerebrum lies a small structure called the **hippocampus,** which plays a crucial role in the formation of long-term memory (see Figure 28.17c). Have you ever tried memorizing a new phone number, only to have someone interrupt you even for an instant? You probably forgot the number because your hippocampus failed to establish it in long-term memory before the interruption occurred. A patient with a damaged hippocampus can recall events that happened before the injury—even decades before—but fails to remember new facts and experiences (including the shapes of objects, phone numbers, and people's names) for more than a few moments. Clearly, we need this area to help us lay down new memories, although the memories are not actually stored there. A patient with a damaged hippocampus may see the same doctor a dozen days in a row, but will greet the physician each time as a stranger. Patients of Alzheimer's disease lose neurons in the hippocampus. Researchers are learning to transplant brain tissue experimentally in the hope of someday helping such patients (see Box 28.1).

Researchers have also made other important discoveries about memory. Stimuli that alter a neuron's impulse activity cause long-lasting changes in the expression of neurotransmitter genes. These changes, in turn, cause alterations in synapse strength and communication to other cells. Essentially, the learning becomes "hard-wired"; that is, a memory is laid down as long-term physical changes take place in the way the neurons interact. These long-lasting changes seem to occur in many different brain regions and suggest that memory involves far more than just the hippocampus.

■ **Language** Complex language distinguishes us from all other animals. It is localized in two areas of the cerebral cortex, one for language structure and the other for language comprehension (see Figure 28.17a). If a stroke (damage to a brain region caused by a blood clot) disrupts the area of language structure, the patient may understand spoken and written language but be unable to speak well. One such patient, talking about a dental appointment, said haltingly, "Monday—Dad and Dick—Wednesday nine o'clock—ten o'clock—doctors—and—teeth." Damage to the area of language comprehension allows the person to speak fluently, but not in a way that is comprehensible to the patient or others. When describing a picture of two boys stealing cookies behind a woman's back, one such patient reported fluently, "Mother is away here working her work to get her better, but when she's looking the two boys looking in the other part."

The two brain areas responsible for the comprehension and structure of language are usually located in the cerebrum's left hemisphere. Corresponding regions of the right hemisphere regulate emotional gesturing and the musical intonation of speech. Thus, mapping shows that the brain is asymmetrical. In most people, language abilities reside mainly on the brain's left side (remember, *l*anguage, *l*eft), as do the underpinnings of analytical thought and fine motor control. The right hemisphere, on the other hand, is mainly responsible for intuitive thought, musical aptitude, the recognition of complex visual patterns, and the expression and recognition of emotion. Almost all right-handers and 75 percent of left-handers have left-brain language areas. Many left-handers, however, have language areas on *both* sides of the cerebral cortex.

■ **Right Cortex Versus Left Cortex** If separate functions are encoded in each hemisphere, how do the two sides of the brain communicate? Once again, neurosurgeons attempting to help epileptics contributed to our understanding of this problem. In epileptics, nerves fire back and forth across the brain in a crescendo of positive electrical feedback. This firing can continue to escalate, like feedback from a loudspeaker, resulting in a seizure. Surgeons thought that they could interrupt the loop by "splitting the brain," or severing the *corpus callosum,* the bridge between the left and right cerebral hemispheres (see Figure 28.15). These "split-brain" patients improved, and they seemed completely normal to friends. But when neurobiologists tested such patients in controlled experiments, they found strange anomalies (Figure 28.18). For example, if the subject saw a key in the left visual field (which projects to the right brain), he or she could point with the left hand to a key in a group of objects. But the patient could not name the object because language ability resides mainly in the left brain, which did not see the key. If the person saw a key in the right visual field (which projects to the left half of each eye and the left brain), the person could both point to the key with the right hand as well as name it, because language resides mainly in the left brain.

Researchers concluded that in a split-brain patient, the right visual cortex could not communicate what it saw to the region of language ability in the left brain. Thus, patients *knew* what they saw but could not *say* it. The split-brain procedure is used much less frequently today to control epilepsy, but it dramatically underscores the principle that specific abilities are localized in particular regions of the brain. Even so, no function is the exclusive property of one brain region; rather, each function requires integrated actions of neurons located in several discrete regions, each region simultaneously carrying out parallel computations that coordinate behavior.

BOX 28.1   P E R S O N A L   I M P A C T

ALZHEIMER'S DISEASE: REPLACEMENT PARTS FOR DAMAGED BRAINS?

A 50-year-old homemaker misplaces her eyeglasses for the third time in one week, then halfway to the grocery store forgets whether or not she turned off the stove and locked the front door. An engineer fails to recall that he has already determined the costs of a project and needlessly recalculates the estimates. A journalist finds it hard to remember the names of familiar objects and how to balance her checkbook. Although these people have common symptoms shared by many of us, their minds are beginning to fail from a silently but relentlessly advancing condition: *Alzheimer's disease.* Within three to ten years, these sufferers will be unable to dress themselves, to care for themselves, and eventually even to walk, sit up, or smile.

Currently, Alzheimer's affects as many as 4 million Americans and accounts for over half of the recorded cases of dementia (mental deterioration). It attacks otherwise healthy middle-aged and older people and causes a progressive loss of mental function. While the behavioral patterns (forgetfulness, inattention to hygiene, and so on) are characteristic, a physician can make a positive diagnosis only by examining brain tissue after the patient's death. The brains of Alzheimer's victims reveal a number of specific features, including loss of neurons from the cerebrum and hippocampus—regions necessary for thought and memory. Alzheimer's patients also show a loss of cells from certain areas of the basal forebrain whose axons synapse with neurons in the cerebrum and hippocampus (see Figure 28.17) and there release the neurotransmitter acetylcholine. In addition to those features, the brain of an Alzheimer's victim has accumulations of twisted filaments within the brain cells and aggregations of protein around blood vessels. Finally, a physician can detect a decrease in the flow of blood to specific regions of the cerebrum in an Alzheimer's patient (Figure 1).

In recent years, researchers have been debating about these protein aggregations, or plaques, made up of β-amyloid proteins. Some think that the buildup of these plaques may cause the nerve degeneration experienced with Alzheimer's, while others argue that something else kills the nerve cells first and that only when the cells degrade does the β-amyloid accumulate. One bit of evidence favoring the former position is that the gene for β-amyloid protein lies on chromosome 21, and researchers are now studying that chromosome in Alzheimer's victims in search of extra copies of the gene, which may be encoding extra protein and explain the buildup. Curiously, adults with Down syndrome also tend to develop β-amyloid plaques, and as we saw in Chapter 12, they have three copies of chromosome 21 (trisomy 21) instead of the usual two.

In 1989, researchers reported finding β-amyloid plaques in the skin, blood vessels, and intestines of Alzheimer's patients, suggesting a possible means of identifying the disease at an earlier stage. In addition, other experimenters in 1990 found that

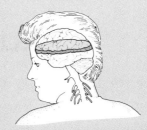

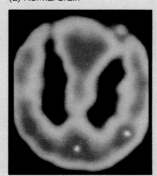   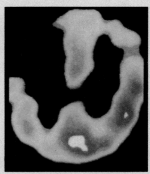

(a) Normal brain   (b) Brain of Alzheimer's victim

FIGURE 1 ■ **Alzheimer's Disease. Comparing a normal brain (a) and the brain of an Alzheimer's victim (b) reveals diminished blood flow to critical brain regions.**

Alzheimer's victims have a defective form of the membrane protein transferrin, which normally binds aluminum and removes it from the bloodstream. Since the defective transferrin fails to remove much of the aluminum, molecules of the metal can circulate to and enter the brain cells of Alzheimer's patients, and there they may block certain neurotransmitters and damage nerves and interfere with nerve transmission. This knowledge may also help provide an earlier means of diagnosing and perhaps treating Alzheimer's.

How can scientists and physicians help alleviate the effects of Alzheimer's and treat those suffering from the disease before it destroys their mental abilities? Although it may sound like science fiction, researchers are now exploring the possibility of transplanting brain tissue. Experiments on rats suggest that cells transplanted from a normal brain into a diseased one may grow and blunt the effects of brain disease. Such experiments provide hope that similar brain transplants in people may someday help alleviate the progressive mental deterioration of the Alzheimer's patient and in so doing ease the pain and disruption they and their families experience. Since the source of cells for the implants might be the brains of stillborn human embryos or aborted fetuses, society must carefully consider the issue, just as we did before passing laws to allow corneas, kidneys, hearts, and livers to be transplanted from dead people to the living.

■ **Association Cortex** While neurobiologists have made considerable progress in mapping functions to different parts of the brain, they have yet to learn how the parts operate. For example, each of the main cerebral lobes contains regions vaguely called *association cortex.* These areas appear to integrate information from other areas and reconstruct it at varying levels of consciousness. But exactly how these areas reassemble information and coordinate it remains unclear. To a large extent, these areas of association cortex contain the physiological underpinnings of emotions, memory, personality, reasoning, judgment, and the conglomeration of traits we call intelligence and remain for future biologists to untangle.

(a) Object in left visual field

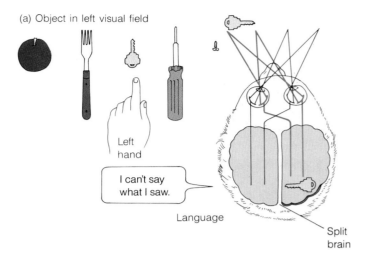

Left hand

I can't say what I saw.

Language

Split brain

(b) Object in right visual field

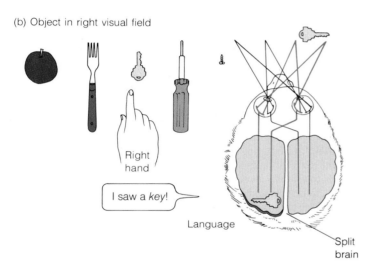

Right hand

I saw a *key*!

Language

Split brain

FIGURE 28.18 ■ **Two Minds in One Brain: Left-Brain and Right-Brain Perception in a Split-Brain Individual.** (a) A patient in which the bridge of nerve tracts linking left and right hemispheres has been surgically severed can perceive a key with the right visual cortex and point to it with the left hand. But since the patient's right brain cannot communicate with the language region in the left brain, the person knows what he or she saw but can't say it. (b) When the patient perceives the key with the left visual cortex, there is no communication block with the language region of the same hemisphere, and thus the person knows and can say what he or she saw.

CONNECTIONS

This chapter is the second in a set of three chapters that explore how electrical communication serves the body of a swooping owl or other animal, allowing the organism to perceive internal (physiological) conditions as well as stimuli from the environment, integrate the information, weigh possible consequences of any response, and issue commands to take appropriate action.

Chapter 27 looked at how individual nerve cells work and how nerve cells communicate with other neurons and cell types. This chapter discussed how neurons and nervous systems sense the environment—allowing us to hear, see, smell, taste, and feel the world around us. It also described how the brain integrates this incoming information and sends out information to the motor neurons of the peripheral nervous system, thus activating and coordinating the work of the body. Chapter 29 takes up at this point to investigate how muscles, glands, and other organs work together to produce the leap of the long jumper and the pounce of the night hunter—and all the other actions that enhance an organism's survival.

Highlights in Review

1. By serving as a communication network, nerves and nervous systems enable an animal to sense its environment and internal states and respond with appropriate behavior.
2. The nervous system integrates and coordinates the activities of all the different systems throughout an organism, so that the organism functions as a unified living being.
3. Sense organs are groups of specialized cells that receive light, odor, touch, or other signals and convert them to a form that can trigger a neural impulse.
4. The cochlea of the inner ear is a highly complex, coiled apparatus that converts sound waves into nerve signals. The cochlea receives sound owing to movement of stereocilia on the hair cells. The fluid-filled semicircular canals are involved in balance and detecting motion.
5. The eye contains two types of photoreceptor cells: rods, which allow us to see in dim light, and cones, which allow us to perceive color when light levels are brighter. In photoreceptor cells, molecules of rhodopsin change shape when light hits them, and this leads to a nerve impulse that the brain interprets as a visual image.
6. Taste and smell are chemical senses based on the direct recognition of flavor and odor molecules coming in from outside.
7. Animals show increasing organization and complexity in their nervous systems, from loosely organized nerve nets to ganglia to a brain that dominates the nerve network.
8. The brain and spinal cord serve as the central nervous system in humans. A peripheral nerve network branches out from the spinal cord, carrying messages to and from outlying muscles, glands, and organs. Peripheral motor nerves can be divided into the somatic nervous system, which primarily controls voluntary muscles, and the autonomic system, which primarily innervates the body's internal organs.

9. The autonomic system is further broken down into the sympathetic and parasympathetic systems, which frequently innervate the same organs, but with opposite effects.

10. The human brain can be divided into three regions: the brain stem, the cerebellum, and the cerebrum. The brain stem helps control autonomic functions. One of the cerebellum's chief roles is the coordination of fine movement. The cerebrum is involved in molding personality and coordinating thought, memory, and learning.

11. The network of neuronal tracts in the reticular formation sends neural processes throughout the brain stem and regulates consciousness.

12. The cerebral cortex is divided into two hemispheres, each with its own area of expertise.

13. Scientists have mapped out several regions of the cortex as either processing incoming information (sensory cortex), programming outgoing signals for movement (motor cortex), or associating the two types of information with higher levels of awareness (association cortex).

Key Terms

brain stem, 580
central nervous system (CNS), 576
cerebellum, 580
cerebral cortex, 581
cerebrum, 580
cochlea, 570
cone, 573
cornea, 573
ganglion, 576
hippocampus, 584
iris, 573
lateral line organ, 572
lens, 573
medulla oblongata, 580
midbrain, 581
motor cortex, 582

nerve net, 576
optic nerve, 573
peripheral nervous system, 576
photoreceptor cell, 573
pons, 581
pupil, 573
reticular formation, 581
retina, 573
rhodopsin, 573
rod, 573
semicircular canal, 572
sense organ, 570
sensory cortex, 582
stereocilia, 572
thalamus, 581

Study Questions

Review What You Have Learned

1. Give an example of how amplifying cells augment the activities of a sensory receptor.
2. What sequence of events in the ear results in sound detection?
3. How do the ear's semicircular canals function?
4. Explain how the eye detects light.
5. What role does rhodopsin play in vision?
6. In what sense is a hydra's nervous system more primitive than that of a flatworm?
7. Describe the reflex pathway at work when you jerk your finger out of a flame.
8. Which body regions are innervated by parasympathetic nerves? Which by sympathetic nerves? Give an example of the action of each.
9. Name the parts of the brain stem and the function(s) of each part.
10. Summarize the major roles of the cerebellum and cerebrum.

11. How are the motor cortex and sensory cortex alike? How are they different?
12. Describe the functions of the right and left hemispheres of the brain.

Apply What You Have Learned

1. Explain how a college student who regularly plays her stereo at top volume may be damaging her sense of hearing.
2. Why is it often harder to taste food when you have a bad head cold?
3. Recent studies suggest that fetal tissue implants may help people suffering from Alzheimer's disease. What problems might be associated with this approach?
4. After a serious automobile accident, a man suffers from emotional instability, poor appetite, and alternating chills and fever. His memory and muscle control seem normal. As students in a medical school, you and your peers must diagnose his illness. One student suggests that the patient has a hormonal imbalance due to a damaged gland. Another argues that the patient has suffered brain stem damage. A third maintains that the patient's cerebral cortex has been injured. A fourth says that the patient's cerebellum has been damaged. Which of your colleagues is more likely to be right? Explain your choice, and discuss why the others are probably wrong.

For Further Reading

ALKON, D. "Memory Storage and Neural Systems." *Scientific American,* July 1989, pp. 42–50.

BROWN, P. "Faulty Protein Holds Clue to Alzheimer's Disease." *New Scientist,* April 7, 1990, pp. 30–31.

CAROLA, R., J. P. HARLEY, and C. R. NOBECK. *Human Anatomy and Physiology.* New York: McGraw-Hill, 1990.

COWLEY, G. "Brain Killer or Bystander? New Clues About the Causes of Alzheimer's Disease." *Newsweek,* March 4, 1991, p. 54.

GAZZANIGA, M. "Organization of the Human Brain." *Science* 245 (1989): 947–952.

KANDEL, E., and J. SCHWARTZ. *Principles of Neural Science.* 2d ed. New York: Elsevier, 1985.

KNUDSEN, E. I. "The Hearing of the Barn Owl." *Scientific American,* December 1981, pp. 113–125.

KUFFLER, S. W., J. NICHOLLS, and A. MARTIN. *From Neuron to Brain.* 2d ed. Sunderland, MA: Sinauer, 1984.

MARX, J. "Alzheimer's Pathology Explored." *Science* 249 (1990): 984–986.

MELZACK, R. "The Tragedy of Needless Pain." *Scientific American,* February 1990, pp. 19–25.

MISHKIN, M., and T. APPENZELLER. "The Anatomy of Memory." *Scientific American,* June 1987, pp. 80–89.

MONTGOMERY, G. "The Mind's Eye." *Discover,* May 1991, pp. 51–56.

SACKS, O. *The Man Who Mistook His Wife for a Hat.* London: Duckworth, 1985.

SCHNAPF, J. L., and D. A. BAYLOR. "How Photoreceptor Cells Respond to Light." *Scientific American,* April 1987, pp. 40–47.

The Dynamic Animal: The Body in Motion

Battle for an Equine Harem

Wild horses roam the Granite Range Mountains of northern Nevada, and these feral descendants of domesticated horses wander many miles each day in search of green grass and fresh water. Biologists have observed that the horses usually live in stable bands with a single stallion protecting and directing a harem of females and young. The leftover bachelors are excluded from the band, but try, from time to time, to wrest control of a harem from the existing leader male. Such challenges can lead to furious contests, and nearly all stallions have wounds and vicious battle scars on their muscular bodies (Figure 29.1).

In May of 1981, researchers watched wild horses battle as the 6-year-old bachelor they called Harry wrangled for three straight days with Moscha, a 21-year-old stallion whose small harem included two mares. Harry, the challenger, constantly circled Moscha and occasionally charged the mares, trying to separate them from his rival. At times, the two combatants engaged in direct hoof-to-hoof combat, kicking at the competitor's head, nipping at his legs, and trying to sever the tendons that join leg muscles to bone and thus cripple his adversary. To defend against a leg bite, a horse would immediately collapse to the ground, and this fall would misdirect the attack. By the time the three-day struggle had ended, the heavily muscled Harry had circled, lunged, and pranced a combined distance of 48 km (30 mi), and the aging Moscha had traveled nearly as far. Finally, the weary Moscha stood with head lowered, too exhausted to protest as Harry escorted his harem off into the desert.

FIGURE 29.1 ▪ **Muscular, Battle-Scarred Stallions Fight Over Leadership of the Harem.**

The clash between Harry and Moscha helps illustrate the topics of this chapter: how the muscles work to drive the skeleton and how the body's physiological systems are integrated in ways that maintain homeostasis during the stress of heavy exercise. Horses are magnificent running machines, designed to move rapidly and continuously over long distances. But nearly all animals employ movement as a key strategy for solving life's most basic problems: capturing sufficient energy and reproducing. The centrality of movement can be seen in the grunion fish wriggling up a California beach to lay eggs, or in a swarm of locusts migrating in search of planted fields, or even in the subtle water-sweeping movements of a stationary coral or sponge filtering food particles from its watery environment. A sponge's collar cells move water using cilia or flagella (see Chapter 18), but most animals generate motion by muscle cells, which are specialized to shorten and transmit force to a skeleton, a rigid internal or external support structure.

Recall the receptor-integrator-effector model of homeostasis mentioned in various chapters in this unit, and recall also that the muscles are the nervous system's main effectors. When a horse's eyes (receptors) detect a combatant's teeth thrusting toward its legs, its brain (integrator) analyzes the information and sends a nerve signal to certain muscles (effectors), which in turn contract, moving the leg and preventing a tendon from being sliced.

Several themes will emerge as we discuss muscles and skeletons. First, muscles and skeletons adjust to stresses both over the lifetime of an individual and over the evolution of the entire species. The skeletons of vertebrate animals share most basic features, but over the course of evolution, a horse's bones have become highly specialized as an adaptation for forward running, while a person's skeleton is less specialized, adapted to running, lifting, carrying, and a range of other activities. Nevertheless, the muscles and skeleton of a horse or a person can adjust to exercise, allowing a given individual's muscles and bones to enlarge and grow stronger (Figure 29.2).

Second, fibers made out of protein provide the strength needed for skeletons and muscles. Crisscrossed fibers of the protein collagen (see Chapter 2) supply the framework for skeletons, and the sliding of fibrous proteins called *actin* and *myosin* allows muscles to contract.

Finally, rigorous physical activity involves not just the muscles and skeleton, but all of the body's physiological systems, and so homeostasis can be maintained even during rad-

FIGURE 29.2 ■ **Muscles and Bones in Action. Stress—in this case, exercise—causes an individual's muscles and bones to adjust by growing larger and stronger.**

ical changes. When horses engage in protracted contests like the titanic battle between Harry and Moscha, or when a person swims the English Channel or runs a marathon, the stress of the event activates heart, lungs, gut, kidneys, nerves, and hormones, in addition to the muscles and skeleton. The activation of these organs and systems requires close integration, enabling the individual to function as one coordinated unit that fights, swims, or runs with power and grace—a dynamic animal.

This chapter will answer several questions about how muscles and skeleton support and move the body:

■ How does the shape and internal structure of a bone allow it to support weight, encase organs, transmit force, store minerals, and make blood cells?
■ How do fibrous proteins, membranes, and energy sources allow muscular activity?
■ How do single bouts of exercise and long-term athletic training affect the body's homeostatic mechanisms?

THE SKELETON: A SCAFFOLD FOR SUPPORT AND MOVEMENT

A primary tactic of fighting horses is to bite at tendons in the legs. This maneuver works because it can sever a muscle's attachment to the skeleton. Like a stretched rubber band attached to a stationary object at one end and a movable object at the other, muscles attach to bones or other hard structures and pull against them. The rigid body support to which muscles attach and apply force is called a **skeleton** in both vertebrates and invertebrates. Skeletons are of two basic types. The most familiar is a *braced framework* that provides solid support; the hard shell of a shrimp or the stiff rods of bone in your arms and legs are examples of braced frameworks. A less familiar but no less effective type of skeleton depends on liquid to transmit force, like the hydraulic brake system in a car. Let's look at this simpler type first.

Water As a Skeletal Support

Animals as diverse as sea anemones, earthworms, and snails have a form of internal support called the **hydroskeleton,** composed of a core of liquid (water or a body fluid such as blood) wrapped in a tension-resisting sheath. A hydroskeleton is like a balloon filled with water: If you squeeze on one end, the fluid transmits force to the other end. Contracting muscles can push against a hydroskeleton, and the transmitted force generates body movement.

The two sea anemones in Figure 29.3 help illustrate how a hydroskeleton works. This denizen of tide pools has a central digestive cavity filled with seawater and surrounded by the body wall. The wall contains two layers of muscles: the *circumferential muscles,* which encircle the wall, and the *longitudinal muscles,* which extend from the animal's base to its tentacles. These two muscle layers act as **antagonistic muscle pairs**—groups of muscles that move the same object in opposite directions. When the longitudinal muscles contract, the animal becomes shorter and wider (as when you push a water-filled balloon from both ends simultaneously). When the circumferential muscles contract, the animal becomes longer and narrower (as when you wrap your hand around a balloon and squeeze).

Hydroskeletons are common among invertebrates, with their lack of bone, but even a system as complex as the human body relies on the principles of hydroskeletons in some situations. People lifting heavy weights off the floor tend to hold their breath and tighten their abdominal muscles. The compressed fluid of the abdominal cavity contributes to the body's rigidity and helps support the load. In many mammals, including humans, a hydroskeleton also stiffens the penis; valves allow blood to flow forcefully into the organ but not to flow out as freely, and the trapped fluid causes the penis to expand and become rigid.

Braced Framework Skeletons

The simple hydroskeleton is perfectly adequate for an aquatic animal like a sea anemone, which has the extra buoyant support of a watery environment, or for certain land animals like the earthworm, whose locomotion requires only moderate versatility, speed, and coordination. A rapidly moving land animal, however, requires a braced framework of solid material made of interlocked proteins, minerals, and polysaccharides. It can be an **exoskeleton**—a stiff sheet or an external covering that surrounds the body, as in shrimp, beetles, scorpions, and other arthropods (Figure 29.4). Or it can be an **endoskeleton**—a set of rigid rods and plates inside the body, as in vertebrates (Figure 29.5).

Mammalian endoskeletons are made of **bone,** a living tissue that contains a matrix of the long, stringy protein collagen hardened by a calcium phosphate salt. Mammals have an **axial skeleton,** which supports the main body axis, and an **appendicular skeleton,** which supports the appendages—the arms and legs (see Figure 29.5). Together, the human axial and appendicular skeletons have 206 individual bones.

■ **The Axial Skeleton**   The skull, vertebral column, ribs, and tailbone make up the axial skeleton. The **skull** is made up of the **cranium,** which encloses and protects the brain, as well as several bones of the face and inner ear (see Figure 28.4). The skull also contains the hyoid bone, or tongue bone, the only human bone not jointed with other bones. (Instead, it is suspended at the back of the mouth by ligaments and muscles.)

The head sits on top of the backbone, or **vertebral column,** which is made up of 33 **vertebrae.** Each vertebra is a bony box that fits together with other vertebrae in a stack,

FIGURE 29.3 ■ A Sea Anemone's Hydroskeleton: Muscle Contraction Against an Incompressible Internal Liquid.   A sea anemone shortens when longitudinal muscles contract (left) and lengthens when circumferential muscles contract (right).

FIGURE 29.4 ■ **A Cicada's Exoskeleton: External Support for the Body and Attachment Sites for Muscles.** A cicada leaves behind its old, tight exoskeleton as it molts. The soft, new, flexible, iridescent one will quickly become more rigid and protective.

the toe will touch only at its tip, with the middle "fingernail" being the hoof. The pressures of natural selection acting over thousands of generations have molded the skeleton of the horse for superb running ability, just as entirely different pressures have led to our adaptations for grasping, carrying, writing, and holding books in our arms.

■ **Bones, a Living Scaffold**   The bones of the skeleton not only support the body and anchor the muscles, but also encase vital organs, form blood cells, and store calcium and phosphate ions. A closer look at a large bone such as the femur shows how these tasks are accomplished. The long shaft, or *diaphysis,* of the femur (Figure 29.7a) allows it to support the body and transmit the weight of the animal to the ground. The femur has at each end an expanded portion called an *epiphysis,* which makes a joint, or *articulation,* with other bones.

encasing and protecting the spinal cord and giving support to the trunk. This movable stack of boxes gives flexibility to the body axis, allowing a shortstop to bend forward and scoop up a line drive, or a gymnast to do a back walkover. The vertebral column, however, is a bit fragile and requires special precautions as you lift, stand, or sit to avoid developing lower back problems (see Box 29.1 on page 597). Curving forward from the vertebrae are 12 pairs of ribs, which meet and attach at the breastbone (sternum) and form a protective compartment around the heart and lungs.

■ **The Appendicular Skeleton**   Other bones are attached to the axial skeleton; these include the **pectoral (shoulder) girdle** and **pelvic (hip) girdle,** the parts of the appendicular skeleton that support the arms and legs and allow them to swing (see Figure 29.5). Attached to the shoulder girdle and the pelvic girdle are the bones of the limbs. If we compare our forelimb (arm) bone to a horse's forelimb (foreleg) bones, it becomes obvious that both species show variations on a common anatomical theme (Figure 29.6). Our forelimbs are specialized for grasping but are not adapted for running (think of a circus performer walking on her hands), while a horse's forelegs are specialized for rapid forward motion but are poorly adapted for movement in other directions (picture the awkwardness of a fallen horse trying to stand up again). The horse's bones that are homologous to our palm and fingers are elongated; this adds an extra functional segment to the limb and means that

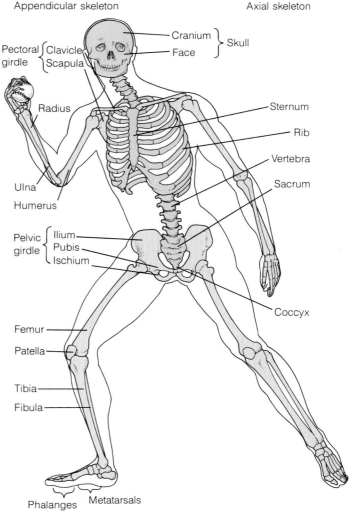

FIGURE 29.5 ■ **The Human Skeleton: Internal Support and Muscle Attachment.** The adult body has 206 bones grouped into two portions: the axial skeleton (80 bones, shown in yellow), and the appendicular skeleton (126 bones, shown in tan).

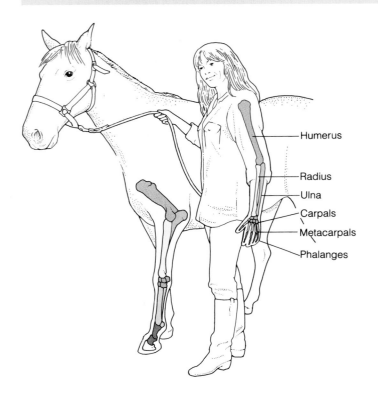

Humerus

Radius

Ulna

Carpals

Metacarpals

Phalanges

FIGURE 29.6 ■ Horse and Human Forelimbs: Variations on a Theme. Although the human forelimb is specialized for grasping and the horse forelimb is specialized for running, the upper bone (humerus) in each is fairly similar, and the human carpal, or wrist bones, are equivalent to the horse's "knee." The human hand, however, consists of 5 metacarpals and 14 phalanges, while the horse has the equivalent of one metacarpal (the cannon bone) and one main phalange with an enlarged nail, its hoof. The horse also has two tiny vestigial phalanges that have helped biologists trace the evolution of the modern horse from three-toed progenitors a bit like the modern tapir.

Projections from the bone called **processes** serve as attachment sites for ligaments and tendons. **Ligaments** are connective bands linking bone to bone, while **tendons** are tissue straps that connect bone to muscle. It is easy to find and feel examples of these skeletal structures. You can feel a large process (a bump) where your head joins the back of your neck

and attaches to muscles that keep your head erect; you can feel the ligament that attaches your kneecap to your shinbone (tibia) just below your knee when you tighten your thigh muscles (quadriceps); and you can easily feel tendons between the muscles in your upper arms and the bone in your forearm by putting your hand beneath the edge of a desk, lifting, and with the other hand, touching the "cables" that now project on the inside of your elbow.

A close-up view of the femur reveals how a bone can anchor, strengthen, store minerals, and perform its other functions (Figure 29.7b). The outer layer of *compact bone* is thick, solid, strong, and resistant to bending; this enables it to provide support. At each end of the femur is a region of *spongy bone,* a looser area crisscrossed by girders that provide strength near the joints. The tissue between girders down the center of the bone's shaft is called the *marrow,* a site of blood cell production. Embedded within the bone are cells that secrete the proteins and minerals of the bone and reabsorb

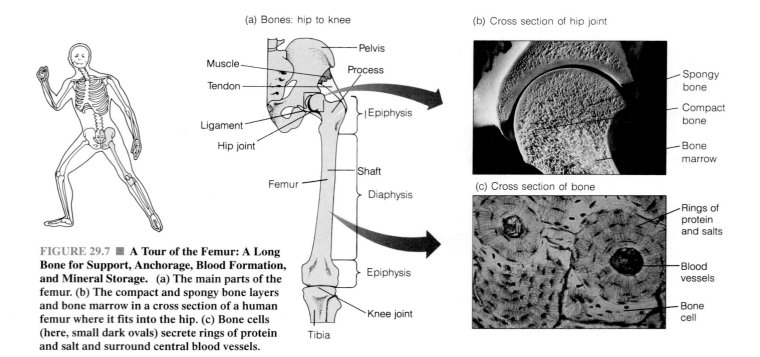

(a) Bones: hip to knee

Muscle

Tendon

Ligament

Hip joint

Femur

Tibia

Pelvis

Process

Epiphysis

Shaft

Diaphysis

Epiphysis

Knee joint

(b) Cross section of hip joint

Spongy bone

Compact bone

Bone marrow

(c) Cross section of bone

Rings of protein and salts

Blood vessels

Bone cell

FIGURE 29.7 ■ A Tour of the Femur: A Long Bone for Support, Anchorage, Blood Formation, and Mineral Storage. (a) The main parts of the femur. (b) The compact and spongy bone layers and bone marrow in a cross section of a human femur where it fits into the hip. (c) Bone cells (here, small dark ovals) secrete rings of protein and salt and surround central blood vessels.

FIGURE 29.8 ■ The Knee. (a) Anatomy of the knee joint. Your knee is the joint between the thighbone (femur) and the shinbone (tibia). Lying between these two bones is a space filled with fluid that acts like a sponge that cushions the pounding of running and jumping. Two pillows of cartilage (each called a meniscus) lie inside the fluid-filled space, and these further muffle shocks. The overuse syndrome called runner's knee results from an injury to these pads, causing severe pain and limiting joint mobility. Your knee resists twisting and extending forward because tough bands of tissue form a pair of crossed braces (the cruciate ligaments) that hold the femur and tibia tightly together. Violent forces (like kicking a soccer ball while off balance) can converge to shred these crossed braces. Finally, straps of tissue (the collateral ligaments) on each side of your knee prevent the joint from bending to the side. A clip on the football field, however, can rip one of these side straps in two. (b) X-ray of the knee as it bends. To see how the knee works, sit on a low chair and fully extend your right leg, then relax your thigh muscles. Now feel the position of your kneecap with your hand; the cap will rest in the position shown, but you can move it around with your fingers. While keeping your leg extended and your hand on your kneecap, tighten your thigh muscles and notice how the ligament lengthens and your kneecap moves up your leg. Finally, while still holding onto your kneecap, bend your leg. You may feel a slight grating as the kneecap slips over the thighbone, perhaps owing to insufficient cartilage or variations in the fluid-filled sac cushioning the two bones.

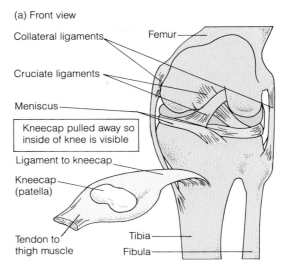

(a) Front view

Collateral ligaments

Femur

Cruciate ligaments

Meniscus

Kneecap pulled away so inside of knee is visible

Ligament to kneecap

Kneecap (patella)

Tendon to thigh muscle

Tibia

Fibula

(b) Side view

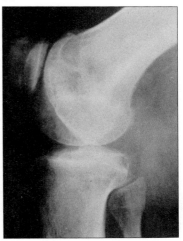

calcium and phosphate from bone for use in other parts of the body (Figure 29.7c). These cells also knit bones back together after a break.

Bone cells are especially active during periods of rapid growth. Bones elongate at a region near their ends called the growth plate, or *epiphysial plate,* that is initially just cartilage. (Recall from Chapter 20 that cartilage is a network of fibrous proteins similar to the stiff sheets that form your outer ear.) As an animal grows, bone is added to this plate, and when the organism matures, the plate disappears and the bones stop growing. If the bones are jarred by too much physical activity during childhood (a result of running marathons, for example, or of carrying heavy loads), the growth plates can be damaged and overall bone growth may be retarded.

The deposition and removal of protein and minerals goes on continuously in the bones. The bumps called processes, for example, grow larger when the muscles that attach to them become stronger. Anthropologists excavating Italian cities buried centuries ago under ash and cinders from the eruption of Mt. Vesuvius are able to distinguish the skeleton of a slave girl from that of a nobleman's daughter simply by comparing the size of the processes on arm and leg bones. If bone cells remove too much calcium, the bones become weak and fragile. This condition, known as *osteoporosis,* is relatively common in women after menopause and in inactive men. Recent research shows that estrogen helps many women absorb calcium from their diets. Some doctors now prescribe small doses of the hormone to help postmenopausal women avoid

osteoporosis. Research also shows that regular exercise stimulates the deposition of calcium in bones, and this, too, can build strong bones and help prevent osteoporosis.

■ Joints: Where Bones Come Together Bones can move with respect to each other because of **joints,** points where adjacent bones nearly touch. Slightly movable joints, such as those between adjacent vertebrae in the spine, have pads of cartilage that absorb shock but allow limited mobility. Freely movable joints, such as the shoulder, hip, and knee (see Figure 29.7a), have pads of cartilage at the ends of the two adjoining bones, but also have a flattened sac (a *bursa*) filled with a fluid called synovial fluid between the bones that cushions the joint and eases the gliding of bones across each other. Temporary inflammation of this joint sac is called *bursitis.* Long-term inflammation of the joint can result in *arthritis* (see Figure 22.15). Figure 29.8 describes how the knee joint works and why it is a source of trouble for joggers and professional athletes alike.

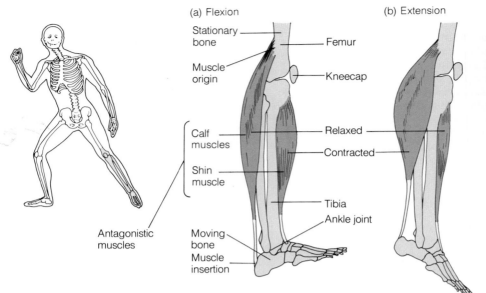

(a) Flexion

Stationary bone

Muscle origin

Calf muscles

Shin muscle

Antagonistic muscles

Moving bone

Muscle insertion

Femur

Kneecap

Relaxed

Contracted

Tibia

Ankle joint

(b) Extension

(c) Extension in action

FIGURE 29.9 ■ Muscles Move Bones to Action. (a) Simultaneous contraction of the shin muscle and relaxation of the calf muscles flex the foot, while (b) contraction of the calf muscles and relaxation of the shin muscle extend the foot. Because the ankle joint acts as a fulcrum, people with longer heels (levers) can jump much higher. (c) Spud Webb, standing a mere 5 ft 6 in., won the National Basketball Association's slam dunk contest one year. Webb, who soared above men more than 1½ ft taller, probably has very efficient muscle and skeleton levers that complement his tough training schedule.

■ How Bones Act As Levers That Transmit Force   Bones can pivot about joints because of the particular way muscles attach to bones, and significantly, <u>muscles are arranged in antagonistic pairs</u>. The shin muscle (anterior tibialis), for example, flexes your foot toward your body (Figure 29.9a), while its antagonists, the calf muscles (gastrocnemius and soleus), extend your foot to point away from the body (Figure 29.9b). Often, when a muscle contracts, its antagonist relaxes. Ordered movements occur because one end of the muscle, the *origin,* attaches to a bone that generally remains stationary during a contraction, while the other end, the *insertion,* attaches across a joint to a bone that moves. Your calf muscles attach to your femur just above the back of your knee, and at the opposite end, to your heel bone (calcaneus). When the calf muscles contract, the shortening forces the foot to rotate around a pivot, the ankle joint, much as a lever rotates around a fulcrum. Because of this arrangement, a small contraction of the muscle transmits a large movement to the bone.

MUSCLES: MOTORS OF THE BODY

A horse can gallop across the prairie and a person can perform aerobic dance exercises because a type of muscle called **skeletal muscle** propels the skeleton (Figure 29.10). Voluntary movement in a skeletal muscle originates when the motor cor-

tex of the cerebrum (review Figure 28.15) sends an action potential down a motor neuron to its synapse with the muscle. When an action potential reaches the synapse, it provokes an action potential in the muscle cell and causes that muscle to contract. But how does an action potential cause the muscle cell to shorten? The answer lies in the unique structure of muscle cells.

Skeletal muscles are made up of **muscle fibers,** giant cells with many nuclei that may extend the full length of the muscle—often several centimeters (Figure 29.11a and b). Each muscle fiber has a system of protein filaments, a system of membranes, and an energy system that act, respectively, as the muscle's engine, electrical ignition switch, and fuel (Figure 29.11c).

Protein Filaments: The Muscular Motor

Within each muscle cell lie threads called **myofibrils** that do the actual job of contracting the muscle (see Figure 29.11b). A longitudinal section of a myofibril reveals repeating units of dark and light bands. Each repeating unit is called a **sarcomere** (from the Greek words for "flesh" and "part of") (see Figure 29.11c). Each sarcomere consists of two types of interdigitating filaments called actin filaments and myosin filaments. Actin filaments are thin, but myosin filaments are

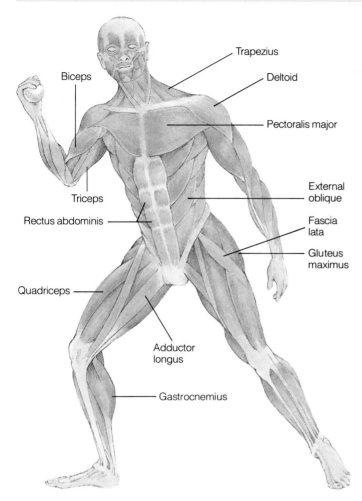

FIGURE 29.10 ■ Skeletal Muscles of the Human Body. The body's 600 or so skeletal muscles attach to and move the bones of the human endoskeleton. If all 600 muscles could contract simultaneously and pull in the same direction, they could lift 25 tons—the weight of a small iceberg nearly 3 m (10 ft) on a side! Of course, this is not actually possible because muscles work in antagonistic pairs. This anatomical chart identifies only the largest muscles of the torso and limbs.

during contraction. To picture this, try thinking of a woman putting her hands into a furry muff; her hands slide past the insides of the muff, her elbows come nearer together, but the muff itself does not decrease in length (Figure 29.12c). Likewise, in a muscle, the actin filaments and the myosin filaments slide past each other as the sarcomere shortens. This remarkable process is called the *sliding filament mechanism* of muscle contraction.

The unique arrangement and behavior of actin and myosin molecules allows the two types of filaments to slide past each other. Each myosin filament is made up of many individual

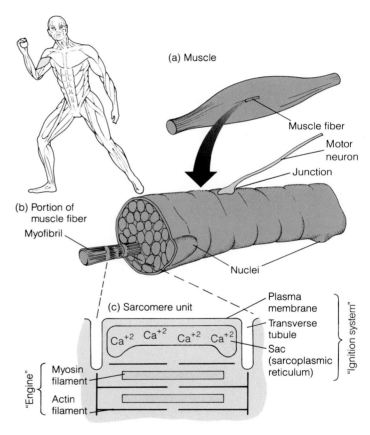

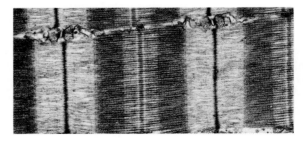

FIGURE 29.11 ■ Structure of Muscle Fibers. The text explains in detail the structure and activities of (a) the muscle, (b) the muscle fiber and myofibrils, and (c) the sarcomere and the myosin and actin filaments. The electron micrograph shows actin and myosin filaments magnified about 20,000 times.

thick, and their overlapping arrangement creates the visible bands, a bit like the pattern you create when you interlace the outstretched fingers of your two hands. Actin and myosin are not exclusive to muscle; in fact, they are found in the cytoskeleton of most eukaryotic cells, where their contraction changes cell shape, localizes materials, and causes dividing cells to pinch in two (review Figure 3.16).

How do these neat arrays of protein filaments cause the individual muscle cells of your biceps to shorten and lift your forearm? By examining contracting muscles with an electron microscope, biologists observed that each sarcomere within a contracting muscle fiber becomes shorter, thus causing the whole fiber to shorten (Figure 29.12a and b). The band of myosin within each sarcomere, however, does not shorten

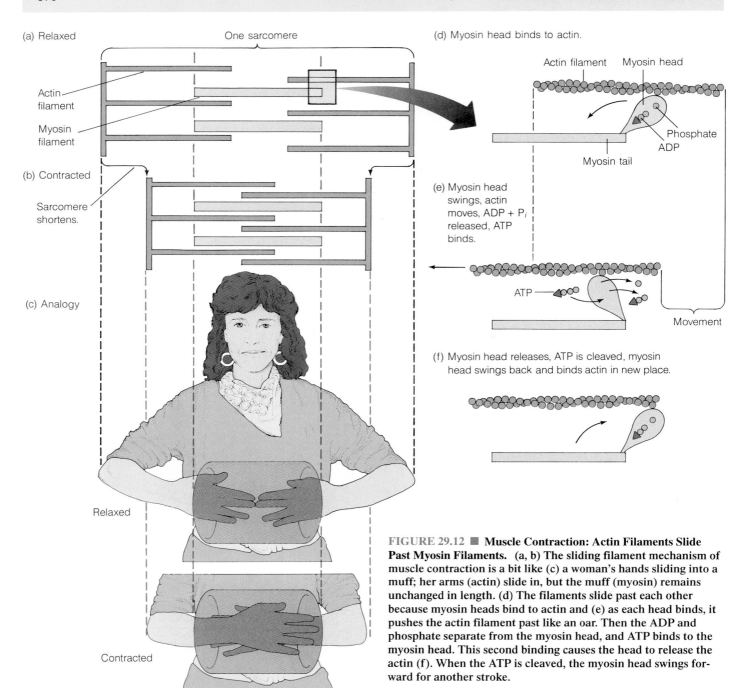

(a) Relaxed

One sarcomere

Actin filament

Myosin filament

(b) Contracted

Sarcomere shortens.

(c) Analogy

Relaxed

Contracted

(d) Myosin head binds to actin.

Actin filament   Myosin head

Phosphate
ADP

Myosin tail

(e) Myosin head swings, actin moves, ADP + P$_i$ released, ATP binds.

ATP

Movement

(f) Myosin head releases, ATP is cleaved, myosin head swings back and binds actin in new place.

FIGURE 29.12 ■ Muscle Contraction: Actin Filaments Slide Past Myosin Filaments. (a, b) The sliding filament mechanism of muscle contraction is a bit like (c) a woman's hands sliding into a muff; her arms (actin) slide in, but the muff (myosin) remains unchanged in length. (d) The filaments slide past each other because myosin heads bind to actin and (e) as each head binds, it pushes the actin filament past like an oar. Then the ADP and phosphate separate from the myosin head, and ATP binds to the myosin head. This second binding causes the head to release the actin (f). When the ATP is cleaved, the myosin head swings forward for another stroke.

myosin protein molecules bound together. Each myosin molecule within the long filament has a long, straight tail at one end and a *head* at the other that reaches out like an oar dipping into the water (see Figure 29.12d). The myosin head swings and moves the actin past it like an oar moves water (see Figure 29.12e). This process of reaching and swinging happens over and over for each myosin molecule as the muscle contracts, literally sliding the actin molecule past it.

Not surprisingly, just as a rowing crew must eat heartily before a regatta, the reaching and swinging of myosin "oars" also requires energy. ATP binds to the myosin head, causing

the head to release the actin. The myosin head then cleaves ATP to ADP and phosphate as the head swings forward and binds actin again for another stroke (see Figure 29.12d). The energy cleavage that propels actin past myosin explains *rigor mortis,* the stiffening that occurs shortly after an animal dies. Since ATP causes the myosin head to release the actin, when the supply of ATP runs out after death, myosin cannot release the actin (the oars get stuck in the water). Thus, myosin cannot glide past actin, so muscles can neither contract nor relax, and they remain stiff until the protein filaments themselves are degraded by enzymatic action. The mechanism of sliding pro-

BOX 29.1   PERSONAL IMPACT

LOW BACK PAIN

Low back pain is the most common medical complaint in America, and while most of us will recover in a few months, about 1 in 100 will suffer the permanent disability of chronic low back pain. By understanding the back's anatomy, however, and by practicing good habits of posture, you stand a much greater chance of escaping an aching back, now or in the future.

The backbone, or vertebral column, running down the center of your back forms three curves (Figure 1a), with the lowest curve the most likely to contribute to low back pain. The stacked vertebrae (Figure 1b) are a bit like round boxes, with a front portion like a small can of tuna and a back portion projecting a series of bumps (processes). The rounded front bears the weight of the body, while the processes serve as sites of muscle attachment and guide the bending of the backbone like a railroad track guiding train wheels.

Between these two parts lies a tunnel, and the spinal cord passes through this opening. Between each pair of adjacent vertebrae lies a cushioning vertebral disc, a rubbery pad with a fluid-filled center and several outer layers of cartilage, like the belts of a tire. Because of this construction, compression in one part of the disc causes another part to bulge out. This squeezable disc cushions the vertebrae on either side, but if the compression is too great, the disc can bulge out temporarily, press on nearby

pain sensors, and cause low back pain, or even blow out like a tire and cause the excruciating pain of a ruptured disc.

For many people, increased compression in the disc begins with an exaggerated curve in the vertebral column's lower bend (a swayback, often due to poor posture). If the overcurving spine puts pressure on nerves carrying sensory neurons from the legs into the spinal cord, the pain may be felt in the legs as well as in the back.

The most extreme compression of the discs occurs during improper lifting: bending forward from the waist and reaching to the floor with legs stiffly locked (Figure 1c), rather than bending at the knees and crouching down to pick up the object.

If swayback is contributing to low back pain, learning to rotate the lower part of the pelvis forward can help relieve the problem. Although it takes practice, you can learn to maintain this forward rotation while walking, standing, sitting in class, or driving. One can even rotate the pelvis during proper lifting to give the discs added protection. Some people have anatomical variations that increase their likelihood of developing back pain no matter how carefully they stand, sit, lift, and carry. But for most people, learning proper posture will substantially improve their chances of being among the small but lucky group that remains free of back problems.

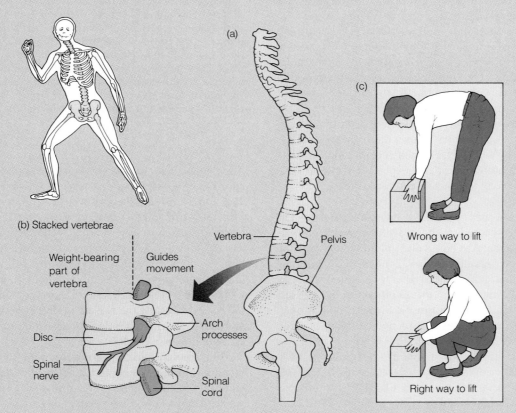

FIGURE 1 ■ The Human Back and Ways of Lifting. (a) The backbone has three curves. (b) Two stacked vertebrae. (c) Wrong and right ways to lift.

tein filaments explains how muscles shorten. It does not, however, explain how muscles know when to contract. That requires signals from the cell membrane.

Membrane System: The Ignition

Muscle contraction enhances an animal's survival only if the organism can control when contraction takes place. This control is exerted by muscle cell membranes, which ignite the muscular motor. Three sets of membranes help stimulate muscle fibers to contract (see Figure 29.11c): the plasma membrane (*sarcolemma*) surrounding the huge fiber cell; the *transverse tubules,* which lead like tunnels from the plasma membrane deep into the cell's interior; and the *sarcoplasmic reticulum* (corresponding to the endoplasmic reticulum of other cells), which forms a sac within the sarcomere and stores large amounts of calcium ions.

When an action potential reaches the synaptic junction between a motor neuron and a muscle cell (see Figure 29.11b), it causes the neuron to release the neurotransmitter acetylcholine into the synaptic cleft and onto the cell's plasma membrane (see Figure 27.10). The transmitter causes an action potential in the muscle cell (similar to the action potential in a neuron; review Figure 27.5), and electrochemical activity on the surface is channeled toward the cell's interior by the transverse tubules. Next, a signal (probably a second messenger; see Chapter 26) passes from the transverse tubules to the sarcoplasmic reticulum, causing the membranous sac to release some of its stored calcium into the cell's cytoplasm.

The jolt of released calcium ions interacts with a set of regulatory proteins (*troponin* and *tropomyosin*). This regulatory interaction allows the myosin heads to bind to actin filaments and the muscle cell to contract. After the cell shortens, an ATP-driven pump quickly returns the calcium ions to the sarcoplasmic reticulum, and the regulatory proteins force myosin away from actin, which allows the muscles to relax. The message here is clear: Muscles can generate force against a skeleton because protein filaments expend energy and slide past each other, causing the muscle cell to shorten.

■ How Cardiac Muscle and Smooth Muscle Move Body Fluids
The mechanisms just discussed apply to skeletal muscles, those muscles that are attached to bones and that move the skeleton (Figure 29.13a). Vertebrates, however, have two additional types of muscles that move fluids rather than bones. **Cardiac muscle,** found only in the heart, drives blood through the circulatory system, while **smooth muscle** propels food through the digestive tract and provides tension in the urinary bladder, uterus, and blood vessels.

Cardiac muscle fibers consist of networks of muscle cells. These are electrically connected to each other via *intercalated discs,* which are specialized junctions like the gap junctions in Figure 3.28 and the electrical synapses in Figure 27.9. Such specialized junctions allow adjacent cells to communicate back and forth (Figure 29.13b). Thus, an impulse initiated any

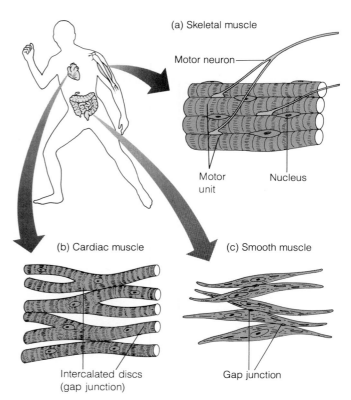

FIGURE 29.13 ■ How Cardiac Muscle, Smooth Muscle, and Skeletal Muscle Differ at the Cellular Level. **(a)** Several skeletal muscle fibers can contract together (a motor unit) when branches of the same motor neuron contact them all. When more motor units fire, the contraction is stronger. **(b)** Cardiac muscle cells have gap junctions organized in structures called intercalated discs. These junctions allow electrical communication between adjacent cardiac muscle cells, coordinating their contraction. **(c)** Smooth muscle cells communicate electrically through gap junctions and tend to contract in sequential waves.

place on the heart muscle propagates through the entire organ, causing the whole fist-sized heart muscle to contract as a unit. The impulse usually begins in the heart's pacemaker (review Figure 21.10), and thus, nerves are not the impetus—a significant advantage during heart transplantations. In contrast to cardiac muscle, skeletal muscle fibers do not communicate with each other directly, although several fibers may be innervated by the same motor neuron (a motor unit) and hence contract together (Figure 29.13a). The more motor units that fire, the more forcefully the skeletal muscle contracts. Like skeletal muscle, cardiac muscle is *striated,* or striped, because its actin and myosin filaments are organized into sarcomeres.

The smooth muscle cells surrounding the digestive tract, bladder, blood vessels, and other hollow body organs lack striations because their actin and myosin filaments are not as well ordered as those in skeletal and cardiac muscle (see Figure 29.13c). Smooth muscle cells usually have a single nucleus and communicate with other smooth muscle cells via gap junctions, as do cardiac muscle cells. This communication allows the rhythmic pushing of food down the digestive tract or the pushing of a baby through the birth canal.

Despite the organizational differences between skeletal, cardiac, and smooth muscle, all three types contribute to an animal's success. Skeletal muscles allow a wild horse to lope across the desert in search of green grass or fight to win mares and mate (thus aiding its reproduction); cardiac muscle propels blood through vessels to every body tissue; and smooth muscle propels food through the digestive tract and helps in various other ways to keep this complex system operating smoothly. All this muscle contraction—so vital to survival—depends on a substantial amount of energy in the form of ATP.

ATP: The Fuel

While actin and myosin filaments cause muscle cells to contract, and membrane systems ignite muscle activity, the fuel for muscle cell contraction is ATP. As we already know, the sources of ATP are the mitochondria and certain enzymes in the cytoplasm. Three energy systems—the *immediate system,* the *glycolytic system,* and the *oxidative system*—supply the ATP that all muscles need. The duration of physical activity dictates which system the body uses (see Figure 5.18a). The immediate energy system is instantly available for a brief explosive action, such as one heave of a heavy shot put. It depends on a muscle cell's stores of ATP plus a high-energy molecule called creatine phosphate, and it can fuel muscle contraction for several seconds. The intermediate energy system is based on the splitting of glucose by glycolysis (see Chapter 5) in the muscles. It can sustain heavy exercise for a few minutes, as in a 200 m swim (see Figure 5.18b). After that, the long-range, or oxidative (aerobic), system takes over for a long period of exercise, such as a run in the mountains (see Figure 5.18c). In the presence of sufficient oxygen, this system can produce energy by breaking down carbohydrates, fatty acids, and amino acids mobilized from other parts of the body. Clearly, anyone interested in melting away body fat should engage in aerobic (oxygen-utilizing) activities like rapid walking, swimming, bicycling, or jogging, which rely primarily on the oxidative energy system and its ability to use fats as fuel.

The immediate, glycolytic, and oxidative energy systems do not contribute equally to the energy budgets of all muscle fibers. *Slow oxidative muscle fibers* (also called slow-twitch muscle fibers) obtain most of their ATP from the oxidative system. Slow-twitch fibers require about one-tenth of a second to contract fully, they are packed with mitochondria, they receive a rich supply of blood, and they have large quantities of the red protein *myoglobin,* which stores oxygen in muscle cells. These characteristics make slow-twitch fibers deep red, like the dark meat of a chicken (Figure 29.14). Slow-twitch fibers are resistant to fatigue and thus able to contract for long periods of time. Athletes trained for endurance sports such as cross-country skiing have a large proportion of slow-twitch muscles. Slow-twitch muscles also provide functions critical for survival, such as maintaining posture. Without the contraction of slow-twitch muscles in your jaw, your mouth would be wide open as you read this sentence.

Whereas slow-twitch fibers bestow endurance, *fast glycolytic fibers* (also called fast-twitch fibers) provide quick power—the kind of power needed for one clean-and-jerk of a heavy barbell, for example (see Figure 29.14). Fast-twitch fibers derive most of their ATP from glycolysis, and they reach maximum contraction twice as quickly as slow-twitch fibers. They soon grow fatigued, however, since they run through their limited stores of glycolytically generated ATP in short order. Fast-twitch fibers are white because they are packed with white actin and myosin proteins (which maximize contractile force) and they contain very little myoglobin for oxygen storage. The white meat in the breast of a chicken is the fast-twitch muscle that powers the wings and enables

Cross section of muscle

 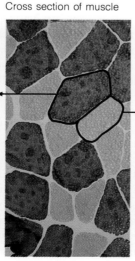 

FIGURE 29.14 ■ Slow-Twitch and Fast-Twitch Muscle Fibers Use Different Energy Sources. With a stain that darkens tissue containing the ATP-splitting myosin head, the myosin-rich fast-twitch fibers in a cross section of muscle tissue appear dark, while the slow-twitch fibers appear as light-colored patches. A person highly trained for and naturally good at weight lifting, shot-putting, or sprinting is more likely to have a high proportion of fast-twitch muscle fibers, while a person trained for aerobic endurance events like cross-country skiing or running in the mountains is more likely to have a high proportion of slow-twitch fibers.

the chicken to suddenly burst away from a fox and fly to safety in a tree. By contrast, a duck's breast is dark meat—red, slow-twitch fibers that can sustain the beating of the wings during long migratory flights without fatigue. A third kind of muscle fiber, *fast oxidative-glycolytic fiber,* has characteristics midway between fast glycolytic and slow oxidative fibers; these fibers are moderately powerful and moderately resistant to fatigue.

Most animals have a combination of fast, slow, and intermediate fibers distributed according to the animal's requirements for survival. American quarter horses are rapid sprinters generally and can outrace thoroughbred horses over short distances owing to genetically determined differences in ratios of fast- to slow-twitch fibers in the two horse breeds. Likewise, human weight lifters, shot-putters, football guards, and sprinters generally have a higher proportion of fast-twitch fibers (up to 63 percent for long jumpers), while cross-country skiers and long-distance runners often have more slow fibers (up to 82 percent for top marathon runners).

Although fiber types are genetically determined to a large degree, training can change the functional characteristics of muscle fibers. For example, endurance training can cause both fast and slow fibers to develop increased oxidative capacity but reduced explosive strength. In contrast, strength training improves immediate energy supply systems and suppresses oxidative capacity.

Training affects much more than the fast and slow muscle types. As the next section explains, it also has dramatic consequences on the homeostatic mechanisms of the entire body.

EXERCISE PHYSIOLOGY AND SURVIVAL

For most animals, survival is linked to locomotion, and in nature, locomotion often must be rapid and coordinated if the animal is to live. Heavy exercise or strenuous work simulates the same fight-or-flight response that electrifies the escape of a field mouse from a diving hawk. The fight-or-flight response is overwhelmingly useful, and it galvanizes nearly every system in the body into action. In fact, the response can really be said to unify our entire discussion of physiology in Chapters 20 through 29. When a threatened animal or an athlete in competition needs to move—and quickly—homeostatic control mechanisms vault into action, allowing the circulatory system to pump blood faster; the respiratory system to deliver oxygen more rapidly; the immune system to suppress inflammatory responses; the digestive system to provide energy compounds of immediate utility; and the kidneys to accommodate sudden differences in ion concentration. All this activity is orchestrated ultimately by the endocrine and nervous systems, which ignite the muscles that move the skeleton—and of course the entire animal—away from danger or toward a needed commodity. An investigation of exercise physiology allows us to integrate many of the principles we have discussed in the past few chapters.

Escape or Skirmish: A Survival Response

A sparrow that has just spotted a cat crouching nearby, a wild horse galloping across the desert, and a student waiting to give an oral report in front of the class are all experiencing stresses that kindle the fight-or-flight response, an automatic set of events managed largely by the hypothalamus. This brain structure serves as a main link between the nervous and endocrine systems and is located at the base of the brain (review Figure 28.15).

In times of stress, as during heavy exercise, the hypothalamus dispatches signals via the sympathetic nervous system to various body tissues (review Figure 28.13). These signals cause chains of biochemical events (Figure 29.15): They raise the amount of vital fuels like glucose, glycerol, and fatty acids in the blood; they elevate heart rate and boost blood pressure, circulating fuel to muscles; they dilate air passages and escalate breathing rate, providing ample oxygen; and they divert blood from the skin and digestive organs to the skeletal muscles, supplying food and oxygen where they are needed the most. The hypothalamus also triggers the secretion of epinephrine from the adrenal medulla, and this nervous system stimulant intensifies and prolongs the various effects (review Figure 26.12). In exercising horses like Moscha and Harry, the activation can be unusually pronounced. The pulse rate of a healthy horse can skyrocket from 25 beats per minute to 250 in the heat of a race. A human athlete, in contrast, might experience a climb from about 45 beats per minute to about 190 in a full-out footrace.

During a fight-or-flight response, the hypothalamus liberates a releasing factor that causes the pituitary to secrete the peptide hormone ACTH (adrenocorticotropic hormone; review Table 26.1). This hormone causes the adrenal gland to secrete the steroid hormone *cortisol,* a so-called stress hormone. Cortisol causes storage proteins to break down and liberate amino acids into the blood, fat cells to release fatty acids, and amino acids or glycerol to form sugar (*gluconeogenesis*). The result of all this activity is that as the fight-or-flight response primes the body for physical activity (perhaps necessary for survival), cortisol fortifies the system with amino acids and energy sources that can help heal tissue that may become damaged. Growth hormone, which also facilitates repair of injured tissues, is also secreted by the pituitary under control of the hypothalamus during stress.

Heavy exercise places special demands on body temperature regulation—a heating and cooling system that relies on the blood vessels to route hot blood to the body surface, and on sweat glands to dump moisture on the skin surface and carry away body heat through evaporation. During stress, the pituitary also secretes antidiuretic hormone (ADH; see Figure 25.14). ADH causes the kidney to retain water, which may be essential in cases of heavy sweating or blood loss. We can summarize this way: Exercise mimics the fight-or-flight response. It calls into action all the major physiological systems of the body, whose combined effects are integrated by the nervous and endocrine systems.

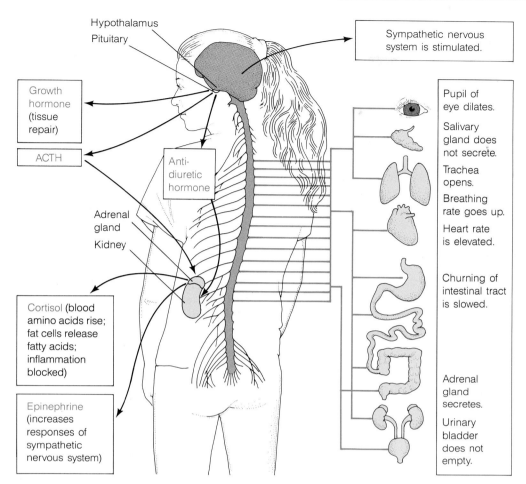

FIGURE 29.15 ■ Stress of Heavy Exercise Stimulates the Fight-or-Flight Response. This flow-chart shows organs (pink) spurred into activity by physical stress, the chemical substances they release (red), and the effects these agents have on the body. Compare to Figure 26.12, which depicts the effects of stress on the hormonal and nervous systems.

How Athletic Training Alters Physiology

Since heavy exercise evokes the fight-or-flight response, people who exercise several times a week enter this state of stress repeatedly. What effects do such repeated periodic challenges have on the body's homeostatic mechanisms? Training works by causing "breakdown" followed by "overshoot"—a breakdown of stored fuel, for example, followed by the increased deposition of fuel molecules; or a slight breakdown of muscle tissues, with an overall strengthening as the tissues are repaired. To increase fitness without injury, therefore, one must carefully, gradually, and progressively augment the intensity, frequency, and duration of workouts.

One goal of human athletic training is to boost the amount of oxygen a person can deliver to working muscles (the so-called *maximal oxygen uptake*). To learn how this uptake occurs, exercise physiologists in Dallas measured subjects' oxygen utilization immediately before and after three solid

weeks of bed rest, then measured it again over an eight-week training period in which each subject ran from 4 to 11 km (2.5 to 7 mi) on 11 different occasions each week. As the graph in Figure 29.16 shows, lying in bed all day long caused a substantial drop in the body's ability to take up and utilize oxygen. After a few days or weeks of training, however, oxygen utilization increased dramatically—especially in the previously sedentary men, who doubled their ability to use oxygen and perform physical work.

What changes in the men's bodies accounted for their increased ability to take up oxygen? Two main factors appeared responsible. First, the men's hearts pumped more blood per minute after training than before. This was due not to a more rapid heart rate but to an increase in the amount of blood pumped per beat (the stroke volume). The more one exercises, in fact, the slower the heart beats at rest: Cross-country skiers and bicycle racers often have resting heart rates of about 45 beats per minute as compared to an average of

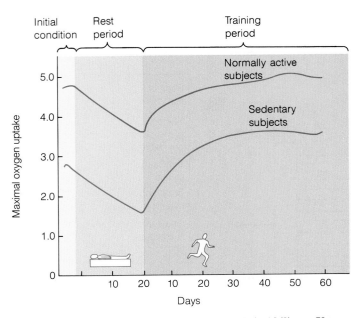

FIGURE 29.16 ■ Training Increases the Body's Ability to Use Oxygen. When experimental subjects stayed in bed for three weeks, the ability of their tissues to take up and use oxygen dipped dramatically. After a period of exercise, however, those who were active before the experiment (red line) regained (or slightly exceeded) their original uptake levels, while those who were initially sedentary (blue line) greatly exceeded their original levels. This experiment suggests that activity improves the body's oxygen utilization, even in the formerly bedridden or sedentary.

about 70 beats per minute in the general population. They have enlarged heart ventricles and thickened ventricle walls, which pump more blood per beat; in other words, they have larger, stronger hearts. Second, physically fit people can use more oxygen because their muscles extract more oxygen from the blood. Energy-supplying mitochondria grow larger, the muscles generate more of the oxygen-storing protein myoglobin, there is an increase in the number of blood capillaries in the muscles, and all these phenomena work together to provide the muscles a larger blood supply from which to extract more oxygen.

Exercise does more than stimulate fight-or-flight mechanisms and increase oxygen intake: It stimulates the release of *endorphins,* a class of naturally occurring morphines, from the pituitary and other brain regions. These peptide hormones act a bit like morphine itself in that they reduce pain and enhance a feeling of well-being. The release of endorphins during strenuous exercise may explain, in part, why many people experience feelings of relaxation and contentment after a workout. In recent years, researchers have found that regular exercise does even more than improve the body's fitness and make you feel good: It also has dramatic positive effects on the heart and vascular system and on a person's chances of living a long and disease-free life.

Exercise and Heart Disease

Many people in industrial nations have adopted sedentary life-styles, riding to work in a car, sitting at a desk or workbench all day, and reading the paper or watching television at night. Our softer life-styles have led to an increase in *hypokinetic* disease, maladies caused by "little motion." Low back pain (see Box 29.1 on page 597), obesity, and heart disease can all result from sedentary life-styles, although family heredity, injuries, and infections can also contribute to these ills. In general, however, the incidence and severity of back, weight, and heart problems can be lessened by activating muscles, skeleton, heart, lungs, nerves, and hormones with regular aerobic exercise.

Available evidence suggests that a physically inactive person is about twice as likely as an active one to die of heart disease. Considerable evidence suggests that heredity, diet, obesity, cigarette smoking, blood pressure, and blood levels of cholesterol may all increase one's risk of heart disease. Nevertheless, regular exercise can lower body weight, decrease one's desire to smoke, lower blood pressure slightly, lower cholesterol levels, and increase blood flow to the heart. While there is no guarantee against heart disease, medical researchers believe that a person can lower his or her risk of developing this killer by remaining smoke-free, eating a balanced diet low in saturated fats, and exercising regularly. It is ironic that many people recognize the importance of diet and exercise to their pets' health, but fail to apply the same principles to their own human bodies. This has led Swedish exercise physiologist Per-Olof Åstrand to make the following suggestion: Walk your dog whether you have one or not.

CONNECTIONS

Self-generated movement is a hallmark of animal life. This chapter investigated how animal movement relies both on organized arrays of protein filaments that slide past each other and on a rigid support, a skeleton, to which the shortening muscles transmit force. Skeletons can be water-filled tubes (as in sea anemones and worms), hard exoskeletons (as in pill bugs and beetles), or bony internal frameworks (as in people and horses). But the muscles that move them contract in the same way in all animals and even share basic features with the structures that allow slime mold cells to move and yeast cells to reorient internal structures. The muscle cells of animals do not contain a special evolutionary invention. Instead, they have simply exploited and specialized an ancient molecular principle—the contraction and sliding of actin and myosin—to generate force. This adaptation allows animals to carry out their special way of life—moving from place to place in search of food and mates.

The ultimate source of the energy that moves the muscles in a horse, rider, or cougar is sunlight. The structures that allow plants to capture the sun's energy, the principles that direct the plant's growth, and the ways that plants cope with environmental stress are all subjects of Part Five.

Highlights in Review

1. Three themes emerge in this chapter: Muscles and skeletons adjust to stress, within single lifetimes and over evolutionary time; protein fibers provide the strength for skeletons and muscles; and rigorous activity involves all of the body's physiological systems.

2. There are three main types of skeletons: the hydroskeleton, which functions by means of incompressible fluid and antagonistic muscle pairs; and two types of braced framework skeletons, the exoskeleton and endoskeleton.

3. Bones typically make a joint, or articulation, with one or more other bones. Processes, or projections, serve as attachment sites for the straplike connectors, tendons and ligaments. Some bone cells secrete hard materials, while others reabsorb calcium and phosphate and help heal breaks.

4. Bones act as levers that transmit force. They pivot about joints and accomplish work because of the way antagonistic muscle groups are attached to the bones.

5. The skeletal muscles that propel the body are made up of giant contractile cells called muscle fibers. Within muscle fibers, repeating units called sarcomeres do the actual contracting. Actin filaments sliding past stationary myosin filaments bring about contraction of the muscle fiber.

6. Three sets of membranes help stimulate muscle fibers to contract: the plasma membrane, the transverse tubules, and the sarcoplasmic reticulum.

7. In addition to skeletal muscle, vertebrates have cardiac, or heart, muscle and smooth muscle. Cardiac muscle cells are electrically connected by junctions, which allows the whole organ to contract as a coordinated unit. Smooth muscle cells lack striations and, like cardiac muscle cells, communicate with each other electrically via junctions; this coordinates waves of contraction in the digestive or reproductive tract.

8. The ATP fuel for muscle contraction comes from three sources—an immediate energy system, a glycolytic system, and the oxidative system—depending on the intensity and duration of activity.

9. Slow-twitch fibers are red, since they are rich in oxygen-storing myoglobin, and they obtain most of their ATP from the oxidative system, whereas fast-twitch fibers are white (owing to high actin and myosin content) and derive most of their ATP from glycolysis.

10. During a fight-or-flight response, the hypothalamus directs the body toward hormonal readiness for instant activity, including the release of cortisol, the stress hormone. Regular exercise causes the body to adjust so that it uses more oxygen and has greater muscle mass and tone, a larger, stronger heart, and stronger bones. Exercise also releases endorphins and helps improve low back pain, heart disease, and obesity.

Key Terms

antagonistic muscle pair, 590
appendicular skeleton, 590
axial skeleton, 590
bone, 590
cardiac muscle, 598
cranium, 590
endoskeleton, 590
exoskeleton, 590

hydroskeleton, 590
joint, 593
ligament, 592
muscle fiber, 594
myofibril, 594
pectoral (shoulder) girdle, 591
pelvic (hip) girdle, 591
process, 592

sarcomere, 594
skeletal muscle, 594
skeleton, 590
skull, 590

smooth muscle, 598
tendon, 592
vertebra, 590
vertebral column, 590

Study Questions

Review What You Have Learned

1. Describe the structure of bone.
2. How do exercise and hormone supplements help reduce the incidence of osteoporosis?
3. How do joints with limited mobility, such as the joints between vertebrae, differ from freely movable joints, such as the knee?
4. Describe the structure of a myofibril.
5. Explain the sliding filament mechanism of muscle contraction.
6. Name the steps of muscle cell contraction.
7. Compare cardiac muscle, smooth muscle, and skeletal muscle.
8. Which energy system most effectively powers sustained muscle contractions? Explain.
9. Name and describe the three types of skeletal muscle fibers.
10. What are the health benefits of regular exercise?

Apply What You Have Learned

1. Using only fossil bones as guides, an anthropologist can recreate what the face of a *Homo erectus* individual may have looked like. Explain how this is possible.
2. A race horse falls, breaks its neck, and dies. A few hours later, the once-limp corpse has stiffened. What has happened?
3. A tired student feels revitalized after playing tennis in the early evening. Explain what brought about this change.
4. A man who suffered general paralysis is rushed to the hospital where you work as a physician. When you examine him, you find that he is alert and his heart is beating normally, but all his skeletal muscles are flaccid. Your colleagues suggest that he suffers from: (a) inadequate ATP production, (b) calcium deficiency, (c) poisoning of the enzymes that degrade acetylcholine, or (d) inhibition of acetylcholine release. Which of these suggested causes is a likely explanation for his paralysis?

For Further Reading

ALBERTS, B., D. BRAY, J. LEWIS, M. RAFF, K. ROBERTS, and J. D. WATSON. *Molecular Biology of the Cell.* 2d ed. New York: Garland, 1989.

ÅSTRAND, P. O., and K. RODAHL. *Textbook of Work Physiology: Physiological Bases of Exercise.* 3d ed. New York: McGraw-Hill, 1986.

BIRKE, L. "Equine Athletes: Blood, Sweat and Biochemistry." *New Scientist,* May 1986, pp. 48–52.

CUNNINGHAM, C., and J. BERGER. "Wild Horses of the Granite Range." *Natural History,* April 1986, pp. 132–138.

HILDEBRAND, M. "Vertebrate Locomotion: An Introduction." *BioScience* 39 (1989): 764–765.

NEWSHOLME, E., and T. LEECH. "Fatigue Stops Play." *New Scientist,* September 22, 1988, pp. 39–43.

PART FIVE

HOW PLANTS
SURVIVE

Plant Architecture and Function

FIGURE 30.1 ■ **Pear Branch in Bloom.** A flowering tree in spring is a reminder of the angiosperm life cycle.

A Blooming Pear Tree

One of the loveliest sights of spring is a mature fruit tree in full bloom. The tall, dark trunk of a pear tree, for example, supports an enormous crown of large and small branches, as well as a cloud of white flowers. Each branch is a study in contrasts, with its delicate, fragrant blossoms growing directly from sturdy twigs (Figure 30.1) and with an entourage of bees flying actively from flower to flower, bristled bodies and legs loaded down with pollen grains.

A blooming pear tree is an apt reminder of how generations alternate in the life cycle of flowering plants (see Chapter 17). The cycle begins with a diploid plant generation (sporophyte), which produces haploid spores by meiosis. In the flowers, these spores develop into tiny haploid plants (the gametophyte generation), which produce gametes—eggs in the ovary, sperm in the pollen grains. When sperm fertilize eggs, new diploid generations develop inside seeds, and wind or animals carry the seeds away. After the seeds germinate and grow into mature plants, the plants flower, and the cycle repeats.

A pear tree can begin producing pear fruits just a decade after germination, when its trunk is still slender and its branches small. It can then live for another 70 years and produce up to 45 kg (99 lb) of fruit annually. That luscious crop is the reason people began cultivating pear plants more than 2000 years ago and why today we harvest more than 7 million tons of pears each year worldwide. The tasty fruit is also the reason birds, deer, and other animals are drawn to pear trees in late summer and early autumn. Each pear seed an animal inadvertently swallows and deposits in a new location is capable of initiating the growth of a new organism—a pear tree, complete with roots, a tall woody trunk, branches, twigs, broad green leaves, fragrant blossoms, and juicy pears.

This chapter picks up the story of the angiosperm life cycle where Chapter 17 left off, and it also provides a foundation for our discussion of plant physiology—how plants survive day to day. Building on the details of pollination and fertilization presented in Chapter 17, we focus here on plant architecture and the development of an embryo to a mature plant. Plants reveal in sharp relief the intimate association of biological form, function, and environment. Most plants, of course, are stationary organisms that cannot move about in pursuit of water, energy, a carbon source, and mineral nutrients. But a plant's anatomy, way of life, and mode of reproduction are all beautifully adapted to meeting its needs while it remains rooted to one spot.

Because sunlight strikes the earth's surface so abundantly, plants can remain stationary and yet still obtain the energy they need to carry out photosynthesis. Moreover, carbon dioxide in air is a universally distributed carbon source that can move into leaves or stems through tiny openings. And plant roots can absorb water and mineral nutrients from supplies in the soil. Water is particularly crucial to plant survival because it transports minerals, sugars, and proteins and is essential for photosynthesis (see Chapter 6). Water also contributes to the rigidity of plant cells and organs that depend on turgor pressure (see Chapter 2), and it helps cool the leaves through evaporation, thereby preventing overheating. Fully 98 percent of the water that enters a plant's roots is lost via transpiration (evaporation in plants), and thus a single mature plant such as corn growing in a farmer's field can require 17 times more water each day than an adult person!

Function also influences form in the flowers, fruits, and seeds of stationary plants (Figure 30.2). Flowers aid pollination by attracting animals or by releasing pollen to wind currents. Fruits aid in the dispersal of the new generation by being fleshy, delectable animal attractors, by clinging to fur or feathers, by floating on water currents, or by sailing on summer breezes. And seeds bearing precious embryonic cargo have tough coats and stores of nutrients that help the offspring survive until they can emerge and obtain nutrients on their own.

Using the pear tree as a central example, we explore the form and function of plants, focusing specifically on flowering plants. As Chapter 17 revealed, not only do angiosperms make up the largest and most diverse of all plant groups, but they have had an immense impact on people. Civilization depends on grains, fruits, vegetables, animal fodder, wood, fiber, spices, beverages, chemicals, drugs, and myriad other plant by-products.

Three unifying themes will emerge in this chapter. First, a plant's form and function strike a compromise between conflicting needs. Plants often have large, flat leaves that can collect maximum sunlight, and the leaves usually have tiny openings that allow gas exchange. Both large surface area and openings, however, conflict with the need for preventing excessive water loss. Second, the anatomical parts that we examine separately make up tissues, organs, and systems that function collectively as whole plants. Transport pipelines, for

FIGURE 30.2 ■ **Pear Slice: Form, Function, and Adaptation.** A pear fruit is more than a delicious snack; it houses and protects seeds and attracts animals. The form and function of a given fruit are adaptations to the plant's stationary growth pattern.

example, are continuous from roots to stems to leaves, and the activities of leaves affect the functioning of roots, and vice versa. Third, plants have open growth based on perpetual embryonic centers that continually produce new organs and larger body size throughout life. This mode of development helps overcome the limitations imposed by rigid cell walls and hardened mature tissues.

Our discussion of plant architecture and function will answer several questions about how a plant's structures develop and enable it to survive in its environment:

■ What are a plant's main tissue types, and how does the rigidity of plant cells influence the organism's overall growth patterns?

■ How do the main structures in flowers, fruits, and seeds contribute to the plant's reproductive success?

■ How does the root anchor the plant, store starch, and channel water and minerals?

■ How does the stem support the plant and transport materials?

■ How does the leaf collect solar energy, store carbohydrates, and supply organic nutrients to the rest of the plant?

THE PLANT BODY: PLANT TISSUES AND GROWTH PATTERNS

Pear trees, with their clouds of white flowers in springtime and their loads of golden pear fruits in autumn, are just one species (*Pyrus communis*) among nearly half a million vascular plants. As a group, vascular plants are distinguished by their vascular tissue, or internal transport tubes. This mammoth group dominates the land and includes ferns, conifers, and flowering plants. A subgroup of vascular plants including gymnosperms and angiosperms produces seeds, and these seed plants have two additional adaptations for life on land: sexual reproduction with internal fertilization and a tough seed coat that protects the developing embryo. These last two characteristics are similar to the innovations of internal fertilization and eggshells we saw in the reptiles (see Chapter 19). Of the seed plants, only the angiosperms produce flowers. The vast majority of vascular plants also have a main axis made up of root and shoot.

The Plant's Main Axis: Root and Shoot

Pear trees, like most typical vascular plants, have roots below the ground and a shoot above the ground. The **shoot** includes the stem, branches, leaves, flowers, and fruits (Figure 30.3a). Lifting the rest of the shoot is the stem, a stiff bundle of internal transport tubes (xylem and phloem) plus surrounding tissues, all enclosed within an external waterproof coating that minimizes water loss. In pear trees as well as other trees and shrubs, the main stem matures into a large, bark-covered trunk with many side branches supporting leaves and flowers.

  Roots are branching organs that grow downward into the soil. Roots help support the plant physically, by spreading out through the soil and attaching the plant in it, and nutritionally, by absorbing and transporting water and mineral nutrients that the plant cannot take in from the air. Pear trees, dandelions, and carrots each have a thick, trunklike **taproot** that develops directly from the embryonic root (radicle) and which grows straight down into the soil. Fine lateral roots branch off the sides of the main taproot (Figure 30.4a). Taproots store water, as well as food in the form of starch. In contrast, in grasses and a few other kinds of plants, the embryonic root dies back early in the organism's life. In these plants, the roots that will support the adult plant grow downward and outward from the plant's stem, branching repeatedly and forming a mass of narrow **fibrous roots** (Figure 30.4b). Fibrous roots store less starch than taproots, but efficiently anchor the plant and absorb water and nutrients. In general, roots that arise from aboveground structures are called **adventitious roots.** Examples are the thick aerial prop roots of a corn plant or banyan tree (Figure 30.4c) and the fine roots that allow ivy plants to cling to walls and fences.

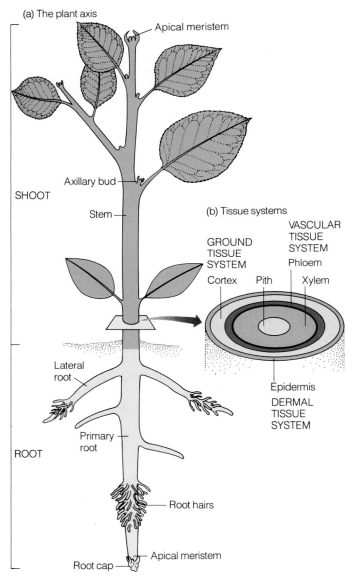

(a) The plant axis

Apical meristem

SHOOT

Axillary bud

Stem

Lateral root

Primary root

ROOT

Root hairs

Apical meristem

Root cap

(b) Tissue systems

GROUND TISSUE SYSTEM

VASCULAR TISSUE SYSTEM

Phloem

Cortex   Pith   Xylem

Epidermis

DERMAL TISSUE SYSTEM

FIGURE 30.3 ■ The Plant Axis: Root and Shoot. **(a) The above- and below-ground portions of the plant have different functions and hence different anatomies. Shoot structure is appropriate for its functions in photosynthesis, support, transport, and sexual reproduction, while root structures anchor the plant and absorb water and minerals. (b) Plants have three tissue systems, each of which carries out a different function.**

Tissue Systems and Tissue Types

Complex flowering plants, such as pear trees, pumpkin vines, and saguaro cactus, have three tissue systems with important physiological roles in the body, and each system contains a few important tissue types (Table 30.1). Each tissue, in turn, is composed of different cell types with distinctive characteristics appropriate to their function (Table 30.2 on page 611). As Figure 30.3b shows, the three main tissue systems in a mature vascular plant are the **dermal tissue system,** which,

(a) Taproot  (b) Fibrous roots  (c) Adventitious roots

FIGURE 30.4 ■ Patterns of Root Growth. (a) A dandelion has a central taproot with fine lateral roots. (b) Grass has numerous fibrous roots that anchor the plant very firmly. (c) A single banyan tree has many adventitious roots that help support the plant and absorb water and minerals.

like skin, protects the plant from water loss and injury to internal tissues; the **ground tissue system,** which provides support and stores starch; and the **vascular tissue system,** which conducts fluids and helps strengthen the roots, stems, and leaves. These tissue systems are continuous throughout the plant.

■ **The Dermal Tissue System: Protection and Waterproofing** Pear trees and other vascular plants have a dermal system that covers every part of the plant and is analogous to an animal's skin (Figure 30.5). In a young seedling, the dermal tissue consists

(a) A petal's dermal tissue

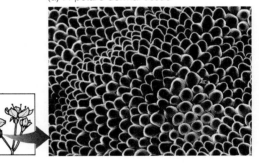

(b) A leaf's dermal tissue

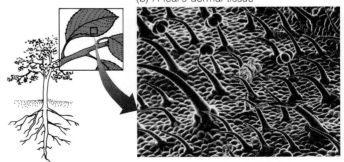

FIGURE 30.5 ■ The Dermal System: A Visible Protective Shield Covering the Root and Shoot. These scanning electron micrographs show (a) the bumpy epidermis covering a flower petal and (b) the waxy cuticle and epidermis of a leaf, with its hairlike, protective trichomes. Trichomes create a surface "hairiness" that helps reflect light and hence retards water loss and also discourages certain predators from eating stems and leaves. Notice that some trichomes are thorn-shaped, while others are globular packets that contain irritating biochemicals.

TABLE 30.1   Tissues and Cell Types in Flowering Plants

Tissue System	Tissues	Cell Types
Dermal	Epidermis (primary growth)	Parenchyma cells Guard cells Trichomes Sclerenchyma cells
	Periderm (secondary growth)	Parenchyma cells Sclerenchyma cells
Ground	Parenchyma Collenchyma Sclerenchyma	Parenchyma cells Collenchyma cells Fibers Sclereids
Vascular	Xylem	Tracheids Vessel members Sclerenchyma cells Parenchyma cells
	Phloem	Sieve tube members Companion cells Parenchyma cells Sclerenchyma cells

of just an *epidermis,* or protective outer covering. Epidermal cells in stems and leaves secrete a waxy waterproof coating, the cuticle. Some epidermal cells form little hairs, or tri-chomes (see Figure 30.5b). In trees and woody shrubs, however, the *periderm,* the outer areas of the bark, replaces the epidermis.

■ **The Ground Tissue System: Storage and Support**   A plant's root and shoot contain a ground (or fundamental) tissue system—a kind of background tissue that packs around the vascular pipelines. The ground tissue system makes up the bulk of most plant organs; it stores the starchy products of photosynthesis, helps keep the plant from collapsing into a formless heap, and has other more specialized roles. There are three tissue types in the ground tissue system, which vary in strength and flexibility: parenchyma, collenchyma, and sclerenchyma.

In both root and shoot, the majority of ground tissue is **parenchyma,** made up of loosely packed, thin-walled, rounded parenchyma cells, such as those that store starch in a potato (Figure 30.6a). Most parenchyma cells have only a thin primary cell wall laid down as the cell grows, and they are often unspecialized, able to give rise to new cells and cell types in a wounded adult plant. Some parenchyma cells, however, are specialized. In the leaves, for example, parenchyma cells photosynthesize but store little starch, while in the roots, they often store large quantities of starch.

Collenchyma tissue is tougher than parenchyma because individual collenchyma cells often have a cylindrical shape and thick primary walls, and they connect end to end in long, stringy fibers such as those just beneath the epidermis of a celery stalk (Figure 30.6b). The word *collenchyma* is based on the Greek word for "glue," and in fact, these shiny cells help hold the plant body together. In cross section, collenchyma cells have uneven, thickened corners (see Table 30.2).

Sclerenchyma tissue (from the Greek word for "hard") is a third tissue type in the ground tissue system. Because of its hardness, sclerenchyma often surrounds and reinforces the tubes of the vascular system. Sclerenchyma cells die at maturity and leave behind a thick secondary cell wall that is deposited after the cell stops growing and is hardened with a tough complex polymer called *lignin.* Sclerenchyma tissue includes stone cells, or *sclereids,* which are hard, crystalline cells (Figure 30.6c). Sclerenchyma cells give pears their grittiness and walnut shells their hardness. The *fibers* that make up the strong, slender threads of hemp and flax and in turn ropes and linen fabrics are a second cell type in sclerenchyma tissue.

These cell types can also occur in other tissue systems, as we shall see.

■ **The Vascular Tissue System: Material Transport and Vertical Strength**   Within a vascular plant, water and materials travel in two kinds of tubular tissues, xylem and phloem, each of which forms a continuous vascular system extending from root tips to leaves. In general, **xylem** transports water and minerals absorbed from the soil up through the roots, stems,

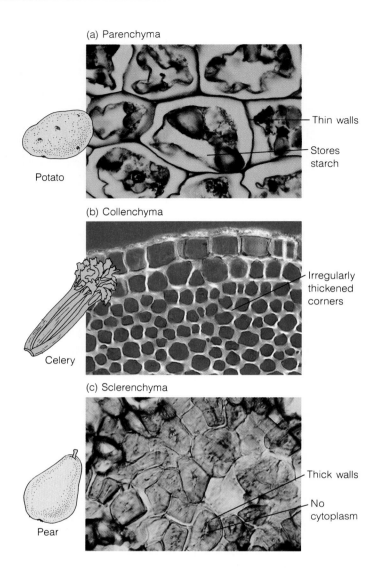

(a) Parenchyma

Potato — Thin walls — Stores starch

(b) Collenchyma

Celery — Irregularly thickened corners

(c) Sclerenchyma

Pear — Thick walls — No cytoplasm

FIGURE 30.6 ■ The Ground Tissue System: Cells That Store, Strengthen, and Lend Hardness to Plant Parts. (a) Most ground tissue cells are parenchyma cells, such as these potato cells with starch grains stained lavender. (b) The stringy fibers just inside the epidermis of a celery stalk are made up of collenchyma cells, with their thickened corners and irregular shapes. (c) The gritty structures in a pear's flesh are a type of sclerenchyma called sclereids, or stone cells; these cells have thick walls but lack cytoplasm when mature.

and leaves. Wood consists of xylem vessels. **Phloem** transports dissolved sugars and proteins from "source to sink"—that is, from cells that produce or store sugars to cells that use sugars rapidly.

The xylem of most flowering plants is composed of several kinds of cells, but principally **tracheids** and **vessel members** (Figure 30.7a and b). Both tracheids and vessel members transport water only after they have died: Each cell type

TABLE 30.2 Characteristics of Cell Types in Flowering Plants

Tissues and Cell Types	Structure	Function
Epidermis		
Guard cells	Pairs of large cells in epidermis surrounding a pore	Regulate gas and water exchange in leaves
Trichomes	Hairs or scales jutting from epidermis	Protection; retard water loss
Parenchyma		
Parenchyma cells	Many-sided, often thin primary wall; alive at maturity; found throughout plant	Photosynthesis, storage, local conduction of materials, wound healing
Collenchyma		
Collenchyma cells	Rectangular; unevenly thickened primary cell wall at corners; alive at maturity	Support of young stems and leaf ribs
Sclerenchyma		
Fibers	Very long and narrow; primary and very thick secondary cell wall; often dead at maturity	Support in stem cortex and vascular system
Sclereids	Often long, but shorter than fibers; primary and thick secondary cell walls; living or dead at maturity	Support and protection throughout plant

Tissues and Cell Types	Structure	Function
Xylem		
Tracheids	Elongated tapering cells; primary and secondary cell walls with pits; dead at maturity	Conduct water in all vascular plants
Vessel members	Long and narrow, but generally shorter than tracheids; primary and secondary cell walls with both pits and perforations; stacked end to end in a vessel; dead at maturity	Main water-conducting cell in xylem of flowering plants
Phloem		
Sieve tube member	Elongated cell; primary cell wall with sieve plates; alive at maturity, but lacks functional nucleus; stacked end to end in sieve tube	Main food-conducting cell in phloem of flowering plants
Companion cells	Somewhat elongated; primary cell wall; alive at maturity and closely connected to sieve tube member	Help regulate activities of sieve tube member

(a) Tracheids

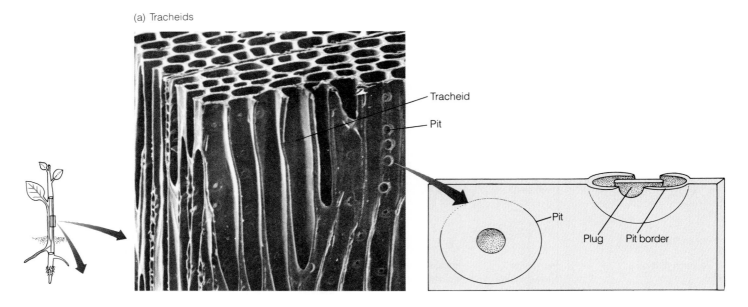

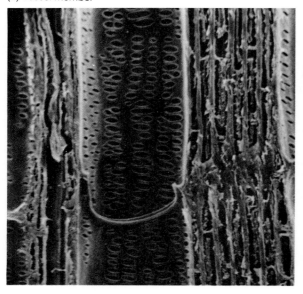

(b) Vessel member

(c) Sieve tube members

Sieve plate

Companion cell

FIGURE 30.7 ■ **The Vascular Tissue System: Tissues That Transport and Support.** (a) Pitted, hollow xylem cells called tracheids transport water and inorganic nutrients in all types of vascular plants. The pits allow lateral movement of water from one pointed tracheid to the neighboring cell. Pressure can force the central plug against the pit borders that prevent water leaks from a high-pressure cell to a low-pressure cell. (b) Vessel members, found only in flowering plants, are also pitted and hollow, but are not pointed like tracheids; their end walls have perforation plates. Vessel members stack end to end, channeling water up the central cavity. (c) The functional cells of phloem are sieve tube members; these lack a nucleus and are adjoined by nucleated companion cells. Sieve plates, which are usually the end walls of sieve tube members, are perforated with tiny pores.

becomes hollow at maturity, when the cell contents disintegrate and leave behind empty cell walls. These hollow cylindrical cells, stacked end to end, form efficient transport pipelines. Tracheids are long, narrow cells that overlap at their slender ends and have in their thick secondary cell walls many *pits,* gaps that allow water to pass from one tracheid cell to an adjoining one (see Figure 30.7a). Nearly all vascular plants have tracheids, but vessel members tend to occur mainly in flowering plants. Like tracheids, vessel members have pits in their walls, and they tend to be stacked directly end to end like the clay pipe sections of a water main. But unlike tracheids,

vessel members have larger holes in their end walls called *perforation plates,* which allow water to flow through the stacked vessel members unimpeded (see Figure 30.7b and Table 30.2 on page 611).

Phloem transport tissue is also composed of cells arranged end to end forming pipelike tubes, but phloem and xylem differ in three important respects: First, phloem cells contain living cytoplasm and are thin-walled, while xylem cells are dead and thick-walled. Second, phloem transports sugars and amino acids dissolved in small quantities of water, while xylem transports minerals and large quantities of water.

Third, while xylem transports materials in one direction, from roots to leaves, phloem can move sugars from photosynthesizing leaves to the roots in summer or from roots to developing leaves in the spring.

The conducting cells of phloem tissue are called **sieve tube members** because **sieve plates,** usually end walls bearing large pores, allow fluids and solutes to pass in and out like a sieve. Sieve tube members remain alive; they have a thin layer of cytoplasm that clings to the cell wall, but they lack a functioning nucleus (Figure 30.7c). Lying next to each living sieve tube member is a small **companion cell** with a nucleus that directs the activities of the nearby sieve tube member. Chapter 32 explains how phloem and xylem transport substances.

Open Growth: A Plant's Pattern of Perpetual Growth and Development

Imagine the chaos that would ensue if an animal—a dog, let's say—were to continue growing throughout life, producing new eyes, ears, legs, feet, livers, and other organs. The result might be a bizarre creature with three tails, seven legs, and four eyes. Conversely, think what would happen if a pear plant grew like an animal, keeping its original root, stem, and two initial leaves, each simply getting bigger and bigger as the plant grew. Clearly, neither growth pattern would suit the other type of organism.

A plant has a growth pattern called **open growth.** Throughout life, the plant adds new organs, such as branches, leaves, and roots, enlarging from the tips of the root and shoot. Plants cannot move toward favorable conditions, but they can grow toward light, water, and mineral nutrients and away from harmful situations. This continual growth is based on **meristems,** tissue that remains perpetually embryonic and that gives rise to new cells and cell types throughout the plant's life. Unlike animals, plants lack a special line of germ cells established during early embryonic development (review Figure 13.23). Instead, meristems allow adult plants to generate reproductive cells and gametes as well as new tissues and organs.

Plants have two types of meristems. The first type, **apical meristems,** are perpetual growth zones at the tips (*apices; singular, apex*) of roots and stems (Figure 30.8a). Apical meristems allow shoots to grow upward toward the light and roots to push ever deeper into the soil in search of water. Growth arising from apical meristems is called **primary growth.** The xylem and phloem produced by apical meristems are called primary xylem and primary phloem. The second type of meristem, **lateral meristems,** are cylinders of dividing cells in the stems and roots that cause these parts to become thicker (Figure 30.8b). The enlarged diameter of a stem or root due to cell divisions in the lateral meristems is called **secondary growth. Vascular cambium** is a type of lateral meristem that produces

wood (secondary xylem) and secondary phloem. Another lateral meristem, **cork cambium,** generates *cork,* the waterproof outer part of the bark. The distinction between primary growth from the apical meristems and secondary growth from the lateral meristems explains why a nail pounded into a young pear tree will never be lifted above its original distance from the ground: Primary growth, which lengthens the tree, occurs only at the apical meristems, which lie at the ends of branches far above the nail. The nail, however, will eventually be buried in bark and wood as secondary growth from lateral meristems adds new tissue to the side of the plant. Although

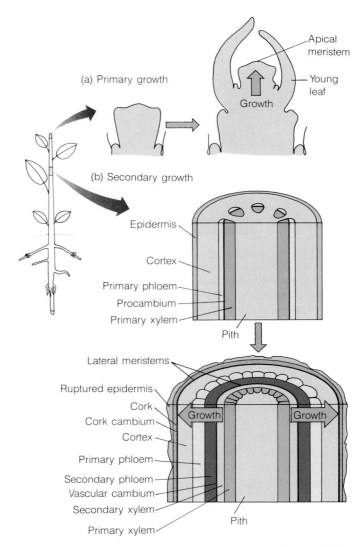

(a) Primary growth

Apical meristem

Young leaf

Growth

(b) Secondary growth

Epidermis

Cortex

Primary phloem

Procambium

Primary xylem

Pith

Lateral meristems

Ruptured epidermis

Cork

Cork cambium

Cortex

Primary phloem

Secondary phloem

Vascular cambium

Secondary xylem

Primary xylem

Growth

Growth

Pith

FIGURE 30.8 ■ Primary and Secondary Growth from Apical and Lateral Meristems. (a) Primary growth from apical meristems at the tips of root and shoot causes the plant to lengthen. **(b)** Secondary growth from lateral meristems causes the shoot and root to increase in diameter.

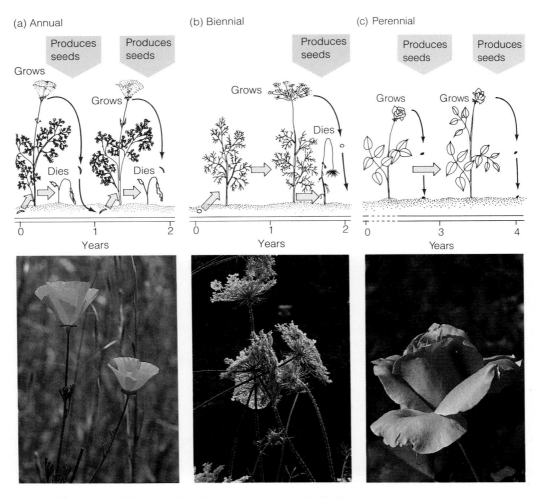

FIGURE 30.9 ■ Annuals, Biennials, and Perennials. (a) Annuals, such as the California poppy shown here, have one-year life cycles and often pass the winter in seed form. (b) Biennials, like this Queen Anne's lace, grow the first year, reproduce the second year, and then die. (c) Perennials, such as this rose, survive and grow year after year.

woody plants, such as pears and pines, have secondary growth, *herbaceous* plants, such as daisies and dandelions, have only primary growth and slender, generally flexible green stems.

While flowering plants have an open growth pattern based on meristems, some live only one season, whereas others live from 2 years to 5000. In a single season, an **annual plant** sprouts from a seed, matures, produces fruits and new seeds, and dies (Figure 30.9a). Marigolds, zinnias, petunias, poppies, and soybeans are familiar annuals. **Biennials** are plants that have a two-year life cycle: They grow from seeds to adults in the first year; then, in the second year, the adults produce flowers, fruits, and seeds and then die (Figure 30.9b). Celery, cabbage, carrots, and Queen Anne's lace are all biennials. Many biennials and most annuals are herbaceous, lacking secondary growth. Finally, **perennials** live for many years and typically bloom and set seeds several times before the

adult plant dies (Figure 30.9c). Some perennials, like tulips and dahlias, are herbaceous; others, like rosebushes and pear trees, are woody.

FLOWERS, FRUITS, SEEDS, AND PLANT EMBRYOS: THE ARCHITECTURE OF CONTINUITY

How many times has each of us chosen a plump, golden pear, eaten the sweet, soft fruit, and tossed out the core without further thought? That pear, however, represents much more than just a luscious snack. At the bottom of a pear sits a small whorl of dry, brown flaps, the sepals and sexual structures from the original pear blossom (Figure 30.10a). The rounded fleshy part, remarkably like a womb in shape, is actually the swollen stalk (receptacle) of the flower. The tough inner ring surrounding the core is the wall of the flower's ovary, and it encircles the small, dark seeds. Each seed contains a tiny living embryo; nourishment that sustains the embryo until it can emerge and take in water, minerals, and solar energy; and all the genetic information needed to generate a tall, productive tree.

The organs of a tree or other plant—the stems, leaves, fruits, and flowers—develop over time, beginning at the fusion of female and male gametes and continuing throughout life. Since the adult plant's tissue systems really get their start through fertilization and development, our discussion must begin at the site of fertilization—the flower.

Flowers: Sex Organs from Modified Leaves

Flowers are highly modified shoots that contain the reproductive organs of trees and other flowering plants. As Chapter 17 explained, flowers produce pollen grains and embryo sacs that are tiny haploid plants (review Figure 17.24). The various parts of the flower help ensure the transfer of pollen, with its sperm nucleus, to the egg inside the embryo sac. Flowers consist of four rings of structures called sepals, petals, stamens, and carpels (Figure 30.10b). The *sepals* in the outer ring and the *petals* in the next ring often have showy colors and shapes and fragrances that attract animal pollinators. A pear blossom's five sepals are green, but its five petals are bright white and attract bees. Nutritious pollen grains and sweet nectar are often the pollinator's reward. The third ring of structures contains the pollen-producing *stamens* (20 in pears), within each

of which the *anther* (containing two pollen sacs) sits high on a *filament* (see Figure 30.10b). This long stalk aids pollen dispersal and is especially helpful in wind-pollinated plants. (Box 30.1 on page 616 explains the "engineering" of wind pollination.) Making up the innermost ring is one or more *carpels,* which may fuse together, as do the five carpels in a pear blossom. The base of the carpel, or *ovary,* houses the *ovules,* and the "neck" of the ovary, or *style,* supports the *stigma,* a sticky surface on which pollen grains germinate. Germinating pollen tubes grow downward through the stigma and style toward the ovules and fertilize the egg (review Figure 17.24).

The numbers and sizes of flower parts vary from species to species; lilies, for example, have flower parts in multiples of three, pears in multiples of five (see Figure 30.10). In both lilies and pears, each flower is *perfect;* that is, each flower contains both stamens and carpels. In other species, such as corn or willows, the flowers are *imperfect;* they are missing either male or female parts. Corn has separate imperfect male and female flowers (the tassel and the ear) on each individual plant, and the species is considered *monoecious* (Greek for "one house"). Willows and American holly have two types of individual plants, some with only imperfect male flowers and others with only imperfect female flowers. Such plants are considered *dioecious* ("two houses").

Pollination, Fertilization, and Seed Formation

Pollination and fertilization are important steps in the angiosperm life cycle, and the pear is a good case study for these events. Pear pollination takes place in spring, when bees carry pollen from the anthers of one blossom to the stigma of

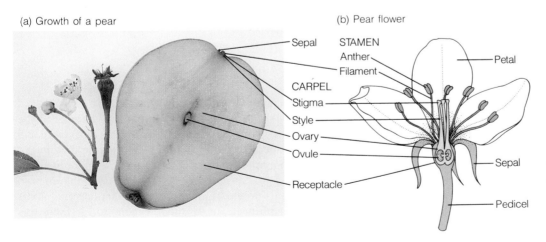

(a) Growth of a pear

(b) Pear flower

Sepal · STAMEN · Anther · Filament · Petal · CARPEL · Stigma · Style · Ovary · Ovule · Receptacle · Sepal · Pedicel

FIGURE 30.10 ■ Pear Fruit and Flower. (a) The remnants of a pear flower—the styles, filaments, and sepals—decorate an inverted pear fruit. The flower stalk (receptacle) expands enormously into the fleshy succulent fruit. (b) Each part of the flower has a specific role in pollination, fertilization, or seed protection. A complete perfect flower like a pear blossom has four concentric rings of flower structures: for pears, 5 sepals, 5 petals, 20 stamens, and 5 carpels. Many kinds of apples also have five fused carpels. The five bumps on the bottom of a red delicious apple reflect that organization.

THE ANSWER WAS BLOWING IN THE WIND

For many observers, wind pollination had always seemed slightly wasteful and even aggravating—a poor cousin to pollen dispersal by animals. After all, for every pollen grain a rose, a tomato, or another animal-pollinated species produces, a wind-pollinated plant produces more than 150—a major expenditure of energy and resources. And it takes 1000 pollen grains, carried aloft from a ragweed or a pine tree, for every grain that eventually reaches the ovule of a target plant. As the rest waft about, at least some land in people's noses and cause the familiar miseries of hay fever.

In his research at Cornell University, botanist Karl Niklas has done a great deal to correct traditional misimpressions about the "inefficiency" of wind pollination. He and his colleagues wondered whether pinecones, for example, are haphazard targets for airborne pollen—no more likely to collect the golden grains than tree leaves, car windshields, or pond surfaces—or whether the compact, whorled, bewinged shape of a female cone somehow "suctions in" pollen via altered air currents.

To test this question, Niklas built a large papier-mâché model of a pinecone, blasted it with air in a wind tunnel, and took stroboscopic photographs of small helium bubbles as they swirled about near the model (Figure 1). He discovered, in this way, that pinecones are uniquely shaped such that they (1) deflect wind (and hence windborne pollen grains) into the core where the ovules develop, (2) swirl air around each ovule's opening (micropyle), and (3) pull pollen grains down into the leeward (downwind) side of the cone. Similarly, he found that whorled pine needles surrounding a cone can draw wind (and pollen) toward a cone's downwind side. And yet other studies revealed that grass influorescences (flower clusters) have aerodynamic properties that tend to warp the flight paths of airborne pollen and in effect channel the grains toward the receptive ovules and eggs.

It's tempting to think that cone and flower designs are evolutionary adaptations for capturing pollen, and this may be true. But, Niklas speculates, the pollen-attracting designs could just as easily be a by-product of adaptations to other unknown envi-

FIGURE 1 ■ Wind Pollination and Cone Shape. Biomechanical formula and diagrams are superimposed on a model of cone and pollen flow.

ronmental pressures. One way or the other, though, he rejects the notion of wind pollination as primitive or inefficient. He points out that gymnosperms, some of the most ancient land plants, have for hundreds of millions of years retained that form of pollination and the general structures of cones. And grasses have evolved aerodynamic architecture only recently in evolutionary history.

---

another. One of the three haploid cells in each pollen grain engineers the growth of a pollen tube down the length of the style to the ovary, while the other two cells, the sperm, travel through that tube and reach the ovule and embryo sac with its eight haploid nuclei (review Figure 17.24). As Chapter 17 explained, angiosperms such as the pear experience a unique double fertilization: One sperm nucleus fuses with the egg nucleus and thus forms the diploid embryo, while the other fuses with two polar nuclei, forming the triploid nutritive tissue, or endosperm, that will nourish the embryo within the seed.

At first, the endosperm inside an ovule receives additional nutrients from the parent tissues and enlarges. Later, as the developing embryo uses the endosperm, that nutritive tissue

may shrink to a thin sheet. As the pear embryo and the rest of the seed develop inside the ovule and ovary, the wall of the ovary and base of the sepals, petals, and stamens enlarge and form a sweet pear fruit (see Figure 30.10), while the ovule's outer covering, or *integument,* gives rise to a hard, brown protective *seed coat.* Surrounded by the seed coat, the embryo produces a tiny embryonic root, or **radicle,** at one end, and a tiny shoot, or **plumule,** at the other. Figure 30.11 shows this in a peanut. These early divisions establish the major axis (root and shoot) that will dominate the pear plant's anatomy for the rest of its life. The tiny radicle will eventually develop into the plant's entire root system, while the plumule will become the shoot with trunk, branches, leaves, blossoms, and fruits.

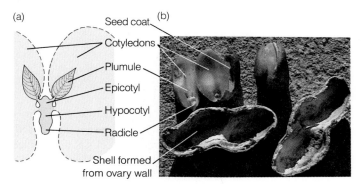

FIGURE 30.11 ■ **Seed of a Peanut Plant.** (a) A peanut embryo's tiny shoot, or plumule, is visible above the primordial root, or radicle. (b) Next time you get some roasted peanuts, examine one before you eat it. Between the two large, tasty cotyledons are the plumule and epicotyl, with the hypocotyl and radicle below.

The embryo's first leaves are thick, oblong structures called **cotyledons,** which grow from the central axis. Many plants, including pears, have embryos with two cotyledons and are therefore called dicotyledons (*dicots*). Others, including grasses, lilies, and palms, have only one cotyledon and are called monocotyledons (*monocots*).

A ballpark peanut is a familiar seed that illustrates typical seed structure more easily than the pear seed (see Figure 30.11). The woody peanut shell is actually the fruit formed from the enlarged ovary wall, while the reddish papery coating around each peanut is the seed coat, the wall of the ovule.

FIGURE 30.12 ■ **Germination: The Seedling Emerges.** In a monocot such as corn (a), the plumule bearing the first true leaf bursts upward and emerges from the soil. In many dicots, such as the bean (b), the whitish hypocotyl arches upward and breaks the soil first, drawing the cotyledons behind it.

The two oval halves of the peanut are the dried, salted remains of the cotyledons. The tiny fleck in the middle is the dried embryo, and close inspection will reveal the rootlike radicle and leafy plumule. Between the nubbin of a radicle and the cotyledons is a short length of stem called the **hypocotyl,** to which the starchy cotyledons are attached. Between the cotyledons and the next leaf is the **epicotyl,** from which the entire shoot will ultimately derive.

It is easy to see how the sweet fragrance and bright color of a fleshy fruit like a pear might attract deer, birds, or other animals and how these hungry animals, in turn, will swallow seeds along with fruits. Later, they inadvertently disperse seeds to new areas when they excrete feces—a rich supply of natural fertilizer. The dispersal of seeds by animals spares immobile plants the energy that they would need for self-propelled motion, and yet their offspring still have a fair chance to become rooted a good distance away from the parent plant and away from direct competition for sunlight, water, and minerals. Other adaptations for seed dispersal include fruits with hooks that attach to passing animals (cockleburs), plumes that lift featherweight seeds on wind currents (dandelions), and hollow shells that roll or float away (coconuts) (review Figure 17.23).

Germination: The New Plant Emerges

To survive in many regions of the world, seeds must withstand the cold temperatures of winter or the lack of water in a dry season. When water is plentiful from rain or snowmelt, however, and sunshine warms the soil, the new embryo begins the process of *germination:* It uses nutrients at a higher rate, resumes growth, and soon cracks out of the seed coat. The tiny radicle of the new seedling can then slip out, and the young root pushes downward into the soil. Soon the shoot pushes upward through the ground and in many plants lifts the cotyledons toward the sun (Figure 30.12).

The plantlet grows as cells divide in the apical meristems of the shoot and root tips, and eventually the cotyledons disappear as their stored starch fuels early growth. As the first normal leaves develop and begin to photosynthesize, the young plant becomes truly independent.

Development from Seedling to Pear Tree

As the spring days lengthen, the shoot of the pear seedling advances skyward and the root burrows deeper into the ground. Growth in both directions is driven by the elongation of cells produced in the rapidly dividing apical meristems. Tiny swellings on the pear shoot called **primordia** flank the

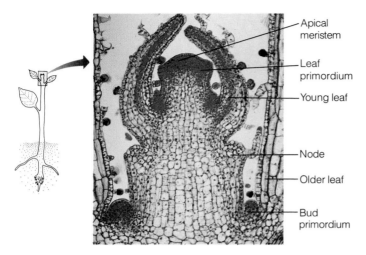

FIGURE 30.13 ■ **Primordia: New Organs Unfold.** The longitudinal section of the growing tip of a pear or other flowering plant (here, a *Coleus*) reveals the primordia that will give rise to new leaves, flowers, and branches.

apical meristem (Figure 30.13). These primordia contain meristematic tissue that will give rise to leaves, new shoots (branches), or flowers. The very first primordia to form, however, are always leaf primordia; this facilitates the seedling's quick development and the rapid unfurling of energy-collecting surfaces.

Tiny bumps called **nodes** form at the level of each leaf primordium. Once a particular leaf has begun to develop, a *bud primordium* forms in the upper angle between the young leaf and the stem and develops a little further into a **bud.** Buds can mature into flowers or new shoots, or they can remain buds. In a young pear tree, buds usually become branches bearing new leaves. Each new branch soon puts out more buds and more side branches. All the while, the root system continues to enlarge from its tips.

About the fourth or fifth year in a pear tree, floral buds appear in many places where buds for new branches would have formed. The floral buds develop into flowers, the blossoms become pollinated, the eggs are fertilized, embryos develop, and a new generation appears.

The rest of the chapter looks at the structure of roots, shoots, and leaves, and Box 30.2 on pages 620 and 621 describes how people use these and other parts from certain exotic plants.

STRUCTURE AND DEVELOPMENT OF ROOTS

Roots are familiar plant organs with important functions: In vascular land plants, the anatomy of roots allows them to anchor the organism firmly to one spot, penetrate the soil and absorb water and minerals, and often store starch. This section tours root anatomy, beginning deep in the soil at the very tip of the root.

The Root: From Tip to Base

■ **Root Cap**   At the lowest tip of a typical root is a dome-shaped **root cap** (Figure 30.14). As the root grows downward, rough soil particles damage and scrape off cells at the surface of the root cap. These damaged cells slough off and cover the rest of the root with a slime that eases penetration through the soil.

■ **Apical Meristem**   Damaged root cap cells are replaced by the apical meristem, a little disc just above the root cap that produces new cells from both its surfaces. The cells from the lower surface of the disc become root cap cells, while those derived from the upper surface elongate, pushing the root tip strongly into the soil. Moving up a root, one can see three general regions (see Figure 30.14): a *zone of cell division,* which includes the apical meristem protected by the root cap cells; a short *zone of elongation,* where individual cells lengthen and force the tip to move through the soil; and a *zone of maturation,* where cells develop their specialized roles as members of the dermal, ground, or vascular tissue systems.

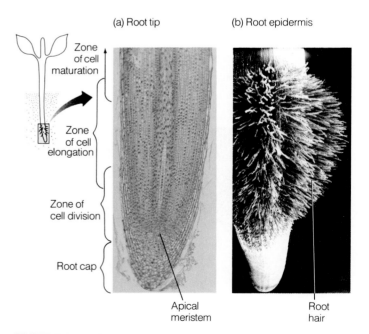

FIGURE 30.14 ■ **The Growing Root Tip.** (a) A longitudinal section of an onion root reveals, from the tip upward, the protective root cap just below the apical meristem; the small new cells of the zone of cell division; the lengthening cells of the zone of cell elongation; and (b) the epidermal cells and root hairs in the zone of maturation.

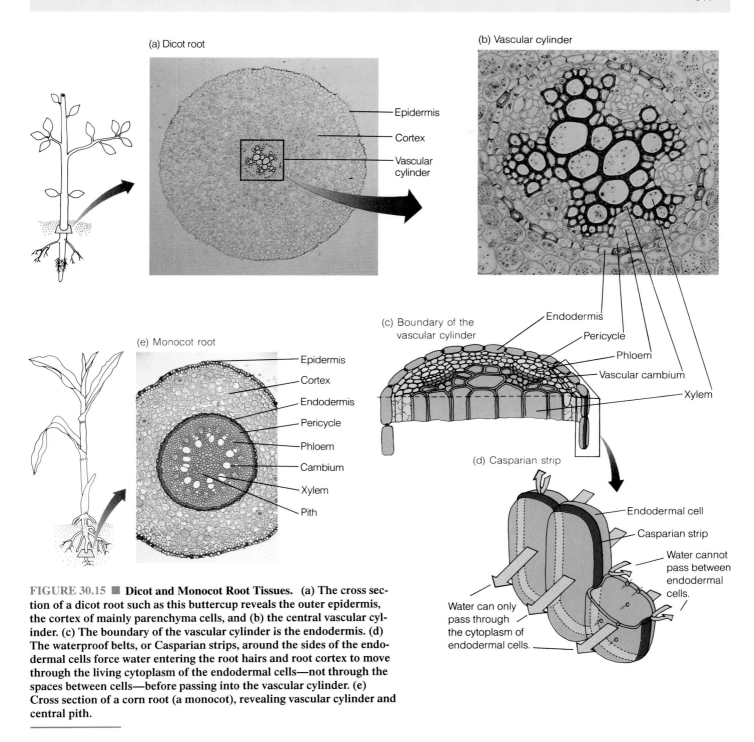

FIGURE 30.15 ■ **Dicot and Monocot Root Tissues.** **(a)** The cross section of a dicot root such as this buttercup reveals the outer epidermis, the cortex of mainly parenchyma cells, and **(b)** the central vascular cylinder. **(c)** The boundary of the vascular cylinder is the endodermis. **(d)** The waterproof belts, or Casparian strips, around the sides of the endodermal cells force water entering the root hairs and root cortex to move through the living cytoplasm of the endodermal cells—not through the spaces between cells—before passing into the vascular cylinder. **(e)** Cross section of a corn root (a monocot), revealing vascular cylinder and central pith.

The Root: From Outside to Inside

■ **The Dermal System: Epidermis and Root Hairs** Looking at a cross section of a mature root cut just above the zone of maturation, one can see an outer protective layer just one cell thick—the epidermis, derived from the outer cells in the zone of maturation (Figure 30.15a). The root epidermis absorbs water and minerals from the soil, and just as microvilli enlarge the absorptive surface area of an animal's intestinal cells (see Figure 24.15), tiny extensions of the root epidermis

BOX 30.2   PERSONAL IMPACT

EXOTIC PLANTS FOR A HUNGRY PLANET

Agriculture is a multibillion-dollar industry, and our lives depend, quite literally, on the products of that ancient enterprise. Despite this, however, our dependence rests on a fairly limited range of plant species. Of the 450,000 plants currently recognized, people have learned to use only about 4000 for food (about a tenth of 1 percent), and of those, we rely heavily on just 30 or so grains, fruits, and vegetables.

Every few decades, American farmers become enthusiastic about a "new" crop. In the 1930s, for example, soybeans were practically unknown outside of Asia, but today, the soybean is the third largest U.S. crop and our major source of vegetable oil. Likewise, in the past 20 years, kiwifruit has risen from obscurity to prominence in American fruit markets.

Many plant researchers are investigating little-known and "forgotten" plant species to recommend new resources for feeding, clothing, and housing the world's burgeoning human population. The suggestions of people like Noel Vietmeyer of the National Academy of Sciences encompass each of the major plant organs discussed in this chapter, and the list that follows, organized by plant part, is just a small sample. Many of these will sound exotic or even bizarre today. But who knows? Some of them may find their way to your table or your garden in the not-so-distant future.

Fruits
- The sweetsop, or sugar apple (*Annona squamosa*), of South America and the carambola (*Averrhoa carambola*) of Malaysia have luscious flavors and highly unusual shapes that many customers would find novel and interesting.
- The pummelo (*Citrus grandis*) is the ancestor of the grapefruit but is hardier and lacks any trace of bitterness.

Seeds
- The seeds of jojoba (*Simmondsia chinensis*), a desert shrub native to Mexico and the southwestern United States, contain a high-quality wax that is already being used in place of sperm whale oil in lotions, shampoos, and other cosmetics, as well as in machine oils and transmission fluids.
- Grain amaranth, or quinoa (species of *Amaranthus;* Figure 1), once a staple for the Aztecs and Incas, produces tiny seeds that, when heated, burst and taste like popcorn. The seeds contain high levels of the essential amino acid lysine as well as high levels of total protein—nutrients often lacking or less abundant in the common grain crops.

Roots and Tubers
- Buffalo gourd (*Cucurbita foetidissima*) produces heavy, starchy roots (up to 40 kg, or 88 lb) in just a few growing seasons, with a flavor and food value much like cassava.
- The groundnut (*Apios americana*) is a common native of eastern North American forests, and the tasty small, round tubers were long a high-protein staple of American Indians.

Leaves
- An Australian tree called leucaena (*Leucaena leucocephala*) is a legume a bit like the American black locust. Its roots associate with another organism that can take nitrogen from air and convert it to amino acids. It grows very rapidly, and the protein-rich leaves make good animal fodder.
- Dry-adapted shrubs of Central America and Mexico, chayas (*Cnidoscolus* species) also grow quickly and serve as an attractive hedge, and people can eat the leaves as a tasty vegetable.

Stems
- An amazingly fast-growing annual from East Africa, kenaf (*Hibiscus cannabinus*) grows 6 m (nearly 20 ft) in a year, and pulp from its thick stem can be used to make good-quality paper and cardboard.

---

called **root hairs** extend the root's absorptive capacity (see Figure 30.14b). Each root hair is an extension of a single epidermal cell, but their numbers can be very great, and their collective surface area can be amazingly large: A single ryegrass plant has about *14 billion root hairs,* with a combined surface area the size of a tennis court!

Root hairs are delicate and short-lived, breaking off as the root tip pushes deeper into the soil. New root hairs continually arise, however, in the region of maturation, and the plant maintains a large absorptive surface area. Since most water absorption occurs in the root hairs near root tips, gardeners must be careful not to tear off young root tips when transplanting flowers or shrubs, and they often get the best results by fertilizing fruit trees several feet out from the trunk, nearest the root tips.

■ **The Root's Ground System: Cortex, Endodermis, and Casparian Strip** A cross section of a root reveals a thick layer of cells called the **cortex** lying just inside the epidermis and surrounding the central core of the vascular system (see Figure 30.15a). You can see this in the kitchen if you cut a slice from a bright orange carrot root. The cortex makes up most of the root's bulk and often stores excess starch.

The innermost layer of the cortex is the **endodermis,** a cylinder of tightly packed cells just one cell thick (Figure 30.15b). Although water moving inward from the soil can pass freely between most cortex cells without entering their cytoplasm, when it reaches the endodermis, the water must pass through the living cytoplasm of endodermal cells before it can reach the vascular bundles and hence the rest of the plant. Each cell in the endodermis is a bit like a brick in a circular brick wall (Figure 30.15b and c). The outside surface of the "bricks" faces the rest of the cortex, and the inside surface faces the vascular tissue in the root's center. The side walls of these cells are impregnated with a waxy water-resistant substance (*suberin*). Called the **Casparian strip,** this

- The sap from the stems of the guayule plant (*Parthenium argenteum*) can be used to make rubber. Guayule rubber is being seriously considered as a substitute for products of the rubber tree and petroleum-based synthetic rubber.

Wood

- A hardwood tree from Australia's tropical rain forest, mangium (*Acacia mangium*) fixes nitrogen, grows as fast as pine, and has the wood quality of walnut.
- Bracatinga (*Mimosa scabrella*) is also a nitrogen-fixing tree, but is native to Brazil, grows up to 15 m tall (nearly 50 ft) in just three years, and is being used to reforest logged areas of Costa Rica.

Multiple Plant Parts

- The winged bean (*Psophocarpus tetragonolobus*), an annual plant from the Philippines, has edible flowers, seeds, seed pods, leaves, tendrils, and tubers. The seeds and tubers are rich in protein, and the ruffled pods can be steamed, stir-fried, or eaten raw. Over the next few years, watch for the winged bean in your grocer's produce section . . . just to the left of the kiwifruit.

Exotic plants such as these relate to one of the biggest issues in modern plant science: preserving plant *germ plasm,* the wealth of genetic information contained in seeds. It is commonplace now for countries to have public and private banks that stock and maintain the seeds of little-studied native plants with the potential genetic resources for new food crops, medicines, fibers, and wood. Controversies have arisen, however, over who should have access to the seeds and at what price. Is plant germ plasm a "common heritage" and "public good" or a "national property" that can be sold at high cost or even withheld from development if a country's price is not met? These questions

FIGURE 1 ■ **Amaranth, or Quinoa, Produces a Tasty Grain.**

remain to be answered. But no one disagrees over the tremendous value of plant germ plasm and the need to preserve genetic resources for future generations. No one knows, after all, where our future food plants will come from next.

narrow water-resistant belt encircles each cell and lies between it and the next endodermal cell (Figure 30.15d). Because of the Casparian strips, water cannot diffuse from one endodermal cell to the next, but can pass only through the cytoplasm of the cells from the cortex to the vascular tissue, and this has an important consequence for the plant: Water and minerals seeping from the soil through the epidermis and into the cortex can reach the central vascular tissue (and from there move throughout the rest of the plant) only by first passing through the living cytoplasm of the endodermal cells (see Chapter 32 for more details).

■ The Vascular System: Xylem, Phloem, and Other Tissues   The

endodermis in a root such as a carrot surrounds yet another cylinder—a central zone of vascular tissue responsible for carrying water and minerals up to the stem and leaves and transporting sugars and amino acids between the leaves,

stems, and roots. Moving inward through the vascular cylinder, one encounters the pericycle, phloem, xylem, and in some plants, pith (see Figure 30.15b).

The **pericycle** lies just inside the endodermis and consists of one or more layers of parenchyma cells. The pericycle gives rise to the lateral roots, which grow out horizontally through the cortex parallel to the surface of the ground.

Encircled by the pericycle is the water- and mineral-conducting xylem and the phloem, which carries the dissolved products of photosynthesis wherever they are needed. The xylem and phloem tubes in plant roots are arrayed in various ways. In the roots of most dicots, the central vascular cylinder contains a star-shaped core of xylem vessels with phloem nestled in the angles between the points of the star (see Figure 30.15a and b). In contrast, most monocots have a ring of phloem inside the pericycle that surrounds a ring of xylem, which in turn encloses a central core of large, thin-walled parenchyma storage cells called **pith** (Figure 30.15e).

■ **Secondary Growth in the Roots**　In perennial plants, such as shrubs and trees, roots experience both primary growth from the apical meristems and secondary growth from lateral meristems. The root tips continually probe downward or outward while the older parts of the root system become woody and thus stronger and more capable of anchoring a large plant such as a pear tree.

As the root begins secondary growth, the pericycle gives rise to the lateral meristems, the *vascular cambium* and *cork cambium* (Figure 30.16). The undifferentiated parenchyma cells of the vascular cambium produce new layers of phloem (to the outside) and new layers of xylem (to the inside). The xylem vessels, being thick-walled and (at maturity) nonliving, create the tissue we call wood. Eventually, the increasing circumference of the root causes the epidermis and cortex layers to rupture (see Figure 30.16b). The meristematic tissue in the cork cambium, however, generates a new waterproof layer, the **cork.** The cork plus the cork cambium form the *periderm,* which protects the root (see Figure 30.16c). The two thin cylinders of vascular cambium and cork cambium in the root continue unbroken into the stem and also generate the wood, bark, and other secondary growth tissues in the shoot.

The root's simple anatomy reflects its relatively simple tasks of anchoring, starch storage, and absorption of water and minerals (described further in Chapter 32). The shoot, however, with its stems, leaves, and flowers, has a more complex structure.

STRUCTURE AND DEVELOPMENT OF THE SHOOT

The tallest California redwood trees have trunks that may reach 100 m (328 ft) into the sky, while the delicate stems of some wildflowers on the ground below may be mere fractions of a centimeter tall. Regardless of size, stems (which are usually a plant's aboveground portions) have two fundamental tasks: supporting the plant's leaves and, if present, flowers or fruit, and acting as a central corridor for the transport of water, minerals, sugars, and other substances. Modified stems can also store starch and allow plants to adhere to vertical surfaces.

(a) Tissue resulting from primary growth

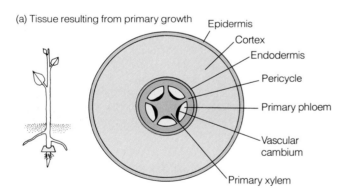

- Epidermis
- Cortex
- Endodermis
- Pericycle
- Primary phloem
- Vascular cambium
- Primary xylem

(b) Early secondary growth

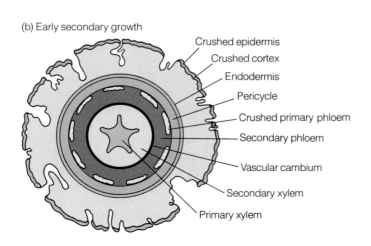

- Crushed epidermis
- Crushed cortex
- Endodermis
- Pericycle
- Crushed primary phloem
- Secondary phloem
- Vascular cambium
- Secondary xylem
- Primary xylem

(c) Secondary growth continues

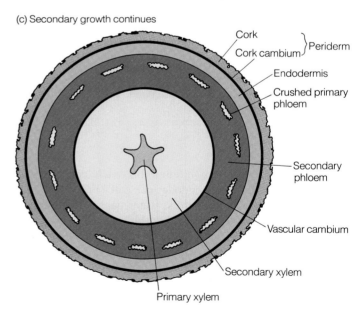

- Cork
- Cork cambium } Periderm
- Endodermis
- Crushed primary phloem
- Secondary phloem
- Vascular cambium
- Secondary xylem
- Primary xylem

FIGURE 30.16 ■ **Secondary Growth in a Dicot Root.** **(a)** At the end of primary growth, the cortex surrounds the vascular cylinder. **(b)** As a plant grows and its root matures, the secondary xylem and phloem form, and the root expands its circumference and crushes the cortex and epidermis. **(c)** As secondary growth continues, the secondary xylem expands and the cork replaces the epidermis.

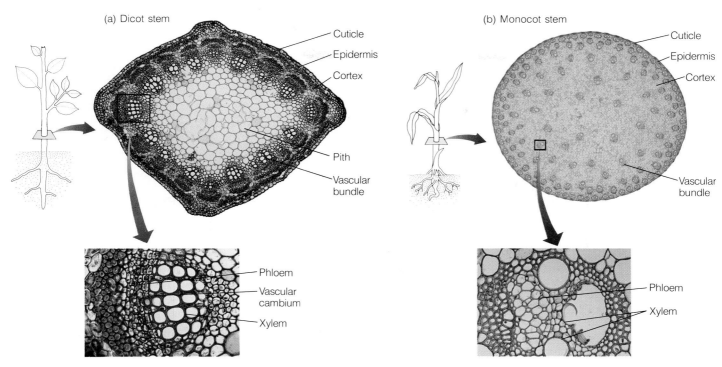

FIGURE 30.17 ■ Primary Growth in Stems. (a) The development of stem tissue in annual dicots such as alfalfa is characterized by a circle of vascular bundles. In perennial dicots, however, a cambium layer within each bundle gives rise to xylem toward the stem's interior and phloem toward its exterior. (b) Annual monocots, such as corn, have scattered vascular bundles.

Stem Structures and Primary Growth

■ **Dermal System: Epidermis**   As in the root, the dermal system of the shoot consists of a layer of epidermal tissue one cell thick (Figure 30.17). While root epidermis absorbs water, shoot epidermis resists water loss. Just as a plastic bag keeps a sandwich from drying out, stem epidermal cells have a water-resistant waxy coating, the *cuticle,* that keeps the plant from desiccating.

■ **Ground System: Cortex and Pith**   As in the root, the ground system of the stem is primarily cortex made up of parenchyma cells (see Figure 30.17). Stems, however, need more structural reinforcement than roots, and many stems also have strands of collenchyma (see Table 30.2 on page 611) around the outer edge of the cortex, just inside the epidermis. Also, a stem's ground system generally has a central core of pith, while a root's ground system may lack it.

■ **Vascular System of Stems During Primary Growth**   In the stem of a young perennial such as a pear seedling or in a herbaceous annual such as a zinnia, xylem and phloem are organized in groupings called *vascular bundles.* Within each bun-

dle, a sheath of parenchyma or sclerenchyma cells surrounds strands of xylem and phloem, with xylem lying inside the phloem. In the stems of most dicots, vascular bundles form a ring around the stem's central core of pith, and a layer of cambium lies between the xylem and phloem (see Figure 30.17a). Cambium from each vascular bundle extends to the adjacent bundles, forming a complete ring of cambium around the pith. In conifers and woody dicots, the arrangement of cambium sandwiched between xylem and phloem is profoundly important to future development, because it can lead to the secondary growth of wood and bark and then to the plant's considerable enlargement. In monocots and a few dicots, by contrast, the vascular bundles are scattered throughout the cortex, there is no vascular cambium, and the stems are incapable of true secondary growth (see Figure 30.17b).

Stem Structures and Secondary Growth

Late in the first summer of a pear seedling's life, lateral meristem cells in the vascular cambium begin dividing and produce additional xylem and phloem cells (Figure 30.18). The new conducting vessels made by the vascular cambium are

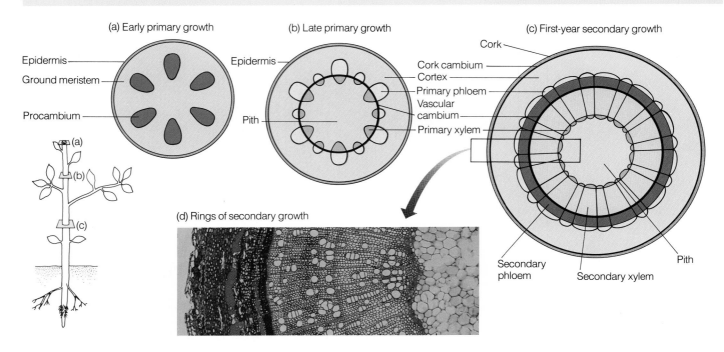

FIGURE 30.18 ■ Secondary Growth in Stems. (a) During early primary growth, the apical meristem lays down the ground meristem, which makes cortex and pith, and the procambium, which makes the primary xylem and phloem. (b) As a plant matures, the primary xylem and primary phloem differentiate from the vascular cambium between them. (c) During the first year of secondary growth, the vascular cambium adds secondary xylem and secondary phloem, the trunk expands, and the cork cambium adds the cork. (d) A cross section of a maple stem reveals several years of secondary growth.

called *secondary xylem* and *secondary phloem.* They transport water and other materials like their primary counterparts. Because the cambium makes far more secondary xylem than secondary phloem, most of the stem tissue in a sapling, a bush, or an older tree is made up of dead xylem cells (or **wood**) in a thick interior rod.

■ **Wood**   A tree cut down in a temperate region—say, in a North American hardwood forest—will have concentric rings of lighter and darker wood (see Figures 30.18d and 30.19). These *growth rings* occur because the cambium produces larger cells in the spring and summer (lighter areas), when water and sunlight are plentiful, and narrower ones in the fall and winter, when water tends to be more scarce. In moist, tropical regions, on the other hand, where a tree can grow year round, the wood may have no obvious growth rings. Scientists use the width of growth rings in ancient trees, such as the 5000-year-old bristlecone pines of eastern California, to accurately reconstruct historic weather patterns. Scientists have dated a massive volcanic eruption on the Greek island of Thera, which must have darkened skies around the world with

ash, to approximately 1625 B.C. They did this by identifying very narrow growth rings in North American bristlecone pines growing at that time.

As a tree grows older, the xylem at the center gradually ceases to conduct water and minerals, and newer xylem nearer the periphery of the trunk takes over. The nonconducting central wood, or *heartwood,* becomes infiltrated with oils, gums, resins, and tannins, all of which make it dark, aromatic, and resistant to rot. The *sapwood* around the outside continues to transport water in the plant.

■ **The Bark**   When the stem expands during secondary growth, it ruptures the seedling's original cortex and epidermal layers. A pear or other sapling still needs a protective waterproof covering, however, and the periderm, with its cork cambium and layer of cork cells, develops here, as it does in the roots. Together, the periderm, the underlying phloem, and many layers of cork cells that die but remain in place make up the outer protective **bark.** Sheets and plugs of cork for flooring, bulletin boards, and bottle stoppers come from the periderm of cork oak trees native to Spain, Portugal, and Algeria.

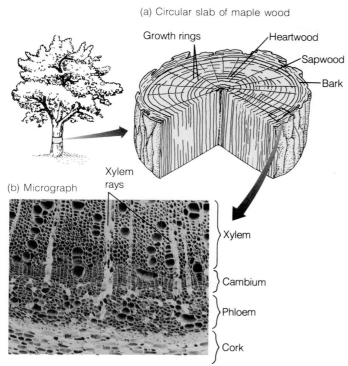

(a) Circular slab of maple wood

Growth rings   Heartwood

Sapwood

Bark

Xylem
rays

(b) Micrograph

Xylem

Cambium

Phloem

Cork

FIGURE 30.19 ■ A Close-Up View of Wood. (a) A circular slab from the trunk of a sugar maple reveals heartwood and sapwood, growth rings, and bark. (b) A small chunk of wood viewed with a scanning electron microscope has easily visible xylem tracheids, phloem sieve tube members, and xylem rays, which allow lateral movement of materials.

Cork cutters are careful to leave the sugar-conducting phloem intact. Without a functioning phloem all the way around the trunk, sugars cannot move downward from the leaves to the roots, and the tree will eventually die. Figure 30.20 shows a simple experiment that proves this point.

Bark does more than keep water in; it helps keep plant-eating insects, fungi, viruses, and other parasites out. Healthy trees also secrete resins that ooze out of holes bored by insects, often forcing the intruders back out. Frankincense and myrrh are fragrant tree resins, and the treasured jewel amber is nothing but fossilized resin.

STRUCTURE AND DEVELOPMENT OF LEAVES

The leaves that grow on plant stems can be smaller than a fingernail, larger than a table top, or shaped like a needle, a knife, a plate, a fan, or a hand. Regardless of form, however, nearly all leaves share the same basic functions: exposing a photosynthetic surface area, usually one that is large, flat, and oriented to catch the maximum amount of sunlight; obtaining carbon dioxide from the atmosphere as a carbon source for photosynthesis, yet retaining as much water vapor as possible; and helping to draw water and nutrients up through the vascular system.

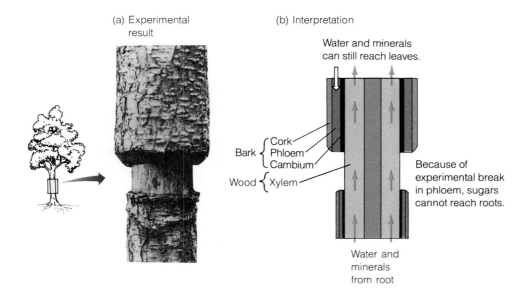

(a) Experimental
result

(b) Interpretation

Water and minerals
can still reach leaves.

Cork
Bark   Phloem
Cambium

Wood   Xylem

Because of
experimental break
in phloem, sugars
cannot reach roots.

Water and
minerals
from root

FIGURE 30.20 ■ What Does Phloem Do? (a) In a simple experiment, researchers stripped the phloem from a black cherry tree (they *girdled* it), but left the xylem intact. After a number of months, the lower portion of the trunk was much smaller than the upper. (b) This diagram explains why. Water and minerals absorbed by the roots could still travel upward to the leaves, but sugars and amino acids could no longer travel from leaves downward through the phloem past the girdled area to lower trunk and roots. Clearly, the phloem is involved in downward transport of organic nutrients. A girdled tree will usually die from the injury.

(a)   (b)

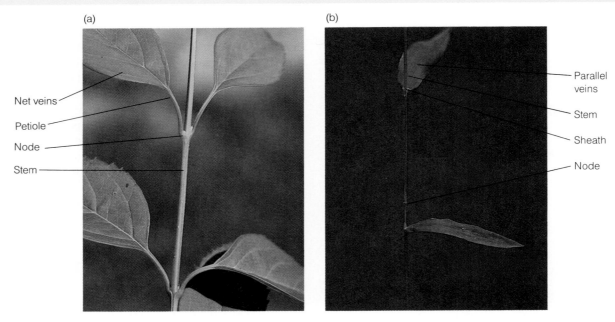

Net veins

Petiole

Node

Stem

Parallel
veins

Stem

Sheath

Node

FIGURE 30.21 ■ **The Architecture of Leaves: Dicot Versus Monocot.** (a) Dicot leaves are often broad and net-veined and have a stalk, or petiole, and a blade. (b) Monocot leaves usually have parallel veins, and often the slender blade is attached to the stem by a sheath rather than by a petiole.

(a) Spiny leaves protect.

FIGURE 30.22 ■ **Specialized Leaves Can Protect, Store, or Attract.** (a) The leaves of this prickly pear cactus are modified into barbed spines that protect the succulent stem. (b) Leaves of a bromeliad, such as this pink-tipped individual, can grow on the branches of tropical trees, and these act as small basins that store water. (c) Brilliantly colored leaves like those of the coleus can help attract insect pollinators to the plant.

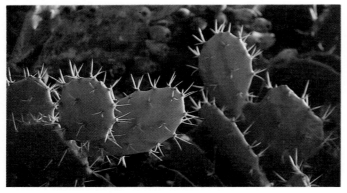

(b) Basin-shaped leaves store water.   (c) Colorful leaves attract.

Leaf Blade and Petiole

Most leaves are composed of a broad, flat portion, called the **blade,** and a **petiole** or *sheath* connecting blade and plant stem (Figure 30.21). Leaf blades are often flat and thin, which maximizes surface area available for absorbing light and carbon dioxide. Structural modifications, however, often help balance the conflicting needs of obtaining carbon dioxide but avoiding water loss. Desert plants such as jojoba and creosote have small leaves covered with a thick, waterproof cuticle and hairy surfaces that reflect light and thus help retard water loss. Cactus leaves are modified into small, dry, needlelike spines that defend rather than photosynthesize (Figure 30.22). Other plants have leaves modified for clinging and climbing, for capturing little pools of water, and for attracting animals toward inconspicuous blossoms. In a few species, including pitcher plants and sundews, the leaves assist predation, trapping a visiting insect, then enzymatically digesting it for its proteins.

Leaves may be *simple,* like those of the pear, and consist of a single blade, or they may be *compound,* like those of the ash tree or pea, and consist of many small leaflets.

Anatomy of a Leaf

Like its external form, a leaf's internal anatomy is intimately associated with its many functions. Most leaves have an outer layer of epidermis protecting internal ground cells that photosynthesize and a vascular system that brings in water and minerals and carries away the products of photosynthesis.

■ **Dermal Tissue**   A leaf's epidermis has a waxy coating, the cuticle. In fact, the leaf is so well sealed that it needs tiny openings called **stomata** (singular, stoma), surrounded by guard cells, to admit carbon dioxide for photosynthesis and to release water vapor and oxygen (Figure 30.23). As Chapter 32 explains, stomata are most numerous on the undersides of leaves, and their opening and closing are based on the plant's conflicting needs to conserve water and take in carbon dioxide.

■ **Ground Tissue**   Sandwiched between the upper and lower epidermal layers is the leaf's **mesophyll** layer, made up of two types of parenchyma. *Palisade parenchyma,* the main photosynthetic tissue, lies just beneath the upper epidermis. It usually consists of one or more rows of vertically oriented, column-shaped cells, each enclosing dozens of chloroplasts. A layer of rounder cells, the *spongy parenchyma,* lies between the palisade parenchyma and the lower epidermis. The loosely packed spongy parenchyma cells provide a huge surface area (analogous to an animal's lung) that absorbs carbon dioxide from the air that enters the stomata. Carbon dioxide entering the stomata can move rapidly through the spongy parenchyma to the palisade parenchyma above, where most of the photosynthesis takes place.

■ **Vascular Tissue**   Leaves need an elaborate ''plumbing'' system to distribute the products of photosynthesis and bring in water and minerals, and a plant's vascular system is continuous between leaves, stem, and roots. Monocot leaves have bundles of xylem and phloem called **veins** running parallel to

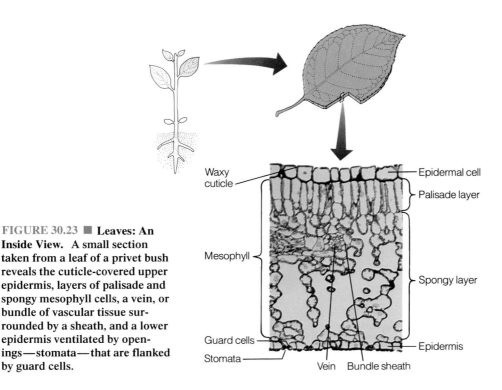

FIGURE 30.23 ■ Leaves: An Inside View. A small section taken from a leaf of a privet bush reveals the cuticle-covered upper epidermis, layers of palisade and spongy mesophyll cells, a vein, or bundle of vascular tissue surrounded by a sheath, and a lower epidermis ventilated by openings—stomata—that are flanked by guard cells.

Waxy cuticle

Mesophyll

Guard cells

Stomata

Epidermal cell

Palisade layer

Spongy layer

Epidermis

Vein   Bundle sheath

each other along the long axis of the leaf (so-called *parallel venation;* see Figure 30.21). Dicot leaves have *netted venation;* that is, the veins form a branching pattern. Surrounding individual veins are *bundle-sheath cells,* which protect the xylem and phloem from direct exposure to air. Collenchyma or sclerenchyma fibers may also stiffen the veins and leaf margins and help each leaf remain flat, which improves solar collection. All summer long, mesophyll cells in the leaf store the sun's energy in sweet sugar molecules, and the phloem transports sugars from the leaves to the developing fruits. The end result is the fruits and the seeds they contain—the plant's evolutionary solution to dispersing the species.

CONNECTIONS

Plant structures—including roots, shoots, parenchyma cells, and root hairs—help reveal the two major distinctions between plants and animals: the source of supplies and movements by and within the organisms. In general, energy and supplies are available to a plant from the sun, the air, and the soil. The rigid walls surrounding each cell collectively provide the strength that supports the leaves and forces roots down into the soil. These rigid cell walls, however, preclude the movement of plant cells and of entire plants. While a rabbit's white blood cell can move to the site of an infection, a plant's cell cannot migrate to the site of a torn-off branch. The rabbit can flee a hungry grizzly bear, but a huckleberry vine cannot.

The realities of limited movement but often-abundant supplies dictate a plant's growth patterns. New plant cells arise only in localized areas as meristems lay down new cells that will become dermal, ground, or vascular tissues. The consequence of this growth from meristems is the continued addition of new organs—new leaves, new branches, and new roots—again in strong contrast with patterns of animal growth, wherein the number of organs is usually fixed early in development and the preestablished organs simply enlarge. Clearly, growth is a central feature of plant survival, and the next chapter explores how environmental factors can trigger the release of chemical regulators inside plants and how these, in turn, can control plant growth.

Highlights in Review

1. Vascular plants have two major parts, the root and the shoot. Roots anchor the plant, store starch, absorb water and minerals, and transport them to the shoot. The shoot typically consists of a branching system of stems that provide support, leaves that capture energy, and flowers that function in reproduction.

2. A vascular plant has a dermal tissue system, the protective outer covering; a ground tissue system, which provides support and storage; and a vascular tissue system, the internal transport tissues.

3. The dermal tissue system contains mostly epidermis during primary growth and periderm during secondary growth. The ground tissue system is mostly squarish parenchyma cells, but can contain hard sclerenchyma cells and thick-cornered collenchyma cells. The vascular tissue system consists of xylem tissue, which conducts water and minerals, and phloem tissue, which conducts photosynthetic products.

4. Plants have open growth based on perpetual growth centers, or meristems. Apical meristems at the tips of roots and stems allow primary growth: extension upward into the air and downward into the soil. Secondary growth depends on lateral meristems (or cambiums); these cylinders of tissue allow root and shoot diameters to thicken.

5. Perfect flowers have rings of sepals, petals, and stamens (male flower parts), as well as carpels (female flower parts). Pollen, the male haploid plant (gametophyte), forms in the stamen, is carried to or lands on the stigma, and grows down through the style and ovary wall to the ovule, which houses the embryo sac, the female haploid gametophyte.

6. The embryo of the diploid sporophyte plant develops within an ovule. An ovule matures into a seed containing embryo and endosperm surrounded by a seed coat. A ripened ovary forms a fruit.

7. At germination, the embryo bursts from the seed. After the first normal leaves appear, growth occurs only at the apical meristem, vascular cambium, and cork cambium. Buds for flowers or branches develop in the notch between a leaf and the stem.

8. The primary root has a protective epidermis, water-absorbing root hairs, and a starch-storing cortex that includes the endodermis. Because of the Casparian strip, water must pass through the cytoplasm of endodermal cells to enter the vascular cylinder.

9. The primary shoot has an epidermis, often tiny surface hairs, and a waxy cuticle that helps prevent water loss.

10. Stem tissues can include cortex and pith inside the vascular cylinder, with its bundles of xylem and phloem.

11. Secondary growth forms the woody parts of shrubs and trees. A ring of meristematic tissue, the vascular cambium lying between the xylem and phloem, gives rise to secondary xylem (the major component of wood) and secondary phloem.

12. As the shoot and root enlarge in circumference, the cortex and epidermis rupture. Tree bark is formed by layers of dead cells from the periderm along with secondary phloem.

13. Leaves have highly photosynthetic palisade parenchyma cells beneath the upper epidermis, as well as spongy parenchyma cells above the lower epidermis. Tiny openings called stomata control the passage of gases into and out of a leaf.

Key Terms

Study Questions

Review What You Have Learned

1. List the types of roots, and give an example of each type.
2. Create a table that lists the types of ground tissue in plants. For each type, give a short description and an example.
3. Describe the structure of xylem and phloem.
4. State three ways in which phloem cells differ from xylem cells.
5. Why is a plant's meristematic tissue so important?
6. What are the parts of a stamen? A carpel?
7. Name the three general regions of a growing root, and briefly describe each region.
8. Where is a plant's pericycle? Where is the endodermis? What is the function of each?
9. Describe the functions of the vascular cambium and the cork cambium.
10. Does most of a leaf's photosynthesis take place in the spongy parenchyma? Explain.

Apply What You Have Learned

1. Why must a gardener transplanting a shrub exercise care in digging up the roots?
2. Botanists were able to correlate unusually narrow growth rings in a bristlecone pine in California with the eruption of a Greek volcano in 1625 B.C. What is the connection between the two?
3. Counselors at a summer camp warn playful campers not to peel bark off the camp's birch trees. What might be the result of this vandalism?
4. How might natural selection bring about the evolution of trees with tasty, fleshy fruits from ancestors bearing hard, unappetizing fruits?

For Further Reading

DARLEY, W. M. "The Essence of 'Plantness.'" *The American Biology Teacher* 52 (1990): 354–357.

ESAU, K. *Plant Anatomy.* 2d ed. New York: Wiley, 1965.

JENSEN, W. A., and F. B. SALISBURY. *Botany.* 2d ed. Belmont, CA: Wadsworth, 1984.

NIKLAS, K. J. "Aerodynamics of Wind Pollination." *Scientific American,* July 1987, pp. 90–95.

RAVEN, P. H., R. F. EVERT, and H. CURTIS. *Biology of Plants.* 4th ed. New York: Worth, 1986.

Regulators of Plant Growth and Development

FIGURE 31.1 ▪ **Cultivating Rice: An Ancient Practice.** People have been cultivating rice (*Oryza sativa*) for about 6000 years, using the traditional methods depicted in this painting by Chiao Ping-chên.

Foolish Seedling Disease

For two out of every five people on this planet, today's meals will be about half rice. This means that 20 percent of all human activity is rice-powered! People began cultivating rice in Thailand nearly 6000 years ago, and the practice spread west to India and Persia and east to China and Japan within a few centuries (Figure 31.1). While farmers in many countries developed traditional methods of growing rice and adhered to them closely, Japanese farmers began, very early on, to breed improved varieties of *Oryza sativa,* the rice plant, and to record and study its diseases.

In 1809, for example, a scholar recorded a particular disease that had plagued rice growers for hundreds of years: Afflicted plants would grow far taller and faster than normal after germination and in the process would become spindly and weak and blow over in the wind and die. The name for this disease was fitting: foolish seedling disease (Figure 31.2).

Japanese scientists continued to study this phenomenon, and in 1926, plant physiologist E. Kurosawa showed that the symptoms of foolish seedling disease were caused by chemicals that the fungus *Gibberella fujikoroi* releases when it infects rice plants. Within a few years, other Japanese researchers had isolated two disease-causing chemicals from that fungus and named them **gibberellins.**

For a few decades, gibberellins were thought to be fungal toxins capable of accelerating plant growth. But in 1956, Western researchers isolated gibberellins from a healthy bean plant. Soon it became clear that the chemicals were molecular messengers, or hormones, that occur naturally in vascular plants. Scientists isolated over 70 kinds of gibberellins and found them to be regulators of plant growth. The *G. fujikoroi* fungus had simply evolved a way to make gibberellins in quantities that can cause the rice plants to literally grow themselves to death.

Gibberellins bring us to a new topic: How do plants regulate their growth and development? In Chapter 30, we saw how a plant grows as the life cycle progresses. It is clear, however, that plants regulate the timing of germination, flowering, and growth and that a plant's daily survival depends on appropriate responses to environmental change. These activities are governed by *plant growth regulators,* also called plant hormones.

A rice plant's intake of water, for example, reflects its rate of photosynthesis and water evaporation, while its nutrient uptake corresponds to metabolic needs. Its roots grow downward and its shoot upward, and not the reverse. The plant does not normally waste energy growing overly large organs (such as the stems of the foolish seedlings). Its leaves tend to grow into areas where they have maximum sunlight. Buds open, flowers bloom, fruits ripen, new leaves sprout, the plant defends itself against microorganisms, insects, and wounds—and all these events take place at appropriate times.

In this chapter, we will see how plant growth, development, and survival depend on *internal regulators*—primarily plant hormones such as gibberellins—as well as on *external regulators,* including temperature, light, gravity, and the number of daylight hours. Molecular messengers in a plant often act like animal hormones (see Chapter 26) in that they are chemicals produced in one tissue that affect growth, development, or metabolism in other tissues, even at low concentrations.

Despite the similarities, however, there are significant differences between plant and animal regulation. Most animals have both nervous and hormonal systems, and the two work together in coordinating activity. Plants, however, lack a nervous system; instead, their hormones trigger responses to environmental cues. Moreover, plant hormones are chemically quite different from animal hormones, and plants have fewer classes of hormones than do animals. Finally, individual plant hormones often have widespread effects in the organism, while an animal's hormones tend to be targeted to a smaller subset of responding cells.

In this chapter, as we consider internal and external plant growth regulators, we encounter four unifying principles. First, plants generally adjust to changing environmental conditions via growth. Unlike animals, plants cannot move away from an environmental problem. A plant, however, can change its shape by localized growth much more radically than most animals can. Thus, depending on conditions, a plant can grow toward high light levels, grow a shorter or longer stem as needed, send roots deeper into soil, or drop its leaves and become dormant during harsh winter weather.

Second, changes in plant morphology (shape) or physiology are often regulated by plant hormones produced in response to environmental factors such as temperature, light, or gravity. For example, plant hormones can promote growth, inhibit growth, or promote maturation, depending on season. Third, plant hormones interact in complex ways; each plant hormone can affect several different processes, and each

FIGURE 31.2 ■ **Foolish Seedling Disease: A Curious Effect of Hormones on Plant Growth.** When a young rice plant (left) is infected by the fungus *Gibberella fujikoroi,* the parasite releases gibberellins, causing the stems to elongate rapidly, grow weak and spindly, and blow over and die in even a gentle breeze. An uninfected healthy, dark green plant is shown at the right.

physiological process can be affected by several different hormones. For example, three kinds of hormones can act either alone or together to cause a plant to grow taller. Fourth, plant hormones act at the level of cells to induce cell division, enlargement, or maturation. One hormone promotes growth, for example, by triggering internal cellular changes that allow cell walls to distend.

This chapter introduces the major types of plant hormones, then follows the life cycle of a flowering plant, answering several questions about how environmental factors and plant growth regulators (hormones) control plant growth and development:

■ What functions does each class of plant hormones perform?

■ How is seed germination regulated?

■ How are plant growth and development controlled?

■ What governs flowering and fruit development?

■ How are dormancy, dying, and the falling of leaves and fruit controlled?

■ How do plants defend themselves from infection and wounds?

PLANT HORMONES: FIVE MAJOR KINDS

The foolish seedling phenomenon brought on by high levels of gibberellins is characteristic of hormonal action in plants: Plants cannot flee from adversity, but they can grow away from the problem. Three of the five main classes of plant hormones, in fact, promote and regulate growth, while the other two classes inhibit growth or promote maturation. The names and activities of all five classes are given in Table 31.1.

In general, plant hormones are small molecules that tend to contain rings of carbon or carbon plus nitrogen. Plant scientists have had difficulty identifying hormone receptors in plant cells, but many suspect that plant hormones act on target cells much the way animal hormones do. Let us survey the five major classes of plant hormones, then see how they interact to regulate a plant's life cycle.

Gibberellins

Gibberellins are a good example of the widespread effects of particular plant hormones. They are made in a variety of organs, such as young leaves, embryos, and roots. Gibberellins can move through the plant's vascular system and are primarily involved in regulating plant height. Too little gibberellin results in dwarf plants, but too much results in long, pale, "foolish" stems. Gibberellins also play an important role in *bolting* (sudden stem lengthening) in rosette plants such as cabbage (Figure 31.3) and in inducing the seeds of rice, barley, and other grasses to germinate. Plant scientists have found that gibberellins are also involved in flowering, fertilization, and the growth of fruits, new leaves, and young branches (see Table 31.1).

Auxins

Like gibberellins, **auxins** are generally growth promoters. The name, in fact, comes from the Greek *auxein,* meaning "to increase, or augment." The most common natural auxin is indoleacetic acid, or IAA. Instead of being produced in various tissues and traveling through the vascular system like gibberellins, auxins are generally made in shoot apical meristems and developing leaves. They diffuse from the site of production downward toward the roots.

Auxins promote growth through cell elongation: They trigger enzyme activities that loosen the tightly woven fibers in cell walls, allowing the cell to expand from within. This action influences the length of cells in stems, leaves, and the wall of the ovary. Auxins also act in other ways that prevent leaves, fruits, or flowers from falling off prematurely and that inhibit growth in lateral buds and roots in favor of growth at the apical meristem. In high concentrations, auxins can cause uncontrolled growth and plant death (Figure 31.4). Some herbicides are based on this principle, including the synthetic auxins 2,4-D and 2,4,5-T, the active ingredients in Agent Orange, which was widely sprayed to defoliate jungle trees and plants during the Vietnam War. Within hours after spraying, the leaves and stems of broad-leaved plants droop, twist, and curl. Within days, the chlorophyll degrades and the plant

---

TABLE 31.1   Internal Regulators of Plant Growth and Development

Regulator	Transport	Action
Hormones		
Gibberellins	Upward and downward in vascular system	*Promote growth:* Promote stem lengthening in dwarf and rosette plants and seed germination in grasses; involved in flowering and fertilization; involved in growth of new leaves, young branches, and fruits
Auxins	From tip of shoot downward in vascular system	*Promote growth:* Augment growth by cell elongation; inhibit growth of lateral buds; foster growth of ovary wall; prevent leaf and fruit drop; orient root and shoot growth
Cytokinins	From root upward in vascular system	*Promote growth:* Stimulate cell division; kindle growth in lateral buds; block leaf senescence
Abscisic acid	Short distances in leaf and fruit	*Inhibits growth:* Opposes the three growth-promoting hormones; induces and maintains dormancy
Ethylene	Through air, as a gas	*Promotes maturation:* Enhances fruit ripening; promotes dropping of leaves, flowers, and fruits
Pigment		
Phytochrome	Not transported; remains in cell that produces it	*Detects light:* Changes form in response to light; mediates flowering, germination, growth, and plant form

FIGURE 31.3 ■ **Bolting. Gibberellin causes the short stem in a rosette plant to greatly elongate. On the right are cabbage plants (*Brassica oleracea*) in normal rosette form. On the left are similar plants after treatment with gibberellins. Bolting takes place naturally in a biennial before flowering in the second year.**

growth inhibitor. It counteracts the growth hormones, apparently by inhibiting translation of mRNA at the ribosomes and thus blocking protein synthesis and new growth. ABA moves only short distances from its site of production. ABA accelerates **abscission,** the dropping, or literally "cutting away," of leaves and fruits, and a drop of ABA on a green leaf also causes a yellow senescent spot. Plant scientists now think that the main role of ABA is to induce and maintain **dormancy,** or metabolic slowdown, especially in buds, and to promote the closing of stomata in leaves, which prevents excess water loss.

Ethylene

Ethylene, the fifth major plant hormone, is a small, simple molecule that exists as a gas at normal temperatures and is dispersed from one plant or plant part to another by air. Ethylene is behind the old adage about one rotten apple spoiling the whole barrel: The hormone is produced by ripening fruits, and it stimulates ripening in nearby fruits. Ethylene also stimulates the aging and dropping of leaves and fruits and may have an important role in plant self-protection (see pages 643 and 644).

Our survey of plant growth regulators has revealed that the five major types of plant hormones are small, mobile molecules that promote or inhibit growth or maturation. Now let us see how the plant hormones interact with each other and with environmental cues to regulate a plant's life cycle.

dries out and dies. These herbicides appear to induce such rapid cell division and tissue proliferation that the phloem becomes plugged with nonconducting cells.

Cytokinins

Cytokinins are the third class of growth-promoting plant hormones, and they generally stimulate cell division, including cytokinesis (see Chapter 7). Cytokinins are less mobile than auxins (and far less so than gibberellins) and appear to move in opposition to auxins—from root upward to shoot, not shoot downward to root. In contrast to auxins, cytokinins promote the growth of lateral buds, not shoot tips. Like most plant hormones, cytokinins also have other effects, including the prevention of leaf aging, or senescence.

Abscisic Acid

If plants had only gibberellins, auxins, and cytokinins, they would grow constantly. Sometimes, however, it is more advantageous for the plant to stop growing—to close its stomata, decrease the level of photosynthesis, drop aging leaves, or become dormant. **Abscisic acid** (ABA; see Table 31.1) is a

FIGURE 31.4 ■ **Synthetic Auxin Causes Some Plants to Grow Themselves to Death. The substance 2,4-D is chemically related to auxin and causes the plant to grow so rapidly that its phloem becomes clogged and it dies. The broad-leaved weeds in the foreground show the effects of the synthetic auxin. The corn in the background was unsprayed.**

HOW HORMONES CONTROL GERMINATION

After animals or air currents carry seeds to new locations, the seeds germinate, and tiny new plants emerge. But what triggers germination at a time when the seedling's chances for survival are greatest? Although specific answers depend on the plant species, environmental cues generally act through plant hormones that in turn regulate the sequential events of germination.

Environmental Cues, Hormones, and Germination

Some seeds germinate shortly after they reach a new location and imbibe enough water for the seed coat to soften and the tiny plantlet to burst out. Rice seeds are in this category, as are willows, poplars, and silver maples. This pattern is especially common in tropical plants, with their relatively mild, moist, and stable environments.

In temperate zones, however, delayed germination is often more advantageous. The seedlings of an annual that emerged in the late summer might not complete an entire life cycle before the short days, cold temperatures, and dry conditions of winter set in. Many temperate plants, therefore, germinate in spring or early summer after surviving winter as a dormant embryo encased and protected within a seed coat. Then in spring, increasing temperatures, lengthening days, and melting snow or rain trigger hormonal activities within the embryo that in turn trigger germination.

For some seeds, especially in dry areas, moisture is the external trigger, counteracting a growth-inhibiting hormone such as abscisic acid. Only when enough moisture is present to leach away most of the inhibitor will growth resume. This chain of events helps guarantee that the ground will be damp enough for the seedling to survive.

For other kinds of seeds, light is a determining factor. Lettuce seeds, for example, are so tiny that seedlings buried deeper than a few millimeters will run out of food reserves before reaching the sunlit surface. Interestingly, lettuce seeds germinate only when they detect light. In contrast, the seeds of some desert plants germinate only in deep, dark, moist soil.

In a few remarkable plant species, seeds can lie dormant for immense periods until conditions are just right. Some seeds from the Japanese lotus (*Nelumbo nucifera*), for example, have germinated successfully after apparently lying dormant for 2000 years. And some lupine seeds frozen in the arctic permafrost have reportedly remained viable for 10,000 years. Nevertheless, most seeds remain dormant and viable for only a few years at best. Even with greatly slowed metabolism, their food reserves are eventually used up and the embryo dies.

How Plant Biologists Learned the Effects of Hormones on Germination

Biologists knew that hormones control germination, but wondered how the growth regulators exert their effect. They noted that when a germinating corn kernel imbibes water, the starch-digesting enzyme amylase appears in the seed's endosperm—the white starchy tissue packed around the embryo (Figure 31.5a). To determine whether the embryo is necessary for the enzyme to appear, researchers removed the embryos from seeds before soaking them in water (Figure 31.5b). They found no increase in enzyme and concluded that the embryo plays a crucial role in the increase in enzyme in the endosperm. The biologists then hypothesized that the embryo might be releasing the hormone gibberellic acid (a type of gibberellin), which acts on endosperm cells and causes them to produce the enzyme. If this hypothesis is true, then treating an embryo-less seed with both water and gibberellic acid

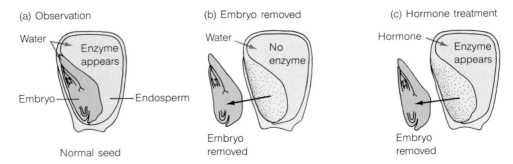

FIGURE 31.5 ■ **Experiments Show that a Hormone Helps Regulate Germination.** As the text explains, (a) water causes the enzyme amylase to appear in the endosperm of a corn kernel. The enzyme digests starch to sugar, which the embryo uses for energy. (b) Removal of the embryo revealed that it is indeed necessary, and (c) further tests confirmed that water causes some tissue in the embryo to secrete a gibberellin hormone called gibberellic acid, which diffuses to the endosperm. The endosperm, in turn, makes the enzyme, which digests the endosperm's starch to sugar. This the embryo uses as a fuel for germination.

should cause the enzyme to appear (Figure 31.5c). This prediction was confirmed by experiment. These results suggest that water causes the embryo to release gibberellic acid, and this hormone acts on endosperm cells, causing them to make the starch-digesting enzyme. The enzyme then digests the starch in the corn kernel, and the embryo uses the released glucose as a fuel during germination.

REGULATION OF PLANT GROWTH AND DEVELOPMENT

Although it might surprise you, farmers in California and along the Gulf Coast grow one-twelfth of the world's rice, and like other rice farmers, they start new crops in springtime. Rice farmers soak rice seeds, wait for the rice grains to germinate, then set the tiny seedlings in rows in the fields. As they grow, rice plants pass through a series of developmental stages, each controlled by external and internal regulators, so that the plants enlarge, develop new leaves and roots, then flower and set seed. This section considers the external and internal triggers of such life cycle events and shows how plants adapt to their surroundings through changing physiology and altered shape and orientation in space.

How Environmental Cues Can Influence a Plant's Orientation

A rice seedling in a field is oriented so that the roots grow down, not up or sideways, and the shoot grows up, not in some other direction. This seems logical enough, but how does a tiny plant manage to grow in the correct orientation? A plant's orientation is based on **tropisms:** bending, turning, or directional growth in response to external and internal stimuli. The term *tropism* comes from the Greek word for "turning toward." A tropism is considered *positive* if the organism's orientation is *toward* the stimulus, and *negative* if the organism's orientation is *away* from the stimulus. Botanists have named a number of tropisms by the external stimulus that causes them. For example, the orientation of roots toward water is called *hydrotropism* (turning toward water). A sweet pea's tendrils exhibit a pressure-sensitive tropism, or *thigmotropism*, as they wrap around any nearby object for support. The orientation of a plant part toward the ground is *gravitropism* (turning toward gravity), and the orientation toward light is *phototropism*. These last two tropisms help explain how a rice seedling's roots grow down and its shoot grows up.

■ **Gravity and Plant Growth** When a farmer sows a rice seed, the little embryo may be upside down or sideways, yet the seedling achieves the proper root-down and shoot-up orien-

tation. This feat is accomplished through the plant's sensitivity to gravity. The root is *positively gravitropic,* since it grows down toward the pull of gravity, and the shoot is *negatively gravitropic,* since it grows up against the pull of gravity (Figure 31.6a).

Investigations of the root tip of plants laid on their sides have revealed that dense starch granules (amyloplasts) rapidly sink to the bottom of the cells (Figure 31.6b and c). Plant physiologists suggest that these starch granules may act like little rocks that respond to the pull of gravity and may induce auxin to move to the lower side of a root and block the elongation of lower cells. As the upper cells continue to elongate, the roots bend and grow downward, thrusting toward a more likely source of moisture and minerals.

Researchers have studied mutant tomato plants with a strange growth habit and found evidence that auxins in plants may bind to receptors in cell membranes, as do many hormones in animals (review Figure 26.2). These mutant tomatoes grow parallel to the ground rather than tall and straight

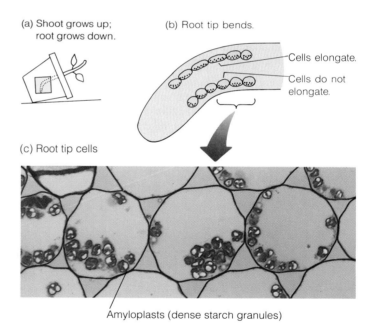

FIGURE 31.6 ■ **Gravitropism: Growth Toward or Away from Gravity. (a)** A plant's shoot will grow upward and its roots downward no matter how the organism is oriented. **(b)** Cells on the upper root surface elongate and cause the root to grow down. **(c)** The downward growth of roots occurs because gravity pulls heavy amyloplasts to the bottom of certain stem and root cells, and the settling of these organelles triggers gravitropism. The photo shows cells from the root of a buttercup plant. After growth reorients the root, the amyloplasts again sink to the bottom of the cell. Amyloplasts appear to work the same way in stem and root cells, but research suggests that in the stem, hormones stimulate cell growth on the organ's downward side, making the stem grow upward.

FIGURE 31.7 ■ **Mutant Tomatoes: Evidence that Plant Hormones May Work Via Receptors.** Certain mutant tomato plants grow parallel to the ground, but they are not deficient in auxin. Research showed them to lack auxin-binding proteins, which may function as auxin receptors on cell surfaces. Here, the mutant stretches horizontally in the foreground while normal tomato plants grow vertically in the background.

like normal garden varieties (Figure 31.7). Plant researchers suspected that the mutant plants might lack the auxin they need for normal orientation to gravity, and so they provided auxin to the prostrate mutant plants. This treatment, however, did not cause the mutant plants to grow upright, suggesting that the problem was not a lack of hormone. By investigating cell membrane proteins, however, the workers found that mutant cells lacked a certain protein that binds auxin in normal cells. They suggested that in normal tomato plants, this protein acts as a receptor that binds to auxin and signals cells to grow normally, thereby resulting in an upright plant.

■ **Light and Plant Growth** Plants need light for photosynthesis, but they may lack sufficient access to it. For example, a wild rice seedling might by chance germinate beneath the bushes at the edge of a forest. But the plant can grow away from the shade and toward light, and this maximizes the plant's exposure to the sun.

The first experiments to probe the mechanisms of phototropism in plants—how the organisms orient toward light—appeared in a book that Charles Darwin and his son Francis published in 1881. The Darwins knew that if a grass plant is illuminated from the side, it will bend toward the light source (Figure 31.8a). Together they performed a series of experiments on oat and canary grass seedlings to find out whether the controller of that bending response is present in the curving portion of the seedling or in its tip. They began by placing a piece of foil around the section of the seedling that actually bends toward the light, to keep the light out, and observed that the seedlings bent toward the light anyway (Figure 31.8b). Then they covered just the tip of a seedling with a foil cap and saw that the seedling did not bend toward the light. Their conclusion: The seedling's tip sends a signal to its lower portion, causing it to bend in the appropriate direction (Figure 31.8c).

Later researchers showed that the growth hormone auxin comes from the seedling's tip and accumulates on the side away from the light. Evidence suggests that the accumulated auxin causes a drop in internal cell pH, the acidic environment allows an enzyme to become active, the enzyme weakens cellulose in the cell walls, and these walls then stretch and enable the entire cell to elongate. Consequently, the seedling bends toward the light.

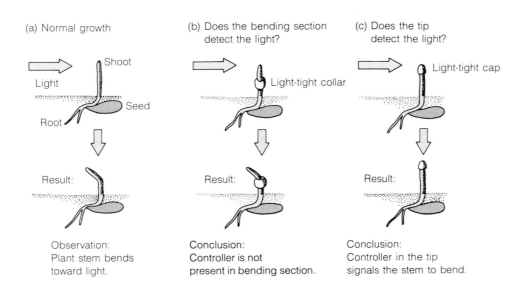

(a) Normal growth

Light — Shoot — Seed — Root

Result:

Observation:
Plant stem bends toward light.

(b) Does the bending section detect the light?

Light-tight collar

Result:

Conclusion:
Controller is not present in bending section.

(c) Does the tip detect the light?

Light-tight cap

Result:

Conclusion:
Controller in the tip signals the stem to bend.

FIGURE 31.8 ■ **The Darwins Prove That the Plant's Tip Controls Phototropic Bending.** (a) By testing grass seedlings, Charles and Francis Darwin noted that the seedlings bend toward a source of light. (b) By shading the bending portion, they deduced that the controller is not located there. (c) Shading the tip revealed that this portion of the plant responds to light by sending a signal to the stem portion, which then bends.

Plant Movements Other Than Tropisms

Although tropisms are immensely important to a plant's orientation in space, not all plant movements are tropisms oriented toward or away from the stimulus that causes them. A leaf of a sensitive plant (*Mimosa pudica;* see Figure 4.20), for example, will droop if touched anywhere along its length, and the leaves of the Venus flytrap snap shut if a small frog or insect triggers any of the little hairs inside the trap (Figure 31.9). Such plant movements whose direction is independent of the direction of the stimulus are called *nastic responses.* These responses are not always slow and irreversible, nor are they based on the growth of cells, as tropisms tend to be; for example, the Venus flytrap snaps shut rapidly, but subsequently reopens.

Some plants exhibit so-called sleep movements, with leaves drooping at night and lifting in the daytime. Even if suddenly kept in complete darkness, these sleep movements will continue for a few days (Figure 31.10). Biologists consider such movements strong evidence that, like animals, plants have internal biological clocks that meter the passage of time and trigger regular responses.

How the Environment Influences a Plant's Shape

Although rooted in place, plants have a tremendous capacity to turn, bend, and grow, which allows them to move part of their body toward or away from light, water, nutrients, gravity, or substrates. Thus, external factors can influence plant growth and shape in ways that adapt the plant to its environment.

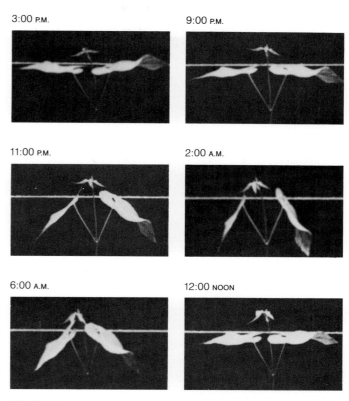

FIGURE 31.10 ■ Nastic Responses: Sleep Movements in Plants. Leaves of the bean (*Phaseolus vulgaris*) droop at night and fully extend by noon, even when kept in the dark. These "sleep movements" are evidence of a biological clock in plants, probably based on hormones.

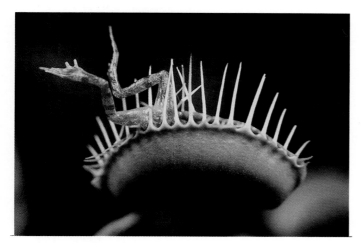

FIGURE 31.9 ■ Venus Flytrap: A Rapid, Reversible Nastic Response. When a small frog or insect lands on a Venus flytrap, an electrical stimulus (rather than a hormonal one like auxin) triggers a substantial change in acidity in the walls of cells where the leaves are hinged. The cell walls weaken, the cells expand, and this eventually forces the leaf to snap shut. Later, other cells on the inside of the hinge enlarge and force the trap open again, leaving it ready for another fly or frog.

■ **Shape Changes in Response to Light** Light can have a profound effect on a plant's shape. A crabgrass plant growing in a dark spot—in the deep shade of a forest or even under a board—is a good example. The plant looks spindly and pale yellow rather than healthy and green, because, deprived of direct light, it expends little energy making chlorophyll and instead grows longer and longer until it reaches the light and slows down (Figure 31.11a). Plants that have grown long and pale in darkness are said to be *etiolated* (from the French word meaning "pale as straw"). Even in less extreme shade, plant survival depends on the plant's ability to detect light levels and adapt by growth to this changing environmental stimulus.

■ **How Plants Detect Light** Plants have a light-sensitive blue-green pigment called **phytochrome** that allows them to detect the quality or quantity of light in the environment. Although it is an internal regulator, phytochrome (meaning "plant color") is not a hormone per se, because it does not diffuse from cell to cell. Phytochrome allows a light-sensitive seed to detect whether light is available and thus whether to germinate. Phytochrome also spurs flowers to emerge at appropriate times for pollination.

(a) Plants grown in darkness

(b) Phytochrome conversions

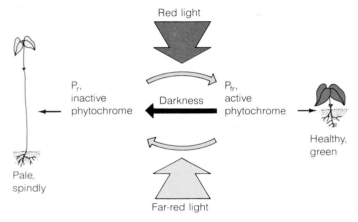

FIGURE 31.11 ■ The Phytochrome Light Timer. (a) Plants grow spindly and pale when deprived of light. In grass under a board, chlorophyll is degraded, and the plants grow toward any light source (here, the edges). **(b)** If a plant is exposed to far-red light or remains in the dark long enough, the phytochrome molecules in its cells convert spontaneously from the active (P_{fr}) to the inactive (P_r) form, and the plant becomes pale and spindly. When exposed to red light, however, phytochrome switches to the active (P_{fr}) form, and the plant stays healthy and green. The ratio of active phytochrome to inactive acts like a biochemical hourglass that times periods of darkness.

Phytochrome helps a plant distinguish between darkness, shade, and sun, as well as the amount of time spent in each, because the chemical changes form, depending on the amount or color of light that strikes it. Visible light consists of a broad spectrum of colors that can be seen individually when they are separated through a prism or water droplets, as in a rainbow (review Figure 6.5). For plants, two of the most important colors in natural light are red and so-called far-red light (Figure 31.11b). Red light is a common sight at sunset and has a longer wavelength than blue, green, yellow, or orange light. Also recall from Chapter 6 that chlorophyll strongly absorbs red light, but reflects green light, explaining why

plants appear green. Far-red light has wavelengths even longer than red light and is barely visible to the human eye.

Phytochrome does in plants what rhodopsin does in human eyes (see Figure 28.7): It responds to light, but specifically to red and far-red light. There are two forms of phytochrome named for the color of light they absorb: P_r absorbs red light, and P_{fr} absorbs far-red light. When P_r absorbs red light, the chemical changes rapidly into the P_{fr} form. When P_{fr} absorbs far-red light, it quickly changes back into P_r. In addition, P_{fr} spontaneously but slowly changes to P_r in the dark (see Figure 31.11b), just as sand in an hourglass pours away over time.

The interconversion of phytochrome affects plant growth in the following way: In normal sunlight, P_r absorbs the substantial amounts of available red light and converts to P_{fr}, the active form of the molecule. The plant responds to this active form by growing green and lush. In darkness, however, P_{fr} changes to P_r, the inactive form. In the absence of P_{fr}, the plants grow long, pale, and spindly. Growth control by phytochrome is just one aspect of how growth regulators affect plant shape. Another control comes from substances at the tips of the plants.

■ **Hormones and Plant Shape**   Plants can take on a wide variety of shapes, and two of these forms are represented by the spruce and maple pictured in Figure 31.12a. Firs usually have a single main trunk with numerous side branches—long at the base and shorter at the top—so that the whole tree has a con-

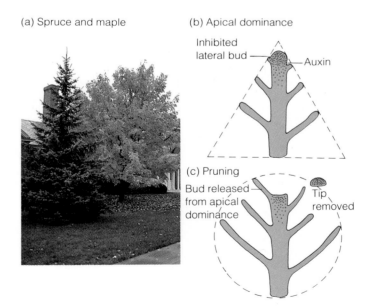

FIGURE 31.12 ■ Apical Dominance: Hormones from Growing Tips Can Affect Overall Plant Shape. (a) A conical spruce tree and a roughly spherical maple tree demonstrate the dramatic effects of apical dominance. **(b)** Auxin produced in the plant's apex travels downward and inhibits the growth of lateral branches. **(c)** If that apex is removed, lower branches are released from inhibition by apical dominance and begin to grow.

ical shape. In contrast, maples often branch low and have several main branches but no central trunk, and the overall shape is roughly spherical, not conical.

A major factor in these plant shapes is **apical dominance** (literally, domination by the tip of the plant), an inhibition of lateral buds by auxin transported down the shoot from the apical bud. Auxin promotes growth in the stem but inhibits growth of lateral buds. Because auxin is continuously broken down as it moves down the stem, its concentration drops off (Figure 31.12b). Thus, buds closest to the tip of the main stem are most inhibited (and smallest), while branches farther from the tip of the plant are the least inhibited (and oldest and largest). The result is a conical shape. Cytokinin also comes upward from the roots, counteracts the effects of auxin, and causes the lateral buds and the lowest branches to grow more strongly. In a plant with a more spherical shape, the apical bud may produce less auxin, and lateral buds and branches may be less sensitive to inhibition by the hormone; thus, the branches both high and low on the stem can grow strongly. If a gardener or a browsing animal removes the apical bud of a plant, the buds farther down the stem are released from apical dominance and begin to grow, producing a bushy appearance (Figure 31.12c). Recent evidence suggests that auxin may cause apical dominance by provoking the release of sugar units called *oligosaccharins* from plant cell walls. Commercial Christmas tree growers take advantage of apical dominance by pruning back the tips of branches. This allows buds toward the inside of the tree to grow, yielding a fuller, more desirable tree.

It is clear from the foregoing information that plants generally respond to environmental hardships or opportunities by growth and change in form and that hormones mediate these responses. Life cycle events are part of this responsiveness, and as in germination and subsequent growth, flowering and fruit formation depend on a mixture of environmental and hormonal triggers.

FIGURE 31.13 ■ Timing Is Everything: Hormones and Environmental Cues Regulate Flowering in *Puya raimondii*. Light, temperature, and internal hormones interact to control flowering in this giant member of the pineapple family. The plant produces a 4.5 m (15 ft) flower spike—the largest floral structure in the plant kingdom—just once every 30 to 150 years.

CONTROL OF FLOWERING AND FRUIT FORMATION

Like germination and growth, flowering and setting fruit are timed to the appropriate seasons. This means that pollination, fertilization, fruit enlargement, seed development, fruit abscission, and seed dispersal all occur in seasons when the species has the greatest potential for success. Environmental factors that change with the seasons—namely, temperature, moisture, and the number of hours of sunlight in a day—act as external regulators of the plant life cycle. Seasonal timing is important for annual plants such as rice, which flower and produce seeds once in a single season, and also for perennials. An exceptional example of the latter is the rare *Puya raimondii,* a large plant of the windswept Andean high plain that can live for nearly two centuries and that produces the world's largest floral structure (Figure 31.13)—but only once every

30 to 150 years. Seasonal timing depends on both environmental cues (such as temperature and light) and internal hormonal triggers.

Temperature and Flowering

Certain plants, including biennials like beets, carrots, and turnips, flower only after a period of exposure to cold, a process called *vernalization.* Such exposure to cold signifies that winter has come and gone. Because flowering comes early in the spring, the plant produces seeds early in its second summer, and the seedlings have plenty of time to grow into hardy plants that can withstand the following winter.

Light and Flowering

For many plants, temperature alone is too risky a trigger for flowering. A cold September followed by a warm October and a harsh November could find a plant covered with flowers but buried in snow. Many plants track the seasons by **photoperiod,** the length of light and dark periods each day, rather than by daily temperature. Early researchers noticed that some plants, like rice, seem to require short days to bloom, and they called these plants *short-day plants.* Other plants seem to require long days and were therefore called *long-day plants.* (They called plants that flower without regard to photoperiod *day-neutral plants.*)

Despite the early terminology, experiments revealed that a short-day or long-day plant is not defined by the absolute length of the day or night, but by whether the day length is longer or shorter than some critical photoperiod. For example, ragweed blooms in the fall and requires *less than* 14½ hours of daylight to blossom; thus, it is a short-day plant. In contrast, spinach needs *more than* 14 hours of light per day to flower, so it is a long-day plant. At 16 hours of light, the spinach will bloom but the ragweed won't, and at 8 hours of light, the ragweed will blossom but the spinach won't (Figure 31.14a and b).

Generally, long-day plants blossom in spring or early summer, when the days are becoming longer, and short-day plants flower in the late summer or fall, when the days are becoming shorter. For example, poinsettias are short-day plants because they will flower only when the days are less than 10 hours long (and the nights are more than 14 hours long), as in autumn. People with indoor potted poinsettias must therefore put them in a dark closet for 14 or more hours each day during the fall in order to enjoy their showy red and yellow displays by Christmas.

To probe how plants measure critical photoperiod, botanists exposed ragweed plants to a day length that should cause flowering, but interrupted their period of darkness with a short blast of light (Figure 31.14c). The result: No flowers formed. Since interrupting the night blocks flowering, plants must measure night length rather than day length; in other words, short-day plants are actually "long-night plants." This explains why even a quick peek at a poinsettia plant in the middle of the night would disrupt the dark period this short-day (long-night) plant needs to set flowers. For a long-day (short-night) plant like spinach, however, a flash of light in the middle of the night promotes flowering (see Figure 31.14c).

Such light flash experiments suggested that plants might measure night length and hence photoperiod via phytochrome. Recall that phytochrome converts spontaneously from the active P_{fr} form to the inactive P_r form in the dark. Flashes of bright light in the middle of the night can cause P_r to change back to P_{fr} and hence reset a plant's night clock. This simple fact helps explain the connection between photoperiods and flowering. In both short-day and long-day plants, P_{fr} gradually disappears in the dark, leaving very little P_{fr} by the end of a long night. In short-day plants, P_{fr} *blocks* flowering; these plants will flower only after a long night allows P_{fr} to disappear. In contrast, in long-day plants, P_{fr} *promotes* flowering; with a short night, enough P_{fr} remains to induce blossoms to form. For these latter plants, a light flash in the

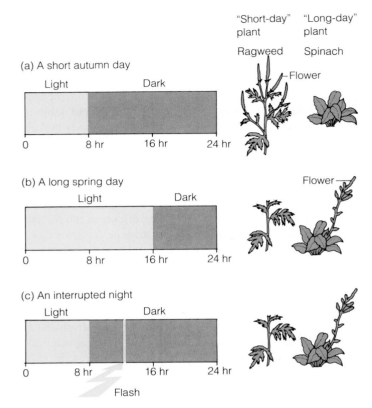

(a) A short autumn day

"Short-day" plant
Ragweed

"Long-day" plant
Spinach

(b) A long spring day

(c) An interrupted night

Flash

FIGURE 31.14 ■ The Light/Dark Cycle Regulates the Onset of Flowering. **(a)** On an 8-hour autumn day, ragweed—a short-day plant that flowers if there are fewer than 14½ hours of light—will form flowers. On the same day, however, spinach—a long-day plant that needs more than 14 hours of light a day—will fail to flower. **(b)** In contrast, on a long spring day, ragweed (the short-day plant) will not flower, but spinach (the long-day plant) will. **(c)** If an experimenter interrupts a long autumn night with a brief flash of light, the plants respond as if the day were long and the night short. This suggests that light resets the "clock" that times the length of the night. Further studies have shown that phytochrome molecules mediate these timing actions.

middle of a long night regenerates high levels of P_{fr}, and thus flowering can take place.

The phytochrome "hourglass" allows such accurate timing of photoperiod that all the individuals of a species can flower within a day or two of each other. This provides a ready source of pollen that wind or animals can convey to other blossoms of the same species. And a plant's photoperiod often coincides with the most active food-seeking portion of an insect's (or other pollinator's) life cycle—evidence of their coevolution.

The Role of Hormones in Flowering

Plant scientists wondered whether phytochrome in the cells of each flower bud measures the day/night cycle for that bud or whether the signal to flower comes from other parts of the plant. To test these possibilities, experimenters exposed the buds of chrysanthemum plants to one light cycle and the leaves to another and checked the plants for flowers (Figure 31.15). The results showed that leaves were responsible for sensing photoperiod, not the bud. If the leaves control flowering, however, then they must send a signal to the tip, which itself subsequently develops into a flower. Plant physiologists suggested that a substance they called **florigen** (literally, "flower generator") might move from the leaves to the bud. Florigen itself has yet to be isolated, however, and many botanists believe that florigen is actually the combined effects of several already familiar plant hormones. Botanists do agree, however, that plants detect changes of seasons by means of a light-sensitive pigment that acts via hormones, causing flowers to appear at favorable times of the year.

Triggers to Fruit Development

As Chapter 30 explained, the fruit—with its cargo of seeds—begins to develop once the flower has been pollinated and seeds start to form. Fruits, as we have seen, help protect the seeds and disperse them. But fruits are also energy-expensive

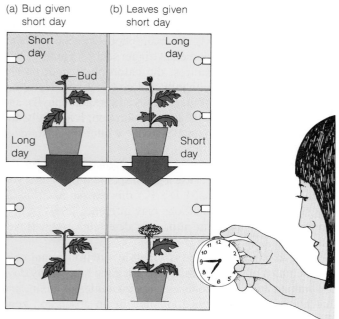

FIGURE 31.15 ■ Does a Bud Control Its Own Destiny? Chrysanthemums normally flower only during the short days of autumn. To determine whether light regulates flowering by acting directly on cells in the bud or on cells in other parts of the plant, researchers put plants in a special light-tight chamber. Researchers exposed the buds of some plants to short days and exposed their leaves to long days (a), while they gave other plants the reverse treatment (b). The results showed that the day length impinging on the leaves was significant but that the day length impinging on the bud did not matter. Researchers suggested that a hormone they called florigen might move from the leaves to the bud and stimulate flowering.

(a) Normal strawberry   (b) Many seeds removed

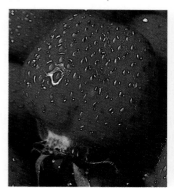

FIGURE 31.16 ■ Seeds Liberate Auxin, Which Triggers Fruit Development. (a) Strawberry seeds secrete auxin, and this promotes fruit development around the seeds. (b) If all but a few seeds are removed, fruit develops just around them.

organs to produce—especially the large, juicy ones—and not surprisingly, some plants ensure that fruits develop only around viable seeds.

The dozens of tiny seeds on the outside of a strawberry, for example, help control the development of a normal fruit (Figure 31.16a). If all the seeds are removed when the fruit is still small and green, it never enlarges and ripens. However, deseeded strawberries treated with auxin will grow normally, suggesting that the seeds promote fruit growth by secreting auxin. Experiments show that strawberry seeds, in fact, do secrete auxin, and if even a few seeds are left on the strawberry, the berry will develop normally around the location of each seed (Figure 31.16b).

Once fruits have developed, the airborne hormone ethylene causes them to ripen. When a peach ripens, it gets softer through the partial breakdown of cell walls; sweeter through the conversion of organic acids and starches to sugars; and more fragrant through the production of aromatic molecules; and it turns from green to rosy yellow through the formation of carotenes and other pigments. All these changes enhance the fruit's appeal to animals that eat it and distribute its seeds, and ethylene is behind most of these changes. Since most fruits release ethylene gas as they begin to ripen, green fruits ripen faster if placed in a plastic bag with ripe fruit.

HOW PLANTS AGE AND HOW THEY PREPARE FOR WINTER

Plant life cycles are strongly synchronized with the seasons. Annuals, for example, set seed and die before winter, while perennials often drop their leaves and become dormant during the cold months. The process of growing old and dying is called **senescence,** while the process that allows mature fruits or dying leaves to fall is called *abscission*.

Senescence and Abscission

Senescence and abscission are technical terms for the events of autumn we all know and enjoy: the turning of green leaves to glorious shades of red, gold, orange, and purple, followed by the fading and finally the falling of the leaves into large fragrant drifts beneath the trees and bushes (Figure 31.17). Several external and internal events control this sequence. Fields of soybeans, for example, grow yellow and die within a few days of each other—evidence that external cues (cold, dryness, diminishing light) trigger mass senescence. Tomatoes, zinnias, and marigold plants in the garden will also die back after they have produced flowers and seeds, each species on its own schedule and dependent on environmental conditions. Flowering itself is a key to this senescence, since in many annuals, the apex of each shoot converts into a flower bud and ends the open growth potential of that branch. When the last branch tip has converted to a flower, the plant stops growing and soon begins to senesce.

In perennials, the plant conserves resources over the dry, cold months of winter, and the **deciduous** growth habit (the dropping of leaves) is very common. In the cold and dark of winter, leaves are much less efficient photosynthetic organs and provide a dangerous avenue for water loss. Leaves may also serve as organs of excretion, storing wastes which then fall from the plant when the leaves drop.

In fall, a deciduous tree, such as a sugar maple, withdraws valuable nutrients—including sugars and amino acids—from the leaves and transports the nutrients to the trunk and roots

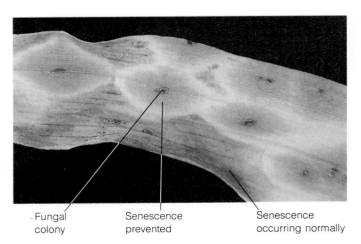

-Fungal colony   Senescence prevented   Senescence occurring normally

FIGURE 31.18 ■ Cytokinins Block the Essence of Senescence. A fungal infection dotting a browned, senescent corn leaf causes islands of green to remain. As in the foolish seedling disease, the fungus releases a plant hormone—in this case, cytokinin, which blocks ABA, the hormone causing senescence. Leaves normally senesce when external cues such as light, temperature, and moisture interact with the relative balance of abscisic acid, gibberellin, and cytokinin.

for storage. One of the last molecules to be broken down and withdrawn is chlorophyll, the main photosynthetic pigment. However, with an environmental trigger such as a night or two of subfreezing temperatures, the chlorophyll, too, breaks down and the other brilliantly colored leaf pigments that remain (red or purple anthocyanins, yellow xanthophylls, and red and orange carotenes) can show through. Eventually, the leaves fall, and the plant survives winter in a state of dormancy.

Researchers have noticed that a fungus that releases the hormone cytokinin will cause a green spot to remain on a yellowing corn leaf (Figure 31.18). For this reason, plant scientists suspect that fall conditions may influence levels of hormones and that these, in turn, may determine when leaves senesce.

Leaf abscission is also governed by environmental cues and the ratios of plant hormones. During the long sunny days of summer, each leaf produces large quantities of auxin, and this hormone inhibits the formation of an abscission zone (Figure 31.19), an area of relatively weak cells at the base of the stem, where the leaf will eventually separate from the branch. In the fall, auxin levels drop, enzymes begin to break down the cell walls in the abscission zone one by one, and eventually, a gust of wind or a few pelting drops of rain will send the leaf fluttering to the ground.

FIGURE 31.17 ■ Senescence Unmasks the Flaming Colors of Fall. A scene in New York State's Catskill Mountains. The breakdown of chlorophyll allows yellow, red, and orange light-gathering pigments to show. Eventually, these too will senesce, and the colors will fade.

Dormancy: The Plant at Rest

Once the leaves have fallen, the plant lives through winter in a state of greatly reduced metabolic activity known as dormancy. Dormancy can take several forms. Some perennial

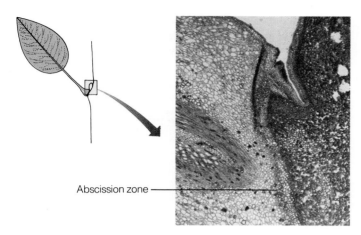

Abscission zone

FIGURE 31.19 ■ Abscission Zone Allows Fruit Detachment and Leaf Fall. Growth and maturation result in an abscission zone, where fruit or leaf will eventually detach. This light micrograph shows the weakened zone, a vertical stripe of lavender cells.

plants like iris and daffodil die back to roots or bulbs that are protected from desiccation and extreme heat or cold by the earth around them, while others like chestnuts and linden trees stand leafless from late autumn until early spring (Figure 31.20). And many plants rest over winter in embryonic form inside dormant seeds. Regardless of exterior form, dormancy is characterized by greatly diminished use of energy and slowed protein synthesis, cell division, and maintenance. Light cycles, perhaps timed by phytochrome, may be involved in the onset of metabolic slowdown, and hormones no doubt relay the message to individual cells and help maintain the state of dormancy.

PLANT PROTECTION: DEFENSIVE RESPONSES TO EXTERNAL THREATS

Plants, as we have seen, usually adjust to changing environmental conditions by growth. One of the biggest environmental threats of all, however, is the onslaught of viruses, bacteria, animal predators, and fungi like the cause of foolish seedling disease in rice. Plants may be stationary, but they are far from defenseless. They can often protect themselves quite effectively by producing physical barriers, such as thorns, leaf hairs, and sticky sap (see Chapter 30); by generating various toxic compounds; and by walling off injured areas.

Chemical Protection

Plant scientists have discovered that in addition to *primary compounds* (biological molecules necessary for growth and regulation), most plants also produce *secondary compounds*, molecules that help ensure survival by repelling, killing, or interfering with the normal activities of plant-eating organ-

isms. Plants can't run from their enemies, but they can resort to chemical warfare! In some tree species, up to 50 percent of the plant matter is secondary defensive compounds. And these are so effective that in a temperate forest in a typical year, predators consume only about 7 percent of the total leaf surface area.

Secondary compounds include such familiar substances as cyanide, camphor, cocaine, caffeine, and nicotine—just a few of the 10,000 plant chemicals that are toxic to animals. Some of these substances, such as camphor and tannins, can discourage insect pests from eating a plant's leaves or laying eggs on them. Others poison attackers. For example, the nicotine in tobacco can kill various insect predators, and cyanide compounds in the bird's-foot trefoil can kill snails that feed on its leaves. Recent evidence also shows that short sugar chains called oligosaccharins from the cell walls of microbes or injured plant cells can cause a plant to make antibiotics that kill attacking fungi and bacteria. Some plants generate compounds that mimic insect hormones and thereby disrupt normal insect metamorphosis.

In response to insect damage, many plants increase their production of secondary compounds, in some cases reaching 10 percent of the total weight of fruits or vegetables. Since these compounds can often harm people as well as insects— many, in fact, can cause cancer—it is best to stay away from diseased or damaged celery, apples, peanuts, and other produce. Noted biologist Bruce Ames suggests that a person who eats damaged organically grown fruit will consume more toxins than one who eats fruit treated at appropriate times with limited quantities of synthetic pesticides.

While most secondary substances are harmful to animals, they have given rise to a large number of human medicines. These include the pain reliever aspirin from willow bark, the heart medicine digitalis from foxglove, the main ingredient in oral contraceptives from yams, and the pain reliever morphine from poppies. Many scientists suspect that thousands more useful medicines will eventually be found in plants. In fact, one reason scientists are fighting for the preservation of the

FIGURE 31.20 ■ Dormancy: A Key to Winter Survival. Many perennial plants, such as this American linden, survive the bitter cold and drought of winter by entering a state of dormancy, triggered by environmental cues and maintained by plant hormones, perhaps including abscisic acid.

vast tracts of unexplored rain forests is so that the plant species—most as yet unstudied—do not become extinct before we can make use of their healing compounds.

Walling Off Injured Areas

Besides producing protective chemicals, plants have a second general growth response that protects them from viruses, bacteria, fungi, animals, or simple physical injury, such as loss of a limb in a storm. In plants, cells cannot migrate into a wounded area and heal the damage. Plants can, however, wall off the damaged area—a response that prevents invaders from gaining access to healthy tissues. This walling-off process is called *compartmentalization,* and it involves the production of toxic chemicals in the invaded area and the plugging of nearby xylem and phloem tubes with thick saps or resins that prevent invaders from spreading to other parts of the plant (Figure 31.21). With the injury walled off, the tree can then continue growing, surviving, and defending itself against new attacks.

CONNECTIONS

A plant's survival depends on its adaptations to the environment. It flowers in seasons appropriate for reproduction. Its seeds germinate when weather favors the tender young sprouts. It becomes dormant before conditions turn harsh. And it grows toward adequate sources of water and light and fends off attackers. The control of all these events involves a combination of external cues (light, temperature, moisture, gravity, seasonal changes) and internal growth regulators. In general, environmental fluctuations trigger internal molecular responses, and the plant then adjusts to its environment. This adjustment can be short- or long-term, but will often involve a change in physical form—an altered shape, the dropping of leaves, or the opening of flowers. Plant regulators—gibberellins, auxins, cytokinins, abscisic acid, and ethylene—have dozens of regulatory roles in plants, and alone or in combination, they control germination, tropisms, plant shape, flowering, fruit development, senescence, leaf drop, and dormancy. Since most of these events involve a change in the plant's physical form, plant hormones are both a practical and

a conceptual bridge between anatomy, environment, and survival. In Chapter 32, we explore that day-to-day survival in more detail by focusing on how plants collect nutrients and transport water.

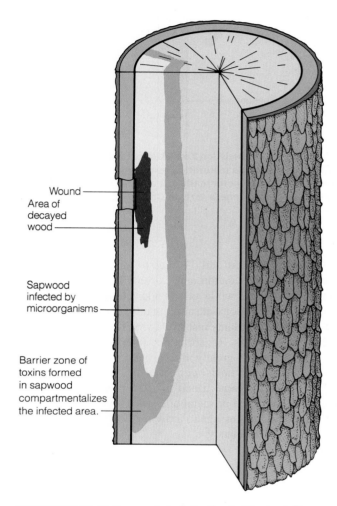

FIGURE 31.21 ■ **Compartmentalization in Plants: A Response to Wounding.** A penetrating wound will inevitably allow microbial and/or fungal infections into the plant (red and pale green), but a zone of parenchymal cells (in lavender) around the infected area begins to produce tannins and other toxic chemicals. These kill the invaders, contain the damage, and allow growth to continue around the walled-off area.

Wound
Area of decayed wood

Sapwood infected by microorganisms

Barrier zone of toxins formed in sapwood compartmentalizes the infected area.

Highlights in Review

1. Plants respond to the environment primarily through modified growth. Growth is modified by hormones, growth regulators that trigger changes in metabolism, development, and other cellular activities.

2. In general, gibberellins, auxins, and cytokinins promote growth. Abscisic acid inhibits growth, maintains dormancy,

and promotes abscission (leaf and fruit fall). Ethylene promotes ripening, senescence, and abscission.

3. Most seeds survive winter in a dormant state and germinate only when suitable conditions of light, temperature, and/or water are present.

4. Plants become oriented in space with respect to light, gravity,

and water by means of tropisms, such as gravitropism and phototropism. In phototropism, auxin from the shoot tip seeps
down the unlighted side of the plant. The cells on that side
grow faster than the cells on the opposite side, and the plant
bends toward the light.

5. Plant movements can be fast or slow, irreversible or reversible,
   and specifically oriented toward or away from a stimulus.
   Movements not oriented with respect to the stimulus are called
   nastic responses.

6. Plants can change shape in response to environmental conditions and internal hormonal triggers. Plants growing in the dark
   become etiolated—long, spindly, and pale—because the light-
   sensitive pigment called phytochrome spontaneously converts
   from the active P_{fr} form to the inactive P_r form in the dark. The
   P_{fr} form promotes normal growth, while the P_r form permits
   etiolated growth.

7. The shoot tip produces auxin; this hormone augments growth
   and helps control cell elongation, but inhibits the growth of
   nearby lateral buds, a process called apical dominance.

8. Cytokinin from the roots acts in opposition to auxin. When
   cytokinin concentrations are high or lateral bud sensitivity to
   auxin is low, the buds and branches throughout the plant are
   released from inhibition and begin to grow.

9. Plants must flower at appropriate times so that pollination and
   subsequent fruit and seed development have the maximum
   potential for success.

10. The regulation of flowering often depends on external cues,
    such as temperature and photoperiod (light/dark cycles), and
    on internal regulators, such as the pigment phytochrome and
    plant hormones.

11. Short-day plants, such as the poinsettia, require nights longer
    than a specific length to flower. Long-day plants, such as spin-
    ach, require nights shorter than a specific length to flower.

12. Plant scientists have hypothesized the existence of a flower-
    inducing hormone (florigen) that moves from the leaves at the
    tips of branches into the flower buds. No one has yet isolated
    this specific chemical, however, and florigen may actually be
    the combined effects of several hormones.

13. Viable seeds strongly promote the growth and development of
    fruit by releasing auxin and in some cases gibberellin. Ethyl-
    ene, produced by maturing fruits, accelerates ripening.

14. In late summer and fall, many plants withdraw starches and
    proteins from their leaves in a process called leaf senescence.
    Abscisic acid promotes senescence; cytokinin inhibits it. When
    the leaves lose chlorophyll, other pigments show through as
    fall colors.

15. Auxin prevents abscission by maintaining an area of weak
    cells, called the abscission zone, at the base of the leaf stalk or
    fruit stem. When auxin levels drop, ethylene and ABA can
    accelerate the dropping of leaves and fruits.

16. Many plants become dormant for the winter after storing
    energy in the form of starches and oils in their roots and stems
    and then dropping their leaves.

17. Plants can contain compounds that repel attackers and/or help
    wall off injured areas.

Key Terms

abscisic acid, 633   auxin, 632
abscission, 633   cytokinin, 633
apical dominance, 639   deciduous, 642

dormancy, 633   photoperiod, 640
ethylene, 633   phytochrome, 637
florigen, 641   senescence, 641
gibberellin, 630   tropism, 635

Study Questions

Review What You Have Learned

1. Name three plant hormones, and describe the activities of each.
2. How does the action of abscisic acid differ from the action of a
   plant's growth-promoting hormones?
3. Explain how moisture, cold, and ground cover affect seed
   germination.
4. How do gibberellins promote germination in a corn seed?
5. Define tropism, and give two examples.
6. What evidence indicates that plants have internal biological
   clocks?
7. If most of a plant's phytochrome has converted to the P_{fr} form,
   is the plant likely to be green and healthy? Explain.
8. How does apical dominance help determine a plant's shape?
9. What is the evidence for florigen?
10. Which hormones are involved in fruit growth, fruit ripening,
    and fruit abscission?
11. Compare the effects of ABA, cytokinin, and auxin on leaf
    senescence in autumn.
12. Describe how secondary compounds help protect plants.

Apply What You Have Learned

1. A gardener sprays the herbicide 2,4-D on a patch of poison ivy.
   What mechanism causes the weed to die?
2. You have been hired by the Christmas Tree Association to
   develop trees that are bushy and densely branched. You decide
   that the best approach is to use genetic engineering to increase
   or decrease the amount of hormones the trees produce. Would
   you increase, decrease, or leave unchanged the production of
   gibberellins, auxins, cytokinins, abscisic acid, and ethylene?
   Explain why you would change (or leave unchanged) the production of each of these hormones.
3. Ragweed is a short-day plant and requires less than 14½ hours
   of light to flower, whereas spinach is a long-day plant and
   requires more than 14 hours of light to flower (review Figure
   31.14a and b). An experimenter exposes a group of ragweed
   plants and a group of spinach plants to a photoperiod of 14¼
   hours of light. Which plants will bloom?
4. The Christmas cactus is a short-day plant. To induce blooming
   in late December, you decide to place the plant in a lightproof
   box in early November and water it only during the day. Will
   this strategy work? If so, how?

For Further Reading

EVANS, M. L., R. MOORE, and K. HASENSTEIN. "How Roots
Respond to Gravity." *Scientific American,* December 1986, pp.
112–119.

RAVEN, P. H., R. F. EVERT, and H. CURTIS. *Biology of Plants.* 4th
ed. New York: Worth, 1986.

ROSENTHAL, G. A. "Chemical Defenses of Higher Plants." *Scientific American,* January 1986, p. 96.

The Dynamic Plant: Transporting Water and Nutrients

FIGURE 32.1 ■ **Tomatoes: America's Favorite Garden Vegetable.** With its deep green leaves, red fruits, and toxic compounds to ward off predators, the tomato is a model of plant survival.

Tomatoes: A Case Study for Plant Transport

On a golden autumn afternoon in 1820, Robert Gibbon Johnson, a young colonel in the U.S. Army, stunned the people of Salem, New Jersey. He stood on the steps of the town's courthouse, and with a crowd gathered before him, he committed what seemed at the time a suicidal act: He ate a tomato. Tomatoes are members of the plant family that includes poisonous mandrake and deadly nightshade, and this connection must have created real skepticism in nineteenth-century New Jersey. Colonel Johnson, of course, survived, and his act—perhaps his bravest!—convinced the townspeople that the scarlet fruits were not only safe but downright delicious.

Today, tomatoes grow in the backyard gardens of nearly half of all American homes (Figure 32.1), and four out of five people prefer tomatoes to any other home-grown food. Americans buy 1.4 million tons of commercially grown fresh tomatoes annually and consume another 6 million tons as juice, sauce, and paste. The once reviled tomato is now a billion-dollar industry, and because we eat so many tomatoes, this single species contributes more vitamins and minerals to our diet than any other vegetable.

The tomato is a good example for many of the plant concepts we have discussed so far. Like virtually all flowering plants, it produces below-ground roots and an aboveground shoot with a main stem, buds, leaves, flowers, and fruit. These organs are made up of dermal, vascular, and ground tissues carrying out all the functions needed for growth and reproduction (see Chapter 30). The tomato plant remains rooted to one spot and must withstand chilly nights, rainstorms, baking sun, hungry cutworms, and a host of viral, bacterial, and fungal diseases. Simultaneously, the plant produces flowers, fruit, and seeds at times when pollination and seed dispersal are most likely to occur. As Chapter 31 explained, certain toxic substances (which tomato leaves and green fruits produce in quantity) help protect a tomato plant against attack by herbivores, while its phytochrome pigments and hormones help regulate the life cycle in response to day length and other external cues.

The tomato is clearly a dynamic plant and a good case study for understanding the physiological mechanisms that make plant survival possible. How does a plant, in general, procure the energy, water, and materials required for growth, development, self-protection, and reproduction? In particular, how do water from the soil, carbon dioxide from the air, and energy from the sun wind up in plump, juicy tomatoes, which are more than 90 percent water and less than 10 percent organic molecules? The answers involve photosynthesis (see Chapter 6) and two topics of central importance to the dynamic plant: material transport and plant nutrition.

The movement of materials through a living plant depends on the anatomy of xylem and phloem tubes, the chemistry of water, and osmotic pressure and related physical phenomena. The underlying mechanisms are the same whether materials are moving through a 1 m (3.28 ft) tomato vine or a 100 m (328 ft) redwood tree, and they represent some of evolution's most remarkable products.

The tomato is also a good example of an exciting trend in modern agriculture: the development of improved crops through the use of tissue culture, gene splicing, artificial seeds, and other special techniques. Traditional plant breeders have altered many characteristics of the wild Latin American vines. By the early 1970s, the fruits of commercial varieties were much larger and ripened more quickly; the plants grew sturdier and more compact; farmers could harvest many types with machines; and the vines bore up to four times more fruits than they did in the 1930s. Now, with the tools of biotechnology and a good understanding of how plants function and survive, plant scientists are making even greater modifications. One company, for example, developed a tomato that is brighter red and far less juicy. This allows canning companies to use less food coloring and buy and ship more tomato solids (and thus less water) from the farmer's field. Another company created a pomato—a plant that produces potatoes below-ground and tomatoes above. The last section of this chapter describes related genetic work and other results of the latest agricultural revolution.

As we consider the uptake and transport of water and nutrients in plants, four unifying themes will emerge: First, plants depend on the special physical properties of water that make it possible for water and nutrients to be pulled from the soil and a continuous flow of materials maintained to individ-

FIGURE 32.2 ■ The Visible Vascular System. Xylem and phloem make up a transport network that carries water and nutrients close to every plant cell.

ual cells throughout the plant body. Second, the avenue for this flow of materials is a network of transport tubes—the vascular system—which integrates all the parts of the plant into a smoothly functioning unit (Figure 32.2). Third, as in animals, plants have mechanisms for maintaining homeostasis, especially a constant fluid environment inside plant cells. Finally, soil quality is crucial to the growth and survival of plants because each individual procures all its inorganic nutrients from the soil in which it germinates and grows.

This chapter will explain how a dynamic plant gets all the materials it needs for daily survival and reproduction. In the process, we will consider several questions:

■ How does a plant take up water, transport it to all body cells, and control excess evaporation from the leaves?

■ What nutrients does a plant need, and where does it get them?

■ How does a plant move the products of photosynthesis from areas of production and storage to areas of use?

■ How are people improving crops through traditional breeding methods and biotechnology?

HOW PLANTS TAKE UP WATER AND RESTRICT ITS LOSS

Consider a typical backyard garden planted with rows of tomatoes, corn, sunflowers, and other common crops. To be successful, the gardener must control weeds and insect pests and apply fertilizers, but the biggest concern must always be water, and the reason is simple: Over the course of the summer, each tomato plant consumes about 120 L (32 gal) of water, which is over half the water a gardener weighing 30 times more will require! Corn and sunflower plants, which are larger than tomato vines, will consume 17 times more than the person, and in each of these plants, 98 percent of the imbibed moisture will evaporate from the leaves.

The explanation for this apparently insatiable "thirst" lies in a fundamental difference between animals and plants: In most animals, fluids tend to recycle through a circulatory system, while in plants, water travels in a one-way path from roots through stems to leaves, then back out into the environment (Figure 32.3). A steady supply of water enables a plant to carry out photosynthesis (see Chapter 6), remain crisp and erect with turgid plant cells (see Chapter 2), stay cool despite a baking sun on outstretched leaves and stems, and transport substances through the plant.

Land plants would quickly lose all the water they take in if it weren't for the waterproof coatings on the aboveground portions of the plant, such as the waxy cuticle or bark layers. But if such layers made an absolutely water- and airtight seal, the plant could not take in the carbon dioxide required for photosynthesis. The evolutionary solution to this dilemma is

Water out

Water and minerals in

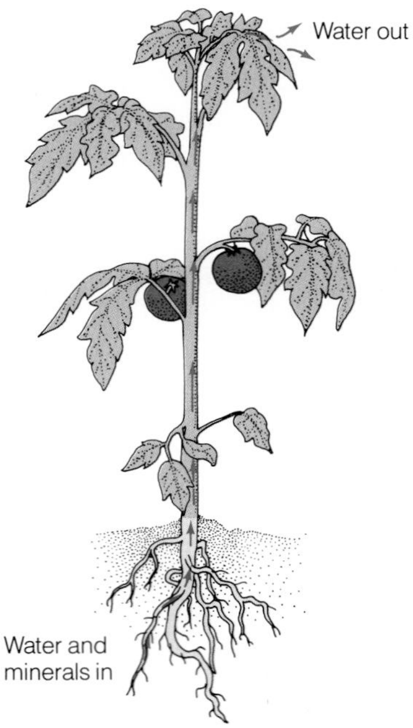

FIGURE 32.3 ■ **One-Way Flow of Water.** Xylem tissue provides a one-way route for the flow of water and inorganic nutrients from the roots through the stem and to the leaves. The transport system and its one-way flow represent an energetically inexpensive way to move materials through a complex organism.

stomata (singular, stoma): tiny access ports in the leaves and stems. The opening and closing of stomata are regulated by adjacent cells called *guard cells,* which allow sufficient air (with its carbon dioxide) to enter the plant while preventing excess water loss through evaporation.

Let us trace the one-way flow of water from soil to roots to leaves to air and see how the physical properties of water and plant tissues—including stomata—make this life-sustaining movement possible.

How Roots Draw Water from the Soil

A garden plant such as a tomato absorbs water from the soil into the roots. Soil is a combination of organic matter and weathered particles of the earth's crust. A film of water and dissolved minerals coats these soil particles, and air fills many of the larger spaces between them. Hundreds of thousands of root hairs (see Figure 30.14) as well as the fungus-root extensions called mycorrhizae (see Figure 17.5) project from the cells of lateral roots into the spaces in soil. These projections have a huge combined surface area and can take up water from the minute reservoirs all around them.

Despite the tremendous importance of water to a plant's survival, water enters roots by passive diffusion. Water diffuses into root hairs and then passes through the root cortex. Some moves through the cytoplasm of root cells, but most passes within the cell walls (Figure 32.4). When it reaches the endodermis (the cell layer that separates the root cortex from the central vascular cylinder), water can no longer follow the cell wall route because of the Casparian strips. These waxy belts surround each endodermal cell and act like gaskets that prevent water from flowing in the spaces between cells, forcing it to pass through the cytoplasm of endodermal cells before it enters the hollow water-conducting tubes of the xylem (review Figure 30.15). Some of the water enters the root by diffusion, but some also enters by bulk flow (review Figure 4.20), replacing water that leaves the xylem in the roots and moves toward the stems and leaves. Passive diffusion and bulk flow can explain how minerals and water enter the root. How, then, do these substances reach the rest of the plant?

Water Transport: Root to Leaf

Simple diffusion can explain how water moves into a plant's roots and enters the xylem vessels. But can this mechanism also explain how water reaches the top of a California redwood, the tallest living thing, more than 100 m (328 ft) above the forest floor?

Plant scientists have proposed four possible mechanisms for water transport: (1) Perhaps the water is pushed up from the roots; (2) maybe pumps in the xylem convey water upward like a bucket brigade; (3) perhaps capillary action in the xylem accounts for the upward rise; or (4) maybe water is pulled upward by the leaves. But which is correct?

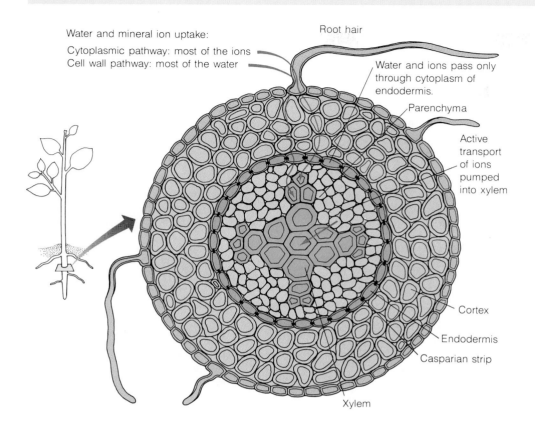

Water and mineral ion uptake:

Cytoplasmic pathway: most of the ions
Cell wall pathway: most of the water

Root hair

Water and ions pass only
through cytoplasm of
endodermis.

Parenchyma

Active
transport
of ions
pumped
into xylem

Cortex

Endodermis

Casparian strip

Xylem

FIGURE 32.4 ■ **How Water Enters Root Hairs and Crosses into the Vascular Cylinder.** Water and mineral nutrients (blue arrows) enter root hairs and move either through the cytoplasm of root cells or within the cell walls of root cells. Most of the water flows in the cell walls, while most of the ions flow through the cell cytoplasm. Materials reaching the endodermis via the cell walls are blocked from entry into the root's center by the Casparian strip. Materials must pass through the cytoplasm of endodermal cells. Once in the xylem, water and nutrients can move up to the rest of the plant. Parenchyma cells load ions into the xylem by active transport.

Root Pressure, Xylem Pumps, and Capillary Action

Farmers have long noticed that in the early morning, the leaves of grass, tomatoes, strawberries, and numerous other plants are sometimes rimmed with tiny droplets of water, the result of a process called *guttation* (Figure 32.5). Guttation occurs when the soil is nearly saturated with water and the leaves are not losing much through evaporation, and it is based on *root pressure,* which in turn is a result of an osmotic gradient. Membrane proteins in the root's parenchyma cells expend energy pumping ions into the xylem; water then follows osmotically. This tends to increase the volume of fluid in the xylem, and this fluid must move somewhere. It cannot flow back between the endodermal cells because the waxy Casparian strip prevents both the solutes and the water from leaking back out of the vascular cylinder (see Figure 32.4). The fluid can move up the xylem, however, where it is forced out of the leaves and accumulates if moisture does not evaporate quickly.

Interestingly, measurements of root pressure show that it is much lower than the force needed to move water to the tops of tall trees, which is about 150 pounds per square inch (psi). In addition, many plants, including pines, do not develop root pressure at all. A simple test confirms that the water in xylem pipelines is not pumped up from below: A pierced plant stem does not spew out water like a punctured garden hose. Instead, air is sucked in. This same test also disproves the idea

FIGURE 32.5 ■ **Guttation and Root Pressure: A Force from Below Helps Transport Water in Some Plants.** A strawberry leaf in early morning is rimmed with water droplets because root pressure forces water through the vascular system faster than it evaporates from the leaves. Root pressure, however, is never high enough to push water all the way to the tops of tall trees.

of xylem pumps relaying water upward, because again, a stem wounded near a "pump" would spew out water, and this does not occur.

Capillary action, the tendency of water to move upward in a thin tube, does occur in narrow xylem vessels. But studies show that water can cling to and creep up the inside of narrow tubes only to a height of about 0.5 m (19½ in.), again not nearly high enough for tall plants. Plant physiologists were left with the possibility that water is somehow pulled upward—an explanation that turned out to be correct.

Transpiration: The Life-Giving Chain of Water

Even though water droplets on tomato leaves prove that root pressure exists in some plants, and capillary action also plays a role in water transport, the major explanation for water transport in plants lies in a fascinating bit of natural engineering: evaporative pull. Tests in plants reveal that when xylem is cut, air moves into the stem and fills the space vacated as water moves up the stem away from the wound, while water moving up from the roots stops at the cut. This observation and numerous other tests led plant scientists to pose the **transpiration pull theory** (also called the **cohesion-adhesion-tension theory**) for water transport in plants. **Transpiration** is the loss of water by evaporation mainly through the stomata on stem and leaves. When stomata are open, water molecules move from a region of high concentration (inside the leaf cells) to a region of lower concentration (the air surrounding the plant) (Figure 32.6).

The physical properties of water help explain how transpiration moves fluid from roots to leaves. As Chapter 2 described, water molecules tend both to *cohere* very strongly to other water molecules and to *adhere* to unlike molecules. Within a given xylem pipeline, hydrogen bonds link water molecules to each other in a long, unbroken liquid chain (cohesion), while additional hydrogen bonding causes the water to adhere to cellulose lining xylem vessels (adhesion). When a water molecule evaporates from an open stoma, the next water molecule in line moves up and takes its place and in turn pulls along the next water molecule. As a result, the entire liquid column moves up in the xylem tube one link, and a new water molecule then moves into the roots below. Because the water molecules pull each other up from above, the entire chain is under constant tension. So long as the water column remains unbroken and solar energy causes evaporation, water will keep rising, and moisture will continue to move in the plant's vascular tissues.

Water Stress: A Break in the Chain

On a hot day, evaporation from leaves can outstrip water absorption by the roots, and tension on the chain of water molecules becomes greater and greater. Under such circumstances, the adhesive forces of water molecules pulling inward against the xylem walls can grow so strong that the diameter of a tree trunk literally shrinks. In hot, dry weather, the tension on the chain of water molecules can become so strained by the rapid upward pull that the column can simply

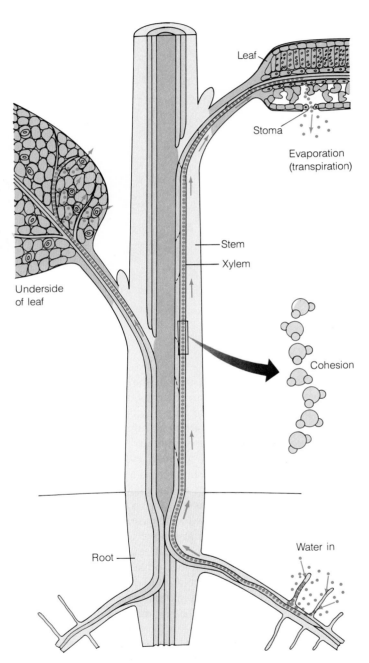

FIGURE 32.6 ■ **The Process of Transpiration.** Water and inorganic nutrients move upward through a plant as a result of transpiration, an upward pull on unbroken chains of water molecules. Water molecules remain strongly linked in chains as a result of cohesion. The evaporation of each water molecule from the leaf helps pull the entire chain upward through the stem and bring a new water molecule into the root below.

snap. In fact, plant physiologists with sensitive microphones can actually hear the snapping and popping inside plants on a hot day!

Once the water column breaks, an air bubble forms inside the xylem tube, and transpiration can no longer pull water from above. If within a short period of time the temperature drops, evaporation slows, or the soil becomes damp again, root pressure from below as well as capillary action within the narrow xylem tubes can help rejoin the ends of the broken water column. If conditions are not reversed quickly enough, however, wilting can destroy part or all of the plant. In a similar way, air enters the xylem vessels of cut flowers in a bou-

quet. But submerging the stems of a bunch of flowers and removing an additional inch or two while they are still underwater will eliminate the section containing the air bubbles and allow the water column to reestablish itself. Transpiration can then continue, and the flowers can remain crisp and beautiful for a longer time.

Stomata and the Regulation of Water Loss

The degree of transpiration is crucial to a plant's daily survival, and water loss is carefully regulated, mainly through the opening and closing of stomata. In the leaf epidermis, each small opening, or stoma, is enclosed by two kidney-shaped guard cells (Figure 32.7a). Like curved water balloons, guard cells arch away from each other when they swell with water, increasing the size of the opening between them (Figure 32.7b). When guard cells lose water and become flaccid, they slump together and close off the stoma, blocking water loss via transpiration (Figure 32.7c).

What causes guard cells to swell or deflate and thus open or close? Experiments show that when a plant begins to dry out, potassium ions are expelled from guard cells and water follows passively by osmosis. Loss of water from guard cells allows them to slump together, close the pore, and inhibit further water loss. After a rain, when the plant has plenty of water, potassium ions enter guard cells, water follows by osmosis, guard cells swell, the stoma opens, and carbon dioxide can enter the leaf, allowing photosynthesis to continue.

Besides desiccation, several other environmental agents affect guard cell function. These include carbon dioxide, light, temperature, and daily rhythms in the plant. The net result is that stomata are generally open when gas exchange can take place without threat of excess water loss, but are closed when there is a danger of dehydration.

Together, the mechanisms of transpiration and guard cell swelling help maintain a constant supply of water to plant cells—a major aspect of homeostasis. For these mechanisms to work, plant cells must absorb appropriate quantities of ions and other mineral nutrients.

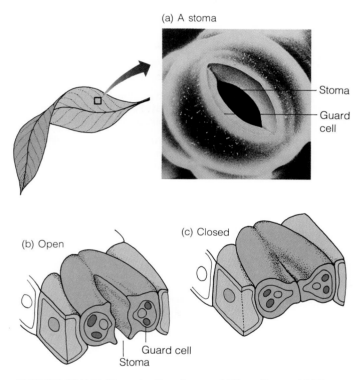

(a) A stoma

Stoma

Guard cell

(c) Closed

(b) Open

Guard cell

Stoma

FIGURE 32.7 ■ Stomata: Regulators of Water Loss. (a) The epidermis on the lower side of a kalanchoe leaf is ventilated by stomata. Each tiny opening, or stoma, is bounded by two guard cells. (b) When the guard cells swell with water, the stoma opens. Specifically, when a leaf's mesophyll cells have plenty of internal moisture and are turgid, concentrations of ethylene and calcium ions cause potassium ions to enter the guard cells, water follows by osmosis, the guard cells swell, and the stoma opens, allowing carbon dioxide to enter the leaf and photosynthesis to resume. (c) When the guard cells lose water, they slump together, closing the stoma. In detail, when mesophyll cells inside a leaf lose turgor pressure because of excess evaporation, they release the hormone abscisic acid (see Chapter 31), which affects the concentration of calcium ions in guard cells. Calcium ions then act as a second messenger, stimulating a series of events that lowers potassium ion concentrations inside guard cells. Water then exits the guard cells by osmosis, allowing the cells to grow flaccid and the stoma to close.

HOW PLANTS ABSORB NEEDED MINERAL NUTRIENTS

Around 1600, a Dutch physician performed what may have been the first quantitative experiment in biology. Jan Baptista van Helmont questioned an assertion made by Aristotle centuries earlier that a plant's body derives most of its substance from soil. To test this idea, he filled a container with 91 kg (200 lb) of dry soil and planted in it a 2.2 kg (5 lb) willow tree. For five years van Helmont watered the tree with rainwater. At the end of that time, he dug the tree out of its pot, removed the soil and dried it, weighed both soil and tree, and discovered that while the tree had gained 75 kg (165 lb), the soil had lost a mere 27.5 g (about an ounce). Thus, van Helmont succeeded in showing quite clearly that the main mass of the willow tree came not from the soil, as Aristotle had suggested, but from some other source.

Today, plant scientists know that plants synthesize organic compounds from the carbon and oxygen in the gas carbon dioxide, from the hydrogen in water, and from small amounts of minerals in soil. Fully 96 percent of a plant's dry weight is made up of carbon, oxygen, and hydrogen. The dozen or so chemical elements that make up the remaining 4 percent, however, are equally essential to plant survival.

What Nutrients Do Plants Need?

Plants require at least 16 different chemical elements, the so-called *essential elements,* which different species require in different amounts. All plants require the **macronutrients** in relatively large amounts and the **micronutrients** in much smaller amounts. By drying many types of plants, weighing the dried plant matter, and analyzing it for chemical content, plant physiologists were able to tabulate information on the nine macronutrients—carbon, oxygen, hydrogen, nitrogen, potassium, calcium, magnesium, phosphorus, and sulfur—and the seven micronutrients—iron, chlorine, manganese, boron, zinc, copper, and molybdenum (Table 32.1). Plants use both macro- and micronutrients in amounts roughly comparable to the levels animals require. And just as animals suffer vitamin and mineral deficiency diseases if deprived of essential nutrients, plants also show specific symptoms relating to specific deficiencies.

■ **Macronutrients** The macronutrients are a plant's fundamental constituents, making up most of the atoms in carbohydrates, proteins, lipids, and nucleic acids (see Chapter 2). Because we have already discussed the roles of carbon, oxygen, and hydrogen in biological molecules and reactions in Chapters 2 through 6, we will concentrate here on the other six macronutrients.

After carbon, oxygen, and hydrogen, nitrogen is the most important macronutrient, being an essential component of amino acids (and hence proteins), chlorophyll, coenzymes, and nucleic acids. Nitrogen is frequently the most important growth-limiting nutrient: The less nitrogen available, the slower the plant grows. And although gaseous nitrogen (N_2) makes up 78 percent of the earth's atmosphere, plants cannot use it directly. Instead, the nitrogen must be *fixed,* or converted from the simple gas N_2 to some other form, such as *ammonia* (NH_3) or *nitrate ions* (NO_3^-)—a process called **nitrogen fixation.** These nitrogen compounds can then be modified further and incorporated into plant proteins.

Farmers and home gardeners often add fixed nitrogen to the soil in the form of nitrate-containing commercial fertilizers. Wild-growing plants, however, do not have such a benefit and rely instead on nitrogen-fixing bacteria, including cyanobacteria, which can convert molecular nitrogen to usable

forms. Many of these microorganisms live independently in the soil and reduce nitrogen gas to ammonia. Some of the most interesting ones, though, live in the root cells of certain vascular plants (Figure 32.8). Peas, beans, alfalfa, clover, lupine, and other **legumes,** or members of the pea family,

TABLE 32.1 Plant Nutrients and Their Functions	
Nutrients (Percent of Dry Weight)*	**Location/Function**
Macronutrients	
Carbon (45.0)	In all organic compounds
Oxygen (45.0)	In most organic compounds, including all sugars and carbohydrates
Hydrogen (6.0)	In all organic compounds
Nitrogen (1.0–4.0)	In proteins, nucleic acids, chlorophyll, and coenzymes
Potassium (1.0)	Involved in activating enzymes, protein synthesis, and regulation of osmotic balance
Calcium (0.5)	In cell walls and starch-digesting enzymes; regulates cell membrane permeability
Magnesium (0.2)	In chlorophyll and many cofactors
Phosphorus (0.2)	In nucleic acids, some coenzymes, ATP, and some lipids
Sulfur (0.1)	In proteins, some lipids, and coenzyme A
Micronutrients	
Iron (0.01)	In electron transport molecules; involved in synthesis of chlorophyll
Chlorine (0.01)	Essential to photosynthesis
Manganese (0.005)	Essential to photosynthesis; activates many enzymes
Boron (0.002)	Unknown
Zinc (0.002)	In some enzymes; involved in protein synthesis
Copper (0.0006)	In chloroplasts and some enzymes
Molybdenum (<0.0001)	Essential to nitrogen fixation and assimilation
Elements Essential to Only Some Plants	
Silicon (0.25–2.0)	In the cell walls of grasses and horsetails
Sodium (Trace)	Essential to a few desert and salt-marsh species
Cobalt (Trace)	Essential for nitrogen fixation

*Concentration of nutrients expressed as percentage of dry weight in a typical plant. Actual proportions vary greatly from species to species.

(a) Roots with nodules   (b) A nodule

FIGURE 32.8 ■ Nitrogen-Fixing Bacteria in Root Nodules Help Nourish Some Plants. Nodules on the roots of an alfalfa plant (a) look granular when magnified about 400 times by a scanning electron microscope (b). Each nodule contains thousands of cells of the *Rhizobium* bacterium, capable of fixing atmospheric nitrogen to ammonia.

have swellings, or **nodules,** on their roots that house nitrogen-fixing bacteria. Certain nonleguminous plants, such as alder trees, also form associations with nitrogen-fixing bacteria.

Figure 32.9 shows how two kinds of plants, a legume, such as the bean plant on the left, and a nonlegume, such as the tomato plant on the right, get enough of the macronutrient nitrogen. When nitrogen-fixing *Rhizobium* bacteria live inside the soybean's root nodules, they can convert nitrogen from the air (trapped in the spaces between soil particles) into ammonia (NH_3). Other bacteria living freely in the soil can make the same conversion. The ammonia molecules quickly pick up hydrogen ions, forming NH_4^+, the biologically useful ammonium ion (see Figure 32.9, step 1).

If root nodules produce ammonium ions in excess of the plant's needs, they are released into the soil, where they join ammonium ions generated by ammonifying bacteria in the soil, which act on decaying organic matter (step 2). Plants such as tomatoes, which lack nodules, can take up ammonium ions in the soil from both these sources (step 3). Ammonium ions remaining in the soil can serve as an energy source for

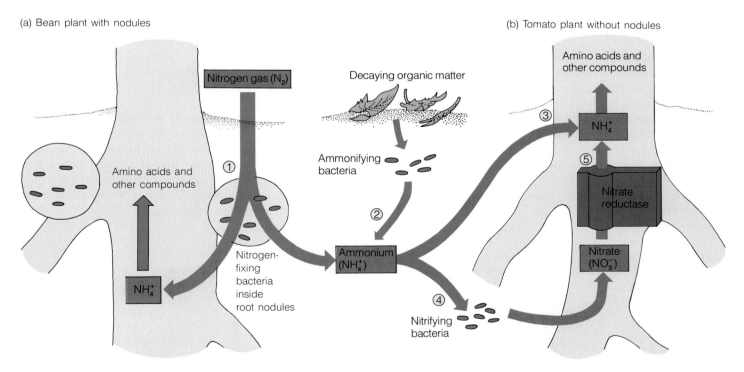

FIGURE 32.9 ■ How Plants Get Needed Nitrogen. Legumes such as the bean (a) and nonlegumes such as the tomato (b) obtain nitrogen in different ways. The bean has root nodules that contain nitrogen-fixing bacteria (1). These bacteria make ammonia from nitrogen gas. Ammonifying bacteria living freely in the soil also make ammonia (2). The tomato takes in fixed nitrogen (NH_4^+) released by the nodules of legumes or by ammonifying bacteria (3). The tomato can also take in nitrates (NO_3^-) released by nitrifying bacteria (4) and convert them via the nitrate reductase enzyme (5) to ammonium ions. Ammonium ions serve in both plants as the starting point for building amino acids and other biological molecules.

species of so-called *nitrifying bacteria,* which convert ammonium ions to nitrate molecules (NO_3^-) (step 4). A tomato plant can take up this nitrate and by means of enzymes convert it to ammonium (step 5). Once in the plant, ammonium is the starting point for the biosynthesis of amino acids and many other nitrogen-containing biological molecules.

Nitrifying bacteria, which convert ammonium to nitrate in the soil, cannot survive in the acidic, waterlogged environment of bogs. Thus, *carnivorous plants,* such as the Venus flytrap (see Figures 1.7 and 31.9) and the pitcher plant, have evolved the ability to gain fixed nitrogen from the proteins of the insects and other small animals they trap and digest.

The formation of root nodules is a classic example of symbiosis (a close association of two dissimilar organisms): The plant supplies the bacteria with high-energy carbohydrates, while the bacteria provide fixed nitrogen produced at a high energy cost. The excess ammonium ions released from nodules is a classic example of community interrelationships among plant species. For centuries, farmers have rotated crops to take advantage of such relationships. Although they may have been unaware of the microbiological basis, they observed that if they grew clover or alfalfa one year, the following year's crop of corn or wheat would grow more luxuriantly. Likewise, rice farmers have encouraged the growth of water ferns in their flooded rice paddies because cyanobacteria living symbiotically in the ferns fix atmospheric nitrogen and enrich the growth of the rice plants (review Figure 16.9).

Plants often require 4 to 40 times more nitrogen than they do the remaining five macronutrients, but these elements are still essential to normal plant growth and development. Calcium acts as an intracellular messenger that controls cell membrane permeability and thus plays a role in the opening and closing of stomata, directional growth in plant cells, phytochrome-triggered responses, and gravitropism (see Chapter 31). Calcium also has a structural role; it is an important component of *pectin,* a substance that glues adjacent cells together, and prevents young plants from making brittle cell walls. People use pectin to put the ''jell'' in jams and jellies.

Potassium regulates osmosis in plant cells such as guard cells and also helps activate enzymes, including those involved in protein synthesis. Potassium deficiency causes mottled, or burnt-edged leaves (Figure 32.10a). Magnesium is a macronutrient because atoms of the element occur in chlorophyll molecules and in the cofactors of many kinds of enzymes, while phosphorus occurs in the backbone of DNA and RNA molecules, in ATP and other high-energy compounds, and in membrane phospholipids. Tomato seedlings deficient in phosphorus have purple leaves (Figure 32.10b). Because sulfur is an important component of two amino acids, plants require it for building most proteins, as well as for manufacturing some fats and coenzymes.

■ **Micronutrients**   Plants require only small amounts of micronutrients for healthy growth (see Table 32.1 on page 652). Because plants require such tiny quantities, deficiencies of the micronutrients are rare. Iron, for example, occurs in several proteins within the energy-harvesting mitochondria and is also involved in the synthesis of chlorophyll (see Chapter 6). A tomato plant with an iron deficiency does not make enough chlorophyll to mask the yellow pigments in leaves (Figure 32.10c). A deficiency of copper (present in chloroplasts and certain enzymes) can result in severe deformation of stems, leaves, and fruits in many plant species. A deficiency of chlorine in a tomato plant will stunt the roots and fruits and wilt the entire plant. Because zinc plays a role in protein synthesis, zinc-deficient apple and peach trees become stunted and grow miniature leaves.

Soil: The Primary Source of Minerals

Soils are the source of all macro- and micronutrients beyond the carbon, oxygen, and hydrogen that plants take in from air and water. Soil is so fundamental and ubiquitous that, aside from farmers and geologists, few of us give it any real thought. But soil composition has a major influence on the kinds of plants (and indirectly, the kinds of animals) that can grow in a particular region of our planet.

(a) Potassium deficiency

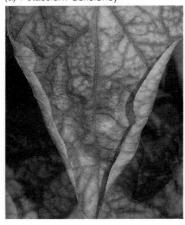

(b) Phosphorus deficiency

(c) Iron deficiency

FIGURE 32.10 ■ **Nutrient Deficiencies in Tomato Plants: Mineral Deficiencies Produce Distinctive Symptoms.** (a) A potassium deficiency causes curled leaves. (b) Phosphorus deficiency leads to purple leaves in a tomato seedling. (c) Iron deficiency renders the fine leaf veins green, but leads to yellowing in the rest of the leaf.

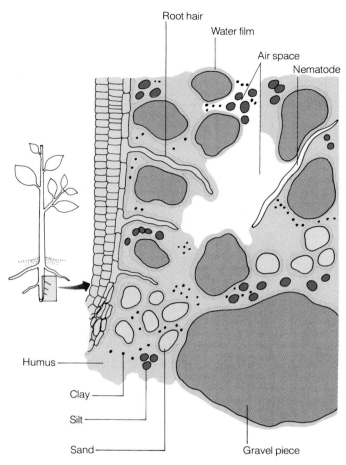

FIGURE 32.11 ■ **Soil: A Source of Air, Water, and Nutrients for Plants.** Soil has inorganic and organic particles, as well as pore spaces filled with air and lined with water. Root hairs probe those spaces and take in water and minerals that dissolve from soil particles.

Soil is a mixture of organic and inorganic material that starts with bedrock, which is relatively unbroken and unweathered rock. Over time, the action of water, wind, heat, and cold disintegrates the bedrock and produces the inorganic parts of soil—particles varying in size from coarse sand and silt to small clay particles (Figure 32.11). Bacteria, fungi, algae, lichens, and plants extract minerals from rocks, sand, silt, and clay and convert them to organic material. When the organisms die, the decomposing organic matter in the soil becomes *humus.*

Air spaces between soil particles are crucial to plant life (see Figure 32.11). In a soil with a mixture of particle sizes, the spaces contain about half water and half air, with the water forming a continuous film over the soil particles. The size of the soil particles, however, determines the soil's water-holding capacity: Soils made up primarily of coarse sand tend to hold water poorly, while soils made up mostly of clay particles hold water tenaciously. Plants such as cacti that grow in rapidly draining sandy soils absorb water quickly and store it,

while plants that grow well in clay soils have roots resistant to dense, soggy, unaerated surroundings. The best agricultural soils are deep layers of *loam* possessing a high mineral and humus content and a mixture of particle sizes.

How Nutrients Enter a Plant

Several things happen before a plant can utilize mineral nutrients from soil. Mineral molecules must first dissolve in the layer of water surrounding soil particles. Then the minerals must move into the root and be distributed to all parts of the plant. Specifically, minerals must pass through the plasma membranes of the root hairs, move through the cells of the root cortex and the endodermis cytoplasm, and be secreted into the xylem (review Figure 32.4). There, the powerful pull of transpiration can lift them (along with water) through the xylem pipelines, and the minerals can be transported to all plant organs. The active transport of ions against a gradient of concentration requires a good deal of energy. Expending this energy to procure necessary inorganic nutrients is simply a cost of staying alive.

The development of millions of fine root hairs provides an enlarged surface area for absorbing water and nutrients. In addition, the symbiotic association of roots and highly specialized mycorrhizal fungi can expand the absorptive surface area still further (review Figure 17.5). When mycorrhizal fungi infect plant roots, they often cover the roots with a spongy mantle of threadlike hyphae. The hyphae can extend outward up to 8 m (26 ft) from the root and even penetrate the roots of nearby plants. In nutrient-poor soils, mycorrhizal fungi provide their hosts with greater concentrations of inorganic nutrients as well as growth-promoting hormones, such as auxins and cytokinins. Some plants, including cultivated citrus and pine, grow far more efficiently with mycorrhizae than without (Figure 32.12).

With fungus   Without fungus

FIGURE 32.12 ■ **Mycorrhizae Allow Faster Growth Through Better Nutrient Absorption.** Even in good soil, a young loblolly pine tree lacking mycorrhizae (right) is half the size of one that has the fungus-root association (left).

Mycorrhizae occasionally interconnect unrelated plants and even allow a few odd plant species to parasitize others. Indian pipe, a ghostly white plant of the forest floor that lacks chlorophyll, has mycorrhizae on its roots that connect it to the roots of nearby trees, absorbing nutrients from the parasitized plant (Figure 32.13).

Since absorbing macro- and micronutrients requires a considerable expenditure of energy, plant roots must have a reliable energy supply. Unable to photosynthesize, root cells must receive energy compounds from the leaves. Thus, the downward transport of sugars is crucial to plant survival.

MOVING ORGANIC MOLECULES THROUGHOUT THE PLANT

Plants have an inherent distribution problem: They generate sugars in their leaves (sources), but tend to store organic nutrients in their roots and use those nutrients in their growing tips, flowers, and fruits (sinks). What's more, sources and sinks can change as the season progresses. In the tarweed, the root is a sink in the active growing season, but in the fall, the root turns from sink to source, as stored carbohydrates are transported up from the root and used to grow a tall stem, flowers,

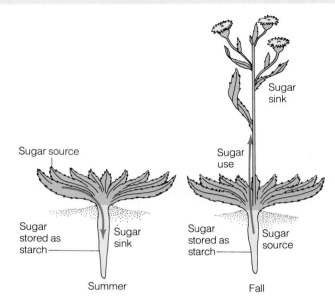

FIGURE 32.14 ■ Seasonal Sugar Transport in the Tarweed. In summer, sugars produced by photosynthesis in the tarweed's leaves are transported downward to the roots and stored as starch. In fall, the transport is reversed: Sugars are chemically cleaved from the stored starch and carried upward, fueling the biosynthesis of materials for an elongating stem, flowers, and seeds.

(a) A mycorrhizal bridge   (b) Indian pipe

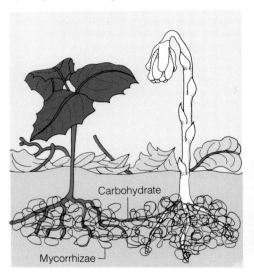

FIGURE 32.13 ■ The Fungal Connection: A Mycorrhizal Bridge Transfers Sugars from Green Plant to Indian Pipe. (a) The Indian pipe (*Monotropa uniflora*) has an obligate relationship with mycorrhizal fungi that are associated with a second plant, in this case a green, actively photosynthesizing angiosperm. The fungus forms a bridge that transfers carbohydrates from the photosynthetic plant to the Indian pipe. (b) Indian pipes pop up on the forest floor and, lacking chlorophyll, look ghostly white.

and seeds (Figure 32.14). Since plants lack a pump and circulatory system, how do they distribute photosynthetic products? The answer is translocation.

Translocation: How Sugar Moves from Source to Sink

Plant scientists devised some clever methods for studying **translocation,** the movement of nutritional materials in the phloem of plants. Phloem tubes are so delicate that puncturing them even with a fine glass probe or needle stops the flow of solutes. Tiny insects called aphids, however, can suck plant sap from phloem tubes without disrupting them, and plant physiologists found that by allowing an aphid to insert its sharp feeding tube into a phloem pipeline, then cutting away the insect's body, the phloem's contents will ooze out of the little tube, whereupon the plant sap can be studied (Figure 32.15). With this technique, they discovered that plant sap contains up to 30 percent sugars (mostly sucrose) and about 70 percent water.

Plant scientists also discovered that they could expose a plant's leaves to radioactive carbon dioxide, then trace the path of the carbon through the phloem once it was incorporated into sugars via photosynthesis. In this way, they found out that sugar flow is often downward from leaves to other plant parts and can be 10,000 times faster than normal diffusion. They therefore postulated that translocation involves a mass movement of phloem fluid based on bulk flow (see Figure 4.20).

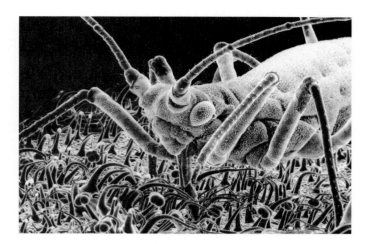

FIGURE 32.15 ■ **Aphids Tap Contents of Phloem.** In this scanning electron micrograph, aphids pierce the veins of a tomato leaf with their hypodermic syringe–like stylets.

Transport of Inorganic Substances in Xylem and Phloem

Nutrients other than carbon, oxygen, and hydrogen—including all the other nutrients listed in Table 32.1 on page 652—can be transported in either of two different ways. Some, such as calcium, move only in the xylem; they are simply carried along in the column of water that moves upward from the roots to the leaves via transpiration. These nutrients are deposited permanently wherever they leave the xylem and enter a living cell, and they cannot be redistributed to the other parts of the plant. Others, such as sulfur or phosphorus,

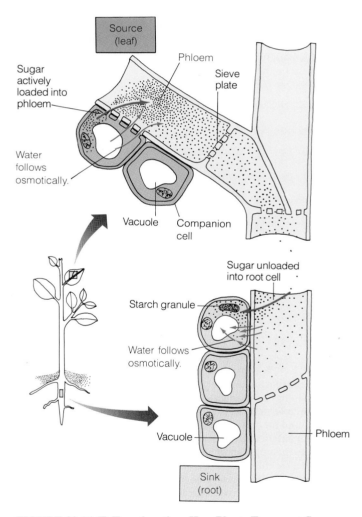

FIGURE 32.16 ■ **Translocation: How Plants Transport Sugars.** In leaves, companion cells load sugars into phloem sieve tube elements, water follows osmotically, and sugar solution is forced through the sieve plates and down a pressure gradient. This gradient is formed because root cells with lower concentrations of sugars and amino acids take in the nutrients, making the solution in the phloem tubes more dilute. In a sink area, as water leaves the phloem and reenters the xylem, the turgor pressure in nearby phloem cells also drops.

Currently, the best hypothesis for phloem transport is the **mass flow theory** (Figure 32.16). It suggests that sugars produced in source regions, such as photosynthesizing leaves, are loaded into the phloem's sieve tube elements by the companion cells (see Chapter 30). This energy-costly active transport increases the solute concentration in the phloem, and water follows by osmosis from the nearby xylem cells. The influx of water plumps up the phloem cells, increasing their turgor pressure and forcing the sugary solution out of the sieve plates at the ends of each sieve tube member and away from the leaf.

Meanwhile, root cells are removing organic solutes from the phloem, and the solute content becomes so low that the osmotic tendency is reversed. Now water flows out of the phloem tubes and back into the xylem tubes, where it is carried upward again by transpirational pull. Ordinary water pressure and the loading activities of companion cells are thus behind the movement of sugars, amino acids, and a few mineral ions, from sources to sinks.

In the early spring, a similar mechanism probably causes plant sap to rise in the xylem of biennials and perennials. At that time, the roots begin breaking down stored starch into sugar and loading it into the phloem's sieve tube members. Water from the xylem follows, and as the roots bring in more water from the soil, enough pressure is created to push the sap up the phloem into the trunk of the tree. New Englanders carefully tap into the phloem of maple trees in early spring, drain off some of the sweet maple sap, and boil it down to make maple syrup.

In a sense, sugar translocates itself, since its production and use create osmotic pressure and mass flow. Other nutrients, however, are moved passively by the flow of fluid in either the xylem or the phloem.

can be redistributed, via the phloem, to new tissues as needed, carried along passively with the sugars and other phloem contents. Figure 32.17 shows an experiment that proves that sulfur goes first from roots to newly forming leaves, then is redistributed to still newer leaves as they form. Plant scientists inferred that sulfur moves first in the xylem, then in the phloem to growing leaves.

These two different modes of transport explain why each mineral deficiency affects a plant in a different way. The symptoms of calcium deficiency, for example, always appear first in new leaves. The old leaves contain sufficient quantities of calcium, but the mineral cannot move out of these leaves— even if they are dying—and be transferred in the xylem to the new leaves. In contrast, the symptoms of phosphorus deficiency always appear first in old leaves; as the plant's environment becomes deficient in the element, the phosphorus in older leaves is mobilized and moved through the phloem to areas of new growth. Thus, in some plants, the old leaves will have the characteristic yellow veins of phosphorus deficiency long before the new leaves show it.

ENGINEERING USEFUL PLANTS

With their knowledge of plant anatomy, development, genetics, and the role of water and nutrients in plant survival, modern plant scientists are approaching—and entering—a new frontier: biological engineering of plant species with dramatically different traits.

Because so many of the world's people are hungry, and because remaining unplanted lands are only marginally useful, plant breeders are constantly working to increase the yield and nutritional content of crops grown under increasingly adverse conditions. They are attempting to combine the hardiness of wild plants that grow in dry, salty, cold, or barren areas with the high-yield traits of the best agricultural crops. The ideal is to develop, for example, wheat, rice, corn, or tomato plants with the drought resistance of cactus, the salt tolerance of marsh grass, the nitrogen-fixing ability of legumes, and the ability to resist insect pests, diseases, and herbicides. Let us explore techniques and early achievements of this new field.

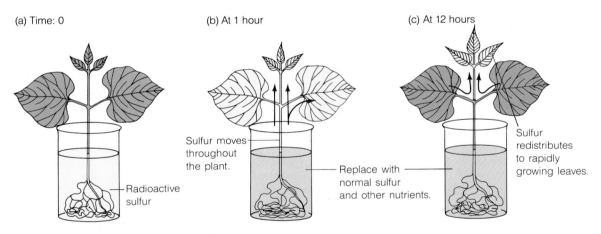

(a) Time: 0

(b) At 1 hour

(c) At 12 hours

Sulfur moves throughout the plant.

Replace with normal sulfur and other nutrients.

Radioactive sulfur

Sulfur redistributes to rapidly growing leaves.

FIGURE 32.17 ■ Movement of Labeled Sulfate Reveals Transport of an Inorganic Ion. In a classic plant physiology experiment, a researcher allowed the roots of a bean plant to absorb a radioactive sulfur compound for an hour (a), then switched the plant to a solution of normal sulfur and other nutrients (b). He then photographed the plant at various times with a technique that reveals the location of the radioactive sulfur (autoradiography). After an hour or so, the radioactive sulfur had been carried throughout both older, larger leaves and newer, smaller ones, presumably via the xylem (c). After 12 hours, however, the sulfur (now presumably incorporated into amino acids and proteins) had been transported, by the phloem, from old to new leaves (d, e). The researcher inferred that xylem initially carried water and nutrients upward to all parts of the plant; then later, phloem redistributed the sulfur, now incorporated in biological molecules, to the rapidly growing parts of the plant.

(c) Autoradiograph at 1 hour

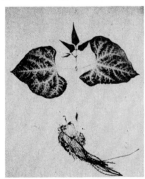

(e) Autoradiograph at 12 hours

(a) Punching out leaf disks

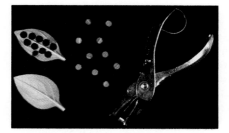

(b) Callus growing from leaf disk   (c) Regenerated plants

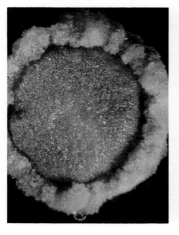 

FIGURE 32.18 ■ Plant Tissue Culture: Growing Whole Plants from Individual Cells or Tissues. **(a) Researchers remove a bit of tissue from a plant (in this case a petunia) with desirable traits by punching out small disks of leaf tissue. (b) Next, they culture clumps of callus. (c) Then they separate individual cells from the clump and induce some of the cells to grow into new petunia plants.**

New Techniques of Artificial Selection

The choice foods we have today—the sweet pears, high-yielding grains, juicy strawberries, and others—are the result of farmers carefully selecting seeds from plants with the best individual traits and planting the seeds for the next genera-tion. New techniques such as tissue culture, however, are speeding up the process of artificial selection.

Plant breeders are often eager to select traits displayed by a particular individual, but these traits can disappear owing to genetic recombination during normal sexual reproduction. **Tissue culture,** or growing new identical plantlets from somatic (body) cells, avoids the problem of genetic recombi-nation. Researchers begin the tissue culture process by punch-ing out a bit of plant tissue from some part of a plant—a leaf or stem, for example (Figure 32.18a)—and placing it in a nutritious culture medium in a warm, well-lighted room. Soon the cells of the tissue lose the distinctive characters that mark them as leaf parenchyma or stem parenchyma, for example, and form an unorganized mass of cells called a *callus* (Figure 32.18b). The plant breeder then breaks up the callus and

places bits of it in a new culture medium with hormones that promote cell differentiation. There, some of the callus cells grow and often develop into a tiny plant (Figure 32.18c). The plants can be raised and studied directly, or the embryos can be grown into plants or packaged as "synthetic seeds" for later use. Either way, since all these plants are derived from the somatic cells of the parent plant, theoretically they all have identical genes and provide numerous copies of the desired original. Some plant nurseries already propagate cer-tain commercial crops using tissue culture—for example, strawberry plants that are free of viral infections.

For unknown reasons, the DNA of many plants grown from the initial tissue-cultured callus contains mutations. Wheat, carrots, celery, and tomatoes produce all sorts of genetic variants in tissue culture—variants that would require years to find by traditional selection techniques. This ten-dency of tissue-cultured plants to display new and heritable traits is called *somaclonal variation* (*soma* referring to the body cells of the plant, and *clonal* because each variant pop-ulation arises from a single initial mutant cell). From such an array of varying cells, breeders select those that express a desired trait. For example, if biologists could develop crop plants that were resistant to salty soils, farmers could exploit many areas not currently open to agriculture. To search for such plants, researchers can treat tissue culture cells with salt. Those cells with genes for salt tolerance live; the rest die. In some species of plants, the surviving salt-tolerant cells can then be treated with hormones so that they grow and differ-entiate into whole plants, from which the experimenters can collect seeds.

In one of the earliest somaclonal variation experiments, researchers grew 230 plants from the cells of a single tomato plant. Among the variants, they found 13 separate mutations, several of which turned out to be commercially useful. Some mutations produced novel bright orange and yellow tomatoes (Figure 32.19), while others affected flower color, chlorophyll

FIGURE 32.19 ■ Retooling the Tomato. **Researchers engi-neered new orange and yellow tomato varieties using somaclonal variation.**

BOX 32.1   PERSONAL IMPACT

RETOOLING THE TASTELESS TOMATO

Sound familiar? You choose the reddest, plumpest-looking tomato you can find in the produce section, take it home, slice into it, and yuck! The inside is light pink, it has no juice, and it tastes like a desk blotter. Depending on your age and appetite, you might mumble something about the good old days when tomatoes had flavor, then toss the bland wedges into a salad anyway, as filler.

Over the past few decades, plant breeders have developed tomato varieties with commercially useful traits like tougher skin for easier harvesting and shipping. Traditional breeding and selection are not, however, the main reasons for the tasteless tomato. In their vine-ripened state, these popular fruits cannot be transported long distances without bruising or splitting. So growers pick tomatoes still green, ship the hard-walled spheres, then gas them with the plant hormone ethylene (see Chapter 31) to induce rapid ripening and reddening. This results in undented produce that is often mealy and insipid as well.

In 1988, plant genetic engineers at Calgene, Inc., in Davis, California, and the University of Nottingham in England set out to solve this problem. They knew that as a tomato ripens on the vine, an enzyme called polygalacturonase breaks down cell walls and allows the fruit to soften. The researchers reasoned that if they could slow down the production of this enzyme, then the fruits would soften more slowly and could be left longer to ripen on the vine and develop juice and flavor.

Team members located the gene for polygalacturonase, as well as the sequence that regulates its activity, and cloned the gene. Then—and this is the novel part—they spliced the gene to the control sequence *backward*. Finally, they inserted this so-called *antisense gene* into a normal tomato chromosome, which also contains the gene in its forward orientation. Apparently, the mRNA produced by the backward, antisense gene somehow counteracts the functioning of the normal gene for polygalacturonase, and the tomato fruit makes only 10 percent as much of the softening enzyme.

Calgene scientists have begun field testing the engineered tomatoes to see whether the fruits with less enzyme do soften more slowly and ship more easily, and whether the gene for the altered enzyme is stably transmitted from generation to generation. If the tests are positive and the engineered fruits reach the marketplace, people may soon be shopping, slicing, and saying, "Now this is a *good* tomato. But I remember when these things tasted like packing crates."

---

production, fruit ripening, and the ratio of juice to solids. Tomatoes with unusually high solid content are valuable to canners of tomato paste. Box 32.1 describes attempts to engineer a better tomato.

Selecting new plant lines from cell cultures has certain limitations. With present technology, biologists cannot grow some kinds of plants from single cells. Also, both traditional selection techniques and modern tissue culture techniques require that some cells in the population already have genes for the desired traits. To select a salt-tolerant tomato from somaclonal variants, for example, requires that some tomato cells already be somewhat salt-tolerant. What happens, though, if tomato cells invariably lack genes for salt tolerance? The answer is that plant researchers can use the same kind of gene transfer techniques that allowed geneticists to develop giant mice (see Chapter 11).

Genetic Engineering in Plants

Genetic engineering offers a pathway to varieties with more profound alterations than traditional techniques are ever likely to produce. In a single generation, a plant geneticist can introduce into a species the extreme expression of a trait or a completely novel trait that might have taken hundreds of generations to develop through standard breeding approaches. Work is now under way on several fronts, and researchers are already reporting initial successes. Belgian plant geneticists,

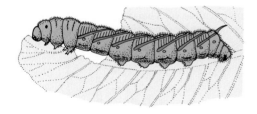

Engineered to resist hornworm   Normal

FIGURE 32.20 ■ **Deadly Dinner: Tobacco Plants Engineered with a Toxin Gene Kill Tobacco Hornworms Before the Worm Kills Them.** An engineered plant (left) containing toxin genes from the *Bacillus thuringiensis* bacterium will poison its predators, while a normal plant lacking the gene (right) merely provides a meal for the hungry hornworms.

for example, have transferred genes from a bacterium into tobacco and tomato plants. The bacterium, *Bacillus thuringiensis,* makes a protein that kills only insects. By removing the gene from the bacterium and inserting it into the plants, researchers conferred upon those plants an immunity to attack by voracious hornworms (Figure 32.20). Similar modifications of important crops are soon to follow. And plant researchers have outfitted petunia plants with a gene for resistance to the widely used herbicide glyphosate. Workers can now spray a field of the resistant plants with the herbicide, and weeds will die off but the petunia crop will remain unharmed.

Ultimately, genetic engineers hope to develop varieties of wheat, corn, and rice that, like meat and dairy products, contain high quantities of all essential amino acids. They also hope to engineer nonleguminous crop species that can form root nodules filled with nitrogen-fixing bacteria, plus dozens of other plant species with increased resistance to insects, diseases, and herbicides, increased nutritional value, and increased tolerance to harsh environments.

Each of the three ways to produce new plant varieties—standard plant breeding, artificial selection from tissue cultures, and genetic engineering—has advantages and disadvantages. Traditional plant breeding is slow, but is well understood, predictable, and controllable. Tissue culturing is much faster, but traits selected for in tissue culture often bring along

undesirable "riders" such as sourness or susceptibility to disease. Genetic engineering has unlimited potential for combining traits that never occur together in nature, but only traits controlled by single genes or very small groups of genes can be added or subtracted. Clearly, the modern tools of plant breeders supplement—but do not replace—traditional breeding techniques.

CONNECTIONS

The tomato plant may look static from the outside, but it is an undeniably dynamic organism: Water and nutrients move the length of the plant from roots to stems and leaves and sometimes back again. Leaves track the sun's movement through the sky. Flowers emerge from buds. Fruits grow from flower bases, becoming plump and red and attracting animals, which disperse the seeds. Toxic compounds form in the leaves and stems and fend off potential predators. And plant hormones coordinate all these activities in response to environmental cues, including light, temperature, day length, and water availability. The next section, which deals with evolution and ecology, fixes our attention more directly on the dynamic interactions of plants, animals, and other organisms with each other and with the physical world.

Highlights in Review

1. In plants, survival depends on absorbing water and inorganic nutrients, as well as on generating sugars through photosynthesis, then transporting these organic nutrients to actively metabolizing cells throughout the plant.

2. Water and minerals enter root hairs, move through or between parenchyma cells of the root cortex, pass through the cell membranes and cytoplasms of endodermal cells, then enter the vascular cylinder. From there, xylem cells pipe the materials upward.

3. The vascular cylinder can build up a low level of root pressure (as evidenced by guttation), but not enough to push water and minerals to the tops of tall trees.

4. Water inside xylem is under tension. Evaporation of water molecules from the leaves draws water molecules out of nearby cells, which is replaced by water molecules from the xylem. Because of the tendency of water molecules to form hydrogen bonds, a chain of water molecules from leaves to roots is pulled up a bit for each molecule that evaporates from a leaf. This is the process of transpiration.

5. Plants lose enormous quantities of water owing to transpiration through tiny holes called stomata. Stomata open when the plant uses additional carbon dioxide but has sufficient water, and they close when the plant has too little water but sufficient amounts of carbon dioxide.

6. Each stoma is flanked by two guard cells whose shape is controlled by an osmotic mechanism; when turgid, these cells pull apart and open the stoma; when flaccid, they slump together and close the stoma.

7. Plants are made up mostly of carbon, oxygen, and hydrogen from carbon dioxide and water. Plants depend on the soil, however, for at least 13 other essential elements, including macronutrients, needed in relatively large quantity, and micronutrients, needed in relatively small quantity.

8. The main growth-limiting element is nitrogen, but plants cannot absorb its plentiful gaseous form directly. Instead, they must take in fixed nitrogen. Legumes absorb ammonium ions from the *Rhizobium* bacteria in their root nodules. Nonlegumes absorb excess ammonium ions released by a legume's roots, as well as ammonium and nitrates fixed by soil bacteria. In both leguminous and nonleguminous plants, ammonium ions are the starting point for biosynthesis of amino acids and other useful biological molecules.

9. Other macronutrients include potassium, calcium, magnesium, phosphorus, and sulfur, and deficiencies of each have characteristic symptoms in plants.

10. Although micronutrient deficiencies are rare, plants deprived of iron, chlorine, manganese, boron, zinc, copper, molybdenum, and other elements also show distinctive symptoms.

11. Soil is composed of a combination of inorganic particles (sand, silt, and/or clay) and organic particles from the decomposing remains of dead organisms (humus).

12. The spaces between the particles of soil are filled with water and air and supply water and oxygen to plant roots.

13. Plants extract nutrients from the soil, but eventually, the nutrients are returned to the soil through nutrient cycling.

14. Plants absorb nutrients from the soil through their own root hairs or with the aid of mycorrhizal fungi, which live on and in the root cells.

15. When leaves photosynthesize, they are a source of sugar, while the developing and growing parts of a plant are sinks that use this sugar. Roots are sinks when they absorb sugar from phloem for their own metabolic needs or when they store it as starch. But roots are sources when they release the sugars from the stored starch to fuel activities in other parts of the plant.

16. Sugars and other organic molecules are translocated throughout the plant. In the leaves, companion cells load sugar into phloem's sieve tube members, and water is pulled into the phloem osmotically. The resulting pressure forms a gradient from source areas—where cells actively load sugars into the phloem—to sink areas—where cells unload sugars from the phloem. This pressure gradient pushes the solution of sugar and water from the source to the sink.

17. Inorganic nutrients can be transported in two ways. Some elements, like calcium, are carried in the xylem to a final destination, where they remain. Others, such as sulfur and phosphorus, move passively through cell membranes and can be redistributed via the phloem.

18. To improve crops, modern plant scientists use traditional plant-breeding techniques, tissue culture and cloning of somatic cells, and the insertion or deletion of specific genes through genetic engineering techniques.

Key Terms

cohesion-adhesion-tension theory, 650	nodule, 653
legume, 653	soil, 655
macronutrient, 652	tissue culture, 659
mass flow theory, 657	translocation, 656
micronutrient, 652	transpiration, 650
nitrogen fixation, 652	transpiration pull theory, 650

Study Questions

Review What You Have Learned

1. What is the function of root hairs? What is the role of the endodermis?
2. How does root pressure cause water to rise in a plant?
3. Explain the transpiration pull theory.
4. What factors affect the opening and closing of stomata?
5. How does a legume get needed nitrogen? How does a nonlegume get nitrogen?
6. Explain the role of each of these macronutrients in a plant: calcium, potassium, magnesium, phosphorus, and sulfur.
7. List four plant micronutrients, and give the major roles of each.

8. How does soil form?
9. Trace the pathway by which minerals enter a plant.
10. What is a source of sugars in a plant? What is a sink of sugars in a plant?
11. Explain how a biologist can use tissue culture to grow genetically identical plants.
12. What are somaclonal variations? How can a plant scientist use them?
13. Describe two accomplishments of genetic engineering in plants.

Apply What You Have Learned

1. A farmer grows a crop of clover and plows it into the ground before planting a crop of corn. What was his purpose in growing clover, a crop he didn't harvest?
2. A Vermont family makes maple syrup early in the spring. What raw material do they gather, and from where?
3. A plant breeder tries using tissue culture to develop salt-tolerant tomato plants. What might cause the experiments to fail?
4. A beginning gardener decides he wants to have the biggest and greenest vegetables possible. Midway through the growing season, he applies chemical fertilizer at a rate ten times greater than the manufacturer recommends. Instead of growing faster, his vegetables wither and die. Why?
5. You own a commercial laboratory that specializes in plant tissue culture. A client asks you to develop a nitrogen-fixing strain of tomato. Should you accept the job? Defend your answer.

For Further Reading

CHAPPIN, F. S., III. "Integrated Responses of Plants to Stress." *BioScience* 41 (1991): 29–36.

CROSSON, P., and N. ROSENBERG. "Strategies for Agriculture." *Scientific American,* September 1989, pp. 128–135.

FISCHHOFF, D. A., K. S. BOWDISH, F. J. PERLAK, P. G. MARRONE, S. M. McCORMICK, J. G. NIEDERMEYER, D. A. DEAN, K. KUSANO-KRETZMER, E. J. MAYER, D. E. ROCHESTER, S. G. ROGERS, and R. T. FRALEY. "Insect Tolerant Transgenic Tomato Plants." *Bio/Technology* 5 (1987): 807–813.

GASSER, C. S., and R. FRALEY. "Genetically Engineering Plants for Crop Improvements." *Science* 244 (1989): 1293–1299.

HITZ, W. D., and R. T. GIAQUINTA. "Sucrose Transport in Plants." *BioEssays* 6 (1987): 217–221.

MOFFAT, A. "Nitrogen-Fixing Bacteria Find New Partners." *Science* 250 (1990): 910–912.

RAVEN, P. H., R. F. EVERT, and H. CURTIS. *Biology of Plants.* 4th ed. New York: Worth, 1986.

ROBERTS, L. "Genetic Engineers Build a Better Tomato." *Science* 241 (1988): 1290.

SOKOLOV, R. "Square, Gassed Tomatoes and Other Modern Myths." *Natural History,* July 1989, pp. 70–72.

STRANGE, C. "Cereal Progress via Biotechnology." *BioScience* 40 (1990): 5–9.

PART SIX

INTERACTIONS: ORGANISMS AND ENVIRONMENT

The Genetic Basis for Evolution

Cheetahs: Sprinting Toward Extinction

Cheetahs of the African savanna are efficient predators that climb to an elevated lookout, such as a rock or fallen tree limb, then search for prey. When a young zebra or Thompson's gazelle strays too close to the predator's perch, the cheetah noiselessly descends and approaches the prey, its buff and brown-spotted coat camouflaged in the long, uneven grassland shadows. The cheetah crouches, then springs toward the startled animal. In just 1 second and with two bounding strides, the cheetah is running 72 km per hour (45 mph), closely pursuing the terrified antelope as it cleaves a zigzag path through the dry grass (Figure 33.1). Chances are good that the feline hunter will soon trip the prey with a powerful swipe of a forepaw, then strangle it with a killing bite to the throat.

Compared to the cheetah, the world's fastest-running animal, even a Corvette seems sluggish. While the lithe cat can be running at its top speed of 116 km per hour (72 mph) within seconds, the Corvette takes three times as long to reach that same speed. The car would eventually overtake a sprinting cheetah, of course, but the animal's unique anatomy allows it to hunt down the fleetest African antelopes, such as impalas and gazelles.

How did the cheetah accumulate the genes necessary for such speed? The answer to that question concerns the chapter's central topic: the genetic basis of evolution. To begin with, virtually every cheetah feature is finely honed toward the single task of sprinting: The head is small and domed; the lungs, blood vessels, and particularly the heart are huge for an animal of its size; the legs are long and slender; the claws are semiextended like spikes on running shoes; the limb muscles are arranged in "high-gear" attachments that allow a large

FIGURE 33.1 ■ **The World's Fastest Runner: Handsome but Endangered.** The cheetah (*Acinonyx jubatus*) evolved with a slender body, camouflage markings, and a high-speed gait. Its claws are semiextended like spikes on running shoes, and it remains airborne half the time during a high-speed chase—here, in pursuit of a Thompson's gazelle.

amount of limb movement from a small amount of muscle contraction; and the backbone is capable of such extreme arching and flexing that the cat can bound in strides up to 7 m (23 ft) long.

Surprisingly, despite the cheetah's near perfection as a sprinter and its enormous success as a hunter, the species is dangerously near to extinction. While the animals once enjoyed a worldwide distribution, they are now limited to a few sites in southern and eastern Africa and to a total population of less than 20,000. What's more, among the remaining cheetahs, every individual is extremely similar to every other, and the result is the kind of inbred depression of genetic vitality one would expect if humans married their cousins for generation after generation.

This lack of genetic diversity has had numerous consequences: Cheetahs are highly susceptible to diseases like cat distemper; male cheetahs produce defective sperm with kinked tails, resulting in a lowering of the species' birthrate; those cubs that do survive tend to be sickly; and while the adults can run like the wind for a mile or less and take down fleet prey, they become so exhausted in the process that nearly any competing predator (lion, leopard, hyena, or even a person with a stick) can drive them off and steal their catch.

The story of how cheetahs became such beautifully adapted hunters and yet have reached the current point of near extinction serves as an excellent case study for the genetic basis of evolution. Evidence suggests that about 7 million years ago, one line of big cats began to diverge into the major feline groups we see today. Over time, mutations and genetic recombinations arose spontaneously in this ancestral lineage, creating new genotypes—and in turn new phenotypes—for anatomical features, biochemistry, and behavior. Subsets of these variations must have been advantageous in different ways, leading to the differential survival and reproduction of cats that were extremely powerful (lions and tigers), that could climb and live in trees (leopards), or that were rapid sprinters (cheetahs). As this divergence progressed, there emerged several distinct *species*—groups of organisms that in nature interbreed and produce healthy, fertile offspring only within the group. *Acinonyx jubatus,* the cheetah, was one such emergent species, and fossil evidence shows that these aerodynamic "running machines" thrived for several million years.

Then, however, perhaps in the last 10,000 years, the cheetah underwent a genetic catastrophe. Some combination of disease, drought, overhunting by people, or other factors caused the near extinction of the species, reducing the remaining numbers of cheetahs worldwide to a tiny group of survivors. New generations descended from these few survivors, and cheetah populations built up once again. All the living descendants, however, were nearly identical genetically. Over time, inbreeding among these "cousins of cousins" led to pairing of like alleles (homozygosity) for negative characteristics and to the current small populations of cheetahs facing extinction.

In this chapter, as we study the role of genetics in the evolution of cheetahs and other living organisms, we will explore four unifying themes. First, a group of interbreeding organisms evolves as its constellation of genes changes over the generations. This chapter explores the mechanisms that cause these genetic changes to occur. Second, most populations possess a remarkable amount of genetic diversity. Some of these diverse genes directly affect the organisms' reproductive ability, while others seem to have little impact on it. Cheetahs once had considerable genetic diversity but today are genetic paupers. Third, genetic diversity is a precondition for evolution: A population lacking a large number of different alleles evolves little or not at all. Modern cheetahs have little opportunity for evolution because of their paucity of genetic variability. Finally, evolution occurs because certain individuals are selected by environmental pressures or chance events to be parents for the next generation. The cheetah's environment necessitates rapid running to catch prey, and historically, cheetahs that ran faster made better parents and left more offspring. However, some past catastrophe eliminated most of the cheetahs, and the few remaining ones, left by chance to be parents, happened to have a few harmful alleles.

Our consideration of the genetic basis of evolution will answer these questions:

- What produces and maintains genetic variation, the raw material of evolution?
- What are the agents of evolution, the factors causing changes in gene frequencies?
- How does natural selection work?
- How do new species arise?
- What evidence supports the evolutionary history of life on earth?
- What trends have occurred in life's evolution?

GENETIC VARIATION: THE RAW MATERIAL OF EVOLUTION

Over several centuries, dog breeders have produced canines as different as Chihuahuas and Great Danes simply by selecting for desired traits among the tremendous amount of heritable variation that existed in the original dogs (Figure 33.2). Charles Darwin, in fact, argued that the numerous varieties of domestic plants and animals prove that most species have extensive amounts of genetic variation. But Darwin had to base his theories on simple observations of organisms and on inferences that their variations were somehow heritable. Although Gregor Mendel was working out the principles of genetics and the mechanisms of inheritance about the same time Darwin was studying evolution, Darwin knew nothing of Mendel's genetic theory (see Chapter 8).

Evolutionary biologists have now combined Darwin's original theory with modern genetics, yielding the *synthetic theory of evolution,* or *modern synthesis,* which explains evolution in genetic terms. A set of principles called **population genetics** clarifies what happens at the genetic level as populations evolve. A population is a group of interacting individuals of the same species that inhabit a defined area. While the sum total of alleles for all the genes carried on an individual's chromosomes is its genotype (see Chapter 8), the sum total of alleles carried in all members of the population is the **gene pool.** Population geneticists study how the frequencies of various alleles in gene pools change as generations pass. Gene pools would never change, however, and evolution could not occur if it were not for genetic variation.

What Is the Extent of Genetic Variation?

Any given gene can exist in several alternative forms, or alleles. For example, the rare albino allele of the human skin color gene (review Table 12.1 on page 248) is evidence of such variation. In heterozygous individuals, one allele is often dominant, the other recessive, and only the dominant allele is expressed in the organism's phenotype. This means that a large degree of genetic variation is hidden, transmitted from generation to generation through recessive genes.

In the early 1960s, molecular biologists developed simple yet sensitive techniques (including gel electrophoresis; see Chapter 10) that enabled researchers to distinguish between proteins that differ by as little as a single amino acid and hence alleles that differ by as little as a single base pair. At first, biologists were shocked to find a huge amount of genetic variation in populations. Each person, for example, has different alleles (is heterozygous) at about 10 percent of the gene sites on his or her chromosomes (Figure 33.3).

Evolutionary biologists have probed genetic variation in human racial groups and found only a small correlation between racial and genetic differences. People have about 100,000 genes, and since each person is heterozygous at about 10 percent, each person has two different alleles for 10,000 genes on average. When biologists compare the distribution of these heterozygous genes *within* a given racial group with the distribution *between* racial groups, they find that nearly all genetic variation (90 percent) exists between individuals of the same race for traits like height, weight, and the amino acid sequences of individual proteins. In contrast, only a small amount (10 percent) of total genetic variation accounts for all the racial differences between European, African, Indian, East Asian, New World, and Oceanic peoples. In other words, it would be easy to find individuals from different races who— except for superficial traits like skin color or facial features— are more genetically similar to you than are many individuals of your own race.

Cheetahs today have one of the lowest rates of genetic diversity among mammals: They are heterozygous for only 0.07 percent of their genes (see Figure 33.3). Thus, they have about 100 times less genetic variability than people and less even than certain highly inbred strains of livestock or laboratory mice. Partly because of this genetic similarity, they face extinction. A single new virus, which the cheetah's immune system lacks sufficiently varied alleles to combat, could wipe

FIGURE 33.2 ■ **Creatures Great and Small: Diversity in Domestic Dogs.** Although a Great Dane weighs 100 times more than a Chihuahua, both are members of the same species, *Canis familiaris,* and descended from the ancient *Tomarctus,* a relative of the wolf.

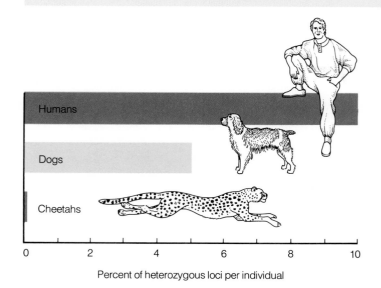

FIGURE 33.3 ■ People and Dogs Have Far More Genetic Variation Than Do Cheetahs. People have about 100,000 genes, and each person is heterozygous (has two different alleles) at 10 percent, or 10,000, of those genes. Dogs are heterozygous at about 5 percent of their genes. By contrast, cheetahs are heterozygous at only about 0.07 percent of their genes.

out their entire population. With such a small amount of genetic variation, cheetahs have little chance for evolutionary change.

Before we can understand how the cheetah came to have such a small amount of genetic variation, we need to consider the sources of genetic variation and what maintains it in populations.

What Are the Sources of Genetic Variation?

Genetic variation can increase in a population by single-gene mutations, gene duplication, exon shuffling, and recombination. The first source, *single-gene mutation*, randomly changes the DNA sequence of a single gene (Figure 33.4a). Some mutations are neutral; their random and spontaneous appearance produces neither harmful nor advantageous effects. Mutations can provide favorable genetic variation in genes that control quantitative traits, like bushels of corn per acre or the size of the heart in cheetahs. Many mutations, however, such as the one that causes hemophilia (review Box 12.1 on page 251), are either harmful or lethal and decrease the gene's efficiency or block its operation altogether. What makes truly new genes with new functions is gene duplication and exon shuffling.

Gene duplication and divergence can produce new genes without destroying the original function (Figure 33.4b). A chance error in DNA replication or recombination can duplicate an original gene, leaving two identical adjacent copies. These two genes are then free to undergo random single-gene mutations independently, so that their sequences will come to

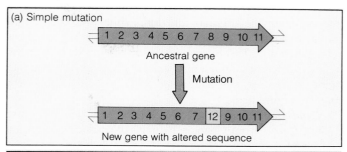

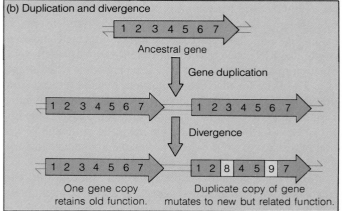

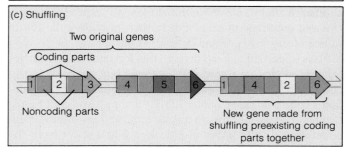

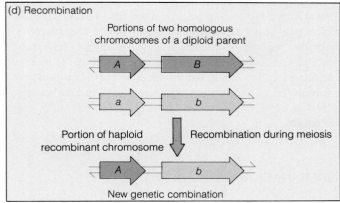

FIGURE 33.4 ■ The Sources of Genetic Variation. (a) In a simple mutation, a nucleotide (here, labeled 8) in a gene can be replaced with an entirely different one (here, labeled 12). (b) In duplication and divergence, an ancestral gene duplicates, and one copy can remain identical (left side), while the other can diverge in sequence by simple mutation (bottom right). (c) In exon shuffling, separate pieces of a gene (here, 1–6) get shuffled into a different set and sequence (1, 4, 2, 7) with a new function. (d) In recombination, random assortment and crossing over lead to chromosomal changes and hence new genetic combinations.

differ, or diverge. Thus, one copy might remain unchanged and maintain the original function, while the other might mutate to a new form that could by chance perform a new but related function. This process has resulted, for example, in the slightly different hemoglobin genes that are active in human embryos, fetuses, and adults and that satisfy their different oxygen requirements.

Exon shuffling is a third source of new genes (Figure 33.4c). Recall from Chapter 10 that eukaryotic genes often consist of *exons,* which can be expressed in proteins, and *introns,* which lie between the exons and are not translated into protein. As a result of chromosome rearrangements, an exon with one function is sometimes duplicated and then positioned by chance next to a different exon. The new combination might evolve into a new gene encoding a novel protein. The gene for the protein that removes cholesterol from the blood, for example, contains an exon from an immune system gene, an exon from a growth factor gene, and exons from genes for blood-clotting proteins. It is as though evolution picked exons off the shelf, pasted them together, and provided a new gene with a new function.

Single-gene mutation, duplication, and exon shuffling can create new alleles and new genes and thus can increase genetic variability. Once these processes produce a large number of alleles, *recombination* in sexually reproducing species can rearrange those alleles further, providing the fourth source of genetic variation (Figure 33.4d).

It is important to keep in mind that recombination shuffles existing alleles into new combinations but does nothing to change the frequency (commonness) of alleles in a population's gene pool (see Chapter 11). By analogy, you could shuffle and deal the 52 cards in a deck into millions of new combinations ("hands"), but the deck would always have the same frequencies of cards—four aces, four queens, and so forth. Only if mutation changed an ace into a king, for example, would the frequencies change. These considerations show that recombination, single-gene mutations, gene duplications, and exon shuffling generate the huge amount of genetic variation found within species, but other principles are required to explain how genetic variation can be maintained once it arises. These principles fooled many early geneticists.

How Is Genetic Variation Maintained?

In 1908, the young Mendelian geneticist R. C. Punnett mulled over a disturbing problem: A mutation causing short, stubby fingers (a defect known as *brachydactyly;* Figure 33.5) is dominant over the allele for normal fingers. Since the mutation is dominant, why doesn't everyone have brachydactyly within a few generations after the mutation arises? One day at lunch, Punnett posed the problem to his friend G. H. Hardy, a famous British mathematician. Hardy thought the brachydactyly problem simple and wrote this equation on a napkin: $p^2 + 2pq + q^2 = 1$. This equation predicts the frequencies of

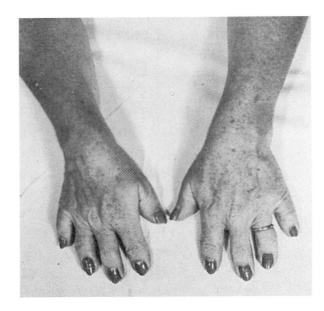

FIGURE 33.5 ■ Brachydactyly: Stubby-Finger Phenotype. The allele for brachydactyly is dominant over the allele for normal fingers. Thus, a person with one or two alleles of the gene for brachydactyly has hands with very stubby fingers; in each finger, the bone nearest the hand is fairly normal in length, but the second and third bones are greatly foreshortened.

three genotypes in a population. Box 33.1 on page 670 explains the equation in detail using the genotypes *AA, Aa,* and *aa,* but the significance of Hardy's equation for our discussion is this: In the absence of any outside forces, the frequency of each allele in a population will not change as generations pass. Hardy's mathematical model is called the **Hardy-Weinberg principle,** because a German physician named W. Weinberg discovered it independently.

Punnett and other Mendelians were puzzled by the brachydactyly problem because they confused the concepts of genotype and phenotype; they were more concerned with the number of stubby-finger phenotypes than with the frequency of the allele that caused the condition. For example, if a person homozygous for the dominant trait of brachydactyly (*BB*) marries a person homozygous for the recessive trait of normal fingers (*bb*), then all their offspring will be heterozygous (*Bb*), and all will show the dominant stubby-finger phenotype. By considering only phenotype, we might conclude that the dominant trait has increased and the recessive trait has disappeared. But by considering the genotype, we can see that the parents (*BB* × *bb*) have four alleles, half of which are *B* and half of which are *b*. Since the children's genotypes are all *Bb* (again, half *B*, half *b*), the allele frequency has not changed. We can continue performing and diagramming theoretical crosses for many generations, but the allele frequencies will stay the same. Geneticists say that such a population is in *Hardy-Weinberg equilibrium.*

The Hardy-Weinberg principle is a useful model for predicting allele frequencies in populations and therefore for determining whether or not a population is evolving. The allele frequencies remain unchanged, and hence the population does not evolve, only so long as the population is free of outside influences. Five conditions must hold for allele frequencies to remain constant over generations and for the Hardy-Weinberg principle to predict the distribution of genotypes in a population: no mutation, no migration, large population size, random mating, and no selection. As long as all five conditions are met, allele frequencies in a population will remain unchanged, and the Hardy-Weinberg principle can accurately predict these frequencies from one generation to the next. A population in equilibrium, however, would be static and unchanging. But as we have seen again and again, populations evolve. In fact, **evolution** can be defined as changes in allele frequencies within a population over time. Thus, the Hardy-Weinberg predictions provide a theoretical standard against which to compare real populations that are interacting in real environments.

In the next section, we will see how violations of the Hardy-Weinberg principle affect populations in nature, and in so doing, we will see the actual agents of evolution at work.

THE AGENTS OF EVOLUTION

Cheetahs and other living cats evolved several million years ago from an ancient population of cats with considerable genetic variation. But while such variation is the raw material of evolution, it could no more change an ancient cat into a

(a) Gene flow
between two
populations

(b) Prairie dogs

FIGURE 33.6 ■ Prairie Dogs and Gene Flow. (a) Migration from one population to another can lead to gene flow, the introduction of new alleles into a population. (b) During most of the year, prairie dogs (*Cynomys ludovicianus*) fend off immigrants from other areas. In the late summer, however, they allow new members to join established populations. Immigrants may introduce new alleles into a population's gene pool.

cheetah, or an ancestral dog into a St. Bernard, than flour could make itself into a bran muffin. Something has to act on the raw materials first. So let's consider each of those "actors" now—the five agents of evolution—mutation, migration, small population size, nonrandom mating, and selection—and see how they work.

Mutation As an Agent of Evolution

Mutation can alter allele frequencies within a population by changing one allele into a different allele. For example, the most common eye tumor in children, retinoblastoma, acts in about 30 percent of all cases as if it were due to a newly arisen dominant mutation. With each new mutation, one normal allele is removed from the gene pool and replaced with the tumor allele. The allele frequency changes a tiny amount and the Hardy-Weinberg equilibrium is disturbed slightly. The primary importance of mutation to evolution, however, is not this small change in allele frequency, but rather the opportunity the new mutation may provide for natural selection.

Migration Alters Allele Frequencies

If individuals migrate from one population to an adjacent population, they may remove alleles from one group and introduce them into a second group (Figure 33.6a). This change in allele frequencies due to immigration or emigration is called **gene flow,** and few populations are so isolated that they escape it entirely. Prairie dogs in Kansas, Nebraska, and other states, for example, live in tight-knit populations separated from one another by both geographical distance and strong social ties to group members (Figure 33.6b). Prairie dogs do not tolerate immigration and drive out any strange prairie dogs that attempt to enter their populations. During the late summer, however, when the pups reach maturity, social restrictions on immigration are relaxed, and dispersing prairie dogs are, for a short time, permitted to establish themselves as breeding members of a new colony. Their alleles are thus added to the gene pool of an existing population and are likewise removed from their parents' population.

Chance Changes in Small Populations

Biologists use the term *genetic drift* to refer to unpredictable changes in allele frequency due to small population size. The Hardy-Weinberg principle can accurately predict genotypic frequencies only for large populations, because as with all probabilities, it is much easier for nonrandom events to occur in small populations. To see why, suppose that an allele *r* exists in one-tenth of the individuals in both a large and a small population of cherry trees. If a chance occurrence like a severe spring ice storm were to strike a large population with 1 million cherry trees and half died, 500,000 would still survive, and the probability is high that one-tenth of the survivors, or 50,000, would still bear the *r* allele. If that same storm

BOX 33.1   H O W   D O   W E   K N O W ?

THE HARDY-WEINBERG PRINCIPLE:
HOW DO WE KNOW
A POPULATION IS EVOLVING?

The Hardy-Weinberg principle provides an idealized standard against which a geneticist can compare a real population and thus detect evolutionary changes. The principle has two main points:

1. If left undisturbed, the frequency of different *alleles* in a population will remain unchanged over time.
2. With no disturbing factors, the frequency of different *genotypes* will not change after the first generation.

We can understand the principle by considering a partially dominant shell color gene in a population of snails that can fertilize either themselves or each other at random (step 1).

① G_0 population

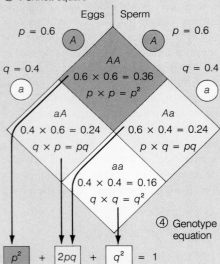

M E I O S I S

② Gametes

③ Punnett square

④ Genotype equation

$$\boxed{p^2} + \boxed{2pq} + \boxed{q^2} = 1$$

⑤ G_1 genotype frequencies

⑥ G_1 allele frequencies

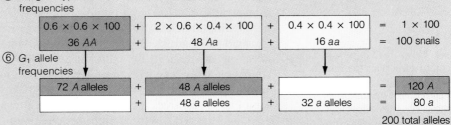

Each of the five snails is a diploid with two copies of the color gene. One allele of the gene (*A*) causes blue color, one (*a*) causes yellow color, and the heterozygote (*Aa*) is green. In this population's gene pool, there are ten alleles: six *A* alleles and four *a* alleles. If the symbol *p* represents the fraction of *A* alleles, then $p = \%_{10}$, or 0.6. If the symbol *q* represents the fraction of *a* alleles, then $q = \frac{4}{10}$, or 0.4.

Since the number of *A* alleles plus the number of *a* alleles represents all of the alleles of this gene for this snail population, $0.6 + 0.4 = 1$, or $p + q = 1$. This is the *allele pool equation*.

To see if allele frequencies change and evolution occurs, we must examine what happens when the snails reproduce, keeping in mind that alleles separate when egg and sperm are formed. Notice that despite meiosis, the frequency of *A* and *a* alleles is the same in the gametes as it was in the original population, $\%_{10}$ *A* and $\frac{4}{10}$ *a* (step 2).

Now what happens to allele frequencies at the time of fertilization? If we assume random mating, then we can write the frequencies into a Punnett square (see Chapter 8 and step 3).

We can call $p^2 + 2pq + q^2 = 1$ the *genotype equation* (step 4). This equation says that the sum of the individuals with *AA* and *Aa* and *aa* genotypes adds up to the entire population (the 1 in the equation).

To determine whether evolution has occurred in the snail population, we must look for a change in allele frequencies between generations. If five snails of the parental G_0 generation produce 100 snails in the G_1 generation, then we can *expect* the genotype frequencies and resulting genotype numbers shown in step 5, barring outside influences. (Thus, the genotype equation predicts the number of each of the three different genotypes in the population.) What, then, are the frequencies of alleles in this new generation? Have they changed?

You can see from step 6 that the frequency of *A* alleles is $^{120}\!/_{200}$, or 0.6, and the frequency of *a* is $^{80}\!/_{200}$, or 0.4—the same as in the original generation. From this generation on, the allele and genotype frequencies will remain the same in the absence of outside influences, as you can prove for yourself by making another Punnett square and filling it in with the new data.

This application of the Hardy-Weinberg equation predicts that in populations without some outside influence, no evolution will occur over the generations. But in real populations, maybe the yellow snails might be less preyed upon by birds, and so leave more progeny than expected under Hardy-Weinberg conditions. These equations provide a benchmark, a point of comparison for measuring any gene changes—or evolution—that might occur.

hit a small population of just 10 trees, however, and only 5 survived, it is much more likely that the single tree bearing the *r* allele would be among the dead. If this happened, the population would lose one allele completely, and the allele frequency will have changed in an unpredictable way.

A natural disaster that drastically reduces a population's size, such as severe weather, widespread disease, or excessive predation, can bring about the phenomenon biologists call a **population bottleneck.** Just as only a small amount of liquid can move through a narrow-necked bottle in a short time, only a small number of organisms survive a population bottleneck (Figure 33.7). For the reason we just discussed, the survivors might have a nonrandom sample of the alleles present in the original population. When the bottlenecked population enlarges once again, it lacks the genetic variability of the original population, and its potential for adapting to further environmental changes may be greatly reduced. Cheetahs probably experienced a population bottleneck due to disease, drought, or overhunting by people about 10,000 years ago, and biologists suspect that the small surviving population had, by chance, alleles for high disease susceptibility and low reproductive rates. Inbreeding among remaining cheetahs must have created the gene pool we see today, with its high numbers of detrimental alleles.

More and more species are experiencing genetic bottlenecks like the one that has endangered the cheetah as humans continue to destroy the conifer forests of America's Pacific Northwest, the tropical forests of Brazil and Borneo, the grasslands of North Africa, and other natural habitats. Alarmingly, biologists warn that the majority of species on earth will become extinct during the next 50 to 100 years—an *extinction crisis* of unprecedented proportions. To help protect endangered species, biologists apply the principles of population genetics and evolution to estimate the *minimum viable population,* the smallest population with enough genetic variability to ensure continued species survival. Armed with this information, conservation biologists can then help citizen groups and government agencies to set aside large enough tracts of natural habitat to sustain populations above their minimum viable levels. This, in turn, can help prevent the chance effects of genetic drift within small populations from leading those groups toward extinction.

Besides population bottlenecks, genetic drift can also be important in a situation called the **founder effect,** which occurs when a few individuals separate from a large population and establish a new one. Since the small group of founders bear only a fraction of the alleles from the original large population, they may represent a nonrandom genetic sample. The 12,000 or more Amish people who live in thriving communities in eastern Pennsylvania are all descendants of about 30 individuals who emigrated from Switzerland beginning about 1720. Some of these "founders" had a recessive allele that causes short forearms and lower legs (Figure 33.8). As a result, in the Amish living around Lancaster, Pennsylvania, the frequency of this allele is 1 per 14, instead of the 1 per 1000 found in other populations. Later in this chapter, we will see how the founder effect can help give rise to entirely new species, proving how important genetic drift is to the process of evolution.

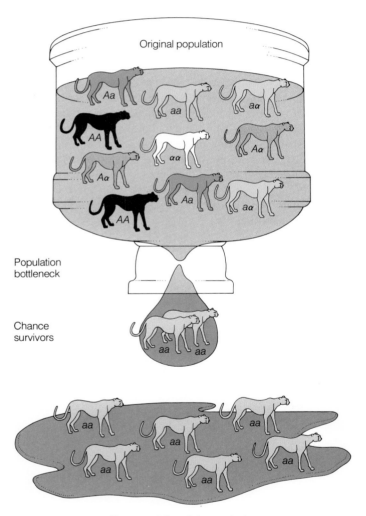

FIGURE 33.7 ■ Bottleneck Effect: A Dramatic Reduction in Population Size Leads to Reduced Genetic Variability. Ancestral cheetahs probably had normal levels of genetic variability, represented in the bottle by the three alleles *A, a,* and α and the six genotypes *AA, Aa, aa, A*α, *a*α, and αα. Ever since some genetic catastrophe perhaps 10,000 years ago, however, each surviving cheetah is homozygous at most genetic loci (here, *aa*), which may well allow the expression of a maladaptive trait.

Nonrandom Mating and Evolution

Besides the processes of mutation, gene flow, and genetic drift, a population's mating tendencies can also alter the frequency of heterozygotes and homozygotes predicted by the Hardy-Weinberg principle. Theoretically, every individual in a population has an equal chance of mating with any other

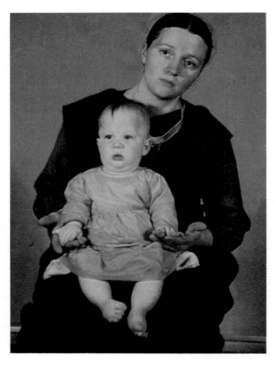

FIGURE 33.8 ■ The Founder Effect. One of the original members of the Amish community in Pennsylvania by chance carried a recessive allele for a rare kind of dwarfism. Inbreeding among members of the colony produced homozygous individuals who show the trait.

member of that population, but often it doesn't turn out this way. Consider a human population in which all short adults chose to mate only with other short adults, and all tall adults with other tall adults. If height were controlled by two separate alleles, one for tall stature and one for short, then such nonrandom mating would lead to few heterozygotes—far fewer than the $2pq$ predicted by the Hardy-Weinberg principle. The proportion of the two alleles would remain the same, but the frequencies of the three genotypes in the population would change.

One type of nonrandom mating in nature is termed **sexual selection,** a type of natural selection in which individuals select a mate on the basis of physical or behavioral characteristics regardless of the effect of that characteristic on general fitness. Take the bowerbird of Australia, for example. The male bowerbird constructs a nestlike "bower" of grass and plant material and then proceeds to decorate it with flowers, shells, teeth, bottle caps, and just about any shiny object it can find (Figure 33.9). Female bowerbirds mate only with males that can lure them into a highly decorated bower. The male's ability to build his nest depends on the specific alleles he inherited for the nest-building instinct. Inheritance affects his reproductive success, and hence, the nest-building genes are selected for through the female's willingness to mate, even

though these alleles may decrease fitness by exposing the bird to predators and decreasing the time he has for feeding.

Selection

Mutation, migration, nonrandom mating, and genetic drift can all cause allele frequencies to change in populations, but probably the most important agent of evolution is selection. Darwin pointed out that some individuals in a population produce more progeny than others, and he theorized that some sort of selection is involved; that is, the environment allows certain individuals with specific traits to survive better and become the parents of the next generation. People can act as agents of *artificial selection.* Swiss dog breeders, for example, selected and mated the largest, strongest animals for many consecutive generations to create the St. Bernard, a breed that historically has helped save skiers lost around Greater St. Bernard Pass in the Swiss Alps. In *natural selection,* the environment plays the role of the breeder, selecting as parents those individuals who can survive, exploit available resources, and reproduce most effectively in their specific environment. An individual's evolutionary *fitness* is its ability to survive and reproduce in a particular environment.

While all five agents of evolution (mutation, migration, genetic drift, nonrandom mating, and selection) can nudge a population away from Hardy-Weinberg equilibrium, evolutionary biologists have been particularly interested, in recent years, in determining what role each agent plays in maintaining genetic variation.

Which Agent of Evolution Is the Most Important?

The first section of this chapter showed that—except for rare exceptions like the cheetah—natural populations have an

FIGURE 33.9 ■ Bowerbird and Sexual Selection. This male, resplendent in his metallic plumage, has built and decorated a nest with bottles, blossoms, and shiny shards that may help attract a female.

immense amount of genetic variation. Some biologists, called *selectionists,* maintain that this variation is based on natural selection. Selectionists contend that different alleles of a gene affect an individual's health, survival, or fertility differently and that natural selection acts on these differences to choose the parents for the next generation.

Neutralists, on the other hand, claim that most variation isn't linked to an organism's survival and reproduction. Neutralists point out that even organisms living in very different environments generally have quite similar rates at which they accumulate mutations in a given gene. For example, in each of the lineages that gave rise to people and goldfish, one amino acid replaced another in some parts of a certain protein every 7 million years on average. This seemingly steady tempo of change has been called a **molecular clock** because it seems to allow a new, mutant allele to replace all the alleles of a given gene in the population over a constant but very long period of time. Neutralists suggest that most alleles are equivalent and that most mutations are "neutral," conferring neither advantages nor disadvantages on an organism's fitness. If a mutation leads to a slightly different protein, that protein need not be identical to the original as long as it works about as well. Neutralists maintain that most genetic variation is the result of neutral mutations, unpredictable genetic drift, and gene flow, while selectionists contend that genetic variation is due to the various forms of natural selection.

Many biologists take a view that combines the best arguments of the selectionists with the sound observations of the neutralists. The intermediate view is that even if most mutations are neutral, there is still plenty of nonneutral variation in every living generation for natural selection to act on and thereby affect the course of adaptive evolution.

So far, we have seen that genetic variation is primarily the result of chance physical processes like mutation and recombination. We have also seen that a number of evolutionary agents can change allele frequencies in breeding populations, again, some influenced by chance and others by the demands of the environment and natural selection. Most biologists agree that natural selection plays a major role in adaptive genetic change. The next section examines this topic in more detail.

NATURAL SELECTION IN ACTION

Question: What do cactus, brown stone plants, poison ivy, and speeding impalas have in common? Answer: All have adaptations that help prevent the organism from being eaten. The cactus has thorns; a stonelike appearance camouflages the stone plant (Figure 33.10a); poison ivy leaves produce a toxic alcohol that poisons and repels hungry animals; and the impala has the speed and agility to stay a few paces ahead of a sprinting cheetah—at least for a while (Figure 33.10b). In each case, nature has selected for these traits among the plants' and animals' ancestors; the thornier, more camouflaged, more poisonous, and faster of the organisms survived to reproductive age more frequently and left more offspring. This exemplifies **natural selection,** the changing of allele frequencies by differential reproduction and survival.

The result of natural selection is **adaptation,** the accumulation of structural, physiological, or behavioral traits that increase an organism's fitness, that is, its ability to survive and reproduce in its environment. Let's look at an example of how people have been adapted by natural selection to live in

(a) Camouflage

(b) Speed and agility

FIGURE 33.10 ■ Strategies to Avoid Being Eaten. (a) The stone plant (*Lithops divergens*) is so well camouflaged that an animal is unlikely to find and consume it. (b) These fleet impalas can escape cheetahs and other attackers on the African savanna by hiding in tall grass and outsprinting the predators over short distances.

particular environments, then examine various mechanisms by which natural selection can act.

Sickle-Cell Anemia and Natural Selection

The inherited human disease sickle-cell anemia is one of the best examples of natural selection at work. In a person with this condition, red blood cells change shape, clog up capillaries, and are removed rapidly from circulation by the spleen (review Figure 9.16). Before the days of modern medicine, the clogged capillaries, enlarged spleen, and other symptoms usually killed victims of sickle-cell anemia before they reached reproductive age.

Affected red blood cells get their sickled shape because of a mutant hemoglobin protein that is encoded by a variant allele *S* of the normal hemoglobin gene (allele *A*). People with sickle-cell anemia have two copies of the *S* allele, while people with one normal *A* allele and one *S* allele have sickle-cell trait, a usually harmless condition in which few cells sickle.

In some areas of Africa, 40 percent of the alleles of this gene in the local human population are *S*, while among black Americans, the figure is 5 percent, and among white Americans, it is only 0.1 percent. What could produce such a high frequency of a harmful allele in specific populations? Biologists began to understand this puzzle when they discovered that the sickle-cell anemia allele was most frequent in areas with a high incidence of malaria, a generally fatal disease (Figure 33.11). They later determined that people with one normal and one *S* allele are less susceptible to malaria than people with two copies of the normal *A* allele. Thus, in malaria-ridden environments, people with the *SS* genotype die of sickle-cell anemia, people with the *AA* normal genotype often die of malaria, but people with the heterozygous *SA* genotype frequently survive both conditions, reproduce, and send more *S* alleles into the next generation. Because of this survival pattern, natural selection altered the frequency of the *S* allele in some African populations to its current high level. In the United States, where malaria does not exert a selective influence, there is no adaptive advantage to the *S* allele, and natural selection seems to be removing it from the black American population.

In this example, malaria is the agent of evolution, and it acts on phenotypes; heterozygotes with sickle-cell trait have a selective advantage. Since phenotypes are due to underlying genotypes, natural selection alters allele frequencies, and it can do so in several ways.

Some Modes of Selection

Natural selection can affect populations in three ways, which biologists refer to as **directional, stabilizing,** and **disruptive selection.**

■ **Directional Selection**   This mode of natural selection favors one extreme form of a trait over all other forms (Figure

(a) Frequency of sickle-cell allele

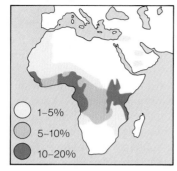

1–5%

5–10%

10–20%

(b) Distribution of malaria

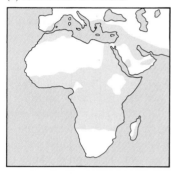

FIGURE 33.11 ■ **Sickle-Cell Anemia and Malaria: Natural Selection in Action.   (a) The *S* allele of the gene that encodes the β chain of hemoglobin is found at surprisingly high frequencies in certain parts of Africa, the Mediterranean, and the Middle East. (b) The map highlights regions where the malarial parasite lives. The nearly overlapping distributions of the sickle-cell allele and malaria provided the first clue that malaria might be an agent of natural selection. In these regions, people that have two doses of the *S* allele usually succumb to sickle-cell anemia, while people with two doses of the normal *A* allele generally die from malaria, leaving the *SA* heterozygotes to survive and reproduce.**

33.12). For example, when the first cheetahs appeared about 4 million years ago, they weighed more than twice as much as modern cheetahs. But over the years, light, fast-running animals reproduced more successfully, and thus, natural selection favored alleles that pushed cheetah weight in one direction—downward.

Another classic case of directional selection includes a little speckled insect called the English peppered moth. An ancient mutation produced two alleles for one of the moth's color genes: a dominant allele for black coloration and a recessive allele for pale gray. During the early 1800s, before the industrial revolution in England, genetically light-colored moths resting on gray tree trunks were so well camouflaged that they often escaped being eaten by birds (Figure 33.13). After industrial soot darkened urban trees, however, light-colored moths became easy targets. Moths bearing the allele for black, on the other hand, blended in perfectly on sooty trees and survived in large numbers, contributing their

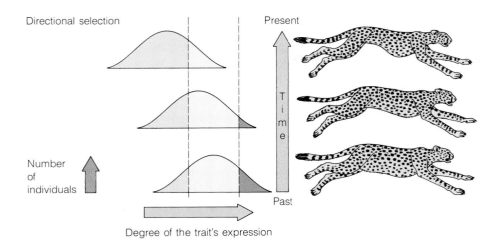

Directional selection

Present

Number
of
individuals

Past

Degree of the trait's expression

FIGURE 33.12 ■ **Directional Selection.**
During the 4-million-year evolution of the
cheetah, there was pressure in the direction
of lighter rather than heavier body weight.
In the original population, cheetahs
expressed a range of weights, but lighter
cheetahs reproduced more successfully and
passed to their offspring alleles causing
lighter weight; thus, the percentage of light
cheetahs increased in the population.

alleles for dark color to succeeding generations. In urban areas, therefore, the frequency of alleles for the extreme trait of dark color increased, and the moth population evolved.

Directional selection is also responsible for the resistance insects and microorganisms acquire to pesticides and antibiotics. A new insecticide may kill all but a few of the insects in a population. The few survivors persist because by chance they already contain mutated genes that render them "immune" to the toxic effects of the poison. These "extremists" then pass the resistance allele on to their offspring, and eventually, a large population of resistant individuals arises. The microorganisms that cause syphilis and gonorrhea have evolved such resistance to penicillin that it now takes ten times as much of the drug to fight these infections as it did just 30 years ago.

■ **Stabilizing Selection**   While directional selection favors an extreme phenotype, stabilizing selection favors intermediate individuals (Figure 33.14a). For example, far more human babies are born weighing about 3.2 kg (7 lb) than any other weight. Very heavy or very light babies have lower chances of surviving, since heavy babies complicate delivery, while lightweight babies tend to be premature and less ready for life outside the womb. Height in adult humans is a similarly stabilized trait, with most people falling toward the center of the bell curve for height (Figure 33.14b). Many biologists believe that stabilizing selection can explain why certain organisms—so-called living fossils—have persisted for millions of years with little or no outward change in form. Examples include coelacanths (see Figure 19.10a), which date back 300 million years; scorpions, which have looked very similar for 350 million years; and redwood trees, which hark back 50 million years (Figure 33.15). Each organism probably arose in an environment that remained unchanged for immense time spans, such as deep oceans or dense rain forests. Alleles causing a particular phenotype especially suited to the stable surroundings would have been selected for, while other alleles would have eventually disappeared.

FIGURE 33.13 ■ **The Peppered Moth:
Directional Selection in Action.** A dark
gray peppered moth on a clean tree trunk
(a) is conspicuous to birds, while a light
gray one on a sooty tree trunk (b) is equally
conspicuous. Experiments showed that
birds can act as an agent of natural selection, preferentially eating poorly camouflaged moths. This pressure of natural
selection is directional—toward the color of
tree trunks in a given locale.

(a) Dark moths on normal tree

(b) Light moths on sooty tree

(a) Stabilizing selection

Present

Number
of
individuals

Past

Degree of the trait's expression

(b) Distribution of heights in a population

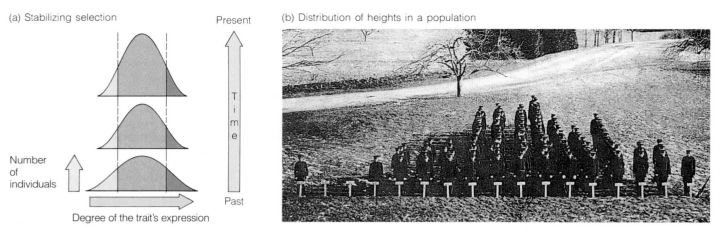

FIGURE 33.14 ■ **Stabilizing Selection.** (a) This form of natural selection favors the middle ground between extreme expressions of a trait. (b) People are more likely to be of average height, and less likely to be very tall or very short, as illustrated by these cadets from a military academy.

■ **Disruptive Selection** This mode is the opposite of stabilizing selection. In disruptive selection the two extremes are selected for (Figure 33.16a), while in stabilizing selection, the two extreme forms are selected against. In parts of Africa, females of the butterfly *Pseudacraea eurytus,* for example, occur in either a blue or an orange form (Figure 33.16b). The blue form looks like one foul-tasting butterfly species (*Bematistes epaea*), and the orange form looks like another species (*B. macaria*) that tastes offensive to birds (Figure 33.16c). *P. eurytus* individuals that are good mimics of either foul-tasting model species will evade predation and be selected for, while the butterflies with alleles for an appearance intermediate

between the two extremes will mimic neither bad-tasting species and will be eaten by predators.

Directional, stabilizing, and disruptive selection can clearly change gene frequencies within populations. Are these changes sufficient, however, to account for actual speciation—for the separation of a population into two separate species? Many kinds of evidence suggest that indeed it is.

HOW NEW SPECIES ARISE

What Is a Species?

Every zoo goer knows that cheetahs, tigers, and lions are three different species of cats, one slender and spotted, one powerful and striped, and one large and tan with a dark mane. Biologists who have tested these three cats have also found that each has its own unique genetic makeup, and in the wild, each breeds and reproduces only with others of its species. Thus, not only do the three species look phenotypically different, but they are genetically different, too, and these criteria place them unambiguously in one species or the other: There are no half tigers/half cheetahs or half cheetahs/half lions in nature, and tiger/lion hybrids are produced only in zoos.

The criteria of unique genetic makeup and absence of interbreeding hold true even for very similar-looking species. Some tropical lakes, for example, have a dozen or more species of frogs that look almost identical. Yet the females of one species respond only to the specific mating calls of their own species' males, never another's. When geneticists examine frog proteins with electrophoresis, they find that the members of an interbreeding group of frogs share a common gene pool that differs from that of other interbreeding groups. In other words, we can be fooled by our perception of phenotype, but not by genotype or ability to interbreed.

(a)   (b)

FIGURE 33.15 ■ **Similarity of Fossil and Living Redwoods Reflects Stabilizing Selection.** (a) This 50-million-year-old fossil of a redwood tree (*Metasequoia* sp.) is very similar to (b) the *Metasequoia glyptostroboides,* or dawn redwood, growing on the University of Oregon campus.

According to modern evolutionary theory, species are groups of populations that interbreed with each other in nature and produce healthy and fertile offspring. Each species, in short, is *reproductively isolated* from (cannot make fertile progeny with) every other species in nature. As we saw with tropical frogs, species that to a human observer look identical can live side by side and not interbreed. Conversely, if very different-looking organisms (like the blue and orange butterflies of Figure 33.16b) can interbreed in nature, they are considered races, or varieties of a single species.

The definition of a species as a shared, reproductively iso-lated gene pool has been the most satisfactory to date, but even it has limitations. The biggest drawback is that it applies only to sexually reproducing organisms. Asexual organisms, such as some bacteria that reproduce only by cell division, must be excluded, because each individual is, in effect, reproductively isolated from all others. With organisms like these, a species must be identified by phenotypic, genetic, and biochemical traits, rather than by the potential for interbreeding.

To see how a species arises, we must first consider the causes of reproductive isolation. In other words, what factors keep separate species like lions and cheetahs or sets of identical-looking tropical frogs from interbreeding and producing fertile offspring?

Reproductive Isolating Mechanisms

Any biological feature that prevents the members of one species from successfully breeding with those of another is considered a **reproductive isolating mechanism.** Some such mechanisms prevent members of different species from mating. For example, dozens of frog species may inhabit the same area, but one species may mate only in ponds, another only in flowing streams, and a third in shallow pools and puddles. Or some may mate only in the spring and others only in the fall. Specific physical structures can also hinder mating. In many spider mites, for example, the male genitals are shaped like "keys" that open the genital plates of their species' females; one species' "keys" cannot open another's "locks."

Sometimes, the sperm of one species can contact the eggs of another species, but reproductive isolating mechanisms prevent them from producing fertile offspring. For example, the gametes of two species can be so different that fertilization is impossible, as with sea urchin eggs that are surrounded by a specific chemical that binds only to sperm from the same species.

Sometimes, matings are partially successful and fertilization does take place, but the **hybrids,** the offspring of the two crossbreeding species, are not viable individuals. With many American frogs, genetic differences between species lead to abnormal hybrid zygotes and freakish individuals.

Occasionally, the cross-mating of two species produces extremely healthy offspring (a result known as *hybrid vigor*), but these offspring are themselves nonreproductive (they experience *hybrid sterility*). Breeders produce mules (see Box 8.2 on page 181) by crossing a male donkey (31 chromosomes in each sperm) with a female horse (32 chromosomes in each egg), and although mules (with their 63 chromosomes per somatic cell) are noted for strength and endurance, they cannot produce further offspring with other mules, with horses, or with donkeys because the chromosomes fail to pair normally during meiosis.

Since reproductive isolating mechanisms separate the gene pools of two related species, we can begin to understand **speciation,** the emergence of new species, by learning how reproductive isolating mechanisms develop.

(a) Disruptive selection

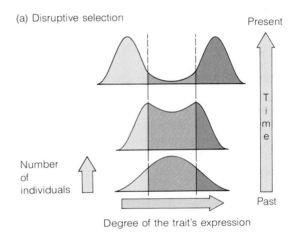

(b) The mimics: two different forms of *P. eurytus*

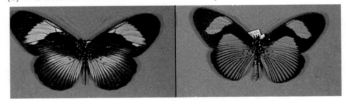

(c) The models: *B. epaea* and *B. macaria*

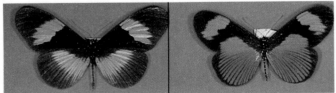

FIGURE 33.16 ■ **Disruptive Selection.** (a) **This form of natural selection favors two extreme expressions of a trait. (b) Members of the species *Pseudacraea eurytus* are found in two forms, one larger with blue fans and the other smaller with orange fans. These two edible forms mimic (c) two other foul-tasting moth species, the blue *Bematistes epaea* and the orange *B. macaria. P. eurytus* individuals intermediate in size or color mimic neither model very well and are hence selected against by predators.**

The Origin of New Species

Most biologists believe that the majority of species arise after populations become geographically isolated and evolve in separate ways. A physical barrier, such as a river, a desert, different vegetation belts, or even a new highway or pipeline, can separate populations of a single species and prevent gene flow between the groups. The split populations slowly diverge as mutation, genetic drift, and adaptation cause different sets of characteristics to accumulate. Eventually, barriers to reproduction emerge and prevent matings even if, in the future, the two populations once again come into contact. This mechanism is called geographical or *allopatric speciation* (*allo* means "different," and *patric* means "native land") (Figure 33.17).

For example, 7 million years ago, the Colorado River began carving out the Grand Canyon in an area inhabited by one squirrel species. Today, the Kaibab squirrel occupies the north rim of the canyon, while the closely related Abert's squirrel inhabits the south rim. The two species most likely arose through allopatric speciation, gradually diverging as the canyon became an uncrossable barrier (see Figure 33.24b). Cheetahs may be an example of a population beginning to be split by allopatric speciation. The only two surviving populations of cheetahs, one in East Africa and the other in South Africa, can no longer experience gene flow because unpassable terrain separates them. Perhaps with time, therefore, enough genetic difference will arise between the two populations that they will no longer be able to produce a fertile hybrid.

The founder effect can bring about a second type of allopatric speciation. Sometimes, a few founding individuals migrate long distances and settle into a new geographical area. Since the founders can be few in number—even a single pregnant female can suffice—the absence of gene flow plus the effects of genetic drift and selection can result in rapid divergence. For example, a species of pygmy mammoths, averaging only 1 m (3.28 ft) in height, once lived on Santa Catalina Island, off the coast of southern California, and biologists suspect that these diminutive pachyderms descended from a few 5-ton mammoths that reached the island some 50,000 years ago.

As we have seen, the splitting of a population into two noninterbreeding groups by a geographical barrier can lead to allopatric speciation. Sometimes, two groups of a population may continue to interbreed at one edge of their range but still diverge anyway—a phenomenon called *parapatric speciation* (meaning "parallel native land"). This type of speciation occurs if, despite the contact, strong reproductive isolating mechanisms develop as the population slowly splits into two separate ecological roles. If the hybrid does not fit well into either ecological role, selection will favor those individuals who mate with members of their own population. Parapatric speciation is common among some plants, snails, flightless insects, mole rats, and other organisms that live in small, isolated groups and travel only short distances. For example,

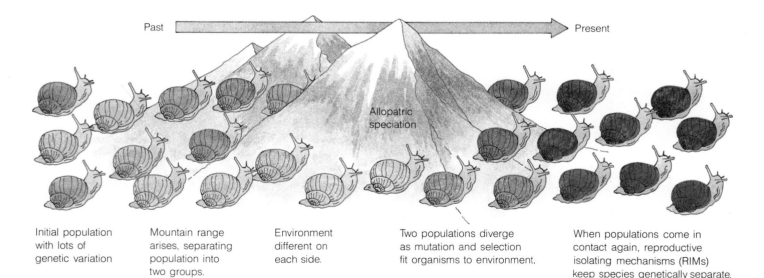

Past   Present

Allopatric speciation

Initial population with lots of genetic variation

Mountain range arises, separating population into two groups.

Environment different on each side.

Two populations diverge as mutation and selection fit organisms to environment.

When populations come in contact again, reproductive isolating mechanisms (RIMs) keep species genetically separate.

FIGURE 33.17 ■ **Geographical Barriers and Allopatric Speciation.** An initial population (here, yellow and green snails) can spread over a wide geographical area and have substantial genetic variation. After a mountain range or other geographical feature separates the initial population into two noninterbreeding subpopulations, natural selection may adapt them to their local environments, and genetic drift may further differentiate the two groups (as seen by their changing colors). If the mountain range erodes and the two groups eventually come into contact again, they may be so dissimilar genetically that they can no longer interbreed, and they will have become two different species (brown and turquoise snails).

populations of some grasses have evolved tolerance to toxic substances in soils contaminated by mining wastes. Simultaneously, they have evolved mechanisms that have decreased their tendency to interbreed with other grass populations that may be growing just a few meters away in the same meadow.

Ecologists have a name for situations in which a previously interbreeding population becomes split into two reproductively isolated groups, even without any spatial separation. They call it *sympatric speciation* ("same native land"). Controversy surrounds most models of sympatric speciation, but many biologists agree that instantaneous speciation can occur via *polyploidy* in plants, an increase in the number of chromosome sets in a cell. A good example of polyploidy is seen in a genus of pretty pink wildflowers called *Clarkia*, first collected in Idaho by members of the Lewis and Clark expedition (Figure 33.18). *Clarkia concinna* has 14 chromosomes in a diploid set, while *C. virgata* has 10. If these two species mate, the hybrid progeny gets 7 haploid chromosomes from one parent and 5 from the other, making a total of 12. This diploid hybrid is sterile because the two different sets of chromosomes do not pair and distribute normally during meiosis and gamete formation. At some time in the past, however, a tetraploid hybrid plant arose in which the chromosome count had doubled from 12 to 24 (two sets of 7 and two sets of 5). In this tetraploid, with four sets of haploid chromosomes, every chromosome had a partner, and so meiosis could occur normally. The new tetraploid hybrid, called *Clarkia pulchella*, was self-fertile, but could not successfully interbreed with either diploid species. A new species was therefore created within one original population in one step without a geographical barrier. Sympatric speciation like this is relatively common in plants and has led to some of our most important crops, including wheat and cotton.

We have seen in this chapter how variation, heritability, differential survival, and chance all set the stage for organisms to evolve. But how well can the modern synthesis of genetics and evolution explain the large-scale evolutionary changes that have produced the immense diversity of life on earth—from worms and beetles to toadstools and tulips—with all its patterns and trends?

PHYSICAL EVIDENCE FOR EVOLUTION

Darwin and Alfred Russel Wallace (see Chapter 1) were correct in suggesting that natural selection could cause evolutionary change in populations of organisms; we have only to look at evidence from cases such as sickle-cell anemia, the peppered moth, moth mimics, and bacteria with penicillin resistance. Evolutionary biologists call these kinds of genetic changes within a species **microevolution.** But what about **macroevolution,** the large-scale evolutionary changes that differentiate taxonomic categories above the species level—the differences between pines and spruce, birds and mammals, fungi and plants? What evidence supports the notion that evolution can account for the full diversity of living

forms on earth? The proof is so pervasive that we could point to examples from every chapter of this book. Here, however, we summarize some of the key evidence from fossil finds, molecular biology, anatomy, and ecology.

Life's History in Stone: Evidence from the Fossil Record

The diversity we see on our planet is a product of billions of years of evolution. Paleontologists (biologists who study

(a) *Clarkia concinna*

(b) *Clarkia* family tree

FIGURE 33.18 ■ Genetic Barriers May Isolate Populations and Lead to Sympatric Speciation. (a) The diploid western wildflower *Clarkia concinna* has two sets of seven chromosomes, while (b) the related diploid species *C. virgata* has two sets of five chromosomes. A diploid hybrid of these two species has one set of five chromosomes and one set of seven chromosomes and is sterile because of mismatched chromosomes during meiosis. But if the chromosome sets double in this diploid hybrid, then a tetraploid results with two sets of five chromosomes and two sets of seven chromosomes. Meiosis occurs normally in this tetraploid *C. pulchella,* because each chromosome can pair with a homologous chromosome. Since the newly formed tetraploid cannot successfully mate with either parental species, and since it arose from the parental populations without a geographical barrier, it represents a case of sympatric speciation.

extinct forms of life through fossils) have shown that the earth formed some 4.6 billion years ago and life appeared about 4 billion years ago (see Chapter 15). The fossil record is the only tangible evidence of what past organisms looked like and when they appeared, but unfortunately, these impressions and mineralized bones occur only in sedimentary rocks, which in turn form only under certain conditions. Episodes of sedimentary rock formation may last only a few days—as when a river overflows its bank and covers a bone with mud—or tens of thousands of years—as when a river deposits a huge delta. Between such episodes there are gaps, times when no sediments are laid down and no fossils formed.

Despite such gaps, paleontologists can examine fossils from many locations and piece together a picture of evolution's overall course. Radiodating of fossils or the rocks in which they are embedded also allows scientists to determine the age of these impressions and hence to determine the order in which fossils appeared over time.

As we saw in Chapter 15 (review Figure 15.13), the oldest fossils are from prokaryotic cells, followed by the fossils of one-celled eukaryotes, then multicellular invertebrate animals and simple plants. Fishes appeared next in the fossil record, preceding amphibians, reptiles, mammals, then birds. This apparent trend from simplicity towards complexity supports the notion of macroevolution. The fossil record, for example,

shows the increasing size of the bony facial bumps on the nose and face of a number of species of hoofed mammals called titanotheres (Figure 33.19). Progressive changes like these—descent with variation—are widely represented in the fossil record and provide clear evidence that evolution has occurred.

Molecular Evidence for Evolution

Darwin and Wallace could not have anticipated the flood of confirming evidence that would someday come from molecular biology. We saw in several earlier chapters how all life shares basic molecular features, such as similar ribosomal RNAs and a single genetic code (review Figure 10.6). These features originated early in life's history and were passed along through each succeeding lineage.

Molecular biologists uncovered one of the strongest proofs of descent with variation by studying the corresponding genes and proteins from different species and comparing their structures. Closely related species share more specific mutations and allele patterns than distantly related species. When molecular geneticists study such gene sequence data, construct family trees, then compare the results with fossil evidence, they find striking consistency. For example, a

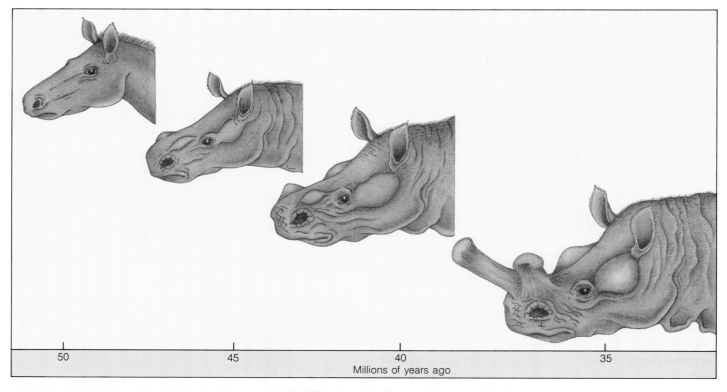

50	45	40	35

Millions of years ago

FIGURE 33.19 ■ **Record in the Rocks: Descent with Modification.** Fossils reveal that over 20 million years or so, several species of hoofed mammals known as titanotheres became progressively larger. Meanwhile, the small bony bumps on their heads evolved into great horns and protrusions.

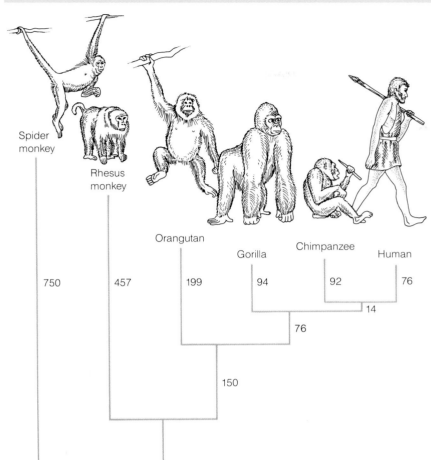

Spider monkey

Rhesus monkey

Orangutan

Gorilla

Chimpanzee

Human

750   457   199   94   92   76

14

76

150

FIGURE 33.20 ■ **Molecular Evolution: A Primate Family Tree.** The base sequences within hemoglobin genes are more similar among closely related species than among distantly related species. The longer the line in this family tree, the more mutations have accumulated in that lineage owing to natural selection and genetic drift. Although the two data sets are totally independent, the molecular tree closely matches fossil analyses (see Figure 19.18).

branching diagram of the number of mutations accumulated in the hemoglobin genes of primates (Figure 33.20) looks very similar to a primate family tree constructed from fossil data (review Figure 19.18). While the two kinds of data are very different, they both support a common pattern of relationships between extinct and living primates.

Evidence from Anatomy: Homology, Vestigial Organs, and Comparative Embryology

Natural selection can only work on genetic variation that is already present in a population. Thus, new structures or functions generally arise as variants in preexisting structures are selected, and the structure becomes modified. Anatomists point to cases of **homology,** wherein structures in related organisms have very different functions now but almost certainly arose from the same structure present in a common ancestor. The forelimbs of various mammals perform different functions: A person's arms and hands can manipulate small and large objects; a cheetah's legs allow it to run swiftly; a whale's flippers allow efficient swimming; a bat's wings allow flight. Yet each type of limb is composed of the same skeletal elements (Figure 33.21). Biologists doubt that

such homologies exist because the best structure for a hand is also the best structure for a wing. They believe, instead, that the forelimbs of various mammals arose from the forelegs of ancestral reptiles as these were modified by natural selection for various tasks.

An especially telling type of homology is seen in **vestigial organs,** rudimentary structures with no apparent utility to the organism, but with a strong resemblance to structures in probable ancestors. People, for example, still have the vertebrae for a tail, muscles that can wiggle the ears, an appendix, and extra back teeth (wisdom teeth), none of which have specific functions today. Likewise, snakes and whales still have a pelvis and femur bones, although neither of these is functional either (Figure 33.22). The fossil record and other evidence suggest that while the ancestors to humans, whales, and snakes had well-developed examples of these structures, changing environments selected against the genes needed for the structures to be fully developed in today's organisms. Natural selection for a smaller appendix might even be operating now when people die from appendicitis.

The third type of anatomical evidence for macroevolution comes from studying embryos. As Figure 14.3 showed, the embryos of vertebrates develop gill arches and gill slits

(a) Human   (b) Cheetah   (c) Whale   (d) Bat

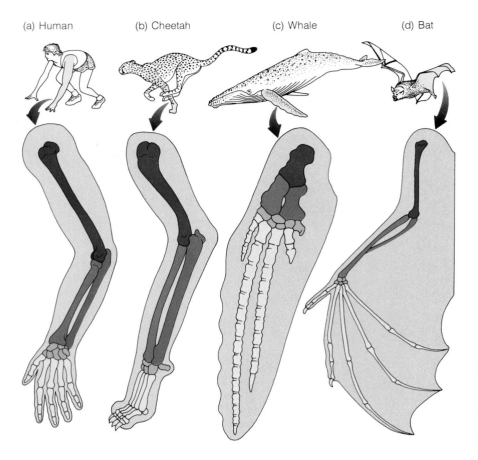

FIGURE 33.21 ■ Comparative Anatomy Suggests Evolution by Descent with Modification. Natural selection has modified the same set of skeletal elements in the forelimbs of a person, a cheetah, a whale, and a bat for different functions: manipulation, running, swimming, and flying.

(a) Vestigial pelvis and femur of a whale

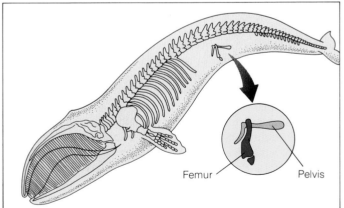

Femur   Pelvis

(b) Vestigial pelvis and femur of a python

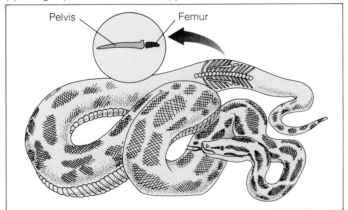

Pelvis   Femur

FIGURE 33.22 ■ Vestigial Organs. A whale (a) and a python snake (b) both have a rudimentary pelvis and femur bones, even though these structures provide no apparent function. They were inherited from ancestors, modified, and remained, even though they don't contribute to the new modes of locomotion.

regardless of whether the adults will have permanent gills. One of the gill slits in human embryos becomes the eustachian tube, and tissues surrounding other gill slits become the thymus and tonsils. What's more, the development of middle-ear bones (see Figure 28.4) in a human embryo suggests that they derive from the jawbones of ancestral reptiles. During evolution, new developmental programs must have been selected that extended and modified programs in ancestral organisms, with the result being homologies, vestigial organs, and other developmental similarities.

(a) Sugar glider   (b) Flying squirrel

FIGURE 33.23 ■ Convergent Evolution. Organisms adapted by natural selection to exploit similar environments evolve similar adaptations even though they are only distantly related. The sugar glider (a) and the flying squirrel (b) belong to two very different groups of mammals, but they have both evolved bushy tails and flaps of skin that act like organic kites and help them exploit their forest environments more efficiently.

Evidence from Ecology: Convergent Evolution and Biogeography

One of Darwin's first clues about evolution was the way organisms are distributed geographically. The world over, groups of organisms that inhabit regions with similar weather and climate tend to look and function similarly, even though they may not be closely related. Evolutionary biologists call this phenomenon **convergent evolution.** Consider, for example, the small mammals with bushy tails that go sailing through the trees of Australian forests, gliding on broad folds of skin that stretch along the side of the body from wrist to ankle (Figure 33.23a). These sugar gliders are marsupials (see Chapter 19), but they look and act like the placental mammals called flying squirrels that inhabit European, Asian, and North American forests (Figure 33.23b). Marsupial counterparts of placental wolves, mice, moles, and anteaters also inhabit parts of Australia. How could such remarkable similarities have come about? Biologists agree that the continent of Australia separated from the rest of the world's landmasses more than 50 million years ago (review Figure 15.13), before placental mammals migrated to that part of the world. Without the competition from placental mammals, marsupials diverged on the Australian continent and came to perform very similar ecological roles to the ones filled by placentals on other continents. Similar environmental pressures selected for similar physical forms and behaviors in the two groups of animals.

The study of where organisms live, or **biogeography,** has also revealed that many distinct species inhabit only particular islands and yet are closely related to species living on the nearby mainland. Darwin saw this as evidence that mainland species invade islands, then diverge into new species as natural selection adapts them to the local environment.

Taken together, data from fossils, molecules, anatomy, and ecology convince nearly all biologists that macroevolution has indeed occurred. As a result, biologists are able to trace the ancestry of today's organisms through the numerous branches of a great family tree—back to an ancient initial origin of life.

TRENDS IN MACROEVOLUTION

A study of **phylogeny,** or the evolutionary history of different groups of organisms, reveals recurring patterns of descent, episodes of extinction, and appearances of evolutionary novelties. These subjects contribute to our understanding of how evolution proceeds.

Patterns of Descent

There are several patterns of evolutionary descent. One is the gradual change in allele frequencies in a population (Figure 33.24a), such as the alterations that little by little led to the diminishing size of the cheetah over geological time. Another is **divergent evolution,** whereby reproductive isolating mechanisms split two populations off from a common ancestral population, after which genetic differences accumulate in the two descendant groups as in the two species of Grand Canyon squirrels (Figure 33.24b).

Sometimes, divergence occurs simultaneously among a number of populations in ways that produce a variety of phenotypes. This fan-shaped branching pattern of evolution is called **adaptive radiation** (Figure 33.24c). Adaptive radiation may be triggered when an ancestral species invades new territories that allow new and different ways of life. The colonization of the Hawaiian Islands by colorful birds called honeycreepers led to a spectacular example of this type of

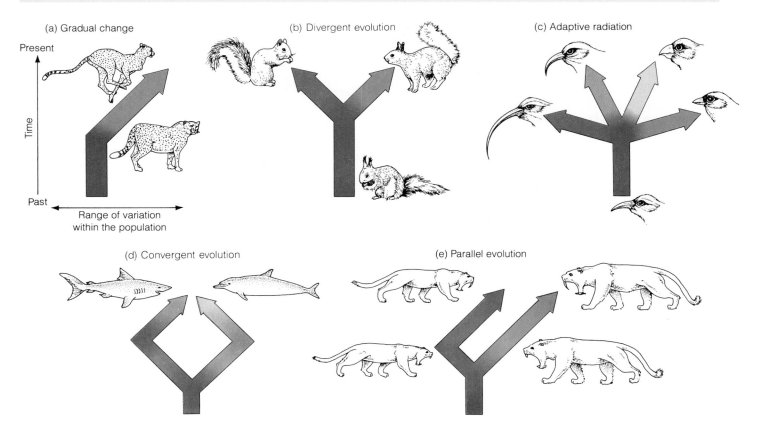

FIGURE 33.24 ■ Patterns of Descent in Evolution. Over time, (a) a population's gene pool may gradually change in a certain direction, as when cheetahs became smaller, or (b) the population may split into two populations, each changing in a different way as time progresses, as when two species of squirrels evolved from one original species when the Grand Canyon formed and caused a reproductive isolation of the two populations. (c) Adaptive radiation results in several new species arising from a single common ancestor, as with the small honeycreepers of Hawaiian forests. (d) In convergent evolution, separate lines evolve similar forms that allow them to exploit similar environments, as with sharks and dolphins. (e) In parallel evolution, two similar and related ancestors lead independently toward new species with similar traits, as in the evolution of different species of saber-toothed cats.

radiation. Descendant species had beaks of various shapes that allowed them to eat diets as different as tiny insects, hard seeds, soft fruits, and the nectar from deep, tubular flowers.

Another evolutionary pattern occurs when two or more dissimilar and distantly related lineages evolve in ways that make them appear more similar superficially. Since the phenotypes "converge," biologists call this phenomenon *convergent evolution* (Figure 33.24d). We have already seen the example of the sugar glider and flying squirrel. Another striking example is the similarity between sharks, ichthyosaurs, and dolphins. All evolved a streamlined body shape because they adapted independently to a similar way of life in the open ocean, not because they inherited the same set of streamlining traits from a common ancestor.

Parallel evolution is a special case of convergence that involves two or more physically similar and genetically related lineages independently changing in the same direction

and away from the ancestral phenotype (both toward larger size and away from smaller size, for example) (Figure 33.24e). Sometime during the last 35 million years, for example, four genera of catlike predators separately evolved saberlike teeth and other adaptations for killing large prey. Since the cats shared a common ancestor, however, their parallel resemblances were far more than superficial.

In the evolutionary pattern called **coevolution,** two species interact so closely that each one's evolutionary fitness depends on the other. An evolutionary change in one, therefore, creates a selective pressure on the other, so that both evolve together. For example, yucca plants and the moths that pollinate them have coevolved to such an extreme that yucca plants can be pollinated solely by the yucca moth, and those animals, in turn, can reproduce only in yucca flowers and seeds. Coevolution is so widespread and important that Chapter 35 discusses it in greater detail.

The Tempo of Evolution

The mechanisms just discussed (adaptive radiation, convergent, divergent, and parallel evolution, and coevolution) imply two basic evolutionary processes. Genetic modification can occur within a single line of descent to make a current population look and act differently from its ancestor population, and genetic changes can split one population into two or more different groups. Traditionally, evolutionary biologists have agreed that after a group splits in two—whether because of geological or biological barriers—both subgroups diverge from each other at about equal rates as natural selection and genetic drift cause modifications within each line. This view is called **phyletic gradualism,** since it assumes that each group gradually becomes different from the original group and from each other (Figure 33.25a).

Adherents of phyletic gradualism suggest that variation in an organism's form occurs gradually during evolution and that steady changes in body form can accrue within an entire population over time without the population necessarily splitting into two distinct species.

If new species do indeed form as one species gradually transforms into the next, then the fossil record should show numerous intermediate species. While Darwin propounded this view, there was little fossil evidence to back it up in his time, and despite more than a century of additional fossil collection, paleontologists still have few records of gradually transformed lineages with numerous intermediate species. Scientists traditionally attribute such gaps in fossil lineages to gaps in the fossil record, explaining that intermediate fossils are missing because the rocks that contained them have eroded away, are still undiscovered, or were never formed.

In 1972, paleontologists Niles Eldredge and Stephen Jay Gould suggested that we should not make the data fit our ideas, but rather make our ideas fit the data. They proposed that the gaps might themselves be telling us how speciation

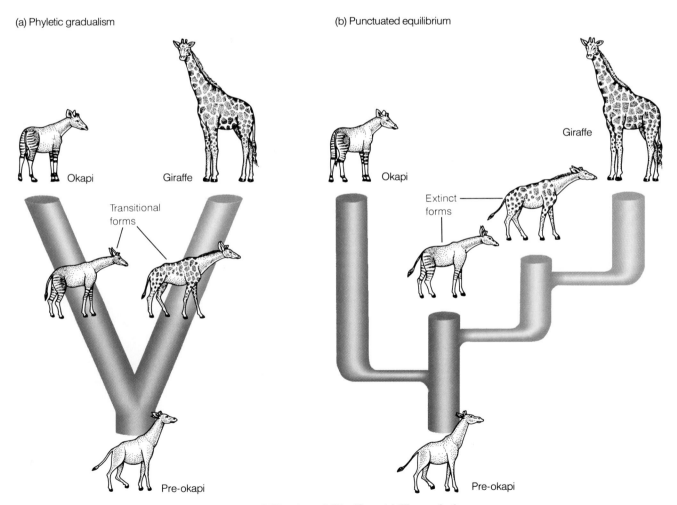

(a) Phyletic gradualism

(b) Punctuated equilibrium

Okapi   Giraffe

Transitional forms

Pre-okapi

Okapi

Giraffe

Extinct forms

Pre-okapi

FIGURE 33.25 ■ Two Hypotheses for the Descent of Okapis and Giraffes. (a) The evolutionary pattern of phyletic gradualism portrays okapi and giraffe evolution as two equal, slowly diverging lines from a common ancestor. (b) The punctuated equilibrium hypothesis suggests that a small population budding off from the ancestral pre-okapi group would lead to modern okapis, while a number of more giraffelike species would arise and became extinct as time passed.

takes place. They pointed out that according to the fossil record, the typical species is not in a constant process of change; instead, it exists for millions of years without significant alteration. These long periods of phenotypic equilibrium are interrupted, or "punctuated," only rarely by great phenotypic changes that result in new species.

This alternative to the phyletic gradualism model is called **punctuated equilibrium** (Figure 33.25b), and it has three tenets: (1) Alterations in body form evolve very rapidly in evolutionary time; (2) during speciation events, changes in form occur almost exclusively in small populations, and the result is that new species are quite different from their ancestral species; and (3) after the burst of change that results in speciation, species retain much the same form until they become extinct, perhaps millions of years later.

To get a feel for the different patterns of descent predicted by punctuated equilibrium and phyletic gradualism, consider two contrasting family trees for the okapi and giraffe. Phyletic gradualism predicts that the two evolutionary lines began to diverge slowly before speciation occurred; then after reproductive isolating mechanisms developed, modifications in form gradually continued until the present-day okapi and giraffe emerged (see Figure 33.25a). In contrast, punctuated equilibrium predicts that offshoots repeatedly emerged as small, isolated populations split from the main group, underwent bursts of evolutionary change, and then maintained their new features for a long or short period of time before the line died out (see Figure 33.25b).

In fact, both processes probably contributed to this evolutionary tree. Evolution may often occur when a small population becomes isolated geographically or ecologically from other members of its species and accumulates genetic changes over many generations (but a short stretch of geological time). Once sufficient change has occurred to fit the new species to its environment, very little additional change may take place. The principles of population genetics can thus explain geologically rapid changes in phenotype followed by long periods of stasis.

Mechanisms of Macroevolution

The principles of macroevolution are intended to explain evolution above the level of species and hence explain the origin of major distinguishing features (reptiles' scales, birds' feathers, mammals' hair). They pose possible answers to the question, Can microevolutionary mechanisms account for the origin of taxa above the species level, or are different kinds of evolutionary mechanisms required for such major differences?

■ **Functional Changes and Mosaic Evolution** Biologists have observed that evolutionary innovations usually follow from a change in size or function in a preexisting body part, rather than from the sudden appearance of something totally new.

For example, horses' high-crowned molars, which they require for chewing silica-laden grasses, arose by gradual increases in a small tooth bump in their ancestors.

Other evolutionary innovations arose when a body part performing one function changed in shape and came to perform a new function. The original structure did not, of course, evolve in anticipation of a future need; but once it had arisen, the structure could be modified by natural selection for a new use. For example, *Archaeopteryx,* the earliest known bird (review Figure 19.12), is distinguished from small, flightless reptiles that lived at the same time solely by its feathers. Mutation and selection changed reptilian scales into lightweight feathers, an evolutionary novelty that allowed better gliding and heat retention. While *Archaeopteryx* had feathers, like a bird, its skeleton still resembled that of a reptile, lacking the specialized breastbone and powerful flight muscles of modern birds. A case like this, in which some characteristics evolve without simultaneous changes in other body parts, is called *mosaic evolution.*

■ **Regulatory Genes and Developmental Changes** Sometimes, the evolution of a dramatic morphological change involving an entire suite of genes can have its roots in a simple developmental event. For example, a single mutation in *Drosophila* can cause four wings to form instead of the usual two (Figure 33.26). Since the mutation is in a regulatory gene that acts early in development and controls the function of many other genes, the result is a startlingly novel effect—four wings, not two. Likewise, simple coordinate changes in the relative growth rates of different body parts during development can cause profound differences in the form of the entire adult organism. As Figure 33.27 shows, modest modifications in the axes of growth can alter a puffer fish to the shape of an ocean sunfish.

FIGURE 33.26 ■ **Development and Evolutionary Novelties.** An evolutionary novelty can arise from a single mutation in a regulatory gene that functions early in development. Here, a single mutation in a fruit fly regulatory gene leads to a mutant fly with four wings instead of two.

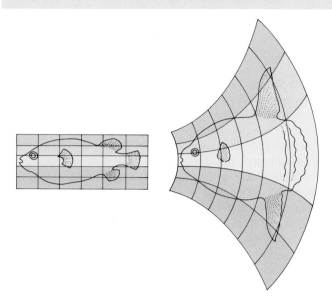

FIGURE 33.27 ■ **Simple Changes in Growth Rate Can Lead to Complex Differences in Body Form.** If a puffer fish (left) experienced accelerated growth in posterior quadrants during development, the adult would resemble an ocean sunfish (right). This kind of model, devised by biologist D'Arcy Thompson, underscores the role of development in evolution.

years. And the number of all living species today—plants, animals, fungi, and microbes included—is probably less than 0.01 percent of the estimated 500 million species that have existed since life began on earth. The most famous extinction event—the demise of the dinosaurs and thousands of other species about 65 million years ago—was probably the result of a meteor impact (review Box 19.1 on page 414). In the most recent mass major extinction, nearly 80 percent of the large mammal species of the Western Hemisphere disappeared about 12,000 years ago. Some paleontologists attribute this extinction to *prehistoric overkill*—overhunting by prehistoric people—and note that extinctions followed human colonization of various regions. The overkill theory, however, is hardly unanimously accepted. Many biologists, in fact, think

Clearly, the principles of microevolution, including preadaptation or developmental mutations, can account for changes that occur during long reaches of time. But can they also account for the patterns of extinction that mark macroevolutionary trends?

■ **Extinction, Species Selection, and Evolutionary Trends**   Modern horses run about on a single toe per foot, but their early ancestors had the primitive condition of four toes per foot. Some macroevolutionists account for such evolutionary trends by invoking the principle of *species selection:* the idea that certain species continue to break up into new species, while other species become extinct. To see how species selection might work, consider the diagrammatic view of evolution in horses shown in Figure 33.28. The change from four toes to one toe did not occur smoothly in a single lineage. Instead, various species arose, gave rise to new species, and then became extinct in a great evolutionary bush with dozens of branches. Our view of this trend is colored because only the genus *Equus* has escaped extinction so far.

Microevolutionary theory—including genetic variation, drift, natural selection, and speciation—can explain the changes in form that took place during horse evolution. But since extinction is not part of microevolution, those models cannot fully explain why living horses, zebras, and donkeys walk about on a single toe per foot. It is left to macroevolutionists to attempt to explain why some lineages branch to form many species and why some lineages become extinct.

Ultimately, extinction is the fate of all species. Most mammal species, for example, survive no longer than 2 to 5 million

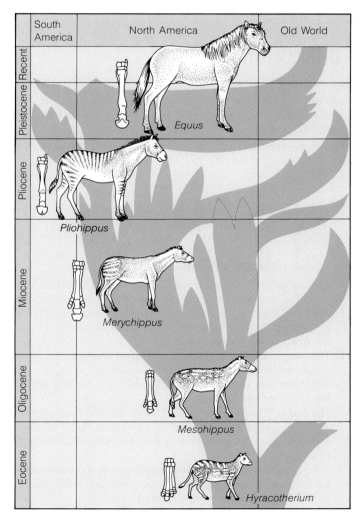

FIGURE 33.28 ■ **Species Selection, Extinction, and Evolutionary Trends.** The tiny, dog-sized progenitor of modern horses, *Hyracotherium* had four toes, but modern *Equus* has one toe. The lineage from *Hyracotherium* to *Equus* was not a gradual change in a single lineage, but rather a bush with many branches, all of which have died out except modern horses.

that rapid climatic and environmental change coming at the end of a long period of glaciation may have caused this period of extinction.

Cheetahs have experienced a number of extinctions. At one time, there were five species ranging over Africa, Europe, Asia, and North America. Such wide distribution is rare among land mammals and shows that cheetahs at one time had the extensive genetic variation necessary to exploit a wide range of environments. Over the last 3 million years, four of these species gradually became extinct. And as we have seen, the final cheetah species, *Acinonyx jubatus* of Africa, suffered a population bottleneck during the Pleistocene and is now hanging on by a thread. Animal conservationists have various plans for breeding cheetahs in zoos and protecting them in the wild to prevent further loss of population. While extinction is the ultimate evolutionary fate of all species, one hopes that the disappearance of the world's fastest-running animal comes later rather than sooner.

CONNECTIONS

Ten thousand years ago, cheetahs bounded after prey in many parts of the globe. Today, however, they are limited to just two parts of Africa, with one population of interbreeding cats in each area. Populations like these constitute the basic units of evolution—the units in which allele frequencies change and evolution occurs. Events that bring about these changes in allele frequencies select the parents for the next generation. Some of these events influence alleles that lead to an organism's reproductive success: Cheetahs bearing alleles for smaller size were more effective at reaching maturity and reproducing than large cheetahs; thus, the frequency of alleles for small size increased in the population, and cheetahs became smaller as a species. Other genetic events occur by chance and have less predictable—but still powerful—consequences. These include migration, mutation, and even environmental catastrophes, such as those that forced cheetah populations through bottlenecks that diminished their genetic variability. Unfortunately, as we have seen, most of the species now alive are currently being forced into this kind of environmental and genetic catastrophe by the rapid expansion of the human population. Only by applying the principles of population genetics and by taking political action to preserve habitats will our species be able to prevent the extinction of half the world's other living species.

Each individual's genetic constellation is just part of the flow of genes in populations. Populations, however, are not islands; they interact daily with other populations of plants, animals, fungi, and microbes, as well as with their physical environments. The next several chapters describe these interactions and address a basic question: What makes organisms live where they do, the way they do, and in the numbers in which they are found?

Highlights in Review

1. Genetic variation among sexually reproducing organisms is so extensive that an almost endless variety of genotypes can be produced. Cheetahs are an example of what happens when a genetic catastrophe eliminates genetic variability; the species is left with maladaptive traits and low viability.

2. The sources of new genes in a population are mutation, gene duplication and divergence, and exon shuffling.

3. Recombination creates variation by shuffling existing alleles into new combinations.

4. The Hardy-Weinberg principle predicts that the frequency of alleles in a population will not change unless the population experiences selection, mutation, gene flow, nonrandom mating, or small population size.

5. Natural selection, or differential reproduction and survival, is the most important agent of evolution. Evolution is the change in allele frequencies in a population over time.

6. Gene flow occurs when individuals enter or leave populations. Genetic drift changes allele frequencies by nonrandom events that occur more frequently in small populations. Genetic drift can lead to higher frequencies of unfavorable alleles after a population bottleneck occurs, and this can contribute to the animal's extinction.

7. Selectionists hold that most alleles impart some selective advantage or disadvantage. Neutralists, however, hold that most alleles confer no advantages or disadvantages for survival.

8. Directional selection favors one extreme form of a trait; stabilizing selection favors the average expression of a trait; and disruptive selection selects two opposite extreme forms of a trait.

9. A species is a group of populations that can interbreed and produce healthy, fertile offspring.

10. Reproductive isolating mechanisms may prevent crosses between species or make those crosses that do occur unsuccessful at producing healthy, fertile young.

11. In allopatric speciation, populations become geographically separated and then diverge along different evolutionary paths. Parapatric speciation occurs in populations that overlap at their edges. Sympatric speciation occurs in populations that overlap for most of their ranges.

12. Microevolution refers to allele changes that take place within populations. Macroevolution refers to evolution above the level of the species.

13. Adaptive radiation occurs when a species simultaneously fragments into several species that exploit different environments. Convergent evolution occurs when two or more distantly

related species come to look alike by adapting to similar ways of life. In parallel evolution, species change following the same general trend.

14. The physical evidence that evolution has occurred includes: the fossil record; genetic similarity between organisms at the molecular level; homology (structures related in shape and embryonic development but unrelated in function); vestigial organs; and the similarity of embryonic structures that develop into separate, distinct organs in adults of different species. The convergent evolution of unrelated species inhabiting similar environments and evidence from biogeography convince most biologists that evolution has occurred.

15. In coevolution, two species interact so closely that each one affects the gene pool of the other.

16. Phyletic gradualism is the notion that evolutionary change is evenly paced and based on the processes of microevolution. Punctuated equilibrium holds that most species remain stable for long periods and then undergo periods of rapid genetic change in small, isolated populations, resulting in new species.

17. Extinction is the ultimate fate of all species and occurs when every individual of a given species has died.

10. How do microevolution and macroevolution differ?
11. Compare phyletic gradualism and punctuated equilibrium.

Apply What You Have Learned

1. What aspects of the cheetah's population genetics contribute to its status as an endangered species?

2. Opponents of evolution point to the theory of punctuated equilibrium as proof that even scientists doubt Darwin's ideas. Do you agree? Explain your answer.

3. The Hardy-Weinberg principle applies only under very special conditions. What is its value, given that real populations rarely fit these conditions?

4. The northern spotted owl is threatened with extinction because of destruction of its habitat, old-growth forests in the Pacific Northwest. Some people argue that a minimum population of several hundred reproductive pairs of owls is necessary to maintain a viable population. Preserving the habitat for these breeding pairs would prevent logging in much of the remaining old-growth forest. Other people argue that we can log the rest of the forest and save a few owl pairs in zoos, thus maintaining a viable population. Compare the underlying assumptions of these two arguments, and discuss their biological validity.

Key Terms

adaptation, 673	macroevolution, 679
adaptive radiation, 683	microevolution, 679
biogeography, 683	molecular clock, 673
coevolution, 684	natural selection, 673
convergent evolution, 683	parallel evolution, 684
directional selection, 674	population bottleneck, 671
disruptive selection, 674	population genetics, 666
divergent evolution, 683	phyletic gradualism, 685
evolution, 669	phylogeny, 683
founder effect, 671	punctuated equilibrium, 686
gene flow, 669	reproductive isolating
gene pool, 666	mechanism, 677
Hardy-Weinberg principle, 668	sexual selection, 672
	speciation, 677
homology, 681	stabilizing selection, 674
hybrid, 677	vestigial organ, 681

Study Questions

Review What You Have Learned

1. Define gene pool.
2. Name four causes of genetic variation.
3. Explain what the following expressions represent in the Hardy-Weinberg equation: p^2, $2pq$, q^2.
4. State five conditions that limit the Hardy-Weinberg principle.
5. Define evolution in genetic terms.
6. List three ways in which natural selection can affect a population, and give an example of each effect.
7. Explain how each of the following serves as an agent of evolution: mutation, gene flow, genetic drift, nonrandom mating.
8. Define species.
9. Distinguish between allopatric, parapatric, and sympatric speciation.

For Further Reading

BEGLEY, S. "A Question of Breeding." *National Wildlife,* February/March 1991, pp. 13–16.

BENTON, M. J. "The Evolutionary Significance of Mass Extinctions." *Trends in Ecology and Evolution* 1 (1986): 127–130.

COOK, L. M., G. S. MANI, and M. E. VARLEY. "Postindustrial Melanism in the Peppered Moth." *Science* 231 (1986): 611–613.

DIAMOND, J. "The Cruel Logic of Our Genes." *Discover,* November 1989, pp. 72–78.

FUTUYMA, D. J. *Evolutionary Biology.* 2d ed. Sunderland, MA: Sinauer, 1986.

KOEHN, R. K., and T. J. HILBISH. "The Adaptive Importance of Genetic Variation." *American Scientist* 75 (1987): 134–141.

LEWIN R. "Molecular Clocks Run Out of Time." *New Scientist,* February 10, 1990, pp. 38–41.

O'BRIEN, S. J., M. E. ROELKE, L. MARKER, A. NEWMAN, C. A. WINKLER, D. MELTZER, L. COLLY, J. F. EVERMANN, M. BUSH, and D. E. WILDT. "Genetic Basis for Species Vulnerability in the Cheetah." *Science* 227 (1985): 1428–1434.

O'BRIEN, S. J., D. E. WILDT, and M. BUSH. "The Cheetah in Genetic Peril." *Scientific American,* May 1986, pp. 84–95.

SLATKIN, M. "Gene Flow and the Geographic Structure of Populations." *Science* 236 (1987): 787–792.

STEBBINS, G. L., and F. J. AYALA. "The Evolution of Darwinism." *Scientific American,* July 1985, pp. 72–82.

TREFIL, J. "Whale Feet." *Discover,* May 1991, pp. 44–48.

WILSON, A. C. "The Molecular Basis of Evolution." *Scientific American,* October 1985, pp. 164–173.

Population Ecology: Patterns in Space and Time

The Rise and Fall of a Desert Population

More than 2000 years ago, a band of Native Americans migrated from Mexico to one of the world's most forbidding deserts, the Salt River valley of central Arizona, where Phoenix sprawls today. Using sharp sticks and manual labor, small teams of the immigrants (and later their descendants) dug an immense network of broad, deep irrigation canals over 1600 km (about 1000 mi) long (Figure 34.1). In winter, these ditches captured water as it tumbled down from the surrounding mountains and channeled it to the Salt River and Gila River valleys, where the Hohokam Indians established their villages and fields. In summer, the canals trapped and carried water from intense local thunderstorms. With this precious redirected resource, the Indians made the barren desert fantastically productive: Each year, they grew two full crops of corn, squash, beans, barley, cotton, and tobacco over an area of more than 25,600 km^2 (about 10,000 mi^2).

Archaeological evidence shows that the Hohokam were probably the first people to irrigate in North America, and as a result of their amazing agricultural success, they flourished in the desert valleys. Their population doubled and redoubled, and their settlements grew from clusters of simple mud huts to at least 22 huge cities of large, multistoried earthen buildings rising from the desert floor like castles. Simply maintaining the canals, once built, must have required a work force of 20,000, and some experts believe that the adobe dwellings housed 1 million people— nearly the population of present-day Phoenix.

How ironic, then, that about 600 years ago, the Hohokam's huge desert culture suddenly vanished. We may never know the reasons for this demise, but archaeologists think that

FIGURE 34.1 ■ **Canals of the Vanished Hohokam Tribe Helped Support a Large Population in the Arizona Desert.** An archaeological expedition in 1964 and 1965 uncovered miles of these 1000-year-old hand-hewn "lifelines," which conveyed water from the surrounding hills to corn, bean, and squash fields in the Salt River valley.

alkali salts may have built up in overirrigated croplands, poisoning the plants and causing agricultural production to plummet. They think that the Hohokam may also have exhausted the surrounding desert of its limited natural resources—firewood in particular. Hardship and starvation must have swept the ancient settlements, and the death rate apparently skyrocketed. Poignantly, half the skeletons from one of the final graveyards are of very young children whose bones show the obvious signs of malnutrition.

Was the rise and fall of the Hohokam unusual? Is a population of human beings independent of the biological principles that regulate population growth in other species, or is it, too, governed by those principles? And if human populations are so governed, then would knowing and applying the principles help us to predict the future of modern human populations and perhaps enable us to take steps to ensure a livable future without catastrophic collapses? Building on the concepts of population genetics and evolution encountered in Chapter 33, this chapter explores questions like these. It focuses on **population ecology:** the study of the distribution and abundance of organisms in space and time (Figure 34.2). As it turns out, no single factor has an absolutely predictable effect on a population's distribution and size, because many factors impinge on every group of living things in a particular area, including climate, availability of food and water, spaciousness or crowding, and influences on reproductive success. We will look at how population ecologists analyze such variables and how the variables affect populations, regardless of species; then we will apply such principles to the Hohokam's success and later disappearance and interpret their meaning for the human populations of today and tomorrow.

Three themes will emerge from our discussion of population ecology. First, closely interacting biological and physical factors govern the abundance and distribution of any species in any area. For example, the amount of rainfall influenced the number of trees available to the Hohokam for fuel, and the interaction of both variables—rainfall and trees—affected the tribe's population size in the Arizona desert valleys. Second, even though populations have complex dynamics, ecologists can predict a population's growth with formal models. Biologists can use these models to construct graphs that correctly predict the size a population might attain in a given period of time. Finally, the ecology of populations is closely tied to the evolution of species. Genetic adaptations determine a population's size, distribution, and overall success, and changes in population size or location can affect the prevalence of specific genes by natural selection.

FIGURE 34.2 ■ **Saguaros at Sunset: A Population of Desert Giants.** **The striking North American native plants called saguaro (*Carnegiea gigantea*) tower stark but majestic above Arizona's Organ Pipe National Monument. The science of ecology helps answer questions like, Why do saguaros grow where they do and nowhere else?**

As the chapter will show, defining relevant variables, estimating a population's future growth, and identifying the selective pressures all allow an observer to understand why a population of birds or plants or fish thrives (or dies out) in a particular place. The other levels of ecology, presented in later chapters, then broaden that understanding to include the interactions of a population's members with other species, with the immediate physical environment, and ultimately with the biosphere, the global setting in which all living things coexist. This chapter will answer several questions about where populations live and how large they grow:

- At what levels do organisms interact with their surroundings?
- Why are particular organisms found in specific places?
- Why are organisms plentiful at one time but not at another?
- How has the population growth of our own species become the most important ecological factor on earth, and where are we headed?

THE SCIENCE OF ECOLOGY: LEVELS OF INTERACTION

Early peoples amassed a simple form of natural history. Their survival depended on knowing the whereabouts of food sources, and they could describe, often through stories or rituals, the seasons when deer was abundant and the valleys where huckleberries grew along with other edible fruits, nuts, and plants. The modern science of ecology has its roots in natural history, but it applies research tools and the scientific method to probe the explanations behind observed phenomena. What factors cause deer populations to increase in the summer, for example, and what allows a particular valley to bristle with huckleberries?

Ecology is the scientific study of how organisms interact with their environment. Ecologists use their science to determine the principles that govern the distribution and abundance of organisms, and the tenets of ecology have become as crucial to human survival as natural history once was to our ancestors. The more people know about the factors that influence plant growth, the better they can manipulate those factors to produce additional food. That is what the Hohokam did when they channeled water to the desert, and it is what modern plant scientists seek to do with plant breeding and genetic engineering.

A key word in the definition of ecology is *interact*. Organisms interact with the other living things that collectively constitute their **biotic** environment, as well as with the nonliving physical surroundings that make up their **abiotic** environment. For a squirrel, the biotic environment includes the acorns it eats, the blue jays that compete for the nuts, the ticks that parasitize and weaken the squirrel, and the siblings that share the squirrel's territory. The abiotic environment includes the rainfall, sunlight, and soil that regulate acorn production and the hot and cold temperatures that the squirrel must endure in summer and winter.

So numerous are each organism's interactions with other living things and the physical environment that biologists organize ecology, the science of interactions, into a hierarchy of four levels: populations, communities, ecosystems, and the biosphere (Figure 34.3).

A **population** is a group of interacting individuals of the same species that inhabit a defined geographical area. The Hohokam Indians of central Arizona, the saguaro cactus of that same region, the saber-toothed cats of Rancho La Brea in prehistoric Los Angeles, and the alligators of a Louisiana swamp are all examples of populations.

A **community** consists of two or more populations of different species occupying the same geographical area. The Hohokam and the giant saguaro cactus whose fruits they gathered and ate made up a community, and so do alligators and the fish they consume and orchids and the bees that pollinate them.

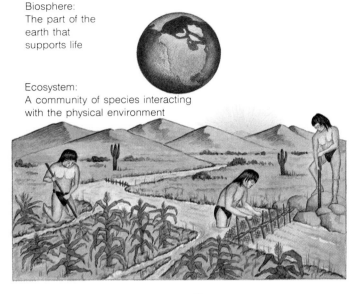

Biosphere: The part of the earth that supports life

Ecosystem: A community of species interacting with the physical environment

Community: Interacting populations of different species

Population: Interacting individuals of the same species

FIGURE 34.3 ■ **Ecological Hierarchy.** The Hohokam Indians were a population of organisms in a desert environment. The Hohokam interacting with cactus, corn, and jackrabbits made up a community. That community, in its dry physical setting, constituted an ecosystem. And all of earth's ecosystems combined make up the biosphere.

Populations and communities include only biotic factors. Such groupings always exist within a physical setting, however, and so ecologists have a third hierarchical level: the **ecosystem,** made up of living things interacting with the physical factors of the environment. The Hohokam's ecosystem included saguaro cactus, corn, and black-tailed jackrabbits, as well as sparse rainfall and the searing summer temperatures of the Arizona desert.

Ecologists recognize that ecosystems are further influenced by global phenomena, such as climate patterns, wind currents, and nutrient cycles. Thus, in the four-tiered hierarchy, groups of ecosystems make up the highest level, the **biosphere**—our entire planet with all its living species, its atmosphere, its oceans, the soil in which living things are found, and the physical and biological cycles that affect them.

The next few chapters examine ecological interactions at increasingly higher levels of organization. Here, we begin with the level of population and discuss the influences on a population's location in time and space, as well as the question of success as measured by a population's size.

DISTRIBUTION PATTERNS: WHERE DO POPULATIONS LIVE?

Space on our planet can be thought of as an ecological vacuum waiting to be filled by living organisms. Yet few organisms are scattered evenly throughout the world; they exist, instead, in only certain spots and are absent entirely from other spots. What limits where an organism lives on a global scale and in individual locales?

Limits to Global Distribution

Why is it that saguaro cactus and mule deer appear naturally in Arizona, where the Hohokam lived, but oak trees and squirrels occur more commonly in the areas where most North Americans live? Questions and answers involving why an organism's range extends to one part of the world but not another are especially important to us when they involve organisms we use for food or materials or when they involve species that cause disease in people and domestic plants and animals. In general, three conditions limit the places where a specific organism might by found: physical factors, other species, and geographical barriers.

■ **Physical Factors**   Organisms may be absent from an area because the region lacks the proper sunlight, water, temperature, mineral nutrients, or any one of a host of physical or chemical requirements. For example, pine trees from the Austrian Alps photosynthesize best at a cool 15°C (59°F), but the *Hammada* bush of the Israeli desert photosynthesizes best at a scorching temperature of 44°C (111°F) (Figure 34.4). Neither plant could survive in the other's environment because genetic adaptations fostered by natural selection have left each plant specialized for a particular set of physical conditions.

■ **Other Species**   Other species may block survival and limit a population's distribution. If certain species are already firmly established in an area, they may prevent the incursion of new species by monopolizing food supplies or acting as predators or parasites. Large regions of Africa, for instance, are nearly uninhabitable to people or cattle because the resident tsetse fly transmits the protist that causes sleeping sickness.

■ **Geographical Barriers**   A species may be absent from even highly favorable areas because a geographical barrier blocks access. Seas, deserts, and mountain ranges can be so wide or high than an organism cannot crawl, swim, fly, or float across the barrier. Europeans artificially bridged such a gap when in 1890 they introduced about 80 starlings into New York City's Central Park (Figure 34.5). Now, millions of the dark speckled birds chatter throughout America's cities and countrysides. North America turned out to be a prime habitat for these organisms; the obstacle to their earlier dispersal was simply a barrier to access.

The factors that limit an organism's global range are only part of what we must know to explain its distribution. The other part is how the organism is dispersed within its range.

(a) A pine that photosynthesizes best at 15°C

(b) A *Hammada* bush that photosynthesizes best at 44°C

FIGURE 34.4 ■ **Temperature: A Physical Factor That Limits Plant Distribution.** (a) Pine trees of the central Alps grow best at a maximum of 15°C (59°F), while (b) dry-adapted *Hammada* bushes of Israel's Negev Desert grow best at 44°C (111°F).

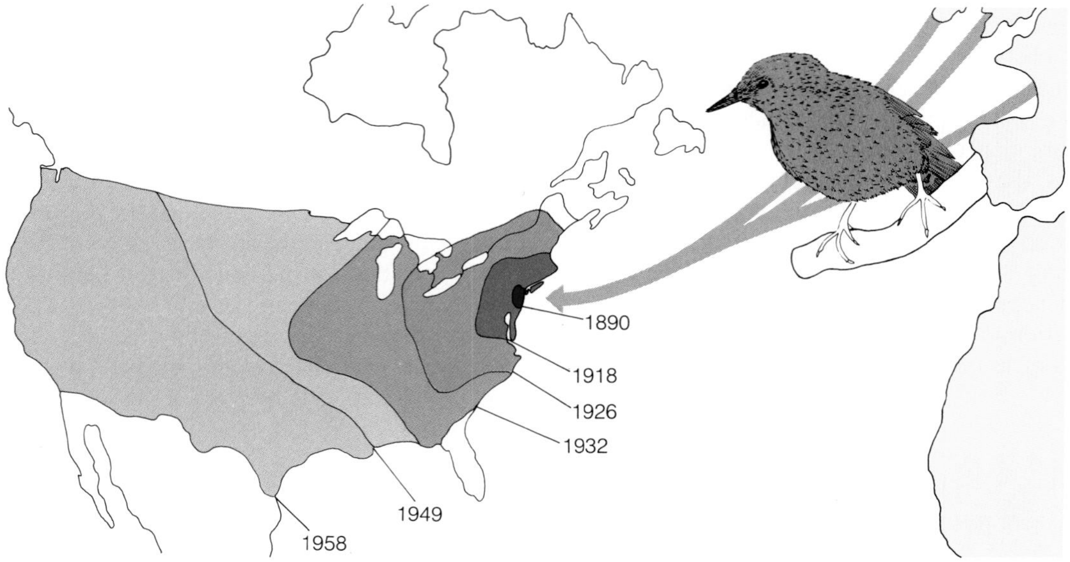

1890
1918
1926
1932
1949
1958

FIGURE 34.5 ■ **Crossing the Atlantic Barrier Allowed Starlings to Spread Throughout North America.** Bird fanciers brought starlings from Europe to New York in 1890, and once the huge, briny geographical barrier had been breached, the birds continually expanded their range westward, reaching the Southwest and Mexico by 1958.

Local Patterns of Distribution

Within their ranges, organisms can be distributed in one of three different patterns: uniform, random, or clumped distribution (Figure 34.6). Organisms have a **uniform distribution** if they are spaced at regular intervals in their range. The apple trees in an orchard have uniform distribution. Such a pattern is rare in natural populations and usually occurs only where physical factors like water and soil quality are also uniform and where organisms compete strongly for some limiting resource. Giant saguaro cactus also have a somewhat uniform distribution on the Arizona desert owing to competition among individuals for water. And creosote bushes have a nearly uniform distribution because their leaf litter gives off toxic substances that inhibit the growth of nearby plants and increase the distance between closest neighbors.

By contrast, organisms will have a **random distribution** if individuals do not influence each other's spacing and if environmental conditions are uniform. Randomly distributed populations include plant species whose fruits are eaten by animals that later drop the seeds haphazardly in their feces. Like uniform distributions, random distributions occur infrequently in nature because resources are seldom uniformly available.

The most common pattern of population distribution in space is a **clumped distribution,** with several members of the population occurring close to each other but a long distance from other groups. Clumping occurs because resources are almost always limited to certain *habitats,* or special areas within a range where an organism can actually live. The Hohokam did not spread out across all of what is now Arizona; instead, they occupied their habitat—river valleys with adequate water and fertile flatlands.

In some species, clumping is tied to social behavior and community activities. This is particularly true of animals that gather in herds, like caribou, or in prides, like lions, or in specific living or nesting areas, like great blue herons. Reproductive processes can also lead to clumping, particularly among plants whose offspring often become established in the parent's immediate vicinity.

FIGURE 34.6 ■ **Distribution Patterns in Local Populations.** (a) Apple trees planted in an orchard have a uniform distribution, but the pattern is uncommon in nature. (b) A random distribution is often found when resources are uniform, but (c) a clumped pattern is the most common distribution, as in live oaks growing on the rolling yellow hills of northern California.

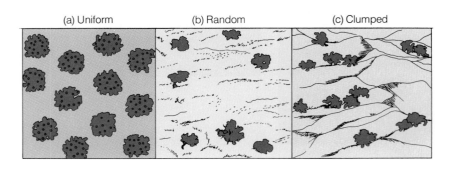

(a) Uniform   (b) Random   (c) Clumped

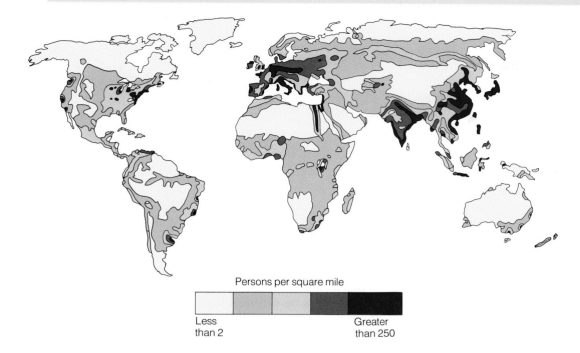

Persons per square mile

Less than 2   Greater than 250

FIGURE 34.7 ■ Worldwide Distribution of Human Population. The crude density of people is less than 2 per square mile over most of the earth's surface. The ecological density, however, is quite high in much of Asia, India, Europe, and eastern North America and around coastlines, rivers, and inland bodies of water the world over.

By describing a species' global range and its local distribution, ecologists define the organism's habitat. Then, by identifying the various biotic and abiotic factors that allow a population to exploit that habitat, they take a big step toward understanding where organisms live and why. The next step is to determine the factors that limit a population's size.

CONSTRAINTS ON POPULATION SIZE

The ancient Hohokam shared a concern with modern-day farmers, fishermen, and foresters: making sure that the populations they harvest remain at optimal levels so that there will be as much grain, protein, or lumber available as possible on a continuing basis. These practical ecologists are concerned with a population's **density,** the number of individuals in a certain amount of space. The number of organisms in an area of a certain size—let's say 100 butterflies per hectare or 50 mice per barn—is the so-called *crude density*. This measure, however, fails to take into account the species' habitat—the fact that the butterflies might cluster around yellow sulfur flowers in the 1-hectare field or that the mice might gather near grain hampers within the barn. A better reflection of an organism's concentration is therefore given by *ecological density,* which counts organisms only where they actually live, in their habitats. The world's human population illustrates the difference between crude density and ecological density quite well: The crude density of the earth's people is less than 3 per 2.6 km² (1 mi²), but as Figure 34.7 shows, the ecological density is quite different from this, because most people live near ocean coasts, riverbanks, and lake shores.

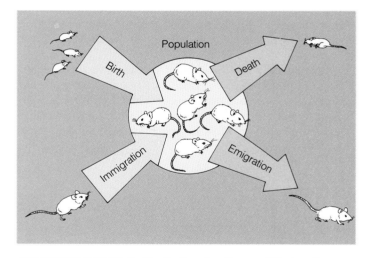

FIGURE 34.8 ■ Birth, Death, Immigration, and Emigration: Factors Affecting Population Size. Birth and immigration tend to increase the size of a population, while death and emigration decrease population size.

Factors Affecting Population Size

The curious biologist is usually not content just to describe a population's ecological density at one point in time, but wants to understand how that population density changes over time. Population size changes when individuals enter or leave the population. Clearly, if more members enter than leave it, the population will grow. Individuals can enter the population by either birth or immigration, and members can leave the population by either death or emigration (Figure 34.8). An equation expresses this concept in shorthand form:

Change in number of individuals in a population
= (number incoming) − (number outgoing)
= (births + immigration) − (deaths + emigration)

The symbol used to represent a change in the number of individuals is *delta N* or ΔN. If the number of individuals gained from births and immigration is exactly equal to the number lost to deaths and emigration, then $\Delta N = 0$, and there is *zero population growth*.

■ **How Survival Varies with Age**   Death and birth are clearly major parameters affecting population size. And for many species, the likelihood that an individual will die depends on its age. The chances that a 90-year-old person will live for one more year are much slimmer than the chances that a 20-year-old will live for one additional year. A **survivorship curve** is a plot of the data representing the proportion of a population that survives to a certain age (Figure 34.9). A table of the numbers used to generate a survivorship curve is called a *life table,* and it shows the **life expectancy** (average time left to live) and probability of death for given ages.

Insurance companies use life tables to determine policy costs for customers of different ages. From the survivorship curve in Figure 34.9, you can understand why insurance companies charge a 70-year-old man more for insurance than a 70-year-old woman: He is more likely to die in the next year than she is.

As one might expect, different species have survivorship curves of different shapes (Figure 34.10). *Late-loss curves* are characteristic of organisms like people and rhinoceroses, with their low mortality in early and middle life and increasing death rate in old age. Most large mammals fit into this cate-

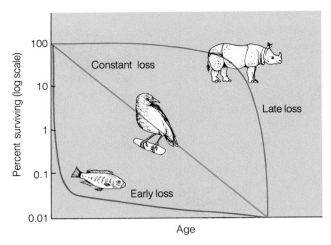

FIGURE 34.10 ■ **Survivorship Curves for Various Species.** With late-loss species, like the rhinoceros, mortality is low in early life but increases for older individuals. For constant-loss species, like most birds, the perils of living are constant; mortality occurs steadily, regardless of age. For early-loss species, like most fishes, surviving early life is quite unlikely, but once past the first months as an easy target for predators, the organism's chances of surviving remain relatively stable until old age.

gory. *Constant-loss curves* reflect a fairly constant death rate at all ages from birth to the end of the life span. Many bird species show constant-loss curves. *Early-loss curves* describe populations in which very young individuals have a high probability of dying, but those that survive this dangerous initial period have a good chance of reaching old age. Many fishes and invertebrates have early-loss curves; in fact, over 90 percent of their mortality can occur during the first 10 percent of the life span. To understand why the shape of the survivorship curve is important, just think what would happen if, in a species with an early-loss curve, most of the young suddenly survived to reproductive age. The numbers of reproducing individuals would be huge, and the population growth rate would soar.

■ **How Fertility Varies with Age**   Survivorship curves help ecologists understand population growth by predicting how many individuals of each age class will leave a population through death. As we saw earlier, birth is the major force that counteracts death. Birthrates, like death rates, also depend on age, and this fact is revealed by *fertility curves,* graphs of reproduction rate versus the age of female population members. (Since only females bear young, ecologists often view populations as females giving rise to more females.) The fertility graph in Figure 34.11 shows that out of every ten American women between the ages of 20 and 30, on average about three will have a little girl in any given year. Women younger than 20 or older than 30 are less likely to reproduce. Birthrates and death rates are frequently the major factors influencing changes in

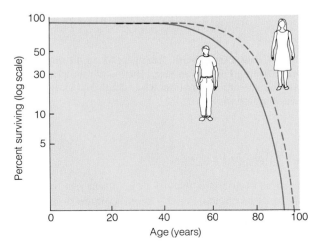

FIGURE 34.9 ■ **Human Survivorship Curve: Most Men Live to Age 70 and Most Women to Age 80, but Succumb Before 90.   For the U.S. population, each year after a person reaches 80, his or her chances of living another year go down substantially. Beyond 90, the drop is precipitous.**

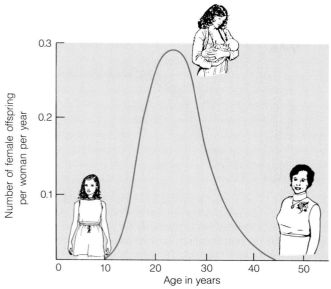

FIGURE 34.11 ■ **Fertility Curve: Younger and Older American Women Are Less Likely to Reproduce Than Are Women Between 20 and 30.** Relatively few American women give birth before age 20 or after age 30. Between those ages, however, a woman's chances are two or three times higher of having a successful pregnancy and of producing a female offspring who will contribute to the population's fertility curve in the future.

of those has one more female per year, there will be 40 female mice at the end of two years, 80 after three years, and so on. The explosive growth results in a **J-shaped curve** representing **exponential growth.**

The speed of population growth is determined by the maximum reproductive rate and the initial population size. If each of the 20 female mice in the preceding example had two female offspring rather than one each year, the population would grow twice as fast, and the *maximum rate of reproduction* (represented by the symbol r_m) would be twice the value calculated for the first example. Moreover, if there are only ten female mice in the initial population producing one female offspring per year, then only ten more will be added by the end of the first year. In other words, the change in the number of individuals in the population depends on both the average reproductive rate and the number of individuals already in the population (N):

Change in number (ΔN)
per unit time (Δt) $=$

$$\text{maximum rate of reproduction } (r_m) \times \text{number already present } (N)$$

or $$\frac{\Delta N}{\Delta t} = r_m \times N$$

population size. By combining the two (and discounting immigration and emigration), population ecologists can make a model for how populations grow.

■ **How Populations Grow** Given plenty of nutrients, space or shelter, water, benign weather, and the absence of predators or agents of disease, every population will expand infinitely, because all organisms have an innate reproductive capability when conditions are ideal. The capacity for reproduction under idealized conditions is called **biotic potential,** and it is amazing. Darwin calculated that a hypothetical pair of impossibly long-lived elephants could give rise to 19 million descendants in 750 years! And most organisms reproduce even faster. If a female cockroach has about 50 surviving daughters each month, the number of female roaches will rise to 800 billion after just seven months. Women seldom approach the maximum human fertility of about 30 children per female. The highest fertility rate recorded for any human population was among the Hutterite communities of Canada's prairie provinces in the early twentieth century, where the average family had 12 children.

Rapid population growth is easier to visualize when plotted on a graph such as the one in Figure 34.12. Let's say that the individuals of a long-lived mouse species grow to reproductive age in one year and that, ignoring males, a population initially consists of ten female mice that each produce one female offspring per year on average. At the end of the first year, the population will include 20 female mice, and if each

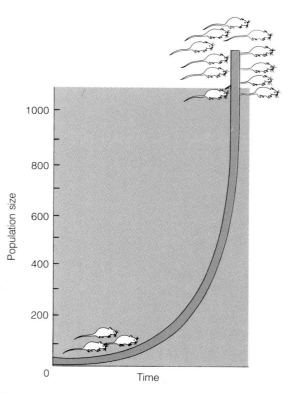

FIGURE 34.12 ■ **Exponential Growth Results in a Population Explosion.** Every species has the innate reproductive potential to found its very own population explosion—a J-shaped curve, heading for an infinity of mice or pear trees or people.

We did not consider the death rate in this formal model, but even with losses due to death, the growth curve would reflect exponential growth; it would simply have taken longer for the population to reach a given number of individuals on its rise to infinity. Under these artificially ideal conditions, any population would follow the J-shaped curve of exponential growth if the birthrate exceeded the death rate by even a small amount.

Fortunately, real organisms in natural situations do not rigidly follow the J-shaped exponential growth pattern. Organisms in nature cannot sustain continued, limitless growth at the full force of their reproductive potential because food supplies and living space are finite. Hence, our planet is not covered with elephants neck-deep in roaches. The realities of supply and demand explain this curb on population growth despite the organism's reproductive capabilities.

Limited Resources Limit Exponential Growth

With the J-shaped curve, ecologists have an idealized standard against which to measure growth in real populations. A classic case was the growth of the sheep population on the island of Tasmania, south of Australia, in the early nineteenth century (Figure 34.13). When English immigrants first introduced the sheep to the new environment, resources were abundant, and the sheep population expanded nearly exponentially for a couple of decades. As the density of sheep on the island rose, competition for limited resources increased,

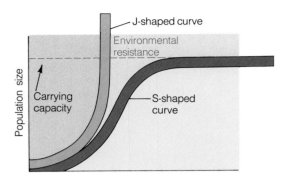

FIGURE 34.14 ■ Environmental Limits on Population Growth: J-Shaped Curves and S-Shaped Curves Compared. A J-shaped curve reflects a species' innate capacity for growth; an S-shaped curve reflects the more realistic growth pattern in a constant but resource-limited environment. The zone between the two curves represents environmental resistance to population growth (limited food and living space, for example), and carrying capacity usually determines the stable population size.

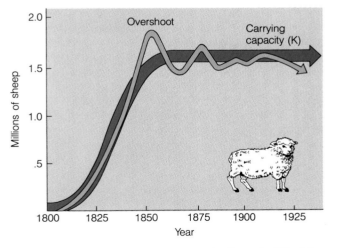

FIGURE 34.13 ■ Sheep on Tasmania: The Growth of Real Populations Often Follows an S-Shaped Curve. Starting in 1800, sheep populations on the island of Tasmania rose exponentially, but by 1855, they overshot *K*, the carrying capacity of the land. Eventually, their numbers oscillated and stabilized, and this plateau is reflected by the S-shaped curve of logistic growth.

and by 1830, each sheep had a smaller share of food and living space. As a result, each individual was less likely to survive and more likely to die, and each had a smaller chance of reproducing. After 1850, the total growth rate decreased, and the population size leveled off at about 1.6 million sheep.

As Figure 34.13 shows, the graph representing population growth began like a J-shaped curve, but flattened into an **S-shaped curve** representing **logistic growth,** a situation in which a large population grows more slowly than a small population would in the same area. The density at which a growing population levels off—1.6 million sheep on Tasmania in the preceding example—is called the **carrying capacity** (represented by the symbol *K*). Carrying capacity is the number of individuals the environment can support for a prolonged period of time. At the carrying capacity, individuals are using all of the resources available to them or the environment. Figure 34.14 contrasts the J-shaped curve, reflecting the innate capacity for growth given limitless resources, with the S-shaped curve, more closely reflecting the growth of real populations in constant but limited environments. After the earliest phases of growth, every point along the J-shaped curve gives a greater population density than do points along the S-shaped curve.

An equation for the S-shaped curve of logistic growth differs from the equation for the J-shaped curve of exponential growth in two ways. First, the growth rate changes from r_m, the innate capacity for increase, which is based solely on birthrates, to r, the rate of population growth per individual, which equals the birthrate minus the death rate. This more accurately reflects natural populations, and r is always smaller than r_m. Second, the model adds a measure of the proportion of resources that remain unused and hence still available for growth:

| Change in number per unit time | = | rate of population growth per individual | × | number of individuals present | × | unused resources |

or

$$\frac{\Delta N}{\Delta t} = r \times N \times \frac{K - N}{K}$$

The part of the equation signifying unused resources can strongly influence changes in population size. If the number of sheep, N, is very small compared with the carrying capacity, K, then the term for unused resources, $(K - N)/K$, will be very close to 1. Another way to say this is that at low population densities, nearly all the resources are unused, so the population can grow at nearly a maximum rate.

As the number of sheep grows to a number that approaches the carrying capacity of the land, the term for unused resources approaches zero. This very low number reflects the fact that only a small proportion of the island's resources are unused. Another way of saying this is that at high densities, almost all of the resources are used, leaving very little left over for further growth. In this way, the formal model of logistic growth can account for the initial population burst of sheep on Tasmania, as well as the period of slower growth and the sheep's eventual stable population size with zero population growth.

Archaeological data show that the Hohokam population probably fit an S-shaped curve, growing from a small initial size 2000 years ago to a relatively stable maximum number and remaining there for two or three centuries before disappearing about 1450 A.D. Examples like sheep on Tasmania and Hohokam in Arizona cause ecologists to ask: What factors prevent unlimited growth?

How the Environment Limits Growth

The limited growth that real populations display comes from a phenomenon called *environmental resistance to growth,* which is due to factors that prevent exponential growth, such as limited food supplies and limited living space. Ecologists call limiting factors like these *population-regulating mechanisms* and classify them as either *density-dependent mechanisms* or *density-independent mechanisms.*

Density-dependent mechanisms become more influential as the population's density increases, and they can have absolutely catastrophic effects. They work by increasing death and limiting reproduction. Disease is a classic example: In dense populations, disease spreads more rapidly; in sparse populations, it spreads more slowly. For instance, in prairie dog towns, as they are called, when the density of the animals is low, the incidence of flea-transmitted bubonic plague is also

FIGURE 34.15 ■ Dense Populations Are More Likely to Be Plagued by Disease. Among these prairie dogs (*Cynomys ludovicianus*), overcrowding allows flea-borne bubonic plague to spread more easily.

very low. But when the dogs are densely packed into their "towns," outbreaks of plague often wipe out entire populations (Figure 34.15). On the other hand, density-dependent factors can stabilize population size, as when a group of squirrels is just large enough to use all the nuts available in a particular forest.

Density-independent mechanisms, as the name implies, exert their effects regardless of population density. These are best illustrated by adverse weather conditions—floods, droughts, or freezing temperatures. These also increase death and lower reproduction, either outright or by affecting food supplies, and take their toll in sparse and dense populations alike. In actual practice, it can be difficult to separate the effects of density-dependent and density-independent mechanisms, because they often work together. For example, a severe drought would kill some individuals in a desert population of any size, but the effects would be far worse if the density were greater. This interaction leads many ecologists to classify population-regulating mechanisms by an alternative criterion: whether the mechanism originates outside or inside the population.

Extrinsic population-regulating mechanisms originate outside the population and include biotic factors, such as food supplies, natural enemies, and disease-causing organisms, as well as physical factors, such as weather, shelter, pollution, and habitat loss (see Box 34.1 on page 700). Rainfall and the availability of saguaro cactus fruits were extrinsic factors that affected Hohokam populations. In contrast, *intrinsic regulating mechanisms* originate in an organism's anatomy, physiology, or behavior. For example, crowded conditions and deple-

BOX 34.1 FOCUS ON THE ENVIRONMENT

AMPHIBIANS: MISSING IN ACTION

Mountain yellow-legged frog. Spade foot toad. Western spotted frog. Tiger salamander. Glass frog. Stomach brooding frog. Goliath frog. Golden toad. If these creatures were characters in a Dr. Seuss story—and they sound remarkably like such characters—the author would probably have them vanishing under mysterious circumstances, one species after another. Dr. Seuss might also convene a congress of puzzled amphibian experts, who would generate a long list of theories and eventually solve the disappearance.

In this case, nature is stranger than fiction. These curious-sounding species are real amphibians, and they are in fact vanishing from California, Colorado, Costa Rica, Cameroon, Haiti, and other places around the world (Figure 1). What's more, a congress did convene recently in Irvine, California, to figure out why. They compared notes on worldwide amphibian populations—the curious as well as the commonplace—and came to a distressing conclusion: we are not just losing a few rare and particularly delicate animals; of the world's 3000 to 4000 amphibian species, all but a handful seem to be fading away. This international group of herpetologists (zoologists who study reptiles and amphibians), meeting in 1990, failed to find a single nefarious Dr. Seuss–like factor behind the worrisome losses. Instead, they suspected that a number of environmental insults may be playing a role, including:

- Destruction of natural habitat, mainly due to the harvesting of timber and the opening of new agricultural land
- Acid rain, which alters the pH of freshwater lakes and streams and can harm amphibian eggs and larvae
- Air, water, and soil pollution (atmospheric gases and toxic materials, such as heavy metals, pesticides, and industrial wastes, can pass through an amphibian's moist skin)
- The stocking of lakes and ponds with sport fish, which consume amphibian eggs
- Global warming, and in some places, unusual drought conditions
- Thinning of the ozone layer and increased penetration of ultraviolet light
- Overhunting, especially in tropical areas, for the exportation of frog legs to European chefs and diners

FIGURE 1 ■ **Why Are Amphibians Like This Haitian Tree Frog (*Hyla vasta*) Vanishing?**

The experts were unable to reach a consensus about the actual causes for dwindling amphibian populations and dispersed to their laboratories and field stations to look for answers. They did agree on one thing, however: A worldwide pattern of vanishing amphibians is an indicator of serious environmental degradation. The earliest land animals, amphibians have survived since long before the dinosaurs and have withstood innumerable climatic shifts and catastrophes. Population losses throughout this entire taxonomic group—with its important place in both aquatic and terrestrial food webs—may well portend continued losses throughout all the kingdoms of life.

If this story had a Dr. Seuss ending, the missing amphibians might turn up in a giant cave somewhere and come hopping, swimming, and crawling back to their old habitats. Truth is sadder than fiction, however, and the animals are simply gone—missing and presumed dead.

---

tion of resources can cause many marsupials, such as kangaroos and koalas, to resorb already developing embryos; this physiological response lowers the rate of population growth. In another example, mouselike creatures called lemmings will migrate away from food-depleted regions, a behavior that lowers the local population density.

■ **Competition** The most important intrinsic regulating mechanism is competition among members of the same species, and competition also depends on population density. As the population grows and resources diminish, competition for

food and space becomes intense among population members. *Scramble competition* occurs when resources are equally available to all members of a population, but individuals must rush for their share or risk losing it. A flock of pigeons descending on a grain lot illustrates scramble competition beautifully: Birds that don't push, peck, and aggressively pursue their morsels risk going hungry.

Contest competition involves literal clashes, usually among males, for social position or territory. The clashing of antlers as male caribou engage in tournaments for leadership of a herd (Figure 34.16) and the jousting of rams in spring are awesome displays of contest competition.

FIGURE 34.16 ■ **Contest Competition: A Clash over Resources.** Male caribou (*Rangifer arcticus*) in arctic Alaska clash over access to mates, a limited resource.

■ **Population Crashes**   While population growth often slows and reaches a plateau, as it did with sheep in Tasmania, there can also be in some species a bust following the boom: a rapid decline following a period of intense population growth. Reindeer introduced onto an island off the southwest coast of Alaska (Figure 34.17) represent a frequently observed growth curve. From an initial population of 25 animals in 1891, the herd grew to about 2000 reindeer in 1938 and then crashed to 8 animals by 1950. The crash can be readily explained on the basis of environmental resistance and carrying capacity. When the reindeer were first introduced, lichens and other food sources were plentiful, having accumulated for centuries

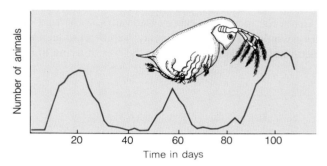

FIGURE 34.18 ■ **Repeated Boom-and-Crash Cycles in Water Flea Populations.**   Populations of *Daphnia magna* have a yo-yo-like existence, booming, vacuuming up available resources, and then crashing—again and again in just 100 days' time.

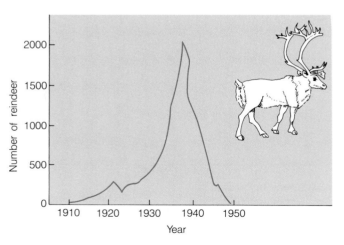

FIGURE 34.17 ■ **Overexploiting Limited Resources Can Lead to a Population Crash: Boom and Crash in Pribilof Island Deer.** Reindeer introduced to a small island in the Aleutian chain southwest of Alaska experienced a boom—exponential J-shaped growth—by eating centuries' worth of slow-growing lichens in a few decades. Once the food ran out, the population crash was rapid and spectacular.

without predation. Thus, the island's carrying capacity was high. After the deer ate the accumulated food, however, new food would appear only as the remaining lichens regrew slowly during each short growing season. The carrying capacity had now changed, and by 1950, the reindeer population had crashed.

Some organisms routinely experience a "boom-and-crash" sequence of population growth. Water flea populations, for example, repeatedly boom and ascend in a roughly J-shaped curve until they overshoot the carrying capacity of their environment, critically deplete resources, and then crash (Figure 34.18).

We have seen that populations in nature usually follow one of two patterns: the S-shaped curve of logistic growth, in which density slowly reaches the carrying capacity of the environment and then remains stable for long periods (as with the Tasmanian sheep), or the boom-and-crash pattern of nearly exponential growth followed by an overshoot of the carrying capacity and a precipitous decline in density (as with the reindeer and water fleas). Individual species clearly respond differently to their environments, but what characteristics of individual species cause them to follow one pattern or the other?

Logistic Growth and Strategies for Survival

Evolution acts on organisms in ways that maximize their individual genetic contributions to future generations. To successfully grow, survive, reproduce, and thus make a genetic contribution to the future, individuals must allocate their limited energy supplies. A very fast growing organism that expends most of its energy enlarging may have little energy left over for reproducing. Conversely, an individual that expends a huge amount of energy attracting a mate or laying thousands of eggs may have little energy remaining for day-to-day survival activities. The way an organism allocates its energy is its **life history strategy.**

Given the inevitable energy trade-offs, what kind of life history would an organism display: rapid reproduction and depletion of stored energy, despite the risk of dying in a brief famine; or delayed reproduction and growth to a larger size, so that temporary food shortages are less of a threat to survival? Species like water fleas, which are tuned by evolution to display rapid rates of reproduction (r) despite risks to survival, are said to show ***r*-selection,** while species like Tasmanian sheep, which have a slow rate of reproduction and can reach and maintain a density close to the carrying capacity (K) of the environment, are said to experience ***K*-selection.**

Organisms that are r-selected show boom-and-crash growth curves and inhabit environments that are generally unoccupied and unpredictable. Weeds, such as dandelions (Figure 34.19a), and small invertebrates, such as fruit flies and water fleas, undergo r-selection. A dandelion's life history strategy is reproduction at as rapid a pace as possible. These plants quickly fill an environment—a newly plowed field, for example—before the environment changes—that is, before winter comes, before the corn grows too large, or before the farmer plows again. Natural selection has caused dandelions to maintain genes that allow rapid embryonic development, early reproduction, and large numbers of small seeds containing few stored nutrients. Dandelions experience an *early-loss* type of survivorship because most of the light,

(a) Dandelions: An *r*-selected species

(b) Rhinoceroses: A *K*-selected species

FIGURE 34.19 ■ **Dandelions and Rhinoceroses: Rapid Reproduction Versus Careful Investment in *r*- Versus *K*-Selected Species.** (a) An r-selected species, such as the dandelion, lives in unpredictable climates and suffers repeated catastrophes owing to weather changes or environmental disturbances. Such species exhibit early-loss survivorship, and their populations grow rapidly and then crash. Organisms that are *r*-selected develop rapidly, usually have small bodies, and tend to reproduce just once early in life. The emphasis is on reproductive productivity that offsets losses induced by the environment. (b) In contrast, *K*-selected species, such as the rhinoceros, live in moderately constant climates and die from factors like competition and disease associated with high population density. *K*-selected species exhibit late-loss survivorship and maintain nearly constant population sizes near the carrying capacity of the environment. They develop slowly, reproduce later in life rather than earlier, have several offspring from sequential reproductive events, and tend to have large bodies. The emphasis is on efficient competition for resources.

windborne seeds die shortly after germination. Our expression "to grow like a weed" reflects the *r*-selected life history strategy.

K-selected species, on the other hand, inhabit more stable environments and tend to maintain their populations near the habitat's carrying capacity. Rhinoceroses (Figure 34.19b), most other mammals, and most large, long-lived plants, such as the saguaro cactus, reflect a *K*-selected life history strategy. In rhinoceroses, embryonic development is slow (gestation takes about 15 months), and reproduction is delayed until the age of about 5 years. Although rhinoceroses have only one calf at a time, the newborns are huge (the weight of an average male college student). Once born, they survive for about 40 years, experiencing a *late-loss* survivorship schedule.

Like other *K*-selected species, rhinoceroses are highly specialized to compete for resources in their environments. Specializations include immense size, thick, armor-plated skin, and a fingerlike extension of the upper lip that pushes grasses and twigs into the mouth. Rhinoceroses are fast approaching extinction because people slaughter them for their nasal horns. While these protuberances are made of the very same protein found in our hair and fingernails, some people have a superstition that the powdered horn bestows aphrodisiacal properties, and they are willing to pay more for the material than for gold, ounce for ounce. Knowing that rhinoceroses are *K*-selected organisms with a slow rate of reproduction, you can appreciate how hard it is for populations of the fascinating beasts to become reestablished once decimated.

In our earlier discussion of the Hohokam's rise and fall, we noted that the science of ecology could analyze our own species' population patterns and make predictions about the future. The final section of the chapter does just that.

THE PAST, PRESENT, AND FUTURE OF THE HUMAN POPULATION

People have achieved unparalleled mastery over their environment through agriculture, medicine, sanitation, transportation, and industrialization. Are we—with our gleaming cities, our gigantic corporate farms, our burgeoning global population—governed by ecological rules? Or have we somehow moved beyond booms, crashes, and growth curves? Stated more formally, to what extent do general ecological principles on the distribution and abundance of organisms apply to populations of people? The answer involves a backward glance at human history, long-term population trends, and some predictions for our future population growth.

Trends in Human Population Growth

You can see the history of human population growth and its staggering current proportions at a glance in Figure 34.20. In the first phase of human history, from our species' origin to about 10,000 years ago, the population grew slowly as people existed by hunting animals and gathering naturally occurring roots and fruits. The worldwide population was probably about 10 million by 8000 B.C. As a species, we seem to fall into the *K*-selection category because of our slow development, long lives, large bodies, our relatively few offspring, our extended and intensive parental care, and our highly specialized brains that help us compete for resources with cunning efficiency.

Population growth accelerated during a second phase beginning about 10,000 years ago, when people started planting and tending crops and domesticating animals in the so-called **agricultural revolution.** The shift to agriculture was rapid and worldwide, perhaps because people are so adaptable and can transmit their culture, or ways of living, to others. As agricultural techniques spread and improved between about 8000 B.C. and 1750 A.D., world population increased from 10 million to about 800 million. Since agriculture allows more efficient use of resources, its practice increases the environment's carrying capacity for humans. The Hohokam, with their intensive irrigation of desert river valleys, exemplify this stage in human cultural development. In its natural state, the Arizona desert has a very low carrying capacity, but irrigation increased the amount of corn the Hohokam could grow and the number of people the desert could support.

A third phase of growth began in eighteenth-century England with the **industrial revolution.** Inventions and scientific advances triggered vast changes that transformed a populace living mainly as farmers, craftspeople, and merchants into a population working mainly in factories and living in crowded cities (see Figure 34.20). In the next 250 years, much of the world would follow this industrialization and social upheaval. The steam engine was a key invention, and its impact was enormous. A farmer with a steam engine attached to a tractor could accomplish the work of dozens of people in a single day and thus increase food production. A steam-driven train or ship could rapidly distribute food and other necessities of life and thus blunt the impact of local famine.

The towering ascension in the graph of human population growth should unnerve you. It is the familiar J-shaped pattern of exponential growth, much like that of the island reindeer just before they overexploited their environment and suffered a population crash. By analyzing the causes of our own population boom, ecologists hope to learn how humans can avert a crash in the future.

Underlying Causes of Change in Human Population Size

How did the agricultural and industrial revolutions quicken the pace of the human population explosion? Ecologists and historians alike have wondered whether the invention of agriculture *allowed* human populations to increase, or whether

FIGURE 34.20 ■ **Human Population Bomb: Our Species' Exponential Growth Rate Is Based on Cultural Advances in the Use of Environmental Resources.** (a) Human population growth since prehistoric times displays a classic J-shaped exponential curve. Currently, our population doubles in about 40 years. (b) Throughout most of human history, we lived as hunter-gatherers like the South African boy shown here. (c) In our agricultural phase, from about 8000 B.C. to about 1750 A.D., populations grew steadily and lived much as these Bolivians do now by farming the land. (d) The industrial phase began in eighteenth-century England with urban sweatshops like this hat factory (which was actually in New England). This phase ushered in the exponential growth that continues today.

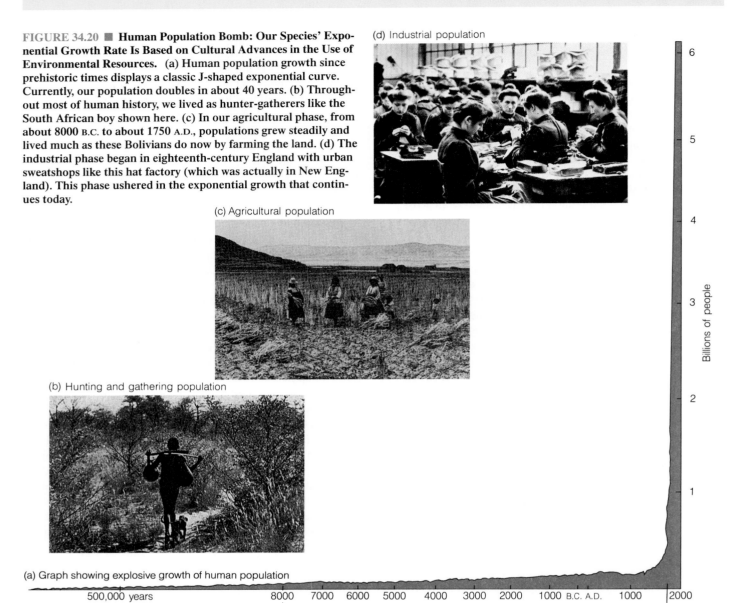

(d) Industrial population

(c) Agricultural population

(b) Hunting and gathering population

(a) Graph showing explosive growth of human population

Billions of people

500,000 years   8000 7000 6000 5000 4000 3000 2000 1000 B.C. A.D. 1000 2000

◀——————Hunting and gathering phase————▶◀————————————Agricultural phase————————————▶ Industrial phase

people were *forced* to invent agricultural practices to help support population densities that were already exceeding the carrying capacity of the land where they lived. Many observers believe the latter and suggest that population growth has been a constant feature of the human experience, continually forcing people to adopt new strategies for increasing the amount of food their land could produce. If true, this necessity has led to some marvelous inventions, including "green revolution" supercrops and genetically engineered hybrid food plants. Carrying capacity, however, cannot be increased forever; the productivity of the land must, at some point, be reached and exceeded.

■ **Birthrates and Death Rates in Developing Nations** To understand the causes of human population increase, particularly the tremendous surge after the industrial revolution, we must recall that for an S-shaped curve, population growth rate equals the birthrate minus the death rate (Figure 34.21). Prior to 1775, the birthrate in developed countries like Sweden was just slightly higher than the death rate, and so the population enlarged at a constant low rate. After 1775, as industry expanded, people enjoyed improved nutrition, better personal and public hygiene, protection of water supplies, and the reduction of communicable diseases such as smallpox. With these factors came a gradual decline in the death rate. As Fig-

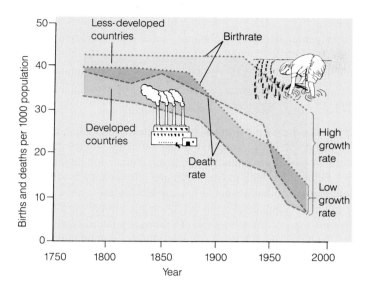

FIGURE 34.21 ■ **The Difference Between Birthrate and Death Rate Detonates the Population Bomb.** Industrialization in developed nations cut the death rate, and in time, a decrease in the birthrate followed. The gap between them has narrowed, and the population growth rate is now very low in the developed countries. In developing nations, the death rate also fell, but a decline in the birthrate did not follow immediately, and the wide gap has allowed a high growth rate and an immense and continuing expansion of population.

ure 34.21 shows, while the death rate began to decline in 1775, the birthrate did not start to drop in developed countries until about a hundred years later. Consequently, each year many more people were born than died, and this translated into an increase in the rate of population growth. By the last decade of the twentieth century, both the birthrates and death rates in industrialized nations had dropped to all-time lows, and the gap between them had once again narrowed. A changing pattern from high birthrate and high death rate to low birthrate and low death rate is called the **demographic transition.**

In industrial Europe and North America, the demographic transition occurred in the first half of the twentieth century and reduced overall growth rates to low levels. The populations of Africa, Asia, and Latin America, however, have continued to grow at immense rates for much of this century (Figures 34.22). The concept of a demographic transition helps explain why. The death rates in those countries remained high until the mid-twentieth century, when the importation of Western medicine and public health technology helped spare lives in record number. Simultaneously, however, the birthrate remained high, based largely on traditional cultural practices. Thus, the gap between the two rates widened. The huge disparity between death rates and birthrates caused an enormous net growth rate and hence an immense population boom in the twentieth century. One of the biggest and most potentially disruptive consequences of this alteration in Third World population size is an increase in the proportion of young people.

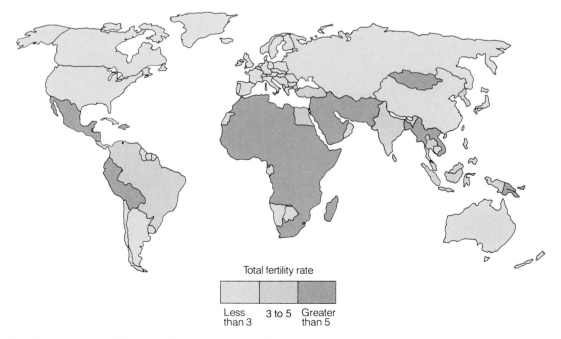

FIGURE 34.22 ■ **Global Patterns of Human Birthrate.** In virtually all developed nations, the total fertility rate is less than 3 children per woman, while in less-developed countries, the rate is 3 to 5 or more. A notable exception is China, where institutionalized policies for late marriage and one-child families have slashed the birthrate in just a few decades.

Growth Rates and Age Structure

A sure sign of a population's growth rate is its **age structure:** the number of people in each age group (Figure 34.23). The age structure of a growing Swedish population in 1900—a time when the death rate had already declined substantially but the birthrate had yet to fall—shows a high percentage of people in the younger age classes. This results in a pyramid-shaped age distribution. The Swedish population's age structure in 1950 shows few teenagers, owing to a low birthrate in the years before World War II, as well as a bulge, the postwar baby boom, of children less than ten years old. By 1977, after the Swedish birthrate had dropped very close to the death rate, each age class was only slightly smaller than the younger one below it. If current trends continue, by 2025, Sweden will have a bullet-shaped age profile with almost no difference between births and deaths until the end of the human life span (the "point" of the "bullet").

Age structure is significant because it can be used to judge a population's growth status. Rapidly growing populations generally have many young individuals and hence a pyramid-shaped age profile, like Sweden in 1900 or Mexico in 1977 (Figure 34.23e). In contrast, stable or declining populations tend to be more bullet-shaped, like Sweden in 1977. With one look at data graphed this way, you can infer the kinds of social services that will be needed by different populations: schools for pyramid-shaped populations, health-care facilities for the elderly in bullet-shaped ones.

What Causes Birthrates to Decline?

By analyzing the demographic transition, it is clear that the key to population control is a low birthrate. At the heart of controversies surrounding population control is a dispute over the factors that lead to low birthrates. Some sociologists feel that birthrates decline because people who have moved to the cities view large families as an economic burden, while those in rural areas continue to see children as potential farmhands and a source of security in old age. Others contend that people have always wanted fewer children, but only widespread knowledge and availability of contraceptives make low birthrates possible.

If improved socioeconomic conditions are a prerequisite for reduced birthrates, then government education programs on contraceptives and the benefits of reduced family size can only be effective after a population experiences an improved standard of living. Most populations with high growth rates, however, cannot afford to wait for improved economic conditions before curtailing their growth, because the added pop-

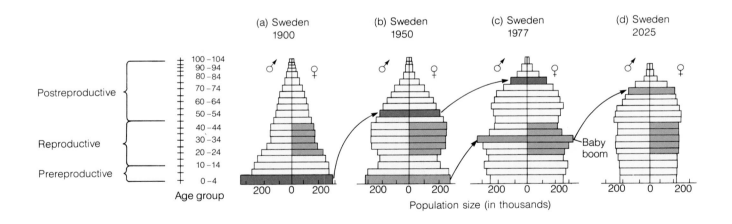

FIGURE 34.23 ■ Age Structure Diagrams Help Reveal Human Population Growth and the Potential for Future Explosions. In these graphs, males are represented on the left, females on the right, and the length of each bar signifies a given age group's proportion of the population. (a) Sweden had approximately 310,000 boys up to four years of age in 1900, compared to about 290,000 girls in this age group. (b) By 1950, when the group members were middle-aged, these figures had dropped to about 210,000 males and 220,000 females; and (c) by 1977, the groups had shrunk still further to 110,000 males and 150,000 females. (d) By 2025, Sweden will have a bullet-shaped age structure. In these graphs, the green bands show reproductive-age females in each population. (e) Note the pyramid-shaped age structure of modern Mexico, with its high birthrate and low death rate and the huge population segment that will soon be reaching reproductive age.

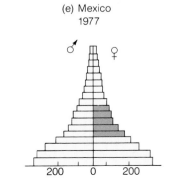

FIGURE 34.24 ■ **A Successful Species?** Will human populations level out from exponential (J-shaped) growth to logistic (S-shaped) growth, or will a crash follow our species' 250-year boom?

ulation actually impedes further development. Hypotheses aside, everyone agrees that even if developing nations immediately instituted stringent regulations over birthrates, the global population would continue to expand for some time.

Population of the Future

The number of people alive today is greater than the total number who have ever lived and died before us. The press of humanity totals over 5 billion people, and if present growth rates continue, another 5 billion will be added in the next 40 years (Figure 34.24). Even if, from this year forward, family size were reduced to the replacement rate of 2.2 children per woman, the global population would continue to grow for decades, because the youthful citizens of countries like Mexico and Nigeria will soon attain reproductive age. Estimates for a stable size for the human population by the year 2040 range from a low of 8 billion, assuming rapidly implemented birth control programs, to a high of 14 billion, with less successful birth control campaigns. Regardless, our planet's human population will either double or triple in the not-so-distant future.

These astounding figures lead us, quite logically, to ask whether the planet can support 8 to 14 billion people at a reasonable standard of living. At present, no one can accurately predict what the earth's ultimate carrying capacity will be. Scientists do suspect that coal, oil, and some of the other resources on which we now depend may well become exhausted early in the twenty-first century. One thing is certain: Without dramatic steps taken immediately and decisively, the crush of humanity will reduce or forever destroy complex and delicate biological systems such as tropical for-

ests, as well as millions of individual species that have taken entire geological eras to evolve on our planet.

CONNECTIONS

The study of ecology, with its organized approach to understanding population sizes, densities, and growth rates, allows us to reconstruct a plausible explanation for the rise and fall of the Hohokam population. These Native Americans flourished in the harsh Southwestern desert, where other peoples could not survive, because they understood some of the interacting factors that limited the abundance and distribution of their crop plants. They introduced corn and squash to the Salt and Gila river valleys (just as Europeans introduced starlings into New York City 2000 years later), and in the process, the tribe overcame the great desert barrier that had prevented those plants from expanding northward on their own.

The Hohokam also identified the abiotic factors (mainly water) that limited the distribution range of their crops and manipulated their environment by irrigation. And they revised the corn genome over generations by selecting seeds from the most drought-resistant plants, altering the plant's evolution, and eventually producing corn that would mature with water from a single irrigation event. After two millennia, modern plant breeders have yet to better this record.

The Hohokam's achievements raised the carrying capacity of their desert environment for both corn and people. Then, however, the density of their settlements grew in a slow but relentless upswinging J-shaped curve. A population biologist taking a census during this time would have found a pyramid-shaped age structure, with many young people at the base and few old people at the tip.

As the population continued to grow, environmental resistance must have caused hardships, increased mortality, a declining birthrate, and a flattening of the growth curve to an S-shaped logistic model of population growth. There simply was not enough good cropland to keep up with an exponential rate of growth. If the tribe experienced a few years of drought, as may have happened around 1450, the desert's carrying capacity would have dropped drastically, leaving the population density well above what the environment would support, like the Alaskan island reindeer. Resources would have vanished. Disease and mortality would have struck the prereproductive class, changing the survivorship curve and causing the death rate to exceed the birthrate. Eventually, the population crashed. Despite great cultural advances, people are clearly constrained by ecological principles.

In this chapter, we focused on the dynamics of populations of individual species—sheep, water fleas, reindeer, dandelions, rhinoceroses, and people. Species, however, do not exist as ecological islands. In the next chapter, we will study communities and see how the distribution and abundance of any particular species depends on interactions with many other species sharing the same environment.

Highlights in Review

1. Biotic and abiotic factors interact in governing the abundance and distribution of species in an area.

2. Biologists identify four levels of ecological interactions: populations of individuals of the same species in an area; communities of interacting species in an area; ecosystems, which are communities in their physical settings; and the biosphere, which consists of all ecosystems and global physical influences combined.

3. Temperature, sunlight, and other physical factors; the activities of other species; and geographical barriers, such as oceans and mountain ranges, can limit where a species lives.

4. Within their ranges, organisms can be uniformly distributed, randomly distributed, or clumped, depending, often, on the distribution of resources.

5. To understand a population's distribution, one must determine crude density, which is the number of organisms in an area, as well as ecological density, which is the number of organisms in the actual habitat within the area.

6. Change in population size, or ΔN, equals births and immigration minus deaths and emigration.

7. The probability of survival varies with an organism's age and can be plotted on a survivorship curve. In a late-loss curve, death usually comes late in life; in a constant-loss curve, there is a constant likelihood of death at every age; and in an early-loss curve, most individuals die young, but the few survivors are often long-lived.

8. The innate capacity for reproduction, or biotic potential, is high in all species. When resources are unlimited, growth follows a J-shaped exponential curve. When environmental resources are limited, population growth often follows an S-shaped logistic curve. The plateau of the S-shaped curve corresponds to the carrying capacity (K), the number of individuals the environment can support for an indefinite period of time.

9. The environment can limit population growth through density-dependent mechanisms like disease or density-independent mechanisms like adverse weather conditions. An alternative classification of environmental limits to growth considers extrinsic mechanisms like limited food supplies versus intrinsic mechanisms like the tendency to migrate when food supplies drop and competition between population members or with other species.

10. Populations that overexploit environmental resources can crash, either once, as did the reindeer on the Alaskan island, or repeatedly, as do water fleas with their boom-and-crash cycles.

11. Individuals allocate their limited energy supplies through r- and K-selected life history strategies. The r-selected species, such as dandelions, reproduce rapidly despite risks to short-term survival, while K-selected species, like rhinoceroses, grow and reproduce slowly, their populations tending to match the habitat's carrying capacity.

12. Our planet's human population has displayed a classic J-shaped exponential growth curve, largely a result of our ability to expand carrying capacity through cultural advances.

13. A demographic transition from high birthrates and high death rates to low birthrates and low death rates took place in developed nations during the twentieth century, and bullet-shaped age structure diagrams reflect the change. Developing nations have yet to complete a similar demographic transition. The result is pyramid-shaped age structures with high reproductive potential.

14. Based on the large Third World populations of reproductive age, our global population is likely to double or triple in the next 50 to 150 years, even with successful birth control campaigns. Ecologists do not know whether the earth's carrying capacity is large enough to support the predicted human populations of 8 to 14 billion people.

Key Terms

abiotic, 692
age structure, 706
agricultural revolution, 703
biosphere, 693
biotic, 692
biotic potential, 697
carrying capacity, 698
clumped distribution, 694
community, 692
demographic transition, 705
density, 695
ecology, 692
ecosystem, 692
exponential growth, 697

industrial revolution, 703
J-shaped curve, 697
K-selection, 702
life expectancy, 696
life history strategy, 702
logistic growth, 698
population, 692
population ecology, 691
random distribution, 694
r-selection, 702
S-shaped curve, 698
survivorship curve, 696
uniform distribution, 694

Study Questions

Review What You Have Learned

1. What is the biotic environment? What is the abiotic environment?

2. Define the following terms, and give an example of each: population, community, ecosystem, biosphere.

3. How do crude density and ecological density differ?

4. The growth of one insect population follows a J-shaped curve, while that of another follows an S-shaped curve. Explain the difference.

5. Do resources influence a population's size? Explain.

6. In the equation for calculating population growth, what do the symbols N, ΔN, K, r, and r_m represent?

7. How does a density-independent mechanism (a population size control) differ from a density-dependent one?

8. Give an example of how a population can "boom" and then "crash."

9. Name a species that shows r-selection and one that shows K-selection, and explain the difference.

10. What main factors have contributed to the J-shaped growth pattern of the human population?

11. Compare the overall demographic changes in industrialized countries with those in Third World nations.

12. According to Figure 34.23, were there more females than males in all age groups in Sweden between 1900 and 1977? Explain.

Apply What You Have Learned

1. Many young Americans are now deferring parenthood until age 30 or later. How will this trend affect the fertility curve shown in Figure 34.11?

2. Some people believe that powdered rhinoceros horn is an aphrodisiac. How does this superstition affect the growth of the world's rhino population?

3. As the human population explosion continues into the twenty-first century, do you believe its J-shaped curve will shift to an S-shaped curve? Support your answer.

4. As a fisheries biologist, you are assigned to determine the level at which a fish population should be maintained to allow the maximum sustainable catch. You have found that the carrying capacity of the environment is 1 million fish. One of your colleagues argues that the population should be maintained at 100,000 individuals. Another biologist says that the catch can be maximized if the population is maintained at 900,000 fish. Another says that the maximum sustainable catch will be achieved from a population of 500,000 fish. Which of these arguments is most likely to be correct? Defend your answer.

For Further Reading

BARINAGA, M. "Where Have All the Froggies Gone?" *Science* 247 (1990): 1033–1034.

BEGON, M., J. L. HARPER, and C. R. TOWNSEND. *Ecology: Individuals, Populations, and Communities.* 2d ed. Boston: Blackwell Scientific Publications, 1990.

BLAKESLEE, S. "Scientists Confront an Alarming Mystery: The Vanishing Frog." *New York Times,* February 20, 1990, p. B7.

CALDWELL, J. C., and P. CALDWELL. "High Fertility in Sub-Saharan Africa." *Scientific American,* May 1990, pp. 118–125.

EHRLICH, P. R., and A. H. EHRLICH. *The Population Explosion.* New York: Simon & Schuster, 1990.

HAURY, E. W. "The Hohokam: First Masters of the American Desert." *National Geographic,* May 1967, pp. 670–701.

KEYFITZ, N. "The Growing Human Population." *Scientific American,* September 1989, pp. 119–126.

KREBS, C. J. *Ecology: The Experimental Analysis of Distribution and Abundance.* 3d ed. New York: Harper & Row, 1985.

TORREY, B. B., and W. W. KINGKADE. "Population Dynamics of the United States and the Soviet Union." *Science* 247 (1990): 1548–1552.

The Ecology of Communities: Populations Interacting

FIGURE 35.1 ▪ **Coevolutionary Partners: The Yucca Moth and Yucca Flower.** *Yucca whipplei,* also called Our Lord's Candle, grows in the Mojave Desert and has an intimate relationship with the yucca moth (inset). The moth lays its eggs in the ovary at the base of a yucca flower and then pollinates the same flower. Moths depend on the plant for food and reproduction; the plant depends on the moth for pollination. This mutually beneficial relationship is called mutualism.

The Intertwined Lives of Flower and Moth

The yucca, a tall, stately member of the lily family, grows in hot, dry regions of the western United States. In California's Mojave Desert, the pointed leaves of a *Yucca whipplei* plant jut upward from the parched ground like a sheaf of swords. Above the leaves and stem rises a single awesome flower stalk, 4 m (13 ft) tall and loaded with more than 1000 white blossoms. Each spring, white female moths flutter about the cream-colored yucca flowers (Figure 35.1) and play out a curious set of behaviors in the desert stillness. These activities might seem purposeful, almost insightful, to an observer unschooled in the principles of natural selection. In fact, though, the interplay of moth and yucca is a classic community interaction that has evolved to the benefit of both parties.

Arriving at the younger flowers near the top of the stalk, the female yucca moth visits several blooms and collects a ball of pollen, which she carries beneath her "chin" with specially modified mouthparts. Bearing this pollen ball, she then flies to another plant and enters an older flower near the base of the stalk. Here she extends a long, sharp drill from the tip of her abdomen, pierces the older flower's ovary, and begins to pump her abdomen. The pumping causes eggs to pass through the drill and into the flower. Egg laying completed, the moth now executes one of nature's most remarkable sequences: She moves up to the flower's green stigma, separates a bit of pollen from the pollen ball with her specialized mouthparts, then rubs the pollen directly across the stigma. This ensures that the flower's eggs will be fertilized and that fruits and seeds will develop and serve as the food source for the moth's offspring as they become caterpillars. Fortunately for the plant, the caterpillars do not eat all of the seeds from fertilized flowers as they grow. The plump mature yucca caterpillars eventually emerge from the fruits, drop to the ground, burrow into the soil, and metamorphose. The next season, adults emerge, mate, and repeat the cycle.

The association of yucca plants and yucca moths is a dramatic indication of the complex interactions between species in nature. To fully understand an organism's ecology—its distribution and abundance—it is not enough to consider the individual species and its birthrate, death rate, and life span in isolation, as in Chapter 34. We must go beyond these concepts to see how different kinds of organisms interact in **communities,** which are assemblies of populations of different species in a particular area at a particular time, such as the yucca plants and yucca moths in the Mojave Desert.

By studying such interactions, ecologists have found, for example, that the yucca moth is the yucca plant's sole agent

FIGURE 35.2 ■ Yucca Community: A Web of Interdependence. The mule deer and sap beetle eat yucca flowers, while the ash-throated flycatcher picks off yucca moths. Below-ground, the yucca roots release a soaplike substance that makes scarce water more available for itself and its nitrogen-fixing neighbors.

of fertilization and that the pollen and seeds of the yucca plant are the sole food source for the larval and adult moths. These two species have coevolved until they are absolutely dependent on each other for their continued existence; the yucca's reproduction depends on the moth, and the moth's daily sustenance comes from the yucca.

Beyond these obligatory dependencies, the yucca and the yucca moth interact with other species in their small desert community (Figure 35.2). Some of these interactions have negative effects: As they fly from flower to flower, yucca moths can be targets for predatory birds called ash-throated flycatchers; and the yucca flowers themselves can serve as a food source for small, dark sap beetles and mule deer. Some interactions, however, can be beneficial. The yucca releases a soaplike substance from its roots, for example, that decreases the surface tension of water. By making water "wetter," this material helps the yucca take in moisture more easily from the dry soil and benefits nearby plants in the same way. And if these neighbors fix nitrogen and release some into the soil, then the yucca gains survival advantage from the association.

This chapter describes the various ways that organisms interact in communities and explains how the associations may have developed. As we study communities, three themes will recur. First, interaction with other species is a major limiting factor in the abundance and distribution of organisms. Beneficial interactions, such as the reproduction-food link between yucca plant and yucca moth, may encourage population growth and a geographical spread of both species, while harmful interactions—competition between the moth, beetle, and deer, for example—may limit a species' success. Second, in extremely tight interactions, two species can influence each other's evolutionary fitness, so that both kinds of organisms evolve together in the process called coevolution. Genes governing the shape of the yucca flower and genes affecting the moth's behavior pattern must have evolved together, gradually adapting each species to the other's habits and life history. Finally, interaction with the human species is the most powerful biological factor in the world today. People encroach upon the desert and destroy the yucca's habitat. In fact, few communities are untouched by our species' all-pervasive influence, and we have begun—and will continue—to strongly shape the earth's ecological communities.

Our exploration of community ecology will address several questions:

■ Where and how do populations live in communities?
■ How does competition between species affect community structure?
■ When one organism consumes another, what is the effect on population densities?
■ How do relationships between two species evolve, resulting in benefit to one or both kinds of organisms?
■ What is the structure of communities, and how do communities change over time and space?

HABITAT AND NICHE: WHERE ORGANISMS RESIDE AND HOW THEY LIVE

To understand the intricate web of relationships between populations of organisms in a community, the observer must discover where the organisms live and how they get needed energy and materials. The general physical place in the environment where a certain kind of organism resides is its **habitat.** A habitat is analogous to an organism's "address," or "home." In describing the general places where aquatic organisms live, for example, an ecologist might speak of an open-water habitat, a shore habitat, a muddy bottom, or a surface-film habitat. Likewise, certain birds may be found in grassland, pinyon-juniper, or tropical habitats—all places where birds live.

Whereas a species' physical home is its habitat, its functional role in the community is its **niche.** The niche is analogous to the organism's "job"—how it gets its supply of energy and materials. The yucca's niche, for example, is that of an autotroph, or *primary producer,* deriving energy directly from the physical environment. The niche of the mule deer or yucca moth is that of *herbivore,* obtaining nourishment from plants like the yucca that grow in its habitat. Within the same community, wolves and ash-throated flycatchers are carnivores, catching, killing, and eating mule deer or yucca moths or other animals. A niche, then, is an organism's functional role: what it does in and for a living community.

Ecologists sometimes represent niches diagramatically as Figure 35.3a does for a warbler's niche in the forests of New England. The bird might conceivably eat insects wherever they occur in trees, at any height and at any distance from the trunk. The bird might also nest any time in June and July. The niches of other species depend on factors such as the temperature and humidity the organisms prefer, the size or type of food they eat, their optimal environmental pH, or the water depth where they find food. The potential range of all biotic and abiotic conditions under which an organism can thrive is called its **fundamental niche.** If a warbler could eat insects any place in the tree, it would be operating in its fundamental niche for prey location.

In nature, however, a warbler cannot obtain insects just anywhere because several species of insect-eating warblers compete for food in eastern forests, each species performing a slightly different but specialized role in the community. The different species obtain insects at different heights in the trees and at different distances from the trunk, and their heavy eating comes at slightly different times during the year, depending on when they nest. The myrtle warbler, for instance, eats insects at the base of trees and in lower branches, while the bay-breasted warbler specializes in insects in middle branches, and the Cape May warbler seeks insects at the outer edges of the top branches (Figure 35.3b). Thus, other com-

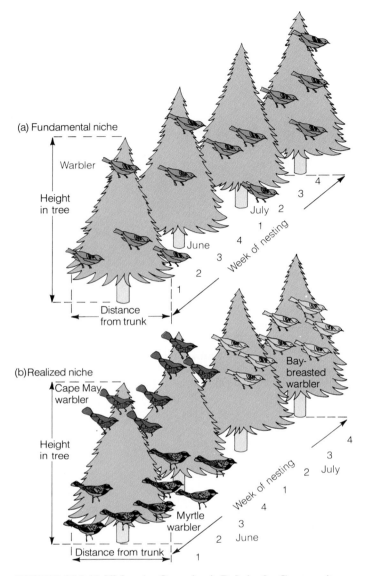

FIGURE 35.3 ■ **Niche: An Organism's Role in the Community.** (a) The fundamental niche is the full range of conditions under which a given organism can operate. As depicted here, a forest warbler can catch insects in a tree at any height and on branches any distance from the trunk, and it can nest any time between early June and late July. (b) The realized niche is the part of the fundamental niche that the organism actually occupies. The Cape May warbler catches insects only in the outer branches toward the top of the tree, and it nests only in June. The bay-breasted and myrtle warblers feed in different places and nest at different times, and thus occupy separate realized niches within the boundaries of the fundamental niche.

munity residents may force a warbler species out of its broader fundamental niche and into its narrower **realized niche:** the part of the fundamental niche that a species actually occupies in nature. The realized niche of the myrtle warbler, as shown in Figure 35.3b, is substantially more limited than its fundamental niche. In general, interactions with other organisms often force a species into a realized niche that is more restricted than its fundamental niche. This restriction is often a major factor in defining a species' distribution and abundance.

The Many Ways Species Interact

Two species can interact in ways that increase, decrease, or leave unchanged the abundance of either or both species. For example, the interaction between yucca plants and yucca moths increases the population of both species. The interaction between a warbler and its insect prey increases the warbler's population but decreases the insect's population size. Ecologists have categorized interactions between species into four general types (Table 35.1). In mutualism and commensalism, neither species is harmed by the interaction, while in competition and predation, one or both of the species suffer. We devote much of this chapter to these four kinds of interactions and to the ways that associations between species and the physical environment affect communities.

TABLE 35.1   Types of Interactions Between Species

Type of Interaction	Species One	Species Two
Competition	Negative	Negative
Predation, parasitism, herbivory, and disease	Beneficial	Negative
Commensalism	Beneficial	Neutral
Mutualism	Beneficial	Beneficial

COMPETITION BETWEEN SPECIES FOR LIMITED RESOURCES

There are never enough good things to go around—good sunny spots in which to germinate and put down roots, good places to build nests, good berry patches, good places to hide, flowers of a certain type, or insects of a particular size. Because of this, different species often compete for the same limited resource, and this interaction restricts the abundance of both species. The key feature of **interspecific competition,** the use of the same resources by two different species, is that one or both competitors have a negative effect on the other's survival or reproduction.

How Species Might Compete in Communities

To see how competition works, imagine a population of red birds that feeds on mosquitoes. The red bird population will continue to expand until competition develops among members of their own species for mosquitoes and other available resources; population growth will then slow as it approaches the carrying capacity of the environment. Now, what would happen if a population of blue birds joined the community and began to compete with the red birds for mosquitoes? With fewer resources available, the carrying capacity will decrease for the red birds, and their population density will fall. How many red birds will eventually survive in the area depends first on how well the red and blue species compete with each other for the limited resources, and second on how crowding affects each species' population growth. If, for example, a high density of one species (say, red birds) affects the growth of the other species (blue birds) more than it affects itself, then the red birds will retain the niche. A situation like this, where one species eliminates another through competition, is called **competitive exclusion.**

In another interesting possibility, competition can affect each species' own population more than it affects the other species. In this case, the density of the red bird population would strongly inhibit its own further growth, and the blue bird population would powerfully slow its own growth. As a result, the two species would end up coexisting.

These possibilities are predicted by a mathematical model and look good on paper or on a computer screen. But do they happen with real organisms? In the 1930s, Russian biologist G. F. Gause designed a laboratory experiment to find out. He began by selecting two species of the slipper-shaped protist *Paramecium,* a larger-celled species and a smaller-celled species that compete for the same food (bacterial cells). When Gause grew each species in its own container, its population growth reflected the standard S-shaped curve (Figure 35.4a and b). When he mixed the two protists, however, and grew them in the same container, interspecific competition occurred, and the smaller-celled species drove the larger-celled species to extinction (Figure 35.4c). In Gause's exper-

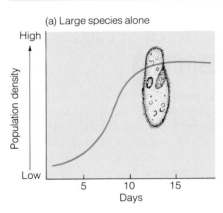

(a) Large species alone

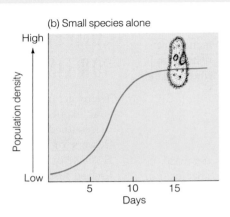

(b) Small species alone

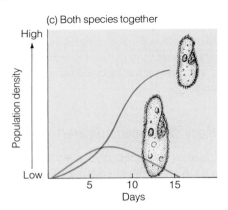

(c) Both species together

FIGURE 35.4 ■ **Competition Between *Paramecium* Species.** (a) A large-celled species (*P. cauda-tum*) experiences logistic (S-shaped) growth in culture. (b) A small-celled species (*P. aurelia*) experiences similar population growth. (c) When the protists compete for the same resources in the same culture, the success of the smaller-celled species spells disaster for the larger-celled species. This is competitive exclusion.

imental system, the smaller-celled species was more resistant to bacterial waste products and hence reproduced more rapidly than the larger-celled species. Gause concluded that competitive exclusion can happen with real species. Experiments with organisms as different as beetles, water fleas, fruit flies, and aquatic plants give similar results and show not only that competition takes place, but that its most likely outcome is competitive exclusion, at least in the laboratory.

Competition in Real Communities: Natural Causes of Competitive Exclusion

It is easy to test the concept of competitive exclusion in laboratory culture dishes and to trace the factors behind the victory of one species and the demise of another. In natural communities, however, it is much harder to pin down the factors critical to each competitor's success because so many biological and physical factors interact. In the western United States, for example, feral burro populations—derived from runaway domesticated animals long ago—have increased dramatically since laws were passed in the 1970s that prohibit their killing. As the burros have multiplied in the arid open lands of Arizona and southern California, however, the number of desert bighorn sheep has rapidly declined (Figure 35.5). Many biologists suspect that the burros, an introduced, or exotic, species to the area, are somehow outcompeting the native bighorn sheep, but they are not sure exactly how.

Sheep and burros eat some of the same desert plant species, but burros may be able to consume or use these plants more efficiently than bighorn sheep. Two kinds of competition may come into play. When two species exploit and have equal access to identical resources, it is called *exploitation competition,* and this may be happening with burros and bighorn sheep. Burros also tend to congregate around desert water holes, however, and often chase other mammals away.

The use of aggressive behavior to keep competitors from a resource is called *interference competition,* and perhaps *this* explains the sheep's demise. Ecologists do not yet know whether the relationship of burros and bighorn sheep will lead to competitive exclusion, with the burros' presence inevitably leading to the sheep's extinction, or whether the relationship will allow a stable coexistence. Experiments with other species suggest possible answers.

The Caribbean island of St. Martin has two lizard species that differ slightly in size but eat the same kinds of insects. To find out whether the two species actually compete with each other, researchers fenced off large squares of land that contained individuals of the smaller-sized species, the larger, or both. They found that where both species coexisted, members of the larger species had less food in their stomachs, grew more slowly, laid fewer eggs, and were forced to perch higher in the bushes than when that species lived alone in an enclosure. Studies like these proved that strong competition does

(a)

(b)

FIGURE 35.5 ■ **Burros Versus Bighorn Sheep: Competitive Exclusion at Work.** Nonnative burros (a) may be eating rings around bighorn sheep (b) and excluding them from their native territories.

Initial
condition

Equilibrium
condition

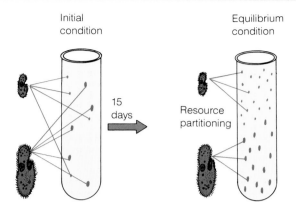

15
days

Resource
partitioning

FIGURE 35.6 ■ Dividing the Spoils: Resource Partitioning in Two *Paramecium* Species. After introduction into a culture tube (left), two species of paramecia can coexist by partitioning resources (right). The smaller species occupies the top territory with greater oxygen penetration, and the larger species inhabits the lower part of the test tube.

exist in natural populations and that the presence of one species can limit another species to its realized (rather than fundamental) niche. Such results make it seem possible that wild burros could indeed restrict bighorn sheep to a smaller realized niche but that the two could continue a stable coexistence.

Competition Can Alter a Species' Realized Niche

The larger lizard in the previous study escaped extinction through an alteration of its realized niche. But is this the way species usually respond to the threat of competitive exclusion? Gause devised a new experiment with *Paramecium* to find out, this time selecting a third species, a challenger, to compete with the previous champion. Since the challenger and champion species consume cells of the same bacterial species, Gause expected them to compete fiercely. But his findings surprised him: Both species coexisted in the same

culture (Figure 35.6). Like the lizards, the paramecia split up the territory, and the challenger occupied a different niche than the champion. While the entire test tube was the fundamental niche for both *Paramecium* species, the realized niche was smaller (only part of the tube) in a mixed culture, and this minimized direct competition. Ecologists use the term **resource partitioning** to indicate the process by which resources are divided, allowing species with similar requirements to use the same resources in different areas, at different times, or in different ways. The warblers of New England forests minimize the harmful effects of competition by resource partitioning—in this case, feeding in different parts of the trees.

In the portions of a species' range where it overlaps the range of a strong competitor, hereditary changes often evolve in both species' physical or behavioral characteristics. Ecologists call such changes *character displacement* and think that it brings about a partitioning of resources by otherwise competing species. Character displacement is an example of *co-evolution*, hereditary changes in two or more species as a consequence of their interactions within a community. For example, imagine an area with huge numbers of elephants, horses, and antelope, as in modern-day Africa or North America some 25,000 years ago. Those animals would serve as an immense food resource for predators like lions, cheetahs, and mountain lions. Cats kill prey with their pointed canine teeth, and animals with larger canines can kill larger prey. Now, if several cat species with approximately equal-sized canines attack and consume prey animals in a given area, they will be limited to prey of the same size, and they will end up in vigorous interspecific competition. Directional selection (see Figure 33.12) would therefore favor different species of cats with different-sized canines. The expected result is indeed the reconstructed history of big cats in North America 25,000 years ago: Huge, saber-toothed cats became specialized for elephants and giant ground sloths; lions became specialized for grazers like buffalo; and the weak-toothed American cheetahs evolved the speed and smaller teeth necessary to kill small gazelles (Figure 35.7). Evolutionary ecologists contend

(a) Specialization for small, fleet prey

(b) Specialization for intermediate prey

(c) Specialization for very large prey

FIGURE 35.7 ■ Character Displacement in Cats Leads to Resource Partitioning. (a) Cheetahs, with their short canines, take down small, fast-running prey—here, a gazelle. (b) Lions, with their medium-sized canines, specialize in larger prey like this Cape buffalo. (c) Extinct saber-toothed cats had huge, daggerlike canine teeth and brought down prey as large as woolly mammoths.

that competition between species and subsequent character displacements probably produce the kind of adaptive radiation seen in honeycreepers (see Figure 33.24c) where one ancestral species diverged into several that specialized on different foods, such as small insects, hard seeds, or soft fruits.

Whether competition leads to competitive exclusion or character displacement and coevolution, it generally decreases the environment's carrying capacity for both species and limits the final density of each population. In contrast, the second major type of community interaction—predation—curtails only prey populations and is thus negative for only one of the two interacting species.

THE HUNTER AND THE HUNTED: PREDATION

Several kinds of interactions between species are beneficial for one species but harmful to the other. Animals like lions or Cape May warblers that kill and eat other animals are **predators,** their food is **prey,** and the act of procurement and consumption is **predation.** Herbivores eat plant parts and often harm the plant without killing it (see Chapter 24). Mule deer and yucca moths are examples of herbivores. Parasites, like tapeworms, feed on a host organism, often without killing it immediately or ever (see Chapter 18). Disease-causing organisms, or *pathogens* (see Chapter 16), are usually fungi, bacteria, or protists that obtain nourishment from a plant or animal host and weaken or kill it. Here we focus on how predation affects the population size of both prey and predator on a short time scale and how hunter and hunted evolve strategies to outwit each other on a longer evolutionary time scale.

Populations of Predator and Prey

In Chapter 16, we encountered a minute but voracious predator called *Didinium* that hunts down *Paramecium* cells and swallows them whole (see Figure 16.1). Laboratory experiments on *Didinium* and *Paramecium* revealed several principles about predator and prey populations. First, as the density of the prey population increases, the predator finds more food to eat, which in turn increases the size of the predator population (Figure 35.8a). Eventually, however, the large number of predators will eat so many prey that the prey population begins to fall. If *Didinium* cells devour all the *Paramecium* cells, then the predator population will crash, too, because its food supply will have been exhausted, and both organisms will finally become extinct in their laboratory setting (see Figure 35.8a). If, however, the experimenter provides the prey with a **refuge,** a safe haven out of the predator's reach, then when the predator has eaten all the accessible prey, it dies out and the prey takes over the environment (Figure 35.8b). Finally, if the experimenter creates a complex environment that offers many partial refuges where prey can survive for a while before predators migrate in and discover them, then the populations of predator and prey tend to oscillate up and down (Figure 35.8c).

For at least 200 years, wild populations of the snowshoe hare and Canadian lynx have periodically risen and fallen on about a ten-year cycle (Figure 35.9). While ecological theory would predict this result, the oscillation does not occur for the reason ecologists originally supposed. Field studies show that shortages of winter food actually cause hare populations to grow smaller, with predators playing only a secondary role. Since the lynx depends on the hare as its main food source, the size of the lynx population drops *along with* that of the hare, instead of the lynx population driving the decline in the prey species' population size.

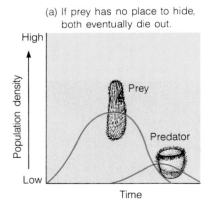
(a) If prey has no place to hide, both eventually die out.

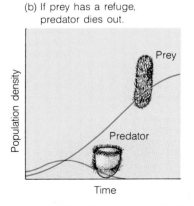
(b) If prey has a refuge, predator dies out.

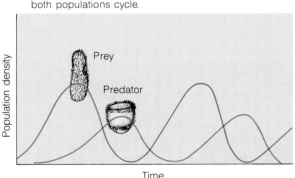
(c) If prey can hide temporarily, both populations cycle.

FIGURE 35.8 ■ **How Populations of Predators and Prey Interact.** **(a) If the predatory protist *Didinium* is grown in culture with its favorite prey species, *Paramecium*, both populations increase until the predator population is so large it eats the prey faster than the prey population can grow. When all the prey have been consumed, the predator also dies out. (b) If the *Paramecium* cells have a refuge inaccessible to the predator, the *Didinium* may starve and die. (c) If the prey species has many partial refuges in which it can grow until "discovered" by the predator, both populations oscillate.**

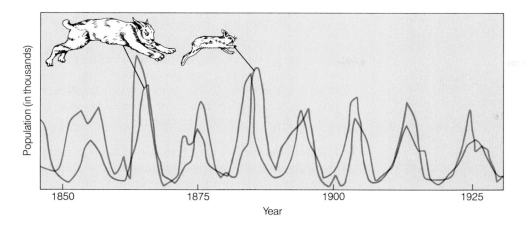

FIGURE 35.9 ■ **The Hare-Lynx Cycle: Responsiveness of Predator and Prey Populations.** In northern Canada, hare populations occasionally "boom" when vegetation is abundant and "crash" when food becomes scarce. When hare populations rise, lynx also increase and devour more of their prey.

Even though predation by the lynx is not the prime force in controlling snowshoe hare populations, other predator species can exert from a modest to a considerable amount of pressure on prey population size. Field observations suggest that predators mainly tend to kill weak animals, especially young or old animals, rather than prey upon animals in their active reproductive years. For example, the wolves that inhabit Isle Royale in Lake Superior mainly kill very young or very old moose, not moose in their prime. Thus, the predator's controlling effect on prey populations is modest.

The case is different, however, for the doglike dingo, Australia's largest carnivore, which feeds on red kangaroos, on the large ground birds called emus, and on other species. More than 20 years ago, sheep ranchers built almost 10,000 km (6200 mi) of dingo-proof fences, then poisoned or shot all the dingos inside the sheep-grazing pastures. Figure 35.10 shows how the prey population responded. With dingos absent, the population density of red kangaroos shot up almost 170 times higher than before. Evidently, a predator like the dingo can strongly limit the density of prey populations.

Since some predators can help regulate populations of their prey, people have used predators to control populations of pest species. Indonesia is the world's fifth most populous country, and its people depend largely on rice for food. In 1985, an outbreak of small insects called brown planthoppers seriously threatened the rice crop, causing it to dry out, fall over, and rot in the fields. The Indonesian government promoted the use of pesticides to control the planthoppers, and this did kill some of them, but it also destroyed the ladybird beetles and spiders that prey upon the planthoppers (Figure 35.11). Without these slow-growing predator species, the planthoppers reproduced rapidly and destroyed even more rice. In 1986, the government began to train farmers to use *integrated pest management* (see Box 35.1 on page 720), which encourages populations of natural predators that can then help control agricultural pests. Farmers using these ecological methods were able to produce more rice per acre more cheaply than farmers who sprayed pesticides indiscriminately. Since then, integrated pest management has become Indonesia's national strategy for protecting rice crops.

As with ladybird beetles and planthoppers or hare and lynx, populations of predator and prey may grow or shrink over the short term, but over the long term, genetic changes can influence the evolutionary balance between hunter and hunted.

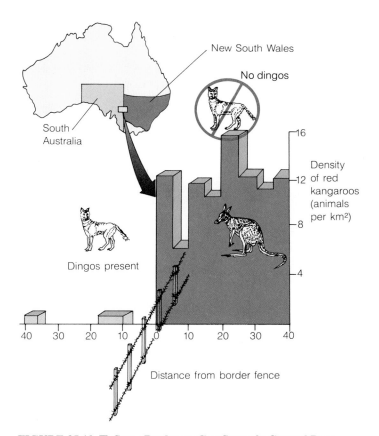

FIGURE 35.10 ■ **Some Predators Can Strongly Control Prey Populations.** Sheep ranchers constructed miles of fences across the border between the states of South Australia and New South Wales, then exterminated all the dingos on their sheep-grazing lands. Because they had eradicated a natural predator that also hunted red kangaroos, these large marsupials proliferated tremendously.

FIGURE 35.11 ■ **Some Predators Help Farmers.** Certain types of ladybird beetles eat brown planthoppers, a pest that threatens Indonesian rice crops. The use of natural predators is part of integrated pest management.

The Coevolutionary Race Between Predator and Prey

Genetic changes in response to natural selection result in a grand coevolutionary race, with predators evolving more efficient ways to catch prey and the prey evolving better ways to escape. It is hard to tell which set of adaptations—the predator's or the prey's—are more fascinating.

■ **Predator Strategies**   Many predators can simply amble, swim, or fly up to a stationary prey target and start to eat. If the prey is mobile, however, then the predator probably needs a way to catch its food, the two main options being *pursuit* and *ambush*. Predators that pursue their prey are selected for speed and often for intelligence, as well. Carnivores store information about the prey's escape strategies and must make quick choices while in pursuit. In keeping with evolutionary pressures, vertebrate predators generally have larger brains (in proportion to their body size) than the prey they catch. Fossils reveal that 60 million years ago, carnivorous predators had larger brains than their hoofed prey (Figure 35.12), and that while selective pressures forced the prey to become more wary, the predators (and their more cunning descendants) always stayed one step ahead.

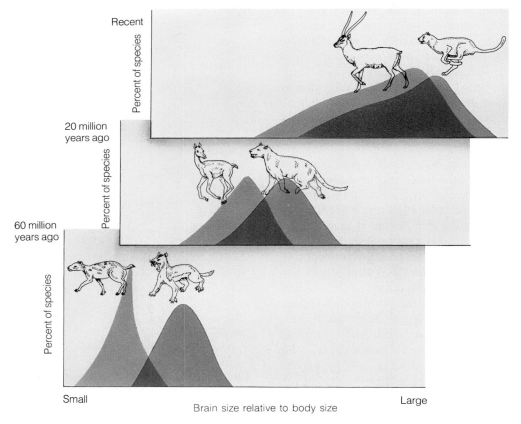

FIGURE 35.12 ■ **The Brain Race: Coevolution of Brain Size in Predator and Prey.** As prey animals' brain size expanded over 60 million years of evolution (green area), brain size of their coevolving predators increased accordingly (red area) and stayed slightly ahead.

(a) A moth that resembles a dead leaf

(b) A *Nemoria* caterpillar emerging in spring resembles a catkin

(c) A *Nemoria* caterpillar emerging in fall resembles a twig

(d) A fish that resembles its surroundings

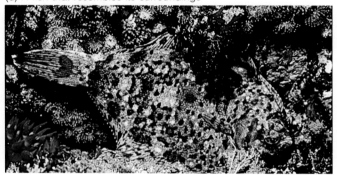

FIGURE 35.13 ■ The Countermeasures of Prey. (a) A moth mimics a dead leaf. (b) Caterpillars of the moth species *Nemoria arizonaria* exhibit a remarkable case of seasonal camouflage. *Nemoria* caterpillars that emerge in the spring and feed on the male flowers (catkins) of oak trees resemble those catkins. (c) Caterpillars that emerge in the fall, however, and feed on leaves look like twigs even though they are apparently genetically identical to the "catkin" caterpillars. Plant chemicals called tannins in the leaves can cause the caterpillar to develop into the twiglike form. When tannins are absent from the diet, the caterpillar develops the catkinlike form. Apparently, predation has selected for two different camouflaged forms, depending on food source, and having two broods each year increases the moth's reproductive potential. (d) This turbot fish blends so well with its background that it is nearly invisible.

For some predators, ambushing is an effective strategy for capturing prey. A familiar example is the frog that ambushes flying insects by snapping out a sticky tongue, hinged at the front of the mouth (see Figure 1.14). Certain mantis insects also ambush their prey by hiding in the open with a camouflaged, plantlike appearance. Those mantises that carry genes for resemblance to the plants they inhabit can be nearly invisible to prey and thus more effective at ambushing their food and surviving to reproduce. An ambush can be even more effective if the predator can lure the prey: The alligator snapping turtle not only blends in with its river-bottom habitat, but has a worm-shaped tongue that entices prey right into its mouth.

■ **The Countermeasures of Prey Species** In addition to rapid running or flying, prey have evolved some remarkably devious tricks that help them avoid being eaten. One defense strategy

avoids confrontation altogether through **camouflage:** the occurrence of shapes, colors, patterns, or even behaviors that enable organisms to blend in with their backgrounds and escape predation. Many insects have evolved shapes that look like twigs, flowers, or leaves—alive or dead—some complete with phony leaf veins (Figure 35.13a). Some even have alternative camouflaged shapes, depending on season and food source (Figure 35.13b and c). Still other insects, and a few amphibians, escape detection by resembling damage or bird excrement on leaves. And behavior plays a role too; if a fish or invertebrate of appropriate color or shape freezes in place, it can resemble the multicolored bottom of a rocky tide pool (Figure 35.13d).

As Chapter 31 explained, chemical warfare is another common defense strategy. Eucalyptus and creosote bushes, for example, produce distasteful oils or toxic substances that kill or harm herbivores. People sometimes plant oleanders (Figure

BOX 35.1   FOCUS ON THE ENVIRONMENT

INTEGRATED PEST MANAGEMENT: COMMUNITY ECOLOGY APPLIED TO AGRICULTURE

Competition is a powerful force in modern agriculture. Species such as corn smut, the cotton bollworm, and the cattle screw-worm compete with us for the bounties of farmland, and while people have had some real victories in controlling "the compe-tition," we have also had some real disasters.

Consider the history of cotton farming in Peru's Cañete Val-ley. Beginning in 1920, farmers planted most of their land with cotton, and in the early years, the few insect pests that bothered their crops were held in check by natural predators and the occa-sional use of mineral insecticides like arsenic and nicotine sul-fate. In 1939, however, a more serious pest appeared quite sud-denly: the cotton bollworm (Figure 1). Within ten years, the predator had so decimated the fields that cotton production plummeted and Peruvian farmers sought more effective ways to control the pest.

In 1949, farmers began using large quantities of DDT and other synthetic chemical pesticides. Cotton production imme-diately soared as populations of the competitor fell, but after the third year of spraying, the insects evolved resistance: Individu-als that happened to already have pesticide resistance alleles survived the insecticide treatment and multiplied, and soon larger and larger doses of pesticide were required. By 1956, despite immense doses of toxic insecticides, resistant boll-worms were again chewing away at the cotton crops, accompa-nied by six new cotton predators with resistance to poisons. That year, the cotton farmers suffered their worst losses ever: Insects ate half their crops. Faced with economic disaster, the Peruvian farmers were forced to reevaluate their pest control strategy, and they decided to ban synthetic pesticides and return to mineral insecticides. In addition, they imported 130 million parasitoid wasps (see Figure 35.17) to attack the eggs of the bollworm. Small numbers of bollworms continued to survive in their fields, but cotton yields reached new highs.

The experience of the Peruvian cotton farmers led to the first success of *integrated pest management,* a system based on the principles of community ecology and aimed at keeping pest populations below economically harmful levels with a mini-mum of chemical pesticides. Integrated pest management

FIGURE 1 ■ A Serious Pest of Cotton.

requires a broad view of all interactions between the crop and other species: weeds that compete for sun and nutrients; insects that eat the crop; predators that eat the insect pests; animals that pollinate the crop; and other wildlife living in the area. Pest managers have found that by carefully controlling the timing, spacing, and intermixing of crops, they can enhance the activi-ties of natural predators. What's more, they can introduce bio-logical control agents from other regions that feed exclusively on the pest, including parasitoid wasps and the bacterium *Bacil-lus thuringiensis* (see Chapter 32). In addition, they can release into the environment millions of sterilized adult male insects, which then mate with wild females and sterilize them. Finally, integrated pest managers use chemical controls only when insects appear poised to do damage above some preestablished level of economic injury. And when spraying is necessary, they choose and apply compounds to hit target species, leaving "innocent bystanders" alone.

Because farmers and ranchers must consider the principles of community ecology if they are to use integrated pest manage-ment, such management is more difficult than simply spraying the fields with chemicals. However, since it avoids a whole host of problems—pests evolving resistance, resurgence of the pest populations, and outbreaks of other pest species—integrated pest management offers an ecologically and economically acceptable alternative to managing crops and saving the 33 per-cent that are lost to pests each year.

35.14a) as decorative shrubs because they resist insect pests, but the leaves are so poisonous that chewing a few can kill a child. Animals are not without their own arsenals: Toads, stinkbugs, and bombardier beetles (Figure 35.14b) produce highly offensive chemicals that repel attackers.

Poisonous prey species usually evolve brightly colored patterns, enabling the experienced predator to recognize and avoid them. This is called **warning coloration** (or *aposematic* coloration). The brilliantly colored but poisonous strawberry frogs of South America (see Figure 1.1) have evolved this strategy. Many nonpoisonous prey species masquerade as

poisonous species; this is one form of the process called **mim-icry,** wherein one species resembles another (review the pairs of blue and orange butterflies in Figure 33.16b, c). Mimicry could arise in the following way. Individuals of a nonpoison-ous butterfly species that by chance contain alleles causing them to resemble the poisonous species even slightly may occasionally escape predation if a hungry animal mistakes them for the poisonous species. A selective pressure like this could, over time, allow the nonpoisonous species to accumu-late more and more alleles for resemblance to the poisonous neighbor. In another type of mimicry, two or more foul-tasting

(a) Poisonous oleander

(b) Defense of the bombardier beetle

FIGURE 35.14 ■ Chemical Warfare in Animals and Plants. (a) The colorful, fast-growing, dry-adapted oleander (*Nerium oleander*) has poisonous leaves and stems. (b) The bombardier beetle (*Brachinus* species) defends itself by spewing a volatile irritant from special glands. When the beetle is threatened, muscles surrounding separate reservoirs of chemicals contract. This mingles the chemicals in another chamber where they react, form a chemical weapon, and generate heat. The hot irritants are forced through the "gun turret" in the beetle's abdomen, and the blast usually repels the attacker.

species come to resemble each other during evolution. To see whether this situation has arisen in the warning coloration of the beautiful viceroy and monarch butterflies (Figure 35.15), experimenters recently removed the wings from seven species of butterflies and presented the insects to redwing blackbirds. The birds took a small nibble from each, but fully consumed only a few of the species, which must have tasted fine. After sampling either a viceroy or a monarch, however, a bird would shake its head, drink excessively, and then strictly avoid *both* types of insects thereafter. Clearly, both the viceroy and monarch are foul-tasting, and birds can recognize desirable prey by taste as well as sight. Since a bird that learns to avoid one species will also avoid the other, this type of mimicry helps both kinds of insects avoid predation.

Plants and animals have also evolved with thorns, spines, sharp spikes, and horns—weapons that discourage predators (Figure 35.16).

Parasites: The Intimate Predators

Parasites are insidious kinds of predators; they are usually smaller than their hosts, often live in close physical association with individual victims, and generally just sap their strength rather than killing them outright. *Ectoparasites,* like fleas, ticks, and leeches, live on the host's exterior, while *endoparasites,* like tapeworms, liver flukes, and some proto-

(a) Butterfly closely related to one from which viceroy evolved

(b) Viceroy

(c) Monarch

(d) Blue jay eating monarch

(e) Jay vomiting after eating monarch

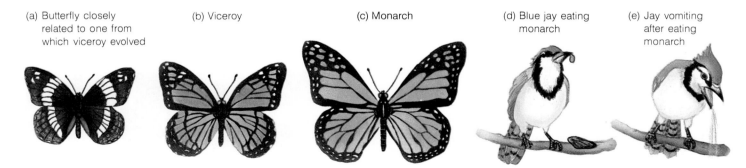

FIGURE 35.15 ■ Warning Coloration and Mimicry: More Defense Strategies. (a) Ancestors of the dull-colored *Limenitis arthemis* butterfly evolved into (b) the flamboyant viceroy (*Limenitis archippus*), mimic of (c) the unrelated monarch butterfly (*Danaus plexippus*). Recent experiments show that monarchs and viceroys are both distasteful to birds, such as redwing blackbirds or blue jays, shown here. The butterflies' nearly identical, bright coloration protects both species from predation, since an individual bird will avoid both species after tasting only one (d) and vomiting (e). Thus, *Limenitis* butterflies that carried alleles for resemblance to monarchs were selected for during evolution.

(a)

(b)

FIGURE 35.16 ■ A Spiny Arsenal for Self-Protection. (a) Porcupines and (b) some kinds of tropical palms have evolved sharp spines that ward off predators.

zoa, inhabit internal organs or the bloodstream. In a special type of parasitic interaction, certain insects develop inside the body of another insect (usually inside a caterpillar or a maggot) and inevitably kill it (Figure 35.17).

Sometimes, parasites and their hosts coevolve in such a way that the parasite becomes less harmful to the host; an especially aggressive parasite that kills its host *before* the parasite reproduces would be selected against. Hosts, in turn, probably coevolve some tolerance to the parasite; native antelope in Africa, for example, suffer little when they harbor trypanosomes, whereas the protists cause lethal cases of sleeping sickness in domestic cattle.

Cuckoos represent the strange phenomenon of *social parasitism,* a special case of parasitic activity in which one species exploits the social behavior of another species during a critical phase of its life cycle. Cuckoos seek out the nest of other bird species, lay an egg in it, then leave the unwitting foster parents to hatch the egg and nurture the young freeloader until it can fend for itself (Figure 35.18). The cuckoo is spared parental duties—a clear benefit—whereas the foster parents expend precious food, energy, and attention on the young of another species while their own may go hungry—a clear disadvantage. This suggests a straightforward case of parasitism. Yet egg-stashing behavior may in some cases actually be more complex than it initially seems.

Like cuckoos, cowbirds sometimes lay their eggs in the nests of other birds, including the bag-shaped nests of large black birds called oropendolas. Experiments showed that the oropendolas in some colonies readily accepted cowbird eggs in their nests. In these oropendola colonies, nests without a cowbird chick were usually infested with the maggots of botflies, parasitic larvae that left the young oropendolas so thoroughly weakened that their survival was unlikely. In contrast, in nests where young cowbirds were raised with oropendolas, the young cowbirds snapped at adult botflies and picked maggots from the skin of oropendola nestlings. The benefit to the oropendolas is great enough to explain how they would

evolve alleles that favor the acceptance of cowbird eggs despite the added cost of an extra mouth to feed. What seemed like a case of parasitism turns out to be an example of both species benefiting from the interaction—a kind of mutually helpful relationship we discuss next.

SHARING AND TEAMWORK: COMMENSALISM AND MUTUALISM

The community relationships we have discussed so far—competition and predation—have involved harm to at least one of the species. Sometimes, however, neither species is harmed by their interactions (see Table 35.1 on page 713). In **commensalism,** one species benefits from the alliance while the other is neither harmed nor helped, whereas in **mutualism,** both species are helped.

FIGURE 35.17 ■ Insect Parasitoid: A Wasp and Its Victim. A parasitic wasp injects eggs into the caterpillar of a tomato sphinx moth. Later, the green caterpillar lies dead (shown above), covered by the white wasp pupae that consumed the host's internal organs.

FIGURE 35.18 ■ **Cuckoo Baby Eating Foster Mother Out of House and Home.** **A female hedge sparrow (left) feeds an enormous freeloading cuckoo chick (right).**

Commensalism

Commensalism is common in tropical rain forests, and the most easily observed examples are the *epiphytes,* or "air plants," that grow on the surfaces of other plants. Epiphytes include Spanish moss, many beautiful orchids, and large and small bromeliads that festoon the branches or decorate the forks of trees (Figure 35.19a). Using the tree merely as a base of attachment, epiphytes take no nourishment from the "host" and do no harm unless their numbers become excessive. Other commensal relationships include birds that nest in trees, algae that harmlessly grow on a turtle's shell, and small fish that live among the stinging tentacles of sea anemones—unharmed and safe from predators (Figure 35.19b).

Mutualism

In a mutualistic interaction, both species benefit; the case of the cowbirds and oropendolas discussed earlier is an example of mutualism. The relationship of the yucca and yucca moth is another example of mutualism. Plants such as the yucca

provide pollinators such as the yucca moth with a high-energy nectar "reward," and as the pollinator visits other individuals of the same plant species in search of more nectar, it unwittingly pollinates the plant (review Figure 35.1). Mutualism between plants and pollinators is so common that one can almost predict that a bizarre flower will have a bizarre pollinator. Charles Darwin once amused his skeptical contemporaries by predicting that a moth with a 30 cm (1 ft) tongue flutters somewhere in the rain forest of Madagascar because a flower lives in those jungles with a 30 cm floral tube. Sure enough, 40 years later, naturalists caught the moth with the longest known tongue.

Sometimes, both species in a mutualistic relationship will benefit, but neither will be wholly dependent on the other for survival. Biologists call this *facultative mutualism.* Many species of aphids, for example, excrete large quantities of sweet, saplike fluid, and ants harvest this sap. The ants receive nutrients, and in return, their presence keeps predators away from the aphids. The ants are not strictly dependent on the sap, however, and the aphids continue to produce it even when the ants are absent.

In contrast, some interacting organisms need each other to survive. A relationship of this type, called *obligatory mutualism,* binds the yucca plant to the yucca moth. The moth receives nourishment only from the yucca, and in turn, the yucca plant is fertilized solely by the yucca moth.

Some tropical acacia trees have equally strong ties to certain ant species. The trees have large, hollow thorns where the ants live, modified leaf tips that the ants consume for protein and oil, and special glands (or "nectaries") at the base of the leaves that dispense carbohydrates that the ants also eat (Figure 35.20a). The acacia benefits, in turn, because the ants attack any herbivore that tries to feed on the tree, they chew up any plant that touches the acacia's branches or leaves, and they clear vegetation from the ground around the trunk (Figure 35.20b). If an experimenter removes all the ants from an acacia, the tree grows much slower and is twice as likely to die. Defense and nutrition bind ants and acacia trees in a coevolutionary system of mutual benefit.

(a) Orchid supported by a tree   (b) Clown fish protected by a sea anemone

FIGURE 35.19 ■ **Commensalism: A Little Harmless Help from a Host.** **(a) In the tropics, orchids often rely on tree trunks for support. (b) On coral reefs, a clown fish can find safe haven within the folds of a sea anemone. In each case, one species benefits from the association, and the other species is neither helped nor harmed.**

(a)    (b)

FIGURE 35.20 ■ **Obligatory Mutualism: A Close Beneficial Interdependence.** The swollen thorns of some Central American acacias, including *Acacia collinsii,* serve as nest sites and safe refuges for ants. (a) Nectar from special nectaries at the base of leaves and nodules at the leaf tips provide the ants with food. (b) The ants clear all vegetation from the base of their host tree. Thus, the tree feeds the ants and the ants protect the tree; both species benefit from the association.

Obligatory mutualism is involved in a minority of existing interspecies interactions, perhaps because it is so risky: No doubt many past cases of obligatory mutualism have disappeared because both species became so specialized that when the environment changed so that one member could no longer survive, the other inevitably became extinct, too.

HOW COMMUNITIES ARE ORGANIZED IN TIME AND SPACE

The principles we have studied—competition, predation, commensalism, and mutualism—affect not just pairs of species, but entire communities consisting of tens to hundreds of species. One of the biggest challenges facing ecologists is to untangle and understand the web of interdependencies in a community and to see how the community itself changes over time and space. One of the first parts of this task was to discover whether communities are fixed groups of species unfailingly bound together or more fluid entities. Research projects undertaken nearly 50 years ago confirmed that a community is not a "superorganism," a package of highly specific groupings of plants and animals (like a vertebrate organism with one heart, two lungs, and a precise assortment of other organs). Rather, communities consist of some species that happened to immigrate into the area and can survive under the available physical conditions and some species that will grow only if other species are also present. Given this premise, a goal of ecologists is to learn how the addition or subtraction of a species will affect the whole community in the short and long term.

Communities Change over Time

Occasionally, a cataclysm will strip an area of its original vegetation. This occurred in 1980 with the volcanic explosion of Mt. Saint Helens in Washington State and again in 1988 when fires swept through much of Yellowstone National Park. A farmer clearing a field can even cause such a change. Left to nature, however, a regular progression of communities will regrow at the site in a process called **succession.** Soon after the denuding, a variety of species begin to colonize the bare ground. These species make up a *pioneer community,* and they modify environmental conditions such as soil quality at the site. These modifications can inhibit or allow additional species to establish themselves and form a *transition community.* The change is rapid at first as more species invade. Then finally, an assemblage of plants and animals becomes relatively stable and tends to perpetuate itself as a *climax community.*

A well-studied example of ecological succession can be seen at Glacier Bay, Alaska, where a glacier that devastated thousands of square kilometers of land has melted back about 100 km (62 mi) in the last 200 years (Figure 35.21), exposing the ground below. The most recently exposed areas are inhospitable piles of rock, sand, and gravel lacking usable nitrogen and essentially devoid of plant or animal life. The first plants to colonize this barren scene form a pioneer community of wind-dispersed species: mosses, horsetails, fireweed, willows, cottonwood, and the matlike rockrose *Dryas.* Most are severely stunted and grow close to the ground as a result of nitrogen deficiency, but *Dryas* has nitrogen-fixing nodules on its roots that provide the growth-limiting nutrient. Within a few years, it crowds out other plants and forms a dense mat over the soil.

In the areas exposed for 20 years or so, alder trees begin to invade from other sites. Alder roots have nitrogen-fixing nodules, so these trees can also grow rapidly. Dead alder leaves add nitrogen to the soil and stimulate the growth of willows and cottonwood. In areas exposed for about 50 years, dense thickets of these plants shade the pioneer species and kill them. Eventually, Sitka spruce invades, and in those areas exposed for 80 years, nitrogen released by the alders enables the spruce to form dense forests that shade out the alders and willows. Finally, after 100 years or so, shade-tolerant hemlock trees invade, grow below the canopy of spruce branches and needles, and become the most frequent tree in the climax community. In low places, however, sphagnum moss invades the forest floor, soaks up large amounts of water, and kills trees by choking off their roots' oxygen supply. This leaves a *mosaic climax* consisting of patches of spruce-hemlock forest intermixed with sphagnum bog.

A succession of communities such as this one may involve *facilitation,* a process in which pioneering species modify the environment in ways that promote the growth of other species. In other cases, species are able to follow pioneers into an area because of their *tolerance* for the prevailing conditions; species that disperse more slowly than the wind-blown pioneers may compete more efficiently for nutrients and are hence more tolerant of shortages. Sometimes, however, pioneers may cause an *inhibition* of other species' growth, and succeeding species can gain a toehold only when the pioneers die or are damaged. Whichever mechanism is operating at a particular site, the general rule is that even after the climax community becomes established, it will probably not remain stable for very long. Local climates change, and a cycle of growth and decay pervades all communities.

Trends in Species Diversity in Communities

Coral reefs and tropical rain forests are dense with interacting species, while arctic tundra and deserts have far fewer. The total number of species found in a community is its **species**

richness, or species diversity. In most communities, there are few common species but many rare types of organisms. In a local area, English ecologists, for example, captured a group of almost 7000 moths and identified individuals of 197 different species. Fully one-quarter of the moths belonged to a single species, and another quarter belonged to just five other species. The remaining half of the moth group fell into the other 191 species—some represented by just one or two individuals.

Both *latitude* (north/south position) and *isolation* (occurrence on peninsulas, island chains, or other out-of-the-way locales) influence the species richness of an area. Some communities in the tropical latitudes, for example, have about 600 types of land birds, while an area of similar size in the arctic tundra may have only 20 to 30 species of land birds at many sites (Figure 35.22). And a single hectare of mainland tropical

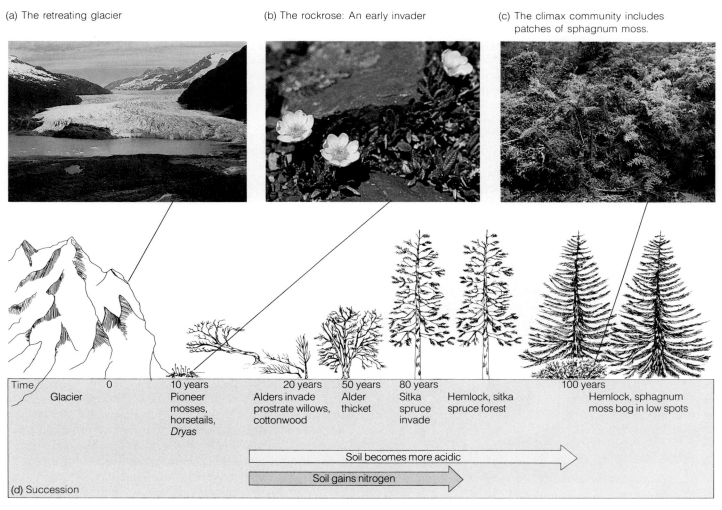

(a) The retreating glacier

(b) The rockrose: An early invader

(c) The climax community includes patches of sphagnum moss.

Time	0	10 years	20 years	50 years	80 years		100 years
	Glacier	Pioneer mosses, horsetails, *Dryas*	Alders invade prostrate willows, cottonwood	Alder thicket	Sitka spruce invade	Hemlock, sitka spruce forest	Hemlock, sphagnum moss bog in low spots

Soil becomes more acidic

Soil gains nitrogen

(d) Succession

FIGURE 35.21 ■ **Succession in the Path of an Arctic Glacier.** (a) As Alaska's massive Mendenhall Glacier receded, a succession of plants followed, from (b) the rockrose (*Dryas* species) to cottonwood and alder and eventually (c) sphagnum moss and hemlock. (d) A time line of the glacier's recession and the plants' succession. Note that the soil gains nitrogen and becomes more acidic with the passing decades.

forest can have 300 different species of trees and tens of thousands of insect species; on a peninsula, however—even a tropical one like Baja California or Florida—the species richness is diminished, as the more limited bird life in those areas illustrates. Species diversity on island chains is limited in such specific ways that ecologists are now applying the principles of island ecology to the design of nature preserves, which are often islands within seas of human development (see Box 35.2).

Recent observations support the conclusion that the more resources available in an area—water and solar energy, for instance—the greater the species richness the area can support. This helps explain the numerous varieties of plants found in tropical forests. Other factors influence species richness, however, including competition and predation. Competition can increase diversity because resource partitioning divides up niches into more and more specialized compartments. For example, a lizard species that has an island all to itself will eat insect prey of any size, but a lizard species that shares an island with four other lizard species will specialize on prey of a particular size. With competition forcing smaller niches, a community can accommodate more species.

Predation can also increase species richness. A sea star along the coast of Washington State, for example, preys on 15 species of barnacles, snails, clams, and mussels. When experimenters removed the sea star from sections of the shore, the diversity dropped to eight species because the mussel population increased and crowded out the other invertebrates. By eating young mussels, the sea star reduces competition for space and so preserves a higher species richness. Species like this sea star, whose activities determine community structure, are called *keystone species*. The abundance of predators and parasites in the tropics may help explain why the lower latitudes have such a high species richness.

Species Diversity, Community Stability, and Disturbances

Some evidence suggests that highly diverse communities, such as those in tropical forests and coral reefs, would be the most *stable* or *resilient*—the best able to return to normal after disturbances like fires, storms, frosts, plagues of insects, or human land-clearing activities. Other analyses, however, suggest that when more species are interwoven into complex networks of competition, predation, and mutualism, the community becomes more fragile. A complex community might be less stable because the loss of each species may affect many other organisms (see Box 35.1 on page 720 for an example from agriculture). For instance, if the yucca moth were exterminated, the yucca plant would die out, too, and so might the sap beetle and many other species whose lives depend on the yucca in unknown ways. Ecologists do not yet know whether diverse or simple communities are more stable, but are convinced that community richness and stability are not related in a very simple way.

Current knowledge suggests that disturbances of intermediate frequency or intensity help to increase species diversity. For example, when tall trees die and fall in a tropical forest, they clear wide gaps of open territory where sunlight can suddenly penetrate to the previously shaded forest floor. Plants requiring strong sunlight will invade the gaps, fruits will become available for different kinds of animals, and species richness will increase.

Perhaps the greatest ecological challenge facing modern civilization is the fact that the immense increase in human population is leading to larger and more frequent disturbances in the species-rich communities of tropical latitudes and is threatening their destruction. Tropical peoples have traditionally cleared land for agriculture by burning the forest, planting crops for a few years, and then moving on when the soil becomes depleted of nutrients (Figure 35.23a). So large are the human populations in these areas now, and so immense are their survival needs, that vast areas of the rain forest are going up in smoke to provide farmland for crops and pastureland on which to graze cattle for export to developed nations. These fires are so huge and so common that they are clearly visible on satellite photos from space (Figure 35.23b). Many ecologists fear that the plant and animal communities of the tropics may not be able to recover from such extreme disturbances and that by the year 2000, no tropical rain forests will remain intact. Still worse, many fear that this disruption may cause up to 20 percent of all living species to become extinct in our lifetime.

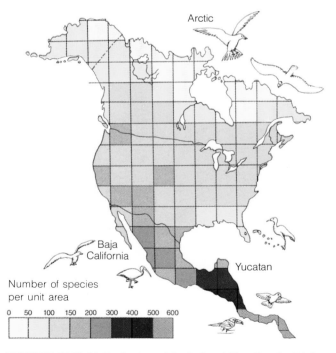

Arctic

Baja California

Yucatan

Number of species per unit area

0   50   100   150   200   300   400   500   600

FIGURE 35.22 ■ Latitude and Isolation Affect Species Richness. Close to the equator, there are 500 to 600 species of breeding land birds in each unit of land (a square about 600 km, or 372 mi, per side). In northern Alaska and Canada, the species count in an equivalent area drops below 20. Even in the tropics, species richness declines along peninsulas. Notice the numbers in Baja California, Florida, and the Yucatan peninsula.

BOX 35.2   FOCUS ON THE ENVIRONMENT

COMMUNITY ECOLOGY OF ISLANDS AND THE DESIGN OF NATURE PRESERVES

Islands are microcosms of evolution and ecology, partially isolated from invasion and relatively unperturbed by the outside world. Many of our nature preserves are islands, too, enveloped not by water but by farms, roads, and towns—hostile environments, zones of human development. These surrounding regions fail to provide proper habitat and present fatal dangers to organisms that stray beyond the safety of the refuge. National parks and other nature preserves represent our best hope for saving many species from extinction. Since a nature preserve is truly an island, community ecologists are now applying the principles of island ecology to their design in the hopes of conserving life's irreplaceable diversity.

At least two factors should affect the species richness on an island: the immigration rate and the extinction rate. The greater the number of species on an island, the lower the rate at which new species enter the island, because it is less likely that any immigrant will represent a new species (the downward-sloping line in Figure 1). At the same time, the greater the number of species on an island, the more likely that some of the species will become extinct, because there are more species that *can* become extinct (the upward-sloping line in Figure 1). At the point where the immigration rate and the extinction rate are about equal, the number of species will remain constant. An ecologist's goal in setting up a nature preserve is to prevent any net loss of species over the years. The equilibrium number of species is therefore a useful parameter for conservationists, and it becomes important for them to understand how an island's size and its distance from sources of colonization affect the equilibrium.

By cataloging species richness on islands—whether reptiles in the West Indies, ants in Melanesia, or vertebrates on Lake Michigan islands—naturalists have found that larger islands have more species than smaller islands. This is probably because population sizes are larger on large islands, and so chance extinctions are less likely. The lesson ecologists draw from this is that large nature preserves are more likely to harbor more species than small ones. Unfortunately, the land area set aside for most preserves today is quite limited.

To determine how the species richness of an island is affected by the distance from the mainland, researchers devised some clever experiments on mangrove islands off the Florida Keys. In addition to low, bushy mangrove trees, these swampy islands, located at various distances from the mainland, are inhabited by insects, spiders, scorpions, and other arthropods. The experimenters sealed the small islands with plastic sheets (Figure 2), fumigated to kill the existing arthropods, then removed the tents and monitored the immigration of arthropod species. They found that the number of species rose for about nine months and then dipped slightly to an equilibrium number near the original level of species richness. The closer an island was to the mainland, the higher its equilibrium number of species. Evidently, the immigration rate is lower on a distant island, and thus it reaches a lower steady-state density of species. Studies like these had an additional message for ecologists who work with nature preserves: If preserves must be separated from each other, they should be as close together as possible to maximize immigration and species diversity.

Based on the principles of island ecology, it is clear that preserves should be large enough to hold sizable populations, but if there must be several small regions, they should be clumped close together so that migration between the preserves can remain high. If human society is to help preserve nature's exquisite diversity, we must act immediately to set aside as many large preserves as possible. Extinctions will no doubt proceed anyway. But if we are too slow to preserve our islands of refuge—if we are more concerned with human developments than evolutionary ones—then a large measure of ecological complexity may be lost forever.

FIGURE 1 ■ Species Richness at Equilibrium.

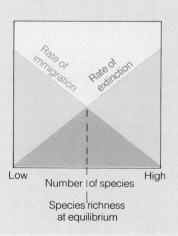

FIGURE 2 ■ Mangrove Tent Experiment.

Ecologists consider it an urgent research priority to learn what makes communities resilient and how they may (or may not) be able to persist in the face of human encroachment. We must solve this problem (discussed in more detail in Chapter 37) if our most diverse and interesting communities are to survive through the twenty-first century.

(a) Clearing the forest

(b) Fires and smoke from burning forests

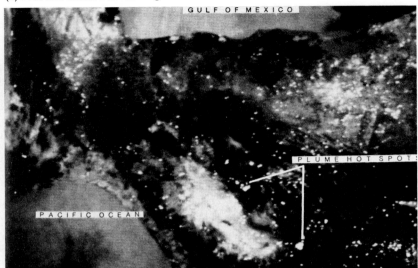

FIGURE 35.23 ■ **Diversity and Destruction in the Tropics.** Disturbances of intermediate intensity or frequency encourage species diversity in the tropics, but massive disturbances destroy, not promote, diversity. (a) Farmers in the tropics often carry out a traditional method of agriculture: "slash and burn" forest clearing, followed by planting, then moving on to new areas when the soil is depleted. (b) A satellite photo of agricultural areas in Mexico taken in 1984 shows the glowing lights of deliberately set fires, as well as white plumes of smoke. Some represent the annual clearing of established fields, but others are the ongoing destruction of virgin forest.

CONNECTIONS

John Muir, the great naturalist whose aggressive advocacy of forest preservation led to the forming of many U.S. national parks, made this observation nearly 100 years ago: "When we try to pick out anything by itself, we find it hitched to everything in the universe."* Our exploration of ecology bears out this wisdom. Populations of individual species, such as those we studied in Chapter 34, are inevitably "hitched" to other species in communities. Each species has its own habitat and niche, but these are impacted by neighboring species. A mutual reproductive dependence, for example, tethers yucca plants to yucca moths, and reciprocal changes in the genomes of coevolutionary partners make such interdependencies permanent.

Despite chains and webs of interconnections, communities are not prepackaged groups of species. The members of

an individual species will live in those places where their special environmental requirements are met. The succession of species that follows disturbances such as glaciers in Alaska reflects such individual needs and the resulting change in community membership over time. The world is in danger of losing the species richness of many communities because as a species, we humans are massively disturbing the globe, not just in the tropics, but in our own climate zone as well, and yet we do not understand how community stability is maintained or how it might bounce back. Biologists do know more about ecological stability than we have so far discussed, however, and our next chapter carries us one level higher in the hierarchy of population, community, ecosystem, and biosphere. There we will see the interactions of communities with sunlight, temperature, rainfall, nutrients, and other physical aspects of the universe to which all organisms are "hitched."

Highlights in Review

1. An organism's habitat is analogous to its "address," the general physical place in the environment where it resides. An

organism's niche is analogous to its job, or role, in the community, such as producer, herbivore, or predator.

2. Communities are assemblies of populations of species in a particular area.

3. In interspecific competition—competition between species for the same limited resources—one species' success can cause

*John Muir, *My First Summer in the Sierras* (Boston: Houghton Mifflin, 1911).

another's extinction. This is competitive exclusion. Over time, species sometimes respond to the threat of competitive exclusion by resource partitioning (dividing up the territory and coexisting) or by character displacement (evolution in physical and behavioral characteristics).

4. The mutual influence that species have on each other's evolution is called coevolution.

5. Predation involves a predator hunting down prey. Other interactions that harm one partner involve herbivores, parasites, parasitoids, and pathogens.

6. Predators use strategies such as pursuit or ambush and generally have larger brains per unit body weight than their prey. The countermeasures of prey species include camouflage, warning coloration, and mimicry.

7. Parasites live in close association with their prey and generally weaken rather than kill them.

8. In commensal relationships, one species benefits, but the other is usually unaffected. In mutualistic relationships, both species benefit, as in the yucca and yucca moth.

9. A community is a haphazard association of individual species that can survive in an area; it is not a superorganism made up of highly specific groups of plants and animals.

10. After a calamity clears an area, there is a succession of communities, including a pioneer community of hardy, wind-dispersed species; a transition community of more slowly dispersed species; and eventually a climax community, a relatively stable assemblage of plants, animals, and other organisms that may change when the climate changes.

11. An area's species richness is dictated partly by latitude (the closer to the equator, the greater the available sunlight, and the richer the assemblage of species), partly by isolation (islands and peninsulas have fewer species than mainlands), and partly by the frequency and intensity of disturbances. The high level of disturbances caused by people is now endangering large numbers of species in the tropics and the overall ecological stability of the zone.

Key Terms

camouflage, 719
commensalism, 722
community, 710
competitive exclusion, 713
fundamental niche, 712
habitat, 712
interspecific competition, 713
mimicry, 720
mutualism, 722
niche, 712

predation, 716
predator, 716
prey, 716
realized niche, 712
refuge, 716
resource partitioning, 715
species richness, 725
succession, 724
warning coloration, 720

Study Questions

Review What You Have Learned

1. Define habitat and niche.
2. Describe the experiment that led G. F. Gause to formulate the principle of competitive exclusion.
3. Does competitive exclusion occur in nature? Explain.

4. Use an example to show how directional selection can lead to resource partitioning.
5. What do biologists mean by coevolution? Give an example.
6. Give an example of a prey population controlled by a predator, and discuss the evidence for your conclusion.
7. Give three examples of predator strategies.
8. Name five adaptations for defense.
9. Do parasites inevitably kill their hosts? Explain.
10. Compare commensalism with mutualism.
11. How does a pioneer community differ from a climax community?
12. How can predation increase species richness?
13. Why do biologists consider a tropical rain forest a fragile community rather than a stable one?

Apply What You Have Learned

1. When wolves and mountain lions living on the Grand Canyon's Kaibab Plateau were exterminated, the area's deer population eventually began to starve. Why?
2. Charles Darwin confidently predicted that investigators would one day find a moth with a 30 cm (1 ft) proboscis in a Madagascar rain forest. Why was he so sure?
3. Some Alaskan glaciers have been receding for 200 years. Recently uncovered land is barren, but trees and shrubs now flourish in the first areas to be freed from the ice two centuries ago. Describe the ecological sequence that has led to this restored plant cover.
4. In a remote valley in South America, cotton growers regularly lost a portion of their crop to bollworms. In the first four years after they started using chemical insecticides, cotton production rose from 490 to 730 kilograms per hectare. The fifth year, it dropped to 390 kg per hectare and dropped even further in subsequent years. The farmers increased the amount of insecticides they applied, but bollworm damage exceeded the levels experienced before they started using the chemicals. What accounts for the temporary rise and ultimate drop in cotton production?

For Further Reading

BEARDSLEY, T. "Recovery Drill: An Old Volcano Teaches Ecologists Some New Tricks." *Scientific American,* November 1990, p. 34.

DAVIES, N. B., and M. BROOKE. "Coevolution of the Cuckoo and Its Hosts." *Scientific American,* January 1991, pp. 92–98.

FULLICK, A., and P. FULLICK. "Biological Pest Control." *New Scientist,* March 9, 1991, pp. 1–4.

KREBS, C. J. *Ecology: The Experimental Analysis of Distribution and Abundance.* 3d ed. New York: Harper & Row, 1985.

MOORE, P. D. "What Makes a Forest Rich?" *Nature* 329 (1987): 292.

NILSSON, L. A. "The Evolution of Flowers with Deep Corolla Tubes." *Nature* 334 (1988): 147–149.

ROMME, W. H., and D. G. DESPAIN. "The Yellowstone Fires," *Scientific American,* November 1989, pp. 37–46.

WALKER, T. "Butterflies and Bad Taste: Rethinking a Classic Tale of Mimicry." *Science News* 139 (1991): 348–349.

Ecosystems: Webs of Life and the Physical World

A Voyage of Discovery

In the spring of 1977, two scientists and a pilot climbed into the *Alvin,* a small spherical submarine. Crew members on *Alvin*'s mother ship then lowered the 16.5-ton submersible into the green Pacific swells 274 km (170 mi) northeast of the Galápagos Islands and due west of Colombia, South America.

Enclosed within the *Alvin,* geochemist John Edmond and biologist John Corliss descended through the zone of light penetration and entered a seldom-seen world of frigid, inky black waters and strange fluorescent fish but little other sea life. After an hour of slow free-fall, 2.5 km (1½ mi) of dark water lay above them. Only *Alvin*'s bright floodlights pierced the abyssal gloom as they started to move slowly forward along the nearly lifeless deep-sea floor.

As their circle of light crept forward, they were surprised to see a pair of large, purple sea anemones loom out of the blackness and were even more perplexed to notice that the water around the animals was shimmering like the air above hot pavement in summer. Not far from that spot, they were transfixed by a fabulous scene that no humans before them had ever witnessed or even suspected.

FIGURE 36.1 ■ **Strange Community of Aquatic Organisms Discovered in the Deep-Ocean Abyss.** In 1977, scientists discovered an entirely new ecosystem based on the heat energy from hydrothermal vents. The living community in this physical setting included yellow vent mussels, crabs (here, looking like ghostly white spiders), large vent clams, and tube worms with red plumes. These plumes take in oxygen, and the worm's body contains chemoautotrophic bacteria, which produce fixed carbon compounds in this ecosystem.

There on the lifeless floor of the ocean abyss lay a virtual oasis of living things (Figure 36.1). Warm water streamed from every bottom crack and fissure over an area larger than four football fields. Clustered all around the deep, warm vents were groups of yellow mussels and fields of giant clams. Pure white crabs scuttled about through the shimmering water. Bizarre-looking worms nearly 1 m (3.28 ft) long lay encased in white tubes with frilly red plumes waving in the slow current. Edmond and Corliss worked quickly to collect samples of the water, sediments, and animals; to measure the temperature, pH, and oxygen content of the environment; and to photograph the amazing spectacle before them.

The team aboard the *Alvin* had inadvertently discovered a new *ecosystem:* a community of organisms interacting with a particular physical environment. This deep-sea vent ecosystem was unlike more familiar ecosystems in several ways. Forests and grasslands, for example, have both plants and animals (not animals alone) and have physical conditions like sunlight and fresh air (not darkness and crushing pressures). As with more familiar ecosystems, however, deep-sea vents must possess an energy source as well as a supply of carbon, nitrogen, and other materials that make up the bodies of living organisms. In the absence of solar energy (and the green plants that capture it), what substitute energy source could be driving the deep-sea vent ecosystem? And how could energy and materials be flowing through the organisms in these bizarre and remote outposts of life? This chapter will answer these questions while also addressing the broader issue of how all communities receive and metabolize energy and materials.

Four major themes will emerge in these discussions. First, energy flows through an ecosystem in a one-way path, entering living things from the physical world, passing from one organism to another, and finally escaping back to the physical environment (Figure 36.2). In most ecosystems, energy comes from nuclear chain reactions in the sun that generate heat energy and light. In contrast, the deep-sea vent ecosystem gets its energy from heat generated by the decay of radioactive elements deep in the earth's core.

Second, materials recycle through an ecosystem, passing from one organism to another, to the physical environment, and then back through the organisms once more (see Figure 36.2). The carbon in carbon dioxide, for example, passes from seawater to bacteria, to the tissues of deep-sea vent clams and tube worms, and then back to the water as the organisms metabolize. Third, because the organisms in an ecosystem have an absolute need for energy and materials, they are dependent on each other and on the physical environment. Plants and certain bacteria capture energy and inorganic compounds from their surroundings; animals, fungi, and protists depend on those autotrophs for their nutrients; and plants depend on bacteria and fungi that decompose dead organisms and return certain of their elements to the soil.

Finally, human activities can drastically affect the health of ecosystems by altering the flow of energy and the recycling of materials. For example, people are destroying entire tropical rain forest ecosystems in many parts of the world. And we are causing more subtle changes in many other areas by replacing certain native species with agricultural crops and

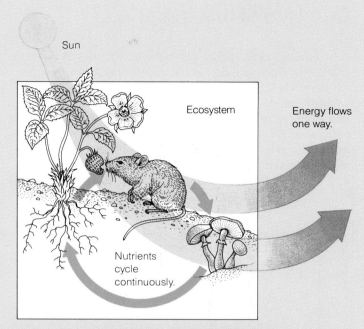

FIGURE 36.2 ■ Pathways of Energy and Materials in Ecosystems. Energy flows in a one-way path, usually from the sun through living things and into the environment, while nutrients and other materials continuously cycle from the physical world to the biological and back again.

livestock and releasing chemicals into the environment. The consequences of our activities to long-term ecosystem function are difficult to predict and are often quite undesirable, as illustrated by the problems of acid rain, the greenhouse effect, and depletion of the ozone layer, as well as by the frightening possibility of a nuclear winter. Ultimately, changes like these endanger us and the other living organisms that share our planet.

This chapter addresses several important questions:

■ How does energy flow, and how do materials cycle through ecosystems?

■ How does the energy in an ecosystem change from one form to another?

■ How do materials like carbon and nitrogen cycle in ecosystems?

■ How have human activities altered the earth's ecosystems?

PATHWAYS FOR ENERGY AND MATERIALS: WHO EATS WHOM IN NATURE?

All organisms need energy and materials for growth, maintenance, and repair and (in the broadest terms) for overcoming the universal tendency toward disorder. In all ecosystems—whether remote ones like the deep-sea world that the *Alvin* first explored or familiar ones like forests and grasslands—organisms have two basic strategies for obtaining energy and materials: autotrophy (self-nourishing) and heterotrophy (other-nourishing). The strategy that each species uses defines its place in feeding levels operating within the community.

Feeding Levels: Strategies for Obtaining Energy

In any ecosystem, autotrophs like plants and certain protists and bacteria are the primary **producers;** they are the only organisms that can use nonliving substances to produce all the biological molecules they need for their growth. Heterotrophs, on the other hand, such as animals, fungi, and many kinds of microbes, obtain their biological molecules by consuming autotrophs or other heterotrophs. Heterotrophs, therefore, are an ecosystem's **consumers.** Ecologists assign every organism in a community to a **trophic level,** or feeding level, depending on whether it is a producer or a consumer and depending on what it eats (Figure 36.3).

At the lowest trophic levels lie the producers—the autotrophic bacteria and plants that support all other organisms directly or indirectly. In most terrestrial ecosystems, green

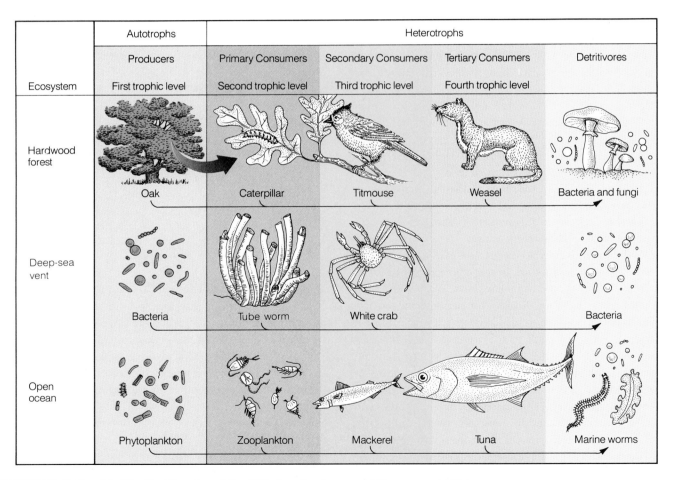

Ecosystem	Autotrophs	Heterotrophs			
	Producers	Primary Consumers	Secondary Consumers	Tertiary Consumers	Detritivores
	First trophic level	Second trophic level	Third trophic level	Fourth trophic level	
Hardwood forest	Oak	Caterpillar	Titmouse	Weasel	Bacteria and fungi
Deep-sea vent	Bacteria	Tube worm	White crab		Bacteria
Open ocean	Phytoplankton	Zooplankton	Mackerel	Tuna	Marine worms

FIGURE 36.3 ■ Trophic (Feeding) Levels and Food Chains: Producers and Consumers in Nature. Organisms vary from one ecosystem to another (shown here are those for forest, deep-sea vent, and open ocean), but in each ecosystem, autotrophic producers form the first trophic level, while consumers form the next three trophic levels, and detritivores utilize the wastes from all levels. Within each ecosystem, one can identify chains of producers and consumers. In the hardwood forest, for example, oak trees are producers, caterpillars consume oak leaves, titmouse birds consume caterpillars, weasels consume the birds, and fungi and soil bacteria live on plant and animal wastes.

plants are the producers. By collecting solar energy and carbon dioxide, they build energy-rich sugar molecules—the main source of carbon atoms in all types of biological molecules. Producers also absorb nitrogen, phosphorus, sulfur, and other needed atoms from the environment and fix them into biological molecules. Ecologists say that producers provide both the *energy fixation base* and the *nutrient concentration base* for the entire ecosystem. In the deep-sea vent ecosystem, chemoautotrophic bacteria (see Chapter 15) are the sole producers, since no light penetrates to these depths and plants cannot survive. Via the process of chemosynthesis (rather than photosynthesis), these bacteria harvest the energy stored in the bonds of hydrogen sulfide (H_2S) molecules dissolved in the water. They use this chemosynthetic energy to fix carbon dioxide from seawater into sugar molecules and then incorporate the carbon from the energy-rich sugars into the other biological molecules they require for growth and reproduction.

At the second trophic level are the **primary consumers,** the organisms that eat the producers. Herbivores, such as caterpillars or cows (see Chapter 24), efficiently digest plant matter for energy and serve as ecological links between the producer level and all other levels (see Figure 36.3). Near deep-sea vents, huge tube worms act as a special kind of primary consumer. Although they lack a mouth or other way of eating, the tube worms have a special internal organ packed with chemoautotrophic bacteria. Red blood coursing through the worm's scarlet plumes delivers hydrogen sulfide to the bacteria, which in turn produce sugars that the tube worms use for energy and carbon. These remarkable worms are related to annelid worms (see Chapter 18) and belong to a small phylum called Vestimentifera.

At the next highest level, **secondary consumers** are carnivores (meat eaters) that consume the herbivores. A titmouse eating a caterpillar is a carnivore, and so is a white crab or deep-sea fish feeding on a tube worm. In the next trophic level, **tertiary consumers** are carnivores that eat other carnivores; a weasel may eat a titmouse, for example, and a tuna may eat a mackerel. Finally, a few ecosystems have one more trophic level containing carnivores that eat tertiary consumers, such as cougars eating weasels and sharks eating other sharks.

A special class of consumers, the **detritivores,** or decomposers, obtain energy and materials from **detritus,** organic wastes and dead organisms that accumulate from all trophic levels. Fungi and bacteria are the most ubiquitous detritivores, but worms, nematodes, many kinds of insects, and carrion feeders like vultures are also important detritus consumers.

The simplified diagram in Figure 36.3 suggests that each trophic level leads directly to the next in a simple chain, and indeed, **food chains** do exist in nature with groups of organisms involved in linear transfers of energy from producer to primary, secondary, and tertiary consumers. More commonly, however, feeding relationships resemble not chains but complex interwoven webs.

Feeding Patterns in Nature

Organisms usually consume more than one other species, and some animals feed at several trophic levels. A high-level carnivore like a cougar may eat herbivores, such as rabbits, as well as other carnivores, such as foxes or hawks. And as an omnivore (see Chapter 24), you yourself eat vegetables (producers), poultry (herbivores), tuna fish (carnivores), and mushrooms (detritivores). Ecologists call complicated interconnected feeding relationships **food webs.** In a hardwood forest ecosystem in New Hampshire, for example, the major producers are sugar maple, beech, and yellow birch trees. These producers support two main food webs, a *grazing food web* that stems directly from the living plants and a *detritus food web* that begins with dead plant parts and animal wastes (Figure 36.4). In the grazing food web, some herbivores, like jays and chipmunks, consume fruits and seeds, while other herbivores, such as mice, deer, and caterpillars, graze on leaves. These primary consumers then fall prey to hawks, skunks, snakes, and other carnivores. In the detritus food web, fungi and bacteria break down dead animals and fallen leaves, twigs, and fruits and are eaten in turn by grubs and earthworms. Grazing and detritus food webs are themselves linked, as detritivores become food for other consumers, like salamanders and shrews.

Trophic levels and food webs describe the general pathways of energy flow and material cycling in ecosystems. They do not, however, portray the *amounts* of energy or materials that pass through each level. To fully understand how ecosystems function, ecologists measure the energy transfers.

HOW ENERGY FLOWS THROUGH ECOSYSTEMS

Whether an organism is a producer or a consumer, it needs energy for movement, for active transport of nutrients and ions, and for synthesis of proteins, nucleic acids, and other large molecules needed for growth and repair (see Chapter 4). Producers obtain their energy directly from the environment in the form of light (in most ecosystems) or inorganic molecules (in deep-sea vents and a few other ecosystems). Consumers, however, can get their energy only from producers. Hence, the activities of a community's producers set a limit for the amount of energy that can be captured and channeled throughout the entire ecosystem. Ecologists have closely studied the hardwood forest ecosystem to precisely measure available energy and how it is spent.

Energy Budget for an Ecosystem

Within the White Mountains of New Hampshire lies the Hubbard Brook Experimental Forest, a fragrant zone of tree-studded rolling hills that has engaged researchers' attention

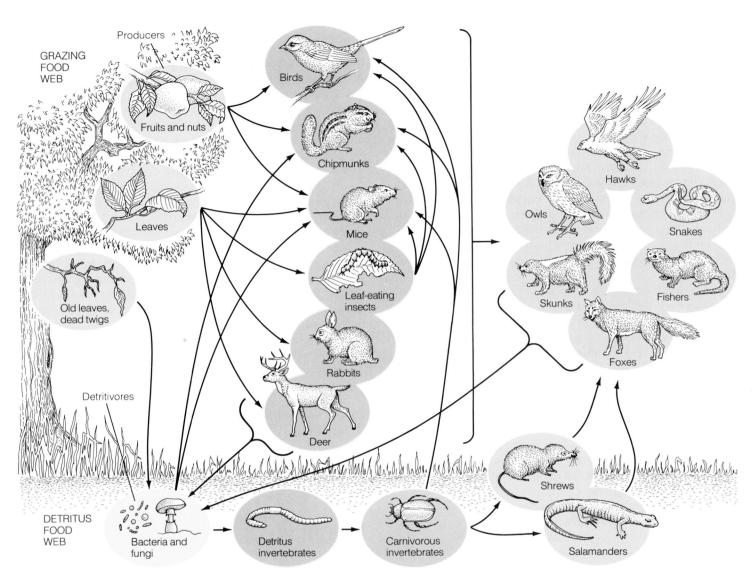

FIGURE 36.4 ■ A Food Web: Cross-Dependencies in the Living World. In the Hubbard Brook Forest ecosystem, dozens of species derive energy from each other in a complex grazing food web, while others participate in the detritus food web. Both webs are linked at numerous points by the activities of specific organisms. Birds, for example, can feed on berries, leaf-eating insects, and insects from the detritus layer. This simplified diagram omits hundreds of additional species and their tangled network of interrelationships.

for more than 25 years (Figure 36.5). Researchers have focused on a few small *watersheds* (regions drained by a single stream or river), because the ridges that separate watersheds provide convenient dividing lines between adjacent ecosystems. Since impermeable bedrock lies below each watershed, researchers can measure all the groundwater leaving the ecosystem (as well as its load of dissolved nutrients) by monitoring the single stream that drains it. By studying the input of energy, water, mineral nutrients, and organic matter

into these watersheds and tracing their incorporation into both living organisms and the physical environment, workers have learned how energy flows through an entire forest.

Over the course of the growing season (June through September), a total of about 500,000 kcal of solar energy, including both heat and light, strike each square meter (10.76 ft^2) in the Hubbard Brook Forest (Figure 36.6a). (To put this in perspective, recall that you burn 100 kcal jogging a mile.) About 15 percent of the sun's radiant energy striking the for-

FIGURE 36.5 ■ **Hubbard Brook Experimental Forest: A Living Laboratory for Studying Energy Flow in Ecosystems.** Ecologists have cut the trees from bands and zones of the Hubbard Brook Forest to determine the effects of logging on soil erosion and nutrient loss, and have monitored energy flow and material cycling through hundreds of specific experiments.

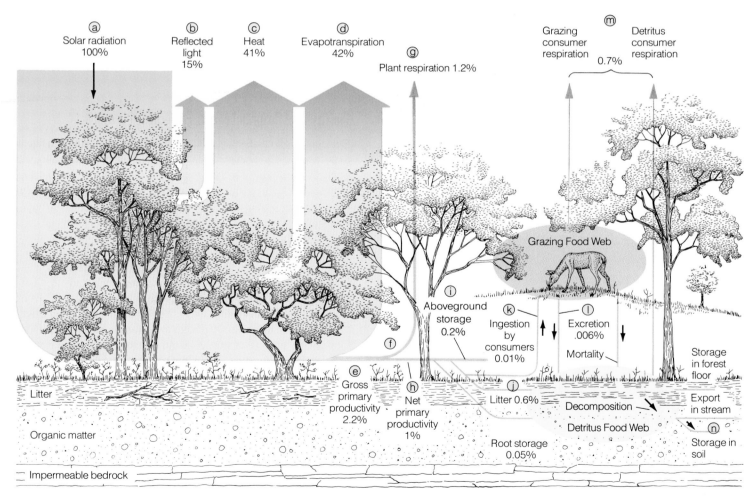

FIGURE 36.6 ■ **Energy Budget of a Hardwood Forest.** This diagram of the Hubbard Brook Experimental Forest traces the flow of energy through an ecosystem. As sunlight strikes the forest, light and heat are returned to the environment, and a small fraction of the original energy is fixed in organisms and their waste products.

estland immediately reflects back into the atmosphere as light (Figure 36.6b). Another large fraction (41 percent) warms the ground and the photosynthesizing plants and eventually radiates back to the atmosphere as heat (Figure 36.6c). Still more of the incoming energy (42 percent) is used in evaporating water from the soil and cells of plant leaves, a combined process called *evapotranspiration* (Figure 36.6d; see also Figure 32.6).

Clearly, over 83 percent of the solar energy that reaches the hardwood forest flows through as heat and another 15 percent as light—for a total of 98 percent that returns rapidly to the physical environment. The 2 percent or so remaining is the amount of energy that producers convert by photosynthesis to chemical energy in the form of sugars and other organic compounds. Ecologists call this small fraction an ecosystem's *gross primary productivity* (Figure 36.6e), and ultimately, it limits an ecosystem's structure, including how many birch trees will grow and how many chipmunks will thrive.

Not all the chemical energy that a plant initially traps will be stored in newly formed leaves, roots, and fruits. Plant cells themselves use a little more than half of this energy to fuel their own cellular respiration, eventually losing 1.2 percent as heat (Figure 36.6f and g). The amount of energy remaining after respiration is called the *net primary productivity,* the amount of chemical energy that is actually stored in new cells, leaves, roots, stems, flowers, and fruits (Figure 36.6h). Of all the energy impinging on the ecosystem, only the net primary productivity—only about 1 percent of the light energy striking the forest—is available to consumers.

During a growing season, plants retain some of the net primary productivity in permanent organs (new stems and roots, for example; Figure 36.6i), but most becomes litter on the forest floor (Figure 36.6j). In fact, nearly twice as much energy is stored in litter and decomposing humus as in the majestic banks of leaves overhead in a forest. Most of the energy contained in the chemical bonds of the organic compounds in the litter fuels the detritus food web. Only a small fraction of the energy stored aboveground in the forest enters the grazing food web (Figure 36.6k), and some of that enters the detritus food web owing to the excretion and mortality of consumers from the grazing food web (Figure 36.6L). Energy dissipation continues as the consumers that graze plants or eat detritus radiate heat via respiration (Figure 36.6m). Only a tiny amount of energy then remains in the soil or exits the ecosystem in the stream that drains the watershed (Figure 36.6n). Clearly, despite the essentially limitless power flowing from the sun to the earth, plants can store only a small fraction of that energy in organic compounds, and that fraction sets an upper limit to the energy available to all other organisms in the ecosystem.

The information in Figure 36.6 allows ecologists to formulate general principles about the energy budget of a hardwood forest like Hubbard Brook. First, even in a lush, leafy green forest, plants and other producers convert only a small fraction (2 percent or less) of the solar energy that enters the ecosystem into stored chemical energy. Second, animals

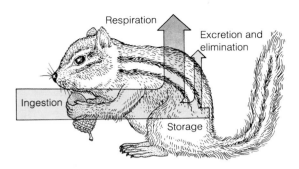

FIGURE 36.7 ■ **Energy Budget of a Forest Herbivore. Energy flows through individual consumers like this chipmunk, as well as through the forest ecosystem it inhabits. Of the energy the chipmunk ingests from seeds and leaves, most is used to fuel its continued respiration; some is lost in the waste products it excretes; and a very small amount (less than 2 percent) is stored in new tissue.**

ingest an even smaller amount (in this case, 0.01 percent) of the energy in the grazing food web. Finally, as energy flows through the trophic levels of the ecosystem, metabolic activities (mostly respiration) release it back to the air, where it ultimately returns to space as heat, a form of energy that does little or no work.

Energy and Trophic Levels: Ecological Pyramids

Ecologists have discovered another important fact in their experiments on energy flow from one trophic level to the next: Food chains rarely have more than four links. To see why, consider the fate of the energy that enters the grazing food web. A chipmunk, for example, one of the most important herbivores of the Hubbard Brook Forest (Figure 36.7), ingests energy in the form of seeds or leaves, assimilates some of the energy, excretes some in liquid wastes, and eliminates some in solid wastes. Most of the assimilated energy is used in aerobic respiration and allows maintenance and repair of the chipmunk's body. Chipmunks store less than 2 percent of ingested food energy in new tissues or offspring. The consequences of a huge loss through respiration and a small net increase in growth is that chipmunks store very little energy in a form that can be used at the next trophic level, say, by a red fox.

Ecologists portray the energy relationships of different trophic levels in an **energy pyramid,** a diagram whose building blocks are proportional in size to the amount of energy available from the level below. Figure 36.8 shows the energy pyramid for a river ecosystem in Silver Springs, Florida, which includes eelgrass and algae (producers); turtles, snails, and caddisflies (herbivores); and beetles, sunfish, and bass (carnivores). At each trophic level, the energy stored by the organisms is substantially less than that of the level below it. Ecologists have constructed similar energy pyramids for hundreds of other ecosystems.

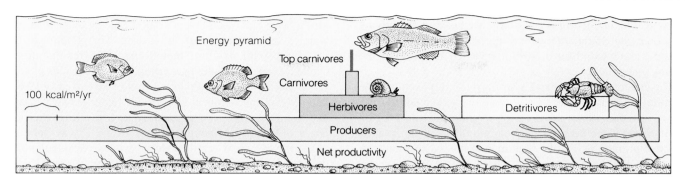

FIGURE 36.8 ■ Energy Pyramid in a River Ecosystem. The energy pyramid for this ecosystem shows that the net productivity of all producers, including eelgrass, is ten times that of all primary consumers (herbivores), including river snails; which is ten times that of all secondary consumers (carnivores), including sunfish; which is ten times that of all top carnivores, including bass. The net productivity of detritivores is larger than that of herbivores.

The stepwise energy decline evident in such pyramid diagrams explains why food chains usually have only four links: The small amount of energy available at the top is too difficult to collect. A handy way to measure the diminishing returns is through **biomass:** the dry weight of organic matter at a particular level. By collecting the eelgrass and other plants from a 1 m² (10.76 ft²) patch on the river bottom in the Silver Springs ecosystem, for example, then drying them to remove their water content and weighing the remaining material, researchers found a biomass of 809 g/m², or about 3 oz/ft² (Figure 36.9).

If they went on to collect all the consumers at different trophic levels in the water above that 1 m² patch and determined their biomasses in a similar way, the data would appear as a **pyramid of biomass,** with each trophic level containing, as a rough approximation, only about 10 percent of the biomass in the level just below it. By moving up four levels (or four links in a given linear food chain), and assuming the 10 percent reduction per link, one finds very little biomass in the top carnivores, such as large, meat-eating bass. Taking it one step further (and another 10 percent reduction in biomass), an angler would need to catch and eat 10 kg (22 lb) of bass to put on 1 kg (2.2 lb) of human tissue. The 10 kg of bass tissue would have come from 100 kg (220 lb) of smaller carnivorous fish, 1000 kg (2200 lb) of insect herbivores, and 10,000 kg (about 10 tons) of plants. All that for 1 kg (2.2 lb) of human being!

A few aquatic ecosystems, such as the waters of Long Island Sound, have inverted pyramids of biomass, with fewer tiny phytoplankton producers than consumers present at any

FIGURE 36.9 ■ Pyramid of Biomass. The biomass—the amount of living tissue—decreases at each trophic level in many ecosystems.

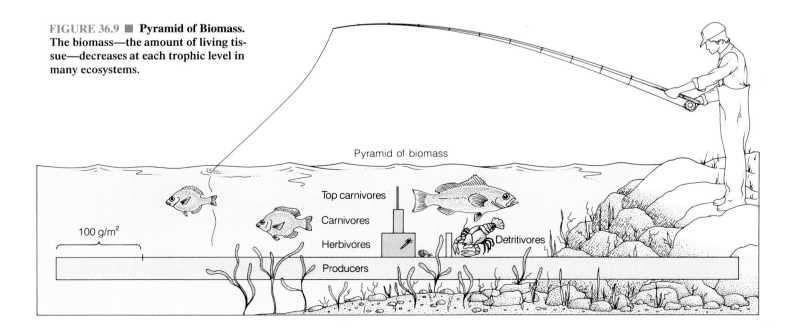

one time. This inversion is possible only because the phytoplankton capture solar energy and reproduce at a fantastically high rate, and most of the offspring are eaten almost immediately by consumers.

The concept of energy and biomass pyramids accounts for the popular 1970s phrase, "Eat low on the food chain." This refers to the fact that it takes 10 kg of grain to build 1 kg of human tissue if the person eats the grain directly, but it takes 100 kg of grain to build 1 kg of human tissue if a cow eats the grain first, and the person eats the beef. Eating lower on the food chain—eating producers, not consumers—saves precious resources on a small planet. (Chapter 24 discusses vegetarian diets in more detail.)

The pyramid pattern of energy flow in ecosystems has an important ramification: **biological magnification,** the tendency for toxic substances to increase in concentration in progressively higher levels of a food chain. Many chemical insecticides, such as DDT and chlordane, resist degradation in the environment and tend to be stored in body fats. If farmers spray DDT on their cabbage plants to control cabbage looper caterpillars, some of the chemical inevitably runs off into streams and lakes. There, instead of breaking down, some of it may enter water plants later eaten by herbivorous fish (Figure 36.10). Fish and other animals cannot break down or excrete the toxin, and it is instead stored in their body fats.

The magnification continues because a carnivorous fish such as a pike must eat many times its weight in herbivorous fish, and the pike stores all the DDT from the many smaller fish it eats. Likewise, when a top carnivore such as an osprey devours many pike, the bird stores the DDT present in all the pike tissue it consumes. The concentration of DDT in the bird's body can reach levels 10,000 times greater than in the water plants that originally took it up. This amount of DDT interferes with calcium metabolism and results in thinshelled, easily broken eggs. During the two decades or so that DDT was commonly used in the United States (early 1940s to late 1960s), the numbers of ospreys, falcons, hawks, eagles, and condors dropped significantly as a result of this lethal chain of events. Although the use of DDT was banned in 1968, the chemical is still widely used in overpopulated and developing nations to control mosquitoes that spread malaria, the greatest killer of humans, and it continues to magnify in the food chain as a consequence of the inefficient energy transfers that also lead to ecological pyramids.

While organisms cannot metabolize many toxic chemicals and recycle their components, they can utilize many compounds that contain nitrogen, sulfur, phosphorus, and other elements. The breakdown and cycling of these nutrients are essential to the long-term health of ecosystems.

HOW MATERIALS CYCLE THROUGH ECOSYSTEMS

In contrast to the essentially inexhaustible stream of solar energy striking our planet, the physical materials on which life depends are limited to what is presently in the soil, water, and air around our globe. Therefore, they must be recycled for

FIGURE 36.10 ■ **Biological Magnification: DDT Concentrates in a Pyramid of Producers and Consumers.** The DDT that a farmer sprays may protect cabbages from caterpillars, but it runs off into nearby streams (red dots), is taken up and incorporated into water plants, and continues to accumulate in each higher level of the food chain, since living things cannot metabolize the hydrocarbon pesticide completely. High accumulations in predatory birds such as osprey result in fragile eggshells and in declining populations of these beautiful animals.

ecosystems to survive. Our planet has several *biogeochemical cycles* (literally, "life-earth-chemical" cycles), global loops of material utilization. In these cycles, substances enter organisms from the atmosphere, water, or soil, reside temporarily in those organisms, then eventually return to the nonliving world when the organisms respire or decompose. Repositories of materials, including the atmosphere, the soil, and living organisms, are called **pools** or reservoirs, and in general, the organismal pool is much smaller than the nonliving pool. Pools remain constant in size as long as a substance's rate of entry equals its rate of departure.

The most important biogeochemical cycles are those for water, nitrogen, phosphorus, and carbon. These underlie the health of all ecosystems, but human activities can alter them in profound and often undesirable ways.

The Water Cycle: Driven by Solar Power

In the global **water cycle** (or *hydrologic cycle*), water moves from the atmosphere to the earth's surface as rain or snow, and back again to the atmosphere by evaporation from puddles, ponds, rivers, and oceans and by transpiration from plants, all driven by energy from the sun (Figure 36.11). In some terrestrial ecosystems, more than 90 percent of the moisture passes through plants and is transpired from their leaves, and only 10 percent evaporates directly from surfaces in the ecosystem. In systems like these, the plants create their own rain: Moisture moves from plants to air to clouds and back to earth in the form of rain wherever the clouds are blown.

The role of vegetation in promoting transpiration, evaporation, and hence rain is especially evident in the tropical rain forests. When people cut down a rain forest in one area, rainwater drains off and eventually reaches the sea instead of rising to the clouds and falling once again on this or another part of the forest. The ongoing massive destruction of the earth's tropical rain forests is changing the environmental conditions needed to support those rain forests, and some ecologists fear the forests may never recover.

The Nitrogen Cycle Depends on Nitrogen-Fixing Bacteria

An organism contains a fairly large quantity of nitrogen in its proteins and nucleic acids. In agricultural ecosystems, where the net primary production helps feed the human race, nitrogen is often the factor that limits productivity. Nitrogen gas makes up 79 percent of our atmosphere, but ironically, most organisms cannot use nitrogen in its gaseous form. Instead, they depend on a few species of nitrogen-fixing bacteria to trap nitrogen in a biologically useful form, while other bacterial species return nitrogen to the atmosphere as nitrogen gas and complete the **nitrogen cycle.** The major steps in nitrogen transformation are nitrogen fixation, ammonification, nitrification, and denitrification (Figure 36.12).

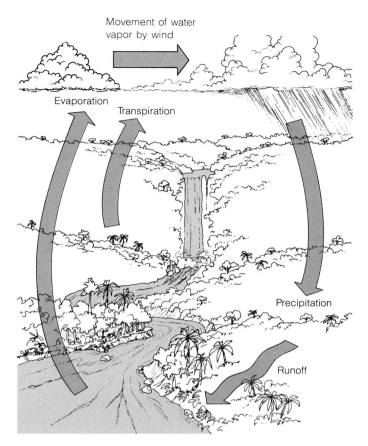

FIGURE 36.11 ■ The Water Cycle. A sun-driven global exchange involving evaporation, precipitation, runoff, and transpiration cycles water from atmosphere to earth's surface to plants and back again.

During **nitrogen fixation,** nitrogen gas (N_2) from the air is transformed into ammonia (NH_3). This colorless gas has an irritating odor and dissolves in water, forming biologically usable ammonium ions (NH_4^+). Nitrogen fixation can occur as a result of lightning or volcanoes or, more importantly, by bacteria in the soil or in root nodules of legumes, aspen trees, or a few other plants (Figure 36.12a; also review Figure 32.8). Plants can absorb these compounds from the soil, change them into nitrate (NO_3^-), and assimilate them into proteins, nucleic acids, and other nitrogen-containing biological molecules (Figure 36.12b). When animals consume plant matter, they break down the plant's nitrogenous compounds and use them to form new animal proteins and other cell components (Figure 36.12c).

After an animal excretes urea or uric acid (see Chapter 25) or after an animal or plant dies, certain bacteria carry out ammonification: They produce ammonium from nitrogen-containing molecules (Figure 36.12d). Plants can then assimilate this ammonium, or still other bacteria can change it to nitrate (NO_3^-) by nitrification (Figure 36.12e). Plants take in some of the nitrate produced in this way (Figure 36.12f). A fourth set of bacterial species acts on the remaining nitrate, and via denitrification, changes fixed nitrogen back to nitrogen gas, thereby completing the cycle (Figure 36.12g).

Since available nitrogen often limits agricultural productivity, farmers have long fertilized their fields to increase the amount of ammonium and nitrate in the soil. Recall that the Pilgrims observed Native Americans burying fish with their corn seeds. Ammonifying soil bacteria produced ammonium from the fish proteins and nucleic acids, the corn assimilated the available nitrogen atoms in the ammonium, and the plants grew taller and faster. And many farmers continue to practice *crop rotation,* as did their predecessors in various parts of the world. They plant leguminous crops such as beans, clover, or alfalfa one year and corn, wheat, or sugar beets the next year to take advantage of the nitrogen that the nitrogen-fixing bacteria in the legumes' root nodules release into the soil.

Today, most large farms around the world depend on nitrogen fertilizers produced through an industrial process rather than a bacterial process. In fact, nitrogen fixed in chem-ical factories may now represent 30 percent of the input to the global nitrogen cycle, a truly great shift in the natural cycle. Unfortunately, industrial nitrogen fixation requires tremendous heat and pressure, and this is usually produced by burning great quantities of fossil fuels. In some areas, people dump more energy into the soil in the form of nitrogen fertilizers than they extract from the soil in food calories. Since reserves of fossil fuels are limited, this enormous reliance on industrially fixed nitrogen fertilizers cannot go on indefinitely. Many biologists think a far better solution would be to genetically engineer plants and soil bacteria to increase natural nitrogen fixation (see Chapter 32).

The nitrogen cycle is sensitive to human activities such as deforestation. To quantify this, ecologists removed all trees from one watershed in the Hubbard Brook Forest and sprayed the area with herbicides to block regrowth (see Figure 36.5). In their test area, 60 times more nitrogen drained away in the stream than in the control watersheds where the forests remained undisturbed. Such severe nitrogen loss drastically limits regrowth of the forest and pollutes groundwater.

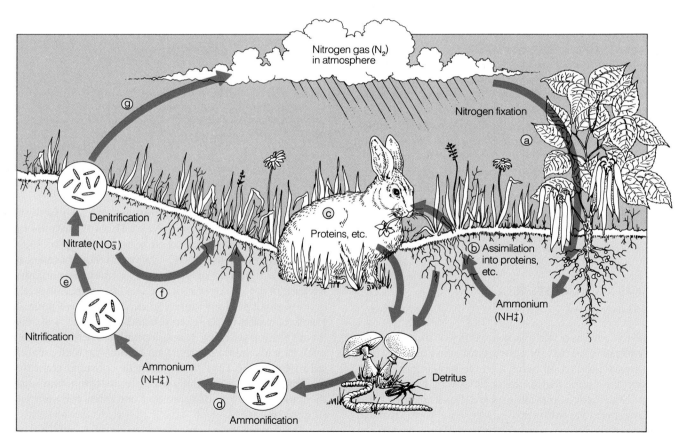

FIGURE 36.12 ■ **The Nitrogen Cycle: From Air to Organisms and Back.** Nitrogen fixation by bacteria in root nodules (a) and in the soil fix atmospheric nitrogen into ammonium or nitrate (d–f). Plants can then use nitrate to make proteins (b), and animals, by eating the plants, can also gain needed nitrogen (c). Denitrifying bacteria release the nitrogen (g), allowing it to return to the atmosphere.

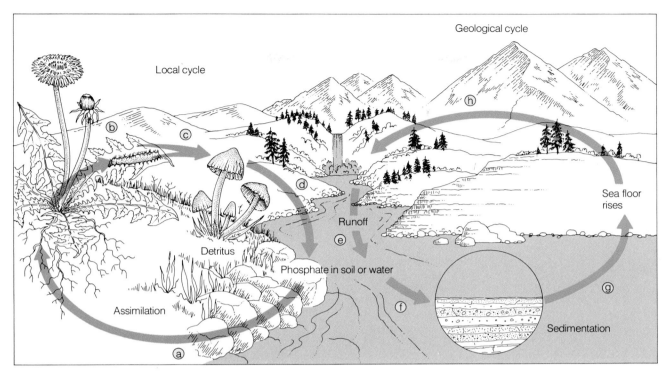

FIGURE 36.13 ■ The Phosphorus Cycle: Short-Term and Local, Long-Term and Global. Plants assimilate phosphates from soil or water (a), animals gain phosphates by eating the plants (b), and when the autotrophs and heterotrophs die or give off wastes (c), the phosphates return to the soil or water in a local cycle (d). Some phosphates become tied up in sedimentary rocks at the bottom of the seas (e, f), but when the seafloor rises after thousands or millions of years (g), the phosphates from this long-term cycle can once again enter the local, short-term cycle (h).

The nitrogen cycle, like some other biogeochemical cycles, has an atmospheric phase. While nitrogen gas, water vapor, and carbon dioxide are airborne, they can be blown anywhere on earth by the wind. This mobility makes the nitrogen, water, and carbon cycles truly planetary. In contrast, some substances, like calcium and phosphorus, lack a gaseous phase and thus cycle locally. Local cycles of these elements can cause great fluctuations in the populations of local organisms. Box 36.1 on page 742 describes such cycles in populations of lemmings in the tundra.

Phosphorus Cycles Locally and in Geological Time

Phosphorus is essential for life. It is a component of cell membranes, nucleic acids, and ATP, a cell's energy currency. The **phosphorus cycle** consists of two interlocking circuits, one that acts locally during short stretches of time, and another that operates more globally over vastly longer time periods. In the local cycle, phosphorus moves from the rocks or soil (where it usually occurs as calcium phosphate) into organisms and back to the soil. Plants assimilate phosphate (PO_4^{-3}) directly from the soil or water (Figure 36.13a), while animals

obtain needed phosphorus from the plants or other animals they consume (Figure 36.13b). When plants and animals die and decay (Figure 36.13c), bacteria convert the organic phosphorus in their tissues into phosphate. This enters the soil once again (Figure 36.13d), and plants may then assimilate it once more, completing the local ecological cycle without a significant atmospheric component or global distribution.

Although phosphorus does not leave local ecosystems by way of the air, it *can* leave terrestrial ecosystems in streams and rivers (Figure 36.13e). Thus, the second phosphorus cycle, if viewed over geological time, does have a global aspect. After phosphate eventually washes from terrestrial waterways down to the sea, it can form insoluble compounds that fall as sediments and become incorporated into rock (Figure 36.13f). Eons later, when the seafloor rises and exposes new land (Figure 36.13g), these phosphorus-containing rocks may form the base of a terrestrial ecosystem, and the phosphorus can once again enter the local ecological cycle (Figure 36.13h).

As with other biogeochemical cycles, human activities have altered the dynamics of the phosphorus cycle—especially in aquatic ecosystems, for which phosphorus is often the factor limiting primary productivity. Phosphates are major

LEMMINGS IN THE TUNDRA: CASE STUDY OF ECOSYSTEM DYNAMICS

Much of the Arctic is windswept and cloaked in near total darkness for several months each year. This desolate region supports only a few major plant species—sedges, grasses, and a knee-high forest of dwarfed birch and willow—and just one major herbivore—the furry tundra mouse called the lemming (Figure 1). Popular wisdom has it that lemmings dash mindlessly over cliffs and into the sea. This is an exaggeration, but one based on a grain of truth: Lemmings do experience periodic population booms—and crashes—triggered by nutrient cycles in the tundra ecosystem.

Observers have noted that every four years, lemmings begin to reproduce and spread rapidly, with each acre of the tundra land suddenly harboring hundreds of the small animals. During these population booms, lemmings grow hungry and migrate in search of food, and it is this phenomenon that gives rise to the popular myth.

High-density populations are possible only when the short grasses and tough tundra herbs are especially rich in phosphorus and calcium, which lemmings need for successful reproduction. When those nutrients are plentiful, however, the tundra vegetation itself takes a terrific beating, since millions of the furry little buzz saws emerge from their burrows and mow the grasses flat. In the process, most of the nutrients transfer from the plants to the lemmings. Lacking additional minerals, the overgrazed tundra plants cannot regenerate new leaves and stems. Now the lemming population boom becomes a population bust: The dwindled food supply cannot sustain large numbers of the hungry animals, and the lemming population crashes to fewer than 1 percent of the animals alive during peak periods.

As the lemmings die, their bodies are slowly decomposed in the detritus food web, but in the cold arctic climate, this process takes a couple of years. As nutrients are gradually returned from the lemmings to the soil, plants start making a comeback from their severely overgrazed state. Eventually, by the fourth year, the plants have completely recovered and produced new leaves and stems, rich in the two limited nutrients, calcium and phos-

FIGURE 1 ■ Collared Lemming Eating Saxifrage Blossoms in the Tundra.

phorus. This boom in nutritious food allows the lemmings to reproduce at high rates again, produce another bumper crop of hungry young, overgraze, and then crash again.

The effects of these nutrient-recovery cycles and population booms ripple throughout the arctic food web. As one might predict, populations of the lemming's main predator, the white arctic fox, also increase and decrease in step with their small furry prey. Curiously, however, so do the populations of the dark-bellied brent goose and certain wading birds. Apparently, if foxes cannot get their fill of lemmings, they eat the eggs and young of these birds and thus cause oscillations in their populations as well.

While other explanations may also help account for the boom and crash of lemming populations, these events drive home the dependence of each species on other species and on the physical nature of the ecosystem.

---

ingredients of agricultural fertilizers and until recently were also main components of detergents. When phosphates from these sources run off into lakes, algae and other aquatic plants grow faster, and the ecosystem becomes **eutrophic** (literally, "well fed"), or overly supplied with nutrients that support primary production. (The opposite term, **oligotrophic,** means "little fed" and describes an ecosystem such as a clear lake, where there is little algae or other vegetation.) As individual algal plants age and die in a eutrophic lake, decomposing bacteria feed on the dead algal cells and use up so much dissolved oxygen that fish and other animals may suffocate.

Ecologists dramatically demonstrated this process with an experiment in a lake in Ontario, Canada. They hypothesized that phosphate is the limiting factor in a lake's primary pro-

ductivity. To test this, they found a lake with a natural hour-glass shape and divided it into two sections by stretching a vinyl curtain across the lake's narrowest part (Figure 36.14). This clever strategy effectively separated the lake into an experimental side and a control side with equivalent conditions at the start of the experiment. The researchers fertilized both halves of the lake with nitrogen and carbon compounds, but in addition, added phosphorus to the experimental half. Without added phosphorus, the control section showed no change in organisms or ecological productivity. Within two months, however, the experimental half, with its added fertilizer, developed a blue-green algal bloom that was visible from an airplane (see Figure 36.14). Only after workers stopped fertilizing the lake with phosphorus did the algal bloom fade

Fertilized with carbon, nitrogen, and phosphorus

Plastic sheet dividing the lake

Fertilized with carbon and nitrogen only

FIGURE 36.14 ■ **A Lake Divided Proves That Phosphorus Is the Limiting Factor in an Aquatic Ecosystem.** A body of fresh water in northwestern Ontario dubbed Lake 226 helped prove the ecological importance of phosphorus. Experimenters divided the lake into two parts and fertilized the near basin with nitrogen and carbon, the far basin with nitrogen, carbon, and phosphorus. After two months, the near basin showed no change, but the far basin experienced an algal bloom, readily visible from the air. Clearly, phosphorus levels were the key to increased productivity. Too much phosphorus from detergents or fertilizers can lead to eutrophication (too much productivity), as in this lake.

and the lake recover its previous condition. This spontaneous recovery suggests that people can decrease the eutrophication of lakes and streams by restricting the use of phosphate-containing detergents. In the mid-1970s, when Lake Erie was becoming seriously eutrophied, regional governments banned the use of phosphate-laden detergents. Since that time, the lake has become much healthier.

The Carbon Cycle: Coupled to the Flow of Energy

Carbon atoms move in a global **carbon cycle** from the physical environment through organisms and back to the nonliving world, just as water moves through the hydrologic cycle. And like the water cycle, the carbon cycle is linked to energy flow because producers—including the photosynthetic plants of the forests and oceans and the chemosynthetic bacteria of deep-sea vents—require environmental energy (either sunlight or inorganic hydrogen compounds) to trap carbon into sugars. The trapped carbon comes from carbon dioxide in the surrounding air or water (Figure 36.15). As the cycle proceeds, consumers devour the organic carbon compounds that producers manufacture. Then, via respiration, both consumers and producers return carbon to the nonliving environment in the form of carbon dioxide. Some carbon accumulates for many years in wood and is eventually returned to the atmosphere in fires or through consumption and respiration by fungi, bacteria, and other detritivores. Organic carbon can leave the cycle for even longer periods of time after sediments bury organic litter, which decomposes only partially and gradually transforms into coal or oil. Carbon also leaves the cycle when cast-off calcium carbonate shells of marine organisms sink to the ocean floor and become covered with sediments that compress them into limestone. Eventually, however, even these carbon deposits are recycled into atmospheric carbon dioxide as the limestone erodes and the fossil fuels are burned in automobiles or in industry. As with the other biogeochemical cycles, human activities are altering the dynamics of the carbon cycle. In some cases, these perturbations have global consequences and affect both our own future and that of the other organisms that share our planet.

HOW HUMAN INTERVENTION ALTERS ECOSYSTEM FUNCTION

Human activities have the potential to drastically modify the nutrient cycles that support life on earth and to alter physical features of the natural environment as well, including air and water temperatures and acidities and the amount of available solar energy. Let us consider four of the most serious current ecological concerns: global warming, acid rain, the dwindling of our natural resources, and the threat of nuclear winter.

Global Warming

The carbon cycle, as we have seen, relies on an atmospheric supply of carbon dioxide, which plants use to make sugars (review Figure 36.15). Although carbon dioxide makes up only about 0.03 percent of our atmosphere, it plays a disproportionately large role in governing the earth's temperature by means of a phenomenon called the *greenhouse effect*. The earth and its envelope of atmospheric gases are a bit like a greenhouse on a sunny day: Light energy passes in, but infrared radiation is trapped inside and the structure warms.

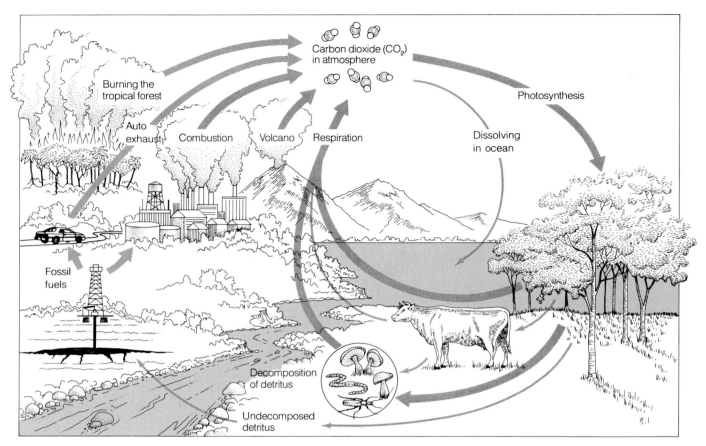

FIGURE 36.15 ■ The Carbon Cycle: From Atmosphere to Plants, Animals, Decomposers, Human Activities, and Back. Carbon cycles, from atmospheric carbon dioxide to biological molecules to organic molecules in the soil to geological deposits of fossil fuels, and back to carbon dioxide. Disruptions of this cycle can lead to increased atmospheric carbon dioxide and to global warming.

The carbon dioxide and other so-called greenhouse gases in the atmosphere are analogous to the glass in a greenhouse or the windshield of a car parked in the sun (Figure 36.16). Light energy passes through atmospheric greenhouse gases, strikes the earth, and warms it. The warm earth then re-irradiates the energy as infrared radiation (see Figure 6.5). While light can pass through greenhouse gases, infrared rays do not, and hence they become trapped near earth's surface and contribute to a buildup of heat. Greenhouse gases include carbon dioxide, methane, chlorofluorocarbons, and nitrous oxide, but carbon dioxide alone contributes about half the human-produced share of the global warming problem. Methane is generated by bacteria that live in flooded rice fields and the guts of cattle and that break down organic matter. *Chlorofluorocarbons* (*CFCs*) are synthetic materials used as coolants in refrigerators and air conditioning as well as in Styrofoam and insulation materials. Nitrous oxide (N_2O_2), or laughing gas, is produced by microbes in soils, and its concentrations increase with the burning of forests and fossil fuels and the production of chemical fertilizers.

Unfortunately, the major greenhouse gas, CO_2, has increased by roughly 25 percent in the past century, and the average atmospheric temperature has risen 0.5°C (about 1°F) along with it. In fact, the National Aeronautics and Space Administration, which has kept available records for the last 140 years, reported that the seven warmest years on record have occurred since 1980, and as this book went to press, the surface temperatures in the year 1990 were the highest to date. Much of the extra carbon dioxide that atmospheric scientists have measured originates with the burning of coal, oil, and gasoline in our factories and automobiles. The burning of tropical rain forests (review Figure 35.23) is another major source of the gas. Some of the excess carbon dioxide dissolves in the ocean. Much of it, however, enters the atmosphere and blocks the escape of infrared radiation.

Many atmospheric scientists predict that sometime between the years 2025 and 2075—well within the expected lifetime of those reading (and writing!) this book—an accumulating blanket of greenhouse gases will trap more and more heat, and global temperatures will climb an average of 2° to

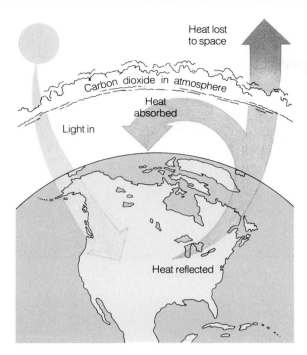

FIGURE 36.16 ■ **The Greenhouse Effect Can Cause Global Warming.** Like sun shining through the glass windows of a greenhouse, light penetrates the atmosphere and warms the earth, but much of the reflected heat becomes trapped in the atmosphere. The more carbon dioxide in the air, the more heat is trapped, and the higher the earth's surface temperatures.

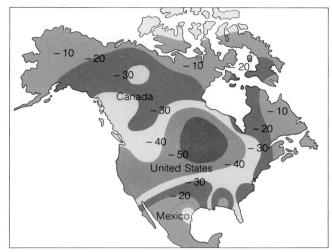

(a) Moisture loss

(b) Dust bowl of the 1930s

4°C (3½° to 7°F). This may seem like a small increase, but its effects could be truly disastrous. According to the predictions, ice at the poles might melt, and sea levels might rise, inundating many of the world's most populous cities—New York, Los Angeles, London, Stockholm, Hong Kong, Tokyo, and others. More importantly, the increased temperatures might change fertile croplands into deserts: The dust bowl conditions of the 1930s could befall the great grain-producing regions of the American Midwest, Canada, and the Soviet Union (Figure 36.17). And irrigation would probably not correct the situation, because groundwater reserves would quickly become depleted. Ominously, even the coldest years of the last decade have been warmer than nearly every year a century ago, and we are beginning to see widespread droughts and resultant forest fires and crop losses.

Most climatologists agree that there is no firm proof that accumulating greenhouse gases are causing the observed warming. Many climatologists suggest, however, that the danger to ecosystems and economies is too great to allow scientific uncertainty to prevent prudent action now. Most ecologists believe that we must decrease our consumption of fossil fuels through an emphasis on energy conservation and a commitment to renewable energy sources such as wind power and solar energy. As individuals, we could, for example, lower the

FIGURE 36.17 ■ **Will the Dust Bowl Return?** (a) If atmospheric levels of carbon dioxide double in the next few decades, as some scientists predict, then the soil will lose precious moisture. This map predicts the relative moisture loss between the years 2025 and 2075: about 10 percent around the coasts, 20 percent farther inland, and in our continent's fertile interior agricultural regions, from 40 to 60 percent drier. (b) Some observers fear that this loss of moisture could lead to another dust bowl like the one in the 1930s, when prolonged drought, overplanting, and agricultural mismanagement allowed soil layers to dry out and topsoil to blow away, driving people from their land.

temperature of our homes in winter, use less air conditioning in summer, and rely on public transportation, bicycles, and our own two feet more often than we do now. Farmers could employ agricultural practices that sustain the levels of organic matter in the soil, and people in tropical regions could step up efforts to replant denuded areas. Finally, biologists could actively develop more drought-resistant and salt-tolerant plants that can substitute for our staple crops if the world's "breadbaskets" become dust bowls. Implementing these solutions will be difficult, but informed citizens must work now to help governments make ecologically sound policy decisions.

When It Rains, It Pours Sulfuric Acid

About 35 years ago, anglers began to note declining fish populations in formerly productive lakes in Sweden, Ontario (Canada), and upper New York State. Swedish ecologists traced the cause to increased acidity in the lake water and in turn to the abnormally low pH of precipitation. In short, they discovered that it is raining sulfuric acid! Observers soon noted that acid rain is also killing trees from Colorado to the Carolinas to Vermont, stunting the growth of lake trout in northern Ontario, and eroding European monuments that are centuries old (Figure 36.18). In Europe alone, acid rain could cost 30 billion dollars each year for the next century in damaged and unusable timber and lumber products.

Clues to the origin of acid rain lay in the distribution of the affected areas: They are all downwind from the industrial centers of Europe and North America. The burning of coal and oil to generate electric power, run factories, and fuel automobile engines creates oxides of nitrogen and sulfur that acidify the rain. (Normal rain is only slightly acidic.) In many instances, airborne pollutants escape from extremely tall industrial smokestacks and can travel hundreds of kilometers, suspended in air currents. Up in the clouds, sulfur and nitrogen oxides combine with water molecules and form sulfuric and nitric acids. When rain precipitates from these clouds, the acids are dissolved in it.

Since the major source of acid rain is pollution from coal-fired plants that generate electricity, the immediate solution would seem to be the removal of sulfur and nitrogen oxides from their exhausts. This can be done by installing scrubbers, devices that use a liquid spray to filter the pollutants. The use of scrubbers raises people's power bills by about 5 percent, however, and so far, electric power utilities have stubbornly resisted their installation. An informed public must decide whether it is willing to pay slightly higher bills for electricity to preserve its forested recreational lands and wood products industry.

Cycle and Recycle: The Sustainable Economy

In nature, carbon, nitrogen, and other fixed resources are cycled and recycled, and human societies must learn to do the same. Americans discard most materials after one use; we take two-thirds of the aluminum we use, three-quarters of the steel and paper, and most of the plastic products and simply throw them away. Yet the amount of energy needed to recycle aluminum is only 5 percent of the energy needed to produce

(a) Trees killed by acid rain

(b) Trout from normal lake (top) and lake damaged by acid rain (bottom)

(c) Statue damaged by acid rain

FIGURE 36.18 ■ Dying Back, Dwindling, Dissolving: The Effects of Acid Rain. (a) These conifer trees in northern Vermont stand like ghostly skeletons, their leaves and tender twigs burned by acid rain. (b) Lake trout in northern Ontario dwindled in size between 1979 and 1982 as they struggled to survive in the acidic water. (c) After gracing a stone cathedral for hundreds of years, this northern European sculpture has begun to dissolve from the highly corrosive effects of acid rain.

it from bauxite, the original raw material. Likewise, recycling steel, newsprint, and glass saves from one-quarter to two-thirds of the energy needed to generate those commodities from scratch. Recycling materials instead of continually producing new ones also cuts air, water, and ground pollution by more than half. But recycled materials can't just be dumped into landfills. Manufacturers must use them to build new consumer products. As individuals, we should not only recycle our newspapers, plastics, and glass, but also encourage the use of recycled materials by asking for and buying new items made from them and packaged in their own recyclable materials. Beyond that, we should learn to compost garbage and garden clippings, to cut down on refuse handling and accumulation, and to generate rich humus for garden and lawn fertilizer. Box 36.2 describes a number of specific ways to recycle and conserve our dwindling natural resources.

Clearly, human activities can unfavorably alter the cycles that support life, as well as help balance and maintain them.

We have an equally awesome —perhaps terminal—power to prevent much of the sun's energy from reaching the earth and driving virtually all life functions.

Nuclear Winter: A Catastrophic Ecological Threat

The flow of energy that supports most living things begins with sunlight striking green plants. Our species now possesses the technology to detonate nuclear weapons, incinerate vast areas, choke the air with dense smoke clouds, block the access of green plants to sunlight, and thus ignite a global ecological disaster.

A large-scale nuclear war employing even a fraction of the warheads now stockpiled by the United States and the Soviet Union would directly kill nearly everyone in North America, Japan, Europe, and the Soviet Union. Most of the survivors would need medical attention that would no longer be available, and thus they, too, would succumb in days or

---

BOX 36.2   FOCUS ON THE ENVIRONMENT

SAVING THE ENVIRONMENT: PERSONAL SOLUTIONS

It's easy to get overwhelmed by environmental news—ozone levels dropping, carbon dioxide rising, acid rain falling, animals disappearing, forests going up in smoke. It's tempting, in fact, to lose interest, resign oneself, and carry on as if these threats were not both real and pressing. We no longer have the luxury of a head-in-the-sand attitude, however, and each of us must develop a personal plan of action. Because the problems are complex and ubiquitous, potential solutions are many. Several recent books list practical conservation tips as well as local and national sources of background information. (See, for example, *50 Simple Things You Can Do to Save the Earth,* by the Earthworks Press, 1989.) What follows are a few beginning suggestions for personal environmental action, and we encourage you to add your own:

- In winter, turn down the furnace, and set the thermostat even lower when you're not at home. In summer, use less air conditioning, and have the furnace pilot light turned off.
- Adjust the water heater to 54°C (130°F), not the normal 60°C (140°F) or higher.
- Leave the car at home and use public transportation, bicycles, or foot power whenever possible.
- Use warm wash and cool rinse when washing clothes, to save gas or electricity.
- Recycle aluminum, tin, glass, newspaper, nonglossy paper (junk mail, old school work, packing materials), and plastic bags and containers.
- Avoid disposable packaging by carrying a string or cloth shopping bag, buying from bins, and reusing plastic produce bags.

- Compost organic wastes when possible, using your own or a community compost pile.
- Plant trees in and around campus, town, and your backyard.
- Refuse to buy ivory and other parts and products of endangered species.
- Refuse to buy teak and other tropical wood products.
- Buy less beef, since cattle are often raised on newly stripped tropical lands, and since the production of animal protein, in general, requires far more energy than the production of plant proteins.
- Cut excess water use by modifying your habits and by using low-flow shower heads and toilet tank modifiers.
- Buy unbleached and recycled paper products.
- Buy more organic produce as a way of encouraging alternative farm practices.
- Avoid CFC propellants and polystyrene cups and containers. If buying a new air conditioner or refrigerator, ask about some of the new alternative coolants now being developed.
- Consider vehicles that use gasohol or methanol instead of gasoline.
- Take leftover toxic materials (used crankcase oil, paints and thinners, oven cleaners, pesticides, etc.) to approved municipal sites instead of dumping them in drains or on the soil, where they can reach surface- and groundwater supplies.
- Switch to integrated pest management techniques in home gardens to avoid using pesticides unnecessarily.

weeks. Horrendous as such a scenario may seem, the longer-term consequences to the earth's ecosystems would be even more devastating. Nuclear warheads detonated at ground level would inject enormous quantities of radioactive dust into the atmosphere, and bombs exploding over urban and forested areas of the Northern Hemisphere would ignite widespread fires that would heave huge clouds of smoke into the air and shroud much of the planet.

What would the climatic effects be from such massive clouds of dust and smoke? Physicists, atmospheric scientists, and ecologists have tried to answer this question by developing computer models that simulate nuclear wars of various magnitude. Simulations taking wind speeds, ocean currents, and seasonal changes into account predict that nuclear smoke clouds would drift from place to place in the Northern Hemisphere, and beneath those clouds, temperatures would plummet to well below freezing, a condition called **nuclear winter.** Most scientists agree that a nuclear war would have catastrophic effects on global climate and have profound consequences for biological systems. Plant species that can normally survive cold winters would probably be killed if the nuclear war occurred during their active growing periods in spring or summer. Monsoon rains would be severely disrupted in the Southern Hemisphere. Rainfall over burned-out forests would trigger massive flooding and erosion. Radiation levels would remain lethal to most animals and many plants for long periods. Finally, poisonous substances released by urban fires would impede the regrowth of vegetation.

Clearly, the conditions facing human survivors of a global nuclear war would be exceedingly grim. Coldness, darkness, starvation, thirst, exposure to radiation and heavy pollution, and the absence of sufficient fuel would destroy civilization

as we know it. The lesson here is straightforward: Even if a nuclear war did not extinguish the human species immediately, any nuclear strike would be suicidal.

Only those organisms that do not depend on sunlight for energy—the bacteria, tube worms, clams, crabs, fishes, and other residents of the deep-sea vent communities—would remain largely unaffected. Independent of light and warmed from within the earth, these assemblages of creatures could survive even if all other living things were destroyed. But what a sad vestige this would be of the fantastic diversity that exists today and has survived over an enormous span of geological time.

CONNECTIONS

Energy flow and nutrient cycles in ecosystems are fundamental biological principles governing life. As history plainly shows, no organisms are exempt from these laws or their implications. Humankind's current polluting activities are changing global environments so fast that organisms have no time to evolve appropriate adaptations. If current trends continue, we will find ourselves the subjects of an unplanned global experiment, and we will not escape its repercussions. We humans are unique among living organisms only in the degree and speed with which we can modify the earth's ecosystems, in our ability to evaluate the future consequences of our actions, and in the capacity to take measures as individuals and as nations to deal with the ecological imbalances and threats we now face.

In the next chapter, we will examine the global forces that link ecosystems into one all-encompassing biosphere.

Highlights in Review

1. The deep-sea vent ecosystem illustrates how a community of interdependent organisms interacts with its physical setting and how energy flows and inorganic materials cycle through the system.

2. Every ecosystem has trophic, or feeding, levels. Autotrophs are the primary producers, while heterotrophs are the primary, secondary, and tertiary consumers. Fungi and decomposing bacteria are the detritivores.

3. The organisms in the various trophic levels are linked in simple food chains and in more complex food webs, such as the grazing and detritus food webs.

4. Most energy flowing into terrestrial ecosystems from the environment is reflected as heat or light or causes evaporation and transpiration of water. The small amount of energy stored in newly formed plant organs becomes the net primary productivity that supports all other life in the ecosystem.

5. Energy pyramid diagrams reveal that the energy available in the producer level is substantially greater than in the next higher level, and so on. The biological magnification of toxic substances is a consequence of energy loss from one trophic level to the next.

6. The physical materials that support life move through massive biogeochemical cycles involving living organisms and their physical environments.

7. The hydrologic, or water, cycle involves evaporation, plant transpiration, and precipitation.

8. The nitrogen cycle involves atmospheric nitrogen, nitrogen fixation by soil bacteria, ammonification and nitrification by other bacteria, and denitrification by still others.

9. The phosphorus cycle is really two interlocking cycles: a short-term, local cycle in which phosphates move from rock and soil into organisms and back to the soil; and a long-term cycle that

operates over geological time, in which phosphates become fixed in ocean bedrocks and are eventually exposed and eroded eons later.

10. The carbon cycle involves carbon dioxide in the air and water, carbon fixation by autotrophs, then a return of the carbon to the air and water through respiration, combustion, and erosion.

11. Carbon dioxide and other greenhouse gases allow light energy to pass toward earth, but can block escape of heat energy. Some ecologists predict that increasing levels of atmospheric carbon dioxide due to human activities may be leading to a global greenhouse effect and global warming.

12. The burning of fossil fuels also releases nitrogen and sulfur oxides, which combine with water in the atmosphere and form acid rain, disrupting lakes and forests downwind from industrial areas.

13. Recycling aluminum, glass, paper, and other materials saves energy, prevents pollution, and protects our dwindling natural resources.

14. The ultimate ecological threat is nuclear winter, in which dust and smoke would block out sunlight, plunge global temperatures, prevent much photosynthesis, lead to the extinction of many species, and permanently disrupt most ecosystems.

Key Terms

biological magnification, 738
biomass, 737
carbon cycle, 743
consumer, 732
detritivore, 733
detritus, 733
energy pyramid, 736
eutrophic, 742
food chain, 733
food web, 733
nitrogen cycle, 739
nitrogen fixation, 739

nuclear winter, 748
oligotrophic, 742
phosphorus cycle, 741
pool, 739
primary consumer, 733
producer, 732
pyramid of biomass, 737
secondary consumer, 733
tertiary consumer, 733
trophic level, 732
water cycle, 739

Study Questions

Review What You Have Learned

1. Define producer and consumer.
2. How do chemosynthesis and photosynthesis differ? Name one type of organism that employs each method.
3. Name several types of detritivores. How do they obtain energy?
4. Create a diagram showing all the organisms in the food web of a forest from your geographical region.
5. True or false: Only 1 percent of the solar energy that reaches the earth is stored via photosynthesis, and yet this provides energy for virtually all other life in the biosphere. Explain your answer.
6. Are there differences in energy content at different levels of an energy pyramid? Explain.
7. What do the various levels in a pyramid of biomass represent?
8. Describe the water cycle.

9. Describe four steps of nitrogen transformation during the nitrogen cycle.
10. Trace the steps in the local phosphorus cycle.
11. What is the greenhouse effect?
12. How does acid rain form?
13. Discuss the ecological hazards of a nuclear war.
14. How can human activity disrupt the carbon cycle?

Apply What You Have Learned

1. Farmers, gardeners, and forest managers are no longer allowed to use DDT as an insecticide in the United States. Why not?
2. Native Americans taught the Pilgrims to plant a fish with each corn seed. Why was this planting strategy successful?
3. The labels of some laundry detergents state that the products contain no phosphates. Why is the absence of phosphates important?
4. A politician proposes that we do nothing about the greenhouse effect, stating: "When global temperatures increased 10,000 years ago, the glaciers retreated toward the poles, followed by climatic and vegetation zones. We don't need to worry about the effects on natural ecosystems, even if the greenhouse effect causes the temperature to rise, because plants and animals will just move northward." How would you analyze and respond to this argument?

For Further Reading

BROWN, L. R. *State of the World 1991.* New York: Norton, 1991.

CHILDRESS, J. J., H. FELBECK, and G. N. SOMERO. "Symbiosis in the Deep Sea." *Scientific American,* May 1987, pp. 114–120.

COHEN, J. E. "Big Fish, Little Fish: The Search for Patterns in Predator-Prey Relationships." *The Sciences* 29 (1989): 37–42.

GOSZ, J. R., R. T. HOLMES, G. E. LIKENS, and F. H. BORMANN. "The Flow of Energy in a Forest Ecosystem." *Scientific American,* March 1978, pp. 93–102.

GRAEDEL, T. E., and P. J. CRUTZEN. "The Changing Atmosphere." *Scientific American,* September 1989, pp. 58–68.

JONES, P., and T. WIGLEY. "Global Warming Trends." *Scientific American,* August 1990, pp. 84–91.

KREBS, C. J. *Ecology: The Experiential Analysis of Distribution and Abundance.* 3d ed. New York: Harper & Row, 1985.

MATTHEWS, S., and J. SUGAR. "Is Our World Warming? Under the Sun." *National Geographic,* October 1990, pp. 66–100.

NEBEL, B. J. *Environmental Science: The Way the World Works.* 2d ed. Englewood Cliffs, NJ: Prentice-Hall, 1987.

ODUM, E. P. *Ecology and Our Endangered Life-Support Systems.* Sunderland, MA: Sinauer, 1989.

PEARCE, F. "Whatever Happened to Acid Rain?" *New Scientist,* September 15, 1990, pp. 57–60.

TURCO, R., O. TOON, T. ACKERMAN, J. POLLACK, and C. SAGEN. "Climate and Smoke: An Appraisal of Nuclear Winter." *Science* 247 (1990): 166–169.

The Biosphere: Earth's Thin Film of Life

The Cathedral Forest

It is extraordinarily still in the cathedral forest (Figure 37.1). The damp air is unmoving and fragrant with the pungence of greenery, wet tree trunks, and dark, spongy leaf litter. Water drips steadily from the mist-enshrouded canopy high overhead, moistening the fern fronds that blanket the forest floor. Only occasionally does an animal's movement break the silence: A Roosevelt elk, huge antlers pointing skyward, picks its way around the fallen trunks of massive evergreens and steps lightly through the tangle of moss-covered branches, loose bark, tiny plants, and rotting logs. A mouselike red-backed vole digs in the damp soil for truffles, the fruiting bodies of fungi growing on and around the forest trees, and moves quickly away with powdery spores clinging to its fur. Hearing the vole's soft scurrying, a spotted owl swoops noiselessly from a high branch, snatches the small rodent, and lifts it to a nest of hungry young in a tall dead tree still standing among the living redwoods, Douglas fir, hemlock, and cedar.

This cathedral grove stands in the world's greatest temperate rain forest, which stretches for more than 1610 km (1000 mi) along the west coast of North America, from northern California to British Columbia. The trees, ferns, and moss growing here require large amounts of precipitation—over 254 cm (100 in.) of rain and snow a year in some places—as well as the mild temperatures influenced by the westerlies blowing in from the Pacific. Worldwide climatic forces set up the wind and water currents that foster the emerald forests of the Pacific Northwest. But in return, these forests affect both the local weather and global climates. Nearly one-quarter of the precipitation that falls in the hundreds of square kilometers of coastal forests starts as fog and clouds; droplets collect on billions of leaves and needles and fall to the ground. And recent evidence shows that old-growth (virgin) forests, such as these in the

FIGURE 37.1 ■ The Cathedral Forest: The Northwest's Great Temperate Rain Forest. A Roosevelt elk stands amid the damp greenery of an old-growth forest in Washington's Olympic National Park.

Pacific Northwest, store a tremendous amount of carbon; when the trees are cut and their wood (either as paper, lumber, or chips) is burned or left to decay, the carbon can be released, contributing additional carbon dioxide to the greenhouse effect and global warming. While young trees planted in place of these virgin forests will take up some carbon dioxide as they grow, it requires many decades, even centuries, before new growth can compensate for the huge release of carbon stored in immense old trees. People are usually motivated to cut ancient, massive trees for economic reasons. But as we will see later in this chapter, environmentalists are battling loggers over the use of old-growth forests, and rare species like the spotted owl are figuring prominently in the arguments over whether to use or preserve the virgin stands.

The temperate rain forest (or moist coniferous forest) is just one example of how closely interconnected are the living and nonliving worlds. To understand these complex interactions, we must study various communities, including forests, deserts, tundra, and grasslands. We must follow the cycling of materials within ecosystems through food webs to nonliving environments and back again. And we must explore how weather and climate dictate where different species of plants and animals will live. The subject of this chapter is the *biosphere,* that portion of our planet, including water, air, and soil, that supports life. The biosphere is the highest level of organization in the natural world (review Figure 1.4), and it includes every body of water—rivers, ponds, lakes, and oceans; the atmosphere to a height of about 10 km (6 mi); the earth's crust to a depth of several meters; and all the living things within this collective zone. As huge as the biosphere may seem, it is actually a delicate veneer at our planet's surface. If you inflated a round balloon to a diameter of 20 cm (8 in.) to represent the earth, then the thickness of its taut rubber skin would be proportional to the biosphere, the thin film of life encircling our world.

Our task in this chapter is to understand how global physical factors—particularly worldwide currents of air and water—produce regional climates, and how these climates, in turn, determine the abundance and distribution of living organisms. The ecologist's goal is to understand why organisms live where and as they do. Only by considering the biosphere, the highest level of ecological organization, can we reach that understanding.

Three main themes underscore our analysis of the biosphere. First, climates originate from currents of air and water on a spinning planet heated by the sun, most strongly at the equator. This heating and spinning generate the moist wind currents that dump rain on the forests of the Pacific North-

■ **Crystalline Stream in a Conifer Forest.** Fresh water, so abundant in the Pacific Northwest, is an important component of the biosphere.

west. Second, as the needle-like leaves of the coniferous forest and the spiny plants of the desert demonstrate so clearly, different climates promote different communities of organisms. Ecologists use the term **biomes** to designate major communities of organisms on land, defined mainly by their vegetation and characterized by specific adaptations to each climate. Temperate and tropical rain forests, deserts, grasslands, and tundra are a few of the world's major biomes. Finally, communities with similar physical environments anywhere in the biosphere contain organisms with similar evolutionary adaptations, even when the individual species are unrelated. Plants inhabiting American deserts tend to have thick body parts, small leaves, and protective spines, and so do many plants from African deserts, even though both plant groups evolved from totally different ancestors.

This chapter explores the global physical forces that determine how the world's major communities of plants and animals are distributed. It addresses several specific questions:

■ How do sunlight and the earth's rotation generate the world's climatic zones?
■ What features characterize each biome, or major terrestrial community of plants and animals?
■ What are the common attributes of organisms inhabiting oceans, lakes, and streams—the world's aquatic ecosystems?
■ How has the biosphere changed in the past, and how might human activities cause it to change in the future?

WHAT GENERATES THE EARTH'S CLIMATIC REGIONS?

After a particularly harsh winter in the Pacific Northwest, elk herds decline substantially in size, and this reveals just how powerfully weather can affect biological systems. **Weather** is the condition of the atmosphere at any particular place and time, including its humidity, wind speed, temperature, and precipitation, whereas **climate** is the accumulation of seasonal weather events over a long period of time. Within the biosphere, weather has temporary, local effects, but climate is the major physical factor determining the abundance and distribution of living things. What, then, causes climate? The answer involves the uneven way that sunlight heats our planet; the behavior of water and air at different temperatures; and the rotation of the earth.

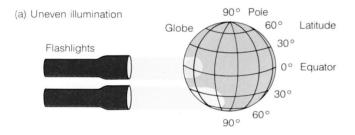

(a) Uneven illumination

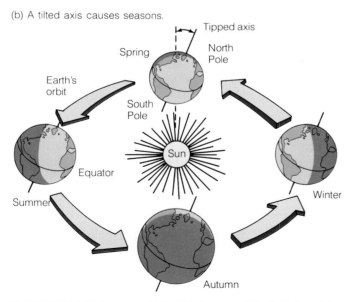

(b) A tilted axis causes seasons.

FIGURE 37.2 ■ Seasons: Light Shining on a Tilted Globe. (a) The sun's angle determines the intensity of solar radiation at a given latitude. Consequently, sunlight is five times stronger at the equator than at the poles. (b) Since the earth spins on a tilted axis, the Northern Hemisphere is tipped closest to the sun in summer and away in winter, and the differences in light intensity generate the seasons.

A Round, Tilted Earth Heats Unevenly

The tropics are warm, the poles are cold, and other regions have their own characteristic climates partly because the earth rotates on a tilted axis and partly because the sun heats the earth's surface unevenly. Like a flashlight beam shining directly down onto a table from above versus one coming in obliquely from the side, sunlight hitting the earth directly is more intense than sunlight striking at an angle, where the same amount of light fans out over a larger area (Figure 37.2a). The sun's angle is therefore the issue, and anyone who has traveled extensively and has also experienced changing seasons knows the importance of *latitude* (the distance between any point and the equator) and *time of year.*

About five times more light and heat energy fall at the earth's equator than at higher latitudes, and the amounts decrease the closer one gets to the poles. These differences in incoming solar power help explain why equatorial regions are warm, polar regions cold, and why tropical forests are usually so productive.

Temperatures grow warmer and colder with the seasons, too, and this is because our spherical planet spins on a tilted axis. This tilt is maintained all year long, even as the earth revolves about the sun (Figure 37.2b); consequently, during the summer, the Northern Hemisphere is maximally inclined toward the sun, while in winter, it is tipped away from the sun and receives less solar energy. The seasons are reversed in the Southern Hemisphere. This tilting of the earth and uneven illumination help explain the warm temperatures and abundant growth of organisms in summer and the cold temperatures and dormancy of living things in winter.

The Formation of Rain

Sunlight striking the earth's tilted sphere at different places and different times heats the air and oceans unequally. Warm air has different properties than cool air, and these account for the formation of rain and snow. Cold air is dense: It weighs more per unit volume and so tends to sink through lighter, warmer air. This is why New England farmers plant cold-sensitive fruit trees on the sides of valleys instead of on the valley floor, where cold, heavy surface air settles at night. Warm air, in contrast, is less dense than cool air and tends to rise. This is why smoke, steam, and hot-air balloons drift upward.

Dense, cool air holds less moisture than light, warm air, and this principle ultimately brings about precipitation. When warm, moist air rises high into the atmosphere, it cools, and its capacity to hold water decreases. The water condenses into droplets, forms clouds, and when the droplets become large and heavy, they fall to the earth as rain (Figure 37.3). This explains why a region like the tropics is so wet. Powerful sunlight at the equator heats the air, which picks up moisture from evaporation and plant transpiration, and rises. Once aloft, the air cools, and the moisture precipitates out, falling onto the lush tropical rain forest.

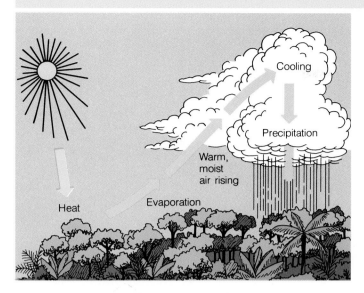

FIGURE 37.3 ■ Let It Rain: The Generation of Precipitation. When heat strikes the land or ocean surface, the warm, moist air rises; the air then cools and releases its moisture content as rain or snow.

Because sunlight strikes the earth unequally, large masses of air warm up, rise, and cool down, then flow to new areas where the cool air falls and flows over the earth in a motion we call **wind.** The earth's natural rotation deflects the wind, influencing climate and weather.

Air and Water Currents: The Genesis of Wind and Weather

The earth's rotation sets up major currents in the atmosphere and oceans. These currents promote rain formation and its precipitation on Pacific Northwest forests. The currents also foster the relatively mild climate of England, and they create the Sahara and the world's other great deserts.

■ **Air Currents: Global Treadmills** Hot air rises, cool air sinks, and the earth turns beneath this mass of mixing atmosphere. Because of the intense sunlight at the equator, hot air rises at the equator and moves north and south at high altitudes (Figure 37.4a). As it moves aloft, the air cools, becomes heavy, and breaks up into six coils of moving air owing to the rota-

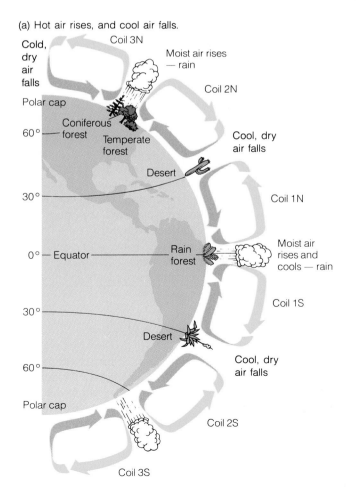

(a) Hot air rises, and cool air falls.

FIGURE 37.4 ■ Air Ascending and Descending Through Massive Coils Creates the Earth's Climatic Zones. (a) Hot air rises at the equator, moves toward the poles, sinks at about 30° N and 30° S, and returns to the equator. (b) The earth's rotation swirls the air into six coils that set up prevailing wind patterns. The air movements create global climatic zones with their characteristic vegetation, such as the tropical rain forests near the equator, where hot, moist air rises, and the deserts of 30° latitude, where dry air descends.

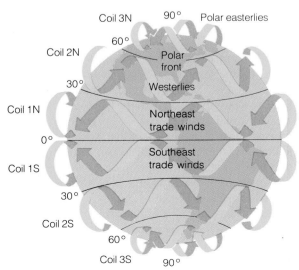

(b) Six coils of air create prevailing winds.

tion of the earth (Figure 37.4b). In each of these coils, the air swirls through a corkscrew pathway. It is as if six "slinkies" of moving air were stretched out around the earth at different latitudes. The movement of air through these coils establishes the direction of prevailing winds at different latitudes and in turn the direction that weather fronts normally travel.

The direction of air flow and the ascent and descent of air masses in their giant coils determine the earth's climatic zones and their vegetation types. Let's consider coil 1 north and coil 1 south, for example. Warm, moist air rises at the equator and drops its moisture in heavy tropical rains as it cools, creating the conditions for tropical rain forests around the equator (see Figure 37.4a). The cooler, drier air now travels at high altitude and descends at latitude 30° (both north and south). This descending dry air creates the great deserts of Australia, North and South Africa, and North America. At ground level, the dry air moves toward the equator, interacts with the rotating earth, and causes surface winds. These predictable breezes, called *trade winds,* propelled traders' sailing ships in past centuries.

In coil 2 north and coil 2 south, dry air descends on the deserts at latitude 30°, moves toward the poles along the earth's surface, and ascends at about latitude 60° (see Figure 37.4a). In the process, it produces winds from west to east (see Figure 37.4b). This flow carries air laden with moisture from the Pacific Ocean over the Coast Range and Cascade Mountains of the Pacific Northwest. There it falls as the rain and snow that encourage the lush growth of temperate rain forests.

Mild air moving toward the poles in coil 2 meets cold air flowing from the poles in coil 3 (see Figure 37.4a). This convergence (the *polar front*) causes the air to rise again at about 60° latitude north and south, cooling and giving up the moisture it picked up as it moved across land and ocean surface. This moisture supports the great temperate forests of the Americas, Europe, and Asia. Once aloft, most of the air moves at high altitude to the poles. There, the dry, frigid air again descends, moves over Canada, and occasionally dips deep into the United States, bringing along polar weather.

The six coils of circulating air help us understand why deserts and forests occur where they do; they also drive the circulation of ocean currents.

■ **Ocean Currents: Patterns of Circulation**   Wind blowing over the ocean surface sets the sea's upper layers moving, and the continents redirect the flow in slow, circular patterns called *gyres* (Figure 37.5). Major gyres occur in the North and South Atlantic and the North and South Pacific. Ocean currents, like air currents, redistribute heat and hence influence climate and the distribution of plants and animals. The Gulf Stream, for example, carries warm water from the tropics up the eastern coast of North America, then eastward across the Atlantic, warming northern Europe (see Figure 37.5a).

Proof that these ocean currents are important to life comes from the *El Niño* event, a temporary reversal of certain currents that occurs about every five years. Under normal conditions, the cold current that flows north along the west coast of South America moves water out to sea near Peru (see Figure

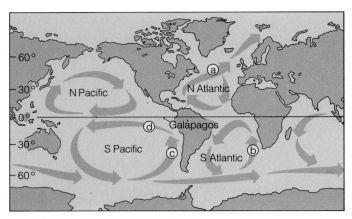

FIGURE 37.5 ■ Ocean Currents Flow in Four Major Gyres, Redistributing Heat. The major warm and cold ocean currents and their directions of flow result from the planet's rotation, the push of prevailing winds across the ocean surfaces, and the position of the continents. These currents, in turn, influence climate, and hence the form of plants and animals, on the landmasses. (a) The Gulf Stream warms Britain and northern Europe. (b) The Benguela Current helps create the coastal desert in southern Africa. (c) The Humboldt Current and (d) the South Equatorial Current cause the upwelling of nutrients that fosters high productivity around the Galápagos Islands.

37.5c and d). To replace the water moving away from the coast, deep waters laden with nutrients well upward and support the prolific growth of plankton (primary producers) and huge populations of anchovies and other fishes (consumers). During an El Niño event, for reasons atmospheric scientists do not fully understand, the trade winds and currents reverse, preventing much of the normal upwelling of nutrient-rich water from the ocean depths. As a result, the ocean's productivity falls, and many organisms die. During the pronounced El Niño of 1983, for example, every Galápagos fur seal pup born that year died because the mother seals could not capture enough fish and squid to feed themselves and their pups.

Despite short-term irregularities such as El Niño, the sun and the earth's tilt and rotation create patterns of rainfall and temperatures that determine the general character of the biomes, the earth's major communities of plants and animals.

BIOMES: LIFE ON THE LAND

Wherever similar climatic conditions exist in the world—the deserts of Australia and Africa, the rain forests of Indonesia and Brazil—plants have evolved similar adaptations that help them exploit the climate's benefits and minimize its drawbacks. Major plant types characterize biomes because plants best reflect adaptations to rain, temperature, light, and other specific climatic conditions, and as primary producers, plants influence the consumers and decomposers that coexist in the biome.

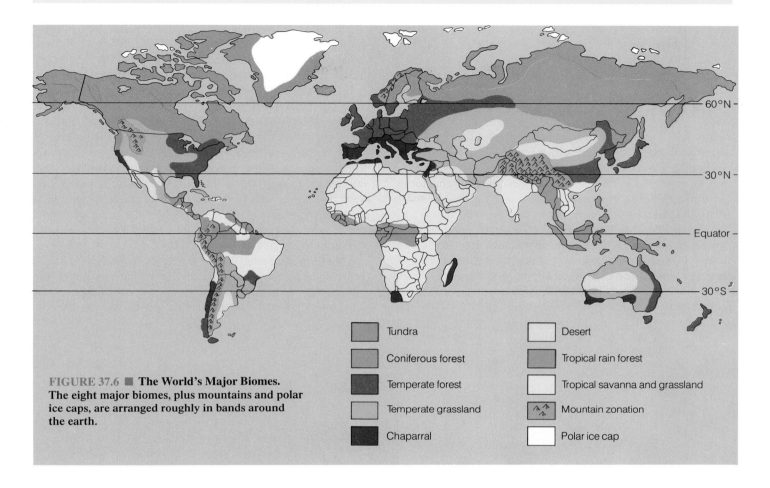

FIGURE 37.6 ■ **The World's Major Biomes.** The eight major biomes, plus mountains and polar ice caps, are arranged roughly in bands around the earth.

Tundra

Coniferous forest

Temperate forest

Temperate grassland

Chaparral

Desert

Tropical rain forest

Tropical savanna and grassland

Mountain zonation

Polar ice cap

FIGURE 37.7 ■ **Biomes and Productivity: Latitude and Climate Determine Life's Bounty.** The net primary productivity of different biomes, measured in terms of grams of biomass per square meter of area each year, varies from one biome to the next, depending on the nearness to the equator (and thus high levels of sunlight), the availability of moisture, and other climatic factors. Notice that the ice caps are the least productive regions; the deserts, tundra, and open ocean are low in productivity; the regions of midlatitude (with their coniferous forests, temperate grasslands, and temperate forests) are intermediate in productivity; and the tropical zones (rain forests and coral reefs) are by far the most productive regions on earth.

Net primary productivity (g/m²/year)

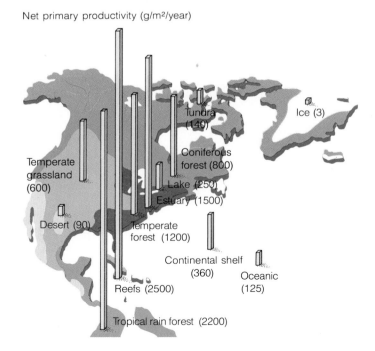

Tundra (140)

Ice (3)

Coniferous forest (800)

Temperate grassland (600)

Lake (250)

Estuary (1500)

Desert (90)

Temperate forest (1200)

Continental shelf (360)

Oceanic (125)

Reefs (2500)

Tropical rain forest (2200)

It becomes obvious that climates determine biomes when we study a biome distribution map (Figure 37.6). Such a map reveals orderly patterns of tropical forests, deserts, deciduous forests, coniferous forests, and tundra lying roughly in bands stacked south to north that correspond with patterns of atmospheric circulation and climate.

Ecologists recognize as few as 7 or as many as 15 or more biomes, depending on how they classify the areas of species mixing between biomes. Here we consider eight biomes, arranged roughly in order from equator to poles—tropical rain forests, savannas, deserts, temperate grasslands, chaparral, temperate forests, coniferous forests, and tundra—plus

mountains and the polar ice caps. This arrangement helps reveal the links between air and water currents, climatic patterns, and biome characteristics. Figure 37.7 compares the net primary productivity (see Chapter 36) in some major biomes and aquatic ecosystems. This figure also reveals how latitude and climate determine life's fruitfulness.

Tropical Rain Forests

Nearly half of all living species reside in the world's warm, wet tropical **rain forests,** even though these lush regions represent only 3 percent of the continental surface area (see Figures 37.6 and 37.8). Tropical forests occur near the equator in Central and South America, Africa, and Southeast Asia, where hot air rises and then dumps its moisture. Rainfall is 200 to 400 cm (78 to 156 in.) per year, and temperatures average about 25°C (77°F). Jungle inhabitants are clearly not limited by water supply or cold temperatures, but light and mineral nutrients can be in short supply, and the dense vegetation blocks the wind. These limitations have shaped the rain forest biome.

The struggle for light has led to three main levels of plant height above the forest floor (Figure 37.8a). Within the upper story, or *emergent layer,* trees up to 50 m (164 ft) rise above the surrounding vegetation, where they capture direct sunlight. Beneath the emergent layer lies the *canopy,* the overlapping tops of shorter forest trees. The canopy is so dense with leaves and branches that only dim light penetrates to the third level, the *understory,* and less than 2 percent of the light eventually reaches the *forest floor.* Because canopy trees do not usually branch in the understory, a person can often walk unimpeded through the humid twilight that filters to ground level in a mature rain forest. The popular image of an impenetrably tangled jungle is accurate only at the edges of cleared forests, along riverbanks, or where a large tree has fallen and the sunlight reaches the ground.

Many jungle plants have evolved specializations for light gathering that allow them to survive in the twilight of the understory. Vines, also called *lianas,* have roots in the soil and long, spindly stems that thrust their leaves into the canopy.

(a) Layers in the rain forest

(b) The lush tropical rain forest

FIGURE 37.8 ■ **Tropical Rain Forest: Lush Equatorial Biome.** (a) Only a few tall trees and vines reach the emergent layer of the rain forest (top), while shorter forest trees form a canopy (upper middle) in which epiphytes and animals thrive. Low trees, shrubs, and underbrush grow in the shady understory (lower middle), and fallen leaves and a few seedlings carpet the scarcely illuminated forest floor (bottom). (b) A tangle of large leaves, vines, and tree trunks shades out the rain forest floor in La Selva Biological Station, Costa Rica.

FIGURE 37.9 ■ Savanna: Tropical Dry Land. Rare Grevy's zebras graze on golden savanna grass beneath flat-topped thorn trees in southern Kenya.

Epiphytes, such as orchids, cling to the trunks and limbs of canopy trees and catch water and nutrient debris in specially shaped leaves. Many plants that dwell in the understory's pale light, like philodendrons, produce huge, dark green leaves capable of capturing the maximum amount of available light, a trait that makes them good houseplants.

Because tropical rain forests have poor soil, roots tend to be shallow, and the trunks of large trees are often expanded into *buttresses,* thick flanges that provide support (see Figure 37.8a). Roots occupy the thin upper layer of soil, where mineral nutrients occur in highest concentrations. Heavy rains tend to leach nutrients from the deeper soils, making the biomass of the living forest itself the biggest source of nutrients. Fallen trees, dropped leaves, and dead animals quickly decay in the warm, damp environment, and their nutrients rapidly recycle.

Because the overhanging canopy is so dense, air in much of the tropical rain forest is deathly still, and a plant that relied on wind pollination might have difficulty reproducing. Still air, however, is not a problem for the many jungle plants that have evolved beautiful and elaborate flowers that attract insects, birds, or bats as animal pollinators. Not surprisingly, a Sumatran rain forest is home to the largest flower on earth: a 1 m (39.4 in.) blossom that weighs 7 kg (15 lb) and smells like rotten meat.

Wind is also undependable for dispersing the seeds of tropical plants, and those plants tend to produce large, succulent, showy fruits that attract animals. Indeed, much of the forest activity takes place in the canopy, where leaves, flowers, and fruits abound.

Despite the timeless beauty and incredible species richness of tropical rain forests, people are rapidly destroying them to help feed their own expanding populations or to harvest hardwood trees or graze cattle for export. This destruction threatens the survival of millions of species (review Box 15.2 on page 333). In addition, it is a threat to all biomes, because tropical forests are the earth's biggest repositories for carbon. Thus, burning the great jungles pours carbon dioxide into the air, and according to some climatologists, this can contribute to global warming and the alteration of climatic patterns all over the globe.

Tropical Savannas

Whereas tropical rain forests grow only where it is constantly wet and warm, year-long warmth with an extended dry season results in a tropical **savanna.** Situated between tropical forests and deserts, savannas contain stunted, widely spaced trees with tall grasses growing in between (Figure 37.9). The world's major savannas lie in Africa, South America, Australia, and parts of Southeast Asia. While the center of life in the rain forest remains high in the canopy, the savanna's productivity is concentrated near the ground, where the leaves and seeds of grasses and short trees are well within reach of grazing animals like Cape buffalo and kangaroos. The herbivores, in turn, support lions, dingoes, and other carnivores.

Trees in the African savanna have peculiar-looking flat tops. This shape is apparently an adaptive compromise—high lower branches to limit predation by animals and a flattened

FIGURE 37.10 ■ Desert: Parched Zone Where Life Is Sparse. The Namibian Desert of southwestern Africa receives almost no rainfall and supports few plants save scrubby drought-tolerant trees and a handful of odd, dry-adapted species like *Welwitschia* (see Chapter 17).

top to prevent excessive energy loss in the dry heat. Many savanna species, including the rhinoceros and the elephant, are threatened with extinction because people kill them for their beautiful coats, ivory tusks, or spectacular horns and destroy their habitats.

Deserts

Desert regions receive less than a tenth the annual rainfall of tropical rain forests; hence, desert plants are widely spaced and cover less than one-third of the ground surface (Figure 37.10). Some desert areas in northern Africa have not seen rain in over a decade and are essentially lifeless tracts of shifting sand.

Some deserts, like those of Mexico, Australia, and North and South Africa, are dry because they lie directly below the zones where dehydrated air—rained out in the tropics—descends back toward earth (see Figure 37.4). Others, like China's Gobi Desert, are dry because they lie at the centers of huge continents, far from the moist sea air. Finally, some deserts form because they are located downwind of tall mountain ranges (Figure 37.11). Before reaching the Great Basin of Nevada and Utah, for example, moist air gains altitude and cools as it approaches and then crosses the Sierra Nevada and White mountains, and the moisture falls as rain or snow on the windward side of the peaks. The air that descends to the basins beyond is thus dry, and the land in the rainshadow becomes a desert.

During the day, deserts are generally hot as well as dry, because on cloudless afternoons, the relentless, blistering sunlight can send temperatures soaring to 49°C (120°F). On cloudless nights, however, heat radiates from the desert back

to space and leaves a biting chill to the air. Many desert plants, including African succulents like ice plant and American cacti like prickly pear, have numerous adaptations that help them cope with searing heat and the potential loss of precious moisture. Such adaptations include thick, waxy cuticles; fleshy or absent leaves; sunken stomata (see Chapter 30); and protective spines. The roots of desert plants may penetrate the sandy soils to a depth of several meters to a source of water, or they may remain shallow and widely spread out, allowing them to

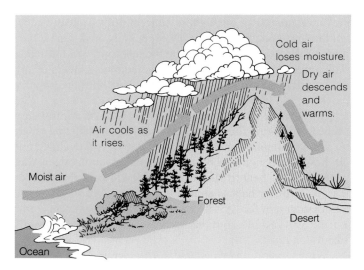

FIGURE 37.11 ■ Rainshadows: How Deserts Form Downwind from Mountains. Moist ocean air loses its moisture as it rises and cools crossing high mountains, and the descending dry air creates a desert.

quickly soak up any rain that falls. Many desert plants also utilize a variant pathway of photosynthesis that conserves water (see Figure 6.11). And others simply avoid harsh desert conditions by germinating, growing, flowering, and setting seed all within the short periods of rainfall. This sets some deserts aglow with dazzling red, gold, and purple flowers for a few brief weeks.

Desert animals have also evolved water conservation strategies. Most are active only in the evening or early morning hours, keeping to their burrows, crevices, or shaded areas during the heat of the day. As Chapter 25 described, the North American kangaroo rat can survive without drinking by producing metabolic water from nutrients in plant seeds and excreting urea in practically crystalline form (see Figure 25.1). By convergent evolution, the jerboa of Saudi Arabia has many similar adaptations. And small crustaceans such as fairy shrimp have evolved the ability to survive for years as dry embryos that hatch quickly when temporary pools form.

Ironically, while human activity is decreasing the areas of tropical rain forests, our actions are causing the world's desert biomes to expand in a process called *desertification*. Deserts enlarge because overgrazing and poor irrigation practices remove grass from the grasslands that rim the deserts (Figure 37.12). The loss of ground cover allows fine, nutrient-bearing soil particles to blow or wash away, leaving behind only sand, gravel, and other coarse materials that cannot hold water.

The fragility of the desert was particularly evident during the Persian Gulf War of 1991. Over thousands of years, the deserts of Kuwait and southern Iraq have accumulated a sur-

FIGURE 37.13 ■ **Temperate Grassland: Treeless Sea of Grass.** Vast herds of bison once grazed the extensive grasslands of central and western North America. Today, just a few protected remnant herds remain, such as these in Yellowstone National Park.

FIGURE 37.12 ■ **The Desert Encroaches: Desertification Claims Precious Grazing Land South of the Sahara.** Overgrazing by goats, cattle, burros, camels, and other livestock can disturb native ground cover and allow precious topsoil to blow away. Through this process, a grazing region can become a desert, such as this spot in Niger.

face layer of pebbles, ranging in size from a pea to a walnut. Since these small rocks are too heavy for even strong desert winds to lift and move, they stabilize the terrain and prevent sandstorms. During the war, however, the movements of thousands of vehicles, the digging of hundreds of kilometers of trenches, and the blasting of countless surface craters disturbed this "desert shield." Experts now fear that sandstorms and the movement of sand dunes will cause the desert to encroach into adjacent agricultural areas.

Temperate Grasslands

Bordering many of the world's deserts are **grasslands,** treeless regions dominated by dozens of grass species (Figure 37.13). Known as *prairies* in North America, the *pampas* in South America, the *steppes* in Asia, and the *veldt* in Africa, these regions are wetter than deserts but drier than forests. The flat expanses of Kansas and the gently rolling hills of Nebraska were typical grassland terrains before the encroachment of farms and ranches. Tall-grass prairies are now quite rare. Unlike savannas, grasslands lack trees and have seasonal extremes of hot and cold rather than wet and dry. Nevertheless, the windswept plains tend to dry out in the summer and fall, so fires are a recurrent feature. As a result, grassland species have evolved the ability to regrow rapidly after a fire. This natural devastation, often touched off by lightning, pre-

FIGURE 37.14 ■ Chaparral: A Biome of Low Bushes and Frequent Fires. The rolling hills and scrubby plants of Spain's chaparral.

vents grasslands from turning into forests. Open grasslands support animals like bison, antelope, prairie dogs, anteaters, armadillos, coyotes, snakes, and hawks.

Grassland soils are richer in organic matter than the soils of other biomes. Grasses have pervasive roots and underground stems, often penetrating to 2 m (6½ ft) and weighing several times more than the aerial leaves and stems. The accumulated debris from generations of these plants makes grass-

land soils thick and fertile. While grasslands make rich agricultural fields and pastures, plowing breaks up the complex soil structure and leaves it vulnerable to erosion. With proper farming techniques such as contour farming (plowing at right angles to the slope) and strip-cropping (leaving broad strips of grass between fields), former grasslands can probably support agriculture indefinitely. Unfortunately, farmers in many parts of the world—including the United States—are often not sufficiently careful to conserve the precious natural resource of topsoil.

Chaparral

A biome called the **chaparral,** or temperate scrublands, borders grasslands and deserts along the shores of the Mediterranean and along the southwest coasts of North and South America, Africa, and Australia (Figure 37.14). Chaparral has hot, dry summers and cool, wet winters. Chaparral plants are generally less than 2 m (6½ ft) tall and have hairy, leathery leaves that stay green year round. Many chaparral plants, like sage and manzanita, have spicy, aromatic odors; the natural alkaloids probably help deter insect herbivores. Because this biome, like the grasslands, experiences frequent fires, the below-ground portions of chaparral perennials have evolved fire resistance, and the seeds of some chaparral annuals must be seared by fire before they germinate.

Temperate Forests

Continuing to move farther north and south from the equator, we leave the grass- and shrub-dominated biomes and enter the **temperate forests,** such as those of eastern North America (Figure 37.15), Europe, China, Japan, New Zealand, Austra-

FIGURE 37.15 ■ Temperate Forest: Deciduous Hardwoods Bordering Grasslands. The new green leaves of spring brighten Tennessee's Great Smoky Mountain National Park.

FIGURE 37.16 ■ Coniferous Forest: Fragrant Evergreens in the Northern Latitudes. A stand of deep-green conifers grows on the shores of Phelps Lake in Grand Teton National Park.

lia, and the tip of South America. These forests are dominated by broad-leaved trees, and as the name suggests, they have intermediate amounts of rainfall and fairly moderate temperatures that fluctuate between summer highs and winter lows. They are about twice as productive as grasslands (review Figure 37.7).

Seasonal temperature changes have strongly shaped the evolution of temperate-forest organisms. The dominant trees, including oak, maple, birch, and hickory, are deciduous (see Chapter 31), dropping their leaves and becoming dormant until spring. The fallen leaves that accumulate on the forest floor allow for the recycling of nutrients and produce an excellent topsoil. In early spring, the leafless trees permit sun to fall unobstructed to the forest floor, and spring wildflowers like wood sorrel, bluebells, and violets bloom. Later, when the canopy leaves enlarge and shade the forest floor, only shade-tolerant plants such as ivy and honeysuckle thrive.

The animals of temperate forests also have adaptations that help them cope with the changing seasons. Many, like bears and snakes, hibernate in winter, while others, like robins, migrate to warmer regions.

Coniferous Forests

At latitudes with cold, snowy winters and short summers, vast forests of conifer trees grow in the **coniferous forest** (also called *taiga* or *boreal forest*) (Figure 37.16). This broad band of mixed pine, fir, spruce, and hemlock trees stretches across much of Canada, northern Europe, and Asia. A few broad-leaved trees such as aspens also survive in these areas around

streams or lakes. The southern border of the coniferous forest tends to fall at the southern limit of the polar front, where cold air sweeping down from the pole meets warm air pushing up from the south (review Figure 37.4). The Southern Hemisphere has little coniferous forest because ocean rather than land occupies appropriate latitudes. Coniferous forests are less productive than temperate forests, but are more productive than grasslands (review Figure 37.7).

During winter in the coniferous forest, ice and snow lock up water, making it unavailable to trees. Conifer leaves, which are needle-shaped and have thick, waxy cuticles, combat water loss. Since these needles are not shed each winter, they can begin collecting sunlight as soon as the short growing season begins. Many conifers have sloping limbs that shed snow easily and tend not to break under the weight. Coniferous forests support bark beetles, elk, porcupines, wolves, and lynx. The *temperate rain forest,* found only in the Pacific Northwest and parts of New Zealand, is a biome closely related to the coniferous forest (review Figure 37.1). Both kinds of coniferous forests contain the world's greatest lumber reserves.

Tundra

At the northern boundary of the coniferous forest, the polar high-pressure center blows cold, dry air down from the pole, even in summer (review Figure 37.4). Here, the fragrant, giant conifers give way to the low vegetation of the **tundra,** a cold, treeless plain (Figure 37.17). The annual temperature in this frosty biome averages −5°C (23°F), and even in the summer,

FIGURE 37.17 ■ **Tundra: Cold, Treeless Plain in the Far North.** The low vegetation in Alaska's Denali National Park erupts with the russets and golds of autumn.

the tundra soil thaws to only about 1 m (39.4 in.). The deeper, permanently frozen soil is called *permafrost,* and this subterranean zone inhibits the establishment of most trees. The soil that does thaw in summer stays soggy and dotted with pools because the land is largely flat and runoff is slow.

Frigid weather and a short growing season have heavily influenced the evolution of tundra organisms. Plants, including grasses, sedges, mosses, lichens, and heathers, grow low, where they can absorb warmth reradiated from the solar-heated ground and minimize the effects of wind and desiccation. These plants support herbivores like caribou, arctic hare, and lemmings, and they, in turn, are prey for foxes, weasels, and snowy owls. During summer, clouds of mosquitoes and other flies fill the air, reproducing in the temporary pools. Many birds migrate to the tundra in the summer, feasting on the insects and breeding during the long summer days.

The tundra is only one-sixteenth as productive as the tropical rain forest; the track of a single vehicle can mar the terrain for decades.

Polar Cap Terrain

The arctic ice cap, a vast frozen ocean covering the planet's North Pole (Figure 37.18), and the continent of Antarctica, a landmass lying beneath a great raft of ice and surrounded by frigid seas, take up millions of square kilometers of land and ocean surface. These icy regions are not considered true

biomes, because they lack major plants; only animals and microbes survive in these regions. In the Arctic, polar bears hunt seals and other aquatic animals, while penguins and seals inhabit the ice shelf at the edge of the Antarctic Ocean. The only true terrestrial animals of Antarctica are a few insects, the largest of which is a wingless mosquito.

Life Zones in the Mountains

Just as ice covers both poles at extremely high latitudes, glaciers and snowfields cap high mountains—even near the equator. In fact, as one goes down in elevation from a high peak, one can often pass through a series of biomes that reverses the order we have just reviewed from the equator to the poles (Figure 37.19). Leaving the snow fields at the top of a mountain in Mexico—let's say the ancient volcanic cone called Citlaltepetl, which is 5754 m (18,878 ft) tall—the traveler descends through an alpine tundra, then a coniferous forest similar to Canada's boreal zones, followed by a deciduous forest like that of the eastern United States, and finally enters a lush tropical rain forest at the mountain's foot. Regions of varying elevation, climate, and organisms in the mountains are called **life zones.** Life zones occur because precipitation and temperature vary with altitude, just as they do with latitude, and each combination of temperature and water influences the kinds of organisms that can survive in those conditions.

FIGURE 37.18 ■ **Polar Caps: Icy Regions at the Highest Latitudes.** A huge polar bear stalks a walrus on a windswept ice floe near the Arctic Circle.

FIGURE 37.19 ■ **Life Zones in the Mountains: Altitude Mimics Latitude.** In descending a high mountain in the tropics, one passes through altitude zones that correspond to latitudes. This diagrammatic approach shows the correspondence between mountaintop and ice cap, alpine regions and tundra, and so on, down to the tropical rain forest shrouding the mountain's base.

Just as climate helps shape terrestrial biomes, it has a similar shaping influence on life in the water—a realm that is less familiar to most of us, but is central to our understanding of the biosphere.

LIFE IN THE WATER

Three-fourths of our planet's surface is covered with water, and most of that area is **marine**—salty oceans and seas. Only 0.8 percent is covered by freshwater lakes, rivers, ponds, streams, swamps, and marshes, but this still represents tens of thousands of square kilometers. Ecologists do not consider marine or freshwater communities as biomes, but their studies show that aquatic ecosystems differ from each other in significant ways.

Properties of Water

As we saw in Chapter 2, water has unique physical characteristics, and these properties strongly influence aquatic organisms. For example, it takes much more energy to heat or chill water than it does to change air temperature. Aquatic organ-

isms therefore suffer fewer rapid temperature changes than their terrestrial counterparts. In addition, air grows steadily denser as it cools, but water has its greatest density at 4°C (39°F). Because of this, ice (at 0°C, or 32°F) floats. Were this not the case, lakes would freeze from the bottom up, only the upper layers would thaw in summer, and bodies of water outside of the tropics would support little life.

Freshwater Communities

Ecologists divide freshwater ecosystems into bodies of running water and bodies of standing water.

■ **Running Water**   A stream that originates from a mountain spring or from melting snow is cold and clear but contains few nutrients. As a consequence, photosynthetic organisms living in the streams contribute less to the biotic system than organic materials washed into the stream from the surrounding watershed. At this point, the stream is heterotrophic overall. As the water tumbles downhill across rocks (Figure 37.20), it picks up oxygen and more nutrients. Species that live in these turbulent zones include algae, mosses, and trout, which thrive only where oxygen is plentiful and water temperatures are low. As nutrients accumulate and the stream widens and is less shaded, the photosynthetic productivity of algae and green aquatic plants increases. The stream is now autotrophic. A river's middle stretch has the greatest species diversity. In the lower section of a large river, the water becomes cloudy and enriched with nutrients and the water moves more sluggishly. The decreased light diminishes the rate of photosynthesis, and the river once again is heterotrophic overall. In these areas, microbes and invertebrates live on detritus in the bottom sediments. Catfish and bass replace trout as the

FIGURE 37.20 ■ **Mountain Stream: Cold, Clear, and Oxygen-Rich.** A narrow stream cascades over rocks and past mossy banks in Michigan's Alger County.

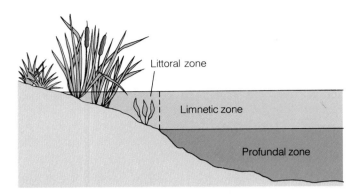

FIGURE 37.21 ■ Freshwater Lakes: Zones of Light Support Life. Cross-sectional view of a lake, showing the littoral, limnetic, and profundal zones with their varying water depths and degrees of light penetration.

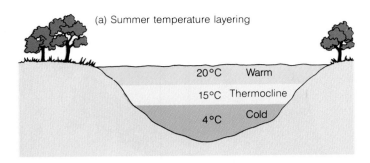

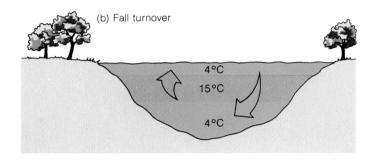

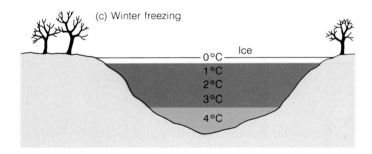

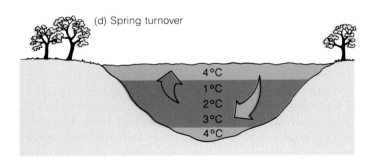

amount of dissolved oxygen decreases in the slower-moving stream. All over the world, people tend to use rivers like "free sewers." Besides pouring wastes into rivers—wastes that kill native species—people have dammed, diked, and channeled rivers in ways that often create worse problems than the ones they sought to avert.

■ **Standing Water**   As a stream enters a lake, the current slows still further, and most suspended particles settle to the bottom. Those particles that remain suspended block light penetration and determine the maximum depth at which photosynthesis can occur. In shallow areas, light can penetrate to the lake's bottom. This is the well-lit *littoral zone,* with abundant producers like water lilies, cattails, and algae and consumers such as insects, snails, amphibians, fishes, and birds (Figure 37.21).

Farther away from the shoreline, the standing water has two main regions: the top layer through which light can penetrate (the *limnetic zone*) and the dark bottom region (the deep, dark *profundal zone*) (see Figure 37.21). The limnetic zone supports huge populations of phytoplankton, such as algae, and zooplankton, such as small crustaceans, while the deep profundal zone supports mainly insect larvae, scavenger fishes, and decomposers.

Dissolved nutrients are as important as light in these biotic zones. For instance, an excess of the nutrient phosphate can cause a bloom of plant growth in a lake or pond (see Figure 36.14). Under normal circumstances, nutrients such as phosphates accumulate in bottom sediments, and in areas where the winters are cold, these nutrients can be stirred up and cycled as the seasons change.

Nutrients cycle in a lake because surface water at 4°C sinks below water that is either warmer or colder. During summer, warm, light layers of water float above cold, heavy ones along a boundary called a *thermocline* (Figure 37.22a).

FIGURE 37.22 ■ Nutrient Cycling in Lakes: Changing Temperatures Mix Water Layers. (a) A thermocline divides warm and cold water layers in summer, and the densest water (4°C) rests at the lake's bottom. (b) During the fall turnover, rapidly chilled top layers sink through warmer midlayers and churn up nutrients. (c) Again in winter, the densest water layers rest at the lake's bottom. (d) During the spring turnover, the top ice melts, water warms to 4°C, sinks once again, and the result is another churning up of nutrients.

As air temperatures drop in autumn, surface water cools to 4°C, becomes heavy, and sinks to the bottom of the lake, carrying oxygen with it (Figure 37.22b). This process, called *fall turnover,* stirs the lower water, and as it rises, it brings nutrients up to the photosynthetic layer of the lake. As winter sets in, ice begins to float on the lake's surface (Figure 37.22c). Below the ice, water becomes gradually warmer with depth, reaching perhaps 4°C at the very bottom. When the lake's ice cover melts in spring, the upper layers once again warm from freezing to 4°C, when they again sink to the bottom through the colder, less dense water below (Figure 37.22d). This again churns up nutrients from the bottom in a *spring turnover* and nourishes a quick bloom of algae.

Saltwater Communities

■ **Estuaries**   Fresh and salt water mingle in **estuaries,** where rivers meet oceans (Figure 37.23). In these areas, temperature and salt concentrations vary widely because of the daily rhythm of the tides and seasonal variations in stream flow. Estuary organisms like brine shrimp can often tolerate wide ranges of salinity (see Chapter 25). Constant water movements stir up nutrients and make estuaries some of the earth's most productive ecosystems and nursery grounds for many species of fish (review Figure 37.7). In many populated areas, people dump urban and industrial effluents into estuaries or fill them with rocks and soil to claim new waterfront property. Such activities are significantly decreasing the great biological productivity of our estuaries.

■ **The Oceans**   Water eventually passes through estuaries and out to sea. Although the ocean may look undifferentiated

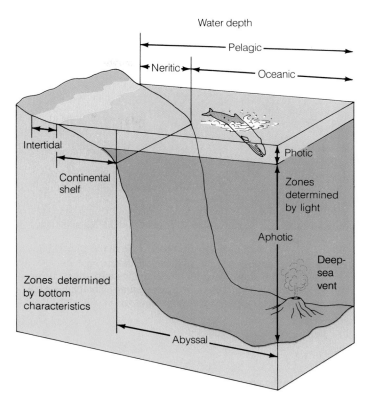

FIGURE 37.24 ■ Ocean Zones: Depth, Bottom, and Light Determine Habitats of the Sea. The ocean is a complex aquatic realm with several zones based on different parameters: Water depth determines the neritic and oceanic zones of the pelagic; bottom characteristics determine the intertidal, continental shelf, and abyssal zones; and light penetration determines the photic and aphotic zones.

FIGURE 37.23 ■ Estuaries: Confluences of Life. In Australia, a silted, richly productive estuary marks the confluence of a freshwater stream and the ocean.

from the surface, various parts receive different amounts of sunlight and nutrients. The bottom can be muddy, sandy, or rocky; the shorelines can vary from smooth, sandy beaches to jagged, rocky cliffs; and the water can be centimeters or kilometers deep. These factors combine to dictate the types of organisms that occupy each region. Ecologists classify ocean habitats and their organisms on the basis of depth, type of bottom, and light level (Figure 37.24).

The *intertidal zone* extends along the coastline between high and low tide levels and is periodically exposed to the air. In many intertidal environments, waves and wind create violent pounding. Most producers (kelp and other algae) have structures that anchor them to rock surfaces (see Chapter 17), while most animals have tough bodies or shells as well as underwater glues. Intertidal organisms also tolerate dry spells or changes in salinity as the tide rolls out each day. A group of burrowing organisms, including clams, snails, and worms, live beneath the extensive tidal mudflats. Many visitors to the seaside enjoy investigating **tide pools,** depressions formed in

FIGURE 37.25 ■ **Tide Pools: Basins of Life in the Intertidal Zone.** A large orange sea star (*Pisaster ochraceus*) cohabits in this shallow tide pool along Washington's Pacific coast with a spiny purple sea urchin (*Strongylocentratus purpuratus*) and aqua blue sea anemones (*Authopleura xarithogrammica*).

rock by wave action, and observing the sea anemones, sea urchins, crabs, sea stars, and fishes that inhabit them (Figure 37.25).

The *neritic* (nearshore) *zone* is the ocean's most productive region, extending from the intertidal zone to the edge of the continental shelf, the submerged part of the continents. Neritic producers include many species of large algae living in extensive beds, as well as huge populations of phytoplankton and zooplankton that serve as a nutrient base for the rest of the ocean's consumers. The largest animals that have ever lived, the blue whale (see Chapter 20) and the whale shark, feed entirely on microscopic plankton.

The shallow tropical neritic zone is home to reef-building corals and the photosynthetic bacteria within them (see Chapter 18), which construct stony mounds with exotic shapes (Figure 37.26). Eventually, these reef builders alter their own environment; the crannies and caves in their stony colonies create sheltered nesting and hiding places for the most diverse communities in the seas. Per unit area, coral reefs are the most productive ecosystems on earth; they are 10 times more productive than freshwater regions and 20 times more so than the open ocean (review Figure 37.7).

The *oceanic zone,* beyond the continental shelf and covering the deep abyssal plane, is nearly deserted, even in the photic zone near the surface (see Figure 37.24). Other zones of the sea include the *pelagic zone,* or open ocean, made up of the water in the neritic and oceanic zones, and the *benthic zone* or seafloor. While the oceanic regions may have plenty of light, oxygen, and carbon dioxide, they lack nutrients such as phosphate and usable nitrogen, and overall they are less productive than the arctic tundra. The ocean's most productive areas occur mainly where currents cause nutrients to well up from very deep waters; a good example is the cold area off the coast of Peru (review Figure 37.5). Although much of the open ocean is not very productive, it does have immense effects on the biosphere, because the oceans modulate climate and help to maintain favorable concentrations of oxygen and carbon dioxide in the atmosphere.

CHANGE IN THE BIOSPHERE

Change is part of nature. Climates shift, continents drift, and organisms evolve, influencing their immediate surroundings and the larger biosphere around them. Recall from Chapter 15 how the evolution of photosynthesis about 2.8 billion years ago decreased atmospheric carbon dioxide and increased oxy-

FIGURE 37.26 ■ **Coral Reef: Fantastic Productivity Beneath the Waves.** A school of bright orange fish darts above a dozen kinds of coral on this reef in the Red Sea.

FIGURE 37.27 ■ **Clear-Cutting the Cathedral Forest.   Denuded hills near Vancouver, Washington, where loggers have stripped the native coniferous forests to bare earth and rock.**

gen gas, causing the oceans to "rust." As we saw in Chapter 36, certain human activities are causing equally pervasive but much faster modifications in the biosphere—modifications that outstrip the speed with which organisms usually evolve adaptations. The degradation of the earth's ozone layer is a prime example of our species' collective impact on the biosphere (see Box 37.1 on page 768).

Recent global phenomena, such as rising carbon dioxide levels and the dwindling ozone layer, contradict our traditional assumption that nature will somehow neutralize all the wastes we release. So great is our impact that some ecologists estimate that humans are currently using about 40 percent of the potential primary productivity on all of the earth's land area! Furthermore, by cutting the world's great forests—tropical rain forests in particular, but the temperate rain forest of the Pacific Northwest to some degree as well (Figure 37.27)—we are destroying habitats for thousands of species.

Harvard ecologist E. O. Wilson points out that we have triggered a new *biodiversity crisis,* during which we as a species could reduce the diversity of life to its lowest level in 65 million years (see Box 15.2 on page 333). For example, loggers have already removed nine-tenths of the old-growth forests in Oregon and Washington, and this habitat loss has threatened the survival of the spotted owl and other species. The remaining 3000 to 5000 breeding pairs of owls are able to live only in old-growth forest, and each pair requires 3000 acres of territory for night hunting. To save the owl, the U.S. Forest Service has considered sparing several million acres of cathedral forest from the loggers' chain saws. This action, on the other hand, could mean the loss of up to 100,000 jobs in the timber-related industries over the next few years. But with no constraints on present rates of cutting, the old-growth forests of the Pacific Northwest will be gone in a dozen years, and the loss of jobs will merely have been postponed. The situation in the tropics is similar: natural habitats versus human jobs and food. Box 37.2 on page 769 describes the important topic of sustainability and some answers to fundamental dilemmas of this kind.

Our species is the only one capable of changing the global environment in so short a time period. But we are also capable of understanding that we are in fact mismanaging the earth's physical and biological resources, and we are capable of foreseeing the consequences. The survival of our species and millions of others depends on the responsibility we take for these capacities and the compromises we work out between preserving our environment and supporting our own needs.

BOX 37.1   FOCUS ON THE ENVIRONMENT

ATTACK ON THE OZONE

Question: What do our long-term health and survival have in common with electric trains, plastic foam, refrigerators, and spray-can propellants?

Answer: Ozone.

The same sharp-smelling ozone gas that floats in the room when a toy train is racing around its track envelops our planet in a layer 15 to 50 km (9 to 31 mi) above ground surface. Ozone forms when ultraviolet light streaming from the sun strikes oxygen molecules in the atmosphere. The earth's ozone layer formed only after photosynthetic organisms evolved and liberated massive quantities of oxygen (see Chapter 15). This ozone formation has had a magnificent consequence: It absorbs about 99 percent of the ultraviolet light that would otherwise strike the planet's surface, and destroys many biological molecules. The shielding effect of the ozone layer probably allowed living things to leave the seas and invade land, and it continues to protect us today from most of the mutational effects of ultraviolet light on DNA (see Chapter 13).

Ominously, in about 1974, atmospheric chemists began to detect decreases in ozone concentrations in the upper atmosphere and increases in the levels of ultraviolet light reaching the earth. Since that time, scientists have detected a huge seasonal hole in the ozone layer over much of Antarctica containing only half as much ozone as it contained a decade ago (Figure 1). They have also discovered a 3 to 5 percent thinning of the entire ozone layer worldwide.

Studies have shown that so-called *chlorofluorocarbons* (CFCs), or chemical compounds containing chlorine and fluorine, can act as catalysts that convert ozone to oxygen gas. Researchers wondered how these compounds got into the ozone layer in the first place. And the answer turned out to be the thousands of tons of CFCs people have used each year. We use CFCs to propel substances from spray cans, to create plastic-foam containers and cushions, and as coolants in refrigerators and air conditioners. These compounds are very long-lived, and they eventually escape into the atmosphere as gases, rise into the ozone layer, and interact with ozone when sunlight powers the reaction. The fact that CFCs act as catalysts means that each molecule in the atmosphere can destroy ozone molecules over and over again for decades.

Scientists predict that for each 1 percent drop in the ozone shield, approximately 2 percent more ultraviolet light will reach us. As a result, they expect to see more cases of skin cancers, more eye cataracts, and more immune system problems, not to mention mutations, stunting, and surface damage to plants, animals and other organisms. They are especially concerned that stronger ultraviolet light will disrupt photosynthesis in Antarctic phytoplankton—a phenomenon that would in turn harm that region's entire food chain, from krill to fishes to birds and mammals.

In the summer of 1990, 93 nations signed an international agreement to phase out the use of CFCs by the year 2000. Some observers still worry that even this remarkably fast and cooperative action will be too little too late, citing recent reports such as the following:

- Antarctic ozone levels dropped to an all-time low in September of 1990, with 95 percent of the gas destroyed in the region

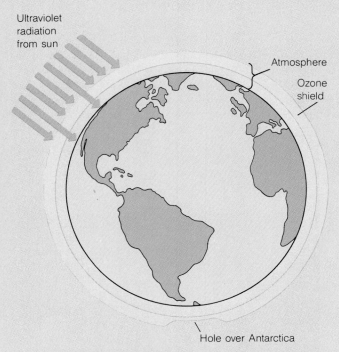

Ultraviolet radiation from sun

Atmosphere

Ozone shield

Hole over Antarctica

FIGURE 1 ■ The Ozone Shield. The ozone layer surrounds the earth and absorbs most of the sun's ultraviolet rays. The thin patch over Antarctica represents the lowered levels of atmospheric ozone, due mainly to human use of chlorofluorocarbons.

between 15 and 23 km (9 and 14 mi) aboveground and 50 percent destroyed at higher altitudes.

- The ozone hole occasionally drifts over major Australian cities, exposing people to increased ultraviolet radiation.
- The same accelerated process of ozone destruction taking place in the frigid air above Antarctica is probably beginning in arctic regions as well. This could be even more disastrous, since northern polar regions have larger human populations than southern polar regions.
- Actual decreases in global ozone levels are already worse than researchers' most careful predictions just a few years ago.
- While manufacturers are racing to develop substitutes for CFCs, they warn that the new products will be less effective and more expensive.

Will the current and future load of CFCs hanging persistently in the stratosphere continue to damage the ozone layer more than scientists have predicted? Will people be willing to curtail their use of plastic foam, smoothly operating air conditioners and refrigerators, and other conveniences in large enough measure to protect the global ozone shield? Only time and the dedication of both atmospheric researchers and an informed public will tell.

CONNECTIONS

The biomes that make up our biosphere can actually be seen as places fashioned by the earth's atmosphere, soil, and water; by the planet's daily rotation on its tilted axis; by its yearly trip around the sun; and by continents drifting and climates changing from hot to cold, dry to moist, and back again. Over vast reaches of time, plants and animals slowly evolve adaptations that allow them to exploit their environments.

Human activities all over the globe are rapidly changing the earth's physical environment to such an extent that entire biomes are affected. We are causing desertification in dry areas and damaging the temperate forests through acid rain.

What's more, some of our activities threaten to affect the entire biosphere in a short time through the greenhouse effect, the loss of ozone, and the terrible possibility of nuclear winter. We must devise and institute solutions with equal speed, so that our all-pervasive influence does not unbalance the complex interdependencies of nature.

All of our foregoing subjects—cell biology, genetics, development, physiology, evolution, and ecology—help us understand how organisms survive in their environments. With this background, we are ready to consider our final topic: a fascinating set of adaptations called animal behavior, which allows animals to respond to each other and to their surroundings in a range of ways from the innate and automatic to the learned and deliberate.

BOX 37.2   FOCUS ON THE ENVIRONMENT

SUSTAINABILITY: A NEW WORLD ORDER

For the past three generations, people have been extraordinarily busy. Since 1900, our species' population has doubled, then doubled again. We are using 30 times more fossil fuel than we used to, we produce 50 times more goods, and our economic activity is 20 times higher. Although the birthrate has leveled off or dropped in some countries, it is still burgeoning in most, and experts expect another doubling of world population during the twenty-first century, from 5 billion to 10 billion. To support these people, even at a minimum level, will require yet another five- to tenfold increase in economic activity—more buying, more selling, more shipping, sewing, building, transporting, financing, using up, burning down, throwing away.

How can we possibly keep this up? How can the earth support twice as many of us? What kind of future will we have? We've seen how a population of bacteria will double and redouble in a closed system like a Petri dish, only to eventually use up most of the available resources, despoil the surroundings with waste products, then crash and die back to a much lower population level. Will our human future be a global version of this boom-and-bust cycle, with ever more hungry, impoverished people in a constantly worsening environment and headed for a series of catastrophes that slash the population? Or will we learn to sustain ourselves in some kind of balance with the biosphere around us?

An important intellectual product of the 1980s was the concept of sustainability—balancing the levels of human population and economic growth against the quantities of available resources and the quality of the physical environment. In the past, economic development has come at the expense of environmental quality, and high prosperity in certain countries has rested, in part, on lower standards of living in other countries. To ensure a stable global environment over the long term, say experts like William Ruckelshaus, former administrator of the Environmental Protection Agency, all people must have a reasonable level of prosperity and security, and this will require a change in human attitudes and actions as great as those that brought about the agricultural and industrial revolutions of the past.

The sustainability revolution will require changes in manufacturing such that industries use fewer materials and less energy and recycle their wastes into other products and processes. Sustainable agriculture will involve rotating crops to increase yields; building up rich soil by preventing erosion and by using natural fertilizers, such as animal manure and blue-green algae; and controlling weeds, insects, and plant diseases through the use of integrated pest management, genetically altered crop plants, and other largely nonchemical methods. Sustainable agriculture will also rely on wind and solar energy instead of fossil fuels wherever possible, and on planting a wide range of traditional and nontraditional crops.

Creative economic solutions will also play a part in global sustainability. Among the many ideas now in circulation are debt-for-nature swaps and conservation easements. In the former, wealthy nations would cancel Third World debts in exchange for the preservation of natural habitats. In the latter, First World nations would pay the equivalent of mineral rights to secure forested areas and nature preserves against destruction.

Besides such approaches, say various experts on sustainability, all nations, rich and poor, will have to provide their citizens with information on sustainability techniques and birth control. They will have to fund international agencies to manage global environments. And they will have to be fully cooperative and committed to the sustainability revolution. The alternative—the human population crash following the present boom—is just too dismal a prospect.

Highlights in Review

1. The biosphere includes the water, air, soil, and all living organisms interacting at a global level.

2. Latitude and time of year influence the amount of sunlight hitting the earth. The earth's tilted axis and uneven heating in turn influence the formation of wind, rain, and ocean currents. The rotation of the earth sets up coils of thermally driven air masses that establish the earth's climatic zones. In turn, climates establish the earth's biomes. Ocean currents affect climatic zones by redistributing heat and moisture.

3. Ecologists identify biomes by their major plant types. We consider eight biomes, roughly in the order they occur from equator to poles.

4. Tropical rain forests contain about half of all living species and occur near the equator, where the climate is warm and moist most of the year. Tropical plants are strongly adapted to compete for light and nutrients.

5. Tropical savannas are warm year round but have a long dry season. Tall grasses and widely spaced flat-topped trees support communities of large grazers and browsers.

6. Deserts are regions that receive little rainfall, either because they lie in nearly constant dry, high-pressure areas; because they lie far from the ocean; or because they lie in the rain-shadows of mountain ranges. Desert plants and animals are strongly adapted to conserving moisture and staying cool.

7. Temperate grasslands are generally flat or rolling, with rich soils and seasonal extremes of hot and cold. Recurrent fires prevent the growth of trees.

8. In the chaparral, summers are hot and dry, winters cool and moist, and the vegetation is scrubby, drought-tolerant bushes with evergreen leaves.

9. Temperate forests occur in the eastern parts of North America and other continents. Temperatures and rainfall are moderate, seasons are pronounced, and broad-leaved hardwood trees are the major plants.

10. Vast coniferous forests are common is northern climes with cold, snowy winters and short summers. A moist version of the forest containing redwoods and firs grows along the west coast of North America.

11. The tundra is a cold, treeless plain that has permafrost 1 m (39.4 in.) below the ground surface, low-growing plants, and animal populations that fluctuate widely based on slow cycling of nutrients.

12. The polar caps are occupied by ice (North Pole) or land and ice (South Pole). Since they lack major plants, they are not true biomes.

13. Mountains have life zones, from low elevation to high, that correspond roughly to the biomes from equator to poles.

14. Three-fourths of the earth's surface is covered by water, and most of that is marine, or salty seas and oceans. The sinking and rising thermal layers in lakes churn up nutrients from the bottom during the spring and fall turnovers.

15. Running water has more oxygen content but fewer nutrients than standing water. Standing water has zones (littoral, limnetic, profundal) where light penetration and nutrients vary.

16. Estuaries are varied and productive areas where rivers flow into oceans.

17. Oceans are varied environments with an intertidal zone periodically exposed to air; a neritic zone of relatively shallow water; and an oceanic zone beyond the continental shelf.

18. Burgeoning human populations are affecting the biosphere in ways that many ecologists predict will be disastrous. These effects include destruction of the protective ozone layer, increased atmospheric carbon dioxide and global warming, and the release of toxic substances into the air, soil, and water.

Key Terms

Study Questions

Review What You Have Learned

1. Define biosphere and biome.
2. Describe the six coils of air currents that circulate around the earth.
3. What factors influence ocean currents?
4. How did oceanic currents change during the 1983 El Niño? How did those changes affect oceanic life?
5. What are the basic features of a tropical rain forest and a tropical savanna?
6. What conditions create a desert?
7. What are the basic features of grassland and chaparral?
8. How do a temperate forest and a coniferous forest differ?
9. What inhibits trees from growing in the tundra?
10. List the living organisms that inhabit the polar ice caps.
11. Prepare a chart that shows the three zones of a lake, the amount of light that reaches each zone, and the types of organisms that inhabit each zone.
12. How do nutrients circulate in a lake?
13. What is an estuary? What is its ecological role?
14. In an ocean ecosystem, where are the intertidal, pelagic, and benthic zones located?

Apply What You Have Learned

1. Philodendrons grow wild in tropical rain forests, but they are also easy to cultivate as houseplants. What makes them so successful indoors?
2. Snow-capped Mt. Kilimanjaro is located at the equator in Africa. Explain why you would or would not expect to find tundra-type plants growing on its slopes.
3. A trout fisherman standing in the rapids of a mountain stream finds the cold, tumbling waters uncomfortable, so he moves downstream to a calmer spot. Have his chances of catching trout improved? Explain.

4. A Brazilian rancher says: "It's okay to clear-cut and burn the tropical rain forest so we can convert it into pasture. In a few years, we'll let the forest grow back." An ecologist counters, "Once the tropical rain forest is clear-cut, it's gone forever." Discuss the validity of these arguments.

 **For Further Reading**

CLOUD, P. "The Biosphere." *Scientific American,* September 1983, pp. 176–189.

DIAMOND, J. M. "Human Use of World Resources." *Nature* 328 (1987): 479–480.

EHRLICH, P. R., and J. ROUGHGARDEN. *The Science of Ecology.* New York: Macmillan, 1987.

EL-BAZ, F. "Do People Make Deserts?" *New Scientist,* October 13, 1990, pp. 41–44.

FROSCH, R. A., and N. E. GALLOPOULOS. "Strategies for Manufacturing." *Scientific American,* September 1989, pp. 144–152.

FURLEY, P. A., W. W. NEWEY, R. P. KIRBY, and J. M. HOTSON. *Geography of the Biosphere: An Introduction to the Nature, Distribution, and Evolution of the World's Life Zones.* London: Butterworth, 1983.

HOLDEN, C. "Kuwait's Unjust Deserts: Damage to Its Desert." *Science* 251 (1991): 1175.

KERR, R. A. "Arctic Ozone Is Poised for a Fall." *Science* 243 (1989): 1007–1008.

KERR, R. A. "Ozone Hole Bodes Ill for the Globe." *Science* 241 (1988): 785–786.

POOL, R. "The Elusive Replacements for CFCs." *Science* 242 (1988): 666–668.

REGANOLD, J. P., R. I. PAPENDICK, and J. F. PARR. "Sustainable Agriculture." *Scientific American,* June 1990, pp. 112–120.

REPETTO, R. "Deforestation in the Tropics." *Scientific American,* April 1990, pp. 36–42.

RUCKELSHAUS, W. D. "Toward a Sustainable World." *Scientific American,* September 1989, pp. 166–174.

VAUGHAN, C. "Streetwise to the Dangers of Ozone." *New Scientist,* May 26, 1990, pp. 56–59.

VOYTEK, M. A. "Addressing the Biological Effects of Decreased Ozone on the Antarctic Environment." *Ambio* 19 (1990): 52–61.

WILSON, E. O. "Threats to Biodiversity." *Scientific American,* September 1989, pp. 108–116.

PART SEVEN

BEHAVIOR AND THE FUTURE

Animal Behavior: Adaptations for Survival

FIGURE 38.1 ■ **A Raucous Male Gull Defends His Territory.** This black-headed gull (*Larus ridibundus*) has carved out a small patch of grass on the beach, and with a raucous call, defends it from other males while awaiting a female's approach.

Menace of the Broken Shell

Each spring, white sea gulls with coal black faces and white-rimmed eyes migrate to the coastlines of northern Europe. Males of this species (the black-headed gull) arrive earliest, select a position on the dunes, and begin stretching forward and calling out raucously in a stereotyped ritual (Figure 38.1). If a newly arrived male tries to claim another male's territory, the occupant rushes the invader, strikes him with a wing, and drives him off.

Males often react quite differently to approaching females, initiating a stylized mating dance in which the pair stretch forward, strut side by side, then, as if by prearranged choreography, turn their faces away from each other (Figure 38.2a). Next the female taps on her suitor's bill with her own, and he regurgitates fish into her mouth. The dancing and dining over, the pair may mate (Figure 38.2b), select a nesting place, then noisily defend it from all comers. The female deposits khaki-colored eggs with dark speckles in the loose mound of grass and straw; then the parents take turns warming and turning the eggs in the nest. After each chick hatches, the parent on duty dumps the empty eggshell far from the nest.

The off-duty parent frequently joins other gulls in noisy flocks as they dive for fish or scratch for worms in freshly plowed fields. When the adult returns to the nest with shredded fish or worms bulging in its crop, the fuzzy, khaki-colored chicks rush up, peck the parent's bill, and receive a regurgitated meal.

Over time, the chicks begin to wander about their parents' territory. Black-headed gulls recognize their own chicks, but harshly peck little ones that wander in from other territories. The chicks, in turn, recognize their parents' attraction calls at mealtimes and ignore those of neighboring adults. Alarm calls, however, are a different matter. Hearing the scream of any adult, the chicks freeze in the nest or quickly dive for cover. Adults respond to an intruding fox or hawk by mobbing it, dive-bombing and defecating until they drive it off.

(a) Facing-away display   (b) Mating

FIGURE 38.2 ■ **A Ritualized Gull Dance Leads to Mating.** (a) A facing-away display is part of the black-headed gull's mating dance. (b) Dancing and dining (with the male regurgitating food into his partner's mouth) often lead to mating.

As summer wears on, the chicks mature, and as winter approaches, the entire colony of black-headed gulls migrates to a warmer winter home in Africa. The following spring they will return once again and enact a new drama of territorial marking, mating, and brooding.

The complex behaviors of the black-headed gull help the organisms obtain energy, reproduce more effectively, and survive more successfully in nature. The scientific study of how animals behave in their natural environments is called **ethology,** and ethologists, those who study animal behavior in nature, ask three main questions: (1) What behaviors do animals display in their normal ecological settings and usual social groupings; (2) how do physiological mechanisms mediate individual behaviors; and (3) why does natural selection favor particular kinds of behavior in one species and different kinds of behavior in another?

An ethologist considering a black-headed gull's tendency to remove broken eggshells from the nest might first observe and record the animal's specific actions: how the bird grasps the eggshell in its beak and flies it to a distant site. Next, the researcher might study the mechanisms that allow a gull to recognize a broken eggshell and remove it. Ethologists call the mechanisms that operate within an individual animal and direct a behavior the **proximate cause** of that behavior. The proximate (or immediate) cause of eggshell removal could be that the ragged edge and shiny white inside of the eggshell trigger the bird to remove the shell. Finally, the ethologist might try to learn why natural selection favors birds that rid their nests of broken eggshells. The answer may reveal the **ultimate cause** of a behavior, the selective advantage that the action confers on the performing individual. Experiments revealed that nests without flashy white eggshells attract fewer predators. Thus, birds possessing genes that foster the recognition and removal of eggshells will produce more sur-

viving chicks than gulls that allow shiny broken eggshells to litter their nests. Thus, the ethologist answers "how" questions with proximate causes and "why" questions with ultimate causes.

As we explore the fascinating range of simple and complex behaviors in the animal kingdom—from mating to migrating, and fighting to befriending—two main themes will recur. First, the ability to perform behaviors originates in genes, arises during development, and may be modified by the environment. The proximate cause for recognition and removal of broken eggshells lies in the patterns of a bird's neural activities. Genes specify these patterns, or programs, which take shape as the embryo develops into a chick and are expressed by the mature adult. Second, natural selection shapes behavior. Behaviors arise because the animals expressing them (and possessing genes that specify them) have more offspring that survive and reproduce than related individuals without those genes and behaviors. Gulls that remove broken eggshells lose fewer chicks to predators than gulls that leave broken eggshells in their nests. The genes that program eggshell removal therefore increase in frequency in the population of black-headed gulls.

Our discussion in this chapter explores how behaviors help animals obtain energy, reproduce effectively, and survive more successfully in nature. It poses several questions:

■ How do genes determine behavior?

■ How does an animal's interaction with the environment modify its behavior?

■ How does natural selection shape behaviors such as avoiding predators or finding a mate, a place to live, and food to eat?

■ Why do some animals live in closely interdependent societies?

GENES DIRECT DEVELOPMENT AND BEHAVIORAL CAPABILITIES

Niko Tinbergen, one of the pioneers of ethology, conducted some of the first experiments into the heritable basis of behavior. He wondered what properties prompt black-headed gulls to remove broken eggshells, so he placed various objects into gull nests alongside the eggs. The birds accepted and incubated round rocks, flashlight batteries, corks, and light bulbs as if they were eggs. But gulls ejected from the nest bottle caps, toy soldiers, seashells, and paper squares. Tinbergen concluded that when the birds saw or felt an object with a jagged edge, they removed it from the nest. Apparently, sensory receptors in the gull's eyes or skin detect sharp edges, and the nervous system directs the bird to remove the object. Since genes determine the development of the bird's sensory and nervous system and set up specific nerve connections, or *neural programs,* it seemed reasonable to suspect that genes direct the behavior.

Other experimenters following Tinbergen's lead conducted breeding experiments with honeybees and garter snakes to test the hypothesis that genes direct behavior. Some honeybees eject diseased bee pupae from their honeycombs. These "hygienic" bees uncap the chambers where the

FIGURE 38.4 ■ Tongue Flicking in Garter Snakes. Both coastal and inland garter snakes (*Thamnophis elegans*) will flick their tongues to "smell" slug juice offered to them by an experimenter. Only coastal garter snakes, however, will eat bits of slug meat. Instinctive slug refusal may help protect the inland snakes from swallowing the leeches that occur in their territory and that might inflict damage to the snake.

infected pupae lie sealed and then expel the diseased animals from the nest (Figure 38.3a). Other strains of honeybees are "nonhygienic": They neither uncap infected cells nor remove diseased animals, even if the experimenter does the uncapping (Figure 38.3b). In nonhygienic colonies, disease can spread rapidly and devastate the hive.

By experimentally crossing the two strains of bees, researchers were able to show that a single gene controls uncapping, while a different, separate gene controls removal. The implication was that individual genes can govern specific behaviors and that a complex behavioral pathway may involve more than one gene.

Studies on garter snakes illuminated the additional point that naturally occurring genetic variation leads to behaviors that adapt individuals to specific environments. Biologists have observed two populations of a single garter snake species, one that inhabits the foggy forests of coastal California and eats mainly slugs, and another that lives in higher, drier areas several miles inland and feeds on fish or frogs from lakes and streams. Researchers wondered whether the distinctly different food choices are due to genetic differences between the two snake populations or whether inland snakes will eat slugs if given the opportunity.

To find out, experimenters captured pregnant snakes from each population, raised the young in the laboratory, and offered them small chunks of fresh banana slug. Baby snakes from inland regions flicked their forked tongues only a few times at a cotton swab soaked in slug juice and then avoided the slimy slug meat (Figure 38.4), while coastal snakes repeatedly flicked their tongues and ate slugs heartily. This result suggests that different genes in the two populations control slug refusal and slug feasting, and crossbreeding showed that alleles for refusal are dominant to alleles for feasting.

(a) Homozygous hygienic bees uncap chambers and remove diseased pupae.

(b) Homozygous nonhygienic bees neither uncap chambers nor remove diseased pupae.

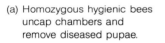 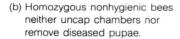

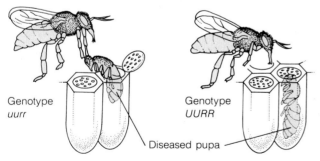

Genotype *uurr*

Genotype *UURR*

Diseased pupa

FIGURE 38.3 ■ Bee Behavior: Two Separate Genes Control Expulsion of Diseased Young from the Hive. (a) Hygienic bees remove the caps from chambers that contain diseased pupae and then throw the sick animals from the hive. **(b)** Nonhygienic bees neither uncap chambers with ill pupae nor remove the afflicted animals. When researchers mated the two strains and obtained an F_2 generation, some were fully hygienic and others were fully nonhygienic. They also found two classes of genetic recombinants; some uncapped diseased chambers but did not remove the sick pupae, while others would not uncap diseased chambers, but would remove sick pupae if the experimenter removed the cap first. The reseachers found the four classes—two parental and two recombinants—in Mendelian ratios, suggesting two independently assorting genes, one for uncapping/not uncapping and one for removal/no removal. (Review Figure 8.12 for a parallel experiment with peas.)

It is easy to see how natural selection would favor preexisting mutations that led snakes of the damp coastal forests to eat slugs: Slug eaters would have an abundant and nutritious food supply that could help them produce more offspring, and through these young, alleles for slug eating would increase in frequency. It is also easy to see how alleles for slug eating would be selected against in snakes of the dry inland mountains: Alleles for slug eating also enhance leech eating, and leeches, although absent from coastal areas, inhabit inland lakes. It would be disadvantageous for garter snakes to swallow potentially damaging leeches on a regular basis. Hence, alleles that allow snakes to accept both slugs and leeches would be selected against and decrease in frequency in inland populations.

The garter snake study shows that proximate causes of behavior sometimes allow us to understand the ultimate causes of behavior. Ethologist Niko Tinbergen, in his studies of broken shell removal by black-headed gulls, was able to build on his knowledge of the proximate causes (jagged edges) to understand ultimate causes (protection from predators). When Tinbergen placed eggshells at various distances from gull nests, he found that nests with eggshells nearby were three times more likely to lose eggs to predators than nests without nearby eggshells. Evidently, natural selection has favored alleles that cause black-headed gulls to remove predator-attracting eggshells from their easily accessible nests. Similar alleles are lacking, however, and appear to be unnecessary in cliff-dwelling kittiwakes (close relatives of black-headed gulls), since foxes, crows, and hawks cannot gain access to the narrow ledges where kittiwakes build their nests (Figure 38.5).

In summary, experiments with bees, gulls, and garter snakes helped identify the proximate causes of animal behavior, such as perceiving environmental signals and responding through neural programs that develop under the direction of genes. They also helped reveal the ultimate causes: selective forces exerted on individuals in their natural environments.

EXPERIENCES IN THE ENVIRONMENT CAN ALTER BEHAVIOR

Animals display a staggering range of behaviors from the instinctive to the highly complex and learned: Birds sing specific songs; gulls perform precise mating dances; rams butt heads in spring; salmon migrate hundreds of kilometers to spawn where they were born; and college students study and remember the intricacies of biology. Some behaviors, as we have seen, have strong genetic determinants and change only over many generations as natural selection winnows out specific alleles but leaves others. For many other behaviors, however, interactions with the environment can lead to modifications during the lifetime of a single individual. There is, in fact, a continuum of behavioral types from **instincts**, which are actions that are innate and unchangeable, to actions with a limited capacity to be changed, to behaviors that can be *learned*, or modified by experience during an individual's lifetime.

Innate Behaviors:
Automatic Responses to the Environment

The very first season a black-headed gull tends a nest, it responds to the jagged edges of broken eggshells. Its behavior is *instinctive:* It is inherited; it does not change with experience; it includes a specific sequence of events; and each member of the species performs it in the same manner.

One of the most thoroughly studied instinctive behaviors is the sequence of actions a graylag goose performs when an egg rolls from the shallow depression on the ground where she nests. When a goose spots an egg outside her nest, she waddles up to it, stretches out her graceful neck, and gently wraps her chin and neck around the egg (Figure 38.6). Once in contact, she pushes the egg along by moving her head from side to side to keep the lopsided object rolling straight, somewhat the way a hockey player advances a puck down the court with both sides of a hockey stick. Ethologists know that this behavior is inborn in graylag geese because even birds raised in isolation do it the first time an egg rolls from the nest. If an experimenter removes the errant egg once the goose has begun rolling it, remarkably, she carries on like a feathered mime, rolling and nesting a nonexistent egg. An innate series of precise physical movements of this sort is called a **fixed-action pattern.**

FIGURE 38.5 ■ **Safe from Predators.** In cliff-dwelling kittiwakes, natural selection does not actively favor genes for removing broken eggshells from the nest, presumably because would-be attackers cannot land on the narrow ledges where the birds nest. While kittiwakes do not share the eggshell removal behavior of black-headed gulls, can you see other behaviors in the photograph that are shared by both species?

FIGURE 38.6 ■ A Graylag Goose Instinctively Retrieves a Stray Egg. When an egg rolls from the nest of a graylag goose (*Anser anser*), the female instinctively retrieves it by arching her neck gracefully around the stray and rolling it back with both sides of her beak like a hockey player advancing a puck. She will retrieve anything "egglike" near the nest, including baseballs and tin cans.

Fixed-action patterns are characterized by five features. First, fixed-action patterns are triggered by specific environmental signals called **sign stimuli,** or *releasers.* Sights, sounds, smells, tastes, or any information picked up by the sense organs may function as releasers. For the graylag goose, a round object near the nest is the releaser for egg-rolling behavior; a goose will retrieve a tennis ball as tenderly as her own egg! For the black-headed gull, the releaser for eggshell removal is an object with a jagged edge.

Second, once a releaser prompts a fixed-action pattern, the behavior proceeds from start to finish in an all-or-none fashion and in a particular sequence. The swallowing reflex is a classic fixed-action pattern. Have you ever tried to stop swallowing a mouthful of food midway down? It can't be done; more than a dozen muscles fire in coordinated sequence, each step unalterably triggered by the preceding step in a classic all-or-none pattern.

Third, since a fixed-action pattern is innate, it is fully formed and functional the first time it is used. Fourth, all species members of the same age and sex perform the behavior the same stereotyped way under similar environmental conditions. And fifth, an animal's instinctive response to a releaser depends on its stage of anatomical and physiological maturation and on physiological states of motivation, or **drives,** which are influenced by hormones or other internal stimuli. Motivation for egg-rolling behavior in the female goose, for example, begins with hormones present only between the time she lays her eggs and the time the eggs hatch. And in the chameleon, a drop in glucose level in the

blood may motivate the animal (physiologically, not consciously) to catch insects by tongue-flicking behavior, another fixed-action pattern (Figure 38.7).

While fixed-action patterns are triggered by environmental cues and depend on the animal's stage of development plus internal states such as blood chemistry, once begun, they are stereotyped and constant. Other behaviors on the continuum from instinct to learning, however, can be modified by information from the environment.

Behaviors with Limited Flexibility

Recall from the chapter introduction that black-headed gulls go on feeding forays, return to their nests, and regurgitate the food for their chicks. Gull parents of many species perform the same sequence, and upon their return, the chicks instinctively peck at the parents' wagging bills. In the herring gull, adults have a distinctive red spot on their bill that serves as the releaser for bill-pecking behavior, a fixed-action pattern (Figure 38.8a). This behavioral sequence, however, has some flexibility, or potential for modification. At first, the chicks will peck at anything resembling the parent's red bill spot and will actually peck more excitedly at a swinging stick with red spots painted on it than at a live gull's bill (Figure 38.8b). As the chicks mature, however, begging behavior becomes more directed. Eventually, they will beg only from their own parents. In these gull chicks, information acquired from the environment (about their parents' physical characteristics) can

FIGURE 38.7 ■ Internal Drive Causes a Chameleon to Flick Its Tongue. This lizard (*Chamaeleo gracilis*) from Kenya's Tsavo National Park can swivel its eyes independently, spot insects or other morsels, and, if hungry (an internal drive), instantly lash out its long, sticky tongue. The programming for such behavior is innate, and chameleons no more than 1 hour old can perform it.

(a) "Begging" behavior

(b) Experimental analysis of the releaser

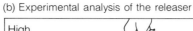
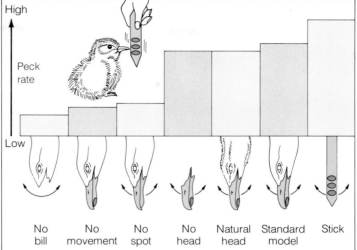

No bill   No movement   No spot   No head   Natural head   Standard model   Stick

FIGURE 38.8 ■ Red Spot Pecking in Young Gull Chicks: A Behavior with Limited Flexibility.
(a) A hungry herring gull chick (*Larus argentatus*) pecks at the red spot on its parent's bill, an action that triggers a regurgitated meal. (b) To identify the releaser, experimenters offered young chicks various artificial objects to peck at and watched for the rate of pecking in response. A head with no bill, a head with a bill but no motion, and a head with a bill and motion but no spot all elicited weak pecking responses. A moving bill with red spot and no head was better at attracting the chick's interest, as were a natural head and a realistic model of a head. Surprisingly, a moving stick with three red spots triggered the highest pecking rate of all.

modify what is initially a fixed-action pattern. The most common categories of behavior with flexibility are imprinting and learning.

Imprinting: Learning with a Time Limit

In the 1930s, Austrian ethologist Konrad Lorenz discovered the phenomenon called **imprinting:** the recognition, response, and attachment of a young animal to a particular adult or object. At the time, Lorenz was studying ducks and geese, and he knew that goslings normally follow their mother from the nest to a nearby pond or lake within a day after hatching. The researcher tried walking away from a group of newly hatched goslings in the nest while he made gooselike honking sounds. To Lorenz's delight, the gaggle of goslings followed him around as readily as they would have their real mother (Figure 38.9). He speculated that to survive, newly hatched graylag geese must quickly learn to recognize and follow their parents, and they are apparently programmed to trail after the first large moving object they see, even if it is a gray-bearded, pipe-puffing ethologist!

Imprinting has an inherent time limit—a **critical period,** or specific developmental stage, during which the imprinting can occur and after which it cannot. In geese, this period lasts from the time of hatching to the second day; imprinting is impossible from the third day on. Imprinting requires specific

releasers, in this case large objects moving away from the nest making goose sounds. Finally, imprinting is irreversible: The imprinted goslings recognized Lorenz as "mother" for the rest of their lives and could not be coaxed to imprint on their real parents later.

FIGURE 38.9 ■ Imprinted for Life. Imprinting has an inherent time limit; geese form parental attachments in the first two days after hatching. The attachments, however, are lifelong. Here, two imprinted geese preen Lorenz's "feathers."

FIGURE 38.10 ■ A Toad Learns a Lesson on Poisoned Prey: Once Sickened, Forever Repulsed. For this toad, eating a poisonous millipede was a big mistake. The nauseated amphibian drops the prey, gags, and learns a permanent lesson about food with this shape and coloration.

Biologists suspect that imprinting can take place in baby birds because development "primes" their nervous systems to change in a defined way based on environmental information shortly after hatching. Imprinting is thus a form of learned behavior, but one limited to a narrow time period. In most forms of learning, behavioral patterns can be repeatedly modified throughout an individual's lifetime.

Learning: Behavior That Changes with Experience

A hungry toad that has never eaten a millipede will, upon seeing the many-legged arthropod, flip out its sticky tongue and snap up the prey. The captured millipede quickly retaliates, however, and exudes a foul-tasting toxin. The poison works quickly, and the nauseated toad spits out the would-be prey unharmed (Figure 38.10). A toad is far from clever, but it will remember this single disagreeable experience forever and will henceforth reject all millipedes, even when hungry. The toad's nervous system has cataloged the shape and markings of a millipede and associated them with a repulsive taste that should be avoided—a one-trial learning event.

The toad's aversion to millipedes encapsulates the essential elements of **learning:** an adaptive and enduring change in an individual's behavior based on personal experiences in the environment. In our example, the toad's learning is adaptive because it allows the animal to avoid poisonous food; the learning is based on a personally experienced event rather than an inherited trait; and the behavioral change endures, in this case the toad's entire life. One-trial learning is just one type of trial-and-error learning. Other kinds of learning

include habituation, classical conditioning, and insight learning. All, as we shall see, can increase an animal's fitness—its ability to survive and reproduce in nature.

■ **Habituation**   **Habituation** is a simple kind of learning in which the animal comes to ignore a repeated stimulus that is never followed by reward or punishment. Most animals, for example, react to a novel stimulus by freezing in place or fleeing. Young black-headed gull chicks instinctively crouch down when objects pass overhead (Figure 38.11a), but this response wastes time and energy if the flying object is a harmless leaf or robin. It is adaptive, therefore, for chicks to habituate and learn to ignore silhouettes of common harmless species (Figure 38.11b), but to continue to cower at the distinctive shapes of predatory birds (Figure 38.11c). Since these predators pass overhead infrequently, habituation for them and other rarely seen objects never occurs, and this, too, is clearly adaptive.

■ **Trial-and-Error Learning**   More complex than habituation is **trial-and-error learning.** Also called **operant conditioning,** or instrumental conditioning, it is a form of learning in which an animal associates an action, or **operant,** with a **reinforcer,** a reward or punishment. The millipede-munching toad, for example, experienced trial-and-error learning: It learned to associate an action, capturing the millipede, with a reinforcer, a foul taste. Likewise, a pet dog will learn to relate the action of begging to a reinforcer, a handout of a dog biscuit.

Behaviorist B. F. Skinner studied operant conditioning under rigorously controlled laboratory settings. He suggested

FIGURE 38.11 ■ Habituation in Gull Chicks: Cowering Ceases with Common Objects, but Continues with Rare Ones. (a) At first, a gull chick will cower no matter what object passes overhead. (b) Eventually, it habituates to the passage of common, harmless objects like robins and leaves, but (c) continues to cower at rare ones like hawks, whether or not they are dangerous predators.

FIGURE 38.12 ■ **Classical Conditioning: Madison Avenue Exploits Our Evolutionary Tendency to Associate Separate Stimuli.** This classic advertisement links security, love, and owning a brand-new 1940 Buick Roadmaster.

that virtually any action can be conditioned, or trained, by a proper schedule of reinforcements. More recent experiments show that operant conditioning works readily only for stimuli and responses that have meaning for animals in nature. Rats rapidly learn to press a lever with their front paws to obtain a food reward, but they do not easily learn to push a lever to avoid an electric shock on their feet. On the other hand, rats quickly learn to jump off the ground to escape a shock to the feet, but they do not readily learn to jump to obtain a food reward. These results make sense when viewed in an ecological context; in nature, rats obtain food by manipulating items with the paws, but this is not how they avoid harmful stimuli. These experiments show that animals are innately prepared to learn some things more easily via operant conditioning than others, and in ways that appear to increase survival and hence reproductive success.

■ **Classical Conditioning**   While operant conditioning involves an association between an action and a reinforcer, **classical conditioning,** or associative learning, involves the association of two separate stimuli. Russian behaviorist Ivan Pavlov developed the concept from what has become the classic case of classical conditioning. He repeatedly presented a dog with meat, a natural stimulus for salivation, and at the same time rang a bell, an arbitrary, unnatural stimulus. Through classical conditioning, the dog learned to associate the natural stimulus and the substitute and began to salivate at the sound of the bell. In nature, a predatory hawk can learn to associate the presence of a broken eggshell with the presence of a defenseless, tender gull chick and act to collect a reward: dinner. Advertising agencies have learned to manipulate our own

species' susceptibility to classical conditioning by associating commercial products with positive images: fancy cars with successful business executives; quarter-pound hamburgers with happy family occasions; toothpaste brand with social popularity (Figure 38.12).

Classical conditioning works best when it mimics natural situations. Experimenters have found that rats quickly learn to avoid food that nauseates them if the food is associated with a particular smell, but not if the food is associated with specific colors. This makes ecological sense because rats usually forage at night, when sight is not a reliable way of evaluating foods.

We can conclude from the study of classical and operant conditioning that an animal's genes program it to learn different things in different ways that fit the animal to its environment. Many animals are adept at both operant and classical conditioning. One final type of learning, however, is common only in primates and reaches its peak in humans.

■ **Insight Learning**   At this moment, as you read, you are acquiring knowledge through **insight learning,** or reasoning. With this type of learning, people and other primates formulate a course of action by understanding the relationships between the parts of a problem. Insight learning allows an animal to encounter a new situation (including a new printed sentence), and based on similar past experiences, to figure out how to respond without actually trying out all possible solutions. A chimpanzee that enters a room where boxes litter the floor and bananas hang from the ceiling out of reach can figure out how to stack the boxes and climb up to grab the food. And experimenters have taught chimpanzees how to use American Sign Language and how to operate computer keyboards—remarkable examples of the chimpanzee's innate capacity to learn (Figure 38.13).

FIGURE 38.13 ■ **Conversations with a Chimp.** Researcher Herbert Terrace taught the chimpanzee "Nim Chimpsky" to use a few American Sign Language gestures. Here, Terrace signs his name "Herb" and Nim copies him.

People often assume that their household pets are capable of reasoning, but in fact, dogs, cats, and other common pets are quite inept at solving even simple insight problems. A dog can wrap its leash around a tree so that it can no longer reach its food and water, but still it cannot conceptualize and address the cause of the problem. A dog could be trained to keep its leash untangled through operant conditioning, but a primate would conceive of the problem beforehand and simply avoid it.

Behavioral ecologists suspect that animals have evolved the particular forms of behavior—from fixed-action patterns to insight learning—that provide the best chance of surviving and propagating. And natural selection seems to have resulted in an immense variety of developmental systems and behavioral strategies, as the next section shows.

HOW NATURAL SELECTION SHAPES BEHAVIORS

A black-headed gull selects a particular nesting site at its summer home on a European beach. The bird feeds itself and its young, detects and drives away predators, selects a mate, and cares for its chicks. As part of all these activities, it communicates to its family and colony mates through a variety of sounds and movements. And it has an innate sense of time and season that allows it to fish at the best time of day and migrate at the right time of year. The gull's complex behavioral repertoire helps illustrate several major categories of animal behavior: choosing a home territory through directional movements such as migration and navigation and then defending the home turf; feeding and avoiding being eaten; reproducing; and communicating. In gulls, as in thousands of other animal species, such behaviors have been shaped by natural selection and markedly influence the organism's chances of obtaining energy and producing progeny.

Locating and Defending a Home Territory

Just as people choose apartments, homes, and building sites, gulls find protected nesting spots on the beach, beavers locate the best site along a stream to build a lodge, and other animals move to territories that suit their needs. For simple animals, the movement can be toward a locally more favorable environment, but for more complex animals, it can involve long migrations.

■ **Simple Movements: Taxes**   Just as a plant's orientation toward light is called a tropism (see Chapter 31), an animal's movement toward light, specific chemicals, or heat is called a **taxis.** Each animal has sensory receptors suitable for detecting such environmental stimuli. A female tick, for example, must locate a mammal and drink its blood before laying her eggs. Light-sensitive receptors in the tick's skin lead her to a well-

FIGURE 38.14 ■ **Tick Taxes: Environmental Cues Lead a Female Tick to a Blood Meal.**   A tick moves toward the smell of butyric acid and the warmth of a mammal's body, bites a hole, and drinks blood. Here a female soft tick (*Ornithoderns monbata*) is enormously swollen with a blood meal and ready to lay her eggs.

lit leaf or branch, and then olfactory receptors help her detect butyric acid with its odor of rancid butter coming from a nearby mammal. Once triggered by smell, the tick drops from the leaf, and if she chances to land on a mammal, her heat detectors enable her to crawl to a warm spot. There she bites a hole, drinks the warm blood, and her eggs begin to mature (Figure 38.14).

The tick's movements involve positive *phototaxis* (movement toward light), positive *chemotaxis* (movement toward a chemical), and positive *thermotaxis* (movement toward heat). Natural selection has favored alleles that direct the development of special sense organs attuned to different physical stimuli; these stimuli, in turn, trigger behaviors that help the animal locate an appropriate home.

■ **Homing and Migration**   Many animals routinely travel hundreds of kilometers in search of an appropriate habitat. Whales, sea turtles, eels, and monarch butterflies travel great distances to their winter habitats or breeding areas (Figure 38.15). Salmon return to breed in the streams where they were hatched. But the champion voyagers are birds. Black-headed gulls trek from Africa to northern Europe and back annually, and the small arctic tern flies 16,000 km (about 10,000 mi), from Greenland and Alaska to Antarctica and back, seemingly in a search for endless summer.

While migrations are seasonal, homing behavior can take place anytime certain kinds of animals are displaced from their home territories. When experimenters transferred an individual shearwater, a long-winged seabird, from its home in Great Britain to a new location in Massachusetts, the remarkable bird was back on its nest in 12 days, having crossed 4800 km (about 3000 mi) of trackless ocean! Likewise, wildlife managers in Yosemite National Park routinely trap troublesome black bears and move them dozens of kilometers away from heavily used campgrounds, only to find the animals have returned in a few days. How do animals find their way, whether homing or migrating? Ethologists are not entirely sure, but tests confirm that some animals have a sense of **orientation,** or direction finding, and of **navigation,** the more exacting ability to move from one map point to another.

To test whether migrating animals orient in a general direction or actually navigate from point to point, researchers studied the homing response of starlings. These shiny black birds normally fly southwest each year from their summer homes in northern Germany to their wintering grounds on the shores of the English Channel (Figure 38.16). Researchers captured birds flying southwest through Holland, then released them in Switzerland almost 800 km (about 500 mi) south of where they were captured. Would the displaced birds continue orienting to the southwest or switch their route and navigate northwest to the English Channel? Interestingly, young starlings who had never made the migration continued orienting to the southwest and flying in the original direction, landing in Spain, far south of their original destination. In contrast, older birds who had flown from Germany to the English Channel the previous year switched their course, flew northwest, and correctly navigated to their proper wintering grounds. These experiments suggest that orientation may be inborn, but navigation is at least partially learned.

Which sensory systems help an animal to navigate? In daytime, birds, ants, and bees orient via the sun's position as

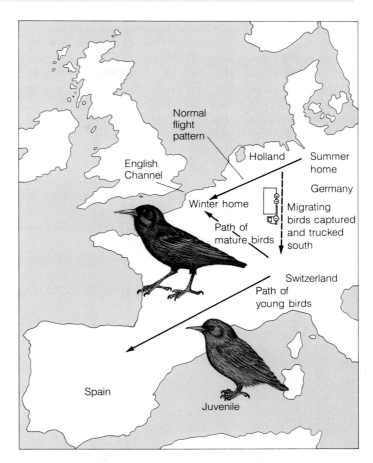

FIGURE 38.16 ■ **Do Starlings Orient During Migration or Actually Navigate?** In autumn, starlings normally fly to the English Channel from their summer home territory in northern Germany. To see if the birds orient or navigate, experimenters trapped starlings midway, and transported them to Switzerland, far south of their normal flight path. When released, novices continued flying southwest, orienting in their usual direction but mistakenly flying to Spain instead of the English Channel. Older birds, however, navigated correctly to their proper wintering grounds. This experiment showed that starlings can navigate, but require experience to do so.

FIGURE 38.15 ■ **A Migration of Monarchs to Mexico.** Each autumn, 100 million monarch butterflies migrate from the United States to the mountains of western Mexico to roost on Oyamel fir trees. Scientists are still studying how they navigate the nearly 2000 km (1240 mi) to this particular region.

it sweeps across the sky relative to their path. For example, a pigeon released west of its birdhouse in the morning will fly toward the sun to get home, while a pigeon released west of its birdhouse in the evening will fly away from the sun to get home. When experimenters shifted the birds' sense of time by altering their exposure to light and dark in a closed room, the birds looked at the sun and flew in the wrong direction. This suggests that pigeons use an internal "timepiece," or *circadian clock,* that helps compute orientation relative to the sun's path.

This is only part of the story, though, because homing pigeons can find their way home even in the dark or when researchers have fitted them with tiny, thickly frosted contact lenses that reduce their vision to a murky haze. Remarkably, birds still orient and navigate under such conditions by detecting the earth's magnetic field. Biologists have discovered a few tiny crystals of magnetic iron oxide in tissues near the brains of birds and porpoises. They suspect that these crystals align with the earth's magnetic field, like the iron needle in a compass, and cause neurons to fire. The researchers concluded that the sun is a pigeon's primary cue for orientation, but the earth's magnetic field acts as a substitute cue in the dark.

These studies on starlings, homing pigeons, and other animals help address the proximate causes of behavior: how animals migrate. But the ultimate causes—why animals migrate in the first place—still must be addressed. After all, migration exacts a high toll. Grizzlies wait in streams and gorge on exhausted salmon migrating home from the sea, and falcons feast on fatigued songbirds arriving at their winter home in Africa. Fuel used by muscles to propel wings, fins, and legs is unavailable for reproductive activities, and time spent on the move is time not spent gathering food.

Clearly, the benefits of migration must outweigh such tremendous costs. Animals have much to gain by moving on to greener pastures after they exhaust an area's seasonal resources. And migration allows animals to fatten themselves in a bountiful but dangerous feeding area, then retire to a more protected region with fewer predators for breeding or birthing. Animal species can travel short distances or long, in isolation or in large groups, but once they reach their destination, they often claim a territory and defend it, even against their own former traveling companions.

■ **Territoriality**   Once a male gull claims a spot on the beach, he rebuffs other male interlopers by assuming stylized body positions and issuing specific calls that proclaim his turf (see Figure 38.1). Such behavior that defends living space from intruders is called **territoriality.** Sleek male impalas of the African savanna are good examples of animals competing for territories rich in food resources. The richer the territory, the more females it entices, and more females mean greater reproductive opportunity for males. The males stake out the richest territories they can find, and when a competing male arrives, they clash, interlocking their graceful horns and thrusting furiously, trying to force each other out. Bands of roving females stay in each territory only so long as the food holds out, and during their stay, the resident male mates with all receptive females. The stakes are clear: Males guarding more food have access to females longer, mate more frequently, and hence tend to have more offspring. A male without fertile territory to defend will have fewer contacts with females and thus far fewer offspring.

The costs of defending a territory are great: Constant battle readiness requires time and energy, and exhausting tournaments leave the combatants vulnerable to predators. The rewards in terms of extra progeny produced, however, are tremendous, just as the principles of natural selection predict.

And the vanquished will often surrender after a brief exchange of threats. Ethologists think that the animal's long-term reproductive success will grow if it survives unscathed and claims its own territory later.

In most territories, the most valuable commodity is a ready source of food, and a second group of behaviors has arisen around obtaining nourishment.

Feeding Behavior: Finding Food While Avoiding Predators

Is it more advantageous for a black-headed gull to expend the energy to fly several kilometers out to sea where herring and other fish are bountiful or to travel a shorter distance but forage where food is less abundant or less nutritious? What fraction of its feeding time should a junco spend scanning the skies for a hungry hawk instead of searching the ground for seeds? To answer such questions, behavioral ecologists have devised and tested the **optimality hypothesis,** the notion that animals behave in ways that allow them to obtain the most food energy with the least effort and the least risk of falling prey to a predator.

Ethologists studied crows, for example, as the birds preyed on whelks—large, snaillike mollusks (Figure 38.17). A crow will typically search the intertidal zone and select a big whelk over several smaller ones, grasp it, fly it to a height of 5 m (16.4 ft), then drop it onto the rocks below. If the shell

FIGURE 38.17 ■ Crow Dropping a Whelk: Minimizing Effort, Maximizing Gain. A crow hunting whelks will choose large prey and smash them on the rocks by dropping the shells from 5 m (16½ ft). If the shell doesn't break, the crow tries again. This precise strategy supports the optimality hypothesis: the idea that natural selection has favored genes causing animals to feed in such a way that feeding effort is minimized while energy gain and safety from predators are maximized.

shatters, the crow consumes the tender flesh; if it doesn't break, the crow lifts and drops it again. Researchers predicted that if natural selection favored genes promoting optimal foraging strategy, then large whelks should be more likely to smash than small ones; a drop of 5 m should be enough to break most whelk shells; and dropping a whelk a second time should be just as likely to lead to a meal as searching for a new whelk. In fact, each of these predictions proved true. Crows waste neither time nor energy dropping small, unbreakable whelks or releasing whelks too low or unnecessarily high for breakage. Considered in evolutionary terms, a crow with optimal feeding behavior spends less time and energy foraging, but has more time and energy left over for producing offspring, and thus sends more genes into the next generation.

The optimal foraging solution is neither conscious and deliberate nor universal. There is no evolutionary advantage to gorging one minute only to be devoured by a predator the next, so animals must often feed in ways that are nutritionally far from optimal in order to avoid the dangers of predation. The dark and light birds called juncos interrupt their search for seeds on the ground about every 4 seconds to scan the sky for hawks. This behavior sacrifices immediate food gain for long-term survival. If food is abundant, juncos form flocks; this cuts the time each bird must spend being watchful by spreading the vigilance over a larger group. Juncos in flocks, however, must share the available seeds with other birds. The optimality hypothesis predicts that individual juncos will alter their behavior to adjust for food availability and imminence of attack and will thus maximize their personal biological fitness.

While territoriality and feeding strategies help animals take in energy, every animal faces a second and equally important challenge: overcoming death by bearing and perhaps caring for young.

Reproductive Behavior: The Central Focus of Natural Selection

Successful organisms reproduce effectively, and different sexes have different strategies for achieving reproductive success. The males of most species tend to make many highly mobile, lightweight gametes and behave in ways that increase the number of eggs they fertilize. In contrast, female animals tend to make fewer but much larger gametes. A woman, for example, produces a few hundred eggs in her lifetime, while a man, in a similar period, produces enough sperm to inseminate all the women in the world. Not surprisingly, such differing reproductive strategies shape the reproductive behaviors of the sexes.

■ **Parental Investment and Mate Selection**   The fact that males and females produce gametes of different sizes and numbers has major consequences, including parental investment and

FIGURE 38.18 ■ **Big Joey Requires Big Parental Investment.** This female gray kangaroo (*Macropus giganteus*) will nurse its large joey for several months after the joey's birth as a tiny, bean-sized marsupial. The mother also teaches her young (largely by example) about grazing and escaping from danger.

mate selection. *Parental investment* is the allocation of a parent's resources, such as food, time, and effort, to each individual gamete and the young that arises from it. Each large egg represents more resources than each small sperm; thus, in a physical sense, the female sacrifices opportunities to make more eggs in favor of increasing the chance that each egg will become a viable descendant. An extreme example of this is the kiwi bird, which lays an egg weighing one-quarter her body weight—the equivalent of a woman having a 13.5 kg (30 lb) baby! Most female mammals shelter and nourish their offspring in the uterus and feed them with milk long after birth (Figure 38.18). Such behaviors extract a heavy cost in physiological stress and increased exposure to predators. Yet the extra care pays dividends in terms of the chances that each offspring will survive and contribute genes to the next generation.

Males usually exhibit much less parental investment. Behaviorists have deduced that male parenting is likely to evolve only in those ecological situations where females cannot provide all the care necessary for raising a brood successfully or where a male who abandons his mate is unlikely to find another partner. In such cases, evolution will favor behavioral mechanisms that ensure paternal proximity and investment, such as establishing and defending a nesting territory, as do black-headed gulls.

The second consequence of gender differences in gamete size is a sex difference in *mate selection*. Typically, males—the sex with the lesser investment—compete among themselves to fertilize many females, while females, with their

FIGURE 38.19 ■ Attempted Mating by Bull Elephant Seal Sets Off the Female's Automatic Alarm Call. A casual advance by an amorous male—a flipper caress—sets the female elephant seal (*Mirounga leonina*) to screaming at high decibel. She calls this way regardless of the suitor's stature in the colony, but if the male is subordinate, the king bull will probably chase him off.

heavier investment, select the male of choice to father their offspring. As a result of competition within a sex and mate selection between sexes, members of the same species exert selective forces on each other—an evolutionary pressure called **sexual selection**.

■ **Sexual Selection in Males**   In many animal species, females, with their greater parental investment per offspring, represent a limited and valuable resource for males. In species where males make little parental investment, therefore, those males compete to inseminate as many females as possible. Biologists call competive pressure between members of the same sex *intrasexual selection*. In some species, this leads to a **dominance hierarchy,** a ranking of group members based on past success in aggressive encounters over access to food or mates. Bull elephant seals, for example, are massive beasts that threaten each other with tremendous roars, and bite and batter competitors. The loser of a skirmish lowers his head and retreats, with the victor in pursuit. A few years ago, an ethologist ranked ten bulls living on a small island in the Atlantic by their track records in winning such encounters, and then counted the successful copulations each achieved during the breeding season. The top-ranking male accomplished nearly 40 percent of all the copulations in the group, the second male less than 20 percent, and the other males only about 5 percent each. Thus, natural selection in elephant seals (and in many other animals as well) favors those behaviors that win dominance contests and hence mates. In other animals, such as chickens, dominance hierarchies may be established for food rather than access to mates.

■ **Sexual Selection in Females**   In many species, females are the sex in demand, and they do the mate choosing. A female generally picks males that provide the "best" genes or material benefits that may enhance her offspring's survival. Females thus engage in *intersexual selection,* wherein one sex (the female) chooses desirable mates, and in so doing, acts as an agent of natural selection on the behavioral and physical traits of the other sex.

By choosing dominant or propertied males, females increase their own biological fitness. Elephant seal cows, for example, create intersexual selective pressure by simply screaming loudly whenever a bull attempts to couple (Figure 38.19). Her cries alert the harem master, and if the courting male is low on the dominance hierarchy, the dominant bull sprints to the site of the tryst and chases away the intruder. The female's automatic cries greatly increase her chances of being inseminated by the dominant male rather than a subordinate and hence of having sons with aggressive behaviors that will favor high social rank.

Elaborate courtship rituals probably also evolved under the pressure of females exercising mate choice. The head turning of the black-headed gull (see Figure 38.2), the strutting dance of the male ptarmigan, and the showy flight of the widowbird flaunting his remarkable tail (Figure 38.20) provide information about these animals' physiological condition, and this, in turn, may reflect gene quality. Experimenters have found that when they cut the tail feathers off a male widowbird, it is less likely to attract a mate, but when they glue on long feathers from another bird, making an extralong

FIGURE 38.20 ■ A Widowbird Flaunts His Tail. The male long-tailed widowbird (*Euplectes progne*) is a jet black African weaverbird with a 60 cm (2 ft) tail that will spread and flutter when he performs a mesmerizing courtship flight for the female. When experimenters cut the tail feathers of these birds to make short tails or glued on long feathers to lengthen the tails, they found that females chose a promising mate on the basis of tail length.

FIGURE 38.21 ■ Luminous Communication: Fireflies on a Summer Night. Fireflies are actually soft-bodied beetles that light up the countryside with glowing abdomens. This drawing depicts the male courtship displays of four species. Each communicates mating interest to waiting females.

tail, the bird can be quite attractive to females. Experiments like these suggest that even a somewhat harmful trait, such as a long tail that hinders flying ability and thus increases the bearer's risk of being injured or eaten, may be selected for if it promotes reproductive success.

Communication: Messages That Enhance Survival

Midwesterners enjoying the fragrant air and trilling insect sounds of a warm summer night can often catch a wonderful aerial display as "lightning bugs," or fireflies, flash their luminous yellow abdomens on and off while they dart about the green lawns, fields, or trees. The flickering streaks they create are based on flashing patterns with a rhythm characteristic for each species (Figure 38.21). Some species give long pulses of continuous light, others blink three times in a row, and still others give a fluttering flicker like a light bulb about to burn out. The airborne flashers are males communicating their availability for mating, and when a female sees the right pattern of flashes, she responds with her own burst of light, revealing her position to the male.

The firefly's blinking light is a form of **communication**, a signal produced by one individual that alters the behavior of a recipient individual and generally improves both participants' biological fitness.

■ **Communication by Sight, Sound, and Scent** Each animal uses one or more channels of communication that fit the species' ecological circumstances. In open areas where individuals live near each other, visual displays such as the courtship dances of ptarmigans and black-headed gulls are common. Animals that hide in thickets or pond vegetation are invisible not only to predators but to potential mates as well, and for them, auditory cues, like the chirp of a cricket or the croak of a bullfrog, are typical. Sound communication, such as the alarm call of a gull or prairie dog, is also advantageous as a warning to colony members who may not see the approach of a predator.

Some animals live such solitary lives that they are unlikely to see or hear signals from other members of their species. For them, odor molecules can convey information. The sex pheromones released into the air by female moths (see Chapter 26) can travel for kilometers on the wind, announcing the females' location and sexual readiness. Chemical signals not only disperse farther than auditory or visual messages, but also can persist longer in the environment. Ants establish durable trails this way and will follow pheromone trails slavishly—even those an experimenter deposits in a circle (Figure 38.22).

■ **Communication by Touch** Groups of animals that live in dark hives, caves, or underground chambers often communicate by touch, as well as by sound and smell. In the 1940s, ethologist Karl von Frisch discovered the classic example of touch communication: the "dances" bees perform in the dark interiors of their hives that communicate the location of rich food sources. If a foraging bee finds flowers that are laden with pollen and nectar and are situated fairly close to the hive (within 80 m, or 262 ft), the bee flies to the hive and imme-

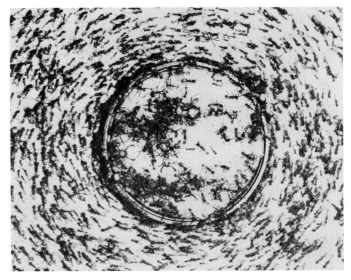

FIGURE 38.22 ■ Scent Circles: Ants Following Pheromone Trails. When an experimenter placed a circular dish amid a colony of army ants, a few explorers traced the periphery and laid down pheromone trails, and the rest followed endlessly like miniature automatons.

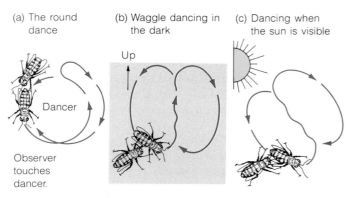

(a) The round dance

Dancer

Observer touches dancer.

(b) Waggle dancing in the dark

Up

(c) Dancing when the sun is visible

FIGURE 38.23 ■ **Honeybee Dances.** Forager bees perform (a) round dances when a discovered food source is close to the hive and (b) waggle dances when the source is more distant. Orientation of the dancing bee on the wall of a dark hive signifies the direction to fly relative to the sun's position, with the convention that a symbol for the sun is straight up. If sunlight shines into the hive (c), the bee orients to the light. A partially blinded bee in a lighted hive dances as if it were dark, but observer bees interpret the dance according to the orientation of the light.

diately performs a *round dance* by walking in tight circles (Figure 38.23a). Observer bees follow the dancer in the dark, sense its movements by touch, and learn that a nearby food source exists. Since worker bees can also smell bits of food clinging to the forager's body, they usually have no trouble locating the specific flowers.

If the forager discovers a food source far from the nest, she returns with a rich load of pollen, climbs onto a vertical wall of the hive, and begins a *waggle dance,* moving in a figure eight while wagging her abdomen from side to side and making an occasional burst of sound (Figure 38.23b). The dancer's animation indicates the amount of energy needed to fly to the food: A lethargic dance with few waggles and little sound indicates a long or uphill trip to the food source; an excited dance with many waggles and frequent bursts of sound suggests a more readily accessible site. Simultaneously, the dancer's orientation on the hive wall reflects the angle of the necessary flight path with respect to the sun. If the hive is dark and the food lies in the sun's direction, the bee waggles while moving straight up (see Figure 38.23b). If the hive is dark and food lies 90° to the right of the sun, the bee waggles while moving horizontally to the right. If sunlight enters the hive, the bee orients her dance directly with respect to the sun's rays (Figure 38.23c).

While von Frisch carefully observed these dances and deduced their meaning, only experiments could confirm that oriented dances guide the observer bees. To test the role of the sun in bee communication, an ethologist captured returning foragers and painted over their three simple eyes but left their two large compound eyes untouched. Under these conditions, a partially blind forager dances as if the hive were dark, and that dance indicates a particular direction for the food. Her

sisters, however, who perceive light shining into the hive, interpret the dance as if the forager, too, could see the light, and so head out in the wrong direction. These experiments proved that sight and touch, not scent, guide the observers' flight path.

Whether an animal communicates via sight, sound, scent, touch, or body positions, the exchange of information has adaptive value: It allows individuals to find homes, food, and mates more successfully or to evade predators and thus to send their alleles on to future generations. While solitary animals like fireflies, moths, and frogs exchange information from time to time, and those living in small family groups, like black-headed gulls, do so more regularly, communication is most complex in animals that live together in large social groupings. By their very complexity, animal societies pose an interesting challenge to evolutionary ecologists: How could such social groups with their roles and rules have evolved?

THE EVOLUTION OF SOCIAL BEHAVIOR

Turtles, beetles, snails, and the vast majority of other animals live solitary lives, interacting with others only to mate and maintaining no long-term contact with parents or offspring. A fascinating minority of animal species, however, live in groups—either as giant colonies, like those of ants, bees, and termites; as large groups, like flocks of gulls; as small, stable packs, like those of wolves (Figure 38.24); or as individuals with only seasonal associations, such as the graceful manatees of Florida. What are the proximate causes of social behavior? What biological mechanisms cause gulls to live in cooperative societies while turtles live as lone individuals? And what

FIGURE 38.24 ■ **Social Contacts.** Many animal species live in social groups. Some are small and intimate, such as this pack of gray wolves. Some are large and permanent, such as colonies of kittiwakes (see Figure 38.5). Some are only seasonal, like the manatee with her offspring of the year (see Figure 19.17d).

FIGURE 38.25 ■ **Helper Scrub Jays Demonstrate Kin Selection.** Older siblings (right) that remain near the nest and assist their parents in raising a new crop of young promote their own alleles. Their behavior is an example of kin selection.

are the ultimate causes of the behavior? How does an individual benefit from, say, cooperative predator mobbing if this behavior puts it in danger? Questions like these are the focus of **sociobiology,** the study of social behavior.

Social Living: Advantages and Disadvantages

What does an individual gain and what does it sacrifice by living in a group? One of the many studies ethologists have designed to answer this question is a comparison of the societies of white-tailed and black-tailed prairie dogs. Regardless of species, a prairie dog has a characteristic response to the approach of a badger, weasel, or other predator: It freezes for a moment, then gives a loud alarm call. This dangerously betrays its location to the predator, but it also warns family members and neighbors to dash off to the safety of their burrows. Animal behaviorists noted that black-tailed prairie dogs live in larger groups than white-tailed prairie dogs. That differential gave them a means of probing the advantages of group living for these animals.

Experimenters pulled a stuffed badger at constant speed through different-sized prairie dog colonies and recorded the time it took before a prairie dog would sound the alarm. They found that among black-tails, with their larger colonies, predators were detected sooner than among the smaller white-tail colonies. With the support of so many watchful colony mates, black-tails could spend less time scanning for predators and more time foraging. Larger colony size is clearly advantageous for providing a means of early predator detection plus more foraging time. So why don't white-tailed prairie dogs live in large colonies too?

The answer to that question lies with the disadvantages of living in larger societies. Observations reveal that prairie dogs in larger groups spend more time aggressively bickering over

territory than those in smaller groups. In addition, plague-carrying flea populations are four times as dense in the larger black-tailed colonies as in the smaller white-tailed groups. Apparently, to avoid predators, black-tails must pay the costs of infighting and disease. One could predict from this that the black-tails' environment exposes them to more predators. Indeed, large colonies of black-tails live exposed on the open plains of the American West, while the smaller colonies of white-tails inhabit shrubland and can hide in the bushes.

From experiments like these, sociobiologists conclude that animals will socialize only when the benefits of group living outweigh the costs. In each case, the benefits require cooperation between individuals, and the cooperation can occasionally cost an individual its feeding time or even its life. How can natural selection explain such seemingly altruistic behaviors?

The Evolution of Altruistic Behavior

Cooperative behavior often favors both helper and helped, as when a lioness chases a wildebeest toward her hunting partners and all share the kill. Sometimes, however, the helper suffers, as when a worker bee stings a raccoon, protecting the hive in the process but ripping out her stinger and subsequently dying. Ethologists have long wondered how such **altruism,** or seemingly unselfish acts for the welfare of others, could evolve, since animals displaying the selfless behavior would tend to reproduce less frequently.

Behaviorists working out the puzzle of altruism have recognized that reproduction multiplies one's own alleles. The obvious way to accomplish this is to reproduce oneself; but there is another way too: encouraging the reproduction of family members that share the same alleles. Evolutionary biologists recognize a specific type of natural selection called **kin selection,** in which an individual increases its genetic contribution to the next generation by helping relatives—who share its alleles—to reproduce.

You can see how kin selection works by considering how many alleles you share with your family members. Because of meiosis and fertilization, you inherited half your alleles from your mother and half from your father. Since you share the same parents, half of your alleles will be carried by each brother or sister. Your aunts, uncles, and grandparents, in turn, carry one-quarter of your alleles, and your cousins one-eighth.

These same degrees of relatedness hold for most other animals and help explain kin selection. Consider, for example, a breeding pair of Florida scrub jays. Some of their offspring serve as helpers at the nest who do not themselves reproduce, but assist their parents to raise more young jays (Figure 38.25). If one of these helper jays were to produce two surviving chicks, each possessing half of its alleles, then the bird's genetic contribution to the next generation would be $\frac{1}{2} \times 2 = 1$. If the helper jay abandoned personal reproduction, however, to help its parents produce three more siblings, all of whom shared half of its alleles, its genetic legacy would be ½ × 3 = 1½, an increased contribution by helping compared to self-reproduction. If certain of the shared alleles promote helping behavior, then kin selection will cause these "altruistic alleles" to increase in frequency in the jay population.

A helper jay may eventually inherit territory when its parents die, and there the jay can attract a mate and set up its own family, complete with new helpers. In such a case, the former helper's biological fitness becomes the sum of its fitness from kin selection as a helper, plus its personal fitness upon reproducing—a sum called **inclusive fitness.**

The proposer of kin selection, W. D. Hamilton, once quipped: "I'd lay down my life for two brothers or eight cousins." By this he meant that an act of altruism toward close relatives promotes survival and transmission of one's own alleles. His theory helps explain how something as drastic as the suicidal sting of a honeybee could have evolved.

■ **Kin Selection and the Evolution of Social Insects**   Before Hamilton proposed the concept of kin selection, the rigid caste system of social bees, wasps, and ants was quite puzzling. Such societies often consist of a single reproducing female, the *queen,* plus a few *drones,* or males, and up to 80,000 sterile female helpers, or *workers* (Figure 38.26). But what advantage comes to the sterile worker who cannot reproduce herself and instead helps the queen reproduce?

Geneticists and animal behaviorists discovered that in bees and wasps, females are diploid, but males are haploid. As a result, daughters share *all* their father's alleles instead of just half, as in most organisms, and so bee sisters are more closely related genetically than human sisters or siblings in other animals. By helping the queen reproduce more sisters (including future queens), they are really favoring the preservation of their own alleles.

FIGURE 38.26 ■ **The Queen and Her Court: How Kin Selection Evolves. A circle of worker bees tend their regal sister, the queen bee, with enlarged abdomen and dark, shiny thorax. By helping the queen (their sister) reproduce, worker bees help perpetuate their own alleles.**

The haploid/diploid sex determination mechanism favored complex caste societies among ants, bees, and wasps. In contrast, many mammalian societies evolved because group hunting provided advantages, as in wolves, or because relatives could assist with the prolonged developmental needs of young ones, as in primates.

Sociobiology and Human Behavior

Biologists can point to thousands of examples from the animal kingdom in which social behavior seems to promote the survival of an animal's alleles through a combination of individual and kin selection. But what about human behaviors? Do similar genetic principles apply to us? In the mid-1970s, Harvard biologist E. O. Wilson stirred a heated controversy by cautiously suggesting that human behavior has evolutionary underpinnings and can be analyzed in terms of individual and kin selection. Some scientists and lay people alike issued strenuous objections to Wilson's ideas, mistakenly thinking that if human behavior is genetically determined and evolutionarily adaptive, our actions cannot and should not be changed. Some critics were justifiably concerned that theories of genetic predeterminism could be twisted to promote fascism, sexism, and racism. Wilson had intended merely to suggest a way of analyzing the biological causes of human behaviors, not justifying them from an ethical standpoint.

Debates continue on the evolution and adaptiveness of specific human behaviors, but it is clear that some human behavioral patterns are innate. The tiny hand of a human baby grasps tightly to a parent's finger, its little lips suck rhythmically at a nipple, and it communicates forcefully with a cry that a parent cannot ignore. A baby born blind still flashes an alluring smile at the sound of its parent's voice—an innate behavior pattern triggered by a releaser. A sociobiologist interpreting such inborn human behavior would suggest that infants inheriting alleles that foster smiling or crying are more likely to solicit parental care and hence more apt to survive and pass on such alleles to the next generation.

While there are many innate behaviors, and while genes underlie at least some aspects of human behavior, many more behaviors are passed along through cultural traditions. Genes determine the native ability to learn language, for example, and this unfolds as neural pathways develop in the growing embryo and make it capable of learning. But whether a child learns to speak English, or Polish, or Swahili after it is born depends on the language its parents teach it.

Sociobiologists propose that this flexibility directs people to accept cultural practices that enhance their inclusive fitness and reject those customs that reduce genetic success. For example, Brazilian tribes living in areas with poor soils and overhunted forests adopt Western agricultural techniques, while clans inhabiting less disturbed areas cling to traditional habits. In both cases, the human nervous system seems to respond to the environment with an innate flexibility that is naturally selected to elevate an individual's economic condition and hence provide for offspring. A few other animals also seem to accept new adaptive cultural practices and transmit them to their offspring. The case of the macaques who seem to have learned to wash sweet potatoes is probably the most famous example (Figure 38.27).

Psychologists have conducted many recent studies on identical and fraternal twins raised together in the same family or apart by adoptive families to explore the contention that genes help shape human personality and behavior (Figure 38.28). These twin studies show that many human personality traits are influenced about equally by genes and the environment. Such traits include social effectiveness, achievement, social closeness, alienation, aggression, respect for authority, and avoidance of harm. A single gene, of course, encodes a protein, not a trait like aggression or achievement. Behavioral geneticists suggest that alleles of hundreds of genes combine to determine a person's range of emotions; the thresholds for anger, hostility, and sexual arousal; and the inclination to learn one type of behavior versus another.

Sociobiologists conclude that the human brain evolved in ways that encourage social behaviors that enhance the bearer's reproductive fitness. Those behaviors evolved when people lived in small, isolated family groups, but in the meantime, we have been enormously—perhaps disastrously—successful as a species. As the biosphere threatens to collapse under the pressure of human population, our gene-encoded

FIGURE 38.28 ■ **Reunited Twins: Something Was Missing.** Identical twins Jerry Levey and Jim Tedesco were separated as infants and long felt something was missing in their lives. Reunited in middle age, they discovered not only a close physical resemblance, but identical vocations—firefighting—and avocations—flirting, telling jokes, and drinking beer.

FIGURE 38.27 ■ **Cultural Transmission in Primates? The Case of the Clean Sweet Potatoes.** A young female macaque accidentally learned to wash the sand off sweet potatoes that she picked up from the beach, and soon others in the troop learned to do the same. This learned behavior became a permanent part of the troop's culture, passed along to subsequent generations, perhaps in the way reading, writing, metallurgy, and agriculture have become parts of human culture.

behavioral plasticity remains. Thus, the hope remains, as well, that we will not destroy the quality of life for the animal called *Homo sapiens,* nor for any of the living species that share our planet, and that the nature of life will remain for us a challenge and a source of perpetual wonder.

CONNECTIONS

The environment selects for genes that direct the development of animal nervous systems in ways that promote specific behaviors. Some behaviors are rigid and stereotyped, like a baby gull pecking at its mother's beak, and others are plastic and variable, like a macaque learning to wash potatoes in the sea. But regardless of flexibility level, behaviors have evolved that increase an individual's survival and genetic perpetuation, as when a gull removes broken eggshells from its nest and in so doing helps to protect its offspring from predators.

Of course, the gull has no awareness of why it removes the broken shells. But those individual gulls who possess alleles promoting eggshell removal tend to lose fewer chicks and have more surviving offspring.

All of the fixed and plastic behaviors that suit an organism to its environment have proximate and ultimate causes, and these help us understand how and why animals act as they do—even in puzzling instances of altruism, such as a scrub jay helping its parents raise another family or a worker bee assisting the queen.

Most sociobiologists believe that the principles of kin selection apply to the evolution of many types of human social behavior, as well as to wolves and prairie dogs. By understanding the evolution of human nature, we stand a better chance of solving our current global crises of population, ecological degradation, disease, starvation, and warlike aggression. Well-informed citizens like the readers of this book are in the best position to apply biological knowledge to these issues and perhaps to help make our world a safer place for life.

Highlights in Review

1. Black-headed gulls display many distinctive behaviors and make a good case study for ethology, the study of how animals behave in their natural environments.
2. Behaviors have proximate causes, or physiological mechanisms that bring them about, and ultimate causes, or selective advantages for the performing individual.
3. Several kinds of experiments helped prove that alleles allowing particular behavioral traits accrue as a result of natural selection.
4. There is a broad continuum of behavioral types, from innate and unchangeable instincts to learned behaviors (behaviors modified by experience).
5. Instinctive sequences of behavior are triggered by sign stimuli, or releasers; proceed in an all-or-none fashion; are innate; are the same in all species members of the same age and sex; and depend on physiological drives.
6. Some behaviors, like the pecking of a red dot by a baby gull chick, can be modified by environmental information.
7. Some behaviors, such as the imprinting of goslings, are flexible, but only within a critical period, a given time frame.
8. Learning reflects truly flexible behavior; it is an adaptive and enduring change in behavior based on personal experiences in the environment.
9. Habituation occurs when an animal learns to ignore a repeated stimulus never followed by reward or punishment.
10. Trial-and-error learning comes about when an animal learns to associate a response, or operant, with a reinforcer, such as a reward or punishment.
11. In classical conditioning, the animal learns to associate two separate stimuli with a response.
12. In insight learning, a trait apparently possessed only by primates, the animal reasons out a solution based on previous experience in similar situations.
13. Choosing and defending a territory can involve a taxis, orientation, navigation, or true territoriality, the defense of a particular space from intruders. The possession of territory tends to confer selective advantage in terms of finding mates or food or hiding from predators.
14. Finding food often involves optimality—the maximizing of energy gain and the minimizing of energy spent collecting food or defending against predators.
15. Numerous behavioral sequences have arisen around the investment of energy for each gamete (and possible offspring) and for mate selection. Members of the same species exert selective forces on each other, called sexual selection. This has led to dominance hierarchies and to courtship rituals by which females can judge male competence.
16. Communication involves a signal (visual, auditory, olfactory, vocal, tactile) produced by one individual that alters the behavior of a recipient and improves both participants' biological fitness.
17. Social behavior, or close group interactions and interdependencies, require communication and evolved because they confer selective advantage to particular species in their particular environments.
18. Altruistic, or selfless, behavior probably evolved via kin selection—the advantage conferred on reproducing individuals by a closely related, often nonreproducing helper.
19. Biologists are still debating the extent of evolutionary preprogramming for specific human behaviors, but twin studies show that many human traits have powerful genetic bases as well as environmental influences.

Key Terms

altruism, 789
classical conditioning, 781
communication, 787
critical period, 779
dominance hierarchy, 786
drive, 778
ethology, 775
fixed-action pattern, 777

habituation, 780
imprinting, 779
inclusive fitness, 790
insight learning, 781
instinct, 777
kin selection, 789
learning, 780
navigation, 783

keys travel in groups of 10 to 20 adult females with their young, and with a single dominant adult male. Other males not associated with female groups try to wrest control of the group from the dominant male. If a takeover is successful, the newly dominant male will attempt to kill all nursing offspring and newborns for 6 months after the takeover. He will not attempt to kill young who are weaned. How might a sociobiologist explain the male's behavior in terms of sociobiology and ultimate causes?

Study Questions

Review What You Have Learned

1. Ethologists distinguish between the proximate cause and the ultimate cause of a behavior. Explain.
2. How does an instinct differ from a learned behavior?
3. List five features of a fixed-action pattern.
4. Name five types of learning.
5. What is a taxis?
6. What factors influence animal navigation?
7. Give an example of territoriality.
8. Which sex generally invests more resources in their offspring, males or females? Explain.
9. List four types of animal communication, and give an example of each type.
10. What is kin selection?
11. Specify the proportion of alleles a person shares with the following family members: father, sister, grandmother, uncle, cousin.

Apply What You Have Learned

1. Will a black-headed gull expel a chicken egg from her nest? Explain.
2. You discover a nest of three-day-old goslings and decide you'd like to have them imprint on you and follow you home. Are you likely to succeed? Why or why not?
3. An advertisement for the Nature Nuggets Company features a hiker eating a bowl of cereal on a mountaintop. What type of learning does the advertisement rely on? Explain.
4. Female langur monkeys have a gestation period of 7 months and usually nurse their offspring for another 8 months. While pregnant and nursing, females will not ovulate. Langur mon-

For Further Reading

ALCOCK, J. *Animal Behavior: An Evolutionary Approach.* 3d ed. Sunderland, MA: Sinauer, 1984.

BREED, M. D. "Genetics and Labour in Bees." *Nature* 333 (1988): 299.

CHENEY, D., and R. SWYFARTH. "In the Minds of Monkeys." *Natural History,* September 1990, pp. 38–47.

COSMIDES, L., and J. TOOBY. "From Evolution to Behavior: Evolutionary Psychology as the Missing Link." In *The Latest on the Best: Essays on Evolution Optimality,* edited by J. Dupre, 277–293. Cambridge, MA: MIT Press, 1987.

GOULD, J. L. *Ethology: The Mechanisms and Evolution of Behavior.* New York: Norton, 1982.

GRIER, J. W. *Biology of Animal Behavior.* St. Louis, MO: Times Mirror/Mosby, 1984.

PECK, J. R., and M. W. FELDMAN. "Kin Selection and the Evolution of Monogamy." *Science* 240 (1988): 1672–1674.

ROSEN, C. M. "The Eerie World of Reunited Twins." *Discover,* September 1987, pp. 36–42.

TINBERGEN, M. "The Shell Menace." *Natural History,* August/ September 1963, pp. 28–35.

WALCOTT, C. "Show Me the Way You Go Home: Researchers Are Still Asking How Homing Pigeons Do It." *Natural History,* November 1989, pp. 40–46.

A P P E N D I X

Appendix A.1

ATOMIC ORBITALS

See pages 29–30.

Cloudlike regions of electron movement give the atom its volume, shape, and reactivity and hence its chemical properties. Diagrams representing the volume "filled" by the rapidly moving electrons are more accurate than the simpler but more static Bohr models. Electrons arrange themselves in atomic orbitals according to a set of basic principles:

1. Electrons are attracted by the nucleus but repelled by each other; thus, they move about as close to the nucleus as possible while remaining as distantly spaced from each other as they can.

2. As a result of this attraction/repulsion, electrons move about the nucleus at nearly the speed of light in cloudlike regions called orbitals. The electron is so small and fast-moving that there is no way to define exactly where it might be at any given moment (a). Thus, an orbital can be thought of as the volume of space where the electron will be found 90 percent of the time. A maximum of two electrons can coexist in the same orbital. Orbitals have various shapes: spheres, dumbbells, and teardrops (b).

3. The more electrons, the more orbitals are needed for the electrons to stay close to the nucleus but as far as possible from each other. Orbitals occur in energy levels, or zones, at given distances from the nucleus: The farther away, the higher the energy. (This is because it takes more centrifugal energy to pull an electron into an orbit farther out from the strongly attractive, positive nucleus than closer to it.) The first energy level is closest to the nucleus; it contains one orbital, the spherical $1s$ orbital, and can accommodate two electrons (a). The second energy level contains four orbitals, the spherical $2s$ and three dumbbell-shaped $2p$

(a) First energy level

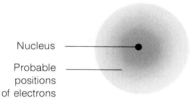

Nucleus

Probable positions of electrons

Spherical orbital, 1s
(can have up to two electrons)

(b) Part of second energy level

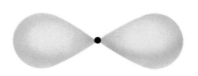

A dumbbell-shaped orbital, 2p,
of second energy level
(can have up to two elections)

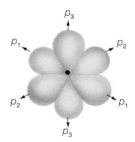

p_3

p_1 p_2

p_2 p_1

p_3

Three-dimensional rendering
of three p orbitals interacting
(each p orbital can have up to two electrons)

(c) Complete second energy level

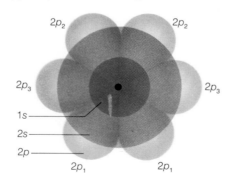

$2p_2$ $2p_2$

$2p_3$ $2p_3$

1s
2s
2p

$2p_1$ $2p_1$

(d) Third energy level

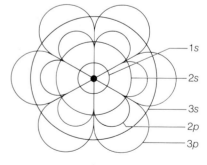

1s

2s

3s

2p

3p

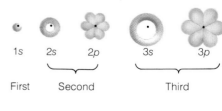

Orbitals:	1s	2s	2p	3s	3p
Energy levels:	First	Second		Third	
Maximum number of electrons:	2	8		8	

orbitals (b, c). Because each orbital can hold up to two electrons, the second energy level can accommodate eight electrons. The third energy level also contains four orbitals, the spherical 3s and three 3p orbitals, which accommodate another eight electrons in total (d). The atoms of heavy elements such as radium (atomic mass 253) and fermium (atomic mass 226) can have as many as seven roughly concentric energy levels.

4. Electrons occupy the orbitals in a predictable way, filling the lowest energy level possible before starting a higher one, and filling simple-shaped orbitals before ones of complex shape (e). Within a given energy level, orbitals of the same shape must have one electron each before any receive the second electron and are thus filled.

As the chart in (e) illustrates, hydrogen, with atomic number 1, has one proton, no neutrons, and one electron. This single electron occupies the 1s orbital, but leaves it unfilled. Carbon, with atomic number 6, has six electrons; two fill the 1s orbital, two fill the 2s orbital, and one each occurs in two of the three 2p orbitals. In accordance with the rules, the lowest energy level is filled first, and in the second energy level, the spherical orbital (a similar shape) is filled before the dumbbell-shaped orbitals (more complex shape), two of which get one electron each. In oxygen, with eight electrons, the 1s and 2s orbitals are filled, and the remaining four electrons are distributed so that each of the three 2p orbitals has one electron before the first 2p orbital gets a second electron. In sodium, the first and second energy levels are filled according to the above principles, and one additional electron orbits alone in the 3s orbital.

(e)

Elements	Atomic Number (Number of Protons)	Number of Neutrons	Electrons in First Level	Electrons in Second Level			Electrons in Third Level		Total Number of Electrons
			1s	2s	$2p_1$	$2p_2$	$2p_3$	3s	$3p_1$	
N	1	0	1							1
C	6	6	2	2	1	1				6
O	8	8	2	2	2	1	1			8
Na	11	12	2	2	2	2	2	1		11

Appendix A.2

MEIOSIS COMPARED TO TWO MITOTIC DIVISIONS

See pages 145–149 and 156–162.

Meiosis and mitosis have major differences in eventual chromosome count, in the presence or absence of genetic recombination, and in where the process takes place in the body. These differences lead to their contrasting evolutionary consequences. This figure illustrates the major differences:

1. During mitosis, the chromosome number remains unchanged; during meiosis, the chromosome number is halved.

2. During mitosis, chromosomes are dupli-

cated exactly; chromosomes are also duplicated exactly in meiosis, but after duplication, they are altered. During mitosis, the chromosomes of the daughter cell are identical to those of the parent cell; during meiosis, independent assortment and crossing over produce new chromosome combinations.

3. Mitosis occurs in many cell types throughout the organism; meiosis occurs only in the gamete-producing reproductive organs.

Small variations in chromosome behavior

account for differences (1) and (2): First, homologous chromosomes align independently during mitosis but align as pairs during meiosis; and second, DNA synthesis always occurs during mitotic interphase, but it does not take place during interphase II of meiosis. Because of these variations, mitosis and meiosis play very different but equally important roles in the life cycles of sexually reproducing eukaryotic organisms.

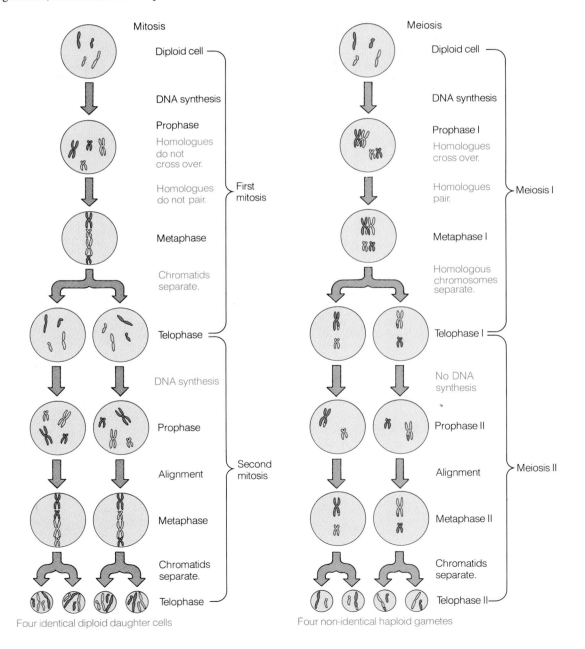

Four identical diploid daughter cells   Four non-identical haploid gametes

Appendix A.3

AIDS: YOU CAN PROTECT YOURSELF FROM INFECTION

See pages 464–465 and 475–476.

Some personal measures are adequate to safely protect yourself and others from infection by the AIDS virus and its complications. Among these are:

- If you have been involved in any of the high-risk sexual activities described below* or have injected illicit intravenous drugs into your body, you should have a blood test to see if you have been infected with the AIDS virus.
- If your test is positive or if you engage in high-risk activities and choose not to have a test, you should tell your sexual partner. If you jointly decide to have sex, you must protect your partner by always using a rubber (condom) during (start to finish) sexual intercourse (vagina or rectum).
- If your partner has a positive blood test showing that he/she has been infected with the AIDS virus or you suspect that he/she has been exposed by previous heterosexual or homosexual behavior or use of intravenous drugs with shared needles and syringes, a rubber (condom) should always be used during (start to finish) sexual intercourse (vagina or rectum).
- If you or your partner is at high risk, avoid mouth contact with the penis, vagina, or rectum.
- Avoid all sexual activities which could cause cuts or tears in the linings of the rectum, vagina, or penis.
- Single teenage girls have been warned that pregnancy and contracting sexually transmitted diseases can be the result of only one act of sexual intercourse. They have been taught to say no to sex. They have been taught to say no to drugs. By saying no to sex and drugs, they can avoid AIDS, which can *kill* them! The same is true for teenage boys who should also not have rectal intercourse with other males. It may result in AIDS.
- Do not have sex with prostitutes. Infected male and female prostitutes are frequently also intravenous drug abusers; therefore, they may infect clients by sexual intercourse and other intravenous drug abusers by sharing their intravenous drug equipment. Female prostitutes also can infect their unborn babies.

Intravenous Drug Users

Drug abusers who inject drugs into their veins are another population group at high risk and with high rates of infection by the AIDS virus. Users of intravenous drugs make up 25 percent of the cases of AIDS throughout the country. The AIDS virus is carried in contaminated blood left in the needle, syringe, or other drug-related implements and the virus is injected into the new victim by reusing dirty syringes and needles. Even the smallest amount of infected blood left in a used needle or syringe can contain live AIDS virus to be passed on to the next user of those dirty implements.

No one should shoot up drugs because addiction, poor health, family disruption, emotional disturbances, and death could follow. However, many drug users are addicted to drugs and for one reason or another have not changed their behavior. For these people, the only way not to get AIDS is *to use a clean, previously unused* needle, syringe, or any other implement necessary for the injection of the drug solution.

Hemophilia

Some persons with hemophilia (a blood-clotting disorder that makes them subject to bleeding) have been infected with the AIDS virus either through blood transfusion or the use of blood products that help their blood clot. Now that we know how to prepare safe blood products to aid clotting, this is unlikely to happen. This group represents a very small percentage of the cases of AIDS throughout the country.

Blood Transfusion

Currently all blood donors are initially screened and blood is *not* accepted from high-risk individuals. Blood that has been collected for use is tested for the presence of antibody to the AIDS virus. However, some people may have had a blood transfusion prior to March 1985 before we knew how to screen blood for safe transfusion and may have been infected with the AIDS virus. Fortunately there are not now a large number of these cases. With routine testing of blood products, the blood supply for transfusion is now safer than it has ever been with regard to AIDS.

Persons who have engaged in homosexual activities or have shot street drugs within the last 10 years should *never* donate blood.

Mother Can Infect Newborn

If a woman is infected with the AIDS virus and becomes pregnant, she is more likely to develop ARC or classic AIDS, and she can pass the AIDS virus to her unborn child. Approximately one-third of the babies born to AIDS-infected mothers will also be infected with the AIDS virus. Most of the infected babies will eventually develop the disease and die. Several of these babies have been born to wives of hemophiliac men infected with the AIDS virus by way of contaminated blood products. Some babies have also been born to women who became infected with the AIDS virus by bisexual partners who had the virus. Almost all babies with AIDS have been born to women who were intravenous drug users.

From *The Surgeon General's Report on Acquired Immune Deficiency Syndrome,* 1986.

---

*These include multiple sex partners and sexual relations with a person infected with AIDS.

Appendix A.4
GUIDELINES FOR GOOD NUTRITION
See pages 496–504.

---

1. Eat a Variety of Foods Daily

- Fresh fruit and vegetables
- Whole grain breads and cereals
- Milk, cheese, or yogurt
- Lean meat, poultry, fish, and eggs
- Legumes (dried peas and beans)

2. Maintain Ideal Weight

- Increase physical activity.
- Prepare smaller portions.
- Avoid "seconds."
- Eat slowly.
- Eat less fat and fatty foods.
- Eat less sugar and sweets.
- Drink less alcohol.

3. Avoid Too Much Fat, Saturated Fat, and Cholesterol

- Choose lean meat, fish, poultry, or dried beans and peas.
- Limit your use of eggs and organ meats (such as liver).
- Limit your use of butter, cream, margarine, shortening, coconut oil, and foods made from these products.
- Trim excess fat off meat.
- Broil, bake, or boil rather than fry.
- Read labels carefully to determine both amount and types of fat contained in food.

4. Eat Foods with Fiber

- Choose foods that are good sources of fiber, such as whole grain breads and cereals, fruits, vegetables, dried beans and peas, nuts, and seeds.

5. Avoid Too Much Sugar

- Use less of all sugars, including white sugar, brown sugar, raw sugar, honey, and syrups.
- Eat less food containing these sugars, such as candy, soft drinks, ice cream, cakes, and cookies.
- Choose fresh fruit, or fruit canned in its own juice without sugar or in a light syrup.

6. Avoid Too Much Salt (Sodium)

- Learn to enjoy the natural flavors of food.
- Cook with only a small amount of added salt.
- Remove salt shaker from the table.
- Limit the amount of salty foods you eat, such as potato chips, pretzels, salted nuts and popcorn, cheese, luncheon meat, pickles, relish, mustard, ketchup, soy sauce, steak sauce, and garlic salt.
- Read labels carefully to determine if salt, sodium, or MSG (monosodium glutamate) has been added.
- Use fresh seasonings and herbs in place of salt.

7. If You Drink Alcohol, Do So in Moderation

- One or two alcoholic drinks a day are considered moderate—this includes beer, wine, or mixed drinks.

8. Reduce Your Cancer Risk

- Limit saturated fats.
- Eat more fiber.
- Avoid foods containing the sweetener saccharin.
- Avoid foods containing nitrates or nitrites.
- Limit alcohol.
- Get plenty of vitamins A, C, and E by eating dark green vegetables, citrus fruits, whole grains, nuts, seeds, and vegetable oils.
- Avoid moldy nuts, grains, or seeds.
- Reduce consumption of smoked foods or foods fried at high temperature.

These recommendations are not meant to prohibit the use of any specific food item or to prevent you from eating a variety of foods.

Adapted from *Nutrition and Cancer Prevention: A Guide to Food Choices* (Berkeley: Northern California Cancer Program, June 1981).

For a more detailed discussion of terms, consult the chapters shown in parentheses after each entry.

Abdomen The posterior part of an arthropod's body; in vertebrates, the abdomen lies between the thorax and the pelvic girdle. (18, 24)

Abiotic (Gr. *a,* without + *bios,* life) Characteristic of the part of an organism's environment made up of nonliving physical surroundings. (34)

Abscisic acid A plant hormone that promotes dormancy. (31)

Abscission (L. *ab,* away, off + *scissio,* dividing) In vascular plants, the dropping of leaves or fruit at a particular time of year, usually the end of the growing season. (31)

Absorption spectrum The wavelengths of light absorbed by a pigment. (6)

Accessory organ One of several organs that aids digestion, including salivary glands, liver, gallbladder, and pancreas. (24)

Acid A proton donor; any substance that gives off hydrogen ions when dissolved in water, causing an increase in the concentration of hydrogen ions. Acidity is measured on a pH scale, with acids having a pH less than 7; the opposite of a base. (2)

ACTH (adrenocorticotropic hormone) A hormone produced by the pituitary in response to stress. ACTH acts on the cortex of the adrenal gland. (26)

Actin (Gr. *actis,* a ray) A major component of microfilaments in contractile cells such as muscle and other cells. (3, 29)

Action potential A temporary all-or-nothing reversal of the electrical charge across a cell membrane, which occurs when a stimulus of sufficient intensity strikes a neuron. (27)

Activation energy The minimum amount of energy that molecules must have in order to undergo a chemical reaction. (4)

Active site A groove or a pocket on an enzyme's surface to which reactants bind. This binding lowers the activation energy required for a particular chemical reaction; thus, the enzyme speeds the reaction. (4)

Active transport Movement of substances against a concentration gradient requiring the expenditure of energy by the cell. (4)

Adaptation (L. *adaptere,* to fit) A particular form of behavior, structure, or physiological process that makes an organism better able to survive and reproduce in a particular environment. (1, 33)

Adaptive radiation Evolutionary divergence of a single ancestral group into a variety of forms adapted to different resources or habitat. (19, 33)

Addiction Physical dependence upon a substance such as alcohol or other drugs, characterized by tolerance of increasingly larger doses, craving the drug, and withdrawal symptoms when use is discontinued. (27)

Adenosine phosphate A class of molecules that are based on modified nucleotide monomers containing the nitrogenous base adenine, the sugar ribose, and a phosphate group. ADP is adenosine diphosphate; ATP is adenosine triphosphate, the major source of chemical energy in metabolism. (2)

Adhesion The tendency of unlike molecules to cling to each other. (2, 32)

Adrenal gland (L. *ad,* to + *renes,* kidney) A hormone-producing endocrine gland located on top of the vertebrate kidney; secretes hormones mainly involved in the body's response to stress, such as epinephrine (adrenaline) and norepinephrine. (26)

Adventitious root A new root that arises from an aboveground structure on a plant. (30)

Aerobic (Gr. *aer,* air + *bios,* life) Conditions where oxygen is present at high levels; also, a biological process that can occur in the presence of oxygen. (5, 29)

Aerobic respiration Oxygen-requiring pathways of fuel breakdown that release energy; takes place within mitochondria in eukaryotic cells. (3, 5)

Age structure The number of individuals in each age group of a given population. (34)

Aging A progressive decline in the maximum functional level of individual cells and whole organs that occurs over time. (14)

Agnathan A jawless fish. (19)

Agricultural revolution The transition of a group of people from an often nomadic hunter-gatherer way of life to a usually more settled life dependent on raising crops, such as wheat or corn, and on livestock. It was under way in the Middle East by 8000 years ago. (34)

Air capillary In insects, the fine, narrow end of an air tube (trachea) through which air flows during respiration. (23)

Air sac A thin-walled extension of the lungs; in birds, air flows through lungs and air sacs during respiration. (19)

Albinism Lack of pigment in body parts of a plant or animal that are normally colored. Albinism in humans is inherited as a recessive. (12)

Aldosterone A steroid hormone released by the cortex of the adrenal gland that regulates salt reabsorption in the kidney. (25)

Alimentary canal A system of tubes and chambers associated with the digestive system; includes the mouth, esophagus, stomach, and intestines. (24)

Alkaline Basic, or having a pH greater than 7. (2)

Allele One of the alternative forms of a gene. (8)

Allopatric speciation The divergence of new species as a result of geographical separation of populations of the same original species. (33)

Allosteric enzyme An enzyme that has two binding sites: a catalytic site and a regulatory site that controls the activity of the enzyme. (5)

Alpha cell An endocrine cell of the pancreas that secretes the hormone glucagon. (26)

Altruism A behavior by one individual that benefits another even if the individual performing the act reduces its own fitness; e.g., a bird giving a warning cry that tells other birds of an approaching predator increases their chances of escape while drawing attention to itself and increases its own chances of being eaten. (38)

Alveolus (L. small cavity) A thin-walled, saclike structure in the vertebrate lung where gas exchange takes place. Each lung contains thousands of alveoli. (23)

Amine hormone A hormone that contains an amino group. (26)

Amino acid A molecule consisting of an amino group (NH_2), an acid group (COOH), and a side group (RCH), where R represents the group of atoms that differs in each amino acid. The 20 amino acids are the basic building blocks of proteins. (2, 10)

Amniocentesis A procedure for obtaining fetal cells from amniotic fluid for the diagnosis of genetic and a few other abnormalities. (12)

Amniote Refers to an egg in which the embryo is encased in water and a series of membranes. Reptiles, birds, and mammals have amniote eggs. (19)

Amniotic cavity A fluid-filled cavity surrounding the developing embryo of reptiles, birds, and mammals that keeps the embryo moist and cushions it from blows. (14)

Amphibian A cold-blooded vertebrate that starts life as an aquatic larva, breathing through gills, and metamorphoses into an air-breathing adult; includes frogs, toads, newts, and salamanders. (19)

Amyloplasts Dense starch-storing granules in plant cells. (31)

Anabolism The building up of complex molecules from simpler ones, usually requiring an input of energy. (4)

Anaerobic (Gr. *an,* without + *aer,* air + *bios,* life) Conditions where oxygen levels are low or absent. Also, biological reactions that do not require oxygen. (5, 15)

Anaphase A period during nuclear division when the chromosomes move toward the poles of the cell. (7)

Anatomy The science of biological structure. (20)

Angiosperm A flowering plant. (17, 30)

Angiotensin A protein hormone that causes blood vessels to constrict and thus increase blood pressure. (25)

Animal A member of the kingdom Animalia; a multicellular organism that can generally move voluntarily, that actively acquires food and often digests it internally, that usually has irritability due to nerve cells and limited growth. (18)

Animalia The kingdom of animals. (15)

Annual plant (L. *annos,* year) A plant that completes its life cycle within one year and then dies. (30)

Antagonistic muscle pair Two muscles that move the same object, such as a limb, in opposite directions. (29)

Antenna One of the paired appendages on the head of arthropods that senses odors and vibrations. (18)

Antenna complex A cluster of pigment molecules acting together during the harvest of light energy in photosynthesis. (6)

Anther In flowers, the terminal portion of a stamen that contains grains of pollen in pollen sacs. (17)

Antheridium The male sperm-producing organ of seedless plants. (17)

Anthropoid A member of the suborder Anthropoidea, order Primates; includes humans, apes, and New and Old World monkeys. (19)

Antibody A protein produced by the immune system in response to the entry of specific antigens (foreign substances) into the body. A specific antibody binds to the antigen that stimulated the immune response. (2, 22)

Anticodon The three bases in a tRNA molecule that are complementary to three bases of a specific codon in mRNA. (10)

Antidiuretic hormone ADH, a posterior pituitary gland hormone that regulates the amount of water passing from the kidney as urine. (25)

Antigen (Gr. *anti,* against + *genos,* origin) Any substance, including toxins, foreign proteins, and bacteria, which when introduced into a vertebrate animal causes antibodies to form. (22)

Aorta In a vertebrate animal, the main artery leading directly from the left ventricle of the heart; supplies blood to most of the animal's body. (21)

Apical dominance The inhibition of lateral buds or meristems by the apical meristem of a plant. (31)

Apical meristem The perpetual growth zone at the tip of a plant's roots and shoots. (30)

Appendicular skeleton That part of the vertebrate skeleton that supports the appendages (arms and legs). (29)

Archegonium The egg-producing organ of ferns and bryophytes, including mosses. (17)

Arteriole A thinner-walled, smaller branch of an artery, which carries blood from arteries to the capillaries in the tissues. (21)

Artery A large blood vessel with thick, multilayered muscular walls, which carries blood from the heart to the body. (21)

Ascending limb In the vertebrate kidney, the arm of the loop of Henle leading back to the cortex. (25)

Ascocarp In certain fungi, a cup-shaped structure that holds sacs (asci) in which nuclear fusion and meiosis occur and spores are formed. (17)

Ascus (plural **asci**) In some fungi, a small sac in which sexually produced spores develop. (17)

Asexual reproduction A type of reproduction in which new individuals arise directly from only one parent. (7)

Atom (Gr. *atomos,* indivisible) The smallest particle into which an element can be broken down and still retain the properties of that element. (2)

Atomic mass The mass of an atom; it is approximately equal to the combined number of protons and neutrons in the nucleus of an atom. (2)

Atomic nucleus The central core of an atom, containing protons and neutrons. (2)

Atomic number The number of protons in the nucleus of an atom, unique to each chemical element. (2)

Atomic theory Set forth by English chemist John Dalton in the early 1820s, the theory that every element is composed of identical particles, called atoms, and that chemical reactions involve rearrangements of atoms, but not their destruction or creation. (2)

ATP Adenosine triphosphate, a molecule consisting of adenine, ribose sugar, and three phosphate groups. ATP can transfer energy from one molecule to another. ATP hydrolyzes to form ADP, releasing energy in the process. (4)

Atrioventricular (AV) node A bundle of modified cardiac muscle cells between the heart's atria and ventricles that relays electrical activity to the ventricles. (21)

Atrium (L. hallway) A chamber of the heart that receives blood from the veins. In mammals, birds, amphibians, and reptiles, the heart has two atria, the right receiving deoxygenated blood from the body and the left receiving oxygenated blood from the lungs. The atria pass blood into the heart's ventricles, which in turn pump it to the body. (21)

Autonomic nervous system Motor neurons of the nervous system that regulate the heart, glands, and smooth muscles in the digestive and circulatory systems. (28)

Autosome A chromosome other than a sex chromosome. (8)

Autotroph (Gr. *auto,* self + *trophos,* feeder) An organism, such as a plant, that can manufacture its own food. (5, 15)

Auxin A plant hormone that stimulates cell elongation, among other activities. (31)

Axial skeleton That part of the vertebrate skeleton that supports the main body axis, consisting of the skull, vertebral column (backbone), sternum, and ribs. (29)

Axon The portion of a nerve cell that carries the impulse away from the cell body. (26, 27)

Bacillus (plural **bacilli**) (L. *baculum,* a rod) A bacterial cell with a rod shape. (16)

Bacteriophage A virus that infects bacterial cells. (9)

Bark Protective, corky tissue of dead cells present on the outside of older stems and roots of woody plants. (30)

Barr body The inactive *X* chromosome in female mammals, visible as a dark-staining body in the nucleus. (12)

Bartholin's glands Lubricating glands surrounding a woman's vagina. (14)

Base Any substance that accepts hydrogen ions when dissolved in water. Basic solutions have a pH greater than 7. (2)

Basidiocarp The reproductive portion of a basidiomycote fungus. (17)

Basidiospore The spore produced by a type of fungus called a basidiomycote. (17)

Basidium (plural **basidia**) In certain fungi, a club-shaped structure that holds the basidiospores. (17)

Basophil A type of white blood cell involved

in inflammation that releases an anticlotting agent at the site of an injury and produces the chemical substance histamine, which delays the spread of invading microorganisms. (21)

B cell (B lymphocyte) A type of white blood cell that makes and secretes antibodies in response to foreign substances (antigens). (22)

Benign Characteristic of a tumor that continues to grow but stays localized and tightly packed. (13)

Beta cell A pancreatic endocrine cell that secretes the hormone insulin. (26)

Beta- (β-) endorphin A naturally occurring protein of the brain and pituitary that has the pain-killing effect of opiate drugs. (26)

Biennial A plant that completes its life cycle in two years and then dies. (30)

Bilateral symmetry A body plan in which the right and left sides of the body are mirror images of each other. (18)

Bile A bitter, alkaline, yellow fluid produced by the liver, stored in the gallbladder, and released into the small intestine that aids in the digestion and absorption of fats. (24)

Binary fission Asexual reproduction by division of a cell or body into two equivalent parts. (7)

Binomial system of nomenclature The system, developed by the Swedish biologist Linnaeus, of giving each type of organism two names, a genus name and a species name. For example, humans are *Homo sapiens*. (15)

Biogeography The study of the geographical location of living organisms. (33)

Biological magnification The tendency for toxic substances to increase in concentration as they move up the food chain. (36)

Biology (Gr. *bios,* life) The scientific study of life. (1)

Biomass The total dry weight of organic matter present at a particular trophic level in a biological community. (36)

Biome A major ecological community type, such as a desert, rain forest, or grassland. (37)

Biosphere (Gr. *bios,* life + *sphaira,* sphere) That part of the planet that supports life; includes the atmosphere, water, and the outer few meters of the earth's crust. (1, 34, 37)

Biotic Characteristic of the part of an organism's environment made up of other living things. (34)

Biotic potential An organism's capacity for reproduction under ideal conditions of growth and survival. (34)

Bivalve A mollusk whose body is enclosed within two valves or shells; includes mussels and clams. (18)

Bladder A storage sac; for example, for urine. (25)

Blade The broad, flat portion of a leaf. (30)

Blastocoel In early embryonic development, a cavity that develops in the mass of cells formed during the blastula stage. (13)

Blastocyst A stage in early embryonic development of mammals consisting of a thin-walled hollow sphere of cells, called the trophoblast, and an inner mass of cells at one side that are destined to become the embryo. (13)

Blastodisc A disc-shaped layer of cells produced by cleavage in a large yolky egg like that of a bird or reptile. (13)

Blastomere A cell that exists in an embryo during the cleavage or blastula stage. (13)

Blastula A stage of embryonic development, at or near the end of cleavage and immediately preceding gastrulation, generally consisting of a hollow ball of cells. (13)

Blubber A thick layer of fat acting as insulation in such mammals as whales, porpoises, and seals. (19)

Body hair Threadlike outgrowths from the skin of mammals. (19)

Bohr effect The phenomenon in which hemoglobin releases oxygen more easily in an acidic environment, as when muscles are working hard. (23)

Bolus A soft, round mass of chewed food, shaped by the tongue, suitable for swallowing. (24)

Bone The main supporting tissue of vertebrates, composed of a matrix of collagen hardened by calcium phosphate. (19, 29)

Book lung A chamber with leaflike plates that exchanges gases, found in spiders and other arthropods. (18)

Bouton The knoblike end of an axon at a synapse. (26)

Bowman's capsule In the vertebrate kidney, the bulbous unit of the nephron that surrounds the blood capillaries of the glomerulus; the region that filters water and solutes from the blood. (25)

Brain stem The most posterior portion of the vertebrate brain, relaying messages to and from the spinal cord and consisting of the medulla, pons, and midbrain. (28)

Bronchiole One of the thousands of small branches from the bronchi that lead to the alveoli in the lungs. (23)

Bronchus (plural **bronchi**) One of two hollow passageways branching off the trachea and entering the lungs. (23)

Bryophytes The division of the plant kingdom comprising mosses, liverworts, and hornworts. (17)

Bud [1] An undeveloped shoot or flower. (30) [2] A daughter cell derived by mitosis in yeast. (17)

Budding A type of asexual reproduction in which some cells differentiate and grow outward from the parent body, eventually breaking off and forming a new individual; occurs in sea anemones and corals. (7, 18)

Buffer A substance that resists changes in pH when acids or bases are added to a chemical solution. (2)

Bulbourethral gland A gland at the base of the human penis that secretes a lubricating substance. (14)

Bulk flow The overall movement of a fluid down a gradient of pressure or gravity. (4)

Bundle-sheath cells Cells in certain plants that surround a vascular bundle. (6)

C_3 plant A plant in which the first stable product of carbon fixation in the light-independent reactions of photosynthesis is a three-carbon compound. Most plants are C_3 plants. (6)

C_4 plant A plant, such as corn, in which the first stable product of carbon fixation is a four-carbon compound. Many C_4 plants are adapted to hot, dry weather. (6)

Calcitonin A thyroid hormone that lowers calcium in the blood, and promotes its deposition in the bones. (26)

Calorie A kilocalorie; the amount of energy needed to raise the temperature of one liter of water 1°C. (24)

Calvin-Benson cycle A series of reactions in the light-independent phase of photosynthesis, during which chloroplasts fix carbon dioxide, form carbohydrates, and regenerate a starting material. (6)

Camouflage Body shapes, colors, or patterns that enable an organism to blend in with its environment and remain concealed from danger. (35)

CAM plant A plant, such as ice plant, in which the C_4 pathway of carbon fixation and the light-independent reactions are in the same cell but occur at different times of day; CAM stands for crassulacean acid metabolism. (6)

Cancer A group of diseases characterized by uncontrolled cell division. (7)

Capillarity The tendency of a liquid substance to move upward against the pull of gravity through a narrow space. (2)

Capillary Tiny blood vessels with walls one cell thick that permeate the tissues and organs of the body. Substances such as oxygen diffuse out of capillaries to the tissues, and waste products diffuse into capillaries from the tissues. (21, 23)

Capillary bed A network of capillaries that links arterial and venous blood vessels. (21)

Capsid The protein coat that encases a virus. (16)

Carapace The exoskeleton covering the cephalothorax of many arthropods. Also refers to the tough outer coverings of the turtle and armadillo. (18, 19)

Carbohydrate (L. *carbo,* charcoal + *hydros,* water) An organic compound composed of a chain or ring of carbon atoms to which hydrogen and oxygen atoms are attached in the ratio of approximately two hydrogens to one oxygen. Carbohydrates include sugars and starch. (2)

Carbon cycle The global flow of carbon atoms from plants through animals to the atmosphere, soil, and water and back to plants. (6, 36)

Carbon fixation The sequence of reactions that converts atmospheric gaseous carbon dioxide into carbohydrates. (6)

Carcinogen A cancer-causing substance. (10)

Cardiac cycle The contraction/relaxation (systolic/diastolic) sequence of atria and ventricles that makes up a single heartbeat. (21)

Cardiac muscle The specialized muscle tissue of the heart. (29)

Carnivore An animal that eats the flesh of other animals. (24)

Carotenoid pigments Certain yellow and orange pigments in plant plastids that can function as accessory pigments in photosynthesis; such pigments are responsible for the color in carrots and in autumn leaves. (6)

Carpel The innermost ring of structures in a flower. A carpel encloses one or more ovules, and together, a flower's carpels form the pistil. (30)

Carrier [1] In genetics, a heterozygous individual not expressing a recessive trait but capable of passing it on to her or his offspring. (12) [2] In biochemistry, a substance, often a protein, that transports another substance. (4)

Carrier-facilitated diffusion The process by which large molecules are carried across cell membranes by carrier proteins. (4)

Carrying capacity The density at which growth of a population ceases due to the limitation imposed by resources. (34)

Casparian strip A water-resistant belt encircling each endodermal cell of a plant, ensuring that any water or minerals entering the plant must flow through the cytoplasm of the endodermal cells. (30)

Caste In a social insect colony, one of several types of individuals; for example, a worker, soldier, or queen in a termite colony. (18)

Catabolism The breaking down of complex molecules to simpler products, with an accompanying release of energy. (4)

Catalyst A substance that speeds up the rate of a chemical reaction by lowering the activation energy without itself being permanently changed or used up during the reaction. (4)

Causality The scientific principle stating that the occurrence of events is due to natural causes. (1)

Cell The basic unit of life; cells are bounded by a lipid-containing membrane and are generally capable of independent reproduction. (3)

Cell cycle The events that take place within the cell between one cell division and the next. (7)

Cell division The splitting of one cell into two cells. (7)

Cell-mediated immune response That part of the immune response in which T cells directly attack invaders. (22)

Cell membrane A biological envelope surrounding a cell; also called the *plasma membrane.* (3)

Cell plate In plant cells, a partition that arises during late cell division from vesicles at the center of the cell and that eventually separates the two daughter cells. (7)

Cell theory The biological doctrine stating that all living things are composed of cells; cells are the basic living units within organisms, and the chemical reactions of life occur within cells; all cells arise from preexisting cells. (3)

Cellular slime mold A type of funguslike protist, usually existing as free-living amoeba-like cells, but aggregating into a multicellular fruiting body before producing reproductive spores. (16)

Cellulose (L. *cellula,* little cell) The chief constituent of the cell wall in green plants, some algae, and a few other organisms. Cellulose is an insoluble polysaccharide composed of straight chains of glucose molecules. (2)

Cell wall A fairly rigid structure that encloses prokaryotic, fungal, and plant cells. (3)

Central cavity A large opening, as in the body of a sponge. (18)

Central nervous system (CNS) The part of the nervous system consisting of the brain and spinal cord, performing the most complex nervous system functions. (28)

Central vacuole A large, single-membraned organelle in plant cells that contains water and storage products. (3)

Centromere The point on the chromosome where the spindle attaches and also where the two chromatids are joined. (7)

Cephalization (Gr. *kephale,* little head) A type of animal body plan or organization in which one end contains a nerve-rich region and functions as a head. (18)

Cephalopod A member of the class Cephalopoda, phylum Mollusca, including squid, octopus, and chambered nautilus. (18)

Cephalothorax The fused head and thorax of many arthropods. (18)

Cerebellum The hindbrain region of the vertebrate brain between the cerebrum and pons; it integrates information about body position and motion, coordinates muscular activities, and maintains equilibrium. (28)

Cerebral cortex The highly convoluted surface layers of the cerebrum, containing about 90 percent of a human brain's cell bodies. It is well developed only in mammals and particularly prominent in humans and is the region involved with conscious sensations and voluntary muscular activity. (28)

Cerebrum (L. brain) The forebrain of the vertebrate brain; the largest part of the human brain. It coordinates and processes sensory input and controls motor responses. (28)

Cervix The narrow necklike junction of the uterus and vagina. (14)

Chaparral A biome that borders deserts and grasslands, characterized by low woody shrubs that are often fragrant and have generally thick, waxy, evergreen leaves. (37)

Chelicera (plural **chelicerae**) (Gr. *chele,* claw + *keras,* horn) One of the first pair of legs of an arachnid (spiders and their close relatives), which possesses poison fangs that kill prey. (18)

Chemical reaction The making or breaking of chemical bonds between atoms or molecules. (4)

Chemiosmotic coupling The coupling of energy capture (ATP production) to gradients of hydrogen ion concentration during electron transfer in mitochondria. (5)

Chemoautotroph An organism that derives energy from a simple inorganic reaction. (15, 16)

Childhood The period of human development from age 2 to about age 12. (14)

Chitin A complex nitrogen-containing polysaccharide that forms the cell walls of certain fungi, the major component of the exoskeleton of insects and some other arthropods, and the cuticle of some other invertebrates. (2)

Chlorophyll (Gr. *chloros*, green + *phyllon*, leaf) Light-trapping pigment molecules that act as electron donors during photosynthesis. (3, 6)

Chloroplast An organelle present in algae and plant cells that contains chlorophyll and is involved in photosynthesis. (3, 6)

Cholecystokinin A hormone secreted by the small intestine that triggers the pancreas to release protein-digesting enzymes. (24)

Chondrichthyes A class of the phylum Chordata that includes cartilaginous fish, such as sharks and rays. (19)

Chordate A member of the animal phylum Chordata, including vertebrates and their relatives. (19)

Chorion A fluid-filled sac that surrounds the embryos of reptiles, birds, and mammals. In placental mammals it gives rise to part of the placenta and produces hCG, a hormone that maintains pregnancy by preventing menstruation. (14)

Chorionic villus sampling A procedure for removing fetal cells from the developing placenta for the diagnosis of genetic defects. (12)

Chromatid A daughter strand of a duplicated chromosome. Duplication of a chromosome gives rise to two chromatids joined together at the centromere. (7)

Chromatin The substance of a chromosome, composed of DNA and proteins. (9)

Chromoplast In plant cells, an organelle that stores yellow, orange, and red pigments, giving color to many fruits and flowers. (3)

Chromosomal mutation A genetic change that affects large regions of chromosomes or entire chromosomes. (10)

Chromosome A self-duplicating body in the cell nucleus made up of DNA and proteins and containing genetic information. It contains genes and a centromere region. Chromosomes are transmitted from one generation to the next via gametes. A human cell contains 23 pairs of chromosomes. In a prokaryote, the DNA circle that contains the cell's genetic information. (3, 7, 9)

Chromosome translocation An instance in which a part of one chromosome becomes attached to a different chromosome. (12)

Chyme The semifluid contents of the stomach consisting of partially digested food and gastric secretions. (24)

Ciliate A protozoan whose cells have rows of cilia that are used in locomotion and in sweeping food particles into the mouth. (16)

Cilium (plural **cilia**) (L. eyelash) A short, centriole-based, hairlike organelle. Rows of cilia propel certain protists. Cilia also aid the movement of substances across epithelial surfaces of animal cells. (3, 6)

Circadian (L. *circa*, about + *dies*, day) Regular rhythms of activity that occur on about a 24-hour cycle. (26)

Circulatory system An organ system, generally consisting of a heart, blood vessels, and blood, that transports substances around the body of an animal. (18, 20, 21)

Class A taxonomic group comprised of members of similar orders. (15)

Classical conditioning A type of learning in which an initially neutral stimulus (like the sound of a bell) evokes a given response (like salivation) by being paired repeatedly with a different stimulus that normally evokes that response (like the presence of food). (38)

Cleavage The early, rapid series of mitotic divisions of a fertilized egg resulting in a hollow sphere of cells known as the blastula. (13)

Climate The accumulation of seasonal weather patterns in an area over a long period of time. (37)

Climax community The most stable community in a habitat and one which tends to persist in the absence of a disturbance. (35)

Clitoris An erectile structure in the female, equivalent to the male penis, which is sensitive to sexual stimulation. (14)

Clone All cells derived from a single parent cell. A replica of a DNA sequence produced by recombinant DNA technologies. (11)

Clumped distribution A pattern in which several members of a population occur close together but are separated by a long distance from other groups. (34)

Cnidarian A member of the phylum Cnidaria, which includes hydras, jellyfish, sea anemones, and corals. (18)

Coagulation The formation of a blood clot. (21)

Coccus (plural **cocci**) A bacterial cell with a spherical shape. (16)

Cochlea A spiral tube in the inner ear that contains sensory cells involved in detecting sound and analyzing pitch. (28)

Codominance The genetic situation in which both alleles in a heterozygotic individual are fully expressed in the phenotype; this is a characteristic of many blood groups. (8)

Codon Three adjacent nucleotide bases in DNA or mRNA that code for a single amino acid. (10)

Coelacanth A type of lobe-finned fish. (19)

Coelom (Gr. *koiloma*, a hollow) The main body cavity of many animals, formed between layers of mesoderm, in which internal organs are suspended. (18)

Coenzyme A nonprotein organic molecule that aids in enzyme-catalyzed reactions, often by acting as an electron donor or acceptor. (4)

Coevolution The evolution of one species in concert with another as a result of their interrelationship within a biological community. (17, 33)

Cohesion The tendency of like molecules to cling to each other. (2)

Cohesion-adhesion tension theory An explanation for the way water is transported upward from plant roots through stem and leaves: Adhesion of water molecules to plant vessel surfaces, cohesion of water molecules with each other, and evaporation of molecules from the water chain moving upward in xylem create the tension on the entire chain that accounts for transpirational pull, also called the *transpirational pull theory*. (32)

Coitus The act of sexual intercourse. (14)

Collagen A fibrous protein found extensively in connective tissue and bone. Collagen also occurs in some invertebrates, as in the cuticle of nematode worms. (3)

Collecting duct Any duct that drains an organ; in the kidney, ducts that drain the renal pelvis of the kidney and carry urine to the ureters. (25)

Collenchyma Plant tissue providing support to young, actively growing structures; it consists of closely packed cells with thickened cell walls, especially at the corners. (30)

Colon The last portion of the large intestine, the wide part of the alimentary canal that leads to the rectum. (24)

Commensalism A relationship between two species in which one species benefits and the other suffers no apparent harm. (35)

Committed In embryology, a characteristic of cells whose developmental fate has become restricted to just one or a few cell types. (13)

Communication A signal produced by one individual that alters the behavior of a recipient individual and generally improves both participants' biological fitness. (38)

Community Two or more populations of different interacting species occupying the same area. (34, 35)

Companion cell In a flowering plant, a specialized cell associated with sieve tube members in phloem. (30)

Competitive exclusion A situation in which

one species eliminates another through competition. (35)

Complementary base pairing In nucleic acids, the hydrogen bonding of adenine with thymine or uracil, and guanine with cytosine; it holds two strands of DNA together, and different parts of RNA molecules in specific shapes and is fundamental to genetic replication, genetic expression, and genetic recombination. (9)

Complement proteins In the immune system, proteins that bind to antibody–invader cell complexes and perforate the invader's membrane, killing the invader. (22)

Compound A substance made up of atoms of two or more elements chemically combined in a fixed ratio. (2)

Condensation The formation of a covalent bond between two smaller molecules in such a way that a water molecule is released and a larger molecule is formed. (2)

Cone One of the two types of photoreceptor (light-sensitive) cells in the retina of the vertebrate eye; cones are sensitive to high levels of light and to color. (*See also* Rod.) (28)

Coniferous forest A biome dominated by conifer trees and usually characterized by cold, snowy winters and short summers. (37)

Conjugation In some prokaryotes, the temporary union of two unicellular organisms of different mating strains, during which time the genetic material is transferred from one to the other. (16)

Connective tissue Animal tissue that connects or surrounds other tissues and whose cells are embedded in a collagen matrix; bones and tendons are mainly connective tissue. (3, 20)

Consumer In ecology, an organism that eats other organisms. (36)

Contact inhibition The process by which cells stop growing and moving when they come into contact with other cells. (7)

Contraception The prevention of conception. (14)

Contractile vacuole In protists and certain animals, an organelle that pumps excess water out of the cell. (3)

Control A check of a scientific experiment based on keeping all factors the same except for the one in question. (1)

Convergent evolution Evolution of similar characteristics in two or more unrelated taxa; often found in organisms living in similar environments. (33)

Copulation (L. *copulare,* to couple) The process by which the male deposits sperm inside the female's body. (13)

Cork An external secondary plant tissue impermeable to water and gases. (30)

Cork cambium A type of lateral meristem that produces cork-forming cells. (30)

Cornea The transparent outer portion of the vertebrate eye, through which light passes. (28)

Corpus luteum The group of follicle cells left behind in the ovary after an egg has been released. If the egg is fertilized, the corpus luteum remains active in the ovary, producing hormones that help to maintain the pregnancy. If the egg is not fertilized, the corpus luteum degenerates. (14)

Cortex [1] The outer surface of an organ. In the kidney, the region where blood filtration takes place. (25) [2] In plants, parenchyma tissue bounded externally by the epidermis and internally by phloem in the stem and pericycle in the root. (30)

Cortisol A steroid hormone secreted by the adrenal cortex that helps an animal respond to stress by speeding the metabolism of sugars, proteins, and fats. (26)

Cotyledon The first leaves of the embryonic plant, they often store food in dicotyledon plants and absorb food in monocotyledon plants. (30)

Countercurrent flow A mechanism in which two different fluids flow in opposite directions and can exchange substances at points of contact along a concentration gradient. (20, 23)

Covalent bond (L. *co,* together + *valere,* sharing) A form of molecular bonding characterized by sharing a pair of electrons between atoms. (2)

Crista (plural **cristae**) Fold(s) formed by the inner membrane of a mitochrondrion. (3)

Critical period A specific developmental stage during which imprinting can occur and after which it cannot. (38)

Crop A specialized region that stores food in the digestive tract of earthworms and other animals. (18)

Crossing over The reciprocal exchange of DNA between homologous chromosomes during meiosis. (7)

Cuticle A waxy or fatty noncellular, waterproof outer layer on epidermal cells in plants and some invertebrates. In parasitic flatworms it prevents the worm from being digested in the gut of the host organism. (17, 18)

Cyanobacterium One of the blue-green algae; a photosynthetic, oxygen-generating and nitrogen-fixing prokaryote. (16)

Cycad A primitive cone-bearing seed plant. (17)

Cyclic phosphorylation Phosphorylation that generates small amounts of ATP during the light-dependent reactions of photosynthesis; involves only photosystem I and the electron transport proteins that link photosystems I and II. (6)

Cytology The study of cells. (8)

Cytokinesis The process of cytoplasmic division accompanying nuclear division. (7)

Cytokinin A growth-promoting plant hormone that regulates cell division, among other things. (31)

Cytoplasm The part of a cell between the outer membrane and the nuclear envelope. (3)

Cytoskeleton Found in the cells of eukaryotes, an internal framework of microtubules, microfilaments, and intermediate filaments that supports and moves the cell and its organelles. (3)

Deciduous (L. *decidere,* to fall off) In plants, the property of shedding leaves at the end of the growing season. (17, 31)

Deductive reasoning Analyzing specific cases on the basis of preestablished general principles. (1)

Demographic transition A changing pattern from high birthrate and high death rate to low birthrate and low death rate. (34)

Dendrite (Gr. *dendron,* a tree) The branching projections of a neuron or nerve cell that transmit nerve impulses to the body of the cell. (27)

Density In relation to population, the number of individuals in a certain amount of space. (34)

Depolarized The loss of electrical polarity in a cell. (27)

Dermal tissue system The protective and waterproofing tissues of a plant, consisting of epidermis and bark. (30)

Descending limb In the vertebrate kidney, the arm of the loop of Henle leading down into the medulla. (25)

Desert A very dry, often barren biome characterized by widely spaced plants with thick waxy leaves and often protective spines. (37)

Desmosome A structure involved in cell adherence. A *belt desmosome* is a ring of actin filaments encircling a cell, helping it adhere to adjacent cells. A *spot desmosome* is a tiny button joining two cells and anchoring intermediate filaments. (3)

Detritivore A consumer organism that obtains energy from dead or waste organic matter. (36)

Detritus A collective term for dead organisms and waste organic matter. (36)

Deuterostome An animal in which the blastophore becomes the anus, while the second opening to develop becomes the mouth; includes echinoderms and chordates. (18)

Development The process by which an offspring increases in size and complexity from a zygote to an adult. (1, 13)

Developmental determinant A chemical in egg cytoplasm that helps direct embryonic development. (13)

Diabetes mellitus A hereditary condition in which insufficient insulin is produced and glucose accumulates in the blood. (26)

Diaphragm The muscle sheet that separates the thoracic cavity from the abdominal cavity and helps draw air into the lungs during breathing. (23)

Diastole The relaxation phase of the heart muscle. (21)

Dicot Short for *dicotyledon,* the larger of the two classes of angiosperms (flowering plants). Dicots are generally characterized by having two cotyledons (seed leaves); floral parts are usually in fours or fives; leaves are typically net-veined. Maple trees and tomato plants are dicots. (17)

Differentiation The process by which a cell becomes specialized for a particular function. (13)

Diffusion (L. *diffundere,* to pour out) The tendency of substances to move from areas of high concentration to areas of low concentration. (4)

Digestion The mechanical and chemical breakdown of food into small molecules that an organism can absorb and use. (24)

Dihybrid cross A mating between two individuals that differ only in two traits. (8)

Dinosaur An extinct giant reptile; the dominant form of land vertebrate during the Jurassic and Cretaceous periods. (19)

Diploid A cell that contains two copies of each type of chromosome in its nucleus (except, perhaps, sex chromosomes). (7)

Directional selection A type of natural selection in which an extreme form of a character is favored over all other forms. (33)

Disaccharide A type of carbohydrate composed of two linked simple sugars. (2)

Disruptive selection A type of natural selection in which the two extreme forms of a trait are favored. (33)

Distal tubule The convoluted tubule in the vertebrate kidney that receives the forming urine after it has passed through the loop of Henle. (25)

Divergent evolution The splitting of a population into two reproductively isolated populations with different alleles accumulating in each one. Over geological time, divergent evolution may lead to speciation. (33)

DNA Deoxyribonucleic acid, the double-helical, ladderlike molecule that stores genetic information in cells and transmits it during reproduction, consists of two very long strands of alternating sugar and phosphate groups. Each "rung" of the ladder is a base (either adenine, guanine, cytosine, or thymine) that extends from the sugar and bonds with a base from the other strand. Genes are made of DNA. (2, 7, 9, 10, 11)

DNA ligase A protein that joins DNA strands together end-to-end during DNA recombination. (10)

Domain A taxonomic group comprised of members of similar kingdoms. (15)

Dominance hierarchy A behavioral phenomenon in which members of a group are ranked from highest to lowest. Ranking is based on past success in aggressive encounters with the other members of the group; the individual winning all fights is the dominant member. Dominant individuals have greater access to food and mates. (38)

Dominant In genetics, an allele or corresponding phenotypic trait that is expressed in the heterozygote. (8)

Dormancy (L. *dormire,* to sleep) A state of reduced physiological activity that occurs in seeds and buds of plants, particularly over the winter. Normal physiological activities such as growth only resume if certain conditions, such as increased temperature, are met. (31)

Double bond A chemical bond created when an atom shares two pairs of electrons with another atom. (2)

Double fertilization In flowering plants, the process in which one sperm from a pollen grain fuses with the small egg cell of the gametophyte, while the second sperm penetrates the adjoining large endosperm cell and fuses with the two polar nuclei. (17)

Double helix The term used to describe the structure of DNA, which resembles a ladder twisted along its long axis. (9)

Down syndrome A genetic condition resulting from having an extra chromosome 21; characterized by mental retardation, heart malformation, and external physical traits. (7)

Drive An organism's physiological state of motivation, influenced by hormones and other internal stimuli. (38)

Drug testing A laboratory technique that checks a person's urine or blood to determine if he or she has taken drugs. (25)

Ecdysone A steroid hormone that regulates the timing of molting of insects and other arthropods. (26)

Echinoderm A member of the phylum Echinodermata, which includes sea stars, brittle stars, sea urchins, sea cucumbers, and sea lilies. (18)

Ecology The scientific study of how organisms interact with their environment and with each other and of the mechanisms that explain the distribution and abundance of organisms. (34)

Ecosystem A community of organisms interacting with a particular environment. (34, 36)

Ectoderm The outer cell layer of an embryo; it gives rise to the skin and nervous system. (13)

Ejaculation The sudden ejection of a fluid from a duct, especially semen from the penis. (14)

Ejaculatory duct The duct connecting the vas deferens and the seminal vesicle in mammals. (14)

Electromagnetic spectrum The full range of electromagnetic radiation in the universe, from highly energetic gamma rays to very low-energy radio waves. (6)

Electron A negatively charged subatomic particle that orbits the nucleus of an atom. The negative charge of an electron is equal in magnitude to the proton's positive charge, but the electron has a much smaller mass. (2)

Electron microscope (EM) An instrument that uses beams of electrons with wavelengths much shorter than visible light to view objects, thus allowing greater magnification. (3)

Electron transport chain The second stage in aerobic respiration in which electrons are passed down a series of molecules, gradually releasing energy that is harvested in the form of ATP. (5, 6)

Electrophoresis A technique used to separate molecules based on their different rates of movement in an electric field. (11)

Element A pure substance that cannot be broken down into simpler substances by chemical means. (2)

Elongation The second main stage of protein synthesis, during which the sequential enzymatic addition of amino acids builds the growing polypeptide chain. (10)

Embryo An organism in the earliest stages of development. In humans this phase lasts from conception to about two months and ends when all the structures of a human have been formed; then the embryo becomes a fetus. (13, 14)

Endergonic A chemical reaction that requires the addition of energy to occur. (4)

Endocrine gland A ductless gland that secretes its hormones into extracellular spaces from where they diffuse into the bloodstream. In vertebrates these glands include the pituitary, adrenal, thyroid, and others. (26)

Endocrine system The collection of endocrine glands of the body. (20)

Endocytosis The process by which a cell membrane invaginates and forms a pocket around a cluster of molecules. This pocket pinches off and forms a vesicle that transports the molecules into the cell. (3)

Endoderm The inner cell layer of an embryo; gives rise to the inner linings of the gut and the organs that branch from it, including the lungs. (13)

Endodermis A unicellular layer of tightly packed cells that surround the vascular tissue of plants. Water must pass through this layer to reach the rest of the plant. (30)

Endometrium The inner membrane lining of the uterus, which thickens in response to estrogens and progesterone and then is shed during menstruation. (14)

Endoplasmic reticulum (ER) A system of membranous tubes, channels, and sacs that form compartments within the cytoplasm of eukaryotic cells; functions in lipid synthesis and in the manufacture of proteins destined for secretion from the cell. (3)

Endoskeleton (Gr. *endos,* within + *skeletos,* hard) An internal supporting structure, such as the bony skeleton of a vertebrate. (18, 29)

Endosperm In flowering plants, triploid nutritive tissue of the developing embryo. (17, 31)

Endospore A heavily encapsulated resting cell formed within many types of bacterial cells during times of environmental stress. (16)

Endosymbiont hypothesis The idea that mitochondria and chloroplasts originated from prokaryotes that fused with a nucleated cell. (15)

Energy The power to perform chemical, mechanical, electrical, or heat-related tasks. (4)

Energy pyramid The energy relationships between different trophic levels in an ecosystem. (36)

Entropy A measure of disorder or randomness in a system. (4)

Enzyme A protein that facilitates chemical reactions by lowering the required activation energy but is not itself permanently altered in the process; also called a *biological catalyst.* (2, 4)

Enzyme-substrate complex In an enzymatic reaction, the unit formed by the binding of the substrate to the active site on the enzyme. (4)

Epicotyl The portion of the plant embryonic axis above the cotyledons but below the next leaf. (30)

Epidermis The outer layer of cells of an organism. (18)

Epididymis A coiled duct leading from the testes to the vas deferens, where sperm accumulate and final sperm maturation takes place. (14)

Epiglottis A flap of tissue just above the larynx that closes during swallowing and prevents food from entering the lungs. (23)

Epinephrine A hormone secreted by the medulla of the adrenal glands, associated with the physiological responses to alarm such as increased concentration of sugar in the blood, raised blood pressure and heart rate, and increased muscle power and resistance to fatigue; also known as *adrenaline.* (26)

Epistasis The interaction between two non-allelic genes in which one gene masks the expression of the other. (8)

Epithelial tissue One of the four main tissue types; covers the body surface and lines the body cavities, ducts, and vessels. (20)

Erectile tissue Spongy tissue within the penis that fills with blood and stiffens during an erection. (14)

Erythrocyte A red blood cell whose main function is transporting oxygen to the tissues. (4, 21)

Esophagus The muscular tube leading from the pharynx to the stomach. (23, 24, 25)

Essential amino acid Any one of the eight amino acids that the human body cannot manufacture and that must be obtained from food. (24)

Essential fatty acid Linoleic acid, the unsaturated fatty acid that cannot be manufactured by the human body and that therefore must be obtained from food. (24)

Estrogen A female sex hormone produced by the ovary; prepares the uterus to receive an embryo in mammals, and causes secondary sex characteristics to develop in mammalian females. (14)

Estuary The area where a river meets an ocean. Estuaries are some of the earth's most productive ecosystems. (37)

Ethology (Gr. *ethos,* habit or custom + *logos,* discourse) The scientific study of animal behavior. (38)

Ethylene A gaseous plant hormone produced by ripening fruits that stimulates ripening in nearby fruits. (31)

Eubacteria The largest, most diverse group of prokaryotes; members of the domain Bacteria; spherical, rodlike, or comma-shaped single-celled organisms. (16)

Eukaryotic cell A cell whose DNA is enclosed in a nucleus and associated with proteins; contains membrane-bound organelles. (3, 16)

Eutrophic An aquatic environment with high nutrient levels, characterized by dense blooms of algae and other aquatic plants. (36)

Evolution Changes in gene frequencies in a population over time. (1, 33)

Excretion The elimination of metabolic waste products from the body; in vertebrates, the main excretory organs are the kidneys. (25)

Exergonic A spontaneously occurring chemical reaction that liberates energy. (4)

Exocrine gland A gland with its own transporting duct that carries its secretions to a particular region of the body; includes the digestive and sweat glands; contrasts with endocrine glands. (26)

Exocytosis The process by which substances are moved out of a cell by cytoplasmic vesicles that merge with the plasma membrane. (3)

Exon The sequence of bases in a eukaryotic gene that ends up in the mature RNA. (10)

Exoskeleton The thick cuticle of arthropods, made of chitin. (18, 29)

Exponential growth Growth of a population without any constraints; hence, the population will grow at an ever-increasing rate. (34)

Extinction The disappearance of an entire taxonomic group. (33)

External fertilization A process taking place in many water-dwelling species, in which eggs and sperm are deposited directly into the water, meet by chance, and fuse. (13)

Extracellular digestion The enzymatic breakdown of nutrient molecules that occurs outside a cell. (24)

Extracellular fluid Fluid within the body but outside the cells. (4)

Extracellular matrix A meshwork of secreted molecules that act as a scaffold and a glue that anchors cells within multicellular organisms. (3)

Eyespot An organelle that is sensitive to light and movement, found in planarians and other organisms. (18)

Facet One of many six-sided units of an arthropod's compound eyes. (18)

Family A taxonomic group comprised of members of similar genera. (15)

Fat An energy storage molecule that contains a glycerol bonded to three fatty acids. Fats in the liquid state are known as oils. (2, 24)

Fatty acid A molecule with a carboxyl group at one end and a long hydrocarbon tail at the other. Fatty acids are components of many lipids. (2)

Fatty acid hormone A hormone derived from straight-chain fatty acids; for example, prostaglandin. (26)

Feather A flat, light, waterproof epidermal structure on a bird; collectively functions as insulation and for flight. (19)

Feces Semisolid undigested waste products, stored in the colon until excreted. (24)

Feedback inhibition The buildup of a metabolic product which in turn inhibits the activity of an allosteric enzyme. As this enzyme is involved in making the original product, the accumulation of the product turns off its own production. (5)

Feedback loop A control system in which the result of a process influences the functioning of the process. (20)

Fermentation Extraction of energy from carbohydrates in the absence of oxygen, generally producing lactic acid or ethyl alcohol as a by-product. (5)

Fertilization The fusion of two haploid gamete nuclei (egg and sperm) which forms a diploid zygote. (7)

Fibrin The activated form of the blood-clotting protein fibrinogen that combines into threads when forming a clot. (21)

Fibrinogen A protein in blood plasma that is converted to fibrin during clotting. (21)

Fibrous root One of a system of many narrow roots in which no one root is more prominent than another. (30)

Filter feeder An organism that strains out and collects food particles suspended in water. (18)

Filtrate In the vertebrate kidney, the fluid in the kidney tubule (nephron) that has been filtered through Bowman's capsule. (25)

Filtration The process in the vertebrate kidney in which blood fluid and small dissolved substances pass into the kidney tubule, but large proteins and blood cells are filtered out and remain in the blood. (25)

Fin A flattened limb found in aquatic animals; used for locomotion. (19)

First filial (F_1) generation In Mendelian genetics, the first generation in the line of descent. (8)

First law of thermodynamics The principle that energy cannot be created or destroyed but can be converted from one form into another, such as potential to kinetic energy, or chemical bond energy to heat. (4)

Fitness Biological fitness; the ability to survive and produce offspring that can in turn reproduce. (33)

Fixed-action pattern In behavioral science, an action that continues to completion even if the stimulus causing the behavior is removed, e.g., a bird that rolls an empty eggshell out of its nest will continue the rolling process to the edge of the nest even if the eggshell is removed during the procedure. (38)

Flagellum (plural **flagella**) Long whiplike organelle protruding from the surface of the cell that either propels the cell acting as a locomotory device, or moves fluids past the cell, becoming a feeding apparatus. (3)

Flame cell A specialized cell containing cilia that functions in excretion in flatworms. (18)

Flatworm A member of the phylum Platyhelminthes, the simplest animal group to display bilateral symmetry and cephalization. (18)

Florigen A plant hormone proposed to stimulate the formation and opening of flowers. (31)

Fluid mosaic model Current model of cell membrane structure in which proteins are embedded in the lipid bilayer and can move freely through the fluidlike lipid structure. (3)

Follicle (L. *folliculus,* a small ball) In the mammalian ovary, an oocyte surrounded by follicular cells. (14)

Follicular cell One of many cells surrounding the oocytes in the ovaries. (14)

Food chain The levels of feeding relationships among organisms in a community; who eats whom. (36)

Food web Complex, interconnected feeding relationships between all the species in a community or ecosystem. (36)

Foot The locomotive organ in mollusks. (18)

Founder effect In evolutionary biology, the principle that individuals founding a new colony carry only a fraction of the total gene pool present in the source population. (33)

Fragile *X* syndrome Mental retardation associated with the breaking of an *X* chromosome near its tip. (12)

Frond The leaflike structure of an individual alga that collects sunlight and produces sugars. Also refers to the large divided leaf on a fern. (17)

Fructose A monosaccharide with the chemical formula $C_6H_{12}O_6$; a sugar found in many fruits. (2)

Fruit In flowering plants, a ripe mature ovary containing seeds. (17)

Fruiting body A spore-producing reproductive structure in many fungi. (16)

Functional group A group of atoms that confers specific behavior to the (usually larger) molecules to which they are attached. (2)

Fundamental niche The potential range of all environmental conditions under which an organism can thrive. (35)

Fungi The kingdom comprising multicellular heterotrophs such as mushrooms or molds; secrete digestive enzymes that decompose other biological tissues. (15, 17)

Fur Dense hair covering the whole body of many mammals. (19)

Furrow In cell division, a dent in the cell surface created by constriction of the contractile ring; the furrow deepens and eventually squeezes the cell in two. (7)

Galactose A monosaccharide with the chemical formula $C_6H_{12}O_6$; a common subunit of many sugars, such as milk sugar. (2)

Gallbladder The sac beneath the right lobe of the liver that stores bile. (24)

Gamete (Gr. wife) A specialized sex cell such as an ovum (egg) or a sperm, which is haploid. A male gamete (sperm) and a female gamete (ovum) fuse and give rise to a diploid zygote, which develops into a new individual. (7)

Gametogenesis The formation of haploid gametes, eggs and sperm, by the process of meiosis and maturation of the cells. (13)

Gametophyte The haploid, gamete-producing phase in the life cycle of plants. (17)

Ganglion (plural **ganglia**) A distinct clump of nerve cells that acts like a primitive brain, found in the head region of many invertebrates; in vertebrates, an aggregation of nerve cell bodies located outside the central nervous system. (28)

Gap junction A group of protein-lined pores that connects adjacent cells and allows small molecules to move from cell to cell. (3)

Gastrin A digestive hormone secreted in the stomach that causes the secretion of other digestive juices. (24)

Gastropod A member of the class Gastropoda in the phylum Mollusca; includes snails, garden slugs, and sea slugs. (18)

Gastrovascular cavity The digestive cavity in cnidarians; also called the *coelenteron.* (18)

Gastrulation The movement of cells in the embryo that generates three cell layers—the ectoderm, mesoderm, and endoderm—each layer in turn giving rise to specific body organs and tissues. (13)

Gene (Gr. *genos,* birth or race) The biological unit of inheritance that transmits hereditary information from parent to offspring and controls the appearance of a physical, behavioral, or biochemical trait. A gene is a specific discrete portion of the DNA molecule in a chromosome that encodes a protein, tRNA, or rRNA molecule. (1, 8)

Gene flow The incorporation into a population's gene pool of genes from one or more other populations through migration of individuals. (33)

Gene mutation A change in the base sequence of DNA that alters individual genes. (1)

Gene pool The sum of all alleles carried by the members of a population; the total genetic variability present in any population. (33)

Gene regulation The process that controls the cell's gene activity. (10)

Genetic code The specific sequence of three nucleotides in mRNA that encodes an individual amino acid in protein. For example, the code for methionine is AUG. (10)

Genetic drift Unpredictable changes in allele frequency occurring in a population due to the small size of that population. (33)

Genetic mapping The process of locating the positions of genes on chromosomes. (8)

Genetic recombination The reshuffling of maternal and paternal chromosomes during meiosis, resulting in new genetic combinations. (7)

Genome The total complement of genes in a single cell. (9)

Genotype The genetic makeup of an individual. (8)

Genus (plural **genera**) In taxonomy, a group of very similar species of common descent. (15)

Geological era A major time span dividing the earth's history; the four geological eras are the Proterozoic, Paleozoic, Mesozoic, and Cenozoic. (15)

Geological period A subdivision of a geological era; the Precambrian period was in the Proterozoic era. (15)

Germ cell A sperm cell or ovum (egg cell), the haploid gametes produced by individuals that fuse to form a new individual. (3, 7)

Germ layers Three layers of cells—ectoderm, mesoderm, and endoderm—formed during embryonic development by the process of gastrulation. These layers develop into all the structures and organs of the body. (13)

Germ line A specialized line of cells, from which come the sperm and egg (ovum). (7)

Germ plasm A region of distinctive granular cytoplasm within the egg cells of some animals from which the germ line develops. (13)

Gibberellins Specific hormones that regulate plant growth, usually by increasing the length of plant stems. (31)

Gill [1] A specialized structure that exchanges gases in water-living animals. (18, 23) [2] The plates under the cap of certain fungi. (17)

Gill filament A fingerlike extension in the gill of a fish through which blood flows and exchanges gases with water. (23)

Gill slit In chordates, an opening through the pharynx to the exterior. (19)

Gizzard A specialized region that grinds food in the digestive tract of earthworms, chickens, and other animals. (18)

Gland A group of cells organized into a discrete secretory organ. (26)

Glial cell A nonneural cell of the nervous system that surrounds the neurons and provides them with protection and nutrients. (27)

Glomerulus (L. a little ball) In the vertebrate kidney, a collection of tightly coiled capillaries enclosed by the Bowman's capsule. (25)

Glucagon A hormone that causes blood glucose levels to rise, thus opposing the effect of insulin. (26)

Glucose The most common type of sugar molecule; broken down to release energy in all living organisms; a simple sugar with the chemical formula $C_6H_{12}O_6$. (2, 5)

Glycerol A molecule of three carbons, each with an OH group. Three fatty acids are bonded to a glycerol molecule in a fat. (2)

Glycogen A polysaccharide made up of branched chains of glucose; an energy-storing molecule in animals, found mainly in the liver and muscles. (2)

Glycolysis The initial splitting of a glucose molecule into two molecules of pyruvate, resulting in the release of energy in the form of two ATP molecules. The series of reactions does not require the presence of oxygen to occur. (5)

Goiter A large lump on the neck caused by an enlarged thyroid. (26)

Golgi apparatus In eukaryotic cells, a collection of flat sacs that process proteins for export from the cell, or for shunting to different parts of the cell. (3)

Gonad An animal reproductive organ that generates gametes. (14)

Gout A painful condition caused by deposits of uric acid crystals in the big toe, elbow, or other joints. (25)

Gradient A change in a quantity, such as pressure or temperature or concentration, with respect to distance. (4)

Granum In chloroplasts, a stack of membrane discs called thylakoids, which contain chlorophylls and are the sites of the light-dependent reactions of photosynthesis. (3)

Grassland A treeless biome dominated by dozens of grass species. (37)

Gray crescent During cleavage, a zone of cytoplasm near the equator of some eggs, on the side opposite where the sperm entered that marks the future dorsal side of the embryo. (13)

Greenhouse effect The result of a buildup of carbon dioxide in the atmosphere (e.g., through the burning of fossil fuels) in which carbon dioxide traps solar heat beneath the atmospheric layers; leads to increased global temperatures, and changes climatic patterns. (6)

Ground tissue system Plant tissues other than the epidermal tissue system and vascular tissue system; provides support and stores starch. (30)

Growth An increase in size. (13)

Growth factor Proteins that can enhance the growth and proliferation of specific cell types. (7)

Guard cells Specialized kidney-shaped cells that enclose the pores (stomata) on the leaf epidermis; their opening and closing regulates water loss from the leaf. (32)

Gymnosperms (Gr. *gymnos,* naked + *sperma,* seed) Conifers and their allies; primitive seed plants whose seeds are not enclosed in an ovary. (17)

Habitat The place within a species' range where the organism actually lives. (35)

Habituation A progressive decrease in the strength of a behavioral response to a constantly applied weak stimulus, when the stimulus has no negative effects. (27, 38)

Halophile A type of archaebacterium that can tolerate extremely high salt concentrations. (16)

Haploid Having only one copy of a chromosome set. A human haploid cell has 23 chromosomes. (7)

Hardy-Weinberg principle In population genetics, the idea that in the absence of any outside forces, the frequency of each allele and the frequency of genotypes in a population will not change over generations. (33)

Head The part of an animal that contains a major concentration of sense organs and generally encounters the environment first as the animal moves along. (18)

Heart   A muscular organ that pumps blood through vessels. (18)

Heat   The random motion of atoms and molecules. (4)

Hemoglobin   A blood protein with iron-containing heme groups that bind and transport oxygen. (23)

Hemolymph   In animals with open circulatory systems, the extracellular fluid in the body cavity that bathes the tissue cells. (21)

Herbivore   An animal that consumes plants as food. (24)

Heredity   The science of inheritance and variation. (8)

Heterotrophs   (Gr. *heteros,* different + *trophos,* feeder) Organisms that take in preformed nutrients; includes animals, fungi, and most prokaryotes and protists. (5, 15)

Heterozygous   (Gr. *heteros,* different + *zygotos,* pair) Having two different alleles for a specific trait. (8)

Hinged jaw   Skeletal elements surrounding the mouth that move about a pivot; found in some arthropods and most vertebrates. (19)

Hippocampus   A structure within the cerebrum of the brain that plays a crucial role in the formation of long-term memory. (28)

Histocompatibility protein   Cell surface substances in vertebrates that can cause the immune rejection of tissues grafted from one individual to another; involved in the recognition of antigens and their presentation from one cell of the immune system to another. (22)

Histone   Protein in the nucleus around which DNA molecules of the chromosomes wind, allowing extremely long DNA molecules to be packed into a cell's nucleus. (9)

Holdfast   A rootlike anchor that attaches an alga to its substrate such as a rock on the ocean floor. (17)

Homeostasis   The maintenance of a constant internal balance despite fluctuations in the external environment. (20)

Homeothermic   "Warm-blooded"; a homeothermic animal can maintain a constant internal body temperature. (19)

Homologous chromosomes   Chromosomes that pair up and separate during meiosis and generally have the same size and shape and genetic information. One member of each pair of homologous chromosomes comes from the mother and the other comes from the father. (7)

Homology   The appearance in related organisms of similar structures, based on the inheritance of similar genetic programs from a common ancestor. (33)

Homo sapiens   (L. *homo,* man + L. *sapiens,* wise) The human species; a bipedal primate with a large brain, capable of language and tool use. (19)

Homozygous   Having two identical alleles for a specific trait. (8)

Hormone   A chemical messenger produced in one part of the body, transported in the blood to another region where it exerts an effect. (25, 26)

Human chorionic gonadotropin (hCG)   A hormone produced by the chorion of a developing embryo that maintains pregnancy by preventing menstruation. (14)

Humoral immune response.   The synthesis and secretion of antibodies by B cells. (22)

Hybrid   An offspring resulting from the mating between individuals of two different genetic constitutions. (8, 33)

Hydrogen bond   A type of weak molecular bond in which a partially negatively charged atom (oxygen or nitrogen) bonds with the partial positive charge on a hydrogen atom when the hydrogen atom is already participating in a covalent bond. (2)

Hydrolysis   The splitting apart of two covalently bound subunits by the insertion of a dissociated water molecule between them. (2)

Hydrophilic   Compounds that dissolve readily in water, such as salt. (2)

Hydrophobic   Compounds that do not dissolve readily in water, such as oil. (2)

Hydroskeleton   A volume of fluid trapped within an animal's tissues that is noncompressible and serves as a firm mass against which opposing sets of muscles can act; for example, in earthworm segments. (18)

Hypertonic   Having a higher concentration of solutes than another solution. (4)

Hypha   (plural **hyphae**)   One of many long thin filaments of cells that makes up a multicellular fungus. (17)

Hypocotyl   The portion of a plant embryo or seedling between the embryonic root (radicle) and the cotyledons. (30)

Hypothalamic release-inhibiting hormone   One of a number of specific hormones released from the hypothalamus that travels to the anterior pituitary and *inhibits* release of a specific pituitary hormone. (26)

Hypothalamic releasing hormone   One of a number of specific hormones released from the hypothalamus that travels to the anterior pituitary and *stimulates* release of a specific pituitary hormone. (26)

Hypothalamus   A collection of nerve cells at the base of the vertebrate brain just below the cerebral hemispheres, responsible for regulating body temperature, many autonomic activities, and many endocrine functions. (26)

Hypothesis   A potential solution to a problem or answer to a question; a provisional working idea that may be accepted for the present but may be rejected in the future if the evidence disproves it. (1)

Hypotonic   Having a lower concentration of solutes than another solution. (4)

Immune system   The network of cells, tissues, and organs that defends the body against invaders. (22)

Immunity   Resistance to infection. (22)

Immunological memory   The capability of memory B cells in vertebrates to serve as remnants of former encounters with antigens. (22)

Implantation   The process by which the fertilized egg burrows into the lining of the uterus in mammals. (14)

Imprinting   The attachment of young animals to a particular adult or object; in nature, the attachment of young to their parents. Animals may imprint on other objects or animals, such as young birds imprinting on humans, if they are exposed to that animal or object at the sensitive stage. (38)

Inclusive fitness   A concept in evolutionary theory in which an individual's total fitness includes both its likelihood of passing on its own genes and the likelihood that it will contribute to the successful passing on of the genes of its relatives. (38)

Incomplete dominance   The genetic situation in which the phenotype of the heterozygote is intermediate between the two homozygotes, e.g., when a cross between homozygous red and homozygous white flowers gives rise to heterozygous pink flowers. (8)

Independent assortment   The random distribution of genes located on different chromosomes to the gametes; Mendel's second law. (7, 8)

Inductive reasoning   Generalizing from specific cases to arrive at broad principles. (1)

Industrial revolution   The replacement of hand tools with power-driven machines (like the steam engine) and the concentration of industry in factories beginning in England in the late eighteenth century. (34)

Inert   Lacking chemical activity. (2)

Infancy   The period of time from birth to about age 2. (14)

Inferior vena cava   One of the two largest veins in the human body; carries blood from the legs and most of the lower body to the right atrium of the heart. (21)

Initiation   The first main stage of protein synthesis, during which tRNA associates with a ribosome and an mRNA, forming a complex of many molecules. (10)

Inner membrane   The folded membrane enclosed by the outer membrane surrounding a mitochondrion. (3)

Inorganic chemistry   The study of substances that do not contain carbon. (2)

Insect   A member of Insecta, arthropods having three main parts—head, thorax, and abdomen—three pairs of legs, and generally two pairs of wings; the largest class of animals. (18)

Insight learning   A type of learning in which individuals formulate a course of action by understanding the relationships between the parts of a problem; common only in higher primates. (38)

Instinct   Any behavioral trait that is innate, i.e., not learned. Instinctive behavior is usually not changed by experience during an animal's lifetime. (38)

Insulin   A protein hormone made in the pancreas that causes cells to remove the sugar glucose from the blood. (2, 26)

Intermediate filaments   Components of the cell's cytoskeleton involved in cell shape, movement, and growth, composed mainly of keratin-like proteins. (3)

Internal fertilization   A process that results when the male deposits sperm inside the female's body in a chamber or tube. (13)

Interneuron   A nerve cell that relays messages between other nerve cells. (27)

Interphase   The period between cell divisions in a cell. During this period, the cell conducts its normal activities and DNA replication takes place in preparation for the next cell division. Interphase is divided into three periods: G_1, S, and G_2. (7)

Interspecific competition   Competition for resources—e.g., food or space—between individuals of different species. (35)

Interstitial cell   A cell embedded in the connective tissue of the seminiferous tubules of the testes; produces testosterone, a male sex hormone; Leydig cell. (14)

Intestine   The long, tubelike section of the digestive tract between the stomach and anus of vertebrates where most food digestion and absorption take place. (18)

Intracellular digestion   The enzymatic breakdown of nutrient molecules that occurs within a cell. (24)

Intracellular fluid   The fluid inside cells. (4)

Intron   A sequence of DNA in a eukaryotic gene that is excised from RNA, and therefore not translated into protein. (10)

Invertebrate   An animal that does not have a backbone. (18)

In vitro fertilization   The fertilization of an egg by a sperm under laboratory conditions. (14)

Ion   An atom that has gained or lost one or more electrons, thereby attaining positive or negative electrical charge. (2)

Ionic bond   A type of molecular bond formed between ions of opposite charge. (2)

Iris   A pigmented ring of tissue in the vertebrate eye that regulates the amount of entering light. (28)

Isomers   Molecules that have the same chemical formula but different molecular shapes and hence different properties; for example, glucose and fructose are isomers. (2)

Isotonic   Two solutions having the same concentration of solutes. (4)

Isotope   An alternative form of an element having the same atomic number but a different atomic mass due to the different number of neutrons present in the nucleus. Some isotopes, such as carbon-14, are unstable and emit radiation. (2)

Joint   The hinge, or point of contact, between two bones. (18)

J-shaped curve   The curve on a graph representing exponential population growth. (34)

Juvenile hormone   In insects, a hormone that, at high levels, prevents metamorphosis. (26)

Kelp   One of the largest members of the algal world, a brown alga. (17)

Kidney   The main excretory organ of the vertebrate body; filters nitrogenous wastes from blood and regulates the balance of water and solutes in blood plasma. (25)

Kinetic energy   The energy possessed by a moving object. (4)

Kingdom   A taxonomic group comprised of members of similar phyla; i.e., Animalia, Plantae, Fungi, etc. (15)

Kin selection   A form of natural selection in which an individual increases its fitness by helping relatives, who share its genes, to reproduce. (38)

Klinefelter syndrome   A genetic condition including sterility exhibited by a human male with two X chromosomes and one Y chromosome (XXY) and two normal sets of autosomes. (12)

Krebs cycle   The first stage of aerobic respiration in which a two-carbon fragment is completely broken down into carbon dioxide

and large amounts of energy are transferred to electron carriers. (5)

K-selection   Selection experienced by populations that typically occur at maximum density; K-selected organisms are often long-lived and have few, large offspring. (34)

Lactose   A disaccharide composed of galactose bonded to glucose; the main sugar found in milk. (2)

Lamella (plural **lamellae**)   A thin, platelike structure; in the gill filaments of fish, lamellae enable respiration in water. (23)

Larynx   A cartilaginous structure containing the vocal cords; it is also known as the "voice box." (23)

Lateral line organ   A system of sense organs in aquatic vertebrates consisting of pores or canals arranged in a line down each side of the body; detects changes in water pressure, including vibrations in the water. (28)

Lateral meristem   An area of actively growing cells in the stems and roots of plants that causes these areas to thicken in secondary growth. (30)

Learning   An adaptive and enduring change in an individual's behavior based on its personal experiences in the environment. (38)

Leg   An appendage that supports or moves an animal. (19)

Legume   A member of the pea family; often contains nitrogen-fixing bacteria in root nodules. (32)

Lens   A circular, crystalline structure in the eye of certain animals that focuses light onto the retina. (28)

Leucoplast   In plant cells, a colorless organelle that stores starch granules; usually found in roots and internal stem tissue. (3)

Leukocyte   A white blood cell; functions in the body's defense against invading microorganisms or other foreign matters. (21)

Lichen   An association between a fungus and an alga, which live symbiotically together. (17)

Life cycle   The events that take place between the birth and death of an organism, or between the reproduction of one generation and that of the next. (7)

Life expectancy   The maximum probable age an individual will reach. (14, 34)

Life history strategy   The way an organism allocates energy to growth, survival, or reproduction. (34)

Life span   An individual's maximum potential age. (14)

Life zone   Any of several mountain regions

of varying elevation, climate, and organisms. (37)

Ligament A band of connective tissue that links bone to bone. (29)

Light-dependent reactions The first phase of photosynthesis, driven by light energy. Electrons that trap the sun's energy pass the energy to high-energy carriers such as ATP or NADPH, where it is stored in chemical bonds. (6)

Light-independent (dark) reactions The second phase of photosynthesis, also called the *Calvin-Benson cycle,* which does not require light. During the six steps of the cycle, carbon is fixed and carbohydrates are formed using the energy trapped in the light-dependent reactions. (6)

Light microscope An instrument containing optical lenses that refract (bend) light rays, magnifying the object viewed. (3)

Linkage group Genes located close to each other on the same chromosome that tend to be transmitted as a single unit. (8)

Lipid bilayer The two-layered sheet of phospholipids that makes up the plasma membrane of cells and is mainly responsible for the membrane's barrier function. (3)

Lipids Biological molecules that consist mainly of hydrocarbons and do not dissolve readily in water, including fatty acids, fats, steroids, and waxes. (2)

Liver A large, lobed gland that destroys blood cells, stores glycogen, disperses glucose to the bloodstream, and produces bile. (24)

Lobe-finned fish A member of the oldest of the two major groups of bony fishes; includes lungfishes and coelacanths. (19)

Locus The location of a gene on a chromosome. (8)

Logistic growth Growth of a population under environmental constraints that set a maximum population size. (34)

Loop of Henle U-shaped region of vertebrate kidney tubule chiefly responsible for reabsorption of water and salts from the filtrate by diffusion. (25)

Lumen The central cavity; of a blood vessel, the central space where the blood flows. (21)

Lung The principal air-breathing organ of most land vertebrates. (19)

Lungfish An air-breathing lobe-finned fish having a lunglike air bladder in addition to gills. (19)

Lymph Fluid forced out of capillaries, occupying spaces between cells, and draining into the lymphatic system. (21)

Lymphatic system In vertebrates, a collective term for a system of vessels carrying lymph—fluid that has been forced out of the capillaries—back to the bloodstream. (21)

Lymphocyte A white blood cell formed in lymph tissue; active in the immune responses that protect the body against infectious diseases. There are two main classes of lymphocyte: B lymphocyte, involved in antibody formation, and T lymphocyte, involved in cell-mediated immunity. (21, 22)

Lysosome Spherical membrane-bound vesicles within the cell containing digestive (hydrolytic) enzymes that are released when the lysosome is ruptured; important in recycling worn-out mitochondria and other cell debris. (3)

Macroevolution Evolutionary changes at levels higher than the species; evolutionary patterns viewed over geological time spans. (33)

Macronucleus The large nucleus governing the activities of a ciliate, containing many sets of chromosomes. (16)

Macronutrient (Gr. *makros,* large + L. *nutrire,* to nourish) An inorganic chemical element, such as nitrogen, potassium, calcium, phosphorus, magnesium, and sulfur, required in large amounts for plant growth. (32)

Macrophage An animal cell that ingests other organisms and substances; in the immune system, a cell that engulfs invaders and consumes debris. (22)

Malignant Characteristic of a tumor that grows without ceasing and spreads throughout the body. (13)

Maltose Two joined glucose subunits. (2)

Mammal A vertebrate animal of the class Mammalia, having the body generally covered with hair, nourishing young with milk from mammary glands, and generally giving birth to live young; lions, whales, rabbits, and kangaroos are some mammals. (19)

Mammary gland Gland that produces milk in female mammals to nourish young. (19)

Mantle A thick fold of tissue found in mollusks that covers the visceral mass and that in some mollusks secretes the shell. (18)

Mantle cavity In mollusks, the space between the mantle and the visceral mass. (18)

Marine Characteristic of oceans and seas. (37)

Marsupial A mammal having a pouch in which it carries its young, which are born in a small and undeveloped state. Found extensively in Australia, with a few representatives in America; includes kangaroos, opossums, and koala bears. (19)

Mass flow theory A model for the large-scale movement of phloem fluid in plants due to the active transport of sugars into the phloem at a sugar source such as the leaves, followed by the passive entry of water, which creates a pressure forcing the fluid to a sugar sink, like the roots, where cells actively remove sugars from the phloem and the water returns to the xylem. (32)

Mate Behavior that brings egg and sperm together during reproduction. (13)

Medulla The central portion of organs; in the kidney, the place that contains the loop of Henle; in the adrenal gland, the place where epinephrine (adrenaline) is made. (25, 26)

Medulla oblongata The lowest region of the brain stem, the most posterior part of the brain; involved in keeping body conditions constant. (28)

Medusa (Gr. mythology, a female monster with snake-entwined hair) A jellyfish, or the free-swimming stage in the life cycle of cnidarians; an inverted umbrella-shaped version of a polyp, with the mouth and tentacles pointing downward. (18)

Megaspore In certain plants, a haploid spore giving rise to the female gametophyte. (17)

Meiosis The type of cell division that occurs during gamete formation; the diploid parent cell divides twice, giving rise to four cells, each of which is haploid. (7)

Melatonin A hormone secreted by the pineal gland in the brain; in some animals, it affects the internal biological rhythms associated with night and day. (26)

Menopause The physiological termination of the menstrual cycle; in human females the onset of menopause normally occurs between the ages 40 and 50. (14)

Menstrual cycle The reproductive cycle of human and other primate females. Generally a 28-day cycle in women, characterized by a gradual thickening of the lining of the uterus (womb) and the maturation of an ovarian follicle from which a mature egg is released. If the egg is not fertilized, the inner lining of the uterus is shed, a process known as menstruation, and the cycle starts again. In humans the menstrual cycle normally starts between ages 12 and 14. (14)

Meristem Plant tissue containing cells that can continually divide throughout the plant's life. (30)

Mesoderm The middle cell layer of an embryo; gives rise to muscles, bones, connective tissue, and reproductive and excretory organs. (13)

Mesoglea A jellylike substance lying between the epidermis and gastrodermis in cnidarians such as jellyfish. (18)

Mesophyll The major photosynthetic tissue in a plant leaf, located between the upper and lower epidermal layers. (30)

Message An mRNA molecule that carries information to make a protein. (*See also* Messenger RNA.) (10)

Messenger RNA (mRNA) An RNA molecule that carries the information to make a specific polypeptide. mRNA is transcribed from structural genes and is translated into protein by the ribosomes. (3, 10)

Metabolic pathway The chain of enzyme-catalyzed chemical reactions that convert energy in cells and in which the product of one reaction serves as the starting substance for the next. (4)

Metabolism (Gr. *metabole*, to change) The sum of all the chemical reactions that take place within the body; includes photosynthesis, respiration, digestion, and the synthesis of organic molecules. (1, 4)

Metamorphosis (Gr. *meta*, after + *morphe*, form + *osis*, state of) The process in which there is a marked change in morphology during postembryonic development; in insects the change in body form that takes place as the individual changes from a larva, such as a caterpillar, and emerges as an adult, such as a butterfly. Also refers to the change from a tadpole to a frog in amphibians. (18, 26)

Metaphase The period during nuclear division (mitosis) when the spindle microtubules cause the chromosomes to line up at the center of the cell. (7)

Metastasis The process by which cancer cells spread throughout the body. (13)

Metazoan An animal; a many-celled eukaryote that obtains its energy from other organisms. (15)

Methanogens Archaebacteria that produce methane as a metabolic by-product. (15)

Microbes (Gr. *mikros*, small) Microscopic organisms; includes members of Monera and Protista kingdoms. (16)

Microbodies Cellular organelles bounded by a single membrane with a granular interior; contain a number of different oxidase-type enzymes like catalase. (3)

Microevolution Evolutionary changes in a species over geologically short time periods, like the accumulation of black forms of certain moths in industrial England. (33)

Microfilaments Components of the cell's cytoskeleton involved in cell shape, movement, and growth, composed mainly of the protein actin. Contraction of microfilaments is the basic mechanism underlying shape changes in eukaryotic cells, including cell motility and muscle contraction. (3)

Micronucleus In ciliates (Protista), one of several nuclei that undergo meiosis and are exchanged during sexual reproduction. (16)

Micronutrient (Gr. *mikros*, small + L. *nutrire*, to nourish) A nutrient required in small amounts for plant growth; includes the minerals iron, copper, zinc, chlorine, manganese, molybdenum, and boron. (32)

Microspore A pollen grain that gives rise to the male gametophyte in certain vascular plants. (17)

Microtubules Components of the cell's cytoskeleton; hollow cylinders of the protein tubulin involved in cell shape, movement, and growth. (3)

Microvillus (plural **microvilli**) One of hundreds of tiny fingerlike projections extending from the surface of cells lining the walls of the intestine that increases the surface area available for absorption. (24)

Midbrain The part of the brain stem between the pons and the thalamus in humans; relays neural signals between different parts of the brain. (28)

Milk A fluid rich in fats and proteins produced in the mammary glands of mammals; nourishes newborns. (19)

Mimicry The evolution of similar appearance in two or more species, which often gives one or all protection; for example, a nonpoisonous species may evolve protection from predators by its similarity to a poisonous model. (35)

Mineral An inorganic element such as sodium or potassium that is essential for survival and is obtained in food. (24)

Mitochondrion (plural **mitochondria**) Organelle in eukaryotic cells that provides energy that fuels the cell's activities. Mitochondria are the sites of oxidative respiration, and almost all of the ATP of nonphotosynthetic eukaryotic cells is produced in the mitochondria. (3)

Mitosis The process of nuclear division in which replicated chromosomes separate and form two daughter nuclei genetically identical to each other and the parent nucleus. Mitosis is usually accompanied by cytokinesis (division of the cytoplasm). (7)

Mitotic spindle Formed during prophase in mitosis, a weblike structure of microtubules that suspends and moves the chromosomes. (7)

Molecular bond The energy that holds atoms together in molecules. (2)

Molecular clock The rate at which mutations accumulate in genes over evolutionary time. (33)

Molecular messenger A chemical that can pass from cell to cell and regulate the activities of other cells. (26)

Molecule A cluster of atoms held together by specific chemical bonds. (2)

Mollusk A member of the phylum Mollusca; includes snails, slugs, clams, oysters, squid, and octopuses. (18)

Molt To shed an animal's skin; in arthropods, molting involves shedding the hard exoskeleton. (18)

Monera In the five-kingdom scheme of classification of all living forms, the kingdom comprising the single-celled prokaryotes, the Eubacteria and Archaebacteria. In the three-domain scheme of classification, Monera is split into the domain Bacteria (eubacteria) and the domain Archaea (archaebacteria). (15)

Monocot Short for *monocotyledon*, the smaller of two classes of angiosperms (flowering plants). Embryos have only one cotyledon; the floral parts are generally in threes; leaves are typically parallel-veined. Corn and lilies are monocots. (17)

Monocyte A type of white blood cell that is attracted to damaged or infected tissue and that consumes bacteria, viruses, and cell debris. (21)

Monohybrid cross A mating between individuals that differ only in one trait. (8)

Monomer A small molecule, many of which can be combined into a polymer. (2)

Monosaccharide A simple sugar that cannot be decomposed into smaller sugar molecules. The most common forms are the hexoses, six-carbon sugars such as glucose, and the pentoses, five-carbon sugars such as ribose. (2)

Monotreme An egg-laying mammal that also has many primitive or reptilian features. The only living forms are the spiny anteater and the duck-billed platypus. (19)

Morphogenesis The growth, shaping, and arrangement of the organs and tissues in a developing embryo. (13)

Morula (L. a little mulberry) An embryo during the process of cleavage; consists of a solid ball of cells. (13)

Motility The self-generated motion of an individual or its parts. (1)

Motor cortex The part of the cerebral cortex of the brain that controls the movement of body parts. (28)

Motor neuron A neuron that sends messages from the brain or spinal cord to muscles or secretory glands. (27)

Mouthpart One of several arthropod head region appendages that chews and sucks. (18)

Mucosa The innermost layer of the lining of many vertebrate canals and organs, such as the uterus, reproductive tracts, and alimentary canal; often consists of mucus-secreting cells, and in the alimentary canal it contains enzyme-secreting cells. (24)

Muscle fiber A muscle cell; a giant cell with many nuclei and numerous myofibrils, capable of contraction when stimulated. (29)

Muscle tissue Tissue that enables an animal to move; the three types are smooth muscle, cardiac muscle, and skeletal muscle. (20)

Mutation Any heritable change in the base sequence of an organism's DNA. (9, 10)

Mutualism A symbiotic relationship in which both species benefit. (34)

Mycelium The mass of filaments (hyphae) that makes up the body of a fungus. (17)

Mycorrhiza An association between the root of a higher plant and a fungus that aids the plant in receiving water and nutrients. (17)

Myelin sheath Lipid membranes made up of Schwann cells, which form an insulating layer around the neurons and speed impulse travel. (27)

Myofibril A fibril composed of actin and myosin found in the cytoplasm of muscle cells that contracts the muscle. (29)

Myosin A muscle protein that interacts with actin and causes muscles to contract. (3)

Nasal cavity A chamber of the nose that opens posteriorly into the pharynx (throat). (23)

Natural selection The increased survival and reproduction of individuals better adapted to the environment. (1, 33)

Navigation The ability to move from one map point to another. (38)

Nematocyst A stinging capsule found in cnidarians, which, when stimulated, shoots out a tiny barb containing a poisonous substance that immobilizes or kills the prey or predator. (18)

Nephridium (plural **nephridia**) An excretory organ in annelids, each body segment having a pair. (18)

Nephron The functional unit of the vertebrate kidney; each of the million nephrons in a kidney consists of a glomerulus enclosed by a Bowman's capsule and a long attached tubule. The nephron removes waste from the blood. (25)

Nerve A group of axons and accompanying cells from many different neurons running in parallel bundles and held together by connective tissue. (27)

Nerve cell A specialized cell that transmits neural impulses; also called a *neuron.* (18)

Nerve cord A strand of nervous tissue that forms part of the central nervous system of invertebrates. (18)

Nerve impulse A change in ion permeabilities in a neuron's membrane that sweeps down the axon to its terminal, where it can excite other cells. (27)

Nerve net A noncentralized, diffuse meshwork of nerve cells interlaced throughout the body. The simplest form of nervous organization in animals. Found in cnidarians, like jellyfish and sea anemones, it permits an equal response to stimuli from any direction in these radially symmetrical animals. (28)

Nervous system An animal's network of nerves; integrates and coordinates the activities of all the bodily systems. (27)

Nervous tissue Tissue containing neurons, cells whose main function is the transmission of electrochemical impulses. (2)

Neural crest Ridge where folds arising from the ectoderm meet during the formation of the neural tube. Neural crest cells migrate and form pigment cells, parts of the face, and certain nerve cells. (13)

Neural tube Embryonic cells that differentiate into the brain and spinal cord. (13)

Neurohormone A hormone secreted by nerve cells. (26)

Neuromodulator A compound occurring in the vicinity of a synapse that influences a receiving neuron's response to neurotransmitters. (27)

Neuron A nerve cell that transmits messages throughout the body; includes dendrites, cell body, and axon. (27)

Neurotransmitter A chemical that transmits a nerve impulse across a synapse. (26)

Neurulation Development of the neural tube. (13)

Neutron (L. *neuter,* either) A subatomic particle without any electrical charge found in the nucleus of an atom. (2)

Neutrophil The most common type of white blood cell, important in combating bacterial infections. (21)

Niche The role, function, or position of an organism in a biological community. (35)

Nitrogen cycle The cycling of nitrogen between organisms and the earth. (36)

Nitrogen fixation The assimilation of atmospheric nitrogen by certain prokaryotes into biologically usable nitrogenous compounds. (32, 36)

Node The position at which a leaf arises on a stem. (30)

Nodule A swelling on the roots of legumes and some other plants that houses nitrogen-fixing bacteria. (32)

Noncyclic photophosphorylation In photosynthesis, the generation of ATP by light-induced electron flow in both photosystems I and II; the Z-scheme. (6)

Nondisjunction Failure of the chromosomes to separate correctly during mitosis or meiosis. Down syndrome results from nondisjunction of chromosome 21. (7)

Nonpolar Not having an asymmetrical distribution of electrical charge; i.e., a nonpolar molecule like most lipids will not dissolve readily in water. (2)

Norepinephrine A hormone secreted by the adrenal medulla; raises blood sugar concentration, blood pressure, and heart rate and increases muscular power and resistance to fatigue; sometimes known as noradrenaline. (26)

Notochord A rod of mesodermal cells in the chordate embryo, marks the location of the backbone in vertebrates. (13)

Nuclear envelope Two bilipid membranes perforated by pores that enclose the nucleus of a eukaryotic cell. (3)

Nuclear pore A hole that perforates the nuclear envelope. (3)

Nuclear winter A condition resulting from the burning and fires that would be generated during a nuclear war. Clouds of smoke and ash would block out sunlight, resulting in long periods of cold and darkness, during which photosynthesis, and with it most life on earth, would be drastically reduced or cease. (6, 36)

Nucleic acid A polymer of nucleotides, e.g., DNA and RNA. (2)

Nucleoid The unbounded region of DNA in a prokaryotic cell. (3)

Nucleolus (plural **nucleoli**) A dark-staining region within the nucleus of a eukaryotic cell where ribosomal RNA is formed. (3)

Nucleosome The basic packaging unit of eukaryotic chromosomes; a histone wrapped with two loops of DNA. (9)

Nucleotide The basic chemical unit of DNA and RNA consisting of one of four nitrogenous bases (A, C, T, or G) linked to a sugar (a ribose or a deoxyribose), which is in turn linked to a phosphate. (2)

Nucleus (atomic) *See* Atomic nucleus.

Nucleus (cell) The membrane-bound region of a eukaryotic cell that contains the cell's DNA. (3)

Nudibranch A sea slug; a marine gastropod mollusk. (18)

Nutrient Any substance that an organism must obtain from the environment in order to survive and reproduce. (24)

Nutrition The science concerned with determining the amounts and kinds of nutrients needed by the body. (24)

Obese Having body weight that is more than 20 percent above ideal. (24)

Oil A fluid lipid that is insoluble in water, often a prime form of energy storage in plants. (2)

Oligotrophic Describes an aquatic environment with low nutrient levels, characterized by small amounts of algae or other aquatic plants. (36)

Omnivore An animal that consumes both plant and animal matter as food. (24)

Oncogene A cancer-causing gene; often a mutant form of a growth-regulating gene that inappropriately causes unrestrained cell replication. (12)

Oocyte The precursor of a mature egg cell (ovum). (13, 14)

Open growth The continuous growth pattern exhibited by a plant throughout its life. (30)

Operant A behavioral response to a stimulus. (38)

Operant conditioning A form of learning in which an animal associates an action (operant; e.g., pressing a lever) with a reward or punishment (reinforcer; e.g., food). (38)

Opercular flap The stiff outer protective structure of a fish's gill. (23)

Operon The collective term for a regulatory protein and a group of genes whose transcription it controls in bacterial cells. (10)

Opposable thumb A first digit or thumb that can be held opposite the other digits; characteristic feature of some primates. (19)

Optic nerve The nerve leading from the eye to the brain. (28)

Optimality hypothesis The idea that animals act in ways that maximize their energy intake and minimize the energy spent on gathering food. (38)

Orbital The path followed by an electron around an atom. (2)

Order A precise arrangement of structural units and activities. (1)

Organ A body structure composed of two or more tissues that together perform a specific function. (18, 20)

Organelle In eukaryotic cells, a complex cytoplasmic structure with a characteristic shape that performs one or more specialized functions. (1)

Organic chemistry The study of carbon compounds. (2)

Organism An individual that can independently carry out all life functions. (20)

Organ system A set of organs with related functions. (18, 20)

Organogenesis The formation of organs during embryonic development. (13)

Orgasm Involuntary contractions of the vaginal and uterine walls or of the muscles lining the seminal vesicles and urethra. (14)

Orientation The ability to sense direction. (38)

Osmoregulation Maintenance of a constant internal salt and water concentration in an organism. (25)

Osmosis The movement of water through a semipermeable membrane from an area of high water concentration (low solute concentration) to one of low water concentration (high solute concentration). (4)

Osteichthyes A class of the phylum Chordata comprised of the bony fishes. (19)

Outer membrane The smooth outer layer of a mitochondrion, surrounding the organelle. (3)

Ovary Egg-producing organ. (14)

Overweight Having body weight that is more than 10 percent above ideal. (24)

Oviduct The tube along which the egg travels from the ovary to the uterus (womb) in women and some other female animals. (13)

Ovulation The release of a mature egg from the ovary. (14)

Ovule In a seed plant, the part of the ovary that develops into a seed. (17)

Ovum An unfertilized egg cell. (13)

Oxidation The removal of electrons from a molecule. (5)

Oxidation-reduction reaction The transfer of electrons from one molecule to another. (5)

Oxytocin A hormone whose release from the pituitary causes the contraction of the uterine muscle and labor. (14)

Pacemaker cells Cells that set the rate of the heartbeat. (21)

Pancreas A gland located behind the stomach that secretes digestive enzymes into the small intestine and the hormones insulin and glucagon into the blood. (24)

Paracrine hormone A primitive hormone that acts on cells immediately adjacent to the ones that secreted it. (26)

Parallel evolution The evolution of two or more physically similar and genetically related lines independently in the same direction and away from the ancestral form. (33)

Parapatric speciation The divergence into separate species of populations living in adjacent areas. (33)

Parasite An organism that lives on or in another living host organism and damages or kills its host as a consequence. (16)

Parasympathetic nerves Neurons in the autonomic system that emanate from the spinal cord and act on the respiratory, circulatory, digestive, and excretory systems as a "housekeeping" system, conserving and restoring body resources. (28)

Parathyroid gland One of a set of four small endocrine glands located on the thyroid gland; it secretes parathyroid hormone, which controls blood calcium levels. (26)

Parathyroid hormone A hormone secreted by the parathyroid gland that regulates blood calcium levels. (26)

Parenchyma A plant tissue composed of living, loosely packed, thin-walled cells; the most abundant plant tissue. (30)

Parental generation In Mendelian genetics, the individuals that give rise to the first filial (F_1) generation. (8)

Parental type An offspring having the characteristics of one of the parents. (8)

Parthenogenesis The process by which an unfertilized egg gives rise to a new individual; occurs in aphids, bees, ants, certain lizards, and some other animals. (13)

Partial pressure The pressure exerted by one gas in a mixture of gases. (23)

Passive transport Movement of substances into or out of a cell from areas of high concentration to areas of low concentration without the expenditure of cellular energy. (4)

Pathogenic (Gr. *pathos,* suffering + *genesis,* beginning) Capable of causing a disease. (16)

Pectoral (shoulder) girdle The skeletal support to which front fins or limbs of vertebrates are attached. (29)

Pedigree An ordered diagram of a family's relevant genetic features. (12)

Pelvic (hip) girdle The skeletal support to which hind fins or limbs of vertebrates are attached. (29)

Pelvis The central cavity of the kidney which collects urine before it passes to the ureter. (*See also* Pelvic girdle.) (25)

Penis In mammals and reptiles, the male organ of copulation and urination. (14)

Pepsin An enzyme secreted by the stomach that digests proteins. (24)

Pepsinogen A precursor to the protein-digesting enzyme pepsin. (24)

Peptide bond A bond joining two amino acids in a protein. (2)

Perennial (L. *per,* through + *annus,* year) A

plant that lives for many years, blooming and setting seeds several times before dying. (30)

Pericycle   In roots of vascular plants, one or more layers of parenchyma tissue lying between the endodermis and the vascular layer; gives rise to lateral roots. (30)

Periderm   In a plant, the outer protective layer including cork and cork cambium that is formed in secondary growth. (30)

Peripheral nervous system   The part of the nervous system consisting of the sensory and motor neurons, connecting the central nervous system with the sense organs, muscles, and glands of the body. (28)

Peristalsis   (Gr. *peristaltikos,* compressing around) In animals, waves of contraction and relaxation of muscles along the length of a tube, such as those in the digestive tract that help move food. (24)

Peritubular capillary   In the vertebrate kidney, one among a network of capillaries surrounding the looped portion of the nephron's tubule. (25)

Petiole   (L. *petiolus,* a little foot) The stalk connecting the leaf to the stem of a plant. (30)

Phagocytosis   The type of endocytosis through which a cell takes in food particles. (3)

Pharynx   [1] In vertebrates, a tube leading from the nose and mouth to the larynx and esophagus; conducts air during breathing and food during swallowing; the *throat.* (24) [2] A short tube connecting the mouth and intestine in flatworms. (18)

Phenotype   The physical appearance of an organism controlled by its genes interacting with the environment. (8)

Pheromone   A compound produced by one individual that affects another individual at a distance; for example, pheromones are secreted by certain female insects and attract males. (26)

Phloem   Food-conducting tissue of plants, consisting of sieve elements with companion cells, phloem parenchyma, and fibers. (30)

Phospholipid   A lipid composed of a phosphate functional group and two fatty acid chains attached to a glycerol molecule; a main component of cell membranes. (2)

Phosphorus cycle   The cycling of phosphorus between organisms and soil, rocks, or water. (36)

Phosphorylation   The transfer of a phosphate group from one organic molecule to another. Phosphorylation often results in the formation of a high-energy bond, as in the formation of ATP. (5)

Photon   A vibrating particle of light radiation that contains a specific quantity of energy. (6)

Photoperiod   The length of light and dark periods each day. (31)

Photoreceptor cells   Light-sensitive cells such as rods and cones in the eye. (28)

Photorespiration   A process that occurs in plants when carbon dioxide levels are depleted but oxygen continues to accumulate and the enzyme RuBP carboxylase (rubisco) fixes oxygen instead of carbon dioxide. (6)

Photosynthesis   The metabolic process in which solar energy is trapped and converted to chemical energy (ATP and NADPH), which in turn is used in the manufacture of sugars from carbon dioxide and water. (3, 6)

Photosystem   A functional light-trapping unit in the thylakoid membrane that includes chlorophyll, pigments, and electron donors and acceptors. (6)

pH scale   A logarithmic scale that measures hydrogen ion concentration; acids range from pH 1 to 7, water is neutral at a pH of 7, and bases range from pH 7 to 14. (2)

Phyletic gradualism   The suggestion that morphological changes occur gradually during evolution and are not always associated with speciation; distinct from *punctuated equilibrium.* (33)

Phylogeny   The study of the evolutionary history of different groups of organisms. (33)

Phylum   A major taxonomic group, lower than the kingdom, comprising members of similar classes, all with the same general body plan. (15)

Physiology   The study of the functioning body of organs. (20)

Phytochrome   Light-sensitive pigment found in plants that absorbs in the red or far-red wavelengths; associated with a number of timing processes, such as flowering, dormancy, leaf formation, and seed germination. (31)

Phytoplankton   Photosynthetic microorganisms that live near the surface of marine and fresh water. (16)

Pineal gland   An endocrine gland located in the vertebrate midbrain that produces the hormone melatonin; probably involved in body rhythms. (26)

Pinocytosis   The type of endocytosis through which a cell takes in fluid. (3)

Pioneer community   The species that are first to colonize a habitat after a disturbance such as fire, plowing, or logging. (35)

Pith   Parenchyma cells occupying the central portion of a stem or root acting as storage cells for the plant. (30)

Pituitary gland   In vertebrates, an endocrine gland connected by a stalk to the hypothalamus at the base of the brain; the anterior lobe secretes growth hormone, prolactin, LH, FSH, ACTH, TSH, and MSH; the posterior lobe stores and releases oxytocin and ADH. Much of the functioning of the pituitary is under the control of the hypothalamus, and pituitary hormones control most of the other endocrine glands. (26)

Placenta   In mammals, the spongy organ rich in blood vessels by which the developing embryo receives nourishment from the mother. (13, 14)

Plantae   The kingdom comprising multicellular autotrophs such as mosses, ferns, and flowering plants. (15)

Plasma   The liquid portion of blood or lymph. (21, 23)

Plasma membrane   The membrane that surrounds all cells, regulating entry and exit of substances; consists of a single lipid bilayer; also called the *cell membrane* and *plasmalemma.* (3)

Plasmid   A circular piece of DNA that can exist either inside or outside bacterial cells but can reproduce only inside a bacterial cell. Plasmids are used extensively in genetic engineering as carriers of foreign genes. (9)

Plasmodesmata   Bridges of cytoplasm that connect plant cells and allow rapid exchange of materials between the cells. (3)

Plastid   An organelle found in plants and some protozoa that harvests solar energy and produces or stores carbohydrates or pigments. (3)

Platelet   A disc-shaped cell fragment in the blood important in blood clotting; also called a *thrombocyte.* (21)

Plate tectonics   The geological building and moving of crustal plates on the surface of the globe. (15)

Pleiotropy   The condition in which a single gene affects two or more distinct and seemingly unrelated traits. (8)

Pleural sac   One of two fluid-filled sacs that encloses the lungs. (23)

Plumule   In an embryonic plant, the portion of the shoot above the cotyledon. (30)

Poikilothermic   "Cold-blooded"; a poikilothermic animal's internal body temperature is the same as that of the environment and therefore fluctuates, rising in warm weather and dropping in cold temperatures. (19)

Poison claw   A modified leg on the first body segment of a centipede, used to kill prey. (18)

Polar   Having an asymmetrical distribution of electrical charge; i.e., a polar molecule like glucose will dissolve readily in water. (2)

Polarized   The resting state of a nerve cell, with the cell's interior negatively charged with respect to the surroundings. (27)

Pole   One end of a cell. (7)

Pollen grain   The male gametophyte of seed plants. (30)

Pollen tube   A tubelike structure extended after pollination by the developing male gametophyte toward the egg cell; when the tube reaches the egg, some of its cytoplasm and the sperm nuclei are transferred into the egg. (17)

Pollinator   An organism that carries pollen from one plant to another; examples are bees and birds. (17)

Polygenic   Characteristic of a trait that varies in quantity (i.e., height, from short to tall and all measurements in between) that depends on the interaction of many genes. (8)

Polymer   A large molecule made up of repeated sequences of similar or identical subunits called monomers. DNA, proteins, and starch are examples of polymers. (2)

Polyp   (L. *polypus,* many-footed) The sedentary stage in the life cycle of cnidarians; a cylindrical organism with a whorl of tentacles surrounding a mouth at one end. Sea anemones and hydras are examples of polyps living alone; corals are examples of colonial polyps. (18)

Polypeptide   (Gr. *polys,* many + *peptin,* to digest) Amino acids joined together by peptide bonds into long chains. A protein consists of one or more polypeptides. (2)

Polypeptide hormone   A hormone consisting of a string of 3 to 200 amino acids; examples include oxytocin and luteinizing hormone. (26)

Polysaccharide   A carbohydrate made up of many simple sugars linked together. Glycogen and cellulose are examples of polysaccharides. (2)

Pons   Part of the brain stem involved in relaying sensory information to other areas of the brain. (28)

Pool   A repository of materials. (36)

Population   A group of individuals of the same species living in a particular area. (34)

Population bottleneck   A situation arising when only a small number of individuals of a population survive and reproduce; therefore only a small percentage of the original gene pool remains. (33)

Population genetics   A set of principles that clarifies what happens at the genetic level as populations evolve. (33)

Pore   A small opening, for example, in the skin or epidermis. (18)

Postsynaptic cell   The nerve cell that receives the message crossing a synapse. (27)

Potassium channel   A protein-lined pore in a cell's plasma membrane that permits potassium ions to flow in and out of the cell. (27)

Potential energy   Energy that is stored and ready to do work. (4)

Precapillary sphincter   A small cuff of smooth muscle near a capillary entrance that contracts, resulting in vasoconstriction (decreased flood flow), or expands, resulting in vasodilation (increased blood flow). (21)

Predation   The act of procurement and consumption of prey by predators. (35)

Predator   An organism, usually an animal, that obtains its food by eating other living organisms. (35)

Prediction   In the scientific method, an experimental result expected if a particular hypothesis is correct. (1)

Pregnancy   Begun by the implantation of the developing embryo in the uterine wall, a series of developmental events involving close cooperation between mother and embryo that transforms a zygote into a baby. (14)

Presynaptic cell   The neuron that sends a message down its axon and across a synapse to another cell. (27)

Prey   Living organisms that act as food for other organisms. (35)

Primary cell wall   The thin, flexible, porous layer immediately outside the plasma membrane of a plant cell deposited as the cell is expanding. (3)

Primary consumer   At the second trophic level in an ecosystem, an organism that eats producers; herbivores (plant eaters like cows and caterpillars) are primary consumers. (36)

Primary embryonic induction   Induction of the embryonic axis during vertebrate development. (13)

Primary growth   Plant growth arising from the apical meristem. (30)

Primary productivity   The amount of energy that producers (plants) convert by photosynthesis to chemical energy in the form of sugars and other organic compounds. *Net* primary productivity is the amount of chemical energy actually stored in new cells, leaves, roots, stems, flowers, and fruits, since much is lost to respiration in the plant. (36)

Primary structure   The level of protein structure referring to the unique linear sequence of amino acids on the polypeptide chain. (2)

Primordium   (plural **primordia**)   One of many tiny swellings flanking the apical meristem, giving rise to leaves, branches, or flowers. (30)

Principle of independent assortment   *See* Independent assortment.

Prion   An intracellular disease-causing entity apparently consisting only of protein and having no genetic material. (16)

Process   A projecting part of an organism or organic structure, such as a bone. (29)

Producer   An organism, usually a plant, that can produce all its nutritional needs from nonbiological substances. (36)

Product   A substance that results from a chemical reaction. (4)

Productive collision   In a chemical reaction, the colliding of molecules with enough kinetic energy to cause the transition state to occur. (4)

Progesterone   A female sex hormone secreted by the corpus luteum in the ovary that stimulates uterine wall thickening and mammary duct growth. (14)

Prokaryotic cell   A cell in which the DNA is not bound by a nuclear envelope such as a eubacterial or archaebacterial cell. Prokaryotic cells generally have no internal membranous organelles, and they evolved earlier than eukaryotic cells. (3)

Prometaphase   An interval during nuclear division when the nuclear membrane disappears and the spindle attaches to the chromosomes. (7)

Propagation   The passing of the action potential from one patch of a nerve cell membrane to another. (27)

Prophase   The first phase of nuclear division in mitosis or meiosis, when the chromosomes condense, the nucleolus disperses, and the spindle forms. (7)

Prosimian   A member of the suborder Prosimii, order Primates; includes lemurs and tarsiers. (19)

Prostaglandin   A fatty acid–derived hormone secreted by many tissues; for example, prostaglandin from the uterus produces strong contractions in the uterine muscle, causing menstrual cramps or inducing labor. (14)

Prostate gland   A reproductive gland in vertebrate males that secretes a milky alkaline fluid that increases motility of sperm and helps neutralize acidity of the vagina. (14)

Protein   A molecule composed of one or more chains of amino acids. Each chain is typically over 100 amino acids long. Enzymes, fingernails, antibodies, and the red protein in the blood are examples of proteins. (2)

Protein synthesis   The assembly of amino acids into a polypeptide chain, in three main stages: initiation, elongation, and termination. (10)

Protista   The kingdom comprising single-celled eukaryotes. (15)

Proton   A positively charged subatomic particle found in the nucleus of an atom. (2)

Protostome   (Gr. *protos,* first + *stoma,*

mouth) Any bilateral animal whose first opening in the embryo (blastopore) becomes the mouth and the second opening the anus; also characterized by spiral cleavage during development; includes annelids, mollusks, and arthropods. (18)

Protozoan Literally a "first animal"; any one of the single-celled eukaryotic organisms that is primarily animal-like in its method of obtaining food. (16)

Proximal tubule In the vertebrate kidney, the convoluted tube between the Bowman's capsule and the loop of Henle. (25)

Proximate cause The mechanisms that operate within an organism, directing a behavior. (38)

Pseudocoelom (Gr. *pseudos,* false + *koiloma,* cavity) A body cavity that is not surrounded by mesoderm-derived cells; found in roundworms and rotifers. (18)

Pseudopod (Gr. *pseudos,* false + *pous,* foot) A limblike cytoplasmic extension used in feeding and locomotion by some protozoans and animal cells. (16)

Puberty The maturation of the sex organs and the development of secondary sex characteristics such as breasts in females and facial hair and a deep voice in males. (14)

Pulmonary artery In amphibians, reptiles, birds, and mammals, the artery that carries blood from the heart's right ventricle to the lungs. (21)

Pulmonary circulation The blood vessel system that carries oxygen-poor blood from the heart to the lungs, where gas exchange occurs, and carries oxygen-rich blood back to the heart. Contrasts to systemic circulation. (21)

Punctuated equilibrium The suggestion that morphological changes evolve rapidly in geological time; during speciation, these changes occur in small populations, with the resulting new species being distinct from the ancestral form. After speciation, species retain much the same form until extinction; distinct from the phyletic gradualism theory. (33)

Punnett square In genetics, a diagrammatic way of presenting the results of random fertilization from a mating. (8)

Pupa In insects with complete metamorphosis, the stage that intervenes between the larva and the adult; in some cases, as in moths, the pupa is encased inside a cocoon. (18)

Pupil The central, shutterlike opening in the iris of the vertebrate eye, through which light passes to the lens and the retina. (28)

Pure-breeding In Mendelian genetics, refer-

ring to organisms that always produce offspring with traits identical to theirs; homozygous. (8)

Pyramid of biomass The relationship between the total masses of various groups of organisms in a food chain, in which there is usually less mass, hence less stored energy, at each successive trophic level. (36)

Quantitative trait *See* Polygenic.

Quaternary structure The level of protein structure referring to the fitting together of two or more polypeptide chains in a three-dimensional configuration. (2)

Radial symmetry A circular body plan having a central axis from which structures radiate outward like the spokes of a wheel; characteristic of cnidarians, like jellyfish and sea anemones. (18)

Radicle The embryonic root of a plant embryo. (30)

Radioactive An unstable isotope that emits energy as it loses neutrons from the nucleus and changes into a more stable form of an element. (2)

Radula A rasping organ in mollusks that shreds plant material by rubbing it against the hardened surface of the mouth. (18)

Rain forest A biome characterized by a warm, wet climate, abundant vegetation, and diverse animal life. (37)

Random distribution A pattern in which members of a population do not influence each other's spacing, if environmental conditions are uniform. (34)

Reactant The starting substance in a chemical reaction. (4)

Reaction center In photosynthesis, the central chlorophyll molecule around which other pigment molecules are arranged. (6)

Reading frame The sequence of three nucleotides in mRNA that corresponds to an amino acid in a protein. There are three possible reading frames for any RNA: ACG GUA AUC, ACGG UAA UC, or AC GGU AAU C. (10)

Realized niche The part of the fundamental niche that a species actually occupies in nature. (35)

Receptor A protein of a specific shape that binds to a particular chemical. (26)

Recessive An allele or corresponding phenotypic trait that is only expressed in the homozygote. (8)

Recombinant DNA technology The techniques involved in excising DNA from one genome and inserting it into a foreign genome. (11)

Recombinant type An offspring with characteristics different from the parents'. (8)

Rectum The portion of the intestine between the colon and the anus that removes solid waste by defecation. (24)

Red tides Dense blooms of certain dinoflagellates that tint water red and produce deadly toxins. (16)

Reduction The addition of electrons to a molecule. (5)

Reflex arc An automatic reaction, involving only a few neurons and requiring no input from the brain, in which a motor response quickly follows a sensory stimulus. (27)

Refractory period The time period following passage of an action potential during which the nerve cell membrane is unable to react to additional stimulation. (27)

Refuge A safe haven out of the reach of predators. (35)

Regeneration The ability of an organism to regrow any lost parts of its body; e.g., a starfish can regenerate a lost arm. (7)

Regulator cell A cell that detects perturbations in the environment and secretes a molecular messenger in response. (26)

Reinforcer In operant conditioning, a reward or punishment. (38)

Renal artery One of two arteries from the aorta to the kidneys. (25)

Renal vein One of two large blood vessels through which blood leaves the kidneys. (25)

Renin An enzyme released in the kidney that converts a blood protein into the hormone angiotensin. (25)

Reproduction The method by which individuals give rise to other individuals of the same type. (1, 7)

Reproductive isolating mechanism Any structural, behavioral, or biochemical feature that prevents individuals of a species from successfully breeding with individuals of another species. (33)

Reptile A cold-blooded, scaly, lung-breathing vertebrate that lays large eggs which usually have a shell; the dominant group of animals in the Mesozoic era; includes crocodiles, lizards, and tortoises. (19)

Resolving power In microscopic viewing, the ability of the human eye to distinguish adjacent objects as distinct and separate. (3)

Resource partitioning The process by which resources are divided, allowing species with similar requirements to use the same resources in different areas, at different times, or in different ways. (35)

Respiration (L. *respirare,* to breathe) [1] The exchange of oxygen and carbon dioxide between cells and the environment. (23) [2] The oxidative breakdown and release of energy from fuel molecules. (5)

Responsiveness The tendency of a living thing to sense and react to its surroundings. (1)

Restriction enzyme A naturally occurring enzyme that cuts DNA at precise points. (11)

Reticular formation A network of neurons extending from the thalamus of the brain stem to the spinal cord; regulates respiration, cardiovascular centers, and awareness. (28)

Retina (L. a small net) A multilayered region lining the back of the vertebrate eyeball, containing light-sensitive cells. (28)

RFLP (restriction fragment length polymorphism) Genetic variation between individuals in the lengths of DNA that can occur when a piece of DNA is cut with a restriction enzyme. (12)

Rhipidistian An extinct lobe-finned fish, probably the first vertebrate to crawl out on land and live for an extended period. (19)

Rhizoid (Gr. *rhiza,* root) In bryophytes, some algae, and fungi, a hairlike structure that anchors the organism to the substrate. (17)

Rhizome An elongated underground horizontal stem. (17)

Rhodopsin The visual pigment in the rods and cones of the vertebrate eye that, activated by light energy, begins a series of events resulting in vision. (28)

Ribonucleotide A base (adenine, cytosine, guanine, or uracil) attached to a ribose sugar and a phosphate; the subunits of RNA. (10)

Ribosomal RNA (rRNA) An RNA molecule that is a structural component of ribosomes and is involved in protein synthesis. (3, 10)

Ribosome A structure in the cell that provides a site for protein synthesis. Ribosomes may lie freely in the cell or attach to the membranes of the endoplasmic reticulum. (3, 10)

Ribozyme An RNA molecule that can catalyze the cleavage or joining of itself and/or of other RNAs. (15)

Rift A juncture between two major rock plates in the earth's crust. (15)

RNA Ribonucleic acid; a nucleic acid similar to DNA except that it is single-stranded and contains the sugar ribose and the base uracil replaces thymine. (3, 7, 9, 10)

RNA polymerase An enzyme that makes RNA by copying the base sequence of a DNA strand. (10)

Rod One of the two types of photoreceptor (light-sensitive) cells in the retina of the vertebrate eye; rods are sensitive to low levels of light and to movement but cannot distinguish color. (*See also* Cone.) (28)

Root The branching structure of a plant that grows downward into the soil. Roots anchor the plant and absorb and transport water and mineral nutrients. (30)

Root cap A mass of cells covering and protecting the growing tissue of the root. (30)

Root hair An extension from the root epidermis that increases the absorptive capacity of a root. (30)

r-selection Selection experienced by populations that typically occurs at levels where competition for resources is unimportant; *r*-selected organisms are often short-lived, with rapid reproduction. (34)

Rough ER The part of the endoplasmic reticulum that is studded with ribosomes; synthesizes lysosomal and membrane proteins and secreted proteins. (3)

Roundworm A member of the phylum Nematoda. (18)

Rubisco Ribulose bisphosphate carboxylase, the enzyme that binds carbon to a biological molecule in photosynthesis; the world's most abundant protein. (6)

Saliva A watery liquid secreted by the salivary glands that moistens food particles for swallowing; contains the starch-digesting enzyme amylase. (24)

Salivary gland A gland that secretes saliva. (24)

Saprobe An organism that lives on decomposing organic matter. (16)

Sarcomere The contractile unit of a skeletal muscle, consisting of repeating bands of actin and myosin. (29)

Saturated fat A lipid containing fatty acids in which no carbon atoms are linked by double bonds, so that the molecule has the maximum possible number of hydrogen atoms. (2)

Savanna A tropical grassland biome containing stunted, widely spaced trees situated between tropical forests and deserts. (37)

Scanning electron microscope (SEM) An instrument used to observe the outer surfaces of cells and organisms at high magnification and great depth of field. (3)

Schistosomiasis A human disease caused by the blood fluke. (18)

Scientific method A series of steps for understanding the natural world based on experimental testing of hypotheses, possible mechanisms for how the world functions. (1)

Sclerenchyma In plants, a supporting tissue composed of cells with lignified cell walls. (30)

Scrotum The sac that contains the testes in male mammals. (14)

Secondary cell wall In certain plant cells, the rigid cell wall between the plasma membrane and the primary cell wall formed after cell elongation has ceased; wood consists of secondary cell walls. (3)

Secondary consumer In an ecosystem, an organism that consumes herbivores; carnivores (meat eaters) are secondary consumers. (36)

Secondary growth The enlarged diameter of a stem or root resulting from cell divisions in the lateral meristem. (30)

Secondary structure The level of protein structure in which the chain is molded into the form of an alpha helix or beta-pleated sheet. (2)

Second filial (F_2) generation In Mendelian genetics, the second generation in the line of descent. (8)

Second law of thermodynamics The second energy law, which states that because energy conversions from one form to another are not 100 percent efficient, all systems tend toward greater states of disorder. (4)

Second messenger A molecule within a cell that causes cellular proteins to change behavior in response to a signal received at the cell surface; cyclic AMP and calcium ions are examples. (26)

Secretin A digestive hormone secreted in the small intestine that triggers the pancreas to secrete bicarbonate, which neutralizes stomach acid. (24)

Seed The product of a fertilized ovule of a seed plant, generally consisting of an embryo with its food reserves enclosed in a protective coat. (17)

Seedless vascular plant Plants with an internal transport system but that require standing water to reproduce because they do not produce seeds; includes horsetails and ferns. (17)

Segmented worm A member of the phylum Annelida. (18)

Segregation principle Mendel's first law of heredity, which states that sexually reproducing diploid organisms have two alleles for each gene, and that during the gamete formation these two alleles segregate from each other so that the resulting gametes have only one allele of each gene. (8)

Selectively permeable membrane A membrane that allows water molecules to pass freely but prevents the passage of large molecules across its surface; also called *semipermeable membrane.* (3, 4)

Self-tolerance The lack of an immune system response to components of one's own body. (22)

Semen (L. *serere*, to sow) Sperm and its surrounding seminal fluid. (14)

Semicircular canal One of several fluid-filled, tubelike channels within the vertebrate inner ear that is sensitive to movement of the head and that maintains balance. (28)

Semiconservative replication The universal mode of DNA replication, in which only one of the two strands of DNA of the parent molecule is passed to a daughter molecule. This inherited strand serves as the template for the synthesis of the other half of the double helix. (9)

Seminal fluid In a vertebrate male, the secretions of the seminal vesicle, prostate gland, and bulbourethral gland in which sperm are suspended. (14)

Seminal vesicle An organ that stores sperm and adds secretions to the semen. (14)

Seminiferous tubules Hollow tubes inside the testes, where sperm formation takes place. (14)

Semipermeable membrane *See* Selectively permeable membrane.

Senescence A condition characterized by a profound decline in cell and organ function that occurs with aging. (14, 31)

Sense organ A group of specialized cells that receives stimulus energy and converts it into neural energy; for example, eyes and ears. (28)

Sensitization An increase in the strength of a response to a stimulus; the opposite of habituation. (27)

Sensory cortex The part of the cerebral cortex of the vertebrate brain that registers and integrates sensations from body parts. (28)

Sensory neuron A nerve cell that receives information from the external or internal environment and transmits this information to the brain or spinal cord. (27)

Septum (plural **septa**) A dividing wall or partition between structures, such as the segments of an earthworm. (18)

Serosa The outermost layer of the alimentary canal, attaching the gut to the inner wall of the body cavity. (24)

Sessile In animals, the quality of being permanently attached to a fixed surface. (18)

Seta (plural **setae**) Attached to segments of earthworms, a bristle that pushes against the ground and enables movement. (18)

Sex chromosomes Pairs of chromosomes where the members of the pair are dissimilar and involved in sex determination, such as the *X* and *Y* chromosomes. (8)

Sex-linked Characteristic of genes that are carried on the sex chromosomes and therefore show different patterns of inheritance between males and females. The gene for color-blindness in humans is sex-linked and is carried on the *X* chromosome. (8)

Sexual reproduction A type of reproduction in which new individuals arise from the mating of two parents. (7)

Sexual selection A type of natural selection in which individuals select a mate on the basis of physical or behavioral characteristics, regardless of the effect of that characteristic on general fitness. (33, 38)

Shoot The main portion of a plant growing above ground. (30)

Sieve plate One of the two perforated end walls of a sieve tube member. (30)

Sieve tube member A conducting cell of phloem tissue in a vascular plant. (30)

Sign stimulus A specific environmental signal that triggers a fixed-action behavioral pattern. (38)

Single bond A chemical bond between two atoms created by the sharing of a pair of electrons. (2)

Sinoatrial (SA) node A lump of modified heart muscle cells that is spontaneously electrically excitable; the pacemaker that governs the basic rate of heart contractions in vertebrates. (21)

Siphon In certain mollusks, the funnel through which water passes on its way through the mantle cavity. (18)

Skeletal muscle Muscle consisting of elongate, striated muscle cells; voluntary muscle. (29)

Skeleton The rigid body support to which muscles attach and apply force, in vertebrates and invertebrates. (29)

Skull The skeleton of the vertebrate head. (19)

Slime mold A protist that derives energy by secreting digestive enzymes outside the cell; the digested material is then absorbed into the cell. (16)

Slime trail The trail along which a snail or slug slowly moves. (18)

Smooth ER The part of the endoplasmic reticulum folded into smooth sheets and tubules, containing no ribosomes; synthesizes lipids and detoxifies poisons. (3)

Smooth muscle Muscle consisting of spindle-shaped, unstriated muscle cells; involuntary muscle found in the digestive, reproductive, and circulatory systems. (29)

Social insect A member of any of the insect groups, such as ants, bees, and termites, that lives in colonies of related individuals and exhibits complex social behaviors. (18)

Sociobiology The study of social behavior. (38)

Sodium channel A protein-lined pore in the plasma membrane through which sodium ions can pass. (26)

Sodium-potassium pump A carrier protein that maintains the cell's osmotic balance by transporting sodium ions out of the cell and potassium ions in. (4)

Soft palate The posterior part of the roof of the mouth. (24)

Soil The portion of the earth's surface consisting of disintegrated rock and decaying organic material. (32)

Solute A substance that has been dissolved in a solvent. (2)

Solvent A substance capable of dissolving other molecules. (2)

Soma The main cell body of a neuron. (27)

Somatic cell (Gr. *soma*, body) A cell in an animal that is not a germ cell. (7)

Somatic cell hybrid A cell derived from the fusion of two or more somatic (normal body) cells (not germ cells). Biologists use somatic cell hybrids between human and mouse cells to facilitate gene mapping. (12)

Somites Blocks of mesoderm that pinch off alongside the neural tube of a vertebrate embryo and develop into muscle blocks or other structures. (14)

Sorus (plural **sori**) A cluster of sporangia on the underside of a fern frond. (17)

Speciation The emergence of a new species. Speciation is thought to occur mainly as a result of populations becoming geographically isolated from each other and evolving in different directions. (33)

Species In taxonomy, a group of organisms whose members have the same structural traits and who can interbreed with each other. (15)

Species richness The total number of species in a community. (35)

Specific immune response The body's most powerful line of defense, in which certain white blood cells produce antibodies that attack invaders; each antibody is specific for a certain invader. (22)

Spermatogenetic cell A sperm-forming cell. (14)

Sphincter A ring of muscle surrounding a tube or tubal opening that controls the size of the opening, for example, the opening from the esophagus to the stomach. (24)

Spicule A slender, spiky rod of silica or calcium carbonate found in sponges that supports the soft wall and provides some protection from predators. (18)

Spider A member of the class Arachnida (phylum Arthropoda) that contains four pairs of legs on the thorax. (18)

Spinal cord   A tube of nerve tissue that runs the length of a vertebrate animal, just above (dorsal to) the notochord. (19)

Spindle   A cluster of microtubules that attach from the centriole to the centromere of a chromosome and move the chromosome toward the pole of the cell during cell division. (3)

Spinneret   A tubular appendage in spiders and some insects that reels out silk threads. (18)

Spiny-finned fish   A member of one of the two major groups of bony fishes; includes trout and zebra fish. (19)

Spiracle   A hole in the body wall of insects that forms the opening of the air tubes (trachea). (23)

Spirillum (plural **spirilla**)   A bacterial cell with a spiral shape. (16)

Sponge   A multicellular animal without true tissues, either radially symmetrical or asymmetrical. Special cells with flagella push water through pores in the body wall from which food particles are filtered. The remains of their skeletal elements make up the common bath sponge. (18)

Sporangium (plural **sporangia**)   A structure in which spores are produced. (17)

Spore   In eukaryotes, a reproductive cell that divides mitotically and produces a new individual. In prokaryotes, a resistant cell capable of surviving harsh conditions and germinating when conditions are once again favorable. (16, 17)

Sporophyte   (Gr. *spora*, seed + *phyton*, plant) The diploid, spore-producing stage in the life cycle of many plants. (17)

S-shaped curve   A curve on a graph representing logistic population growth. (34)

Stabilizing selection   Type of natural selection in which the intermediate form of a trait is favored over extreme forms. (33)

Stamen   The pollen-producing structure within a flower. (30)

Starch   A polysaccharide composed of long chains of glucose subunits; the principal energy source of plants. (2)

Start codon   In DNA and mRNA, the codon that signals where the translation of a portion begins. (10)

Stem   The central, often elongated part of a plant composed mainly of vascular tissue, that supports leaves and transports water and nutrients. (17)

Stem cell   A normal body (somatic) cell that can continue to divide, replacing cells that die (e.g., in the skin or blood) during an animal's life. (13)

Stereocilia   Threadlike projections on the hair cells of the cochlea of the inner ear. (28)

Stereoscopic vision   Overlapping visual fields with good depth perception, allowing animals to discriminate distances well. (19)

Sternum   The breastbone; in birds, a blade-shaped anchor for the pectoral muscles that enable flight. (19)

Steroid   (Gr. *stereos*, solid + L. *ol*, from + *oleum*, oil) A major class of lipids based on a 17-carbon-atom ring system and often a hydrocarbon tail. Cholesterol and sex hormones are steroids. (2)

Steroid hormone   A hormone synthesized from cholesterol containing four joined rings of carbon atoms; e.g., estrogen, testosterone, cortisol. (26)

Stigma   [1] The tiny, light-sensitive eyespot of a euglenoid. (16) [2] The sticky top of a flower that serves as a pollen receptacle. (17)

Stipe   The stemlike structure that provides vertical support to an alga. (17)

Stoma (plural **stomata**)   A tiny hole surrounded by two guard cells in the epidermis of plant leaves and stems, through which gases can pass. (30, 32)

Stomach   An expandable, elastic-walled sac of the gut that receives food from the esophagus. (24)

Stop codon   In DNA and mRNA, the codon that signals where the translation of a protein ends. (10)

Style   A structure leading from the ovary of flowering plants to the stigma (pollen receptacle). (17)

Submucosa   A connective tissue outside the mucosa of the digestive tract that is richly supplied with blood and lymph vessels and nerves. (24)

Substrate   A reactant in an enzyme-catalyzed reaction; fits into the active site of an enzyme. (4)

Succession   A progression of communities in a particular habitat, each one replacing the other until a persistent, stable, climax community is established. (35)

Sucrose   A disaccharide composed of glucose bonded to fructose, common in many plants; table sugar. (2)

Summation   The cumulative effect of individual postsynaptic potentials in a neuron. (27)

Superior vena cava   One of the two largest veins in the human body, carrying blood from the head, neck, and arms to the right atrium of the heart. (21)

Supporting cell   A cell in the seminiferous

tubule walls of the vertebrate testes that nourishes the developing sperm. (14)

Surface tension   The tendency of molecules at the surface of a liquid to cohere to each other rather than to the molecules of air above them. (2)

Surface-to-volume ratio   The ratio of the surface area of a cell to its volume; imposes limits on cell size because as the cell's linear dimensions grow, its surface area increases less than its volume. (3)

Survivorship curve   A plot of the data representing the proportion of a population that survives to a certain age. (34)

Symbiont   An organism that lives in a close relationship with an organism of another species. (16)

Symbiosis   A close interrelationship of two different species. (16)

Sympathetic nerves   Neurons of the autonomic nervous system that emanate from the central spinal cord and act on the respiratory, circulatory, digestive, and excretory systems in response to stress or emergency, preparing for flight or fight. (28)

Sympatric speciation   A situation in which a population diverges into two species after a genetic, behavioral, or ecological barrier to gene flow arises between subgroups of the population inhabiting the same region. (33)

Synapse   The region of communication between two neurons or between a neuron and a muscle. (27)

Synaptic vesicle   At the tip of a nerve cell's axon, a small round packet containing neurotransmitter molecules. (27)

Systemic circulation   The blood vessel system that carries oxygen-rich blood from the heart to the body and returns oxygen-poor blood to the heart. Contrasts to pulmonary circulation. (21)

Systole   The contraction phase of the heart muscle. (21)

Tail   In chordates, a structure that protrudes beyond the anus. (19)

Taproot   In plants, a root system with a prominent main root developing directly from the embryonic root, growing vertically downward and bearing lateral roots, e.g., the root of a carrot plant. (30)

Target cell   Any cell that responds to specific hormones. (26)

Taxis (plural **taxes**)   An animal's movement toward light, heat, or specific chemicals. (38)

Taxonomic group   A category based on shared characteristics; the groups are species, genus, family, order, class, phylum, kingdom, and domain. (15)

Taxonomy (Gr. *taxix,* arrangement + *nomos,* law) The science of classifying organisms into different categories. (15)

T cell (T lymphocyte) A type of white blood cell that kills foreign cells directly and also regulates the activities of other lymphocytes. (22)

Teleost A modern bony fish; includes most living fish; characterized by the presence of a swim bladder and spiny fins. (19)

Telophase The final phase of nuclear division when the chromosomes are at opposite poles of the cell, the nuclear membrane and nucleolus reappear, and the spindle disappears. (7)

Temperate forest A biome that occurs north or south of subtropical latitudes, characterized by generally mild climate and varied populations of evergreen and deciduous trees. (37)

Template A strand of DNA or RNA that serves as a pattern or model for the construction of a complementary strand of DNA or RNA. (9)

Tendon A connective tissue strap that connects bone to muscle. (29)

Termination The third main stage of protein synthesis, during which the growth of the polypeptide chain comes to a halt, when the ribosome reaches the stop codon. (10)

Territoriality A form of behavior in which an individual defends its living or feeding space against any intruders. (38)

Tertiary consumer In an ecosystem, a carnivore that eats other carnivores. (36)

Tertiary structure The level of protein structure referring to the three-dimensional folding of the entire polypeptide chain. (2)

Testcross The mating of an individual of unknown genotype with a homozygous recessive in order to determine the unknown genotype. (8)

Testis (plural **testes**) The male reproductive organ; produces sperm and sex hormones. (14)

Testosterone A hormone produced by the testis in vertebrate males; stimulates embryonic development of male sex organs, sperm production, male secondary sex characteristics, and male behaviors. (14)

Thalamus Located beneath the cerebrum and above the hypothalamus of the vertebrate brain; relays and processes sensory impulses. (28)

Thallus The flat, leaflike body of some algae and other lower plants. (17)

Theory A broad general hypothesis that is repeatedly tested but never disproved. (1)

Thoracic cavity The region within the rib cage directly over the heart, in which the lungs are suspended. (23)

Thorax The central region of the body of an arthropod or vertebrate between the head and the abdomen. (18)

Thrombocyte *See* Platelet.

Thylakoid (Gr. *thylakos,* sac + *oides,* like) A stack of flattened membranous discs containing chlorophyll found in the chloroplasts of eukaryotic cells. (3, 6)

Thymus A gland in the neck or thorax of many vertebrates; makes and stores lymphocytes in addition to secreting hormones. (21)

Thyroid gland A large endocrine gland in the neck of vertebrates regulating the body's growth and use of energy by secreting thyroxine. (26)

Thyroxine The most abundant thyroid hormone, which governs metabolic and growth rates and stimulates nervous system function. (26)

Tidal ventilation The in-and-out flow of air that characterizes mammalian respiration. (23)

Tide pool Along the ocean shore, a depression formed in rock by wave action. (37)

Tight junction A band around some eukaryotic cells where the plasma membrane touches the membrane of an adjacent cell, forming a tight seal and preventing passage of substances in the space between adjacent cells. (3)

Tissue A group of cells of the same type performing the same function within the body. (20)

Tissue culture The growth of tissue cells "in the test tube" in a way that often permits further differentiation and maturation of the cells. In plant breeding, the growing of new identical plants from somatic (body) cells. (32)

Tongue A muscular organ on the floor of the mouth of most higher vertebrates that carries taste buds and manipulates food. (24)

Trachea The "windpipe," the major airway leading into the lungs of vertebrates. (23)

Tracheae Branching networks of hollow air passages used for gas exchange in insects. (18, 23)

Tracheid An elongated hollow cell with thick, rigid, pitted walls; a basic unit of vascular tissue in most vascular plants. (30)

Transcription (L. *trans,* across + *scribere,* to write) The formation of RNA from a single strand of a DNA molecule; the process is catalyzed by the enzyme RNA polymerase. (10)

Transfer RNA (tRNA) A small RNA molecule that translates a codon in mRNA into an amino acid during protein synthesis. (10)

Transformation The process of transferring an inherited trait by incorporating a piece of foreign DNA into a prokaryotic or eukaryotic cell. (9)

Transition state In a chemical reaction, the intermediate springlike state between reactant and product. (4)

Translation The conversion of the information on a strand of RNA into a sequence of amino acids in a protein; occurs on ribosomes. (10)

Translocation In plants, the transport of solutes in phloem cells. (32)

Transmission electron microscope (TEM) An instrument used to observe the inner structures of thin stained sections of cells at high magnification. (3)

Transpiration The loss of water from plants by evaporation, mainly through the stomata on stems and leaves. (32)

Transpiration pull theory The theory that water is transported in plants by the sun's energy evaporating water from leaves via transpiration and, due to the cohesion of water molecules, pulling the entire column of water up from the soil; also called *cohesion-adhesion-tension* theory. (32)

Transposon A gene that can move around the genome, within or between chromosomes; also called a *jumping gene.* (11)

Trial-and-error learning *See* Operant conditioning.

Triglyceride A fat or oil composed of three fatty acids joined to the three carbons of glycerol. (2)

Trimester Any of the following three-month periods of pregnancy: first to third, fourth to sixth, or seventh to ninth months. (14)

Trophic level A particular feeding level in a community or ecosystem. (35)

Tropism (Gr. *trope,* turning) The movement of a plant in response to external or internal stimuli; e.g., phototropism, the response of plants to light. (31)

True hormone A chemical produced by nonneural cells in one part of the body that has an effect on another part of the body. (26)

True slime mold A type of funguslike protist characterized by the plasmodium, a mass of continuous cytoplasm surrounded by one plasma membrane containing many diploid nuclei. (16)

Tubular reabsorption In the vertebrate kidney, the process whereby water, salts, and nutrients are returned from the kidney tubule to the blood. (25)

Tubular secretion In the vertebrate kidney, the process whereby ions and drugs are secreted from the blood into the kidney tubule. (25)

Tubulin   A globular protein that makes up the microtubules. (3)

Tumor   An uncontrolled or abnormal growth of cells. (7)

Tundra   A biome characterized by cold, treeless plains with grasses, lichens, and occasional small shrubs; deep tundra soil that remains frozen permanently is called permafrost. (37)

Tunic   The outer envelope of a sea squirt, enclosing the pharynx. (19)

Turgor pressure   (L. *turgor,* a swelling) Internal pressure on a cell wall caused by osmotic movement of water into the cell. (4)

Turner syndrome   Features exhibited by a human female with only one *X* chromosome (*XO*); includes sterility due to rudimentary or missing ovaries. (12)

Ultimate cause   The selective advantage of a particular action or behavior. (38)

Umbilical cord   A cord joining the embryo and the placenta. (14)

Uniform distribution   A pattern in which members of a population are spaced at regular intervals in their range. (34)

Uniformity   The scientific principle that the laws of nature always have operated and always will operate the same way. (1)

Unsaturated fat   A lipid containing fatty acids in which two or more carbon atoms are linked by double bonds, so that there are fewer than the maximum possible number of hydrogen atoms. (2)

Urea   A water-soluble nitrogenous waste product formed during the breakdown of proteins, nucleic acids, and other substances, excreted by mammals and some fishes. (25)

Ureter   The tube that carries urine from the kidney to the bladder. (25)

Urethra   (Gr. *ouerin,* to urinate) The tube that carries urine and releases it to the outside; in males this tube also carries sperm. (14, 25)

Uric acid   A water-insoluble nitrogenous waste product excreted by birds, land reptiles, and insects. (25)

Urination   The process by which urine exits the body through the urethra. (25)

Urine   A fluid that washes uric acid or urea from the body. (25)

Uterus   (L. womb) The thick-walled chamber where the embryo develops. (14)

Vagina   A hollow muscular tube that receives the penis during copulation and through which the fetus passes during birth. (14)

Valve   A tonguelike flap extending into the internal space of a vein that helps regulate the flow of blood. (21)

Vasoconstriction   Contraction of blood vessel walls; regulates blood flow. (21)

Vasodilation   Relaxation of blood vessel walls; regulates blood flow. (21)

Vascular cambium   A type of lateral meristem that produces secondary phloem and secondary xylem, or wood. (30)

Vascular plants   Plants that possess an internal transport system for water and food in the form of xylem and phloem cells. (17)

Vascular tissue system   Plant tissue that conducts fluid throughout the plant and helps strengthen roots, stems, and leaves; consists of xylem and phloem cells. (30)

Vas deferens   (L. *vas,* a vessel + *defere,* to carry down) A tube that carries sperm from the epididymis to the ejaculatory duct in male vertebrates. (14)

Vegetative reproduction   A process in which new individuals genetically identical to each other and to the parent emerge from the parent's body. (17)

Vein   [1] A large thin-walled blood vessel that brings blood from the body to the heart. (21) [2] In plant leaves, bundles of xylem and phloem cells that form a branching pattern in dicots and run parallel to each other in monocots. (30)

Ventilation   The breathing in and out of the lungs. (23)

Ventricle   A muscular chamber of the heart that pumps blood to the lungs or to the rest of the body. (21)

Venule   A small thin-walled blood vessel that arises from capillaries and carries blood from the tissues to the veins. (21)

Vertebra   One of a series of interlocking bones that makes up the backbone of vertebrate animals. (19, 29)

Vertebral column   The backbone. (29)

Vertebrate   An animal that possesses a backbone made of bony segments known as vertebrae. (18)

Vesicle   A small membranous container in a cell. (3)

Vessel   [1] A tubelike structure that carries blood. (18) [2] A tubelike structure of xylem that conducts water in plants. (30)

Vessel member   In the xylem of flowering plants, a hollow cell with pitted walls, stacked end to end, through which water flows from the roots to the rest of the plant. (30)

Vestigial organ   A rudimentary structure with no apparent utility but bearing a strong resemblance to structures in probable ancestors. (33)

Vibrio   A bacterial cell with a curved rod shape. (16)

Villus (plural **villae**)   (L. a tuft of hair) A fingerlike projection of the intestinal wall that increases the surface area for absorption of nutrients. (24)

Viroid   An intracellular parasite that affects plants, consists only of small RNA molecules without any protein coat. (16)

Virus   An infectious agent consisting of RNA or DNA encased in a protein coat (capsid); incapable of metabolism or reproduction outside a host cell. (16)

Vitamin   (L. *vita,* life + *amine,* of chemical

origin) An organic compound needed in small amounts for growth and metabolism obtained in food. (24)

Warning coloration   Bright colors or striking patterns of prey species that warn predators to stay away; also known as aposematic coloration. (35)

Water cycle   A sun-driven global exchange involving evaporation, precipitation, runoff, and transpiration that cycles water from the atmosphere to the earth's surface through organisms and back again. (36)

Water mold   A type of funguslike protist containing several nuclei within a common cytoplasm and forming relatively large immobile egg cells; *Oomycota.* (16)

Water vascular system   In echinoderms, a system of fluid-filled canals that includes

hundreds of short branches called tube feet that can attach to objects and thus aid in locomotion or feeding. (18)

Wax   A sticky, solid, waterproof lipid that forms the comb of bees and waterproofing of plant leaves. (2)

Weather   The condition of the atmosphere at any particular place and time, including its temperature, humidity, wind speed, and precipitation. (37)

Wind   Air flowing over the earth. (37)

Wood   The hard, cellulose-containing dead xylem cells of a perennial plant. (30)

X **chromosome**   The sex chromosome found in two doses in female mammals, fruit flies, and many other species. (8)

X-**linked**   Characteristic of a heritable trait that occurs on the *X* chromosome. (8)

Xylem   (Gr. *xylon,* wood) Water- and mineral-transporting tissue of plants composed of tracheids, vessels, parenchyma cells, and fibers; dead xylem cells form wood. (30)

Y **chromosome**   The sex chromosome found in a single dose in male mammals, fruit flies, and many other species. (8)

Yolk   Material consisting of proteins, carbohydrates, nucleic acids, and lipids stored in an animal egg as a nutrient supply for a developing embryo. (13)

Zygote   (Gr. *zygotos,* paired together) The diploid cell that results from the fusion of an egg and a sperm cell. A zygote may either form a line of diploid cells by a series of mitotic cell divisions or undergo meiosis and develop into haploid cells. (7)

Chapter 9
Figure 9.1: R. Langridge/D. McCoy/Rainbow. Figure 9.2: Photo Researchers, Inc. Figure 9.3: (a) Maclyn McCarty, Rockefeller University. Figure 9.4: (b) K. Kleinschmidt et al., *Biochimica et Biophysica Acta* 61 (1962): 857. Figure 9.5: Hank Morgan/Photo Researchers, Inc. Figure 9.6: C. W. Schwartz/Animals Animals. Box 9.1, Figure 1: John Postlethwait. From *Science* 250 (1990). Copyright 1990 by the American Association for the Advancement of Science. Figure 9.9: Ruth Kavenoff/©Designer Genes Posters Ltd. Figure 9.10: (e) Ulrich K. Laemmli, *Cell* 55 (1988): 937–944. Figure 9.12: Sheldon Wolff and Judy Bodgcote, Laboratory of Radiobiology and Environmental Health, University of California, San Francisco. Box 9.2, Figure 1: (b) Peter Magubane/Gamma-Liaison. Figure 9.13: (c) W. L. Nyhan and N. A. Sakati, *Diagnostic Recognition of Genetic Disease* (Lea & Febiger, Philadelphia, 1987). Figure 9.15: Jackie Lewin/Photo Researchers, Inc. Box 9.3, Figure 1: Jonathan T. Wright/Bruce Coleman, Inc.

Chapter 10
Figure 10.1: Cystic Fibrosis Foundation. Figure 10.8: (b) Elena Kiselava. Figure 10.9: (a) S. J. Krasemann/Photo Researchers, Inc.; (b) Richard Kolar/Animals Animals. Figure 10.12: *Oxford Textbook of Medicine*, D. J. Weatherall, ed. (Oxford University Press, Oxford, 1987). Figure 10.13: (a) James E. Cleaver, Ph.D., and Mary C. Curry, M.D., University of California, San Francisco. Figure 10.16: (c) Brian Matthews, University of Oregon. Figure 10.19: Conrad E. Firling, University of Minnesota. Box 10.1, Figure 1: Alexander Lowry/Photo Researchers, Inc. Figure 10.20: Howard Sochurek.

Chapter 11
Figure 11.1: Dr. R. L. Brinster, Laboratory of Reproductive Physiology, University of Pennsylvania. Box 11.1, Figure 1: Wide World Photos, Inc. Figure 11.6: Ted Spiegel. Box 11.2, Figure 1: Fred McDonald. Figure 11.7: Monsanto Company. Figure 11.8: (a, b) Steven A. Rosenberg, *Scientific American* (May 1990): 65. Figure 11.9: Thomas W. Sasek, Duke University.

Chapter 12
Figure 12.1: Milupa Corporation. Figure 12.3: (a–c) D. D. Whitney, University of Nebraska. Reprinted from *Journal of Heredity* 30 (1969): F1; (d) C. Stern, W. R. Wall, S. S. Sarkow, "New Data on the Problem of Y-Linkage of Hairy Pinnae," *American Journal of Human Genetics* 16 (1964): 467. Figure 12.6: J. D. Spillane. Reprinted from *The Metabolic Basis of Inherited Disease*, John B. Stanbury, ed., 2:1262, Figure 52.2. Figure 12.7: (a, b) Wide World Photos, Inc. Figure 12.8: CNRI/SPL/Photo Researchers, Inc. Box 12.1, Figure 1: The Bettmann Archive. Figure 12.9: (a) Mike and Moppet Reed/Animals Animals. Figure 12.10: (b) *Science* 231 (1986): 920. Copyright 1986 by the American Association for the Advancement of Science. Figure 12.11: (a) G. Turner, Prince of Wales Children's Hospital; (b) Christine J. Harrison, *Journal of Medical Genetics* 20 (1983): 280. Figure 12.12: American Psychological Association. Box 12.1, Figure 1: From Nancy Wexler, photo by Steve Uzzell. *Science* 232 (1986): 317. Copyright 1986 by the American Association for the Advancement of Science. Box 12.3, Figure 1: Stephen Ferry/Gamma-Liaison.

Chapter 13
Figure 13.1: Dr. Adam Felsenfeld. Figure 13.2: (a–h) Don Kane, University of Oregon. Figure 13.3: (a) Gregory Dimijian/Photo Researchers, Inc.; (b) Scott Camazine/Photo Researchers, Inc. Figure 13.4: G. Schatten/Photo Researchers, Inc. Figure 13.5: (b, c) Jorge Blaquier, *Fertility and Sterility* 47: 302. Figure 13.6: (a) G. Schatten/Photo Researchers, Inc. Figure 13.8: John Gerhart, from S. D. Black and J. C. Gerhart, *Development Biology* 116 (1986): 228–240. Figure 13.9: (c) Charles Mayer/Photo Researchers, Inc. Figure 13.12: (b) R. Fink, Mt. Holyoke College; (c) Neil G. McDaniel/Photo Researchers, Inc. Figure 13.14: (c) N. K. Wessels, Stanford University. Box 13.1, Figure 1: (a, b) John Postlethwait, University of Oregon; (c) C. K. Tuggle, et al., *Genes and Development* 4 (1990): 180–189. Figure 13.16: (bottom left) Science Source/Photo Researchers, Inc.; (bottom right) John Ciannicehi/Photo Researchers, Inc. Figure 13.18: (b) Photo Researchers, Inc. Figure 13.20: Kristine S. Vogel, University of Oregon. Figure 13.21: (a) Mark McKenna; (b) Julian Heath, Baylor College of Medicine, Houston. From *Discover* (Oct. 1989): 26.

Chapter 14
Figure 14.1: Hank Morgan/Time Magazine. Figure 14.2: Wide World Photos, Inc. Figure 14.7: (d) Manfred Kage/Peter Arnold, Inc. Figure 14.9: Science Source/Photo Researchers, Inc. Box 14.1, Figure 1: Carl Djerassi, Stanford University. Figure 14.11: Science Source/Photo Researchers, Inc. Box 14.2, Figure 1: G. W. Willis/Biological Photo Service. Figure 14.14: Leonard McCombe/Time, Inc. Figure 14.15: Charles Belinsky/Photo Researchers, Inc. Figure 14.16: Wide World Photos, Inc. Figure 14.19: Wide World Photos, Inc. Figure 14.21: Richard Hutchings/Info Edit.

Chapter 15
Figure 15.1: Larry Lefever/Grant Heilman Photography, Inc. Figure 15.2: Alexander Zehnder, Agricultural University, The Netherlands. Figure 15.3: Comstock. Figure 15.5: Chesley Bonestell. Figure 15.6: (a) Andrew Fracknoi/Photo Researchers, Inc. Figure 15.7: Roger Ressmeyer/Starlight. Figure 15.9: Pieter Cullis and Michael Hope, University of British Columbia. From *Scientific American* 256 (1987): No. 1, p. 20. Figure 15.10: Fred Bavendam/Peter Arnold, Inc. Box 15.1, Figure 1: Wide World Photos, Inc. Figure 15.14: Adrienne T. Gibson/Animals Animals. Box 15.2, Figure 1: Nigel Stork, Department of Entomology, Natural History Museum, London.

Chapter 16
Figure 16.1: Biophoto Associates/Science Source/Photo Researchers, Inc. Figure 16.3: L. J. LeBeau, University of Illinois Hospital/Biological Photo Service. Figure 16.4: (a) Science Source/Photo Researchers, Inc.; (b) John Cunningham/Visuals Unlimited; (c) David M. Phillips, Visuals Unlimited; (d) Science Source/Photo Researchers, Inc.; (e) Ralph Robinson/Visuals Unlimited. Figure 16.5: Ken Graham. Figure 16.6: Alfred Pasieka/Photo Researchers, Inc. Figure 16.7: H. Vali, *Nature* 343 (1990): 213. Copyright © 1990 Macmillan Magazines Ltd. Figure 16.8: Helen Carr/Biological Photo Service. Figure 16.9: Biological Photo Service. Figure 16.10: Manfred Kage/Peter Arnold, Inc. Figure 16.11: Arthur J. Olson, The Scripps Research Institute. Figure 16.12: Kichara Homata/PhotoNats. Box 16.1, Figure 1: Arthur J. Olson, The Scripps Research Institute. Figure 16.14: (a) J. Paulin/Biological Photo Service; (b) John Cunningham/Visuals Unlimited. Figure 16.15: Eric Gravé/Photo Researchers, Inc. Figure 16.16: (a) Robert Gwadz, *Science* 247 (1990). Copyright 1990 by the American Association for the Advancement of Science; (b) Science Source/Photo Researchers, Inc. Figure 16.17: Manfred Kage/Peter Arnold, Inc. Figure 16.18: (a) F. J. R. Taylor and G. Gaines, Institute of Oceanography, University of British Columbia, Vancouver; (b) Science Source/Photo Researchers, Inc. Figure 16.19: Manfred Kage/Peter Arnold, Inc. Figure 16.20: (b) Stephen P. Parker/Photo Researchers, Inc. Figure 16.21: Nichlos Devore III/Bruce Coleman, Inc.

Chapter 17
Figure 17.1: J. Baylor Roberts/National Geographic Society. Figure 17.2: (b) Gwen Fidler/Comstock. Figure 17.5: (a) D. H. Marx/Visuals Unlimited. Figure 17.6: S. N. Postlethwait. Figure 17.7: Doug Wechsler/Earth Scenes. Figure 17.8: (a) W. H. Hodge/Peter Arnold, Inc.;

(b) G. H. Harrison/Grant Heilman Photography, Inc. Figure 17.12: Breck P. Kent/Animals Animals. Figure 17.13: Robert B. Evans/Peter Arnold, Inc. Box 17.1, Figure 1: Steven D. Aust, Utah State University. Figure 17.14: Biophoto Associates/Photo Researchers, Inc. Figure 17.15: Photo Researchers, Inc. Figure 17.17: S. N. Postlethwait. Figure 17.18: (b) J. R. Waaland, University of Washington/Biological Photo Service; (c) J. N. A. Lott, McMaster University/Biological Photo Service. Figure 17.20: (a) S. N. Postlethwait; (b) Barry Runk and Stanley Schoenberger/Grant Heilman Photography, Inc. Figure 17.21: Photo Researchers, Inc. Figure 17.23: (a) John Serraro/Visuals Unlimited; (b) John Colwell/Grant Heilman Photography, Inc.; (c) Tom J. Ulrich/Visuals Unlimited. Figure 17.25: A. Nelson/Animals Animals.

Chapter 18
Figure 18.1: E. S. Ross. Figure 18.4: (top) S. Conway Morris; (bottom) Marianne Collins. Figure 18.5: Nancy Seton/Photo Researchers, Inc. Figure 18.6: (a) Tom Branch/Photo Researchers, Inc.; (b) Glenn Oliver/Visuals Unlimited; (c) Steve Early/Animals Animals; (d) Claudia Mills, University of Washington/Biological Photo Service. Figure 18.8: Joe Ghiold, *New Scientist* (June 2, 1990). Figure 18.10: (a) Michael Abbey/Photo Researchers, Inc. Figure 18.11: Armed Forces Institute of Pathology. Figure 18.12: Ed Reschke/Peter Arnold, Inc. Figure 18.14: Ken Lucas/Biological Photo Service. Figure 18.16: (a) Breck P. Kent/Animals Animals; (b) David J. Wrobel/Biological Photo Service. Figure 18.17: Joy Spurr/Bruce Coleman, Inc. Figure 18.18: (a) Oxford Scientific Films/Animals Animals; (b) David J. Wrobel/Biological Photo Service. Figure 18.19: (a) Fred McConnaughey/Photo Researchers, Inc.; (b) Seung K. Kim. Figure 18.22: (a) K. E. Lucas/Biological Photo Service; (b) John Gerlach/Visuals Unlimited. Figure 18.24: Oxford Scientific Films/Animals Animals. Figure 18.25: (a) D. Ellis/Visuals Unlimited; (b) Michael Fogden/Animals Animals. Figure 18.26: Robert Redden/Animals Animals. Figure 18.27: Karl Maslowski/Visuals Unlimited. Figure 18.28: Peter J. Bryant/Biological Photo Service. Figure 18.29: (a) Robert F. Myers/Visuals Unlimited; (b) David Hall, Photo Researchers, Inc. Figure 18.30: (b) G. I. Bernard/Animals Animals.

Chapter 19
Figure 19.1: Stephen Dalton/Animals Animals. Figure 19.5: (c) John Cunningham/Visuals Unlimited. Figure 19.7: Tom Stack and Associates. Figure 19.9: (a) Zig Leszczynski/Animals Animals; (b) L. L. T. Rhodes/Animals Animals. Figure 19.10: (a) Tom McHugh/

Photo Researchers, Inc.; (b) Zig Leszczynski/Animals Animals; (c) Gary Retherford/Photo Researchers, Inc. Figure 19.11: (a) Zig Leszczynski/Animals Animals; (b) Photo Researchers, Inc. Figure 19.13: (a) John Nees/Animals Animals; (b) San Diego Zoo; (c) John Nees/Animals Animals; (d) Frans Lanting/Photo Researchers, Inc. Box 19.1, Figure 1: Field Museum of Natural History Neg. No. GN 84985.11c. Figure 19.15: (a) Bruce Coleman, Inc.; (b) San Diego Zoo. Figure 19.17: (a) Science Source/Photo Researchers, Inc.; (b) John Chellman/Animals Animals; (c) Leonard Rue III/Visuals Unlimited; (d) Ted Levin/Animals Animals. Figure 19.19: Michael Dick/Animals Animals. Figure 19.20: M. Austerman/Animals Animals. Figure 19.21: Mickey Gibson/Animals Animals. Figure 19.22: (a, b) San Diego Zoo; (c) Lawrence Migdale/Photo Researchers, Inc. Figure 19.25: David L. Brill, National Geographic Society. Figure 19.26: (a) Anthropology Archives, Smithsonian Institution; (b) Musée de l'Homme.

Chapter 20
Figure 20.1: François Gohier/Photo Researchers, Inc. Figure 20.3: Tom J. Ulrich/Visuals Unlimited. Figure 20.6: (c) Ed Reschke/Peter Arnold, Inc. Figure 20.7: Ed Reschke/Peter Arnold, Inc. Figure 20.8: (a) G. W. Willis/Biological Photo Service; (b) Manfred Kage/Peter Arnold, Inc. Figure 20.9: G. C. Kelley/Photo Researchers, Inc. Box 20.1, Figure 1: Larry Mulvehill/Photo Researchers, Inc. Figure 20.13: J. Weber/Visuals Unlimited. Box 20.2, Figure 1: Stephen J. Krasemann/Photo Researchers, Inc.

Chapter 21
Figure 21.1: E. S. Ross. Figure 21.3: David M. Phillips/Visuals Unlimited. Figure 21.7: (c) CBS/Visuals Unlimited. Figure 21.8: Custom Medical Stock Photo. Figure 21.11: Jerry L. Ferrara/Photo Researchers, Inc. Box 21.1, Figure 1: Howard Sochurek. Figure 21.12: (a) Karen Pruess; (b) Alan Carey/The Image Works. Figure 21.13: David M. Phillips/Visuals Unlimited.

Chapter 22
Figure 22.1: NIBSC/Science Photo Library/Photo Researchers, Inc. Figure 22.2: Gad Gross/JB Pictures. Figure 22.5: Lennart Nilsson/Boehringer Ingelheim International, GMBH. Figure 22.8: (c) Ian Wilson, Research Institute of Scripps Clinic. From Stanfield et al., *Science* 248 (1990): 712–719. Copyright 1990 by the American Association for the Advancement of Science. Box 22.1, Figure 1: Custom Medical Stock Photo. Figure 22.11: Lennard Nilsson/Boehringer Ingelheim International, GMBH. Figure 22.13: (b) David

Weintraub/Photo Researchers, Inc. Figure 22.15: James Stevenson/Photo Researchers, Inc.

Chapter 23
Figure 23.1: Kjell B. Sandved. Figure 23.2: Tom McHugh/Photo Researchers, Inc. Figure 23.3: Carl Roessler/Bruce Coleman, Inc. Figure 23.4: Victor H. Hutchison/Visuals Unlimited. Figure 23.6: Harry Howard, University of Oregon. Figure 23.9: (a) Michael Gabridge/Visuals Unlimited; (b) Martin Rotker. Figure 23.11: Jonathan T. Wright/Bruce Coleman, Inc. Figure 23.13: Bill O'Connor/Peter Arnold, Inc.

Chapter 24
Figure 24.1: Keith Gillett/Animals Animals. Figure 24.3: Kenneth C. Balcomb/EarthViews. Figure 24.6: Gerard Vandystadt/Photo Researchers, Inc. Box 24.1, Figure 1: David M. Phillips. Figure 24.8: (a) Tim Rock/Animals Animals; (b) Kim Taylor/Bruce Coleman, Inc. Figure 24.10: John Postlethwait, University of Oregon.

Chapter 25
Figure 25.1: Tom McHugh/Photo Researchers, Inc. Figure 25.5: John Cunningham/Visuals Unlimited. Figure 25.6: Stouffer Productions/Animals Animals. Figure 25.10: Peter Andrews, George Washington University Medical Center. Figure 25.12: Heinz Kluetmeier/Sports Illustrated. Box 25.1, Figure 1: Richard Hutchings/Photo Researchers, Inc. Figure 25.15: (a) Richard LaVal/Animals Animals; (b) R. Ingo Riepl/Animals Animals. Figure 25.18: John Postlethwait, University of Oregon.

Chapter 26
Figure 26.1: (a) William D. Griffin/Animals Animals; (b, c) Wallace Kirkland/Animals Animals; (d) Breck P. Kent/Animals Animals. Figure 26.3: Mark Jenike/Anthro-Photo. Figure 26.10: Elizabeth Crews/The Image Works. Figure 26.11: Lester Bergman and Associates. Box 26.1, Figure 1: Laurence Frank, University of California, Berkeley. Figure 26.14: Breck P. Kent/Animals Animals. Box 26.2, Figure 1: David Madison/DUOMO.

Chapter 27
Figure 27.1: Chemistry and Medical Departments, Brookhaven National Laboratory. Figure 27.2: CBS/Visuals Unlimited. Figure 27.7: (a) Steinhart Aquarium/Photo Researchers, Inc.; (b) A. L. Hodgkin and R. D. Keynes, "Experiments on the Injection of Substances into Squid Giant Axons by Means of a Microsyringe," *Journal of Physiology* 131 (1956):

615. Figure 27.8: (c) Carol Readhead et al., "Expression of a Myelin Basic Protein Gene in Transgenic Shiverer Mice: Correction of the Dismyelinating Phenotype," *Cell* 48 (1987): 703. Figure 27.10: (a) From Stuart Ira Fox, *Human Physiology,* 2d ed. (William C. Brown Publishers, Dubuque, Iowa, 1987). Figure 27.11: Frank R. George/National Institute of Drug Abuse. Figure 27.13: Daniel W. Gotshall/ Visuals Unlimited.

Chapter 28
Figure 28.1: Stephen Dalton/Photo Researchers, Inc. Figure 28.3: Milton H. Tierney, Jr./ Visuals Unlimited. Figure 28.5: *Discover* (Nov. 1982): 95. Figure 28.8: Lennart Nilsson. Figure 28.9: (a, b) Richard G. Kessel and Randy H. Kardon, *Tissues and Organs: A Text-Atlas of Scanning Electron Microscopy,* p. 135, Fig. 5 and p. 133, Fig. 2 (W. H. Freeman and Co., San Francisco, 1979). Figure 28.10: C. S. Goodman et al., *Science* 225. Copyright by the American Association for the Advancement of Science. Figure 28.14: Jonathan Wright/Bruce Coleman, Inc. Figure 28.15: (b) Science Source/Photo Researchers, Inc. Box 28.1, Figure 1: (a, b) Visuals Unlimited.

Chapter 29
Figure 29.1: Skylar Hansen. Figure 29.2: Lawrence Migdale/Photo Researchers, Inc. Figure 29.3: Roger Jackman/Animals Animals. Figure 29.4: F. Collet/Photo Researchers, Inc. Figure 29.7: (b) Andreas Feininger, Life Magazine © Time Warner, Inc.; (c) Fred Hossler/Visuals Unlimited. Figure 29.8: (b) Martin M. Rotker. Figure 29.9: (b) John McDonough/Sports Illustrated. Figure 29.11: (c) C. Franzini-Armstrong, University of Pennsylvania. Figure 29.14: (a) David Madison/DUOMO; (b) J. D. MacDougall, McMaster University, Hamilton, Ontario, Canada; (c) Bonnie Kamin/Comstock.

Chapter 30
Figure 30.1: Grant Heilman/Grant Heilman Photography, Inc. Figure 30.2: Jerome Wexler/ Photo Researchers, Inc. Figure 30.4: (a) Barry Runk and Stanley Schoenberger/Grant Heilman Photography, Inc.; (b) John Cunningham/ Visuals Unlimited; (c) George Harrison/Grant Heilman Photography, Inc. Figure 30.5: (a, b) Eric Schabtach, University of Oregon. Figure 30.6: (a–c) George Wilder/Visuals Unlimited. Figure 30.7: (a) From *Three-Dimensional Structure of Wood,* B. A. Meylan and B. G. Butterfield, p. 65 (Syracuse University Press, 1972); (b) G. Shih and R. Kessel/Visuals Unlimited; (c) From *Biology,* 3d ed., by H. Curtis (Worth Publishers, New York, 1979). Figure 30.9: (a) R. Humbert/Biological Photo Service; (b) Dick Thomas/Visuals Unlimited;

(c) William Ferguson. Figure 30.10: (a) Harry Howard, University of Oregon; (b) Barry Runk and Stanley Schoenberger/Grant Heilman Photography, Inc. Box 30.1, Figure 1: Peter Angelo Simon/Phototake. Figure 30.11: (b) S. N. Postlethwait. Figure 30.12: (a) William J. Weber/Visuals Unlimited; (b) Breck P. Kent/ Earth Scenes. Figure 30.13: J. R. Waaland, University of Washington/Biological Photo Service. Figure 30.14: (a) Barry Runk and Stanley Schoenberger/Grant Heilman Photography, Inc.; (b) From Gene Shih and Richard Kessel, *Living Images,* p. 41, Fig. 3 (Science Books International, 1982; present publishers, Jones and Bartlett). Figure 30.15: (a, b) Harry Howard, University of Oregon; (e) Jim Solliday/Biological Photo Service. Box 30.2, Figure 1: Grant Heilman Photography, Inc. Figure 30.17: (a, b) Ray F. Evert, University of Wisconsin, Madison. Figure 30.18: (d) John Cunningham/Visuals Unlimited. Figure 30.19: (b) R. Gregory/Visuals Unlimited. Figure 30.20: (a) U.S. Forest Service. Figure 30.21: (a, b) S. N. Postlethwait. Figure 30.22: (a) S. N. Postlethwait; (b) John Cunningham/Visuals Unlimited; (c) William Ferguson. Figure 30.23: S. N. Postlethwait.

Chapter 31
Figure 31.1: Library of Congress. Figure 31.2: International Rice Research Institute, Manila, Philippines. Figure 31.3: Sylvan H. Wittwer, Michigan State University. Figure 31.4: David Newman/Visuals Unlimited. Figure 31.6: (c) S. N. Postlethwait. Figure 31.7: Terri Lomax, Oregon State University. Figure 31.9: Oxford Scientific Films/Earth Scenes. Figure 31.10: Frank B. Salisbury, Utah State University. Figure 31.11: (a) Mark McKenna. Figure 31.12: (a) David Newman/Visuals Unlimited. Figure 31.13: Loren McIntyre. Figure 31.16: (a) David M. Doody/Tom Stack and Associates; (b) University of California Statewide IPM Project. Figure 31.17: Larry Lefever/Grant Heilman Photography, Inc. Figure 31.18: *Trends in Genetics* 3 (1987): No. 6. Figure 31.19: Ed Reschke. Figure 31.20: Ray Ellis/ Photo Researchers, Inc.

Chapter 32
Figure 32.1: G. I. Bernard/Animals Animals/ Earth Scenes. Figure 32.2: John Cunningham/Visuals Unlimited. Figure 32.5: (a) Jeremy Burgess/Photo Researchers, Inc. Figure 32.8: (a) C. P. Vance/Visuals Unlimited; (b) Jeremy Burgess/Photo Researchers, Inc. Figure 32.10: (a–c) University of California Statewide IPM Project. Figure 32.12: C. R. Berry/Visuals Unlimited. Figure 32.13: (b) S. N. Postlethwait. Figure 32.15: Eric Schabtach, University of Oregon. Figure 32.17: (c, e) G. Biddulph et al., *Plant Physiology* 33 (1958): 295.

Figure 32.18: (a–c) Monsanto Chemical Company. Figure 32.19: DNAP. Figure 32.20: Lynn Riddiford, Monsanto Chemical Company.

Chapter 33
Figure 33.1: Gregory G. Dimijian/Photo Researchers, Inc. Figure 33.2: Reynolds Photography. Figure 33.5: Victor McKusick, *Human Genetics,* 2d ed., p. 49 (Prentice Hall, Englewood Cliffs, NJ, 1969). Figure 33.6: (b) G. Perkins/Visuals Unlimited. Figure 33.8: Victor McKusick, Johns Hopkins University. Figure 33.9: Patti Murray/Animals Animals. Figure 33.10: (a) John Trager/Visuals Unlimited; (b) John Cunningham/Visuals Unlimited. Figure 33.13: (a) Breck P. Kent/Photo Researchers, Inc.; (b) M. W. Tweedie/Photo Researchers, Inc. Figure 33.14: (b) *Journal of Heredity* 5 (1914): No. 11. Figure 33.15: (a) Ken Lucas/Biological Photo Service; (b) Biological Photo Service. Figure 33.16: (b, c) Animals Animals. Figure 33.18: (a) Edward S. Ross, California Academy of Sciences. Figure 33.23: (a) Natural History Photographic Agency; (b) Richard Alan Wood/Animals Animals. Figure 33.26: David Scharf.

Chapter 34
Figure 34.1: Arizona State Museum, University of Arizona, Tucson. Figure 34.2: John Gerlach/Visuals Unlimited. Figure 34.4: (a) A. Link/Visuals Unlimited; (b) M. Zohary, *Plants of the Bible* (Cambridge University Press, 1982). Figure 34.15: Albert Copley/Visuals Unlimited. Box 34.1, Figure 1: Jeff Lepore/ Photo Researchers, Inc. Figure 34.16: W. E. Ruth/Bruce Coleman, Inc. Figure 34.19: (a) Renee Purse/Photo Researchers, Inc.; (b) R. S. Virdee/Grant Heilman Photography, Inc. Figure 34.20: (b) B. M. Fagen; (c) C. B. Hieser, from *Science* 232 (1986): 1381. Copyright 1986 by the American Association for the Advancement of Science; (d) The Bettmann Archive. Figure 34.24: Van Bucher/Photo Researchers, Inc.

Chapter 35
Figure 35.1: Brooking Tatum/Visuals Unlimited; (inset) Gilbert Grant/Photo Researchers, Inc. Figure 35.5: Rick McIntyre. Figure 35.7: (a) David L. Pearson/Visuals Unlimited; (b) Artmus-Bertrand/Photo Researchers, Inc.; (c) B. H. Brattstrom/Visuals Unlimited. Figure 35.11: Food and Agriculture Organization of the United Nations. Figure 35.13: (a) J. A. Alcock/Visuals Unlimited; (b, c) Erick Greene, *Science* 243 (1989). Copyright 1989 by the American Association for the Advancement of Science; (d) Daniel W. Gotshall/Visuals Unlimited. Box 35.1, Figure 1: E. R. Degginger/Animals Animals. Figure 35.14: (a) Susan

Leavines/Photo Researchers, Inc.; (b) Thomas Eisner and D. Aneshansley. Figure 35.16: (a) William Weber/Visuals Unlimited; (b) John Cunningham/Visuals Unlimited. Figure 35.17: Alexander Klotz, from Peter Farb, *The Insects* (Life Nature Library, Time, 1962). Figure 35.18: Michael Leach/Animals Animals. Figure 35.19: (a) Raymond A. Mendez/Earth Scenes; (b) Richard Gross. Figure 35.20: (a) Oxford Scientific Films/Animals Animals; (b) Daniel H. Janzen. Figure 35.21: (a) Steve McCutcheon/Visuals Unlimited; (b) Steve Coombs/Photo Researchers, Inc.; (c) Kevin and Betty Collins/Visuals Unlimited. Box 35.2, Figure 1: Daniel Simberloff, Florida State University. Figure 35.23: (a) R. T. Domingo/Visuals Unlimited; (b) George Stephens/National Oceanographic and Atmospheric Administration.

Chapter 36

Figure 36.1: Robert R. Hessler, Scripps Institution of Oceanography, La Jolla, CA. Figure 36.5 Robert Pierce/U.S. Department of Agriculture. Box 36.1, Figure 1: Tom McHugh/Photo Researchers, Inc. Figure 36.14: D. W. Schindler, *Science* 184 (1974): 897. Copyright 1974 by the American Association for the Advancement of Science. Figure 36.17: (b) The Bettmann Archive. Figure 36.18: (a) Gary Randay/Visuals Unlimited; (b) D. W. Schindler, "Effects of Years of Experimental Acid," *Science* 228 (1985). Copyright 1985 by the

American Association for the Advancement of Science; (c) John Cunningham/Visuals Unlimited.

Chapter 37

Figure 37.1: Tom and Pat Leeson. Page 751: Bruce Matheson/PhotoNats. Figure 37.8: (b) Gregory C. Dimijian/Photo Researchers, Inc. Figure 37.9: Michael Rogers. Figure 37.10: Arthur Gloor/Earth Scenes. Figure 37.12: Victor Englebert/Photo Researchers, Inc. Figure 37.13: C. C. Lockwood/Animals Animals. Figure 37.14: John Cunningham/Visuals Unlimited. Figure 37.15: James R. Fisher/Photo Researchers, Inc. Figure 37.16: John Lemker/Earth Scenes. Figure 37.17: Johnny Johnson/Earth Scenes. Figure 37.18: Bill Curtsinger/Photo Researchers, Inc. Figure 37.20: John Lemker/Earth Scenes. Figure 37.23: Joel Arrington/Visuals Unlimited. Figure 37.25: Anne Wertheim/Animals Animals. Figure 37.26: Carl Roessler/Animals Animals. Figure 37.27: Gar Braasch.

Chapter 38

Figure 38.1: Gordon Langsbury/Bruce Coleman, Inc. Figure 38.2: Monica Impeckoven. Figure 38.4: W. A. Banaszewski/Visuals Unlimited. Figure 38.5: Glenn Oliver/Visuals Unlimited. Figure 38.7: Zig Leszczynski/Animals Animals. Figure 38.8: Louis Darling. Figure 38.9: Life Picture Service. Figure 38.10:

Thomas Eisner. Figure 38.12: The Bettmann Archive. Figure 38.13: Susan Kuklin/Photo Researchers, Inc. Figure 38.14: Oxford Scientific Films/Animals Animals. Figure 38.15: Ron Austing/Photo Researchers, Inc. Figure 38.17: R. H. Armstrong/Animals Animals. Figure 38.18: Janet Hopson. Figure 38.19: William E. Townsend, Jr./Photo Researchers, Inc. Figure 38.20: Bruce Coleman, Inc. Figure 38.22: American Museum of Natural History. Figure 38.24: Charles Palek/Animals Animals. Figure 38.26: Scott Camazine/Photo Researchers, Inc. Figure 38.27: Tom Cajacob, Minnesota Zoo. Figure 38.28: Bob Sacha.

Part Openers

Part One: Manfred Kage/Peter Arnold, Inc. Part Two: Frans Lanting/Minden Pictures. Part Three: Lefever/Grushow/Grant Heilman Photography, Inc. Part Four: Jim Brandenberg/Minden Pictures. Part Five: Frans Lanting/Minden Pictures. Part Six: Grant Heilman/Grant Heilman Photography, Inc. Part Seven: Frans Lanting/Minden Pictures.

Feature Box Photo Logos

Biology: A Human Endeavor: Flip Chalfant/The Image Bank. *Focus on the Environment:* Galen Rowell/Peter Arnold, Inc. *How Do We Know?:* Will McIntyre/Photo Researchers, Inc. *Personal Impact:* Hank Delespinasse/The Image Bank.

INDEX

Pages on which definitions or main discussions of topics appear are indicated by **boldface**.
Pages containing illustrations or tables are indicated by *italics*.

Metric–English and English–Metric Conversions

Prefixes Used with Metric Units

Prefix	Symbol	Value	Equivalent	A Biological Example
Kilo-	k	1000, or 10^3	1 kilometer (km) = 1×10^3 m	$^6/_{10}$ of a mile
Centi-	c	$^1/_{100}$, or 10^{-2}	1 centimeter (cm) = 0.01 m	$^1/_{100}$ m (about the width of your little finger)
Milli-	m	$^1/_{1000}$, or 10^{-3}	1 millimeter (mm) = 0.001 m	$^1/_{1000}$ m (about the length of a fruit fly)
Micro-	μ	$^1/_{1,000,000}$, or 10^{-6}	1 micrometer (μm) = 1×10^{-6} m	$^1/_{1,000,000}$ m ($^1/_{100}$ the thickness of a human hair)
Nano-	n	$^1/_{1,000,000,000}$, or 10^{-9}	1 nanometer (nm) = 1×10^{-9} m	$^1/_{1,000,000,000}$ m ($^1/_3$ the size of a small protein; myoglobin is about 3 nm across)
Pico-	p	$^1/_{1,000,000,000,000}$, or 10^{-12}	1 picomol (pmol)	6×10^{11} molecules of any substance
Femto-	f	$^1/_{1,000,000,000,000,000}$, or 10^{-15}	1 femtomol (fmol)	600,000,000 molecules of any substance

Length: Abbreviations and Conversions*

Metric	=	English (USA)
millimeter (0.001 m)	=	0.0394 in.
centimeter (0.01 m)	=	0.394 in.
meter (1 m)	=	3.28 ft, 39.4 in.
kilometer (1×10^3 m)	=	0.621 mi, 1094 yd, 3281 ft

English (USA)	=	Metric
inch (in.)	=	2.54 cm
foot (ft)	=	0.305 m, 30.5 cm
yard (yd)	=	0.914 m, 91.4 cm
mile (5280 ft)	=	1.61 km, 1609 m

*Some metric equivalents in the text have been rounded to whole numbers.

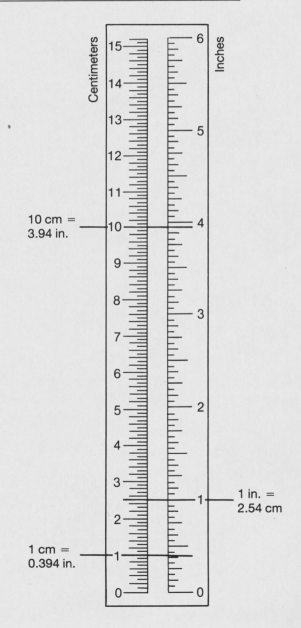

10 cm = 3.94 in.

1 in. = 2.54 cm

1 cm = 0.394 in.

Practice Examples:

A foot-long tennis shoe is 30.5 cm long.
  (30.5 cm = 12 in. × 2.54 cm/1 in.)
A 32-inch softball bat is 0.8 m long.
  (0.8 m = 32 in. × 1 m/39.4 in.)
A 100-yard long football field is 91 m long.
  (91 m = 100 yd × 36 in./1 yd × 1 m/39.4 in.)
When driving at 55 mph, you will travel 88.6 km in an hour.
  (88.6 km = 55 mi × 1 km/0.621 mi)